Birkhäuser

Static & Dynamic Game Theory: Foundations & Applications

Static & Dynamic Game Theory: Foundations & Applications aims to publish top quality state-of-the-art textbooks and research monographs at the graduate and post-graduate levels in game theory and its applications in a variety of fields, including biology, communications, computer and computational science, ecology, economics, environmental science, engineering, management science, networks, operations research, political science, security, and the social sciences. The scope of the series includes both cooperative and non-cooperative games; static, repeated, and dynamic games, defined in discrete time as well as in continuous time (so-called differential games); deterministic as well as stochastic games (such as Markov games); and complete or incomplete information games. The series will also include titles that study learning, evolution, and asymptotics in high player population regimes such as mean-field analysis.

The series will appeal to a broad interdisciplinary readership of mathematically-oriented engineers, physical and social scientists, computer and computational scientists, and economists, as well as to mathematicians at the graduate and post-graduate levels. Students and researchers in these fields will benefit from the comprehensive treatment of topical areas, groundbreaking theoretical exploration and development of computational tools, and applications covering the entire interdisciplinary spectrum where multi-agent decision-making is at the forefront. Theoreticians and practitioners in industry, government, and academia will find the titles in the series accessible across the varied terrain of game-theoretical research.

Tamer Başar • Boualem Djehiche •
Hamidou Tembine

Mean-Field-Type Game Theory I

Foundations and New Directions

Birkhäuser

Tamer Başar
Dept. Electrical & Computer Engineering
University of Illinois at Urbana-Champaign
Urbana, IL, USA

Boualem Djehiche
Department of Mathematics
KTH Royal Institute of Technology
Stockholm, Sweden

Hamidou Tembine
Electrical Engineering and Computer
Science, School of Engineering
Quebec University
Trois-Rivieres, QC, Canada

ISSN 2363-8516 ISSN 2363-8524 (electronic)
Static & Dynamic Game Theory: Foundations & Applications
ISBN 978-3-032-07026-5 ISBN 978-3-032-07027-2 (eBook)
https://doi.org/10.1007/978-3-032-07027-2

Mathematics Subject Classification: 91, 93, 65, 35, 37, 82, 90, 94

This book is published under the imprint Birkhäuser, www.birkhauser-science.com by the registered company Springer Nature Switzerland AG
The registered company address is: Gewerbestrasse 11, 6330 Cham, Switzerland

To my wife Tangül (T.B.),
To my grandchildren Adam, Selma, Bilal (B.D.),
To Yandai & Pama Tembine, Florence & Jean-Pierre & Marie-Claire Tixier and Anna Netzhammer for their unconditional support (H.T.).

Preface

While there are certainly excellent resources available that analyze the subject of dynamic game theory with a high degree of mathematical rigor and precision, there seems to be a distinct need for a book that offers a consolidated exploration of mean-field-type game theory. This book is intended to cater to a wider audience, ranging from undergraduate students to professionals in various scientific and industrial fields. The primary goal is to address this gap by presenting both discrete and continuous-time variations of basic mean-field-type game theory in a coherent manner. A particular emphasis is placed on the semi-explicit solutions of Mean-Field-Type Games, equipping readers to apply the techniques and methodologies to specific problems of interest.

Mean-Field-Type Games, featuring multiple decision-makers and diverse types, can be conceptualized as a sequence of interconnected interdependent decision-making problems. When the population profile involves numerous decision-makers for each type, a common strategy involves aggregating individual decision-makers and employing continuous variables to represent the collective averages of type-specific actions. Various authors, including Cournot, Roos, Borel, von Neumann, von Stackelberg, Nash, and Berge, have proposed equilibrium criteria. The aggregate models, referred to as non-atomic or population games, were initially examined by Wardrop in 1952 within a deterministic and stationary framework involving indistinguishable decision-makers. In the context of games with infinite populations, an equilibrium is characterized by a fixed set inclusion: the population profile's support is encompassed within the maximizers' set of the payoff function. Essentially, Mean-Field-Type Games extend beyond dependence solely on state-action pairs; they also consider the distributions of state-action pairs and/or their marginals. This concept applies both to infinite population games and finite games, featuring both von Neumann and non-von Neumann type payoffs.

The book is composed of two volumes, with the objective of this first volume being to offer an introductory exploration of fundamental concepts in mean-field-type game theory. The volume endeavors to provide a well-balanced perspective on key components of Mean-Field-Type Game Theory. Its organization ensures that key pertinent topics are covered at a foundational level of detail. This first volume is structured into six parts, facilitating a comprehensive exploration of the subject matter.

Part I: Discrete State Markov Games of Mean-Field Type

Chapter 1: One-Shot Mean-Field-Type Games We start with games with a single state or no state, called one-shot games. We begin with a clear definition of one-shot Mean-Field-Type Games. These games lay the foundation for understanding how decision-makers, each making individual decisions, collectively shape the dynamics of an interactive system. We will examine the essential elements that define these games and their relevance in diverse variance-aware, quantile-aware, and other risk-aware scenarios.

Chapter 2: Markov Games
Chapter 2 focuses on Markov Games. These games introduce the element of time and one-step memory into the strategic interactions, allowing us to capture basic state transition dynamics. We will explore how Markov processes and game theory intertwine to create a rich framework for modeling dynamic systems.

Chapter 3: Mean-Field-Type Games with Discrete State Spaces Building on the concepts introduced in the previous chapters, Chap. 3 navigates through Mean-Field-Type Games with discrete state spaces. We will investigate how the discrete nature of states influences the dynamics of these games and the unique challenges they present. Key foundational elements are introduced: equilibrium systems and master adjoint systems (MASS).

Part II: Equilibrium Principles

Chapter 4: Stochastic Maximum Principle In Part II, we shift our focus to the Stochastic Maximum/Minimum Principle. This principle plays a crucial role in optimizing strategies within Mean-Field-Type Games. We will unravel the mathematics behind this principle and explore its applications in various scenarios.

Chapter 5: Dynamic Programming Principle Dynamic Programming is a powerful tool in understanding and solving complex sequential decision-making problems. In this chapter, we present risk-sensitive and risk-neutral Dynamic Programming Principle and its significance in the context of Mean-Field-Type Games. We will explore how this principle helps us analyze and optimize strategies in the space of probability measures.

Part III: Classes of Mean-Field-Type Games

Chapter 6: Non-asymptotic Mean-Field-Type Games Moving forward, we explore non-asymptotic Mean-Field-Type Games. These games deviate from traditional asymptotic approaches and introduce stimulating new perspectives on strategic interactions and their approximations in terms of equilibrium payoffs as well as in terms of equilibrium strategies. We will present the nuances of these games and their implications in real-world scenarios.

Chapter 7: Linear-Quadratic Mean-Field and Mean-Field-Type Differential Games Risk awareness is a critical aspect of decision-making in uncertain environments. In this chapter, we introduce linear-quadratic-induced interactions in Mean-Field-Type Games. We will explore how risk considerations shape strategic decisions and outcomes.

Chapter 8: Mean-Field-Type Games with Jump and Regime Switching Mean-Field-Type Games are not limited to linear dynamics; they can encompass jump and regime-switching models. In this chapter, we explore Mean-Field-Type Games with jump and regime switching, offering insights into more complex and realistic scenarios.

Chapter 9: MASS: Master Adjoint Systems The concept of Master Adjoint Systems (MASS) is a powerful tool for understanding the dynamics of Mean-Field-Type Games. This chapter takes an in-depth look at MASS and its role in analyzing and optimizing strategies in dynamic environments.

Chapter 10: Semi-Explicit Solutions in Non-quadratic Mean-Field-Type Games Semi-explicit solutions provide valuable insights into solving Mean-Field-Type Games. In this chapter, we explore techniques for finding semi-explicit solutions in non-quadratic Mean-Field-Type Games, shedding light on practical applications.

Chapter 11: Stackelberg Mean-Field-Type Games Stackelberg games introduce hierarchy into strategic interactions with variance-awareness. In Chap. 11, we explore Stackelberg Mean-Field-Type Games, where leaders and followers make strategic decisions. We delve into the dynamics and solution concepts within this framework. These games extend the concept of leadership and followership, revealing the complexities of risk-aware multi-level decision-making. We then focus on distributed planning and investigate how Mean-Field-Type Games can be used to model and analyze distributed planning scenarios, offering practical insights into decentralized decision-making.

Chapter 12: Mean-Field-Type Games Driven by Rosenblatt Noises Rosenblatt noises introduce randomness and uncertainty which are non-Gaussian, non-Poisson, and non-Markov, and with fractional features into Mean-Field-Type Games. In this chapter, we present the dynamics of games driven by Rosenblatt noises, exploring their impact on strategic interactions and decision-making.

Chapter 13: Mean-Field-Type Games with Asymmetric Information Information asymmetry is a common feature in many real-world scenarios. In this chapter, we examine Mean-Field-Type Games with asymmetric information, shedding light on how incomplete information shapes strategic interactions.

Chapter 14: Difference Games of Mean-Field Type In this chapter, we start with linear-quadratic discrete-time infinite-population mean-field games. We then continue with linear-quadratic difference Mean-Field-Type Games. These classes of games provide a unique feature as the type of noises involved are relatively general, without restricting to specific distributions. Then, discrete-time adversarial Mean-Field-Type Games are introduced. These games examine adversarial elements into strategic interactions, revealing the complexities of competitive decision-making. Our journey concludes with a connection to Q-values, MinMax-Q-values in finite horizon Mean-Field-Type Games.

Part IV: Wrap-up

Chapter 15: Conclusions and New Directions In this chapter, we present some summary and new directions.

Appendix The Appendix has two chapters A and B. Chapter A focuses on inequalities and backward stochastic differential equations. Chapter B provides some mean-field convergence results under indistinguishability properties.

Volume II

The second volume of the book focuses on applications of mean-field-type games in engineering, computer science, evolutionary biology, chemistry, and network economics.

Urbana, IL, USA — Tamer Başar
Stockholm, Sweden — Boualem Djehiche
Trois-Rivieres, QC, Canada — Hamidou Tembine
June 2025

Acknowledgments We would like to express our sincere gratitude to everyone who contributed to the completion of the two volumes of this book. Some of the research described in the book has emerged from collaborations with others, and the authors are indebted to all of their co-authors on collaborative work cited in the bibliography sections. The authors would like to thank Xin Luo, Ahmed Farhan Hanif, Jian Gao, Yida Xu, Julian Barreiro-Gomez, Massa N'Dong, Salah Choutri, Tyrone E. Duncan, Bozenna Pasik-Duncan, Dario Bauso, and Alain Bensoussan and four anonymous reviewers of this book for their interesting questions and comments on mean-field-type models and approaches.

Over the last ten years, we received partial support from different grants and research projects:

- The Verg Foundation, Stockholm
- AFOSR: NonCooperative Dynamic Games with Large Population of Players
- Tommo-So Dogon: Audio-Centric Machine Intelligence for Tommo-So Dogon Using Mean-Field-Type Game Theory
- TIMADIE: MFTG for Risk-Aware Textless Audio-to-Audio Machine Intelligence
- GRABALTON: Risk-Aware Graph-Chain for Livestock & Poultry
- GUINAGATON: Graph-Chain for Agriculture: From Farm Sources to Consumers' Table
- TIMADIE: Risk-Aware Blockchain and Last Mile Integration
- MFTG4MI: Mean-Field-Type Game Theory for Machine Intelligence
- MCF: Data-Driven Mean-Field-Type Game Theory Modelling of SARS-CoV-2
- AFOSR: MFTG: Mean-Field-Type Games: Foundations
- SITE: Center on Stability, Instability and Turbulence
- MFTG4Evacuation: Mean-Field-Type Games for Multi-Level Building Evacuation
- MFTG4Water: Mean-Field-Type Games for Water Distribution Networks
- MFTG: Mean-Field-Type Games for Intelligent Transportation Systems
- MELANGE: Mean-Field-Type Games: Learning, Explicit Solutions and Implementation
- 5G: Learning and Incentives in Large Scale Millimeter Wave Wireless communications
- GOLD: Game Theory, Optimization, Learning and Dynamics
- Center for Uncertainty Quantification in Computational Science & Engineering
- MEAN-MACCS: MEAN-field games for Multi-Agent Allocation and Control in Communication Systems
- MORE: Advanced Mathematical Tools for Complex Network Engineering
- LEG: Learning, Evolution and Games
- RMG: Random Matrix Games in Network Engineering, Power Grid and Transportation Science
- COWINET: Evolutionary Coalitional Games in Wireless Networks
- TF: Evolutionary Games for Audio-Literacy

Artwork

The scientific graphs in the book are generated using the MATLAB® software by Mathworks Inc. and the Mean-Field-Type Package for Simulation of Large Population Games.

Competing Interests The authors have no competing interests to declare that are relevant to the content of this manuscript.

Contents

Abbreviations

ESS	Evolutionarily Stable Strategy
MFTG	Mean-Field-Type Game
MFG	Mean-Field Game
ODE	Ordinary Differential Equation
PDE	Partial Differential Equation
PIDE	Partial-Integro Differential Equation
MP	Maximum (Minimum) Principle
SMP	Stochastic Maximum (Minimum) Principle
DPP	Dynamic Programming Principle

Part I
Discrete State Markov Games of Mean-Field Type

Chapter 1
One-Shot Mean-Field-Type Games

In this chapter, we introduce one-shot Mean-Field-Type Games (MFTG) in which the decisions are taken once. These games are distinguished by their unique characteristics, where the (instantaneous) payoff functionals not only hinge on the action profiles adopted by the decision-makers but also on the distribution of these action profiles [102, 330, 331, 333]. This added feature results in the emergence of nonlinear expected payoffs in the product measure encompassing the strategies made by all decision-makers involved.

We start with fundamental constrained one-shot MFTGs, where multiple decision-makers interact within a stationary environment. In these scenarios, the payoff functionals extend their influence beyond the mere action profile, reaching into the realm of the first moments, encapsulating the expected values of the actions taken by the decision-makers. This novel static game approach serves as the breeding ground for innovative risk-aware solution concepts, a critical aspect of game theory in practical contexts.

We introduce several solution concepts to one-shot MFTGs. These include mean-field-type best-response strategies, mean-field-type constrained Nash equilibria, and mean-field-type variational equilibria, each offering a unique perspective on equilibrium solutions within this dynamic framework. To provide a more comprehensive understanding, we examine the structure of payoffs with a decomposition that incorporates variance-aware cost considerations, enriching our understanding of risk within these games. Furthermore, we present learning algorithms, exploring methods that enable us to reach mean-field-type best Nash equilibria while respecting constraints imposed by the game environment. This marks a pivotal point in our exploration, as it underscores the practical applicability of these concepts in real-world scenarios (see Volume II of the book on applications). We also incorporate migration costs and migration incentives, factors that can significantly impact the behavior of decision-makers and their choice components within the MFTG. This extension of our framework adds a layer of realism, making our analysis more representative of decision-making in interactive environments. We

T. Başar et al., *Mean-Field-Type Game Theory I*, Static & Dynamic Game Theory: Foundations & Applications, https://doi.org/10.1007/978-3-032-07027-2_1

then focus on one-shot mean-variance games and distributionally robust mean-variance MFTGs. We turn our attention to MFTGs, featuring coherent risk measures as loss functionals, with a particular emphasis on the entropic value at risk (evar), and its significance within MFTGs [333, 339, 342].

The rest of the chapter is organized as follows. Section 1.1 introduces generic one-shot MFTGs. We start by presenting a motivational example, followed by a class of action-constrained MFTGs with loss functions that are dependent on the moments. We provide diverse learning algorithms which would converge to the solutions of such games under various solution concepts. Section 1.2 develops transformation of variance-dependent loss functions using Legendre–Fenchel, f-divergence, and Bregman discrepancies. Section 1.3 considers the entropic value at risk (evar) and provides an example of a MFTG with coherent risk measures. Finally, we present expectile-based MFTGs.

1.1 A Class of One-Shot Mean-Field-Type Games

In this section, we define a class of one-shot MFTGs. These games present an intriguing departure from traditional game theory, as they introduce a novel dimension of interaction feature by incorporating not only the actions of decision-makers but also the distribution of these actions into their (instantaneous) payoff functions. This unique characteristic gives rise to nonlinear expected payoffs with respect to own mixed strategies, which add a layer of richness and depth to the ensuing analysis.

Definition 1.1 The basic ingredients of a one-shot MFTG are given by:

- The set of decision-makers $\mathscr{I}$, with cardinality $I = |\mathscr{I}| \geq 2$,
- For every decision-maker $i \in \mathscr{I}$,, there is a set of actions $\mathscr{A}_i$ which is non-empty.
- Every decision-maker i has a preference structure that can be represented by an instant performance functional $L_i : \prod_{j\in\mathscr{I}} \mathscr{A}_j \times \mathscr{P}\left(\prod_{j\in\mathscr{I}} \mathscr{A}_j\right) \rightarrow \mathbb{R}$, $(a, D_a) \mapsto L_i(a, D_a)$, with $a = (a_1, \ldots, a_I)$. The expected performance functional of decision-maker i is given by $E[L_i(a, D_a)] := \int_{b\in\prod_{j\in\mathscr{I}} \mathscr{A}_j} L_i(b, D_a)D_a(db)$.

  The collection $\mathscr{G} = (\mathscr{I}, (\mathscr{A}_i, E[L_i])_{i\in\mathscr{I}})$, is called a MFTG in strategic form.

We have the following basic observations: Assume that the functional L_i is not trivially zero for any $i \in \mathscr{I}$. Then, the expected performance function of decision-maker i :

$$D_a \mapsto \hat{L}_i(D_a) := \int L_i(b, D_a)D_a(db),$$

is not necessarily linear [102, 331]. Let $(D_{a_1}, \ldots, D_{a_I})$ be the mixed strategies of all the decision-makers. Then,

$$(D_{a_1}, \ldots, D_{a_I}) \mapsto \int L_i(a'_1, \ldots a'_I, D_{a_1}, \ldots, D_{a_I}) \prod_{j \in \mathcal{I}} D_{a_j}(da'_j),$$

is not necessarily linear. For each decision-maker $i \in \mathcal{I}$, and $(D_{a_j})_{j \neq i}$,, the expected functional

$$D_{a_i} \mapsto \int L_i(a'_1, \ldots a'_I, D_{a_1}, \ldots, D_{a_I}) \prod_{j \in \mathcal{I}} D_{a_j}(da'_j),$$

is not necessarily linear. This latter property creates a departure from classical game theory with expected von Neumann utility. For this reason, $E[L_i(a, D_a)]$ will be called non-von Neumann expected (dis)utility.

Below we denote the probability distribution of the random variable over the action a_i of decision-maker i by $D_{a_i} =: x_i$. In this context of time-independent interaction, $x_i \in \mathcal{P}(\mathcal{A}_i)$ is interpreted as a mixed strategy of decision-maker i. The vector $x = (x_1, \ldots, x_I)$ is a strategy profile of all decision-makers. We denote $(x_j)_{j \neq i} \in \prod_{j \neq i} \mathcal{P}(\mathcal{A}_j)$ by x_{-i}.

In addition to the strategies of the decision-makers, we add an extra randomness in the instantaneous loss function via a set Ω.

Static MFTGs with Nature Uncertainty

A game $\mathcal{G}_{nature} = \left(\mathcal{I}, (\mathcal{A}_i, \hat{L}_i)_{i \in \mathcal{I}}; \tilde{m} \in \mathcal{P}(\Omega)\right)$, with

$$\begin{cases} I = |\mathcal{I}| \geq 2, \\ L_i : \prod_{j \in \mathcal{I}} \mathcal{A}_j \times \left(\prod_{j \in \mathcal{I}} \mathcal{P}(\mathcal{A}_j)\right) \times \Omega \to \mathbb{R}, \\ (a_1, \ldots, a_I, x_1, \ldots, x_I, \omega) \mapsto L_i(a_1, \ldots, a_I, x_1, \ldots, x_I, \omega), \\ \hat{L}_i(x, \tilde{m}) = E_{x \otimes \tilde{m}}[L_i(a_1, \ldots, a_I, x_1, \ldots, x_I, \omega)] \\ = \int L_i(a_1, \ldots, a_I, x_1, \ldots, x_I, \omega) \left(\prod_j x_j(da_j)\right) \tilde{m}(d\omega), \end{cases} \tag{1.1}$$

is a MFTG in the presence of the uncertainty $\omega \sim \tilde{m}$ due to nature, where $(x \otimes \tilde{m})(da d\omega)$ denotes product measure $\left(\prod_j x_j(da_j)\right) \tilde{m}(d\omega)$.

Definition 1.2 (Mean-Field-Type Additive ϵ-Best Response) Let $\epsilon_i \geq 0$. A strategy $x_i \in \mathcal{P}(\mathcal{A}_i)$ of decision-maker i is a mean-field-type additive ϵ_i-best response strategy to $(x_j)_{j \neq i} =: x_{-i} \in \prod_{j \neq i} \mathcal{P}(\mathcal{A}_j)$ if

$$x_i \in \text{aBR}_{i,\epsilon_i}(x_{-i}) = \left\{x_i \in \mathscr{P}(\mathscr{A}_i)|\ \hat{L}_i(x,\tilde{m}) - \epsilon_i \leq \inf_{x_i' \in \mathscr{P}(\mathscr{A}_i)} \hat{L}_i(x_i', x_{-i}, \tilde{m})\right\}$$

Definition 1.3 (Mean-Field-Type Additive ϵ-Equilibrium) The strategy profile $(x_1, \ldots, x_I) \in \prod_{j \in \mathscr{I}} \mathscr{P}(\mathscr{A}_j)$ is an additive $(\epsilon_1, \ldots, \epsilon_I)-$equilibrium of the MFTG $\mathscr{G}_{nature}$ if for each decision-maker $i \in \mathscr{I}$, one has $x_i \in \text{aBR}_{i,\epsilon_i}(x_{-i})$. A mean-field-type additive 0-equilibrium is obtained for $\epsilon_i = 0,\ \forall i \in \mathscr{I}$.

The value of ϵ in the additive epsilon-equilibrium is relative, and this definition needs more attention. If we consider a game with payoff r_j with bounded equilibrium set in a finite-dimensional space, then, we can choose a positive real number c_1 as the supremum of the absolute value of the equilibrium payoffs (which is finite). Now we use the fact that the set of equilibria is invariant by conic transformation. Without loss of generality, we can choose $c_1 > 0$. Hence, the game where the payoffs are $\frac{r_j}{c_1 10^{c_2}}$, $c_2 > 1$, has the same set of equilibria as the original game. However, the equilibrium payoff of this new game are all less than 10^{-c_2} which can be arbitrary small if c_2 is large. In particular, for games with bounded payoffs, the number c_1 will be chosen to be the infinite norm of the payoff, and every strategy is an additive $\epsilon-$equilibrium in the novel game which has the same set of equilibria as the original game. In conclusion, the notion of additive $\epsilon-$equilibrium does not capture the $0-$equilibrium structure as we see that it is not invariant by conic transformation (multiplication by a positive scalar).

Alternatively, one can introduce a multiplicative $(\epsilon_1, \ldots, \epsilon_I)$—equilibrium is a strategy profile that approximately satisfies the condition of Nash mean-field-type equilibrium with a ratio $1 - \epsilon_i$ for decision-maker i.

Definition 1.4 (Mean-Field-Type Multiplicative ϵ-Equilibrium) What is a (multiplicative) epsilon-equilibrium or near equilibrium?

Let $\epsilon_i \geq 0$. A strategy $y_i \in \mathscr{P}(\mathscr{A}_i)$ of decision-maker i is a mean-field-type multiplicative ϵ_i-best response strategy to $x_{-i} \in \prod_{j \neq i} \mathscr{P}(\mathscr{A}_j)$ if

$$y_i \in \text{mBR}_{i,\epsilon_i}(x_{-i}) = \left\{x_i \in \mathscr{P}(\mathscr{A}_i)|\ \hat{L}_i(x,\tilde{m}) \leq (1+\epsilon_i) \inf_{x_i' \in \mathscr{P}(\mathscr{A}_i)} \hat{L}_i(x_i', x_{-i}, \tilde{m})\right\}$$

The strategy profile $(x_1, \ldots, x_I) \in \prod_{j \in \mathscr{I}} \mathscr{P}(\mathscr{A}_j)$ is a multiplicative $(\epsilon_1, \ldots, \epsilon_I)$-equilibrium of $\mathscr{G}_{nature}$ if for every decision-maker $i \in \mathscr{I}$, one has $x_i \in \text{mBR}_{i,\epsilon_i}(x_{-i})$.

Below we provide some examples of static MFTGs.

Example 1.1 (Variance-Awareness) Variance-aware games constitute extensions of standard games with expected payoffs, where we not only consider what decision-makers do and what they receive as payoffs but also take into account how uncertain those payoffs are. This adds a layer of complexity and realism to the interaction, making it particularly relevant in situations where risk management plays a crucial role. In essence, these games challenge us to think not only about maximizing expected payoffs but also about minimizing the associated variance.

We start with a generic formulation: Consider $I \geq 2$ decision-makers. Decision-maker $i \in \{1, \ldots, I\}$ has a non-empty action space $\mathscr{A}_i$. For each decision-maker i, there is a real-valued square integrable function $l_i : (\prod_{k=1}^{I} \mathscr{A}_k) \times \Omega \rightarrow \mathbb{R}$, and possibly an extra randomness $\omega \sim \tilde{m}$ such that the variance-aware loss function of i is

$$\hat{L}_i(x, \tilde{m}) = f_i(E_{x,\tilde{m}}[l_i(a, \omega)], SD_{x,\tilde{m}}[l_i(a, \omega)]),$$

where f_i is a real-valued function for the single objective, or vectorial function in the multi-objective case, $E_{x,\tilde{m}}[l_i(a, \omega)] = \int_{(\prod_{k=1}^{I} \mathscr{A}_k)\times\Omega} l_i(a, \omega)(\prod_{k=1}^{I} x_k(da_k))\tilde{m}(d\omega)$, denotes the expectation operator $SD_{x,\tilde{m}}[l_i(a, \omega)] = \sqrt{\mathrm{var}_{x,\tilde{m}}[l_i(a, \omega)]}$ denotes the standard deviation operator with the respect to the probability measure x and $\tilde{m}$, $\mathrm{var}_{x,\tilde{m}}[l_i(a, \omega)]$ is the variance $\mathrm{var}_{x,\tilde{m}}[l_i(a, \omega)] = E_{x,\tilde{m}}\{[l_i(a, \omega) - E_{x,\tilde{m}}[l_i(a, \omega)]]^2\}$ denoting the variance operator. The variance-aware game $\mathscr{G} = \left(\mathscr{I}, (\mathscr{P}(\mathscr{A}_i), \hat{L}_i)_{i\in\mathscr{I}}; \tilde{m} \in \mathbb{P}(\Omega)\right)$, is a MFTG in the sense of (1.1).

We have the following observations:

- Clearly, $(x, \tilde{m}) \mapsto E_{x,\tilde{m}}[l_i(a, \omega)]$ is nonlinear. However, $x \otimes \tilde{m} \mapsto E_{x,\tilde{m}}[l_i(a, \omega)]$ is linear.
- $x \otimes \tilde{m} \mapsto \mathrm{var}_{x,\tilde{m}}[l_i(a, \omega)]$ is nonlinear.
- $x \otimes \tilde{m} \mapsto SD_{x,\tilde{m}}[l_i(a, \omega)]$ is nonlinear.
- Given $(x_{-i}, \tilde{m})$, the function $x_i \mapsto SD_{x,\tilde{m}}[l_i(a, \omega)]$ is nonlinear.
- Given $(x_{-i}, \tilde{m})$,, the function $x_i \mapsto \hat{L}_i(x, \tilde{m})$ is nonlinear.

Non-convexity with Respect to Self Strategies

In classical game theory, it is well known that any finite game (with a finite and non-empty set of choices, a finite number of decision-makers, and finite payoff entries) has at least one equilibrium in mixed strategies. This is known as Nash's theorem and can be easily proven using Kakutani's fixed-point theorem. You can simplify the proof by using the Brown-von Neumann-Nash map or the Smith map. Upon careful examination of the proof, it becomes evident that having convex-valued reactions or convex cost functions in self's mixed strategy is crucial. This is satisfied, because the (mixed extended) payoff functions here are multilinear.

Now, how does this relate to MFTG? The MFTG cost function is not necessarily linear and not necessarily convex, even in a 2×2 MFTG. The mapping

(continued)

$$\begin{cases} i \in \mathscr{I}, \\ x_i \mapsto E_{x\otimes\tilde{m}}[L_i(a_1,\ldots,a_I,x_1,\ldots,x_I,\omega)] \\ = \int L_i(a_1,\ldots,a_I,x_1,\ldots,x_I,\omega)\left(\prod_j x_j(da_j)\right)\tilde{m}(d\omega), \end{cases} \tag{1.2}$$

is not necessarily convex as the instantaneous cost $L_i(a_1,\ldots,a_I,x_1,\ldots,x_I,\omega)$ depends on x_i. As we will see below in the mean-variance case, the non-convexity of the cost function with respect to self-strategy may create a nonexistence of mean-field-type equilibria in 2×2-induced MFTGs.

Example 1.2 (Mean-Variance) A mean-variance game [102] entails interactive decision-making under uncertainty, where individuals or entities seek to optimize their choices based on both the mean (average) and the scaled variance (risk or variability) of possible outcomes [333, 339, 342]. The scale of i is $\frac{r_i}{\epsilon+|l_i(a^0)|}$ where a^0 is an action profile, $\epsilon > 0$, and r_i is the scaling factor (risk-sensitivity index) for decision-maker i. The objective of decision-maker i in the mean-variance game is to optimize their choices by minimizing the following cost function

$$\hat{L}_i(x) = E_x[l_i(a)] + \frac{r_i}{\epsilon+|l_i(a^0)|}\cdot var_x[l_i(a)],$$

subject to any other constraints or requirements specific to the game at hand.

If there is an extra randomness $\omega \sim \tilde{m}$ in the cost function, the formulation becomes:

$$\hat{L}_i(x,\tilde{m}) = E_{x,\tilde{m}}[l_i(a,\omega)] + \frac{r_i}{\epsilon+|l_i(a^0,\omega^0)|}\cdot var_{x,\tilde{m}}[l_i(a,\omega)],$$

and the resulting mean-variance game $\mathscr{G}_{mv} = \left(\mathscr{I},(\mathscr{P}(\mathscr{A}_i),\hat{L}_i)_{i\in\mathscr{I}};\tilde{m}\right)$, is a MFTG.

Example 1.3 (Mean-Variance) For a payoff function p_i to be maximized, the mean-variance formulation becomes

$$\hat{P}_i(x,\tilde{m}) = E_{x,\tilde{m}}[p_i(a,\omega)] - \frac{r_i}{\epsilon+|p_i(a^0,\omega^0)|}\cdot var_{x,\tilde{m}}[p_i(a,\omega)],$$

and the resulting mean-variance game $\mathscr{G}_{mv} = \left(\mathscr{I},(\mathscr{P}(\mathscr{A}_i),\hat{P}_i)_{i\in\mathscr{I}};\tilde{m}\right)$, is a MFTG.

Example 1.4 Consider the following basic game. Consider two decision-makers. Decision-maker 1 chooses a strategy $(x, 1-x),\ x \in [0, 1]$, and decision-maker 2 chooses $(y, 1-y),\ y \in [0, 1]$. The expected payoff of decision-maker 1 is $m_1 = (2y-1)(4x-3)$. The variance of her payoff is $v_1 = x + 9(1-x) - m_1^2 = -8x+9-(2y-1)^2(4x-3)^2$. The mean-variance payoff of decision-maker 1 is given by $mv_1 = (2y-1)(4x-3)+8x-9+(2y-1)^2(4x-3)^2$. The mean-variance payoff of decision-maker 2 is given by $mv_2 = -(2y-1)(4x-3)+8x-9+(2y-1)^2(4x-3)^2$. These are polynomials and hence smooth functions.

We note that at the pure strategy $(x, y) = (0, 1),$, one has $(m_1, m_2) = (-3, 3)$, $(v_1, v_2) = (0, 0)$, and $(mv_1, mv_2) = (-3, 3)$. At another pure strategy $(x, y) = (1, 1),$, one has $(m_1, m_2) = (1, -1)$, $(v_1, v_2) = (0, 0)$, and $(mv_1, mv_2) = (1, -1)$.

For every given $y \in [0, 1],\ y \neq \frac{1}{2},$, the function $x \mapsto mv_1(x, y)$, is strictly convex and is to be maximized (and not minimized) by decision-maker 1 over $[0, 1]$. The critical points of the gradient are not useful here, as we are not minimizing mv. We therefore look at the extreme points of the domain. At $y = \frac{1}{2},\ mv_1(x, \frac{1}{2}) = 8x - 9$, is maximized at $x = 1$.

Note that $(x, y) = (0, 1)$ is not an equilibrium: At $x = 0$, we obtain $mv_1(0, y) = -9 - 3(2y-1) + 9(2y-1)^2$. At $x = 1$, we obtain $mv_1(1, y) = -1 + (2y-1) + (2y-1)^2$ which means $x = 1$ is better than $x = 0$ against $y = 1$.

Similarly, $(x, y) = (1, 1)$ is not an equilibrium because $mv_2(1, 0) = 1 > -1 = mv_2(1, 1)$, and the same holds for the two other extreme points. Thus, there is no mean-variance equilibrium at the extreme points.

Nonexistence of Mean-Variance Equilibria in Mixed Strategies
Let $\mathcal{G}_{mv} = (\mathcal{I}, (\mathcal{P}(\mathcal{A}_i), \mathrm{mv}_i)_{i\in\mathcal{I}})$, where the pure action space is $\mathcal{A}_i = \{0, 1\}$, the mixed strategy space $\mathcal{P}(\mathcal{A}_i) = \{(x, 1-x) \mid x \in [0, 1]\}$ is replaced by $[0, 1]$ which is a convex, non-empty, compact set. From the structure of the functions, $mv_1 = (2y-1)(4x-3) + 8x - 9 + (2y-1)^2(4x-3)^2$ and $mv_2 = -(2y-1)(4x-3) + 8x - 9 + (2y-1)^2(4x-3)^2,$, there is no mean-variance equilibrium in the game $\mathcal{G}_{mv}$. This means that the non-concavity of the payoff mv_i with respect to the decision-maker strategy x_i plays an important role in the nonexistence of mean-variance equilibrium.

Remark 1.1 (Mixture of Mixed Strategies) We have seen that the previous mean-variance game with action profile space $[0, 1]^2$ does not have an equilibrium. However, the action space of each decision-maker is a non-empty, convex, and compact set, and the mean-variance functions are all continuous over $[0, 1]^2$. Therefore we can define a second-level mixed extension over Borel sets. We define Δ_i to be the set of Borel probability measures over $[0, 1]$. Following the same procedure as above, we define the mixed extension of the mean-variance game over $\Delta_1 \times \Delta_2$. For $(\sigma_1, \sigma_2) \in \Delta_1 \times \Delta_2,$, the payoffs are

$$\int_{[0,1]}\int_{[0,1]} mv_i(x,y)\sigma_1(dx)\sigma_2(dy) = E_{\sigma_1\otimes\sigma_2}[mv_i], \quad i \in \{1,2\}.$$

By Glicksberg fixed-point theorem [218], the mixed extension of the mean-variance game has at least one (mixed of mixed) equilibrium as the mean-variance game is a continuous game with continuous mv utility functions over compact action spaces $[0,1]^2$.

1.1.1 Mean-Variance Blockchain Security Games

A blockchain is a decentralized and transparent digital ledger technology that securely records transactions across a network of machines, computers, and devices using cryptographic techniques to ensure immutability and fostering trust without the need for a central authority. In the context of blockchain security, consider a scenario where two entities, say mobile nodes A and B, are involved in reaching a consensus on a transaction. Each mobile node has an option to either validate the transaction or reject it. The outcome of the validation process influences the overall state of the blockchain. If the hash is even, Node A wins by validating while B loses by rejecting. Conversely, if the hash is odd, Node B wins by validating and A loses by rejecting. The validation winning values are different and captured by the matrix payoffs (1.3) given by

$$G_1 := \begin{pmatrix} (-1,1) & (1,-1) \\ (3,-3) & (-3,3) \end{pmatrix} \tag{1.3}$$

Risk-Neutral G_1

The expected payoff of decision-maker 1 is given by

$$\begin{cases} m_1(x,y) := (x, 1-x)G_1\begin{pmatrix} y \\ 1-y \end{pmatrix} = (x,1-x)\begin{pmatrix} -2y+1 \\ 3(2y-1) \end{pmatrix} \\ = -x(2y-1) + 3(1-x)(2y-1) = -(4x-3)(2y-1), \\ 0 \le x \le 1, \ 0 \le y \le 1, \end{cases} \tag{1.4}$$

where $(x, 1-x)$ is the mixed strategy of decision-maker 1 and $(y, 1-y)$ is the mixed strategy of decision-maker 2. The expected payoff of decision-maker 2 is $m_2(x,y) = (4x-3)(2y-1)$. The risk-neutral game G_1 is a zero-sum game and has no pure Nash equilibrium (saddle point). The game has a unique risk-neutral mixed Nash equilibrium given by $(\frac{3}{4}, \frac{1}{4}), (\frac{1}{2}, \frac{1}{2})$.

Variance Payoffs from G_1

Let $S_1 = (G_{11}^2(i,j), G_{12}^2(i,j))_{i,j}$ be the matrix whose entries are the squared entries of G_1.

$$S_1 := \begin{pmatrix} (1,1) & (1,1) \\ (9,9) & (9,9) \end{pmatrix}. \tag{1.5}$$

The variance payoff [102] of decision-maker 1 is given by

$$\begin{cases} v_1(x,y) := (x, 1-x)S_{11}\begin{pmatrix} y \\ 1-y \end{pmatrix} - m_1^2(x,y) \\ = (x, 1-x)\begin{pmatrix} 1 \\ 9 \end{pmatrix} - m_1^2(x,y) = -8x + 9 - m_1^2(x,y) \\ = -8x + 9 - (4x-3)^2(2y-1)^2. \end{cases} \tag{1.6}$$

Similarly, the variance payoff of decision-maker 2 is given by $v_2(x,y) := -8x + 9 - (4x-3)^2(2y-1)^2$.

Note that, by definition, $v_1(x,y) \geq 0$ and $v_2(x,y) \geq 0$ for every $(x,y) \in [0,1]^2$. As expected, these two functions are zero at the pure strategies in $\{(1,0),(1,0)\}, \{(1,0),(0,1)\}, \{(0,1),(1,0)\}, \{(0,1),(0,1)\}$. This is because the variance is zero for non-stochastic objects. On the other hand, the variances $v_1(x,y)$ and $v_2(x,y)$ are maximized at $\{(0,1), (\frac{1}{2}, \frac{1}{2})\}$.

Mean-Variance Payoffs from G_1

The mean-variance payoff of decision-maker 1 with risk-sensitive index λ_1 is given by

$$\begin{cases} mv_1(x,y) := m_1(x,y) - \lambda_1 v_1(x,y) \\ = -(4x-3)(2y-1) - \lambda_1(-8x + 9 - (4x-3)^2(2y-1)^2). \end{cases} \tag{1.7}$$

The mean-variance payoff of decision-maker 2 with risk-sensitive index λ_2 is given by

$$\begin{cases} mv_2(x,y) := m_2(x,y) - \lambda_2 v_2(x,y) \\ = (4x-3)(2y-1) - \lambda_2(-8x + 9 - (4x-3)^2(2y-1)^2). \end{cases} \tag{1.8}$$

The mean-variance game $\mathcal{G}_{mv,\lambda} = (\mathcal{I}, (\mathcal{P}(\mathcal{A}_i), \text{mv}_i)_{i\in\mathcal{I}})$, is given by

- Decision-makers $\mathcal{I} = \{1, 2\}$.
- Decision-makers 1 and 2 action space is $\mathcal{A}_i = \{(1,0), (0,1)\}$, and their strategy space is $\mathcal{P}(\mathcal{A}_i) = \{(x, 1-x),\ 0 \leq x \leq 1\}$.
- Decision-makers 1 and 2 payoffs are $mv_1(x,y), mv_2(x,y)$, respectively. The parameter λ_i is the risk-sensitivity index of decision-maker i.

The game $\mathcal{G}_{mv,\lambda}$ is clearly a MFTG. The variance v_1 is nonlinear x, and the variance v_2 is nonlinear in y. As we will see below, the sign of risk-sensitivity index of decision-makers will play a crucial role in the existence and structure of mean-field-type equilibria. The general combination depending on the sign of $\lambda_1 \in \mathbb{R}, \lambda_2 \in \mathbb{R}$, is discussed below.

The mean-variance equilibrium set is given by

$$\mathcal{E}(\mathcal{G}_{mv,\lambda}) = \left\{ \begin{array}{l} ((x^*, 1-x^*), (y^*, 1-y^*)) : \\ x^* \in \arg\max_{0\leq x\leq 1} mv_1(x, y^*), \\ y^* \in \arg\max_{0\leq y\leq 1} mv_2(x^*, y) \end{array} \right\}, \tag{1.9}$$

where

$$\left\{ \begin{array}{l} \lambda_1 \in \mathbb{R}, \lambda_2 \in \mathbb{R},\ 0 \leq x \leq 1,\ 0 \leq y \leq 1, \\ mv_1(x, y) := -(4x-3)(2y-1) - \lambda_1(-8x+9-(4x-3)^2(2y-1)^2), \\ mv_2(x, y) := (4x-3)(2y-1) - \lambda_2(-8x+9-(4x-3)^2(2y-1)^2). \end{array} \right. \tag{1.10}$$

We compute the following payoffs when one of the decision-makers uses a pure strategy:

$$\left\{ \begin{array}{l} mv_1(0, y) := 3(2y-1) - 9\lambda_1(1-(2y-1)^2), \\ mv_1(1, y) := -(2y-1) - \lambda_1(1-(2y-1)^2), \\ mv_2(0, y) := -3(2y-1) - 9\lambda_2(1-(2y-1)^2), \\ mv_2(1, y) := (2y-1) - \lambda_2(1-(2y-1)^2), \end{array} \right. \tag{1.11}$$

and

$$\left\{ \begin{array}{l} mv_1(x, 0) := (4x-3) - \lambda_1(-8x+9-(4x-3)^2), \\ mv_1(x, 1) := -(4x-3) - \lambda_1(-8x+9-(4x-3)^2), \\ mv_2(x, 0) := -(4x-3) - \lambda_2(-8x+9-(4x-3)^2), \\ mv_2(x, 1) := (4x-3) - \lambda_2(-8x+9-(4x-3)^2). \end{array} \right. \tag{1.12}$$

These equations will be used to compute the sign of $mv_1(0, z) - mv_1(1, z)$ which determines the best response of decision-maker 1 who uses a pure strategy. Similarly, the sign of $mv_2(z, 0) - mv_2(z, 1)$ will be used to check if $(z, 0)$ is a partially mixed equilibrium.

We provide equilibrium conditions when one of the decision-makers uses a pure strategy:

$$\begin{cases} (0, y_0^*) : \\ mv_1(0, y_0^*) - mv_1(1, y_0^*) = 4((2y_0^* - 1) - 2\lambda_1(1 - (2y_0^* - 1)^2)) > 0, \\ y_0^* \in \arg\max_y - (2y - 1) - 3\lambda_2(1 - (2y - 1)^2), \\ (1, y_1^*) : \\ 4((2y_1^* - 1) - 2\lambda_1(1 - (2y_1^* - 1)^2)) < 0, \\ y_1^* \in \arg\max_y (2y - 1) - \lambda_2(1 - (2y - 1)^2), \end{cases} \tag{1.13}$$

and

$$\begin{cases} (x_0^*, 0) : \\ mv_2(x_0^*, 0) - mv_2(x_0^*, 1) = -2(4x_0^* - 3) > 0, \\ x_0^* \in \arg\max_x (4x - 3) - \lambda_1(-8x + 9 - (4x - 3)^2), \\ (x_1^*, 1) : \\ mv_2(x_1^*, 0) - mv_2(x_1^*, 1) = -2(4x_1^* - 3) < 0, \\ x_1^* \in \arg\max_x - (4x - 3) - \lambda_1(-8x + 9 - (4x - 3)^2). \end{cases} \tag{1.14}$$

Case 1: Both Are Risk-Averse $\lambda_1 > 0$ and $\lambda_2 > 0$
We demonstrate that when both decision-makers are risk-averse, the resulting mean-variance game lacks an equilibrium in mixed strategies. Initially, we note that the unique risk-neutral equilibrium is not mean-variance equilibrium. Subsequently, we establish that due to the strict convexity of the mean-variance payoff (except at one point), the best response of decision-maker 1 leads to a pure strategy. However, the pure strategies yield zero variance, and the risk-neutral game has no pure equilibrium.

Proposition 1.1 *Consider two risk-averse decision-makers with $\lambda_1 > 0$ and $\lambda_2 > 0$. Then, the mean-variance game obtained from the 2×2-matrix payoff G_1 has no pure Nash equilibrium in $\{(1, 0), (0, 1)\}^2$.*

The mean-variance game obtained from the 2×2-matrix payoff G_1 has no mixed Nash equilibrium in $\{(z, 1 - z),\ 0 \le z \le 1\}^2$. The mean-variance game obtained from the 2×2-matrix payoff G_1 has at least one Nash equilibrium in $(\mathscr{B}[0, 1])^2$ where $\mathscr{B}([0, 1])$ is the set of Borel measures over the interval $[0, 1]$.

Proof We check the four pure strategies.

$\{(1, 0), (1, 0)\}$: is not an equilibrium as $mv_1(1, 1) = -1 < mv_1(0, 1) = 3$. Decision-maker 1 has incentive to deviate unilaterally.

$\{(1, 0), (0, 1)\}$: is not an equilibrium as $mv_2(1, 0) = -1 < mv_2(1, 1) = 3$. Decision-maker 2 has incentive to deviate unilaterally.

$\{(0, 1), (1, 0)\}$: is not an equilibrium as $mv_2(0, 1) = -3 < mv_2(1, 1) = 1$. Decision-maker 2 has incentive to deviate unilaterally.

$\{(0, 1), (0, 1)\}$: is not an equilibrium as $mv_1(0, 0) = -3 < mv_1(1, 1) = 1$. Decision-maker 2 has incentive to deviate unilaterally.

This proves that independently of (λ_1, λ_2), the mean-variance game obtained from the 2×2-matrix payoff G_1 has no pure Nash equilibrium in $\{(1,0),(0,1)\}^2$.

We now look for mean-variance Nash equilibria in mixed strategies in $\{(z, 1-z),\ 0 \le z \le 1\}^2$. For $y = \frac{1}{2}$, we have $mv_1(x, \frac{1}{2}) = \lambda_1(8x - 9)$ is maximized at $x = 1$. Hence, $(\frac{3}{4}, \frac{1}{4}), (\frac{1}{2}, \frac{1}{2})$ is not an equilibrium.

For $y \neq \frac{1}{2}$,, the function $x \mapsto mv_1(x, y)$ is strictly convex. This means that $\arg\max_{0 \le x \le 1} mv_1(x, y) \subset \{0, 1\}$. The response of decision-maker 1 is a pure strategy in this case.

For $x = \frac{3}{4}$, $mv_2(\frac{3}{4}, y) = -3\lambda_2$. Every strategy $(y, 1-y)$ is best response. $mv_1(\frac{3}{4}, 0) = mv_1(\frac{3}{4}, 1)$.

For $x \neq \frac{3}{4}$,, the function $y \mapsto mv_2(x, y)$ is strictly convex. The best response of decision-maker 2 satisfies $\arg\max_{0 \le y \le 1} mv_2(x, y) \subset \{0, 1\}$.

It remains to check the partially mixed strategy:

$mv_1(\frac{3}{4}, 0) = -3\lambda_1 < 1 = mv_1(1, 0)$ which implies that $(\frac{3}{4}, \frac{1}{4}), (0, 1)$ is not an equilibrium. $mv_1(\frac{3}{4}, 1) = -3\lambda_1 < 3 = mv_1(0, 1)$ which implies that $(\frac{3}{4}, \frac{1}{4}), (1, 0)$ is not an equilibrium.

We conclude that there is no fixed point in mixed strategies in $\{(z, 1-z),\ 0 \le z \le 1\}^2$. We now focus on the last statement. By identifying the mixed strategy set $\{(z, 1-z),\ 0 \le z \le 1\}$ to the set $\{z,\ 0 \le z \le 1\} = [0, 1]$. We will work on $\mathscr{B}[0, 1]$ is the set of Borel measures over the interval $[0, 1]$. The payoff functions $(x, y) \mapsto mv_1(x, y)$ and $(x, y) \mapsto mv_2$ are continuous over the convex, compact, and non-empty set $[0, 1]^2$. Then, independently of the values of (λ_1, λ_2),, the mean-variance game obtained from the 2×2-matrix payoff G_1 has at least one second-order mixed Nash equilibrium in $(\mathscr{B}[0, 1])^2$. This completes the proof. ∎

Case 2: Risk-Averse and Risk-Neutral $\lambda_1 > 0$ and $\lambda_2 = 0$

The same rationale of pure strategies applies when decision-maker 1 is risk-averse and decision-maker 2 is risk-neutral.

Proposition 1.2 *Assume that decision-maker 1 is risk-averse $\lambda_1 > 0$ and decision-maker 2 is risk-neutral $\lambda_2 = 0$. Then, there is no mean-variance equilibrium in mixed strategies.*

Proof We first examine the case where $y = \frac{1}{2}$.

$$\begin{cases} \text{At } (x, \frac{1}{2}) : \\ mv_1(x, \frac{1}{2}) = \lambda_1(8x - 9) \text{ is maximized } x = 1 \\ (\frac{3}{4}, \frac{1}{2}) \text{ is not an equilibrium} \\ mv_2(x, \frac{1}{2}) = 0, \\ (1, \frac{1}{2}) \text{ is not an equilibrium because} \\ mv_2(1, y) = (2y - 1) \text{ is maximized at } y = 1. \end{cases} \tag{1.15}$$

We now examine the case where $y \neq \frac{1}{2}$. In this case, $x \mapsto mv_1(x, y)$ is strictly convex as $\lambda_1 > 0$. The best response set $\arg\max_{0 \le x \le 1} mv_1(x, y) \subset \{0, 1\}$. This means that x is a pure strategy. We therefore concentrate the search to the partially mixed strategies $(0, y)$ and $(1, y)$ for $y \neq \frac{1}{2}$. As $mv_2(0, y) = -3(2y - 1)$ is maximized at $y = 0$ and $mv_2(1, y) = (2y - 1)$ is maximized at $y = 1$,, we conclude that the best responses $\arg\max_{0 \le y \le 1} mv_2(0, y) = \{0\}$ and $\arg\max_{0 \le y \le 1} mv_2(1, y) = \{1\}$. But these pure strategies are not equilibria.

The mean-variance game obtained from the 2×2-matrix payoff G_1 has no mixed Nash equilibrium in $\{(z, 1 - z),\ 0 \le z \le 1\}^2$ when decision-maker 1 is risk-averse and decision-maker 2 risk-neutral. ■

Case 3: Risk-Averse and Risk-Seeking $\lambda_1 > 0$ **and** $\lambda_2 < 0$

When decision-maker 1 is risk-averse but decision-maker 2 risk-seeking, the dynamics is different. The risk-neutral equilibrium is still not viable due to the variance contributions. Decision-maker 1 exhibits a pure-strategy best response due to strict convexity. The mean-variance equilibrium seeking reduces to the existence of an interior best response of decision-maker 2, which depends on its risk-sensitivity index $\lambda_2 < 0$.

Proposition 1.3 *Assume that* $\lambda_1 > 0$ *and* $\lambda_2 < 0$. *The game has a mean-variance equilibrium if* $\lambda_2 < -\frac{1}{2}$ *and* $\frac{1-2\lambda_2}{-4\lambda_2} \le \frac{4\lambda_1 - 1 + \sqrt{(4\lambda_1-1)^2 + 8\lambda_1}}{8\lambda_1}$, *in which case the equilibrium is unique and it is given by* $(1, 0), (y_1^*, 1 - y_1^*)$ *with* $y_1^* = \frac{1}{2} - \frac{1}{4\lambda_2}$.

Proof

$$\begin{cases} \lambda_1 > 0,\ \lambda_2 < 0, \\ 0 \le x \le 1,\ 0 \le y \le 1, \\ x \mapsto mv_1(x, y) \text{ is strictly convex for } y \neq \frac{1}{2}, \\ mv_1(x, \frac{1}{2}) = \lambda_1(8x - 9) \text{ maximized at } x = 1, \\ y \mapsto mv_2(x, y) \text{ is strictly concave for } x \neq \frac{3}{4}, \end{cases} \tag{1.16}$$

The equilibrium conditions are given by

$$\begin{cases} (0, y_0^*),\ y_0^* \in (0, 1) : \\ mv_1(0, y_0^*) - mv_1(1, y_0^*) = 4((2y_0^* - 1) - 2\lambda_1(1 - (2y_0^* - 1)^2)) > 0, \\ y_0^* \in \arg\max_y -(2y - 1) - 3\lambda_2(1 - (2y - 1)^2), \\ (1, y_1^*),\ y_1^* \in (0, 1) : \\ 4((2y_1^* - 1) - 2\lambda_1(1 - (2y_1^* - 1)^2)) < 0, \\ y_1^* \in \arg\max_y (2y - 1) - \lambda_2(1 - (2y - 1)^2). \end{cases} \tag{1.17}$$

$$\begin{cases} mv_2(0,y) = -9\lambda_2 - 3(2y-1) + 9\lambda_2(2y-1)^2, \\ \partial_y mv_2(0,y) = 6(-1-6\lambda_2+12\lambda_2 y) \text{ vanishes at } y_0^* = \frac{1+6\lambda_2}{12\lambda_2}\mathbb{I}_{\lambda_2<-\frac{1}{6}}, \\ mv_1(0,y_0^*) - mv_1(1,y_0^*) = 4(2y_0^*-1) - 8\lambda_1(1-(2y_0^*-1)^2) > 0 \\ mv_2(1,y) = -\lambda_2 + (2y-1) + \lambda_2(2y-1)^2, \\ \partial_y mv_2(1,y) = 2(1+2\lambda_2(2y-1)) \text{ vanishes at } y_1^* = (\frac{1-2\lambda_2}{-4\lambda_2})\mathbb{I}_{\lambda_2<-\frac{1}{2}}, \\ mv_1(0,y_1^*) - mv_1(1,y_1^*) = 4((2y_1^*-1) - 2\lambda_1(1-(2y_1^*-1)^2)) < 0. \end{cases} \tag{1.18}$$

We examine the sign of $mv_1(0,z) - mv_1(1,z)$.

$$\begin{cases} (2z-1) - 2\lambda_1(1-(2z-1)^2) \\ < 0 \text{ on } \left(\frac{4\lambda_1-1-\sqrt{(4\lambda_1-1)^2+8\lambda_1}}{8\lambda_1}, 0\right) \cup \left(0, \frac{4\lambda_1-1+\sqrt{(4\lambda_1-1)^2+8\lambda_1}}{8\lambda_1}\right), \\ = 0 \text{ on } \left\{\frac{4\lambda_1-1-\sqrt{(4\lambda_1-1)^2+8\lambda_1}}{8\lambda_1}, \frac{4\lambda_1-1+\sqrt{(4\lambda_1-1)^2+8\lambda_1}}{8\lambda_1}\right\}, \\ > 0 \text{ on } \left(-\infty, \frac{4\lambda_1-1-\sqrt{(4\lambda_1-1)^2+8\lambda_1}}{8\lambda_1}\right) \cup \left(\frac{4\lambda_1-1+\sqrt{(4\lambda_1-1)^2+8\lambda_1}}{8\lambda_1}, +\infty\right). \end{cases} \tag{1.19}$$

For $0 \le z \le 1$,, one has

$$(2z-1) - 2\lambda_1(1-(2z-1)^2) \begin{cases} < 0 \text{ on } \left(0, \frac{4\lambda_1-1+\sqrt{(4\lambda_1-1)^2+8\lambda_1}}{8\lambda_1}\right), \\ = 0 \text{ if } z = \frac{4\lambda_1-1+\sqrt{(4\lambda_1-1)^2+8\lambda_1}}{8\lambda_1}, \\ > 0 \text{ on } \left(\frac{4\lambda_1-1+\sqrt{(4\lambda_1-1)^2+8\lambda_1}}{8\lambda_1}, 1\right). \end{cases} \tag{1.20}$$

The mean-variance equilibrium conditions reduce to the feasibility of the following system:

$$\begin{cases} (0,y_0^*), \text{ mv equilibrium if } y_0^* = \frac{1+6\lambda_2}{12\lambda_2}\mathbb{I}_{\lambda_2<-\frac{1}{6}} > \frac{4\lambda_1-1+\sqrt{(4\lambda_1-1)^2+8\lambda_1}}{8\lambda_1}, \\ (1,y_1^*), \text{ mv equilibrium if } y_1^* = (\frac{1-2\lambda_2}{-4\lambda_2})\mathbb{I}_{\lambda_2<-\frac{1}{2}} < \frac{4\lambda_1-1+\sqrt{(4\lambda_1-1)^2+8\lambda_1}}{8\lambda_1} \end{cases}. \tag{1.21}$$

As $y_0^* < 1/2 < \frac{4\lambda_1-1+\sqrt{(4\lambda_1-1)^2+8\lambda_1}}{8\lambda_1}$,, the first branch is eliminated for $\lambda_1 > 0, \lambda_2 < 0$,

Hence if $\lambda_2 < -\frac{1}{2}$, $(1, y_1^*)$, is a mean-variance equilibrium if

$$y_1^* = \frac{1-2\lambda_2}{-4\lambda_2} = \frac{1}{2} + \frac{1}{-4\lambda_2} < \frac{4\lambda_1 - 1 + \sqrt{(4\lambda_1-1)^2+8\lambda_1}}{8\lambda_1}.$$

‘

If $\frac{1}{2} + \frac{1}{-4\lambda_2} = \frac{4\lambda_1-1+\sqrt{(4\lambda_1-1)^2+8\lambda_1}}{8\lambda_1}$ for a certain $\lambda_2 < -\frac{1}{2}, \lambda_1 > 0$ then

$\frac{1}{-4\lambda_2} = \frac{-1+\sqrt{(4\lambda_1-1)^2+8\lambda_1}}{8\lambda_1}$ and
$mv_1(0, y_1^*) = mv_1(1, y_1^*)$ for a strictly convex function to be maximized in x.

$(1, y_1^*)$ is still an equilibrium. However $(0, y_1^*)$ is not an equilibrium as $mv_2(0, z)$ is strictly concave and maximized y_0^* with $y_0^* = \frac{1+6\lambda_2}{12\lambda_2} = \frac{1}{2}+\frac{1}{12\lambda_2} < \frac{1}{2}+\frac{1}{-4\lambda_2} = y_1^*$. This completes the proof. ■

Note that

- For $(\lambda_1, \lambda_2) = (1, -1)$,, the profile $(1, 0), (y_1^*, 1 - y_1^*) = (\frac{3}{4}, \frac{1}{4})$ is the unique mean-variance equilibrium.
- For $\lambda_1 > 0, \lambda_2 \in (-\frac{1}{2}, +\infty)$, there is no mean-variance equilibrium.

Case 4: Risk-Neutral and Risk-Averse $\lambda_1 = 0$ and $\lambda_2 > 0$

Proposition 1.4 *Let $\lambda_1 = 0$ and $\lambda_2 > 0$. Then, the game has a unique mean-variance equilibrium given by $(\frac{3}{4}, \frac{1}{4}), (\frac{1}{2}, \frac{1}{2})$.*

Proof For $\lambda_1 = 0$ and $\lambda_2 > 0$, the mean-variance equilibrium problem is given by

$$\begin{cases} \lambda_1 = 0, \lambda_2 > 0, \\ 0 \le x \le 1, \ 0 \le y \le 1, \\ mv_1(x, y) := -(4x-3)(2y-1), \\ mv_2(x, y) := (4x-3)(2y-1) - \lambda_2(-8x+9-(4x-3)^2(2y-1)^2). \end{cases} \tag{1.22}$$

We compute the best responses. We start with the case where $x \neq \frac{3}{4}$. This means that the function $y \mapsto mv_2(x, y)$ is strictly convex leading to pure best response set for decision-maker 2 : $\arg\max_{0\le y\le 1} mv_2(x, y) \subset \{0, 1\}$. We evaluate for decision-maker 1's payoff $mv_1(x, 0) = (4x - 3)$ and $mv_1(x, 1) = -(4x - 3)$. Both lead to a pure best response by strict monotonicity. These pure outcomes are not equilibria.

We now examine the case where $x = \frac{3}{4}$. As a result, decision-maker 2's payoff is $mv_2(\frac{3}{4}, y) = -3\lambda_2, \ \forall y,$, i.e., $\arg\max_{0\le y\le 1} mv_2(\frac{3}{4}, y) = [0, 1]$ and

$$\begin{cases} mv_1(\frac{3}{4}, y) = 0 < mv_1(0, y) = 3(2y-1) \ \text{ if } \ y > \frac{1}{2}, \\ mv_1(\frac{3}{4}, y) = 0 < mv_1(1, y) = -(2y-1) \ \text{ if } \ y < \frac{1}{2}, \\ \arg\max_{0\le x\le 1} mv_1(x, \frac{1}{2}) = [0, 1]. \end{cases} \tag{1.23}$$

Hence the fully mixed strategy $(\frac{3}{4}, \frac{1}{4}), (\frac{1}{2}, \frac{1}{2})$ is a mean-variance equilibrium. The equilibrium condition for any other type leads to pure outcomes (which are not equilibria.) ■

Case 5: Both Are Risk-Neutral $\lambda_1 = 0$ and $\lambda_2 = 0$

In this case, there is a unique equilibrium which is the unique risk-neutral fully mixed equilibrium $(\frac{3}{4}, \frac{1}{4}), (\frac{1}{2}, \frac{1}{2})$. At this equilibrium, the expected equilibrium payoff is zero for both decision-makers.

Case 6: Risk-Neutral and Risk-Seeking $\lambda_1 = 0$ and $\lambda_2 < 0$

Proposition 1.5 *Let $\lambda_1 = 0, \lambda_2 < 0$. Then the fully mixed strategy $(\frac{3}{4}, \frac{1}{4}), (\frac{1}{2}, \frac{1}{2})$ is the unique mean-variance equilibrium.*

Proof

$$\begin{cases} \lambda_1 = 0, \lambda_2 < 0, \\ 0 \leq x \leq 1, \ 0 \leq y \leq 1, \\ mv_1(x, y) = -(4x-3)(2y-1), \\ mv_2(x, y) = (4x-3)(2y-1) - \lambda_2(-8x+9-(4x-3)^2(2y-1)^2), \\ y \mapsto mv_2(x, y) \text{ is strictly concave for } x \neq \frac{3}{4}, \\ mv_2(\frac{3}{4}, y) = -3\lambda_2, \\ (\frac{3}{4}, \frac{1}{4}), (\frac{1}{2}, \frac{1}{2}) \text{ is a mean-variance equilibrium }, \\ y < \frac{1}{2} \implies x = 1 \\ mv_2(1, y) = (2y-1) - \lambda_2(1-(2y-1)^2), \\ \arg\max_{0 \leq y \leq 1} mv_2(1, y) = \{\frac{1-2\lambda_2}{-4\lambda_2}\}, \\ \frac{1-2\lambda_2}{-4\lambda_2} > \frac{1}{2}, \\ (1, \frac{1-2\lambda_2}{-4\lambda_2}) \text{ is not a mean-variance equilibrium }, \\ y > \frac{1}{2} \implies x = 0, \\ mv_2(0, y) = -3(2y-1) - 9\lambda_2(1-(2y-1)^2), \\ \arg\max_{0 \leq y \leq 1} mv_2(0, y) = \{\frac{-1-6\lambda_2}{-12\lambda_2}\}, \ \lambda_2 < -\frac{1}{6} \\ \frac{-1-6\lambda_2}{-12\lambda_2} < \frac{1}{2}, \\ (0, \frac{-1-6\lambda_2}{-12\lambda_2}) \text{ is not a mean-variance equilibrium.} \end{cases} \tag{1.24}$$

This completes the proof. ∎

Case 7: Both Are Risk-Seeking $\lambda_1 < 0$ and $\lambda_2 < 0$

Proposition 1.6 *Let $\lambda_1 < 0$ and $\lambda_2 < 0$ such that*

$$\begin{cases} (\frac{1}{\lambda_1} + \frac{1}{\lambda_2}) < -2, \\ \frac{1}{\lambda_2}(\frac{1}{\lambda_1} + \frac{1}{\lambda_2}) < 12. \end{cases} \tag{1.25}$$

Then the strategy profile $(\frac{1}{2} + \frac{\lambda_1\lambda_2}{(\lambda_1+\lambda_2)}, \frac{3}{4} - \frac{(\lambda_1+\lambda_2)}{16\lambda_1\lambda_2^2})$ is a mean-variance equilibrium. Otherwise, there is no mean-variance equilibrium.

Note that the existence of mean-variance equilibrium restrict the risk-sensitivity values of both decision-makers.

Proof We check the equilibrium conditions for interior solution due to strict concavity.

$$\begin{cases} \lambda_1 < 0, \lambda_2 < 0, \\ 0 \le x \le 1,\ 0 \le y \le 1, \\ mv_1(x,y) := -(4x-3)(2y-1) - \lambda_1(-8x+9-(4x-3)^2(2y-1)^2), \\ mv_2(x,y) := (4x-3)(2y-1) - \lambda_2(-8x+9-(4x-3)^2(2y-1)^2), \\ mv_1(x,\frac{1}{2}) = \lambda_1(8x-9),\ \text{is maximized}\ x=0 \\ mv_2(\frac{3}{4},y) = -3\lambda_2, \\ (\frac{3}{4},\frac{1}{4}), (\frac{1}{2},\frac{1}{2})\ \text{is not an equilibrium} \\ x \mapsto mv_1(x,y)\ \text{is strictly concave for}\ y \neq \frac{1}{2}, \\ y \mapsto mv_2(x,y)\ \text{is strictly concave for}\ x \neq \frac{3}{4}, \\ \text{For}\ x \neq \frac{3}{4}, y \neq \frac{1}{2}: \\ \partial_x mv_1(x,y) := 4(-(2y-1)+2\lambda_1(1+(4x-3)(2y-1)^2)), \\ \partial_y mv_2(x,y) := 2(4x-3)(1+2\lambda_2(4x-3)(2y-1)), \end{cases} \tag{1.26}$$

This leads to the following system:

$$\begin{cases} -(2y-1)+2\lambda_1(1+(4x-3)(2y-1)^2)=0, \\ 1+2\lambda_2(4x-3)(2y-1)=0 \end{cases} \tag{1.27}$$

We make the following change of variables

$$\begin{cases} Y = 2y-1 \implies y = \frac{Y+1}{2}, \\ X = 4x-3 \implies x = \frac{X+3}{4} \end{cases} \tag{1.28}$$

Then $-1 \le Y \le 1, -3 \le X \le 1$.
The system becomes

$$\begin{cases} -Y+2\lambda_1(1+XY^2)=0, \\ 1+2\lambda_2 XY = 0 \end{cases} \tag{1.29}$$

By multiplying by Y, we obtain

$$\begin{cases} -Y+2\lambda_1(1+XY^2)=0, \\ Y+2\lambda_2 XY^2 = 0 \end{cases} \tag{1.30}$$

It follows that

$$\begin{cases} -1 \le Y \le 1, -3 \le X \le 1, \\ XY^2 = -\frac{\lambda_1}{(\lambda_1+\lambda_2)}, \\ Y = \frac{2\lambda_1\lambda_2}{(\lambda_1+\lambda_2)}, \\ X = -\frac{(\lambda_1+\lambda_2)}{4\lambda_1\lambda_2^2}, \end{cases} \tag{1.31}$$

Using the change of variables, we obtain

$$\begin{cases} Y = 2y - 1 \implies y = \frac{Y+1}{2} = \frac{1}{2} + \frac{\lambda_1\lambda_2}{(\lambda_1+\lambda_2)}, \\ X = 4y - 3 \implies x = \frac{X+3}{4} = \frac{3}{4} - \frac{(\lambda_1+\lambda_2)}{16\lambda_1\lambda_2^2} \end{cases} \tag{1.32}$$

provided the latter terms are between 0 and 1. This completes the proof. ■

Note that for $\lambda_1 = \lambda_2 = -1/2$,

$$\begin{cases} y = \frac{1}{2} + \frac{\lambda_1\lambda_2}{(\lambda_1+\lambda_2)} = \frac{1}{4}, \\ x = \frac{3}{4} - \frac{(\lambda_1+\lambda_2)}{16\lambda_1\lambda_2^2} = \frac{3}{4} - \frac{1}{2} = \frac{1}{4} \end{cases} \tag{1.33}$$

which means that $(\frac{1}{4}, \frac{3}{4}), (\frac{1}{4}, \frac{3}{4})$ is the mean-variance equilibrium.

Case 8: Risk-Seeking and Risk-Neutral $\lambda_1 < 0$ and $\lambda_2 = 0$

Proposition 1.7 *Let* $\lambda_1 < 0$ *and* $\lambda_2 = 0$. *If* $\lambda_1 < -\frac{1}{2}$, *the strategy* $(x_0^*, 1 - x_0^*), (0, 1)$ *with* $x_0^* = \frac{1}{2} + \frac{1}{-8\lambda_1}$. *is an equilibrium. Otherwise, there is no mean-variance equilibrium.*

Proof

$$\begin{cases} \lambda_1 < 0, \lambda_2 = 0, \\ 0 \le x \le 1, \ 0 \le y \le 1, \\ mv_1(x, y) := -(4x-3)(2y-1) - \lambda_1(-8x + 9 - (4x-3)^2(2y-1)^2), \\ mv_2(x, y) := (4x-3)(2y-1), \\ mv_1(x, \frac{1}{2}) = \lambda_1(8x - 9), \text{ is maximized } x = 0, \\ mv_2(\frac{3}{4}, y) = 0, \\ (\frac{3}{4}, \frac{1}{4}), (\frac{1}{2}, \frac{1}{2}) \text{ is not an equilibrium} \\ x \mapsto mv_1(x, y) \text{ is strictly concave for } y \ne \frac{1}{2}, \\ x < \frac{3}{4} \implies y = 0 : mv_1(x, 0) = (4x-3) - \lambda_1(-8x + 9 - (4x-3)^2), \\ \partial_x mv_1(x, 0) = 4(1 + 4\lambda_1(-1 + 2x)), \\ x_0^* = \frac{1-4\lambda_1}{-8\lambda_1}, \\ \lambda_1 < -\frac{1}{2} : \ (x_0^*, 0) \text{ is an equilibrium} \\ x > \frac{3}{4} \implies y = 1 : \\ mv_1(x, 1) = -(4x-3) - \lambda_1(-8x + 9 - (4x-3)^2), \\ \partial_x mv_1(x, 1) = 4(-1 + 4\lambda_1(-1 + 2x)), \\ x_1^* = \frac{-1-4\lambda_1}{-8\lambda_1} = \frac{1}{2} + \frac{1}{-8\lambda_1} < \frac{1}{2} < \frac{3}{4}, \ \lambda_1 < -\frac{1}{4}, \\ (x_1^*, 1) \text{ is not an equilibrium.} \end{cases} \tag{1.34}$$

This completes the proof. ■

Case 9: Risk-Seeking and Risk-Averse $\lambda_1 < 0$ **and** $\lambda_2 > 0$

Proposition 1.8 *We consider a risk-averse $\lambda_1 < 0$ and a risk-seeking $\lambda_2 > 0$. If $\lambda_1 < -\frac{1}{2}$ then $(x_0^*, 1 - x_0^*), (0, 1)$ with $x_0^* = \frac{1}{2} + \frac{1}{-8\lambda_1}$ is a mean-variance equilibrium. Otherwise, there is no mean-variance equilibrium.*

Note that the existence of mean-variance equilibrium restrict the risk-sensitivity value of decision-maker 1 to $\lambda_1 < -\frac{1}{2}$.

Proof We recall the partially mixed equilibrium conditions:

$$\begin{cases} (x_0^*, 0): \\ mv_2(x_0^*, 0) - mv_2(x_0^*, 1) = -2(4x_0^* - 3) > 0 \text{ if } x_0^* < \frac{3}{4}, \\ x_0^* \in \arg\max_x (4x - 3) - \lambda_1(-8x + 9 - (4x - 3)^2), \\ (x_1^*, 1): \\ mv_2(x_1^*, 0) - mv_2(x_1^*, 1) = -2(4x_1^* - 3) < 0 \text{ if } x_1^* > \frac{3}{4}, \\ x_1^* \in \arg\max_x - (4x - 3) - \lambda_1(-8x + 9 - (4x - 3)^2). \end{cases} \tag{1.35}$$

We check the above conditions when $\lambda_1 < 0, \lambda_2 > 0$. By computing $mv_1(x, \frac{1}{2}) = \lambda_1(8x - 9)$, we observe that the latter is decreasing and maximized $x = 0$. The best response of decision-maker 1 to $y = \frac{1}{2}$ is 0. This implies that $(\frac{3}{4}, \frac{1}{4}), (\frac{1}{2}, \frac{1}{2})$ is not an equilibrium in this case. We have $mv_2(\frac{3}{4}, y) = -3\lambda_2$ for all y. Any strategy y of decision-maker 2 is best response to $x = \frac{3}{4}$.

Now, for $x \neq \frac{3}{4}$,, the mapping $y \mapsto mv_2(x, y)$ is strictly convex. From the maximization of convex function, this means that the best responses of decision-maker are pure strategies (at the extreme points). With this, the equilibrium condition is reduced to

$$\begin{cases} x^* \in \arg\max_{0 \leq x \leq 1} mv_1(x, y^*), \\ y^* \in \arg\max_{0 \leq y \leq 1} mv_2(x^*, y) \subset \{0, 1\}. \end{cases} \tag{1.36}$$

We evaluate the payoffs $mv_1(x, 0)$ and $mv_1(x, 1)$ when decision-maker 2 chooses pure strategies.

The payoff of 1 at $y = 0$ yields $mv_1(x, 0) = (4x - 3) - \lambda_1(-8x + 9 - (4x - 3)^2)$. It follows $y \mapsto mv_1(x, 0)$ is strictly concave for $\lambda_1 < 0$. The derivative in x is given by $\partial_x mv_1(x, 0) = 4(1 + 4\lambda_1(-1 + 2x))$. For the maximization of concave functions, the first optimality conditions are sufficient. The derivative vanishes at $x = \frac{1-4\lambda_1}{-8\lambda_1} = \frac{1}{2} + \frac{1}{-8\lambda_1}$. For such point to be eligible, we check the interior condition: $0 < \frac{1}{2} + \frac{1}{-8\lambda_1} < 1$. The first branch $0 < \frac{1}{2} + \frac{1}{-8\lambda_1}$ is satisfied as $\lambda_1 < 0$. The second inequality $x_0^* = \frac{1}{2} + \frac{1}{-8\lambda_1} < 1$ is satisfied if $\lambda_1 < -\frac{1}{4}$.

For $(x_0^*, 0)$ to be an equilibrium, it is necessary to have $mv_2(x_0^*, 0) \geq mv_2(x_0^*, 1)$. One has $mv_2(x_0^*, 0) - mv_2(x_0^*, 1) = -2(4x_0^* - 3)$ which is positive if $x_0^* < \frac{3}{4}$. The condition $x_0^* < \frac{3}{4}$ holds if $\frac{1}{-8\lambda_1} < \frac{1}{4}$ which $\lambda_1 < -\frac{1}{2} < -\frac{1}{4}$. This means that if $\lambda_1 < -\frac{1}{2}, \lambda_2 > 0$, then $(x_0^*, 1 - x_0^*), (0, 1)$ is an equilibrium.

We now check the other partially mixed profile $(x, 1)$; The mean-variance payoff of 1 is $mv_1(x, 1) : -(4x - 3) - \lambda_1(-8x + 9 - (4x - 3)^2)$, which is strictly concave.

The derivative in x is $\partial_x mv_1(x,1) = 4(-1+4\lambda_1(-1+2x))$,, which the critical point $x_1^* = \frac{-1-4\lambda_1}{-8\lambda_1} = \frac{1}{2} + \frac{1}{8\lambda_1} < \frac{1}{2}$, $\lambda_1 < -\frac{1}{4}$. The computation $mv_2(x_1^*, 0) - mv_2(x_1^*, 1) = -2(4x_1^* - 3)$, which is not negative. Thus, $(x_1^*, 1-x_1^*), (1,0)$ is not an equilibrium.

This completes the proof. ■

Based on Propositions 1.2 and 1.4, we observe that the risk-sensitivity index is not permutable. If decision-maker 1 is risk-averse and decision-maker 2 is risk-neutral, the resulting game has no equilibrium in mixed strategies. But if decision-maker 1 is risk-neutral and decision-maker 2 is risk-averse, the resulting game has a unique equilibrium.

Note that for any value of $(\lambda_1, \lambda_2) \in \mathbb{R}^2$, by Glicksberg fixed-point theorem [218], the mixed extension of the mean-variance game has at least one (mixed of mixed) equilibrium as the mean-variance game is a continuous game with continuous payoff functions over the convex, compact and non-empty space $[0,1]^2$.

Example 1.5 (Mean-Standard Deviation Trade-off) This is a similar formulation as the mean-variance trade-off where the scaled variance is replaced by the standard deviation. $SD_x[l_i(a)]$ is the standard deviation of the cost $l_i(a)$ under x at hand.

The objective of decision-maker i in the mean-standard deviation game is to optimize their choices by maximizing the following utility function:

$$\hat{L}_i(x) = E_x[l_i(a)] + r_i \cdot SD_x[l_i(a)],$$

subject to any other constraints or requirements specific to the game at hand, where r_i is the scaling factor (risk-sensitivity index) for decision-maker i. Here, the units of the two additive terms are similar, and thus r_i is a dimensionless scalar.

If there is an extra randomness $\omega \sim \tilde{m}$ in the payoff, the formulation becomes

$$\hat{L}_i(x, , \tilde{m}) = E_{x,\tilde{m}}[l_i(a,\omega)] + r_i \cdot SD_{x,\tilde{m}}[l_i(a,\omega)]$$

This mean-standard deviation game $\mathcal{G}_{msd} = \left(\mathcal{I}, (\mathcal{P}(\mathcal{A}_i), \hat{L}_i)_{i\in\mathcal{I}}; \tilde{m}\right)$ is indeed a MFTG.

Example 1.6 (Mean to Standard Deviation Ratio) The objective of decision-maker i in the mean to standard deviation ratio game is to maximize the following utility function:

$$\hat{P}_i(x) = \frac{E_x[p_i(a)]}{\epsilon + SD_x[p_i(a)]},$$

subject to any other constraints or requirements specific to the game at hand.

Example 1.7 (Dispersion Index) The objective of decision-maker i in the dispersion index game is to minimize the following cost function:

$$\hat{L}_i(x) = \frac{SD_x[p_i(a)]}{\epsilon + E_x[p_i(a)]},$$

where $\epsilon > 0$ and $p_i(a)$ is assumed to be nonnegative.

Example 1.8 A variance-constrained game is a type of strategic interactive decision-making problem where multiple decision-makers participate, each with their own individual payoff objectives. In this game, the focus is on self-preferences subject to managing and limiting the variability or dispersion (variance) of the decision-makers' payoffs. Each decision-maker in a variance-constrained game aims to optimize her objectives while adhering to a specified constraint on the variance of her payoff. This constraint sets a limit on how much variability or risk the decision-makers would be willing to tolerate in their outcomes. The decision-makers' strategies will need to balance maximizing their expected payoffs with the requirement to keep the variance of their payoffs below the specified level. The variance constraint adds an additional layer of interaction to the decision-making process, as participants must consider not only the potential gains and losses associated with their choices but also the extent to which those choices contribute to the variance of their payoffs. The interactions and strategies employed by each decision-maker will influence the overall outcomes and equilibrium of the variance-constrained game [333, 339, 342].

Specifically, for $\rho_i > 0,$, the decision-maker problem is formulated as below, for each i.

$$\begin{array}{ll} \text{Sup}_{a_i \in \mathscr{A}_i} & E_{x,\tilde{m}}[p_i(a,\omega)] \\ \text{Subject to:} & \text{var}_{x,\tilde{m}}[p_i(a,\omega)] \leq \rho_i \end{array}. \tag{1.37}$$

This is a MFTG with coupled action constraints. The mean-field term here is through the variance $\text{var}_{x,\tilde{m}}$.

Example 1.9 (Variance-Reduction in Satisfaction Games) A satisfaction game entails a type of strategic interaction where each decision-maker's objective is to achieve a payoff that exceeds a certain satisfaction level. In this game, participants strive to make choices that result in outcomes surpassing their predetermined levels of satisfaction. A variance-reduction in satisfaction game involves a strategic scenario where multiple decision-makers engage in a game where their main goal is twofold: First, they aim to reduce the variance (uncertainty or variability) of their individual payoffs, and second, they are subject to a constraint that their payoffs must exceed a specified satisfaction levels.

In a variance-reduction in satisfaction game, decision-makers need to strike a balance between minimizing the variability of their outcomes (variance reduction) while ensuring that their payoffs remain above their satisfaction threshold. This type of a game introduces a dual concern for participants, who must manage risk by reducing variance, while also maintaining payoffs at or above their desired satisfaction levels. The strategic choices made by each decision-maker influence the overall dynamics and outcomes of the variance reduction in satisfaction game.

In case no extra randomness is involved, the game becomes the following in mixed strategies. The objective of each decision-maker i remains to minimize the variance of their payoff $var_x[p_i(a)]$ subject to the constraint that their expected payoff exceeds a prespecified satisfaction level S_i:

$$\begin{array}{ll} \text{Inf}_{a_i} & var_x[p_i(a)] \\ \text{Subject to:} & E_x[p_i(a)] \geq S_i. \end{array} \tag{1.38}$$

In case an extra randomness is involved, the game becomes

$$\begin{array}{ll} \text{Inf}_{a_i} & \text{var}_{x,\tilde{m}}[p_i(a,\omega)] \\ \text{Subject to:} & E_{x,\tilde{m}}[p_i(a,\omega)] \geq S_i \end{array} \tag{1.39}$$

This is a MFTG with coupled action constraints. The mean-field term is through the variance $\text{var}_{x,\tilde{m}}$

1.1.2 Action-Coupled Constrained Mean-Field-Type Games

One-shot MFTGs with action-coupled constraints constitute a subclass of MFTGs that combine payoffs with joint action-based limitations or rules. These games involve decision-makers taking actions within a specified space, and the constraints are directly linked to these actions in a coupled manner.

Static Action-Coupled Constrained MFTGs
A game $\mathcal{G}_c = (\mathcal{I}, \mathcal{X}, (\mathcal{A}_i, \mathcal{X}_i, E[L_i])_{i\in\mathcal{I}})$, with the constraints

$$\begin{cases} |\mathcal{I}| \geq 2, \ \mathcal{X} \subset \prod_{j\in\mathcal{I}} \mathcal{A}_j, \\ \mathcal{X}_i : \prod_{j\neq i} \mathcal{A}_j \to 2^{\mathcal{A}_i}, \\ a_{-i} = (a_j)_{j\in\mathcal{I}\setminus\{i\}} \mapsto \mathcal{X}_i(a_{-i}), \\ a_i \in \mathcal{X}_i(a_{-i}) = \{a_i' \in \mathcal{A}_i \mid (a_i', a_{-i}) \in \mathcal{X}\}, \end{cases} \tag{1.40}$$

and the mean-field-type instantaneous payoff functionals

$$\begin{cases} L_i : \mathcal{X} \times \mathcal{P}(\mathcal{X}) \to \mathbb{R}, \\ (a, D_a) \mapsto L_i(a, D_a), \\ E[L_i(a, D_a)] = \int_{b\in\prod_{j\in\mathcal{I}} \mathcal{A}_j} L_i(b, D_a) D_a(db), \end{cases} \tag{1.41}$$

is called a static MFTG with action-coupled constraints.

Here, the action spaces $\mathscr{A}_j$ can be seen as orthogonal constraints (independent of the other decision-maker choices). $\mathscr{X}_i$ is a coupled constraint set of decision-maker i, and $\mathscr{X}$ is the set of (joint) action constraints. The sets are assumed to be non-empty.

Remark 1.2 As the MFTG works with the probability distribution D,, we can also define distributional constraints such as

- Orthogonal distributional constraints: $\{D_i|\ \hat{g}_i(D_i) \leq 0\}$, with a certain function $\hat{g}_i$. Another example is $\{D_i|\ d_{g_i}(D_i, \tilde{m}_i) \leq \rho_i\}$, where $\rho_i \geq 0$, $\tilde{m}_i$ is a measured data distribution and d_g could be a Wasserstein distance of a certain order, a g_i-divergence, a g_i-Bregman discrepancy between two measures.
- Coupled distributional constraints: $\{D|\ g(D) \geq 0\}$ or $\hat{X}_i(D_{-i}) = \{D_i' \in \mathscr{P}(\mathscr{X}_i) \mid (D_i', D_{-i}) \in \hat{X}\}$ for a certain distribution set $\hat{X}$.

See Sect. 1.2.1 on distributional robustness for more details.

1.1.2.1 A Basic Constrained Game

In order to highlight the differences between the standard constrained deterministic static games, and the constrained static MFTGs as formulated above, we present an illustrative example showing the mean-variance terms appearing under the MFTG framework.

- There are $I \geq 2$ decision-makers. The set of decision-makers is denoted by $\mathscr{I} = \{1, \ldots, I\}$.
- Decision-maker $i \in \mathscr{I}$ has an action space which is the set of nonnegative real numbers.
- The actions are coupled through the constraints:

$$x_i \geq 0,\ \forall\, i \in \mathscr{I} \text{ and } \sum_{j \in \mathscr{I}} x_j \geq 1. \tag{1.42}$$

- Decision-maker $i \in \mathscr{I}$ has a cost function $l_i : \mathbb{R}^I \to \mathbb{R}$ given by

$$l_i(x) = (x_i + 1)^2 + \sum_{j \in \mathscr{I}\setminus\{i\}} x_j^2. \tag{1.43}$$

Decision-maker $i \in \mathscr{I}$ is seeking to minimize the cost function $l_i(x)$ subject to some constraints as follows:

$$\begin{array}{l} \min_{x_i \in \mathbb{R}_+} l_i(x) \\ \text{s.t. } x_i \geq 0, \quad \text{and} \quad \sum_{j \in \mathscr{I}} x_j \geq 1. \end{array} \tag{1.44}$$

We would like to investigate the best reaction of a decision-maker given the strategy of the other decision-makers.

Observe that the set of constraints (1.42) is not empty as it contains the simplex of dimension $\mathscr{I}-1$. It is a convex set. However, it is not a compact set as it is unbounded. The non-compactness of the constraint set does not create any difficulty, because the cost function l_i is coercive, i.e., $l_i(x) \to +\infty$ as $x_i \to +\infty$.

We compute the best-response strategy for decision-maker $i \in \mathscr{I}$ under the constraints in (1.42).

The constrained best-response set for decision $i \in \mathscr{I}$ is

$$\mathrm{BR}_{i,0}\left(\{x_j\}_{j\in\mathscr{I}\setminus\{i\}}\right) = \left\{\max\left(0, 1 - \sum_{j\in\mathscr{I}\setminus\{i\}} x_j\right)\right\}. \tag{1.45}$$

Using the constrained best-response strategy, one can introduce the concept of constrained Nash equilibrium (CNE), defined next.

Definition 1.5 A constrained Nash equilibrium (CNE) is a profile $(x_i)_{i\in\mathscr{I}}$ such that the constraints in (1.42) are satisfied, and $x_i \in \mathrm{BR}_{i,0}\left(\{x_j\}_{j\in\mathscr{I}\setminus\{i\}}\right)$, for all $i \in \mathscr{I}$.

The constrained Nash equilibrium is also referred to as generalized Nash equilibrium (GNE). According to Definition 1.5 and the solution obtained in (1.45), the set of CNE is the $(I-1)$–dimensional simplex given by

$$\Delta_{I-1} = \left\{(x_1, \ldots, x_I) \mid x_i \geq 0,\ \forall\, i \in \mathscr{I},\ \sum_{j\in\mathscr{I}} x_j = 1\right\},$$

which is a non-empty, convex, and compact set. In particular, there is a continuum of CNE of the game (1.44), which calls for selection of an efficient equilibrium as discussed next.

1.1.2.2 A Constrained Mean-Field-Type Game

In order to point out the fundamental differences between standard deterministic static games and MFTGs in the stationary regime, we modify the example presented above by incorporating a random variable $w = (w_i)_{i\in\mathscr{I}}$ in the cost function of each decision-maker, with the following mean and variance: $E[w_i] = 1$, and $\mathrm{var}(w_i) = \sigma^2$. Then, the problem in (1.44) becomes a stochastic game described as follows:

There are $I \geq 2$ decision-makers. The set of decision-makers is denoted by $\mathscr{I} = \{1, \ldots, I\}$. Let $w_i \in L^2(\mathbb{R})$ be a random variable with certain probability distribution $D_{w_i} \in \mathscr{P}(\mathbb{R})$, for all $i \in \mathscr{I}$. Decision-maker $i \in \mathscr{I}$ has an action space which is the set of random variables with finite mean and second moments. The pure strategy x_i of decision-maker $i \in \mathcal{I}$ could be correlated with the random variable w_i. The expected value of the actions are coupled through the constraints:

$$E[x_i] \geq 0, \ \sum_{j \in \mathcal{I}} E[x_j] \geq 1.$$

Decision-maker $i \in \mathcal{I}$ has a cost function $L_i : (L^2(\mathbb{R}))^{I+1} \times \mathscr{P}(\mathbb{R})^{I+1} \to \mathbb{R}$ given by

$$L_i(x, w_i, D_x, D_{w_i}) = (x_i + w_i)^2 + \sum_{j \in \mathcal{I} \setminus \{i\}} x_j^2 + \sum_{j \in \mathcal{I}} \varepsilon_{ij}(x_j - E[x_j])^2.$$

Decision-maker $i \in \mathcal{I}$ is seeking to minimize the expected cost function $E[L_i]$ subject to some constraints as follows:

$$\begin{array}{l} \min_{x_i} E\left[L_i(x, w_i, D_x, D_{w_i})\right] \\ E[x_i] \geq 0, \ \sum_{j \in \mathcal{I}} E[x_j] \geq 1. \end{array} \tag{1.46}$$

Remark 1.3 If $\varepsilon_{ij} = 0$ for all $i, j \in \mathcal{I}$, and $\sigma = 0$, then we recover the deterministic problem in (1.44).

Proposition 1.9 *The stochastic game problem presented in* (1.46) *is equivalent to Problem* (1.47) *given by*

$$\begin{array}{l} \min_{x_i} E\left[L_i(x, w_i, D_x, D_{w_i})\right] = (1 + \varepsilon_{ii})\text{var}(x_i) + \text{var}(w_i) + 2\text{cov}(x_i, w_i) \\ +E[x_i]^2 + E[w_i]^2 + 2E[x_i]E[w_i] + \sum_{j \in \mathcal{I} \setminus \{i\}}(1 + \varepsilon_{ij})\text{var}(x_j) + E[x_j]^2, \end{array} \tag{1.47}$$

subject to the same constraints (1.46).

$$E[x_i] \geq 0, \ i \in \mathcal{I}, \ \sum_{j \in \mathcal{I}} E[x_j] \geq 1.$$

Proof We use the decompositions $x_i = x_i - E[x_i] + E[x_i]$ and $w_i = w_i - E[w_i] + E[w_i]$ in the cost function (1.46). Then,

$$\begin{array}{l} E[(x_i - w_i)^2] = E\Big[(x_i - E[x_i])^2 + E[x_i]^2 + (w_i - E[w_i])^2 \\ +E[w_i]^2 + 2E[x_i]E[w_i] + 2(x_i - E[x_i])(w_i - E[w_i])\Big], \end{array} \tag{1.48}$$

noticing that $E[E[x_i](x_i - E[x_i])] = 0$ and $E[E[w_i](w_i - E[w_i])] = 0$ since $(x_i - E[x_i]) \perp E[x_i]$ as well as $(w_i - E[w_i]) \perp E[w_i]$; and $E[E[x_i](w_i - E[w_i])] = 0$, $E[E[w_i](x_i - E[x_i])] = 0$, thus arriving at the problem in (1.47). ∎

Let

$$L_i^1(y, w_i - E[w_i]) = (1+\varepsilon_{ii})E[y_i^2] + 2E[y_i(w_i - E[w_i])] \\ + \sum_{j\in\mathscr{I}\backslash\{i\}} (1+\varepsilon_{ij})E[y_j^2],$$

$$L_i^2(E[x]) = (E[x_i]+1)^2 + \sum_{j\in\mathscr{I}\backslash\{i\}} E[x_j]^2, \tag{1.49}$$

The stochastic problem can be decomposed into two subproblems as follows:

$$\min_{y_i} L_i^1(y, w), \text{ s.t. } E[y_i] = 0. \tag{1.50}$$

and

$$\begin{aligned} &\min_{E[x_i]} L_i^2(E[x]), \\ &\text{s.t. } E[x_i] \geq 0,\ i \in \mathscr{I},\ \textstyle\sum_{j\in\mathscr{I}} E[x_j] \geq 1. \end{aligned} \tag{1.51}$$

Note that the optimization in the first problem (1.47) can be performed by using completion of squares. For the sake of simplification in the notation, let $\xi_{ij} = (1+\varepsilon_{ij})$, arriving at

$$\begin{aligned} &\xi_{ii}E[y_i^2] + 2E[y_i(w_i - E[w_i])] + \textstyle\sum_{j\in\mathscr{I}\backslash\{i\}} \xi_{ij}E[y_j^2] \\ &= \xi_{ii}\left(E[y_i^2] + 2E\left[y_i\frac{(w_i-E[w_i])}{\xi_{ii}}\right]\right) + \textstyle\sum_{j\in\mathscr{I}\backslash\{i\}} \xi_{ij}E[y_j^2], \end{aligned} \tag{1.52}$$

where

$$E\left[y_i + \frac{w_i-E[w_i]}{1+\varepsilon_{ii}}\right]^2 = E[y_i^2] + 2E\left[y_i\frac{(w_i-E[w_i])}{\xi_{ii}}\right] + \left(\frac{w_i-E[w_i]}{1+\varepsilon_{ii}}\right)^2. \tag{1.53}$$

Then, we conclude that

$$\begin{aligned} &\xi_{ii}E[y_i^2] + 2E[y_i(w_i - E[w_i])] + \textstyle\sum_{j\in\mathscr{I}\backslash\{i\}} \xi_{ij}E[y_j^2] \\ &= \xi_{ii}E\left[y_i + \frac{w_i-E[w_i]}{\xi_{ii}}\right]^2 + E\textstyle\sum_{j\in\mathscr{I}\backslash\{i\}} \xi_{ij}y_j^2 - \frac{\sigma^2}{\xi_{ii}}. \end{aligned} \tag{1.54}$$

And by minimizing in y_i, the optimal decision variable is obtained as

$$x_i^* = E[x_i^*] - \frac{w_i-E[w_i]}{1+\varepsilon_{ii}}. \tag{1.55}$$

Since $E[x^*] \in \Delta_{I-1}$ according to the solution of the deterministic game approach in (1.45), note that the set of CNE of the MFTG is a simplex perturbed by a random noise, i.e., $x^* \in \Delta_{n-1} - \{\vec{N}_{\text{CNE}}\}$, with $N_i = -\frac{w_i-E[w_i]}{1+\varepsilon_{ii}},\ i \in \mathscr{I}$.

Remark 1.4 Notice that the cost is reduced by $\sigma^2/(1+\varepsilon_{ii})$ when the MFTG approach is adopted.

1.1.2.3 Constrained Variational Equilibrium

In the previous two problems, i.e., the deterministic and MFTGs, the set of CNE is an infinite set. It is desirable to refine the set of equilibria by introducing the concept of variational equilibrium, which is defined below.

Definition 1.6 The strategy profile $x^* = E[x^*] + y^*$ is a variational equilibrium of the MFTG in (1.46) if

$$E\left[\langle \nabla_{y_i} L_i^1(y^*, w), y_i - y_i^* \rangle\right] + E\left[\langle \nabla_{E[x_i]} L_i^2(E[x^*]), E[x_i] - E[x_i^*] \rangle\right] \geq 0, \forall i \in \mathcal{I}, \tag{1.56}$$

for every $E[y_i] = 0$, $E[x_i] \geq 0$, and $\sum_{j\in\mathcal{I}} E[x_j] \geq 1$.

Proposition 1.10 *There is at most one mean-field-type variational equilibrium x^{VI} of the game above.*

Proof We can restrict the search domain to be bounded by $[0, 10]^I$. Hence, the intersection set will be convex, compact, and non-empty. Noting that $\left(\nabla_{y_i} L^1(y, w), \nabla_{E[x_i]} L^2(E[x])\right)$ is strongly monotone operator, the variational inequality in (1.56) has at at most one solution in the intersection set. ■

Proposition 1.11 *The profile $x^{\text{VE}} = (\frac{1}{I}, \ldots, \frac{1}{I}) - \vec{N}_{\text{CNE}}$ is a variational equilibrium of the MFTG.*

Proof We directly verify that the strategic profile given by $x^{\text{VE}} = (1/I, \ldots, 1/I) - \vec{N}_{\text{CNE}}$ is a variational equilibrium, since x^* satisfies the inequality in (1.56). ■

We define a most efficient CNE as a CNE with the minimum total sum-cost $\sum_{i\in\mathcal{I}} E[L_i(x^*)]$.

Proposition 1.12 *The variational equilibrium of the MFTG is the most efficient CNE.*

Proof Observe that

$$\begin{cases} \inf_{x^*\in\text{CNE}} \sum_{i\in\mathcal{I}} E[L_i(x^*)] \\ = \inf_{y\in\{\vec{N}_{\text{CNE}}\}} \sum_{i\in\mathcal{I}} L_i^1(y, w - E[w]) + \inf_{E[x]\in\Delta_{n-1}} \sum_{i\in\mathcal{I}} L_i^2(E[x^*]). \end{cases} \tag{1.57}$$

Taking the second term from the right-hand side in (1.57) yields

$$\begin{aligned} &\inf_{E[x]\in\Delta_{I-1}} \sum_{i\in\mathcal{I}} \left\{\sum_{j\in\mathcal{I}} E[x_j]^2 + 2E[x_i] + 1\right\} \\ &= \inf_{E[x]\in\Delta_{I-1}} I \sum_{j\in\mathcal{I}} E[x_j]^2 + 2 + I, \end{aligned} \tag{1.58}$$

and $E[x^*] \in \arg\min_{E[x]\in\Delta_{I-1}} \sum_{j\in\mathscr{I}} E[x_j]^2$. Now, for a concave function f, applying Jensen's inequality, we have $f\left(\frac{1}{I}\sum_{i=1}^{I} x_i\right) \geq \frac{1}{I}\sum_{i=1}^{I} f(x_i)$. Let $f(x) = -x^2$; then $-\left(\frac{1}{I}\right)^2 \geq -\frac{1}{I}\sum_{i=1}^{I} x_i^2$, and $\frac{1}{I} \leq \sum_{i=1}^{I} x_i^2$, showing that $E[x^{\mathrm{VI}}] = \left(\frac{1}{I}, \ldots, \frac{1}{I}\right)$ minimizes $\sum_{i\in\mathscr{I}} L_i^2(E[x^*])$. ■

Proposition 1.13 *The most efficient MFT-CNE does not coincide with the global optimum [102] in the stochastic case as $\vec{N}_{\mathrm{global}} \neq \vec{N}_{\mathrm{CNE}}$, where*

$$\begin{aligned} \left(\vec{N}_{\mathrm{global}}\right)_i &= -\frac{w_i - E[w_i]}{\delta_i}, \\ \left(\vec{N}_{\mathrm{CNE}}\right)_i &= -\frac{w_i - E[w_i]}{1+\varepsilon_{ii}}, \end{aligned} \tag{1.59}$$

and $\delta_i = \sum_{j\in\mathscr{I}}(1+\varepsilon_{ji})$.

Proof Taking the first term from the right-hand side in (1.57) yields

$$\begin{aligned} &\inf_{y\in\{\vec{N}_{\mathrm{CNE}}\}} \sum_{i\in\mathscr{I}} \left\{\sum_{j\in\mathscr{I}}(1+\varepsilon_{ij})E[y_j^2] + 2E[y_i(w_i - E[w_i])]\right\}, \\ &= \inf_{y\in\{\vec{N}_{\mathrm{CNE}}\}} \sum_{j\in\mathscr{I}} \left(\sum_{i\in\mathscr{I}}(1+\varepsilon_{ij})\right) E[y_j^2] + 2\sum_{i\in\mathscr{I}} E[y_i(w_i - E[w_i])], \\ &= \inf_{y\in\{\vec{N}_{\mathrm{CNE}}\}} \sum_{j\in\mathscr{I}} \delta_j \left(E[y_j] + \frac{w_i - E[w_i]}{\delta_j}\right)^2 - \sum_{j\in\mathscr{I}} \frac{\sigma^2}{\delta_j}, \end{aligned} \tag{1.60}$$

completing the proof. ■

1.1.2.4 Potential Mean-Field-Type Games

Figure 1.1 illustrates the gap between $\mathrm{var}\left(x^{\mathrm{VI}} = x^{\mathrm{best-CNE}}\right)$ and $\mathrm{var}\left(x^{\mathrm{Global}}\right)$ as different parameters $\varepsilon_{ij} = \varepsilon$, for all $(i, j) \in \mathscr{I}^2$, and number of decision-makers I change for a fixed variance $\sigma = 1$.

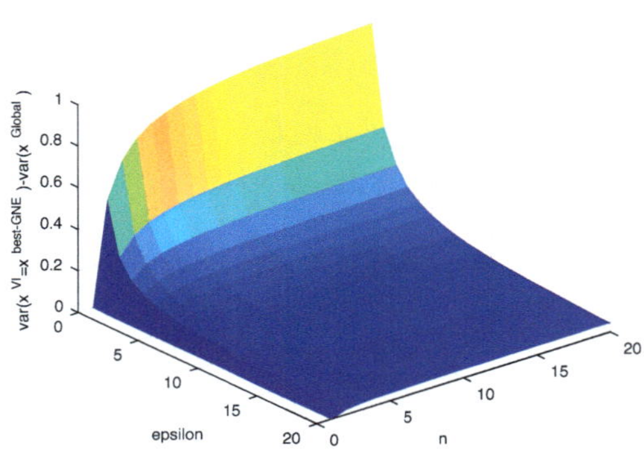

Fig. 1.1 Gap between $\mathrm{var}\left(x^{\mathrm{VI}} = x^{\mathrm{best-CNE}}\right)$ and $\mathrm{var}\left(x^{\mathrm{Global}}\right)$ as different parameters $\varepsilon_{ij} = \varepsilon$, for all $(i, j) \in \mathscr{I}^2$, and number of decision-makers I change for a fixed variance $\sigma = 1$

Extending the potential games definition from the deterministic counterpart, the constrained MFTG is potential if the objectives of all the decision-makers can be represented by means of a unique potential function. Thus, let us consider the following constrained potential game problem:

$$\begin{array}{ll} \min_{x_i} & E\left[P^1(x - E[x], w)\right] + P^2(E[x]), \\ \text{s.t.} & E[x_i] \geq 0, \text{ and } \sum_{j \in \mathcal{N}} E[x_j] \geq 1, \end{array} \tag{1.61}$$

where $P^1(x - E[x], w - E[w]) = \sum_{i \in \mathcal{I}} (1 + \varepsilon_{ii}) \text{var}(x_i) + 2 \sum_{i \in \mathcal{I}} \text{cov}(x_i, w_i)$, and $P^2(E[x], E[w]) = \sum_{i \in \mathcal{I}} (E[x_i] + E[w_i])^2$. One can verify that the functions P^k solve

$$\begin{array}{ll} \nabla_{E[x_i]} P^2(E[x], \bar{w}) & = \nabla_{E[x_i]} L_i^2(E[x], \bar{w}), \\ \nabla_{y_i} P^1(x - E[x], w - E[w]) & = \nabla_{y_i} L_i^1(x - E[x], w - E[w]). \end{array} \tag{1.62}$$

Therefore, the MFTG can be solved by optimizing the single potential functions $P^1(x - E[x], w)$ and $P^2(E[x])$ satisfying the given constraints.

1.1.2.5 Efficiency Analysis

According to Proposition 1.13, none of the mean-field-type equilibrium provides the same efficiency of the global optimum in the stochastic case [102]. Therefore, we are interested in analyzing the efficiency of the CNE with respect to the global optimum as different parameters change in the game. Then, we analyze how the efficiency changes under four specific scenarios, i.e., Variation in the variance, Variation in the ε-parameters, Variation in the number of decision-makers, and Variation in connectivity of the graphs. Let Φ denote a risk key performance index (KPI) that computes the gap between the variance for the best CNE and the variance under the global solution as follows:

$$\begin{aligned} \Phi &= \text{var}\left(x^{\text{VI}} = x^{\text{best-CNE}}\right) - \text{var}\left(x^{\text{Global}}\right), \\ &= \sigma^2 \sum_{i \in \mathcal{I}} \left(\frac{1}{1+\varepsilon_{ii}} - \frac{1}{1+\varepsilon_{ii} + \sum_{j \in \mathcal{I} \setminus \{i\}} (1 + \varepsilon_{ij})} \right). \end{aligned}$$

Next, we analyze how the risk KPI is affected by different parameters in the strategic game according to (1.63):

1.1.2.6 Variations in the Variance

It can be seen from (1.63) that the risk key performance index Φ is directly proportional and unbounded with respect to the variance σ of the random variable, i.e., the gap between the variance obtained with the variational equilibrium and the global solution increases as the variance of the noisy signals w_i, for all $i \in \mathcal{I}$, increase.

1.1.2.7 Variations in the ε-Parameters

To analyze the risk key performance index Φ under the variation of the ε-parameter, let us fix the variance of the random variable, i.e., $\sigma = 1$, and let $\varepsilon_{ij} = \varepsilon$, for all $i, j \in \mathscr{I}$. Figure 1.1 shows the behavior of Φ. It can be seen that the existing gap between the variance for the best CNE and under the global solution vanish as ϵ increasing, independently of the number I of decision-makers.

1.1.2.8 Variations in the Number of Decision-Makers

To analyze the risk key performance index Φ as the number of decision-makers changes, let us fix $\sigma = 1$ and $\varepsilon_{ij} = \varepsilon$, for all $i, j \in \mathscr{I}$. It can be seen in Fig. 1.1 that, different from the trend of the changes over ε, the key performance index Φ increases as the number of decision-makers increases and has a limit given by 1.

1.1.2.9 Variations in Connectivity Under Graphs

Let $\mathscr{G} = (\mathscr{I}, \mathscr{E})$ be a directed graph where $\mathscr{E} \subseteq \{(i, j)|\ i, j \in \mathscr{I}\}$ denotes the set of links representing how decision-makers are connected to others, more precisely, (i, j) means that the decision-maker $i \in \mathscr{I}$ incorporates the variance of the decision-maker $j \in \mathscr{I}$ within its cost function $L_i^1(y, w)$. Thus, the set of links are $\mathscr{E} = \{(i, j)|\varepsilon_{ij} \neq 0\}$. Therefore, notice that the existing gap between the variance for the best CNE and under the global solution behaves similarly as with the number of decision-makers I but now with the number of links. Indeed, notice that the parameter δ_i in $\vec{N}_{\text{global}} = -(w_i - E[w_i])/\delta_i$ becomes $\delta_i = (1 + \varepsilon_{ii}) + \sum_{j \in \mathscr{N}_i}(1 + \varepsilon_{ji})$ being $\mathscr{N}_i = \{j|\varepsilon_{ji} \neq 0\}$.

1.1.2.10 Learning Variational Equilibria

We design a learning algorithm that converges to a MFT-CNE. More precisely, it is desirable to guarantee converge to the efficient CNE given by the MFT-VE. To this end, consider the following learning dynamics:

$$\begin{cases} i \in \mathscr{I}, \\ y_i(0) \text{ given} \\ \dot{y}_i = -\frac{\partial}{\partial y_i} P^1(y, w - E[w]), \\ E[\dot{x}_i] = -E[x_i]\left[\frac{\partial}{\partial E[x_i]} P^2(E[x], E[w]) - \sum_{j \in \mathscr{I}} E[x_j] \frac{\partial}{\partial E[x_i]} P^2(E[x], E[w])\right], \\ E[x_i(0)] > 0, \forall i \in \mathscr{I}, \text{ and } \sum_{j \in \mathscr{I}} E[x_j(0)] = 1, \end{cases} \tag{1.63}$$

which converges to the variational equilibrium given by $x^* = y^* + E[x^*]$. Let us consider the MFTG in (1.46) for two decision-makers. Figure 1.2 shows the evolution of the strategies x_1 and x_2 under the learning algorithm presented in (1.63). We can see the convergence of the trajectories to the MFT-VE given by $x^{\text{best}-\text{CNE}} = x^{\text{VI}} = (1/2, 1/2)$.

1.1.3 Constrained Mean-Field-Type Game Model

Based on the motivating example of the previous section, we propose a class of constrained MFTGs in stationary regime. We present some analysis and some learning algorithms for the variational mean-field-type equilibria.

- Decision-maker $i \in \mathcal{I}$ has a cost function $L_i : \mathbb{R}^I \times L^2(\mathbb{R}^I) \times \mathcal{P}(\mathbb{R}^I) \to \mathbb{R}$ given by

$$L_i(x, w, D_x, D_w) = L_i^1(x - E[x], w - E[w]) + L_i^2(E[x]),$$

Each decision-maker is interested in minimizing its own mean-field-type cost function denoted by $EL_i(x, w, D_x, D_w)$, for all $i \in \mathcal{I}$, which can be split into two parts, i.e.,

$$\begin{cases} L_i(x, w, D_x, D_w) = L_i^1(y, w) + L_i^2(E[x]), \\ x = y + E[x], \end{cases} \tag{1.64}$$

where $y, x, w \in \mathbb{R}^I$, with x being the strategy profile of all decision-makers, and w a random variable with certain distribution D_w. We consider different types of constraints:

$$\begin{cases} \min_{y_i, E[x_i]} E\left[L_i^1(y, w) + L_i^2(E[x])\right], \\ \text{s.t. } G \cdot E[x] \leq g, \ x = y + E[x], \end{cases} \tag{1.65}$$

where $G \in \mathbb{R}^{b \times n}$ and $g \in \mathbb{R}^b$ are used to describe b inequality constraints.

Alternatively, one can consider the following related constraints of mean-field-type:

$$\begin{cases} \min_{y_i, E[x_i]} E\left[L_i^1(y, w) + L_i^2(E[x])\right], \\ \text{s.t. } \sum_{j \in \mathcal{I}} E[x_j] = m, \\ 0 \leq \underline{x}_j \leq E[x_j] \leq \bar{x}_j, \forall j \in \mathcal{I}, \end{cases} \tag{1.66}$$

where $m \in \mathbb{R}_{>0}$.

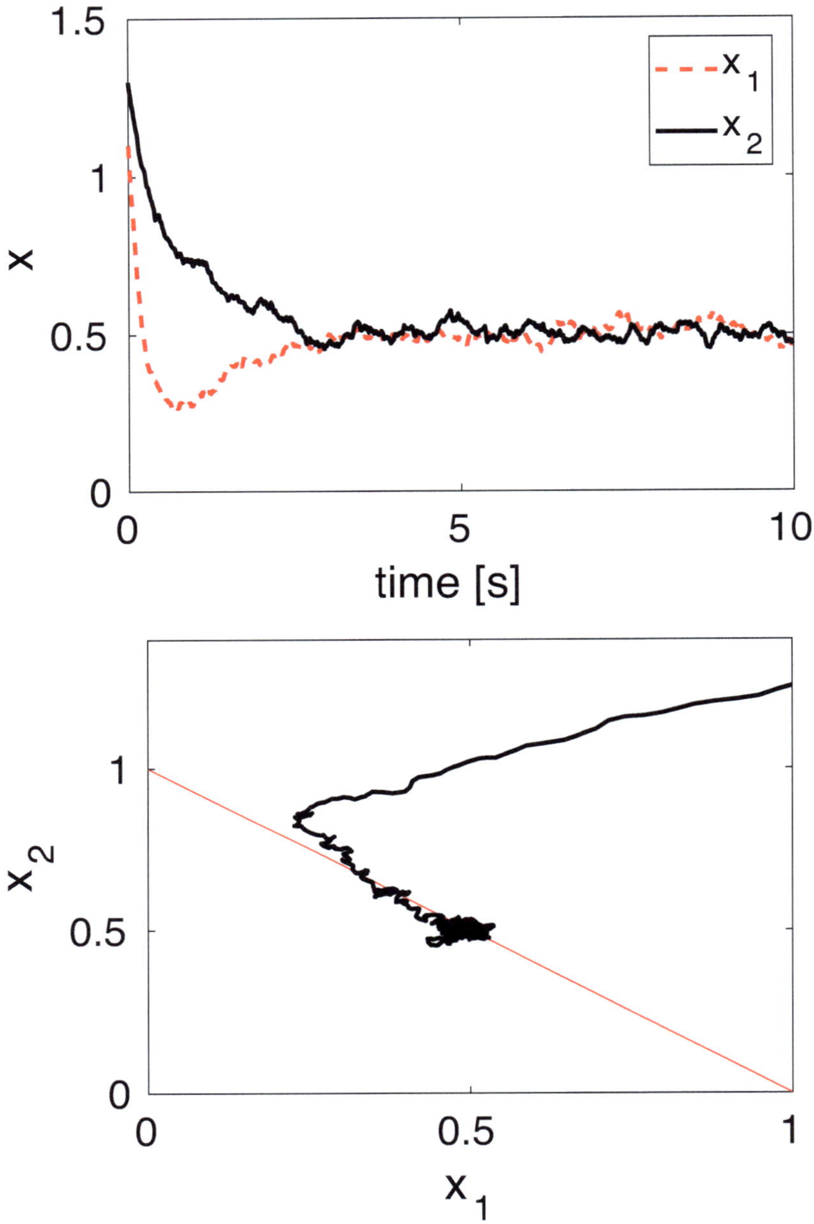

Fig. 1.2 Evolution of the variable x under the learning algorithm in (1.63) for the constrained MFTG presented in (1.46) with $I = 2$

Finally, the MFTG problem has the same constraints as in (1.46), i.e.,

$$\begin{cases} \min_{y_i, E[x_i]} E\left[L_i^1(y, w) + L_i^2(E[x])\right], \\ \text{s.t. } \sum_{j\in\mathscr{I}} E[x_j] = m, \\ E[x_j] \geq 0, \quad j \in \mathscr{I} \end{cases} \tag{1.67}$$

In Sect. 1.1.2, we introduced the concepts of mean-field-type constrained Nash equilibrium and mean-field-type variational equilibrium for the specific stochastic game in (1.46). Next, we present the definitions of the same concepts, but in general terms corresponding to problem (1.65).

Definition 1.7 (Mean-Field-Type Best-Response Strategy (MTG-BR)) A strategy profile is said to be feasible if it satisfies the constraints. Any feasible strategy

$$\left(x_1, \ldots, x_{i-1}, x_i^* = E[x_i^*] + y_i^*, x_{i+1}, \ldots, x_I\right)$$

such that

$$\begin{cases} y_i^* \in \arg\min_{y_i} E[L_i^1(y, w)], \\ E[x_i^*] \in \arg\min_{E[x_i]} L_i^2(E[x]), \end{cases} \tag{1.68}$$

satisfying constraints in either (1.65), (1.66), or (1.67), for all $i \in \mathscr{I}$, is a mean-field-type best response of the decision-maker $i \in \mathscr{I}$, denoted by $\text{BR}_i(\{x_j\}_{j\in\mathscr{I}\setminus\{i\}})$, against the strategies $x_{-i} = E[x_{-i}] + y_{-i}$ selected by the other decision-makers $\mathscr{I} \setminus \{i\}$.

The set of mean-field-type constrained Nash equilibrium (MFT-CNE) is defined by using the MFT-BR as shown next.

Definition 1.8 (Mean-Field-Type Constrained Nash Equilibrium) Any feasible strategy $x^* = E[x^*] + y^*$ such that $x_i^* \in \text{BR}_i(\{x_j^*\}_{j\in\mathscr{I}\setminus\{i\}})$, for all $i \in \mathscr{I}$, is a mean-field-type constrained Nash equilibrium.

As observed in Sect. 1.1.2, there can be multiple MFT-CNEs. One way to refine this set in the differentiable case is with variational MFTG equilibrium, defined next.

Definition 1.9 (Mean-Field-Type Variational Equilibrium) A strategy profile $x^* = E[x^*] + y^*$ is a mean-field-type variational equilibrium (MFT-VE) of the MFTG in (1.46) if it is feasible and

$$\begin{aligned} &E\left[\langle \nabla_{y_i} L_i^1(y^*, w), y_i - y_i^* \rangle\right] \\ &+E\left[\langle \nabla_{E[x_i]} L_i^2(E[x^*]), E[x_i] - E[x_i^*] \rangle\right] \geq 0, \end{aligned} \tag{1.69}$$

for all $i \in \mathscr{I}$ for every $E[y_i] = 0, \; G \cdot E[x] \geq g$.

Definition 1.10 Let $I \geq 1$. An operator $F : \mathbb{R}^I \to \mathbb{R}^I$ is strictly monotone if the following inequality holds

$$\langle x - y, F(x) - F(y) \rangle > 0, \forall x \neq y.$$

Note that in one dimension, strict monotonicity is equivalent to $x > y$ implying $F(x) > F(y)$ for all $x \neq y$.

Strictly monotone operators are of interest, because they provide a very important result in equilibrium seeking theory. The variational inequality $\langle x - x^*, F(x^*)\rangle \geq 0, \forall x$ has at most one solution if F is a strictly monotone operator. To prove this statement, first observe that if two elements x_1 and x_2 are solutions, then $\langle y - x_1, F(x_1)\rangle \geq 0, \ \forall y$ and $\langle y - x_2, F(x_2)\rangle \geq 0$. By swapping the roles of x_1 and x_2, one arrives at

$$\begin{aligned} &\langle (x_1 - x_2), -F(x_1)\rangle \geq 0, \\ &\langle x_1 - x_2, F(x_2)\rangle \geq 0 \end{aligned} \tag{1.70}$$

By summing up the two inequalities, one obtains

$$\langle x_1 - x_2, F(x_1) - F(x_2)\rangle \leq 0.$$

By strictly monotonicity of F, this says that $x_1 = x_2$.

A function $f : \mathbb{R}^I \to \mathbb{R}$ is called *quasi-convex* if for all $x, y \in \mathbb{R}^I$ and for all $\lambda \in [0, 1]$, the following inequality holds

$$f(\lambda x + (1 - \lambda)y) \leq \max\{f(x), f(y)\}.$$

Proposition 1.14 *Let L_i^1, L_i^2 be quasi-convex functions in their i-th component and jointly continuous. Assume, in addition, that these functions are coercive. Then, there exists at least one CNE of mean-field type under constraints of type (1.65) (1.66) and (1.67). Moreover, there is at most one MFT-VE if the gradients generate a strictly monotone operator.*

Proof We observe that the set of constraints is non-empty, convex, but not compact. We can reduce it to a compact set, because the objective functions are coercive. The problem is then reduced to a mean-field type game problem a under non-empty, convex, and compact constraint set. Under the assumption above, there exists at least one point $E[x]$ for the deterministic part and one for the stochastic part. Combining together, we obtain the desired result. ■

Next we introduce a class of constrained games known as constrained potential games, for which we study several learning algorithms to seek constrained Nash equilibria.

1.1.3.1 Learning Algorithms

Definition 1.11 (Constrained Potential MFTGs) A MFTG whose cost functional is given by

$$L_i(x, w_i, D_x, D_{w_i}) = L_i^1(x - E[x], w_i - E[w_i]) + L_i^2(E[x], E[w_i]),$$

is a potential MFTG if there exist functions $P^1(x - E[x], w - E[w])$ and $P^2(E[x], E[w])$ such that $\nabla_{v^k} P^k = \nabla_{v^k} L_i^k$, for all $i \in \mathscr{I}$, $k \in \{1, 2\}$ with $(v^1, v^2) = (x_i - E[x_i], E[x_i])$.

According to Definition 1.11, some of the equilibria of the MFTG can be obtained by means of the optimization of the corresponding potential functions. Thus, several learning algorithms can be implemented in the solution for this type of games with convergence guarantees when there is a strictly convex potential function. Below, we discuss the use of learning algorithms to seek mean-field-type variational equilibria [102].

Considering potential MFTGs with strictly monotone gradient potential functions, several learning algorithms can be implemented depending on the imposed constraints. We discuss here some possible algorithms that can be used in seeking MFT-VE (see Definition 1.9). Note that if the potential function is strongly convex and the domain is unconstrained, then any gradient-based learning algorithm is suitable to seek the MFTG equilibrium [102]. The issue here comes from the constraints in (1.65). Then, the constrained optimization problem in (1.65) can be solved by means of the dual decomposition algorithm with separable objectives. To this end, let us introduce the Lagrangian function, i.e., $\mathscr{L}(y, w, E[x], \mu) = P^1(y, w - E[w]) + P^2(E[x], E[w]) + \mu^\top (GE[x] - g)$. The dual decomposition algorithm then consists of

$$\begin{aligned} &\hat{y} \in \arg\min_y P^1(y, w - E[w]), \\ &\hat{x} \in \arg\min_{E[x]} P^2(E[x], E[w]) + \mu^\top G E[x], \\ &\mu^+ = \min\left\{\mu + \varrho\left(G\hat{x} - g\right), 0\right\}, \end{aligned} \tag{1.71}$$

with $\varrho > 0$, which converges to the mean-field-type variational equilibrium (MFT-VE), since the potential functions $P^1(y, w - E[w])$ and $P^2(E[x], E[w])$ are assumed to be strictly convex.

On the other hand, when considering an equality constraint, there are some suitable alternative algorithms that have been mainly designed for solving resource allocation problems. Thus, the problem in (1.66) can be addressed by using the constrained evolutionary dynamics, e.g., using the Smith-replicator dynamics given by

$$\begin{aligned} &\dot{y}_i = -\frac{\partial}{\partial y_i} P^1(y, w - E[w]), \\ &E[\dot{x}_i] = \\ &-[\hat{x}_i]_+ \textstyle\sum_{j\in\mathscr{I}} [\check{x}_j]_+ \left[\frac{\partial P^2(E[x],E[w])}{\partial E[x_i]} - \frac{\partial P^2(E[x],E[w])}{\partial E[x_j]}\right]_+ \\ &+[\check{x}_i]_+ \textstyle\sum_{j\in\mathscr{I}} [\hat{x}_j]_+ \left[\frac{\partial P^2(E[x],E[w])}{\partial E[x_j]} - \frac{\partial P^2(E[x],E[w])}{\partial E[x_i]}\right]_+, \\ &\underline{x} \le E[x(0)] \le \bar{x}, \text{ and } \textstyle\sum_{j\in\mathscr{I}} E[x_j(0)] = m, \end{aligned}$$

where $[\cdot]_+ = \min(0, \cdot)$, $\hat{x} = \bar{x} - E[x]$, $\check{x} = E[x] - \underline{x}$. This algorithm converges to the MFT-VE, since the potential functions $P^1(y, w - E[w])$ and $P^2(E[x], E[w])$

are assumed to be strictly convex. Finally, the example in (1.63) consists of a gradient system for an unconstrained optimization problem together with the replicator dynamics, which is a gradient algorithm that evolves within an invariant set forcing satisfaction of the constraints imposed in (1.67). In fact, the problem in (1.67) can be addressed by means of any evolutionary dynamics algorithm either centralized or distributed, e.g., the distributed Smith dynamics

$$
\begin{aligned}
&\dot{y}_i = -\frac{\partial}{\partial y_i} P^1(y, w - E[w]),\\
&E[\dot{x}_i] = -\sum_{j\in\mathscr{S}} E[x_j]\left[\frac{\partial P^2(E[x],E[w])}{\partial E[x_i]} - \frac{\partial P^2(E[x],E[w])}{\partial E[x_j]}\right]_+\\
&+E[x_i]\sum_{j\in\mathscr{S}}\left[\frac{\partial P^2(E[x],E[w])}{\partial E[x_j]} - \frac{\partial P^2(E[x],E[w])}{\partial E[x_i]}\right]_+,\\
&E[x_i(0)] > 0, \forall i \in \mathscr{I}, \text{ and } \sum_{j\in\mathscr{I}} E[x_j(0)] = m,
\end{aligned}
$$

where $[\cdot]_+ = \min(0, \cdot)$, which converges to the MFT-VE, since the potential functions $P^1(y, w - E[w])$ and $P^2(E[x], E[w])$ are assumed to be strictly convex.

1.1.3.2 Equilibrium Under Migration Constraints

In this section, we assume that each decision-maker i has a migration incentive or cost $\eta_{i,ll'}$ to move from component (i, l) to (i, l') with l' within $\mathscr{N}_{il}$. When $\eta_{i,ll'} > 0$, there is a migration incentive for i from l and l'. When $\eta_{i,ll'} < 0$, it becomes a migration cost. When $\eta = 0$ and $\mathscr{N}_{il}$ is the set of all choices of i, then we recover the classical mean-field-type equilibrium. Next we define a mean-field-type equilibrium with migration incentive/cost. A feasible strategic profile $x^* = E[x^*] + y^*$ is an equilibrium under migration incentive/cost η of the MFTG in (1.46) if

$$
\begin{aligned}
&E[x^*_{il}] > 0 \implies\\
&\begin{cases}
E\left[\nabla_{y_{il}} L^1_i(y^*, w)\right] = \min_{l'\in\mathscr{N}_{il}} E\left[\nabla_{y_{il'}} L^1_i(y^*, w)\right],\\
\nabla_{E[x_{il}]} L^2_i(E[x^*]) + \eta_{i,ll} = \min_{l'\in\mathscr{N}_{il}} \nabla_{E[x_{il'}]} L^2_i(E[x^*]) + \eta_{i,ll'},
\end{cases}
\end{aligned}
$$

for every decision-maker $i \in \mathscr{I}$. This type of solution concepts has plenty of applications. It includes the following:

- Cost with $\eta_{ij} > 0$ (i) electricity transmission and distribution cost from the point of production to the end user, (ii) pricing of electric vehicle charging station in smart cities, (iii) transportation and Uber pricing, and (iv) cost sharing between cities in water distribution networks,
- Benefit with $\eta_{ij} < 0$: (i) smart city planning with incentive/benefit for the citizens to join specific areas and (ii) mechanism design for cloud servers in blockchain-based networks (validation task delegation).

This equilibrium has three types of constraints: strategy constraint on the $E[x]$, choice movement constraint $\mathcal{N}_{i,l}$, and incentive/cost for a change of choice η.

Note that this equilibrium can be reformulated as an equilibrium of a modified game with cost functional $\tilde{L}_i^2 = L_i^2 + \sum_{l,l'} \eta_{i,ll'}\tilde{x}_{i,ll'}$ where $x_{il} = \sum_{l'} \tilde{x}_{i,l'l}$. It then follows that

$$\nabla_{E[\tilde{x}_{i,ll}]}\tilde{L}_i^2(E[x^*]) = \min_{l' \in \mathcal{N}_{il}} \nabla_{E[\tilde{x}_{ill'}]}\tilde{L}_i^2(E[x^*]),$$

leading to an equilibrium similar to that in Definition 1.9 with the neighborhood constraint $\mathcal{N}_{il}$.

To recap, we have thus introduced here a class of constrained MFTGs under the stationary regime. These kinds of games entail distribution-dependent cost through the first moment and could take into account both expected value and variance terms in the payoffs of the decision-makers. We have presented concepts such as mean-field-type equilibrium, mean-field-type variational equilibrium, and mean-field-type potential games, and we have introduced some learning algorithms to converge to the best Nash equilibrium.

1.2 Transformation of Variance-Dependent Loss Functions

Consider a set $\mathcal{I}$ of decision-makers with cardinality $|\mathcal{I}| \geq 2$. Decision-maker $i \in \mathcal{I}$ has its own strategy space $\mathcal{X}_i$ (which is non-empty). Let $(\Omega, \mathcal{F}, \tilde{m})$ be a probability space. Decision-maker $i \in \mathcal{I}$ has its own loss functional $E_{\tilde{m}}[L_i]$ with $L_i : \prod_{j \in \mathcal{I}} \mathcal{X}_j \times \Omega \to \mathbb{R}$ expressed as $L_i(x, \omega)$. It is assumed that $E_{\tilde{m}}[L_i] < +\infty$ for every $i \in \mathcal{I}$.

A corresponding risk-neutral game is given by

$$G_{neu} = (\mathcal{I}, (\mathcal{X}_i, E[L_i])_{i \in \mathcal{I}}, \tilde{m}).$$

The risk-neutral equilibrium problem of G_{neu} is formulated as follows:

$$\begin{aligned} & i \in \mathcal{I}, \\ & \inf_{x_i \in \in \mathcal{X}_i} E_{\tilde{m}}[L_i(x, \omega)] \end{aligned} \tag{1.72}$$

Note that $\tilde{m} \mapsto E_{\tilde{m}}[L_i(x, \omega)]$ is linear, for every x and for $i \in \mathcal{I}$. This linearity (in the measure) property will play a crucial role in transforming the performance functional. One of the challenges with G_{neu} is when the exact measure $\tilde{m}$ is unknown to the decision-makers. In this case, we propose alternative approaches based on a distributional uncertainty set around known data.

1.2.1 Distributionally Robust Risk-Neutral

In many cases, the true distribution of the underlying data over Ω is not available or unknown to the decision-maker. We formulate the problem from the variations $\tilde{m}$ around the data-driven measured or observed distribution m.. For convex and proper function f,, we define f-divergence as

$$D_f(\tilde{m}||m) = \int_{\omega} f\left(\frac{d\tilde{m}}{dm}\right) dm - f(1).$$

The term $f(1)$ is used for controlling the sign. The f-divergence is not a distance metric as it is not symmetric. However, it can be used to capture discrepancies between two probability measures.

For $\rho_i > 0$,, the distributional uncertainty set driven by f-divergence ball is

$$B_{\rho_i}(m) = \{\tilde{m}| \int_{\Omega} \tilde{m}(d\omega) = 1, \ D_f(\tilde{m}||m) \leq \rho_i\}.$$

Given a convex function $f : \mathscr{X} \to (-\infty, +\infty]$, its Legendre–Fenchel conjugate $f^* : \mathscr{X}^* \to (-\infty, +\infty]$ is defined as

$$f^*(y) = \sup_{x\in\mathscr{X}} \{\langle x, y\rangle - f(x)\} = -\inf_{x\in\mathscr{X}} \{f(x) - \langle x, y\rangle\}.$$

Let $(\Omega, \mathscr{F}, \tilde{m})$ is a measure space, a measurable function f, and ν an absolutely continuous measure; i.e., there exists $\rho(.) > 0$ such that $\nu(dx) = \rho(x)\tilde{m}(dx)$. Let l be a function such that e^l is $\tilde{m}$-integrable. The Legendre–Fenchel transform of the relative entropy yields

$$\sup_{\nu}\left\{\int l(x)\nu(dx) - \int \log(\rho(x))\nu(dx)\right\} = \log\left(\int_{\Omega} e^{l(\omega)}\tilde{m}(d\omega)\right)$$

$$\begin{aligned}&\sup_{\nu}\left\{\int l_i(x)\nu(dx) - \tfrac{1}{\epsilon_i}\int \log(\rho(x))\nu(dx)\right\}\\ &= \sup_{\nu}\left\{\int l_i(x)\nu(dx) - \tfrac{1}{\epsilon_i}\int \log(\tfrac{\nu(dx)}{\tilde{m}(dx)})\nu(dx)\right\} = \tfrac{1}{\epsilon_i}\log\left(\int_{\Omega} e^{\epsilon_i l_i(\omega)}\tilde{m}(d\omega)\right),\end{aligned} \tag{1.73}$$

which is the log of the expected value of exponentiated loss. For a very small ϵ_i,, one can connect the latter expression to mean variance through

$$\frac{1}{\epsilon_i}\log\left(\int_{\Omega} e^{\epsilon_i l_i(\omega)}\tilde{m}(d\omega)\right) = E_{\tilde{m}} l_i + \frac{\epsilon_i}{2}\text{var}_{\tilde{m}}[l_i] + O(\epsilon_i^2).$$

It follows from the previous observation that minimizing the log of the expected value of exponentiated loss can be seen as a distributionally robust optimization

$$\begin{aligned}&\inf_{a_i\in\mathscr{A}_i}\frac{1}{\epsilon_i}\log\left(\int_\Omega e^{\epsilon_i l_i(a,\omega)}\tilde{m}(d\omega)\right)\\&=\inf_{a_i\in\mathscr{A}_i}\sup_\nu\left\{\int l_i(x)\nu(dx)-\frac{1}{\epsilon_i}\int\log\left(\frac{\nu(dx)}{\tilde{m}(dx)}\right)\nu(dx)\right\},\end{aligned}\tag{1.74}$$

with the distributional constraint that the measure μ is absolutely continuous with respect to $\tilde{m}$.

Distributionally Robust Risk-Neutral Game
Consider a set $\mathscr{I}$ of decision-makers with $|\mathscr{I}|\geq 2$. Each decision-maker $i\in\mathscr{I}$ has its own loss functional $E_{\tilde{m}}L_i$ with $L_i:\prod_{j\in\mathscr{I}}\mathscr{X}_j\times\Omega\rightarrow\mathbb{R}$ expressed as $L_i(x,\omega)$. Further, the distributionally robust risk-neutral game is given by

$$G_{dr}=\left(\mathscr{I},(\mathscr{X}_i\times B_{\rho_i}(m),E_{\tilde{m}}L_i)_{i\in\mathscr{I}},(\rho_i)_{i\in\mathscr{I}},m\right).$$

The distributionally robust equilibrium problem of G_{dr} is formulated as follows:

$$\begin{aligned}&i\in\mathscr{I},\\&\inf_{x_i\in\mathscr{X}_i}\sup_{\tilde{m}\in B_{\rho_i}(m)}E_{\tilde{m}}L_i(x,\omega)\end{aligned}\tag{1.75}$$

We transform the distributionally robust term as follows. The distributionally robust term is given by

$$\begin{aligned}&\sup_{\tilde{m}\in B_{\rho_i}(m)}E_{\tilde{m}}L_i(x,\omega)\\&=\inf_{\lambda_i\geq 0,\mu}E_m\left\{\lambda_i(\rho_i+f(1))+\mu+\lambda_i f^*\left(\frac{L_i(x,\omega)-\mu}{\lambda_i}\right)\right\}\end{aligned}\tag{1.76}$$

Proposition 1.15 (Triality Theory) *Let l_3 a real-valued function $(\omega_1,\omega_2,\omega_3)\mapsto l_3(\omega_1,\omega_2,\omega_3)$ defined on $\prod_{k=1}^3\Omega_k$. Then, the following inequalities holds*

$$\begin{aligned}&\sup_{\omega_2\in\Omega_2}inf_{\omega_1\in\Omega_1,\omega_3\in\Omega_3}l_3(\omega_1,\omega_2,\omega_3)\leq\\&inf_{\omega_3\in\Omega_3}\sup_{\omega_2\in\Omega_2}inf_{\omega_1\in\Omega_1}l_3(\omega_1,\omega_2,\omega_3)\leq\\&inf_{\omega_1\in\Omega_1,\omega_3\in\Omega_3}\sup_{\omega_2\in\Omega_2}l_3(\omega_1,\omega_2,\omega_3),\end{aligned}\tag{1.77}$$

and similarly

$$\begin{aligned}&\sup_{\omega_1\in\Omega_1,\omega_3\in\Omega_3}inf_{\omega_2\in\Omega_2}l_3(\omega_1,\omega_2,\omega_3)\leq\\&\sup_{\omega_3\in\Omega_3}inf_{\omega_2\in\Omega_2}\sup_{\omega_1\in\Omega_1}l_3(\omega_1,\omega_2,\omega_3)\leq\\&inf_{\omega_2\in\Omega_2}\sup_{\omega_1\in\Omega_1,\omega_3\in\Omega_3}l_3(\omega_1,\omega_2,\omega_3).\end{aligned}\tag{1.78}$$

A proof of Proposition 1.15 can be obtained from the duality inequality. More details on triality theory can be found in [54, 164, 165, 167, 170, 322].

We now use triality theory (Proposition 1.15) to derive the following expression:

$$\begin{aligned}
&\inf_{x_i \in \mathscr{X}_i} \sup_{\tilde{m} \in B_{\rho_i}(m)} E_{\tilde{m}} L_i(x, \omega) \\
&= \inf_{a_i \in \mathscr{X}_i} \inf_{\lambda_i \geq 0, \mu} E_m \left\{ \lambda_i(\rho_i + f(1)) + \mu + \lambda_i f^* \left(\tfrac{L_i(x,\omega) - \mu}{\lambda_i} \right) \right\} \\
&= \inf_{(x_i, \lambda_i, \mu) \in \mathscr{X}_i \times \mathbb{R}_+ \times \mathbb{R}} E_m \hat{L}_i(x, \lambda_i, \mu, \omega),
\end{aligned} \tag{1.79}$$

where

$$\hat{L}_i(x, \lambda_i, \mu, \omega) = \lambda_i(\rho_i + f(1)) + \mu + \lambda_i f^* \left(\frac{L_i(x, \omega) - \mu}{\lambda_i} \right).$$

Based on the above, the distributionally robust game can be transformed into the following game $\hat{G}_{rdr}$

Reduced Distributionally Robust Risk-Neutral Game

Consider a set $\mathscr{I}$ of decision-makers with $|\mathscr{I}| \geq 2$. Each decision-maker $i \in \mathscr{I}$ has its own loss functional $E_m \hat{L}_i$ with $\hat{L}_i : \prod_{j \in \mathscr{I}} \mathscr{X}_j \times \mathbb{R}_+ \times \mathbb{R} \times \Omega \to \mathbb{R}$ expressed as $\hat{L}_i(x, \lambda_i, \mu, \omega)$.

The reduced distributionally robust risk-neutral game is given by

$$\hat{G}_{rdr} = \left(\mathscr{I}, (\mathscr{X}_i \times \mathbb{R}_+ \times \mathbb{R}, E_m \hat{L}_i)_{i \in \mathscr{I}}, \rho_i, m \right).$$

Further, the reduced distributionally robust equilibrium problem of $\hat{G}_{rdr}$ is formulated as follows:

$$\begin{aligned}
&i \in \mathscr{I}, \\
&\inf_{(x_i, \lambda_i, \mu) \in \mathscr{X}_i \times \mathbb{R}_+ \times \mathbb{R}} E_m \hat{L}_i(x, \lambda_i, \mu, \omega)
\end{aligned} \tag{1.80}$$

The advantage of the reduced distributionally robust game $\hat{G}_{rdr}$ is that the action space $\mathscr{X}_i \times B_{\rho_i}(m)$ is reduced to the space $\mathscr{X}_i \times \mathbb{R}_+ \times \mathbb{R}$.

1.2.2 Mean-Variance Loss

MFTG with Mean-Variance Loss
Consider a set $\mathscr{I}$ of decision-makers with $|\mathscr{I}| \geq 2$. Each decision-maker $i \in \mathscr{I}$ has its own loss functional $\text{mv}_i : \prod_{j\in\mathscr{I}} \mathscr{X}_j \times \mathbb{P}(\Omega) \to \mathbb{R}$ expressed as

$$\text{mv}_i(x, \tilde{m}) = E_{\tilde{m}}[L_i(x, \omega)] + \hat{r}_i \cdot var_{\tilde{m}}[L_i(x, \omega)].$$

Here $\hat{r}_i = \frac{r_i}{\epsilon+|L_i(x^0,\omega^0)|}$, $r_i > 0, \epsilon > 0$ with r_i being the risk-sensitivity index of decision-maker i.

The mean-variance game is given by

$$G_{mv} = (\mathscr{I}, (\mathscr{X}_i, \text{mv}_i)_{i\in\mathscr{I}}, \tilde{m}).$$

Further, the mean-variance equilibrium problem of G_{mv} is formulated as follows:

$$\begin{cases} i \in \mathscr{I}, \\ x_i \in \arg\min_{x_i'\in\mathscr{X}_i} \text{mv}_i(x_i', x_{-i}, \tilde{m}) \end{cases} \tag{1.81}$$

Note that $\tilde{m} \mapsto \text{mv}_i(x, \tilde{m})$ is nonlinear. This nonlinearity in the measure $\tilde{m}$ creates an extra challenge in the search of the best-response strategy.

The Mean-Variance as the Best Case of an Expected Loss
Let mv be a mean-variance trade-off of L, $\hat{\lambda} > 0$. Using the equality $-\sup(f) = +\inf(-f)$,, we have

$$\begin{aligned} mv &= E[L] + \hat{\lambda} \cdot \text{var}[L] \\ &= E[L] + \hat{\lambda}(E[L^2]) - \hat{\lambda}(E[L])^2 \\ &= E[L] + \hat{\lambda}E[L^2] - \hat{\lambda}\sup_{z\in\mathbb{R}}\left\{2\langle z, E[L]\rangle - z^2\right\} \\ &= E[L] + \hat{\lambda}E[L^2] + \hat{\lambda}\inf_{z\in\mathbb{R}}\left\{-2\langle z, E[L]\rangle + z^2\right\} \\ &= \inf_{z\in\mathbb{R}}\left[E[L] + \hat{\lambda}E[L^2] + \hat{\lambda}\left\{-2\langle z, E[L]\rangle + z^2\right\}\right] \\ &= \inf_{z\in\mathbb{R}}\left\{(1 - 2\hat{\lambda}z)E[L] + \hat{\lambda}z^2\right\} + \hat{\lambda}E[L^2] \\ &= \inf_{z\in\mathbb{R}}E\left[\left\{(1 - 2\hat{\lambda}z)L + \hat{\lambda}z^2\right\} + \hat{\lambda}L^2\right] \\ &= \inf_{z\in\mathbb{R}}E[\tilde{L}(z)], \end{aligned} \tag{1.82}$$

where $\tilde{L}(z) = (1 - 2z\hat{\lambda})L + \hat{\lambda}z^2 + \hat{\lambda}L^2$.

Based on the above, we transform the mean-variance game G_{mv} as follows:

$$\begin{aligned} &\inf_{x_i \in \in \mathscr{X}_i} E_{\tilde{m}}[L_i(x,\omega)] + \hat{r}_i \cdot var_{\tilde{m}}[L_i(x,\omega)] \\ &= \inf_{(x_i,z_i)\in \mathscr{X}_i \times \mathbb{R}} E_{\tilde{m}} \tilde{L}_i(x, z_i, \omega), \end{aligned} \tag{1.83}$$

where

$$\tilde{L}_i(x, z_i, \omega) = (1 - 2z_i\hat{r}_i)L_i + \hat{r}_i z_i^2 + \hat{r}_i L_i^2.$$

MFTG with Measure-Linearized Mean-Variance Loss
Consider a set $\mathscr{I}$ of decision-makers with $|\mathscr{I}| \geq 2$. Each decision-maker $i \in \mathscr{I}$ has own loss functional $E_{\tilde{m}} \tilde{L}_i(x, z_i, \omega)$ with $\tilde{L}_i : \prod_{j\in\mathscr{I}} \mathscr{X}_j \times \mathbb{R} \times \Omega \to \mathbb{R}$ expressed as $\tilde{L}_i(x, z_i, \omega)$.

The reduced mean-variance game is given by

$$\tilde{G}_{rmv} = \left(\mathscr{I}, (\mathscr{X}_i \times \mathbb{R}, E_{\tilde{m}} \tilde{L}_i)_{i\in\mathscr{I}}, \tilde{m}\right).$$

The reduced mean-variance equilibrium problem of $\tilde{G}_{rmv}$ is formulated as follows:

$$\begin{aligned} &i \in \mathscr{I}, \\ &\inf_{(x_i,z_i)\in \mathscr{X}_i \times \mathbb{R}} E_{\tilde{m}} \tilde{L}_i(x, z_i, \omega) \end{aligned} \tag{1.84}$$

The advantage of $\tilde{G}_{rmv}$ formulation is that $\tilde{m} \mapsto E_{\tilde{m}} \tilde{L}_i(x, z_i, \omega)$ is linear. The price to pay is that the action space of decision-maker i is now augmented from $\mathscr{X}_i$ to $\mathscr{X}_i \times \mathbb{R}$.

1.2.3 Distributional Robust Mean Variance of a Loss Functional

The exact probability measure $\tilde{m}$ over Ω is known to the decision-makers. Given a data set with size D,, a certain data-driven distribution m is estimated. The decision-makers consider the distributional robustness of their estimation m.

Distributionally Robust Mean-Variance MFTG
Consider a set $\mathscr{I}$ of decision-makers with $|\mathscr{I}| \geq 2$. Each decision-maker $i \in \mathscr{I}$ has own loss functional $\text{drmv}_i : (\prod_{j\in\mathscr{I}} \mathscr{X}_j) \times \mathbb{P}(\Omega) \to \mathbb{R}$ expressed as $\text{drmv}_i(x, \tilde{m})$. The distributionally robust mean-variance game is given by

(continued)

$$G_{drmv} = \left(\mathscr{I}, (\mathscr{X}_i \times B_{\rho_i}(m), \text{drmv}_i)_{i\in\mathscr{I}}, \rho_i, m\right).$$

The distributionally robust mean-variance equilibrium problem of G_{drmv} is formulated as follows:

$$\begin{array}{l} i \in \mathscr{I}, \\ \inf_{x_i\in\mathscr{X}_i} \sup_{\tilde{m}\in B_{\rho_i}(m)} \text{mv}_i(x, \tilde{m}) \end{array} \tag{1.85}$$

The challenge of G_{drmv} is that the function $\tilde{m} \mapsto \text{mv}_i(x, \tilde{m})$ is nonlinear (in the measure) and the distributional uncertainty set $B_{\rho_i}(m)$ is infinite dimensional (in general).

$$\begin{array}{l} \inf_{x_i\in\mathscr{X}_i} \sup_{\tilde{m}\in B_{\rho_i}(m)} mv(L_i(x,\omega); \tilde{m}||m) \\ = \inf_{x_i\in\mathscr{X}_i} \sup_{\tilde{m}\in B_{\rho_i}(m)} E_{\tilde{m}} L_i(x,\omega) + \hat{r}_i \text{var}_{\tilde{m}}[L_i(x,\omega)] \\ = \inf_{x_i\in\mathscr{X}_i} \sup_{\tilde{m}\in B_{\rho_i}(m)} \inf_{z_i\in\mathbb{R}} E_{\tilde{m}}\left[(1-2z_i\hat{r}_i)L_i(x,\omega) + \hat{r}_i z_i^2 + \hat{r}_i L_i^2(x,\omega)\right] \\ = \inf_{(x_i,z_i)\in\mathscr{X}_i\times\mathbb{R}} \sup_{\tilde{m}\in B_{\rho_i}(m)} E_{\tilde{m}} \tilde{L}_i(x, z_i, \omega) \\ = \inf_{(x_i,z_i)\in\mathscr{X}_i\times\mathbb{R}} \inf_{\lambda_i\geq 0,\mu} E_m \left\{\lambda_i(\rho_i + f(1)) + \mu + \lambda_i f^*\left(\frac{\tilde{L}_i(x,z_i,\omega)-\mu}{\lambda_i}\right)\right\} \\ = \inf_{(x_i,z_i,\lambda_i,\mu)\in\mathscr{X}_i\times\mathbb{R}\times\mathbb{R}_+\times\mathbb{R}} E_m \hat{\text{mv}}_i(x, z_i, \lambda_i, \mu, \omega), \end{array} \tag{1.86}$$

where

$$\hat{\text{mv}}_i(x, z_i, \lambda_i, \mu, \omega) = \lambda_i(\rho_i + f(1)) + \mu + \lambda_i f^*\left(\frac{\tilde{L}_i(x, z_i, \omega) - \mu}{\lambda_i}\right).$$

$$\tilde{L}_i(x, z_i, \omega) = (1 - 2z_i\hat{r}_i)L_i + \hat{r}_i z_i^2 + \hat{r}_i L_i^2,$$

and where we have used (1.79).

Distributionally Robust Mean-Variance MFTG: Linearized Measure and f-Divergence

Consider a set $\mathscr{I}$ of decision-makers with $|\mathscr{I}| \geq 2$. Each decision-maker $i \in \mathscr{I}$ has its own loss functional $E_m \hat{\text{mv}}_i(x, z_i, \lambda_i, \mu, \omega)$ with

$$\hat{\text{mv}}_i : (\prod_{j\in\mathscr{I}} \mathscr{X}_j) \times \mathbb{R} \times \mathbb{R}_+ \times \mathbb{R} \times \Omega \to \mathbb{R}.$$

(continued)

The reduced distributionally robust mean-variance game is given by

$$\hat{G}_{drmv} = \left(\mathscr{I}, (\mathscr{X}_i \times \mathbb{R} \times \mathbb{R}_+ \times \mathbb{R}, E_m \hat{\text{mv}}_i)_{i\in\mathscr{I}}, \rho_i, m\right).$$

Further, the reduced distributionally robust mean-variance equilibrium problem of $\hat{G}_{drmv}$ is formulated as follows:

$$\begin{array}{l} i \in \mathscr{I}, \\ \inf_{(x_i, z_i, \lambda_i, \mu)\in \mathscr{X}_i\times\mathbb{R}\times\mathbb{R}_+\times\mathbb{R}} E_m \hat{\text{mv}}_i(x, z_i, \lambda_i, \mu, \omega). \end{array} \tag{1.87}$$

Thanks to the previous transformation, the function $\text{mv}_i(x, \tilde{m})$ is transformed into infimization over a linearized function in the measure $\tilde{m}$ resulting in an augmented action space. As a consequence of this step, the action space $\mathscr{X}_i \times B_{\rho_i}(m)$ is augmented to be $\mathscr{X}_i \times \mathbb{R} \times B_{\rho_i}(m)$ for decision-maker i. As a second step, the action space $\mathscr{X}_i \times \mathbb{R} \times B_{\rho_i}(m)$ is reduced to $\mathscr{X}_i \times \mathbb{R} \times \mathbb{R}_+ \times \mathbb{R}$ thanks to the Fenchel transformation above.

1.2.4 Variance-Aware Bregman Learning

Let g be a strictly convex and twice continuously differentiable function $g : \mathscr{X} \to \mathbb{R}$. The Bregman discrepancy $d_g : \mathscr{X} \times \text{reint}(\mathscr{X}) \to \mathbb{R}$ is given by

$$d_g(y, x) = g(y) - g(x) - \langle g_x(x), y - x\rangle,$$

with reint($\mathscr{X}$) being the relative interior of $\mathscr{X}$.

The Bregman discrepancy $d_g(y, x)$ is the difference between the value of g at the destination point y and the first-order Taylor expansion of g around x. Since g is a convex function, $d_g(y, x) \geq 0$. Moreover, $d_g(y, x) > 0$, $\forall x \neq y$ (by strict convexity of g). Let

$$\hat{\text{mv}}_i(x, y, \lambda, \mu) = E_m \lambda(\rho + f(1)) + \mu + \lambda f^*\left(\frac{\tilde{L}(x, y, \omega) - \mu}{\lambda}\right),$$

where

$$\tilde{L}_i(x, y, \omega) = (1 - 2y\hat{r}_i)L_i(x, y, \omega) + \hat{r}_i y^2 + \hat{r}_i L_i^2(x, y, \omega).$$

Proposition 1.16 (Regret Bound, [54, 167, 170]) *Let* $z = (x, y, \lambda, \mu)$, *and consider the following dynamics*

$$
\begin{aligned}
& z(0) = z_{00},\ \dot z(0) = z_{01},\ \ddot z(0) = z_{02}, \\
& \dddot z = -\frac{3}{t}\ddot z - (e^{\alpha} - \dot\alpha)(\ddot z + \frac{2}{t}\dot z) \\
& \quad - \frac{e^{2\alpha+\beta}}{t} g_{zz}^{-1}(z + [t + 2e^{-\alpha}]\dot z + te^{-\alpha}\ddot z) l_z(t\dot z + z),
\end{aligned} \tag{1.88}
$$

Then, we have the following bound:

$$
0 \le \hat{mv}(z(t)) - \hat{mv}(z^*) \le \frac{c_0}{t}\int_0^t e^{-\beta(s)} ds,
$$

where $c_0 = e^{\beta(0)}[\hat{mv}(z(0)) - \hat{mv}(z^*)] + d_g(z^*, z_0 + e^{-\alpha(0)}\dot z_0) > 0$ *is the initial error plus the initial Bregman discrepancy.*

1.2.4.1 Particle Swarm Variance-Aware Bregman Learning

We now focus on the accuracy bounds over the sample data $\{\tilde\omega_k \sim m, k \in \{1, \dots, N\}\}$ with $N \ge 1$ points.

Proposition 1.17 (Sample Regret bound) *Let*

$$
\hat{mv}(x, y, \lambda, \mu, \omega) = \lambda(\rho + f(1)) + \mu + \lambda f^*\left(\frac{\tilde L(x, y, \omega) - \mu}{\lambda}\right),
$$

and particle swarm-based stochastic Bregman dynamics given by

$$
\begin{aligned}
& z(0) = z_{00},\ \dot z(0) = z_{01} \\
& \tfrac{d}{dt}[g_z(z + e^{-\alpha}\dot z)] = -e^{\alpha+\beta}\tfrac{1}{N}\textstyle\sum_{k=1}^N \hat{mv}_z(x, y, \lambda, \mu, \tilde\omega_k), \\
& \tilde\omega_k \sim m, k \in \{1, \dots, N\}
\end{aligned} \tag{1.89}
$$

For each sample $(\tilde\omega_k)_{k\in\{1,\dots,N\}} \in \Omega^N$*, and convex function* $\tilde L$*, one has*

$$
0 \le \frac{1}{N}\sum_{k=1}^N \hat{mv}(z(t), \omega_k) - \hat{mv}(z_N^*, \omega_k) \le \frac{c_{0,N}}{t}\int_0^t e^{-\beta(s)} ds,
$$

where $c_{0,N} = e^{\beta(0)}[\hat{mv}(z_{0,N}) - \hat{mv}(z_N^*)] + d_g(z_N^*, z_{0,N} + e^{-\alpha(0)}\dot z_0) > 0$ *is the initial error and* $z_N^* \in \arg\min_z \frac{1}{N}\sum_{k=1}^N \hat{mv}(z, \omega_k)$.

Proof The proof follows from the inequalities in Proposition 1.16. ■

1.3 Coherent Risk Quantification

In this section, we introduce three other classes of risk measures: entropic value at risk, expectile, and extremile. They satisfy the properties of coherent risk measures defined next.

Definition 1.12 A risk measure of a random loss functional assigns a real value to it in order to determine its risk degree. A risk measure R is coherent if it satisfies the following properties:

- Translation principle: $R(L_1 + \lambda) = R(L_1) + \lambda$ for all $\lambda \in \mathbb{R}$.
- Subadditivity: $R(L_1 + L_2) \leq R(L_1) + R(L_2)$ for all L_1, L_2
- Monotonicity: If $L_1 \leq L_2$, then $R(L_1) \leq R(L_2)$
- Positive homogeneity: $R(\lambda L) = \lambda R(L)$, for all $\lambda \geq 0$,

Noncoherent Risk Measures
The expected value of L, denoted by $E[L]$ which is also the arg min of $l \mapsto E[(L-l)^2]$ is a coherent risk measure. It is risk-neutral. The variance $var(L)$ is not coherent as $var(L + \lambda) = var(L)$ for all $\lambda \in \mathbb{R}$ and $L \in L^2$. Also, the variance is not positively homogeneous. Similarly, the standard deviation is not coherent. The mean-variance, mean-standard deviation, variance-to-mean ratio, mean-to-variance ratio, Sharpe ratio, and dispersion index are not coherent risk measures. The quantile and the value at risk are not coherent risk measures. Similarly, the exponentiated return and the entropic risk measure given by

$$RS(L_i) = \frac{1}{\eta_i} \log\left(\frac{1}{\alpha_i} E_D[e^{\eta_i L_i(a_1,\ldots,a_I,D)}]\right)$$

are not coherent as they do not satisfy positive homogeneity.

Example 1.10 (Do Not Use Value-at-Risk for Diversification) The α-level value-at-risk (VaR_α) for a loss variable L is

$$\text{VaR}_\alpha(L) = \inf\{l \in \mathbb{R} : P(L > l) \leq 1 - \alpha\}$$

The subadditivity property is $\rho(X+Y) \leq \rho(X)+\rho(Y) \quad \forall X, Y$ where X and Y are random loss variables. This property ensures that **diversification reduces risk**.

The value at risk violates subadditivity. As an illustration, we consider two independent, identically distributed loans X and Y with the following loss distribution:

$$X, Y = \begin{cases} 0 & \text{with probability } 0.96 \\ 100 & \text{with probability } 0.04 \end{cases}$$

Next we compute $\text{VaR}_{0.95}$ for $\alpha = 0.95$. The individual VaR are

$$\begin{aligned} P(X \leq 0) = 0.96 \geq 0.95 &\implies \text{VaR}_{0.95}(X) = 0 \\ P(Y \leq 0) = 0.96 \geq 0.95 &\implies \text{VaR}_{0.95}(Y) = 0 \end{aligned} \tag{1.90}$$

The portfolio VaR is as follows: The combined loss $S = X + Y$ has a distribution:

$$\begin{aligned} P(S = 0) \quad &= (0.96)^2 = 0.9216 \\ P(S = 100) &= 2 \times 0.96 \times 0.04 = 0.0768 \\ P(S = 200) &= (0.04)^2 = 0.0016 \end{aligned} \tag{1.91}$$

The cumulative probabilities are

$$\begin{aligned} P(S \leq 0) \quad &= 0.9216 < 0.95 \\ P(S \leq 100) &= 0.9216 + 0.0768 = 0.9984 \geq 0.95 \end{aligned} \tag{1.92}$$

Thus

$$\text{VaR}_{0.95}(S) = 100 \quad (\text{smallest } l \text{ where } P(S \leq l) \geq 0.95)$$

By comparing portfolio VaR to sum of individual VaRs:

$$\underbrace{\text{VaR}_{0.95}(X + Y)}_{100} > \underbrace{\text{VaR}_{0.95}(X) + \text{VaR}_{0.95}(Y)}_{0+0=0}$$

This violates subadditivity $\rho(X + Y) \not\leq \rho(X) + \rho(Y)$

It follows that

Property	Expression	Result for VaR
Subadditivity	$\rho(X + Y) \leq \rho(X) + \rho(Y)$	Fails

This means that VaR ignores loss severity beyond the quantile threshold. Merging MI portfolios can appear riskier than individual positions. It discourages diversification when tail events are correlated.

VaR is not a coherent risk measure. Examples of coherent risk measures that can be used in generative MI include expected shortfall or conditional value at risk, expectile value at risk, extremile value-at-risk, and entropic value at risk.

1.3.1 Entropic Value-at-Risk MFTGs

Entropic Value at Risk Is a Coherent Risk Measure
The entropic value-at-risk function of i with confidence level $1-\alpha_i,\ \alpha_i \in (0,1]$ is given by

$$\hat{L}_i(D) = \inf_{\eta_i>0}\left\{\frac{1}{\eta_i}\log\left(\frac{1}{\alpha_i}E_D[e^{\eta_i L_i(a_1,\ldots,a_I,D)}]\right)\right\}.$$

Clearly, $D_{a_i} \mapsto \hat{L}_i(D_{a_1},\ldots,D_{a_I})$ is nonlinear. This holds even if $L_i(a_1,\ldots,a_I,D)$ is independent of D_{a_i}. For $\alpha_i = 1$, we recover a minimization over the risk-sensitivity index in (1.74).

Proposition 1.18 *The entropic value-at-risk function $\hat{L}_i$ is a coherent risk measure.*

Proof To prove the statement, we can verify the properties one by one: The translation principle follows from the exponential property. The subadditivity property follows from the convexity. The monotonicity follows from composite functions exponential and logarithm and positivity of η. The infimization preserves the inequality. The positive homogeneity follows from the infimization on positive η. ■

Remark 1.5 On the existence of equilibria:

- Mixed strategy equilibria under exponential cost Let

$$\mathcal{G}_{ec} = \left(\mathcal{I}, (\mathcal{P}(\mathcal{A}_i), E_{x_1\otimes\ldots\otimes x_I} e^{\eta_i l_i})_{i\in\mathcal{I}}; \{\eta_i\}_{i\in\mathcal{I}} \in \mathbb{R}^I_{++}\right),$$

  where $\mathcal{A}_i$ is non-empty and finite and the exponential cost function of i is given by

$$E_{x_1\otimes\ldots\otimes x_I}[e^{\eta_i l_i(a_1,\ldots,a_I)}].$$

  Then, $\mathcal{G}_{ec}$ has at least one equilibrium in $\prod_{i\in\mathcal{I}}\mathcal{P}(\mathcal{A}_i)$.
- Entropic risk measure mixed strategy equilibria: Let

$$\mathcal{G}_{erm} = \left(\mathcal{I}, (\mathcal{P}(\mathcal{A}_i), \hat{l}_i)_{i\in\mathcal{I}}; \{\eta_i\}_{i\in\mathcal{I}} \in \mathbb{R}^I_{++}, \{\alpha_i\}_i \in (0,1]^I\right),$$

  where $\mathcal{A}_i$ is non-empty and finite and the entropic risk measure function of i with confidence level $1-\alpha_i \in [0,1)$ is given by

$$\hat{l}_i(x,\eta_i,\alpha_i) = \frac{1}{\eta_i}\log\left(\frac{1}{\alpha_i}E_{x_1\otimes\ldots\otimes x_I}[e^{\eta_i l_i(a_1,\ldots,a_I)}]\right).$$

  Then, $\mathcal{G}_{erm}$ has at least one equilibrium in $\prod_{i\in\mathcal{I}}\mathcal{P}(\mathcal{A}_i)$.

Indeed, one has

$$\begin{cases} \alpha_i e^{\eta_i \hat{l}_i(x,\eta)} = E_{x_1\otimes\ldots\otimes x_I}[e^{\eta_i l_i(a_1,\ldots,a_I)}], \\ i \in \mathcal{I}, \ x_i \in \mathcal{P}(\mathcal{A}_i). \end{cases} \tag{1.93}$$

The existence follows from the result of the game $\mathcal{G}_{ec}$. This result does not extend to the general case where l_i depends on x_i as it can lead to non-quasi-convexity of the cost function $\hat{l}_i$ in x_i.

- The existence result extends to games with payoff $g^{-1}\big(E_{x_1\otimes\ldots\otimes x_I}[g(l_i(a_1,\ldots, a_I))]\big)$ where g and its inverse are monotone real-valued functions and l_i independent of x_i.
- Taking the infimization over $\eta_i > 0$, the game with the entropic value-at-risk function of i with confidence level $1-\alpha_i$, $\hat{L}_i(x) = \inf_{\eta_i>0}\hat{l}_i(x,\eta_i,\alpha_i)$ has a mixed strategy ϵ-equilibrium for $\epsilon_i > 0$.

One-Shot MFTG with Entropic Value-at-Risk

For a payoff function p to be maximized, we set $\tilde{R}(p) = R(-p)$. For maximization problems, a coherent risk measure satisfies the following:

- Translation principle: $\tilde{R}(p_1+\lambda) = \tilde{R}(p_1) - \lambda$ for all $\lambda \in \mathbb{R}$.
- Subadditivity: $\tilde{R}(p_1+p_2) \le \tilde{R}(p_1) + \tilde{R}(p_2)$ for all p_1, p_2
- Monotonicity: If $p_1 \le p_2$,, then $\tilde{R}(p_1) \ge \tilde{R}(p_2)$
- Positive homogeneity: $\tilde{R}(\lambda p) = \lambda \tilde{R}(p)$, for all $\lambda \ge 0$,

MFTG with Entropic Value-at-risk

The MFTG with entropic value at risk (evar) is given by $\mathcal{G}_{evar} = \left(\mathcal{I}, (\mathcal{P}(\mathcal{A}_i), \hat{R}_i)_{i\in\mathcal{I}}\right)$, where the entropic value-at-risk function $\hat{R}_i$ of decision-maker i with confidence level $1-\alpha_i$ is given by

$$\begin{cases} |\mathcal{I}| \ge 2, \\ p_i : \prod_{j\in\mathcal{I}} \mathcal{A}_j \times \prod_{j\in\mathcal{I}} \mathcal{P}(\mathcal{A}_j) \to \mathbb{R}, \\ (a_1,\ldots,a_I, D_{a_1},\ldots,D_{a_I}) \mapsto p_i(a_1,\ldots,a_I, D_{a_1},\ldots,D_{a_I}), \\ \hat{R}_i : \prod_{j\in\mathcal{I}} \mathcal{P}(\mathcal{A}_j) \to \mathbb{R}, \\ \hat{R}_i(D_{a_1},\ldots,D_{a_I}) = \inf_{\eta_i>0}\left\{\frac{1}{\eta_i}\log\left(\frac{1}{\alpha_i}\int e^{-\eta_i p_i(a_1',\ldots,a_I',D)} \prod_{j\in\mathcal{I}} D_{a_j}(da_j')\right)\right\} \end{cases} \tag{1.94}$$

The risk $\hat{R}_i$ is to be minimized by decision-maker i. We can also define the evar as a supremum as follows. The payoff $\hat{P}_i$ is to be maximized by decision-maker i:

(continued)

$$\begin{cases} i \in \mathcal{I}, \\ \hat{P}_i(x_1, \ldots, x_I) = \sup_{z_i > 0} \left\{ -\frac{1}{z_i} \log \left(\frac{1}{\alpha_i} \int e^{-z_i p_i(a'_1, \ldots, a'_I, x)} \prod_{j \in \mathcal{I}} x_j(da'_j) \right) \right\} \end{cases} \tag{1.95}$$

An ϵ-equilibrium of $\mathcal{G}_{evar}$ is called an entropic value-at-risk ϵ-equilibrium. For every $(\epsilon_1, \ldots, \epsilon_I)$ with $\epsilon_i > 0, \forall i$, the existence of ϵ-equilibrium follows from Remark 1.5 for finite games.

1.3.2 Expectile MFTGs

Expectiles are a more generalized concept that allows for a nonlinear division of the data. Instead of dividing the data into two parts based on a fixed proportion, expectiles divide the data such that a certain proportion of the expected values of the data falls below the expectile. The expectile is law-invariant. For any $\frac{1}{2} \leq \alpha < 1$, the α-expectile is a coherent risk measure. Expectiles are often used in risk engineering and financial modeling when dealing with asymmetric and non-normal distributions, where traditional quantiles may not provide an accurate representation of the data. They can be particularly useful in situations where you want to focus on a specific range of expected outcomes rather than just looking at a fixed percentile.

Let $(\Omega, \mathcal{F}, \mathbb{P})$ be an atomless probability space and $L^q(\Omega, \mathcal{F}, \mathbb{P})$ be the L^q space with $q \geq 1$. Let α-expectile of an integrable random variable L, with $\alpha \in (0, 1)$ is defined by

$$\begin{cases} \rho_\alpha : L^2(\Omega, \mathcal{F}, \mathbb{P}) \to \mathbb{R}, \\ \rho_\alpha(L) \in \arg\min_{l \in \mathbb{R}} E\{\alpha[(L-l)_+]^2 + (1-\alpha)[(l-L)_+]^2\} \\ = \arg\min_{l \in \mathbb{R}} E\{W_\alpha(L-l)\}, \\ W_\alpha : \mathbb{R} \to \mathbb{R}, \\ W_\alpha(x) = \alpha[x_+]^2 + (1-\alpha)[(-x)_+]^2 \\ = \alpha x^2 \mathbb{I}_{\{x \geq 0\}} + (1-\alpha) x^2 \mathbb{I}_{\{x<0\}} = |\alpha - \mathbb{I}_{\{x<0\}}| x^2 \end{cases} \tag{1.96}$$

For $\alpha = \frac{1}{2}$,, the above definition yields $\arg\min_{l \in \mathbb{R}} E[(L-l)^2]$, we know that $\{E[L]\} = \arg\min_{l \in \mathbb{R}} E[(L-l)^2]$ which means that $\rho_{\frac{1}{2}}(L)$ the half-expectile of L coincides with the expected value of L.

Lemma 1.1

(a) *For each* $\alpha \in (0, 1)$, *and a square integrable* $L \in L^2$,, *the mapping* $l \mapsto E\{W_\alpha(L-l)\}$ *is convex.*

(b) *For each* $\alpha \in (0,1)$, *and a square integrable* $L \in L^2$,, *the mapping* $l \mapsto E\{W_\alpha(L-l)\}$ *is differentiable, and the derivative is well-defined in* L^1, *and it is a one to one mapping, given by* $-2E\{U_\alpha(L-l)\}$ *with*

$$\begin{cases} U_\alpha : \mathbb{R} \to \mathbb{R}, \\ U_\alpha(x) = \alpha x \mathbb{I}_{\{x\geq 0\}} + (1-\alpha)x\mathbb{I}_{\{x<0\}} = |\alpha - \mathbb{I}_{\{x<0\}}|x \end{cases} \tag{1.97}$$

(c) *Denote by* $\mathbb{U}_\alpha(l) = E\{U_\alpha(L-l)\}$. *For each* $\alpha \in (0,1)$, *and a square integrable* $L \in L^2$,, *the mapping* $l \mapsto E\{W_\alpha(L-l)\}$ *has a unique minimizer, and it is given by* $\mathbb{U}_\alpha^{-1}(0) = \rho_\alpha(L)$. *This means that the* α*-expectile of the loss* L *is the number* l *uniquely determined by such that*

$$\alpha E[(L-l)_+] = (1-\alpha)E[(l-L)_+].$$

(d) *For each* $\frac{1}{2} \leq \alpha < 1$,, $\rho_\alpha : L^2 \to \mathbb{R}$ *is a coherent risk measure.*
(e) *For each* $0 < \alpha < \frac{1}{2}$, $\tilde{\rho}_\alpha : L^2 \to \mathbb{R}$, *defined by* $\tilde{\rho}_\alpha(L) = -\rho_\alpha(-L)$ *is a coherent risk measure.*
(f) *For each* $0 < \alpha < 1$, $\rho_\alpha : L^1 \to \mathbb{R}$, *is law-invariant, i.e.,* $\mathcal{R}_\alpha(\mathbb{P}_{L_1}) = \mathcal{R}_\alpha(\mathbb{P}_{L_2})$ *if* $\mathbb{P}_{L_1} = \mathbb{P}_{L_2}$, *where* $\mathcal{R}_\alpha(\mathbb{P}_L) = \rho_\alpha(L) = \mathbb{U}_\alpha(L)^{-1}(0)$, *and* $\mathbb{U}_\alpha(L)(l) = E\{U_\alpha(L-l)\} = \int U_\alpha(y-l)\mathbb{P}_L(dy)$ *for* $l \in \mathbb{R}$.
(g) $\mu \mapsto \mathcal{R}_\alpha(\mu)$ *is Frechet-differentiable*
(h) *The* α*-expectile of Gaussian random variable satisfies* $\mathcal{R}_\alpha(\mathcal{N}(m,\sigma^2)) = m + (\alpha - \frac{1}{2})|\sigma|\sqrt{\frac{8}{\pi}} + O((\alpha-\frac{1}{2})^3)$. *For* $\frac{1}{2} < \alpha < 1$,, *set* $\alpha - \frac{1}{2} = \frac{\sqrt{\beta}}{2}$, *then*

$$\mathcal{R}_{\frac{1+\sqrt{\beta}}{2}}(\mathcal{N}(m,\sigma^2)) = m + |\sigma|\sqrt{\frac{2\beta}{\pi}} + O(\frac{\beta\sqrt{\beta}}{8}).$$

(i) *Let* $x(t) = x(t_0) + \int_{t_0}^t b(t', z(t'))dt' + \int_{t_0}^t \sigma(t', z(t'))dB(t')$, *to an stochastic differential equation driven by standard Brownian motion with*

$$z(t') = (x(t'), u(t')).$$

Then,

$$f(t, x(t)) = f(t_0, x(t_0)) + \int_{t_0}^t [f_t + f_x.b + \frac{\sigma^2}{2} f_{xx}]dt' + \int_{t_0}^t \sigma f_x dB(t').$$

One has
$\frac{1}{\epsilon}\mathcal{R}_{\frac{1+\sqrt{\beta}}{2}}(\int_t^{t+\epsilon} c(t', z(t'))dt' + f(t+h, x(t+h)) - f(t,x)|x(t) = x)$ *goes to*

$$c + f_t + f_x.b + \frac{\sigma^2}{2} f_{xx} + |\sigma f_x|\sqrt{\frac{2\beta}{\pi}}$$

as ϵ goes zero. In particular, the $(\frac{1+\sqrt{\beta}}{2})$*-expectile dynamic programming principle differs from the the standard dynamic principle.*

(j) *Computing expectile as fixed point:*

Let $\alpha \in (0,1)$ *and* L *an integrable real-valued random variable. The expectile of* L *with level* α*, denoted by* $\rho_\alpha(L)$ *is the unique solution of the equation*

$$\begin{array}{l} x = E[\max((1+\beta)L - \beta x, L)], \\ \beta := \frac{2\alpha-1}{1-\alpha}. \end{array} \tag{1.98}$$

We refer to [271] for a proof ■.

Expectile Mean-Field-Type Games

Decision-maker i aims to look at an expectile of the loss L_i with risk-sensitivity index $\alpha_i := \frac{1+\sqrt{\beta_i}}{2}$ as

$$\begin{cases} i \in \mathscr{I}, \\ \inf_{a_i \in \mathscr{A}_i} \mathscr{R}_{\alpha_i}(L_i(a, s, \mathbb{P}_s, \mathbb{P}_a)), \end{cases} \tag{1.99}$$

1.3.3 Extremile Mean-Field-Type Games

The extremile of order α of a loss function L with cumulative function F is

$$E_\alpha(F) \in \arg\min_{l \in \mathbb{R}} E\{k'_\alpha(F(L))\left|L - l\right|^2\} \tag{1.100}$$

where

$$k_\alpha(x) = (1 - (1-x)^{s_\alpha})\mathbb{I}_{\{0 \le \alpha \le \frac{1}{2}\}} + x^{r_\alpha}\mathbb{I}_{\{\frac{1}{2} < \alpha < 1\}}$$

and

$$r_\alpha = \frac{\log(\frac{1}{2})}{\log(\alpha)} = s(1-\alpha).$$

Denote by $y_l = \inf\{y \mid F(y) > 0\}$, $y_u = \sup\{y \mid F(y) < 1\}$, the lower and upper endpoints of the support of F. The extremile is explicitly given by

$$E_\alpha(F) = \begin{cases} \{-\int_{y_l}^0 \{1 - (1 - F(y))^{s_\alpha}\} dy + \int_0^{y_u} (1 - F(y))^{s_\alpha}\, dy\}\mathbb{I}_{\{0 \le \alpha \le \frac{1}{2}\}} \\ +\{-\int_{y_l}^0 (F(y))^{r_\alpha} dy + \int_0^{y_u} (1 - F(y))^{r_\alpha}\, dy\}\mathbb{I}_{\{\frac{1}{2} < \alpha \le 1\}} \end{cases} \tag{1.101}$$

If the support of F is the entire set of real numbers, then

$$E_\alpha(F) = \int_{-\infty}^{0} (g_\alpha(1 - F(y)) - 1)dy + \int_{0}^{+\infty} g_\alpha(1 - F(y))dy,$$

where

$$g_\alpha(x) = x^{s_\alpha}\mathbb{I}_{\{0\leq\alpha\leq\frac{1}{2}\}} + (1 - (1 - x)^{r_\alpha})\mathbb{I}_{\{\frac{1}{2}<\alpha\leq 1\}}.$$

Let L have a finite absolute first moment and L^i independent observations drawn from the distribution of F of L: Then,

$$E_\alpha(F) = \begin{cases} E[\max(L^1, \ldots, L^r)]\mathbb{I}_{\{\alpha=(\frac{1}{2})^{\frac{1}{r}},\ r\geq 1\}} \\ E[\min(L^1, \ldots, L^s)]\mathbb{I}_{\{\alpha=1-(\frac{1}{2})^{\frac{1}{s}},\ s\geq 1\}} \end{cases} \quad (1.102)$$

One-Shot Quantile Mean-Field-Type Games

Notice that the quantile is not a coherent risk measure. The quantile can be recovered from the following optimization: The quantile of order α of a loss functional L of mean-field type and with cumulative F is given by

$$Q_\alpha(F) \in \arg\min_{l\in\mathbb{R}} E\{k'_\alpha(F(L))\Big|L - l\Big|\} \quad (1.103)$$

The resulting game with payoffs $Q_\alpha(F_i)$ is called a quantile MFTG.

In this chapter, we have introduced one-shot MFTGs with and without coupled-action constraints. We have also presented risk measures and connection with mean-field-type payoffs.

1.4 Notes

For more details on constrained one-shot games, see [34, 38, 45, 49, 55, 164, 165, 169, 170, 170, 320, 322, 329, 341, 350, 376]. For more recent developments on mean-field-type games, see [15, 15–17, 27–29, 33, 33, 33, 35–37, 39, 40, 43, 46–48, 50–52, 62, 99, 102, 124, 125, 127, 132, 146, 157, 158, 166, 168, 185, 194, 245, 247, 281, 282, 293, 294, 296, 307, 308, 311, 320, 330–332, 334, 336–338, 350, 359, 367] and the references therein. Mean-field learning algorithms can be found in [213, 237, 317, 328, 335, 344–346].

Chapter 2
Markov Games

2.1 Markov Chains

A Markov chain, named after Andrey Markov, is a mathematical system that undergoes transitions from one state to another, between a finite and infinite number of possible states. A first-order Markov chain is a random process characterized as memoryless: The next state depends only on the current state and not on the sequence of events that preceded it. This specific kind of "memorylessness" is called the Markov property.

A first-order Markov chain is a sequence of random variables $X_1, X_2, X_3, \ldots$ with the (first-order) Markov property, namely, that, given the present state, the future and past states are independent. Formally,

$$\mathbb{P}(X_{t+1} = x' \mid X_1 = x_1, X_2 = x_2, \ldots, X_t = x_t) = \mathbb{P}(X_{t+1} = x' | X_t = x_t). \tag{2.1}$$

The possible values of X_t form a set $\mathscr{X}$ called the state space of the chain.

A first-order Markov chain is described by a set of states $\mathscr{X}$ and a probability transition from one state to the next. The process starts in one of these states and moves successively from one state to another. Each move is called a step. If the chain is currently in state x at step t, then it moves to state x' at the next step with a probability $q_{t,x'x} = \mathbb{P}(X_{t+1} = x' \mid X_t = x)$., and this probability does not depend upon which states the chain was in before the current state. The probabilities $\mathbb{P}(X_{t+1} = x' \mid X_t = x)$ are called transition probabilities. The process can remain in the state it is in, and this occurs with probability $\mathbb{P}(X_{t+1} = x \mid X_t = x)$.

T. Başar et al., *Mean-Field-Type Game Theory I*, Static & Dynamic Game Theory: Foundations & Applications, https://doi.org/10.1007/978-3-032-07027-2_2

A Markov chain of order m (or a Markov chain with memory m), where m is finite, is a process satisfying

$$\begin{aligned}&\mathbb{P}(X_t = x_t | X_{t-1} = x_{t-1}, X_{t-2} = x_{t-2}, \dots, X_1 = x_1)\\&= \mathbb{P}(X_t = x_t | X_{t-1} = x_{t-1}, X_{t-2} = x_{t-2}, \dots, X_{t-m} = x_{t-m}) \text{ for } t > m.\end{aligned} \tag{2.2}$$

In other words, the future state depends on the past m states.

2.1.1 Chapman-Kolmogorov Relation

If a Markov chain is time-homogeneous, then the transition matrix P is the same after each step, so the t-step transition probability can be computed as the t-th power of the transition matrix, P^t. This is called *Chapman-Kolmogorov relation.*

Let $p_{t,xx'}$ denote the t-step transition probability, $\mathbb{P}(X_t = x' \mid X_0 = x)$,, and let P_t denote the matrix whose (x,x')th entry is $p_{t,xx'}$. Then, $P_{t+t'} = P_t P_{t'}$, i.e., $p_{t+t',xx'} = \sum_{y\in\mathcal{X}} p_{t,xy} p_{t',yx'}$. In other words, the probability of going from state x to state x' in $t + t'$ steps is the sum over all y of the probability of going from x to y in t steps and then from state y to x in t' steps. In particular, $P_t = P_{t-1}P_1 = P_{t-2}P_1^2 = ... = P_1^{t-1}P_1 = P_1^t$, where P_1 is the just the one-step transition probability matrix denoted as P. Thus, $P_t = P^t$. For nonhomogeneous Markov chains, the Chapman-Kolmogorov relation is $P_t = Q_{t-1}Q_{t-2}\cdots Q_2Q_1$, where Q_i is the transition matrix in the step $(i, i+1)$.

A stationary distribution ω is a (row) vector, whose entries are nonnegative and sum to 1, satisfying the equation:

$$\omega = \omega P.$$

In other words, a stationary distribution ω is a normalized (meaning that the sum of its entries is (1) left eigenvector of the transition matrix associated with the eigenvalue 1. Alternatively, ω can be viewed as a fixed point of the linear (hence continuous) transformation on the unit simplex associated with the matrix P. As any continuous transformation in the unit simplex has a fixed point, a stationary distribution always exists (by, the Brouwer fixed-point theorem) but is not guaranteed to be unique, in general.

There is a simple class of Markov chains for which the fixed-point algorithm (Picard iterates) converges. A Markov chain is called an ergodic or irreducible Markov chain if it is possible to eventually get from every state to every other state with positive probability. A state x is said to have a period of $d_x \geq 1$ if d_x is the greatest integer such that $(P^t)_{xx} = 0$ if t is not multiple of d_x. If $(P^t)_{xx} = 0, \forall t$, we say that $d_x = +\infty$. A state x of a Markov chain is aperiodic if its period is 1. A Markov chain is said to be aperiodic if all its states are aperiodic. If the Markov chain is irreducible and aperiodic, then there is a unique stationary distribution ω. Additionally, in this case, P^t converges to a rank-one matrix in which each row is the stationary distribution ω, that is,

$$\lim_{t \to \infty} \mathbf{P}^t = \mathbf{1}\omega.$$

where $\mathbf{1}$ is the column vector with all entries equal to 1. Moreover, the convergence is an exponential decay: There exists $\alpha > 0, \beta > 0$ such that

$$\max_{x,x'} \left| (\mathbf{P}^t)_{x'x} - \omega_x \right| \leq \alpha e^{-\beta t}, \quad t \geq 1.$$

This is stated as the Perron-Frobenius theorem.

The limit $\lim_{t \longrightarrow +\infty} P^t$ may not exist, as illustrated by the transition matrix P given by

$$P = \begin{pmatrix} 0 & 1 \\ 1 & 0 \end{pmatrix}.$$

To see this, note that the stationary distribution obtained by solving $\omega P = \omega$ gives $\omega = (\frac{1}{2}, \frac{1}{2})$. However, the system does not converge to this stationary distribution starting from any initial condition. Starting from $(1, 0)$ its moves to $(0, 1)$, then $(1, 0)$, and then $(0, 1)$, and so on. Therefore, $\lim_t P^t$ cannot be $(\frac{1}{2}, \frac{1}{2})$, and there is no limit to a point, but there is a "limit cycle." The reason that this Markov chain does not converge to the stationary distribution is due to the fact that the state periodically alternates between the two states.

The evolution of the distribution of the process X_t is determined by the so-called Kolmogorov forward equation:

$$m_{t+1} = \mathscr{L}(X_{t+1}) = m_t P = m_0 P^{t+1},$$

where m_0 is the initial distribution of the starting state $x_0 \sim m_0$.

2.1.2 From Markov Chains to Graph Theory

One can associate with a matrix P a certain directed graph. It has exactly n vertices, where n is the size of P, and there is an edge from vertex x' to vertex x precisely when $p_{xx'} > 0$. Then, the matrix P is irreducible if and only if its associated graph G is strongly connected. The Perron-Frobenius theorem describes the properties of the leading eigenvalue and of the corresponding eigenvectors when P is a nonnegative real square matrix. Early results were due to Oskar Perron (1907) and concerned positive matrices. Later, Georg Frobenius (1912) found their extension to certain classes of nonnegative matrices.

Theorem 2.1 (Perron-Frobenius Theorem) *Let P be an irreducible nonnegative square matrix with spectral radius $\rho(P)$. Then the following statements hold.*

- *The number $\rho(P)$ is a positive real number, and it is an eigenvalue of the matrix P, called the Perron-Frobenius eigenvalue.*
- *The Perron-Frobenius eigenvalue is simple. Both right and left eigenspaces associated with $\rho(P)$ are one-dimensional.*
- *P has a left eigenvector with eigenvalue $\rho(P)$ whose components are all positive.*
- *Likewise, P has a right eigenvector with eigenvalue $\rho(P)$ whose components are all positive.*
- *The only eigenvectors whose components are all positive are those associated with the eigenvalue $\rho(P)$.*
- *The Perron-Frobenius eigenvalue satisfies the inequalities*

$$\min_{x'} \sum_{x} p_{xx'} \leq \rho(P) \leq \max_{x'} \sum_{x} p_{xx'}.$$

In the next subsection, we will associate a decision process to a Markov chain and formulate a dynamic Markov optimization problem: Markov decision process.

2.2 Markov Decision Process

Many processes we encounter can be described by action and state-space systems. These include processes in engineering, computer science, economics, physics, chemistry, biology, and many other fields. If we can derive a mathematical model for a process, then we can use the tools of mathematics to control the process, calculate the associated risks, and obtain information about the process. This is why action and state-space systems are so important, particularly to engineers and industrial researchers. If we know the state of a system at the present time, and we know all of the present and future inputs, then we can deduce the values of all future outputs of the system for an associated action or control. State-space models can be linear or nonlinear. Although most real processes are nonlinear, the mathematical tools that are available for estimation and control are much more accessible and well understood for linear systems. That is why nonlinear systems are often approximated as linear systems plus second-order perturbations. That way we can use the tools that have been developed for linear systems to derive estimation, adaptive, iterative, filtering, and control algorithms.

Markov decision processes (MDPs) provide a mathematical framework for modeling decision-making in situations where outcomes are partly random and partly under the control of a decision-maker. MDPs were known at least as early as the 1950s (Bellman 1952–57). Much research in the area was spawned due to Ronald A. Howard's book, Dynamic Programming and Markov Processes, in 1960. More precisely, a Markov decision process is a stochastic control process. At each time step, the process is in some state x, and the decision-maker may choose an action a that is available in state x. The process responds at the next time step by randomly moving into a new state x', and giving the decision-maker a corresponding payoff $r(t, x, a, x')$. The probability with which the process moves to its new state x' is influenced by the chosen action (decision). Specifically, it is given by the state transition function

$$q_{t,xax'} = P(X_{t+1} = x' \mid X_t = x, A_t = a).$$

Thus, the next state x' depends on the current state x and the decision-maker's action a. But given x and a, it is conditionally independent of all previous states and actions; in other words, the state transitions of an MDP possess the Markov property.

Markov decision processes are an extension of Markov chains; the difference here is the addition of actions (allowing choice) and payoffs (giving motivation). Conversely, if only one action exists at each state and all rewards are constant, a Markov decision process reduces to a Markov chain. As we will see later, in a broader context, an MDP is a single-decision-maker stochastic game.

An MDP is described as follows:

- A single decision-maker (single decision-maker),
- A time set $\mathcal{T}$.
- $\mathcal{X}$ a (non-empty) state space,
- $\mathcal{A}(x)$ (non-empty) action space in state $x \in \mathcal{X}$,
- The set of all possible choices, given as the union $\bigcup_{x \in \mathcal{X}} \mathcal{A}(x)$,
- A transition function at time step $t \in \mathcal{T} : (x, a) \longmapsto q_t(.|x, a) \in \Delta(\mathcal{X})$ satisfying

$$q_t(x'|x, a) := q_{t,xax'} \geq 0, \sum_{x' \in \mathcal{X}} q_{t,xax'} = 1,$$

(continued)

where $\Delta(\mathscr{X})$ denotes the set of probability measures over $\mathscr{X}$.
Letting

$$\mathscr{X}\mathscr{A} = \{(x, a) \mid x \in \mathscr{X},\ a \in \mathscr{A}(x)\},$$

be the state-action space, then

$$q_t : \mathscr{X}\mathscr{A} \longrightarrow \Delta(\mathscr{X}).$$

$$q_{t,xax'} = \mathbb{P}\left(x_{t+1} = x' \,|x_t = x, a_t = a\right)$$

$$= \mathbb{P}\left(x_{t+1} = x' \,|x_t = x, a_t = a,\ x_{t-1}, a_{t-1}, x_{t-2}, a_{t-2}, \ldots, x_0\right).$$

- an instantaneous payoff function $r_t : \mathscr{X}\mathscr{A} \longrightarrow \mathbb{R}$
- A long-term payoff that could be cumulative, multiplicative, or a terminal payoff.

2.3 From Markov Decision Processes to Markov Games

A branch of *game theory* that takes into consideration the state evolution of the system and the behavior of the decision-makers in a dynamic environment is that of *dynamic game theory*.

Dynamic game theory deals with sequential situations of several decision-makers (often called decision-makers) where the objective for each one of the decision-makers may be a function of not only its own preference and decision but also of the decisions of other decision-makers. Dynamic games allow modeling of sequential decision-making, time-varying interactions, and uncertainty and randomness in the interactions of the decision-makers. They allow modeling of situations in which the parameters defining the games vary in time and decision-makers can adapt their strategies (or policies) according to changes in the environment. At some given time, a decision-maker takes a decision (also called an action) according to some strategy. A strategy of a decision-maker is a collection of history-dependent maps that dictates at each time the choice (which can be probabilistic) of that decision-maker. The vector of actions chosen by decision-makers at a given time (called action profile) may determine not only the payoff for each decision-maker at that time but also the state evolution [333, 339, 342]. A particular class of dynamic games widely studied in the literature is the class of *stochastic games*. These are dynamic games with probabilistic state transitions (stochastic state evolution) controlled by one or more decision-makers. The discrete time state evolution with kernel descriptions is often modeled as interactive Markov decision processes, while the continuous time state evolution is referred to stochastic differential games. Discounted stochastic games have been introduced in the 1950s by Shapley [289].

Stochastic Games with Common State
Stochastic games, first introduced by Shapley [289], study dynamic interactions in which the environment changes in response to behavior of the decision-makers.

A stochastic game with a common state space is a collection

$$\Gamma = (\mathcal{I}, \mathcal{S}, (\tilde{\mathcal{A}}_j, \mathcal{A}_j, r_j)_{j\in\mathcal{I}}, \mathcal{T}, \{q_t\}_{t\in\mathcal{T}})$$

where

- $\mathcal{I}$ is the set of decision-makers. The set $\mathcal{I}$ can be finite or infinite depending on the model. Typical number of decision-makers can be $1, 2, 10, 10^2, 10^4, 10^6$, etc.
- $\mathcal{T}$ is the time horizon of the game.
- $\mathcal{S}$ is a common state space. A state may represent the environment state or the system state.
- For every decision-maker $j \in \mathcal{I}$, $\tilde{\mathcal{A}}_j$ is the set of all possible actions of that decision-maker. $\mathcal{A}_j : \mathcal{S} \longrightarrow 2^{\tilde{\mathcal{A}}_j}$ is a set-valued map (correspondence) that assigns to each state $s \in \mathcal{S}$ the set of actions $\mathcal{A}_j(s)$ that are available to decision-maker j. We denote the state-action space by

$$\text{SA} = \{(s, a) : s \in \mathcal{S}, a = (a_j)_{j\in\mathcal{I}}, a_j \in \mathcal{A}_j(s), \forall j \in \mathcal{I}\}$$

- For every decision-maker $j \in \mathcal{I}$, $r_{j,t} : \text{SA} \longrightarrow \mathbb{R}$ is an instantaneous payoff function for decision-maker j.
- $\forall t,\ q_t : \text{SA} \longrightarrow \Delta(\mathcal{S})$ is a transition function where $\Delta(\mathcal{S})$ is the space of probability distributions over $\mathcal{S}$.
- The long-term payoff functional for each decision-maker.
- A discount factor $\delta_j \in (0, 1)$ is associated with decision-maker $j \in \mathcal{I}$.

2.3.1 How Is the Stochastic Game Played?

The stochastic game starts at an initial state $s \in \mathcal{S}$ and is played as follows. Time is discrete and time space is a subset of natural numbers $\mathbb{N}$. The current state s_t is known by all decision-makers. At each time $t \in \mathbb{N}$,, each decision-maker $j \in \mathcal{I}$ chooses an action $a_{j,t} \in \mathcal{A}_j(s_t)$, and receives an instantaneous payoff $r_{j,t}(s_t, a_t)$, where $a_t = (a_{j,t})_{j \in \mathcal{J}}$, and the game move to a new state according to a probability distribution given by $q_t(. \mid s_t, a_t, s_{t-1}, a_{t-1}, \ldots, s_0) \in \Delta(\mathcal{S})$, where

$$q_t(s' \mid s_t, a_t, s_{t-1}, a_{t-1}, \ldots, s_0) = \mathbb{P}\left(s_{t+1} = s' | s_t, a_t, s_{t-1}, a_{t-1}, \ldots, s_0\right), \; s' \in \mathcal{S}.$$

The transition structure is said to be Markovian (one-step-Markov) if the transition kernel satisfies

$$q_t(. \mid s_t, a_t, s_{t-1}, a_{t-1}, \ldots, s_0) = q_t(. \mid s_t, a_t).$$

In this case, the transition kernel q_t is completely determined by

$$q_t(\mathcal{S}' \mid s_t, a_t) = \mathbb{P}\left(s_{t+1} \in \mathcal{S}' \mid s_t, a_t\right),$$

where $\mathcal{S}'$ describes a measurable subset of $\mathcal{S}$.

This class of stochastic games is referred to (time-inhomogeneous) *Markov games* or *interactive Markov decision process* [153].

2.3.2 Repeated Games as a Subclass of Stochastic Games

Standard repeated games are games in which the decision-makers play the same one-shot game several times. This model of stochastic games is related to repeated games and Markov decision processes as described below.

- Stochastic games are generalization of standard repeated games. If the state space $\mathcal{S}$ is reduced to a singleton, one recovers repeated games.
- Stochastic games formulated over an infinite horizon encapsulates formulation of finite repeated games of horizon T by assuming that the decision-makers move, at time T, to an absorbing state with instantaneous payoff of zero to all the decision-makers after the horizon T.
- In particular, stochastic games are generalization of one-shot games which are played only once.
- Markov decision processes [143] are stochastic games with a single decision-maker (equivalently a team decision-maker).
- Markov decision teams are stochastic games where all team members have the same payoff. This is also referred to as common interest stochastic games.

The use of multiple discount factors appeared with the seminal paper of Shapley [289]. The framework of [289] considers two-decision-maker zero-sum stochastic games with uncontrolled stopping probabilities known by both decision-makers. For more details on stochastic games, we refer to [302] and the references therein.

2.3.3 *Histories and Strategies*

2.3.3.1 Complete Information of Past Play

We say the decision-makers have complete information of past play if, at each time-slot t, they know the collection $\{s_0, a_0, s_1, a_1, \ldots, s_{t-1}, a_{t-1}, s_t\}$ of states that were visited in the past and the actions that were chosen by all decision-makers. We say that the decision-maker j has perfect recall if for any time t, he/she remembers all its history up to t. Note that this requires memory build-up.

The assumption of complete information and perfect recall are too strong for most of the engineering and computer science applications, and we will discuss later how this assumption can be relaxed.

2.3.3.2 Histories

Under complete information of past play, a history of length $t + 1$ corresponds to the sequence $h_t = (s_0, a_0, s_1, a_1, \ldots, s_{t-1}, a_{t-1}, s_t)$.

2.3.3.3 Strategies

A pure strategy σ_j of decision-maker j is a mapping that assigns to every finite history h_t an element of $\mathscr{A}_j(s_t)$. A mixed strategy is a probability distribution on the set of pure strategies. The set of mixed strategies is denoted by Σ_k.

2.3.3.4 Behavioral Strategies

A behavioral strategy σ_j of decision-maker j is a mapping that assigns to every finite history h_t an element of $\Delta(\mathscr{A}_j(s_t))$, which is the set of probability measures over $\mathscr{A}_j(s_t)$.

2.3.3.5 (One-Step Memory) Markov Strategies

At each time t, the strategy $\sigma_{j,t}$ depends only on t, s_t but not on the past.

2.3.3.6 Stationary Strategies

A simple class of strategies is the class of stationary strategies. A strategy profile $(\sigma_j)_{j\in\mathscr{J}}$ is stationary if $\forall j,\ \sigma_j(s_0, a_0, s_1, a_1, \ldots, s_{t-1}, a_{t-1}, s_t)$ depends only on the current state s_t. A stationary strategy of decision-maker j can be identified with an element of the product-space $\prod_{s\in\mathscr{S}} \Delta(\mathscr{A}_j(s))$.

2.3.3.7 Extension to the σ-Algebra of Infinite Plays

Every profile $\sigma = (\sigma_j)_{j\in\mathscr{I}} \in \prod_j \Sigma_j$ of mixed strategies, together with the initial state s_0,, induces a probability distribution $\mathbb{P}_{s_0,\sigma}$ over the space of infinite plays $\text{SA}^{\mathbb{N}}$. We denote the corresponding expectation operator by $E_{s_0,\sigma}$.

2.3.4 *Different Types of Payoffs*

We examine four classes of cumulative payoff functions to evaluate the performance of the sequences of payoffs that the decision-makers receive in a stochastic game. Let σ be a mixed strategy profile.

- The finite-horizon payoff, say of duration T, where a decision-maker considers the cumulative payoff during the first T stages. For every finite $T \in \mathbb{N}$,, the finite-horizon payoff for decision-maker j is

$$F_{j,T}(\sigma) = E_{s_0,\sigma}\left[\sum_{t=0}^{T-1} r_j(s_t, a_t)\right]. \tag{2.3}$$

  We denote the associated stochastic game by Γ_T.
- The discounted payoff, where a decision-maker considers the discounted sum of his stage payoffs: For a discount factor δ_j, $0 \le \delta_j < 1$, the δ_j-discounted payoff under the strategy σ for decision-maker j is

$$F_{j,\delta}(\sigma) = (1-\delta_j)E_{s_0,\sigma}\left[\sum_{t=0}^{+\infty} \delta_j^t r_j(s_t, a_t)\right], \tag{2.4}$$

  where δ_j could be different for different decision-makers. We denote the associated stochastic game by Γ_δ.
- The liminf (or limsup) payoff, in which a decision-maker considers the liminf of his long-run average payoff. For decision-maker j, the liminf payoff is

$$F_{j,\infty}(\sigma) = E_{s_0,\sigma}\left[\underline{\lim}_{T\longrightarrow+\infty} \frac{1}{T}\sum_{t=0}^{T-1} r_j(s_t, a_t)\right]. \tag{2.5}$$

  We denote the associated stochastic game by Γ_∞.

- A stochastic game with terminal payoff. Here, there is no stage payoff. Only a terminal payoff is received by the decision-maker at the end.

A (Nash) equilibrium of a stochastic game is a strategy profile such that no decision-maker can benefit from a unilateral deviation. As we will see below, this corresponds to a 0-Nash equilibrium in a broader class of equilibria.

Definition 2.1 Let $\epsilon_j \geq 0,\ j \in \mathscr{I}$.

- A strategy profile $\sigma^* = (\sigma_j^*)_{j\in\mathscr{I}}$ is a T-stage additive $(\epsilon_1, \ldots, \epsilon_I)$-equilibrium if

$$F_{j,T}(s_0, \sigma^*) \geq F_{j,T}(s_0, \sigma_j, \sigma_{-j}^*) - \epsilon_j,\ \forall s_0 \in \mathscr{S},\ \forall j \in \mathscr{I},\ \forall \sigma_j \in \Sigma_j \quad (2.6)$$

- A strategy profile $\sigma^* = (\sigma_j^*)_{j\in\mathscr{I}}$ is a δ-discounted additive ϵ-equilibrium if

$$F_{j,\delta}(s_0, \sigma^*) \geq F_{j,\delta}(s_0, \sigma_j, \sigma_{-j}^*) - \epsilon_j,\ \forall s_0 \in \mathscr{S},\ \forall j \in \mathscr{I},\ \forall \sigma_j \in \Sigma_j \quad (2.7)$$

- A strategy profile $\sigma^* = (\sigma_j^*)_{j\in\mathscr{I}}$ is a $\underline{\lim}$ additive ϵ-equilibrium if

$$F_{j,\infty}(s_0, \sigma^*) \geq F_{j,\infty}(s_0, \sigma_j, \sigma_{-j}^*) - \epsilon_j,\ \forall s_0 \in \mathscr{S},\ \forall\, j \in \mathscr{I},\ \forall \sigma_j \in \Sigma_j \quad (2.8)$$

- The cumulative payoff that corresponds to an ϵ-equilibrium is called ϵ-equilibrium payoff.

2.3.5 Existence of Equilibria

Theorem 2.2 *Assume that the state and action spaces are discrete and finite and the action spaces are non-empty. Then,*

- *the stochastic game Γ_T has at least one T-stage 0-equilibrium.*
- *the stochastic game Γ_δ has at least one δ-discounted 0-equilibrium.*

Under complete information and perfect monitoring, the existence result for two-decision-maker zero-sum discounted stochastic game has been proved by Shapley [289]. Takahashi [309] and Fink [154] extended Shapley's result to discounted 0-equilibria in nonzero-sum stochastic games. The proof is based on dynamic programming [56] and fixed-point theorems. The following Lemma characterizes the stationary discounted 0-equilibria and the associated equilibrium payoffs.

Lemma 2.1 *A stationary strategy σ is a discounted 0-equilibrium if and only if the corresponding payoff $\{v_{j,\delta}(s)\}_{(j,s)\in\mathscr{I}\times\mathscr{S}}$ satisfies the dynamic programming recursion:*

$$v_{j,\delta}(s,\sigma_{-j}) = \sup_{y_j \in \Delta(\mathscr{A}_j(s))} \Big\{(1-\delta_j)r_j(s,y_j,\sigma_{-j}) + \delta_j \sum_{s' \in \mathscr{S}} q(s' \mid s, y_j, \sigma_{-j}) v_{j,\delta}(s',\sigma_{-j})\Big\}$$

for all $s \in \mathscr{S},\ j \in \mathscr{I}$.

A proof of this lemma can be found in [70] in conjunction with Theorem 6 there. ∎

Theorem 2.3 (Takahashi [309], Fink [154]) *Every I-decision-maker stochastic game with finite state and finite (and non-empty) action spaces has a δ-discounted 0-equilibrium in stationary strategies.*

For the T-stage stochastic game, the dynamic programming recursion associated with the value functions for the time-average payoff is

$$T v_{j,T}(s,\sigma_{-j}) = \sup_{y_j \in \Delta(\mathscr{A}_j(s))} \Big\{ r_j(s,y_j,\sigma_{-j}) + (T-1) \sum_{s' \in \mathscr{S}} q(s'|s,y_j,\sigma_{-j}) v_{j,T-1}(s',\sigma_{-j}) \Big\}$$

The backward induction principle of the cumulative payoff leads to the recursion

$$v_{j,t}(s,\sigma_{-j}) = \sup_{y_j \in \Delta(\mathscr{A}_j(s))} \Big\{ r_j(s,y_j,\sigma_{-j}) + \sum_{s' \in \mathscr{S}} q(s' \mid s, y_j, \sigma_{-j}) v_{j,t+1}(s',\sigma_{-j}) \Big\},$$

$$v_{T+1} = 0.$$

Regarding the existence of the liminf equilibrium, little is known. One of the most significant results in this direction is by Vieille [355–357], who proved that every two-decision-maker stochastic game with finite state and action spaces has an ϵ-equilibrium for every $\epsilon > 0$. The existence of equilibria has been proved for a specific class of stochastic games under ergodicity conditions or stochastic games on product state spaces [155].

It is known from a counterexample by Gillette [175] that even in the two-decision-maker-zero-sum case, a 0-equilibrium may not exist in the average payoff case.

The existence of 0-equilibria in Γ_∞ is known for the class of irreducible Markov decision processes. Next we define the notion of irreducible stochastic games.

Definition 2.2 A stochastic game is irreducible if for each decision-maker k, and for each pair of states $(s,s') \in \mathscr{S}^2$, there is a finite horizon $T > 0$ and a sequence

$(s_1, \ldots, s_T)$ of states such that $s_1 = s$ and $s_T = s'$, and for each $t < T$,, there is a profile α such that $q(s_{t+1}|s_t, a_j, \alpha_{-j}) > 0$ for all $a_j \in \mathscr{A}_j$.

Theorem 2.4 (Sobel [298, 299]) *Consider a finite and irreducible stochastic game. Then,*

- *a* 0*-equilibrium exists for the average payoff.*
- *there exists a stationary strategy* $(\sigma(s))_s$ *and for each decision-maker* $j \in \mathscr{I}$ *a constant* g_j *and a bounded function* $v_j(.)$ *such that*

$$
\begin{aligned}
& g_j + v_{j,\infty}(s) \\
& = \sup_{y_j \in \Delta(\mathscr{A}_j(s))} \left\{ r_j(s, y_j, \sigma_{-j}(s)) + \sum_{s' \in \mathscr{S}} q(s'|s, y_j, \sigma_{-j}(s)) v_{j,\infty}(s') \right\},
\end{aligned}
\tag{2.9}
$$

for all $s \in \mathscr{S}$, *where* $\sigma_j(s)$ *achieves the maximum on the right-hand side of (2.9) for all* $s \in \mathscr{S}$. *Moreover,* $F_{j,\infty}(s, \sigma) = g_j$.

A refined notion of Nash equilibrium is the subgame perfect equilibrium and the ϵ-subgame perfect equilibrium. They can be seen as the requirement for the credibility of an equilibrium strategy. In the context of stochastic games, the notion subgame perfect equilibrium has been studied in [349].

Definition 2.3 A strategy profile σ^* is said to provide a subgame perfect equilibrium if $\sigma^*(h_t)$ is an equilibrium strategy profile for every finite history h_t.

Theorem 2.5 ([242, 301]) *The* I*-decision-maker discounted stochastic game* Γ_δ *has a subgame perfect* 0*-equilibrium.*

Next, we present a result on the existence of 0-equilibrium in pure strategies for a special class of stochastic games where the decision-makers have the same payoffs, i.e., stochastic dynamics teams, are strategically equivalent to a stochastic dynamic teams.

Theorem 2.6

- *If, in an* I*-decision-maker discounted stochastic game the payoff functions of all the decision-makers are the same, then the discounted stochastic game has at least one pure (Nash)* 0*-equilibrium.*
- *If, in an* I*-decision-maker discounted stochastic game, the transitions between the states are independent of the actions played and for each state* $s \in \mathscr{S}$, *the finite game* $G(s) = (\mathscr{I}, \mathscr{A}_j(s), r_j(s, .))$ *is a potential game in the sense of [248], then the discounted stochastic game has at least one pure (Nash)* 0*-equilibrium.*

Proof The first result follows from the fact that the discounted stochastic game can be seen as a discounted team problem. By the use of Bellman optimality, we associate a dynamic programming principle to the common instantaneous payoff function r. Then, we use the result of Blackwell [69] which gives the existence of optimal stationary pure strategies. The recursion equation is

$$v \in \mathbb{R}^{|\mathcal{S}|},\ v_s = \sup_{a_j \in \mathcal{A}_j(s)} \left\{ r(s,a) + \delta \sum_{s'} q(s' \mid s,a) v_{s'} \right\}.$$

The latter system corresponds to global optimality [102] which is much stronger than Nash equilibrium, which corresponds to person-by-person notion.

We next prove the second result. Since the decision-makers have no influence in the transition probabilities, each decision-maker does its best in each of the visited states. At each state s, one has a pure Nash equilibrium a_s^* of the auxiliary game with

$$r_j(s,a) + \delta_j \sum_{s'} q(s' \mid s) v_{j,s'}$$

by using Monderer & Shapley. The collection of the pure equilibria $a^* = (a_s^*)_{s \in \mathcal{S}}$ is a stationary pure equilibrium of the discounted stochastic game. This completes the proof. □

2.3.6 *Characterization of* 0*-Equilibrium Payoffs*

We next focus on the characterization of the set of 0-equilibrium payoffs in stochastic games. The folk theorem studies the set of 0-equilibrium payoffs. Generically, the folk theorem identifies conditions under which the set of 0-equilibrium payoffs of the stochastic game is related to the the set of feasible and individually rational payoffs, which will be specified in Sect. 2.3.6.2.

2.3.6.1 Zero-Sum Stochastic Games

Zero-sum stochastic games with complete information have been studied in [153, 303]. A two-decision-maker zero-sum stochastic game with imperfect monitoring has been studied in [303]. Neyman & Sorin have studied zero-sum stochastic games with a public uncertain horizon [260].

2.3.6.2 Irreducible Stochastic Games

Dutta [140, 141] identifies a class of stochastic games with infinite horizon for which the folk theorem holds. The proof of this result raises an additional difficulty compared to the single-state repeated games. One of the reason is that the decision-makers may want to deviate not only in order to improve their current payoffs but also to affect the probability distribution on the states. For this reason, asymptotic state invariance of feasible long-run payoff and additional assumptions are needed.

Dutta [140] constructed a punishment strategy such that for all individually rational payoffs, there is a subgame perfect strategy whose payoff approaches the former. The equilibrium punishment strategy consists of minimaxing the deviator during several periods and then playing a strategy that provides payoffs to decision-makers who were punished during the previous phase. Note that the strategy constructed by Dutta [140] may not be subgame perfect in the finite horizon case. As an extension of the result of Dutta [140], the folk theorem for irreducible stochastic games with imperfect public monitoring has been established by Fudenberg and Yamamoto [160] and by Hörner et al. [193]. However, the folk theorem remains an open question even for finite stochastic games with complete information and full monitoring when the chain is not irreducible.

Below we describe in detail the result of Fudenberg and Yamamoto [160]. Denote by $\tilde{\mathscr{Y}}$ the set of public signals. In each period t, decision-makers observe the state s_t and then move simultaneously, with decision-maker k choosing an action a_j from a finite set $\mathscr{A}_j$. Given an action profile a, decision-makers observe a public signal y from a finite set $\tilde{\mathscr{Y}}$,, and the state in period $t+1$ is chosen from the set $\mathscr{S}$. Let $\pi(s', \tilde{y} \mid s, a)$ denote the probability that decision-makers observe a signal $\tilde{y}$, and the state for the next period is s' when today's state is s and decision-makers play action profile a. (Note that the distributions of $\tilde{y}$ and s' may be correlated.) decision-maker k's realized payoff is $\bar{r}_j(s, a_j, \tilde{y})$, so that her expected payoff conditional on (s, a) is

$$r_j(s, a) = \sum_{s' \in \mathscr{S}} \sum_{\tilde{y} \in \tilde{\mathscr{Y}}} \pi(s', \tilde{y} \mid s, a) \bar{r}_j(s, a_j, \tilde{y}).$$

Denote by E_{δ, s_0} the set of feasible payoffs when the initial state is s_0 and the discount factor is δ. Dutta [140] has shown that for an *irreducible stochastic game* with public signal, the set of feasible payoffs is independent of the initial state in the limit as δ goes to one. The minimax payoff to decision-maker j in a stochastic game with initial state s_0 is defined to be

$$v^*_{j,\delta,s_0} = \inf_{\sigma_{-j}} \sup_{\sigma_j} F_{j,\delta}(s_0, \sigma).$$

Dutta [140] has shown that for irreducible stochastic games, the limit of the minimax payoff as δ goes to one is independent of the initial state from s_0. We define below the notion of irreducibility in the presence of public signals.

Definition 2.4 A stochastic game is irreducible if for each decision-maker j, and for each pair of states $(s, s') \in \mathscr{S}^2$, there is a finite horizon $T > 0$ and a sequence $(s_1, \ldots, s_T)$ of states such that $s_1 = s$ and $s_T = s'$, and for each $t < T$,, there is a profile α such that $\sum_{\tilde{y} \in \mathscr{Y}} \pi(s_{t+1}, \tilde{y} \mid s_t, a_j, \alpha_{-j}) > 0$ for all $a_j \in \mathscr{A}_j$.

Define a matrix $M_j^{s,s'}(\alpha)$ with rows $\pi(s', \tilde{y} \mid s, a_j, \alpha_{-j})_{\tilde{y} \in \tilde{\mathscr{Y}}}$ for all $a_j \in \mathscr{A}_j$.

Definition 2.5 A profile α has full rank for the triplets (j, s, s') if the rank of $M_j^{s,s'}(\alpha)$ is of rank $\sharp\mathscr{A}_j$ where $\sharp\mathscr{A}_j$ denotes the cardinality of $\mathscr{A}_j$.

Definition 2.6 For each $j \neq j'$ and (s, s'),, the profile α a has *pairwise full rank* for (j, j', s, s') if the rank of the matrix obtained by stacking the two matrices $M_j^{s,s'}(\alpha)$ and $M_{j'}^{s,s'}(\alpha)$ is of rank $\sharp\mathscr{A}_j + \sharp\mathscr{A}_{j'} - 1$.

Theorem 2.7 (Fudenberg and Yamamoto [160]) *Assume that*

- ***Condition 1:*** *For each $(s, s') \in \mathscr{S}^2$,, there is an integer $T > 0$ and a sequence $(s_1, \ldots, s_T)$ of states such that $s_1 = s$, $s_T = s'$, and for each $t < T$,, every pure action profile has individual full rank for (s_t, s_{t+1}) and every decision-maker j.*
- ***Condition 2:*** *For each $(s, s') \in \mathscr{S}^2$,, there is an integer $T > 0$ and a sequence $(s_1, \ldots, s_T)$ of states such that $s_1 = s$, $s_T = s'$, and for each (j, j') and $t < T$,, there is a profile α_t that has pairwise full rank for (s_t, s_{t+1}, j, j') and decision-maker k.*

  Denote by V^* the limit of the set of feasible and individually rational payoffs.
- ***Condition 3:*** *The dimension of V^* is exactly the cardinality of $\mathscr{I}$.*

Then,

$$\lim_{\delta \longrightarrow 1} E_{\delta,s_0} = V^*$$

Note that the full dimensionality condition (Condition 3) may not be satisfied in some important classes of games such as (anti)coordination games and team problems, and irreducibility conditions fail in the presence of absorbing states.

Note also that when $\tilde{\mathscr{Y}} = \prod_j \mathscr{A}_j$, one arrives at the perfect monitoring case, and then the folk theorem holds under Conditions 1, 2, and 3. A similar result has been proved by Hörner et al. [192].

2.4 Risk-Sensitive Stochastic Games

Mixed (or randomized) actions have been widely considered in game theory and its applications (economics, finance, biology, ecology, engineering, etc.). When at least one of the decision-makers randomizes its action, the resulting payoff of the decision-makers becomes a random variable. The classical approach resulting from von Neumann & Morgernstern utility theory suggests using the mathematical expectation of the payoff function also called *expected payoff*. The expected payoff has natural interpretation in statistics and learning theory based on the basic observation from the law of large numbers and ergodic theory.

If the game is played infinitely many times and the decision-makers always implement a fixed randomized action profile, then the time-average payoff obtained by the decision-makers converge with probability one to the expected payoff.

Now, consider the following two scenarios:

Scenario 1: Each decision-maker gets 0 surely.

Scenario 2: Each decision-maker gets a -10^6 with probability 1 if he/she looses and he/she gets $+10^6$ with probability 1 if he/she wins. Each decision-maker wins or looses with probability $1/2$.

Using the classical approach with the expected payoff criterion, the scenario 1 and the scenario 2 lead to the *same outcome*. It is clear that in scenario 2, the decision-maker faces a big risk which can be reflected by the variance (or higher moments) of the outcome. A risk-neutral decision-maker ignores the risk. However, it is clear that there is always a risk whenever the action profiles are selected at random.

The first step to take into consideration this phenomenon is to modify the expected payoff criterion to incorporate also variance, and possibly higher moments as well.

A criterion that takes into consideration all the moments of the random payoff could be

$$r_{j,\mu_j} = \frac{1}{\mu_j} \log \left[E e^{\mu_j r_j} \right]$$

where $\mu_j \neq 0$ is the so-called risk sensitivity index.

Using Jensen's inequality for $\mu_j < 0$, we have $r_{j,\mu_j} \leq E[r_j]$, and when $\mu_j > 0$,, the opposite inequality holds. In the first case ($\mu_j < 0$), the decision-maker is referred to as risk-averse and in the second case ($\mu_j > 0$) it is referred to as risk-seeking.

It should be transparent that the risk-sensitive payoff r_{j,μ_j} takes into consideration all the moments of the payoff r_j. Its Taylor expansion for μ_j close to zero yields

$$r_{j,\mu_j} \underbrace{\approx}_{\mu_j \sim 0} Er_j + \frac{\mu_j}{2} var(r_j) + o(\mu_j),$$

where $o(\mu_j)$ includes terms of higher order. A strategic form game with risk-sensitive decision-makers (one index μ_j per decision-maker) is called a *risk-sensitive game*.

Next we examine a specific class of stochastic games in which each decision-maker has its own state which he/she controls and makes a decision based on its own state.

2.5 Stochastic Games with Additional Individual States

In this section, we discuss the class of stochastic games with individual states and random numbers of interacting decision-makers, which is closely related to the mean-field model that we will study in the following chapters. We consider a stochastic game in discrete time in which each decision-maker has a finite number of states. For each individual state, the decision-maker has a finite number of actions.

In contrast to the classical formulation, here we describe explicitly the individual state evolution and the variability of the set of interacting decision-makers, leading to a stochastic game with *random number of interacting decision-makers*.

A stochastic game with additional individual states and a random set of interacting decision-makers is a collection

$$\Gamma = \left(\mathcal{I}, \mathcal{T}, \mathcal{S}, (\mathcal{X}_j, \tilde{\mathcal{A}_j}, \mathcal{A}_j, r_j)_{j\in\mathcal{I}}, q, \mathcal{B}^I\right)$$

where

- $\mathcal{I}$ is the set of decision-makers. The cardinality of $\mathcal{I}$ is I, assumed to be $I \geq 2$.
- $\mathcal{T}$ is the time horizon of the game.
- $\mathcal{S}$ is a set of environment states.
- For every decision-maker j, $\mathcal{X}_j$ is its own state space. A state has two components as follows: the type of the decision-maker and the internal state. The type is a constant during the course of the game. The state of decision-maker j at time t is denoted by $x^I_{j,t} = (\theta_j, y^I_{j,t})$ where θ_j is the type. The set of possible states $\mathcal{X}_j = \{1, 2, ..., \Theta\} \times \mathcal{Y}_j$ is finite. The set $\mathcal{Y}_j$ may include other parameters, such as space location, current direction, and so on.
- For every decision-maker $j \in \mathcal{I}$, $\tilde{\mathcal{A}_j}$ is the set of actions of that decision-maker. $\mathcal{A}_j : \mathcal{S} \times \mathcal{X}_j \longrightarrow 2^{\tilde{\mathcal{A}_j}}$ is a set-valued map (correspondence) that assigns to each state $(s, x_j) \in \mathcal{S} \times \mathcal{X}_j$ the set of actions $\mathcal{A}_j(s, x)$ that are available to decision-maker j, which is assumed to be non-empty. We denote the state-action space by

$$\begin{aligned} \text{SXA} = \Big\{&(s, x, a) : (s, x_1, \ldots, x_I) \in \mathcal{S} \times \textstyle\prod_{j\in\mathcal{I}} \mathcal{X}_j, \\ &a = (a_j)_{j\in\mathcal{I}}, \ a_j \in \mathcal{A}_j(s, x), \ j \in \mathcal{I}\Big\}. \end{aligned} \tag{2.10}$$

- For every decision-maker $j \in \mathcal{I}$, r_j : SXA $\longrightarrow \mathbb{R}$ is an instant payoff function for decision-maker j.
- q : SXA $\longrightarrow \Delta(\mathcal{S} \times \prod_j \mathcal{X}_j)$ is a transition function where $\Delta(\mathcal{S} \times \prod_j \mathcal{X}_j)$ is the space of probability distributions over $\mathcal{S} \times \prod_j \mathcal{X}_j$. The marginal of q is denoted by q_j, which represents the transition probabilities of state of decision-maker j.

(continued)

- $\mathcal{B}^I$ is a random set of decision-makers: $\mathcal{B}^I \subseteq \mathcal{I}$ denotes the set of decision-makers interacting at the current time. The payoff function will be denoted by $r_j^{\mathcal{B}^I}$.
- A long-term payoff functional for each decision-maker

How does such a dynamic game evolve?

The dynamic game starts at an initial $(s_0, x_0) \in \mathcal{S} \times \prod_{j\in\mathcal{I}} \mathcal{X}_j$ and is played as follows. Time is discrete, and time space is $\mathbb{N}$. A random set $\mathcal{B}_t^I$ of decision-makers are randomly selected for a one-shot interaction. The current state $s_t, x_t, \mathcal{B}_t^I$ is known by all the selected decision-makers. At each time $t \in \mathbb{N}$,, each decision-maker $j \in \mathcal{B}_t^I$, chooses an action $a_{j,t} \in \mathcal{A}_j(s_t, x_t)$, and receives an instant payoff $r_j(s_t, x_t, a_t)$, where $a_t = (a_{j,t})_{j\in\mathcal{I}}$, and the game moves to a new state according to a probability distribution given by $q(. \mid s_t, x_t, a_t) \in \Delta(\mathcal{S} \times \prod_{j'\in\mathcal{I}} \mathcal{X}_{j'})$. The payoffs of those decision-makers not selected are zero, and the payoff function $r_j^{\mathcal{B}_t^I}$ of a selected decision-maker j does not depend on the state actions of the nonselected decision-makers.

Note that stochastic game with individual states is not necessarily a particular case of the standard stochastic games (with full monitoring), because some components of the states may not be observed by the decision-maker. For example, the case where each decision-maker observes its own individual state but not the individual state of the other decision-makers leads to a partial monitoring problem. The tools for such problems need to account for partial observation in stochastic games. The associated dynamic programming principle leads to state space on the beliefs.

2.5.1 Policies and Strategies

We start with the case where the actions and the value of the past states are observed, that is, at each time t, the collection $\{s_0, x_0, \mathcal{B}_0^I, a_0, s_1, x_1, \mathcal{B}_1^I, a_1, \ldots, s_t, x_t, \mathcal{B}_t^I\}$ of states actions that were visited in the past and the actions that were chosen by all the decision-makers is known at time t. This assumption is too strong for most of network applications, and it needs to be relaxed to, for example, the scenario where the set of active decision-makers is not necessarily known by all the decision-makers and the states of the other decision-makers may not be known.

Histories

Under complete information of past play, a history of length $t + 1$ corresponds to the sequence

$$h_t = (s_0, x_0, \mathcal{B}_0^I, a_0, s_1, x_1, \mathcal{B}_1^I, a_1, \ldots, s_t, x_t, \mathcal{B}_t^I).$$

Some Classes of Strategies

A **pure** strategy σ_j of decision-maker j is a mapping that assigns to every finite history h_t an element of $\mathscr{A}_j(s_t, x_t)$.

A **mixed** strategy is a probability distribution on the set of the pure strategies. We denote the set of mixed strategies by Σ_j. A **behavioral** strategy of j is a collection of mappings $(\sigma_{j,t})_{t\geq 0}$ with $\sigma_{j,t}$ from the space of histories to a probability distribution over $\mathscr{A}_j(s_t, x_t) \subseteq \tilde{A}_j$.

Stationary strategies A simple class of strategies is the class of stationary strategies. A strategy profile $(\sigma_j)_{j\in\mathscr{I}}$ is stationary if $\forall\ j,\ \sigma_j(h_t)$ depends only on the current pair (s_t, x_t). A stationary strategy of decision-maker j can be identified as an element of the product space of $\prod_{(s,x)\in\mathscr{S}\times\mathscr{X}} \Delta(\mathscr{A}_j(s, x))$.

Every profile $\sigma = (\sigma_j)_{j\in\mathscr{I}} \in \prod_j \Sigma_j$ of mixed strategies, together with the initial state s_0, x_0 induces a probability distribution $\mathbb{P}_{s_0,x_0,\sigma}$ over the space of infinite plays $\text{SXA}^{\mathbb{N}}$. We denote the corresponding expectation operator by $E_{s_0,x_0,\sigma}$.

2.5.2 *Different Types of Payoffs*

We examine three classes of payoff functions to evaluate the sequences of payoffs that the decision-makers receive in a stochastic game. Let σ be a mixed strategy profile.

- In the case of a finite-horizon payoff, a decision-maker considers the cumulative payoff during the first T times. For every finite $T \in \mathbb{N}$, the finite-horizon payoff for decision-maker j is

$$F_{j,T}(s_0, x_0, \sigma) = E_{s_0,x_0,\sigma}\left[\sum_{t=0}^{T-1} r_j^{\mathscr{B}_t^l}(s_t, x_t, a_t)\mathbb{1}_{\{j\in\mathscr{B}_t^l\}}\right]. \tag{2.11}$$

  We denote the associated stochastic game by Γ_T.
- In the case of a discounted payoff, a decision-maker considers a discounted sum of his instantaneous payoffs: For different discount factors $0 \leq \beta_j < 1$, the β_j-discounted payoff for decision-maker j under the strategy σ is

$$F_{j,\beta}(s_0, x_0, \sigma) = (1-\beta_j)E_{s_0,x_0,\sigma}\left[\sum_{t=0}^{+\infty} \beta_j^t r_j^{\mathscr{B}_t^l}(s_t, x_t, a_t)\mathbb{1}_{\{j\in\mathscr{B}_t^l\}}\right]. \tag{2.12}$$

  The term $(1-\beta_j)$ is a normalization factor: Since $\sum_{t=0}^{+\infty} \beta_j^t = \frac{1}{1-\beta_j}$, by multiplying both sides by $(1-\beta_j)$, one gets $(1-\beta_j)\sum_{t=0}^{+\infty} \beta_j^t = 1$. We denote the associated stochastic game by Γ_β.
- In the case of a liminf (or limsup) payoff, in which a decision-maker considers the liminf of his long-run average payoffs. For decision-maker j,, the liminf payoff

is

$$F_{j,\infty}(s_0, x_0, \sigma) = E_{s_0,x_0,\sigma}\left[\underline{\lim}_{T\longrightarrow+\infty} \frac{\sum_{t=0}^{T-1} r_j^{\mathscr{B}_t^I}(s_t, x_t, a_t)\mathbb{1}_{\{j\in\mathscr{B}_t^I\}}}{\sum_{t'=0}^{T-1} \mathbb{1}_{\{j\in\mathscr{B}_{t'}^I\}}}\right]. \tag{2.13}$$

We will consider the liminf or the limsup when the limit of Cesaro-mean payoff does not exist. We denote the associated stochastic game by Γ_∞.

Next we introduction additive ϵ-solution concepts (see Chap. 1 for the multiplicative ϵ-solution concepts).

Definition 2.7 Let $\epsilon \geq 0$.

- A strategy profile $\sigma^* = (\sigma_j^*)_{j\in\mathscr{I}}$ is a T-stage ϵ-(Nash) equilibrium if

$$\forall(s_0, x_0) \in \mathscr{S} \times \prod_j \mathscr{X}_j,\ \forall\, j \in \mathscr{I},\ \forall\sigma_j \in \Sigma_j$$

$$F_{j,T}(s_0, x_0, \sigma^*) \geq F_{j,T}(s_0, x_0, (\sigma_j, \sigma_{-j}^*)) - \epsilon, \tag{2.14}$$

  where σ_{-j} means $(\sigma_i)_{i\neq j}$ and $(\sigma_j, \sigma_{-j}^*)$ denotes the profile

$$(\sigma_1^*, \ldots, \sigma_{j-1}^*, \sigma_j, \sigma_{j+1}^*, \ldots, \sigma_I^*).$$

- A strategy profile $\sigma^* = (\sigma_j^*)_{j\in\mathscr{I}}$ is a β-discounted ϵ-equilibrium if

$$\forall(s_0, x_0) \in \mathscr{S} \times \prod_j \mathscr{X}_j,\ \forall j \in \mathscr{I},\ \forall\sigma_j \in \Sigma_j$$

$$F_{j,\beta}(s_0, x_0, \sigma^*) \geq F_{j,\beta}(s_0, x_0, (\sigma_j, \sigma_{-j}^*)) - \epsilon, \tag{2.15}$$

- A strategy profile $\sigma^* = (\sigma_j^*)_{j\in\mathscr{I}}$ is a $\underline{\lim}$ ϵ-equilibrium if

$$\forall(s_0, x_0) \in \mathscr{S} \times \prod_{j'} \mathscr{X}_{j'},\ \forall\, j \in \mathscr{I},\ \forall\sigma_j \in \Sigma_j,$$

$$F_{j,\infty}(s_0, x_0, \sigma^*) \geq F_{j,\infty}(s_0, x_0, (\sigma_j, \sigma_{-j}^*)) - \epsilon, \tag{2.16}$$

- The payoff of an ϵ-equilibrium is called ϵ-equilibrium payoff.
- A 0-equilibrium (simply called—Nash—equilibrium) of the stochastic game is a strategy profile such that no decision-maker can benefit from by any unilateral deviation. A refined notion of (Nash) equilibrium (resp., ϵ-equilibrium) is subgame perfect equilibrium (resp., ϵ-subgame perfect equilibrium). This can be seen as the requirement for the credibility of an equilibrium strategy. A strategy profile σ^* is called a subgame perfect ϵ-equilibrium if $\sigma(h_t)$ is an ϵ-equilibrium strategy profile for every finite history h_t.

Note that the random set $\mathscr{B}_t^I$ can be seen as a state component. When $j \in \mathscr{B}_t^I$ its action space is $A_j(s_t, x_t)$ otherwise his action space is a set of dummy actions in terms of payoffs, i.e.,

$$B_j(s_t, x_t, \mathscr{B}_t^I) = \begin{cases} \mathscr{A}_j(s_t, x_t) \text{ if } j \in \mathscr{B}_t^I \\ \tilde{B}_j \text{ otherwise} \end{cases}$$

where the actions in $\tilde{B}_j$ do not affect the payoffs. Then, under complete information and perfect monitoring, the stochastic game with random number of interacting decision-makers can be transformed in the classical formulation with $(s_t, x_t, \mathscr{B}_t^I)$ as the *augmented state* of the dynamic game. Therefore, the following result follows:

Lemma 2.2

- *For every $I \in \mathbb{N}$,, the stochastic games with a random number of interacting decision-makers, Γ_T, Γ_β, have at least one* 0*-equilibrium.*
- *The stochastic game Γ_∞ has an equilibrium under* ***ergodicity conditions*** *[301]*
- *The discounted stochastic game Γ_β has at least one stationary* 0*-equilibrium. Moreover, Γ_β has at least one subgame perfect ϵ-equilibrium.*

The first and second statements are classical results in stochastic games with individual states; see [324] and the references therein. The proof of the last statement can be found in [301]. □

2.5.3 Computation of Equilibria

Under discount factors, complete information and perfect recall and fixed set of decision-makers during all the interaction, we can write the Bellman-Shapley operator to compute the equilibria (the solutions obtained by dynamic programming are also subgame perfect equilibria and lead to Markovian strategies.) The operator $\mathscr{O} = (\mathscr{O}_j)_{j \in \mathscr{I}}$ is given by

$$\mathscr{O}_j(V)[s, x] = \max_{u_j} \left\{ (1 - \beta_j) r_j(s, x, u) + \beta_j \sum_{s', x'} q(s', x' | s, x, u) V_j(s', x') \right\}$$

where $u = (u_1, \ldots, u_I)$ is a stationary strategy profile. Let $\mathscr{O}_j(V) = (\mathscr{O}_j(V)[s, x])_{s,x}$. A fixed point solution of $\mathscr{O}$ gives the equilibrium payoff as well as the β-discounted 0-equilibrium obtained by collecting the maximizers of

$$(1 - \beta_j) r_j(s, x, a) + \beta_j \sum_{s', x'} q(s', x' | s, x, a) V_j(s', x').$$

Under the same assumptions, one can write a similar dynamic programming principle (by omitting the factor β_j) for the finite horizon case (backward induction).

With the variable set of interacting decision-makers, the dynamic programming principle becomes

$$\mathscr{O}_j(V)[s, x, \tilde{\mathscr{B}}_0^I, u_{-j}^*] = \max_{u_j} \Bigg\{ (1-\beta_j) r_j^{\tilde{\mathscr{B}}_0^I}(s, x, u)$$

$$+ \beta_j \sum_{s', x', \tilde{\mathscr{B}}_1^I} L(s', x', \tilde{\mathscr{B}}_1^I | s, x, \tilde{\mathscr{B}}_0^I, u) V_j(s', x', \tilde{\mathscr{B}}_1^I, u_{-j}^*) \Bigg\}$$

where $L(s', x', \tilde{\mathscr{B}}_1^I | s, x, \tilde{\mathscr{B}}_0^I, u) = q(s', x' | s, x, u) q_2(\tilde{\mathscr{B}}_1^I | \tilde{\mathscr{B}}_0^I)$, $q_2(\tilde{\mathscr{B}}_1^I | \tilde{\mathscr{B}}_0^I) = \mathbb{P}(\mathscr{B}_t^I = \tilde{\mathscr{B}}_1^I)$ is the probability that $\tilde{\mathscr{B}}_1^I$ is the set of active decision-makers at the current time. The set $\tilde{\mathscr{B}}_1^I$ describes all possible subsets of $\mathscr{I}$ (except the empty set). So, one has $2^I - 1$ elements in addition to the set $S \times \prod_{j' \in \mathscr{I}} \mathscr{X}_{j'}$. Naturally, several questions arise on the computational complexity of this system. The first question is how to reduce the complexity of solving the dynamic programming equations. The second question is whether the dynamic programming principle still applies when we have less information and relaxed observation assumptions.

These questions are crucial in terms of computation of the solutions and the application of stochastic games in experimental interactive scenarios. In next subsection, we examine the case where the knowledge of the transition probabilities can be relaxed via Q-learning and stochastic approximations.

2.5.4 *Q-Values*

While an optimal strategy can, in principle, be obtained by the methods of dynamic programming, policy iteration, and value iteration, such computations are often prohibitively time-consuming, in particular if the size of the state space grows exponentially with the number of state variables, a phenomenon referred to by Bellman as the *curse of dimensionality*. Similarly, the size of the action space can also lead to computational intractability. In the single decision-maker case, Q-learning [364] perhaps the most well-known example of reinforcement learning, is a stochastic-approximation-based solution approach to solving the fixed-point equation

$$Q^*(s, x, a) = (1-\beta) r(s, x, a) + \beta \sum_{s', x'} q(s', x' \mid s, x, a) \left(\sup_{b \in \mathscr{A}(s', x')} Q^*(s', x', b) \right)$$

One nice feature here is to learn the Q-values without any knowledge of the transition probabilities $q(s', x' \mid s, x, a)$. The iterative version is given by

$$\begin{aligned} Q_{t+1}(s,x,a) = Q_t(s,x,a) + \lambda_t(s,x)\,[(1-\beta)r(s,x,a)+ \\ \beta\left(\sup_{b\in\mathscr{A}(s',x')} Q_t(s',x',b)\right) - Q_t(s,x,a)\big] \end{aligned} \tag{2.17}$$

where $\lambda_t(s,x) \geq 0$ is a learning rate function satisfying

$$\sum_t \lambda_t(s,x) = +\infty, \ \sum_t \lambda_t(s,x)^2 < +\infty,$$

which is a standard assumption in stochastic approximation. In the single decision-maker case, the Q-learning algorithms show good properties. However, the convergence of the algorithm in the multi-decision-maker stochastic game case is still an open and challenging issue. In particular, when there are multiple equilibria, it is not clear what convergence properties the Q-values have.

2.5.5 Classes of Stochastic Games

In this section, we provide a brief overview of some important classes of stochastic games and the most important issues that arise in such games [9, 187, 341, 343]. A two-decision-maker stochastic game is called

- *a unichain stochastic game,* if with respect to any profile of strategies there is only one ergodic set of states.
- *a perfect information stochastic game*, if the state space $\mathscr{S}$ can be partitioned into $\mathscr{S}_j,\ j \in \{1,2\}$ such that $|\mathscr{A}_j| = 1$ for all $s \in S_j,\ j \in \{1,2\}$.
- *a switching control stochastic game*, if $\mathscr{S}$ can be partitioned into $\mathscr{S}_j,\ j \in \{1,2\}$ such that $q_j(.|s, a(s))$ is independent of $a_{-j}(s)$ for all $a_j \in \mathscr{A}_j(s)$; $s \in \mathscr{S}_j$; $j \in \{1,2\}$.
- *a stochastic game with additive reward and additive transition (ARAT) structure*, if the instantaneous payoff functions and transitions probabilities are separable and additive, i.e., $r_j(s,a) = \sum_j \tilde{r}_j(s,a_j),\ q_j(.|s,a) = \sum_j q_j(.|s,a_j)$ for $a_j \in \mathscr{A}_j(s),\ s \in \mathscr{S}$.
- *a stochastic game with state independent transitions (SIT)*, if the cardinality of the action spaces is independent of the state and, assuming that $\mathscr{A}_j(s) = \mathscr{A}_j, \forall j, s$, it holds that $q_j(.|s,a) = q(.|s,a) = q(.|a) = \bar{q}$, for all $s \in \mathscr{S}$.
- *a stochastic game with separable rewards and state independent transitions (SER- SIT),* if it is a SIT stochastic game and, assuming that $\mathscr{A}_j(s) = \mathscr{A}_j \forall\ s$, $r_j(s,a) = r_j^1(s) + r_j^2(a),\ \forall\ (s,a,j)$
- *a repeated game with absorbing states*, if all the states but one are absorbing.
- *a recursive stochastic game*, if the payoffs are equal to zero in all nonabsorbing states.

- *a stochastic game over a product state space* if the action space of decision-maker j only depends on the j-th coordinate of the state w, and the jth coordinate of the transitions from any state s only depends on the j-th coordinate s_j of the state and on the action $a_j(s)$ chosen by decision-maker j,

Theorem 2.8 *Consider a general-sum stochastic game.*

- *In unichain stochastic games, stationary equilibria exist ([142, 278, 298]).*
- *In perfect information stochastic games, equilibria exist ([229, 349]).*
- *In switching control stochastic games, ϵ-equilibria exist for all $\epsilon > 0$ [349].*
- *In ARAT stochastic games, equilibria exist (see [349]).*
- *In SIT stochastic games, ϵ-equilibria exist for all $\epsilon > 0$ (see [348]).*
- *In SER-SIT stochastic games, stationary equilibria exist such that the prescribed mixed actions are state independent (see [273]).*
- *In all two-decision-maker stochastic games, [357] proved that for all $\epsilon > 0$, ϵ-equilibria exist.*
- *In stochastic games on product space, 0-equilibria exist [155]*
- *Constrained discounted stochastic games with finite number of decision-makers and finite number of actions per state, 0-equilibrium exists under Slater conditions*

2.5.6 Dynamic Potential Games

We say that a game $G = (\mathcal{I}, \mathcal{A}_j, \tilde{r}_j)$ is a potential game if there exists a global function P such that for all decision-maker j, action profiles a and a', one has

$$\tilde{r}_j(a_j, a_{-j}) - \tilde{r}_j(b_j, a_{-j}) = P(a_j, a_{-j}) - P(b_j, a_{-j}).$$

In the case of *action-independent transitions*, we show the existence of pure 0-equilibrium when the auxiliary games are potential games [333, 339, 342].

Below, we present sufficient conditions for the existence of pure 0-equilibrium in the game Γ_β defined in Sect. 2.3.4.

Proposition 2.1 *The following holds:*

- *If in the discounted stochastic game Γ_β the payoff function is the same for all the decision-makers, i.e., $r_j(.) = r(.)$,, then the discounted stochastic game is a team problem or a coordination game and has at least one pure (Nash) 0-equilibrium.*
- *If in the discounted stochastic game Γ_β, the transitions between the states are independent of the actions played and for each state (s, x), the finite game $G(s, x) = (\mathcal{I}, \mathcal{A}_j(s, x), \tilde{r}_j(s, x, .))$ is a potential game, then the discounted stochastic game has at least one pure (Nash) 0-equilibrium, where $\tilde{r}_j(s, x, .) = (1 - \beta_j)r_j(s, x, .) + \beta_j \sum_{s',x'} q(s', x'|s, x)V_j(s', x')$.*

Proof The first result follows from the fact that the discounted stochastic game can be seen as a discounted team problem. By the use of Bellman optimality, we associate a dynamic programming principle to the common payoff function r. Then, we use the result of Blackwell [69] which gives the existence of optimal stationary pure strategies. $V \in \mathbb{R}^{|\mathcal{S}\times\mathcal{X}|}$, $V(s,x) = \max_{u_j\in\Delta(\mathcal{A}_j(s,x))}\{r(s,x,u) + \beta\sum_{s',x'} q(s',x' \mid s,x,u)V(s',x')\}$.

We now prove the second result. Since the decision-makers have no influence on the transition probabilities, each decision-maker does its best in the visited states. For each state (s,x),, one has a pure (Nash) equilibrium $a^*_{s,x}$ of the auxiliary game with

$$r_j(s,x,a) + \beta_j \sum_{s',x'} q(s',x' \mid s,x)V_j(s',x'),$$

by using Monderer and Shapley [248]. The collection of the pure equilibria $a^* = (a^*_{s,x})_{(s,x)\in\mathcal{S}\times\mathcal{X}}$ is a stationary pure equilibrium of the discounted stochastic game. This completes the proof. □

2.6 Asymptotics of Markov Games

We consider a class of stochastic games with a finite number of resource/environment states, individual states, and actions per state. At each time stage of the game, a random set of decision-makers interact. The states and actions of all the interacting decision-makers determine together the instantaneous payoffs and the transitions to the next states. We study the convergence of the stochastic game with a variable set of interacting decision-makers when the total number of possible decision-makers grows without bound. Within such a setting, the optimal payoffs and the mean-field equilibrium payoffs are solution of a system of coupled backward-forward equations which can be rewritten as one global Bellman system using an extended state space. The limiting games are closely related to the so-called discrete time anonymous sequential games or to differential population games. Using (multidimensional) diffusion processes and the indistinguishability per type assumption, a generic mean-field convergence to stochastic differential equation is given.

Stochastic games and interactive Markov decision processes are widely used for modeling sequential decision-making problems that arise in engineering, computer science, operations research, and social sciences. However, it is well known that many real-world problems modeled by stochastic games have huge state and/or action spaces, leading to the well-known *curse of dimensionality* that makes solution of the resulting models intractable. In addition, if the size of the system grows, the number of parameters (states, actions, transitions) explodes exponentially.

In the mean-field stochastic Markov game modeling in discrete time, there must be an equation to express the dynamic optimization problem of each decision-maker. Usually this involves one equation for each decision-maker. If decision-makers are classified together by similar decision-maker classes, there is one equation per class. This equation is generally a Bellman-Shapley equation, since a large proportion of optimization problems fall within the framework of dynamic programming. Hence, the Bellman-Shapley equations will be used to compute optimal behavioral strategies. An equation is also needed to express the class' behavior, the mean-field behavior of each class. The dynamics of the distribution is governed by a Kolmogorov forward equation. In this Kolmogorov forward equation, the optimal behaviors of the decision-makers occur as data, since it is the infinite collection of individual behaviors that is aggregated and constitutes collective behavior by consistency. Thus, the modeling of a the behavior of a group of decision-makers naturally leads to a BS-K (Bellman-Shapley and Kolmogorov) system of equations. The discrete BS-K have been studied by Jovanovic & Rosenthal in the eighty's. The novelty in their study is that the mean-field games formalism involves *the density of decision-makers on the state space can enter the Bellman-Shapley equation*. Thus, the mean-field equilibrium is defined by a BS-K system in which the Bellman-Shapley equations are doubly coupled: Individual behaviors are given for the Kolmogorov forward equation, and, at the same time, the distribution of decision-makers in the state space enters in the Bellman equation which is completely innovative. This means that decision-makers can incorporate into their preferences the density of states/actions of other decision-makers at the anticipated equilibrium. Therefore each decision-maker can construct his strategy by taking into account the anticipated distribution of strategies and of the actions of other decision-makers. In order to examine the limiting behavior of Markov games when the asymptotic is taken on the number of decision-makers, one takes into account a crucial assumption. This assumption is that the state transitions and the cost functions are asymptotically invariant by permutation within the same class of decision-makers. This property is referred to as *Asymptotic Indistinguishability Per Class*. Under this setting, mean-field equilibria can be examined. We develop a mean-field limit for stochastic games with a variable number of interacting decision-makers. Mean-field solutions are obtained by identifying a consistency relationship between the individual-state-mass interaction such that in the population limit, each individual optimally responds to the mass effect and these individual strategies also collectively produce the same mass effect presumed initially. This leads to a coupled system of forward/backward equations. The system can be rewritten as one single backward equation known as master Bellman equation and its master adjoint system (MASS).

2.6.1 Mean-Field Interaction Without Decision

Mean-field interactions with spatially distributed decision-makers and types can be described as a sequence of dynamic games. Since the population profile involves many decision-makers for each type or class, a common approach is to ignore individual decision-makers and to use continuous variables to represent the aggregate average of *type-location-secondary actions*. The validity of this method has been proven only under specific time-scaling techniques and regularity assumptions. The *mean-field limit* is then modeled by a state and location-dependent process. The large population approach has been suggested in [361], and [258, 259] proposed the population mass action interpretation. This type of aggregate models, also known as nonatomic or population games, has been studied by [363] in the deterministic and stationary setting of identical decision-makers. We describe below some recent advances in the population limit context.

In the infinite population games, an equilibrium is characterized by a fixed inclusion: The support of the population profile is included in argmax of the payoff function,

$$\overline{\{x,\ m_x > 0\}} = \text{support}(m) \subseteq \arg\max_{x'} r_{x'}(m) \tag{2.18}$$

In other words, if the fraction of decision-makers under a specific action is nonzero, then the payoff of the corresponding action is maximized. This large-scale methodology has inherent connections with evolutionary game theory where one has a large number of interacting decision-makers in different subpopulations. Different solution concepts such as evolutionarily stable states or strategies, neutrally stable strategies, and invadable states have been proposed, and several applications can be found in evolutionary biology, ecology, control design, networking, economics, and transportation.

2.6.2 A Basic Mean-Field Interaction Model

We consider interacting decision-makers in discrete time, with a finite number of states. The decision-makers share local resources, which have a finite number of states. The decision-makers are observable only through their states. In the limit, when the number of decision-makers goes to infinity, we show that, under suitable conditions, the system can be approximated by a deterministic, usually nonlinear dynamical system called the mean-field limit. The mean-field is in discrete or continuous time, depending on how the model scales with the number of decision-makers. If the expected number of transitions per decision-maker per time slot vanishes when n grows, then the limit is in continuous time. Otherwise, the limit is in discrete time. One of the main advantages of the mean-field approach is that it could drastically reduce the number of parameters in the analysis of large systems by using an aggregate dynamical system description.

The first assumption is that x_t^I is a homogeneous Markov chain. The second one is that the transition kernel L^I of x_t^I is invariant under any permutation of the labeling of the decision-makers. That is,

$$L^I(x_1, .., x_I, s; x_1', .., x_I', s') = L^I(x_{\sigma(1)}, .., x_{\sigma(I)}, s; x_{\sigma(1)}', .., x_{\sigma(I)}', s')$$
$$= \mathbb{P}(x_{1,t+1}^I = x_1', .., x_{I,t+1}^I = x_I', s_{t+1} = s' \mid x_{1,t}^I = x_1, .., x_{I,t}^I = x_I, s_t = s), \tag{2.19}$$

for any permutation σ of the index set $\{1, 2, \ldots, I\}$. Then, x_t^I is called a *mean-field interaction model* with n decision-makers. The occupancy measure is defined as

$$M_t^I(x) = \frac{1}{I}\sum_{j=1}^{I} \mathbb{1}_{\{x_{j,t}^I = x\}}.$$

By the invariance assumption, (s_t, M_t^I) is a homogeneous Markov chain. Since we are interested in the asymptotic results when I grows to infinity, we need to set the proper time scaling. Assume that there exists a vanishing δ_I, called intensity (or step-size), which can be understood as the expected number of decision-makers that do a transition in one time slot. Also define the drift $f^I(m, s) = \mathbb{E}(M_{t+1}^I - M_t^I \mid M_t^I = m, s_t = s)$, which is the expected change in the occupancy measure in one time slot. The main scaling assumptions in this model are the following:

H1 The resource does not scale with I; that is, there exists L such that

$$\lim_{I\to\infty} L_{s;s'}^I(m) = L_{s;s'}(m), \quad \forall s, s' \in \mathcal{S}$$

where

$$L_{s;s'}^I(m) = \mathbb{P}(s_{t+1} = s' \mid M_t^I = m, s_t = s).$$

Moreover, the transition matrix $L(m)$ is ergodic; i.e., it is possible to eventually reach from every state s to every other state s' with positive probability.

H2 The intensity vanishes at a rate δ_I; that is, there exists $\delta_I > 0$, such that $\lim_{I\to\infty} \delta_I = 0$, and

$$\lim_{I\to\infty} \frac{f^I(m, s)}{\delta_I} = f(m, s), \quad \forall\, s \in \mathcal{S}.$$

H3 Second moment of the number of decision-maker transitions per time slot is bounded; that is, $\forall s \in \mathcal{S}$,

$$\int_{m'} \mathbb{1}_{\{\|m-m'\|\leq 2\}}(m_x - m_x')(m_{x'} - m_{x'}')\frac{1}{\delta_I}\mathcal{L}^I(dm'; m, s) \to 0$$

where

$$\mathscr{L}^I(m'; m, s) := \sum_{\substack{(x'_1,\ldots,x'_I) \\ \frac{1}{I}\sum_{j=1}^{I} \delta_{x'_j} = m'}} L^I(x'; x, s).$$

H4 $L^I_{s;s'}(m)$ is a smooth function of δ_I and m.

H5 $f^I(m, s)$ is a smooth function of δ_I and m.

Later, we will see that these assumptions are indeed relaxed in our approach. The following theorem can be used to approximate M^I_t by the solution $m(\delta_I t)$ of an ODE with the same initial condition.

Theorem 2.9 ([324]) *If $M^I(0) \to m_0$ in probability [resp., in mean square] as $n \to \infty$ then $\sup_{0 \le \tau \le T} \| M^I(\tau) - m(\tau) \| \to 0$ in probability [resp., in mean square], $\tau \in [t\delta_I, (t+1)\delta_I]$, where $m(\tau)$ satisfies*

$$\begin{cases} \dot{m} &= \tilde{f}(m) \\ m(0) &= m_0, \end{cases}$$

where

$$\tilde{f}(m) = \sum_{s \in \mathscr{S}} w_s(m) f(m, s)$$

and $w(m)$ is the invariant probability of the transition matrix $L(m)$.

2.6.3 Mean-Field Interactions with Decisions

We now include a decision component in the above model. We consider large populations of decision-makers in which frequent interactions occur between small numbers of chosen individuals. Each interaction in which a decision-maker is involved can be described as one stage of a dynamic game. The state and actions of the decision-makers at each stage determine an immediate payoff for each decision-maker as well as the transition probabilities of a controlled Markov chain. Each decision-maker wants to optimize its expected payoff averaged over time. This model with a finite number of decision-makers is in general difficult to analyze because of the huge state space required to describe actions and dynamics of all the decision-makers. Then taking the asymptotics as the number of decision-makers grows to infinity, the whole behavior of the population is replaced by a deterministic limit (under the assumptions H1–H5) that represents the system's state, which is fraction of the population at each individual state that use a given action [333, 339, 342].

In [313, 324, 335], the authors study the asymptotic dynamic behavior of the system in which the population profile evolves in time. For large I, under mild assumptions, the mean-field converges to a deterministic measure that satisfies a nonlinear ordinary differential equation for any stationary strategy. This mean-field interaction is asymptotically equivalent to a Markov decision evolutionary game. The mean-field asymptotic calculations for large n for given choices of strategies allow one to compute the equilibrium of the game in the asymptotic regime.

Theorem 2.10 *If $M_0^I \to m_0$ in probability as $I \to \infty$ then, for any stationary strategy u, and any time t, M_t^I converges in distribution to the solution of*

$$\begin{cases} \dot{m} &= \tilde{f}(u, m), \\ m(0) &= m_0 \end{cases}$$

A proof of Theorem 2.10 can be derived from [324]. □

One of the limitations of the classical mean-field approach is the approximation of an inherently stochastic system with a *deterministic* representation (ordinary differential equation or inclusion). We propose an approach in which we obtain the main advantage of the mean-field approach, that is, the reduction of the parameters in the analysis of large systems but adding a random or "noisy" component. This new addition could lead to a more realistic mathematical model of the original situation, and many local and simultaneous interactions may occur at the same time making the moments unbounded. In this context, the above results are not applicable anymore (because the assumption H3 below may not be satisfied). A typical scenario where these assumptions are not satisfied is when many decision-makers do a transition without bound and the second order term does not vanish when the number of decision-makers goes to infinity.

The idea of the proposed analysis is that, if the third order term in the Taylor approximation is bounded, a convergence to a stochastic mean-field limit can be established even if the noise is not negligible. Inspired from the work of [305] based on multidimensional diffusion process, we are able to establish mean-field convergence to *non-deterministic* differential equations and extend the previous works in mean-field interaction models (with and without controls). This novel mean-field limit, which is stochastic, is called *"noisy" mean-field limit* and is applied to malware propagation in opportunistic networks. Using multidimensional diffusion process approximation, we provide a general mean-field convergence to stochastic differential equations. Based on Itô's formula, we characterize the payoff evolutions as well as the mean-field optimality conditions.

In next subsection, we formulate the stochastic game with a random set of interacting decision-makers. We first provide a general convergence result to mean-field which is characterized by a stochastic differential equation, and the payoff evolution is a solution to a partial differential equation. Following that, we focus on deterministic mean-field convergence under suitable conditions. We characterize the mean-field optimality by a coupled system of

- Bellman-Shapley optimality equations combined with a discrete time mean-field (forward) equation,
- Hamilton-Jacobi-Bellman equation combined with a mean-field differential equation.

We examine the existence of solutions of these systems and show the existence of *mean-field equilibrium*. We establish a connection to differential population games and anonymous sequential games. Finally, we apply mean-field to opportunistic wireless large networks.

We assume throughout that a decision-maker does not have access to the states of the other decision-makers. Each decision-maker has then its own *private history* $h_{j,t} = (s_0, x_{j,0}, O_0, a_{j,0}, \ldots, s_t, x_{j,t}, O_t, a_{j,t})$ where O_t is a random variable indicating if j is active or not, $O_t = \mathbb{1}_{\{j \in \mathscr{B}_t^I\}}$. In particular, a decision-maker does not observe the past actions of other decision-makers and does not know with whom he is currently interacting (the subset $\mathscr{B}_t^I$ is unknown to the decision-makers). We assume that $\mathscr{A}_j(s, x) = \mathscr{A}_j(s, x_j),\ \forall\, x_{-j} := (x_{j'})_{j' \neq j} \in \prod_{j' \neq j} \mathscr{X}_{j'}$.

2.6.4 Basic Mean-Field Interaction with Decisions

We introduce a decision and control-dependent mean-field interaction model. The finite version of this model is a particular case of stochastic games with individual states. We restrict our attention to a particular class of behavioral strategies for which we are able to establish the mean-field convergence. This restriction is due to the fact that when the number of decision-makers goes to infinity, the dimension of the set of stationary strategies goes to infinity as well. The mean-field approach offers an alternative to direct computation of infinite systems known as curse of dimensionality. By letting the size of the system go to infinity, the discrete stochastic game problem is replaced by a limit of a system of Hamilton-Jacobi-Bellman equations coupled with a mean-field limit ordinary differential equation (ODE) or coupled system of Bellman-Shapley optimality equations and discrete mean-field evolution, which are deterministic and where the dimensionality of the original system has been transformed into the mass behavior of the system.

Time $t \in \mathbb{N}$ is discrete. The *global state* of the system at time t is $(s_t, x_t^I) = (s_t, x_{1,t}^I, ..., x_{I,t}^I)$. Denote by $a_t^I = (a_{1,t}^I, \ldots, a_{I,t}^I)$ the action profile at time t. The system (s_t, x_t^I) is Markovian once the action profile a_t^I are drawn under Markovian strategies. We denote the set of Markovian strategies by $\mathscr{U}$. Define M_t^I to be the current population profile, i.e.,

$$M_t^I(x) = \frac{1}{I} \sum_{j=1}^{I} \mathbb{1}_{\{x_{j,t}^I = x\}}.$$

At each time t,, M_t^I is in the finite set $\{0, \frac{1}{I}, \frac{2}{I}, \ldots, 1\}^{|\mathscr{X}|}$, and $M_t^I(x)$ is the fraction of decision-makers which belong to population of individual state x. For a subset $X_1 \subseteq \mathscr{X}$,, define

$$M_t^I[u](X_1) := \frac{1}{I}\sum_{j=1}^{I}\delta_{\{x_{j,t}^I[u]\in X_1\}}.$$

Similarly, we associate the process $U_a^I(t) = \frac{1}{I}\sum_{j=1}^{I}\mathbb{1}_{\{a_{j,t}^I=a\}}$ with the fraction of actions and

$$\tilde{M}_t^I(x,a)) = \frac{1}{I}\sum_{j=1}^{I}\mathbb{1}_{\{x_{j,t}^I=x, a_{j,t}^I=a\}}.$$

Strategies and Random Set of Interacting Decision-Makers In the time slot t,, an ordered list $\mathscr{B}_t^I$ of decision-makers in $\{1, 2, \ldots, I\}$, without repetition, is selected randomly as follows. First we draw a random number of decision-makers k_t such that $\mathbb{P}(|\mathscr{B}_t^I| = k \mid M_t^I = m) =: J_k^I(m)$ where the distribution $J_k^I(m)$ is given for any $n, m \in \{0, \frac{1}{I}, \frac{2}{I}, \ldots, 1\}^{|\mathscr{X}|}$. Second, we set $\mathscr{B}_t^I$ to an ordered list of k_t decision-makers drawn uniformly at random among the $I(I-1)...(I-k_t+1)$ possible ones.

Each decision-maker such that $j \in \mathscr{B}_t^I$ takes part in a one-shot interaction at time t, as follows. First, each selected decision-maker $j \in \mathscr{B}_t^I$ chooses an action $a_{j,t} \in \mathscr{A}(s, x_j)$ with probability $u(a_j \mid s, x_j)$ where (s, x_j) is the current decision-maker state. The stochastic array u is the strategy profile of the population.

Denote the current set of interacting decision-makers by $\mathscr{B}_t^I = \{j_1, \ldots, j_k\}$. Given the actions $a_{j_1}, ..., a_{j_k}$ drawn by the k decision-makers, we draw a new set of individual states $(x'_{j_1}, ..., x'_{j_k})$ and resource state s' with probability $L_{s;s'}^I(k, m, a)$, where a is the vector of the selected actions by the interacting decision-makers.

We assume that for any given Markovian strategy, the transition kernel L^I is invariant under any permutation of the index of the decision-makers within the same type. This implies in particular that the decision-makers are only distinguishable through their individual states. Moreover, this means that the process M^I is also Markovian once the sequence of strategies is given. Denote by $w_{s,s'}^I(u, m)$ the marginal transition probability between the resource states. Given any Markov strategy and any vector m of $\Delta(\mathscr{X})$, the resource state generates an irreducible Markov decision process with limiting invariant measure $w_s(u, m)$. Then, we can simplify the analysis by fixing the resource state $S_t = s$ without losing generality.

The model is entirely determined by the probability distributions J^I, the transition kernels L^I, and the strategy profile u. The next theorem provides the mean-field limit as the population size grows to infinity.

Theorem 2.11 *Let* $\mathscr{M}_I^d = \{m \mid Im \in \mathbb{N}^I\}$. *Suppose that*

- *For every* s,, *the function* $w_s(u, m)$ *is continuously differentiable in* (m, u).

- *$\exists\, 0 < \delta_I, \epsilon_I \searrow 0$, and a continuously differentiable function $f : \mathbb{R}^d \times \mathscr{U} \times \mathscr{S} \longrightarrow \mathbb{R}^d$ such that*

$$\lim_I \sup_{u\in\mathscr{U}} \sup_{\|m\|\leq 1} \| \frac{f^I(m,u,s)}{\delta_I} - f(m,u,s) \| = 0,$$

where $x \in \mathscr{X}$ and $f^I_x(m,u,s) =$

$$\int_{m'\in\mathscr{M}^d_I} \mathbb{1}_{\|m'-m\|\leq 2}(m'_x - m_x)\mathscr{L}^I(dm';m,u,s),$$

- $$\sup_I \sup_{u\in\mathscr{U}} \frac{1}{\delta_I} \int_{m'\in\mathscr{M}^d_I} \| m' - m \| \mathscr{L}^I(dm';m,u,s) < +\infty$$

- $\lim_I \sup_{u\in\mathscr{U}} \frac{1}{\delta_I} \int_{m'\in\mathscr{M}^d_I} \mathbb{1}_{\|m'-m\|>\epsilon_I} \| m' - m \| \mathscr{L}^I(dm';m,u,s) = 0,$
- *$M^I(0) = m^I_0$ converges to $m_0 \in \Delta(\mathscr{X})$.*

Then, for all $\epsilon > 0$, $T < +\infty$,

$$\lim_I \mathbb{P}\left(\sup_{t\in[0,T]} \| M^I_{\frac{t}{\delta_I}} - m(t,u,m_0) \| > \epsilon \right) = 0,$$

where $m(t,u,m_0)$ is the unique solution of the ordinary differential equation $\dot{m} = \tilde{f}(u,m)$ starting from $m_0 \in \Delta(\mathscr{X})$, where

$$\tilde{f}(u,m) := \sum_{s\in\mathscr{S}} w_s(u,m) f(u,m,s)$$

A proof can be obtained from [324]. □

2.6.4.1 Big Step Size

We analyze the case where the step size is not vanishing when I goes to infinity. In that case, the mean-field limit is in discrete time and driven by the probability transition $L_t(u,m)$. Given an initial population profile m_0 and a terminal payoff, the sequence of population profiles $\{m_t\}_t$ is driven by the transition probabilities $\{L_{t,s,x,s',x'}(u,m_t)\}_t$.

$$m_{t+1}(x) = \sum_{x'\in\mathscr{X}} m_t(x') L_{t,x,x'}(u,m_t), \tag{2.20}$$

where $L_{t,x,x'}(u,m) = \sum_{k\geq 0}\sum_s w_s(u,m)L_{t,s,x,x'}(u,m;k)J_k(m)$, $L_{t,s,x,x'}(u,m;k)$ is the limiting probability transition from x to x' when the resource state is s and the number of interacting decision-makers is k. Combining with the Bellman-Shapley optimality criterion, one arrives at the following system in the finite horizon case:

$$\begin{cases} v_{j,t}(s,x) = \max_{u_j}\{r(s,x,u,m_t) + \sum_{s',x'} L(s',x'|s,x,u,m_t)v_{j,t+1}(s',x')\} \\ m_{t+1}(x) = \sum_{x'\in\mathcal{X}} m_t(x')L_{t,x',x}(u,m_t) \end{cases}$$

For the discounted case, one gets

$$\begin{cases} v_{j,\beta}(s,x) = \max_{u_j}\ \{(1-\beta_j)r(s,x,u,m_t) + \beta_j \sum_{s',x'} L(s',x'|s,x,u,m_t)v_{j,\beta}(s',x')\} \\ m_{t+1}(x) = \sum_{x'\in\mathcal{X}} m_t(x')L_{t,x',x}(u,m_t), \end{cases}$$

and for the average case

$$\begin{cases} h_j + v_j(s,x) = \max_{u_j}\ \{r(s,x,u,m) + \sum_{s',x'} L(s',x'|s,x,u,m)v_j(s',x')\} \\ m_{t+1}(x) = \sum_{x'\in\mathcal{X}} m_t(x')L_{t,x',x}(u,m_t) \end{cases}$$

where h_j is a constant.

Definition 2.8 (Anonymous Sequential Population Game) To a game as defined in Sect. 2.6.3, we associate a macroscopic population game, defined as follows. Each member j of the population has state $x_{j,t}$ and a population profile $m_t[u]$. The initial conditions of the game are $s_0 = s$, $X_{j,0} = x$, $m_0[u] = m_0$. The population profile is solution to the discrete time evolution, and $s_t, x_{j,t}$ evolves as a jump process given by the marginal of L which depends on m_t and u. Further, let $F_T(s,x,u',u,m)$ be the T-stage payoff of decision-maker j in this game, given that $X_{j,0} = x$ and m_0, i.e.,

$$F_T(s,x,u',u_t,m) = E\left(g(m_T) + \sum_0^T r(s_t,x_{j,t},u',u_t,m_t) \mid s_0 = s, x_{j,0} = x, m_0\right), \tag{2.21}$$

where m_t evolves as in (2.39). The discounted F_β and limiting average payoff F_∞ are defined in the same way.

2.6.4.2 Connection to Anonymous Sequential Game

Under the assumptions of Definition 2.8, the discrete-time mean-field stochastic game is equivalent to a large population anonymous sequential game. Also, the discounted case is an anonymous sequential population game in the sense of [210]: The instantaneous payoff of an individual with strategy v, state x, when the population profile (mass-behavior) is m driven by u is defined by $r(s,x,v,u,m)$. The transition between the states is given by $L(s',x'|s,x,v,u,m)$.

If (v, u, m) satisfies the following system:

$$\begin{cases} m_{t+1}(x') = \sum_x m_t(x) L_{x;x'}(u^*, m) \\ \forall t, x, a \text{ such that } m_x u(a|s, x) > 0 \Longrightarrow \\ a \in \arg\max_b (1-\beta) r(s, x, b, m) + \beta \sum_{s',x'} v_{t+1}(s', x', m) L(s', x'|s, x, b, u, m), \end{cases}$$

then we say that (v, u, m) is a *mean-field equilibrium*.

Proposition 2.2 *Under the assumptions of Definition 2.8, the finite horizon (resp., the discounted) anonymous sequential population game has at least one mean-field equilibrium.*

Proof The proof uses the Fan-Glicksberg-Debreu (1952) fixed-point theorem. Under the above assumptions, the actions spaces are non-empty and discrete, and the payoff functions are continuous and bounded. We extend the discrete action spaces to the behavioral strategies sets by considering the simplex-valued measurable mapping. Hence, the best-response correspondence (set-valued mapping) in the space of measures "m" is non-empty, closed-valued, convex-valued, and upper semicontinuous. The best response is coupled with compatibility with the intersection of the measures generated by the Kolmogorov forward equations on the finite-dimensional simplex. Then, we use Fan-Glicksberg-Debreu (1952) fixed-point theory to conclude. □

2.6.4.3 Mean-Field Coordination Games

In this subsection, we provide sufficient conditions for mean-field convergence in team problems and global optimal control problems [102]. Consider the case of a single team. A payoff of a decision-maker in the team is $\tilde{r}(x, a)$ when the team action is $a \in \mathscr{A}$ and x is the state $\in \mathscr{X}$. The payoff function is the same for each member of the team. For a measure $m \in \Delta(\mathscr{X})$,, we define the expected payoff of the team as $(r_a(m))_{a \in \mathscr{A}}$.

$$r_a(m) = \sum_{x \in \mathscr{X}} \tilde{r}(x, a) m_x$$

A history of length t is now a sequence of $(A^I_{t'}, M^I_{t'})_{t' \leq t}$. Let σ be a behavioral strategy. The payoff defined above becomes

$$F^I_T(m_0, \sigma) = E_{m_0,\sigma}\left[\sum_{t=0}^{T-1} r_{a_t}(M^I_t)\right]. \tag{2.22}$$

$$F_{\beta}^{I}(m_0, \sigma) = (1 - \beta) E_{m_0,\sigma} \left[\sum_{t=0}^{+\infty} \beta^t r_{a_t}(M_t^I) \right]. \tag{2.23}$$

$$F_{\infty}^{I}(m_0, \sigma) = E_{m_0,\sigma} \left[\underline{\lim}_{T \longrightarrow +\infty} \frac{1}{T} \sum_{t=0}^{T-1} r_{a_t}(M_t^I) \right]. \tag{2.24}$$

We keep the index I in the payoff functions, because the strategy σ takes a sequence of occupancy measures or population profiles M_t^I and yields $A^I(t)$, which depends on I. Our aim is to prove the convergence of this finite coordination game to a mean-field coordination game for (i) strategies, (ii) population profile, and (iii) payoffs.

Proposition 2.3 *Assume that*

- $\lim_I \sup_{m^I(x_1)>0} \parallel \bar{L}^I(.|x_1, u, m^I) - \bar{L}(.|x_1, u, m^I) \parallel = 0$,
- *There exists* $c_1 > 0$ *such that for all* u, x_1, m, m'*, one has* $\parallel \bar{L}^I(.|x_1, u, m) - \bar{L}(.|x_1, u, m') \parallel \leq c_1 \parallel m - m' \parallel$,
- $M^I(0) \longrightarrow m_0$.

Then, the following holds

- *Convergence of payoff functions:* $\lim_I F_{\beta}^I(M^I(0)) = F_{\beta}(m_0)$.
- *Characterization of limiting optimal strategies: For all* $I' \in \mathbb{N}$,*, there exists* $I > I'$ *such that*

  $$BR_{\beta}^I(\theta, M^I(0)) \subseteq BR_{\beta}(\theta, m_0)$$

  where

  $$v_{\beta}^I(\theta, m) = \sup_a \{(1-\beta) r_{\theta,a}(m) + \beta \sum_{m'} \bar{L}^I(m'|m, \theta, a) v_{\beta}^I(\theta, m')\}$$

  and

  $$BR_{\beta}^I(\theta, m) = \{a,\ v_{\beta}^I(\theta, m) = (1-\beta) r_{\theta,a}(m) + \beta \sum_{m'} \bar{L}^I(m'|m, \theta, a) v_{\beta}^I(\theta, m')\}$$

  is the set of actions reaching the maximum, and

  $$BR_{\beta}(\theta, m_*) = \{a,\ v_{\beta}(\theta, m_*) = (1-\beta) r_{\theta,a}(m_*) + \beta v_{\beta}(\theta, \langle m_*, \bar{L}(.|m_*, \theta, .)\rangle)\}$$

- *A mean-field limit solves* $m_* = \langle m_*, \bar{L}(.|u_*, m_*, \theta, .)\rangle$ *and* $u_* \in BR(m_*)$.

Proof of Proposition 2.3 Let $\mathscr{F}_t = \sigma(x_{t'}^I, a_{t'}^I)_{0 \leq t' \leq t}$ be the filtration associated with the processes $\{x^I\}$, and $\{a^I\}$ up the time t. The evolution of the system is described by the transition probabilities. We first note that the sequence $\{F_{\beta}^I(\theta, M^I(0))\}_{I \geq 0}$ is bounded by the norm sup of $\tilde{r}$. Assume that the transition probabilities are regular enough:

(i)

$$\lim_{I} \sup_{m^I(x_1)>0} \| \bar{L}^I(.|x_1, u, m^I) - \bar{L}(.|x_1, u, m^I) \| = 0.$$

(ii) There exists $c_1 > 0$ such that for all u, x_1, m, m', one gets

$$\| \bar{L}^I(.|x_1, u, m) - \bar{L}(.|x_1, u, m') \| \leq c_1 \| m - m' \|$$

(iii) $M^I(0) \longrightarrow m_0$. Then, for an increasing mapping $\phi : \mathbb{N} \longrightarrow \mathbb{N}$,, one has

$$\lim_{I} F_{\beta}^{\phi(I)}(\theta, M^{\phi(I)}(0)) = F_{\beta}(\theta, m_0)$$

Let (u_*^I, m_*^I) be the discounted optimal pair of processes with starting population profile m_0^I. The process (u_*^I, m_*^I) is tight, because $\mathscr{A}$ is finite. This leads to a convergence in law. Denote by \$ the law of the limit. Then, the equality $F_{\beta}(\theta, m_0) = v_{\beta}(\theta, m_0)$ holds \$-almost surely. Thus, $\lim_{\phi(I)\longrightarrow+\infty} F_{\beta}^{\phi(I)}(\theta, m_0) = F_{\beta}(\theta, m_0)$ and the mapping

$$(u, m) \longmapsto F_{\beta}(\theta, u, m)$$

are continuous for any θ.

A similar reasoning gives the convergence of the optimal strategies and the announced inclusions. This completes the proof. □

2.6.5 *Vanishing Step Size*

We define the rescaled processes with step size δ_I :

$$\widetilde{X}_j^I(\delta_I t) = x_{j,t}^I, \ j \in \mathscr{I}, \ \widetilde{M}_j^I(\delta_I t) = M_t^I, \ t \in \mathbb{N}.$$

$\widetilde{M}^I(\delta_I t + t') = M_t^I + \frac{t'}{\delta_I}(M_{t+1}^I - M_t^I), \ t' \in (0, \delta_I)$
Define the drift as

$$f^I(u, m, s)$$
$$= E\left(\widetilde{M}^I(\delta_I(t+1)) - \widetilde{M}^I(\delta_I t) \mid \widetilde{M}^I(\delta_I t) = m, \tilde{A}^I(\delta_I t) \sim u, S(\delta_I t) = s\right).$$

We assume

(i) $\frac{1}{\delta_I} f^I(u, m, s) \longrightarrow f(u, m, s)$ where f is locally Lipschitz,
(ii) The first and second moments of $\mathscr{B}^I$ are finite when $I \longrightarrow +\infty$.

Under these assumptions, if the initial process $\tilde{M}^I(0)$ converges in distribution to $m_0 \in \Delta(\mathscr{S} \times \prod_j \mathscr{X}_j)$,, then the process $\tilde{M}^I$ converges in distribution to a deterministic measure (mean-field limit) m_t which is solution of the ordinary differential equation $\dot{m} = \tilde{f}(m)$.

Proposition 2.4 ([324]) *Under the same assumptions as above, the process $\tilde{M}^I[u]$ (which is the $\tilde{M}^I$ under the strategy u) converges in distribution to a deterministic measure $m_t[u]$ which is a solution of the ordinary differential equation $\dot{m} = \tilde{f}(u_s, m)$.*

Definition 2.9 (Individual Optimization Framework in Population Game) To the game as defined in Sect. 2.6.3, we associate a macroscopic population game, defined as follows. Each member j of the population has state X_j and a population profile m. The initial condition of the game is $S(t) = s, X_j(0) = x, m_0[u] = m_0$. The population profile is a solution to the discrete-time evolution equation, and (s, x_j) evolves as a jump process given by the marginal of q which depends on m, u and the strategy u' of j. Further, let $F_T(s, x, u', u, m)$ be the T-stage payoff of decision-maker j in this game, given $x_{j,0} = x$ and m_0,

$$F_T(s, x, u', u, m) = E\left(g(m(T)) + \int_0^T r(s, x_j, u', u, m) |\; s_0 = s, x_{j,0} = x,\; m_0\right)$$

where m is a solution of the ODE $\dot{m} = f(u, m)$. The discounted F_β and limiting average payoff F_∞ are defined in the same way.

Equivalence to Differential Population Game of [324] Under the above assumptions, the continuous-time mean-field stochastic game is equivalent to a *differential population game*. The instantaneous payoff of an individual with strategy v, state x when the population profile (mass-behavior, [333, 339, 342]) is m driven by u is defined by $r(s, x, v, u, m)$. The transition between the state is given by $L(s', x'|s, x, v, u, m)$.

2.6.5.1 Centralized Mean-Field-Type Control

We now provide the feedback optimality system for the global expected mean-field type payoff $\bar{r}(u, m)$. The T-stage mean-field-type optimization problem can be written as

$$\sup_u \; g(m(T)) + \int_0^T \bar{r}(u(t), m(t))\, dt$$

subject to $\dot{m} = f(u, m), \;\; m_0 = m$.

A set of strategies $u_*(t) = \phi(t, m)$ constitutes an optimal mean-field solution if there exist continuously differentiable function $v(t, m)$ defined on $[0, T] \times \mathbb{R}^d \longrightarrow \mathbb{R}$ satisfying the following Hamilton-Jacobi-Bellman equation combined with the mean-field ODE.

$$0 = v_t(t, m) + \max_u \{\bar{r}(u, m) + \langle \nabla_m v(t, m), f(u, m) \rangle\}, \qquad (2.25)$$

$$v(T, m) = g(m) \qquad (2.26)$$

In many situations, the horizon of the game is not known (e.g., the lifetime of a user in the system is unknown). One way to handle this problem is to consider the infinite horizon game. The optimality equation for discounted payoff

$$\sup_u \int_{t_0}^{+\infty} e^{-\beta'(t-t_0)} \bar{r}(u, m)\, dt$$

subject to $\dot{m} = f(u, m)$ where $\beta' > 0$, is

$$0 = -\beta' v(t, m) + \max_u \{\bar{r}(u, m) + \langle \nabla_m v(t, m), f(u, m) \rangle\} \qquad (2.27)$$

2.6.6 *Mean-Field Q/H Learning*

Mean-Field Q-Learning

Q-learning is a technique used to compute an optimal policy for a controlled Markov chain based on observations of the system controlled using a nonoptimal policy. Many interesting results have been obtained for models with finite state and action spaces.

While an optimal strategy $u(t, s, x, m)$ can, in principle, be obtained using dynamic programming, policy iteration, or value iteration, such computations are often prohibitively time-consuming. In particular, the size of the state space grows exponentially with the number of state variables, a phenomenon referred to by Bellman as the *curse of dimensionality*. Similarly, the size of the action space can also lead to computational intractability. Following the idea of Q-learning [364], we define the Q-value

$$Q(s, x_j, a_j, m_t) = r_j(s, x_j, a, m_t) + \sum_{s', x'} L(s', x_j' \mid s, x_j, a_j, u_t, m_t) v(s', x_j').$$

Let $\sigma = (\sigma_t)_{t \geq 0}$, where σ_t is a mapping that assigns to every finite history $h_{j,t}$ an element of $\Delta(\mathscr{A}_j(s_t, x_{j,t}))$. h_t is a collection of history up to t that are available to decision-maker j. Every strategy σ, together with the initial state $(s_0, x_{j,0})$ and the mean-field m induces a probability distribution $\mathbb{P}_{s_0,x_{j,0},\sigma}$ over the space of infinite plays. We denote the corresponding expectation operator by $E_{s_0,x_{j,0},\sigma}$. For the discounted payoff case with discount factor β_j,, the payoff is

$$F^I_{j,\beta_j}(s_0, x_{j,0}, \sigma) = E_{s_0,x_{j,0},\sigma}\left[\sum_{t\in\delta_I\mathbb{N}} \beta_j^t r_j^{\mathscr{B}_t^I}(s_t, x_t, a_t)\mathbb{1}_{\{j\in\mathscr{B}_t^I\}}\right] \tag{2.28}$$

Assuming that $j \in \mathscr{B}^I(t), t \geq 0$, one gets

$$Q^*(s, x_j, a_j, m)=r_j(s, x_j, a_j, m)$$

$$+\beta_j \sum_{s',x_j'} L(s', x_j' \mid s, x_j, a_j, m)\left(\sup_{b_j\in\mathscr{A}_j(s',x_j')} Q^*(s', x_j', b_j, m)\right) \tag{2.29}$$

One nice feature here is to be able to learn the Q-values without the knowledge of the transition probabilities. The iterative version is given by

$$Q_{t+1}(s, x_j, a_j, m)=Q_t(s, x_j, a_j, m)+\lambda_t(s, x_j)\left[r_j(s, x_j, a_j, m)+\right.$$

$$\left.\beta_j\left(\sup_{b_j\in\mathscr{A}_j(s',x_j')} Q_t(s', x_j', b_j, m)\right)-Q_t(s, x_j, a_j, m)\right] \tag{2.30}$$

where $\lambda_t(s, x_j) \geq 0$ is a learning rate function satisfying

$$\sum_t \lambda_t(s, x_j) = +\infty, \ \sum_t \lambda_t(s, x_j)^2 < +\infty,$$

which is a standard assumption in stochastic approximation.

2.6.7 Mean-Field H-Learning

What does Q-learning have to do with the Hamilton-Jacobi-Bellman equation and the Pontryagin maximum principle?

The answers to this question have been examined in detail in [241] in a deterministic setting. The authors have shown that the Hamiltonian appearing in control theory is essentially the same as the Q-function that is the quantity of interest in Q-learning. They established a close connection between Q-learning and differential dynamic programming [205]. This remarkable result allows us to address the H-learning (learning under the Hamiltonian function) in mean-field limit. The analogue of the Q-function is the H-function defined by

$$H(u, m, p, G) = r(u, m) + \langle p, \tilde{f} \rangle + \frac{1}{2} trace(\tilde{a} G),$$

where p is a d-dimensional vector and $G \in \mathscr{R}^{d \times d}$. Denote by W_l an approximated value at iteration l. The algorithm is then described as follows:

- Initialize m_0
- At iteration l use m_l to compute the payoff $r(., m_l)$.
- Use the Hamilton-Jacobi-Bellman-Fleming optimality to obtain W_l
- Compute the optimal control u_l^* via the H-function
- Use u_l^* to get m_l^*, solution of the Fokker-Planck-Kolmogorov forward equation

2.6.8 Mean-Field Interaction with Regime Switching

In this section, we provide the main result of mean-field game dynamics where the resource state changes on a slower time scale than the individual dynamics, that is, a mean-field convergence to noisy mean field with switching. The key generalization is the relaxation of condition **A1** to let the resource to switch between different regimes:

A1' $S(t)$ is a finite state space Markov chain with time space $\mu_I \mathbb{N}$, $\mu_I > 0$. Note that $S(t)$ does not scale with the intensity of the interaction δ_I, that is, $\frac{\delta_I}{\mu_I} \to 0$. We may think of μ_I as the intensity of the availability of the resource. This says that the individual states change faster than the resource states. That is, the process $S(t)$ is constant for $t \in [lk, (l+1)k)$, $l \in \mathbb{N}$.. Let

$$w_{s's} := \mathbb{P}\left(S(lk) = s \mid S(l(k-1)) = s'\right)$$

be the transition probability of the regime switching process. The next theorem identifies the mean-field limit under regime switching.

Theorem 2.12 *Let* $M_0^I \longrightarrow m_0$ *in law where* m_0 *is a probability measure. Under A1–A4 [resp., A1'–A4'], the process* M_t^I *converges in law to a weak [resp., strong] solution of the Itô stochastic differential with switching equation given by*

$$\begin{aligned} dm(s(t)) &= f(m, u, s(t))dt + \sigma(m, u, s(t))d\mathscr{B}_t \\ S(t) &= S(lk) = s \ \ \text{if } t \in [lk, (l+1)k), \ l \in \mathbb{N} \end{aligned} \tag{2.31}$$

The proof of this result follows the same line as for the mean-field convergence applied to $(s(t), M^I(t))$. □

2.6.8.1 Existence and Uniqueness

Theorem 2.13 *Let* $m \in \mathbb{R}^{|\mathcal{X}|}$, $\mathcal{S} = \{1, \ldots, |\mathcal{S}|\}$, $|\mathcal{S}| < +\infty$. *Let* $w_{ss'}$ *be a* $|\mathcal{S}| \times |\mathcal{S}|$ *matrix. Consider the pair process*

$$y(t) = (m(t), s(t))$$

given by (2.31) with initial condition $(m(0), s(0)) = (m_0, s_0)$. *Assume that the functions* $f(\cdot, \cdot, \cdot)$ *and* $\sigma(\cdot, \cdot, \cdot)$ *satisfy*

$$|f(m, u, s)| + |\sigma(m, u, s)| \leq C(1 + \|m\|)$$

for some constant $C > 0$, *and* $\forall\, u \in \mathcal{U}, s \in \mathcal{S}$. *Moreover, we assume that* $\forall t \in [0, T], s \in \mathcal{S}, (m, m') \in \mathbb{R}^{|\mathcal{X}|}$

$$|f(m, u, s) - f(m', u, s)| + |\sigma(m, u, s) - \sigma(m', u, s)| \leq C\|m - m'\|$$

Then there exists a unique solution $y(t)=(m(t), s(t))$ *to the hybrid SDE with the initial condition* (m_0, s_0) *in which the evolution of the piecewise constant jump process is given by (2.31).*

Proof See [297] pages 103–104. □

2.6.8.2 Stability

Definition 2.10 The stationary point m^* is said to be stable in probability under strategy u if for any $\epsilon > 0$ and any $s \in \mathcal{S}$

$$\lim_{m_0 \to m^*} \mathbb{P}\left(\sup_{t \geq 0} \|m^{m_0, s_0}[u](t) - m^*\| > \epsilon\right) = 0$$

and m^* is said to be unstable in probability if it is not stable in probability.

Definition 2.11 The mean-field m^* is asymptotically stable in probability under the strategy u if it is stable in probability and satisfies

$$\lim_{m_0 \to m^*} \mathbb{P}\left(\lim_{t \to +\infty} m^{m_0, s_0}[u](t) = m^*\right) = 1$$

for any m_0, s_0.

Proposition 2.5 (Lyapunov Stability for Controlled Hybrid Systems) *Let* $D \subset \mathbb{R}^{|\mathcal{X}|}$ *be a neighborhood of* m^*. *Suppose that for each* $s \in \mathcal{S}$, *there exists a function* $\omega_s(m, u)$, $\omega : \mathbb{R}^{|\mathcal{X}|} \times \mathcal{U} \to \mathbb{R}$,, *and let* $\bar{\omega}_s(m, u) = \omega_s(m, u) - \omega_s(m^*, u)$ *be nonnegative such that*

- *(1)* $\bar{\omega}_s(m,u)$ *is continuous in* $\mathbb{R}^{|\mathscr{X}|} \times \mathscr{U}$
- *(2)* $\bar{\omega}_s(m,u) = 0 \Rightarrow m = m^*$
- *(3)* $\bar{\omega}_s(m,u)$ *is twice continuously differentiable in* $D\backslash\{m^*\}$ *with respect to* m *and*

$$\langle f, \nabla_m \bar{\omega}_s\rangle + \tfrac{1}{2} tr\left(\sigma(m,u,s)\sigma'(m,u,s)\partial^2_{m_x m_{x'}}\bar{\omega}_s(m,u)\right) \\ + \textstyle\sum_{s'\in\mathscr{S}} w_{ss'}\bar{\omega}_{s'}(m,u) \le 0, \quad \forall m \neq m^* \tag{2.32}$$

then m^* *is stable in probability under* u.

Proof Let $\epsilon > 0$ be such that the open ball centered at m^* with radius ϵ, denoted by $B(m^*,\epsilon) = \{m \in \mathbb{R}^{|\mathscr{X}|} \mid \ \| m - m^* \| < \epsilon\}$ is included in D and its boundary $\partial B(m^*,\epsilon) := \{m \in \mathbb{R}^{|\mathscr{X}|} \mid \ \| m - m^* \| = \epsilon\}$ is in D. Let $i_\epsilon(u) = \inf\{\bar{\omega}_{s'}(m',u),\ m' \in D\backslash B(m^*,\epsilon),\ s' \in \mathscr{S}\}$. The continuity and compactness, and 1),2) implies $i_\epsilon(u) > 0$. Using 3) and Dynkin formula

$$E_{m,s}\bar{\omega}_{s(t\wedge\tau_\epsilon)}(m(t\wedge\tau_\epsilon),u) \\ = \bar{\omega}_s(m,u) + E_{m,s}\int_0^{t\wedge\tau_\epsilon} O\bar{\omega}_{s(t'\wedge\tau_\epsilon)}(m(t'\wedge\tau_\epsilon),u)ds' \le \bar{\omega}_s(m,u), \tag{2.33}$$

where

$$O\bar{\omega}_{s(t'\wedge\tau_\epsilon)}(m(t'\wedge\tau_\epsilon),u) \\ = \langle f, \nabla_m \omega_s\rangle + \frac{1}{2}\mathrm{tr}\left(\sigma(m,u,s)\sigma'(m,u,s)\partial^2_{m_x m_{x'}}\omega_s(m,u)\right) \\ + \sum_{s'\in\mathscr{S}} w_{ss'}[\omega_{s'}(m,u) - \omega_s(m^*,u)] \le 0, \tag{2.34}$$

and

$$\tau_\epsilon = \tau_\epsilon(u) = \inf\{t \ge 0,\ m^{m_0,u,s}(t) \notin B(m^*,\epsilon)\}.$$

It is clear that $\tau_\epsilon \le t \Longleftrightarrow \sup_{0\le t'\le t} \| m(t') - m^* \| > \epsilon$. Thus,

$$E i_\epsilon(u)\mathbb{1}_{\{\tau_\epsilon \le t\}} \le E_{m_0,s_0}\left(\bar{\omega}_{s(\tau_\epsilon)}(m(\tau_\epsilon),u)\mathbb{1}_{\{\tau_\epsilon\le t\}}\right)$$

The LHS of the above inequality is precisely $i_\epsilon(u)\mathbb{P}(\tau_\epsilon \le t) = i_\epsilon(u)\mathbb{P}\left(\sup_{0\le t'\le t} \| m(t') - m^* \| > \epsilon\right)$. The RHS is less than $\bar{\omega}_s(m,u) - \bar{\omega}_s(m^*,u)$. Thus,

$$\mathbb{P}_{m_0,s_0}\left(\sup_{0\le t'\le t} \| m(t') - m^* \| > \epsilon\right) \le \frac{\omega_s(m,u) - \bar{\omega}_s(m^*,u)}{i_\epsilon(u)}$$

By the continuity assumption, the difference $\omega_s(m,u)-\bar{\omega}_s(m^*,u) \geq 0$ is arbitrarily small in the considered neighborhood, and this completes the proof. □

2.6.9 Stability Conditions for Hybrid Mean-Field Limit

In this subsection, we provide a sufficient condition for stability in probability of the hybrid mean-field limit in the linearized case.

Proposition 2.6 *Suppose that for each resource state (switching regime) $s \in \mathcal{S}$,, there exists a square matrix $A(m^*,u,s)$ and a unique stationary distribution w_s such that*

$$f(m,u,s) = f(m^*,u,s) + A(m^*,u,s)(m-m^*) + o(\|m-m^*\|).$$

Then,

- *If $\sum_{s\in\mathcal{S}} w_s\lambda_{\max}\left(A(m^*,u,s)+A'(m^*,u,s)\right) < 0$, t, the linearized system in m^* with switching is **stable** in probability.*
- *If $\sum_{s\in\mathcal{S}} w_s\lambda_{\min}\left(A(m^*,u,s)+A'(m^*,u,s)\right) > 0$,, the linearized system in m^* with switching is **unstable** in probability.*

where $\lambda_{\max}(.)$ denotes the maximum eigenvalue and $\lambda_{\min}(.)$ denotes the minimum eigenvalue.

Remark 2.1 Note that Proposition 2.6 does not provide any insight into what happens between the two sufficiency conditions: There may be a gap between the $\lambda_{\max}$ and $\lambda_{\min}$. However, in the case of two states per decision-maker, the mean-field limit can be reduced to a one-dimensional ODE with switching. Thus, if $\sum_{s\in\mathcal{S}} w_s\bar{A}(m^*,u,s) < 0$, then stability in probability holds, and if $\sum_{s\in\mathcal{S}} w_s\bar{A}(m^*,u,s) > 0$, then instability in probability holds.

Proof See Theorem 2.14. □

2.6.10 Stability Conditions for Hybrid Noisy Mean Field

Theorem 2.14 *Suppose that for each resource state (switching regime) $s \in \mathcal{S}$,, there exists square matrices $A(m^*,u,s)$, $B(m^*,u,s)$ and a unique stationary distribution w_s such that*

$$f(m,u,s) = f(m^*,u,s) + A(m^*,u,s)(m-m^*) + o(\|m-m^*\|), \tag{2.35}$$

$$\sigma(m,u,s) = \sigma(m^*,u,s) + \left[B_1(m^*,u,s)(m-m^*)\Big|\ldots\Big|B_{|\mathcal{X}|}(m^*,u,s)(m-m^*)\right]$$

$$+ o(\|m - m^*\|). \tag{2.36}$$

Then,

- *The linearized system with switching is asymptotically stable in probability under u if*

$$\sum_{s \in \mathcal{S}} w_s \lambda_{\max} \left(A(m^*, u, s) + A'(m^*, u, s) + \sum_{x \in \mathcal{X}} B_x(m^*, u, s) B'_x(m^*, u, s) \right) < 0$$

- *m^* is asymptotically unstable in probability if*

$$\sum_{s \in \mathcal{S}} w_s [\lambda_{\min}(A(m^*, u, s) + A'(m^*, u, s) + \sum_{x \in \mathcal{X}} B_x(m^*, u, s) B'_x(m^*, u, s)) \\ - \frac{1}{2} \sum_{x \in \mathcal{X}} \rho[B_x(m^*, u, s) + B'_x(m^*, u, s)]^2] > 0, \tag{2.37}$$

where $\rho(B_x(m^*, u, s))$ is the spectral radius the supremum among the absolute values of the elements in the spectrum of $B_x(m^*, u, s)$.

Note that when $\sigma = 0$, (and hence $B_x(.) = 0$), Theorem 2.14 reduces to Proposition 2.6.

As a corollary, we have the following result in the two individual states case $|\mathcal{X}| = 2$,. If $\sum_{s \in \mathcal{S}} w_s[\bar{A}(m^*, u, s) + \frac{\bar{B}_1(m^*, u, s)^2}{2}] < 0$, then stability in probability holds and if $\sum_{s \in \mathcal{S}} w_s[\bar{A}(m^*, u, s) - \frac{\bar{B}_1(m^*, u, s)^2}{2}] > 0$,, then instability in probability holds.

2.6.11 Differential Population Game

We provide in this section a mean-field equilibrium characterization of the *differential population game* where each generic decision-maker reacts to the mean field within a finite horizon $[0, T]$.

2.6.12 Global Optimization Under Symmetric per Type Strategies

We first start by a payoff function of the form $\bar{r}(t, u, m)$.

$$(*) \ \sup_u \left[\bar{g}(m(T)) + \int_t^T \bar{r}(t', u, m)\, dt' \right]$$

subject to the population mean-field dynamics

$$m(t) = m_0 + \int_0^t \tilde{f}(t', u, m)\, dt'.$$

Note that here, the control u is a collective control.

We say the pair of trajectories $(u^*(t), m^*(t))_{t\geq 0}$ constitutes a consistent mean-field response if u_t^* is an optimal strategy to be above problem (*) where m_t^* is the mean field at time t and u_t^* produces the population mean-field $m_t[u^*, m_0] = m_t^*$

A consistent mean-field response is characterized by a backward-forward equation

$$\begin{cases} \bar{v}_t + \sup_u \left\{ \bar{r}(t, u, m) + \langle \nabla_m \bar{v}, \tilde{f}(t, u, m) \rangle \right\} = 0, \\ \bar{v}(T, m) = \bar{g}(m) \\ \\ m(t) = m_0 + \int_0^t \tilde{f}(t', u^*, m)\, dt' \end{cases}$$

where u^* is in the argmax of $u \mapsto \bar{r}(t, u, m) + \langle \nabla_m \bar{v}, \tilde{f}(t, u, m) \rangle$.

2.6.13 Individual Response to Population Mean Field

Next, we consider an individual state-dependent payoff $r(t, x, u, m)$. Define

$$F^1(x, u, m) = g(x(T), m(T)) + \int_t^T r(t', x, u, m)\, dt',$$

where g is a terminal payoff. Consider the individual response problem as follows:

$$(**) \sup_u \ [g(x(T), m(T)) + \int_t^T r(t', x, u, m)\, dt'],$$

subject to the population mean-field dynamics

$$m(t) = m_0 + \int_0^t \tilde{f}(t', u, m)\, dt',$$

where the individual state $x = x[u]$ is a continuous time Markov jump process under u. We denote by $\bar{q}$ the infinitesimal generator of $x[u]$.

We say that the pair of trajectories $(u^*(t), m^*(t))_{t\geq 0}$ constitutes a mean-field equilibrium if $\{u^*(t)\}_{t\geq 0}$ is a mean-field best response to the above problem (**) where $m^*(t)$ is the population mean field at time t and $u^*(t)$ produces the population mean-field $m[u^*, m_0] = m^*$

Theorem 2.15 *Consider a differential population game with a single type. Assume that there exists a unique pair (u^*, m^*) such that*

(i) there exists a bounded, continuously differentiable function $\tilde{v}_x : [0, T] \times \mathbb{R}^{|\mathcal{X}|}$, $\tilde{v}_x^*(t, m) = W(t, x, m)$ and a differentiable function $m^* : [0, T] \longrightarrow \mathbb{R}^{|\mathcal{X}|}$, $m^* = m[u^*, m_0]$ as solution to the master system

$$\begin{cases} W_t(t,x,m) + \sup_u \left\{ r(t,x,u,m) + \langle \nabla_m W(t,x,m), \tilde{f}(t,u^*,m)\rangle + \sum_{x'\in\mathcal{X}} \bar{q}_{xux',t}(m) W(t,x',m) \right\} = 0 \\ W(T,x,m) = g(x,m), \end{cases}$$

(ii) $u^*(t, x, m)$ maximizes the function

$$r(t, x, u, m) + \sum_{x'\in\mathcal{X}} \bar{q}_{xux',t}(m) W(t, x', m)$$

where $\bar{q}_{xux',t}(m)$ is the transition of the infinitesimal generator of x under the strategy u and m, $\sum_{x'} \bar{q}_{xux',t}(m) = 0$, and the term $\sum_{x'\in\mathcal{X}} \bar{q}_{xux',t}(m) W(t, x', m)$ is

$$\sum_{x'\neq x} \bar{q}_{xux',t}(m)(W(t, x', m) - W(x, m)),$$

Then, $(u^*(t), m^*(t))_{t\geq 0}$ with $m^* = m[u^*, m_0]$ constitutes a mean-field equilibrium, and the equilibrium payoff is $F(T, x, u^*, m^*)$

Proof We first identify the pair (x, m) as an augmented state of the interaction. This new process is Markovian when the strategies are Markovian. Then the result follows by applying the standard dynamic programming principle. □

We have presented mean-field Markov games with control-dependent regime switching. The result opens new questions for discontinuous mappings (viz., the drift and the variance). In that case, one may expect to have a stochastic differential inclusion for which the existence of solution and the well-posedness need more attention.

2.6.14 Mean-Field Games with Discrete State Space

In this section, we address (population) mean-field games with discrete time and discrete state spaces.

2.6.14.1 Problem Description

Consider a (population) mean-field game described by the collection

$$\Gamma_{MFG} = (\mathscr{I}, \mathscr{X}, \mathscr{M}, \tilde{\mathscr{A}}, \mathscr{A}, \mathscr{T}, \{q_t\}_{t\in\mathscr{T}})$$

where

- $\mathscr{T}$ is the horizon of the game.
- $\mathscr{I}$ is the set of decision-makers (infinite).
- $\mathscr{X}$ is the state space of a generic decision-maker.
- For every decision-maker $j \in \mathscr{I}$, $\tilde{\mathscr{A}}$ is the set of all possible actions of that decision-maker. $\mathscr{A} : \mathscr{X} \longrightarrow 2^{\tilde{\mathscr{A}}}$ is a set-valued map (correspondence) that assigns to each state $x \in \mathscr{X}$ the set of actions $\mathscr{A}(x)$ that are available to decision-maker j. We denote the state-action space of a generic decision-maker by

$$\text{XA} = \{(x, a) : x \in \mathscr{X},\ \ a \in \mathscr{A}(x)\}.$$

- $\mathscr{M}$ is the set of probability measures of the state-action pair.
- For every decision-maker $j \in \mathscr{I}$, r : XA $\times \mathscr{M} \longrightarrow \mathbb{R}$ is an instantaneous payoff function for decision-maker j.
- $\forall t \in \mathscr{T}$, q_t : XA $\times \mathscr{M} \longrightarrow \Delta(\mathscr{X})$ is a transition function where $\Delta(\mathscr{X})$ is the space of probability distributions over $\mathscr{X}$.
- A long-term payoff functional for each individual decision-maker.

2.6.14.2 How Is the Population Mean-Field Game Played?

Decision-makers' initial states are generated according to an initial distribution m_0. The generic decision-maker starts at an initial state $x \in \mathscr{X}$. The population mean-field game is played as follows: Time is discrete, and time space is the set $\mathscr{T}$. The current state x_t of the generic decision-maker j is realized according to the marginal distribution with the respect to state in m_t and is observed by all the decision-makers. At each time $t \in \mathscr{T}$,, each decision-maker $j \in \mathscr{I}$ chooses a control action $a_t \in \mathscr{A}(x_t)$, and receives an instantaneous payoff $f(x_t, u_t, m_t)$, and the game moves to a new state according to a probability distribution given by $q_t(. \mid x_t, u_t, m_t, x_{t-1}, u_{t-1}, m_{t-1}, \ldots, x_0) \in \Delta(\mathscr{X})$, where

$$\begin{aligned} &q_t(x' \mid x_t, u_t, m_t, x_{t-1}, u_{t-1}, m_{t-1}, \ldots, x_0) \\ &= \mathbb{P}\left(x_{t+1} = x' \mid x_t, u_t, m_t, x_{t-1}, u_{t-1}, m_{t-1}, \ldots, x_0\right),\ x' \in \mathscr{X}, \end{aligned} \tag{2.38}$$

and a new distribution m_{t+1} will be generated from the choices. The transition structure is Markovian (one-step-Markov) if

$$q_t(. \mid x_t, u_t, m_t, x_{t-1}, u_{t-1}, \ldots, x_0) = q_t(. \mid x_t, u_t, m_t).$$

The kernel q_t is completely determined by

$$q_t(\mathscr{X}' \mid x_t, u_t, m_t) = \mathbb{P}\left(x_{t+1} \in \mathscr{X}' \mid x_t, u_t, m_t\right),$$

where $\mathscr{X}'$ describes a measurable subset of $\mathscr{X}$.

2.6.14.3 Bellman-Shapley Equation

The evolution of the population distribution of states is given by

$$m_{t+1}(S) = \mathbb{P}(x_{t+1} \in S). \tag{2.39}$$

Given a population mean-field trajectory $\{m_t\}_{\{0,\ldots,T\}}$,, the best response in the finite horizon with length T is given by

$$\begin{cases} \inf_u \ E[h(x_T, m_T) + \sum_{t=0}^{T-1} f(t, x_t, u_t, m_t)] \\ \text{subject to} \\ x_{t+1} \sim q_t(. \mid x_t, u_t, m_t), \quad x_0, \end{cases} \tag{2.40}$$

Note that this formulation differs from the one in Chap. 1, where the expected payoff is nonlinear in the measure $P_{x_{t+1}}$.

Bellman-Shapley optimality criterion for the finite horizon problem (2.40) yields

$$\begin{cases} v_T(x) = h(x, m), \\ v_t(x) = \min_u\{f(t, x, u, m_t) + \sum_{x'} q_t(x'|x, u, m_t)v_{t+1}(x')\} \\ m_{t+1}(x) = \sum_{x' \in \mathscr{X}} m_t(x')L_{t,x'x}(u_t^*, m_t), \\ m_0 \end{cases}$$

Let $\beta \in (0, 1)$ be a discount factor, and consider the following best-response problem under discounted cost:

$$\begin{cases} \inf_u \ (1-\beta) \sum_{t=0}^{+\infty} \beta^t f(t, x_t, u_t, m_t) \\ \text{subject to} \\ x_{t+1} \sim q_t(. \mid x_t, u_t, m_t), \ x_0. \end{cases} \tag{2.41}$$

The Bellman's equation associated with (2.41) is

$$\begin{cases} v_\beta(x) = \min_{u_t} \ \{(1-\beta)f(t, x, u_t, m_t) + \beta \sum_{x'} q_t(x'|x, u_t, m_t)v_\beta(x')| \ x_t = x\} \\ m_{t+1}(x) = \sum_{x' \in \mathscr{X}} m_t(x')L_{t,x'x}(t, u_t^*, m_t), \end{cases}$$

In the homogenous and stationary case, the mean-field equilibrium system becomes

$$\begin{cases} v_\beta(x) = \min_u \{(1-\beta)f(x,u,m) + \beta \sum_{x'} q(x'|x,u,m)v_\beta(x')\} \\ \qquad\qquad m(x) = \sum_{x'\in\mathcal{X}} m(x')L_{x'x}(u^*,m). \end{cases}$$

Now, consider the limiting time-average cost with homogenous and ergodic processes:

$$\begin{cases} J_\infty(u,m) := \liminf_T E\frac{1}{T}\sum_{t=0}^{T-1} f(x_t,u_t,m_t), \\ \inf_u \ J_\infty(u,m) : \\ \text{subject to} \\ x_{t+1} \sim q(. \mid x_t,u_t,m_t), \ x_0. \end{cases} \tag{2.42}$$

$$\begin{cases} v + \hat{v}(x) = \min_u \ \{f(x,u,m) + \sum_{x'} q(x'|x,u,m)\hat{v}(x')\} \\ \qquad\qquad m(x) = \sum_{x'\in\mathcal{X}} m(x')L_{x'x}(u^*,m), \end{cases}$$

where v is the equilibrium cost.

We have presented dynamic programming principle for population mean-field game with discrete state space. These dynamic programming equations do not necessarily hold when the payoffs and the state dynamics involve the individual mean field such as the distribution of self-state. In that case, one needs to introduce a new state that includes both the individual mean field and the population mean field in order to get a new dynamic programming principle (see Chap. 3).

In this chapter, we have examined interactions in asymptotic regime and established connection with anonymous games. The methodology presented above is easily extended with not only the population mean field of states m but also the population mean field of actions $\tilde{m}$ which is the distribution of other decision-makers' states. These are particular cases of the mean-field-type games. The regime switching is also covered in Chaps. 8 and 10.

2.6.15 Notes

Discrete-time mean-field games with a continuum of decision-makers have been studied by Jovanovic and Rosenthal [210] under the name *anonymous sequential games*. The authors considered the evolution of the mean-field limit in the Bellman dynamic programming equation. The work in [210] showed the existence of such mean-field equilibria in the case where the mean-field limit of decision-makers' characteristics evolves nonstochastically. Bergin and Bernhardt [65] showed how *stochastic mean-field limit* can be introduced into the model (so the mean-field limit evolves stochastically), extending the Jovanovic-Rosenthal's existence theorem to the stochastic mean-field limit. The paradigm of mean-field dynamics has been to associate a relative growth rate to actions according to the expected payoff they achieved and then study the asymptotic trajectories of the state of the system, i.e., the fraction of decision-makers per individual state action. The authors in [314, 315, 324, 329, 341] have introduced evolving stochastic games with a finite number decision-makers, in which each decision-maker in the population interacts with other randomly selected decision-makers (the number of decision-makers in interaction evolves in time). The states and actions of each decision-maker in an interaction together determine the instantaneous payoff for all involved decision-makers. They also determine the transition probabilities to move to the next state. Each individual wishes to maximize the total expected payoff over an infinite horizon. Under the restricted class of Markovian strategies, the random process consisting of one specific decision-maker and the remaining population converges weakly to a jump process driven by the solution of a system of differential equations. The authors proved that the large population asymptotic of the microscopic model is equivalent to a (macroscopic) stochastic population game in which a local interaction is described by a single decision-maker against an evolving population profile. Another characterization of the long run mean-field game is described by using the so-called differential population games. Here we extend the work in [324] which is limited to deterministic limit under strong conditions (bounded second moment, stationary strategies). Our work here extends also [57] in which the decision of the decision-makers is not examined. See also [4, 25, 284–286, 353]. In [179], mean-field models in discrete time with a finite number of states and a very large number of decision-makers are considered. The authors studied existence, uniqueness, and exponential convergence properties of the mean-field Nash equilibrium system. The Bellman systems and the master adjoint system for mean-field-type games can be found in [44, 296, 333, 338].

Chapter 3
Mean-Field-Type Games With Discrete State Spaces

In Sect. 2.6.14, a mean-field game problem with an infinite number of players was introduced. The mean-field term which was a population mean-field (distribution of states of all other players) was not affected by a single agent's individual state. This chapter examines the case where an individual state of a player may affect a mean-field term: The probability distribution of own-state appears in the dynamics and/or the payoff functional. This is a continuation of the formulation in Chap. 1 where the expected payoff is nonlinear in the measure $P_{x_{t+1}}$. We analyze the existence of optimal strategies in games associated with payoff functionals of mean-field type, under a dynamics driven by a class of Markov chains of mean-field type.

A Markov chain of mean-field type (also known as a nonlinear Markov chain) is a pure jump process with a discrete state space whose jump intensities further depend on the marginal law of the process. The marginal law of the nonlinear process, obtained as a deterministic limit of a sequence of empirical distribution functions representing the states of the finite system, satisfies a "nonlinear" Fokker–Planck–Kolmogorov equation of *McKean–Vlasov type*. In a sense, it represents the law of a typical trajectory in the underlying collection of interacting jump processes. In particular, games based on the nonlinear process dynamics would give an insight into the effect of the design of game strategies for large systems of interacting jump processes [333, 339, 342].

This chapter provides a proof of existence and uniqueness of optimal strategies for Markov chains of mean-field type using a fixed point argument based on a Girsanov-type change of measure and the Csiszár-Kullback-Pinsker inequality. Moreover, games associated with payoff functionals of mean-field type, when the nonlinear Markov chain is controlled through its jump intensities, are considered. The state is a pure jump processes x whose jump intensities at time t depend on the whole path over the time interval $[0, T]$, $T > 0$ and also on the marginal law of $x(t)$, as long as they are predictable. In a sense, this way of constructing a nonlinear jump process is a generalization of the classical thinning procedure of a point process.

T. Başar et al., *Mean-Field-Type Game Theory I*, Static & Dynamic Game Theory: Foundations & Applications, https://doi.org/10.1007/978-3-032-07027-2_3

After a short section of preliminaries, we introduce the class of Markov chains of mean-field type and prove their existence and uniqueness under rather weak conditions on the underlying unbounded jump intensities. After that, we consider an optimal control problem of mean-field type and prove the existence of an optimal control. We then consider a related zero-sum game and show the saddle point existence under the so-called Isaacs' condition. In both cases, the main results are derived using techniques involving Markov chain backward stochastic differential equations. Section 3.5 introduces multiple decision-makers interacting over a non-empty discrete state space. We then focus on Bellman system and master adjoint systems for mean-field-type games with discrete state spaces and finitely many agents.

Preliminaries

A σ-field (or σ-algebra) is a collection of subsets of a given set Ω that satisfies three key properties: It includes the set Ω itself, it is closed under complementation (meaning if a set is in the σ-field, its complement is also in the σ-field), and it is closed under countable unions (meaning if a countable collection of sets is in the σ-field, their union is also in the σ-field). Let $I = \{0, 1, 2, \ldots\}$ equipped with its discrete topology and σ-field, and let $\Omega := \mathscr{D}([0, T], I)$ be the space of functions from $[0, T]$ to I that are right continuous with left limits at each $t \in [0, T)$ and are left continuous at time T. Denote by $\mathscr{F}$ the Borel σ-field over Ω which is the σ-field generated by the open sets of Ω. Given $t \in [0, T]$ and $\omega \in \Omega$, put $x(t, \omega) \equiv \omega(t)$, and denote by $\mathscr{F}_t^0 := \sigma(x(s),\ s \leq t),\ 0 \leq t \leq T$, the filtration generated by x. Let $|x|_t := \sup_{0 \leq s \leq t} |x_s|,\quad 0 \leq t \leq T$.

To $x_{\cdot}$, we associate the indicator process $I_i(t) = \mathbf{1}_{\{x(t)=i\}}$ whose value is 1 if the chain is in state i at time t and 0 otherwise, and the counting processes $N_{ij}(t),\ i \neq j$, such that

$$N_{ij}(t) = \#\{\tau \in (0, t] : x(\tau^-) = i, x(\tau) = j\}, \quad N_{ij}(0) = 0,$$

which counts the number of jumps from state i into state j during the time interval $(0, t]$. Naturally, x being right continuous with left limits implies that both I_i and N_{ij} are right continuous with left limits. Moreover, by the relationship

$$x(t) = \sum_i i I_i(t), \quad I_i(t) = I_i(0) + \sum_{j:\, j \neq i} \left(N_{ji}(t) - N_{ij}(t)\right),$$

the state process, the indicator processes, and the counting processes carry the same information, which is represented by the natural filtration $\mathbb{F}^0 := (\mathscr{F}_t^0,\ 0 \leq t \leq T)$ of x.

Below, C and c_1 denote generic positive constants which may change from line to line.

Markov Chains

The time-dependent matrice $G(t) = (g_{ij}(t),\ i,\ j \in I),\ 0 \le t \le T$, is a Q-matrix if

$$g_{ij}(t) \ge 0,\ \ i \neq j, \quad g_{ii}(t) = -\sum_{j:\, j\neq i} g_{ij}(t), \quad \sum_{i,j:\, j\neq i} \int_0^T g_{ij}(t)\, dt < +\infty. \tag{3.1}$$

In view of, e.g., Theorem 7.3 in [147], or Theorem 20.6 in [279] (for the finite state-space and time independent case), given the Q-matrix G and a probability measure ξ_0 over I, there exists a unique probability measure P on $(\Omega, \mathscr{F})$ under which the coordinate process x is a time-inhomogeneous Markov chain with intensity matrix G and starting distribution ξ_0, i.e., such that $P \circ x^{-1}(0) = \xi$. Equivalently, P solves the martingale problem for G with initial probability distribution ξ meaning that, for every f on I, the process defined by

$$M_t^f := f(x(t)) - f(x(0)) - \int_{(0,t]} (G(s)f)(x(s))\, ds \tag{3.2}$$

is a local martingale relative to $(\Omega, \mathscr{F}, \mathbb{F}^0)$, where

$$G(s)f(i) := \sum_j g_{ij}(s) f(j) = \sum_{j:\, j\neq i} g_{ij}(s)(f(j) - f(i)),\ \ i \in I,$$

and

$$G(s)f(x(s)) = \sum_{i,j:\, j\neq i} I_i(s) g_{ij}(s)(f(j) - f(i)). \tag{3.3}$$

By Lemma 21.13 in [279], the compensated processes associated with the counting processes N_{ij}, defined by

$$M_{ij}(t) = N_{ij}(t) - \int_{(0,t]} I_i(s^-) g_{ij}(s^-)\, ds, \quad M_{ij}(0) = 0, \tag{3.4}$$

are zero mean, square integrable, and mutually orthogonal P-martingales whose predictable quadratic variations are

$$\langle M_{ij}\rangle_t = \int_{(0,t]} I_i(s^-) g_{ij}(s^-)\, ds. \tag{3.5}$$

Moreover, at jump times t, we have

$$\Delta M_{ij}(t) = \Delta N_{ij}(t) = I_i(t^-) I_j(t). \tag{3.6}$$

Thus, the optional variation of M is

$$[M](t) = \sum_{0<s\leq t} |\Delta M(s)|^2 = \sum_{0<s\leq t} \sum_{i,j:\, j\neq i} |\Delta M_{ij}(s)|^2 = \sum_{0<s\leq t} \sum_{i,j:\, j\neq i} I_i(s^-)I_j(s). \tag{3.7}$$

We call $M := \{M_{ij},\ i \neq j\}$ the accompanying martingale of the counting process $N := \{N_{ij},\ i \neq j\}$ or of the Markov chain x.

We denote by $\mathscr{F}f := (\mathscr{F}_t)_{0\leq t\leq T}$ the completion of $(\mathscr{F}_t^0)_{t\leq T}$ with the P-null sets of Ω. Hereafter, a process from $[0, T] \times \Omega$ into a measurable space is said to be predictable (resp., progressively measurable) if it is predictable (resp., progressively measurable) with respect to the predictable σ-field on $[0, T] \times \Omega$ (resp., $\mathbb{F}$).

For a real-valued matrix $m := (m_{ij},\ i, j \in I)$ indexed by $I \times I$, we let

$$\Big|m\Big|_g^2(t) := \sum_{i,j:\, i\neq j} |m_{ij}|^2 g_{ij} \mathbf{1}_{\{w(t^-)=i\}} < \infty. \tag{3.8}$$

If m is time-dependent, we simply write $\Big|m(t)\Big|_g^2$.

Let $(Z_{ij},\ i \neq j)$ be a family of predictable processes, and set

$$\Big|Z(t)\Big|_g^2 := \sum_{i,j:\, i\neq j} Z_{ij}^2(t) I_i(t^-) g_{ij}(t^-), \quad 0 < t \leq T, \tag{3.9}$$

$$\sum_{0<s\leq t} Z(s)\Delta M(s) := \sum_{0<s\leq t} \sum_{i,j:\, i\neq j} Z_{ij}(s)\Delta M_{ij}(s). \tag{3.10}$$

Consider the local martingale

$$W(t) = \int_0^t Z(s)dM(s) := \sum_{i,j:\, i\neq j} \int_0^t Z_{ij}(s)dM_{ij}(s). \tag{3.11}$$

Then, the optional variation of the local martingale W is

$$[W](t) = \sum_{0<s\leq t} |Z(s)\Delta M(s)|^2 = \sum_{0<s\leq t} \sum_{i,j:\, i\neq j} |Z_{ij}(s)\Delta M_{ij}(s)|^2 \tag{3.12}$$

and its compensator is

$$\langle W\rangle_t = \int_{(0,t]} \Big|Z(s)\Big|_g^2 ds. \tag{3.13}$$

Provided that

$$E\left[\int_{(0,T]} \Big|Z(s)\Big|_g^2 ds\right] < \infty, \tag{3.14}$$

W is a square-integrable martingale, and its optional variation satisfies

$$E\left[[W](t)\right] = E\left[\sum_{0<s\leq t} |Z(s)\Delta M(s)|^2\right] = E\left[\int_{(0,t]} \left|Z(s)\right|_g^2 ds\right]. \tag{3.15}$$

Moreover, the following Doob's inequality holds

$$E\left[\sup_{0\leq t\leq T}\left|\int_0^t Z(s)dM(s)\right|^2\right] \leq 4E\left[\int_{(0,T]} \left|Z(s)\right|_g^2 ds\right]. \tag{3.16}$$

If $\widetilde{Z}$ is another predictable process that satisfies (3.14), setting

$$\langle Z(t), \widetilde{Z}(t)\rangle_g := \sum_{i,j:i\neq j} Z_{ij}(t) I_i(t^-) g_{ij}(t^-), \quad 0 \leq t \leq T, \tag{3.17}$$

and considering the martingale

$$\widetilde{W}(t) = \int_0^t \widetilde{Z}(s)dM(s) := \sum_{i,j:i\neq j} \int_0^t \widetilde{Z}_{ij}(s)dM_{ij}(s),$$

it is easy to see that

$$E\left[[W, \widetilde{W}](t)\right] = E\left[\int_{(0,t]} \langle Z(s), \widetilde{Z}(s)\rangle_g ds\right]. \tag{3.18}$$

Since the filtration $\mathbb{F}$ generated by the chain x is the same as the filtration generated by the family of counting processes $\{N_{ij},\ i \neq j\}$, we state the following martingale representation theorem (see, e.g., [74], Theorem T11 or [279], IV-21, Theorem 21.15).

Theorem 3.1 *If $\mathscr{L}$ is a (right-continuous) square-integrable $\mathbb{F}$-martingale, there exists a unique ($dP \times g_{ij}(s^-)I_i(s^-)ds$-almost everywhere) family of predictable processes Z_{ij}, $i \neq j$, satisfying*

$$E\left[\int_{(0,T]} \left|Z(s)\right|_g^2 ds\right] < +\infty, \tag{3.19}$$

where

$$\left|Z(t)\right|_g^2 := \sum_{i,j:i\neq j} Z_{ij}^2(t) I_i(t^-) g_{ij}(t^-), \quad 0 \leq t \leq T, \tag{3.20}$$

such that

$$\mathscr{L}_t = \mathscr{L}_0 + \int_0^t Z(s)dM(s), \quad 0 \le t \le T, \tag{3.21}$$

where

$$\int_0^t Z(s)dM(s) := \sum_{i,j:\, i \neq j} \int_0^t Z_{ij}(s)dM_{ij}(s).$$

In particular, at jump times t, we have

$$\Delta\mathscr{L}_t = \sum_{i,j:\, i\neq j} Z_{ij}(t)\Delta M_{ij}(t) = \sum_{i,j:\, i\neq j} Z_{ij}(t)I_i(t^-)I_j(t).$$

Next, we provide an important application of Theorem 3.1 to the local martingale M^f given by (3.2) where an explicit form of the process Z can be displayed in terms of the function f. At jump times t, we have

$$\begin{aligned}\Delta M_t^f = M_t^f - M_{t^-}^f &= \sum_i I_i(t^-) \sum_{j:\, j\neq i} I_j(t)(f(j) - f(i)) \\ &= \sum_{i,j:\, j\neq i} (f(j) - f(i))\Delta M_{ij}(t),\end{aligned}$$

since, by (3.6), at a jump time t,

$$I_i(t^-)I_j(t) = \Delta M_{ij}(t) \text{ and } I_i(t^-)I_i(t) = I_{\{x(t^-)=i, x(t)=i\}} = 0.$$

We may now define $Z^f(t) = (Z_{ij}^f(t))_{ij}$ by

$$Z_{ij}^f(t) := f(j) - f(i), \quad i, j \in I, \tag{3.22}$$

to obtain

$$M_t^f = \int_{(0,t]} Z^f(s)dM(s) = \sum_{i,j:\, j\neq i} (f(j) - f(i))M_{ij}(t), \quad 0 \le t \le T, \tag{3.23}$$

provided that $\sum_{i,j:\, j\neq i} (f(j) - f(i))^2 \int_0^T g_{ij}(t)\, dt < +\infty$.

Probability Measures and Metrics on Discrete Spaces

Let $\mathscr{P}(I)$ denote the set of probability measures on I. For $\mu, \nu \in \mathscr{P}(I)$, the total variation distance is defined by the formula

$$d(\mu, \nu) = 2 \sup_{A \subset I} |\mu(A) - \nu(A)| = \sum_{i \in I} |\mu(\{i\}) - \nu(\{i\})|. \tag{3.24}$$

Furthermore, let $\mathscr{P}_2(\Omega)$ be the space of probability measures P on Ω such that

$$\left|P\right|_2^2 := \int_\Omega |w|_T^2 P(dw) = E[|x|_T^2] < +\infty,$$

where $|x|_t := \sup_{0 \le s \le t} |x(s)|, \quad 0 \le t \le T$. Define on $\mathscr{F}$ the total variation metric

$$d(P, Q) := 2 \sup_{A \in \mathscr{F}} |P(A) - Q(A)|. \tag{3.25}$$

Similarly, on the filtration $\mathbb{F}$, we define the total variation metric between two probability measures P and Q as

$$D_t(P, Q) := 2 \sup_{A \in \mathscr{F}_t} |P(A) - Q(A)|, \quad 0 \le t \le T. \tag{3.26}$$

It satisfies

$$D_s(P, Q) \le D_t(P, Q), \quad 0 \le s \le t. \tag{3.27}$$

Endowed with the total variation metric D_T, $\mathscr{P}_2(\Omega)$ is a complete metric space. Moreover, D_T carries the usual topology of weak convergence.

For $P, Q \in \mathscr{P}_2(\Omega)$ with time marginals $P_t := P \circ x^{-1}(t)$ and $Q_t := Q \circ x^{-1}(t)$, the total variation distance between P_t and Q_t satisfies

$$d(P_t, Q_t) \le D_t(P, Q), \quad 0 \le t \le T. \tag{3.28}$$

Indeed, we have

$$\begin{aligned} d(P_t, Q_t) &:= 2 \sup_{B \subset I} |P_t(B) - Q_t(B)| = 2 \sup_{B \subset I} |P(x^{-1}(t)(B)) - Q(x^{-1}(t)(B))| \\ &\le 2 \sup_{A \in \mathscr{F}_t} |P(A) - Q(A)| = D_t(P, Q). \end{aligned}$$

3.1 Jump Processes of Mean-Field Type

In this section, we prove the existence of a unique probability measure $\widetilde{P}$ on $(\Omega, \mathscr{F})$ under which the coordinate process x is a jump process with intensities $\lambda_{ij}(t, x, \widetilde{P} \circ x^{-1}(t)),\ i, j \in I$, where we allow the jump intensities at time t depend on the whole path x over the time interval $[0, T]$ and also on the marginal law of $x(t)$, as long as the intensities are predictable. Because of the dependence of its jump intensities on the marginal law, we call it *jump process of mean-field type*. If the intensities are deterministic functions of t and of the marginal law of $x(t)$,, i.e., they are of the form $\lambda_{ij}(t, \widetilde{P} \circ x^{-1}(t)),\ i, j \in I$, we call x a *Markov chain of mean-field type* or simply a *nonlinear Markov chain.*

The probability measure $\widetilde{P}$ is constructed as follows. We start with the probability measure P which solves the martingale problem associated with $G = (g_{ij})$, where the intensities g_{ij} are assumed to be time-independent, making the coordinate process x a time-homogeneous Markov chain. Then, using a Girsanov-type change of measures in terms of a Dolean-Dade exponential martingale for jump processes, which involves the intensities λ_{ij} and g_{ij}, we obtain our probability measure $\widetilde{P}$. It is also possible to choose G time-dependent. But it is easier to deal with time-independent intensities.

Below, we will set $\left|a\right|^2 := \sum_{i,j:\, j\neq i} |a_{ij}|^2$ for $a = (a_{ij},\ i, j \in I,\ j \neq i)$.

Let λ be a measurable function from $[0, T] \times I \times I \times \Omega \times \mathscr{P}(I)$ into $(0, +\infty)$ such that

(A1) For every $Q \in \mathscr{P}_2(\Omega)$, the process $((\lambda_{ij}(t, x, Q \circ x^{-1}(t)))_t$ is predictable.

(A2) There exists a positive constant c_1 such that for every $(t, i, j) \in [0, T] \times I \times I;\ i \neq j,\ w \in \Omega$ and $\mu, \nu \in \mathscr{P}(I)$,

$$\lambda_{ij}(t, w, \mu) \geq c_1 > 0.$$

(A3) For every $t \in [0, T]$, $w \in \Omega$ and $\mu, \nu \in \mathscr{P}(I)$,

$$\left|\lambda(t, w, \mu) - \lambda(t, w, \nu)\right| \leq Cd(\mu, \nu).$$

(A4) For $p \in \{1, 2\}$ and for every for $t \in [0, T]$, $w \in \Omega$ and $\mu \in \mathscr{P}_2(I)$,

$$\sum_{i,j:\, j\neq i} |j - i|^p \lambda_{ij}(t, w, \mu) \leq C(1 + |w|_t^p + \int |y|^p \mu(dy)).$$

(A5) The probability measure ξ on I has finite second moment: $\left|\xi\right|_2^2 := \int |y|^2 \xi(dy) < \infty$.

Remark 3.1 Assumption (A4) is needed to guarantee that the chain has finite second moment. The constants C may be depend on the parameters but all bounded with the interval of interest $[0, T]$.

Example 3.1 (A Mean-Field Schlögl Model) In the mean-field version of the Schlögl model (cf. [108, 151] and [150]), the intensities are

$$\lambda_{ij}(w, \mu) := \begin{cases} \nu_{ij} & \text{if } j \neq i + 1, \\ \nu_{ii+1} + \left|\mu\right|_1 & \text{if } j = i + 1, \end{cases} \tag{3.29}$$

where $\left|\mu\right|_1 = \int |y|\mu(dy)$ is the first moment of the probability measure μ on I and (ν_{ij}) is a Q-matrix satisfying $\inf_{i\in I}\nu_{ii+1} > 0$ and there exists $N_0 \geq 1$ such that $\nu_{ij} = 0$, for $|i-j| \geq N_0$. The martingale problem formulation states that, for every f on I, the process defined by

$$M_t^f := f(x(t)) - f(x(0)) - \int_{(0,t]} (\widetilde{G}(s)f)(x(s))\,ds$$

is a local martingale relative to $(\Omega, \mathscr{F}, \mathbb{F})$, where

$$\widetilde{G}(s)f(i) = \sum_{j:\,j\neq i} \nu_{ij}(f(j)-f(i)) + \sum_j jP(x(s)=j)(f(i+1)-f(i)). \tag{3.30}$$

Let P be the probability under which x is a time-homogeneous Markov chain such that $P\circ x^{-1}(0) = \xi$ and with Q-matrix $(g_{ij})_{ij}$ satisfying (3.1), with g_{ij} time-independent. Assume further that

$$c_2 := \inf_{i,j:i\neq j} g_{ij} > 0 \quad \text{if and only if} \quad \lambda_{ij}(t,w,\mu) \geq c_1 > 0. \tag{3.31}$$

We impose this condition, because we are going to use a Girsanov-type change of measure between two probability measures under which the chain has jump intensities of λ_{ij} and g_{ij}, respectively. This amounts to only taking into account nonzero jump intensities.

To ease notation, we set, for $(t,i,j) \in [0,T]\times I\times I$ and $Q \in \mathscr{P}_2(\Omega)$,

$$\lambda_{ij}^Q(t) := \lambda_{ij}(t,x,Q\circ x^{-1}(t)), \quad \lambda_{ii}^Q(t) := -\sum_{j\neq i}\lambda_{ij}^Q(t). \tag{3.32}$$

Let P^Q be the measure on $(\Omega,\mathscr{F})$ defined by

$$dP^Q := L^Q(T)dP, \tag{3.33}$$

where

$$L^Q(T) := \prod_{\substack{i,j\\ i\neq j}} \exp\left\{\int_{(0,T]} \ln\frac{\lambda_{ij}^Q(t)}{g_{ij}}dN_{ij}(t) - \int_0^T (\lambda_{ij}^Q(t) - g_{ij})I_i(t)dt\right\}, \tag{3.34}$$

is the Dolean-Dade exponential. It is the solution of the following linear stochastic integral equation

$$L^Q(t) = 1 + \int_{(0,t]} L^Q(s^-)\sum_{i,j:i\neq j} I_i(s^-)\ell_{ij}^Q(s)dM_{ij}(s), \tag{3.35}$$

where

$$\ell_{ij}^{Q}(s) = \begin{cases} \lambda_{ij}^{Q}(s)/g_{ij} - 1 \textbf{ if } i \neq j, \\ 0 \textbf{ if } i = j, \end{cases} \tag{3.36}$$

and $(M_{ij})_{ij}$ is the P-martingale given in (3.4).

By Girsanov theorem, if L^Q is a P-martingale, then P^Q is a probability measure on $(\Omega, \mathscr{F})$ under which the coordinate process x is a jump process with intensity matrix $\lambda^Q := (\lambda_{ij}^Q(t))_{i,j}$ and starting distribution ξ, i.e., $P^Q \circ x^{-1}(0) = \xi$. In particular, the compensated processes associated with the counting processes N_{ij} defined by

$$M_{ij}^{Q}(t) := N_{ij}(t) - \int_{(0,t]} I_i(s^-)\lambda_{ij}^{Q}(s)ds, \quad M_{ij}^{Q}(0) = 0, \tag{3.37}$$

are zero mean, square integrable and mutually orthogonal P^Q-martingales whose predictable quadratic variations are

$$\langle M_{ij}^{Q}\rangle_t = \int_{(0,t]} I_i(s^-)\lambda_{ij}^{Q}(s)ds. \tag{3.38}$$

Using (3.36), we may write M_{ij}^Q in terms of M_{ij} as follows.

$$M_{ij}^{Q}(t) = M_{ij}(t) - \int_{(0,t]} \ell_{ij}^{Q}(s)I_i(s^-)g_{ij}ds. \tag{3.39}$$

Now, since L^Q is a positive P-local martingale, it is a supermartingale. Thus, $E[L_T^Q] \leq 1$. In order to show that it is a P-martingale, we need to show that $E[L_T^Q] = 1$. We note that the imposed conditions (A1)–(A4) on the intensity matrix λ^Q do not fit with the assumptions displayed in the literature ranging from [74], Theorem T11, to [300], Theorem 2.4, to guarantee that L^Q is a P-martingale.

To show that L^Q is a P-martingale, we will use the following *a priori* estimate.

Lemma 3.1 *Let $Q \in \mathscr{P}_2(\Omega)$ and assume that λ^Q and ξ satisfy (A1)–(A5). If P^Q given by* (3.33) *is a probability measure on $(\Omega, \mathscr{F})$, then*

$$\left|P^Q\right|_2^2 = E_{P^Q}[|x|_T^2] \leq Ce^{CT}(1 + \left|\xi\right|_2^2 + \left|Q\right|_2^2) < +\infty. \tag{3.40}$$

In particular,

$$\left|P\right|_2^2 = E[|x|_T^2] \leq Ce^{CT}(1 + \left|\xi\right|_2^2) < +\infty. \tag{3.41}$$

Proof Since, by Girsanov theorem, under P^Q, x has jump intensity $\lambda_{ij}(t, x, Q \circ x^{-1}(t))$, applying a similar formula as (3.2) to $f(x) = x$, where instead of G we use the matrix λ^Q, we obtain

$$x(t) = x(0) + \int_{(0,t]} \lambda^Q(s)x(s)\, ds + M_t^x, \tag{3.42}$$

where

$$M^x(t) = \int_{(0,t]} Z^x(s)dM(s) = \sum_{i,j:\, j\neq i} (j-i)M_{ij}(t), \tag{3.43}$$

with

$$\left|Z^x(s)\right|^2_{\lambda^Q} = \sum_{i,j:\, j\neq i} (j-i)^2 I_i(s^-)\lambda^Q_{ij}(s),$$

and

$$\lambda^Q(s)x(s) = \sum_{i,j:\, j\neq i} (j-i)I_i(s)\lambda^Q_{ij}(s),$$

which, in view of (A4), satisfies

$$|\lambda^Q(s)x(s)|^2 + \left|Z^x(s)\right|^2_{\lambda^Q} \le C\big(1 + |x|_s^2 + \int w_s^2 Q(dw)\big), \;\; 0 \le s \le T. \tag{3.44}$$

Therefore, applying the Cauchy-Schwarz inequality together with (3.16) to (3.42), we obtain

$$E_{P^Q}[|x|_T^2] \le C E_{P^Q}\left[|x(0)|^2 + \int_{(0,T]}\left(|\lambda^Q(s)x(s)|^2 + \left|Z^x(s)\right|^2_{\lambda^Q}\right) ds\right],$$

and by (3.44), we get

$$E_{P^Q}[|x|_T^2] \le C\Big(1 + \left|\xi\right|_2^2 + \int_{(0,T]}\Big(E_{P^Q}[|x|_s^2] + \int w_s^2 Q(dw)\Big)\, ds\Big). \tag{3.45}$$

Using $\int w_s^2 Q(dw) \le \int |w|_s^2 Q(dw) \le \left|Q\right|_2^2$ and applying Grönwall's inequality, we finally get (3.40):

$$\left|P^Q\right|_2^2 = E_{P^Q}[|x|_T^2] \le Ce^{CT}(1 + \left|\xi\right|_2^2 + \left|Q\right|_2^2).$$

∎

Proposition 3.1 *Let $Q \in \mathscr{P}_2(\Omega)$ and assume λ^Q and ξ satisfy (A1)–(A5). Then, L^Q is a P-martingale.*

Proof The proof is inspired by the proof of Proposition (A.1) in [144]. As mentioned above, it suffices to prove that $E[L^Q_T] = 1$. For $I \geq 0$, let λ^I be the predictable intensity matrix given by $\lambda^I_{ij}(t) := \lambda^Q_{ij}(t)\mathbf{1}_{\{\omega,\ \left|x(\omega)\right|_t \leq I\}}$, and let L^I be the associated Dolean-Dade exponential and P^I the positive measure defined by $dP^I = L^I_T dP$. Note that, for $i, j \in I$, $i \neq j$, $|i - j| \geq 1$, by (A4), we have

$$\lambda_{ij}(t, w, \mu) \leq C(1 + |w|_t + \int |y|\mu(dy)).$$

Thus, for every $I \geq 1$, $\lambda^I_{ij}(t) \leq C(1 + I + \left|Q\right|_2)$, i.e., λ^I_{ij} is bounded. In view of [74], Theorem T11, L^I is a P-martingale. In particular, $E[L^I_T] = 1$ and P^I are a probability measure. By (3.41), $\left|x\right|_T < \infty$, P-a.s. Therefore, on the set $\{\omega,\ \left|x(\omega)\right|_T \leq I_0\}$, for all $I \geq I_0$, $L^I_T(\omega) = L^Q_T(\omega)$. This in turn yields that $L^I_T \to L^Q_T$, P-a.s., as $I \to +\infty$. Now, if $(L^I_T)_{I \geq 1}$ is uniformly integrable, the P-a.s. convergence implies $L^1(P)$-convergence of L^I_T to L^Q_T, yielding $E[L^Q_T] = 1$. It remains to show that $(L^I_T)_{I \geq 1}$ is uniformly integrable:

$$\lim_{a\to\infty} \sup\nolimits_{I \geq 1} \int_{\{L^I_T > a\}} L^I_T \, dP = 0.$$

For $m \geq 1$, set $\theta_m = \inf\{t \leq T,\ \left|x\right|_t \geq m\}$ if the set is non-empty and $\theta_m = T + 1$ if it is empty. Denoted by E^I the expectation with respect to P^I, we have

$$\int_{\{\theta_m \leq T\}} L^I_T \, dP = P^I(\theta_m \leq T) = P^I(\left|x\right|_T \geq m) \leq E^I[\left|x\right|_T]/m \leq C/m, \tag{3.46}$$

where, by (3.40), C does not depend on I.

Let $\eta > 0$. Choose $m_0 \geq 1$ such that $C/m_0 < \eta$. We have, for all $I \geq m_0$, $L^I_{T\wedge\theta_{m_0}} = L^{m_0}_{T\wedge\theta_{m_0}}$. This entails that

$$\sup_{I\geq 1} \int_{\{L^I_{T\wedge\theta_{m_0}} > a\}} L^I_{T\wedge\theta_{m_0}} \, dP = \max_{I \leq m_0} \int_{\{L^I_{T\wedge\theta_{m_0}} > a\}} L^I_{T\wedge\theta_{m_0}} \, dP \to 0,\ a \to \infty.$$

So there exists $a_0 > 0$ such that whenever $a > a_0$,

$$\max_{I \leq m_0} \int_{\{L^I_{T\wedge\theta_{m_0}} > a\}} L^I_{T\wedge\theta_{m_0}} \, dP < \eta. \tag{3.47}$$

We have

$$\begin{aligned}
&\sup_{I\geq 1}\int_{\{L_T^I>a\}} L_T^I\,dP \leq \sup_{I\geq 1}\int_{\{L_T^I>a,\,\theta_{m_0}\leq T\}} L_T^I\,dP + \sup_{I\geq 1}\int_{\{L_T^I>a,\,\theta_{m_0}>T\}} L_T^I\,dP\\
&\leq \sup_{I\geq 1}\int_{\{\theta_{m_0}\leq T\}} L_T^I\,dP + \sup_{I\geq 1}\int_{\{L^I_{T\wedge\theta_{m_0}}>a,\,\theta_{m_0}>T\}} L^I_{T\wedge\theta_{m_0}}\,dP\\
&\leq \sup_{I\geq 1}\int_{\{\theta_{m_0}\leq T\}} L_T^I\,dP + \max_{I\leq m_0}\int_{\{L^I_{T\wedge\theta_{m_0}}>a\}} L^I_{T\wedge\theta_{m_0}}\,dP \leq C/m_0+\eta<2\eta,
\end{aligned} \tag{3.48}$$

in view of (3.46) and (3.47). This completes the proof, since η is arbitrary. ■

Next, we will show that there is $\widehat{Q}$ such that $P^{\widehat{Q}}=\widehat{Q}$, i.e., $\widehat{Q}$ is a fixed point. It is the probability measure under which the coordinate process is a jump process of mean-field type.

Theorem 3.2 *The map*

$$\begin{aligned}
&\Phi:\mathscr{P}_2(\Omega)\longrightarrow\mathscr{P}_2(\Omega)\\
&\qquad Q\mapsto\Phi(Q):=P^Q;\quad dP^Q:=L^Q(T)dP,\quad P^Q\circ x^{-1}(0)=\xi,
\end{aligned}$$

admits a unique fixed point, $\widehat{Q}$*, which satisfies*

$$\left|\widehat{Q}\right|_2^2\leq Ce^{CT}(1+\left|\xi\right|_2^2)<+\infty. \tag{3.49}$$

Proof First, we note that if $Q\in\mathscr{P}_2(\Omega)$ then by (3.40),

$$\left|\Phi(Q)\right|_2^2=E_{\Phi(Q)}[|x|_T^2]\leq Ce^{CT}(1+\left|\xi\right|_2^2+\left|Q\right|_2^2)<+\infty,$$

which implies that $\Phi(Q)\in\mathscr{P}_2(\Omega)$, since $\left|\xi\right|_2^2<+\infty$ by (A5).

Next, we show the contraction property of the map Φ. To this end, given $Q,\widetilde{Q}\in\mathscr{P}_2(\Omega)$, we use an estimate of the total variation distance $D_T(\Phi(Q),\Phi(\widetilde{Q}))$ in terms of the relative entropy $H(\Phi(Q)|\Phi(\widetilde{Q}))$ between $\Phi(Q)$ and $\Phi(\widetilde{Q})$ given by the celebrated Csiszár-Kullback-Pinsker inequality:

$$D_T^2(\Phi(Q),\Phi(\widetilde{Q}))\leq 2H(\Phi(Q)|\Phi(\widetilde{Q})), \tag{3.50}$$

where

$$H(\Phi(Q)|\Phi(\widetilde{Q}))=E_{\Phi(Q)}\Big[\log\frac{L^Q(T)}{L^{\widetilde{Q}}(T)}\Big].$$

right-hand side In view of (3.34), we have

$$\log\frac{L^Q(T)}{L^{\widetilde{Q}}(T)}=\sum_{i,j:i\neq j}\int_{(0,T]}\ln\frac{\lambda_{ij}^Q(t)}{\lambda_{ij}^{\widetilde{Q}}(t)}dN_{ij}(t)-\int_0^T(\lambda_{ij}^Q(t)-\lambda_{ij}^{\widetilde{Q}}(t))I_i(t)dt.$$

Taking expectation with respect to $\Phi(Q)$, using (3.37), we obtain

$$H(\Phi(Q)|\Phi(\widetilde{Q})) = \sum_{i,j:i\neq j} E_{\Phi(Q)}\left[\int_0^T \left(\tau(\lambda_{ij}^{Q}(t)) - \tau(\lambda_{ij}^{\widetilde{Q}}(t)) \right.\right. \\ \left.\left. -(\lambda_{ij}^{Q}(t) - \lambda_{ij}^{\widetilde{Q}}(t)) \log \lambda_{ij}^{\widetilde{Q}}(t)\right) I_i(t)dt\right],$$

where $\tau(x) := x \log x - x + 1,\ x > 0$ is a convex function. We note that the right-hand side of this last equality is nonnegative, since, by convexity, we have

$$\tau(x) \geq \tau(y) + (x - y)\tau'(y),$$

where $\tau'(y) = \log y$. Using Taylor expansion, we get

$$\tau(x) = \tau(y) + (x - y)\tau'(y) + \frac{1}{2}(x - y)^2\tau''(z),$$

for some $z \in \{tx + (1 - t)y\ :\ t \in (0, 1)\}$, where $\tau''(z) = 1/z$. Taking x, y such that $x, y \geq c_1 > 0$, as in (A2), we obtain

$$\tau(x) \leq \tau(y) + (x - y)\tau'(y) + \frac{1}{2c_1}(x - y)^2. \tag{3.51}$$

Applying (3.51) to the entropy (3.6), we obtain

$$H(\Phi(Q)|\Phi(\widetilde{Q})) \leq \frac{1}{2c_1} \sum_{i,j:i\neq j} E_{\Phi(Q)}\left[\int_0^T (\lambda_{ij}^{Q}(t) - \lambda_{ij}^{\widetilde{Q}}(t))^2\, dt\right].$$

Combining this inequality with (3.50), we obtain

$$D_T^2(\Phi(Q), \Phi(\widetilde{Q})) \leq \frac{1}{c_1} \sum_{i,j:i\neq j} E_{\Phi(Q)}\left[\int_0^T (\lambda_{ij}^{Q}(t) - \lambda_{ij}^{\widetilde{Q}}(t))^2\, dt\right]. \tag{3.52}$$

We may use (A3) to obtain

$$\sum_{i,j:i\neq j} E_{\Phi(Q)}\left[(\lambda_{ij}^{Q}(t) - \lambda_{ij}^{\widetilde{Q}}(t))^2\right] \leq Cd^2(Q_t, \widetilde{Q}_t) \leq CD_t^2(Q, \widetilde{Q}).$$

Therefore,

$$D_T^2(\Phi(Q), \Phi(\widetilde{Q})) \leq 2H(\Phi(Q)|\Phi(\widetilde{Q})) \leq \frac{C}{c_1}\int_0^T D_t^2(Q, \widetilde{Q})\, dt.$$

Iterating this inequality, we obtain, for every $N > 0$,

$$D_T^2(\Phi^I(Q), \Phi^I(\widetilde{Q})) \leq C^I \int_0^T \frac{(T-t)^{N-1}}{(N-1)!} D_t^2(Q, \widetilde{Q})dt \leq \frac{C^I T^I}{N!} D_T^2(Q, \widetilde{Q}),$$

where Φ^I denotes the N-fold composition of the map Φ. Hence, for N large enough, Φ^I is a contraction which implies that Φ admits a unique fixed point.

Finally, using (3.45) with $\Phi(\widehat{Q}) = \widehat{Q}$, and noting that $\int w_s^2 \widehat{Q}(dw) \leq \int |w|_s^2 \widehat{Q}(dw) = E_{\widehat{Q}}[|x|_s^2]$, we get

$$E_{\widehat{Q}}[|x|_T^2] \leq C\Big(1 + \Big|\xi\Big|_2^2 + 2\int_{(0,T]} E_{\widehat{Q}}[|x|_s^2]\, ds\Big).$$

Applying Grönwall's inequality, we obtain the estimate (3.49):

$$\Big|\widehat{Q}\Big|_2^2 \leq Ce^{2CT}(1 + \Big|\xi\Big|_2^2) < \infty.$$

■

Corollary 3.1 *The mapping $t \mapsto P \circ x^{-1}(t)$ is continuous. More precisely, we have*

$$d(P_t, P_s) \leq C(1 + \Big|P\Big|_2)(t-s), \quad 0 \leq s \leq t \leq T. \tag{3.53}$$

Proof The inequality (3.53) follows by applying the estimates obtained earlier to the martingale (3.2) with $f(x) = I_{\{x \in A\}}$, $A \subset I$, where we use the matrix λ instead of G. ■

Markov Chain Backward Stochastic Equation

An important consequence of Theorem 3.1 is solutions (Y, Z) of Markov chain backward stochastic differential equations (BSDEs) defined on $(\Omega, \mathscr{F}, \mathbb{F}, P)$ by

$$-dY(t) = f(t, \omega, Y(t^-), Z(t))dt - Z(t)dM(t), \quad Y(T) = \zeta. \tag{3.54}$$

It is easily seen that if (Y, Z) solves (3.54), then it admits the following representation:

$$Y(t) = E\Big[\zeta + \int_t^T f(s, \omega, Y(s^-), Z(s))\, ds\Big|\mathscr{F}_t\Big], \quad t \in [0, T].$$

Moreover, $t \mapsto Y(t)$ is right-continuous with left limits. Therefore, $Y(t^-) = Y(t)$, dt-a.e. Hence, we may write

$$Y(t) = E\Big[\zeta + \int_t^T f(s, \omega, Y(s), Z(s))\, ds\Big|\mathscr{F}_t\Big], \quad t \in [0, T].$$

Existence and uniqueness results of solutions of Markov chain BDSEs (3.54) based on the martingale representation theorem (L^2-theory) have been recently studied in a series of papers by Cohen and Elliott (see, e.g., [103] and the references therein). Their approach essentially adapts the method for solving Brownian motion driven BSDEs established first in [267]. Recently, Confortola *et al.* [104] derived existence and uniqueness results for more general classes of BSDEs driven by marked point processes under only L^1-integrability conditions. In this section, we use the L^2-theory as we want to use the martingale representation theorem in our optimal control problem.

Below, we establish existence of an optimal control and existence of a saddle point for the zero-sum game using some properties of the following class of BSDEs which is a special case of (3.54).

$$-\,dY(t) = \phi(t, x, Z(t))dt - Z(t)dM(t), \quad Y(T) = \zeta, \tag{3.55}$$

where $(M_{ij})_{ij}$ is the P-martingale given in (3.4) and the driver ϕ is essentially of the form

$$\phi(t, x, p) := f(t, x) + \langle \ell(t, x), p\rangle_g, \tag{3.56}$$

where $p := (p_{ij},\, i, j \in I)$ is a real-valued matrix indexed by $I \times I$ and $\ell = (\ell_{ij},\, i, j \in I)$ is given by

$$\ell_{ij}(s, x) := \begin{cases} \lambda_{ij}(s, x)/g_{ij} - 1 \text{ if } i \neq j, \\ 0 \text{ if } i = j, \end{cases} \tag{3.57}$$

where the predictable process $\lambda(t, x) = (\lambda_{ij}(t, x),\, i, j \in I)$ is the intensity matrix of the chain x under a probability measure $\widetilde{P}$ on $(\Omega, \mathscr{F})$ given by a similar formula as (3.33)–(3.34). In particular, as in (3.39), the processes

$$\widetilde{M}_{ij}(t) = M_{ij}(t) - \int_{(0,t]} \ell_{ij}(s, x) I_i(s^-) g_{ij} ds \tag{3.58}$$

are zero mean, square integrable, and mutually orthogonal $\widetilde{P}$-martingales whose predictable quadratic variations are

$$\langle \widetilde{M}_{ij}\rangle_t = \int_{(0,t]} I_i(s^-)\lambda_{ij}(s, x) ds. \tag{3.59}$$

Moreover, ϕ satisfies a "stochastic" Lipschitz condition. More precisely, we make the following assumptions on the driver ϕ and the terminal value ξ.

(H1) P-a.s., for all $(t, x) \in [0, T] \times \Omega$, $z^1 = (z^1_{ij}), z_2 = (z^2_{ij})$, $z^1_{ij}, z^2_{ij} \in \mathbf{R}$,

$$|\phi(t,x,z_1)-\phi(s,x,z_2)| \le a(t)\left|z^1-z^2\right|_g(t),$$

(see notation 3.8) where $(a(t))_t$ is a nonnegative and progressively measurable process which belongs to $L^2([0,T]\times\Omega, dt\otimes dP)$.

(H2) $\phi(t,\omega,0)$ is bounded.

(H3) ζ is an $\mathscr{F}_T$-measurable and bounded random variable.

First we establish a comparison result for solutions of the BSDE (3.55). This result is a key argument in the proof of existence and uniqueness of solutions of our BSDE.

Proposition 3.2 *Let, for $i\in\{1,2\}$, (Y^i,Z^i) be the solutions of the BSDE* (3.55) *associated with (ϕ^i,ζ^i) where (ϕ^i,ζ^i) satisfy (H1) to (H3). Assume that*

(H4) $\zeta^1\ge\zeta^2$, P-a.s.;

(H5) for any $(t,\omega)\in[0,T]\times\Omega$, $\phi^1(t,\omega,Z^2(t))\ge\phi^2(t,\omega,Z^2(t))$ $P-a.s.$

Then, $Y^1\ge Y^2$ on $[0,T]$ P-a.s.

Proof Set

$$\widehat{Y}:=Y^1-Y^2,\quad \widehat{Z}:=Z^1-Z^2,\quad \widehat{\zeta}=\zeta^1-\zeta^2.$$

Using (H4) and (H5), we obtain

$$\begin{aligned}
\widehat{Y}_t &= \widehat{\zeta}+\textstyle\int_t^T(\phi^1(s,x,Z^1(s))-\phi^1(s,x,Z^2(s)))ds\\
&+\textstyle\int_t^T(\phi^1(s,x,Z^2(s))-\phi^2(s,x,Z^2(s)))ds-\int_t^T\widehat{Z}(s)dM(s)\\
&\ge \textstyle\int_t^T(\phi^1(s,x,Z^1(s))-\phi^1(s,x,Z^2(s)))ds-\int_t^T\widehat{Z}(s)dM(s)\quad P\text{-a.s.}
\end{aligned}$$

Since

$$\phi^1(s,x,Z^1(s))-\phi^1(s,x,Z^2(s))=\langle\ell(s,x),\widehat{Z}(s)\rangle_g,$$

by (3.58)

$$\int_0^t\widehat{Z}(s)d\widetilde{M}(s)=\int_0^t\widehat{Z}(s)dM(s)-\int_0^t\langle\ell(s,x),\widehat{Z}(s)\rangle_g ds$$

is a zero-mean $\widetilde{P}$-martingale. Since $\widetilde{P}$ is absolutely continuous with respect to P, we also have

$$\widehat{Y}_t\ge-\int_t^T\widehat{Z}(s)d\widetilde{M}(s)\quad \widetilde{P}\text{-a.s.}$$

Hence, taking conditional expectation with respect to $\mathscr{F}_t$, we obtain

$$\widehat{Y}_t \geq -\widetilde{E}\left[\int_t^T \widehat{Z}(s)d\widetilde{M}(s)|\mathscr{F}_t\right] = 0 \quad P \text{ and } \widetilde{P}\text{−a.s.}$$

■

Theorem 3.3 *Let* (ϕ, ζ) *satisfies the assumptions (H1) to (H3). Then, the BSDE* (3.55) *associated with* (ϕ, ζ) *admits a solution* (Y, Z) *consisting of an adapted process* Y *which is right-continuous with left limits and a predictable process* Z *which satisfy*

$$E\left[\sup_{t\in[0,T]}|Y(t)| + \int_{(0,T]} \left|Z(s)\right|_g^2 ds\right] < +\infty.$$

This solution is unique up to indistinguishability for Y *and equality* $dP \times g_{ij}(s^-)I_i(s^-)ds$*-almost everywhere for* Z.

Using Proposition 3.2, the proof of the theorem is similar to that of the Brownian motion driven BSDEs derived in [184], Theorem I-3, using an approximation scheme by an increasing sequence of standard Markov chain BSDEs for which existence, uniqueness, and comparison results are similar to that of the Brownian motion driven BSDEs derived in [267] and [145], along with the properties (3.15) and (3.16) related to the martingale W displayed in (3.11) together with Itô's formula for semimartingales driven by counting processes. We do not include the details.

3.2 Mean-Field-Type Optimal Control of Jump Processes

Let (U, δ) be a compact metric space with its Borel field $\mathscr{B}(U)$ and $\mathscr{U}$ the set of $\mathbb{F}$-progressively measurable processes $u = (u(t),\ 0 \leq t \leq T)$ with values in U. We call $\mathscr{U}$ the set of admissible controls. In this section, we consider a control problem of the jump process of mean-field type introduced above, where the control enters the jump intensities.

For $u \in \mathscr{U}$, let P^u be the probability measure on $(\Omega, \mathscr{F})$ under which the coordinate process x is a jump process with intensities

$$\lambda_{ij}(t, u(t)) := \lambda_{ij}(t, x, P^u \circ x^{-1}(t), u(t)),\quad i, j \in I,\ 0 \leq t \leq T, \tag{3.60}$$

satisfying the following assumptions similar to (A1)–(A5).

(B1) For any $u \in \mathscr{U}$, $i, j \in I$, the process $((\lambda_{ij}(t, x, P^u \circ x^{-1}(t)))_t$ is predictable.
(B2) There exists a positive constant c_1 such that for every $(t, i, j) \in [0, T] \times I \times I$; $i \neq j$, $w \in \Omega$, $u \in U$ and $\mu, \nu \in \mathscr{P}(I)$

$$\lambda_{ij}(t, w, \mu, u) \geq c_1 > 0.$$

(B3) There exists a positive constants C such that for every $t \in [0, T]$, $w \in \Omega$, $u, v \in U$ and $\mu, \nu \in \mathscr{P}(I)$,

$$\left|\lambda(t, w, \mu, u) - \lambda(t, w, \nu, v)\right| \leq C\left(d(\mu, \nu) + \delta(u, v)\right).$$

(B4) There exists a positive constants C such that for $p \in \{1, 2\}$ and for every $t \in [0, T]$, $w \in \Omega$, $u \in U$ and $\mu \in \mathscr{P}_2(I)$,

$$\sum_{i,j:\, j\neq i} |j - i|^p \lambda_{ij}(t, w, \mu, u) \leq C(1 + |w|_t^p + \int |y|^p \mu(dy)).$$

(B5) The probability measure ξ on I has finite second moment:

$$\left|\xi\right|_2^2 := \int |y|^2 \xi(dy) < \infty.$$

Existence of P^u such that $P^u \circ x^{-1}(0) = \xi$ is obtained as a fixed point of Φ^u defined in the same way as in Theorem 3.2 except that the intensities $\lambda_{ij}(\cdot)$ further depend on u, which does not raise any major issues.

Let P be the probability measure on $(\Omega, \mathscr{F})$ under which x is a time-homogeneous Markov chain such that $P \circ x^{-1}(0) = \xi$ and with Q-matrix $(g_{ij})_{ij}$ satisfying (3.1) and (3.31). We have

$$dP^u := L^u(T)dP, \tag{3.61}$$

where, for $0 \leq t \leq T$,

$$L^u(t) := \prod_{\substack{i,j \\ i\neq j}} \exp\left\{\int_{(0,t]} \ln \frac{\lambda_{ij}(s, u(s))}{g_{ij}} dN_{ij}(s) - \int_0^t (\lambda_{ij}(s, u(s)) - g_{ij}) I_i(s) ds\right\}, \tag{3.62}$$

which satisfies

$$L^u(t) = 1 + \int_{(0,t]} L^u(s^-) \sum_{i,j:\, i\neq j} I_i(s^-) \ell_{ij}^u(s) dM_{ij}(s), \tag{3.63}$$

where $\ell_{ij}^u(s) := \ell_{ij}(t, x, P^u \circ x^{-1}(s), u(s))$ is given by the formula

$$\ell_{ij}^u(s) = \begin{cases} \lambda_{ij}(s, u(s))/g_{ij} - 1 \ \textbf{if} \ i \neq j, \\ 0 \ \textbf{if} \ i = j, \end{cases} \tag{3.64}$$

and $(M_{ij})_{ij}$ is the P-martingale given in (3.4). Moreover, in a similar way as in (3.39), the accompanying martingale $M^u = (M^u_{ij})_{ij}$ satisfies

$$M^u_{ij}(t) = M_{ij}(t) - \int_{(0,t]} \ell^u_{ij}(s) I_i(s^-) g_{ij} ds. \tag{3.65}$$

We first derive continuity of the map $u \mapsto P^u$ and then state the optimal control problem we want to solve.

Let E^u denote the expectation with respect to P^u. Using (3.49), we have for every $u \in \mathscr{U}$,

$$\left|P^u\right|_2^2 = E^u[|x|_T^2] \le Ce^{CT}(1 + \left|\xi\right|_2^2) < \infty. \tag{3.66}$$

We further have the following estimate of the total variation between P^u and P^v.

Lemma 3.2 *For every $u, v \in \mathscr{U}$, it holds that for some $C > 0$,*

$$D_T^2(P^u, P^v) \le C \sup_{\{0 \le t \le T\}} E^u[\delta^2(u(t), v(t))]. \tag{3.67}$$

In particular, the function $u \mapsto P^u$ from U into $\mathscr{P}_2(\Omega)$ is Lipschitz continuous: for every $u, v \in U$,

$$D_T(P^u, P^v) \le C\delta(u, v). \tag{3.68}$$

Moreover,

$$K_T := \sup_{u \in U} \left|P^u\right|_2 \le C < \infty, \tag{3.69}$$

for some constant $C > 0$ that depends only on T and ξ.

Proof A similar estimate as (3.52) yields

$$D_T^2(P^u, P^v) \le \frac{1}{c_1} \sum_{i,j: i \ne j} E^u \left[\int_0^T (\lambda^u_{ij}(t) - \lambda^v_{ij}(t))^2 \, dt \right]. \tag{3.70}$$

Using (B3), we obtain

$$D_T^2(P^u, P^v) \le \frac{C}{c_1} E^u \left[\int_0^T d^2(P^u(t), P^v(t)) + \delta^2(u(t), v(t)) dt \right]$$

By (3.28) and Gr'onwall's inequality, we finally obtain

$$D_T^2(P^u, P^v) \le C \sup_{\{0\le t\le T\}} E^u[\delta^2(u(t), v(t))].$$

Inequality (3.68) follows from (3.67) by letting $u(t) \equiv u \in U$ and $v(t) \equiv v \in U$. It remains to show (3.69). But this follows from (3.66) and the continuity of the function $u \mapsto P^u$ from the compact set U into $\mathscr{P}_2(\Omega)$. ∎

Let f be a measurable function from $[0, T] \times \Omega \times \mathscr{P}_2(I) \times U$ into $\mathbf{R}$ and h be a measurable functions from $I \times \mathscr{P}_2(I)$ into $\mathbf{R}$ such that

(B5) For any $u \in \mathscr{U}$ and $Q \in \mathscr{P}_2(\Omega)$, the process $(f(t, x, Q \circ x^{-1}(t), u(t)))_t$ is progressively measurable. Moreover, $h(x(T), Q\circ x^{-1}(T))$ is $\mathscr{F}_T$-measurable.
(B6) There exists $C > 0$ such that for every $t \in [0, T]$, $w \in \Omega$, $u, v \in U$, and $\mu, \nu \in \mathscr{P}_2(I)$, $|\phi(t, w, \mu, u) - \phi(t, w, \nu, v)| \le C(d(\mu, \nu) + \delta(u, v))$. for $\phi \in \{f, h\}$.
(B7) f and h are uniformly bounded.

The cost functional $J(u)$, $u \in \mathscr{U}$ associated with the jump process controlled through the intensities $\lambda_{ij}(t, u(t))$ is

$$J(u) := E^u\left[\int_0^T f(t, x, P^u \circ x^{-1}(t), u(t))dt + h(x(T), P^u \circ x^{-1}(T))\right], \tag{3.71}$$

where f and h satisfy (B5), (B6), and (B7) above.

Any $\widehat{u} \in \mathscr{U}$ satisfying

$$J(\widehat{u}) = \min_{u\in\mathscr{U}} J(u) \tag{3.72}$$

is called optimal control. The corresponding optimal dynamics is given by the probability measure $\widehat{P}$ on $(\Omega, \mathscr{F})$ defined by

$$d\widehat{P} = L^{\widehat{u}}(T)dP, \tag{3.73}$$

where $L^{\widehat{u}}$ is given by the same expression as (3.108) and under which the coordinate process x is a jump process with intensities

$$\lambda_{ij}(t, \widehat{u}(t)) := \lambda_{ij}(t, x, P^{\widehat{u}} \circ x^{-1}(t), \widehat{u}(t)), \quad i, j \in I, \ 0 \le t \le T.$$

We want to prove the existence of such an optimal control and further characterize the optimal cost functional $J(\widehat{u})$.

For $(t, w, \mu, u) \in [0, T] \times \Omega \times \mathscr{P}_2(I) \times U$ and a matrix $p = (p_{ij})$ on $I \times I$ with real-valued entries, we introduce the Hamiltonian associated with the optimal control problem (3.72) with (3.71)

$$H(t, w, \mu, p, u) := f(t, w, \mu, u) + \langle \ell(t, w, \mu, u), p\rangle_g, \tag{3.74}$$

where

$$\langle \ell(t, w, \mu, u), p\rangle_g := \sum_{i,j:\, i\neq j} p_{ij}\ell_{ij}(t, w, \mu, u) g_{ij}\mathbf{1}_{\{w(t^-)=i\}},$$

Recalling that $\ell_{ij}(t, w, \mu, u)g_{ij} = \lambda_{ij}(t, w, \mu, u) - g_{ij}$ for $i \neq j$, we have

$$\begin{aligned}\langle \ell(t, w, \mu, u) - \ell(t, w, \nu, v), p\rangle_g = \sum_{i,j:\, i\neq j} & p_{ij}(\lambda_{ij}(t, w, \mu, u) \\ & - \lambda_{ij}(t, w, \nu, v))\mathbf{1}_{\{w(t^-)=i\}}.\end{aligned}$$

Using (B3), we obtain

$$|\lambda_{ij}(t, w, \mu, u) - \lambda_{ij}(t, w, \nu, v)| \le C(d(\mu, \nu) + \delta(u, v)), \quad j \neq i.$$

Therefore, provided that $\Big|p\Big|_g(t) < +\infty$ (see the notation (3.8)),

$$\begin{aligned}|\langle \ell(t, w, \mu, u) - \ell(t, w, \nu, v), p\rangle_g| &\le \tfrac{1}{c_2}\Big|p\Big|_g(t)\Big|\lambda(t, w, \mu, u) - \lambda(t, w, \nu, v)\Big|_g \\ &\le C\Big|p\Big|_g(t)(d(\mu, \nu) + \delta(u, v))\Big(\sum_{i,j:\, i\neq j} g_{ij}\Big)^{1/2}.\end{aligned}$$

where $\sum_{i,j:\, i\neq j} g_{ij} < +\infty$, by (3.1).

Hence, in view of (B3), the Hamiltonian H satisfies

$$|H(t, w, \mu, p, u) - H(t, w, \nu, p, v)| \le C(1 + \Big|p\Big|_g(t))(d(\mu, \nu) + \delta(u, v)). \tag{3.75}$$

Moreover, note that for $i \neq j$, $|i - j| \ge 1$, we may use (B4) to obtain

$$\ell_{ij}(t, w, \mu, u) \le 1 + \frac{\lambda_{ij}(t, w, \mu, u)}{g_{ij}} \le 1 + \frac{C}{c_2}(1 + |w|_t + \int |y|\mu(dy)),$$

that is

$$\ell_{ij}(t, w, \mu, u) \le C(1 + |w|_t + \int |y|\mu(dy)), \quad i, j \in I, \ i \neq j.$$

Therefore,

$$|H(t, w, \mu, p, u) - H(t, w, \mu, p', u)| \le C(1 + |w|_t + \int |y|\mu(dy))\Big|p - p'\Big|_g(t). \tag{3.76}$$

Next, we show that the cost functional $J(u),\ u \in \mathscr{U}$ can be expressed by means of solutions of a linear BSDE.

Proposition 3.3 *For every $u \in \mathscr{U}$, the BSDE*

$$\begin{cases} -dY^u(t) = H(t, x, P^u \circ x^{-1}(t), Z^u(t), u(t))dt - Z^u(t)dM(t), \\ Y^u(T) = h(x(T), P^u \circ x^{-1}(T)), \end{cases} \tag{3.77}$$

admits a solution (Y, Z) which consists of an $\mathbb{F}$-adapted process Y which is right-continuous with left limits and a predictable process Z which satisfy

$$E\left[|Y^u|_T^2 + \int_0^T \left|Z^u(s)\right|_g^2 ds\right] < +\infty. \tag{3.78}$$

This solution is unique up to indistinguishability for Y and equality $dP \times g_{ij} I_i(s^-)ds$-almost everywhere for Z.

Moreover, $Y_0^u = J(u)$.

Proof Since the Hamiltonian $H(t, w, \mu, p, u)$ is linear in p, by Theorem 3.3, existence and uniqueness of solutions of the BSDE (3.77) satisfying (3.78) follow from (3.76), the boundedness of $h(x(T), P^u \circ x^{-1}(T))$ and the boundedness of $H(t, x, P^u \circ x^{-1}(t), 0, u(t))$ which follows from (B7).

It remains to show that $Y_0^u = J(u)$. Indeed, in terms of the $(\mathbb{F}, P^u)$-martingale

$$M_{ij}^u(t) = M_{ij}(t) - \int_{(0,t]} \ell_{ij}(s, u(s)) I_i(s^-) g_{ij} ds.$$

the process (Y^u, Z^u) satisfies, for $0 \le t \le T$,

$$Y^u(t) = h(x(T), P^u \circ x^{-1}(T)) + \int_t^T f(s, x, P^u \circ x^{-1}(s), u(s))ds - \int_t^T Z^u(s)dM^u(s).$$

Therefore,

$$Y^u(t) = E^u\left[\int_t^T f(s, x, P^u \circ x^{-1}(s), u(s))ds \right.$$
$$\left. + h(x(T), P^u \circ x^{-1}(T)) \middle| \mathscr{F}_t \right] \quad P^u - \text{a.s.}$$

In particular,

$$Y_0^u = E^u\left[\int_0^T f(s, x, P^u \circ x^{-1}(s), u(s))dt + h(x(T), P^u \circ x^{-1}(T))\right] = J(u).$$

■

Existence of an Optimal Control

We want to find $\widehat{u} \in \mathscr{U}$ such that $\widehat{u} \in \arg\min_{u\in\mathscr{U}} J(u)$. A way to find such an optimal control is to proceed as in Proposition 3.3 and introduce a linear BSDE whose solution Y^* satisfies $Y_0^* = \inf_{u\in\mathscr{U}} J(u)$. Then, by comparison (cf. Proposition 3.2), the problem can be reduced to minimizing the corresponding Hamiltonian with respect to the control u.

We may use (3.28), (3.68) and (3.75) to see that the function $u \mapsto H(t, w, P^u \circ w^{-1}(t), p, u)$ is continuous on the compact set U. Thus, for each (t, w, p), $H^*(t, w, p) := \inf_{u\in U} H(t, w, P^u \circ w^{-1}(t), p, u)$ is finite. Moreover, the set of minima of H,

$$U(t, w, p) := \{u \in U,\ H^*(t, w, p) = H(t, w, P^u \circ w^{-1}(t), p, u)\}, \tag{3.79}$$

is not empty.

Furthermore, by (3.28), (3.68) and (B6), the function $u \mapsto h(w_T, P^u \circ w^{-1}(T))$ is continuous. Therefore, $g^*(w_T) := \inf_{u\in U} h(w_T, P^u \circ w^{-1}(T))$ is finite, and the set of minima of h

$$\widetilde{U}(w_T) := \{u \in U,\ h^*(w_T) = h(w_T, P^u \circ w^{-1}(T))\} \tag{3.80}$$

is not empty.

We have the following

Lemma 3.3 *For every $(t, w) \in [0, T] \times \Omega$, the function $p \mapsto H^*(t, w, p)$ from the set of $I \times I$-matrices with real entries into $\mathbf{R}$ is Lipschitz continuous. More precisely, we have*

$$|H^*(t, w, p) - H^*(t, w, p')| \le C(1 + |w|_t + \sup_{u\in U} \left|P^u\right|_2)\left|p - p'\right|_g(t). \tag{3.81}$$

Proof We have

$$\begin{aligned}
&|H^*(t, w, p) - H^*(t, w, p')| \\
&= \left|\inf_{u\in U} H(t, w, P^u \circ w_t^{-1}, p, u) - \inf_{u\in U} H(t, w, P^u \circ w_t^{-1}, p', u)\right| \\
&\le \sup_{u\in U} \left|H(t, w, P^u \circ w_t^{-1}, p, u) - H(t, w, P^u \circ w_t^{-1}, p', u)\right| \\
&\le C(1 + |w|_t + \sup_{u\in U} \left|P^u\right|_2)\left|p - p'\right|_g(t),
\end{aligned}$$

by (3.76), where by the continuity of $u \mapsto P^u$, $K_T := \sup_{u\in U} \left|P^u\right|_2$ is finite. ∎

By the Beneš selection theorem ([58]) applied to the sets of minima $U(t, w, p)$ and $\widetilde{U}(w_T)$, there exists a progressively measurable function u_1^* from $[0, T) \times \Omega \times l^2(I)$

into U such that

$$u_1^*(t, w, p) \in U(t, w, p). \tag{3.82}$$

Moreover, there exits an $\mathscr{F}_T$-measurable process u_2^* from I into U such that

$$u_2^*(w_T) \in \widetilde{U}(w_T). \tag{3.83}$$

Combining (3.82) and (3.83), it is easily seen that the progressively measurable function u^* defined by

$$\widehat{u}(t, w, p) := \begin{cases} u_1^*(t, w, p), & t < T, \\ u_2^*(w_T), & t = T, \end{cases} \tag{3.84}$$

satisfies

$$H^*(t, w, p) = H(t, w, P^{\widehat{u}} \circ w^{-1}(t), p, \widehat{u}), \;\; h^*(w_T) = h(w_T, P^{\widehat{u}} \circ w^{-1}(T)). \tag{3.85}$$

In the next theorem, we characterize the set of optimal controls associated with (3.72).

Theorem 3.4 *The BSDE*

$$\begin{cases} -dY^* = H^*(t, x, Z_t^*)dt - Z_t^* dM(t), & 0 \leq t < T, \\ Y_T^* = h^*(x(T)), \end{cases} \tag{3.86}$$

admits a solution (Y^*, Z^*) *which consists of an* $\mathbb{F}$*-adapted process* Y^* *which is right-continuous with left limits and a predictable process* Z *which satisfy*

$$E\left[|Y^*|_T^2 + \int_0^T \left|Z^*(s)\right|_g^2 ds \right] < +\infty. \tag{3.87}$$

This solution is unique up to indistinguishability for Y^* *and equality* $dP \times g_{ij} I_i(s^-)ds$*-almost everywhere for* Z^*.

Moreover, for every $t \in [0, T]$, $Y_t^* = \text{ess}\inf_{u \in \mathscr{U}} Y_t^u$, *where* Y^u *is solution of the BSDE (3.77), and the process* $u_t^* := \widehat{u}(t, x, Z_t^*)$ *given by (3.84) is an optimal control for the problem (3.72).*

Proof Existence and uniqueness of solutions (Y^*, Z^*) of (3.86) satisfying (3.87) follow from the estimates (3.81), (3.41) and the boundedness of the function $H^*(t, x, 0)$ which follows from (B7) and the boundedness of $g^*(x(T))$ which also follows from (B7). Furthermore, since $g^*(x(T)) \leq g(x(T), P^u \circ x^{-1}(t))$ and $H^*(t, x., p) \leq H(t, x, P^u \circ x^{-1}(t), p, u)$, by Theorem 3.3, it holds that $Y_t^* \leq Y_t^u$ a.s.. Hence, $Y_t^* \leq \text{ess}\inf_{u \in \mathscr{U}} Y_t^u$, a.s.. In view of (3.84), (3.85) and the uniqueness of the solution of (3.86), we have $Y_t^* = Y_t^{u^*} \geq \text{ess}\inf_{u \in \mathscr{U}} Y_t^u$.

Therefore,

$$Y_t^{u^*} = Y_t^* = \text{ess inf}_{u \in \mathcal{U}} Y_t^u.$$

In particular, $Y_0^* = \inf_{u \in \mathcal{U}} J(u) = J(u^*)$. This concludes the proof of the theorem. ■

3.3 Two-Player Zero-Sum Game

In this section, we consider a two-player zero-sum game. Let $\mathcal{U}$ (resp., $\mathcal{V}$) be the set of admissible U-valued (resp., V-valued) control strategies for the first (resp., second) player, where (U, δ_1) and (V, δ_2) are compact metric spaces.

For $(u, v), (\bar{u}, \bar{v}) \in U \times V$, we set

$$\delta((u, v), (\bar{u}, \bar{v})) := \delta_1(u, \bar{u}) + \delta_2(v, \bar{v}). \tag{3.88}$$

The distance δ defines a metric on the compact space $U \times V$.

As in the previous section, let P be the probability measure on $(\Omega, \mathcal{F})$ under which x is a time-homogeneous Markov chain such that $P \circ x^{-1}(0) = \xi$ and with Q-matrix $(g_{ij})_{ij}$ satisfying (3.1) and (3.31).

For $(u, v) \in \mathcal{U} \times \mathcal{V}$, let $P^{u,v}$ be the measure on $(\Omega, \mathcal{F})$ defined by

$$dP^{u,v} := L_T^{u,v} dP, \tag{3.89}$$

where

$$L^{u,v}(t) := \prod_{\substack{i,j \\ i \neq j}} \exp\left\{ \int_{(0,t]} \ln \frac{\lambda_{ij}^{u,v}(s)}{g_{ij}} dN_{ij}(s) - \int_0^t (\lambda_{ij}^{u,v}(s) - g_{ij}) I_i(s) ds \right\}, \tag{3.90}$$

$$\lambda_{ij}^{u,v}(t) := \lambda_{ij}(t, x, P^u \circ x^{-1}(t), u(t), v(t)), \quad i, j \in I, \ 0 \le t \le T, \tag{3.91}$$

satisfying the following assumptions similar to (A1)–(A5).

(C1) For any $(u, v) \in \mathcal{U} \times \mathcal{V}$, $i, j \in I$, the process $((\lambda_{ij}(t, x, P^u \circ x^{-1}(t), u(t), v(t)))_t$ is predictable.

(C2) There exists a positive constant c_1 such that for every $(t, i, j) \in [0, T] \times I \times I$; $i \neq j$, $w \in \Omega$, $u \in U$, $v \in V$ and $\mu, \nu \in \mathcal{P}(I)$

$$\lambda_{ij}(t, w, \mu, u, v) \ge c_1 > 0.$$

(C3) For every $t \in [0, T]$, $w \in \Omega$, $(u, v), (\bar{u}, \bar{v}) \in U \times V$ and $\mu, \nu \in \mathcal{P}(I)$,

$$\left| \lambda(t, w, \mu, u, v) - \lambda(t, w, \nu, \bar{u}, \bar{v}) \right| \le C \left(d(\mu, \nu) + \delta((u, v), (\bar{u}, \bar{v})) \right).$$

(C4) For $p = 1, 2$ and for every $t \in [0, T]$, $w \in \Omega$, $u \in U$, $v \in V$ and $\mu \in \mathcal{P}_2(I)$,

$$\sum_{i,j:\, j \neq i} |j-i|^p \lambda_{ij}(t, w, \mu, u, v) \leq C(1 + |w|_t^p + \int |y|^p \mu(dy)).$$

(C5) The probability measure ξ on I has finite second moment: $\left|\xi\right|_2^2 := \int |y|^2 \xi(dy) < \infty$.

By Proposition 3.1, these assumptions guarantee that $P^{u,v}$ is a probability measure on $(\Omega, \mathcal{F})$ under which the coordinate process x is a chain with intensity matrix $\lambda^{u,v}$. Let $E^{u,v}$ denote the expectation with respect to $P^{u,v}$.

Let f be a measurable function from $[0, T] \times \Omega \times \mathcal{P}_2(I) \times U \times V$ into $\mathbf{R}$ and h be a measurable function from $I \times \mathcal{P}_2(I)$ into $\mathbf{R}$ such that

(C6) For any $(u, v) \in \mathcal{U} \times \mathcal{V}$ and $Q \in \mathcal{P}_2(\Omega)$, the process is $(f(t, x, Q \circ x^{-1}(t), u(t), v(t)))_t$ which is progressively measurable. Moreover, $h(x(T), Q \circ x^{-1}(T))$ is $\mathcal{F}_T$-measurable.

(C7) There exists a positive constant C such that for every $t \in [0, T]$, $w \in \Omega$, $(u, v), (\bar{u}, \bar{v}) \in U \times V$, and $\mu, \nu \in \mathcal{P}_2(I)$,

$$|\phi(t, w, \mu, u, v) - \phi(t, w, \nu, \bar{u}, \bar{v})| \leq C(d(\mu, \nu) + \delta((u, v), (\bar{u}, \bar{v})),$$

for $\phi \in \{f, h\}$.

(C8) There exists a positive constant C such that for every $t \in [0, T]$, $w \in \Omega$, $(u, v) \in U \times V$, and $\mu \in \mathcal{P}_2(I)$,

$$|f(t, w, \mu, u, v)| \leq C(1 + |w|_t + \int |y| \mu(dy)).$$

(C9) f and h are uniformly bounded.

The performance functional $J(u, v)$, $(u, v) \in \mathcal{U} \times \mathcal{V}$, associated with the controlled Markov chain, is

$$J(u, v) := E^{u,v}\left[\int_0^T f(t, x, P^{u,v} \circ x^{-1}(t), u(t), v(t))dt + h(x(T), P^{u,v} \circ x^{-1}(T))\right]. \tag{3.92}$$

The zero-sum game we consider is between two players, where the first player (with control u) wants to minimize the payoff (3.92), while the second player (with control v) wants to maximize it. The zero-sum game boils down to showing existence of a saddle point for the game, i.e., to show the existence of a pair of strategies $(\widehat{u}, \widehat{v})$ such that

$$J(\hat{u}, v) \le J(\widehat{u}, \widehat{v}) \le J(u, \widehat{v}) \tag{3.93}$$

for each $(u, v) \in \mathscr{U} \times \mathscr{V}$.

The corresponding optimal dynamics is given by the probability measure $\widehat{P}$ on $(\Omega, \mathscr{F})$ defined by

$$d\widehat{P} = L_T^{\widehat{u},\widehat{v}} dP \tag{3.94}$$

under which the chain has intensity $\lambda^{\widehat{u},\widehat{v}}$.

For $(t, w, \mu, u) \in [0, T] \times \Omega \times \mathscr{P}_2(I) \times U \times V$ and matrices $p = (p_{ij})$ with real-valued entries, we introduce the Hamiltonian associated with the optimal control problem (3.92)

$$H(t, w, \mu, p, u, v) := f(t, w, \mu, u, v) + \langle \ell(t, w, \mu, u, v), p\rangle_g, \tag{3.95}$$

where we recall that $\ell_{ij}(t, w, \mu, u, v)g_{ij} = \lambda_{ij}(t, w, \mu, u, v) - g_{ij}$ for $i \neq j$.

In a similar way as for (3.75) and (3.76), whenever $\left|p\right|_g(t)$ and $\left|p'\right|_g(t)$ are finite, the Hamiltonian H satisfies, for some constant C,

$$\begin{aligned}|H(t, w, \mu, p, u, v) - H(t, w, \nu, p, \bar{u}, \bar{v})| \le C(1 + \left|p\right|_g(t))(d(\mu, \nu)\\ + \delta((u, \bar{u}), (v, \bar{v})),\end{aligned} \tag{3.96}$$

and

$$|H(t, w, \mu, p, u, v) - H(t, w, \mu, p', u, v)| \le C(1 + |w|_t + \int |y|\mu(dy))\left|p - p'\right|_g(t). \tag{3.97}$$

Definition 3.1 (Isaacs' Condition) We say that the Isaacs' condition holds for the game if, for $(t, w) \in [0, T] \times \Omega$ and matrices $p = (p_{ij})$ with real-valued entries,

$$\begin{aligned}&\min_{u\in\mathscr{U}}\max_{v\in V} H(t, w, P^{u,v} \circ w^{-1}(t), p, u, v)\\ &\quad = \max_{v\in V}\min_{u\in U} H(t, w, P^{u,v} \circ w^{-1}(t), p, u, v),\end{aligned} \tag{3.98}$$

$$\min_{u\in U}\max_{v\in V} g(w(T), P^{u,v} \circ w^{-1}(T)) = \max_{v\in V}\min_{u\in U} g(w(T), P^{u,v} \circ w^{-1}(T)).$$

In view of (3.28), (3.68), (3.96), and (C7), the functions

$$(u, v) \mapsto H(t, w, P^u \circ w^{-1}(t), p, u, v),\ \ h(w(T), P^{u,v} \circ w^{-1}(T))$$

are continuous on the compact set $U \times V$. Thus, both sides of (3.98) are finite.

Using the Beneš selection theorem ([58]), with a similar construction that yielded (3.84), Isaacs' condition is equivalent to the following condition.

(S) There exist progressively measurable functions u^* from $[0, T] \times \Omega \times \mathbf{R}^{I\times I}$ into U and v^* from $[0, T] \times \Omega \times \mathbf{R}^{I\times I}$ into V such that

$$\begin{cases} \widetilde{H}(t, w, p) \le H(t, w, P^{u,v^*} \circ w^{-1}(t), p, u, v^*(t, w, p)), \\ \widetilde{h}(w(T)) \le h(w(T), P^{u,v^*} \circ w^{-1}(T)), \end{cases} \quad \forall u \in U,$$

$$\begin{cases} \widetilde{H}(t, w, p) \ge H(t, w, P^{u^*,v} \circ w^{-1}(t), p, u^*(t, w, p), v), \\ \widetilde{h}(w(T)) \ge g(w(T), P^{u^*,v} \circ w^{-1}(T)), \end{cases} \quad \forall v \in V,$$

where

$$\widetilde{H}(t, w, p) := H(t, w, P^{u^*,v^*} \circ w^{-1}(t), p, u^*(t, w, p), v^*(t, w, p))$$

and

$$\widetilde{h}(w(T)) := h(w(T), P^{u^*,v^*} \circ w^{-1}(T)).$$

Equivalently,

$$\begin{aligned} &\widetilde{H}(t, w, p) = \min_{u\in U} \max_{v\in V} H(t, w, P^{u,v} \circ w^{-1}(t), p, u, v), \\ &\widetilde{g}(w(T)) = \min_{u\in U} \max_{v\in V} g(w(T), P^{u^*,v^*} \circ w^{-1}(T)). \end{aligned} \tag{3.99}$$

Existence of a Saddle Point

In the remaining part of this section, we want to find $(\widehat{u}, \widehat{v}) \in \mathcal{U} \times \mathcal{V}$ which satisfies the pair of inequalities (3.93). A way to find such a saddle point is to proceed as in Proposition 3.3 and introduce a linear BSDE associated with each of the Hamiltonians and end values involved in the inequality (S). Then, by the comparison theorem for BSDEs (cf. Proposition 3.2), we show that (u^*, v^*) satisfies (3.93), i.e., it is a saddle point for the game.

We start by showing that the performance functional $J(u, v)$, $(u, v) \in \mathcal{U} \times \mathcal{V}$ can be expressed in terms of solutions of a linear BSDE.

Proposition 3.4 *For every* $(u, v) \in \mathcal{U} \times \mathcal{V}$, *the BSDE*

$$\begin{cases} -dY^{u,v}(t) = H(t, x, P^{u,v} \circ x^{-1}(t), Z^{u,v}(t), u(t), v(t))dt \\ \qquad\qquad -Z^{u,v}(t)dM_t, \quad 0 \le t < T, \\ Y^{u,v}(T) = g(x(T), P^{u,v} \circ x^{-1}(T)), \end{cases} \tag{3.100}$$

admits a solution $(Y^{u,v}, Z^{u,v})$ *which consists of an* $\mathbb{F}$*-adapted process* $Y^{u,v}$ *which is right-continuous with left limits and a predictable process* $Z^{u,v}$ *which satisfy*

$$E\left[|Y^{u,v}|_T^2 + \int_0^T \left|Z^{u,v}(s)\right|_g^2 ds\right] < +\infty. \tag{3.101}$$

This solution is unique up to indistinguishability for $Y^{u,v}$ and equality $dP \times g_{ij} I_i(s^-)ds$-almost everywhere for $Z^{u,v}$. Moreover, $Y_0^{u,v} = J(u, v)$.

Proof Since the Hamiltonian $H(t, w, \mu, p, u)$ is linear in p, existence and uniqueness of solutions of the BSDE (3.100) satisfying (3.101) follow from (3.97), (3.41), and the boundedness of $H(t, x, P^{u,v} \circ x^{-1}(t), 0, u(t), v(t))$ which follow from (C7) and (C8) and the boundedness of h. The remaining proof is similar to the one of Proposition 3.3. ■

We have the following lemma whose proof is similar to that of Lemma 3.3 and therefore is not included.

Lemma 3.4 *For every $(t, w) \in [0, T] \times \Omega$, the function*

$$p \mapsto H(t, w, P^{u^*,v^*} \circ w^{-1}(t), p, u^*(t, w, p), v^*(t, w, p))$$

from the set of matrices on $I \times I$ with real-valued entries into **R** *is Lipschitz continuous.* ■

In the next theorem, we characterize the set of saddle points of the zero-sum game associated with (3.92).

Theorem 3.5 *The BSDE*

$$\begin{cases} -d\widehat{Y}(t) = \widehat{H}(t, x, \widehat{Z}(t))dt - \widehat{Z}(t)dM(t), \quad 0 \le t < T, \\ \widehat{Y}(T) = \widehat{h}(x(T)), \end{cases} \tag{3.102}$$

admits a solution $(\widehat{Y}, \widehat{Z})$ which consists of an $\mathbb{F}$-adapted process $\widehat{Y}$ which is right-continuous with left limits and a predictable process $\widehat{Z}$ which satisfy

$$E\left[|\widehat{Y}|_T^2 + \int_0^T \left|\widehat{Z}(s)\right|_g^2 ds\right] < +\infty, \tag{3.103}$$

where

$$\widehat{H}(t, x, \widehat{Z}(t)) := H(t, x, P^{\widehat{u},\widehat{v}} \circ x^{-1}(t), \widehat{Z}_t, \widehat{u}(t), \widehat{v}(t)),$$

$$\widehat{h}(x(T)) := h(x(T), P^{\widehat{u},\widehat{v}} \circ x^{-1}(t))$$

and

$$(\widehat{u}(t), \widehat{v}(t)) := (u^*(t, x, \widehat{Z}(t)), v^*(t, x, \widehat{Z}(t))).$$

This solution is unique up to indistinguishability for $\widehat{Y}$ and equality $dP \times g_{ij} I_i(s^-)ds$-almost everywhere for $\widehat{Z}$.

Moreover, $\widehat{Y}_0 = J(\widehat{u}, \widehat{v})$.

Furthermore, the process $(\widehat{u}, \widehat{v})$ is a saddle point for the zero-sum game associated with (3.92).

Proof Existence and uniqueness of solutions $(\widehat{Y}, \widehat{Z})$ of (3.102) satisfying (3.103) follow from Lemma 3.4 and the boundedness of $\widehat{H}(t, x, 0)$ which follows from (C7) and (C8), and the boundedness of h.

We will now show that $(\widehat{u}, \widehat{v})$ is a saddle point. Let $u \in \mathcal{U}$, and consider the following BSDE

$$\begin{cases} -d\bar{Y}^u(t) = H(t, x, P^{u,\widehat{v}} \circ x^{-1}(t), \bar{Z}^u(t), u(t), \widehat{v}(t))dt \\ \qquad\qquad -\bar{Z}^u(t)dM(t), \quad 0 \leq t < T, \\ \bar{Y}^u(T) = h(x(T), P^{u,\widehat{v}} \circ x^{-1}(T)). \end{cases}$$

By Proposition 3.4, this BSDE has a unique solution $(\bar{Y}^u, \bar{Z}^u)$ for which $\bar{Y}^u_0 = J(u, \widehat{v})$. Since, by (S), $H(t, x, P^{u,\widehat{v}} \circ x^{-1}(t), \widehat{Z}(t), u(t), \widehat{v}(t)) \geq \widehat{H}(t, x, \widehat{Z}(t))$ and $h(x(T), P^{u,\widehat{v}} \circ x^{-1}(T)) \geq \widehat{h}(x(T))$ P-a.s., we may apply the comparison theorem for BSDEs (cf. Theorem (3.3)) to obtain that $\bar{Y}^u(t) \geq \widehat{Y}(t)$, P-a.s. Hence, $J(u, \widehat{v}) \geq J(\widehat{u}, \widehat{v})$.

Now, let $v \in \mathcal{U}$, and consider the following BSDE

$$\begin{cases} -d\underline{Y}^v(t) = H(t, x, P^{\widehat{u},v} \circ x^{-1}(t), \underline{Z}^v(t), \widehat{u}(t), v(t))dt \\ \qquad\qquad -\underline{Z}^v(t)dM(t), \quad 0 \leq t < T, \\ \underline{Y}^v(T) = h(x(T), P^{\widehat{u},v} \circ x^{-1}(T)). \end{cases}$$

Then, in a similar fashion, the inequalities $H(t, x, P^{\widehat{u},v} \circ x^{-1}(t), \widehat{Z}(t), \widehat{u}(t), v(t)) \leq \widehat{H}(t, x, \widehat{Z}(t))$ and $h(x(T), P^{\widehat{u},v} \circ x^{-1}(T)) \leq \widehat{h}(x(T))$ yield $J(\widehat{u}, v) \leq J(\widehat{u}, \widehat{v})$. ■

It is possible to characterize the optimal controls $\hat{u}$ and the equilibrium points $(\hat{u}, \hat{v})$ in terms of a stochastic maximum principle.

3.4 Stochastic Maximum Principle for Discrete State Spaces

We consider controls with values in some subset U of $\mathbf{R}^d$ and let $\mathcal{U}$ the set of $\mathbb{F}$-progressively measurable processes $u = (u(t), 0 \leq t \leq T)$ with values in $U \subset \mathbf{R}^d$. $\mathcal{U}$ is the set of admissible controls.

For $u \in \mathcal{U}$, let P^u be the probability measure on (Ω, Φ) under which the coordinate process x is a jump process with intensities

$$\lambda^u_{ij}(t) := \lambda_{ij}(t, x, E^u[\kappa(x(t))]), u(t)), \quad i, j \in I, \ 0 \leq t \leq T, \tag{3.104}$$

The cost functional associated with P^u is of the form

$$J(u) := E^u\left[\int_0^T f(t, x, E^u[\kappa_f(x(t))], u(t))dt + h\left(x(T), E^u[\kappa_h(x(T))]\right)\right]. \tag{3.105}$$

In this section, we propose to characterize minimizers $\bar{u}$ of J, i.e., $\bar{u} \in \mathscr{U}$ satisfying

$$J(\bar{u}) = \min_{u \in \mathscr{U}} J(u) \tag{3.106}$$

in terms of a stochastic maximum principle (SMP). We first state and prove the sufficient optimality conditions. Then, we state the necessary optimality conditions.

Let P be the probability measure on $(\Omega, \mathscr{F})$ under which x is a time-homogeneous Markov chain such that $P \circ x^{-1}(0) = \xi$ and with Q-matrix $(g_{ij})_{ij}$ satisfying (3.1). Then, by a Girsanov-type result for pure jump processes (see, e.g., [74, 279]), it holds that

$$dP^u := L^u(T)dP, \tag{3.107}$$

where, for $0 \leq t \leq T$,

$$L^u(t) := \prod_{\substack{i,j \\ i \neq j}} \exp\left\{\int_{(0,t]} \ln \frac{\lambda^u_{ij}(s)}{g_{ij}} dN_{ij}(s) - \int_0^t (\lambda^u_{ij}(s) - g_{ij}) I_i(s) ds\right\}, \tag{3.108}$$

which satisfies

$$L^u(t) = 1 + \int_{(0,t]} L^u(s^-) \sum_{i,j:\, i \neq j} I_i(s^-) \ell^u_{ij}(s) dM_{ij}(s), \tag{3.109}$$

where $\ell^u_{ij}(s) := \ell_{ij}(t, x, E^u[\kappa(x(s))], u(s))$ is given by the formula

$$\ell^u_{ij}(s) = \begin{cases} \lambda^u_{ij}(s)/g_{ij} - 1 \text{ if } i \neq j, \\ \qquad\qquad 0 \text{ if } i = j, \end{cases} \tag{3.110}$$

and $(M_{ij})_{ij}$ is the P-martingale given in (3.4). Moreover, the accompanying martingale $M^u = (M^u_{ij})_{ij}$ satisfies

$$M^u_{ij}(t) = M_{ij}(t) - \int_{(0,t]} \ell^u_{ij}(s) I_i(s^-) g_{ij} ds. \tag{3.111}$$

Note that

$$J(u) = E\left[L^u(T)\int_0^T f(t,x,E^u[\kappa_f(x(t))],u(t))dt\right.$$
$$\left.+L^u(T)h(x(T),E^u[\kappa_h(x(T))])\right].$$

Integrating by parts and taking expectation, we obtain

$$J(u) := E\left[\int_0^T L^u(t)f(t,x,E[L^u(t)\kappa_f(x(t))],u(t))dt\right.$$
$$\left.+L^u(T)h(x(T),E[L^u(T)\kappa_h(x(T))])\right]. \tag{3.112}$$

We have recast our problem of controlling a Markov chain through its intensity matrix to a standard control problem which aims at minimizing the cost functional (3.112) under the dynamics given by the density process L^u which satisfies (3.109), to which the mean-field stochastic maximum principle can be applied. The corresponding optimal dynamics is given by the probability measure $\bar{P}$ on (Ω,Φ) defined by

$$d\bar{P} = L^{\bar{u}}(T)dP, \tag{3.113}$$

where $L^{\bar{u}}$ is the associated density process. $(L^{\bar{u}},\bar{u})$ is called optimal pair associated with (3.106).

For $w = y,\bar{y},u$, ψ_w denotes the partial derivative of the function $\psi(y,\bar{y},u)$ with respect to w.
for $\alpha = \ell, f, h$, we set

$$\alpha(t) := \alpha(t,E[L^u(t)\kappa_\alpha(x(t))],u(t)),\quad \bar{\alpha}(t) := \alpha(t,E[L^{\bar{u}}(t)\kappa_\alpha(x(t))],\bar{u}(t)).$$

To the admissible pair of processes $(L^{\bar{u}},\bar{u})$, we associate the solution (p,q) (if it exits) of the following linear BSDE of mean-field type, known as first-order adjoint equation [102]:

$$\begin{cases} dp(t) = -\left\{\langle\bar{\ell}(t),q(t)\rangle_g - \bar{f}(t) + \kappa(x(t))E[L^{\bar{u}}(t)(\langle\bar{\ell}_{\bar{y}}(t),q(t)\rangle_g]\right. \\ \left.-\kappa_f(x(t))E[L^{\bar{u}}(t)\bar{f}_{\bar{y}}(t)\right\}dt + q(t)dM(t), \\ \\ p(T) = -\bar{h}(T) - \kappa_h(x(T))E[L^{\bar{u}}(T)\bar{h}_{\bar{y}}(T)]. \end{cases} \tag{3.114}$$

In the next proposition, we give sufficient conditions on f,h,ℓ,κ,κ_f and κ_h that guarantee existence of a unique solution to the BSDE (3.114).

Proposition 3.5 *Assume that*

(A1) The function ℓ *is differentiable in* $\bar{y}$*. Moreover,* $\left|\ell\right|_g$ *and* $\left|\ell_{\bar{y}}\right|_g$ *are bounded.*
(A2) The functions f, h, κ_f *and* κ_h *are bounded.* f *and* h *are differentiable in* $\bar{y}$ *with bounded derivatives.*

Then, the BSDE (3.114) *admits a solution* (p, q) *consisting of an adapted process* p *which is right-continuous with left limits and a predictable process* q *which satisfy*

$$E\left[\sup_{t\in[0,T]} |p(t)| + \int_{(0,T]} \left|q(s)\right|_g^2 ds\right] < +\infty. \tag{3.115}$$

This solution is unique up to indistinguishability for p *and equality* $dP \times g_{ij} I_i(s^-)ds$*-almost everywhere for* q*.*

Proof Assumptions (A1) and (A2) make the driver of the BSDE (3.114) Lipschitz continuous in q. The proof is similar to that of Theorem 3.1 for the Brownian motion driven mean-field BSDE derived in [76, 77, 272] by considering the following norm

$$\left|(p,q)\right|_\beta^2 := E\int_0^T e^{\beta t}(|p(t)|^2 + \left|q(t)\right|_g^2)dt,$$

where $\beta > 0$, along with Itô's formula for purely discontinuous semimartingales. We do not include the details. ■

Remark 3.2

(i) The boundedness on f and h and their derivatives is strong and can be considerably weakened using standard truncation techniques.
(ii) If $\ell_{\bar{y}} \doteq 0$, i.e., the intensity does not contain any mean-field coupling, the BSDE (3.114) becomes standard. Thanks to Theorem 3.10 in [98, 99], it is solvable only by imposing conditions similar to (H1)–(H3) therein.
(iii) If $\ell_{\bar{y}} \neq 0$, i.e., the intensity is of mean-field type, we do not know whether we can relax the imposed boundedness of ℓ, κ and $\ell_{\bar{y}}$, because without this condition, the standard comparison theorem for Markov chain BSDEs simply does not, in general, apply for such drivers.

Let $(L^{\bar{u}}, \bar{u})$ be an admissible pair and (p, q) be the associated first-order adjoint process solution of (3.114).

For $v \in U$, we introduce the Hamiltonian associated with our control problem

$$H(t,v) := L^{\bar{u}}(t)\left(\langle \ell(t, E[L^{\bar{u}}(t)\kappa(x(t))], v), q(t)\rangle_g - f(t, E[L^{\bar{u}}(t)\kappa_f(x(t))], v)\right). \tag{3.116}$$

Next, we state the SMP sufficient and necessary optimality conditions.

In the next two theorems, we assume that (A1) and (A2) of Proposition 3.5 hold.

Theorem 3.6 (Sufficient Optimality Conditions) *Let $(L^{\bar{u}}, \bar{u})$ be an admissible pair and (p, q) be the associated first-order adjoint process which satisfies* (3.114)–(3.115). *Assume that:*

(A4) The set of controls U is a convex body (i.e., U is convex and has a non-empty interior) of $\mathbf{R}^d$, and the functions ℓ and f are differentiable in u.
(A5) The functions $(y, \bar{y}, u) \mapsto y\ell(\cdot, \bar{y}, u)$ and $(y, \bar{y}, u) \mapsto -yf(\cdot, \bar{y}, u)$ are concave in $(y, \bar{y}, u)$ for a.e., $t \in [0, T]$, P-almost surely,
(A6) The function $(y, \bar{y}) \mapsto yh(\cdot, \bar{y})$ is convex.

If the admissible control $\bar{u}$ satisfies

$$H(t, \bar{u}(t)) = \max_{v \in U} H(t, v), \qquad a.e.\ t \in [0, T], \qquad P\text{-}a.s. \tag{3.117}$$

then, the pair $(L^{\bar{u}}, \bar{u})$ is optimal.

Proof We want to show that if the pair $(L^{\bar{u}}, \bar{u})$ satisfies (3.117), then

$$J(u) - J(\bar{u}) = E\left[\int_0^T (L^u(t)f(t) - L^{\bar{u}}(t)\bar{f}(t))dt + L^u(T)h(T) - L^{\bar{u}}(T)\bar{h}(T)\right] \geq 0.$$

Since $(y, \bar{y}) \mapsto yh(\cdot, \bar{y})$ is convex, we have

$$\begin{aligned} E[L^u(T)h(T) - L^{\bar{u}}(T)\bar{h}(T)] &\geq E[(\bar{h}(T) + \kappa(T)E[L^{\bar{u}}(T)\bar{h}_{\bar{y}}(T)])(L^u(T) - L^{\bar{u}}(T))] \\ &= -E[p(T)(L^u(T) - L^{\bar{u}}(T))]. \end{aligned}$$

Integrating by parts, using (3.114), we obtain

$$\begin{aligned} &E[p(T)(L^u(T) - L^{\bar{u}}(T))] \\ &= E\left[\int_0^T (L^u(t^-) - L^{\bar{u}}(t^-))dp(t) + p(t^-)d(L^u(t) - L^{\bar{u}}(t)) + d[L^u - L^{\bar{u}}, p](t)\right] \\ &= -E\Big[\int_0^T \Big(\big\{\langle \bar{\ell}(t), q(t)\rangle_g - \bar{f}(t) + \kappa(x(t))E[L^{\bar{u}}(t)(\langle \bar{\ell}_{\bar{y}}(t), q(t)\rangle_g] \\ &\quad -\kappa_f(x(t))E[L^{\bar{u}}(t)\bar{f}_{\bar{y}}(t)\big\}(L^u(t) - L^{\bar{u}}(t)) - \langle L^u(t)\ell(t) - L^{\bar{u}}(t)\bar{\ell}(t), q(t)\rangle_g\Big)dt\Big]. \end{aligned}$$

We introduce the following "Hamiltonian" function:

$$\mathscr{H}(t, y, \bar{y}, u, z) := y\langle \ell(t, \bar{y}, u), z\rangle_g - yf(t, \bar{y}, u). \tag{3.118}$$

Furthermore, for u and $\bar{u}$ in $\mathscr{U}$, we set

$$\begin{cases} \mathscr{H}(t) := \\ L^u(t)\left(\langle \ell(t, E[L^u(t)\kappa(x(t))], u(t)), q(t)\rangle_g - f(t, E[L^u(t)\kappa_f(x(t))], u(t))\right), \\ \bar{\mathscr{H}}(t) := \\ L^{\bar{u}}(t)\left(\langle \ell(t, E[L^{\bar{u}}(t)\kappa(x(t))], \bar{u}(t)), q(t)\rangle_g - f(t, E[L^{\bar{u}}(t)\kappa_f(x(t))], \bar{u}(t))\right). \end{cases} \tag{3.119}$$

Since $(y, \bar{y}, u) \mapsto y\ell(\cdot, \bar{y}, u)$ and $-\ yf(\cdot, \bar{y}, u)$ are concave, we have

$$\begin{aligned}&\mathscr{H}(t) - \bar{\mathscr{H}}(t) \leq \bar{\mathscr{H}}_y(t)(L^u(t) - L^{\bar{u}}(t))\\&+\bar{\mathscr{H}}_{\bar{y}}(t)(E[\kappa(x(t))(L^u(t) - L^{\bar{u}}(t))]) + \bar{\mathscr{H}}_u(t) \cdot (u(t) - \bar{u}(t)).\end{aligned}$$

Since, by (3.117), $\bar{\mathscr{H}}_u(t) = H_u(t, \bar{u}(t)) = 0$ a.e. $t \in [0, T]$, we obtain

$$\begin{aligned}&E[\mathscr{H}(t) - \bar{\mathscr{H}}(t)] \leq E\left[\left\{\langle\bar{\ell}(t), q(t)\rangle_g - \bar{f}(t) + \kappa(x(t))E[L^{\bar{u}}(t)(\langle\bar{\ell}_{\bar{y}}(t), q(t)\rangle_g]\right.\right.\\&\left.\left.-\kappa_f(x(t))E[L^{\bar{u}}(t)\bar{f}_{\bar{y}}(t)\right\}(L^u(t) - L^{\bar{u}}(t))\right]\end{aligned}$$

for a.e. $t \in [0, T]$.

Therefore,

$$\begin{aligned}&E[L^u(T)h(T) - L^{\bar{u}}(T)\bar{h}(T)] \geq\\&E\left[\int_0^T \left(\mathscr{H}(t) - \bar{\mathscr{H}}(t) - \langle L^u(t)\ell(t) - L^{\bar{u}}(t)\bar{\ell}(t), q(t)\rangle_g\right) dt\right].\end{aligned}$$

Hence,

$$\begin{aligned}J(u) - J(\bar{u}) \geq E\Big[\int_0^T \big(&\mathscr{H}(t) - \bar{\mathscr{H}}(t) + L^u(t)f(t) - L^{\bar{u}}(t)\bar{f}(t)\\&-\langle L^u(t)\ell(t) - L^{\bar{u}}(t)\bar{\ell}(t), q(t)\rangle_g\big) dt\Big] = 0.\end{aligned}$$

■

Theorem 3.7 (Necessary Optimality Conditions (Verification Theorem)) *If* $(L^{\bar{u}}, \bar{u})$ *is an optimal pair for the control problem* (3.106) *and there is a unique pair of* $\mathscr{F}$*-adapted processes* (p, q)*, associated with* $(L^{\bar{u}}, \bar{u})$*, which satisfies* (3.114)–(3.115)*, then*

$$H(t, \bar{u}(t)) = \max_{v\in U} H(t, v), \qquad a.e.\ t \in [0, T], \quad P\text{-}a.s.$$

Remark 3.3 Unfortunately, the sufficient optimality conditions are quite rare to obtain, due to the fact that the convexity conditions imposed on the involved coefficients are not always satisfied, even for the simplest examples: Assume ℓ and f without mean-field coupling and linear in the control u. Then none of the functions $(y, u) \mapsto y\ell(\cdot, u)$ and $(y, u) \mapsto -yf(\cdot, u)$ is concave in (y, u). However, the verification theorem in terms of necessarily optimality conditions holds for a fairly general class of functions with sufficient smoothness. Hence, if we can solve the associated BSDEs, the necessary optimality conditions result can be useful.

We first solve the adjoint equation associated with an optimal control problem related to a standard two-state Markov chain, and then we extend the problem to a two-state Markov chain of mean-field type. As mentioned in Remark (3.3), whether sufficient or necessary conditions apply depending of course on the smoothness of the involved functions. Not all the functions involved in the next examples satisfy the convexity conditions imposed in Theorem (3.6).

Two-State Markov Decision Process
We study the optimal control of a simple Markov chain x whose state space is $\mathscr{X} = \{a, b\}$, where $(0 \leq a < b)$ are integers and its jump intensity matrix is

$$\lambda^u(t) = \begin{pmatrix} -\alpha & \alpha \\ u(t) & -u(t) \end{pmatrix}, \tag{3.120}$$

where α is a given positive constant intensity and u is the control process we assume to be nonnegative, bounded, and predictable. Let P the probability measure under which the chain x has intensity matrix

$$G = \begin{pmatrix} -g_{ab} & g_{ab} \\ g_{ba} & -g_{ba} \end{pmatrix}, \quad g_{ab}, \ g_{ba} > 0. \tag{3.121}$$

Further, let $L^u(t) = \frac{dP^u}{dP}|_{\Phi_t}$ be the density process given by (3.109), where ℓ is defined by

$$\ell^u_{ij}(t) = \begin{cases} \lambda^u_{ij}(t)/g_{ij} - 1 \textbf{ if } i \neq j, \\ \qquad\qquad\quad 0 \textbf{ if } i = j. \end{cases}$$

The control problem we want to solve consists of finding the optimal control $\bar{u}$ that minimizes the linear-quadratic cost functional

$$J(u) = E^u \left[\frac{1}{2} \int_0^T u^2(t)dt + h(x(T)) \right], \quad h(b) \geq h(a). \tag{3.122}$$

Given a control $v \in U$, consider the Hamiltonian

$$H(t, L^{\bar{u}}(t), q(t), v) := L^{\bar{u}}(t)(\langle \ell^v(t), q(t) \rangle_g - \frac{1}{2} v^2) =: H(t, v),$$

where

$$\langle \ell^v(t), q(t) \rangle_g = q_{ab}(t)(\alpha - g_{ab}) I_a(t^-) + q_{ba}(t)(v - g_{ba}) I_b(t^-).$$

By the first-order optimality conditions, an optimal control $\bar{u}$ is solution of the equation $\frac{\partial H(t,v)}{\partial v} = 0$, which implies

$$0 = \langle \frac{\partial \ell^v}{\partial v}(t), q(t) \rangle_g - v = q_{ba}(t) I_b(t^-) - v. \tag{3.123}$$

The optimal control is thus

$$\bar{u}(t) = q_{ba}(t)I_b(t^-). \tag{3.124}$$

where, for each t, $q_{ba}(t) \geq 0$, since $\bar{u}(t) \geq 0$.

It remains to identify $q_{ba}(t)$. Consider the associated adjoint equation given by

$$\begin{cases} 0 \leq t < T: \\ dp(t) = -\left\{\langle \ell^{\bar{u}}(t), q(t)\rangle_g - \frac{1}{2}\bar{u}^2(t)\right\} dt + q_{ab}(t)dM_{ab}(t) + q_{ba}(t)dM_{ba}(t), \\ p(T) = -h(x(T)). \end{cases}$$

In view of (3.124), the driver reads

$$\begin{aligned} &\langle \ell^{\bar{u}}(t), q(t)\rangle_g - \tfrac{1}{2}\bar{u}^2(t) \\ &= q_{ab}(t)I_a(t^-)(\alpha - g_{ab}) + q_{ba}(t)I_b(t^-)\left\{\tfrac{1}{2}q_{ba}(t)I_b(t^-) - g_{ba}\right\}. \end{aligned} \tag{3.125}$$

The adjoint equation becomes

$$\begin{aligned} dp(t) &= q_{ab}(t)\left\{-(\alpha - g_{ab})I_a(t^-)dt + dM_{ab}(t)\right\} \\ &+ q_{ba}(t)\left\{-(\tfrac{1}{2}q_{ba}(t) - g_{ba})I_b(t^-)dt + dM_{ba}(t)\right\}. \end{aligned}$$

Now, considering the probability measure $\widetilde{P}$ under which x is a Markov chain whose jump intensity matrix

$$\widetilde{G}(t) = \begin{pmatrix} -\alpha & \alpha \\ \frac{1}{2}q_{ba}(t) & -\frac{1}{2}q_{ab}(t) \end{pmatrix}, \tag{3.126}$$

the processes defined by

$$\begin{cases} d\widetilde{M}_{ab}(t) = dM_{ab}(t) - (\alpha - g_{ab})I_a(t^-)dt, \\ d\widetilde{M}_{ba}(t) = dM_{ba}(t) - (\frac{1}{2}q_{ba}(t) - g_{ba})I_b(t^-)dt, \end{cases}$$

are $\widetilde{P}$-martingales having the same jumps as the martingales M_{ij}:

$$\Delta\widetilde{M}_{ab}(t) = \Delta M_{ab}(t) = I_a(t^-)I_b(t), \;\; \Delta\widetilde{M}_{ba}(t) = \Delta M_{ba}(t) = I_b(t^-)I_a(t) \tag{3.127}$$

and

$$dp(t) = q_{ab}(t)d\widetilde{M}_{ab}(t) + q_{ba}(t)d\widetilde{M}_{ba}(t). \tag{3.128}$$

This yields

$$\Delta p(t) = q_{ab}(t)I_a(t^-)I_b(t) + q_{ba}(t)I_b(t^-)I_a(t). \tag{3.129}$$

Integrating (3.128) and then taking conditional expectation yield

$$p(t) = -\widetilde{E}[h(x(T))|\mathscr{F}_t].$$

Therefore,

$$\Delta p(t) = -\Delta \widetilde{E}[h(x(T))|\mathscr{F}_t]. \tag{3.130}$$

Under the probability measure $\widetilde{P}$

$$\begin{aligned} &h(x(T)) = h(x(t)) \\ &+\int_t^T \left\{\alpha(h(b)-h(a))I_a(s^-) + \tfrac{1}{2}q_{ba}(s)(h(a)-h(b))I_b(s^-)\right\} ds \\ &+\int_t^T (h(b)-h(a))d\widetilde{M}_{ab}(s) + \int_t^T (h(a)-h(b))d\widetilde{M}_{ba}(s). \end{aligned} \tag{3.131}$$

Taking conditional expectation, we obtain

$$\begin{aligned} &\widetilde{E}[h(x(T))|\mathscr{F}_t] = h(x(t)) + \\ &\int_t^T \widetilde{E}\left[\alpha(h(b)-h(a))I_a(s^-) + \tfrac{1}{2}q_{ba}(s)(h(a)-h(b))I_b(s^-)|\mathscr{F}_t\right] ds, \end{aligned}$$

and

$$\begin{aligned} \Delta \widetilde{E}[h(x(T))|\mathscr{F}_t] = \Delta h(x(t)) &= -(h(b)-h(a))I_a(t^-)I_b(t) \\ &\quad - (h(a)-h(b))I_b(t^-)I_a(t), \end{aligned} \tag{3.132}$$

which in view of (3.130) implies that

$$q_{ab}(t) = h(a) - h(b), \quad q_{ba}(t) = h(b) - h(a). \tag{3.133}$$

Therefore,

$$\bar{u}(t) = (h(b)-h(a))I_b(t^-) = h(b)I_b(t^-) - h(a) + h(a)I_a(t^-) = h(x(t^-)) - h(a), \tag{3.134}$$

which yields the following explicit form of the optimal control

$$\bar{u}(t) = h(x(t^-)) - h(a).$$

In the next two examples, we highlight the effect of the mean-field coupling in both the jump intensity and the cost functional on the optimal control.

Two-State Mean-Field-Type Markov Decision Process

We consider the same chain as in the first example but with the following mean-field type jump intensities, ($t \in [0, T]$):

$$\lambda^u(t) = \begin{pmatrix} -\alpha & \alpha \\ u(t) + E^u[x(t^-)] & -u(t) - E^u[x(t^-)] \end{pmatrix}, \quad \alpha > 0, \quad u(t) + E^u[x(t^-)] \geq 0, \tag{3.135}$$

and want to minimize the cost functional

$$J(u) = E^u\left[\frac{1}{2}\int_0^T u^2(t)dt\right] + Var^u(x(T)), \tag{3.136}$$

where $Var^u(x(T))$ denotes the variance of $x(T)$ under the probability P^u defined by

$$Var^u(x(T)) := E^u\left[\left(x(T) - E^u[x(T)]\right)^2\right].$$

Given a control $v \in U$, consider the Hamiltonian

$$H(t, v) := L^{\bar{u}}(t)(\langle \ell^v(t), q(t)\rangle_g - \frac{1}{2}v^2),$$

where

$$\langle \ell^v(t), q(t)\rangle_g = q_{ab}(t)(\alpha - g_{ab})I_a(t^-) + q_{ba}(t)(v + E^{\bar{u}}[x(t^-)] - g_{ba})I_b(t^-).$$

Performing similar calculations as in Example 3.1, we find that the optimal control is given by

$$\bar{u}(t) = q_{ba}(t)I_b(t^-). \tag{3.137}$$

We will now identify q_{ba}. The associated adjoint equation is given by

$$\begin{cases} dp(t) = -\left\{\langle \ell^{\bar{u}}(t), q(t)\rangle_g - \frac{1}{2}\bar{u}^2(t) + x(t)E^{\bar{u}}[\bar{H}_{\bar{y}}(t)]\right\}dt + q_{ab}(t)dM_{ab}(t) \\ \qquad\qquad +q_{ba}(t)dM_{ba}(t), \\ p(T) = -\left(x(T) - E^{\bar{u}}[x(T)]\right)^2. \end{cases}$$

In view of (3.137), the driver reads

$$\begin{aligned} &\langle \ell^{\bar{u}}(t), q(t)\rangle_g - \frac{1}{2}\bar{u}^2(t) + x(t)E^{\bar{u}}[\bar{H}_{\bar{y}}(t)] \\ &= q_{ba}(t)\left(\frac{1}{2}q_{ba}(t) + E^{\bar{u}}[x(t^-)] - g_{ba}\right)I_b(t^-) \\ &\quad + q_{ab}(t)(\alpha - g_{ab})I_a(t^-) + x(t)E^{\bar{u}}[q_{ba}(t)I_b(t^-)] \end{aligned}$$

The adjoint equation becomes

$$\begin{aligned}dp(t) &= q_{ab}(t)\left\{dM_{ab}(t) - (\alpha - g_{ab})I_a(t^-)dt\right\} - x(t)E^{\bar{u}}[q_{ba}(t)I_b(t^-)]dt\\ &+q_{ba}(t)\left\{dM_{ba}(t) - \left(\tfrac{1}{2}q_{ba}(t) + E^{\bar{u}}[x(t^-)] - g_{ba}\right)I_b(t^-)dt\right\}.\end{aligned} \tag{3.138}$$

Consider the probability measure $\widetilde{P}$, under which x is a Markov chain whose jump intensity matrix is

$$\widetilde{G}(t)=\begin{bmatrix} -\alpha & \alpha \\ \frac{1}{2}q_{ba}(t) + E^{\bar{u}}[x(t^-)] & -\frac{1}{2}q_{ba}(t)-E^{\bar{u}}[x(t^-)]\end{bmatrix}, \quad \frac{1}{2}q_{ba}(t)+E^{\bar{u}}[x(t^-)]\geq 0. \tag{3.139}$$

This change of measure yields the $\widetilde{P}-$martingales

$$\begin{cases} d\widetilde{M}_{ab}(t) = dM_{ab}(t) - (\alpha - g_{ab})I_a(t^-)dt,\\ d\widetilde{M}_{ba}(t) = dM_{ba}(t) - (\frac{1}{2}q_{ba}(t) + E^{\bar{u}}[x(t^-)] - g_{ba})I_b(t^-)dt\end{cases}$$

and

$$dp(t) = -x(t)E^{\bar{u}}[q_{ba}(t)I_b(t^-)]dt + q_{ab}(t)d\widetilde{M}_{ab}(t) + q_{ba}(t)d\widetilde{M}_{ba}(t). \tag{3.140}$$

This yields

$$\Delta p(t) = q_{ab}(t)I_a(t^-)I_b(t) + q_{ba}(t)I_b(t^-)I_a(t). \tag{3.141}$$

Integrating (3.140) and then taking conditional expectation yield

$$p(t) = -\widetilde{E}[\left(x(T) - E^{\bar{u}}[x(T)]\right)^2 |\mathscr{F}_t] + \widetilde{E}[\int_t^T x(s)E^{\bar{u}}[q_{ba}(t)I_b(s^-)]ds|\mathscr{F}_t]$$

Therefore,

$$\Delta p(t) = -\Delta\widetilde{E}[\left(x(T) - E^{\bar{u}}[x(T)]\right)^2 |\mathscr{F}_t]. \tag{3.142}$$

Next, we compute the right-hand side of (3.142) and then identify q_{ba} by matching.

Set $\bar{\mu}(t) := E^{\bar{u}}[x(t)]$and $\phi(t, x(t)) := (x(t) - \bar{\mu}(t))^2$. Under $\widetilde{P}$, Dynkin's formula yields

$$\phi(T, x(T)) = \phi(t, x(t)) + \int_t^T \left(\frac{\partial\phi}{\partial s} + \widetilde{G}\phi\right)(s, x(s))ds + \widetilde{M}_T^{\phi} - \widetilde{M}_t^{\phi}.$$

Taking conditional expectation yields

$$\widetilde{E}[\phi(T, x(T))|\mathscr{F}_t] = \phi(t, x(t)) + \int_t^T \widetilde{E}\left[\left(\frac{\partial\phi}{\partial s} + \widetilde{G}\phi\right)(s, x(s))\,\Big|\,\mathscr{F}_t\right]ds,$$

and

$$\Delta \widetilde{E}[\phi(T, x(T))|\mathscr{F}_t] = \Delta\phi(t, x(t)), \tag{3.143}$$

where

$$\begin{aligned}\Delta\phi(t, x(t)) &= \sum_{i,j:\, i\neq j} (\phi(t, j) - \phi(t, i))\, I_i(t^-)I_j(t) \qquad (i, j\ \in \{a, b\}) \\ &= \Big((b^2 - a^2) - 2\bar{\mu}(t)(b - a)\Big)\, I_a(t^-)I_b(t) \\ &\quad + \Big((a^2 - b^2) - 2\bar{\mu}(t)(a - b)\Big)\, I_b(t^-)I_a(t).\end{aligned} \tag{3.144}$$

Therefore,

$$\begin{aligned}\Delta p(t) &= -\Delta\widetilde{E}[(x(T) - \bar{\mu}(T))^2\, |\mathscr{F}_t] \\ &= \Big((a^2 - b^2) + 2\bar{\mu}(t)(b - a)\Big)\, I_a(t^-)I_b(t) \\ &\quad + \Big((b^2 - a^2) + 2\bar{\mu}(t)(a - b)\Big)\, I_b(t^-)I_a(t).\end{aligned}$$

Matching (3.141) with (3.6) yields

$$\begin{cases} q_{ab}(t) = (a^2 - b^2) + 2\bar{\mu}(t)(b - a) \\ q_{ba}(t) = (b^2 - a^2) + 2\bar{\mu}(t)(a - b). \end{cases}$$

Hence,

$$\bar{u}(t) = \Big((b^2 - a^2) + 2\bar{\mu}(t)(a - b)\Big)\, I_b(t^-).$$

Note that $a \leq \bar{\mu}(t) \leq b$ to guarantee that both $\lambda^{\bar{u}}(t)$ and $\widetilde{G}(t)$ above are indeed intensity matrices, it suffices to impose that

$$0 \leq \bar{\mu}(t) \leq \frac{1}{2}(a + b). \tag{3.145}$$

We further characterize the optimal control $\bar{u}(t)$ by finding $\bar{\mu}(t)$ which satisfies (3.145). Indeed, under $P^{\bar{u}}$, x has the representation

$$\begin{aligned}x(t) = x(0) + \textstyle\int_0^t \big\{\alpha(b - a)I_a(s^-) + \bar{\mu}(s)(a - b)I_b(s^-) + \bar{u}(s)(a - b)I_b(s^-)\big\}\, ds \\ + \textstyle\int_0^t (b - a)dM^{\bar{u}}_{ba}(s) + \int_0^t (a - b)dM^{\bar{u}}_{ab}(s).\end{aligned} \tag{3.146}$$

Taking the expectation under $P^{\bar{u}}$ yields

$$\bar{\mu}(t) = \bar{\mu}(0) + E^{\bar{u}}[\int_0^t \{\alpha(b-a)I_a(s^-) + \bar{\mu}(s)(a-b)I_b(s^-) \\ +(a-b)\bar{u}(s)I_b(s^-)\} ds]. \tag{3.147}$$

In particular, the mapping $t \to \mu(t)$ is absolutely continuous.

Using the fact that $(a-b)I_b(t^-) = a - x(t^-)$ and $(b-a)I_a(t^-) = b - x(t^-)$, Eq. (3.147) becomes

$$\begin{aligned}
\bar{\mu}(t) &= \bar{\mu}(0) + \int_0^t \{\alpha(b-\bar{\mu}(s)) + \bar{\mu}(s)(a-\bar{\mu}(s))\} ds \\
&\quad + \int_0^t \left\{E^{\bar{u}}[(a-b)\bar{u}(s)I_b(s^-)]\right\} ds \\
&= \bar{\mu}(0) + \int_0^t \{\alpha(b-\bar{\mu}(s)) + \bar{\mu}(s)(a-\bar{\mu}(s))\} ds \\
&\quad + \int_0^t \left\{E^{\bar{u}}\left[(a-b)I_b(s^-)\left((b^2-a^2) + 2\bar{\mu}(s)(a-b)\right)\right]\right\} ds \\
&= \bar{\mu}(0) + \int_0^t \{\alpha(b-\bar{\mu}(s)) + \bar{\mu}(s)(a-\bar{\mu}(s))\} ds \\
&\quad + \int_0^t \left\{(b^2-a^2)(a-\bar{\mu}(s)) + 2(a-b)\bar{\mu}(s)(a-\bar{\mu}(s))\right\} ds \\
&= \bar{\mu}(0) + \int_0^t \left\{\left(\alpha b + a(b^2-a^2)\right) + (2(b-a)-1)\,\bar{\mu}^2(s)\right. \\
&\quad \left. +(3a^2 + a(1-2b) - b^2)\bar{\mu}(s)\right\} ds,
\end{aligned} \tag{3.148}$$

with

$$\begin{cases} A := 2(b-a) - 1, \\ B := 3a^2 + a(1-2b) - b^2, \\ C := \alpha b + a(b^2 - a^2). \end{cases}$$

Thus, in view (3.145), $\bar{\mu}$ should satisfy the following constrained Riccati equation

$$\begin{cases} \dot{\bar{\mu}}(t) = A\bar{\mu}^2(t) + B\bar{\mu}(t) + C, \\ \bar{\mu}(0) = m_0, \\ 0 \le \bar{\mu}(t) \le \frac{1}{2}(a+b), \end{cases} \tag{3.149}$$

where m_0 is a given initial value. As it is well known, without the imposed constraint on $\bar{\mu}$, the Riccati equation admits an explicit solution that may explode in finite time unless the involved coefficients a, b, α and m_0 belong to certain ranges. With the imposed constraint on $\bar{\mu}$, these ranges become even tighter. Below we illustrate this through a few cases. See Tables 3.1 and 3.2 on the numerical estimates of these conditions.

As shown in the tables below, for low values of α, the ODE (3.149) can be solved for any time. How low the intensity should be mainly depends on the sizes of b and $b-a$; the larger is b, the wider is the range for α for which the ODE is solvable. In particular, when $a=0$ and $b=1$, (3.149) is solvable for any time when $\alpha \in \{0.1, 0.2\}$. For greater values of α, the ODE violates the constraint proportionally "faster."

These results also show that the initial conditions may affect the time horizon T. Starting with values reasonably close to $\frac{a+b}{2}$, the ODE (3.149) is solvable only for relatively shorter time horizon than when we start with values reasonably close to zero (Table 3.3).

Mean-Field-Type Schlögl Model

We now address the control problem associated with a mean-field type version of the Schlögl model where the intensities are of the form

$$\lambda_{ij}^{u}(t,x,u(t)) := \begin{cases} \nu_{ij}(t) & \text{if } j \neq i-1, \\ u(t)+\beta E^{u}[x(t)] & \text{if } j = i-1, \end{cases} \tag{3.150}$$

for some predictable and positive control process u, where $\beta > 0$ and $(\alpha_{ij})_{ij}$ are a deterministic Q-matrix for which there exists $N_0 \geq 1$ such that $\alpha_{ij} = 0$ for $|j-i| \geq N_0$ and $\alpha_{ij} > 0$ for $|j-i| < N_0$.

We consider the following mean-field-type cost functional

Table 3.1 Scenario 1: Impact of a, b on the horizon T

a	b	α	$T_{m_0=0}$	$T_{m_0=0.25}$	a	b	α	$T_{m_0=0.25}$	$T_{m_0=1}$
0	1	0.1	.	.	1	2	0.1	.	.
0	1	0.2	.	.	1	2	0.2	.	.
0	1	0.3	5.145	3.762	1	2	0.3	.	.
0	1	0.4	2.355	1.481	1	2	0.4	2.644	2.153
0	1	0.5	1.571	0.928	1	2	0.5	1.429	1.001
0	1	0.6	1.870	0.676	1	2	0.6	1.073	0.692
0	1	0.7	0.955	0.532	1	2	0.7	0.878	0.535
0	1	0.8	0.800	0.439	1	2	0.8	0.750	0.438
0	1	0.9	0.689	0.373	1	2	0.9	0.659	0.371
0	1	1	0.605	0.325	1	2	1	0.589	0.322
0	1	5	0.104	0.053	1	2	5	0.121	0.053
0	1	10	0.051	0.026	1	2	10	0.062	0.026

Table 3.2 Scenario 2: Impact of a, b on the horizon T

a	b	α	$T_{m_0=0.25}$	$T_{m_0=2}$	a	b	α	$T_{m_0=0}$	$T_{m_0=0.75}$
2	3	0.1	.	.	0	2	0.1	.	.
2	3	0.2	.	.	0	2	0.2	.	.
2	3	0.3	.	.	0	2	0.3	.	.
2	3	0.4	.	.	0	2	0.4	.	.
2	3	0.5	1.206	0.761	0	2	0.5	.	.
2	3	0.6	0.899	0.494	0	2	0.6	.	.
2	3	0.7	0.746	0.373	0	2	0.7	5.593	1.433
2	3	0.8	0.648	0.302	0	2	0.8	2.227	0.636
2	3	0.9	0.578	0.254	0	2	0.9	1.470	0.418
2	3	1	0.524	0.220	0	2	1	1.111	0.312
2	3	5	0.131	0.035	0	2	5	0.112	0.029
2	3	10	0.070	0.018	0	2	10	0.053	0.014

Table 3.3 Scenario 3: Impact of a, b on the horizon T

a	b	α	$T_{m_0=0}$	$T_{m_0=1}$
0	3	0.1	.	.
0	3	0.2	.	.
0	3	0.3	.	.
0	3	0.4	.	.
0	3	0.5	.	.
0	3	0.6	.	.
0	3	0.7	.	.
0	3	0.8	.	.
0	3	0.9	.	.
0	3	1	.	.
0	3	5	0.126	0.043
0	3	10	0.056	0.019

$$J(u) = E^u\left[\int_0^T \frac{1}{2}u^2(t)dt + x(T)\right]. \tag{3.151}$$

Given a control $v > 0$, the associated Hamiltonian reads

$$H(t, v) := L^{\bar{u}}\left(\sum_{i,j,j\neq i,i-1} \{I_i(t^-)(\alpha_{ij} - g_{ij})q_{ij}(t)\}\right.$$

$$\left.+\sum_i \{I_i(t^-)(v + \beta_1 E^{\bar{u}}[x(t)] - g_{ii-1})q_{ii-1}(t)\} - \frac{v^2}{2}\right). \tag{3.152}$$

The first-order optimality conditions yield

$$\bar{u}(t) = \sum_i I_i(t^-) q_{ii-1}(t) \tag{3.153}$$

Next, we write the associated adjoint equation and identify q_{ii-1}.

$$\begin{cases} dp(t) = -\sum\limits_i q_{ii-1}(t) \left\{ I_i(t^-)(\frac{1}{2}q_{ii-1}(t) + \beta E^{\bar{u}}[x(t)] - g_{ii-1})dt - dM_{ii-1}(t) \right\} \\ - \sum\limits_{ij, j \neq i, i-1} q_{ij}(t) \left\{ I_i(t^-)(v_{ij} - g_{ij})dt - dM_{ij}(t) \right\} - \beta x(t) E^{\bar{u}} \left[\sum\limits_i q_{ii-1}(t) I_i(t^-) \right] dt, \\ p(T) = -x(T). \end{cases}$$

Consider the probability measure $\widetilde{P}$, under which x is a pure jump process whose jump intensity matrix is

$$\widetilde{G}_{ij}(t) = \begin{cases} \alpha_{ij} & \text{if } j \neq i-1, \\ \frac{1}{2}q_{ij}(t) + \beta E^{\bar{u}}[x(t)] & \text{if } j = i-1. \end{cases}$$

The adjoint equation becomes

$$\begin{cases} dp(t) = -\beta x(t) E^{\bar{u}} \sum\limits_i [q_{ii-1}(t) I_i(t^-)] dt + \sum\limits_i q_{ii-1}(t) d\widetilde{M}_{ii-1}(t) \\ \qquad + \sum\limits_{i \neq j, j \neq i-1} q_{ij}(t) d\widetilde{M}_{ij}(t), \\ p(T) = -x(T), \end{cases}$$

where $\widetilde{M}_{ij}, \; i \neq j$ are mutually orthogonal $\widetilde{P}$-martingales.

Thus,

$$p(t) = -\widetilde{E}[x(T)|\mathscr{F}_t] + \beta \int_t^T \widetilde{E}\left[x(s) E^{\bar{u}} \left[\sum_i q_{ii-1} I_i(s^-) \right] |\mathscr{F}_t \right] ds,$$

and

$$\Delta p(t) = \Delta \widetilde{E}[-x(T)|\mathscr{F}_t]. \tag{3.154}$$

Following the same steps leading to (3.133), from 3.154, we obtain $q_{ij}(t) = i - j$, thus $q_{i,i-1}(t) = 1, \; i \in \{1, 2, \ldots\}$

Therefore,

$$\bar{u}(t) = \sum_{i \geq 1} I_i(t^-) = 1 - I_0(t^-).$$

3.5 MASS: Master Adjoint System for Discrete State Spaces

In this section, we established new dynamic programming principle on the space of measures for mean-field-type games with discrete state spaces. From these optimality equations, we can derive equations in the dual space due the functional differentiability. These are called Master adjoint systems (MASS). MASS also play also important roles in finding equilibrium strategies in mean-field-type games [102, 330, 331, 333]. We establish such systems for generic rate transitions of mean-field type.

A Basic Model of Mean-Field-Type Game

Consider $I \geq 2$ number decision-makers interacting over a non-empty discrete state space $\mathcal{S}$. Assume that the cardinality of $\mathcal{S}$ is finite. The transition rates are given by the $Q-$matrix function with entries $q_{s,s'}(t, a_1, \ldots a_I, m),\ (s, s') \in \mathcal{S}^2$, with

$$\begin{cases} q_{s,s'}(t, a_1, \ldots, a_I, m) > 0,\ s' \neq s, \\ q_{s,s}(t, a_1, \ldots, a_I, m) = -\sum_{s' \neq s} q_{s,s'}(t, a_1, \ldots, a_I, m), \\ a_i \in \mathcal{A}_i. \end{cases} \tag{3.155}$$

The probability law m of the state $s(t)$ satisfies the Kolmogorov forward equation

$$\begin{cases} \dot{m}_s = \sum_{s' \neq s} m_{s'} q_{s',s}(t, a_1, \ldots, a_I, m) - m_s \sum_{s' \neq s} q_{s,s'}(t, a_1, \ldots, a_I, m), \\ m(0) = m_0 \in \Delta(\mathcal{S}), \\ s \in \mathcal{S}. \end{cases} \tag{3.156}$$

A Mean-Field-Type Game with Discrete State Space

The mean-field-type game problem is given by

$$\begin{cases} i \in \mathcal{I}, \\ \inf_{a_i \in \mathcal{A}_i} \sum_{s \in \mathcal{S}} h_i(s, m(T)) m_s(T) + \int_0^T \sum_{s \in \mathcal{S}} l_i(t, s, a, m) m_s dt, \\ \dot{m}_s = \sum_{s' \neq s} m_{s'} q_{s',s}(t, a_1, \ldots, a_I, m) - m_s \sum_{s' \neq s} q_{s,s'}(t, a_1, \ldots, a_I, m), \\ m(0) = m_0, \\ s \in \mathcal{S} \end{cases} \tag{3.157}$$

Introduce the following I functions

$$\begin{cases} \hat{V}_i(t,m) = \inf_{a_i \in \mathscr{A}_i} \sum_{s\in\mathscr{S}} h_i(s, m(T))m_s(T) + \int_t^T \sum_{s\in\mathscr{S}} l_i(t,s,a,m)m_s dt, \\ \dot{m}_s = \sum_{s'\neq s} m_{s'}q_{s',s}(t, a_1, \ldots, a_I, m) - m_s \sum_{s'\neq s} q_{s,s'}(t, a_1, \ldots, a_I, m), \\ m(t) = m, \\ s \in \mathscr{S}. \end{cases} \tag{3.158}$$

Applying the standard dynamic programming in finite dimensions with the (new) state m yields

$$\begin{cases} \partial_t \hat{V}_i(t,m) + \sum_{s\in\mathscr{S}} m_s \inf_{a_i} \left[l_i(t,s,a,m) + \sum_{s'\in\mathscr{S}} [\nabla_{m_{s'}} \hat{V}_i(t,m)] q_{s,s'} \right] = 0, \\ \hat{V}_i(T,m) = \sum_{s\in\mathscr{S}} m_s h_i(t,s,m), \\ i \in \mathscr{I} \end{cases} \tag{3.159}$$

This leads to the following Bellman system:

$$\begin{cases} \partial_t \hat{V}_i(t,m) + \sum_{s\in\mathscr{S}} m_s \hat{H}_i(t,s,m,\nabla\hat{V}) = 0, \\ \hat{V}_i(T,m) = \sum_{s\in\mathscr{S}} m_s h_i(t,s,,m), \\ i \in \mathscr{I} \end{cases} \tag{3.160}$$

where

$$\begin{cases} \hat{H}_i(t,s,m,\nabla\hat{V}) := \inf_{a_i} \left[l_i(t,s,a,m) + \sum_{s'\in\mathscr{S}} [\nabla_{m_{s'}} \hat{V}_i(t,m)] q_{s,s'}(t,a,m) \right], \\ i \in \mathscr{I} \end{cases} \tag{3.161}$$

Master Adjoint Systems (MASS)
The MASS yields

$$\begin{cases} \partial_t \hat{U}_i(t,k,m) + \hat{H}_i(t,k,m,\hat{U}) + \sum_{s\in\mathscr{S}} m_s \partial_{m_k} \hat{H}_i(t,s,m,\hat{U}) = 0, \\ \hat{U}_i(T,k,m) = h_i(t,k,m) + \sum_{s\in\mathscr{S}} m_s \partial_{m_k} h_i(t,s,m), \\ i \in \mathscr{I}, \end{cases} \tag{3.162}$$

where

$$\begin{cases} \partial_{m_k} \hat{H}_i(t,s,m,\nabla\hat{V}) = \hat{H}_{i,m_k}(t,s,m,\nabla\hat{V}) \\ \qquad + \sum_{j,l} \hat{H}_{i,p_{jl}}(t,s,m,\nabla\hat{V}) \partial_{m_k m_l} \hat{V}_j(t,m) \\ = \hat{H}_{i,m_k}(t,s,m,\hat{U}) + \sum_{j,l} \hat{H}_{i,p_{jl}}(t,s,m,\hat{U}) \partial_{m_k} \hat{U}_j(t,l,m), \\ i \in \mathscr{I}, \end{cases} \tag{3.163}$$

If there is a smooth solution to the Bellman system (3.160), then the gradient $\hat{U}_i(t,k,m) = \partial_{m_k} \hat{V}_i(t,m)$ satisfies the MASS (3.162).

Unnormalized Master Adjoint Systems

Unbalanced dynamics with birth and death processes are critical components in various models, measurements, and observations. These processes capture the natural fluctuations in populations, accounting for births and deaths, input-output, inflow-outflow, and in-out, and enable the development of more accurate and relevant models. We follow the same notation as in Sect. 3.5. In addition to the movement between states, there is a birth rate process $\tilde{b}(t, s, a_1, \ldots, a_I, m)$ and a death rate process $\tilde{d}(t, s, a_1, \ldots, a_I, m)$. The Kolmogorov forward equation of the unnormalized mean-field is given by

$$\begin{cases} \dot{m}_s = \sum_{s' \neq s} m_{s'} q_{s',s}(t, a_1, \ldots, a_I, m) - m_s \sum_{s' \neq s} q_{s,s'}(t, a_1, \ldots, a_I, m) \\ +\tilde{b}_s(t, a_1, \ldots, a_I, m) - \tilde{d}_s(t, a_1, \ldots, a_I, m), \\ m(0) = m_0, \\ s \in \mathcal{S}. \end{cases} \tag{3.164}$$

A Mean-Field-Type Game with Discrete State Space and Death and Birth

The unnormalized mean-field-type game problem is given by

$$\begin{cases} \inf_{a_i} \sum_{s \in \mathcal{S}} h_i(s, m(T)) m_s(T) + \int_0^T \sum_{s \in \mathcal{S}} l_i(t, s, a, m) m_s dt, \\ \dot{m}_s = \sum_{s' \neq s} m_{s'} q_{s',s} - m_s \sum_{s' \neq s} q_{s,s'} + \tilde{b}_s - \tilde{d}_s, \\ m(0) = m_0, \\ s \in \mathcal{S}, i \in \mathcal{I}. \end{cases} \tag{3.165}$$

Introduce the I functions, $\hat{V}_i(t, m) : [0, T] \times \mathbb{R}^{\sharp \mathcal{S}} \to \mathbb{R}$, defined by

$$\begin{cases} \hat{V}_i(t, m) = \inf_{a_i} \sum_{s \in \mathcal{S}} h_i(s, m(T)) m_s(T) + \int_t^T \sum_{s \in \mathcal{S}} l_i(t, s, a, m) m_s dt, \\ \dot{m}_s = \sum_{s' \neq s} m_{s'} q_{s',s} - m_s \sum_{s' \neq s} q_{s,s'} + \tilde{b}_s - \tilde{d}_s, \\ s \in \mathcal{S}, \\ i \in \mathcal{I}. \end{cases} \tag{3.166}$$

Applying the standard dynamic programming in finite dimensions with the state m yields

$$\begin{cases} \partial_t \hat{V}_i(t, m) + \inf_{a_i} \Big\{ \sum_{s \in \mathcal{S}} m_s \Big[l_i(t, s, a, m) + \sum_{s' \in \mathcal{S}} [\partial_{m_{s'}} \hat{V}_i(t, m)] q_{s,s'} \Big] \\ + \sum_{s' \in \mathcal{S}} [\partial_{m_{s'}} \hat{V}_i(t, m)](\tilde{b}_{s'} - \tilde{d}_{s'}) \Big\} = 0, \\ \hat{V}_i(T, m) = \sum_{s \in \mathcal{S}} m_s h_i(t, s, m), \\ i \in \mathcal{I} \end{cases} \tag{3.167}$$

Bellman System for (3.165)
This leads to the following Bellman system

$$\begin{cases} \partial_t \hat{V}_i(t, m) + H_i(t, m, \nabla \hat{V}) = 0, \\ \hat{V}_i(T, m) = \sum_{s \in \mathcal{S}} m_s h_i(t, s, m), \\ i \in \mathcal{I} \end{cases} \tag{3.168}$$

where

$$\begin{cases} H_i(t, m, \nabla \hat{V}) := \inf_{a_i} \left\{ \sum_{s \in \mathcal{S}} m_s \left[l_i(t, s, a, m) + \sum_{s' \in \mathcal{S}} [\partial_{m_{s'}} \hat{V}_i(t, m)] q_{s,s'} \right] \right. \\ \left. + \sum_{s' \in \mathcal{S}} [\partial_{m_{s'}} \hat{V}_i(t, m)](\tilde{b}_{s'} - \tilde{d}_{s'}) \right\}, \\ i \in \mathcal{I} \end{cases} \tag{3.169}$$

Unnormalized Master Adjoint Systems
Unnormalized master adjoint systems (uMASS) are highly valuable tools in mean-field-type games, as they offer an efficient and structured approach to finding equilibrium strategies and optimizing decisions in scenarios involving many interacting entities. In this case, the uMASS yields

$$\begin{cases} \partial_t \hat{U}_i(t, k, m) + \partial_{m_k} H_i(t, m, \hat{U}) = 0, \\ \hat{U}_i(T, k, m) = h_i(t, k, m) + \sum_{s \in \mathcal{S}} m_s \partial_{m_k} h_i(t, s, m), \\ i \in \mathcal{I}, \end{cases} \tag{3.170}$$

where

$$\begin{cases} H_i(t, m, \hat{U}) := \inf_{a_i} \left\{ \sum_{s \in \mathcal{S}} m_s \left[l_i(t, s, a, m) + \sum_{s' \in \mathcal{S}} \hat{U}_i(t, s', m) q_{s,s'} \right] \right. \\ \left. + \sum_{s' \in \mathcal{S}} \hat{U}_i(t, s', m)(\tilde{b}_{s'} - \tilde{d}_{s'}) \right\}, \\ \partial_{m_k} H_i(t, m, \hat{U}) = H_{i,m_k}(t, m, \hat{U}) + \sum_{j,l} \hat{H}_{i,p_{jl}}(t, m, \hat{U}) \partial_{m_k} \hat{U}_j(t, l, m), \\ i \in \mathcal{I}, \end{cases} \tag{3.171}$$

If there is a smooth solution to the Bellman system (3.168), then the gradient $\hat{U}_i(t, k, m) = \partial_{m_k} \hat{V}_i(t, m)$ satisfies the uMASS (3.170).

Reduced Bellman System
In the absence of birth-and-death process or when $\tilde{b} = \tilde{d}$,, the Bellman system (3.168) reduces to the following system:

$$\begin{cases} \partial_t \hat{V}_i(t,m) + \sum_{s\in\mathscr{S}} m_s \hat{H}_i(t,s,m,\nabla\hat{V}) = 0, \\ \hat{V}_i(T,m) = \sum_{s\in\mathscr{S}} m_s h_i(t,s,,m), \\ i\in\mathscr{I} \end{cases} \tag{3.172}$$

where

$$\begin{cases} \hat{H}_i(t,s,m,\nabla\hat{V}) := \inf_{a_i}\left[l_i(t,s,a,m) + \sum_{s'\in\mathscr{S}}[\nabla_{m_{s'}}\hat{V}_i(t,m)]q_{s,s'}(t,a,m)\right], \\ i\in\mathscr{I} \end{cases} \tag{3.173}$$

Normalized MASS
In the absence of birth-and-death process or when $\tilde{b} = \tilde{d}$ uMASS (3.170) reduces to the following normalized master adjoint system (MASS, [333, 339, 342]):

$$\begin{cases} \partial_t \hat{U}_i(t,k,m) + \hat{H}_i(t,k,m,\hat{U}) + \sum_{s\in\mathscr{S}} m_s \partial_{m_k}\hat{H}_i(t,s,m,\hat{U}) = 0, \\ \hat{U}_i(T,k,m) = h_i(t,k,m) + \sum_{s\in\mathscr{S}} m_s \partial_{m_k} h_i(t,s,m), \\ i\in\mathscr{I}, \end{cases} \tag{3.174}$$

where

$$\begin{cases} \hat{H}_i(t,s,m,\hat{U}) := \inf_{a_i}\left[l_i(t,s,a,m) + \sum_{s'\in\mathscr{S}}\hat{U}_i(t,s',m)q_{s,s'}(t,a,m)\right], \\ \partial_{m_k}\hat{H}_i(t,s,m,\hat{U}) = \hat{H}_{i,m_k}(t,s,m,\hat{U}) + \sum_{j,l}\hat{H}_{i,p_{jl}}(t,s,m,\hat{U})\partial_{m_k}\hat{U}_j(t,l,m), \\ i\in\mathscr{I}, \end{cases} \tag{3.175}$$

In this chapter, we have examined mean-field-type games over discrete state spaces and established master adjoint systems (MASS). The regime switching process is also covered in Chaps. 8 and 10 of this volume.

3.6 Notes

Mean-field-type games over discrete state spaces have been examined in [98]. The authors established existence of Markov chains of mean-field type with unbounded jump intensities by means of a fixed point argument using the total variation distance. The authors have also provided conditions for existence of a saddle point for a zero-sum differential game associated with payoff functionals of mean-field type, under dynamics driven by such Markov chains of mean-field type. Master adjoint systems (MASS) for discrete state spaces can be found in [336, 338]. In general, the master adjoint systems differ from the master equations [10, 63, 67, 68, 85–87, 89, 93–95, 113, 114, 161–163, 171, 180, 182, 206, 240, 254, 255, 277, 354]. The master adjoint system is extended in Chap. 8, Sect. 8.3, to finitely many decision-makers under jump-diffusion processes.

Part II
Equilibrium Principles

Chapter 4
Stochastic Maximum Principle

We start a basic optimal control problem that is mean-field free.

4.1 Pontryagin's Principle for Mean-Field-Free Optimal Control

4.1.1 A Heuristic Derivation of Pontryagin's Principle

Consider the following optimal control problem: Minimize

$$J(u) = \int_0^T f(x(t), u(t))dt + h(x(T)), \tag{4.1}$$

with respect to $u : [0, T] \longrightarrow U$, subject to

$$\begin{cases} \dot{x}(t) = b(x(t), u(t)), & 0 < t \leq T, \\ x(0) = x_0, \end{cases} \tag{4.2}$$

where U is a given non-empty set of controls.

Using the Lagrange multiplier method, solving the constrained Problem (4.1)–(4.2) is equivalent to minimizing the Lagrangian functional

$$\mathscr{L}(x, u, p) := \int_0^T (f(x(t), u(t)) + p(t) \cdot (\dot{x}(t) - b(x(t), u(t))\, dt + h(x(T)), \tag{4.3}$$

where $'\cdot'$ denotes the scalar product.

T. Başar et al., *Mean-Field-Type Game Theory I*, Static & Dynamic Game Theory: Foundations & Applications, https://doi.org/10.1007/978-3-032-07027-2_4

Introducing the so-called Hamiltonian

$$H(x, u, p) := p \cdot b(x, u) - f(x, u), \tag{4.4}$$

the Lagrangian functional becomes

$$\mathscr{L}(x, u, p) := \int_0^T (-H(x(t), u(t), p(t)) + p(t) \cdot \dot{x}(t))\, dt + h(x(T)). \tag{4.5}$$

If $(\bar{x}, \bar{u}, p)$ is a minimizer of $\mathscr{L}$, performing a first-order Taylor expansion, the infinitesimal variation of $\mathscr{L}$ with perturbation $(\delta x, \delta u, \delta p)$, such that $\delta x(0) = 0$, reads

$$\begin{aligned}\delta\mathscr{L} &:= \mathscr{L}(\bar{x} + \delta x, \bar{u} + \delta u, p + \delta p) - \mathscr{L}(\bar{x}, \bar{u}, p)\\ &\approx \textstyle\int_0^T \left(-H_x \cdot \delta x - H_u \cdot \delta u - H_p \cdot \delta p + p \cdot \frac{d}{dt}(\delta x) + \delta p \cdot \dot{\bar{x}}\right) dt + h_x \cdot \delta x(T),\end{aligned}$$

where $H_j(t) := \nabla_j H\Big(\bar{x}(t), \bar{u}(t), p(t)\Big)$, $j \in \{x, u, p\}$ and $h_x := h_x(\bar{x}(T))$.

Integrating by parts, we obtain

$$\begin{aligned}\delta\mathscr{L} \approx \textstyle\int_0^T &\left((-H_x(t) - \dot{p}) \cdot \delta x - H_u(t) \cdot \delta u + \left(-H_p(t) + \dot{\bar{x}}\right) \cdot \delta p + \frac{d}{dt}(p \cdot \delta x)\right) dt\\ &+ h_x \cdot \delta x(T).\end{aligned}$$

Requiring $\delta\mathscr{L} = 0$ and noting that $\delta x(0) = 0$ yield

$$\begin{aligned}0 = \textstyle\int_0^T &\left((-H_x(t) - \dot{p}) \cdot \delta x - H_u(t) \cdot \delta u + \left(-H_p(t) + \dot{\bar{x}}\right) \cdot \delta p\right) dt\\ &+ (h_x + p(T)) \cdot \delta x(T).\end{aligned}$$

This in turn suggests that the extremal $(\bar{x}, \bar{u}, p)$ satisfies the so-called *Hamiltonian system* associated with the optimal control problem (4.1)–(4.2):

$$\begin{cases}\bar{x}(0) = x_0, & & \text{(Initial value)}\\ \dot{\bar{x}}(t) = H_p(\bar{x}(t), \bar{u}(t), p(t)), & 0 < t \le T, & \text{(Controlled dynamics)}\\ \dot{p}(t) = -H_x(\bar{x}(t), \bar{u}(t), p(t)), & 0 \le t < T, & \text{(Adjoint equation)}\\ p(T) = -h_x(\bar{x}(T)), & & \text{(Adjoint terminal value).}\end{cases}$$

and

$$H_u(\bar{x}(t), \bar{u}(t), p(t)) = 0, \quad 0 \le t \le T, \tag{4.6}$$

which means that $\bar{u}$ is extremal for the function $u \longmapsto H(\bar{x}(t), \cdot, p(t))$. Noting that

$$H_p(x, u, p) = b(x, u),$$

the first two equations in (4.6) are nothing but the dynamics (4.2). The new element in this heuristic approach is a backward ordinary differential equation called the *adjoint equation* satisfied by the Lagrange multiplier p. We call p the *first-order adjoint function*. It satisfies the adjoint equation

$$\begin{cases} \dot{p}(t) = -H_x(\bar{x}(t), \bar{u}(t), p(t)), & 0 \le t < T, \\ p(T) = -h_x(\bar{x}(T)). \end{cases} \tag{4.7}$$

Hence, solving the optimal control problem amounts to solving the Hamiltonian system (4.6), which is a system of forward-backward ordinary differential equations. Pontryagin's maximum principle can be simply stated as follows.

Pontryagin's Maximum Principle

If $(\bar{x}, \bar{u})$ is an optimal solution of the control problem (4.1)–(4.2), then there exists a function p solution of the adjoint equation (4.7) and

$$\bar{u}(t) \in \arg\max_{v \in \mathcal{U}} H(\bar{x}(t), v, p(t)), \quad 0 \le t \le T. \quad \text{(Maximum Principle)} \tag{4.8}$$

The perturbation method used to derive the maximum principle implicitly assumes that the set U of controls is *linear*, i.e., if $u \in U$ then also $u + \delta u \in U$. But, in many practical examples, the set/space U is not linear. It is rather convex or a general metric space. Therefore, a crucial question is whether it is possible to extend the abovementioned maximum principle to more general sets/spaces of controls. This, in turn, would suggest to device a new perturbation method which should be compatible with the structure of the set of controls.

If the set of controls U is convex, then a natural perturbation method would be a convex perturbation of a given optimal control $\bar{u}$:

$$\bar{u} + \delta u := \bar{u} + \eta(v - \bar{u}), \quad v \in U,$$

where η will tend to 0. Hence, if $\bar{u}$ is optimal for J, we have

$$J(\bar{u} + \eta(u - \bar{u})) - J(\bar{u}) \ge 0.$$

Thus, we can write

$$(J^{\psi}(\bar{u}), v - \bar{u}) \ge 0, \qquad \forall v \in U,$$

provided that J is Gateaux differentiable. Mimicking the previous formal computations, we would expect the following (local) version of the maximum principle.

$$\frac{dH}{du}(\bar{x}(t), \bar{u}(t), p(t)) \cdot (u - \bar{u}(t)) \le 0, \qquad 0 \le t \le T, \quad \forall u \in U. \tag{4.9}$$

If the set U is not necessarily convex, Pontryagin's approach suggests that the following perturbation method, called *spike variation*, would be appropriate.

For $\eta > 0$, pick a subset $E_\eta \subset [0, T]$ such that $|E_\eta| = \eta$. The control process u^η is a spike variation of u if

$$u^\eta(t) := \begin{cases} u, & t \in E_\eta, \\ \bar{u}(t), & t \in E_\eta^c, \end{cases} \tag{4.10}$$

where $u \in U$ is an arbitrary control.

Using the spike variation technique, Pontryagin's main contribution is the following key relation between the performance functional (4.1) and the associated Hamiltonian (4.4):

$$J(u^\eta) - J(\bar{u}) = -\int_0^T \delta H(s) I_{E_\eta}(s)\, ds + o(\eta), \tag{4.11}$$

for arbitrary $v \in U$, where

$$\delta H(s) := H(\bar{x}(s), v, p(s)) - H(\bar{x}(s), \bar{u}(s), p(s)).$$

We will show in the next section how to arrive at (4.8), using (4.11).

In this chapter, we will rigorously derive this principle for both the deterministic systems (4.1)–(4.2) and for stochastic systems driven by diffusion processes of Itô type. As we will see throughout the next few sections, the main ingredients needed to arrive at the maximum principle are the Taylor expansion and the so-called duality relation, which permit to arrive at (4.11), using (4.7).

Relation to the Hamilton–Jacobi–Bellman's Equation

Again for the same optimal control problem (4.1)–(4.2), an alternative approach is to use dynamic programming. This leads to the Hamilton–Jacobi–Bellman (HJB) equation, which is a nonlinear backward PDE

$$\begin{cases} -\frac{\partial V}{\partial t}(t, x) + \max_u H(x, u, -V_x(t, x)) = 0, \\ V(T, x) = h(x). \end{cases} \tag{4.12}$$

satisfied by the value function $V(t, x)$ defined by

$$V(t, x) = \min_{u: x(t)=x} \int_t^T f(x(s), u(s))dt + h(x(T)) \tag{4.13}$$

The associated optimal control is then of the *state-feedback* type:

$$u^*(t, x) \in \arg\max_u H(x, u, -V_x(t, x)), \tag{4.14}$$

which depends on $V_x(t, x)$ but not on V.

An informal comparison between (4.8) and (4.14) suggests that

$$p(t) = -V_x(t, x), \tag{4.15}$$

provided that the HJB equation admits a smooth solution V. Unfortunately, HJB equations admit smooth solutions in very few cases. Most real-world applications exhibit non-smooth solutions of the HJB equation, on which we will say more later in the chapter.

4.2 Rigorous Derivation of Pontryagin's Principle

In this section, we derive the Pontryagin's Maximum (Minimum) Principle for the deterministic control problem [333, 339, 342].

Consider (4.1)–(4.2), and let $d_0 : U \times U \to \mathbb{R}_+$ be a metric, $d \geq 1$ be the dimension of the state x. Let us assume the following:

- (2-H1) The space of controls (U, d_0) is a separable metric space.
- (2-H2) The maps $b : [0, T] \times \mathbb{R}^d \times U \longrightarrow \mathbb{R}^d$, $f : [0, T] \times \mathbb{R}^d \times U \longrightarrow \mathbb{R}$, and $h : \mathbf{R}^d \longrightarrow \mathbb{R}$ are measurable and satisfy the following conditions: There exists a constant $L > 0$ such that

$$\begin{cases} |\varphi(t, x, u) - \varphi(t, \hat{x}, u)| \leq L|x - \hat{x}|, & t \in [0, T], x, \hat{x} \in \mathbb{R}^d, u \in U, \\ |\varphi(t, x, u)| \leq C(1 + |x|), & t \in [0, T], u \in U, \end{cases}$$

  for $\varphi(t, x, u) \in \{b(t, x, u), f(t, x, u), h(x)\}$.
- (2-H3) The maps b, f and h are C^1 in x and are such that

$$|\varphi_x(t, x, u) - \varphi_x(t, \hat{x}, u)| \leq L|x - \hat{x}|, \quad t \in [0, T], x, \hat{x} \in \mathbb{R}^d, u \in U,$$

  for $\varphi(t, x, u) \in \{b(t, x, u), f(t, x, u), h(x)\}$.

Let $\mathscr{U}[0, T] := \{u : [0, T] \longrightarrow U, u \text{ measurable}\}$. A control u is called admissible if it belongs to $\mathscr{U}[0, T]$.

The classical result due to Carathèodory (see Warga (1972), Theorem II.4.1 and II.4.2, or Davis (1993), Theorem 41.4) guarantees that under (2-H1)–(2-H2), for any $v \in \mathscr{U}[0, T]$, Eq. (4.1) admits a unique solution and (4.2) is well defined. The point here is that b needs only be *measurable* in t.

The deterministic optimal control problem can then equivalently be stated as follows:

Problem (DC) Characterize $\bar{u} \in \mathscr{U}[0, T]$ such that

$$\bar{u} \in \underset{u \in \mathscr{U}[0,T]}{\arg\min}\, J(u). \tag{4.16}$$

Denote by $\bar{x}(\cdot) = \bar{x}^{\bar{u}}(\cdot)$ the solution of (4.1) corresponding to $\bar{u}$, and consider (4.4), the Hamiltonian H associated with (4.1)–(4.2). Taking the gradient of H with respect to x,

$$H_x(t, x, u, p) = b_x^*(t, x, u)p - f_x(t, x, u), \tag{4.17}$$

where A^* denotes the transpose of the matrix A.

4.2.1 The Maximum Principle

The following theorem states the Pontryagin's Maximum Principle, which gives a set of first-order necessary optimality conditions for the pair $(\bar{x}, \bar{u})$.

Theorem 4.1 *Let assumptions (H1)–(H2) hold. Let $\bar{u}$ be an optimal control for Problem (DC) and $\bar{x}$ be the corresponding trajectory. Then, there exists an absolutely continuous function $p : [0, T] \longrightarrow \mathbf{R}^d$ satisfying*

$$\begin{cases} \dot{p}(t) = -H_x(t, \bar{x}(t), \bar{u}(t), p(t)), & \text{a.e.} t \in [0, T), \\ p(T) = -h_x(\bar{x}(T)), \end{cases} \tag{4.18}$$

and

$$H(t, \bar{x}(t), \bar{u}(t), p(t)) = \max_{v \in U} H(t, \bar{x}(t), v, p(t)), \quad \text{a.e.} t \in [0, T]. \tag{4.19}$$

Proof of the Maximum Principle
Let $(\bar{x}, \bar{u})$ be an optimal solution of the control problem (4.1)–(4.2). We derive (4.19), in several steps, starting with the fact that

$$J(u^\eta) - J(\bar{u}) \geq 0,$$

where u^η is the spike variation of $\bar{u}$, defined as follows (cf. 4.10).

For $\eta > 0$, pick a subset $E_\eta \subset [0, T]$ such that $|E_\eta| = \eta$, and consider the control process (spike variation of u)

$$u^\eta(t) := \begin{cases} v, & t \in E_\eta, \\ \bar{u}(t), & t \in E_\eta^c, \end{cases}$$

where $v \in U$ is an arbitrary control. Let $x^\eta := x^{u^\eta}$ denote the corresponding state process which satisfies (4.1).

We will arrive at (4.19) by establishing the following key relation between the performance functional (4.2) and the associated Hamiltonian (4.4):

$$J(u^\eta) - J(\bar{u}) = -\int_0^T \delta H(s) I_{E_\eta}(s)\, ds + o(\eta), \tag{4.20}$$

for arbitrary $v \in U$ and $t \in [0, T]$, where

$$\delta H(s) = H(s, \bar{x}(s), v, p(s)) - H(s, \bar{x}(s), \bar{u}(s), p(s)).$$

We will use the following notation throughout this section. For $\varphi \in \{b, f, h\}$, we define

$$\delta\varphi(t) := \varphi(t, \bar{x}(t), u) - \varphi(t, \bar{x}(t), \bar{u}(t)), \quad \varphi_x(t) := \varphi_x(t, \bar{x}(t), \bar{u}(t)). \tag{4.21}$$

Let y^η be the solutions of the following ODE:

$$\begin{cases} \dot{y}^\eta(t) = b_x(t) \cdot y^\eta(t) + \delta b(t) I_{E_\eta}(t), & \text{a.e.} t \in (0, T], \\ y^\eta(0) = 0, \end{cases} \tag{4.22}$$

Equation (4.22) is called the first-order variational equation.

4.2.2 *Duality and Taylor Expansion*

The duality relation between p and y^η displayed in the next lemma follows immediately from the differential chain rule.

Lemma 4.1 *We have*

$$p(T) \cdot y^\eta(T) = \int_0^T \Big(f_x(t) \cdot y^\eta(t) + p(t) \cdot \delta b(t) I_{E_\eta}(t) \Big) dt. \tag{4.23}$$

Recall the equations satisfied by $x^\eta := x^{u^\eta}$, $\bar{x} := \bar{x}^{\bar{u}}$:

$$\dot{x}^\eta(t) = b(t, x^\eta(t), u^\eta(t)), \quad \text{a.e.} t \in (0, T], \quad x^\eta(0) = x_0, \tag{4.24}$$

$$\dot{\bar{x}}(t) = b(t, \bar{x}(t), \bar{u}(t)), \quad \text{a.e.} t \in (0, T], \quad \bar{x}(0) = x_0. \tag{4.25}$$

The estimates displayed in the next result are very important for the proof of Theorem 4.1. If not specified, we will denote by C some appropriate positive constants, which are independent of η, that appear in the estimates below and which may differ from line to line.

Lemma 4.2 *Assume that conditions (2-H2)–(2-H3) hold. Then,*

$$\sup_{t\in[0,T]} |x^\eta(t) - \bar{x}(t)| \leq C\eta, \tag{4.26}$$

$$\sup_{t\in[0,T]} |y^\eta(t)| \leq C\eta, \tag{4.27}$$

$$\sup_{t\in[0,T]} |x^\eta(t) - \bar{x}(t) - y^\eta(t)| \leq C\eta. \tag{4.28}$$

Proof The estimates (4.26), (4.27) and (4.28) follow from Gröonwall's inequality, using Assumptions (2-H2)–(2-H3). For a proof, see Lemma 3.2.2. in Yong and Zhou (1999). ■

Using these estimates, we obtain the following useful relation. Assume $\phi(t, x, u)$ satisfies Assumptions (2-H2) and (2-H3). Then, we have

$$\begin{aligned}\varphi(t, x^\eta(t), u^\eta(t)) - \varphi(t, \bar{x}(t), \bar{u}(t)) - \varphi_x(t, \bar{x}(t), \bar{u}(t))y^\eta(t)\\ +\delta\varphi(t)I_{E_\eta}(t) + o(\eta).\end{aligned} \tag{4.29}$$

Indeed, set $\xi^\eta(t) := x^\eta(t) - \bar{x}(t)$, and write

$$\varphi(t, x^\eta(t), u^\eta(t)) - \varphi(t, \bar{x}(t), \bar{u}(t)) = J_1(t) + J_2(t),$$

where

$$\begin{aligned}J_1(t) &:= \varphi(t, x^\eta(t), u^\eta(t)) - \varphi(t, \bar{x}(t), u^\eta(t)),\\ J_2(t) &:= \varphi(t, \bar{x}(t), u^\eta(t)) - \varphi(t, \bar{x}(t), \bar{u}(t)) = \delta\varphi(t)I_{E_\eta}(t).\end{aligned}$$

We have

$$\begin{aligned}J_1(t) &= \int_0^1 \varphi_x(t, \bar{x}(t) + \theta\xi^\eta(t), u^\eta(t))\, d\theta\, \xi^\eta(t)\\ &= \int_0^1 \Big(\varphi_x(t, \bar{x}(t) + \theta\xi^\eta(t), u^\eta(t)) - \varphi_x(t, \bar{x}(t), u^\eta(t))\Big)\, d\theta\, \xi^\eta(t)\\ &+\Big(\varphi_x(t, \bar{x}(t), u^\eta(t)) - \varphi_x(t, \bar{x}(t), \bar{u}(t))\Big)\xi^\eta(t)\\ &+\varphi_x(t, \bar{x}(t), \bar{u}(t))(\xi^\eta(t) - y^\eta(t)) + \varphi_x(t, \bar{x}(t), \bar{u}(t))y^\eta(t)\\ &:= I_1(t) + I_2(t) + I_3(t) + \varphi_x(t, \bar{x}(t), \bar{u}(t))y^\eta(t).\end{aligned}$$

In view of (2-H3) and (4.26), we have

$$\sup_{t\in[0,T]} |I_1(t)| + \sup_{t\in[0,T]} |I_2(t)| \leq C\eta^2 + C\eta.$$

Moreover, by (4.28), we have

$$\sup_{t\in[0,T]} |I_3(t)| \leq C\eta,$$

where we have used the estimate $|\varphi_x(t,x,u)| \leq L,\ t \in [0,T], x \in \mathbb{R}^d, u \in U$, which follows from the Lipschitz condition (2-H2).

Thus,

$$\sup_{t\in[0,T]} |\varphi(t,x^\eta(t),u^\eta(t)) - \varphi(t,\bar{x}(t),\bar{u}(t)) - \varphi_x(t,\bar{x}(t),\bar{u}(t))y^\eta(t) \\ -\delta\varphi(t)I_{E_\eta}(t)| \leq C\eta.$$

This completes the proof of (4.29).

Proof of Theorem 4.1 Applying (4.29) to f and h, we obtain

$$\begin{aligned} 0 \leq J(u^\eta) - J(\bar{u}) &= \int_0^T (f(t,x^\eta(t),u^\eta(t)) - f(t,\bar{x}(t),\bar{u}(t)))\,dt \\ &\quad +h(x^\eta(T)) - h(\bar{x}(T)) \\ &= \int_0^T \left(f_x(t)\cdot y^\eta(t) + \delta f(t)I_{E_\eta}(t)\right) dt + h_x(T)\cdot y^\eta(T) + \mathrm{o}(\eta) \\ &= \int_0^T (-p(t)\cdot \delta b(t) + \delta f(t))\, I_{E_\eta}(t)\,dt + \mathrm{o}(\eta) \\ &= -\int_0^T \delta H(t)I_{E_\eta}(t)\,dt + \mathrm{o}(\eta), \end{aligned}$$

where the third equality follows from the duality relation (4.23), by noting that $p(T) = -h_x(\bar{x}(T))$. Now, since the set E_η is arbitrary, we can choose it of the form $E_\eta := [\bar{t}, \bar{t}+\eta]$, for arbitrarily chosen $\bar{t} \in [0,T]$. This yields

$$0 \leq J(u^\eta) - J(\bar{u}) = -\int_{\bar{t}}^{\bar{t}+\eta} \delta H(t)\,dt + \mathrm{o}(\eta).$$

Dividing by η and then letting η go to zero, by the separability of U, we obtain (4.19). ∎

4.2.3 Sufficient Optimality Conditions

Theorem 4.2 *Let assumptions (2-H2) and (2-H3) hold. Let $(\bar{x},\bar{u})$ be an admissible pair, and let p be the associated adjoint process. Assume*

- *(2-H4) The set of controls U is a convex body (i.e., U is convex and has a non-empty interior) of $\mathbb{R}^k$, and the maps b and f are differentiable in u,*
- *(2-H5) The Hamiltonian $H(t,\cdot,\cdot,p(t))$ is concave in (x,v) for* a.e.$t \in [0,T]$,
- *(2-H6) The function h is convex.*

Then, the pair $(\bar{x},\bar{u})$ is optimal if

$$H(t,\bar{x}(t),\bar{u}(t),p(t)) = \max_{v\in U} H(t,\bar{x}(t),v,p(t)), \quad \text{a.e.}\, t \in [0,T]. \tag{4.30}$$

Proof Set

$$\psi(t) := \psi(t, x(t), u(t)), \quad \bar{\psi}(t) := \psi(t, \bar{x}(t), \bar{u}(t))$$

for $\psi \in \{b, f\}$, and

$$H(t) = b(t) \cdot p(t) - f(t), \quad \bar{H}(t) = \bar{b}(t) \cdot p(t) - \bar{f}(t).$$

We want to show that if the pair $(\bar{x}, \bar{u})$ satisfies (4.30) where the adjoint process p satisfies (4.18), then

$$J(u) - J(\bar{u}) = \int_0^T (f(t) - \bar{f}(t))dt + h(x(T)) - h(\bar{x}(T)) \geq 0.$$

Since H is concave in (x, u), we have

$$H(t) - \bar{H}(t) \leq \bar{H}_x(t) \cdot (x(t) - \bar{x}(t)) + \bar{H}_u(t) \cdot (u(t) - \bar{u}(t)).$$

Moreover, by (4.30), we have $\bar{H}_u(t) = 0$ a.e.$t \in [0, T]$.
Hence

$$H(t) - \bar{H}(t) \leq \bar{H}_x(t) \cdot (x(t) - \bar{x}(t)), \quad \text{a.e.} t \in [0, T]. \tag{4.31}$$

By the convexity of h, we have

$$h(x(T)) - h(\bar{x}(T)) \geq h_x(\bar{x}(T)) \cdot (x(T) - \bar{x}(T)) = -p(T) \cdot (x(T) - \bar{x}(T)).$$

Integrating by parts and using (4.18), we obtain

$$\begin{aligned} -p(T) \cdot (x(T) - \bar{x}(T)) &= -\int_0^T (x(t) - \bar{x}(t)) \cdot dp(t) + p(t) \cdot d(x(t) - \bar{x}(t)) \\ &= \int_0^T \Big(\bar{H}_x(t) \cdot (x(t) - \bar{x}(t)) - p(t) \cdot (b(t) - \bar{b}(t))\Big)dt. \end{aligned}$$

Hence, in view of (4.31), we obtain

$$h(x(T)) - h(\bar{x}(T)) \geq \int_0^T \Big(H(t) - \bar{H}(t) - p(t) \cdot (b(t) - \bar{b}(t))\Big)dt. \tag{4.32}$$

Therefore,

$$J(u) - J(\bar{u}) \geq \int_0^T \Big(f(t) - \bar{f}(t) - p(t) \cdot (b(t) - \bar{b}(t)) + H(t) - \bar{H}(t)\Big)dt = 0.$$

■

4.3 Pontryagin's Principle for Stochastic Differential Equations

In this section, we derive the Pontryagin's principle for systems that are governed by stochastic differential equations (SDEs). In a first step, we consider the following control system

$$\begin{cases} dx(t) = b(t, x(t), u(t))dt + \sigma(t, x(t))dW_t, \\ x(0) = x_0. \end{cases} \tag{4.33}$$

for some functions b and σ, and a Brownian motion W. For every t, the control $u(t)$ is allowed to take values in some action space U.

The objective of the control problem is to minimize a performance functional of the form

$$J(u) = E\left[\int_0^T f(t, x(t), u(t))\, dt + h(x(T))\right], \tag{4.34}$$

for given functions f and h.

The system (4.33) can be seen as a random perturbation of the ODE studied in the previous section in the sense that the control enters only the drift term b and does not enter the diffusion term σ. Mimicking the duality relation derived in the previous section together with the first-order Taylor expansion of the spike variation, we expect to obtain a *stochastic* Hamiltonian system which consists of a *forward-backward* SDE and a maximum principle of a similar form as (4.19).

We will see in the next section that controlling the diffusion coefficient exhibits an extra complication. Roughly speaking, the main difficulty lies in the fact that the Itô integral $\int_t^{t+\eta} \sigma dW(s)$ is only of order $\sqrt{\eta}$ rather than η as in the Lebesgue integral. This means that the first-order variation method would not suffice characterizing the optimal control. We should rather consider both the first and *second* order terms in the Taylor expansion of the spike variation and device a *stochastic* Hamiltonian system which would consist of *two forward-backward* SDEs and a Maximum Principle with an additional *quadratic* term of the diffusion coefficient.

4.3.1 State-Dependent Diffusion

Let $T > 0$ be a fixed time horizon and $(\Omega, \mathscr{F}, \mathbb{F}, \mathbb{P})$ be a given filtered probability space, whose filtration $\mathbb{F} = (\mathscr{F}_t)_{t\geq 0}$, satisfies the usual conditions of right-continuity and completeness, on which a d-dimensional standard Brownian motion $W = \{W_t\}_{t\geq 0}$ is given. We assume that $\mathbb{F}$ is the natural filtration of W augmented by $\mathbb{P}-$null sets of $\mathscr{F}$.

Consider the stochastic control system (4.33) with

$$b : [0, T] \times \mathbb{R}^d \times U \longrightarrow \mathbb{R}^d, \quad \sigma : [0, T] \times \mathbb{R}^d \longrightarrow \mathbb{R}^{d\times d}.$$

For convenience, we have chosen to assume σ to be a square matrix. The results below hold even for non-square matrices.

An admissible control u is an $\mathbb{F}$-adapted process with values in a separable metric space (U, d). We denote the set of all admissible controls by $\mathscr{U}[0, T]$.

$$\mathscr{U}[0, T] := \{u : \ [0, T] \times \Omega \longrightarrow U, \ u \text{ is } \mathbb{F}\text{-adapted}\}.$$

The optimal control problem we are concerned with is to minimize the cost functional (4.34) over $\mathscr{U}[0, T]$: with

$$f : \ [0, T] \times \mathbb{R}^d \times U \longrightarrow \mathbb{R}, \qquad h : \ \mathbb{R}^d \longrightarrow \mathbb{R}.$$

Any $\bar{u} \in \mathscr{U}[0, T]$ satisfying

$$J(\bar{u}) = \inf_{u\in\mathscr{U}[0,T]} J(u)$$

is called an optimal control. The corresponding state process, which is the solution to (4.33), is denoted by $\bar{x} := x^{\bar{u}}$. Our stochastic optimal control problem can then be stated as follows.

Problem (SC) Characterize $\bar{u} \in \mathscr{U}[0, T]$ such that

$$\bar{u} \in \arg\min_{u\in\mathscr{U}[0,T]} J(u). \tag{4.35}$$

4.3.2 *The Maximum Principle: Smooth Coefficients*

In this section, we will establish the maximum principle when the coefficients b, σ, f and h are smooth, i.e., satisfy the following assumptions:

- (3-H1) The maps b, σ, f and h are measurable and satisfy the following conditions: There exists a constant $L > 0$ such that

$$\begin{cases} |\varphi(t, x, u) - \varphi(t, \hat{x}, u)| \leq L|x - \hat{x}|, & t \in [0, T], x, \hat{x} \in \mathbb{R}^d, u \in U, \\ |\varphi(t, x, u)| \leq L(1 + |x|), & t \in [0, T], u \in U, \end{cases}$$

  for $\varphi(t, x, u) \in \{b(t, x, u), \sigma(t, x), f(t, x, u), h(x)\}$.
- (3-H2) The maps b, σ, f and h are C^1 in x and are such that

$$|\varphi_x(t, x, u) - \varphi_x(t, \hat{x}, u)| \leq L|x - \hat{x}|, \quad t \in [0, T], x, \hat{x} \in \mathbb{R}^d, u \in U,$$

for $\varphi(t, x, u) \in \{b(t, x, u), \sigma(t, x), f(t, x, u), h(x)\}$.

Under these assumptions, for each $u \in \mathscr{U}[0, T]$, the SDE (4.33) admits a unique strong solution. Moreover, for every $p \geq 0$,

$$E\left[\sup_{0\leq t\leq T} |x(t)|^p\right] \leq C_p, \tag{4.36}$$

where C_p is a positive constant depending only on p and the Lipschitz constant L.

We will see below that these conditions can be made much weaker.

Recall the following notation that will be used throughout this section. For $\varphi \in \{b, \sigma, f, h\}$ and $v \in U$, we introduce

$$\delta\varphi(t) := \varphi(t, \bar{x}(t), v) - \varphi(t, \bar{x}(t), \bar{u}(t)), \quad \varphi_x(t) := \varphi_x(t, \bar{x}(t), \bar{u}(t)). \tag{4.37}$$

Consider the Hamiltonian associated with the control system

$$H(t, x, v, p) := b(t, x, v) \cdot p + \mathrm{tr}\,(q^*\sigma(t, x)) - f(t, x, v), \tag{4.38}$$

and let

$$\delta H(t) := p(t) \cdot \delta b(t) - \delta f(t) \tag{4.39}$$

and

$$H_x(t) := b_x^*(t, \bar{x}(t), \bar{u}(t))p(t) + \sum_{j=1}^{d} \sigma_x^j(t, \bar{x}(t))^* q_j(t) - f_x(t, \bar{x}(t), \bar{u}(t)). \tag{4.40}$$

4.3.2.1 The Adjoint Equations

In this subsection, we introduce the adjoint equation involved in the stochastic maximum principle for our control problem.

The first-order adjoint equation is the following linear backward SDE.

$$\begin{cases} dp(t) = -H_x(t)dt + q(t)dW_t, \\ p(T) = -h_x(T). \end{cases} \tag{4.41}$$

Under assumptions (3-H1)–(3-H2), by Lemma A.3 in the Appendix, Eq. (4.41) admits a unique $\mathbb{F}$-adapted solution (p, q) such that

$$E[\sup_{t\in[0,T]} |p(t)|^2 + \int_0^T |q(t)|^2\, dt] < \infty. \tag{4.42}$$

Note that the adjoint equation (4.41) is a random perturbation (by the semimartingale $\bar{M}_t := \int_0^t q(s)dW_s - \sum_{j=1}^d \sigma_x^j(s, \bar{x}(s))^T q_j(s)ds$) of the adjoint equation (4.18) associated with the deterministic control system discussed in the previous section.

Since Eq. (4.41) is linear in the variable p, in view of the martingale representation theorem for Brownian functionals, by Lemma A.4, in the Appendix, it admits the following explicit solution.

$$p(t) = -E\left[\int_t^T \Phi^*(s,t) f_x(s, \bar{x}(s), \bar{u}(s))ds + \Phi^*(T,t) h_x(\bar{x}(T)) \Big| \mathscr{F}_t\right], \quad (4.43)$$

where $\Phi(s,t)$, $s \geq t$, is the fundamental solution of the linear equation

$$\begin{cases} d\Phi(s,t) = b_x(s, \bar{x}(s), \bar{u}(s))\Phi(s,t)ds + \sum_{j=1}^d \sigma_x^j(s, \bar{x}(s))\Phi(s,t)dW_s^j, \\ \Phi(t,t) = I_d, \end{cases}$$

where Φ^* is the transpose of the matrix Φ.

4.3.3 The Maximum Principle

The following theorem states a stochastic version of the Pontryagin's maximum principle, which provides a set of first-order necessary optimality conditions for the pair $(\bar{x}, \bar{u})$.

Theorem 4.3 (Stochastic Maximum Principle) *Let assumptions (3-H1)–(3-H2) hold. If $(\bar{x}, \bar{u})$ is an optimal solution of the control problem (4.33)–(4.35), then there is a unique pair of $\mathbb{F}$-adapted processes (p, q) that satisfies (4.41)–(4.42) such that*

$$H(t, \bar{x}(t), \bar{u}, p(t)) = \max_{v \in U} H(t, \bar{x}(t), v, p(t)), \text{ a.e.} t \in [0,T], \ \mathbb{P}\text{a.s.} \quad (4.44)$$

Proof of the Maximum Principle

To streamline the presentation, we only consider the one-dimensional case. The extension to the multidimensional case is by now straightforward.

Let $(\bar{x}, \bar{u})$ be an optimal solution of the control problem (4.33)–(4.35). We derive (4.44) in several steps, starting with the fact that

$$J(u^\eta) - J(\bar{u}) \geq 0,$$

where u^η is the spike variation of $\bar{u}$ defined as follows (cf. (4.10)):

For $\eta > 0$, pick a subset $E_\eta \subset [0,T]$ such that $|E_\eta| = \eta$,, and consider the control process (spike variation of $\bar{u}$)

$$u^\eta(t) := \begin{cases} u(t), & t \in E_\eta, \\ \bar{u}(t), & t \in E_\eta^c, \end{cases} \tag{4.45}$$

where $u \in \mathscr{U}[0, T]$ is an arbitrary admissible control. Let $x^\eta := x^{u^\eta}$ denote the corresponding state process which satisfies (4.33).

We will arrive at (4.44) by establishing the following key relation between the performance functional (4.34) and the associated Hamiltonian (4.38):

$$J(u^\eta) - J(\bar{u}) = -E\left[\int_0^T \delta H(s) I_{E_\eta}(s)\, ds\right] + o(\eta), \tag{4.46}$$

for arbitrary $u \in \mathscr{U}[0, T]$.

Let y^η be the solution of the following SDE:

$$\begin{cases} dy^\eta(t) = \left\{b_x^*(t) y^\eta(t) + \delta b(t) I_{E_\eta}(t)\right\} dt + \sigma_x^*(t) y^\eta(t) dW_t, \\ y^\eta(0) = 0. \end{cases} \tag{4.47}$$

Equation (4.47) is called the first-order variational equation.

4.3.4 Duality and Taylor Expansions

The duality relation between p and y^η displayed in the next lemma follows immediately from Itô's formula.

Lemma 4.3 *We have*

$$E\left[p(T) \cdot y^\eta(T)\right] = E\left[\int_0^T \left(f_x(t) \cdot y^\eta(t) + p(t) \cdot \delta b(t) I_{E_\eta}(t)\right) dt\right]. \tag{4.48}$$

Recall the equations satisfied by $x^\eta := x^{u^\eta}$, $\bar{x} := x^{\bar{u}}$.

$$dx^\eta(t) = b(t, x^\eta(t), u^\eta(t))dt + \sigma(t, x^\eta(t), u^\eta(t))dW_t, \quad x^\eta(0) = x_0; \tag{4.49}$$

$$d\bar{x}(t) = b(t, \bar{x}(t), \bar{u}(t))dt + \sigma(t, \bar{x}(t), \bar{u}(t))dW_t, \quad \bar{x}(0) = x_0. \tag{4.50}$$

The estimates displayed in the next result are very important for the proof of Theorem 4.3. If not specified, we will denote by C the positive constants, which are independent of η, that appear in the estimates below, and these may differ from line to line.

Proposition 4.1 *Let assumptions (3-H1)–(3-H3) hold. Then, for any $k \geq 1$,*

$$E\left[\sup_{t\in[0,T]}|x^\eta(t)-\bar{x}(t)|^{2k}\right]\leq C\eta^k, \tag{4.51}$$

$$E\left[\sup_{t\in[0,T]}|y^\eta(t)|^{2k}\right]\leq C\eta^k, \tag{4.52}$$

$$E\left[\sup_{t\in[0,T]}|x^\eta(t)-\bar{x}(t)-y^\eta(t)|^{2k}\right]\leq C\eta^{2k}. \tag{4.53}$$

Proof Estimates (4.51), (4.52) and (4.53) follow from standard arguments using Grönwall's inequality and the Burkholder–Davis–Gundy inequalities for diffusion processes (see, e.g., Revuz and Yor (2004)). For a proof, see Theorem 4.4 in Yong and Zhou (1999). ■

Proof of Theorem 4.3 We perform the proof in the one-dimensional case, the multidimensional case being straightforward. We have

$$\begin{aligned}
J(u^\eta)-J(\bar{u}) &= E\left[\int_0^T (f(t,x^\eta(t),u^\eta(t))-f(t,\bar{x}(t),\bar{u}(t)))dt\right]\\
&\quad +E[h(x^\eta(T))-h(\bar{x}(T))]\\
&= E\left[\int_0^T f_x(t)y^\eta(t)+\delta f(t)I_{E_\eta}(t)\,dt\right]+E[h_x(T)y^\eta(T)]+o(\eta).
\end{aligned}$$

Hence, noting that $p(T)=-h_x(\bar{x}(T))$, in view of (4.48), we obtain

$$0\leq J(u^\eta)-J(\bar{u})=-E\Big[\int_0^T \delta H(t)I_{E_\eta}(t)\,dt\Big]+o(\eta).$$

Now, choose $E_\eta=[\bar{t},\bar{t}+\eta]$, for an arbitrary $\bar{t}\in[0,T]$, and let A be an arbitrary set of $\mathscr{F}_{\bar{t}}$. Consider the control

$$u(t):=vI_A+\bar{u}(t)I_{\Omega\setminus A},\quad t\in E_\eta.$$

for an arbitrary $v\in U$. Obviously, $u(t)$ is an admissible control. Moreover, we have, for every $t\in E_\eta$,

$$\delta H(t)=\Big(H(t,\bar{x}(t),v,p(t))-H(t,\bar{x}(t),\bar{u}(s),p(t))\Big)I_A.$$

Therefore, we obtain

$$\begin{aligned}
J(u^\eta)-J(\bar{u}) = -E\Big[\int_{\bar{t}}^{\bar{t}+\eta} (&H(t,\bar{x}(t),v,p(t))\\
&-H(t,\bar{x}(t),\bar{u}(s),p(t)))\,dtI_A]+o(\eta).
\end{aligned}$$

Dividing by η and then letting η go to zero, by the separability of U, we obtain (4.44). ∎

4.3.5 The Maximum Principle: Non-smooth Coefficients

In this section, we will establish the maximum principle with considerably relaxed conditions on the smoothness of the coefficients b, σ, f and h. We assume them to be only Lipschitz continuous instead of differentiable in x and satisfying (3-H2). Using an embedding result by Bouleau and Hirsch (1988), the idea is to define a slightly different stochastic differential equation defined on an enlarged probability space, where the initial condition $x(0)$ will be taken as a random element.

To this end, we let $\Omega = \mathscr{C}_0(\mathbb{R}_+, \mathbb{R}^d)$ be the space of continuous functions ω such that $\omega(0) = 0$, endowed with the topology of uniform convergence on compact subsets of $\mathbb{R}_+$. $\mathscr{F}$ is the Borel σ-algebra over Ω, $\mathbb{P}$ the Wiener measure on $(\Omega, \mathscr{F})$, and $\mathbb{F} = (\mathscr{F}_t)_{t\geq 0}$, the filtration of the coordinate process augmented with $\mathbb{P}$-null sets of $\mathscr{F}$. We make the coordinate process ω a $\mathbb{P}$-Brownian motion, by setting $W_t(\omega) := \omega(t)$, $t \geq 0$, so that $\big(\Omega, \mathscr{F}, \mathbb{F}, \mathbb{P}, (W_t)_{t\geq 0}\big)$ defines a Brownian motion. To keep the exposition of the results less technical, we let the set of controls U be a compact subset of $\mathbb{R}^m$ for some $m \geq 1$.

- (3-A1) The maps b, σ, f and h are measurable and satisfy the following conditions: There exists a constant $L > 0$ such that

$$\begin{cases} |\varphi(t,x,u) - \varphi(t,\hat{x},\hat{u})| \leq L|x-\hat{x}|, & t \in [0,T], x, \hat{x} \in \mathbb{R}^d, u, \in U, \\ |\varphi(t,x,u)| \leq L(1+|x|), & t \in [0,T], u \in U. \end{cases}$$

  for $\varphi(t,x,u) \in \{b(t,x,u), \sigma(t,x), f(t,x,u), h(x)\}$.

Since b, σ, f and h are Lipschitz continuous in x, by Rademacher's theorem, they are differentiable almost everywhere in the sense of the Lebesgue measure: For $\varphi(t,x,u) \in \{b(t,x,u), \sigma^j(t,x), f(t,x,u), h(x)\}$, $j \in \{1,\ldots,d\}$, we have

$$\partial_x \varphi(t,x,u) = \varphi_x(t,x,u) \quad dx\text{-a.e.},$$

where $\varphi_x(t,x,u) \in \{b_x(t,x,u), \sigma_x^j(t,x), f_x(t,x,u), h_x(x)\}$, with b_x, σ_x, f_x and h_x being Borel measurable functions. It should be clear that these almost everywhere derivatives are bounded by the Lipschitz constant L.

Finally, we assume that

- (3-A2) $b_x(t,x,u)$ and $f_x(t,x,u)$ are continuous in u, uniformly in (t,x).

To describe the Bouleau and Hirsch embedding, let ψ be a continuous positive function on $\mathbb{R}^d$ satisfying $\int \psi(x)dx = 1$ and $\int \psi^2(x)dx < \infty$, and consider the space of functions

$$D := \{f \in L^2(\psi dx) \text{ such that } \frac{\partial f}{\partial x_i} \in L^2(\psi dx),\ i \in \{1, \dots, d\}\},$$

where $\frac{\partial f}{\partial x_i}$ denotes the derivative of f in the sense of distributions. Endowed with the norm

$$\|f\|_D := \Big[\int f^2(x)\psi(x)dx + \sum_{i=1}^{d} \Big(\frac{\partial f}{\partial x_i}(x)\Big)^2 \psi(x)dx\Big]^{1/2},$$

D is a Hilbert space.

Let $\widetilde{\Omega} := \mathbb{R}^d \times \Omega$, $\widetilde{\mathcal{F}}$ be the Borel σ-algebra over $\widetilde{\Omega}$, $\widetilde{\mathbb{P}} := \psi dx \otimes \mathbb{P}$, $\widetilde{W}_t(x, \omega) := W_t(\omega)$ and $(\widetilde{\mathcal{F}}_t)_{t\geq 0}$ denote the natural filtration of $\widetilde{W}$ augmented with $\widetilde{\mathbb{P}}$-negligible sets. It can be easily seen that on $\big(\widetilde{\Omega}, \widetilde{\mathcal{F}}, (\widetilde{\mathcal{F}}_t)_{t\geq 0}, \widetilde{\mathbb{P}}\big)$, $\widetilde{W}$ is a Brownian motion.

Let $\widetilde{x}$ solve the extended SDE:

$$\begin{cases} d\widetilde{x}(t) = b(t, \widetilde{x}(t), \widetilde{u}(t))dt + \sigma(t, \widetilde{x}(t))d\widetilde{W}_t, \\ \widetilde{x}(0) = \xi, \end{cases} \tag{4.54}$$

associated with the control

$$\widetilde{u}(t, x, \omega) := u(t, \omega)$$

on the enlarged space $\big(\widetilde{\Omega}, \widetilde{\mathcal{F}}, (\widetilde{\mathcal{F}}_t)_{t\geq 0}, \widetilde{\mathbb{P}}\big)$. Since the coefficients b and σ are Lipschitz continuous, with linear growth, (4.54) admits a unique $\widetilde{\mathcal{F}}_t$ adapted continuous solution. Compared with (4.33), uniqueness of the solution of (4.54) is slightly weaker, due to the enlargement of the underlying probability space. This will allow us to perform operations for $\widetilde{x}$ which are not defined for x. In fact, the uniqueness of the solution of (4.54) implies that for each $t \geq 0$, $x(t) = \widetilde{x}(t)$, $\widetilde{\mathbb{P}}$a.s.

4.3.6 The Maximum Principle

The following theorem states a stochastic version of the Pontryagin's maximum principle, which provides a set of first-order necessary optimality conditions for the pair $(\bar{x}, \bar{u})$.

Theorem 4.4 (Stochastic Maximum Principle) *Let Assumptions (3-A1)–(3-A2) hold. If $(\bar{x}, \bar{u})$ is an optimal solution of the stochastic control problem (4.33)–(4.35), then there is a unique process p that satisfies*

$$p(t) = -E\Big[\int_t^T \Phi^*(s,t) f_x(s, \bar{x}(s), \bar{u}(s))ds + \Phi^*(T,t)h_x(\bar{x}(T))\big|\widetilde{\mathcal{F}}_t\Big], \tag{4.55}$$

where $\Phi(s,t), s \geq t$ is the fundamental solution of the linear equation

$$\begin{cases} d\Phi(s,t) = b_x(s,\bar{x}(s),\bar{u}(s))\Phi(s,t)ds + \sum_{j=1}^{d} \sigma_x^j(s,\bar{x}(s))\Phi(s,t)d\widetilde{W}_s^j, \\ \Phi(t,t) = I_d. \end{cases}$$

Moreover, the following maximum holds.

$$H(t,\bar{x}(t),\bar{u},p(t)) = \max_{v\in U} H(t,\bar{x}(t),v,p(t)), \text{ a.e.} t \in [0,T], \ \widetilde{\mathbb{P}}\text{a.s.} \tag{4.56}$$

Proof of the Maximum Principle

The proof of Theorem 4.4 is based on a result by Bouleau and Hirsch (1988) on differentiability in the sense of distributions of the solution of an SDE with Lipschitz continuous coefficients with respect to the initial data and on Ekeland's variational principle 4.13, discussed in Sect. 4.6. We first recall Bouleau and Hirsch result and then proceed with the proof of the theorem.

Lemma 4.4 *Let $\widetilde{x}$ be a solution of (4.54) on $\left(\widetilde{\Omega}, \widetilde{\mathscr{F}}, (\widetilde{\mathscr{F}}_t)_{t\geq 0}, \widetilde{\mathbb{P}}\right)$. Then, for $\widetilde{\mathbb{P}}$-a.e.ω*

- *For all $t \geq 0$, $\xi \mapsto \widetilde{x}(t,\omega) := \widetilde{x}^{\xi}(t,\omega)$ is in D^d.*
- *There exists an $\widetilde{\mathscr{F}}_t$-adapted $GL_n(\mathbb{R})$-valued continuous process $(\widetilde{\Phi}_t)_{t\geq 0}$ such that for $\widetilde{\mathbb{P}}$-a.e.ω and every $t \geq 0$,*

$$\frac{\partial}{\partial \xi}(\widetilde{x}^{\xi}(t,\omega)) = \widetilde{\Phi}_t(\xi,\omega) \quad dx\text{-a.e.}$$

  where the differentiation is in the sense of distributions. Here, $GL_n(\mathbb{R})$ denotes the linear group of $d \times d$-invertible matrices with real entries.
- *For every $t \geq 0$, the image measure of $\widetilde{\mathbb{P}}$ through the map $\widetilde{x}(t)$ is absolutely continuous with respect to the Lebesgue measure.*
- *The distributional derivative $\widetilde{\Phi}_t$ is the fundamental solution of the linear SDE*

$$\begin{cases} d\widetilde{\Phi}(s,t) = b_x(s,\bar{x}(s),\bar{u}(s))\widetilde{\Phi}(s,t)ds + \sum_{j=1}^{d} \sigma_x^j(s,\bar{x}(s))\widetilde{\Phi}(s,t)d\widetilde{W}_s^j, \\ \widetilde{\Phi}(t,t) = I_d, \end{cases}$$

  where b_x and σ_x^j are versions of the almost everywhere derivatives of b and σ^j.

4.3.7 *A Maximum Principle for a Family of Perturbed Control Problems*

Let ψ be a mollifier function, i.e., a $\mathscr{C}^\infty$, nonnegative function defined on $\mathbb{R}^d$, with support in the unit ball such that $\int \psi(y)dy = 1$. For $n \geq 1$,, introduce the following smooth functions

$$\varphi^n(t,x,u) = n^d \int \varphi(t,x-y,u)\psi(ny)dy, \tag{4.57}$$

for $\varphi \in \{b, \sigma^j, f, h\}$, $j = 1, \ldots, d$. In the next lemma, we collect some properties satisfied by these smooth functions that we will use later.

Lemma 4.5 *Let $\varphi \in \{b, \sigma^j, f, h\}$, $j = 1, \ldots, d$ and φ^n given by (4.57). Then,*

(a) φ^n are Borel measurable functions and Lipschitz continuous in x with the Lipschitz constant L.

(b) There exists a constant $C > 0$ such that, for every $t \in [0, T]$,

$$|\varphi^n(t,x,u) - \varphi(t,x,u)| \le C/n,$$

(c) φ^n is $\mathscr{C}^1$ in x, and for every $t \in [0, T]$,

$$\lim_{n\to\infty} \varphi^n_x(t,x,u) = \varphi_x(t,x,u), \quad dx\text{-a.e.}$$

(d) For every $p \ge 1$ and $r > 0$,

$$\lim_{n\to\infty} \int_{[0,T]\times B(0,r)} \sup_{u\in U} |\varphi^n_x(t,x,u) - \varphi_x(t,x,u)|^p dtdx = 0,$$

where $B(0,r)$ denotes the ball in $\mathbb{R}^d$ of radius r.

Proof Statements (a), (b), and (c) are classical facts. We therefore only prove (d). Since φ_x and φ^n_x are bounded by the Lipschitz constant L, by the Lebesgue dominated convergence theorem, it suffices to prove that, for every $t \in [0, T]$,

$$\lim_{n\to\infty} \sup_{u\in U} |\varphi^n_x(t,x,u) - \varphi_x(t,x,u)| = 0, \quad dx\text{-a.e.}$$

In view of (c), for every $t \in [0, T]$ and $u \in U$, there exists a dx-negligible subset $N(u)$ in $\mathbb{R}^d$ such that for all $x \notin N(u)$,

$$\lim_{n\to\infty} \sup_{u\in U} |\varphi^n_x(t,x,u) - \varphi_x(t,x,u)| = 0.$$

Set $N := \cup_{u\in\hat{U}} N(u)$, where $\hat{U}$ is a countable dense subset of U (recall that U is separable). Pick $u \in U$ and a sequence $(u_p)_{p\ge0}$ in $\hat{U}$ such that $\lim_{p\to\infty} u_p = u$. We have

$$\begin{aligned}|\varphi^n_x(t,x,u) - \varphi_x(t,x,u)| &\le |\varphi^n_x(t,x,u) - \varphi^n_x(t,x,u_p)| \\ &\quad + |\varphi^n_x(t,x,u_p) - \varphi_x(t,x,u_p)| + |\varphi_x(t,x,u_p) - \varphi_x(t,x,u)|.\end{aligned}$$

But, the first and third terms on the right-hand side tend to 0 as $p \to \infty$ uniformly in (t, x) and n, and the second term tends to 0 for each $x \notin N$, which achieves the proof. ■

Now, consider the family $(\mathscr{P}^n)$ of perturbed control problems obtained by replacing the data b, σ, f and h with their respective mollified functions b^n, σ^n, f^n and h^n on the enlarged probability space $\left(\widetilde{\Omega}, \widetilde{\mathscr{F}}, (\widetilde{\mathscr{F}}_t)_{t\geq 0}, \widetilde{\mathbb{P}}, \widetilde{W}\right)$. Let y^n be the solution of the controlled SDE

$$\begin{cases} dy^n(t) = b^n(t, y^n(t), u(t))dt + \sigma^n(t, y^n(t))d\widetilde{W}_t, \\ y^n(0) = \xi, \end{cases} \tag{4.58}$$

and consider the performance functional

$$J^n(u) = \widetilde{E}\left[\int_0^T f(t, y^n(t), u(t))dt + h(y^n(T))\right]. \tag{4.59}$$

In view of Lemma 4.5 and standard arguments, based on Grönwall's and Burkholder–Davis–Gundy's inequalities, it is easy to derive the following estimates which relate the original stochastic control problem (4.33)–(4.34) and the perturbed one (4.58)–(4.59).

Lemma 4.6 *Consider (4.33)–(4.34) and (4.58)–(4.59). There exist positive constants C_1 and C_2 such that*

(i)

$$\widetilde{E}\left[\sup_{0\leq t\leq T} |y^n(t) - x(t)|^2\right] \leq C_1/n^2,$$

(ii)

$$|J^n(u) - J(u)| \leq C_2/n.$$

Let $(\bar{u}, \bar{x})$ be an optimal pair for stochastic control problem (4.33)–(4.34). We note that $\bar{u}$ is not necessarily optimal for the perturbed control problem $(\mathscr{P}^n)$. However, in view of Lemma 4.6, there exists a positive constant $\delta_n = 2C_2/n$, such that

$$J^n(\bar{u}) \leq \inf\{J^n(u), \ u \in \mathscr{U}[0, T]\} + \delta_n.$$

that is $\bar{u}$ is δ_n-optimal for the problem $(\mathscr{P}^n)$. In view of Lemma 4.8 of Sect. 4.6, the performance functional J^n being continuous with respect to the topology induced by the metric d_E defined in (4.187), by applying Ekeland's variational principle, Lemma 4.13, to $\bar{u}$ with $\lambda = \delta_n^{1/2}$, there exists an admissible control u^n such that

- $d(u^n, \bar{u}) \leq \delta_n^{1/2}$,
- $J^n(u^n) \leq J^n(\bar{u})$,
- u^n is optimal for the performance functional $J^n(u) + \delta_n^{1/2} d(u, u^n)$.

Denote by x^n the unique solution of (4.58) corresponding to u^n, and let $\Phi^n(s, t)$, $(s \geq t)$ be the fundamental solution of the linear equation

$$\begin{cases} d\Phi_n(s,t) = b_x^n(s, x^n(s), u^n(s))\Phi_n(s,t)ds + \sum_{j=1}^d \sigma_x^{j,n}(s, \bar{x}(s))\Phi_n(s,t)d\widetilde{W}_s^j, \\ \Phi_n(t,t) = I_d. \end{cases}$$

The adjoint process p^n and the Hamiltonian associated with (x^n, u^n) are then defined as follows:

$$\begin{aligned} p^n(t) = -\widetilde{E}\Big[\textstyle\int_t^T \Phi_n^*(s,t) f_x^n(s, x^n(s), u^n(s))ds \\ +\Phi_n^*(T,t)h_x^n(x^n(T))\Big|\widetilde{\mathscr{F}}_t\Big], \end{aligned} \tag{4.60}$$

where Φ_n^* is the transpose of the matrix Φ_n, and

$$H^n(t, x, u, p) := p \cdot b^n(t, x, u) - f^n(t, x, u). \tag{4.61}$$

Proposition 4.2 *For each $\delta_n > 0$,, there exists an admissible control u^n and an $(\mathscr{F}_t)$-adapted p^n given by (4.60), and a Lebesgue null set N, such that for $t \notin N$*

$$\widetilde{E}\left[H^n(t, x^n(t), u^n(t), p^n(t))\right] \geq \widetilde{E}\left[H^n(t, x^n(t), v, p^n(t))\right] - \delta_n^{1/2}, \tag{4.62}$$

for every U-valued $\mathscr{F}_t$-measurable random variable v.

Proof Since u^n is optimal for the cost functional $J_\delta^n := J^n(u) + \delta_n^{1/2} d(u, u^n)$, we proceed as in the case of smooth coefficients to derive the maximum principle for u^n. Let $\bar{t} \in (0, T)$, and let v be a U-valued $\mathscr{F}_t$-measurable random variable. Consider the spike variation u_θ^n of u^n on the interval $(\bar{t}, \bar{t}+\theta)$. The optimality of u^n for J_δ^n implies that

$$J^n(u_\theta^n) - J^n(u^n) \geq -\delta_n^{1/2}\theta.$$

Since the coefficients associated with the controlled problem $(\mathscr{P}^n)$ are differentiable, using the results of the first part of this section, we obtain

$$\begin{aligned} &\widetilde{E}\left[H^n(t, x^n(t), u^n(t), p^n(t)) - H^n(t, x^n(t), v, p^n(t))\right] \\ &\qquad = \lim_{\theta \to 0} \tfrac{1}{\theta}(J^n(u_\theta^n) - J^n(u^n)) \geq -\delta_n^{1/2}, \end{aligned}$$

for every U-valued $\mathscr{F}_t$-measurable random variable v. This completes the proof. ■

Let $\bar{x}^n$ denote the unique solution of (4.58) corresponding to $\bar{u}$:

$$\begin{cases} d\bar{x}^n(t) = b^n(t, \bar{x}^n(t), \bar{u}(t))dt + \sigma^n(t, \bar{x}^n(t))d\widetilde{W}_t, \\ \bar{x}^n(0) = \xi, \end{cases} \tag{4.63}$$

and let $\Phi^n(s,t)$, $(s \geq t)$ be the fundamental solution of the linear equation

$$\begin{cases} d\bar{\Phi}_n(s,t) = b_x^n(s, \bar{x}^n(s), \bar{u}(s))\bar{\Phi}_n(s,t)ds + \sum_{j=1}^d \sigma_x^{j,n}(s, \bar{x}^n(s))\bar{\Phi}_n(s,t)d\widetilde{W}_s^j, \\ \bar{\Phi}_n(t,t) = I_d, \end{cases}$$

The adjoint process $\bar{p}^n$ associated with $(\bar{x}^n, \bar{u})$ is then defined as follows:

$$\begin{aligned} \bar{p}^n(t) = -\widetilde{E}\Big[\int_t^T \bar{\Phi}_n^*(s,t) f_x^n(s, \bar{x}^n(s), \bar{u}(s))ds \\ + \bar{\Phi}_n^*(T,t) h_x^n(\bar{x}^n(T)) \Big| \widetilde{\mathscr{F}}_t\Big], \end{aligned} \tag{4.64}$$

where $\bar{\Phi}_n^*$ is the transpose of the matrix $\bar{\Phi}_n$.

For the processes $(\bar{x}^n, \bar{u}, \bar{p}^n)$, the conclusions of Proposition (4.2) hold:

Corollary 4.1 *There exists an $(\widetilde{\mathscr{F}}_t)$-adapted process $\bar{p}^n$ given by (4.64) and a Lebesgue null set N such that for $t \notin N$,*

$$\widetilde{E}\left[H^n(t, \bar{x}^n(t), \bar{u}(t), \bar{p}^n(t))\right] - \widetilde{E}\left[H^n(t, \bar{x}^n(t), \upsilon, \bar{p}^n(t))\right] \geq -\delta_n^{1/2}, \tag{4.65}$$

for every U-valued $\mathscr{F}_t$-measurable random variable υ.

The next lemma relates the adjoint process p, the fundamental solution Φ, and the associated Hamiltonian given in Theorem 4.4, and $\bar{p}^n$, $\bar{\Phi}_n$ and the corresponding Hamiltonian associated with $(\bar{x}^n, \bar{u})$.

Lemma 4.7 *We have*

$$\lim_{n\to\infty} \widetilde{E}\left[\sup_{t\leq s\leq T} |\bar{\Phi}_n(s,t) - \Phi(s,t)|\right] = 0, \tag{4.66}$$

$$\lim_{n\to\infty} \widetilde{E}\left[\sup_{0\leq t\leq T} |\bar{p}^n(t) - p(t)|\right] = 0, \tag{4.67}$$

$$\lim_{n\to\infty} \widetilde{E}\left[|H^n(t, \bar{x}^n(t), \bar{u}(t), \bar{p}^n(t)) - H(t, \bar{x}(t), \bar{u}(t), p(t))|\right] = 0. \tag{4.68}$$

The proof of the lemma follows from the previous results, the Grönwall's and the Burkholder–Davis–Gundy inequalities.

Proof of Theorem 4.4 In view of Corollary 4.1 and Lemma 4.7, we obtain

$$\widetilde{E}\left[H(t,\bar{x}(t),\bar{u}(t),p(t))\right] \geq \widetilde{E}\left[H(t,\bar{x}(t),w,p(t))\right], \quad dt\text{-a.e.} \tag{4.69}$$

for every U-valued $\mathscr{F}_t$-measurable random variable w. Let v be an arbitrarily chosen element of U, and set

$$w = \bar{u}(t)I_{A^c} + vI_A, \quad A \in \mathscr{F}_t.$$

We then get

$$\begin{aligned} H(t,\bar{x}(t),\bar{u}(t),p(t)) &- H(t,\bar{x}(t),w,p(t)) \\ &= \left(H(t,\bar{x}(t),\bar{u}(t),p(t)) - H(t,\bar{x}(t),v,p(t))\right) I_A. \end{aligned}$$

Therefore,

$$\widetilde{E}\left[\left(H(t,\bar{x}(t),\bar{u}(t),p(t)) - H(t,\bar{x}(t),v,p(t))\right) I_A\right] \geq 0, \quad A \in \mathscr{F}_t.$$

Hence,

$$H(t,\bar{x}(t),\bar{u}(t),p(t)) = \min_{v\in U} H(t,\bar{x}(t),v,p(t)), \quad dt\text{-a.e.}, \quad \widetilde{\mathbb{P}}\text{a.s.}$$

■

4.4 Pontryagin's Principle for Control-Dependent Diffusions

In this section, we derive the Pontryagin's principle for systems that are governed by stochastic differential equations (SDEs), where the control enters both the drift and the diffusion coefficients. As mentioned in the previous section, we will show in fact that when controlling the diffusion coefficient to establish a maximum principle for the related system, we need to consider both the first- and second-order terms in the Taylor expansion of the spike variation. This suggests that the associated *stochastic* Hamiltonian system would consist of *two forward-backward* SDEs. Moreover, the stochastic maximum principle would have an additional *quadratic* term of the diffusion coefficient. This stochastic maximum principle was first derived by Peng (1990) and since then it is called Peng's stochastic maximum principle.

Consider the following control system

$$\begin{cases} dx(t) = b(t,x(t),u(t))dt + \sigma(t,x(t),u(t))dW_t, \\ x(0) = x_0. \end{cases} \tag{4.70}$$

where

$$b : [0,T] \times \mathbb{R}^d \times U \longrightarrow \mathbb{R}^d, \quad \sigma : [0,T] \times \mathbb{R}^d \times U \longrightarrow \mathbb{R}^{d\times d}.$$

For convenience, we have chosen to assume σ to be a square matrix. The results below hold even for non-square matrices.

Notice that the main difference between (4.33) and (4.70) is the dependence of diffusion coefficient function σ on the control $u(t)$.

An admissible control u is an $\mathbb{F}$-adapted with values in a separable metric space (U, d_0). We denote the set of all admissible controls by $\mathscr{U}[0, T]$.

The objective of the optimal control problem is to minimize a cost functional over $\mathscr{U}[0, T]$.

$$J(u) = E\left[\int_0^T f(t, x(t), u(t))\,dt + h(x(T))\right], \tag{4.71}$$

where

$$f : [0, T] \times \mathbb{R}^d \times U \longrightarrow \mathbb{R}, \qquad h : \mathbb{R}^d \longrightarrow \mathbb{R}.$$

Any $\bar{u} \in \mathscr{U}[0, T]$ satisfying

$$J(\bar{u}) = \inf_{u \in \mathscr{U}[0,T]} J(u)$$

is called an optimal control. The corresponding state process, i.e., the solution to (4.70), is denoted by $\bar{x}(\cdot) := x^{\bar{u}}(\cdot)$. Our stochastic optimal control problem can be stated as follows.

Problem (SC) Characterize $\bar{u} \in \mathscr{U}[0, T]$ such that

$$\bar{u} \in \arg\min_{u \in \mathscr{U}[0,T]} J(u). \tag{4.72}$$

4.4.1 The Stochastic Maximum Principle

In this section, we will establish the stochastic maximum principle when the coefficients b, σ, f and h are smooth; i.e., satisfy the following assumptions.

- (4-H1) The maps b, σ, f and h are measurable and satisfy the following conditions: There exists a constant $L > 0$ such that

$$\begin{cases} |\varphi(t, x, u) - \varphi(t, \hat{x}, u)| \le L|x - \hat{x}|, & t \in [0, T], x, \hat{x} \in \mathbb{R}^d, u, \in U, \\ |\varphi(t, x, u)| \le C(1 + |x|), & t \in [0, T],\ u \in U. \end{cases}$$

  for $\varphi(t, x, u) \in \{b(t, x, u), \sigma(t, x, u), f(t, x, u), h(x)\}$.
- (4-H2) The maps b, σ, f and h are C^2 in x and are such that

$$|\varphi_x(t,x,u)-\varphi_x(t,\hat{x},u)|+|\varphi_{xx}(t,x,u)-\varphi_{xx}(t,\hat{x},u)|\le L|x-\hat{x}|,$$
$$t\in[0,T], x,\hat{x}\in\mathbb{R}^d, u\in U,$$

for $\varphi(t,x,u)\in\{b(t,x,u),\sigma(t,x,u),f(t,x,u),h(x)\}$.

Under these assumptions, for each $u\in\mathscr{U}[0,T]$, the SDE (4.70) admits a unique strong solution. Moreover, for every $p\ge 0$,

$$E\left[\sup_{0\le t\le T}|x(t)|^p\right]\le C_p, \tag{4.73}$$

where C_p is a positive constant depending only on p and the Lipschitz constant L.

As in the previous sections, we will use the following notation. For $\varphi\in\{b,\sigma,f,h\}$ and $v\in U$, we define

$$\begin{cases}\delta\varphi(t):=\varphi(t,\bar{x}(t),v)-\varphi(t,\bar{x}(t),\bar{u}(t));\\ \varphi_x(t):=\varphi_x(t,\bar{x}(t),\bar{u}(t)), \quad \varphi_{xx}(t):=\varphi_{xx}(t,\bar{x}(t),\bar{u}(t)),\\ \mathscr{L}_t(\varphi,y):=\frac{1}{2}\varphi_{xx}(t)y^2.\end{cases} \tag{4.74}$$

The Hamiltonian associated with the control system is

$$H(t,x,u,p,q):=b(t,x,u)\cdot p+\operatorname{tr}(\sigma(t,x,u)q)-f(t,x,u), \tag{4.75}$$

and denote

$$\delta H(t):=\delta b(t)\cdot p(t)+\operatorname{tr}(\delta\sigma(t)q(t))-\delta f(t) \tag{4.76}$$

Let

$$\begin{aligned}H_x(t)&=b_x^*(t)p(t)+\textstyle\sum_{j=1}^d\sigma_x^j(t)^*q_j(t)-f_x(t),\\ H_{xx}(t)&=b_{xx}^*(t)p+\textstyle\sum_{j=1}^d\sigma_{xx}^j(t)^*q_j(t)-f_{xx}(t).\end{aligned} \tag{4.77}$$

4.4.1.1 The Adjoint Equations

In this subsection, we introduce the adjoint equations involved in the stochastic maximum principle for our control problem.

The first-order adjoint equation is the following linear backward SDE

$$\begin{cases}dp(t)=-H_x(t)dt+q(t)dW_t,\\ p(T)=-h_x(T).\end{cases} \tag{4.78}$$

Under the assumptions (4-H1)–(4-H2), Eq. (4.78) admits a unique $\mathbb{F}$-adapted solution (p,q) such that (cf. Lemma A.3)

$$E[\sup_{t\in[0,T]} |p(t)|^2 + \int_0^T |q(t)|^2\,dt] < \infty. \tag{4.79}$$

The second-order adjoint equation is also a linear backward SDE but involves the second derivative of the Hamiltonian, since, as we will see below, we will have to consider the Taylor expansion up to the second-order term.

$$\begin{cases} dP(t) = -\Big(b_x^*(t)P(t) + P(t)b_x(t) + \sum_{j=1}^d \sigma_x^j(t)^* P(t)\sigma_x^j(t) + \sigma_x^j(t)^* Q_j(t) \\ \qquad + \sum_{j=1}^d Q_j(t)\sigma_x^j(t) + H_{xx}(t)\Big)dt + \sum_{j=1}^d Q_j(t)\,dW_t^j, \\ P(T) = -h_{xx}(\bar{x}(T)). \end{cases} \tag{4.80}$$

This is a standard linear backward SDE; by Lemma A.3, it admits a unique $\mathbb{F}$-adapted solution (P, Q) satisfying the following estimate

$$E\Big[\sup_{t\in[0,T]} |P(t)|^2 + \int_0^T |Q(t)|^2\,dt\Big] < \infty. \tag{4.81}$$

Theorem 4.5 (Stochastic Maximum Principle) *Let Assumptions (4-H1)–(4-H2) hold. If $(\bar{x}, \bar{u})$ is an optimal solution of the control problem (4.70)–(4.72), then there are pairs of $\mathbb{F}$-adapted processes (p, q) and (P, Q) that satisfy (4.78)–(4.79) and (4.80)–(4.81), respectively, such that (cf. the notation (4.76))*

$$\delta H(t) + \tfrac{1}{2}\mathrm{tr}\,\{\delta\sigma(t)^* P(t)\delta\sigma(t)\} \le 0, \ \forall v \in U, \ \text{a.e.} t \in [0, T], \ \mathbb{P}\text{a.s.} \tag{4.82}$$

Remark 4.1 Define the so-called $\mathscr{H}$-function associated with the process $(\bar{x}(t), \bar{u}(t),$
$p(t), q(t), P(t))$:

$$\begin{aligned}\mathscr{H}^{(\bar{x}(t),\bar{u}(t))}(t, x, u) := H(t, x, u, p(t), q(t) - P(t)\sigma(t, \bar{x}(t), \bar{u}(t))) \\ +\tfrac{1}{2}\mathrm{tr}\,\{\sigma^*(t, x, u)P(t)\sigma(t, x, u)\}.\end{aligned} \tag{4.83}$$

Then it is easily checked that

$$\begin{aligned}\delta H(t) + \tfrac{1}{2}\mathrm{tr}\,\{\delta\sigma(t)^* P(t)\delta\sigma(t)\} = \mathscr{H}^{(\bar{x}(t),\bar{u}(t))}(t, \bar{x}(t), v) \\ -\mathscr{H}^{(\bar{x}(t),\bar{u}(t))}(t, \bar{x}(t), \bar{u}(t)).\end{aligned}$$

Therefore, the inequality (4.82) is equivalent to

$$\mathscr{H}^{(\bar{x}(t),\bar{u}(t))}(t, \bar{x}(t), \bar{u}(t)) = \max_{v\in U} \mathscr{H}^{(\bar{x}(t),\bar{u}(t))}(t, \bar{x}(t), v), \tag{4.84}$$

a.e.$t \in [0, T]$, $\mathbb{P}$a.s.

For all practical purposes, it would be nice to find or characterize equilibrium points, through only maximizing the Hamiltonian H, which amounts to only solving the first-order adjoint equation (4.78). In fact, we know from the previous section that this happens in the special case where the diffusion coefficient does not contain the control variable, i.e.,

$$\sigma(s, x, u) \equiv \sigma(s, x), \quad (s, x, u) \in [0, T] \times \mathbb{R}^d \times U,$$

whence, manifestly, the inequality (4.82) is equivalent to

$$H(t, \bar{x}(t), \bar{u}(t), p(t), q(t))) = \max_{v \in U} H(t, \bar{x}(t), v, p(t), q(t))), \tag{4.85}$$

a.e.$t \in [0, T]$, $\mathbb{P}$a.s.

Another very useful case is described in the following proposition whose proof follows, e.g., from Lemma 5.1. in Yong and Zhou (1999).

Proposition 4.3 *Let Assumptions (4-H1)–(4-H2) hold. Furthermore, assume that the action space U is a convex body in $\mathbb{R}^k$ and the coefficients b, σ, f and h are locally Lipschitz in v. If $(\bar{x}, \bar{u})$ is optimal, then there is a pair of $\mathbb{F}$-adapted processes (p, q) that satisfies (4.78)–(4.79) and for which*

$$H(t, \bar{x}(t), \bar{u}(t), p(t), q(t)) = \max_{v \in U} H(t, \bar{x}(t), v, p(t), q(t)), \tag{4.86}$$

a.e.$t \in [0, T]$, $\mathbb{P}$a.s.

In the following counterexample, we will show that if the set of controls U is *not convex*, the first-order adjoint equation does not suffice to establish the stochastic maximum principle. One needs the second-order adjoint equation. This means that the stochastic maximum principle remains essential.

4.4.1.2 A Counterexample

The controlled process satisfies the one-dimensional SDE:

$$x^u(t) = x_0 + \int_0^t u(s) dW(s), \quad t \in [0, T]$$

where the control process u takes values in $U = \{a, b\}$, with $a < 0 < b$ and $|a| < |b|$. Consider the performance function to be minimized over $L^2([0, T], U)$

$$J(u) = \frac{1}{2} E[x^2(T)].$$

Obviously, for each admissible control $u \in L^2([0, T], U)$, we have

$$E[\int_0^T u(s)^2 ds] \geq \int_0^T a^2 ds = a^2 T.$$

Therefore,

$$J(u) = \frac{x_0^2}{2} + \frac{1}{2}E[\int_0^T u(s)^2 ds] \geq \frac{x_0^2}{2} + \frac{1}{2}\int_0^T a^2 ds.$$

Hence, $\bar{u}(t) = a$ is an optimal control, and the corresponding path is $\bar{x}(t) = x_0 + aW(t)$.

Now, the associated Hamiltonian being

$$H(x, u, p, q) = qu,$$

the adjoint processes (p, q) satisfy the following BSDE.

$$dp(t) = q(t)dW(t), \;\; p(T) = -h_x(T) = -x_0 - aW(T).$$

It follows that

$$p(t) = E[p(T)|\mathscr{F}_t] = -x_0 - aW(t).$$

This implies that $q(t) = -a > 0$. Now, note that

$$\max_{v \in \{a,b\}} H(v, q(t)) = -ab > -a^2 = q(t)\bar{u}(t) = H(\bar{u}(t), q(t)).$$

Thus, if the stochastic maximum principle (4.5) through the first-order adjoint equation was true, we would have a contradiction.

Now, let us consider the second-order adjoint equation. For our model, it reads

$$dP(t) = Q(t)dW(t), \;\; P(T) = -h_{xx}(T) = -1.$$

We then get

$$P(t) = E[P(T)|\mathscr{F}_t] = -1.$$

According to stochastic maximum principle (4.5)

$$\max_{v \in U} \left(H(\bar{x}(t), u, p(t), q(t)) + \frac{1}{2}P(t)(\sigma(\bar{x}(t), \bar{u}(t)) - \sigma(\bar{x}(t), u))^2 \right)$$

should be equal to $H(\bar{x}(t), \bar{u}(t), p(t), q(t))$ which in our case reads

$$\max_{u\in\{a,b\}}\left(q(t)u+\frac{1}{2}P(t)(u-\bar{u}(t))^2\right)=q(t)\bar{u}(t),$$

Indeed, noting that $q(t)=-\bar{u}(t)=-a$ and $P(t)=-1$, we easily check that

$$\max_{u\in\{a,b\}}\left(-au-\frac{1}{2}(u-a)^2\right)=-a^2.$$

Hence, the stochastic maximum principle (4.5) holds, while the stochastic maximum principle based only on the first-order adjoint equation does not.

Proof of the Stochastic Maximum Principle
Let $(\bar{x},\bar{u})$ be an optimal solution of the control problem (4.70)–(4.72). We shall derive the variational inequality (4.82), in several steps, from the fact that

$$J(u^\eta)-J(\bar{u})\geq 0,$$

where u^η is the spike variation (4.45) of $\bar{u}$. Let $x^\eta(\cdot) := x^{u^\eta}(\cdot)$ denote the corresponding state process which satisfies (4.70).

We will arrive at (4.82) by establishing the following key relation between the performance functional J and the associated Hamiltonian (4.75) (see (4.1)).

Proposition 4.4

$$\begin{aligned}J(u^\eta)-J(\bar{u})=-E\Big[\textstyle\int_0^T\Big(&\mathcal{H}^{(\bar{x}(s),\bar{u}(s))}(s,\bar{x}(s),u(s))\\&-\mathcal{H}^{(\bar{x}(s),\bar{u}(s))}(s,\bar{x}(s),\bar{u}(s))\Big)I_{E_\eta}(s)\,ds\Big]+o(\eta),\end{aligned}\tag{4.87}$$

for arbitrary $u\in\mathcal{U}[0,T]$ *and* $t\in[0,T]$.

We will carry out the proof of Proposition 4.4 and derive the stochastic maximum principle only in the one-dimensional case.

Let y^η and z^η be, respectively, the solutions of the following SDEs:

$$\begin{cases}dy^\eta(t)=\left\{b_x(t)y^\eta(t)+\delta b(t)I_{E_\eta}(t)\right\}dt\\ \qquad\qquad+\left\{\sigma_x(t)y^\eta(t)+\delta\sigma(t)I_{E_\eta}(t)\right\}dW_t,\\ y^\eta(0)=0,\end{cases}\tag{4.88}$$

and

$$\begin{cases}dz^\eta(t)=\left\{b_x(t)z^\eta(t)+\mathcal{L}_t(b,y^\eta(t))+\delta b_x(t)y^\eta(t)I_{E_\eta}(t)\right\}dt\\ \qquad+\left\{\sigma_x(t)z^\eta(t)+\mathcal{L}_t(\sigma,y^\eta(t))+\delta\sigma_x(t)y^\eta(t)I_{E_\eta}(t)\right\}dW_t,\\ z^\eta(0)=0.\end{cases}\tag{4.89}$$

Equation (4.88) (resp., (4.89)) is called the first- (resp., second-) order variational equation.

4.4.2 Duality and Taylor Expansions

The duality relations between p, y^η and z^η displayed in the next lemma, follow immediately from Itô's formula.

Lemma 4.8 *We have*

$$E[p(T)y^\eta(T)] = E\Big[\int_0^T \Big(y^\eta(t) f_x(t) + (p(t)\delta b(t) + q(t)\delta\sigma(t))\, I_{E_\eta}(t)\Big)dt\Big] \tag{4.90}$$

and

$$E[p(T)z^\eta(T)]=E\Big[\int_0^T (z^\eta(t) f_x(t)+(p(t)\delta b_x(t)+q(t)\delta\sigma_x(t)) y^\eta(t) I_{E_\eta}(t)+(p(t)\mathscr{L}(b, y^\eta(t))+q(t)\mathscr{L}(\sigma, y^\eta(t)))\, dt\Big]. \tag{4.91}$$

Recall the equations satisfied by $x^\eta := x^{u^\eta}$, $\bar{x} := \bar{x}^{\bar{u}}$.

$$\begin{cases} dx^\eta(t) = b(t, x^\eta(t), u^\eta(t))dt + \sigma(t, x^\eta(t), u^\eta(t))dW_t, \\ x^\eta(0) = x_0; \end{cases} \tag{4.92}$$

$$\begin{cases} d\bar{x}(t) = b(t, \bar{x}(t), \bar{u}(t))dt + \sigma(t, \bar{x}(t), \bar{u}(t))dW_t, \\ \bar{x}(0) = x_0. \end{cases} \tag{4.93}$$

The estimates displayed in the next result are very important for the proof of Theorem 4.5. If not specified, we will denote by C the appropriate positive constants that appear in the estimates below, which may differ from line to line.

Proposition 4.5 *Assume that conditions (4-H1) and (4-H2) hold. Then, for any $k \geq 1$,*

$$E\left[\sup_{t\in[0,T]} |x^\eta(t) - \bar{x}(t)|^{2k}\right] \leq C\eta^k, \tag{4.94}$$

$$E\left[\sup_{t\in[0,T]} |y^\eta(t)|^{2k}\right] \leq C\eta^k, \tag{4.95}$$

$$E\left[\sup_{t\in[0,T]} |z^\eta(t)|^{2k}\right] \le C\eta^{2k}, \tag{4.96}$$

$$E\left[\sup_{t\in[0,T]} |x^\eta(t) - \bar{x}(t) - y^\eta(t)|^{2k}\right] \le C\eta^{2k}, \tag{4.97}$$

$$E\left[\sup_{t\in[0,T]} |x^\eta(t) - \bar{x}(t) - y^\eta(t) - z^\eta(t)|^{2k}\right] \le C\eta^{2k}\rho_k(\eta), \tag{4.98}$$

where $\rho_k : (0,\infty) \longrightarrow (0,\infty)$ *is such that* $\rho_k(\eta) \to 0$ *as* $\eta \to 0$.

Proof Estimates (4.94), (4.95), (4.96) and (4.97) follow from standard arguments using Grönwall's and the Burkholder–Davis–Gundy inequalities for diffusion processes (see, e.g., Revuz and Yor (2004)). For a proof, see Theorem 4.4 in Yong and Zhou (1999). Estimate (4.98) will derived for a more general model in Sect. 4.5. ∎

Using these estimates, we obtain the following useful relation. Assume that $\phi(t,x,u)$ satisfies Assumptions (4-H1) and (4-H2). Then, following the same argument used to derive (4.29), we have

$$\begin{aligned}\phi(t,x^\eta(t),u^\eta(t)) - \phi(t,\bar{x}(t),\bar{u}(t)) = \delta\phi(t)I_{E_\eta}(t) + \phi_x(t)(y^\eta(t)+z^\eta(t))\\ + \mathscr{L}_t(\phi, y^\eta(t)) + o(\eta),\end{aligned} \tag{4.99}$$

where we recall that $\mathscr{L}_t(\phi, y^\eta(t)) := \frac{1}{2}\phi_{xx}(t,\bar{x}(t),\bar{u}(t))(y^\eta(t))^2$.

Below we will use the following estimates related to the solutions of the adjoint processes (p,q) and (P,Q) given by (4.79) and (4.81), respectively.

Lemma 4.9

$$E\Big[\int_0^T \Big|\Big(p(t)\delta b_x(t) + q(t)\delta\sigma_x(t)\Big)y^\eta(t)I_{E_\eta}(t)\Big|\,dt\Big] \le \eta\rho(\eta),\ \ \eta > 0, \tag{4.100}$$

where $\rho : (0,\infty) \longrightarrow (0,\infty)$ *are such that* $\rho(\eta) \to 0$ *as* $\eta \to 0$.

Proof Set $\|y^\eta\| := \Big(E[\sup_{t\in[0,T]} |y^\eta(t)|^2]\Big)^{1/2}$. In view of (4.79) and (4.95), we have

$$\begin{aligned}E\Big[\int_0^T \Big|q(t)\delta\sigma_x(t)y^\eta(t)I_{E_\eta}(t)\Big|\,dt\Big] &\le C\|y^\eta\|\Big(E\Big[\Big(\textstyle\int_0^T |q(t)|I_{E_\eta}(t)\,dt\Big)^2\Big]\Big)^{1/2}\\ &\le C\eta^{1/2}\Big(E\Big[\textstyle\int_0^T |q(t)|^2 I_{E_\eta}(t)\,dt\Big]\Big)^{1/2}\eta^{1/2}\\ &\le \eta\rho_1(\eta),\end{aligned}$$

where

$$\rho_1(\eta) := C\Big(E\Big[\int_0^T \Big|q(t)I_{E_\eta}(t)\Big|^2 dt\Big]\Big)^{1/2} \longrightarrow 0, \quad \eta \downarrow 0.$$

Similarly,

$$E\Big[\int_0^T \Big|\Big(p(t)\delta b_x(t)y^\eta(t)I_{E_\eta}(t)\Big| dt\Big] \le C\|y^\eta\|\Big(E\Big[\Big(\int_0^T |p(t)|I_{E_\eta}(t)\,dt\Big)^2\Big]\Big)^{1/2} \\ \le \eta\rho_2(\eta),$$

where

$$\rho_2(\eta) := C\Big(E\Big[\int_0^T \Big|p(t)I_{E_\eta}(t)\Big|^2 dt\Big]\Big)^{1/2} \longrightarrow 0, \quad \eta \to 0.$$

Setting, $\rho(\eta) = \rho_1(\eta) + \rho_2(\eta)$, we get the estimate (4.100). ■

Using Lemma 4.9 and the duality relations (4.90)–(4.91), we get

$$\begin{aligned} E[p(T)(y^\eta(T) + z^\eta(T))] = E\Big[\int_0^T (f_x(t)(y^\eta(t) + z^\eta(t)) + (p(t)\delta b(t) \\ +q(t)\delta\sigma(t))I_{E_\eta} + p(t)\mathscr{L}_t(b, y^\eta(t)) + q(t)\mathscr{L}_t(\sigma, y^\eta(t)))\,dt\Big] + o(\eta) \end{aligned} \tag{4.101}$$

Proof of Theorem 4.5 Using (4.99), we have

$$\begin{aligned} 0 \le J(u^\eta) - J(\bar u) &= E\Big[\int_0^T \Big(f(t, x^\eta(t), u^\eta(t)) - f(t, \bar x(t), \bar u(t))\Big)\,dt\Big] \\ &\quad + E[h(x^\eta(T)) - h(\bar x(T))] \\ &= E\Big[\int_0^T \Big(f_x(t)(y^\eta(t) + z^\eta(t))\Big)\,dt\Big] \\ &\quad + E[h_x(T)(y^\eta(T) + z^\eta(T)) + \mathscr{L}_T(h, y^\eta(T))] \\ &\quad + E\Big[\int_0^T \Big(\delta f(t)I_{E_\eta}(t) + \mathscr{L}_t(f, y^\eta(t))\Big)\,dt)\Big] + o(\eta). \end{aligned}$$

Hence, in view of (4.101) and using the notation (4.76) and (4.77), we have

$$\begin{aligned} J(u^\eta) - J(\bar u) = -E[\int_0^T \delta H(t)I_{E_\eta}(t)\,dt\Big] + \frac{1}{2}E[h_{xx}(T)(y^\eta(T))^2] \\ -\frac{1}{2}E\Big[\int_0^T H_{xx}(t)(y^\eta(t))^2\,dt\Big] + o(\eta). \end{aligned}$$

Now, applying Itô's formula to $P(t)(y^\eta(t))^2$ and taking expectation, we get

$$E[P(T)(y^\eta(T))^2] = E\Big[\int_0^T \Big(-H_{xx}(t)(y^\eta(t))^2 + P(t)\delta\sigma(t)^2 I_{E_\eta}\Big)\,dt\Big] + o(\eta).$$

Noting that $P(T) = -h_{xx}(T)$, we obtain

$$J(u^\eta)-J(\bar{u})= - E\Big[\int_0^T \Big(\delta H(t)+\frac{1}{2}P(t)(\delta\sigma(t))^2\Big)I_{E_\eta}\,dt\Big]+ \circ(\eta). \quad (4.102)$$

It follows that

$$\begin{aligned} H(t,\bar{x}(t),v,p(t),q(t)) - H(t,\bar{x}(t),\bar{u}(t),p(t),q(t)) \\ +\tfrac{1}{2}P(t)\Big(\sigma(t,\bar{x}(t),v) - \sigma(t,\bar{x}(t),\bar{u}(t))\Big)^2 \le 0 \\ \forall v \in U, \quad \text{a.e.} t \in [0,T], \ \mathbb{P}\text{a.s.} \end{aligned}$$

■

4.4.3 *Sufficient Optimality Conditions*

Theorem 4.6 *Let Assumptions (4-H1) and (4-H2) hold. Let $(\bar{x},\bar{u})$ be an admissible pair, and let (p,q) be the associated first-order adjoint processes. Assume*

- *(4-H3) The set of controls U is a convex body (i.e., U is convex and has a nonempty interior) of $\mathbb{R}^k$, and the maps b,σ and f are differentiable in u,*
- *(4-H4) The Hamiltonian $H(t,\cdot,\cdot,p(t),q(t))$ is concave in (x,u) for* a.e. $t \in [0,T]$ *almost surely,*
- *(4-H5) The function h is convex.*

Then the pair $(\bar{x},\bar{u})$ is optimal if

$$H(t,\bar{x}(t),\bar{u}(t),p(t),q(t)) = \max_{v\in U} H(t,\bar{x}(t),v,p(t),q(t)),$$

a.e.$t \in [0,T]$, $\mathbb{P}$a.s.

Proof The proof is similar to the proof of Theorem 4.2. We leave it as an exercise.

■

4.5 Pontryagin's Principle for Stochastic Differential Equations of the Mean-Field Type

We consider, in this section, stochastic control problems for state processes governed by a stochastic differential equation (SDE) of the mean-field type, in the sense that the coefficients of the SDE are allowed to depend on the state of the process as well as its expected value. More specifically, the SDE is defined as

$$\begin{cases} dx(t) = b(t, x(t), E[x(t)], u(t))dt + \sigma(t, x(t), E[x(t)], u(t))dW_t, \\ x(0) = x_0. \end{cases} \tag{4.103}$$

for some functions b and σ, and a Brownian motion W. For every t, the control u_t is allowed to take values in some action space U. This mean-field SDE is obtained as the mean-square limit, when $n \to \infty$, of a system of interacting particles of the form

$$dx_t^{i,n} = b\left(t, x_t^{i,n}, \frac{1}{n}\sum_{j=1}^{n} x_t^{j,n}, u(t)\right) dt + \sigma\left(t, x_t^{i,n}, \frac{1}{n}\sum_{j=1}^{n} x_t^{j,n}, u(t)\right) dW_t^i,$$

where $(W^i, i \geq 1)$ is a collection of independent Brownian motions (see, e.g., Sznitman (1989) and Jourdain *et al.* (2008) and the references therein).

The objective of the control problem is to minimize a cost functional of the form

$$J(u) = E\left[\int_0^T f(t, x(t), E[x(t)], u(t))\, dt + h(x(T), E[x(T)])\right], \tag{4.104}$$

for given functions f and h. This cost functional is also of mean-field type, as the functions f and h depend on the marginal law of the state process through its expected value.

The fact that J through h may be nonlinear in the expected value makes the control problem time inconsistent in the sense that Bellman's optimality principle, based on applying the law of iterated conditional expectations on the cost functional, does not hold[102].

In this section, we obtain a stochastic maximum principle, specifying the necessary conditions for optimality, for a general action space, using the classical spike perturbation. This stochastic maximum principle differs from the classical one discussed earlier in this chapter in the sense that here the first-order adjoint equation turns out to be a linear mean-field backward SDE, but the second-order adjoint equation remains the same as in the stochastic maximum principle without expected value terms. The result of this section can be extended to include $E[u(t)]$ in the coefficient functions f, b, σ.

Statement of the Problem

Let $T > 0$ be a fixed time horizon and $(\Omega, \mathcal{F}, \mathbb{F}, \mathbb{P})$ be a given filtered probability space, whose filtration $\mathbb{F} = \{\mathcal{F}_s,\ 0 \leq s \leq T\}$ satisfies the usual conditions of right-continuity and completeness, on which a k-dimensional Brownian motion $W = \{W_s\}_{s\geq 0}$ is given. We assume that $\mathbb{F}$ is the natural filtration of W augmented by $\mathbb{P}$-null sets of $\mathcal{F}$.

Consider the stochastic control system:

$$\begin{cases} dx(t) = b(t, x(t), E[x(t)], u(t))dt + \sigma(t, x(t), E[x(t)], u(t))dW_t, \\ x(0) = x_0, \end{cases} \tag{4.105}$$

where

$$b : [0, T] \times \mathbb{R}^d \times \mathbb{R}^d \times U \longrightarrow \mathbb{R}^d, \quad \sigma : [0, T] \times \mathbb{R}^d \times \mathbb{R}^d \times U \longrightarrow \mathbb{R}^{d\times k}.$$

An admissible control u is an $\mathbb{F}$-adapted process with values in a separable metric space (U, d_0) with d_0 being a distance on $U \times U$. We denote the set of all admissible controls by $\mathscr{U}[0, T]$.

The optimal control problem we are concerned with is to minimize the following cost functional over $\mathscr{U}[0, T]$:

$$J(u) = E\left[\int_0^T f(t, x(t), E[x(t)], u(t))\, dt + h(x(T), E[x(T)])\right], \tag{4.106}$$

where

$$f : [0, T] \times \mathbb{R}^d \times \mathbb{R}^d \times U \longrightarrow \mathbb{R}, \qquad h : \mathbb{R}^d \times \mathbb{R}^d \longrightarrow \mathbb{R}.$$

Any $\bar{u} \in \mathscr{U}[0, T]$ satisfying

$$J(\bar{u}) = \inf_{u\in\mathscr{U}[0,T]} J(u) \tag{4.107}$$

is called an optimal control. The corresponding state process, which is the solution to (4.105), is denoted by $\bar{x} := x^{\bar{u}}$.

The following assumptions will be in force throughout this section.

- (5-H1) The maps b, σ, f and h are measurable and satisfy the following conditions: There exists a constant $L > 0$ such that

$$\begin{cases} |\varphi(t, x_1, y_1, u) - \varphi(t, x_2, y_2, u)| \le L(|x_1 - x_2| + |y_1 - y_2|), \\ \qquad t \in [0, T], x_1, x_2, y_1, y_2 \in \mathbb{R}^d, u, \in U, \\ |\varphi(t, \hat{x}, \hat{y}, u)| \le L(1 + |\hat{x}| + |\hat{y}|), \quad t \in [0, T], \hat{x}, \hat{y} \in \mathbb{R}^d, u \in U, \end{cases}$$

  for $\varphi \in \{b, \sigma, f, h\}$.
- (5-H2) The maps b, σ, f and h are C_b^2 in x and y and are such that, for $w = x, y, xx$,

$$|\varphi_w(t, x_1, y_1, u) - \varphi_w(t, x_2, y_2, u)| \le L(|x_1 - x_2| + |y_1 - y_2|),$$

  for all $t \in [0, T], x_1, x_2, y_1, y_2 \in \mathbb{R}^d, u \in U$, where $\varphi \in \{b, \sigma, f, h\}$.

Under assumption (5-H2), for each $u \in \mathscr{U}[0, T]$, the SDE (4.105) admits a unique strong solution (see, e.g., Sznitman (1989) and Jourdain *et al.* (2008)).

For convenience, we recall the following notation. For $\varphi \in \{b, \sigma, f, h\}$, we define

$$\begin{cases} \delta\varphi(t) = \varphi(t, \bar{x}(t), E[\bar{x}(t)], v) - \varphi(t, \bar{x}(t), E[\bar{x}(t)], \bar{u}(t)), \\ \varphi_w(t) := \varphi_w(t, \bar{x}(t), E[\bar{x}(t)], \bar{u}(t)), \quad w \in \{x, y, xx, yy, xy\}, \\ \mathscr{L}_t(\varphi, y) := \frac{1}{2}\varphi_{xx}(t) y^2. \end{cases} \tag{4.108}$$

Let us introduce the Hamiltonian associated with the integrable random variable X:

$$\begin{aligned} H(t, X, u, p, q) := {} & b(t, X, E[X], u) \cdot p + \operatorname{tr}(\sigma(t, X, E[X], u) q) \\ & - f(t, X, E[X], u), \end{aligned} \tag{4.109}$$

and let

$$\delta H(t) := p(t) \cdot \delta b(t) + \operatorname{tr}(q(t) \delta\sigma(t)) - \delta f(t) \tag{4.110}$$

and note that

$$H_w(t) = b_w^*(t) p + \textstyle\sum_{j=1}^d \sigma_w^j(t)^* q_j - f_w(t), \qquad w \in \{x, y, xx\}. \tag{4.111}$$

4.5.1 The Adjoint Equations and the Stochastic Maximum Principle

In this subsection, we introduce the adjoint equations involved in the stochastic maximum principle for the stochastic control problem formulated above.

The first-order adjoint equation is the following linear backward SDE of mean-field type

$$\begin{cases} dp(t) = -\{H_x(t) + E\left[H_y(t)\right]\} + q(t) dW_t, \\ p(T) = -h_x(T) - E[h_y(T)]. \end{cases} \tag{4.112}$$

This equation reduces to the standard one, when the coefficients do not explicitly depend on the expected value of the underlying diffusion process. Thanks to Theorem 3.1. in Buckdahn *et al.* (2009) (cf. Lemma A.3), under assumptions (5-H1) and (5-H2), Eq. (4.112) admits a unique $\mathbb{F}$-adapted solution (p, q) such that

$$E[\sup_{t \in [0,T]} |p(t)|^2 + \int_0^T |q(t)|^2 dt] < \infty. \tag{4.113}$$

The second-order adjoint equation is the classical linear backward SDE that appears in the stochastic maximum principle:

$$\begin{cases} dP(t) = -\Big(b_x^T(t)P(t)+P(t)b_x(t)+\sum_{j=1}^d \sigma_x^j(t)^T P(t)\sigma_x^j(t)+\sigma_x^j(t)^T Q_j(t) \\ \qquad +\sum_{j=1}^d Q_j(t)\sigma_x^j(t)+H_{xx}(t)\Big)dt + \sum_{j=1}^d Q_j(t)\, dW_t^j, \\ P(T) = -h_{xx}(\bar{x}(T)). \end{cases} \tag{4.114}$$

This is a standard linear backward SDE, whose unique $\mathbb{F}$-adapted solution (P, Q) satisfies the following estimate

$$E\Big[\sup_{t\in[0,T]} |P(t)|^2 + \int_0^T |Q(t)|^2\, dt\Big] < \infty. \tag{4.115}$$

The following theorem is main the result of the section.

Theorem 4.7 (Stochastic Maximum Principle) *Let assumptions (5-H1) and (5-H2) hold. If $(\bar{x}, \bar{u})$ is an optimal solution of the control problem (4.105)–(4.107), then there are uniques pairs of $\mathbb{F}$-adapted processes (p, q) and (P, Q) satisfying (4.112)–(4.113) and (4.114)–(4.115), respectively, such that (cf. the notation (4.110))*

$$\delta H(t)+\tfrac{1}{2}\mathrm{tr}\,\{\delta\sigma(t)^T P(t)\delta\sigma(t)\} \le 0,\ \ \forall v \in U,\ \ \text{a.e.}t \in [0,T],\ \ \mathbb{P}\text{a.s.} \tag{4.116}$$

Proof of the Stochastic Maximum Principle of the Mean-Field Type
Let $(\bar{x}, \bar{u})$ be an optimal solution of the control problem (4.105)–(4.107). We derive the variational inequality (4.116) in several steps from the fact that

$$J(u^\eta) - J(\bar{u}) \ge 0,$$

where u^η is the spike variation of $\bar{u}$. Let $x^\eta := x^{u^\eta}$ denote the corresponding state process which satisfies (4.105).

We will arrive at (4.116), by establishing the following key relation between the performance functional (4.104) and the associated Hamiltonian (4.109):

$$J(u^\eta) - J(\bar{u}) = -E\left[\int_0^T \left(\delta H(s) + \tfrac{1}{2}P(s)(\delta\sigma(s))^2\right) I_{E_\eta}(s)\, ds\right] + R(\eta),$$

for arbitrary $u \in \mathscr{U}[0,T]$ and $t \in [0,T]$, where

$$|R(\eta)| \le \eta\bar{\rho}(\eta),$$

for some function $\bar{\rho} : (0,\infty) \to (0,\infty)$ such that $\bar{\rho}(\eta) \to 0$ as $\eta \to 0$.

Let y^η and z^η be, respectively, the solutions of the following SDEs:

$$\begin{cases} dy^\eta(t) = \big\{b_x(t)y^\eta(t) + b_y(t)E\left[y^\eta(t)\right] + \delta b(t)I_{E_\eta}(t)\big\}\, dt \\ \qquad + \big\{\sigma_x(t)y^\eta(t) + \sigma_y(t)E\left[y^\eta(t)\right] + \delta\sigma(t)I_{E_\eta}(t)\big\}\, dW_t, \\ y^\eta(0) = 0, \end{cases} \tag{4.117}$$

and

$$\begin{cases} dz^\eta(t) = \{b_x(t)z^\eta(t) + b_y(t)E\left[z^\eta(t)\right] + \delta b_x(t)y^\eta(t)I_{E_\eta}(t) \\ \qquad\qquad + \mathscr{L}_t(b, y^\eta)\}\, dt + \{\sigma_x(t)z^\eta(t) + \sigma_y(t)E\left[z^\eta(t)\right] \\ \qquad\qquad\qquad\qquad + \mathscr{L}_t(\sigma, y^\eta) + \delta\sigma_x(t)y^\eta(t)I_{E_\eta}(t)\}\, dW_t, \\ z^\eta(0) = 0. \end{cases} \tag{4.118}$$

Equation (4.117) (resp. (4.118)) is called the first- (resp., second-) order variational equation.

4.5.2 Duality

The duality relations between p y^η and z^η displayed in the next lemma follow immediately from Itô's formula.

Lemma 4.10 *We have*

$$\begin{aligned} E\left[p(T)y^\eta(T)\right] = E&\left[\int_0^T y^\eta(t)\left(f_x(t) + E[f_y(t)]\right) dt\right] \\ &+ E\left[\int_0^T \left(p(t)\delta b(t) + q(t)\delta\sigma(t)\right) I_{E_\eta}(t) dt\right] \end{aligned} \tag{4.119}$$

and

$$\begin{aligned} E\left[p(T)z^\eta(T)\right] = E&\left[\int_0^T z^\eta(t)(f_x(t) + E[f_y(t)])dt\right] \\ +E&\left[\int_0^T \left(p(t)\delta b_x(t) + q(t)\delta\sigma_x(t)\right) y^\eta(t)I_{E_\eta}(t)dt\right] \\ +E&\left[\int_0^T \left(p(t)\mathscr{L}_t(b, y^\eta(t)) + q(t)\mathscr{L}_t(\sigma, y^\eta(t))\right) dt\right]. \end{aligned} \tag{4.120}$$

4.5.3 Taylor Expansions

Recall the equations satisfied by $x^\eta := x^{u^\eta}$ and $\bar{x} := \bar{x}^{\bar{u}}$:

$$\begin{cases} dx^\eta(t) = b(t, x^\eta(t), E\left[x^\eta(t)\right], u_t^\eta)dt + \sigma(t, x^\eta(t), E\left[x^\eta(t)\right], u_t^\eta)dW_t, \\ x^\eta(0) = x_0; \end{cases}$$

$$\begin{cases} d\bar{x}(t) = b(t, \bar{x}(t), E\left[\bar{x}(t)\right], \bar{u}_t)dt + \sigma(t, \bar{x}(t), E\left[\bar{x}(t)\right], \bar{u}_t)dW_t, \\ \bar{x}(0) = x_0. \end{cases}$$

The estimates displayed in the next result are very important for the proof of Theorem 4.7. If not specified, we will denote by C the positive constants that appear in the estimates below, which may differ from line to line. The estimates displayed in Proposition (4.5) constitute a special case.

Proposition 4.6 *Assume that conditions (5-H1) and (5-H2) hold. Then, for any* $k \geq 1$,

$$E\left[\sup_{t\in[0,T]} |x^\eta(t) - \bar{x}(t)|^{2k}\right] \leq C\eta^k, \tag{4.121}$$

$$E\left[\sup_{t\in[0,T]} |y^\eta(t)|^{2k}\right] \leq C\eta^k, \tag{4.122}$$

$$\sup_{t\in[0,T]} |E\left[y^\eta(t)\right]|^2 \leq \eta\rho(\eta), \quad \eta > 0, \tag{4.123}$$

for some function $\rho : (0, \infty) \to (0, \infty)$ *such that* $\rho(\eta) \to 0$ *as* $\eta \to 0$.

$$E\left[\sup_{t\in[0,T]} |z^\eta(t)|^{2k}\right] \leq C\eta^{2k}, \tag{4.124}$$

$$E\left[\sup_{t\in[0,T]} |x^\eta(t) - \left(\bar{x}(t) + y^\eta(t)\right)|^{2k}\right] \leq C\eta^{2k}, \tag{4.125}$$

$$E\left[\sup_{t\in[0,T]} |x^\eta(t) - \left(\bar{x}(t) + y^\eta(t) + z^\eta(t)\right)|^{2k}\right] \leq C\eta^{2k}\rho_k(\eta), \tag{4.126}$$

where $\rho_k : (0, \infty) \to (0, \infty)$ *is such that* $\rho_k(\eta) \to 0$ *as* $\eta \to 0$.

Estimate (4.123) in Proposition 4.6 will be derived as a consequence of the following result.

Lemma 4.11 *For any progressively measurable process* $(\Phi(t))_{t\in[0,T]}$ *for which, for all* $p \geq 1$, *there exists a positive constant* C_p, *such that*

$$E[\sup_{t\in[0,T]} |\Phi(t)|^p] \leq C_p, \tag{4.127}$$

there exists a function $\tilde{\rho} : (0, \infty) \to (0, \infty)$ *with* $\tilde{\rho}(\eta) \to 0$ *as* $\eta \to 0$, *such that*

$$|E[\Phi(T)y^\eta(T)]|^2 + \int_0^T |E[\Phi(s)y^\eta(s)]|^2\, ds \leq \eta\tilde{\rho}(\eta), \quad \eta > 0. \tag{4.128}$$

Proof Recall the dynamics (4.117) of y^η and set, for $t \in [0, T]$,

$$\eta_t := \exp\{-\int_0^t \sigma_x(s)dW_s - \int_0^t (b_x(s) - \frac{1}{2}|\sigma_x(s)|^2)ds\} \tag{4.129}$$

and

$$\rho_t := \eta_t^{-1} := \exp\{\int_0^t \sigma_x(s)dW_s + \int_0^t (b_x(s) - \frac{1}{2}|\sigma_x(s)|^2)ds\}. \tag{4.130}$$

Applying Itô's formula to $\eta_t y^\eta(t)$, we get

$$\begin{aligned}\eta_t y^\eta(t) = \textstyle\int_0^t \eta_s\{&(b_y(s) - \sigma_x(s)\sigma_y(s))E[y^\eta(s)] \\ &+(\delta b(s) - \sigma_x(s)\delta\sigma(s))I_{E_\eta}(s)\}ds \\ &\quad + \textstyle\int_0^t \eta_s\{\sigma_y(s)E[y^\eta(s)] + \delta\sigma(s)I_{E_\eta}(s)\}dW_s.\end{aligned}$$

Since b_x, σ_x are bounded, for all $p \geq 1$, there exists a positive constant C_p, such that $E[\sup_{t\in[0,T]}|\rho_t|^p] \leq \tilde{C}_p$. Moreover, in view of (4.127), since

$$E[\sup_{t\in[0,T]}(|\Phi(t)|^p + |\rho_t|^p)] \leq C_p,$$

it also follows that

$$E[\sup_{t\in[0,T]}|\Phi(t)\rho_t|^p] \leq C_p.$$

From the Itô Representation Theorem (since $\mathbb{F} = \mathbb{F}^W$), for every $t \in [0, T]$, there exists a unique $\mathbb{F}$-adapted and square-integrable (over $[0, t]$) process $\gamma_{\cdot,t}$ such that

$$\Phi(t)\rho_t = E[\Phi(t)\rho_t] + \int_0^t \gamma_{s,t}dW_s, \qquad \mathbb{P}\text{a.s.}$$

We note that, for every $p \geq 1$, there exists another constant (renamed C_p) independent of t such that

$$\sup_{t\in[0,T]} E\left[\left(\int_0^t |\gamma_{s,t}|^2\, ds\right)^{p/2}\right] \leq C_p. \tag{4.131}$$

Indeed, since $\gamma_{\cdot,t}$ is square-integrable, by the concavity of the function $x \to \sqrt{x}$, (4.131) holds true for $p = 1$. For $p > 1$, using the Burkholder–Davis–Gundy and Doob's Inequalities, we get

$$
\begin{aligned}
E\left[\left(\int_0^t |\gamma_{s,t}|^2\, ds\right)^{p/2}\right] &\leq C_p E\left[\sup_{s\in[0,t]} \left|\int_0^s \gamma_{r,t}\, dW_r\right|^p\right]\\
&\leq C_p\left(\tfrac{p}{p-1}\right)^p E\left[\left|\int_0^t \gamma_{s,t}\, dW_s\right|^p\right]\\
&\leq C_p E[|\Phi(t)\rho_t - E[\Phi(t)\rho_t]|^p]\\
&\leq C_p E[|\Phi(t)\rho_t|^p] + |E[\Phi(t)\rho_t]|^p\\
&\leq C_p E\left[|\Phi(t)\rho_t|^p\right]\\
&\leq C_p E\left[\sup_{t\in[0,T]} |\Phi(t)\rho_t|^p\right] \leq C_p.
\end{aligned}
$$

whence the inequality (4.131).

Consider

$$
\Phi(t) y_t^\eta = J_1^\eta(t) + J_2^\eta(t),
$$

where

$$
\begin{aligned}
J_1^\eta(t) = \Phi(t)\rho_t \int_0^t \eta_s\{(b_y(s) - \sigma_x(s)\sigma_y(s))E[y^\eta(s)]\\
+(\delta b(s) - \sigma_x(s)\delta\sigma(s))I_{E_\eta}(s)\}ds,
\end{aligned}
$$

$$
J_2^\eta(t) = \Phi(t)\rho_t \int_0^t \eta_s\{\sigma_y(s)E[y^\eta(s)] + \delta\sigma(s)I_{E_\eta}(s)\}dW_s.
$$

The following estimate of $E[J_1^\eta(t)]$ holds.

$$
\left|E[J_1^\eta(t)]\right|^2 \leq C\left(\eta^2 + \int_0^t |E[y^\eta(s)]|^2\, ds\right), \qquad t \in [0, T]. \tag{4.132}
$$

Indeed, in view of (5-H2) and (4.131), we have

$$
\begin{aligned}
\left|E[J_1^\eta(t)]\right| &= \Big|\Phi(t)\rho_t \int_0^t \eta_s\{(b_y(s) - \sigma_x(s)\sigma_y(s))E[y^\eta(s)]\\
&\quad +(\delta b(s) - \sigma_x(s)\delta\sigma(s))I_{E_\eta}(s)\}ds\Big|,\\
&\leq CE\left[\sup_{t\in[0,T]} |\Phi(t)\rho_t| \sup_{t\in[0,T]} |\eta_t| \left(\int_0^t \left|E[y^\eta(s)]\right| ds + \eta\right)\right]\\
&\leq C\left(E\left[\sup_{t\in[0,T]} |\Phi(t)\rho_t|^2\right]\right)^{1/2}\left(E[\sup_{t\in[0,T]} |\eta_t|^2]\right)^{1/2}\\
&\quad \left(\int_0^t |E[y^\eta(s)]|\, ds + \eta\right)\\
&\leq C\eta + C\int_0^t |E[y^\eta(s)]|\, ds,
\end{aligned}
$$

which yields (4.132).

To estimate $E[J_2^\eta(t)]$, we note that

$$
\begin{aligned}
E[J_2^\eta(t)] &= E\left[(\Phi(t)\rho_t)\int_0^t \eta_s\{\sigma_y(s)E[y^\eta(s)] + \delta\sigma(s)I_{E_\eta}(s)\}dW_s\right]\\
&= E\left[\int_0^t \gamma_{s,t}\eta_s\sigma_y(s)E[y^\eta(s)]\, ds\right] + E\left[\int_0^t \gamma_{s,t}\eta_s\delta\sigma(s)I_{E_\eta}(s)\}\, ds\right]\\
&:= K_1^\eta(t) + K_2^\eta(t).
\end{aligned}
$$

Now, for every $t \in [0, T]$,

$$\begin{aligned} |K_1^\eta(t)|^2 &\le CE\left[\sup_{t\in[0,T]} |\eta_t|^2] \int_0^t |\gamma_{s,t}|^2 \, ds\right] \int_0^t |E[y^\eta(s)]|^2 \, ds \\ &\le C \int_0^t |E[y^\eta(s)]|^2 \, ds. \end{aligned} \tag{4.133}$$

Furthermore,

$$\begin{aligned} |K_2^\eta(t)| &\le CE\left[\sup_{s\in[0,T]} |\eta_s| \int_0^t |\gamma_{s,t}| I_{E_\eta}(s) \, ds\right] \\ &\le C\sqrt{\eta} E\left[\sup_{s\in[0,T]} |\eta_s| \left(\int_0^t |\gamma_{s,t}|^2 I_{E_\eta}(s) \, ds\right)^{1/2}\right]. \end{aligned} \tag{4.134}$$

Therefore,

$$\int_0^T |K_2^\eta(t)|^2 \, dt \le C\eta\Big(E\Big[\Big(\int_0^T \int_0^t |\gamma_{s,t}|^2 I_{E_\eta}(s) \, dsdt\Big)^2\Big]\Big)^{1/2}. \tag{4.135}$$

Setting

$$\rho_1(\eta) := C\Big(E\Big[\Big(\int_0^T \int_0^t |\gamma_{s,t}|^2 I_{E_\eta}(s) \, dsdt\Big)^2\Big]\Big)^{1/2}, \tag{4.136}$$

we note that, since

$$E\Big[\Big(\int_0^T \int_0^t |\gamma_{s,t}|^2 I_{E_\eta}(s) \, dsdt\Big)^2\Big] \le C_T \sup_{t\in[0,T]} E\Big[\Big(\int_0^t |\gamma_{s,t}|^2 \, ds\Big)^2\Big] < \infty,$$

and $I_{E_\eta} \to 0$ in measure $dt \otimes d\mathbb{P}$, by the dominated convergence theorem,

$$\rho_1(\eta) \to 0, \quad \epsilon \to 0.$$

Hence,

$$\int_0^T |K_2^\eta(t)|^2 \, dt \le \eta\rho_1(\eta), \tag{4.137}$$

with $\rho_1(\eta) \to 0$ as $\eta \downarrow 0$.

In view of (4.132), (4.133) and (4.137), we have

$$\begin{aligned} |E[\Phi(t)y^\eta(t)]|^2 &\le 2|E[J_1^\eta(t)]|^2 + 2|E[J_2^\eta(t)]|^2 \\ &\le C\Big(\eta^2 + |K_2^\eta(t)|^2 + \int_0^t |E[y^\eta(s)]|^2 \, ds\Big), \quad t \in [0, T], \end{aligned} \tag{4.138}$$

which yields

$$\int_0^t |E[\Phi(s)y^\eta(s)]|^2\,ds \le C\Big(\eta^2+\eta\rho_1(\eta)+\int_0^t\int_0^s |E[y^\eta(r)]|^2\,dr\,ds\Big), \quad (4.139)$$

where $C=C(\Phi)$ depends on the semimartingale Φ.

Now, taking $\Phi=1$ and $C_1:=C(1)$ in (4.139), we get, for every $t\in[0,T]$,

$$\int_0^t |E[y^\eta(s)]|^2\,ds \le C_1\left(\eta^2+\eta\rho_1(\eta)+\int_0^t\int_0^s |E[y^\eta(r)]|^2\,dr]\,ds\right) \quad (4.140)$$

which, by Grönwall's Lemma, yields

$$\int_0^t |E[y^\eta(s)]|^2\,ds \le C_1\left(\eta^2+\eta\rho_1(\eta)\right),\ \eta>0. \quad (4.141)$$

Consequently, from (4.139), we get

$$\int_0^t |E[\Phi(s)y^\eta(s)]|^2\,ds \le C\Big(\eta^2+\eta\rho_1(\eta)\Big), \quad \eta>0,\quad t\in[0,T]. \quad (4.142)$$

Setting $t=T$ in the estimate (4.134), we get

$$\begin{aligned}|K_2^\eta(T)|^2 &\le C\eta E\left[\sup_{s\in[0,T]}|\eta_s|^2\int_0^T|\gamma_{s,T}|^2 I_{E_\eta}(s)\,ds\right]\\ &\le C\eta\Big(E\Big[\Big(\int_0^T|\gamma_{s,T}|^2 I_{E_\eta}(s)\,ds\Big)^2\Big]\Big)^{1/2} := \eta\rho_T(\eta),\ \eta>0,\end{aligned}$$

where, by the dominated convergence theorem,

$$\rho_T(\eta)=C\Big(E\Big[\Big(\int_0^T|\gamma_{s,T}|^2 I_{E_\eta}(s)\,ds\Big)^2\Big]\Big)^{1/2}\to 0,\quad \eta\to 0.$$

Hence, in view of (4.139) and (4.141), we have

$$|E[\Phi(T)y_T^\eta]|^2 \le C\left(\eta^2+\eta\rho_T(\eta)+\eta\rho(\eta)\right),\quad \eta>0. \quad (4.143)$$

Finally, setting

$$\tilde\rho(\eta):=C\left(\eta+\rho_T(\eta)+\rho(\eta)\right),\quad \eta>0.$$

and summing (4.142) and (4.143), we obtain the estimate (4.128). ■

Proof of Proposition 4.2 Estimates (4.121), (4.122), (4.124) and (4.125) follow from standard arguments using Grönwall's inequality and the moments inequalities for diffusion processes (see, e.g., Revuz and Yor (2004)). ■

Proof of Estimate (4.123) Since

$$E[y^\eta(t)] = \int_0^t \left(E[b_x(s)y^\eta(s)] + E[b_y(s)]E\left[y^\eta(s)\right] + E[\delta b(s)I_{E_\eta}(s)]\right) ds,$$

in view of Proposition 4.11, setting $\Phi(t) = b_x(t)$, we obtain

$$\left|\int_0^t E[b_x(s)y^\eta(s)]\, ds\right|^2 \le t\int_0^t |E[b_x(s)y^\eta(s)]|^2\, ds \le T\eta\tilde{\rho}(\eta), \qquad t \in [0, T].$$

Thus, in view of (5-H1) and (5-H2), it holds that

$$|E[y^\eta(t)]| \le (T\eta\tilde{\rho}(\eta))^{1/2} + C\int_0^t (\eta + |E[y^\eta(s)]|)\, ds, \qquad t \in [0, T].$$

From Grönwall's Lemma, we finally get (4.123), with $\rho(\eta) = C(\tilde{\rho}(\eta) + \eta), \quad \eta > 0$. ∎

Proof of Estimate (4.126) Set

$$\eta_t^\eta := x^\eta(t) - (\bar{x}(t) + y^\eta(t) + z^\eta(t)).$$

We have

$$d\eta_t^\eta = \alpha_t^\eta(b)dt + \alpha_t^\eta(\sigma)dW_t, \tag{4.144}$$

where, for $\phi \in \{b, \sigma\}$,

$$\begin{aligned}\alpha_t^\eta(\phi) := {}&\phi(t, x^\eta(t), E[x^\eta(t)], u_t^\eta) - \phi(t, \bar{x}(t), E[\bar{x}(t)], \bar{u}_t)\\ &-\phi_x(t)(y^\eta(t) + z^\eta(t)) - b_y(t)E[y^\eta(t) - z^\eta(t)]\\ &-\mathscr{L}_t(\phi, y^\eta) - \delta b_x(t)y^\eta(t)I_{E_\eta}(s).\end{aligned}$$

∎

Estimates of $\alpha^\eta(\phi)$ We have, for each $t \in [0, T]$,

$$\begin{aligned}\phi(t, x^\eta(t), E[x^\eta(t)], u_t^\eta) - \phi(t, \bar{x}(t), E[\bar{x}(t)], u_t^\eta) = {}&\textstyle\int_0^1 d\theta\, \{\phi_x^\theta(t)(x^\eta(t) - \bar{x}(t))\\ &+\phi_y^\theta(t)E[x^\eta(t) - \bar{x}(t)]\},\end{aligned}$$

where, for $j \in \{x, y, xx, xy\}$, and for real λ,

$$\phi_j^\lambda(t) := \phi_j(t, \bar{x}(t) + \lambda(x^\eta(t) - \bar{x}(t)), E[\bar{x}(t) + \lambda(x^\eta(t) - \bar{x}(t))], u_t^\eta).$$

Moreover,

$$\begin{aligned}&\phi(t,x^\eta(t),E[x^\eta(t)],u_t^\eta)-\{\phi(t,\bar{x}(t),E[\bar{x}(t)],\bar{u}_t)+\phi_x(t)(y^\eta(t)+z^\eta(t))\}\\&\quad+b_y(t)E[y^\eta(t)+z^\eta(t)]=\int_0^1 d\theta\,\{\phi_x^\theta(t)\eta_t^\eta+\phi_y^\theta(t)E[\eta_t^\eta]\\&\qquad+(\phi_x^\theta(t)-\phi_x(t))(y^\eta(t)+z^\eta(t))+(\phi_y^\theta(t)-\phi_y(t))E[y^\eta(t)+z^\eta(t)]\}.\end{aligned}$$

Similarly,

$$\begin{aligned}\phi_x^\theta(t)-\phi_x(t)&=\theta\int_0^1 d\gamma\,\{\phi_{xx}^{\theta\gamma}(t)(x^\eta(t)-\bar{x}(t))+\phi_{xy}^{\theta\gamma}(t)E[x^\eta(t)-\bar{x}(t)]\}\\&\quad+\delta\phi_x(t)I_{E_\eta}(t)=\theta\int_0^1 d\gamma\,\{\phi_{xx}^{\theta\gamma}(t)\eta_t^\eta+\phi_{xy}^{\theta\gamma}(t)E[\eta_t^\eta]\}\\&\quad+\theta\int_0^1 d\gamma\,\{\phi_{xx}^{\theta\gamma}(t)(y^\eta(t)+z^\eta(t))\}\\&\quad+\theta\int_0^1 d\gamma\,\{\phi_{xy}^{\theta\gamma}(t)E[y^\eta(t)+z^\eta(t)]\}+\delta\phi_x(t)I_{E_\eta}(t),\end{aligned}$$

and

$$\begin{aligned}&\phi_y^\theta(t)-\phi_y(t)=\theta\int_0^1 d\gamma\,\{\phi_{xy}^{\theta\gamma}(t)\eta_t^\eta+\phi_{yy}^{\theta\gamma}(t)E[\eta_t^\eta]\}\\&\quad+\theta\int_0^1 d\gamma\,\{\phi_{xy}^{\theta\gamma}(t)(y^\eta(t)+z^\eta(t))+\phi_{yy}^{\theta\gamma}(t)E[y^\eta(t)+z^\eta(t)]\}+\delta\phi_x(t)I_{E_\eta}(t).\end{aligned}$$

Introduce the following notation:

$$\begin{cases}x(t)^{1,\eta}(\phi)=\int_0^1 d\theta\,\theta\int_0^1 d\gamma\,\{\phi_{xx}^{\theta\gamma}(t)\eta_t^\eta(y^\eta(t)+z^\eta(t))\\\qquad+\phi_{xy}^{\theta\gamma}(t)(y^\eta(t)+z^\eta(t))E[\eta_t^\eta]+\eta_t^\eta E[y^\eta(t)+z^\eta(t)])\}\\\qquad+\int_0^1 d\theta\,\theta\int_0^1 d\gamma\,\{\phi_{yy}^{\theta\gamma}(t)E[\eta_t^\eta]E[y^\eta(t)+z^\eta(t)]\},\\x(t)^{2,\eta}(\phi)=\int_0^1 d\theta\,\theta\int_0^1 d\gamma\,\{\phi_{xx}^{\theta\gamma}(t)((y^\eta(t)+z^\eta(t))-(y^\eta(t))^2)\\\qquad+2\phi_{xy}^{\theta\gamma}(t)(y^\eta(t)+z^\eta(t))E[y^\eta(t)+z^\eta(t)]\}\\\qquad+\int_0^1 d\theta\,\theta\int_0^1 d\gamma\,\{\phi_{yy}^{\theta\gamma}(t)(E[y^\eta(t)+z^\eta(t)])^2\},\\x(t)^{3,\eta}(\phi)=\int_0^1 d\theta\,\theta\int_0^1 d\gamma\,\{(\phi_{xx}^{\theta\gamma}(t)-\phi_{xx}(t))(y^\eta(t))^2\},\\x(t)^{4,\eta}(\phi)=\int_0^1 d\theta\,\theta\int_0^1 d\gamma\,\{\delta\phi_x(t)z^\eta(t)+\delta\phi_y(t)E[y^\eta(t)+z^\eta(t)]\}I_{E_\eta}(t).\end{cases}$$

We have

$$\begin{aligned}\alpha_t^\eta(\phi)&=x(t)^{1,\eta}(\phi)+x(t)^{2,\eta}(\phi)+x(t)^{3,\eta}(\phi)+x(t)^{4,\eta}(\phi)\\&\quad+\int_0^1 d\theta\,\{\phi_x(t)\eta_t^\eta+\phi_y(t)E[\eta_t^\eta]\}.\end{aligned}\tag{4.145}$$

From the definition of η^η, and the estimates (4.124) and (4.125), we have

$$E[\sup_{t\in[0,T]}|\eta_t^\eta|^{2k}]\le C\eta^{2k},\quad k\ge 1,\tag{4.146}$$

which, together with (4.122) and (4.124), yields

$$E[\sup_{t\in[0,T]}|x(t)^{1,\eta}(\phi)|^{2k}]\le C\eta^{3k},\quad \eta>0,\ k\ge 1.\tag{4.147}$$

Again, using (4.122) and (4.124), we have

$$E\Big[\sup_{t\in[0,T]}\Big|(y^\eta(t)+z^\eta(t))^2-(y^\eta(t))^2\Big|^{2k}\Big]\le C\eta^{3k},\quad k\ge 1;$$

Also, by (4.122), (4.123) and (4.124), we have

$$E\Big[\sup_{t\in[0,T]}\Big|(y^\eta(t)+z^\eta(t))E[y^\eta(t)+z^\eta(t)]\Big|^{2k}]\le \eta^{2k}\bar\rho_k(\eta),\ \eta>0,\ \ k\ge 1;$$

where $\bar\rho_k(\eta):=\eta^k+\eta^{2k}+(C\eta^k+1)\rho^k(\eta)$, which satisfies $\bar\rho_k(\eta)\to 0$ as $\eta\downarrow 0$.

Consequently,

$$\sup_{t\in[0,T]}|E[y^\eta(t)+z^\eta(t)]|^{4k}\le\eta^{2k}\bar\rho_k(\eta),\quad \eta>0,\ \ k\ge 1. \tag{4.148}$$

Hence,

$$E[\sup_{t\in[0,T]}|x(t)^{2,\eta}(\phi)|^{2k}]\le\eta^{2k}\bar\rho_k(\eta),\quad \eta>0,\ \ k\ge 1. \tag{4.149}$$

Using Hölder's inequality, we get

$$\begin{aligned}
&E\Big[\Big(\int_0^T|x(t)^{3,\eta}(\phi)|^2\,dt\Big)^k\Big]\\
&\le E\Big[\Big\{\int_0^T\Big(\int_0^1\int_0^1|\phi_{xx}^{\theta\gamma}(t)-\phi_{xx}(t)||y^\eta(t)|^2\,d\theta\,d\gamma\Big)^2dt\Big\}^k\Big]\\
&\le\Big(E\Big[\int_0^T\int_0^1\int_0^1|\phi_{xx}^{\theta\gamma}(t)-\phi_{xx}(t)|^{4k}\,dt\,d\theta\,d\gamma\Big]E[\sup\nolimits_{t\in[0,T]}|y^\eta(t)|^{8k}]\Big)^{1/2}.
\end{aligned}$$

Therefore, in view of (4.122), it holds that

$$E\Big[\Big(\int_0^T|x(t)^{3,\eta}(\phi)|^2\,dt\Big)^k\Big]\le\eta^{2k}\hat\rho_k(\eta),\quad \eta>0,\ \ k\ge 1. \tag{4.150}$$

where

$$\hat\rho_k(\eta):=C\Big(E\Big[\int_0^T\int_0^1\int_0^1|\phi_{xx}^{\theta\gamma}(t)-\phi_{xx}(t)|^{4k}\,dt\,d\theta\,d\gamma\Big]\Big)^{1/2}\eta^{2k},$$

which satisfies $\hat\rho_k(\eta)\to 0$ as $\eta\to 0$.

We will finally obtain estimates of $X^{4,\eta}(\phi)$. Indeed, in view of (4.122)–(4.124), we have, for every $\eta>0,\ k\ge 1$,

$$E\Big[\Big(\int_{E_\eta}|\delta\phi_x(t)z^\eta(t)|^2\,dt\Big)^k\Big]\le CE[\sup_{t\in[0,T]}|z^\eta(t)|^{2k}]\eta^k\le C\eta^{3k},$$

$$E\Big[\Big(\int_{E_\eta}|\delta\phi_y(t)E[z^\eta(t)]|^2\,dt\Big)^k\Big]\le C\eta^{3k},$$

$$E\Big[\Big(\int_{E_\eta}|\delta\phi_y(t)E[y^\eta(t)]|^2\,dt\Big)^k\Big]\le CE[\sup_{t\in[0,T]}|y^\eta(t)|^{2k}]\eta^k\le C\eta^{3k}\rho(\eta)^k,$$

Therefore,

$$E\Big[\Big(\int_0^T|x(t)^{4,\eta}(\phi)|^2\,dt\Big)^k\Big]\le\eta^{2k}\breve{\rho}_k(\eta),\tag{4.151}$$

where $\breve{\rho}_k(\eta):=C(2\eta^k+\rho^k(\eta))$ which is such that $\breve{\rho}_k(\eta)\to 0$ as $\eta\to 0$.

Finally, by (4.147), (4.149), (4.131), and (4.151), we have

$$E\Big[\Big(\int_0^t|\alpha_t^\eta(\phi)|^2\,dt\Big)^k\Big]\le\eta^{2k}\bar{\rho}_k(\eta)+C\int_0^t E[|\eta_s^\eta|^{2k}]\,ds,\quad t\ge 0.\tag{4.152}$$

where

$$\rho_k(\eta):=C\eta+\bar{\rho}_k(\eta)+\tilde{\rho}(\eta)+\breve{\rho}_k(\eta),$$

which satisfies $\rho_k(\eta)\to 0$ as $\eta\to 0$. Hence, recalling the dynamics (4.144) of η^η and using (4.152), with $\phi=b,\sigma$, Grönwall's inequality yields

$$E\Big[\sup_{t\in[0,T]}|\eta_t^\eta|^{2k}\Big]\le C\eta^{2k}\rho_k(\eta),\quad \eta>0,\ \ k\ge 1,\tag{4.153}$$

which is (4.126). ■

In the next proposition, we collect useful estimates for $\alpha^\eta(\phi)$ given by (4.145) obtained in view of (4.152) and (4.153).

Proposition 4.7 *The following estimates hold*

$$E\Big[\Big(\int_0^t|\alpha_t^\eta(\phi)|^2\,dt\Big)^k\Big]\le C\eta^{2k}\rho_k(\eta),\quad \eta>0,\ \ k\ge 1,\tag{4.154}$$

$$E\Big[|\alpha_T^\eta(\phi)|\Big]\le C\eta\rho(\eta),\quad \eta>0,\tag{4.155}$$

where $\rho(\eta)$ and $\rho_k(\eta)\to 0$ as $\eta\to 0$.

Proof of Theorem 4.7

We first prove the following estimates related to the solutions of the adjoint processes (p,q) and (P,Q) given, respectively, by (4.113) and (4.115).

Lemma 4.12

$$E\Big[\int_0^T \Big|\Big(p(t)\delta b_x(t) + q(t)\delta\sigma_x(t)\Big)y^\eta(t)I_{E_\eta}(t)\Big|\, dt\Big] \le \eta\rho_1(\eta), \qquad (4.156)$$

and

$$E[\int_0^T \Big|\Big(P(t)b_y(t) + P(t)\sigma_x(t)\sigma_y(t)\Big)y^\eta(t)E[y^\eta(t)]\Big|\, dt] \le \eta\rho_2(\eta), \ (4.157)$$

where for $i = 1, 2$, $\rho_i : (0, \infty) \to (0, \infty)$ *are such that* $\rho_i(\eta) \to 0$ *as* $\eta \to 0$.

Proof We only prove (4.156). Estimate (4.157) follows in a similar fashion. Set

$$\|y^\eta\| := \Big(E[\sup_{t\in[0,T]} |y^\eta(t)|^2]\Big)^{1/2}.$$

In view of (4.113) and (4.122), we have

$$\begin{aligned} E\Big[\int_0^T \Big|q(t)\delta\sigma_x(t)y^\eta(t)I_{E_\eta}(t)\Big|\, dt\Big] &\le C\|y^\eta\|\Big(E\Big[\Big(\int_0^T |q(t)|I_{E_\eta}(t)\, dt\Big)^2\Big]\Big)^{1/2} \\ &\le C\eta^{1/2}\Big(E\Big[\int_0^T |q(t)|^2 I_{E_\eta}(t)\, dt\Big]\Big)^{1/2}\eta^{1/2} \\ &\le \eta\rho_3(\eta), \end{aligned}$$

where

$$\rho_3(\eta) := C\Big(E\Big[\int_0^T \Big|q(t)I_{E_\eta}(t)\Big|^2 dt\Big]\Big)^{1/2} \to 0, \quad \eta \to 0.$$

Similarly,

$$\begin{aligned} E\Big[\int_0^T \Big|p(t)\delta b_x(t)y^\eta(t)I_{E_\eta}(t)\Big|\, dt\Big] &\le C\|y^\eta\|\Big(E\Big[\Big(\int_0^T |p(t)|I_{E_\eta}(t)\, dt\Big)^2\Big]\Big)^{1/2} \\ &\le \eta\rho_4(\eta), \end{aligned}$$

where

$$\rho_4(\eta) := C\Big(E\Big[\int_0^T \Big|p(t)I_{E_\eta}(t)\Big|^2 dt\Big]\Big)^{1/2} \to 0, \quad \eta \to 0.$$

Setting, $\rho_1(\eta) = \rho_3(\eta) + \rho_4(\eta)$, we get the estimate (4.156).

Using Lemma 4.12 and the duality relations (4.119)–(4.120), we get

$$
\begin{aligned}
E[p(T)(y^\eta(T)+z^\eta(T))] = & E\Big[\int_0^T(p(t)\delta b(t)+q(t)\delta\sigma(t))I_{E_\eta}(t)\,dt\Big] \\
& +E\Big[\int_0^T(y^\eta(t)+z^\eta(t))(-f_x(t)-E[f_y(t)])\,dt\Big] \\
& +E\Big[\int_0^T(p(t)\mathscr{L}_t(b,y^\eta)+q(t)\mathscr{L}_t(\sigma,y^\eta)\,dt\Big]+\circ(\eta)
\end{aligned}
\tag{4.158}
$$

We will now proceed to the proof of Theorem 4.7. We have

$$
\begin{aligned}
J(u^\eta)-J(\bar u) = & E\Big[\int_0^T\Big(f(t,x^\eta(t),E[x^\eta(t)],u_t^\eta)-f(t,\bar x(t),E[\bar x(t)],\bar u_t\Big)\,dt\Big] \\
& +E[h(x^\eta(T),E[x^\eta(T)])-h(\bar x(T),E[\bar x(T)])] \\
= & E\Big[\int_0^T\Big(f_x(t)(y^\eta(t)+z^\eta(t))+f_y(t)E[y^\eta(t)+z^\eta(t)]\Big)\,dt\Big] \\
& +E[h_x(T)(y^\eta(T)+z^\eta(T))]+E[h_y(T)]E[y^\eta(T)+z^\eta(T)] \\
& +E\Big[\int_0^T\Big(\delta f(t)I_{E_\eta}(t)+\mathscr{L}_t(f,y^\eta)\Big)\,dt)\Big]+\circ(\eta) \\
= & E\Big[\int_0^T\Big(\delta f(t)I_{E_\eta}(t)+\mathscr{L}_t(f,y^\eta)\Big)\,dt\Big] \\
& +E\Big[\int_0^T(y^\eta(t)+z^\eta(t))(f_x(t)+E[f_y(t)])\,dt\Big] \\
& +E\Big[\Big(h_x(T)+E[h_y(T)]\Big)(y^\eta(T)+z^\eta(T))\Big]+\circ(\eta).
\end{aligned}
$$

Hence, in view of (4.158), (4.110) and (4.111), we have

$$
\begin{aligned}
J(u^\eta)-J(\bar u) = & E[\int_0^T-\delta H(t)I_{E_\eta}(t)\,dt\Big]+\tfrac{1}{2}E\Big[h_{xx}(T)(y^\eta(T))^2 \\
& -\int_0^T H_{xx}(t)(y^\eta(t))^2\,dt\Big]+\circ(\eta).
\end{aligned}
$$

Now, applying Itô's formula to $P(t)(y^\eta(t))^2$, noting that $P(T)=-h_{xx}(T)$, and taking expectation, we get

$$
\begin{aligned}
E[P(T)(y^\eta(T))^2] = & E\Big[\int_0^T-H_{xx}(t)(y^\eta(t))^2dt\Big] \\
& +E\Big[\int_0^T P(t)\Big(\sigma_y^2(t)(E[y^\eta(t)])^2+\delta\sigma(t)^2I_{E_\eta}\Big)\,dt\Big] \\
& +2E\Big[\int_0^T\Big(P(t)b_y(t)+P(t)\sigma_x(t)\sigma_y(t)+Q(t)\sigma_y(t)\Big)y^\eta(t)E[y^\eta(t)]\,dt\Big] \\
& +\circ(\eta).
\end{aligned}
$$

In view of the estimates (4.156) and (4.157), we get

$$
E[P(T)(y^\eta(T))^2]=-\int_0^T\Big(P(t)\delta\sigma(t)^2I_{E_\eta}+H_{xx}(t)(y^\eta(t))^2\Big)\,dt+\circ(\eta).
$$

Therefore,

$$
0\le J(u^\eta)-J(\bar u)=-E\Big[\int_0^T\Big(\delta H(t)+\frac{1}{2}P(t)(\delta\sigma(t))^2\Big)I_{E_\eta}\,dt\Big]+\circ(\eta).
$$

It follows that

$$
\begin{aligned}
&H(t, \bar{x}(t), u, p(t), q(t)) - H(t, \bar{x}(t), \bar{u}(t), p(t), q(t)) \\
&\quad + \tfrac{1}{2} P(t)\Big(\sigma(t, \bar{x}(t), E[\bar{x}(t)], u) - \sigma(t, \bar{x}(t), E[\bar{x}(t)], \bar{u}(t))\Big)^2 \leq 0 \\
&\qquad\qquad\qquad \forall u \in U, \quad \text{a.e.} t \in [0, T], \quad \mathbb{P}\text{a.s.}
\end{aligned}
$$

■

4.5.4 Sufficient Optimality Conditions

Theorem 4.8 *Let Assumptions (5-H1) and (5-H2) hold. Let $(\bar{x}, \bar{u})$ be an admissible pair, and let (p, q) be the associated first-order adjoint processes. Assume*

- *(5-H3) The set of controls U is a convex body (i.e., U is convex and has a nonempty interior) of $\mathbb{R}^k$, and the maps b, σ and f are differentiable in u,*
- *(5-H4) The functions b, σ and f are convex in (x, y, u) for* a.e. $t \in [0, T]$ *almost surely,*
- *(5-H5) The function $(x, y) \mapsto h(x, y)$ is convex.*

Then the pair $(\bar{x}, \bar{u})$ is optimal if

$$
H(t, \bar{x}(t), \bar{u}(t), p(t), q(t)) = \max_{v \in U} H(t, \bar{x}(t), v, p(t), q(t))), \tag{4.159}
$$

a.e.$t \in [0, T]$, $\mathbb{P}$a.s. *where, for $v \in U$,*

$$
\begin{aligned}
H(t, \bar{x}(t), v, p(t), q(t)) := {}& b(t, \bar{x}(t), E[\bar{x}(t)], v) \cdot p + \operatorname{tr}(\sigma(t, \bar{x}(t), E[\bar{x}(t)], v) q) \\
& - f(t, \bar{x}(t), E[\bar{x}(t)]), v).
\end{aligned}
$$

Proof Set

$$
\psi(t) := \psi(t, x(t), E[x(t)], u(t)), \quad \bar{\psi}(t) := \psi(t, \bar{x}(t), E[\bar{x}(t)], \bar{u}(t))
$$

for $\psi \in \{b, \sigma, f, h\}$ and

$$
\begin{aligned}
H(t) &= b(t) \cdot p(t) + \operatorname{tr}(\sigma(t) q(t)) - f(t), \\
\bar{H}(t) &= \bar{b}(t) \cdot p(t) + \operatorname{tr}(\bar{\sigma}(t) q(t)) - \bar{f}(t).
\end{aligned}
$$

We want to show that if the pair $(\bar{x}, \bar{u})$ satisfies (4.159), where the adjoint process (p, q) satisfies (4.112), then

$$
J(u) - J(\bar{u}) = E\Big[\int_0^T (f(t) - \bar{f}(t)) dt + h(T) - \bar{h}(T)\Big] \geq 0.
$$

Since H is concave in (x, u), we have

$$H(t) - \bar{H}(t) \le \bar{H}_x(t) \cdot (x(t) - \bar{x}(t)) + \bar{H}_y(t) \cdot (E[x(t)] - E[\bar{x}(t)]) \\ + \bar{H}_u(t) \cdot (u(t) - \bar{u}(t)).$$

Since, in view of (4.159), $\bar{H}_u(t) = 0$ a.e.$t \in [0, T]$, we obtain

$$E[H(t) - \bar{H}(t)] \le E[(\bar{H}_x(t) + E[\bar{H}_y(t)]) \cdot (x(t) - \bar{x}(t))], \tag{4.160}$$

a.e.$t \in [0, T]$. By the convexity of h, we have

$$E[h(T) - \bar{h}(T)] \ge E[(\bar{h}_x(T) + E[\bar{h}_y(T)]) \cdot (x(T) - \bar{x}(T))] \\ = -E[p(T) \cdot (x(T) - \bar{x}(T))].$$

Integrating by parts and using (4.18), we obtain

$$E[p(T) \cdot (x(T) - \bar{x}(T))] = E\Big[\int_0^T (x(t) - \bar{x}(t)) \cdot dp(t) + p(t) \cdot d(x(t) - \bar{x}(t))\Big] \\ = -E\Big[\int_0^T \Big((\bar{H}_x(t) + E[\bar{H}_y(t)]) \cdot (x(t) - \bar{x}(t)) \\ -p(t) \cdot (b(t) - \bar{b}(t)) - \operatorname{tr}((\sigma(t) - \bar{\sigma}(t))q(t))\Big)dt\Big].$$

In view of (4.160), we obtain

$$E[h(T) - \bar{h}(T)] \ge E\Big[\int_0^T \Big(H(t) - \bar{H}(t) - p(t) \cdot (b(t) - \bar{b}(t)) \\ -\operatorname{tr}((\sigma(t) - \bar{\sigma}(t))q(t))\Big)dt\Big].$$

Hence,

$$J(u) - J(\bar{u}) \ge E\Big[\int_0^T \Big(f(t) - \bar{f}(t) - p(t) \cdot (b(t) - \bar{b}(t)) \\ -\operatorname{tr}((\sigma(t) - \bar{\sigma}(t))q(t)) + H(t) - \bar{H}(t)\Big)dt\Big] = 0.$$

■

Mean-Variance Portfolio Selection

In this section, we will illustrate the stochastic maximum principle by solving (in terms of a verification theorem) the optimal mean-variance portfolio problem.

We consider a market with a risky asset and a risk free bank account and denote the prices at time t by S_t^1 and S_t^0, respectively. The price process evolves according to the equations

$$\begin{cases} dS_t^0 = \rho_t S_t^0 dt, \\ dS_t^1 = \alpha_t S_t^1 dt + \sigma_t S_t^1 dW_t, \end{cases}$$

where $\alpha_t, \sigma_t, \rho_t$ are bounded deterministic functions. If $u(t)$ denotes the amount of money invested in the risky asset at time t, we can write down the value $x(t)$ of a self-financing portfolio consisting of the risky and the risk-free assets, as

$$dx(t) = (\rho_t x(t) + (\alpha_t - \rho_t)\, u(t))\, dt + \sigma_t u(t) dW_t, \quad x_0 = x(0). \tag{4.161}$$

A widely discussed version of the mean-variance portfolio problem is to find an admissible control u which minimizes the variance

$$\mathrm{Var}(x(T)) = E[x^2(T)] - (E[x(T)])^2,$$

under the condition that

$$E[x(T)] = a,$$

where a is a given real number. This problem has been formulated and solved using both the dynamic programming principle and the stochastic maximum principle; see Yong and Zhou (1999).

The portfolio selection problem, in which one minimizes the variance while maximizing the expected return, differs from the original one, since the Bellman optimality principle does not hold due to the time inconsistency. It has been solved with different methods such as in Zhou and Li (2000), by embedding it into a stochastic LQ control problem. We will apply the stochastic maximum principle stated above to get a candidate for a feedback optimal control. This control coincides with the optimal control obtained by Zhou and Li (2000) through their sufficient optimality condition (Theorem 4.1 in that paper).

The cost functional to be minimized is given by

$$J(u) = \frac{\gamma}{2}\mathrm{Var}(x(T)) - E[x(T)]. \tag{4.162}$$

By rewriting it as

$$J(u) = E\left[\frac{\gamma}{2}x^2(T) - x(T)\right] - \frac{\gamma}{2}\left(E[x(T)]\right)^2,$$

We first write down the associated Hamiltonian for this system:

$$H(t, x, u, p, q) = (\rho_t x + (\alpha_t - \rho_t)\, u)\, p + \sigma_t u q.$$

The adjoint equation becomes

$$\begin{cases} dp(t) = -\rho_t p(t) dt + q(t) dW_t, \\ p_T \quad = -\gamma(x_T - \mu_T) + 1, \end{cases}$$

where $\mu_t = E[x(t)]$. Looking at the terminal condition of $p(t)$, it is reasonable to try a solution of the form $p(t) = A_t(x(t) - \mu_t) + C_t$, with A_t, C_t deterministic functions such that $A_T = -\gamma, C_T = 1$. Noting that from (4.161), we have

$$d\mu_t = (\rho_t \mu_t + (\alpha_t - \rho_t)E[u(t)])\,dt, \tag{4.163}$$

we may apply Itô's formula on $p(t)$ to get

$$\begin{aligned} dp(t) = \big(&A_t\,(\rho_t\,(x_t - \mu_t) + (\alpha_t - \rho_t)\,(u_t - E[u(t)])) + \dot{A}_t(x(t) - \mu_t) \\ &+\dot{C}_t\big)\,dt + A_t\sigma_t u_t dW_t, \end{aligned} \tag{4.164}$$

where $\dot{A}_t$ and $\dot{C}_t$ denote the derivatives with respect to t. By comparing (4.163) with (4.164), we get

$$\begin{aligned} &A_t\,(\rho_t(x(t) - \mu_t) + (\alpha_t - \rho_t)(u_t - E[u(t)])) + \dot{A}_t(x(t) - \mu_t) + \dot{C}_t \\ &\quad = -\rho_t A_t(x(t) - \mu_t) - \rho_t C_t, \end{aligned} \tag{4.165}$$

$$q(t) = A_t\sigma_t u(t). \tag{4.166}$$

Since H is linear in the control variable u, we consider the first-order condition for minimizing the Hamiltonian, which yields

$$(\alpha_t - \rho_t)p(t) + \sigma_t q(t) = 0.$$

Inserting (4.166) into the latter expression gives us the following candidate of feedback form for the optimal control:

$$\bar{u}(t) = -\frac{(\alpha_t - \rho_t)}{\sigma_t^2}\,(\bar{x}(t) - \bar{\mu}_t) - \frac{(\alpha_t - \rho_t)}{\sigma_t^2 A_t} C_t, \tag{4.167}$$

which is square integrable since $\bar{x}(t)$ is. The expected value of $\bar{u}(t)$ is

$$E[\bar{u}(t)] = \frac{(\rho_t - \alpha_t)C_t}{\sigma_t^2 A_t}. \tag{4.168}$$

Moreover, by inserting (4.168) into (4.165), we also get

$$\begin{aligned} \bar{u}(t) = \tfrac{1}{A_t(\rho_t - \alpha_t)}\,\big(&(2\rho_t A_t + \dot{A}_t)(\bar{x}(t) - \bar{\mu}_t) + \dot{C}_t \\ &+ \left(\rho_t + \tfrac{(\alpha_t - \rho_t)^2}{\sigma_t^2}\right) C_t\big). \end{aligned} \tag{4.169}$$

Identifying (4.167) and (4.169) as two equal regressions on $\bar{x}(t) - \bar{\mu}_t$, we can compare the coefficients to obtain the following equations for A_t and C_t.

$$\begin{cases} (\rho_t - \alpha_t)^2 A_t - \left(2\rho_t A_t + \dot{A}_t\right)\sigma_t^2 = 0, \ A_T = -\gamma, \\ \rho_t C_t + \dot{C}_t = 0, \ C_T = 1. \end{cases}$$

The solutions to these equations are

$$\begin{cases} A_t = -\gamma e^{\int_t^T (2\rho_s - \Lambda_s)ds}, \\ C_t = e^{\int_t^T \rho_s ds}, \end{cases}$$

where we have defined

$$\Lambda_t = \frac{(\rho_t - \alpha_t)^2}{\sigma_t^2}.$$

Hence, in view of (4.167), the candidate for the optimal control is given by

$$\begin{aligned} \bar{u}(t) =& -\frac{\alpha_t - \rho_t}{\sigma_t^2}\left(C_t A_t^{-1} + (\bar{x}(t) - \bar{\mu}_t)\right) \\ =& \frac{\alpha_t - \rho_t}{\sigma_t^2}\left(\frac{1}{\gamma}e^{\int_t^T (\Lambda_s - \rho_s)ds} + (\bar{x}(t) - \bar{\mu}_t)\right), \end{aligned} \tag{4.170}$$

with expected value

$$E[\bar{u}(t)] = \frac{\alpha_t - \rho_t}{\sigma_t^2}\frac{1}{\gamma}e^{\int_t^T (\Lambda_s - \rho_s)ds}. \tag{4.171}$$

By inserting (4.171) into (4.165), we get the following equation for $\bar{\mu}_t$.

$$d\bar{\mu}_t = \left(\rho_t \bar{\mu}_t + \frac{1}{\gamma}\Lambda_t e^{\int_t^T (\Lambda_s - \rho_s)ds}\right)dt, \quad \bar{\mu}_0 = x_0. \tag{4.172}$$

The solution to (4.172) is

$$\bar{\mu}_t = x_0 e^{\int_0^t \rho_s ds} + \frac{1}{\gamma}e^{\int_t^T (\Lambda_s - \rho_s)ds}\left(e^{\int_0^t \Lambda_s ds} - 1\right). \tag{4.173}$$

Finally, by inserting (4.173) into (4.170), we get the solution candidate for the mean-variance portfolio selection problem (4.162), when x_t obeys (4.161), given in feedback form by

$$\bar{u}(t, \bar{x}(t)) = \frac{\alpha_t - \rho_t}{\sigma_t^2}\left(x_0 e^{\int_0^t \rho_s ds} + \frac{1}{\gamma}e^{\int_0^T \Lambda_s ds - \int_t^T \rho_s ds} - \bar{x}(t)\right),$$

which is identical to the optimal control found in Zhou and Li (2000) (cf. (5.12), (6.7) and the subsequent comments).

4.5.5 Open-Loop Adapted Mean-Field-Type Nash Equilibria

In this subsection, we apply the stochastic maximum principle to an $I-$ person *noncooperative* stochastic differential game, where the i-th player in the game would like to minimize her own performance functional regardless of the other players' performance functionals [102].

Let $u = (u_1, \ldots, u_I)$, where $u_i \in \mathcal{U}_i[0, T]$, is an open-loop adapted strategy. and denote

$$(u_{-i}, v) := (u_1, \ldots, u_{i-1}, v, u_{i+1}, \ldots, u_I), \quad i \in \{1, \ldots, I\}.$$

Then, Player i selects $u_i \in \mathcal{U}_i[0, T]$ to minimize the functional

$$v \mapsto J^i(u_{-i}, v) := J^i(u_1, \ldots, u_{i-1}, v, u_{i+1}, \ldots, u_I),$$

where for $i \in \{1, \ldots, I\}$,

$$J^i(u) = E\left[\int_0^T f^i(t, x(t), E[x(t)], u(t))\,dt + h^i(x(T), E[x(T)])\right], \tag{4.174}$$

subject to the dynamics that couples all the players through the vector of controls u given by the following SDE:

$$\begin{cases} dx(t) = b(t, x(t), E[x(t)], u(t))dt + \sigma(t, x(t), E[x(t)], u(t))dW_t, \\ x(0) = x_0. \end{cases} \tag{4.175}$$

where x, b is $\mathbb{R}^d$-valued, σ is $\mathbb{R}^{d\times k}$-valued, and W is k-dimensional Brownian motion. We note that $J^i(u)$ not only depends on u_i but also depends on the other controls $u_j,\ j \neq i$. Therefore, the optimal strategy of Player i depends on the strategies of the other players. This suggests that we use the notion of *equilibrium*, which we recall here.

Definition 4.1 (Nash Equilibrium) An N-tuple $\bar{u} = (\bar{u}_1, \ldots, \bar{u}_I)$ of strategies from $\Pi_{j=1}^I \mathcal{U}_j[0, T]$ is a Nash equilibrium for the game if the following holds:

$$J^i(\bar{u}) \le J^i(\bar{u}_{-i}, v), \qquad \forall v \in \mathcal{U}_i[0, T], \quad 1 \le i \le I. \tag{4.176}$$

To characterize Nash equilibria through a Pontryagin's stochastic maximum principle, we introduce the Hamiltonian and the adjoint processes associated with the dynamics of the i-th player:

$$\begin{aligned} H^i(t, X, u, p^i, q^i) &:= b(t, X, E[X], u) \cdot p^i + \operatorname{tr}(\sigma(t, X, E[X], u)q^i) \\ &\quad - f^i(t, X, E[X], u), \end{aligned}$$

and denote, for $1 \le i \le I$,

$$\delta H^i(t) := p^i(t) \cdot \delta b(t) + \mathrm{tr}\,(\delta\sigma(t) q^i(t)) - \delta f^i(t), \tag{4.177}$$

and

$$\begin{aligned} H^i_w(t) &= b^T_w(t) p^i(t) + \textstyle\sum_{j=1}^d \sigma^j_w(t)^T q^i_j(t) - f^i_w(t), \quad w = x, y, \\ H^i_{xx}(t) &= b^T_{xx}(t) p^i(t) + \textstyle\sum_{j=1}^d \sigma^j_{xx}(t)^T q^i_j(t) - f^i_{xx}(t), \end{aligned} \tag{4.178}$$

where $1 \le i \le I$,

$$\delta\varphi^i(t) = \varphi(t, \bar{x}(t), E[\bar{x}(t)], (\bar{u}_{-i}(t), v(t))) - \varphi(t, \bar{x}(t), E[\bar{x}(t)], \bar{u}(t)). \tag{4.179}$$

The first-order adjoint equation is the following linear backward SDE of mean-field type

$$\begin{cases} dp^i(t) = -\{H^i_x(t) + E\left[H^i_y(t)\right]\} + q^i(t) dW_t, \\ p^i(T) = -h^i_x(T) - E[h^i_y(T)]. \end{cases} \tag{4.180}$$

The unique solution of this equation satisfies

$$E[\sup_{t\in[0,T]} |p^i(t)|^2 + \int_0^T |q^i(t)|^2\, dt] < \infty, \quad 1 \le i \le I. \tag{4.181}$$

The second-order adjoint equation reads

$$\begin{cases} dP^i(t) = -\Big(b^T_x(t) P^i(t) + P^i(t) b_x(t) + \sum_{j=1}^d \sigma^j_x(t)^T P^i(t) \sigma^j_x(t) \\ \qquad + \sigma^j_x(t)^T Q^i_j(t) + \sum_{j=1}^d Q^i_j(t) \sigma^j_x(t) + H^i_{xx}(t)\Big) dt \\ \qquad + \sum_{j=1}^d Q^i_j(t)\, dW^j_t, \\ P^i(T) = -h^i_{xx}(\bar{x}(T)). \end{cases} \tag{4.182}$$

The unique solution satisfies the following estimate

$$E\Big[\sup_{t\in[0,T]} |P^i(t)|^2 + \int_0^T |Q^i(t)|^2\, dt\Big] < \infty, \quad 1 \le i \le I. \tag{4.183}$$

We then have the following Pontryagin's stochastic maximum principle for Nash equilibria of the I-person stochastic differential game.

Theorem 4.9 (*Stochastic Maximum Principle*) *Let Assumptions (5-H1) and (5-H2) hold. Suppose* $\bar{u} = (\bar{u}_1, \ldots, \bar{u}_I) \in \Pi_{j=1}^I \mathcal{U}_j[0, T]$ *is a Nash equilibrium of the stochastic differential game, and let* $\bar{x}$ *denote the solution of (4.175). Then there* *exist pairs of $\mathbb{F}$-adapted processes* $\{(p^j, q^j), \ , 1 \leq j \leq I\}$ *and* $\{(P^j, Q^j), \ , 1 \leq j \leq I\}$ *that satisfy (4.180)–(4.181) and (4.182)–(4.183), respectively, such that (cf. the notation(4.179)), for every* $1 \leq i \leq I$,

$$\delta H^i(t) + \frac{1}{2}\mathrm{tr}\,(\delta\sigma^i(t)^* P^i(t)\delta\sigma^i(t)) \ \forall v \in U_i, \quad \text{a.e.} t \in [0, T], \ \mathbb{P}\text{a.s.} \quad (4.184)$$

The proof of this theorem is similar to that of Theorem 4.7.

Sufficient conditions for Nash equilibria which can be formulated is a similar fashion as in Theorem 4.8.

Extensions to Other Mean-Field Couplings

Extensions of the stochastic maximum principle (SMP) to mean-field couplings of the form $E[\varphi(x(t))]$ have been derived in Andersson and Djehiche (2010). Hosking (2012) has extended the SMP to mean-field couplings, which includes the law of the control itself, of the form $E[\varphi(x(t), u(t))]$. Extensions to mean-field couplings of the form $\varphi(t, x(t), \mathbb{P} \circ x^{-1}(t), u(t))$, where $\mathbb{P} \circ x^{-1}(t)$ is the marginal law of the process x at time t, have been considered in Buckdahn, Li, and Ma (2016).

4.6 Stochastic Near-Optimal Control

In this section, we introduce near-optimality and briefly account for the necessary and sufficient conditions of near-optimality for the control of SDEs of mean-field type.

For a given $\eta > 0$, an admissible control u^η is called near or η-optimal for the cost functional F over the set of controls V if

$$F(u^\eta) - \eta \leq \inf\{F(v); \ v \in V\}.$$

Near-optimal controls enjoy many useful and desirable properties that optimal controls do not have. In fact, thanks to Ekeland's variational principle that we will include below, under very mild conditions on the control set V and the cost functional F, near-optimal controls always exist while optimal controls may not. Moreover, there are many candidates for near-optimal controls which make it possible to select among them appropriate ones that are easier to implement and handle both analytically and numerically. For stochastic systems where uncertainties are intrinsic and inherent, robustness to perturbations and disturbances is also another desirable feature of near-optimal controls: while, by its very nature, an optimal control is usually very sensitive to perturbations in the sense that it loses its optimality as soon as it is disturbed, a near-optimal control is insensitive to small perturbations.

We first present Ekeland's variational principle, Lemma 4.13, along with the metric (4.187) on the space $\mathscr{U}[0, T]$ of admissible controls and some continuity lemmas. Then we derive necessary and sufficient conditions for near-optimality.

4.6.1 Ekeland's Variational Principle

The classical variational principle states that if a differentiable cost functional F attains its minimum at some point $\bar{u}$, then $F'(\bar{u}) = 0$. Ekeland's principle states that if a differentiable cost functional F has a lower bound, although it need not attain it, then for every $\eta > 0$, there exists some control $u_\eta \in V$ such that the derivative $F'(\bar{u}_\eta)$ can be made arbitrarily small, i.e., $|F'(\bar{u}_\eta)| \leq \eta$.

Lemma 4.13 (Ekeland's Variational Principle) *Let (V, ρ) be a complete metric space and $F : V \longrightarrow \mathbb{R} \cup \{+\infty\}$ be lower-semicontinuous and bounded from below (i.e., $\inf F > -\infty$). Given $\eta > 0$, let $u^\eta \in V$ be **η-optimal** for F. Then, for any $\lambda > 0$, there exists $u^* \in V$ such that*

- $F(u^*) \leq F(u^\eta)$;
- $\rho(u^\eta, u^*) \leq \lambda$;
- $]\, F(u^*) < F(w) + \frac{\eta}{\lambda}\rho(w, u^*), \quad w \neq u^*.$

This principle states that even if we do not know whether the cost functional F admits a minimum or not, the cost functional

$$\widetilde{F}(w) := F(w) + \frac{\eta}{\lambda}\rho(w, u^*)$$

always admits a minimum. Moreover, this minimum is reached at u^*. This principle is the basis for our characterization of η-optimal controls.

4.6.2 Near-Optimal Controls for SDEs of Mean-Field Type

Consider the stochastic control system

$$\begin{cases} dx(t) = b(t, x(t), E[x(t)], u(t))dt + \sigma(t, x(t), E[x(t)], u(t))dW_t, \\ x(0) = x_0, \end{cases} \tag{4.185}$$

where

$$b : [0, T] \times \mathbb{R}^d \times \mathbb{R}^d \times U \longrightarrow \mathbb{R}^d, \quad \sigma : [0, T] \times \mathbb{R}^d \times \mathbb{R}^d \times U \longrightarrow \mathbb{R}^{d\times k},$$

and W is a k-dimensional Brownian motion. We recall that an admissible control u is an $\mathbb{F}$-adapted process with values in a separable metric space (U, d_0). We denote the set of all admissible controls by $\mathscr{U}[0, T]$.

The cost functional over $\mathscr{U}[0, T]$ is

$$J(u) = E\left[\int_0^T f(t, x(t), E[x(t)], u(t))\,dt + h(x(T), E[x(T)])\right], \tag{4.186}$$

where

$$f:\ [0, T] \times \mathbb{R}^d \times \mathbb{R}^d \times U \longrightarrow \mathbb{R}, \quad h:\ \mathbb{R}^d \times \mathbb{R}^d \longrightarrow \mathbb{R}.$$

We assume that (U, d_0) is closed and that the conditions (5-H1) and (5-H2) of the previous section are in force.

In this section, we derive necessary and sufficient conditions of near-optimality for the control problem (4.185)–(4.186). It turns out that this is possible, thanks to the following metric, introduced by Ekeland (1974), Lemma 7.2, in the deterministic case and extended to the stochastic case in Elliott and Kohlmann (1980).

Definition 4.2 Define the metric d_E on the space $\mathscr{U}[0, T]$ of admissible controls as follows. For $u, v \in \mathscr{U}[0, T]$,

$$d_E(u, v) := \widehat{\mathbb{P}}\{(\omega, t) \in \Omega \times [0, T],\ d_0(u(\omega, t), v(\omega, t)) > 0\}, \tag{4.187}$$

where $\widehat{\mathbb{P}}$ is the product measure of $\mathbb{P}$ and the Lebesgue measure on $[0, T]$.

Lemma 4.14 *We have*

(i) d_E *is a distance.*
(ii) $(\mathscr{U}[0, T], d_E)$ *is a complete metric space.*

Proof

(i) We only show the transitivity property of d_E. Take any u_1, u_2, u_3 in $\mathscr{U}[0, T]$. From

$$\begin{aligned}\{(\omega, t),\ d(u_1(\omega, t), u_3(\omega, t)) > 0\} \subset \{(\omega, t),\ d(u_1(\omega, t), u_2(\omega, t)) > 0\}\\ \cup\{(\omega, t),\ d(u_2(\omega, t), u_3(\omega, t)) > 0\}\end{aligned}$$

, we obtain

$$\begin{aligned}\widehat{\mathbb{P}}\{(\omega, t),\ d(u_1(\omega, t), u_3(\omega, t)) > 0\} \le \widehat{\mathbb{P}}\{(\omega, t),\ d(u_1(\omega, t), u_2(\omega, t)) > 0\}\\ +\widehat{\mathbb{P}}\{(\omega, t),\ d(u_2(\omega, t), u_3(\omega, t)) > 0\}\end{aligned}$$

(ii) Let $(u_n)_{n\in\mathbb{N}}$ be a Cauchy sequence in $\mathscr{U}[0, T]$. We can extract a subsequence $(u_{n_k})_{k\in\mathbb{N}}$ such that $d_E(u_{n_k}, u_{n_{k+1}}) \le 1/2^k$. This sequence converges. Indeed, set

$$E_k := \bigcup_{p \geq k} \{(\omega, t),\ d(u_{n_p}(\omega, t), u_{n_{p+1}}(\omega, t)) > 0\}.$$

We have $E_{k+1} \subset E_k$ and

$$\widehat{\mathbb{P}}(E_k) = \sum_{p=k}^{\infty} \frac{1}{2^p} = \frac{1}{2^{k-1}}.$$

Define $\hat{u} \in \mathscr{U}[0, T]$ by

$$\hat{u}(\omega, t) = u_{n_k}(\omega, t), \quad (\omega, t) \notin E_k.$$

By definition, the subsequence $(u_{n_k})_{k \in \mathbb{N}}$ converges to $\hat{u}$, and since $(u_n)_{n \in \mathbb{N}}$ is Cauchy, the whole sequence converges to $\hat{u}$. Now, since d_E-convergence implies $\widehat{\mathbb{P}}$-a.e. convergence, it follows that $\hat{u} \in \mathscr{U}[0, T]$.

■

The next continuity results for both the solutions of SDE (4.185) and the cost functional (4.186) with respect to d_E are the main ingredients to the necessary and sufficient near-optimality for our system.

Proposition 4.8

(*i*) *For any* $0 < \alpha < 1$ *and* $p \geq 0$ *satisfying* $\alpha p < 1$, *there exists a positive constant* $C = C(\alpha, p)$ *such that, for any* $u, v \in \mathscr{U}[0, T]$, *it holds that*

$$E\Big[\sup_{0 \leq t \leq T} |x^u(t) - x^v(t)|^{2p}\Big] \leq C\Big[d_E(u, v)\Big]^{\alpha p}, \tag{4.188}$$

where x^u *and* x^v *are solutions of (4.70), corresponding to the controls* u *and* v.

(*ii*) *The performance functional* $J : \Big(\mathscr{U}[0, T], d_E\Big) \longrightarrow \mathbb{R}$ *is continuous. More precisely, for any* $u, v \in \mathscr{U}[0, T]$, *it holds that*

$$|J(u) - J(v)| \leq C\Big[d_E(u, v)\Big]^{\alpha/2}. \tag{4.189}$$

The proof employs the standard Grönwall and Burkholder–Davis–Gundy inequalities. See Zhou (1998), Lemma 3.1 for the details.

We have the following stability lemmas for the pairs of adjoint processes (p, q) and (P, Q) solutions of (4.112) and (4.114), respectively.

Lemma 4.15 *For any* $0 < \alpha < 1$ *and* $1 < \beta < 2$ *satisfying* $(1 + \alpha)\beta < 2$, *there exists a constant* $C = C(\alpha, \beta) > 0$ *such that for any admissible controls* u *and* u' *along with the corresponding dynamics* x *and* x' *and adjoint processes*

(p,q), (P,Q), and (p',q'), (P',Q') solutions of (4.112) and (4.114) associated with (x,u) and (x',u'), the following estimates hold:

$$E\left[\int_0^T \left(\left|p(t)-p'(t)\right|^\beta + \left|q(t)-q'(t)\right|^\beta\right) dt\right] \le C\left[d_E(u,u')\right]^{\frac{\alpha\beta}{2}} \quad (4.190)$$

and

$$E\left[\int_0^T \left(\left|P(t)-P'(t)\right|^\beta + \left|Q(t)-Q'(t)\right|^\beta\right) dt\right] \le C\left[d_E(u,u')\right]^{\frac{\alpha\beta}{2}}. \quad (4.191)$$

The proof uses usual arguments from the theory of backward stochastic differential equations (see Zhou (1998), Lemma 3.2, for the details).

Definition 4.3 (Near-Optimality) For a given $\eta > 0$, an admissible control $u^\eta \in \mathscr{U}[0,T]$ is called η-optimal for (4.185)–(4.186) if

$$J(u^\eta) \le \inf_{u\in\mathscr{U}[0,T]} J(u) + \eta. \quad (4.192)$$

We denote by x^η the solution of (4.185) corresponding to u^η. (x^η, u^η) is called η-optimal pair.

Necessary conditions for near-optimality in terms of a stochastic integrated near-maximum of the underlying $\mathscr{H}$-function (4.83) are displayed in the next result.

Recall the Hamiltonian associated with (4.185)–(4.186).

$$\begin{aligned} H(t,X,u,p,q) &:= b(t,X,E[X],u)\cdot p + \mathrm{tr}\,(\sigma(t,X,E[X],u)q) \\ &\quad - f(t,X,E[X],u), \end{aligned}$$

and

$$\begin{aligned} \mathscr{H}^{(x^\eta(t),u^\eta(t))}(t,X,u) := H(t,X,u,&\,p^\eta(t),q^\eta(t) \\ &-P^\eta(t)\sigma(t,x^\eta(t),E[x^\eta(t)],u^\eta(t))) \\ &+\tfrac{1}{2}\mathrm{tr}\,\{\sigma^T(t,X,E[X],u)P^\eta(t)\sigma(t,X,E[X],u)\}, \end{aligned}$$

where (p^η, q^η) and (P^η, Q^η) are the adjoint processes associated with (x^η, u^η).

Theorem 4.10 (Necessary Conditions for Near-Optimality) *For any $0 \le \gamma < 1/3$, there exists a positive constant $C = C(\gamma)$ such that for any $\eta > 0$ and any η-optimal pair (x^η, u^η), we have*

$$\begin{aligned} &E\left[\int_0^T \mathscr{H}^{(x^\eta(t),u^\eta(t))}(t,x^\eta(t),u^\eta(t))dt\right] \\ &\qquad \ge \sup_{u\in U} E\left[\int_0^T \mathscr{H}^{(x^\eta(t),u^\eta(t))}(t,x^\eta(t),u^\eta(t))dt\right] - C\eta^\gamma. \end{aligned} \quad (4.193)$$

The proof is based on the relation (4.5.1) and the estimates derived in the previous section. For a proof of the mean-field free case, see Zhou (1998), Theorem 4.1.

Under some convexity conditions of the underlying Hamiltonian and the terminal cost function h, the integrated near-maximum condition of the $\mathscr{H}$-function is sufficient for near-optimality.

Theorem 4.11 *Let Assumptions (5-H3) to (5-H5) hold. For some $\eta > 0$, consider an admissible pair (x^η, u^η),, and let (p^η, q^η) be the associated first order adjoint processes. If*

$$E\left[\int_0^T \mathscr{H}^{(x^\eta(t),u^\eta(t))}(t, x^\eta(t), u^\eta(t))dt\right] \geq \sup_{u\in U} E\left[\int_0^T \mathscr{H}^{(x^\eta(t),u^\eta(t))}(t, x^\eta(t), u^\eta(t))dt\right] - C\eta,$$

then

$$J(u^\eta) \leq \inf_{u\in\mathscr{U}[0,T]} J(u) + C\eta. \tag{4.194}$$

where $C > 0$ is a constant independent of η, i.e., (x^η, u^η) is an η-optimal pair.

The proof is based on the relation (4.5.1) and the estimates derived in the previous section. For a proof of the mean-field free case, see Zhou (1998), Theorem 5.1.

4.7 Pontryagin's Maximum Principle for Risk-Sensitive Mean-Field Type Control

In this section, we study mean-field type control problems with risk-sensitive performance functionals. To ease the presentation of the main results, we only consider dynamics in one dimension. Extension to the multidimensional case is straightforward.

Statement of the Problem

Let $T > 0$ be a fixed time horizon and $(\Omega, \mathscr{F}, \mathbb{F}, \mathbb{P})$ be a given filtered probability space on which a one-dimensional standard Brownian motion $W = \{W_s\}_{s\geq 0}$ is given, and the filtration $\mathbb{F} = \{\mathscr{F}_s,\ 0 \leq s \leq T\}$ is the natural filtration of B augmented by $\mathbb{P}-$null sets of $\mathscr{F}$.

We consider the stochastic control system

$$\begin{cases} dx^u(t) = b(t, x^u(t), E[x^u(t)], u(t))dt + \sigma(t, x^u(t), E[x^u(t)], u(t))dW_t, \\ x^u(0) = x_0, \end{cases} \tag{4.195}$$

where

$$b, \sigma : [0, T] \times \mathbb{R} \times \mathbb{R} \times U \longrightarrow \mathbb{R}.$$

An admissible control u is an $\mathbb{F}$-adapted and square-integrable process with values in a separable metric space (U, d_0) where d_0 is a distance on U^2. We denote the set of all admissible controls by $\mathscr{U}[0, T]$.

Given $u \in \mathscr{U}$, Eq. (4.195) is a stochastic differential equation (SDE) with random coefficients.

The risk-sensitive cost functional associated with (4.195) is given by

$$J^\theta(u) = Ee^{\theta\left[\int_0^T f(t, x^u(t), E[x^u(t)], u(t))\,dt + h(x^u(T), E[x^u(T)])\right]}, \tag{4.196}$$

where θ is the risk sensitivity index,

$$f : [0, T] \times \mathbb{R} \times \mathbb{R} \times U \longrightarrow \mathbb{R}, \quad h : \mathbb{R} \times \mathbb{R} \longrightarrow \mathbb{R},$$

Any $\bar{u} \in \mathscr{U}$ satisfying

$$J^\theta(\bar{u}) = \inf_{u \in \mathscr{U}} J^\theta(u) \tag{4.197}$$

is called a risk-sensitive optimal control. The corresponding state process, solution of (4.195), is denoted by $\bar{x}(\cdot) := x^{\bar{u}}(\cdot)$.

The optimal control problem we are concerned with is to characterize the pair $(\bar{x}, \bar{u})$ solution of the problem (4.284).

Let $\Psi_T := \int_0^T f(t, x(t), E[x(t)], u(t))dt + h(x(T), E[x(T)])$. Then the risk-sensitive loss functional is given by

$$\Psi_\theta := \frac{1}{\theta} \log Ee^{\theta\left[\int_0^T f(t, x(t), E[x(t)], u(t))dt + h(x(T), E[x(T)])\right]} = \frac{1}{\theta} \log\left[Ee^{\theta \Psi_T}\right].$$

When the risk-sensitive index θ is small, the loss functional Ψ_θ can be expanded as

$$E[\Psi_T] + \frac{\theta}{2}\mathrm{var}(\Psi_T) + O(\theta^2),$$

where $\mathrm{var}(\Psi_T)$ denotes the variance of Ψ_T. If $\theta < 0$, the variance of Ψ_T, as a measure of risk, improves the performance Ψ_θ, in which case the optimizer is called *risk seeker*. But, when $\theta > 0$, the variance of Ψ_T worsens the performance Ψ_θ, in which case the optimizer is called *risk averse*. The risk-neutral loss functional $E[\Psi_T]$ can be seen as a limit of risk-sensitive functional Ψ_θ when $\theta \to 0$.

For $\phi \in \{b, \sigma, f, h\}$ and $u \in \mathscr{U}$, we define

$$\begin{cases} \delta\phi(t) = \phi(t, \bar{x}(t), E[\bar{x}(t)], u(t)) - \phi(t, \bar{x}(t), E[\bar{x}(t)], \bar{u}(t)); \\ \phi_x(t) = \frac{\partial \phi}{\partial x}(t, \bar{x}(t), E[\bar{x}(t)], \bar{u}(t)), \quad \phi_{xx}(t) = \frac{\partial^2 \phi}{\partial x^2}(t, \bar{x}(t), E[\bar{x}(t)], \bar{u}(t)), \\ \phi_y(t) = \frac{\partial \phi}{\partial y}(t, \bar{x}(t), E[\bar{x}(t)], \bar{u}(t)). \end{cases} \tag{4.198}$$

We define the "risk-neutral" Hamiltonian associated with random variables $X \in L^1(\Omega, \mathscr{F}, \mathbb{P})$ as follows. For $(p, q) \in \mathbb{R} \times \mathbb{R}$,

$$\begin{aligned} H(t, X, u, p, q) := &\, b(t, X, E[X], u)p + \sigma(t, X, E[X], u)q \\ &- f(t, X, E[X], u), \end{aligned} \tag{4.199}$$

We also introduce the risk-sensitive Hamiltonian. For $\theta \in \mathbb{R}$ and $(p, q, \ell) \in \mathbb{R} \times \mathbb{R} \times \mathbb{R}$,

$$\begin{aligned} H^\theta(t, X, u, p, q, \ell) := &\, b(t, X, E[X], u)p + \sigma(t, X, E[X], u)(q + \theta \ell p) \\ &- f(t, X, E[X], u). \end{aligned} \tag{4.200}$$

We have $H = H^0$.

Moreover, we denote

$$\begin{aligned} \delta H(t) &:= p(t)\delta b(t) + q(t)\delta\sigma(t) - \delta f(t), \\ \delta H^\theta(t) &:= p(t)\delta b(t) + (q + \theta \ell p)\delta\sigma(t) - \delta f(t), \\ H_k(t) &:= b_k(t)p + \sigma_k(t)q - f_k(t), \\ H_k^\theta(t) &:= p(t)b_k(t) + (q + \theta \ell p)\sigma_k(t) - f_k(t), \end{aligned} \tag{4.201}$$

for $k \in \{x, y, xx\}$.

We will only consider the case where the risk-sensitive parameter is positive, $\theta > 0$. The case $\theta < 0$ can be treated in a similar fashion by considering $\theta = -\bar{\theta}, \bar{\theta} > 0$ and $\bar{f} = -f, \bar{h} = -h$ in the performance functional (4.196).

Throughout this section, we will make the following assumption.

- (H1)The functions b, σ, f, h are twice continuously differentiable with respect to (x, y). Moreover, b, σ, f, h and all their derivatives up to second-order with respect to (x, y) are continuous in (x, y, u) and bounded.

Under Assumption (H1), for each $u \in \mathscr{U}$, the SDE (4.195) admits a unique strong solution x^u.

We now provide an example of mean-field-type control problem that satisfies Assumption (H1).

Example 4.1 (Effect of the Mean in Cooperative Dynamics) Consider the following mean-field dynamics

$$dx(t) = \{\sin(-x^3(t) + x(t)) - \mu \sin(x(t) - E[x(t)])\}dt + \sigma dW_t, \tag{4.202}$$

where $\mu > 0, \sigma \in \mathbb{R}$. The first term in the drift, $(\sin(-x^3(t) + x(t)))$, is often replaced by a control action $u(t) \in [-1, 1]$ to get

$$dx(t) = \{u(t) - \mu \sin(x(t) - E[x(t)])\}dt + \sigma dW_t \tag{4.203}$$

which clearly is in the framework of Assumption (6-H1). This type of mean-field SDE models has been used to understand muscle contraction. Other similar models have been widely studied in chemical kinetics, statistical mechanics, and economics to capture cooperative behavior of a generic particle, oscillator, or an agent [102].

We now introduce the first- and second-order adjoint processes involved in the risk-sensitive SMP for the mean-field-type control problem.

The (risk-sensitive) first-order adjoint processes (p, q, v^θ, ℓ) (all depending on θ, but we suppress the labeling in p, q and ℓ to ease notation) solve the following system of backward SDEs of mean-field type:

$$\begin{cases} d\bar{p}(t) = -\left\{H_x^\theta(t) + \frac{1}{v^\theta(t)}E[v^\theta(t)H_y^\theta(t)]\right\}dt + \bar{q}(t)(-\theta\ell(t)dt + dW_t), \\ dv^\theta(t) = \theta\ell(t)v^\theta(t)dW_t, \\ v^\theta(T) = \phi^\theta(T), \\ \bar{p}(T) = -h_x(T) - \frac{1}{\phi^\theta(T)}E[\phi^\theta(T)h_y(T)], \end{cases} \tag{4.204}$$

where

$$\phi^\theta(T) := e^{\theta[h(\bar{x}(T),E[\bar{x}(T)]) + \int_0^T f(t,\bar{x}(t),E[\bar{x}(t)],\bar{u}(t))dt]}. \tag{4.205}$$

Note that the Hamiltonian terms in (4.204) are evaluated at the optimal state and optimal control $(\bar{x}(\cdot), \bar{u}(\cdot))$, i.e., $H_k^\theta(t, \bar{x}(t), E[\bar{x}(t)], \bar{u}(t), \bar{p}(t), \bar{q}(t), \ell(t)), \quad k \in \{x, y\}$.

We have

Proposition 4.9 *Under Assumption (H1), the system of BSDEs (4.204) admits a unique $\mathbb{F}$-adapted solution $(\bar{p}, \bar{q}, v^\theta, \ell)$ such that*

$$E\left[\sup_{t\in[0,T]}|\bar{p}(t)|^2 + \sup_{t\in[0,T]}|v^\theta(t)|^2 + \int_0^T \left(|\bar{q}(t)|^2 + |\ell(t)|^2\right)dt\right] < \infty. \tag{4.206}$$

The (risk-sensitive) second-order adjoint processes $(\bar{P}, \bar{Q})$ are solutions of the following backward SDE:

$$\begin{cases} d\bar{P}(t) = -\left\{\left(2b_x(t) + \sigma_x^2(t) + 2\theta\ell(t)\sigma_x(t)\right)\bar{P}(t) + 2\sigma_x(t)\bar{Q}(t) \right. \\ \left. -\theta(\bar{q}(t) + \sigma_x(t)\bar{p}(t))^2 + H_{xx}^\theta(t)\right\}dt + \bar{Q}(t)(-\theta\ell(t)dt + dW_t), \\ \bar{P}(T) = -h_{xx}(T). \end{cases} \tag{4.207}$$

Equation (4.207) is a standard linear backward SDE (in $(\bar{P}, \bar{Q})$). Under assumption (H1), (4.207) has a unique $\mathbb{F}$-adapted solution $(\bar{P}, \bar{Q})$ satisfying

$$E\left[\sup_{t\in[0,T]}|\bar{P}(t)|^2 + \int_0^T |\bar{Q}(t)|^2 dt\right] < \infty. \tag{4.208}$$

Remark 4.2 The main difference between the risk-sensitive first-order adjoint equation (4.204) and the mean-field-free case is the presence of the terms θ, v^θ and ℓ. As θ goes to zero, $v^\theta \to 1$, $\mathbb{P}$a.s. and the system reduces to

$$\begin{cases} d\bar{p}(t) = -\left\{H_x(t) + E[H_y(t)]\right\} dt + \bar{q}(t)dB_t, \\ \bar{p}(T) = -h_x(T) - E[h_y(T)], \end{cases} \tag{4.209}$$

which is the risk-neutral first-order adjoint equation for the corresponding mean-field-type control problem derived in Theorem 4.116. Similarly, we retrieve the risk-neutral second-order adjoint equation as θ vanishes.

We show below that

$$\frac{v^\theta(t)}{v^\theta(0)} = \exp\left(\int_0^t \theta\ell(s)dW_s - \frac{\theta^2}{2}\int_0^t |\ell(s)|^2 ds\right) := L_t^\theta, \quad 0 \le t \le T, \tag{4.210}$$

is a uniformly integrable martingale which defines a probability measure $\mathbb{P}^\theta$ equivalent to $\mathbb{P}$:

$$d\mathbb{P}^\theta = L_t^\theta d\mathbb{P}, \quad \text{on} \quad \mathscr{F}_t.$$

By Girsanov's Theorem,

$$W_t^\theta := W_t - \theta \int_0^t \ell(s)ds$$

is a $\mathbb{P}^\theta$-Brownian motion. This makes the adjoint equation in (4.204) satisfied by $(\bar{p}, \bar{q})$ becomes

$$\begin{cases} d\bar{p}(t) = -\left\{H_x^\theta(t) + \frac{1}{L^\theta(t)} E^\theta[H_y^\theta(t)]\right\} dt + \bar{q}(t)dW_t^\theta, \\ \\ \bar{p}(T) = -h_x(T) - \frac{1}{L^\theta(T)} E^\theta[h_y(T)], \end{cases} \tag{4.211}$$

where the expectation $E^\theta[\,\cdot\,]$ is with respect to the probability measure $\mathbb{P}^\theta$. We note that (4.211) is quite similar to the risk-neutral adjoint equation (4.209).

We now have the following main result:

Theorem 4.12 (Risk-Sensitive Maximum Principle) *Let Assumption (H1) hold. If $(\bar{x}(\cdot), \bar{u}(\cdot))$ is an optimal solution of the risk-sensitive control problem (4.195)–(4.284), then there are three pairs of $\mathbb{F}$-adapted processes (v^θ, ℓ), $(\bar{p}, \bar{q})$ and $(\bar{P}, \bar{Q})$ that satisfy (4.204)–(4.206) and (4.207)–(4.208), respectively, such that*

$$\delta H^\theta(t) + \frac{1}{2}\left(\bar{P}(t) - \theta\bar{p}^2(t)\right)(\delta\sigma(t))^2 \le 0, \tag{4.212}$$

for all $u \in U$, almost every $t \in [0,T]$ and $\mathbb{P}$-almost surely.
In particular, if $\sigma(t,x,u) := \sigma(t,x)$ is independent of u, then

$$H^\theta(t,\bar{x}(t),\bar{u}(t),\bar{p}(t),\bar{q}(t),\ell(t)) = \max_{u\in U} H^\theta(t,\bar{x}(t),u,\bar{p}(t),\bar{q}(t),\ell(t)).$$

Remark 4.3 If the model is mean-field free, i.e., for which $\sigma_y = h_y = b_y = f_y = 0$, and when $\ell(t) := -\sigma(t,\bar{x}(t),\bar{u}(t))\bar{p}(t)$, then the generic martingale v^θ becomes the smooth value function of a Markovian or feedback control dynamics, whose gradient is the adjoint process.

Proof of the Risk-Sensitive Stochastic Maximum Principle
The proof of Theorem 4.13 is carried out in the next few subsections.

4.7.1 An Intermediate SMP for Mean-Field Type Control

We first reformulate the risk-sensitive control problem (4.195)–(4.284) in terms of an augmented state process and terminal payoff problem. An intermediate stochastic maximum principle is then obtained by applying the SMP of Theorem 4.7 for loss functionals without running cost. Then, we transform the intermediate first- and second-order adjoint processes to a more simpler form. We introduce an auxillary state process $\xi(t)$ which is solution of the following forward SDE:

$$d\xi(t) = f(t,x(t),E[x(t)],u(t))dt, \quad \xi(0) = 0.$$

The mean-field-type control problem (4.284) under the dynamics (4.195) is equivalent to

$$\begin{cases} \inf_{u\in\mathcal{U}} Ee^{\theta[h(x(T),E[x(T)])+\xi(T)]}, \\ \text{subject to} \\ dx(t)=b(t,x(t),E[x(t)],u(t))dt+\sigma(t,x(t),E[x(t)],u(t))dW_t, \\ d\xi(t) = f(t,x(t),E[x(t)],u(t))dt, \\ x(0) = x_0, \quad \xi(0) = 0. \end{cases} \tag{4.213}$$

Recall that

$$\phi^\theta(T) := e^{\theta[h(\bar{x}(T),E[\bar{x}(T)])+\int_0^T f(t,\bar{x}(t),E[\bar{x}(t)],\bar{u}(t))dt]}.$$

Under Assumption (H1), we may apply the SMP for risk-neutral mean-field type control, Theorem 4.7, to the augmented state dynamics (x,ξ) to derive the first-order adjoint equation: There exists a unique $\mathbb{F}$-adapted pair of processes $(\vec{p},\vec{q})$, which solves the BSDE

$$
\begin{cases}
d\vec{p}(t) = -\left\{\begin{pmatrix} b_x(t) & 0 \\ f_x(t) & 0 \end{pmatrix}' \vec{p}(t) + \begin{pmatrix} \sigma_x(t) & 0 \\ 0 & 0 \end{pmatrix}' \vec{q} \right. \\
\qquad\qquad \left. +E\left[\begin{pmatrix} b_y(t) & 0 \\ f_y(t) & 0 \end{pmatrix}' \vec{p} + \begin{pmatrix} \sigma_y(t) & 0 \\ 0 & 0 \end{pmatrix}' \vec{q}(t)\right]\right\} dt + \vec{q}(t) dW_t, \\
\vec{p}(T) = -\theta\phi^{\theta}(T)\begin{pmatrix} h_x(T) \\ 1 \end{pmatrix} - \theta E\left[\phi^{\theta}(T)\begin{pmatrix} h_y(T) \\ 0 \end{pmatrix}\right],
\end{cases}
\tag{4.214}
$$

with

$$
E\left[\sup_{0\le t\le T} |\vec{p}(t))|^2 + \int_0^T |\vec{q}(t)|^2 dt\right] < \infty. \tag{4.215}
$$

Let $\tilde{H}^{\theta}$ be the Hamiltonian associated with the optimal state dynamics $(\bar{x}(\cdot), \bar{\xi}(\cdot))$ and the pair of adjoint processes $(\vec{p}(t), \vec{q}(t))$:

$$
\begin{aligned}
\tilde{H}^{\theta}(t, \bar{x}(t), u, \vec{p}(t), \vec{q}(t)) := &\begin{pmatrix} b(t, \bar{x}(t), E[\bar{x}(t)], u) \\ f(t, \bar{x}(t), E[\bar{x}(t)], u) \end{pmatrix} \cdot \vec{p}(t) \\
&+ \begin{pmatrix} \sigma(t, \bar{x}(t), E[\bar{x}(t)], u) \\ 0 \end{pmatrix} \cdot \vec{q}(t),
\end{aligned}
\tag{4.216}
$$

where "$\cdot$" denotes the usual scalar product in $\mathbb{R}^2$. The dependence of the Hamiltonian on θ stems from the dependence of the adjoint processes $(\vec{p}, \vec{q})$ of θ through the end condition in (4.214).

Again, by Theorem 4.7, there exists a unique $\mathbb{F}$-adapted pair of processes (P, Q), which solves the second-order adjoint equation

$$
\begin{cases}
dP(t) = -\left\{\begin{pmatrix} b_x(t) & 0 \\ f_x(t) & 0 \end{pmatrix} P(t) + P(t)\begin{pmatrix} b_x(t) & 0 \\ f_x(t) & 0 \end{pmatrix}' \right. \\
\quad + \begin{pmatrix} \sigma_x(t) & 0 \\ 0 & 0 \end{pmatrix} Q(t) + Q(t)\begin{pmatrix} \sigma_x(t) & 0 \\ 0 & 0 \end{pmatrix}' \\
\quad \left. + \begin{pmatrix} \sigma_x(t) & 0 \\ 0 & 0 \end{pmatrix} P(t)\begin{pmatrix} \sigma_x(t) & 0 \\ 0 & 0 \end{pmatrix}' + \begin{pmatrix} \tilde{H}^{\theta}_{xx}(t) & 0 \\ 0 & 0 \end{pmatrix}\right\} dt + Q(t) dW_t, \\
P(T) = -\theta\phi^{\theta}(T)\begin{pmatrix} \theta h_x^2(T) + h_{xx}(T) & \theta h_x(T) \\ \theta h_x(T) & \theta \end{pmatrix},
\end{cases}
\tag{4.217}
$$

and satisfies

$$
E\left[\sup_{0\le t\le T} ||P(t))||^2 + \int_0^T ||Q(t)||^2 dt\right] < \infty, \tag{4.218}
$$

where $||\cdot||$ denotes the norms of the corresponding matrices.

We have the following

Proposition 4.10 *Let Assumption (H1) hold; if* $(\bar{x}, \bar{\xi}, \bar{u})$ *is an optimal solution of the risk-neutral control problem (4.213), then there are two pairs of* $\mathbb{F}$*-adapted processes* $(\vec{p}, \vec{q})$ *and* (P, Q) *that satisfy (4.214)–(4.215) and (4.217)–(4.218), respectively, such that*

$$\delta\tilde{H}^{\theta}(t) + \frac{1}{2}\begin{pmatrix}\delta\sigma(t)\\ 0\end{pmatrix}' P(t)\begin{pmatrix}\delta\sigma(t)\\ 0\end{pmatrix} \leq 0, \tag{4.219}$$

for all $u \in U$*, almost every* t *and* $\mathbb{P}$*-almost surely,*

where

$$\delta\tilde{H}^{\theta}(t) := \tilde{H}^{\theta}(t, \bar{x}(t), u, \vec{p}(t), \vec{q}(t)) - \tilde{H}^{\theta}(t, \bar{x}(t), \bar{u}(t), \vec{p}(t), \vec{q}(t)).$$

4.7.2 Transformation of the First-Order Adjoint Process

Although the result of Proposition 4.10 is a good SMP for the risk-sensitive mean-field type control, the fact that augmenting the state process with the second component ξ yields a system of two adjoint equations that appears to make it complicated to solve in concrete situations. We will introduce a transformation (through a generic square-integrable martingale) of the pair of adjoint processes $(\vec{p}, \vec{q})$ in such a way so as to get rid of the second component (p_2, q_2) in (4.214) and express the SMP in terms of only one adjoint process that we denote by $(\tilde{p}_1, \tilde{q}_1)$ which solves a backward SDE whose driver is quadratic in $\tilde{p}_1$. This generic martingale allows to transform the pair (p_2, q_2) into the adjoint process $(\tilde{p}_2, 0)$, where the process $\tilde{p}_2$ is still a square-integrable martingale, which would mean that $\tilde{p}_2(t) = \tilde{p}_2(T)$ and is equal to the constant $E[\tilde{p}_2(T)]$. But, this generic martingale need not be related to the pair of adjoint process $\vec{p}$. Instead, it will be part of the adjoint equation associated with the risk-sensitive SMP (see Theorem 4.13, above). Roughly, noting that $dp_2(t) = q_2(t)dW_t$ and $p_2(T) = -\theta\phi^{\theta}(T)$,, the explicit solution of this backward SDE is

$$p_2(t) = -\theta E[\phi^{\theta}(T) \mid \mathscr{F}_t] = -\theta v^{\theta}(t), \tag{4.220}$$

where

$$v^{\theta}(t) := E[\phi^{\theta}(T) \mid \mathscr{F}_t], \qquad 0 \leq t \leq T. \tag{4.221}$$

In view of (4.220), it would be natural to choose a transformation of $(\vec{p}, \vec{q})$ into an adjoint process $(\tilde{p}, \tilde{q})$, where

$$\tilde{p}(t) := \begin{pmatrix} \tilde{p}_1(t) \\ \tilde{p}_2(t) \end{pmatrix}, \quad \tilde{q}(t) := \begin{pmatrix} \tilde{q}_1(t) \\ \tilde{q}_2(t) \end{pmatrix}, \quad 0 \le t \le T,$$

such that

$$\tilde{p}_2(t) = \frac{p_2(t)}{\theta v^\theta(t)} = -1, \qquad 0 \le t \le T, \tag{4.222}$$

which would imply that, for almost every $0 \le t \le T$, $\tilde{q}_2(t) = 0$, $\quad \mathbb{P}$a.s.

We consider the following transform

$$\tilde{p}(t) = \begin{pmatrix} \tilde{p}_1(t) \\ \tilde{p}_2(t) \end{pmatrix} := \frac{1}{\theta v^\theta(t)} \vec{p}(t), \qquad 0 \le t \le T. \tag{4.223}$$

In view of (4.214) and (4.307), we have

$$\tilde{p}(T) = - \begin{pmatrix} h_x(T) + \frac{1}{v^\theta(T)} E[v^\theta(T) h_y(T)] \\ 1 \end{pmatrix}. \tag{4.224}$$

The following properties of the generic martingale v^θ are essential in order to investigate the properties of these new processes $(\tilde{p}, \tilde{q})$.

First, we note that since, by Assumption (H1), f and h are bounded by some constant $C > 0$, we have

$$0 < e^{-(1+T)C\theta} \le \phi^\theta(T) \le e^{(1+T)C\theta}. \tag{4.225}$$

Therefore, v^θ is a uniformly bounded $\mathbb{F}$-martingale satisfying

$$0 < e^{-(1+T)C\theta} \le v^\theta(t) \le e^{(1+T)C\theta}, \qquad 0 \le t \le T. \tag{4.226}$$

Furthermore, the martingale v^θ enjoys the following useful *logarithmic transform*:

$$v^\theta(t) = \exp\left(\theta Y_t + \theta \int_0^t f(s, \bar{x}(s), E[\bar{x}(s)], \bar{u}(s)) ds\right), \quad 0 \le t \le T, \tag{4.227}$$

and

$$v^\theta(0) = E[\phi^\theta(T)] = \exp(\theta Y_0), \tag{4.228}$$

where in view of (4.226) and the boundedness of f,

$$\sup_{0 \le t \le T} |Y_t| \le C_T, \tag{4.229}$$

where C_T is a positive constant that depends only on T and the bounds of f and h. Moreover, the process Y is the first component of the $\mathbb{F}$-adapted pair of processes (Y, ℓ) which is the unique solution to the following quadratic BSDE (see Appendix):

$$\begin{cases} dY_t = -\{f(t, \bar{x}(t), E[\bar{x}(t)], \bar{u}(t)) + \frac{\theta}{2}|\ell(t)|^2\}dt + \ell(t)dW_t, \\ Y_T = h(\bar{x}(T), E[\bar{x}(T)]), \end{cases} \tag{4.230}$$

where

$$E\left[\int_0^T |\ell(t)|^2 dt\right] < \infty. \tag{4.231}$$

In particular, v^θ is the unique solution of the following linear backward SDE

$$dv^\theta(t) = \theta\ell(t)v^\theta(t)dW_t, \quad v^\theta(T) = \phi^\theta(T). \tag{4.232}$$

Hence,

$$\frac{v^\theta(t)}{v^\theta(0)} = \exp\left(\int_0^t \theta\ell(s)dB_s - \frac{\theta^2}{2}\int_0^t |\ell(s)|^2 ds\right) := L_t^\theta, \quad 0 \le t \le T, \tag{4.233}$$

is a uniformly bounded $\mathbb{F}$-martingale.

We wish to identify the processes $\tilde{\alpha}$ and $\tilde{q}$ such that

$$d\tilde{p}(t) = -\tilde{\alpha}(t)dt + \tilde{q}(t)dW_t. \tag{4.234}$$

We may apply Itô's formula to the process $\vec{p}(t) = \theta v^\theta \tilde{p}(t)$, using the expression of v^θ in (4.316), to obtain

$$d\vec{p}(t) = \theta v^\theta(t)d\tilde{p}(t) + \theta^2\ell(t)v^\theta(t)\tilde{q}(t)dt + \theta^2\ell(t)v^\theta(t)\tilde{p}(t)dW_t.$$

Thus,

$$d\tilde{p}(t) = \frac{d\vec{p}(t)}{\theta v^\theta(t)} - \theta\ell(t)\tilde{q}(t)dt - \tilde{p}(t)\theta\ell(t)dW_t.$$

Substituting the expression of $d\vec{p}$ in (4.214) and identifying the coefficients, we get the diffusion term

$$\tilde{q}(t) = \frac{1}{\theta v^\theta(t)}\vec{q}(t) - \theta\ell(t)\tilde{p}(t), \quad 0 \le t \le T, \tag{4.235}$$

and the drift term of the process $\tilde{p}$

$$
\begin{aligned}
\tilde{\alpha}(t) &= \begin{pmatrix} b_x(t) & 0 \\ f_x(t) & 0 \end{pmatrix}' \frac{1}{\theta v^\theta(t)} \vec{p}(t) + \begin{pmatrix} \sigma_x(t) & 0 \\ 0 & 0 \end{pmatrix}' \frac{1}{\theta v^\theta(t)} \vec{q}(t) \\
&+ \frac{1}{\theta v^\theta(t)} E\left[\begin{pmatrix} b_y(t) & 0 \\ f_y(t) & 0 \end{pmatrix}' \vec{p}(t) + \begin{pmatrix} \sigma_y(t) & 0 \\ 0 & 0 \end{pmatrix}' \vec{q} \right] + \theta \ell(t) \tilde{q}(t).
\end{aligned}
$$

Now using the relations

$$
\vec{p}(t) = \theta v^\theta(t) \tilde{p}(t), \quad \vec{q}(t) = \theta v^\theta(t) \tilde{q}(t) + \theta \tilde{p}(t) \theta \ell(t) v^\theta(t),
$$

we finally arrive at

$$
\begin{aligned}
\tilde{\alpha}(t) &= \begin{pmatrix} b_x(t) & 0 \\ f_x(t) & 0 \end{pmatrix}' \tilde{p}(t) + \begin{pmatrix} \sigma_x(t) & 0 \\ 0 & 0 \end{pmatrix}' \{\tilde{q}(t) + \theta \ell(t) \tilde{p}(t)\} + \theta \ell(t) \tilde{q}(t) \\
&+ \frac{1}{v^\theta(t)} E\left[v^\theta(t) \begin{pmatrix} b_y(t) & 0 \\ f_y(t) & 0 \end{pmatrix}' \tilde{p}(t) + v^\theta(t) \begin{pmatrix} \sigma_y(t) & 0 \\ 0 & 0 \end{pmatrix}' \{\tilde{q}(t) + \theta \ell(t) \tilde{p}(t)\} \right].
\end{aligned}
\tag{4.236}
$$

It is easily verified that

$$
d\tilde{p}_2(t) = \tilde{q}_2(t)[-\theta \ell(t) dt + dW_t], \quad \tilde{p}_2(T) = -1.
$$

In view of (4.317), we may use the Girsanov Theorem to claim that

$$
d\tilde{p}_2(t) = \tilde{q}_2(t) dW_t^\theta, \quad \mathbb{P}^\theta \text{a.s.} \qquad \tilde{p}_2(T) = -1,
$$

where

$$
W_t^\theta := W_t - \int_0^t \theta \ell(s) ds
$$

is a $\mathbb{P}^\theta$-Brownian motion, where

$$
\frac{d\mathbb{P}^\theta}{d\mathbb{P}}|_{\mathscr{F}_t} := L_t^\theta = \exp\Big(\int_0^t \theta \ell(s) dW_s - \frac{\theta^2}{2} \int_0^t |\ell(s)|^2 ds \Big), \qquad 0 \le t \le T.
$$

In view of (4.317) and (4.225), the probability measures $\mathbb{P}$ and $\mathbb{P}^\theta$ are in fact equivalent. Hence, noting that $\tilde{p}_2(t) := [\theta v^\theta(t)]^{-1} p_2(t)$ is square-integrable, we get that $\tilde{p}_2(t) = E^{\mathbb{P}^\theta}[\tilde{p}_2(T)|\mathscr{F}_t] = -1$. Thus, its quadratic variation is zero: $\int_0^T |\tilde{q}_2(t)|^2 dt = 0$. This implies that, for almost every $0 \le t \le T$, $\tilde{q}_2(t) = 0$, $\mathbb{P}^\theta$ and $\mathbb{P}$a.s.

Therefore, the first component of $\tilde{\alpha}$ given by (4.236) reads

$$\tilde{\alpha}_1(t) = b_x(t)\tilde{p}_1(t) + \sigma_x(t)\left(\tilde{q}_1(t) + \theta\ell(t)\tilde{p}_1(t)\right) - f_x(t) + \tilde{q}_1(t)\theta\ell(t) \\ + \tfrac{1}{v^\theta(t)} E\left[v^\theta(t)\left(b_y(t)\tilde{p}_1(t) + \sigma_y(t)\left(\tilde{q}_1(t) + \theta\ell(t)\tilde{p}_1(t)\right) - f_y(t)\right)\right]. \tag{4.237}$$

Hence, the risk-sensitive first-order adjoint equation satisfied by $(\tilde{p}_1, \tilde{q}_1)$ and (v^θ, ℓ) becomes

$$\begin{cases} d\tilde{p}_1(t) = -\tilde{\alpha}_1(t)dt + \tilde{q}_1 dW_t, \\ dv^\theta(t) = \theta\ell(t)v^\theta(t)dW_t, \\ v^\theta(T) = \phi^\theta(T), \\ \tilde{p}_1(T) = -h_x(T) - \tfrac{1}{\phi^\theta(T)}E[\phi^\theta(T)h_y(T)]. \end{cases} \tag{4.238}$$

In view of the uniqueness of $\mathbb{F}$-adapted pairs $(\vec{p}, \vec{q})$, solution of (4.214) and the pair (v^θ, ℓ) obtained by (4.230) and (4.316), the solution of the system of backward SDEs (4.238) is unique. This achieves the proof of Proposition 4.9.

4.7.3 Transformation of the Hamiltonian

In view of (4.309) and (4.235), the Hamiltonian $\tilde{H}^\theta$, associated with (4.213), given by

$$\tilde{H}^\theta(t, \bar{x}(t), u, \vec{p}(t), \vec{q}(t)) = \begin{pmatrix} b(t, \bar{x}(t), E[\bar{x}(t)], u) \\ f(t, \bar{x}(t), E[\bar{x}(t)], u) \end{pmatrix} \cdot \vec{p}(t) \\ + \begin{pmatrix} \sigma(t, \bar{x}(t), E[\bar{x}(t)], u) \\ 0 \end{pmatrix} \cdot \vec{q}(t),$$

satisfies

$$\tilde{H}^\theta(t, \bar{x}(t), u, \vec{p}(t), \vec{q}(t)) = [\theta v^\theta(t)]H^\theta(t, \bar{x}(t), u, \tilde{p}_1(t), \tilde{q}_1(t), \ell(t)) \tag{4.239}$$

where H^θ is the risk-sensitive Hamiltonian given by (4.200).

Using the notation (4.198), we have the following relation between the drift term $\tilde{\alpha}_1$ in (4.236) and the gradient of the risk-sensitive Hamiltonian H^θ:

$$\tilde{\alpha}_1(t) = H_x^\theta(t) + \frac{1}{v^\theta(t)}E[v^\theta(t)H_y^\theta(t)] + \theta\ell(t)\tilde{q}_1(t). \tag{4.240}$$

Hence, the risk-sensitive first-order adjoint equation (4.238) becomes

$$\begin{cases} d\tilde{p}_1(t) = -\left\{H_x^\theta(t) + \frac{1}{v^\theta(t)}E[v^\theta(t)H_y^\theta(t)]\right\}dt + \tilde{q}_1(t)(-\theta\ell(t)dt + dW_t), \\ dv^\theta(t) = \theta\ell(t)v^\theta(t)dW_t, \\ v^\theta(T) = \phi^\theta(T), \\ \tilde{p}_1(T) = -h_x(T) - \frac{1}{\phi^\theta(T)}E[\phi^\theta(T)h_y(T)]. \end{cases} \tag{4.241}$$

4.7.4 Transformation of the Second-Order Adjoint Process

For the second-order adjoint equation, we let, for $0 \le t \le T$,

$$\tilde{P}(t) = \frac{P(t)}{\theta v^\theta(t)} + \theta\tilde{p}(t)\tilde{p}'(t), \tag{4.242}$$

and

$$\tilde{Q}(t) = \frac{Q(t)}{\theta v^\theta(t)} + \theta\tilde{q}(t)\tilde{p}'(t) + \theta\tilde{p}(t)\tilde{q}'(t) - \theta\ell(t)\left(\tilde{P}(t) - \theta\tilde{p}(t)\tilde{p}'(t)\right). \tag{4.243}$$

In view of (4.217) satisfied by (P, Q), lengthy (but straightforward) calculations yield that

$$\tilde{P}(t) = \begin{pmatrix} \tilde{P}_1(t) & 0 \\ 0 & 0 \end{pmatrix}, \quad \tilde{Q} = \begin{pmatrix} \tilde{Q}_1(t) & 0 \\ 0 & 0 \end{pmatrix}, \quad 0 \le t \le T, \tag{4.244}$$

where $(\tilde{P}_1, \tilde{Q}_1)$ is a pair of one-dimensional processes that uniquely solve the risk-sensitive second-order adjoint equation:

$$\begin{cases} d\tilde{P}_1 = -\Big\{\left(2b_x(t) + \sigma_x^2(t) + 2\theta\ell(t)\sigma_x(t)\right)\tilde{P}_1(t) + 2\sigma_x(t)\tilde{Q}_1(t) \\ \qquad\quad -\theta(\tilde{q}_1(t) + \sigma_x(t)\tilde{p}_1(t))^2 + H_{xx}^\theta(t)\Big\}dt \\ \qquad\quad +\tilde{Q}_1(t)(-\theta\ell(t)dt + dW_t), \\ \tilde{P}_1(T) = -h_{xx}(T). \end{cases} \tag{4.245}$$

4.7.5 Risk-Sensitive Stochastic Maximum Principle

To arrive at a risk-sensitive SMP expressed in terms of the adjoint processes $(\tilde{p}_1, \tilde{q}_1)$, (v^θ, ℓ) and $(\tilde{P}_1, \tilde{Q}_1)$, which solve (4.241) and (4.245), respectively, we note that in view of (4.242), (4.243) and (4.244), the second term in the variational inequality (4.219) satisfies

$$\begin{pmatrix} \delta\sigma(t) \\ 0 \end{pmatrix}' P(t) \begin{pmatrix} \delta\sigma(t) \\ 0 \end{pmatrix} = [\theta v^\theta(t)] \begin{pmatrix} \delta\sigma(t) \\ 0 \end{pmatrix}' \left[\tilde{P}(t) - \theta\tilde{p}(t)\tilde{p}'(t)\right] \begin{pmatrix} \delta\sigma(t) \\ 0 \end{pmatrix}$$
$$= \theta v^\theta(t) \left(\tilde{P}_1(t) - \theta\tilde{p}_1^2(t)\right) (\delta\sigma(t))^2 .$$

Combining this relation with (4.239), we obtain

$$\delta\tilde{H}^\theta(t) + \frac{1}{2} \begin{pmatrix} \delta\sigma(t) \\ 0 \end{pmatrix}' P(t) \begin{pmatrix} \delta\sigma(t) \\ 0 \end{pmatrix}$$

$$= \theta v^\theta(t) \left[\delta H^\theta(t) + \frac{1}{2}\left(\tilde{P}_1(t) - \theta\tilde{p}_1^2(t)\right)(\delta\sigma(t))^2\right].$$

Hence, since $v^\theta > 0$, the variational inequality (4.219) translates into

$$\delta H^\theta(t) + \frac{1}{2}\left(\tilde{P}_1(t) - \theta\tilde{p}_1^2(t)\right)(\delta\sigma(t))^2 \leq 0,$$

for all $u \in U$, almost every $t \in [0, T]$ and $\mathbb{P}$—almost surely.

This completes the proof of Theorem 4.13.

Illustrative Example: Linear-Quadratic Risk-Sensitive Model

As an illustration, we consider in the next two subsections two one-dimensional cases with linear state dynamics and exponential (at most) quadratic cost functionals. The linear-quadratic setup does not satisfy the boundedness condition of Assumption (H1). However, existence and uniqueness of risk-sensitive linear-quadratic optimal control problems have been widely investigated in the mean-field-free case, and their optimality can be checked by a direct method in the some time interval. Here, we incorporate a small modification by adding $\mu E[x(T)]$ in the cost functional. It is well known that in the absence of mean-field coupling, the optimal control is a linear feedback control whose feedback gain is obtained from the solution of a risk-sensitive Riccati equation which has an additional term when compared to the (classical) Riccati equation for the quadratic cost problem. In the examples below, we will show that this feature is still valid in the LQ risk-sensitive problem (with and without the mean-field coupling).

We apply the optimality equations derived in Theorem 4.13 even if the coefficients are unbounded. Since we do not have bounded coefficients, $\bar{p}(t)$ can blow up in finite time. The existence of the optimal control is therefore related to the existence of the adjoint process $\bar{p}(t)$ which in turn is conditioned on the existence of solution to a risk-sensitive Riccati equation. Thus, the linear-quadratic risk-sensitive mean-field-type control has a unique solution whenever the associated risk-sensitive Riccati equation solution does not blow up in the time interval $[0, T]$. The range of values of the risk-sensitivity parameter θ for which the Riccati equation solution remains well defined can be tuned to get a larger interval in the form $(t_*(\theta), T]$.

4.7.6 LQ Without Mean-Field Coupling

In order to show the relationship between our SMP result and the available results in the literature, we consider the following linear-quadratic risk-sensitive control problem:

$$\begin{cases} \inf_{u\in\mathcal{U}} Ee^{\theta\left[\frac{1}{2}\int_0^T u^2(t)dt+\frac{1}{2}x^2(T)\right]}, \\ \text{subject to} \\ dx(t) = (ax(t) + bu(t))\,dt + \sigma dW_t, \\ x(0) = x_0, \end{cases} \tag{4.246}$$

where a, b and σ are real constants.

An admissible pair $(\bar{x}(\cdot), \bar{u}(\cdot))$ that satisfies the necessary optimality conditions of Theorem 4.13 can be obtained by solving the following system of forward-backward SDEs:

$$\begin{cases} d\bar{x}(t) = (a\bar{x}(t) + b\bar{u}(t))\,dt + \sigma dW_t, \\ d\bar{p}(t) = -\left\{a\bar{p}(t) + \theta\ell(t)\bar{q}(t)\right\}dt + \bar{q}(t)dW_t, \\ dv^\theta(t) = \theta\ell(t)v^\theta(t)dW_t, \\ v^\theta(T) = \phi^\theta(T), \\ \bar{x}(0) = x_0, \quad \bar{p}(T) = -\bar{x}(T), \end{cases} \tag{4.247}$$

where $\phi^\theta(T) := e^{\theta\left[\frac{1}{2}\int_0^T \bar{u}^2(t)dt+\frac{1}{2}\bar{x}^2(T)\right]}$.

This system involves only the first adjoint equation, because the diffusion coefficient in the state dynamics is independent of the control.

The associated risk-sensitive Hamiltonian is

$$H^\theta(t, x, u, p, q, \ell) := (ax + bu)p - \frac{1}{2}u^2 + \sigma(q + \theta\ell p).$$

We have

$$H_x^\theta = ap, \qquad H_u^\theta = bp - u.$$

Maximizing the Hamiltonian yields

$$\bar{u}(t) = bp(t). \tag{4.248}$$

The associated state dynamics $\bar{x}$ then solves the SDE

$$d\bar{x}(t) = \left(a\bar{x}(t) + b^2\bar{p}(t)\right)dt + \sigma dW_t. \tag{4.249}$$

Thanks to the martingale representation theorem (4.313), the BSDE satisfied by (v^θ, ℓ) is equivalent to the solution (Y, ℓ) of the BSDE

$$\begin{cases} dY_t = -\{\frac{1}{2}\bar{u}^2(t) + \frac{\theta}{2}\ell^2(t)\}dt + \ell(t)dW_t, \\ Y_T = \frac{1}{2}\bar{x}^2(T). \end{cases} \tag{4.250}$$

We try a solution of the form

$$Y_t = \frac{\alpha(t)}{2}\bar{x}^2(t) + \beta(t), \tag{4.251}$$

where α and β are deterministic functions such that $\alpha(T) = 1$, $\beta(T) = 0$. In view of the SDE satisfied by $\bar{x}$ in (4.247), applying Itô's formula and identifying the coefficients, we obtain

$$\ell(t) = \sigma\alpha(t)\bar{x}(t) \tag{4.252}$$

and

$$\begin{aligned} &\tfrac{1}{2}\left(\bar{u}(t) + b\alpha(t)\bar{x}(t)\right)^2 + \tfrac{1}{2}\left(\dot{\alpha}(t) + (\theta\sigma^2 - b)\alpha^2(t) + 2a\alpha(t)\right)\bar{x}^2(t) \\ &\quad + \left(\dot{\beta}(t) + \tfrac{1}{2}\sigma^2\alpha(t)\right) = 0. \end{aligned} \tag{4.253}$$

Hence,

$$\bar{u}(t) = -b\alpha(t)\bar{x}(t) = -\frac{b}{\sigma}\ell(t), \tag{4.254}$$

and

$$\begin{cases} \dot{\alpha}(t) + (\theta\sigma^2 - b)\alpha^2(t) + 2a\alpha(t) = 0, \quad \alpha(T) = 1, \\ \dot{\beta}(t) + \frac{\sigma^2}{2}\alpha(t) = 0, \quad \gamma(T) = 0. \end{cases} \tag{4.255}$$

where we note that α solves the standard risk-sensitive Riccati equation, as expected.

Now, in view of (4.8.4) and (4.254), we obtain the relationship between the adjoint process p and ℓ,

$$\ell(t) = -\sigma\bar{p}(t). \tag{4.256}$$

Thus,

$$\bar{p}(t) = -\alpha(t)\bar{x}(t). \tag{4.257}$$

Identifying the coefficients in the BSDE satisfied by p in (4.247) and (4.257), we obtain

$$\bar{q}(t) = -\sigma\alpha(t). \quad (4.258)$$

Summing up, the optimal control is

$$\bar{u}(t) = -b\alpha(t)\bar{x}(t), \quad (4.259)$$

where α solves the risk-sensitive Riccati equation

$$\dot{\alpha}(t) + (\theta\sigma^2 - b)\alpha^2(t) + 2a\alpha(t) = 0, \quad \alpha(T) = 1, \quad (4.260)$$

and the optimal dynamics solves the linear SDE

$$d\bar{x}(t) = \left(a - b^2\alpha(t)\right)\bar{x}(t)dt + \sigma dW_t, \quad \bar{x}(0) = x_0. \quad (4.261)$$

4.7.7 LQ Risk-Sensitive Control with a Mean-Field Coupling

We keep the same functions f, σ, b as in (4.323) but we modify the terminal cost h to be

$$h(x(T), E[x(T)]) = \frac{1}{2}x^2(T) + \mu E[x(T)],$$

for some given constant μ, where the only mean-field coupling is $E[x(T)]$.

The associated risk-sensitive Hamiltonian is

$$H^\theta(t, x, u, p, q, \ell) := (ax + bu)p - \frac{1}{2}u^2 + \sigma(q + \theta\ell p).$$

We have

$$H_x^\theta = ap, \qquad H_u^\theta = bp - u.$$

Maximizing the Hamiltonian yields

$$\bar{u}(t) = b\bar{p}(t). \quad (4.262)$$

The first-order adjoint equation remains the same as (4.247), but the terminal condition becomes

$$\bar{p}(T) = -\bar{x}(T) - \frac{\mu}{\phi^\theta(T)}E[\phi^\theta(T)], \quad (4.263)$$

where $\phi^\theta(T) := e^{\theta\left[\frac{1}{2}\int_0^T \bar{u}^2(t)dt + \frac{1}{2}\bar{x}^2(T) + \mu E[\bar{x}(T)]\right]}$.

In view of (4.313), we have

$$L_t^\theta := \frac{v^\theta(t)}{E[\phi^\theta(T)]} = \frac{v^\theta(t)}{v^\theta(0)} = \exp\Big(\theta \int_0^t \ell(s)dB_s - \frac{\theta^2}{2}\int_0^t |\ell(s)|^2 ds\Big). \tag{4.264}$$

which satisfies the linear SDE

$$dL_t^\theta = \theta \ell(t) L_t^\theta dW_t, \qquad L_0^\theta = 1. \tag{4.265}$$

Hence, the end-value (4.263) becomes

$$L_T^\theta \bar{p}(T) = -L_T^\theta \bar{x}(T) - \mu. \tag{4.266}$$

In view of the form (4.266), we try a solution $p(t)$ such that

$$\bar{p}(t)L_t^\theta = -\alpha(t)\bar{x}(t)L_t^\theta - \mu\beta(t), \tag{4.267}$$

where $\alpha(t)$ and $\beta(t)$ are deterministic functions such that $\alpha(T) = 1$ and $\beta(T) = 1$. Proceeding as above, in view of (4.262), we apply Itô's formula to the process $L_t^\theta \bar{p}(t)$ using (4.247) and (4.265), and then (4.267) and (4.8.4), and identifying the coefficient, we obtain

$$\bar{q}(t) = \frac{\theta\ell(t)}{L_t^\theta}\mu\beta(t) - \sigma\alpha(t), \tag{4.268}$$

and

$$\begin{aligned}
&-\mu\dot{\beta}(t) + \mu(b^2\alpha(t) - a)\beta(t) \\
&- \big(\dot{\alpha}(t) + 2a\alpha(t) + (\theta\sigma^2 - b^2)\alpha^2(t)\big)\,\bar{x}(t)L_t^\theta \\
&+ \big(\sigma\theta\alpha(t)\ell(t) - \theta\sigma^2\alpha^2(t)\bar{x}(t)\big)\,L_t^\theta = 0.
\end{aligned} \tag{4.269}$$

Therefore,

$$\begin{cases} \dot{\alpha}(t) + 2a\alpha(t) + (\theta\sigma^2 - b^2)\alpha^2(t) = 0, \quad \alpha(T) = 1, \\ -\dot{\beta}(t) + (b^2\alpha(t) - a)\beta(t) = 0, \quad \beta(T) = 1, \end{cases} \tag{4.270}$$

and

$$\ell(t) = \sigma\alpha(t)\bar{x}(t). \tag{4.271}$$

Now, the relationship between the adjoint processes p and ℓ reads

$$\ell(t) = -\sigma\bar{p}(t) - \mu\sigma\beta(t)(L_t^\theta)^{-1}. \tag{4.272}$$

Another way to check that ℓ satisfies (4.271) is to solve the BSDE satisfied by (v^θ, ℓ) or equivalently, by the martingale representation theorem (4.313), the BSDE satisfied by (Y, ℓ):

$$\begin{cases} dY_t = -\{\frac{1}{2}\bar{u}^2(t) + \frac{\theta}{2}\ell^2(t)\}dt + \ell(t)dW_t, \\ Y_T = \frac{1}{2}\bar{x}^2(T) + \mu E[\bar{x}(T)], \end{cases} \tag{4.273}$$

where by (4.262) and (4.267),

$$\bar{u}(t) = b(\alpha(t)\bar{x}(t) + \mu\beta(t)(L_t^\theta)^{-1}).$$

Set

$$\bar{y}(t) = \frac{\alpha(t)}{2}\bar{x}^2(t) + \beta(t)E[\bar{x}(t)] + \gamma(t), \tag{4.274}$$

where

$$\dot{\gamma}(t) + \frac{\sigma^2}{2}\alpha(t) - \mu^2 b^2\lambda(t)\beta(t)E[(L_t^\theta)^{-1}] = 0, \quad \gamma(T) = 0, \tag{4.275}$$

and (α, β) satisfy (4.270).

It is easily verified that

$$\begin{aligned} &\bar{y}(t) = \bar{y}(T) \\ &+ \int_t^T \tfrac{1}{2}(b\alpha(s)\bar{x}(s))^2 + b^2\mu\alpha(s)\beta(s)\bar{x}(t)(L^\theta)^{-1}(s)ds \\ &+ \int_t^T \tfrac{\theta}{2}(\sigma\alpha(s)\bar{x}(s))^2 ds - \int_t^T (\sigma\alpha(s)\bar{x}(s))dB_s. \end{aligned} \tag{4.276}$$

Moreover,

$$\begin{cases} \ell(t) = \sigma\alpha(t)\bar{x}(t), \\ dY_t = d\bar{y}(t) - \frac{1}{2}\left(b\mu\beta(s)(L_t^\theta)^{-1}\right)^2 dt, \\ Y_T = \bar{y}(T) = \frac{1}{2}\bar{x}^2(T) + \mu E[\bar{x}(T)]. \end{cases} \tag{4.277}$$

solves (4.273).

Summing up, the optimal control of the LQ-problem with mean-field coupling is

$$\bar{u}(t) = -b\alpha(t)\bar{x}(t) + \mu\beta(t)(L_t^\theta)^{-1}, \tag{4.278}$$

where

$$dL_t^\theta = \theta\sigma\alpha(t)\bar{x}(t)L_t^\theta dW_t, \quad L_0^\theta = 1, \tag{4.279}$$

and (α, β) solves

$$\begin{cases} \dot{\alpha}(t) + 2a\alpha(t) + (\theta\sigma^2 - b^2)\alpha^2(t) = 0, \quad \alpha(T) = 1, \\ -\dot{\beta}(t) + (b^2\alpha(t) - a)\beta(t) = 0, \quad \beta(T) = 1, \end{cases} \tag{4.280}$$

and the optimal dynamics solves the linear SDE

$$d\bar{x}(t) = \{\left(a - b^2\alpha(t)\right)\bar{x}(t) + \mu\beta(t)(L^{\theta})^{-1}(t)\}dt + \sigma dW_t, \ x(0) = x_0. \tag{4.281}$$

We note that if we set $\mu = 0$ in (4.278), we obtain the optimal control related to the mean-field free case given in the previous subsection.

4.8 Risk-Sensitive Mean-Field Type Control Under Partial Observation

The purpose of this section is to establish a stochastic maximum principle for a class of risk-sensitive mean-field type control problems under partial observation. We present the model and state the partially observable risk-sensitive SMP.

Statement of the Problem
Let $T > 0$ be a fixed time horizon and $(\Omega, \mathscr{F}, \mathbb{F}, \mathbb{P})$ be a given filtered probability space on which there are defined two independent standard one-dimensional Brownian motions $W = \{W_s\}_{s\geq 0}$, and $Y = \{Y_s\}_{s\geq 0}$. Let $\mathscr{F}_t^W$ and $\mathscr{F}_t^Y$ be the $\mathbb{P}$-completed natural filtrations generated by W and Y, respectively. Set $\mathbb{F}^Y := \{\mathscr{F}_t^Y, \ 0 \leq s \leq T\}$ and $\mathbb{F} := \{\mathscr{F}_s, \ 0 \leq s \leq T\}$, where $\mathscr{F}_t = \mathscr{F}_t^W \vee \mathscr{F}_t^Y$.

The model is defined as follows:

(i) An admissible control u is an $\mathbb{F}^Y$-adapted process with values in a non-empty subset (not necessarily convex) U of $\mathbb{R}$ and satisfies $E[\int_0^T |u(t)|^2 dt] < \infty$. We denote the set of all admissible controls by $\mathscr{U}$. The control u is called partially observable.

(ii) Given a control process $u \in \mathscr{U}$, we consider the signal-observation pair (x^u, Y) which satisfies the following SDE of mean-field type

$$\begin{cases} dx^u(t) = b(t, x^u(t), E[x^u(t)], u(t))dt + \sigma(t, x^u(t), E[x^u(t)])dW_t \\ \qquad\qquad +\alpha(t, x^u(t), E[x^u(t)])d\widetilde{W}_t^u, \quad x^u(0) = x_0, \\ dY_t = \beta(t, x^u(t))dt + d\widetilde{W}_t^u, \quad Y_0 = 0, \end{cases} \tag{4.282}$$

where

$$b(t, x, m, u): \ [0, T] \times \mathbb{R} \times \mathbb{R} \times U \longrightarrow \mathbb{R},$$

$$\alpha(t, x, m), \ \sigma(t, x, m): \ [0, T] \times \mathbb{R} \times \mathbb{R} \longrightarrow \mathbb{R}$$

and $\beta(t, x) : [0, T] \times \mathbb{R} \longrightarrow \mathbb{R}$ are Borel measurable function.

In this model, the observation process Y, which carries out the controls u, is assumed to be a given Brownian motion independent of W and is supposed to admit a decomposition as a trend $\int_0^{\cdot} \beta(t, x^u(t))dt$ (a functional of the state process x^u) corrupted by a process $\widetilde{W}^u$ which are a priori not observable.

A more general model of the function β would be to let it depend on the control u and be of mean-field-type [333, 339, 342]. To keep the presentation simpler, we skip this case in this section. But the main results do extend to this case.

Before we formulate the control problem, we show that the system (4.282) has a weak solution. Introduce the density process defined on $(\Omega, \mathscr{F}, \mathbb{F}, \mathbb{P})$ by

$$\rho^u(t) := \exp\left\{\int_0^t \beta(s, x^u(s))dY_s - \frac{1}{2}\int_0^t |\beta(s, x^u(s))|^2 ds\right\},$$

which solves the linear SDE

$$d\rho^u(t) = \rho^u(t)\beta(t, x^u(t))dY_t, \quad \rho^u(0) = 1.$$

Assuming the function β bounded (see Assumption 4.287), ρ is a uniformly integrable martingale such that, for every $p \geq 2$,

$$E[\sup_{0 \leq t \leq T} \rho_t^p] \leq C, \tag{4.283}$$

where C is a constant which depends only on the bound of β, p, and T. Define $d\mathbb{P}^u = \rho^u(T)d\mathbb{P}$. By Girsanov's theorem, $\mathbb{P}^u$ is a probability measure. Moreover, $\widetilde{W}^u$ is a $\mathbb{P}^u$-standard Brownian motion independent of W. This in turn entails that $(\mathbb{P}^u, x^u, Y, W, \widetilde{W}^u)$ is a weak solution of (4.282).

The objective is to characterize admissible controls which minimize the risk-sensitive cost functional given by

$$J^\theta(u(\cdot)) = E^u\left[\exp\left(\theta\left[\int_0^T f(t, x^u(t), E^u[x^u(t)], u(t))\, dt + h(x^u(T), E^u[x^u(T)])\right]\right)\right],$$

where θ is the risk-sensitivity index,

$$\begin{aligned} &f(t, x, m, u): \ [0, T] \times \mathbb{R} \times \mathbb{R} \times U \longrightarrow \mathbb{R}, \\ &h(x, m): \ \mathbb{R} \times \mathbb{R} \longrightarrow \mathbb{R}. \end{aligned}$$

Any $\bar{u}(\cdot) \in \mathscr{U}$ which satisfies

$$J^\theta(\bar{u}(\cdot)) = \inf_{u(\cdot) \in \mathscr{U}} J^\theta(u(\cdot)) \tag{4.284}$$

is called a risk-sensitive optimal control under partial observation.

Let $\Psi_T = \int_0^T f(t, x(t), E^u[x(t)], u(t))dt + h(x(T), E^u[x(T)])$, and consider the payoff functional given by

$$\widetilde{\Psi}_\theta := \frac{1}{\theta} \log E^u e^{\theta \Psi_T}.$$

When the risk-sensitive index θ is small, the loss functional $\widetilde{\Psi}_\theta$ can be expanded as

$$E^u[\Psi_T] + \frac{\theta}{2}\text{var}_u(\Psi_T) + O(\theta^2),$$

where $\text{var}_u(\Psi_T)$ denotes the variance of Ψ_T w.r.t. $\mathbb{P}^u$. If $\theta < 0$, the variance of Ψ_T, as a measure of risk, improves the performance $\widetilde{\Psi}_\theta$, in which case the optimizer is called *risk seeker*. But, when $\theta > 0$, the variance of Ψ_T worsens the performance $\widetilde{\Psi}_\theta$, in which case the optimizer is called *risk averse*. The risk-neutral loss functional $E^u[\Psi_T]$ can be seen as a limit of risk-sensitive functional $\widetilde{\Psi}_\theta$ when $\theta \to 0$.

Since $d\mathbb{P}^u = \rho^u(T)d\mathbb{P}$, the associated risk-sensitive cost functional becomes

$$J^\theta(u(\cdot)) = E\left[\rho^u(T)e^{\theta\left[\int_0^T f(t,x^u(t),E[\rho^u(t)x^u(t)],u(t))\,dt + h(x^u(T),E[\rho^u(T)x^u(T)])\right]}\right], \tag{4.285}$$

where, on $(\Omega, \mathscr{F}, \mathbb{F}, \mathbb{P})$, the process (ρ^u, x^u) satisfies the following dynamics:

$$\begin{cases} d\rho^u(t) = \rho^u(t)\beta(s, x^u(s))dY_t, \\ dx^u(t) = \{b(t, x^u(t), E[x^u(t)], u(t)) - \alpha(t, x^u(t), E[x^u(t)])\beta(t, x^u(t))\}\,dt \\ \qquad\qquad +\sigma(t, x^u(t), E[x^u(t)])dW_t + \alpha(t, x^u(t), E[x^u(t)])dY_t, \\ \rho^u(0) = 1, \;\; x^u(0) = x_0. \end{cases} \tag{4.286}$$

We have recast the partially observable control problem into the following completely observable control problem: Minimize $J^\theta(u(\cdot))$ defined by (4.285) subject to (4.286).

The main result of this section is a stochastic maximum principle (SMP) in terms of necessary optimality conditions for the problem (4.284) subject to (4.286).

We will only consider the case where the risk-sensitive parameter is positive, $\theta > 0$. The case $\theta < 0$ can be treated in a similar fashion by considering $\theta = -\bar{\theta}, \bar{\theta} > 0$, and $\bar{f} = -f, \bar{h} = -h$ in the performance functional (4.285).

We will make the following assumption.

$$\begin{cases} \text{The functions } b, \sigma, \alpha, \beta, f, h \text{ are twice continuously differentiable with respect to} \\ (x, m). \text{ Moreover, these functions and their first derivatives with respect to } (x, m) \\ \text{are bounded and continuous in } (x, m, u). \end{cases} \tag{4.287}$$

To keep the presentation less technical, we impose these assumptions, although they are restrictive and can be made weaker.

Under these assumptions, for each $u \in \mathscr{U}$, the SDE (4.286) admits a unique weak solution (ρ^u, x^u).

We now state an SMP to characterize optimal controls $\bar{u}(\cdot) \in \mathscr{U}$ which minimize (4.285), subject to (4.286). Let $(\bar{\rho}, \bar{x}) := (\rho^{\bar{u}}, x^{\bar{u}})$ denote the corresponding state process, defined as the solution of (4.286).

We introduce the following notation.

$$
\begin{aligned}
& X := \begin{pmatrix} \rho \\ x \end{pmatrix}, \quad \bar{X} := \begin{pmatrix} \bar{\rho} \\ \bar{x} \end{pmatrix}, \quad X_0 = \bar{X}_0 := \begin{pmatrix} 1 \\ x_0 \end{pmatrix}, \quad B_t := \begin{pmatrix} Y_t \\ W_t \end{pmatrix}, \\
& F(t, X, m, u) := \begin{pmatrix} 0 \\ c(t, x, m, u) \end{pmatrix}, \quad G(t, X, m) := \begin{pmatrix} \rho\beta(t, x) & 0 \\ \alpha(t, x, m) & \sigma(t, x, m) \end{pmatrix}, \\
& c(t, x, m, u) := b(t, x, m, u) - \alpha(t, x, m)\beta(t, x), \ \phi(X) := x, \ \tilde{\phi}(X) := \rho x, \\
& \phi(\bar{X}) := \bar{x}, \ \tilde{\phi}(\bar{X}) := \bar{\rho}\bar{x}.
\end{aligned}
\tag{4.288}
$$

We define the risk-neutral Hamiltonian as follows. For $(p, q) \in \mathbb{R}^2 \times \mathbb{R}^{2\times 2}$,

$$
H(t, X, m, p, q, u) := \langle F(t, X, m, u), p\rangle + \mathrm{tr}(G^*(t, X, m)q) - f(t, x, m, u),
\tag{4.289}
$$

where $'*'$ denotes the transposition operation of a matrix or a vector.

We also introduce the risk-sensitive Hamiltonian: $(p, q, \ell) \in \mathbb{R}^2 \times \mathbb{R}^{2\times 2} \times \mathbb{R}^2$,

$$
\begin{aligned}
H^\theta(t, X, m, u, p, q, \ell) := & \langle F(t, X, m, u), p\rangle - f(t, x, m, u) \\
& + \mathrm{tr}(G^*(t, X, m)(q + \theta\ell p^*)).
\end{aligned}
\tag{4.290}
$$

We have $H = H^0$.

Setting

$$
\ell := \begin{pmatrix} \ell_1 \\ \ell_2 \end{pmatrix}, \quad p := \begin{pmatrix} p_1 \\ p_2 \end{pmatrix}, \quad q := \begin{pmatrix} q_{11} & q_{12} \\ q_{21} & q_{22} \end{pmatrix},
$$

the explicit form of the Hamiltonian (4.290) reads

$$
\begin{aligned}
H^\theta(t, X, m, u, p, q, \ell) := & c(t, x, m, u)p_2 - f(t, x, m, u) + \rho\beta(t, x)(q_{11} + \theta\ell_1 p_1) \\
& + \alpha(t, x, m)(q_{21} + \theta\ell_2 p_1) + \sigma(t, x, m)(q_{22} + \theta\ell_2 p_2).
\end{aligned}
$$

Setting $\theta = 0$, we obtain the explicit form of the Hamiltonian (4.289):

$$
\begin{aligned}
H(t, X, m, u, p, q) := & c(t, x, m, u)p_2 - f(t, x, m, u) + \rho\beta(t, x)q_{11} \\
& + \alpha(t, x, m)q_{21} + \sigma(t, x, m)q_{22}.
\end{aligned}
$$

With the obvious notation for the derivatives of the functions $b, \alpha, \beta, \sigma, f, h$ with respect to the arguments x and m, we further set

$$\begin{cases} H_x^\theta(t, X, m, u, p, q) := c_x(t, x, m, u)p_2 - f_x(t, x, m, u) \\ \qquad\qquad +\rho\beta_x(t, x)(q_{11} + \theta\ell_1 p_1) \\ \qquad\qquad +\alpha_x(t, x, m)(q_{21} + \theta\ell_2 p_1) + \sigma_x(t, x, m)(q_{22} + \theta\ell_2 p_2), \\ \check{H}_m^\theta(t, X, m, u, p, q) := c_m(t, x, m, u)p_2 + \alpha_m(t, x, m)(q_{21} + \theta\ell_2 p_1) \\ \qquad\qquad +\sigma_m(t, x, m)(q_{22} + \theta\ell_2 p_2), \\ H_\rho^\theta(t, X, m, u, p, q) := \beta(t, x)(q_{11} + \theta\ell_1 p_1). \end{cases}$$

With this notation, the system (4.286) can be rewritten in the following compact form

$$\begin{cases} dX(t)=F(t, X(t), E[\phi(X(t))], u(t))dt+G(t, X(t), E[\phi(X(t))])dB_t, \\ X(0)=X_0. \end{cases} \tag{4.291}$$

We define the risk-neutral Hamiltonian associated with random variables X such that $\phi(X)$ and $\tilde{\phi}(X)$ are $L^1(\Omega, \mathscr{F}, \mathbb{P})$ as follows (with the obvious abuse of notation): For $(p, q) \in \mathbb{R}^2 \times \mathbb{R}^{2\times 2}$,

$$\begin{aligned} H(t, X, p, q, u) &:= \langle F(t, X, E[\phi(X)], u), p\rangle - f(t, x, E[\tilde{\phi}(X)], u) \\ &\quad + \mathrm{tr}(G^*(t, X, E[\phi(X)])q). \end{aligned} \tag{4.292}$$

We also introduce the risk-sensitive Hamiltonian: $(p, q, \ell) \in \mathbb{R}^2 \times \mathbb{R}^{2\times 2} \times \mathbb{R}^2$,

$$\begin{aligned} H^\theta(t, X, u, p, q, \ell) &:= \langle F(t, X, E[\phi(X)], u), p\rangle - f(t, x, E[\tilde{\phi}(X)], u) \\ &\quad + \mathrm{tr}(G^*(t, X, E[\phi(X)])(q + \theta\ell p^*)). \end{aligned} \tag{4.293}$$

For $g \in \{b, c, \sigma, \alpha, \beta\}$ and $u \in U$, we set

$$g_x(t) := g_x(t, \bar{x}(t), E[\bar{x}(t)], \bar{u}(t)), \quad g_m(t) := g_m(t, \bar{x}(t), E[\bar{x}(t)], \bar{u}(t)) \tag{4.294}$$

and

$$\begin{cases} f_x(t) := f_x(t, \bar{x}(t), E[\bar{\rho}(t)\bar{x}(t)], \bar{u}(t)), \quad f_m(t) := f_m(t, \bar{x}(t), E[\bar{\rho}(t)\bar{x}(t)], \bar{u}(t)), \\ h_x(t) := h_x(\bar{x}(t), E[\bar{\rho}(t)\bar{x}(t)]), \quad h_m(t) := h_m(\bar{x}(t), E[\bar{\rho}(t)\bar{x}(t)]). \end{cases} \tag{4.295}$$

Let

$$\psi_T^\theta := \bar{\rho}(T)\exp\theta\left[\int_0^T f(t, \bar{x}(t), E[\bar{\rho}(t)\bar{x}(t)], \bar{u}(t))dt + h(\bar{x}(T), E[\bar{\rho}(T)\bar{x}(T)])\right].$$

We introduce the adjoint equations involved in the risk-sensitive SMP for our control problem.

$$\begin{cases} d\hat{p}(t) = -\begin{pmatrix} H_\rho^\theta(t) + \frac{1}{v^\theta(t)}E[v^\theta(t)\check{H}_m^\theta(t)] - \frac{\bar{x}(t)}{v^\theta(t)}E[v^\theta(t)f_m(t)] \\ H_x^\theta(t) + \frac{1}{v^\theta(t)}E[v^\theta(t)\check{H}_m^\theta(t)] - \frac{\bar{\rho}(t)}{v^\theta(t)}E[v^\theta(t)f_m(t)]] \end{pmatrix} dt \\ \qquad\qquad +\hat{q}(t)(-\theta\ell(t)dt + dB_t), \\ dv^\theta(t) = \theta v^\theta(t)\langle \ell(t), dB_t\rangle, \\ \hat{p}(T) = -\begin{pmatrix} (\theta\bar{\rho}(T))^{-1} \\ h_x(T) \end{pmatrix} - \begin{pmatrix} \bar{x}(T) \\ \bar{\rho}(T) \end{pmatrix} \frac{1}{\psi_T^\theta} E[\psi_T^\theta h_m(T)], \\ v^\theta(T) = \psi_T^\theta, \end{cases} \tag{4.296}$$

where, in view of (4.8) and (4.295), for $k \in \{\rho, x\}$,

$$\begin{aligned} H_k^\theta(t) := \langle F_k(t, \bar{X}(t), E[\phi(\bar{X}(t))], \bar{u}(t)), \hat{p}(t)\rangle - f_k(t, \bar{x}(t), E[\tilde{\phi}(\bar{X}(t))], \bar{u}(t)) \\ +\text{tr}(G_k^*(t, \bar{X}(t), E[\phi(\bar{X}(t))])(\hat{q}(t) + \theta\ell\hat{p}^*(t)) \end{aligned}$$

and

$$\begin{aligned} \check{H}_m^\theta(t) := \langle F_m(t, \bar{X}(t), E[\phi(\bar{X}(t))], \bar{u}(t)), \hat{p}(t)\rangle \\ +\text{tr}(G_m^*(t, \bar{X}(t), E[\phi(\bar{X}(t))])(\hat{q}(t) + \theta\ell\hat{p}^*(t)). \end{aligned}$$

We note that the processes $(\hat{p}, \hat{q}, \ell)$ may depend on the sensitivity index θ. To ease notation, we omit to make this dependence explicit.

Below, we will show that, under Assumption 1, (4.296) admits a unique $\mathbb{F}$-adapted solution $(\hat{p}, \hat{q}, v^\theta, \ell)$ such that

$$E\left[\sup_{t\in[0,T]} |\hat{p}(t)|^2 + \sup_{t\in[0,T]} |v^\theta(t)|^2 + \int_0^T \left(|\hat{q}(t)|^2 + |\ell(t)|^2\right) dt\right] < \infty. \tag{4.297}$$

Moreover,

Lemma 4.16 *The process defined on* $(\Omega, \mathcal{F}, \mathbb{F}, \mathbb{P})$ *by*

$$L_t^\theta := \frac{v^\theta(t)}{v^\theta(0)} = \exp\left(\int_0^t \theta\langle \ell(s), dB_s\rangle - \frac{\theta^2}{2}\int_0^t |\ell(s)|^2 ds\right), \quad 0 \le t \le T, \tag{4.298}$$

is a uniformly integrable $\mathbb{F}$*-martingale.*

The process L^θ defines a new probability measure $\mathbb{P}^\theta$ equivalent to $\mathbb{P}$ by setting $L_t^\theta := \left.\frac{d\mathbb{P}^\theta}{d\mathbb{P}}\right|_{\mathcal{F}_t}$. By Girsanov's theorem, the process $B_t^\theta := B_t - \theta\int_0^t \ell(s)ds,\ 0 \le t \le T$ is a $\mathbb{P}^\theta$-Brownian motion.

The following theorem is the main result of the section. Let $E^\theta[\,\cdot\,]$ denote the expectation w.r.t. $\mathbb{P}^\theta$.

Theorem 4.13 (Risk-Sensitive Maximum Principle) *Let Assumption 4.287 hold. If the process $(\bar{\rho}(\cdot), \bar{x}(\cdot), \bar{u}(\cdot))$ is an optimal solution of the risk-sensitive control problem (4.284)–(4.286), then there are two pairs of $\mathbb{F}$-adapted processes (v^θ, ℓ) and $(\hat{p}, \hat{q})$ which satisfy (4.296)–(4.297), such that*

$$E^\theta[H^\theta(t, \bar{\rho}(t), \bar{x}(t), \hat{p}(t), \hat{q}(t), \ell(t), u) - H^\theta(t, \bar{\rho}(t), \bar{x}(t), \hat{p}(t), \hat{q}(t), \ell(t), \bar{u}(t))|\mathscr{F}_t^Y] \leq 0,$$

for all $u \in U$, almost every t and $\mathbb{P}^\theta$—almost surely.

Remark 4.4 The boundedness assumption of the involved coefficients and their derivatives imposed in Assumption 1, in Theorem 4.13, guarantees the solvability of the system of forward-backward SDEs (4.286) and (4.296). In fact Theorem 4.13 applies provided we can solve system of forward-backward SDEs (4.286) and (4.296). A typical example of such a situation is the classical linear-quadratic (LQ) control problem, in which the involved coefficients are at most quadratic, but not necessarily bounded.

4.8.1 Proof of the Main Result

In this section, we give a proof of Theorem 4.13 here presented in several steps.

4.8.2 An Intermediate SMP for Mean-Field-Type Control

In this subsection, we first reformulate the risk-sensitive control problem associated with (4.285)–(4.286) in terms of an augmented state process and terminal payoff problem. An intermediate stochastic maximum principle is then obtained by applying the SMP for loss functionals without running cost. Then, we transform the intermediate first-order adjoint processes to a simpler form. The mean-field type control problem for the cost functional (4.285) under the dynamics (4.286) is equivalent to

$$\inf_{u(\cdot)\in\mathscr{U}} E\left[\rho(T)e^{\theta[h(x(T),E[\rho(T)x(T)])+\xi(T)]}\right], \tag{4.299}$$

subject to

$$\begin{cases} d\rho(t) = \rho(t)\beta(t, x(t))dY_t, \\ dx(t) = \{b(t, x(t), E[x(t)], u(t)) - \alpha(t, x(t), E[x(t)])\beta(t, x(t))\}\, dt \\ \qquad\qquad +\sigma(t, x(t), E[x(t)])dW_t + \alpha(t, x(t), E[x(t)])dY_t, \\ d\xi = f(t, x(t), E[\rho(t)x(t)], u(t))dt, \\ \rho(0) = 1,\ x(0) = x_0,\ \xi(0) = 0. \end{cases} \tag{4.300}$$

We introduce the following notation.

$$R := \begin{pmatrix} \rho \\ x \\ \xi \end{pmatrix} = \begin{pmatrix} X \\ \xi \end{pmatrix}, \quad \bar{R} := \begin{pmatrix} \bar{\rho} \\ \bar{x} \\ \bar{\xi} \end{pmatrix} = \begin{pmatrix} \bar{X} \\ \bar{\xi} \end{pmatrix}, \quad R_0 = \bar{R}_0 := \begin{pmatrix} X_0, \\ 0 \end{pmatrix},$$

$$\Gamma(t, R, m) := \begin{pmatrix} G(t, X, m, u) \\ 0 \end{pmatrix},$$

$$\phi(R)=\phi(X)=x, \;\; \tilde{\phi}(R)=\tilde{\phi}(X)=\rho x, \;\; \phi(\bar{R})=\phi(\bar{X})=\bar{x}, \;\; \tilde{\phi}(\bar{R})=\tilde{\phi}(\bar{X}) = \bar{\rho}\bar{x}.$$

With this notation, the system (4.300) can be rewritten in the following compact form

$$\begin{cases} dR(t) = \begin{pmatrix} F(t, R(t), E[\phi(R(t))], u(t)) \\ f(t, R(t), E[\tilde{\phi}(R(t))], u(t)) \end{pmatrix} dt + \Gamma(t, R(t), E[\phi(R(t))])dB_t, \\ R(0) = R_0 \end{cases}$$

and the risk-sensitive cost functional (4.285) becomes

$$J^{\theta}(u(\cdot)) := E[\Phi\left(R(T), E[\tilde{\phi}(R(T))]\right)],$$

where

$$\Phi\,(R(T), E[\phi(R(T))]) := \rho(T)\exp\left(\theta h(x(T), E[\rho(T)x(T)]) + \theta\xi(T)\right).$$

We define the Hamiltonian associated with random variables R such that $\phi(R) \in L^1(\Omega, \mathscr{F}, \mathbb{P})$ as follows. For $(p, q) \in \mathbb{R}^3 \times \mathbb{R}^{3\times 3}$,

$$\begin{aligned} H^e(t, R, p, q, u) := \langle \begin{pmatrix} F(t, R(t), E[\phi(R(t))], u(t)) \\ f(t, R(t), E[\tilde{\phi}(R(t))], u(t)) \end{pmatrix}, p\rangle \\ +\mathrm{tr}(\Gamma^*(t, R, E[\phi(R)])q), \end{aligned} \tag{4.301}$$

where Γ^* denotes the transpose of the matrix Γ.

Setting

$$p := \begin{pmatrix} p_1 \\ p_2 \\ p_3 \end{pmatrix}, \quad q := \begin{pmatrix} q_{11} & q_{12} \\ q_{21} & q_{22} \\ q_{31} & q_{32} \end{pmatrix}, \tag{4.302}$$

the explicit form of the Hamiltonian (4.301) reads

$$\begin{aligned} H^e(t, \rho, x, \xi, p, q, u) := H^e(t, R, p, q, u) &= c(t, x, E[x], u)p_2 \\ &+f(t, x, E[\rho x], u)p_3 \\ &+\sigma(t, x, E[x])q_{22} + \rho\beta(t, x)q_{11} + \alpha(t, x, E[x])q_{21}. \end{aligned}$$

In view of (4.294), we set

$$\begin{cases} H_x^e(t) := c_x(t)p_2(t) + f_x(t)p_3(t) + \sigma_x(t)q_{22}(t) + \bar{\rho}(t)\beta_x(t)q_{11}(t) \\ \qquad\qquad +\alpha_x(t)q_{21}(t), \\ \check{H}_m^e(t) := c_m(t)p_2(t) + f_m(t)p_3(t) + \sigma_m(t)q_{22}(t) + \alpha_m(t)q_{21}(t), \\ H_\rho^e(t) = \beta(t,\bar{x}(t))q_{11}(t). \end{cases}$$

We may apply the SMP for risk-neutral mean-field-type control ([78], Theorem 2.1) to the augmented state dynamics (ρ, x, ξ) to derive the first-order adjoint equation:

$$\begin{cases} dp(t) = -\begin{pmatrix} H_\rho^e(t) + E[\check{H}_m^e(t)] + \bar{x}(t)E[f_m(t)p_3(t)] \\ H_x^e(t) + E[\check{H}_m^e(t)] + \bar{\rho}(t)E[f_m(t)p_3(t)] \\ 0 \end{pmatrix} dt + q(t)dB_t, \\ p(T) = -\theta\psi_T^\theta \begin{pmatrix} (\theta\bar{\rho}(T))^{-1} \\ h_x(T) \\ 1 \end{pmatrix} - \theta \begin{pmatrix} \bar{x}(T) \\ \bar{\rho}(T) \\ 0 \end{pmatrix} E[\psi_T^\theta h_m(T)]. \end{cases} \tag{4.303}$$

This is a system of linear backward SDEs of mean-field type which, in view of ([77], Theorem 3.1), under Assumption 4.287, admits a unique $\mathbb{F}$-adapted solution (p, q) satisfying

$$E\left[\sup_{t\in[0,T]} |p(t)|^2 + \int_0^T |q(t)|^2 dt\right] < \infty, \tag{4.304}$$

where $|\cdot|$ denotes the usual Euclidean norm with appropriate dimension.

We may apply the SMP for SDEs of mean-field-type control from ([78], Theorem 2.1) together with the SMP for risk-neutral partially observable SDEs to obtain the following SMP.

Proposition 4.11 *Let Assumption 4.287 hold. If $(\bar{R}(\cdot), \bar{u}(\cdot))$ is an optimal solution of the risk-neutral control problem (4.299) subject to the dynamics (4.300), then there is a unique pair of $\mathbb{F}$-adapted processes (p, q) which satisfies (4.303)–(4.304) such that*

$$E[H^e(t, \bar{R}(t), p(t), q(t), u) - H^e(t, \bar{R}(t), p(t), q(t), \bar{u}(t))|\mathscr{F}_t^Y] \le 0,$$

for all $u \in U$, almost every t and $\mathbb{P}$−almost surely.

4.8.3 Transformation of the First-Order Adjoint Process

Although the result of Proposition 4.11 is a good SMP for the risk-sensitive mean-field-type control with partial observations, augmenting the state process with the third component ξ yields a system of three adjoint equations that appears complicated to solve in concrete situations. In this subsection, we apply the transformation of the adjoint processes (p, q) introduced in [122, 123] in such a way to get rid of the third component (p_3, q_{31}, q_{32}) in (4.303) and express the SMP in terms of only two adjoint processes that we denote $(\hat{p}, \hat{q})$, where

$$\hat{p} := \begin{pmatrix} \hat{p}_1 \\ \hat{p}_2 \end{pmatrix}, \quad \hat{q} := \begin{pmatrix} \hat{q}_1 \\ \hat{q}_2 \end{pmatrix}, \quad \hat{q}_i := (\hat{q}_{i1}, \hat{q}_{i2}), \ \ i = 1, 2. \tag{4.305}$$

Indeed, noting that from (4.303), we have $dp_3(t) = \langle q_3(t), dB_t \rangle$ and $p_3(T) = -\theta \psi_T^\theta$, the explicit solution of this backward SDE is

$$p_3(t) = -\theta E[\psi_T^\theta \mid \mathscr{F}_t] = -\theta v^\theta(t), \tag{4.306}$$

where

$$v^\theta(t) := E[\psi_T^\theta \mid \mathscr{F}_t], \qquad 0 \le t \le T.$$

In particular, we have $v^\theta(0) = E[\psi_T^\theta]$. Therefore, in view of (4.306), it would be natural to choose a transformation of (p, q) into an adjoint process $(\hat{p}, \hat{q})$, where

$$\hat{p} := \begin{pmatrix} \hat{p}_1 \\ \hat{p}_2 \\ \hat{p}_3 \end{pmatrix}, \quad \hat{q} := \begin{pmatrix} \hat{q}_{11} \ \hat{q}_{12} \\ \hat{q}_{21} \ \hat{q}_{22} \\ \hat{q}_{31} \ \hat{q}_{32} \end{pmatrix},$$

such that

$$\hat{p}_3(t) = \frac{p_3(t)}{\theta v^\theta(t)} = -1, \qquad 0 \le t \le T. \tag{4.307}$$

This would imply that, for almost every $0 \le t \le T$,

$$\hat{q}_3(t) = (\hat{q}_{31}(t), \hat{q}_{32}(t)) = 0, \ \ \mathbb{P} - \text{a.s.}, \tag{4.308}$$

which in turn would reduce the number of adjoint processes to those of the form given by (4.305).

We consider the following transform:

$$\hat{p}(t) := \frac{1}{\theta v^\theta(t)} p(t), \qquad 0 \le t \le T. \tag{4.309}$$

In view of (4.303), we have

$$\hat{p}(T) = -\begin{pmatrix} (\theta \bar{\rho}(T))^{-1} \\ h_x(T) \\ 1 \end{pmatrix} - \begin{pmatrix} \bar{x}(T) \\ \bar{\rho}(T) \\ 0 \end{pmatrix} \frac{1}{\psi_T^\theta} E[\psi_T^\theta h_m(T)]. \tag{4.310}$$

We should identify the processes $\hat{\alpha}$ and $\hat{q}$ such that

$$d\hat{p}(t) = -\hat{\alpha}(t)dt + \hat{q}(t)dB_t, \tag{4.311}$$

for which (4.307) and (4.308) are satisfied.

In order to investigate the properties of these new processes $(\hat{p}, \hat{q})$, the following properties of the generic martingale v^θ, used in [122, 123], are essential. We reproduce them here for the sake of completeness. Since, by Assumption 4.287, f and h are bounded by some constant $C > 0$, we have

$$0 < e^{-(1+T)C\theta}\rho(T) \le \psi_T^\theta \le e^{(1+T)C\theta}\rho(T).$$

Therefore, v^θ is a uniformly integrable $\mathbb{F}$-martingale satisfying

$$0 < e^{-(1+T)C\theta}\rho(t) \le v^\theta(t) \le e^{(1+T)C\theta}\rho(t), \qquad 0 \le t \le T.$$

Hence, in view of (4.283), we have

$$E[\sup_{0\le t\le T} |v^\theta(t)|^2] \le C. \tag{4.312}$$

Furthermore, the martingale v^θ enjoys the following useful *logarithmic transform* established in [122]:

$$v^\theta(t)=\exp\left(\theta Z_t+\theta\int_0^t f(s, \bar{x}(s), E[\bar{\rho}(s)\bar{x}(s)], \bar{u}(s))ds\right), \; 0 \le t \le T, \tag{4.313}$$

and

$$v^\theta(0) = E[\psi_T^\theta] = \exp(\theta Z_0).$$

Moreover, the process Z is the first component of the $\mathbb{F}$-adapted pair of processes (Z, ℓ) which is the unique solution to the following quadratic BSDE:

$$\begin{cases} dZ_t = -\{f(t, \bar{x}(t), E[\bar{\rho}(s)\bar{x}(s)], \bar{u}(t)) + \frac{\theta}{2}|\ell(t)|^2\}dt + \langle \ell(t), dB_t\rangle, \\ Z_T = \frac{1}{\theta}\ln\bar{\rho}(T) + h(\bar{x}_T, E[\bar{\rho}(T)\bar{x}(T)]). \end{cases} \tag{4.314}$$

where $\ell(t) = (\ell_1(t), \ell_2(t))$ satisfies

$$E\left[\int_0^T |\ell(t)|^2 dt\right] < \infty. \tag{4.315}$$

In particular, v^θ solves the following linear backward SDE

$$dv^\theta(t) = \theta v^\theta(t)\langle \ell(t), dB_t\rangle, \quad v^\theta(T) = \psi_T^\theta. \tag{4.316}$$

Hence,

Proof of Lemma 4.16 In view of (4.312),

$$\frac{v^\theta(t)}{v^\theta(0)} = \exp\left(\int_0^t \theta\langle \ell(s), dB_s\rangle - \frac{\theta^2}{2}\int_0^t |\ell(s)|^2 ds\right) := L_t^\theta,\ 0 \le t \le T, \tag{4.317}$$

is a uniformly integrable $\mathbb{F}$-martingale. ■

To identify the processes $\tilde{\alpha}$ and $\tilde{q}$ such that

$$d\hat{p}(t) = -\hat{\alpha}(t)dt + \hat{q}(t)dB_t,$$

we may apply Itô's formula to the process $p(t) = \theta v^\theta \tilde{p}(t)$, use (4.303) and (4.316) and identify the coefficients. We obtain

$$\begin{cases} \hat{\alpha}(t) = \frac{1}{\theta v^\theta(t)}\begin{pmatrix} H_\rho^e(t) + E[\check{H}_m^e(t)] + \bar{x}(t)E[f_m(t)p_3(t)] \\ H_x^e(t) + E[\check{H}_m^e(t)] + \bar{\rho}(t)E[f_m(t)p_3(t)] \\ 0 \end{pmatrix} + \theta\hat{q}(t)\ell(t), \\ \hat{q}(t) = \frac{1}{\theta v^\theta(t)}q(t) - \theta\hat{p}(t)\ell(t). \end{cases} \tag{4.318}$$

Therefore,

$$\begin{cases} d\hat{p}(t) = -\frac{1}{\theta v^\theta(t)}\begin{pmatrix} H_\rho^e(t) + E[\check{H}_m^e(t)] + \bar{x}(t)E[f_m(t)p_3(t)] \\ H_x^e(t) + E[\check{H}_m^e(t)] + \bar{\rho}(t)E[f_m(t)p_3(t)] \\ 0 \end{pmatrix} dt + \hat{q}(t)dB_t^\theta, \\ \hat{q}(t) = \frac{1}{\theta v^\theta(t)}q(t) - \theta\hat{p}(t)\ell(t), \\ dv^\theta(t) = \theta v^\theta(t)\langle \ell(t), dB_t\rangle, \\ \hat{p}(T) = -\begin{pmatrix} (\theta\bar{\rho}(T))^{-1} \\ h_x(T) \\ 1 \end{pmatrix} - \begin{pmatrix} \bar{x}(T) \\ \bar{\rho}(T) \\ 0 \end{pmatrix} \frac{1}{\psi_T^\theta} E[\psi_T^\theta h_m(T)], \\ v^\theta(T) = \psi_T^\theta, \end{cases} \tag{4.319}$$

where $B_t^\theta := B_t - \theta\int_0^t \ell(s)ds,\ 0 \le t \le T$, which is, in view of (4.317) and Girsanov's Theorem, a $\mathbb{P}^\theta$-Brownian motion, where $\left.\frac{d\mathbb{P}^\theta}{d\mathbb{P}}\right|_{\mathscr{F}_t} := L_t^\theta$.

In particular,

$$d\hat{p}_3(t) = \langle \hat{q}_3(t), -\theta\ell(t)dt + dB_t\rangle, \quad \hat{p}_3(T) = -1.$$

Therefore, noting that $\hat{p}_3(t) := [\theta v^\theta(t)]^{-1} p_3(t)$ is square-integrable, we obtain

$$\hat{p}_3(t) = E^{\mathbb{P}^\theta}[\hat{p}_3(T)|\mathscr{F}_t] = -1.$$

Thus, its quadratic variation becomes $\int_0^T |\hat{q}_3(t)|^2 dt = 0,\ \mathbb{P}^\theta$ – a.s. This implies that, for almost every $0 \le t \le T$, $\hat{q}_3(t) = 0,\ \mathbb{P}^\theta$ and $\mathbb{P}$ – a.s.

Hence, we can drop the last components from the adjoint processes $(\hat{p}, \hat{q})$ and only consider (keeping the same notation)

$$\hat{p} := \begin{pmatrix} \hat{p}_1 \\ \hat{p}_2 \end{pmatrix}, \quad \hat{q} := \begin{pmatrix} \hat{q}_{11} & \hat{q}_{12} \\ \hat{q}_{21} & \hat{q}_{22} \end{pmatrix},$$

for which (4.319) reduces to the risk-sensitive adjoint equation:

$$\begin{cases} d\hat{p}(t) = -\frac{1}{\theta v^\theta(t)} \begin{pmatrix} H^e_\rho(t) + E[\check{H}^e_m(t)] - \bar{x}(t)E[f_m(t)] \\ H^e_x(t) + E[\check{H}^e_m(t)] - \bar{\rho}(t)E[f_m(t)] \end{pmatrix} dt + \hat{q}(t)dB^\theta_t, \\ \hat{q}(t) = \frac{1}{\theta v^\theta(t)} q(t) - \theta \hat{p}(t)\ell(t), \\ dv^\theta(t) = \theta v^\theta(t)\langle \ell(t), dB_t \rangle, \\ \hat{p}(T) = -\begin{pmatrix} (\theta\bar{\rho}(T))^{-1} \\ h_x(T) \end{pmatrix} - \begin{pmatrix} \bar{x}(T) \\ \bar{\rho}(T) \end{pmatrix} \frac{1}{\psi^\theta_T} E[\psi^\theta_T h_m(T)], \\ v^\theta(T) = \psi^\theta_T. \end{cases} \tag{4.320}$$

In view of the uniqueness of $\mathbb{F}$-adapted pairs (p, q), solution of (4.303), and the pair (v^θ, ℓ) obtained satisfying (4.315) and (4.316), the solution of the system of backward SDEs (4.320) is unique and satisfies (4.297).

4.8.4 Risk-Sensitive Stochastic Maximum Principle

We may use the transform (4.309) and (4.318) to obtain the explicit form (4.293) of the risk-sensitive Hamiltonian H^θ defined by

$$H^\theta(t, \bar{X}(t), \hat{p}(t), \hat{q}(t), \ell(t), u) := \frac{1}{\theta v^\theta(t)} H^e(t, \bar{R}(t), p(t), q(t), u), \tag{4.321}$$

where H^e is defined by (4.301).

Let

$$\delta H^e(t) := H^e(t, \bar{R}(t), p(t), q(t), u) - H^e(t, \bar{R}(t), p(t), q(t), \bar{u}(t))$$

and

$$\delta H^\theta(t) = H^\theta(t, \bar{X}(t), \hat{p}(t), \hat{q}(t), \ell(t), u) - H^\theta(t, \bar{X}(t), \hat{p}(t), \hat{q}(t), \ell(t), \bar{u}(t)).$$

We have

$$E[\delta H^e(t)|\mathscr{F}_t^Y] = \theta E[v^\theta(t)\delta H^\theta(t)|\mathscr{F}_t^Y] = \theta v^\theta(0)E^\theta[\delta H^\theta(t)|\mathscr{F}_t^Y],$$

where we recall that $v^\theta(t)/v^\theta(0) = L_t^\theta = d\mathbb{P}^\theta/d\mathbb{P}|_{\mathscr{F}_t}$.

Now, since $\theta > 0$ and $v^\theta(0) = E[\psi_T^\theta] > 0$, the variational inequality (4.219) translates into

$$E^\theta[H^\theta(t, \bar{p}(t), \bar{x}(t), \hat{p}(t), \hat{q}(t), \ell(t), u) - H^\theta(t, \bar{p}(t), \bar{x}(t), \hat{p}(t), \hat{q}(t), \ell(t), \bar{u}(t))|\mathscr{F}_t^Y] \leq 0.$$

for all $u \in U$, almost every t and $\mathbb{P}^\theta$—almost surely. This finishes the proof of Theorem 4.13.

Linear-Quadratic Risk-Sensitive Model Under Partial Observation

To illustrate our approach, we consider a one-dimensional linear diffusion with exponential quadratic cost functional. Perhaps, the easiest example of a linear-quadratic (LQ) risk-sensitive control problem with mean-field coupling is

$$\begin{cases} \inf_{u(\cdot)\in\mathscr{U}} E^u e^{\theta\left[\frac{1}{2}\int_0^T u^2(t)dt + \frac{1}{2}x^2(T) + \mu E^u[x(T)]\right]}, \\ \text{subject to} \\ dx(t) = (ax(t) + bu(t))\,dt + \sigma dW_t + \alpha d\widetilde{W}_t^u, \\ dY_t = \beta x(t)dt + d\widetilde{W}_t^u, \\ x(0) = x_0,\ Y_0 = 0, \end{cases} \tag{4.322}$$

where a, b, α, β, μ and σ are real constants.

We consider the linear-quadratic risk-sensitive control problem:

$$\begin{cases} \inf_{u(\cdot)\in\mathscr{U}} E^u e^{\theta\left[\frac{1}{2}\int_0^T u^2(t)dt + \frac{1}{2}x^2(T)\right]}, \\ \text{subject to} \\ dx(t) = (ax(t) + bu(t))\,dt + \sigma dW_t + \alpha d\widetilde{W}_t^u, \\ dY_t = \beta x(t)dt + d\widetilde{W}_t^u, \\ x(0) = x_0,\ Y_0 = 0, \end{cases} \tag{4.323}$$

where a, b, α, β and σ are real constants.

An admissible process $(\bar{p}(\cdot), \bar{x}(\cdot), \bar{u}(\cdot))$ satisfying the necessary optimality conditions of Theorem 4.13 is obtained by solving the following system of forward-backward SDEs (cf. (4.286) and (4.296)) (see Remark 4.4, above).

$$\begin{cases} d\bar{\rho}(t) = \beta\bar{\rho}(t)\bar{x}(t)dY_t, \\ d\bar{x}(t) = \{c\bar{x}(t) + b\bar{u}(t)\}\,dt + \sigma dW_t + \alpha dY_t, \\ dp(t) = -\begin{pmatrix} H_\rho^\theta(t) \\ H_x^\theta(t) \end{pmatrix} dt + q(t)(-\theta\ell(t)dt + dB_t), \\ dv^\theta(t) = \theta v^\theta(t)\langle \ell(t), dB_t\rangle, \\ p(T) = -\begin{pmatrix} (\theta\bar{\rho}(T))^{-1} \\ \bar{x}(T) \end{pmatrix}, \\ v^\theta(T) = \psi_T^\theta, \\ \bar{\rho}(0) = 1,\ \bar{x}(0) = x_0, \end{cases} \tag{4.324}$$

where

$$c := a - \alpha\beta,\ B_t := \begin{pmatrix} Y_t \\ W_t \end{pmatrix},\ \ell := \begin{pmatrix} \ell_1 \\ \ell_2 \end{pmatrix},\ p := \begin{pmatrix} p_1 \\ p_2 \end{pmatrix},\ q := \begin{pmatrix} q_{11}\ q_{12} \\ q_{21}\ q_{22} \end{pmatrix},$$

$$\psi_T^\theta := \bar{\rho}(T)e^{\theta\left[\frac{1}{2}\int_0^T \bar{u}^2(t)dt + \frac{1}{2}\bar{x}^2(T)\right]},$$

and the associated risk-sensitive Hamiltonian is

$$\begin{aligned} H^\theta(t,\rho,x,u,p,q,\ell) := (cx+bu)p_2 - \tfrac{1}{2}u^2 + \rho\beta x(q_{11} + \theta\ell_1 p_1) \\ +\alpha(q_{21} + \theta\ell_2 p_1) + \sigma(q_{22} + \theta\ell_2 p_2). \end{aligned} \tag{4.325}$$

In general, the solution (v^θ, ℓ) primarily gives the correct form of the process ℓ which may be a function of the optimal control $\bar{u}$. Inserting ℓ in the BSDE satisfied by (p, q) in the system (4.324) and solving for (p, q), we arrive at the characterization the optimal control of our problem.

For the LQ-control problem, it turns out that by considering the BSDE satisfied by (v^θ, ℓ), we will find an explicit form of the optimal control $\bar{u}$. Indeed, by (4.313), this is equivalent to consider the BSDE satisfied by (Z, ℓ):

$$\begin{cases} dZ_t = -\{\frac{1}{2}\bar{u}^2(t) + \frac{\theta}{2}|\ell(t)|^2\}dt + \langle \ell(t), dB_t\rangle, \\ Z_T = \frac{1}{\theta}\ln\bar{\rho}(T) + \frac{1}{2}\bar{x}_T^2. \end{cases}$$

Since $\bar{u}$ is $\mathscr{F}_t^Y$, the form of Z_T suggests that we characterize $\bar{u}$ and ℓ such that

$$E^\theta[Z_t|\mathscr{F}_t^Y] = E^\theta[\frac{\gamma(t)}{2}\bar{x}^2(t) + \frac{1}{\theta}\ln\bar{\rho}(t) + \eta(t)|\mathscr{F}_t^Y], \quad 0 \le t \le T,$$

where γ and η are deterministic functions such that $\gamma(T) = 1$ and $\eta(T) = 0$. In view of the SDEs satisfied by $(\bar{\rho}, \bar{x})$ in (4.324), applying Itô's formula and identifying the coefficients, we get

$$\ell_1(t) = (\alpha\gamma(t) + \beta/\theta)\bar{x}(t), \quad \ell_2(t) = \sigma\gamma(t)\bar{x}(t) \tag{4.326}$$

and

$$E^{\theta}[\tfrac{1}{2}\left(\dot{\gamma}(t)+2(c+\alpha\beta)\gamma(t)+(\theta(\sigma^2+\alpha^2)-b^2)\gamma^2(t)\right)\bar{x}^2(t)|\mathscr{F}_t^Y] \\ +E^{\theta}[\dot{\eta}(t)+\tfrac{1}{2}(\sigma^2+\alpha^2)\gamma(t)+(\bar{u}(t)+b\gamma(t)\bar{x}(t))^2|\mathscr{F}_t^Y]=0.$$

Hence,

$$\begin{cases}\dot{\gamma}(t)+2(c+\alpha\beta)\gamma(t)+(\theta(\sigma^2+\alpha^2)-b^2)\gamma^2(t)=0, \quad \gamma(T)=1,\\ \dot{\eta}(t)+\tfrac{1}{2}(\sigma^2+\alpha^2)\gamma(t)=0, \quad \eta(T)=0,\end{cases} \tag{4.327}$$

where the first equation is the risk-sensitive Riccati equation and

$$E^{\theta}[(\bar{u}(t)+b\gamma(t)\bar{x}(t))^2|\mathscr{F}_t^Y]=0.$$

By the conditional Jensen's inequality, we have

$$\left|E^{\theta}[\bar{u}(t)+b\gamma(t)\bar{x}(t)|\mathscr{F}_t^Y]\right|^2\le E^{\theta}[(\bar{u}(t)+b\gamma(t)\bar{x}(t))^2|\mathscr{F}_t^Y].$$

Therefore, the optimal control is

$$\bar{u}(t)=-b\gamma(t)E^{\theta}[\bar{x}(t)|\mathscr{F}_t^Y], \tag{4.328}$$

and the optimal dynamics solves the linear SDE

$$d\bar{x}(t)=\left(c\bar{x}(t)-b^2\gamma(t)E^{\theta}[\bar{x}(t)|\mathscr{F}_t^Y]\right)dt+\sigma dW_t+\alpha dY_t, \ \bar{x}(0)=x_0, \tag{4.329}$$

where $\pi_t(\bar{x}):=E^{\theta}[\bar{x}(t)|\mathscr{F}_t^Y]$ is the solution of the SDE on $(\Omega,\mathscr{F},\mathbb{F},\mathbb{P}^{\theta})$:

$$\pi_t(\bar{x})=x_0+\int_0^t(c-b^2\gamma(s))\pi_s(\bar{x})ds+\int_0^t(\alpha+(\theta\alpha\gamma(t)+\beta) \\ \times\left[\pi_s(\bar{x}^2)-\pi_s^2(\bar{x})\right])d\bar{Y}_s^{\theta},$$

where $\bar{Y}_t^{\theta}=Y_t-\int_0^t(\theta\alpha\gamma(s)+\beta)\pi_s(\bar{x})ds$ is an $(\Omega,\mathscr{F},\mathbb{F}^Y,\mathbb{P}^{\theta})$-Brownian motion.

Inserting the form (4.326) of ℓ in the BSDE satisfied by (p,q) in the system (4.324) and solving for (p,q), we arrive at the same characterization the optimal control of our problem, obtained as a maximizer of the associated H^{θ} given by (4.325). We sketch the main steps and omit the details.

We have

$$H_u^{\theta}=bp_2-u, \quad H_{\rho}^{\theta}=\beta x(q_{11}+\theta\ell_1p_1), \quad H_x^{\theta}=cp_2+\beta\rho(q_{11}+\theta\ell_1p_1).$$

The BSDE satisfied by (p,q) then reads

$$\begin{cases} dp_1(t) = -\{q_{11}(t)(\beta\bar{x}(t) + \theta\ell_1(t)) + \theta(\ell_1(t)p_1(t)\bar{x}(t) + q_{12}(t)\ell_2(t))\}dt \\ \qquad\qquad +q_{11}(t)dY_t + q_{12}(t)dW_t, \\ dp_2(t) = -\{cp_2(t) + \beta\rho(t)(q_{11}(t) + \theta\ell_1(t)p_1(t))\}dt \\ \qquad\qquad +\theta(q_{21}\ell_1(t) + q_{22}\ell_2(t))dt + q_{21}(t)dY_t + q_{22}(t)dW_t, \\ p_1(T) = -\frac{1}{\theta\bar{\rho}(T)}, \quad p_2(T) = -\bar{x}(T). \end{cases} \tag{4.330}$$

In view of Theorem 4.13, if $\bar{u}$ is an optimal control of the system (4.323), it is necessary that

$$E^\theta[bp_2(t) - \bar{u}(t)|\mathscr{F}_t^Y] = 0.$$

This yields

$$\bar{u}(t) = bE^\theta[p_2(t)|\mathscr{F}_t^Y].$$

The associated state dynamics $\bar{x}$ solves then the SDE

$$d\bar{x}(t) = \left\{c\bar{x}(t) + b^2E^\theta[p_2(t)|\mathscr{F}_t^Y]\right\}dt + \sigma dW_t + \alpha dY_t.$$

It remains to compute $E^\theta[p_2(t)|\mathscr{F}_t^Y]$. Indeed, inserting the form (4.326) of ℓ in the BSDE satisfied by (p, q) in the system (4.330), by Itô's formula and identifying the coefficients, it is easy to check that $(p_1(t), q_{11}(t), q_{12}(t))$ given by

$$p_1(t) := -\frac{1}{\theta\bar{\rho}(t)}, \quad q_{11}(t) := \frac{\beta}{\theta}\frac{\bar{x}(t)}{\bar{\rho}(t)}, \quad q_{12}(t) := 0$$

solves the first adjoint equation in (4.330). Furthermore, since $p_2(T) = -\bar{x}(T)$, setting

$$E^\theta[p_2(t)|\mathscr{F}_t^Y] = -\lambda(t)E^\theta[\bar{x}(t)|\mathscr{F}_t^Y],$$

where λ is a deterministic function such that $\lambda(T) = 1$, and identifying the coefficients, we find that λ satisfies the risk-sensitive Riccati equation in (4.327). Moreover,

$$q_{21}(t) = -\sigma\lambda(t), \quad q_{22}(t) = -\alpha\lambda(t).$$

By uniqueness of the solution of the risk-sensitive Riccati equation in (4.327), it follows that $\lambda = \gamma$. Therefore,

$$E^\theta[p_2(t)|\mathscr{F}_t^Y] = -\gamma(t)E^\theta[\bar{x}(t)|\mathscr{F}_t^Y], \quad q_{21}(t) = -\sigma\gamma(t), \quad q_{22}(t) = -\alpha\gamma(t).$$

Summing up: The optimal control of the LQ-problem (4.324) is

$$\bar{u}(t) = -b\gamma(t)E^{\theta}[\bar{x}(t)|\mathscr{F}_t^Y], \tag{4.331}$$

where γ solves the risk-sensitive Riccati equation

$$\dot{\gamma}(t) + 2(c + \alpha\beta)\gamma(t) + (\theta(\sigma^2 + \alpha^2) - b^2)\gamma^2(t) = 0, \quad \gamma(T) = 1. \tag{4.332}$$

The optimal dynamics solves the linear SDE

$$d\bar{x}(t) = \left(c\bar{x}(t) - b^2\gamma(t)E^{\theta}[\bar{x}(t)|\mathscr{F}_t^Y]\right)dt + \sigma dW_t + \alpha dY_t, \quad \bar{x}(0) = x_0, \tag{4.333}$$

and the filter $\pi_t(\bar{x}) := E^{\theta}[\bar{x}(t)|\mathscr{F}_t^Y]$ is solution of the SDE on $(\Omega, \mathscr{F}, \mathbb{F}, \mathbb{P}^{\theta})$:

$$\pi_t(\bar{x}) = x_0 + \int_0^t (c - b^2\gamma(s))\pi_s(\bar{x})ds + \int_0^t (\alpha + (\theta\alpha\gamma(t) + \beta) \\ \times \left[\pi_s(\bar{x}^2) - \pi_s^2(\bar{x})\right]) d\bar{Y}_s^{\theta},$$

where $\bar{Y}_t^{\theta} = Y_t - \int_0^t (\theta\alpha\gamma(s) + \beta)\pi_s(\bar{x})ds$ is an $(\Omega, \mathscr{F}, \mathbb{F}^Y, \mathbb{P}^{\theta})$-Brownian motion.

Chapter 5
Dynamic Programming Principle

Dynamic programming is a computational approach that helps decision-makers find optimal strategies in sequential or dynamic games by breaking down the decision-making process into smaller steps and solving for the best course of action at each stage of the interactions. The key issue here is that the standard dynamic programming principle may not apply as the state of mean-field-type game involves both state and mean-field terms. Therefore, the dynamic programming principle needs to be extended to the space of measures.

This chapter presents the dynamic programming principle (DPP) for payoff functionals of mean-field type. We start with a generic setup with measurement noise. The drift and the diffusion functions of the instantaneous cost and the state are both taken to be dependent on time, state, control, and a positive measure [333, 339, 342].

The chapter is organized into three main sections. Section 5.1 introduces Gâteaux differentiability. Section 5.2 risk-sensitive mean-field-type games, and Sect. 5.3 covers its counterpart for risk-sensitive mean-field games. The chapter ends with a section on bibliographic notes.

5.1 Gâteaux Differentiability

In order to establish a dynamic programming principle in the space of measures over the states, it is useful to have an integration formula, which involves variations of the functionals involved. Toward that end, we introduce here a directional derivative (may not be unique) as follows.

Let $T > 0$, $\alpha > 1$ and $(\Omega, \mathscr{X}, \mathbb{P})$ be a measure space. Let $b \in L^{\alpha}(\Omega \times [0, T] \times \mathscr{X} \times \mathscr{P}(\mathscr{X}), \mathbb{R})$, $O \subset L^{\alpha^*}$ be a non-empty open set and α^* the conjugate of α such that $\frac{1}{\alpha} + \frac{1}{\alpha^*} = 1$.

T. Başar et al., *Mean-Field-Type Game Theory I*, Static & Dynamic Game Theory: Foundations & Applications, https://doi.org/10.1007/978-3-032-07027-2_5

Definition 5.1 (Gâteaux Differential [173, 174]) The Gâteaux differential $b_m[\tilde{m}]$ of b at $m \in O$ in the direction $\tilde{m}$ is defined as

$$\lim_{\epsilon\to 0_+} \frac{d}{d\epsilon} b(., t, x, m + \epsilon\tilde{m}) = \int b_m(., t, x, m)(\xi)\, \tilde{m}(d\xi).$$

If the limit exists for all directions $\tilde{m}$, then b is Gâteaux differentiable at m.

The limit appearing in Definition 5.1 is taken relative to the weak topology of L^{α^*}.
We provide Gâteaux differential of some basic functions.

Example 5.1 Mean state: Consider the functional $g(m) := \int ym(dy)$. Then, one has

$$\lim_{\epsilon\to 0_+} \frac{d}{d\epsilon} g(m + \epsilon\tilde{m}) = \int \xi\, \tilde{m}(d\xi).$$

Thus, the Gâteaux differential $g_m[\tilde{m}]$ of g at m in the direction $\tilde{m}$ is given by the identity function.

Example 5.2 Second moment: If $g(m) = \frac{1}{2}\int y^2 m(dy)$,, then

$$\lim_{\epsilon\to 0_+} \frac{d}{d\epsilon} g(m + \epsilon\tilde{m}) = \frac{1}{2}\int \xi^2\, \tilde{m}(d\xi).$$

Hence, $g_m(m)(\xi) = \frac{1}{2}\xi^2$.

Example 5.3 Squared mean: Let $g(m) = \frac{1}{2}(\int ym(dy))^2$, then

$$\lim_{\epsilon\to 0_+} \frac{d}{d\epsilon} g(m + \epsilon\tilde{m}) = (\int ym(dy)) \int \xi\, \tilde{m}(d\xi).$$

Hence the Gâteaux differential is $g_m(m)(\xi) = \xi(\int ym(dy))$,

Example 5.4 α-th moment: $g(m) = \int y^\alpha m(dy)$. Then, $\lim_{\epsilon\to 0_+} \frac{d}{d\epsilon} g(m + \epsilon\tilde{m}) = \int \xi^\alpha\, \tilde{m}(d\xi)$. Thus, $g_m(m)(x) = x^\alpha$,

Example 5.5 Let b be a nonzero functional and $\alpha > 1$. The L^α-norm of b with respect to the measure m is given by

$$\bar{b} = \left(\int_{y\in\mathcal{X}} |b|^\alpha(., t, x, y, u) m(dy)\right)^{\frac{1}{\alpha}}.$$

'The Gâteaux differential of $\bar{b}$ is

$$\bar{b}_m(., t, x, u, m)(\xi) := \frac{b^\alpha(., t, x, \xi, u)}{\alpha\bar{b}^{\alpha-1}}.$$

As we will see below, the form $\partial_x v_m$ for functional v appears in the structure of the optimal control strategy. Thus, Gâteaux differentiation allows for a direct computation of the functional behind the random variables.

5.2 DPP for Risk-Sensitive Mean-Field-Type Games

Let $T > 0$ be a fixed time horizon, $\alpha \geq 1$ and $(\Omega, \mathscr{F}, \mathbb{F}, \mathbb{P})$ be a given filtered probability space on which independent one-dimensional standard Brownian motions $B = \{B_s\}_{s\geq 0}$, and $B_i = \{B_{i,s}\}_{s\geq 0}$, $i \in \{1, 2, \ldots\}$ are given, and the filtration $\mathbb{F} = \{\mathscr{F}_s,\ 0 \leq s \leq T\}$ is the natural filtration of B, B_i augmented by $\mathbb{P}$-null sets of $\mathscr{F}$. Introduce the random (and noisy) payoff functional of decision-maker i as

$$\begin{cases} \Psi_i = \int_0^T f_i(t, x^u(t), m^u(t), u(t))\, dt + h_i(x^u(T), m^u(T)) + \int_0^T \tilde{\sigma}_i(t, x^u(t), \\ \quad m^u(t), u(t))\, dB_i,\ i \in \{1, 2, \ldots\} \end{cases} \tag{5.1}$$

and the risk-sensitive payoff of decision-maker i as

$$\bar{J}_i^{\theta}(u_i(\cdot), u_{-i}(\cdot)) = \frac{1}{\theta_i} \log\left[E e^{\theta_i \Psi_i}\right], \tag{5.2}$$

The underlying risk-sensitive best-response problem of decision-maker i is

$$\begin{cases} \inf_{u_i \in \mathscr{U}_i} \bar{J}_i^{\theta}(u_i, u_{-i}) \text{ subject to} \\ dx^u(t) = b(t, x^u(t), m^u(t), u(t))\, dt + \sigma(t, x^u(t), m^u(t), u) dB(t), \\ x^u(0) = x_0, \\ m^u(t, .) := \mathscr{L}(x^u(t)), \end{cases} \tag{5.3}$$

where u_{-i} denotes $(u_1, \ldots, u_{i-1}, u_{i+1}, \ldots, u_I)$, $I \geq 2$, and by abuse of notation, $u = (u_i, u_{-i})$. $\mathscr{L}(x^u(t)) = \mathbb{P}_{x^u(t)}$ is the probability measure of $x^u(t)$. The Brownian motions B_i and B are independent. U_i denotes the control set of decision-maker i. The realized control of decision-maker i is $u_i(t, .) \in U_i$. A strategy u_i of decision-maker i is progressively measurable process. We have

$$f_i(t, x, m, u), \tilde{\sigma}_i(t, x, m, u) : [0, T] \times \mathscr{X} \times \mathscr{P}(\mathscr{X}) \times \prod_i U_i \longrightarrow \mathbb{R},$$

$$b(t, x, m, u),\ \sigma(t, x, m, u), \tilde{\sigma}_i(t, x, m, u) : [0, T] \times \mathscr{X} \times \mathscr{P}(\mathscr{X}) \times \prod_i U_i \longrightarrow \mathbb{R},$$

$$h_i(x, m) : \mathscr{X} \times \mathscr{P}(\mathscr{X}) \longrightarrow \mathbb{R},$$

The control strategy u_i is chosen by decision-maker i. An admissible control strategy u_i is an $\mathbb{F}$-adapted process with values in a non-empty subset U_i of $\mathbb{R}^d$. We denote the set of all admissible strategies of the decision-maker i by $\mathscr{U}_i$. In view of (5.3), up to a change of the parameter θ_i into $-\theta_i$, the optimization of $\bar{J}_i^\theta$ is the same as optimizing

$$J_i^\theta(u) = Ee^{\theta_i \Psi_i}. \tag{5.4}$$

Any $u_i \in \mathscr{U}_i$ satisfying

$$J_i^\theta(u_i, u_{-i}) = \inf_{u_i' \in \mathscr{U}_i} J_i^\theta(u_i', u_{-i}), \tag{5.5}$$

is called a risk-sensitive best response strategy of decision-maker i to u_{-i}. The corresponding state process, solution of the SDE in (5.3), is denoted by $x(\cdot) := x^u(\cdot)$. Then the risk-sensitive loss functional is given by

$$\bar{J}_i^\theta = \frac{1}{\theta_i} \log J_i^\theta = \frac{1}{\theta_i} \log \left[Ee^{\theta_i \Psi_i} \right].$$

When the risk-sensitive index θ_i is small, the loss functional $\bar{J}_i^\theta$ can be expanded in terms of θ_i as

$$E[\Psi_i] + \frac{\theta_i}{2} \mathrm{var}(\Psi_i) + O(\theta_i^2),$$

where $\mathrm{var}(\Psi_i)$ denotes the variance of the random quantity Ψ_i. If $\theta_i < 0$, the variance of Ψ_i, as a measure of risk, improves the performance, in which case the optimizer is called *risk seeker*. But, when $\theta_i > 0$, the variance of Ψ_i worsens the performance $\bar{J}_i^\theta$, in which case the optimizer is called *risk averse*. The risk-neutral loss functional $E[\Psi_i]$ can be seen as a limit of risk-sensitive functional $\bar{J}_i^\theta$ when $\theta_i \to 0$. This is one of the reasons why this criterion has attracted considerable of attention. The criterion has also interesting connections with H_∞-mean-field-type optimization. This can easily be deduced from the Donsker–Varadhan formula:

$$\frac{1}{\theta} \log \left(\int e^{\theta \phi} \, d\nu \right) = \sup_{\mu \in \mathscr{P}(\Omega)} \left[\int \phi \, d\mu - \frac{1}{\theta} \tilde{H}(\mu | \, \nu) \right], \tag{5.6}$$

for any measurable bounded function ϕ on Ω, where ν a probability measure on Ω. Moreover, the supremum is *uniquely* achieved by the *imitative* Boltzmann-Gibbs distribution μ_*, widely used in distributed strategic learning, which is given by

$$d\mu_* = \frac{e^{\theta \phi} \, d\nu}{\int_\Omega e^{\theta \phi} \, d\nu}.$$

In (5.6), the function $\tilde{H}(.|.)$ is the relative entropy from Ω to $\bar{\mathbb{R}}$ given by

$$\tilde{H}(\mu|\nu) := \int \log\left(\frac{d\mu}{d\nu}\right) d\mu = \int \frac{d\mu}{d\nu} \log\left(\frac{d\mu}{d\nu}\right) d\nu,$$

whenever $\mu \in \mathscr{P}(\Omega)$ is absolutely continuous with respect to ν; otherwise we set $\tilde{H}(\mu|\nu) = +\infty$. The problem of infimization of $\bar{J}_i^\theta$ then becomes equivalent to

$$\inf_{u_i} \bar{J}_i^\theta(u) = \inf_{u_i} \sup_{\mu \in \mathscr{P}(\Omega)} \left[E_\mu[\Psi_i] - \frac{1}{\theta_i}\tilde{H}(\mu|\,\mathbb{P}) \right]. \tag{5.7}$$

The measurement noise term $\tilde{\sigma}_i$ does not play any role in the risk-neutral case [333, 338] because

$$E\left[\int_0^T \tilde{\sigma}_i(t, x^u(t), m^u(t), u(t))\, dB_i\right] = 0,$$

by the martingale property. However, $\tilde{\sigma}_i$ plays an important role in the risk-sensitive case when $\theta_i \neq 0$ as demonstrated below.

The problem is equivalent to optimization of the functional

$$e^{\theta_i \bar{J}_i^\theta(u)} = E\left[e^{\theta_i \Psi_i}\right].$$

The sign of θ_i plays an important role as it determines whether the optimization problem is one of maximization or minimization.

5.2.1 Augmented Finite-Dimensional State

Introduce the following augmented state (x, z) such that $z_i(0) = 0$, and $dz_i^u(t) = f_i(t, x^u(t), m^u(t), u(t))\, dt + \tilde{\sigma}_i(t, x^u(t), m^u(t), u(t))\, dB_i$. Then,

$$e^{\theta_i \bar{J}_i^\theta(u)} = E e^{\theta_i z_i(T) + \theta_i h_i(x^u(T), m^u)}.$$

Thus, we obtain the following mean-field-type control problem for decision-maker i:

$$\begin{cases} \inf_{u_i} E e^{\theta_i [z_i(T) + h_i(x^u(T), m^u(T))]} \\ dz_i^u(t) = f_i(t, x^u(t), m^u(t), u(t))\, dt + \tilde{\sigma}_i(t, x^u(t), m^u(t), u(t))\, dB_i, \\ z_i^u(0) = 0, \\ dx^u(t) = b(t, x^u(t), m^u(t), u(t))dt + \sigma(t, x^u(t), m^u(t), u(t))dB(t) \\ x^u(0) = x_0. \end{cases} \tag{5.8}$$

The dynamics of the augmented state (x, z) can be rewritten in a vector form:

$$d\begin{pmatrix} x \\ z_1 \\ z_2 \\ z_3 \\ \vdots \\ z_I \end{pmatrix} = \begin{pmatrix} b \\ f_1 \\ f_2 \\ f_3 \\ \vdots \\ f_I \end{pmatrix} dt + \begin{pmatrix} \sigma & 0 & 0 & 0 & \dots & 0 \\ 0 & \tilde{\sigma}_1 & 0 & 0 & \dots & 0 \\ 0 & 0 & \tilde{\sigma}_2 & 0 & \dots & 0 \\ 0 & 0 & 0 & \tilde{\sigma}_3 & \dots & 0 \\ \vdots & & & \dots & \dots & \vdots \\ 0 & 0 & 0 & 0 & 0 & \tilde{\sigma}_I \end{pmatrix} d\begin{pmatrix} B \\ B_1 \\ B_2 \\ B_3 \\ \vdots \\ B_I \end{pmatrix}$$

with $I \geq 2$.

5.2.2 Fokker–Planck–Kolmogorov Forward Equation

Let us introduce $\mu(t, dxdz) = \mathscr{L}(x^u(t), z^u(t))$, the probability measure associated with the joint process $(x^u(t), z^u(t))$. Generically, the measure μ solves the Fokker–Planck–Kolmogorov forward equation in the distributional (weak) sense.

$$\mu_t(t, x, z) = -\partial_x[b\mu] - \sum_{i=1}^{I} \frac{\partial}{\partial z_i}[f_i\mu] + \frac{1}{2}\sum_{i,j} \frac{\partial^2}{\partial \tilde{z}_i\, \partial \tilde{z}_j}\left[D_{ij}\mu\right],$$

with $\tilde{z}_i \in \{x, z_i\}$, drift vector (b, f) and the diffusion

$$D_{ij}(t, x, z, m, u) = \sum_{k=1}^{I+1} \hat{\sigma}_{ik}\hat{\sigma}_{jk},$$

where $\hat{\sigma}$ is a diagonal matrix consisting of σ and $(\tilde{\sigma}_1, \dots, \tilde{\sigma}_I)$. Thus,

$$\mu_t = -\partial_x[b\mu] - \text{div}_z(f\mu) + \frac{1}{2}\partial_{xx}(\sigma'\sigma\mu) + \frac{1}{2}\partial_{zz}\text{trace}(\hat{\sigma}'\hat{\sigma}\mu) =: \tilde{b}, \tag{5.9}$$

with initial distribution $\mu(0, dx, dz) = m_0(dx)\delta_0(dz)$ and $\hat{\sigma}'$ is the transpose of $\hat{\sigma}$.

5.2.3 Risk-Sensitive DPP in Infinite Dimensions

The cost functional $e^{\theta_i \bar{J}_i^{\theta}(u)}$ can be rewritten in a deterministic manner as

$$\int \mu(T, dx, dz)\; e^{\theta_i(z_i + h_i(x, \int_{\tilde{z}} \mu(T,.,d\tilde{z})))}\,.$$

This is a terminal cost in the sense that it is evaluated only at $\mu(T,.)$. μ is an infinite dimensional quantity that serves as a state in this problem. The advantage now is that $\mu(.)$ is a deterministic quantity. Since there is no running cost, one can write directly the HJB equation using classical calculus of variations for

$$V_i(t,\mu) = \inf_{u_i\in\mathscr{U}_i} \int \mu(T,dx,dz)\, e^{\theta_i(z_i+h_i(x,\int_{\tilde z}\mu(T,.,d\tilde z)))},$$

starting from $\mu(t,.) = \mu$ at time t :

$$0 = V_{i,t}(\mu) + \inf_{u_i\in\mathscr{U}_i}[\langle \tilde b, V_{i,\mu}\rangle], \tag{5.10}$$

where

$$\langle \tilde b, V_{i,\mu}\rangle = \int \Big[b\partial_x V_{i,\mu} + \langle f, \partial_z V_{i,\mu}\rangle + \frac{\sigma^2}{2}\partial_{xx}V_{i,\mu} + \frac{1}{2}\sum_{i=1}^n \tilde\sigma_i^2 \partial_{z_i z_i} V_{i,\mu}\Big]\mu(t,dx,dz)$$

The risk-sensitive HJB system under loss measurement noise yields

$$\begin{aligned} 0 &= V_{i,t}(t,\mu) + \int \inf_{u_i}\Big[b\partial_x V_{i,\mu} + \langle f, \partial_z V_{i,\mu}\rangle + \frac{\sigma^2}{2}\partial_{xx}V_{i,\mu} \\ &\quad + \frac{1}{2}\sum_{k=1}^n \tilde\sigma_k^2 \partial_{z_k z_k} V_{i,\mu}\Big]\mu(t,dx,dz), \\ V_i(T,\mu) &= \int \mu(T,dx,dz)\, e^{\theta_i(z_i+h_i(x,\int_{\tilde z}\mu(T,.,d\tilde z)))}, \\ i &\in \{1,2,\ldots,n\} \end{aligned} \tag{5.11}$$

This is an infinite-dimensional PIDE system. Note that

$$V_i(T,\mu) == V_i(T,\mu_i) \tag{5.12}$$

where the measure $\mu_i(t,.)$, given by $\mu_i(t,dx,dz_i) = \int_{z_{-i}}\mu(t,dxdz_idz_{-i})$, is the marginal of μ with respect to (x,z_i) at time t. The above calculation shows that the value function of i depends on the measure μ_i. The partial derivatives of $V_{i,\mu_i}(T,.)$ do not depend on $z_j,\ j\neq i$. Moreover, at any time $t,$, the functions $b, f, \sigma, \tilde\sigma$ do not depend explicitly on $z,$, and the terminal cost of decision-maker i depends only on (x,z_i), the required working state for decision-maker i. Therefore, one can look for a solution in the form $V_i(t,\mu) = \tilde V_i(t,\mu_i)$, for all time t.

$$\begin{cases} 0 = \tilde{V}_{i,t}(t,\mu_i) + \int \inf_{u_i} \left[b\partial_x \tilde{V}_{i,\mu_i} + f_i \partial_{z_i} \tilde{V}_{i,\mu_i} \rangle + \frac{\sigma^2}{2}\partial_{xx}\tilde{V}_{i,\mu_i} + \frac{\tilde{\sigma}_i^2}{2}\partial_{z_i z_i}\tilde{V}_{i,\mu_i} \right] \\ \quad \mu_i(t,dx,dz_i), \\ \tilde{V}_i(T,\mu_i) = \int \mu_i(T,dx,dz_i)\, e^{\theta_i(z_i + h_i(x, \int_{\tilde{z}} \mu_i(T,.,d\tilde{z})))}, \\ \mu_i(t,dx,dz_i) = \int_{z_{-i}} \mu(t, dx dz_i dz_{-i}), \\ \mu_t = -\partial_x[b\mu] - \mathrm{div}_z(f\mu) + \frac{1}{2}\partial_{xx}(\sigma^2\mu) + \frac{1}{2}\partial_{zz}(\tilde{\sigma}'\tilde{\sigma}\mu), \\ \mu(t,dx,dz) = m_0(dx)\delta_0(dz), \\ i \in \{1,2,\ldots,n\} \end{cases} \tag{5.13}$$

Differentiating weakly (5.13) with respect to μ_i,, we obtain

$$0 = \partial_t \tilde{V}_{i,\mu_i}(\mu_i)(t,x,z_i) + \hat{H}_i(\mu_i)(t,x,z_i) + \int \hat{H}_{i,\mu_i}(\mu_i)(t,\tilde{x},\tilde{z}_i)\mu_i(t,d\tilde{x},d\tilde{z}_i) \tag{5.14}$$

where

$$\hat{H}_i = b\partial_x \tilde{V}_{i,\mu_i} + f_i \partial_{z_i}\tilde{V}_{i,\mu_i} + \frac{\sigma^2}{2}\partial_{xx}\tilde{V}_{i,\mu_i} + \frac{\tilde{\sigma}_i^2}{2}\partial_{z_i z_i}\tilde{V}_{i,\mu_i}. \tag{5.15}$$

Definition 5.2 (Dual Function) The function $\tilde{v}_i^*(t,x,z_i) := \tilde{V}_{i,\mu_i}(\mu_i)(t,x,z_i)$ is the dual function associated with the risk-sensitive best response value of decision-maker i.

The dual function $\tilde{v}_i^*(t,x,z_i) := \tilde{V}_{i,\mu_i}(\mu_i)(t,x,z_i)$ solves the following PDE system:

$$\begin{aligned} 0 = \tilde{v}^*_{i,t} + \tilde{v}^*_{i,z_i} H_i(t,x,z_i,\mu_i, \frac{\tilde{v}^*_{i,x}}{\tilde{v}^*_{i,z_i}}, [\sigma\tilde{v}^*_{i,xx}, \tilde{\sigma}_i\tilde{v}^*_{i,z_i z_i}]\frac{1}{2\tilde{v}^*_{i,z_i}}) \\ + \int \tilde{v}^*_{i,\tilde{z}_i} H_{i,\mu_i}(t,\tilde{x},\tilde{z}_i, \frac{\tilde{v}^*_{i,\tilde{x}}}{\tilde{v}^*_{i,\tilde{z}_i}}, [\sigma\tilde{v}^*_{i,\tilde{x}\tilde{x}}, \tilde{\sigma}_i\tilde{v}^*_{i,\tilde{z}_i\tilde{z}_i}]\frac{1}{2\tilde{v}^*_{i,\tilde{z}_i}})(t,x,z_i)\mu_i(t,d\tilde{x},d\tilde{z}_i), \end{aligned} \tag{5.16}$$

where $H_i(t,x,z_i,\mu_i,p_i,q_{1i},q_{2i}) = f_i + bp_i + \sigma q_{1i} + \tilde{\sigma}_i q_{2i}$.

The terminal condition for this PDE is

$$\begin{aligned} \tilde{v}_i^*(T,x,z_i) = e^{\theta_i(z_i + h_i(x,m(T)))} + \theta_i \int h_{i,m}(\tilde{x},m(T))(x) e^{\theta_i(\tilde{z}_i + h_i(\tilde{x},m(T)))} \\ \times \mu_i(T,d\tilde{x},d\tilde{z}_i). \end{aligned}$$

Note that, in general, the dual function $\tilde{v}^*(t,x,z)$ is different from the value function $\tilde{V}(\mu_i)$ or $V(\mu)$.

Lemma 5.1 *If $\theta_i > 0$,, then $\eta_i(t) = \tilde{v}^*_{i,z_i}(t, x(t), z(t)) > 0$. Moreover, the process η_i satisfies the backward SDE*

$$d\eta_i = \eta_i[l_1 dB + l_2 dB_i], \quad \eta_i(T) = \theta_i e^{\theta_i[z_i(T)+h_i(x(T),m(T))]}, \tag{5.17}$$

where $l_1 = \sigma\frac{\eta_{i,x}}{\eta_i} = \sigma\partial_x[\log\eta_i]$ and $l_2 = \tilde{\sigma}_i\frac{\eta_{i,z_i}}{\eta_i} = \tilde{\sigma}_i\partial_{z_i}[\log\eta_i]$. Furthermore, by Itô's formula in multiple dimension, the function $(t, x, z_i) \mapsto \eta(t, x, z_i)$ solves the following partial differential equation, whenever the weak derivatives are well defined:

$$\begin{cases} 0 = \eta_{i,t} + b\eta_{i,x} + f_i\eta_{i,z_i} + \frac{\sigma^2}{2}\eta_{i,xx} + \frac{\tilde{\sigma}_i^2}{2}\eta_{i,z_iz_i}, \\ \eta_i(T, x, z_i) = \theta_i e^{\theta_i[z_i+h_i(x,m(T,.))]}, \\ i \in \{1, 2, \ldots\}. \end{cases} \tag{5.18}$$

Note that for $\theta_i \neq 0$,, the process $\frac{1}{\eta_i(t)}$ satisfies

$$d(\frac{1}{\eta_i}) = \frac{1}{\eta_i}[l_1^2 + l_2^2]dt - \frac{d\eta_i}{\eta_i} = \frac{1}{\eta_i}[l_1^2 + l_2^2]dt - (l_1 dB + l_2 dB_i). \tag{5.19}$$

5.2.4 Risk-Sensitive DPP to Risk-Sensitive SMP

5.2.4.1 Perfect Measurement Case

This subsection examines the perfect measurement case; i.e., $\tilde{\sigma}_i$ is identically zero for all i. The risk-sensitive value function of (5.14) becomes

$$\begin{cases} 0 = \tilde{V}_{i,t}(\mu) + \int \inf_{u_i\in\mathcal{U}_i}\Big[b\partial_x\tilde{V}_{i,\mu_i}(\mu_i)(t, x, z_i) + f_i\partial_z\tilde{V}_{i,\mu_i}(\mu_i)(t, x, z_i) \\ +\frac{\sigma^2}{2}\partial_{xx}\tilde{V}_{i,\mu_i}(\mu_i)(t, x, z_i)\Big]\mu_i(t, dx, dz_i), \end{cases} \tag{5.20}$$

which is a partial differential equation with state μ_i which is a measure on $\mathcal{X} \times \mathbb{R}$. If we introduce $\tilde{v}_i^*(t, x, z_i) := \tilde{V}_{i,\mu_i}(t, \mu_i)(t, x, z_i)$ as a dual function (because of the Gâteaux differential with respect to μ_i), then $(\tilde{v}_1^*, \ldots, \tilde{v}_I^*)$ is finite-dimensional and solves the dual system

$$0 = \tilde{v}^*_{i,t} + \tilde{v}^*_{i,z_i}H_i(t, x, z_i, \mu_i, \frac{\tilde{v}^*_{i,x}}{\tilde{v}^*_{i,z_i}}, [\sigma\tilde{v}^*_{i,xx}]\frac{1}{2\tilde{v}^*_{i,z_i}}) \tag{5.21}$$

$$+ \int \tilde{v}^*_{i,\tilde{z}_i}H_{i,\mu_i}(t, \tilde{x}, \tilde{z}_i, \frac{\tilde{v}^*_{i,\tilde{x}}}{\tilde{v}^*_{i,\tilde{z}_i}}, [\sigma\tilde{v}^*_{i,\tilde{x}\tilde{x}}]\frac{1}{2\tilde{v}^*_{i,\tilde{z}_i}})(t, x, z_i)\mu_i(t, d\tilde{x}, d\tilde{z}_i),$$

where $H_i(t, x, z_i, \mu_i, p_i, q_i) = f_i + bp_i + \sigma q_i$, $u_i \in \arg\min H_i$, $H_i^*(t, x, z_i, \mu_i, p_i, q_i) = \inf_{u_i}[f_i + bp_i + \sigma q_i]$, $m(t, dx) = \int \mu_i(t, dx, dz_i)$.

We define the risk-neutral Hamiltonian associated with random variable $X \in L^\alpha(\Omega, \mathscr{F}, \mathbb{P})$ as follows. For $(p, q) \in \mathbb{R} \times \mathbb{R}$

$$H_i(t, X, m, u, p, q) := b(t, X, m, u)p + \sigma(t, X, m, u)q - f_i(t, X, m, u). \tag{5.22}$$

We also introduce the risk-sensitive Hamiltonian: for $\theta_i \in \mathbb{R}$ and $(p_i, q_i, \ell_i) \in \mathbb{R} \times \mathbb{R} \times \mathbb{R}$,

$$H_i^\theta(t, X, m, u, p_i, q_i, \ell_i) := bp_i + \sigma(q_i + \theta_i \ell_i p_i) - f_i. \tag{5.23}$$

Clearly $H = H^0$. The sign $(-f)$ is used here for the only purpose of stating the maximum principle instead of the minimum principle and does not fundamentally change the methodology. Moreover, we let

$$H_{i,k}^\theta(t) := p_i(t)b_k(t) + (q_i + \theta_i \ell_i p_i)\sigma_k(t) - f_{i,k}(t), \tag{5.24}$$

for $k \in \{x, m\}$, where the subscript stands for derivative.

Lemma 5.2 *Consider the following mean-field backward SDE*

$$p(t) = p(T) + \int_t^T \tilde{E}[\hat{f}(s, \tilde{p}(s), \tilde{q}(s), p(s), q(s))]\, ds - \int_t^T q(s)dB(s),$$

where $p(T)$ is a progressively measurable, square integrable random variable. Let $\hat{f}(t, ., ., ., .)$ be Lipschitz for all time $t \in [0, T]$ and $t \mapsto f(t, 0, 0, 0, 0)$ be square integrable over $[0, T]$. Then, the mean-field backward SDE has a unique adapted solution satisfying

$$E\left[\sup_{t\in[0,T]} |p(t)|^2 + \int_0^T |q(t)|^2 dt\right] < \infty. \tag{5.25}$$

Further, by choosing

$$\hat{f}(t, \tilde{p}(t), \tilde{q}(t), p(t), q(t))$$

$$= a_0(t, .) + a_1(t, .)\tilde{p}(t) + a_2(t, .)\tilde{q}(t) + a_3(t, .)p + a_4(t, .)q(t),$$

where $a_i(t, .)$ are measurable bounded coefficient functions, one gets a backward equation in the form of the adjoint equations.

Let $(p_i^*, q_i^*, \eta_i^*, l_i)$ satisfy the (risk-sensitive) first-order adjoint system given by

$$\begin{cases} dp_i^* = -\left[H_{i,x}^* + \frac{1}{\eta_i^*}E[\eta_i^* \partial_x H_{i,m}^*] + \sigma_x(\frac{\sigma v_{i,xx}^*}{\eta_i^*}) - p_i^*(l_i^*)^2 + \frac{\sigma l_i^* v_{i,xx}^*}{\eta_i^*}\right]dt \\ \quad +(-p_i^* l_i^* + \frac{\sigma v_{i,xx}^*}{\eta_i^*})dB_i \\ d\eta_i^* = \eta_i^* l_i^* dB_i, \\ \eta_i^*(T) = \theta_i e^{\theta_i[z_i(T)+h_i(x(T),m(T))]}, \\ q_i = (-p_i^* l_i^* + \frac{\sigma v_{i,xx}^*}{\eta_i^*}) \\ p_i^*(T) = h_{i,x}(T) + \frac{1}{\phi_i^\theta(T)}E[\phi_i^\theta(T)\partial_x h_{i,m}(T)], \\ i \in \{1, 2, \ldots, n\} \end{cases} \tag{5.26}$$

where

$$\phi_i^\theta(T) := e^{\theta_i[h_i(\bar{x}(T),\bar{m}(T))+\int_0^T f_i(t,\bar{x}(t),\bar{m}(t),\bar{u}(t))dt]}. \tag{5.27}$$

Proposition 5.1 *If the functions b, σ, f_i, h_i, are twice continuously differentiable with respect to (x, m), and b, σ, f_i, h_i and all their first-order derivatives with respect to (x, m) are continuous in (x, m, u_i) and bounded, then (5.26) admits a unique $\mathbb{F}$-adapted solution $(\bar{p}, \bar{q}, \eta, \ell)$ such that*

$$E\left[\sup_{t\in[0,T]} |\bar{p}(t)|^2 + \sup_{t\in[0,T]} |\eta(t)|^2 + \int_0^T \left(|\bar{q}(t)|^2 + |\ell(t)|^2\right) dt\right] < \infty. \tag{5.28}$$

Proof Under the assumptions of Proposition 5.1, the processes p, q solve a backward SDE coupled with the process v^θ. Moreover, these equations can be transformed into linear SDEs of mean-field-type, which involve $(p, q, E[p], E[q])$ and random coefficients involving $(x, \frac{\eta}{E[\phi^\theta(T)]}, l)$. We now check that the coefficients of the linear SDEs do not blow up within the horizon $[0, T]$. For the functions σ, f_i, h_i, and their derivatives, boundedness follows from the assumption. Then, for each direction chosen in the sub-differential, the assumptions of Lemma 5.2 are fulfilled, and hence the existence of solution to the first-order risk-sensitive adjoint equations follows. Moreover, b_x and $\partial_x b_m$ are (uniquely) well defined and bounded. Using Lemma 5.2 again, we confirm the existence and uniqueness of the solution.■

Note that the boundedness and differentiation conditions can be weakened using Rademacher's theorem, which states that Lipschitz functions admit almost everywhere weak Gâteaux differential. The following Proposition states a stochastic maximum principle.

Proposition 5.2 (Risk-Sensitive Stochastic Maximum Principle) *Let the Assumptions of Proposition 5.1 hold. If $(\bar{x}(\cdot), \bar{u}(\cdot))$ is an equilibrium of the risk-sensitive mean-field-type game problem (5.3) with $\bar{\sigma}_i = 0$,, then there are two pairs of $\mathbb{F}$-adapted processes (η, ℓ), $(\bar{p}, \bar{q})$ for each decision-maker that satisfy (5.26), such that*

$$
\begin{aligned}
&H_i^{\theta}(t, \bar{x}(t), \bar{m}(t), \bar{u}(t), \bar{p}(t), \bar{q}(t), \ell(t)) \\
&\quad = \max_{u_i} H_i^{\theta}(t, \bar{x}(t), \bar{m}(t), u_i, \bar{u}_{-i}(t), \bar{p}(t), \bar{q}(t), \ell(t)),
\end{aligned}
$$

for almost every $t \in [0, T]$ and $\mathbb{P}$-almost surely.

To prove Proposition 5.2, we use a logarithmic transformation as treated above. The proof is therefore omitted.

5.2.5 Risk-Neutral Case

We now seek for a DPP equation when θ goes to zero in the approach above. The noise terms $\tilde{\sigma}_i$ in (5.13) are irrelevant in this case, and the infinite-dimensional PDE becomes

$$
\begin{cases}
0 = \tilde{V}_{i,t}(t, \mu_i) + \int \inf_{u_i} \left[b\partial_x \tilde{V}_{i,\mu_i} + f_i \partial_{z_i} \tilde{V}_{i,\mu_i}\rangle + \frac{\sigma^2}{2}\partial_{xx}\tilde{V}_{i,\mu_i} \right] \mu_i(t, dx, dz_i), \\
\tilde{V}_i(T, \mu_i) = \int \mu_i(T, dx, dz_i)\, [z_i + h_i(x, \int_{\tilde{z}} \mu_i(T, ., d\tilde{z}))], \\
\mu_i(t, dx, dz_i) = \int_{z_{-i}} \mu(t, dx dz_i dz_{-i}), \\
\mu_t = -\partial_x[b\mu] - \mathrm{div}_z(f\mu) + \frac{1}{2}\partial_{xx}(\sigma^2\mu) + \frac{1}{2}\partial_{zz}(\tilde{\sigma}'\tilde{\sigma}\mu), \\
i \in \{1, 2, \ldots, I\}
\end{cases}
$$

The dual functions $\{\tilde{v}_i^*,\ i \in \{1, 2, \ldots, I\}\}$ satisfy the following system:

$$
\begin{aligned}
0 = {}& \tilde{v}_{i,t}^* + \tilde{v}_{i,z_i}^* H_i^*(t, x, z_i, \mu_i, \frac{\tilde{v}_{i,x}^*}{\tilde{v}_{i,z_i}^*}, \frac{\sigma \tilde{v}_{i,xx}^*}{2\tilde{v}_{i,z_i}^*}) \\
& + \int \tilde{v}_{i,\tilde{z}_i}^* H_{i,\mu_i}^*(t, \tilde{x}, \tilde{z}_i, \frac{\tilde{v}_{i,\tilde{x}}^*}{\tilde{v}_{i,\tilde{z}_i}^*}, \frac{\sigma \tilde{v}_{i,\tilde{x}\tilde{x}}^*}{2\tilde{v}_{i,\tilde{z}_i}^*})(t, x, z_i)\mu_i(t, d\tilde{x}, d\tilde{z}_i),
\end{aligned}
\tag{5.29}
$$

where $H_i^*(t, x, z_i, \mu_i, p_i, q_i) = f_i + bp_i + \sigma q_i$.

The terminal condition for this PDE is

$$
\tilde{v}_i^*(T, x, z_i) = z_i + h_i(x, m(T, .)) + \int h_{i,m}(\tilde{x}, m(T))(x)\, \mu_i(T, d\tilde{x}, d\tilde{z}_i).
$$

This implies that $\tilde{v}_{i,z_i}^*(T, x, z_i) = 1 = \eta_i(T, x, z_i)$ and for all $t \in (0, T)$ one has

$$
0 = \eta_{i,t} + b\eta_{i,x} + f_i \eta_{i,z_i} + \frac{\sigma^2}{2}\eta_{i,xx}.
$$

Then, the terminal function is constant and solves the PDE. Hence, $\eta_i(t, x, z_i) = 1 = \tilde{v}^*_{i,z_i}(t, x, z_i)$

Hence, the dual functions solve

$$0 = \tilde{v}^*_{i,t} + H^*_i(t, x, z_i, \mu_i, \tilde{v}^*_{i,x}, \frac{\sigma \tilde{v}^*_{i,xx}}{2}) \tag{5.30}$$
$$+ \int H^*_{i,\mu_i}(t, \tilde{x}, \mu_i, \tilde{z}_i, \tilde{v}^*_{i,\tilde{x}}, \frac{\sigma}{2}\tilde{v}^*_{i,\tilde{x}\tilde{x}})(t, x, z_i)\mu_i(t, d\tilde{x}, d\tilde{z}_i),$$

The terminal condition for this PDE is

$$\begin{aligned} \tilde{v}^*_i(T, x, z_i) &= z_i + h_i(x, m(T, .)) + \textstyle\int h_{i,m}(\tilde{x}, m(T))(x)\ \mu_i(T, d\tilde{x}, d\tilde{z}_i) \\ &= z_i + h_i(x, m(T, .)) + \textstyle\int h_{i,m}(\tilde{x}, m(T))(x)\ m(T, d\tilde{x}). \end{aligned}$$

By differentiating with respect to x,, we arrive at

$$0 = \tilde{v}^*_{i,tx} + H_{i,x} + \int \partial_x H_{i,m}(t, \tilde{x}, m, \tilde{v}^*_{i,\tilde{x}}, \frac{\sigma}{2}\tilde{v}^*_{i,\tilde{x}\tilde{x}})(t, x)m(t, d\tilde{x}),$$

The adjoint processes $\{p_i(t) = \tilde{v}^*_{i,x}(t, x^u(t), z^u(t)),\ \ q_i(t) = \sigma\tilde{v}^*_{i,xx}(t, x^u(t), z^u(t))\}$ solve the following backward SDE system:

$$\begin{aligned} dp_i &= [\tilde{v}^*_{i,tx} + b\tilde{v}^*_{i,xx} + f_i\tilde{v}^*_{i,z_ix} + \tfrac{\sigma^2}{2}\tilde{v}^*_{i,xxx}]dt + \sigma\tilde{v}^*_{i,xx}dB \\ &= [-H_{i,x} - \partial_x \textstyle\int H_{i,m}(t, \tilde{x}, m, p_i, q_i)(t, x)m(t, d\tilde{x})]dt + q_i dB \\ p_i(T) &= h_{i,x}(x, m(T, .)) + \textstyle\int \partial_x h_{i,m}(\tilde{x}, m(T))(x)\ m(T, d\tilde{x}). \end{aligned}$$

Note that in order to establish a stochastic maximum principle in the control-dependent diffusion case, one may need second-order processes (P, Q) in addition to (p, q) specially when the control set U is non-convex.

5.2.5.1 Aggregative Mean-Field-Type Games

In this subsection, we take b, σ, f_i, h_i as functions of m^u only through aggregative terms $\int \phi_.(x)m^u(t, dx) = E[\phi_.(x^u(t))]$, and $\int \phi_.(u')\mathbb{P}_u(t, du') = E[\phi_.(u(t))]$. In that case, the Gâteaux differentiation with respect to m can be reduced to a finite-dimensional differentiation.

$$\begin{cases} \bar{J}^0_i = E\Big[\int_0^T f_i(t, x^u(t), E[\phi_{f_i}(x^u(t))], u(t), E[\phi_{f_i}(u(t))])\, dt \\ +\int_0^T \bar{\sigma}_i(t, x^u(t), E[\phi_{\bar{\sigma}_i}(x^u(t))], u(t), E[\phi_{f_i}(u(t))])\, dB_i \\ +h_i(x^u(T), E[\phi_{h_i}(x^u(t))])\Big], \\ \inf_{u_i\in\mathcal{U}_i} \bar{J}^0_i(u_i, u_{-i}) \ \text{ subject to} \\ dx^u(t) = b(t, x^u(t), E[\phi_b(x^u(t))], u(t), E[\phi_b(u(t))])\, dt \\ +\sigma(t, x^u(t), E[\phi_\sigma(x^u(t))], u(t), E[\phi_\sigma(u(t))])dB(t), \\ x^u(0) = x_0. \end{cases} \tag{5.31}$$

If b, σ, f_i are only functions of $(t, x^u, E[\phi_.(x^u(t))], u, E[\phi_.(u(t))])$,, then the partial derivatives of the dual functions $p_i(t) = \tilde{v}^*_{i,x}(t, x^u(t), z^u(t))$ and $q_i(t) = \sigma\tilde{v}^*_{i,xx}(t, x^u(t), z^u(t))$ solve backward SDE system:

$$\begin{cases} dp_i = \{-H_{i,x} - \phi_{f_i,x}E[f_{i,y}] - \phi_{b,x}E[b_y p_i] - \phi_{\sigma,x}E[\sigma_y q_i]\}dt + q_i dB, \\ p_i(T) = h_{i,x}(x^u(T), E[\phi_{h_i}(x^u(T))]) + \phi_{h,x}(T)E[h_{i,y}(x^u(T), E[\phi_{h_i}(x^u(T))])] \end{cases} \tag{5.32}$$

where $H_{i,x} = f_{i,x} + b_x.p_i + \text{trace}(\sigma_x' q_i)$.

Example 5.6 (First Moment Risk-Neutral Mean-Field-Type Games) We take b, σ, f_i as functions of the first moments $\int x m^u(t, dx) = E[x^u(t)]$, and $\int u'\mathbb{P}_u(t, du') = E[u(t)]$. Then the aggregative functions are reduced to the identity function $\phi_.(y) = y$. Then, $\phi_{.,y} = 1$. Hence, the first-order risk-neutral adjoint processes $p_i(t) = \tilde{v}^*_{i,x}(t, x^u(t), z^u(t))$ and $q_i(t) = \sigma\tilde{v}^*_{i,xx}(t, x^u(t), z^u(t))$ solve a simpler backward SDE system and (5.32) reduces to

$$\begin{aligned} dp_i &= \{-H_{i,x} - E[H_{i,y}(t, x^u, E[x^u], p_i, q_i)]\}dt + q_i dB \\ p_i(T) &= h_{i,x}(x^u(T), E[x^u(T)]) + E[h_{i,y}(x^u(T), E[x^u(T)])]. \end{aligned}$$

5.3 DPP for Risk-Sensitive Mean-Field Games

In this section, we consider the case where the mean-field term $m^*(t, .)$ affecting a generic decision-maker i is generated by an infinite population of other decision-makers' states or control actions. Therefore, the individual control action does not affect $m^*(t, .)$. Consider the minimization of the risk-sensitive cost functional

$$J_\theta(u) = \frac{1}{\theta}\log\left(Ee^{\theta\int_0^T f(t,x(t),u(t),m^*(t,.))dt+\theta h(x(T),m^*(T))}\right), \tag{5.33}$$

subject to

$$\begin{cases} dx(t) = b(t, x(t), u(t), m^*(t))dt + \sigma(t, x(t), u(t), m^*(t))dW(t), & 0 < t \le T, \\ x(0) = x_0, \end{cases} \tag{5.34}$$

where U is a given control set.

This is equivalent to optimization of the functional

$$e^{\theta J_\theta(u)} = E\left[e^{\int_0^T \theta f(t,x(t),u(t),m^*(t,.))dt+\theta h(x(T),m^*(T))}\right].$$

Introduce the following augmented state (x, z) such that $z(0) = 0$, and $dz(t) = \theta f(t, x(t), u(t), m^*(t, .))dt$. Then,

$$e^{\theta J_\theta(u)} = E e^{z(T)+\theta h(x(T),m^*(T))}.$$

Thus, we obtain the following standard stochastic optimization problem:

$$\begin{cases} \inf_u E e^{z(T)+h(x(T),m^*(T))} \\ dz(t) = \theta f(t, x(t), u(t), m^*(t, .))dt \\ z(0) = 0, \\ dx(t) = b(t, x(t), u(t), m^*(t))dt + \sigma(t, x(t), u(t), m^*(t))dW(t) \\ x(0) = x_0. \end{cases} \tag{5.35}$$

The problem (5.35) can be seen as a standard stochastic optimal problem *without running cost*. There is only a final cost given by $e^{z(T)+\theta h(x(T),m^*(T))}$). We apply DPP to (5.35) to obtain

$$\begin{cases} \hat{v}_t(t, x, z) + \inf_u \Big[\hat{v}_x \cdot b(t, x, u, m^*) + \hat{v}_z \cdot \theta f(t, x, u, m^*) \\ \qquad\qquad + \frac{1}{2} trace(\sigma' \sigma \tilde{v}_{xx}(t, x, z))\Big] = 0, \\ \hat{v}(T, x, z) = e^{z+\theta h(x,m^*(T,.))} = e^z.e^{\theta h(x,m^*(T,.))} \end{cases} \tag{5.36}$$

We try a solution in the following form: $\hat{v}(t, x, z) = e^z \tilde{v}(t, x)$. Then, v solves

$$\begin{cases} e^z \tilde{v}_t(t, x) + \inf_u \Big[e^z \tilde{v}_x \cdot b(t, x, u, m^*) + e^z \tilde{v}(t, x) \cdot \theta f(t, x, u, m^*) \\ \qquad\qquad + e^z \frac{1}{2} \text{trace}(\sigma' \sigma \tilde{v}_{xx}(t, x, z))\Big] = 0, \\ \tilde{v}(T, x) = e^{\theta h(x,m^*(T,.))} \end{cases} \tag{5.37}$$

Simplifying by e^z,, we get the dynamic programming equation for (5.35):

$$\begin{cases} \tilde{v}_t(t, x) + \inf_u \Big[\tilde{v}_x \cdot b(t, x, u, m^*) + \theta \tilde{v}(t, x) f(t, x, u, m^*) \\ \qquad\qquad + \frac{1}{2} trace(\sigma' \sigma \tilde{v}_{xx}(t, x, z))\Big] = 0, \\ \tilde{v}(T, x) = e^{\theta h(x,m^*(T,.))} \end{cases} \tag{5.38}$$

It is clear that $\tilde{v}(t, x) > 0$. By making the transformation $v(t, x) = \frac{1}{\theta} \log[\tilde{v}(t, x)]$,, one arrives at the HJB equation associated with (5.33):

$$\begin{cases} (\theta e^{\theta v}) v_t(t, x) + \inf_u \Big[(\theta e^{\theta v}) v_x \cdot b(t, x, u, m^*) + \theta e^{\theta v} f(t, x, u, m^*) \\ + \frac{1}{2} trace(\sigma' \sigma \tilde{v}_{xx}(t, x, z))\Big] = 0, \\ \tilde{v}(T, x) = e^{\theta h(x,m^*(T,.))} = e^{\theta v(T,x)} \end{cases} \tag{5.39}$$

Computing

$$\tilde{v}_{xx} = \partial_x[(\theta e^{\theta v}) v_x] = \theta[e^{\theta v} v_{xx} + \theta e^{\theta v} (v_x)'.v_x],$$

we obtain

$$\begin{cases} (\theta e^{\theta v})v_t + \inf_u \Big[(\theta e^{\theta v})v_x \cdot b(t,x,u,m^*) + \theta e^{\theta v} f(t,x,u,m^*) \\ +\frac{1}{2}trace(\sigma'\sigma\theta e^{\theta v} v_{xx} + \sigma'\sigma\theta^2 e^{\theta v}(v_x)'.v_x)\Big] = 0, \\ v(T,x) = h(x, m^*(T,.)). \end{cases} \tag{5.40}$$

$$\begin{cases} v_t + \inf_u \left[v_x \cdot b + f + \frac{1}{2}trace(\sigma'\sigma v_{xx}) + \theta\left|\sigma'.v_x\right|^2 \right] = 0, \\ v(T,x) = h(x, m^*(T,.)). \end{cases} \tag{5.41}$$

The risk-sensitive (optimized) Hamiltonian is

$$\tilde{H}_\theta(t,x,m^*,p,q) := \inf_u \left[p\cdot b + f + \frac{1}{2}trace(\sigma'\sigma q) + \theta\left|\sigma p\right|^2 \right].$$

and

$$\begin{cases} v_t + \tilde{H}_\theta(t,x,m,v_x,v_{xx}) = 0, \\ v(T,x) = h(x, m^*(T,.)). \end{cases} \tag{5.42}$$

As θ approaches 0,, the function $\tilde{H}_\theta$ approaches the risk-neutral Hamiltonian $\tilde{H}_*(t,x,m,p,q) := \inf_u \left[p\cdot b + f + \frac{1}{2}trace(\sigma'\sigma q)\right]$.

5.3.1 DPP for Second-Order Risk-Sensitive Mean-Field Games

Consider a representative decision-maker in an infinite population with distribution $m^*(t,.)$. Consider the problem of minimization of the risk-sensitive functional

$$J^\theta(u) = \frac{1}{\theta}\log\left[E e^{\int_0^T \theta f(t,x(t),u(t),m^*(t,.))dt + \theta h(x(T),m^*(T))} \right], \tag{5.43}$$

with respect to $u : [0,T] \longrightarrow U$, subject to

$$\begin{cases} dx(t) = b(t,x(t),u(t),m^*(t.))dt + \sigma(t,x(t),u(t),m^*(t))dW(t), \quad 0 < t \le T, \\ x(0) = x_0, \end{cases} \tag{5.44}$$

where U is a given control set.

If (u^*, m^*, v) is solution of the following backward-forward equation:

$$\begin{cases} v_t + \tilde{H}_\theta(t, x, m^*, v_x, v_{xx}) = 0, \quad 0 \le t < T, \\ v(T, x) = h(x, m^*(T, .)), \\ m_t^* + \partial_x[m^* b(t, x, m^*, u^*)] - \frac{1}{2}\partial_{xx}(m^*\sigma^2(t, x, u^*, m^*)) = 0, \quad 0 < t \le T, \\ m^*(0, x) = m_0(x), \\ u^*(t, .) \in \arg\min_u \left\{ v_x \cdot b(t, x, u, m^*) + f(t, x, u, m^*) + \frac{1}{2}(\sigma^2 v_{xx}) + \theta(\sigma v_x)^2 \right\}, \end{cases} \tag{5.45}$$

then (u^*, m^*) constitutes a mean-field equilibrium of the risk-sensitive mean-field game and $v^* = v$ is the equilibrium cost at (u^*, m^*).

5.3.2 DPP for Risk-Neutral Second-Order Mean-Field Games

We now examine (5.45) when θ vanishes. Consider a representative decision-maker in an infinite population with distribution $m^*(t, .)$. Consider the problem of minimizing the risk-neutral functional

$$J(u) = E \int_0^T f(t, x(t), u(t), m^*(t, .))dt + h(x(T), m^*(T)), \tag{5.46}$$

with respect to $u : [0, T] \longrightarrow U$, subject to

$$\begin{cases} dx(t) = b(t, x(t), u(t), m^*(t.))dt + \sigma(t, x(t), u(t), m^*(t))dW(t), \quad 0 < t \le T, \\ x(0) = x_0, \end{cases} \tag{5.47}$$

where U is a given control set. Introduce the function $v(t, x)$

$$\inf_u \left\{ \int_t^T f(s, x(s), u(s), m^*(s))ds + h(x(T), m^*(T)) \mid x(t) = x, m^*(.) \right\}. \tag{5.48}$$

Given the trajectory of m^*, $v(t, x)$ corresponds to the optimal value of the stochastic optimization problem (5.46) starting from x at time t.

If $v(t, x)$ is of class $C^{1,2}(\mathscr{T} \times \mathscr{X}, \mathbb{R})$,, i.e., continuously differentiable in t and twice continuously differentiable in x,, then it solves the HJB equation

$$v_t + \tilde{H}_*(t, x, m^*, v_x, v_{xx}) = 0, \; v(T, x) = h(x, m^*(T, .)). \tag{5.49}$$

If (u^*, m^*, v) is solution of the following backward-forward partial differential equation:

$$\begin{cases} v_t + \tilde{H}_*(t, x, m^*, v_x, v_{xx}) = 0, \quad 0 \le t < T, \\ v(T, x) = h(x, m^*(T, .)), \\ m_t^* + \mathrm{div}_x[m^* b(t, x, m^*, u^*)] \\ \quad -\frac{1}{2}\, trace\partial_{xx}(m^*\sigma'\sigma(t, x, u^*, m^*)) = 0, \quad 0 < t \le T, \\ m^*(0, x) = m_0(x), \\ u^*(t, .) \in \arg\min_u \left\{ v_x \cdot b(t, x, u, m^*) + f(t, x, u, m^*) + \frac{1}{2} trace(\sigma'\sigma v_{xx}) \right\}, \end{cases} \tag{5.50}$$

then (u^*, m^*) constitutes a risk-neutral mean-field equilibrium and $v^* = v$ is the equilibrium cost at (u^*, m^*).

5.3.3 DPP for First-Order Mean-Field Games

We say that (u^*, m^*) is a risk-neutral mean-field equilibrium if u^* minimizes the cost functional given m^* and m^* coincides with the population state distribution reproduced from u^*. If (u^*, m^*, v) is solution of the following backward-forward partial differential equation:

$$\begin{cases} v_t + H_*(t, x, m^*, v_x) = 0, \quad 0 < t \le T, \\ v(T, x) = h(x, m^*(T, .)), \\ m_t^* + \mathrm{div}_x[m^* b(t, x, m^*, u^*)] = 0, \quad 0 < t \le T, \\ m^*(0, x) = m_0(x), \\ u^*(t, .) \in \arg\min_u \{p \cdot b(t, x, u, m^*) + f(t, x, u, m^*)\}, \end{cases} \tag{5.51}$$

then the pair (u^*, m^*) constitutes a mean-field equilibrium and v^* is the equilibrium cost at (u^*, m^*).

The equilibrium is in feedback form, i.e., u^* is a function of time t, individual state $x(t)$, and the population mean-field measure $m^*(t, .)$. It is an individual state-and-mean-field feedback strategy.

5.3.4 On the Ill-Posedness of the Initial-Terminal Problems

Note that the system (5.51) is a backward-forward system of ordinary differential equations with initial-terminal values. We ask if such a system has always a solution whenever the coefficients are smooth. There are many examples of systems of backward-forward equations which do not admit solutions. As a very simple example, consider the system

$$\begin{cases} \dot{v} = m, \\ v(T,.) = -m(T,.), \\ \dot{m} = -v, \\ m(0) = m_0 \end{cases} \tag{5.52}$$

It is obvious that the coefficients of this pair of backward-forward differential equations are all uniformly bounded and smooth. We show below that, depending on T, this may not be solvable for $m_0 \neq 0$.

Let $T = k\pi + 3\pi/4$, with k being a nonnegative integer. Then, the above two-point boundary value problem does not admit a solution for any $m_0 \neq 0$,, and it admits infinitely many solutions for $m_0 = 0$.

Suppose $m_0 = 0$,, then

$$\frac{d^2}{dt^2}m + m = 0.$$

This is a second-order ordinary differential equation with constant coefficients. The explicit solutions have the form

$$m(t) = a.\cos t + b.\sin t,$$

for some $(a, b) \in \mathbb{R}^2$. Since $m_0 = 0$,, one has $a = 0$ and hence $m(t) = b\sin(t)$ and $v(t) = -b\cos(t)$. We check that

$$m(T) = b\sin(k\pi + 3\pi/4) = -b\cos(k\pi + 3\pi/4),$$

for any real number k. This leads to an infinite number of solutions of the system of backward-forward ODEs.

Suppose now that $m_0 \neq 0$. Then, $a \neq 0$. Let $T = 7\pi/4$. For this horizon, we claim that the system has no solution. To see this, we compute

$$m_T = a\cos(-\pi/4) + bsin(-\pi/4) \tag{5.53}$$

$$= a\cos(\pi/4) - b\sin(\pi/4) \tag{5.54}$$

$$= \frac{1}{\sqrt{2}}(a - b) \tag{5.55}$$

and the terminal value of v

$$v_T = -a\sin(T) + b\cos(T) = \frac{a}{\sqrt{2}} + \frac{b}{\sqrt{2}} = \frac{a+b}{\sqrt{2}}.$$

It follows that $m_T = -v_T$ if and only if $a = 0$. On the other hand, $m_0 \neq 0$ implies that $a \neq 0$ which is incompatible with $m_T = -v_T$. Hence the system has no solution.

5.3.5 Explicit Solution to the HJB Equation

Below we provide some particular cases of the HJB equation that can be solved explicitly or that has an explicit representation.

Proposition 5.3 (Explicit Solution: Hopf–Lax–Oleinik [191, 224, 263, 264]) *Let* $v(t,x) = \inf_{y\in\mathbb{R}^k}[h(y) + tH^*(\frac{x-y}{t})]$, H^* *be the Legendre–Fenchel transform of a convex function* H *that satisfies* $\lim_{|p|\to+\infty} \frac{H(p)}{|p|} = 0$. *Then,* v *solves the HJB equation*

$$v_t + H(v_x) = 0, \quad v(0,x) = h(x). \tag{5.56}$$

$v(t,x) = \sup_{p\in\mathbb{R}^d}[\langle p, x\rangle - h^*(p) - tH(p)]$ *where* h^* *is the Legendre–Fenchel transform of the convex function* h.

Let H be in the form $H(x,p) = \Phi(H_0(x,p))$ where Φ is convex, nondecreasing, $\Phi(0) = 0$, H_0 is a continuous function on $\mathbb{R}^{2k}$, $p \mapsto H_0(x,p)$ is convex, $H_0(x,p) \geq 0$, $H_0(x,\lambda p) = \lambda H_0(x,p)$, $|H_0(x,p) - H_0(y,p)| \leq \omega(|x-y|(1+|p|))$ for all $x, y, p, \lambda > 0$ and for some modulus ω such that $\lim_{s\to 0_+}\omega(s) = 0$.

Under the assumptions made on H_0,, the solution d of the stationary PDE

$$H_0(x, d_x) = 1 \text{ on } \mathbb{R}^k\backslash\{y\}, \tag{5.57}$$

is of eikonal type. For each $y \in \mathbb{R}^k$,, the stationary PDE (5.57) has a unique viscosity solution $d = d(x,y)$ for all x, y and $d(y,y) = 0$.

Proposition 5.4 (Explicit Solution: State-Dependent Case) *If* h *is lower semicontinuous,* $g(x) \geq -c(1+|x|)$ *for some* $c > 0$,, *then the function*

$$v(t,x) = \inf_{y\in\mathbb{R}^k}[h(y) + t\Phi^*(\frac{d(x,y)}{t})],$$

is the unique lower semicontinuous viscosity solution to the HJ equation

$$v_t + \Phi(H_0(x, v_x)) = 0, \quad v(0,x) = h(x), \tag{5.58}$$

which is bounded below by linear growth and $\liminf_{(y,t)\to(x,0_+)} v(t,y) = h(x)$.

Proposition 5.5 (Explicit Solution: Local Mean-Field Case) *Consider the one-dimensional case* $k = 1$. *Let the Hamiltonian be in the form* $H_*(x, m^*(x), v_x)$ *with* $p \mapsto H_*(x, m^*(x), p)$ *convex and* $|H_*(x, m^*(x), p) - H_*(m^*(y), y, p)| \leq \omega(|x -$

$y|(1+|p|))$. Then, the HJB equation is

$$v_t + H_*(x, m^*(x), v_x) = 0, \quad v(0, x) = h(x, m(x)) = \tilde{h}(x), \tag{5.59}$$

which has an explicit representation given by

$$v(t, x) = \inf_y \sup_{c > c_*} \{\tilde{h}(y) + d^c(x, y) - ct\},$$

where $d^c(x, y)$ is the Mané potential at level c given by

$$d^c(x, y) = \inf_{\xi, t>0}\{\int_0^t c + H^*(\xi(s), \dot{\xi}(s))\, ds, \ \xi(0) = y, \xi(t) = x\}, \ (x, y) \in \mathbb{R}^2.$$

The infimum is taken over all absolutely continuous functions $\xi : [0, +\infty) \to \mathbb{R}$ and $t > 0$, c_* is a critical value. $y \mapsto d^c(x, y)$ is a viscosity subsolution of the stationary HJB

$$H_*(y, m(y), d_y) = c, \ \ y \in \mathbb{R}.$$

To date, the extension of Proposition 5.5 to two or higher dimensions is an open problem.

5.3.6 *Sufficiency Conditions for Having at Most One Solution*

In this subsection, we provide sufficient conditions under which the mean-field system has at most one solution. Each decision-maker state evolves according to an SDE which is influenced not only by its own control but also by the states of the other decision-makers. The corresponding reduced mean-field game system can be written in the form

$$\begin{cases} v_t + H_*(x, v_x, m) + \frac{1}{2}\sigma^2(x, m) v_{xx} = 0, \\ m_t + \partial_x (m_t b(x, g^{-1}[H_{*,p}(x, v_x, m)], m) - \frac{1}{2}\partial_{xx}\left[\sigma^2(x, m) m\right] = 0, \\ m_0(.) \text{ fixed}, \\ v(T, .) \text{ fixed}, \\ v, m \ \text{normalized} \end{cases} \tag{5.60}$$

Assumption W0 The drift term $b(x, u, m)$ is sufficiently smooth and invertible with respect to u,, i.e., the function $g : u \longmapsto b(x, u, m)$ is a one-to-one mapping, and g^{-1} is smooth for any fixed (x, m).

The generic expression for the optimal control is given by

$$u^* = g_{x,m}^{-1}[H_{*,p}(x, v_x, m)],$$

where $g^{-1} = g_{x,m}^{-1}$ denotes the inverse of the function g.

Suppose that there exist two smooth solutions $(\hat{v}_1, \hat{m}_1)$, $(\hat{v}_2, \hat{m}_2)$ of the system (5.60). Now, consider the function

$$MONO : t \longmapsto \int_{x \in \mathcal{X}} (\hat{v}_2(t,x) - \hat{v}_1(t,x))(\hat{m}_2(t,x) - \hat{m}_1(t,x))dx.$$

Note that this function takes the value 0 at $t = 0$, because the measures coincide initially, and the function MONO is also 0 at the final time $t = T$, because the final values coincide. Therefore, the function MONO will be identically 0 in $[0, T]$ if we can show that it is monotone. This will imply that the integrand is zero, and hence one of the two terms in $(\hat{v}_2(t,x) - \hat{v}_1(t,x))(\hat{m}_2(t,x) - \hat{m}_1(t,x))$ should be 0. Now, if the measures coincide, we look at the HJB equation to arrive at a conclusion. If the values coincide, we look at the FPK equation for the uniqueness of the measure. Thus, it remains to find a sufficient condition for monotonicity of $MONO$.

We are looking for a sufficient condition under which the quantity

$$\int_{x \in \mathcal{X}} (\hat{v}_2(t,x) - \hat{v}_1(t,x))(\hat{m}_2(t,x) - \hat{m}_1(t,x))dx$$

is monotone in time. For sake of conciseness, we suppress the dependence on time t. To compute the time derivative of the quantity above, we use Fubini's rule to swap differentiation and integration over a compact set. This leads to

$$S := \frac{d}{dt}\left[\int_{x \in \mathcal{X}} (\hat{v}_2 - \hat{v}_1)(\hat{m}_2(x) - \hat{m}_1(x))dx\right] \tag{5.61}$$

$$= \int_{x \in \mathcal{X}} (\partial_t \hat{v}_2 - \partial_t \hat{v}_1)(\hat{m}_2(x) - \hat{m}_1(x))dx \tag{5.62}$$

$$+ \int_{x \in \mathcal{X}} (\hat{v}_2 - \hat{v}_1)(\partial_t \hat{m}_2(x) - \partial_t \hat{m}_1(x))dx \tag{5.63}$$

Next we expand the first term $A := \int_{x \in \mathcal{X}} (\partial_t \hat{v}_2 - \partial_t \hat{v}_1)(\hat{m}_2(x) - \hat{m}_1(x))dx$. To do so, we take the difference between the two HJB equations and multiply by $\hat{m}_2 - \hat{m}_1$:

$$\partial_t \hat{v}_1 + H_*(x, \partial_x \hat{v}_1, \hat{m}_1) + \frac{1}{2}\sigma^2(x, \hat{m}_1)\partial_{xx}^2 \hat{v}_1 = 0, \tag{5.64}$$

$$\partial_t \hat{v}_2 + H_*(x, \partial_x \hat{v}_2, \hat{m}_2) + \frac{1}{2}\sigma^2(x, \hat{m}_1)\partial_{xx}^2 \hat{v}_2 = 0 \tag{5.65}$$

which yields

$$\partial_t \hat{v}_2 - \partial_t \hat{v}_1 = -H_*(x, \partial_x \hat{v}_2, \hat{m}_2) + H_*(x, \partial_x \hat{v}_1, \hat{m}_1)$$

$$-\frac{1}{2}\sigma^2(x,\hat{m}_2)\partial^2_{xx}\hat{v}_2 + \frac{1}{2}\sigma^2(x,\hat{m}_1)\partial^2_{xx}\hat{v}_1$$

$$\begin{aligned}
A &:= \int_x [\partial_t \hat{v}_2 - \partial_t \hat{v}_1](\hat{m}_2(x) - \hat{m}_1(x))dx \\
&= -\int_x H_*(x, \partial_x \hat{v}_2, \hat{m}_2)(\hat{m}_2(x) - \hat{m}_1(x))dx \\
&\quad + \int_x H_*(x, \partial_x \hat{v}_1, \hat{m}_1)(\hat{m}_2(x) - \hat{m}_1(x))dx \\
&\quad - \int_x \frac{1}{2}\sigma^2(x,\hat{m}_2)\partial^2_{xx}(\hat{v}_2)(\hat{m}_2(x) - \hat{m}_1(x))dx \\
&\quad + \int_x \sigma^2(x,\hat{m}_1)\frac{1}{2}\partial^2_{xx}(\hat{v}_1)(\hat{m}_2(x) - \hat{m}_1(x))dx
\end{aligned} \tag{5.66}$$

Next we expand the second term $B := \int_{x\in\mathscr{X}}(\partial_t \hat{m}_2 - \partial_t \hat{m}_1)(\hat{v}_2 - \hat{v}_1)dx$.

$$\partial_x \hat{m}_1 + div(\hat{m}_1 b(x, g^{-1}_{(x,\hat{m}_1)}[H_{*,p}(x, \partial_x \hat{v}_1, \hat{m}_1)], \hat{m}_1)) - \frac{1}{2}\partial^2_{xx}[\sigma^2(x,\hat{m}_1)\hat{m}_1] = 0,$$

$$\partial_x \hat{m}_2 + div(\hat{m}_2 b(x, g^{-1}_{(x,\hat{m}_2)}[H_{*,p}(x, \partial_x \hat{v}_2, \hat{m}_2)], \hat{m}_2)) - \frac{1}{2}\partial^2_{xx}\sigma^2(x,\hat{m}_2)\hat{m}_2 = 0$$

Taking the difference between the two FPK equations, and multiplying by $\hat{v}_2 - \hat{v}_1$,, we get

$$\begin{aligned}
\partial_t \hat{m}_2 - \partial_t \hat{m}_1 &= -div(\hat{m}_2 b(x, g^{-1}_{(x,\hat{m}_2)}[H_{*,p}(x, \partial_x \hat{v}_2, \hat{m}_2)], \hat{m}_2) \\
&\quad + div(\hat{m}_1 b(x, g^{-1}_{(x,\hat{m}_1)}[H_{*,p}(x, \partial_x \hat{v}_1, \hat{m}_1)], \hat{m}_1)) \\
&\quad + \frac{1}{2}\partial^2_{xx}[\sigma^2(x,\hat{m}_2)\hat{m}_2] - \frac{1}{2}\partial^2_{xx}[\sigma^2(x,\hat{m}_1)\hat{m}_1]
\end{aligned}$$

$$\begin{aligned}
B &:= \int_x (\partial_t \hat{m}_2 - \partial_t \hat{m}_1)(\hat{v}_2 - \hat{v}_1)dx \qquad (5.67)\\
&= -\int_x div(\hat{m}_2 b(x, g^{-1}_{(x,\hat{m}_2)}[H_{*,p}(x, \partial_x \hat{v}_2, \hat{m}_2)], \hat{m}_2))(\hat{v}_2 - \hat{v}_1)dx \\
&\quad + \int_x div(\hat{m}_1 b(x, g^{-1}_{(x,\hat{m}_1)}[H_{*,p}(x, \partial_x \hat{v}_1, \hat{m}_1)], \hat{m}_1))(\hat{v}_2 - \hat{v}_1)dx \\
&\quad + \int_x \frac{1}{2}\partial^2_{xx}(\sigma^2(x,\hat{m}_2)\hat{m}_2)(\hat{v}_2 - \hat{v}_1)dx
\end{aligned}$$

$$- \int_x \frac{1}{2} \partial^2_{xx} (\sigma^2(x, \hat{m}_1)\hat{m}_1)(\hat{v}_2 - \hat{v}_1) dx$$

By integration by parts,

$$\int_x \text{div}(k)\phi = - \int_x k.\partial_x \phi,$$

where ϕ is a test function defined over $\mathscr{X}$. Thus, B becomes

$$\begin{aligned} B := & \int_x (\partial_t \hat{m}_2 - \partial_t \hat{m}_1)(\hat{v}_2 - \hat{v}_1) dx \\ = & + \int_x \hat{m}_2(x) b(x, g^{-1}_{(x,\hat{m}_2)}[H_{*,p}(x, \partial_x \hat{v}_2, \hat{m}_2)], \hat{m}_2)(\partial_x \hat{v}_2 - \partial_x \hat{v}_1) dx \\ & - \int_x \hat{m}_1(x) b(x, g^{-1}_{(x,\hat{m}_1)}[H_{*,p}(x, \partial_x \hat{v}_1, \hat{m}_1)], \hat{m}_1)(\partial_x \hat{v}_2 - \partial \hat{v}_1) dx \\ & + \int_x \frac{1}{2} \partial^2_{xx} (\sigma^2(x, \hat{m}_2)\hat{m}_2)(\hat{v}_2 - \hat{v}_1) dx \\ & - \int_x \frac{1}{2} \partial^2_{xx} (\sigma^2(x, \hat{m}_1)\hat{m}_1)(\hat{v}_2 - \hat{v}_1) dx \end{aligned} \tag{5.68}$$

Now, reorganizing the terms in $A + B$,, we obain

$$\begin{aligned} A + B = & - \int_x H_*(x, \partial_x \hat{v}_2, \hat{m}_2)(\hat{m}_2(x) - \hat{m}_1(x)) dx \\ & + \int_x H_*(x, \partial_x \hat{v}_1, \hat{m}_1)(\hat{m}_2(x) - \hat{m}_1(x)) dx \\ & + \int_x \hat{m}_2(x) b(x, g^{-1}_{(x,\hat{m}_2)}[H_{*,p}(x, \partial_x \hat{v}_2, \hat{m}_2)], \hat{m}_2)(\partial_x \hat{v}_2 - \partial_x \hat{v}_1) dx \\ & - \int_x \hat{m}_1(x) b(x, g^{-1}_{(x,\hat{m}_1)}[H_{*,p}(x, \partial_x \hat{v}_1, \hat{m}_1)], \hat{m}_1)(\partial_x \hat{v}_2 - \partial_x \hat{v}_1) dx \\ & + \text{the Laplacian terms} \\ & - \int_x \frac{1}{2} [\sigma^2(x, \hat{m}_2) \partial^2_{xx} \hat{v}_2 - \sigma^2(x, \hat{m}_1) \partial^2_{xx} \hat{v}_1](\hat{m}_2 - \hat{m}_1) \\ & + \int_x \frac{1}{2} [\partial^2_{xx} \sigma^2(x, \hat{m}_2)\hat{m}_2 - \partial^2_{xx} \sigma^2(x, \hat{m}_1)\hat{m}_1](\hat{v}_2 - \hat{v}_1) dx \end{aligned} \tag{5.69}$$

$$S_t = A + B$$

We next introduce

$$\hat{m}_\lambda := (1-\lambda)\hat{m}_1 + \lambda \hat{m}_2 = \hat{m}_1 + \lambda(\hat{m}_2 - \hat{m}_1).$$

The measure $\hat{m}_\lambda$ starts with $\hat{m}_1$ for the parameter $\lambda = 0$ and ends up at the measure $\hat{m}_2$ for $\lambda = 1$. Define

$$\hat{v}_\lambda := (1-\lambda)\hat{v}_1 + \lambda \hat{v}_2.$$

Introduce an auxiliary integral

$$\begin{aligned}
C_\lambda := &-\int_x H_*(x, \partial_x \hat{v}_\lambda, \hat{m}_\lambda)(\hat{m}_\lambda(x) - \hat{m}_1(x))dx \\
&+\int_x H_*(x, \partial_x \hat{v}_1, \hat{m}_1)(\hat{m}_\lambda(x) - \hat{m}_1(x))dx \\
&+\int_x \hat{m}_\lambda(x) b(x, g^{-1}_{(x,\hat{m}_\lambda)}[\partial_p H^*_*(x, \partial_x \hat{v}_\lambda, \hat{m}_\lambda)], \hat{m}_\lambda)(\partial_x \hat{v}_\lambda - \partial_x \hat{v}_1)dx \\
&-\int_x \hat{m}_1(x) b(x, g^{-1}_{(x,\hat{m}_1)}[\partial_p H^*_*(x, \partial_x \hat{v}_1, \hat{m}_1)], \hat{m}_1)(\partial_x \hat{v}_\lambda - \partial_x \hat{v}_1)dx \\
&-\int_x \frac{1}{2}[\sigma^2(x, \hat{m}_\lambda)\partial^2_{xx}\hat{v}_\lambda - \sigma^2(x, \hat{m}_1)\partial^2_{xx}\hat{v}_1](\hat{m}_\lambda - \hat{m}_1) \\
&+\int_x \frac{1}{2}[\partial^2_{xx}\sigma^2(x, \hat{m}_\lambda)\hat{m}_\lambda - \partial^2_{xx}\sigma^2(x, \hat{m}_1)\hat{m}_1](\hat{v}_\lambda - \hat{v}_1)dx
\end{aligned} \tag{5.70}$$

By making the substitutions $\hat{v}_\lambda - \hat{v}_1 = \lambda(\hat{v}_2 - \hat{v}_1)$ and $\hat{m}_\lambda - \hat{m}_1 = \lambda(\hat{m}_2 - \hat{m}_1)$,, we obtain

$$\begin{aligned}
\frac{C_\lambda}{\lambda} := &-\int_x H_*(x, \partial_x \hat{v}_\lambda, \hat{m}_\lambda)(\hat{m}_2(x) - \hat{m}_1(x))dx \\
&+\int_x H_*(x, \partial_x \hat{v}_1, \hat{m}_1)(\hat{m}_2(x) - \hat{m}_1(x))dx \\
&+\int_x \hat{m}_\lambda(x) b(x, g^{-1}_{(x,\hat{m}_\lambda)}[H_{*,p}(x, \partial_x \hat{v}_\lambda, \hat{m}_\lambda)], \hat{m}_\lambda)(\partial_x \hat{v}_2 - \partial_x \hat{v}_1)dx \\
&-\int_x \hat{m}_1(x) b(x, g^{-1}_{(x,\hat{m}_1)}[\tilde{H}_{*,p}(x, \partial_x \hat{v}_1, \hat{m}_1)], \hat{m}_1)(\partial_x \hat{v}_2 - \partial_x \hat{v}_1)dx \\
&-\int_x \frac{1}{2}[\sigma^2(x, \hat{m}_\lambda)\partial^2_{xx}\hat{v}_\lambda - \sigma^2(x, \hat{m}_1)\partial^2_{xx}\hat{v}_1](\hat{m}_2 - \hat{m}_1)dx \\
&+\int_x \frac{1}{2}[\partial^2_{xx}\sigma^2(x, \hat{m}_\lambda)\hat{m}_\lambda - \partial^2_{xx}\sigma^2(x, \hat{m}_1)\hat{m}_1](\hat{v}_2 - \hat{v}_1)dx.
\end{aligned} \tag{5.71}$$

Using the continuity of the terms on the RHS and the compactness of $\mathscr{X}$,, we deduce that

$$\lim_{\lambda \longrightarrow 0} \frac{C_\lambda}{\lambda} = 0,$$

because all the remaining expressions are in m_1.

We next seek a condition under which the one-dimensional function $\lambda \longmapsto \frac{C_\lambda}{\lambda}$ is monotone in λ. We need to compute the variations of $\frac{d}{d\lambda}\left(\frac{C_\lambda}{\lambda}\right)$:

$$\begin{aligned}\frac{d}{d\lambda}\left(\frac{C_\lambda}{\lambda}\right) = &-\int_x \big[H_{*,p}(x, \partial_x\hat{v}_\lambda, \hat{m}_\lambda)(\partial_x\hat{v}_2 - \partial_x\hat{v}_1) + \\ &H_{*,z}(x, \partial_x\hat{v}_\lambda, \hat{m}_\lambda)(\hat{m}_2(x) - \hat{m}_1(x))\big]\,(\hat{m}_2(x) - \hat{m}_1(x))dx \\ &+ \int_x \partial_\lambda\left(\hat{m}_\lambda(x) b(x, g^{-1}_{(x,\hat{m}_\lambda)}[H_{*,p}(x, \partial_x\hat{v}_\lambda, \hat{m}_\lambda)], \hat{m}_\lambda)\right) \\ &\times(\partial_x\hat{v}_2 - \partial_x\hat{v}_1)dx \\ &-\frac{1}{2}\int_x \partial_\lambda\left(\sigma^2(x, \hat{m}_\lambda)\partial^2_{xx}\hat{v}_\lambda\right)(\hat{m}_2 - \hat{m}_1)dx \\ &+\frac{1}{2}\int_x \partial_\lambda\left(\partial^2_{xx}[\sigma^2(x, \hat{m}_\lambda)\hat{m}_\lambda]\right)(\hat{v}_2 - \hat{v}_1)dx. \end{aligned} \tag{5.72}$$

$$\begin{aligned}\frac{d}{d\lambda}\left(\frac{C_\lambda}{\lambda}\right) = &-\int_x H_{*,p}(x, \partial_x\hat{v}_\lambda, \hat{m}_\lambda)(\partial_x\hat{v}_2 - \partial_x\hat{v}_1)(\hat{m}_2(x) - \hat{m}_1(x))dx \\ &-\int_x H_{*,z}(x, \partial_x\hat{v}_\lambda, \hat{m}_\lambda)(\hat{m}_2(x) - \hat{m}_1(x))^2 dx \\ &+\int_x (\hat{m}_2 - \hat{m}_1) b(x, g^{-1}_{(x,\hat{m}_\lambda)}[H_{*,p}(x, \partial_x\hat{v}_\lambda, \hat{m}_\lambda)], \hat{m}_\lambda)(\partial_x\hat{v}_2 - \partial_x\hat{v}_1)dx \\ &+\int_x \hat{m}_\lambda \partial_\lambda b(x, g^{-1}_{(x,\hat{m}_\lambda)}[H_{*,p}(x, \partial_x\hat{v}_\lambda, \hat{m}_\lambda)])(\partial_x\hat{v}_2 - \partial_x\hat{v}_1)dx \\ &-\frac{1}{2}\int_x \partial_\lambda\partial^2_{xx}\left(\sigma^2(x, \hat{m}_\lambda)\hat{v}_\lambda\right)(\hat{m}_2 - \hat{m}_1)dx \\ &+\frac{1}{2}\int_x \partial_\lambda\partial^2_{xx}\left(\sigma^2(x, \hat{m}_\lambda)\hat{m}_\lambda\right)(\hat{v}_2 - \hat{v}_1)dx \end{aligned} \tag{5.73}$$

We now compute the term

$$\hat{m}_\lambda(x)\partial_\lambda\left(b(x, g^{-1}_{(x,\hat{m}_\lambda)}[H_{*,p}(x, \partial_x\hat{v}_\lambda, \hat{m}_\lambda)], \hat{m}_\lambda)\right).$$

Let $\eta(x, y, m) = g^{-1}_{(x,m)}(y)$. Then,

$$
\begin{aligned}
D_\lambda &= \partial_\lambda \left(b(x, g^{-1}_{(x,\hat{m}_\lambda)}[H_{*,p}(x, \partial_x \hat{v}_\lambda, \hat{m}_\lambda)], \hat{m}_\lambda)\right) \\
&= \partial_u b \partial_\lambda g^{-1}_{(x,\hat{m}_\lambda)}[H_{*,p}(x, \partial_x \hat{v}_\lambda, \hat{m}_\lambda)] + (\hat{m}_2 - \hat{m}_1)\partial_m b \\
&= (\partial_u b)\left[(\partial_y \eta).\partial_\lambda (H_{*,p}) + (\hat{m}_2 - \hat{m}_1)\partial_z \eta\right] + (\hat{m}_2 - \hat{m}_1)\partial_m b \\
&= (\partial_u b)\left[(\partial_y \eta).\left(H_{*,pp}.(\partial_x \hat{v}_2 - \partial_x \hat{v}_1) + H_{*,pz}.(\hat{m}_2 - \hat{m}_1)\right) + (\hat{m}_2 - \hat{m}_1)\partial_z \eta\right] \\
&\quad + (\hat{m}_2 - \hat{m}_1)\partial_m b.
\end{aligned} \tag{5.74}
$$

$$
\frac{d}{d\lambda}\left(\frac{C_\lambda}{\lambda}\right) \tag{5.75}
$$

$$
\begin{aligned}
&= \int_x \hat{m}_\lambda (\partial_u b(x, u, m))(\partial_z \eta)(\hat{m}_2 - \hat{m}_1)(\partial_x \hat{v}_2 - \partial_x \hat{v}_1)dx \\
&+ \int_x \hat{m}_\lambda (\partial_u b(x, u, m)).(g^{-1})'_{(x,\hat{m}_\lambda)}[H_{*,p}(x, \partial_x v, \hat{m}_\lambda)] H_{*,pp}(x, \partial_x v, \hat{m}_\lambda) \\
&\times (\partial_x \hat{v}_2 - \partial_x \hat{v}_1)^2 dx \\
&+ \int_x \hat{m}_\lambda (\partial_u b(x, u, m)).(g^{-1})'_{(x,\hat{m}_\lambda)}[H_{*,p}(x, \partial_x v, \hat{m}_\lambda)] H_{*,pz}(x, \partial_x v, \hat{m}_\lambda) \\
&\times (\hat{m}_2 - \hat{m}_1)(\partial_x \hat{v}_2 - \partial_x \hat{v}_1)dx \\
&+ \int_x \hat{m}_\lambda (\partial_m b)_{(x,u^*,m)} + \frac{b(x, g^{-1}\left(H_{*,p}(x, \partial v, \hat{m}_\lambda)\right), m_\lambda)}{m_\lambda} \\
&\times (\hat{m}_2 - \hat{m}_1)(\partial_x \hat{v}_2 - \partial_x \hat{v}_1)dx \\
&- \int_x \hat{m}_\lambda [\frac{H_{*,p}(x, \partial v, m_\lambda)}{m_\lambda}](\hat{m}_2 - \hat{m}_1)(\partial_x \hat{v}_2 - \partial_x \hat{v}_1)dx \\
&- \int_x \hat{m}_\lambda [\frac{\partial_z H_*(x, \partial v, m_\lambda)}{m_\lambda}](\hat{m}_2 - \hat{m}_1)^2 dx \\
&- \frac{1}{2}\int_x \partial_\lambda \left(\sigma^2(x, \hat{m}_\lambda)\partial^2_{xx}\hat{v}_\lambda\right)(\hat{m}_2 - \hat{m}_1)dx \\
&+ \frac{1}{2}\int_x \partial_\lambda \left(\partial^2_{xx}[\sigma^2(x, \hat{m}_\lambda)\hat{m}_\lambda]\right)(\hat{v}_2 - \hat{v}_1)dx
\end{aligned} \tag{5.76}
$$

where

$$
\eta = g^{-1}_{(x,m_\lambda)}\left(H_{*,p}(x, \partial v_\lambda, m_\lambda)\right).
$$

Now, we need to compute

$$\partial_z \eta = \partial_z (g^{-1}_{(x,z)}(H_{*,p}(x, \partial_x v_\lambda, z))) = (\partial_y \eta) H_{*,mp} + (\partial_x \eta) \tag{5.77}$$

We want to express this as a quadratic form plus noise terms:

$$\begin{aligned} \frac{d}{d\lambda}\left(\frac{C_\lambda}{\lambda}\right) = &= \int m_\lambda (\partial_x \hat{v}_2 - \partial_x \hat{v}_1, \hat{m}_2 - \hat{m}_1) \begin{pmatrix} a_{11} & a_{12} \\ a_{12} & a_{22} \end{pmatrix} \begin{pmatrix} \partial_x \hat{v}_2 - \partial_x \hat{v}_1 \\ \hat{m}_2 - \hat{m}_1 \end{pmatrix} + \\ & -\frac{1}{2}\int_x \partial_\lambda \left(\sigma^2(x, \hat{m}_\lambda)\partial^2_{xx}\hat{v}_\lambda\right)(\hat{m}_2 - \hat{m}_1)dx \\ & +\frac{1}{2}\int_x \partial_\lambda \left(\partial^2_{xx}[\sigma^2(x, \hat{m}_\lambda)\hat{m}_\lambda]\right)(\hat{v}_2 - \hat{v}_1)dx \end{aligned} \tag{5.78}$$

where

$$a_{11} := (\partial_u b)(g^{-1}_{(x,m_\lambda)})'(H_{*,p}(x, \partial_x v_\lambda, m_\lambda))H_{*,pp} \tag{5.79}$$

$$a_{22} := -\frac{H_{*,z}}{m_\lambda} \tag{5.80}$$

$$\begin{aligned} a_{12} := \frac{1}{2}\Bigg(& (\partial_u b)(\partial_z \eta) + (\partial_u b)(g^{-1})'_{(x,m_\lambda)}(H_{*,p}(x, \partial_x v_\lambda, m_\lambda))H_{*,pz} \\ & +(\partial_m b) + \frac{b(x, g^{-1}(H_{*,p}), m_\lambda)}{m_\lambda} - \frac{H_{*,p}}{m_\lambda}\Bigg) \end{aligned} \tag{5.81}$$

Note that the last two terms in (5.78) vanish as a result of integration by parts.

$$-\int_x \partial_\lambda \left(\sigma^2(x, \hat{m}_\lambda)\partial^2_{xx}\hat{v}_\lambda\right)(\hat{m}_2 - \hat{m}_1)dx \tag{5.82}$$

$$+\int_x \partial_\lambda \left(\partial^2_{xx}[\sigma^2(x, \hat{m}_\lambda)\hat{m}_\lambda]\right)(\hat{v}_2 - \hat{v}_1)dx = 0 \tag{5.83}$$

A sufficient condition for uniqueness of regular solution is then given by the positivity of the matrix in the first expression of (5.78), whose elements are (5.79)–(5.81) for any (x, p, m).

Theorem 5.1 *Let*

$$\begin{cases} v_t + H_*(x, v_x, m) + \frac{1}{2}\sigma^2(x, m)\partial^2_{xx} v = 0, \\ m_t + div(mb(x, g^{-1}[\partial_p H_*(x, v_x, m)], m) - \frac{1}{2}\partial^2_{xx}[\sigma^2(x, m)m] = 0, \\ m_0(.) \textit{ fixed}, \\ v_T(.) \textit{ fixed}, \\ v, m \ \textit{normalized}, \\ x \in (0, 1)^d := \mathscr{X}. \end{cases}$$

where H_* is the following transform (with respect to the control):

$$H_*(x,p,m) = \inf_u\{\langle p.b(x,u,m)\rangle + f(x,u,m)\}$$

Suppose that $(x,p,m) \longmapsto H_*(x,p,m)$ is twice continuously differentiable with respect to (p,m) (eventually with respect to the operator) and for all $(x,p,m) \in \mathscr{X} \times \mathbb{R}^d \times \mathscr{M}_{+},$, the following matrix is positive definite

$$\begin{pmatrix} a_{11} & a_{12} \\ a_{12} & a_{22} \end{pmatrix} \succ 0.$$

Then, there exists at most one regular solution to the mean-field system.

For $b(x,u,m) = u$ and H_* in the form $H_*(x, \nabla v, m(x))$, one has

$$\partial_u b \equiv 1, g^{-1}_{(x,m)}(y) = id(y), (g^{-1}_{(x,m)})'(y) \equiv 1, \partial_z \eta = \partial_z(id(u)) \equiv 0, \partial_m b \equiv 0$$

- $a_{11} := (\partial_u b)(g^{-1})'_{(x,m_\lambda)}(H_{*,p}(x, \partial_x v_\lambda, m_\lambda))H_{*,pp} = H_{*,pp}$
- $a_{22} = -\frac{H_{*,z}}{z}$,
- $a_{12} = \frac{1}{2}\left((\partial_z \eta) + H_{*,pz} - \frac{H_{*,p}}{m_\lambda}\right) = \frac{1}{2}\left(1.0 + 1.1.H_{*,pz} + 0 + \frac{\partial_p H_*}{m_\lambda} - \frac{H_{*,p}}{m_\lambda}\right) = \frac{1}{2}H_{*,pz}$.

Remark 5.1

- Theorem 5.1 extends to the discounted case by considering the expression

$$\frac{d}{dt}\left[e^{-\beta t}\int_x (v_1(x) - v_2(x))(m_1(x) - m_2(x))\,dx\right].$$

- For multiple type mean-field games, similar conditions as in Theorem 5.1 are obtained by considering

$$\sum_{\theta\in\Theta}\frac{d}{dt}\left[e^{-\beta t}\int_x (v_{1,\theta}(x) - v_{2,\theta}(x))(m_{1,\theta}(x) - m_{2,\theta}(x))\,dx\right].$$

  where Θ denotes the set of types.

5.4 Notes

In this chapter, we have considered finite horizon risk-sensitive mean-field games with exponentiated mean-field payoff functionals. The reader is referred to a series of studies for the infinite horizon case with long-run time average payoff or discounted payoff. Stationary solutions to mean-field games have been examined in [220] which also relies on long-time behavior of the solutions of Hamilton–Jacobi–Bellman equations [257] and on large-time behavior for a nonlocal diffusion equation with absorption and bounded initial data [73, 79, 188, 347].The work in [84] examines a class of long-time average, as the time horizon tends to infinity, of the solution of a mean-field game with a nonlocal coupling. They show an exponential convergence to the solution of the associated stationary ergodic mean-field game. Ergodic mean-field games with Hörmander diffusions have been examined in [134]. Markov–Nash equilibria of mean-field games with discounted cost have been studied in [283]. The work [2] studies a class of infinite horizon mean-field games over networks. See also the books [92] and the references therein. Long-run-time average linear quadratic risk-sensitive and robust mean-field games can be found in [250]. Variational formulation of some stationary second-order mean-field game systems can be found in [243]. The authors in [32, 81–83, 88] show that the asymptotic behavior of the mean-field game system is strongly related with the long-time behavior of the so-called master equation and with the vanishing discount limit of the discounted master equation. Solutions of mean-field games with ergodic cost are examined in [13]. Existence of classical solutions to stationary mean-field games have been investigated in [8, 100, 176–178], and their algorithms can be found in [75]. The derivation of ergodic mean-field game equations for several populations of decision-makers can be found in [149]. Master adjoint systems (MASS) can be found in [338]. Recent developments on master equations can be found in [10, 63, 67, 68, 85–87, 89, 93–95, 113, 114, 161–163, 171, 180, 182, 206, 240, 254, 255, 277, 354]. Other equilibrium principles have been developed to solve MFTGs. These include Wiener Chaos Expansions [132, 331, 333] and Direct Methods of Sect. 7.2.

Part III
Classes of Mean-Field-Type Games

Chapter 6
Non-asymptotic Mean-Field Games

Mean-field games have been studied under the assumption of a very large number of interacting decision-makers. For such large systems, the basic idea entails approximating large games by a stylized game model with a continuum of decision-makers. The approach has been shown to be useful in many applications in economics and engineering. However, the stylized game model with a continuum of decision-makers is rarely observed in practice, and the approximation proposed in the asymptotic regime is generally not appropriate for networks with just a few interacting decision-makers. In this chapter, we propose a mean-field framework that is suitable not only for large systems but also for a small world with few number of decision-makers.

The classical works provide rich mathematical foundations and equilibrium concepts in the asymptotic regime but relatively little in the way of computational and representational insights that would allow for few number of decision-makers. Below we give some limitations of the *asymptotic* mean-field approach in engineering, computer science, and economics.

- In wireless networks, the number of interacting nodes at the same slot in the same range is finite, and currently the capacity/bandwidth of the system is limited. Therefore, a mean-field model for infinite capacity and infinite number of nodes is not plausible. The result obtained under the infinite system assumption may not capture the real system with only a small number of nodes.
- In most of current markets, the number of traders is *finite*. In that context, it is known that the Bayesian Cournot game may not have an ex post equilibrium whenever the number of traders is finite. However, the infinite game with a continuum of traders has a pure (static mean-field) equilibrium. In what sense, the (static) mean-field ex post equilibrium captures the finite system is not clear.

Our primary goal is to provide a simple and easy-to-check condition such that mean-field theory can be used at finite scale which we call *non-asymptotic mean-field* approach. We investigate two basic conditions in the present chapter:

T. Başar et al., *Mean-Field-Type Game Theory I*, Static & Dynamic Game Theory: Foundations & Applications, https://doi.org/10.1007/978-3-032-07027-2_6

- The first condition is indistinguishability (or interchangeability) of the payoff functions. This indistinguishability property is easy to verify.
- The second condition is the (regularity) smoothness of the payoff functions. This regularity property is relatively easy to check as well.

Based on these two conditions, we present a simple approximation framework for finite horizon mean-field systems. The framework can be easily extended to the infinite horizon case. The non-asymptotic mean-field approach is based on a simple observation that the many effects of different actions cancel out when the payoff is indistinguishable. Nevertheless, it can lead to a significant simplification of mathematical mean-field models in finite-population regime. The approach presented here is non-asymptotic and is unrelated to the mean-field convergence that originates from the law of large numbers (and its generalization to de Finetti–Hewitt–Savage functional mean-field convergence) in large populations. The non-asymptotic mean-field approach holds even when there is only a small number of decision-makers in a game, or just a few nodes in a network.

One of the motivations of the asymptotic mean-field game approach is that it may reduce the complexity analysis of large systems. The coverage of this chapter goes beyond that. We believe that if the complexity of the infinite system can be reduced easily, then the finite system can also be studied using a non-asymptotic mean-field approach.

In order to apply the mean-field approach to a system with an arbitrary number of decision-makers, we will exploit more the structure of the payoff function and the main assumption on the model which is the indistinguishability property; i.e., the performance index remains unchanged if one permutes the labels of the decision-makers.

The aggregative structure of the problem and the indistinguishability property of the decision-makers are used to derive an error bound for any number of decision-makers. Interestingly, this result holds not only for large numbers of decision-makers but for just few numbers as well. For example, for $I = 5$ decision-makers, there is no systematic way to apply the existing theory for games with an infinite number of decision-makers, but the non-asymptotic mean-field result of this chapter could be applied. The non-asymptotic mean-field result does not impose additional assumptions on the payoff functions of the decision-makers. We further show that the indistinguishability property provides an accurate error bound for any system size. We show that the total sum equilibrium payoff of all decision-makers with heterogeneous parameters can be approximated by the symmetric payoff where the symmetry is with respect to the average of those parameters. These parameters can be real numbers, vectors, matrices, or infinite dimensional functionals. The proof of the approximation error is essentially based on a Taylor expansion which cancels out the first-order terms due to indistinguishability. The methodology extends to mean-field-type games [124, 166, 311, 318, 333, 335–338] following the payoff structures presented in Chap. 1.

Table 6.1 Summary of notation

Symbol	Meaning
$\mathscr{I}$	Set of potential decision-makers
I	Cardinality of $\mathscr{I}$
$\mathscr{A}$	Action space $\mathbb{R}$
a_j	Action of decision-maker j
$r_g(a_1, \ldots, a_I)$	Global payoff function
$\mathbb{1}_{\{.\}}$	Indicator function
$\bar{m}^{\bigotimes I}$	$= (\bar{m}, \ldots, \bar{m})$, there are I such $\bar{m}$
τ_j	Strategy of decision-maker j
$R_{j,T}$	Long-term payoff of decision-maker j with horizon $[0, T-1]$

We provide examples where non-asymptotic mean-field interaction is required and the indistinguishability property could be exploited more efficiently. We present an example of a queueing system with only a few servers where the closed-form expression of the waiting time is not available and the use of the present framework provides appropriate bounds.

The chapter is structured as follows. In Sect. 6.1, we present a mean-field system with an arbitrary number of interacting entities and propose a non-asymptotic static mean-field framework. In Sect. 6.2, we extend the basic results to a dynamic setup.

We list some of the notation using in the chapter in Table 6.1.

6.1 Mean-Field System for an Arbitrary Number of Entities

Consider an interactive system with $I \geq 2$ entities (decision-makers). The global system performance is captured by a certain function $r_g(a_1, \ldots, a_I) \in \mathbb{R}$. Each decision variable a_j belongs to a Polish space $\mathscr{A}$.

Assumption A1 Indistinguishability. The general payoff function is invariant to permutation the indices of the decision-makers, i.e.,

$$r_g(a_1, \ldots, a_i, \ldots, a_j, \ldots, a_I) = r_g(a_{\pi(1)}, \ldots, a_{\pi(i)}, \ldots, a_{\pi(j)}, \ldots, a_{\pi(I)}),$$

for every permutation $\pi : \mathscr{S}_I \longrightarrow \mathscr{S}_I$, where $\mathscr{S}_I$ is the set $\mathscr{I} := \{1, 2, \ldots, I\}$. Note that, it suffices to check for pairwise interchangeability, i.e., permutation of any two of the coordinates.

Assumption A2 Smoothness. We assume that the objective function r_g is twice differentiable with respect to the variables.

It is important to notice that assumption A1 can be easily checked. In practice, A1 will be satisfied by functions that can be expressed in terms of the mean $\bar{m} = \frac{1}{I}\sum_{j=1}^{I} a_j, \frac{1}{I}\sum_{j=1}^{I} a_j^2, \frac{1}{I}\sum_{j=1}^{I}\phi(a_j,\bar{m})\ldots, \left(\prod_{j=1}^{I} a_j\right)^{\frac{1}{I}}$, etc.

These types of payoff functions have a wide range of applications:

In economics and financial markets, the market price (of products, good, phones, laptops, etc.) is influenced by the total demand and total supply. Examples include

- Public good: Goods are called public if one person's consumption of them does not preclude consumption by others. Typical examples are television programs and uncongested roads. Public good provisioning game with total payoff $g(\bar{m})$ where $\bar{m}$ is the average contribution of the participants.
- Beauty contest games with payoff $R- \parallel a_j - p\bar{m} \parallel, 0 < p < 1$
- Cournot oligopoly model with payoff $a_j p(\frac{\sum_{j'=1}^{I} a_{j'}}{I}) - c(a_j)$.

In queueing theory, the task completion of data centers or servers is influenced by the mean of how much the other data centers/servers can serve. In resource sharing problems, the utility/disutility of a decision-maker depends on the demand of the other decision-makers. Examples include cost sharing in coalitional system and capacity and bandwidth sharing in cloud networking. In wireless networks, the performance of a wireless node is influenced by the interference created by the other transmitters. In congestion control, the delay of a network depends on the aggregate (total) flow and the congestion level of the links/routes.

Basic Results

We now provide the basic results that hold for both non-asymptotic and asymptotic static systems.

Theorem 6.1 *Assume that A1 and A2 hold. Then the following hold:*

- $\partial_{a_j} r_g(\bar{m}^{\bigotimes I}) = \partial_{a_1} r_g(\bar{m}^{\bigotimes I})$ *and*

$$\bar{m} = \frac{1}{I}\sum_{j=1}^{I} a_j = \int bm(db), \tag{6.1}$$

with

$$m(db) = \frac{1}{I}\sum_{j=1}^{I} \delta_{a_j}(db) \tag{6.2}$$

and δ_{a_j} *is the Dirac measure concentrated at the point* a_j.
- *The structure of the payoff function implies that the first-order term in the Taylor expansion is cancelled out.*

- *The cross-derivatives are independent of the labels:* $\partial^2_{a_i a_j} r_g(\bar{m}^{\bigotimes I}) = \partial^2_{a_1 a_2} r_g(\bar{m}^{\bigotimes I}),\ i \neq j$

Proof The first item is immediately proved by using the indistinguishability property in the definition of directional derivative. For the second item, we use the first item and the relation (6.1).

The following gradient term $\langle a - \bar{m}^{\bigotimes I}, \nabla_a r_g \rangle$ vanishes:

$$D^1 := \sum_{j=1}^{I} (a_j - \bar{m}) \partial_{a_j} r_g(a) = \sum_{j=1}^{I} (a_j - \bar{m}) \partial_{a_1} r_g(a) \tag{6.3}$$

$$= \partial_{a_1} r_g(a) \left[\sum_{j=1}^{I} (a_j - \bar{m}) \right] = \partial_{a_1} r_g(a) \left[\sum_{j=1}^{I} a_j - \sum_{j=1}^{I} a_j \right] = 0 \tag{6.4}$$

The third item uses another exchange of positions between i and j. ■

Theorem 6.2 *Suppose that the payoff function r_g satisfies the assumptions A1 and A2. Assume that a is in a small neighborhood of the mean vector $\bar{m}^{\bigotimes I} := (\bar{m}, \ldots, \bar{m})$; i.e., there is a small positive number $c_{\bar{m}, r_g}$ which may depend on $\bar{m}$ and the function r_g such that $\| (a_1, \ldots, a_I) - \bar{m}^{\bigotimes I} \| \leq c_{\bar{m}, r_g}$. Then, $r_g(a) - \bar{r}(\bar{m}) = O\left(\| a - \bar{m}^{\bigotimes I} \|_2^2 \right)$, and*

$$\| r_g(a) - \bar{r}(\bar{m}) \| \leq \delta c^2_{\bar{m}, r_g}$$

where $\bar{r}(\bar{m}) := r_g(\bar{m}, \ldots, \bar{m}) = r_g(\bar{m}^{\bigotimes I}), \delta > 0$.

The proof of 6.2 follows from the following result which gives an explicit error bound for δ and for $c_{\bar{m}, r_g}$

Theorem 6.3 *Assume that A1 and A2 hold. Then, the explicit error bound for arbitrary number of decision-makers is $r_g(a) - \bar{r}(\bar{m})$ is of the order of*

$$\delta_{\bar{m}, \bar{r}} \sum_{j=1}^{I} (a_j - \bar{m})^2,$$

where

$$\delta_{\bar{m}, \bar{r}} = \left| \frac{I}{2(I-1)} \left(-\frac{1}{I^2} \bar{r}''(\bar{m}) + \partial^2_{a_1 a_1} r_g(\bar{m}^{\bigotimes I}) \right) \right|$$

Proof Let O_2 be the error term (second order).

$$
\begin{aligned}
2O_2 &= \sum_{j,j'=1}^{I} (a_j - \bar{m})(a_{j'} - \bar{m})\partial^2_{a_j a_{j'}} r(\bar{m}^{\otimes I}) \\
&= \sum_{j,j',\, j\neq j'}^{I} (a_j - \bar{m})(a_{j'} - \bar{m})\, \underbrace{\partial^2_{a_j a_{j'}} r(\bar{m}^{\otimes I})}_{=\partial^2_{a_1 a_2}} + \sum_{k=1}^{I}(a_k - \bar{m})^2\, \underbrace{\partial^2_{a_k a_k} r}_{=\partial^2_{a_1 a_1}} \\
&= (-\partial^2_{a_1 a_2} r + \partial^2_{a_1 a_1} r) \sum_{k=1}^{I}(a_k - \bar{m})^2
\end{aligned}
$$

On the other hand, one has

$$
r(\bar{m}^{\otimes I} + \varepsilon \mathbb{1}) = \bar{r}(\bar{m} + \varepsilon) \tag{6.5}
$$

This implies that

$$
2O_2 = \varepsilon^2 \bar{r}'' = I(I-1)\varepsilon^2(\partial^2_{a_1 a_2} r) + I\varepsilon^2(\partial^2_{a_1 a_1} r) \tag{6.6}
$$

Hence, $(\partial^2_{a_1 a_2} r) = \frac{\bar{r}''(\bar{m}) - I\partial^2_{a_1 a_1} r}{I(I-1)}$. It follows that

$$
\begin{aligned}
O_2 &= \frac{1}{2}\left[\frac{-\bar{r}''(\bar{m}) + I\partial^2_{a_1 a_1} r}{I(I-1)} + \partial^2_{a_1 a_1} r\right] \sum_{j=1}^{I}(a_j - \bar{m})^2 \\
&= \frac{1}{2}\left[\frac{-\bar{r}''(\bar{m}) + I^2\partial^2_{a_1 a_1} r}{I(I-1)}\right] \sum_{j=1}^{I}(a_j - \bar{m})^2 \\
&= \frac{I}{2(I-1)}\left[-\frac{1}{I^2}\bar{r}''(\bar{m}) + \partial^2_{a_1 a_1} r\right] \sum_{j=1}^{I}(a_j - \bar{m})^2,
\end{aligned}
$$

and we get the exact error for arbitrary system size. ■

Remark 6.1 In order to compute the error bound, one needs only $\bar{r}$ and $\partial^2_{a_1 a_1} r_g(\bar{m}^{\otimes I})$. The expression of the function $r_g(a)$ is not required for vectors with nonsymmetric components.

If $\bar{r}''$ is bounded by β and $\frac{\sum_{j=1}^{I}(a_j - \bar{m})^2}{I} \leq \sigma$, then

$$
r_g(a) - \bar{r}(\bar{m}) \leq \frac{\beta}{2(I-1)} \frac{\sum_{j=1}^{I}(a_j - \bar{m})^2}{I} \leq \frac{\beta\sigma}{2(I-1)}.
$$

In particular, if the finite regime has a solution in a certain sense, that is ε-close to a vector with symmetric component, then the solution of infinite regime provides automatically an $O(\varepsilon^2)$-solution for any number of decision-makers $I \geq 2$. This is a non-asymptotic result in the sense that it holds for all range of system sizes $I \geq 2$.

Collaborative Effort

Consider the geometric aggregate as payoff function given by

$$r_j(a) = \left(\prod_{j'=1}^{I} a_{j'} \right)^{\frac{1}{I}} = e^{\frac{1}{I} \sum_{j'=1}^{I} \log(a_{j'})}$$

over $[0, 1]^I$.

Lemma 6.1 *The following statements hold:*

- *The payoff functions $r_j(a)$ are indistinguishable.*
- *0-effort and 1-effort as equilibrium. 1 is strong-equilibrium (resilience to any deviation of any size $k > 1$).*
- *Define the analogue of the price of anarchy (PoA) for maximization problem as the ratio between the worst equilibrium payoff and the social optimum. It follows that $PoA = 0$. The price of stability (the ratio between the best equilibrium payoff and the social optimum) is $PoS = 1$.*
- *0-effort is not an evolutionarily stable strategy in the infinite-population regime.*

Proof The payoff functions satisfy $r_j(a) = r_i(a) = r(a) = \left(\prod_{j'=1}^{I} a_{j'}\right)^{\frac{1}{I}}$, which remains invariant to interchange of indices

If all the decision-makers do the maximum effort, i.e., $a_j = 1$, then every decision-maker receives the maximal payoff $r_j = 1$ and no decision-maker has incentive to deviate. This is clearly a pure Nash equilibrium. Suppose now that a subset of decision-makers (a coalition) deviates and jointly chooses an action that is different than $(1, \ldots, 1)$, then the payoff of all the decision-makers is less than 1. In particular, the members of the coalition get a lower payoff. A resilience to any coalition of any size is also called strong Nash equilibrium. Since this analysis holds for any coalition of any size, the action profile $(1, \ldots, 1)$ is a strong Nash equilibrium. Clearly, $PoS = 1$.

If one of the decision-makers chooses 0-effort (no effort), then the payoff of the every decision-maker will be zero. This then makes $(0, \ldots, 0)$ is a pure Nash equilibrium. Remark that the equilibrium payoff at this equilibrium is the lowest possible payoff that a decision-maker can receive. $(0, \ldots, 0)$ is the worst equilibrium in terms of payoffs. Hence, $PoA = 0$.

We now show that the action with 0-effort (i.e., the action profile $(0, \ldots, 0)$) is not an evolutionarily stable strategy in the infinite-population regime. Let a population state where all the other decision-makers do no effort. This then makes the mean action to be 0. If all the other decision-makers deviate and do a very small effort $\varepsilon > 0$, then a best response is nonzero, and the payoff would be ε. ∎

The non-asymptotic mean-field approach allows to link the geometric mean with the arithmetic mean action. We note that the geometric mean as a payoff satisfies the indistinguishability property, and it is smooth in the positive orthant. Here $\bar{r}$ is the identity function, because when all the actions are identical, the geometric mean coincides with the arithmetic mean. In order to illustrate the error bound in the non-asymptotic mean field, let us consider the case of two decision-makers. We expand the payoff for an asymmetric input level of size ε.

$$\begin{array}{l} r(a_1,a_2) = \sqrt{a_1 a_2},\ \bar{m} = \frac{a_1+a_2}{2} \\ r = \sqrt{(\bar{m}+\varepsilon\gamma_1)(\bar{m}+\varepsilon\gamma_2)} = \sqrt{\bar{m}^2 + \varepsilon \underbrace{(\gamma_1+\gamma_2)}_{=0} + \varepsilon^2 \underbrace{\gamma_1\gamma_2}_{abs.\leq 1}} \\ r^2 - \bar{m}^2 \leq \varepsilon^2 \end{array} \tag{6.7}$$

If $a_1 = 1$ and $a_2 = (1-\varepsilon)^2 < 1$. then $r = 1-\varepsilon$ and $\bar{m} = 1-\varepsilon+\frac{\varepsilon^2}{2}$. This leads to

$$\bar{m} - r = \frac{\varepsilon^2}{2} < \varepsilon^2.$$

The above computation illustrates that if the system is indistinguishable, then we can work directly with the mean of the mean-field with error $O(\varepsilon^2)$ where ε captures the asymmetry level of the system.

Queueing Mean-Field Games

Consider I servers and an $M/M/I$ system with parameters λ and μ_i. Assume that the customers are indistinguishable. Each customer will be assigned to one of the non-busy servers (if any is available). If not, the customer joins a queue. Our goal is to investigate the delay, i.e., the propagation delay and the expected waiting time (WT) in the queue. Let the expected propagation delay be d_p.

We can obtain an expression for the waiting time $WT(\lambda, \mu_i = \bar{m})$ for the case of similar service rates $\mu_i = \bar{m}$.

The probability that all servers are busy is

$$C = \frac{\left(\frac{(I\rho)^I}{I!}\right)\left(\frac{1}{1-\rho}\right)}{\sum_{k=0}^{I-1}\frac{(I\rho)^k}{k!} + \left(\frac{(I\rho)^I}{I!}\right)\left(\frac{1}{1-\rho}\right)},\quad \rho = \frac{\lambda}{I\bar{m}}.$$

Hence the waiting response time for the symmetric case is $WT(\lambda, \mu_i = \bar{m}) = \frac{C}{I\bar{m}-\lambda}$

Next, let us address the nonasymmetric case, where the computation of $WT(\lambda, \mu_1, \ldots, \mu_I)$ is highly complex.

Clearly WT is regular (in $\mu_1, \ldots, \mu_I$ for $\lambda < \sum_{j=1}^{I} \mu_j$), and satisfies the indistinguishability property. Then, using a non-asymptotic mean-field approach,

$$WT(\lambda, (\mu_i = \bar{m} + \varepsilon\gamma_i)_i) = WT(\lambda, \bar{m}) + O(\varepsilon^2),\ \ \gamma_i = \frac{\mu_i - \bar{m}}{\varepsilon}$$

6.2 Non-asymptotic Approximation Bound

In this section, we provide approximation results for dynamic interactive systems. We consider a finite horizon with length $T \geq 1$.

6.2.1 Non-asymptotic Mean-Field Optimization

Consider I users and a designer in a network. Assume that the designer aims to achieve a certain goal with objective function given by the choice variables that the designer dictates to the users. The objective function is $R_{g,T}(\tau) = \sum_{t=0}^{T-1} r_g(a_t)$ where r_g satisfies assumptions $A1-2$ and $a_t = (a_{1,t}, \ldots, a_{I,t})$ is the choice variable at time t. Let $\bar{m}_t = \frac{1}{I}\sum_{j=1}^{I} a_{j,t}$ be the sequence of mean actions.

Theorem 6.4 *An explicit error bound with an arbitrary number of decision-makers for $r_g(a) - \bar{r}(\bar{m})$ is*

$$\delta_{T,\bar{m},\bar{r}} \parallel a - m \parallel^2_{l^2_T},$$

where

$$\delta_{T,\bar{m},\bar{r}} = |\frac{I}{2(I-1)}\left(-\frac{1}{I^2}\bar{r}''(\bar{m}_t) + \partial^2_{a_1a_1} r_g(\bar{m}_t^{\otimes I})\right)|$$

See Theorem 6.5 for a proof.

6.2.2 Non-asymptotic Dynamic Mean-Field Games

Consider a stochastic game with I decision-makers, defined over the discrete time interval. Time space is $\mathcal{T} = \{0, 1, \ldots, T-1\}$. Each decision-maker j has its individual state $s_{j,t},\ t \in \mathcal{T}$ which evolves according to a Markovian process. The action space a decision-maker depends on its current state, $\mathcal{A}(s_{j,t}) \subset \mathbb{R}$. A strategy of a decision-maker at time t is a mapping from history up to t to the current action space $\mathcal{A}(s_{j,t})$. The instantaneous payoff function of decision-maker j is $r(s_{j,t}, a_{j,t}, m_t)$ where m_t is the state-action distribution, which satisfies the indistinguishability property with the respect to the other decision-makers. We assume the payoff function is smooth. Denote by τ_j the strategy of decision-maker j. The state transition law of decision-maker j is given by $q(.|s_{j,t}, a_{j,t}, m_t, \tau_t)$. The long-term payoff of decision-maker j is

$$R_{j,T}(s_0, \tau) = E\left[\sum_{t\in\mathcal{T}} r(s_{j,t}, a_{j,t}, m_t) \mid (s_{j,0},\ \tau_j)_{j\in\mathcal{I}}\right]$$

A strategy profile τ is a (Nash) equilibrium if no decision-maker can improve her payoff by unilateral deviation. Let $V_{j,T}(s_0, \tau_{-j})$ be the value function of decision-maker j; i.e., it is the supremum, over all possible strategies, of the expectation of the payoff $R_{j,T}$ starting from an initial s_0 when the other strategy profile is τ_{-j}.

Fixing decision-maker j and taking expectation over the other decision-makers' states, the recursive Bellman-Kolmogorov equation is given by

$$V_{j,t}(s_j) = \sup_{a_j}\left[r(s_j, a_j, m_t) + E_{s'}V_{j,t+1}(s'_j \mid s_j, a_j, \tau)\right],$$
$$m_{t+1} \sim \mathbb{P}(.\mid m_t, \tau_t)$$
$$s_{j,t+1} \sim q(.\mid s_{j,t}, a_{j,t}, m_t, \tau_t) = \mathbb{P}(.\mid s_{j,t}, a_{j,t}, m_t, \tau_t)$$

Theorem 6.5 *Let $a_{j,t}(h_{j,t}) = \bar{m}_t(h_{j,t}) + \varepsilon\gamma_{j,t}(h_{j,t})$, where*

$$\gamma_{j,t}(h_{j,t}) = \frac{a_{j,t}(h_{j,t}) - \bar{m}_t(h_{j,t})}{\varepsilon},$$

$$\bar{m}_t(h_{j,t}) = \frac{1}{I}\sum_{j'=1}^{I} a_{j',t}(h_{j',t}),$$

and

$$\varepsilon = \max_j \sup_{t\in\mathcal{T}} \sup_{h_{j,t}\in\mathcal{H}_{j,t}} |a_{j,t}(h_{j,t}) - \bar{m}_t(h_{j,t})|,$$

Assume the state transition q is continuous. Then $R_g(s_0; \tau_1, \ldots, \tau_I)$ is of the order of $R_g(s_0; \bar{m}, \ldots, \bar{m}) + O(\varepsilon^2)$ for any $I \geq 2$.

Proof Using similar lines as in 6.3 for any time $t \in \mathcal{T}$, $|r_j(s_t, a_t) - r_j(s_t, \bar{m}_t)|$ is bounded by $\delta_{t,\bar{m}_t,\bar{r}}(s_t)\left|a_t - \bar{m}_t\right|^2$. Now, a small change in the action may change the state, and hence the term $\delta_{t,\bar{m}_t,\bar{r}}$ is changed. Using the continuity of the state transition q,, we take a uniform bound by considering the supremum $\bar{\delta}_{t,\bar{m}_t,\bar{r}} = \sup_s \delta_{t,\bar{m}_t,\bar{r}}(s)$. Then, the global error is bounded by $\varepsilon^2\left(\sup_{t\in\mathcal{T}}\bar{\delta}_{t,\bar{m}_t,\bar{r}}\right)$. ■

Many research works have delved into mean-field games with the assumption of a vast number of decision-makers interacting. When dealing with such extensive systems, the main concept involves simplifying large games into a stylized model featuring a continuous spectrum of decision-makers. This approach has proven beneficial across various applications in economics, computer science, and engineering. However, in practical scenarios, we seldom encounter the stylized game model with a continuous spectrum of decision-makers, and the proposed approximation in the asymptotic regime isn't generally suitable for networks with only a handful of interacting decision-makers.

In this chapter, we have presented a mean-field approximation framework that not only fits well with large systems but also accommodates small-scale scenarios with a limited number of decision-makers. The approximation methods presented here share payoff properties akin to infinite-population mean-field games but now cater to a finite number of decision-makers. The result reveals that these games entail an approximation error, for which we have furnished a bound. This broadens the scope of the theory's applicability to a range of scenarios where explicit payoff expressions are unavailable and the indistinguishability property can be confirmed through the given definitions. The method proposed here can be extended to mean-field-type games with arbitrary number of decision-makers.

6.3 Notes

Non-asymptotic mean-field games have been studied in [316, 318]. Several open issues on asymptotic indistinguishability are discussed in [128, 129, 132, 310, 333] for the form

$$r_{i,I}\left(x_i, a_i, \mathbb{P}_{x_i}, \mathbb{P}_{a_i}, \frac{1}{I-1}\sum_{j\neq i}\delta_{x_j}, \frac{1}{I-1}\sum_{j\neq i}\delta_{a_j}\right)$$

in payoffs as well as in the state dynamics, where $\mathbb{P}_Y$ denotes the probability measure of the random variable Y. An asymptotic version of this model is examined in Chap. 9 for both atomic and nonatomic decision-makers. See Appendix B for more details.

Chapter 7
Linear-Quadratic Mean-Field and Mean-Field-Type Differential Games

We undertake in this chapter a comprehensive study of mean-field and mean-field-type differential games with linear dynamics and quadratic (as well as exponential of quadratic) cost functions for the decision-makers. The chapter is composed of ten main sections.

Section 7.1 covers basic linear-quadratic mean-field differential games where the stochastic state dynamics are driven by standard Brownian motions. Three different methods are discussed: (i) direct method, (ii) stochastic maximum principle, (iii) dynamic programming principle, for finding expressions for both equilibrium strategies and equilibrium costs. Section 7.1.1 collects some preliminary results on risk-neutral linear-quadratic mean-field differential games using three different approaches. In Sect. 7.1.2, we address the risk-sensitive differential game problem, elucidating the difference between risk-sensitive and risk-neutral Riccati equations. Section 7.1.3 is devoted to the study of robust mean-field differential games. In Sect. 7.1.4, we discuss linear exponential-quadratic robust mean-field differential games. In Sect. 7.1.5, we characterize equilibria of backward linear-quadratic mean-field differential games. Section 7.1.6 focuses on social optima of linear-quadratic mean-field differential games using mean-field-type control.

Section 7.2 considers basic linear-quadratic mean-field-type games. The key difference with the formulation in Sect. 7.1 is in the mean-field term. In the mean-field-type framework:

- The decision of a single decision-maker may affect directly the mean-field term.
- The number of decision-makers is not necessarily large.
- The payoff and/or the control actions are decision-maker specific and do not require indistinguishability, interchangeability, or symmetry (in distribution) between the decision-makers.

Section 7.2.1 presents a basic risk-neutral game problem. In Sect. 7.2.2, we consider the risk-neutral robust differential game problem.

T. Başar et al., *Mean-Field-Type Game Theory I*, Static & Dynamic Game Theory: Foundations & Applications, https://doi.org/10.1007/978-3-032-07027-2_7

In Sect. 7.3, a multi-person, mean-field-type game is formulated and solved that is described by a linear jump-diffusion system of conditional mean-field type where the conditioning is with respect to common noise and a quadratic cost functional involving the second moment, the square of the conditional expectation of the state, and the control actions of the decision-makers.

In Sect. 7.4, we formulate and solve mean-field-type games described by a linear jump-diffusion system of conditional mean-field type where the conditioning is with respect to common noise and a quadratic empathetic cost functional involving the second moment, the square of the expected values of the state, and the control actions for the decision-makers.

The price of simplicity, which is the gap between the solution of the original problem and the solution achieved by the mean-field game problem, is unbounded. This suggests a careful design of the decomposition of the original problem instead of freezing the mean-field terms. Section 7.5 presents the price of simplicity of mean-field-type games in large population regime through basic examples with linear state dynamics and variance-aware cost functions. We show that the game solution obtained using the large population approach (which freezes the population mean-field term) may lead to poor performance compared to the original optimization problem (see also Chap. 15 of Volume II of the book).

Section 7.6 focuses on linear-quadratic mean-field-type game under a jump-diffusion-regime switching state dynamics that include mean-field terms which are:

- conditional expectation of the state with respect to the filtration of the regime switching, in the drift, diffusion, and jump coefficient functions,
- conditional expectation of the control actions, in the drift, diffusion, and jump coefficient functions.

The model considered has one-dimensional state process. The instantaneous cost function includes:

- conditional variance of the state,
- square of the conditional expected value of the state,
- conditional variance of the control-action process,
- square of the conditional expected value of the control-action process,
- covariance between the state and the control,
- product term between the conditional expected value of the control action and the conditional expected value of the state.

Section 7.7 extends the problem of Sect. 7.6. We consider a generic linear-quadratic mean-field-type game problem under a jump-diffusion-regime switching state dynamics that include common noise and mean-field terms which are:

- the conditional expectation of the state with respect to the filtration of the regime switching, in the drift, diffusion, and jump coefficient functions,
- the conditional expectation of the control actions, in the drift, diffusion, and jump coefficient functions.

The model still has one-dimensional state process, and the common noise is of two types: a Brownian motion and a Poisson random process. The instantaneous cost function includes:

- conditional variance of the state,
- square of the conditional expected value of the state,
- conditional variance of the control-action process,
- square of the conditional expected value of the control-action process,
- cross-term such as covariance between the state and the control,
- product term between the conditional expected value of the control-action and the conditional expected value of the state,
- a noisy cost measurement, which is a jump-diffusion-regime switching and common noise dependent.

Section 7.8 extends the problem of Sect. 7.7 to adversarial games. A generic adversarial linear-quadratic mean-field-type game problem under a jump-diffusion-regime switching state dynamics that include common noise and mean-field terms which includes:

- the conditional expectation of the state with respect to the filtration of the regime switching, in the drift, diffusion, and jump coefficient functions,
- the conditional expectation of the control actions, in the drift, diffusion, and jump coefficient functions.

The state process is still one dimensional and the common noise is again of two types: a Brownian motion and a Poisson random process. The instantaneous cost function includes:

- the conditional variance of the state,
- square of the conditional expected value of the state,
- conditional variance of the control-action process,
- square of the conditional expected value of the control-action process,
- cross-term such as covariance between the state and the control,
- product term between the conditional expected value of the control action and the conditional expected value of the state
- noisy cost measurement, which is a jump-diffusion-regime switching and common noise dependent.

It is shown that a mean-field-type saddle point equilibrium can be obtained in a semi-explicit way for the adversarial linear-quadratic game problem with jump-diffusion process and regime switching, where all drift, diffusion, and jump terms are state, control input, conditional expected value state, conditional expected value control, and switching dependent. In addition, the common noise includes not only a diffusion term but also a jump process, which is crucial in many application areas. The presented method is fairly straightforward, not requiring either integrated partial integral-differential equations with switching terms or the backward–forward stochastic maximum principle.

Section 7.9 focuses on a special class of risk-sensitive linear-quadratic mean-field-type game problems under a jump-diffusion-regime switching state dynamics. In the study of time-dependent mean-field-type games, we have thus far restricted the analysis to the scalar case. The multidimensional extension entails several challenges, stemming from noncommutativity and nonsymmetric structures. One such extension is provided in Sect. 7.10 where we revisit the problem examined in Sect. 7.4 in the multidimensional case. The chapter ends with a bibliographical notes in section.

7.1 Large Population Games with Measure-Linear Cost

Notations and Preliminaries

Let $T > 0$ be a fixed time horizon and $(\Omega, \mathscr{F}, \mathbb{F}, \mathbb{P})$ be a given filtered probability space on which a one-dimensional standard Brownian motion $B = \{B_t\}_{t\geq 0}$ is defined, and the filtration $\mathbb{F} = \{\mathscr{F}_t,\ 0 \leq t \leq T\}$ is the natural filtration of B augmented by $\mathbb{P}$-null sets of $\mathscr{F}$. A set is $\mathbb{P}$-null (or zero measure) if its measure is zero under $\mathbb{P}$.

We introduce the following notations.

- Let $k \geq 1$. $L^k(0, T; \mathbb{R})$ is the set of measurable functions $f : [0, T] \to \mathbb{R}$ such that $\int_0^T |f(t)|^k dt < \infty$.
- $\mathscr{L}_{\mathbb{F}}^k(0, T; \mathbb{R})$ is the set of $\mathbb{F}$-adapted $\mathbb{R}$-valued processes $X(\cdot)$ such that $E\left[\int_0^T |X(t)|^k dt\right] < \infty$.
- For $t \in [0, T]$, $|f|_t := \sup_{0\leq s\leq t} |f(s)|$.

Next, we introduce the class of admissible controls, which plays an important role to obtain well-defined moments of the state process.

Definition 7.1 (Admissible Control Strategy) An admissible control strategy u is an $\mathbb{F}$-adapted and square-integrable process with values in a non-empty subset U of $\mathbb{R}$. We denote the set of all admissible controls by $\mathscr{U}$:

$$\mathscr{U} = \{u(\cdot) \in \mathscr{L}_{\mathbb{F}}^2(0, T; \mathbb{R});\ \ u(t) \in U\ \ a.e.\ \ t \in [0, T],\ \ \mathbb{P} - a.s.\}$$

Given $u \in \mathscr{U}$, consider the following one-dimensional controlled linear stochastic differential equation:

$$\begin{cases} dx(t) = [ax(t) + \bar{a}m(t) + bu(t)]dt + \sigma dB(t), \\ x(0) = x_0 \in \mathbb{R}, \end{cases} \tag{7.1}$$

where $a, \bar{a}, b, \sigma$ are real numbers and m is a deterministic function such that $|m|_T < \infty$.

Definition 7.2 (Liouville Dynamics) A dynamical system is of Liouville type if the transition kernel functional (drift, diffusion, jump coefficient functions) involves not only the states and the control actions but also an independent measure over the state, control actions, and/or other related parameters' space.

Note that here $m(t)$ is not necessarily equal to $\int y\mathbb{P}_{x(t)}(dy)$ unless if we are at an equilibrium (to be defined later). The case where m is replaced by $\int y\mathbb{P}_{x(t)}(dy)$ is discussed in the next section.

Lemma 7.1 *Consider (7.1) with $u \in \mathscr{U}$. Then, the following holds:*

(1) *There exists a constant $C_T > 0$ depending on $T, a, \bar{a}, b, \sigma$ and $|m|_T$ such that*

$$E[|x|_T^2] \leq C_T E\left(1 + |x(0)|^2 + E\left[\int_0^T |u(t)|^2 dt\right]\right). \tag{7.2}$$

(2) *If f is a Borel function from $[0, T] \times \mathbb{R}$ into $\mathbb{R}$ and is of linear growth, i.e., there exists a constant $C > 0$ such that $|f(t, x)| \leq C(1 + |x|)$, for any $(t, x) \in [0, T] \times \mathbb{R}$, then the process defined, for $0 \leq t \leq T$, by*

$$\mathscr{E}_t(x) := \exp\left(\int_0^t f(s, x(s))dB(s) - \frac{1}{2}\int_0^t |f(s, x(s))|^2 ds\right), \tag{7.3}$$

is an $\mathbb{F}$-martingale.
In particular,

$$E\left[\mathscr{E}_T(x)\right] = 1. \tag{7.4}$$

Proof The first assertion follows from the Grönwall and Burkholder–Davis–Gundy's inequalities. The second assertion follows from Corollary 3.5.16 in [212]. ∎

From (7.2), we deduce that if $x(0) \in L^2$ and u is admissible, then $x \in \mathscr{L}_{\mathbb{F}}^2(0, T; \mathbb{R})$. The second assertion (ii) with (7.3) and (7.4) allows us to use Girsanov type of transforms when dealing with risk-sensitive decision-makers. Since the coefficients are bounded, we will see below that the optimal control and the equilibrium strategy structures are such that the drift function of the state dynamics is of linear growth.

7.1.1 Risk-Neutral Large Population Regime with Measure-Linear Cost

In this subsection, we formulate and solve a mean-field game described by a linear stochastic dynamics and a quadratic cost functional for each generic decision-maker. The optimal strategies for the decision-makers will be determined explicitly using three different methods: (i) a simple and direct method based on completion of squares and a Girsanov-type change of measure, (ii) a stochastic maximum principle, and (iii) dynamic programming. We will show that there is unique best response strategy to the mean of the state and provide a simple sufficient condition of existence and uniqueness of mean-field equilibrium.

Consider a very large number of risk-neutral decision-makers. Given $m(.)$, the best response of a decision-maker is the following risk-neutral linear-quadratic stochastic control problem where the mean-field term is through the mean of all the states of all the decision-makers:

$$\begin{cases} \inf_{u(\cdot)\in\mathscr{U}} \frac{1}{2} E\left[q(T)x^2(T) + \bar{q}(T)[x(T)-m(T)]^2 \right. \\ \qquad \left. + \int_0^T q(t)x^2(t) + \bar{q}(t)(x(t)-m(t))^2 + r(t)u^2(t)dt\right], \\ \text{subject to} \\ dx(t) = [ax(t) + \bar{a}m(t) + bu(t)]dt + \sigma dB(t), \\ x(0) \in \mathbb{R}, \end{cases} \tag{7.5}$$

where $q(t), \bar{q}(t), r(t) \geq \epsilon > 0, \forall\, t$ and $a, \bar{a}, b, \sigma$ are real numbers and where $m(t)$ is the mean state trajectory created by all decision-makers at an equilibrium (if it exists).

Definition 7.3 (Best Response to Mean-Field) Any $\bar{u}(\cdot) \in \mathscr{U}$ satisfying the infimum in (7.5) is called a risk-neutral best-response strategy of a generic decision-maker to the mean-field term $(m(t))_{t\in[0,T]}$. The corresponding state process, solution of (7.5), is denoted by $\bar{x}(\cdot) := x^{\bar{u}}[m, x(0)](\cdot)$.

The risk-neutral mean-field equilibrium problem we are concerned with is characterization of the triplet $(\bar{x}, \bar{u}, m)$ solution of the problem (7.5), and characterization of mean state created by all the decision-makers coincides with m, i.e.,

$$m(t) = E[x^{\bar{u}}[m](t)],$$

which leads to a fixed-point system.

Definition 7.4 A mean-field equilibrium is a triplet $(\bar{x}, \bar{u}, m)$ such that $\bar{u}$ is a minimizer in (7.5) and $E[\bar{x}] = m$.

Note that we define a mean-field equilibrium through the mean state and not the distribution itself.

Next, we employ three different methods to determine the best response of a generic decision-maker: (i) direct method, (ii) stochastic maximum (minimum) principle (Pontryagin's approach) , and (iii) dynamic programming (Bellman's approach).

7.1.1.1 Direct Method

In this subsection, we employ a very simple and efficient method, called *direct method*, which essentially involves a completion of squares in a Hilbert space, to solve the best-response problem (7.5). Let

$$L(u,m)=\tfrac{1}{2}[q(T)x^2(T)+\bar{q}(T)[x(T)-m(T)]^2 \\ +\int_0^T\{q(t)x^2(t)+\bar{q}(t)(x(t)-m(t))^2+r(t)u^2(t)\}dt]. \tag{7.6}$$

Then, the risk-neutral cost functional in (7.5) is $E[L(u,m)]$, the expected value of L.

The direct method consists of writing a generic structure of the cost functional, with unknown deterministic functions to be identified. Inspired by the structure of the terminal cost function, we try a generic solution in a quadratic form. Let $f(t,x(t))=\frac{1}{2}\beta(t)x^2(t)+\alpha(t)x(t)+\gamma(t)$, where α, β and γ are deterministic and measurable functions of time, such that

$$f(T,x(T))=\frac{1}{2}\left[q(T)x^2(T)+\bar{q}(T)[x(T)-m(T)]^2\right],$$

i.e., $\beta(T)=q(T)+\bar{q}(T),\ \alpha(T)=-\bar{q}(T)m(T),\ \gamma(T)=\frac{\bar{q}(T)}{2}m^2(T)$.

Since the state process is an Itô diffusion process with twice differentiable scalar function $f(t,x(t))$ of two real variables t and $x(t)$, we can apply Itô's formula to $f(t,x(t))$ to obtain

$$\begin{aligned}
&f(T,x(T))-f(0,x(0))\\
&=\int_0^T\{\frac{\dot{\beta}(t)}{2}+a\beta(t)\}x^2(t)+(\dot{\alpha}(t)+a\alpha(t)+\bar{a}\beta(t)m(t))x(t)dt\\
&+\int_0^T(\dot{\gamma}(t)+\bar{a}\alpha(t)m(t)+\frac{\sigma^2}{2}\beta(t))+bu(\beta(t)x(t)+\alpha(t))dt\\
&+\int_0^T\sigma(\beta(t)x(t)+\alpha(t))dB(t).
\end{aligned}\tag{7.7}$$

We evaluate $L(u,m)-\frac{1}{2}\beta(0)x^2(0)-\alpha(0)x(0)-\gamma(0)$:

$$\begin{aligned}
&L(u,m)-\frac{1}{2}\beta(0)x^2(0)-\alpha(0)x(0)-\gamma(0)=\int_0^T\{\frac{\dot{\beta}(t)}{2}+a\beta(t)+\frac{q(t)+\bar{q}(t)}{2}\}x^2(t)dt\\
&+\int_0^T(\dot{\alpha}(t)+a\alpha(t)+\bar{a}\beta(t)m(t)-\bar{q}(t)m(t))x(t)dt\\
&+\int_0^T(\dot{\gamma}(t)+\bar{a}\alpha(t)m(t)+\frac{\sigma^2}{2}\beta(t)+\frac{\bar{q}(t)}{2}m^2(t))dt
\end{aligned}$$

$$+\int_0^T [bu(t)(\beta(t)x(t)+\alpha(t))+\frac{r(t)}{2}u^2(t)]+\sigma(\beta(t)x(t)+\alpha(t))dB(t).$$

We now use the following simple completion of squares relationship:

$$bu(\beta x+\alpha)+\frac{r}{2}u^2=\frac{r}{2}\left(u+\frac{b}{r}(\beta x+\alpha)\right)^2-\frac{b^2\beta^2}{2r}x^2-\frac{b^2\alpha\beta}{r}x-\frac{b^2\alpha^2}{2r}. \tag{7.8}$$

Hence,

$$\begin{aligned}
&L(u,m)-\frac{1}{2}\beta(0)x^2(0)-\alpha(0)x(0)-\gamma(0) \qquad (7.9)\\
&=\int_0^T\{\frac{\dot\beta(t)}{2}+a\beta(t)+\frac{q(t)+\bar q(t)}{2}-\frac{b^2\beta^2(t)}{2r(t)}\}x^2(t)\\
&+\int_0^T(\dot\alpha(t)+a\alpha(t)+\bar a\beta(t)m(t)-\bar q(t)m(t)-\frac{b^2\alpha(t)\beta(t)}{r(t)})x(t)dt\\
&+\int(\dot\gamma(t)+\bar a\alpha(t)m(t)+\frac{\sigma^2}{2}\beta(t)+\frac{\bar q(t)}{2}m^2(t)-\frac{b^2\alpha^2(t)}{2r(t)})dt\\
&+\int_0^T\frac{r(t)}{2}\left[u(t)+\frac{b}{r(t)}(\beta(t)x(t)+\alpha(t))\right]^2+\sigma(\beta(t)x(t)+\alpha(t))dB(t).
\end{aligned}$$

Since the expected value of the last term (7.10) is zero and $r>0$, we have

$$E[L(u,m)]\geq\frac{1}{2}\beta(0)x^2(0)+\alpha(0)x(0)+\gamma(0),\quad u\in\mathcal{U}, \tag{7.10}$$

if and only if

$$\begin{cases}
\dot\beta(t)+2a\beta(t)-\frac{b^2}{r(t)}\beta^2(t)+q(t)+\bar q(t)=0,\\
\beta(T)=q(T)+\bar q(T)\geq 0,\\
\dot\alpha(t)+a\alpha(t)+(\bar a\beta(t)-\bar q(t))m(t)-\frac{b^2}{r(t)}\alpha(t)\beta(t)=0,\\
\alpha(T)=-\bar q(T)m(T),\\
\dot\gamma(t)+\bar a\alpha(t)m(t)+\frac{\sigma^2}{2}\beta(t)+\frac{\bar q(t)}{2}m^2(t)-\frac{b^2}{2r(t)}\alpha^2(t)=0,\\
\gamma(T)=\frac{\bar q(T)}{2}m^2(T).
\end{cases} \tag{7.11}$$

with equality if and only if

$$u(t)=-\frac{b}{r(t)}(\beta(t)x(t)+\alpha(t)).$$

Since $b \neq 0, r > 0$ by assumption, the risk-neutral Riccati equation in (7.11) has a unique positive solution β. Incorporating it into the equation satisfied by α, one gets a solution $\alpha[m]$ by direct integration. Thus, the unique best-response strategy is

$$\bar{u}(t) = -\frac{b}{r(t)}(\beta(t)\bar{x}(t) + \alpha[m](t)), \tag{7.12}$$

where α, β are solutions of (7.11). Moreover, the associated performance is

$$E[L(\bar{u})] = \frac{1}{2}\beta(0)x^2(0) + \alpha(0)x(0) + \gamma(0). \tag{7.13}$$

Risk-Neutral Mean-Field-Type Equilibrium: Large Population Regime

We now look for an equilibrium of the risk-neutral linear-quadratic game with measure-linear cost, which we call risk-neutral mean-field equilibrium. If $(\bar{x}, \bar{u}, m)$ is a mean-field equilibrium, then m solves the following fixed-point equation:

$$m(t) = m(0) + \int_0^t [(a+\bar{a})m(s) - \frac{b^2}{r(s)}(\beta(s)m(s) + \alpha[m](s))]\, ds =: \Phi[m](t), \tag{7.14}$$

which can be rewritten as $m = \Phi[m]$. The Banach–Picard fixed-point theorem (also known as the contraction mapping theorem) is an important tool in the theory of complete metric spaces. It guarantees the existence and uniqueness of fixed points of certain self-maps of complete metric spaces and provides a constructive method to find those fixed points by Banach–Picard iterates.

From the inequality in Lemma 7.1, we deduce that the operator Φ maps $L^2(0, T; \mathbb{R})$ into itself. $L^2(0, T; \mathbb{R})$ is a complete metric space. Therefore, a sufficient condition for having a unique fixed point of Φ is given by a strict contraction of Φ, which is ensured if its Lipschitz constant L_Φ is strictly less than one. Since

$$(\alpha[m] - \alpha[m'])(t) = -\bar{q}(T)(m - m')(T)$$

$$+ \int_t^T (a - \frac{b^2}{r(t')}\beta(t'))(\alpha - \alpha')(t') + (\bar{a}\beta(t') - \bar{q}(t'))(m - m')(t')\, dt',$$

a standard Grönwall inequality yields

$$|\alpha[m] - \alpha[m']|_T \leq [\bar{q}(T) + |\bar{a}\beta - \bar{q}|_T]e^{T|a - \frac{b^2}{r}\beta|_T}|m - m'|_T$$

$$\Phi[m](t) - \Phi[m'](t) = \int_0^t [(a + \bar{a} - \frac{b^2}{r(t')}\beta(t'))(m - m')(t')]dt'$$

$$- \int_0^t \frac{b^2}{r(t')}(\alpha[m] - \alpha[m'])(t')]\, dt'.$$

It follows that

$$
\begin{aligned}
&|\Phi[m]-\Phi[m']|_T \\
&\leq T\left[|a+\bar{a}-\frac{b^2}{r}\beta|_T+|\frac{b^2}{r}|_T[\bar{q}(T)+|\bar{a}\beta-\bar{q}|_T]e^{T|a-\frac{b^2}{r}\beta|_T}\right]|m-m'|_T.
\end{aligned}
$$

Thus,

$$
L_\Phi \leq T\left[|a+\bar{a}-\frac{b^2}{r}\beta|_T+|\frac{b^2}{r}|_T[\bar{q}(T)+|\bar{a}\beta-\bar{q}|_T]e^{T|a-\frac{b^2}{r}\beta|_T}\right].
$$

This means that

$$
L_\Phi \leq T[g_1+|\frac{b^2}{r}|_T[\bar{q}(T)+\epsilon_1 e^{T|a-\frac{b^2}{r}\beta|_T}]], \tag{7.15}
$$

where

$$
g_1=|a+\bar{a}-b^2\beta/r|_T, \quad \epsilon_1=|\bar{a}\beta-\bar{q}|_T. \tag{7.16}
$$

Thus, we have proved the following:

Proposition 7.1 *Suppose that* $r>0, s>0, q+\bar{q}>0$. *Then there exists a unique best-response strategy* $\bar{u}=-\frac{b}{r}(\beta x+\alpha)$, *where* α *and* β *solve the Riccati equations (7.11).*

In addition, if $T[g_1+|\frac{b^2}{r}|_T[\bar{q}(T)+\epsilon_1 e^{T|a-\frac{b^2}{r}\beta|_T}]]<1$, *then there is a unique risk-neutral mean-field equilibrium.*

We now refine the equation satisfied by α by setting $\alpha=\eta m$:

$$
\begin{cases}
\dot{\alpha}(t)+a\alpha(t)+(\bar{a}\beta(t)-\bar{q}(t))m(t)-\frac{b^2}{r(t)}\alpha(t)\beta(t)=0, \\
\alpha(T)=-\bar{q}(T)m(T).
\end{cases} \tag{7.17}
$$

Then,

$$
\begin{cases}
\dot{\eta}+[2a+\bar{a}-2\frac{b^2}{r}\beta]\eta-\frac{b^2}{r}\eta^2+\bar{a}\beta-\bar{q}=0, \\
\eta(T)=-\bar{q}(T).
\end{cases} \tag{7.18}
$$

Closed-Form Expression of Mean-Field Equilibrium

Whenever α and β do not blow up within $[0, T]$, the explicit mean-field equilibrium is given by

$$\begin{cases} m(t) = m(0)e^{\int_0^t (a+\bar{a}-\frac{b^2}{r}(\beta+\eta))dt} \\ \dot{\beta}(t) + 2a\beta(t) - \frac{b^2}{r(t)}\beta^2(t) + q(t) + \bar{q}(t) = 0, \\ \beta(T) = q(T) + \bar{q}(T), \\ \dot{\eta}(t) + [2a + \bar{a} - 2\frac{b^2}{r(t)}\beta(t)]\eta(t) - \frac{b^2}{r(t)}\eta^2(t) + (\bar{a}\beta(t) - \bar{q}(t)) = 0, \\ \eta(T) = -\bar{q}(T) \\ \bar{u} = -\frac{b}{r(t)}(\beta(t)x(t) + \eta(t)m(t)). \end{cases} \tag{7.19}$$

Note that the equilibrium strategy $\bar{u}$ is a linear combination of the state $x(t)$, and the mean of the mean-field $m(t)$.

Summary of the Direct Method

- Write the structure of the cost as a polynomial of degree 2 in the form:

$$\frac{\beta(t)}{2}x^2(t) + \eta(t)x(t)m(t) + \gamma(t)m^2(t)$$

- Use Itô's formula and compute the difference between the two cost functionals.
- Derive the structure of the best response u as in (7.19).
- Use fixed-point argument to derive the equilibrium strategy $\bar{u}$.
- Use fixed-point argument to derive the equilibrium mean-field $\bar{m}$.
- Rewrite the equilibrium state trajectory $\bar{x}$ by incorporating $(\bar{u}, \bar{m})$.

7.1.1.2 Stochastic Minimum Principle

This subsection describes how to efficiently solve risk-neutral linear-quadratic mean-field-type games in large population regime employing stochastic minimum/maximum principle (SMP) . We solve the best-response problem in (7.5).

The risk-neutral (non-optimized) Hamiltonian (also called Pontryagin function) is

$$H(t, x, m, u, p, z) := (ax + \bar{a}m + bu)p + \frac{1}{2}[qx^2 + \bar{q}(x-m)^2 + ru^2] + \sigma z,$$

where p and z are adjoint processes. H is smooth in (x, u) and we have

$$H_x = ap + qx + \bar{q}(x-m), \qquad H_u = bp + ru.$$

Minimizing the Hamiltonian yields

$$\bar{u}(t) = -\frac{b}{r}p(t). \tag{7.20}$$

The associated state dynamics $\bar{x}$ solves the SDE:

$$d\bar{x}(t) = \left(a\bar{x}(t) + \bar{a}m(t) - \frac{b^2}{r}p(t)\right)dt + \sigma dB(t). \tag{7.21}$$

The first-order adjoint equation yields

$$dp = -H_x dt + zdB(t) = -[ap + qx + \bar{q}(x - m)]dt + zdB, \tag{7.22}$$

$$p(T) = q(T)x(T) + \bar{q}(T)(x(T) - m(T)). \tag{7.23}$$

We try a solution of the form

$$p(t) := \beta(t)\bar{x}(t) + \alpha(t), \tag{7.24}$$

where $\beta(t)$, $\alpha(t)$ are deterministic functions to be determined. From (7.23), it follows that

$$\beta(T) \ = q(T) + \bar{q}(T), \text{ and } \quad \alpha(T) = -\bar{q}(T)m(T). \tag{7.25}$$

Applying Itô's formula to (7.24) yields

$$d[\beta x + \alpha] = \{\dot{\beta} + a\beta - \frac{b^2\beta^2}{r}\}xdt + [\dot{\alpha} - \frac{b^2\alpha\beta}{r} + \bar{a}\beta m]dt \tag{7.26}$$

$$+\sigma\beta dB(t). \tag{7.27}$$

Equating (7.22) and (7.27), we get

$$\dot{\beta} + a\beta - \frac{b^2\beta^2}{r} = -(a\beta + q + \bar{q}), \tag{7.28}$$

$$\dot{\alpha} - \frac{b^2\alpha\beta}{r} + \bar{a}\beta m = -a\alpha + \bar{q}m, \tag{7.29}$$

$$z = \sigma\beta. \tag{7.30}$$

Reorganizing the above, we get the following system of the Riccati equations:

$$\begin{cases} \dot{\beta} + 2a\beta - \frac{b^2\beta^2}{r} + q + \bar{q} = 0, \\ \beta(T) \ = q(T) + \bar{q}(T), \\ \dot{\alpha} + a\alpha - \frac{b^2\alpha\beta}{r} + (\bar{a}\beta - \bar{q})m = 0, \\ \alpha(T) = -\bar{q}(T)m(T), \\ z = \sigma\beta, \\ \bar{u} = -\frac{b}{r}(\beta x + \alpha). \end{cases} \tag{7.31}$$

Note that (7.31) is the same system as the one obtained by the direct method in (7.11). Moreover, the best-response strategy is a linear feedback of the state, and the coefficient α depends on m. To prove the optimality of the system via stochastic minimum principle, we use the fact that the (unoptimized) Hamiltonian function is strictly convex in (x, u).

Summary of the Stochastic Minimum Principle

- Introduce the (non-optimized) Hamiltonian function and write the adjoint equation:

$$dp = -H_x dt + z dB$$

- Derive the structure of the best response.
- Based on the terminal value $p(T)$, write a generic form of the adjoint process $p = \beta x + \eta m$.
- Compute $dp = d[\beta x + \eta m]$ and $dp = -H_x dt + z dB$.
- Check if the necessary condition of optimality is sufficient.
- Derive the structure of the best response u.
- Use fixed-point argument to derive the equilibrium strategy $\bar{u}$.
- Use fixed-point argument to derive the equilibrium mean-field $\bar{m}$.
- Rewrite the equilibrium state trajectory $\bar{x}$ by incorporating $(\bar{u}, \bar{m})$.

7.1.1.3 Dynamic Programming Principle

This subsection describes how to efficiently solve risk-neutral linear-quadratic mean-field-type games with measure-linear cost by employing the dynamic programming principle (DPP) . Given the mean-field m, we introduce the following function:

$$v(t, x) = \inf_u \left\{ \frac{1}{2}[q(T)x^2(T) + \bar{q}(T)[x(T) - m(T)]^2] \right.$$

$$+\frac{1}{2}\int_t^T q(t')x^2(t')+\bar{q}(t')(x(t')-m(t'))^2+r(t')u^2(t')dt',\ x(t)=x\Bigg\}. \quad (7.32)$$

The best response to the mean-field m at time t is

$$\bar{u} \in \arg\min_u H(t, x, m, u, v_x, \frac{\sigma}{2}v_{xx}),$$

and v solves the Hamilton–Jacobi–Bellman equation:

$$v_t + H^*(t, x, m, v_x, \frac{\sigma}{2}v_{xx}) = 0,$$

where $H^*(t, x, m, p, z) = H(t, x, m, \bar{u}, p, z)$. The terminal condition is $v(T, x) = \frac{q}{2}x^2 + \frac{\bar{q}}{2}(x - m)^2$. The DPP equation can be written as

$$\begin{cases} v_t + (ax + \bar{a}m)v_x - \frac{b^2}{2r}v_x^2 + \frac{1}{2}[qx^2 + \bar{q}(x - m)^2] + \frac{\sigma^2}{2}v_{xx} = 0, \\ v(T, x) = \frac{q(T)}{2}x^2 + \frac{\bar{q}(T)}{2}(x - m)^2. \end{cases} \quad (7.33)$$

We try a solution in the form

$$v(t, x) = \frac{\beta(t)}{2}x^2(t) + \alpha(t)x(t) + \gamma(t).$$

The partial differential equation (7.33) becomes

$$\begin{cases} \frac{\dot{\beta}}{2}x^2 + \dot{\alpha}x + \dot{\gamma} \\ +(ax + \bar{a}m)\beta x + (ax + \bar{a}m)\alpha \\ -\frac{b^2}{2r}\beta^2x^2 - \frac{b^2}{r}\alpha\beta x - \frac{b^2}{2r}\alpha^2 \\ +\frac{1}{2}[qx^2 + \bar{q}(x - m)^2] + \frac{\sigma^2}{2}\beta = 0. \end{cases} \quad (7.34)$$

Reorganizing the terms in terms of the basis $\{1, x, x^2\}$, we obtain

$$\begin{cases} \{\frac{\dot{\beta}}{2} + a\beta - \frac{b^2}{2r}\beta^2 + \frac{q+\bar{q}}{2}\}x^2 \\ +\{\dot{\alpha} + a\alpha - \frac{b^2}{r}\alpha\beta + (\bar{a}\beta - \bar{q})m\}x \\ +\dot{\gamma} + \bar{a}m\alpha - \frac{b^2}{2r}\alpha^2 + \frac{\bar{q}}{2}m^2 + \frac{\sigma^2}{2}\beta = 0. \end{cases} \quad (7.35)$$

Recall that in a field $(\mathbb{R}, +, \times)$, a polynomial in the form $a_0 + a_1x + \ldots + a_nx^n$ is the null polynomial iff $a_0 = a_1 = \ldots = a_n = 0$. This means that all three coefficients are zero, and

$$\begin{cases} \dot{\beta} + 2a\beta - \frac{b^2}{r}\beta^2 + q + \bar{q} = 0 \\ \beta(T) = q(T) + \bar{q}(T), \\ \dot{\alpha} + a\alpha - \frac{b^2}{r}\alpha\beta + (\bar{a}\beta - \bar{q})m = 0, \\ \alpha(T) = -\bar{q}(T)m(T), \\ \dot{\gamma} + \bar{a}m\alpha - \frac{b^2}{2r}\alpha^2 + \frac{\bar{q}}{2}m^2 + \frac{\sigma^2}{2}\beta = 0, \\ \gamma(T) = \frac{\bar{q}(T)}{2}m^2(T), \\ \bar{u} = -\frac{b}{r}v_x = -\frac{b}{r}(\beta x + \alpha), \\ v(0,x) = \frac{\beta(0)}{2}x^2 + \alpha(0)x + \gamma(0). \end{cases} \tag{7.36}$$

Again, we retrieve (7.11) and (7.31).

Summary of the Dynamic Programming Principle

- Introduce the Hamiltonian function and write the Hamilton–Jacobi–Bellman (HJB) equation.
- Derive the best response to the mean of the other decision-makers state as a function of v_x, and possibly v_{xx}.
- Based on the terminal value $v(T, x)$, write a generic form of the value function $v(t, x) = \frac{\beta(t)}{2}x^2 + \alpha(t)x + \gamma(t)$.
- Identify the coefficient function α, β, γ via HJB.
- Derive the structure of the best response u.
- Use fixed-point argument to derive the equilibrium strategy $\bar{u}$.
- Use fixed-point argument to derive the equilibrium mean-field $\bar{m}$.
- Rewrite the equilibrium state trajectory $\bar{x}$ by incorporating $(\bar{u}, \bar{m})$.

From these simple calculations, we observe that the three methods, (i) direct method of Sect. 7.1.1.1, (ii) stochastic minimum principle of Sect. 7.1.1.2, (iii) dynamic programming principle of Sect. 7.1.1.3, provide the same best-response strategy to the mean field in the large population regime case with measure-linear cost. We will see that this observation does not hold in the general linear-quadratic mean-field-type games where open-loop adapted solutions differ from state-mean-field-type feedback solutions.

We can also work directly with the individual state distribution $\mu = \mathbb{P}_x$. Let us define $V(t, \mu) = \int v(t, x)\mu(dx)$. We retrieve that the value function $v(t, x)$ coincides with $V(t, \delta_x)$. In this particular case, by linearity in μ, the master adjoint equation obtained $U(x, \mu) = V_\mu(\mu)[x]$ provides also the value function of the problem. This latter observation does not hold in the general mean-field-type game case.

7.1.2 Risk-Sensitive Large Population Regime with Measure-Linear Cost

We now consider a very large number of risk-sensitive decision-makers with risk-sensitivity index $\theta > 0$. The best response of a decision-maker is then obtained by solving the following risk-sensitive linear-quadratic stochastic control problem where the mean-field term is through the mean of all the states of all the decision-makers [325, 327]:

$$\begin{cases} \inf_{u(\cdot)\in\mathscr{U}} Ee^{\theta L(u,m)}, \\ \text{subject to} \\ dx(t) = [ax(t) + \bar{a}m(t) + bu(t)]dt + \sigma dB(t), \\ x(0) \in \mathbb{R}. \end{cases} \tag{7.37}$$

Any $\bar{u}(\cdot) \in \mathscr{U}$ delivering the infimum in (7.37) is a risk-sensitive best response of a generic decision-maker to the mean-field term $(m(t))_t$. The corresponding state process, solution of (7.37), will be denoted by $\bar{x}(\cdot) := x^{\bar{u}}[m](\cdot)$.

The risk-sensitive mean-field equilibrium problem we are concerned with is characterization of the triplet $(\bar{x}, \bar{u}, m)$ solution of the problem (7.37), and the state created by all the decision-makers agrees with m, i.e.,

$$m(t) = E[x^{\bar{u}}[m](t)],$$

which is a fixed-point equation.

7.1.2.1 Direct Method

We complete with the term $-\frac{1}{2}\int_0^T \theta^2\sigma^2(\beta(t)x(t) + \alpha(t))^2 dt + \theta\int_0^T [\frac{1}{2}\theta\sigma^2]$ $(\beta(t)x(t) + \alpha(t))^2 dt$ in $L(u, m)$ to obtain

$$\theta[L(u,m) - \frac{1}{2}\beta(0)x^2(0) - \alpha(0)x(0) - \gamma(0)] \tag{7.38}$$

$$\begin{aligned}
&= \theta\int_0^T \{\frac{\dot{\beta}}{2}(t) + a\beta(t) + \frac{q(t)+\bar{q}(t)}{2} + (-\frac{b^2}{2r(t)} + \frac{1}{2}\theta\sigma^2)\beta^2(t)\}x^2(t)dt \\
&+\theta\int_0^T (\dot{\alpha}(t) + a\alpha(t) + \bar{a}\beta(t)m(t) - \bar{q}(t)m(t) + (-\frac{b^2}{r(t)} + \theta\sigma^2(t))\alpha(t)\beta(t))x(t)dt \\
&+\theta\int (\dot{\gamma}(t) + \bar{a}\alpha(t)m(t) + \frac{\sigma^2}{2}\beta(t) + \frac{\bar{q}(t)}{2}m^2(t) - \frac{b^2\alpha^2(t)}{2r(t)} + \frac{1}{2}\theta\sigma^2\alpha^2(t))dt \\
&+\theta\int_0^T \frac{r}{2}\left[u(t) + \frac{b}{r(t)}(\beta(t)x(t) + \alpha(t))\right]^2 dt \\
&+\int_0^T \theta\sigma(\beta(t)x(t) + \alpha(t))dB(t) - \frac{1}{2}\int_0^T \theta^2\sigma^2(\beta(t)x(t) + \alpha(t))^2 dt.
\end{aligned} \tag{7.39}$$

The process of exponentiation followed by taking the expectation yields

$$
\begin{aligned}
Ee^{\theta[L(u,m)-\frac{1}{2}\beta(0)x^2(0)-\alpha(0)x(0)-\gamma(0)]} &= E\left[\mathscr{E}_T(x)e^{\theta\int_0^T \frac{r(t)}{2}\left[u(t)+\frac{b}{r(t)}(\beta(t)x(t)+\alpha(t))\right]^2 dt}\right] \\
&\geq E\left[\mathscr{E}_T(x)\right]
\end{aligned}
\tag{7.40}
$$

if and only if (β, α, γ) satisfies the following system of equations, where we call the equation satisfied by β the risk-sensitive Riccati equation:

$$
\begin{cases}
\dot{\beta}(t) + 2a\beta(t) - (\frac{b^2}{r(t)} - \theta\sigma^2)\beta^2(t) + q(t) + \bar{q}(t) = 0 \\
\dot{\alpha}(t) + a\alpha(t) + (\bar{a}\beta(t) - \bar{q}(t))m(t) + (-\frac{b^2}{r(t)} + \theta\sigma^2)\alpha(t)\beta(t) = 0 \\
\dot{\gamma}(t) + \bar{a}\alpha(t)m(t) + \frac{\sigma^2}{2}\beta(t) + \frac{\bar{q}(t)}{2}m^2(t) - \frac{b^2\alpha^2(t)}{2r(t)} + \frac{\theta\sigma^2}{2}\alpha^2(t) = 0, \\
\gamma(T) = \frac{\bar{q}(T)}{2}m^2(T),
\end{cases}
\tag{7.41}
$$

with equality if and only if

$$
\bar{u}(t) = -\frac{b}{r(t)}(\beta(t)\bar{x}(t) + \alpha(t)), \tag{7.42}
$$

where $\bar{x}$ solves the dynamics in (7.37) associated with $\bar{u}$, and

$$
\mathscr{E}_t(x) := \exp\left(\int_0^t \theta\sigma(\beta(s)x(s)+\alpha(s))dB(s) - \frac{1}{2}\int_0^t \theta^2\sigma^2(\beta(s)x(s)+\alpha(s))^2 ds\right).
$$

Now, the risk-sensitive Riccati equation in β above has a unique positive solution if $\frac{b^2}{r} - \theta\sigma^2 > 0$, i.e., $\theta < \frac{b^2}{r\sigma^2}$, in which case, in view of Lemma 7.1, we have

$$
E[\mathscr{E}_T(x)] = 1, \quad u \in \mathscr{U}.
$$

Hence,

$$
Ee^{\theta L(u,m)} \geq e^{\theta\left[\frac{1}{2}\beta(0)x^2(0)+\alpha(0)x(0)+\gamma(0)\right]}, \quad u \in \mathscr{U},
$$

with equality for $u = \bar{u}$, which constitutes the unique best response to the mean-field process.

Hence, the cost functional associated with the best response $\bar{u}$ given by (7.42) is

$$
Ee^{\theta L(\bar{u})} = e^{\theta\left(\frac{1}{2}\beta(0)x^2(0)+\alpha(0)x(0)+\gamma(0)\right)}. \tag{7.43}
$$

Note that the presence of the term $\theta\sigma^2$ in the risk-sensitive Riccati equation (7.41) comes from the completion of the exponential martingale. One gets the risk-neutral Riccati equation as θ vanishes.

Using the estimate in (7.15) with $\theta < \frac{b^2}{r\sigma^2}$, then the Banach–Picard fixed-point theorem applies for T sufficiently small enough such that

$$L_{rs} := T\left[g_1 + |\frac{b^2}{r}|_T[\bar{q}(T) + \epsilon_1 e^{T|a+(-\frac{b^2}{r}+\theta\sigma^2)\beta|_T}]\right] < 1.$$

In this case, the existence of the fixed-point m follows. For a larger T, the bound L_{rs} may exceed 1 and is not useful for the strict contraction argument. Note that for $\theta > \frac{b^2}{r\sigma^2}$, the solution β may blow up in finite time and the control $-\frac{b}{r}(\beta x + \alpha)$ becomes ill-defined.

Thus, we have proved the following:

Proposition 7.2 *Consider the linear-quadratic risk-sensitive mean-field game associated with (7.37). Suppose that* $r > 0, s > 0, q + \bar{q} > 0$ *and* $\theta < \frac{b^2}{r\sigma^2}$. *Then, there exists a unique best-response strategy* $\bar{u} = -\frac{b}{r}(\beta x + \alpha)$, *where* α *and* β *solves the risk-sensitive Riccati equation (7.41). In addition, if* $T\left[g_1 + |\frac{b^2}{r}|_T[\bar{q}(T) + \epsilon_1 e^{T|a+(-\frac{b^2}{r}+\theta\sigma^2)\beta|_T}]\right] < 1$, *then there is a unique risk-sensitive mean-field equilibrium, where* g_1 *and* ϵ_1 *are given by (7.16).*

7.1.2.2 Stochastic Minimum Principle

We use the risk-sensitive maximum principle equations to derive a best-response strategy. The risk-sensitive minimum principle is given by

$$dp = -H_x dt + z[-\theta\sigma p dt + dB], \quad p(T) = qx + \bar{q}(x - m),$$

where

$$H(t, x, m, u, p, z) := (ax + \bar{a}m + bu)p + \frac{1}{2}[qx^2 + \bar{q}(x - m)^2 + ru^2] + \sigma z$$

$$\begin{aligned} d[\beta x + \eta m] = &(\dot{\beta} + a\beta - \frac{b^2}{r}\beta^2)x dt \\ &+(\dot{\eta} + [a + \bar{a} - \frac{b^2}{r}(\beta + \eta)]\eta + \bar{a}\beta - \frac{b^2}{r}\eta\beta)m dt \\ &+\sigma\beta dB. \end{aligned} \tag{7.44}$$

On the other hand, the risk-sensitive first-order adjoint equation solves

$$dp = -\{(a\beta + q + \bar{q})x + (a\eta - \bar{q})m\}dt + z[-\theta\sigma(\beta x + \eta m)dt + dB].$$

Then

$$z = \sigma\beta, \tag{7.45}$$

$$dp = -\{a\beta+q+\bar{q}+\theta\sigma^2\beta^2\}xdt+-\{(a+\theta\sigma^2\beta)\eta-\bar{q}\}mdt+\sigma\beta dB. \tag{7.46}$$

By identifying the coefficients, we obtain

$$\dot{\beta} + 2a\beta + (-\frac{b^2}{r} + \theta\sigma^2)\beta^2 + q + \bar{q} = 0$$

and

$$\left\{\dot{\eta} + [2a + \bar{a} + (-2\frac{b^2}{r} + \theta\sigma^2)\beta]\eta - \frac{b^2}{r}\eta^2 + \bar{a}\beta - \bar{q}\right\} m = 0.$$

The pair (η, β) provides (p, z), which in turn provides a best response to the infinite-population mean-field m. In order to find a risk-sensitive equilibrium, we use the mean state dynamics that solves the fixed-point

$$m(t) = m(0) + \int_o^t \left(a + \bar{a} - \frac{b^2}{r}(\beta + \eta)\right) m(t')dt', \tag{7.47}$$

which provides the mean-field

$$m(t) = m(0)e^{\int_0^t (a+\bar{a}-\frac{b^2}{r}(\beta+\eta))dt}.$$

7.1.2.3 Dynamic Programming Principle

We use the dynamic programming principle (DPP) to find the best-response strategy of risk-sensitive decision-makers. The risk-sensitive Hamilton–Jacobi–Bellman equation associated with $\frac{1}{\theta}\log E[e^{\theta L(u,m)}]$ is given by

$$v_t + H^*(t, x, v_x, \frac{\sigma}{2}v_{xx}) + \frac{\theta}{2}\sigma^2 v_x^2 = 0,$$

with $v(T, x) = \frac{q(T)}{2}x^2 + \frac{\bar{q}(T)}{2}(x - m)^2$, which can be expanded to arrive at

$$\begin{cases} v_t + (ax + \bar{a}m)v_x - \frac{b^2}{2r}v_x^2 + \frac{1}{2}[qx^2 + \bar{q}(x - m)^2] + \frac{\sigma^2}{2}v_{xx} + \frac{\theta}{2}\sigma^2 v_x^2 = 0, \\ v(T, x) = \frac{q(T)}{2}x^2 + \frac{\bar{q}(T)}{2}(x - m)^2. \end{cases} \tag{7.48}$$

We try a solution in the form

$$v(t,x)=\frac{\beta(t)}{2}x^2(t)+\alpha(t)x(t)+\gamma(t).$$

Reorganizing the terms in function of the basis $(1,x,x^2)$, we obtain

$$\begin{cases}\{\frac{\dot\beta}{2}+a\beta+(-\frac{b^2}{2r}+\frac{\theta}{2}\sigma^2)\beta^2+\frac{q+\bar q}{2}\}x^2\\ +\{\dot\alpha+a\alpha+(-\frac{b^2}{r}+\theta\sigma^2)\alpha\beta+(\bar a\beta-\bar q)m\}x\\ +\dot\gamma+\bar a m\alpha+(-\frac{b^2}{2r}+\frac{\theta}{2}\sigma^2)\alpha^2+\frac{\bar q}{2}m^2+\frac{\sigma^2}{2}\beta=0.\end{cases}\tag{7.49}$$

By substituting $\alpha=\eta m$, we get

$$\begin{cases}\{\frac{\dot\beta}{2}+a\beta+(-\frac{b^2}{2r}+\frac{\theta}{2}\sigma^2)\beta^2+\frac{q+\bar q}{2}\}x^2\\ +\{\dot\eta+[2a+\bar a+(-2\frac{b^2}{r}+\theta\sigma^2)\beta]\eta+(\bar a\beta-\bar q)\}mx\\ +\dot\gamma+\bar a m^2\eta+(-\frac{b^2}{2r}+\frac{\theta}{2}\sigma^2)\eta^2m^2+\frac{\bar q}{2}m^2+\frac{\sigma^2}{2}\beta=0,\end{cases}\tag{7.50}$$

which gives the same optimal strategy as in (7.41). Similarly, the mean field is given by

$$m(t)=m(0)e^{\int_0^t(a+\bar a-\frac{b^2}{r}(\beta+\eta))dt}$$

provided that the term $\frac{b^2}{r}(\beta+\eta)$ is integrable over $(0,T)$. The risk-sensitive equilibrium is given by $(\bar x,\bar u,\bar m)$:

$$\begin{cases}\bar x(t)==\left\{(a-\frac{b^2}{r(t)}\beta(t))\bar x+(\bar a-\frac{b^2}{r(t)}\eta(t))m(t)\right\}dt+\sigma dB(t),\bar x(0)=x(0).\\ \bar u(t)=-\frac{b}{r(t)}(\beta(t)\bar x(t)+\eta(t)m(t)),\\ \bar m(t)=m(0)e^{\int_0^t(a+\bar a-\frac{b^2}{r(t')}(\beta(t')+\eta(t')))dt'}.\end{cases}\tag{7.51}$$

The equilibrium cost of a generic decision-maker is $v(0,x(0))=\frac{\beta(0)}{2}x^2(0)+\alpha(0)x(0)+\gamma(0)$, where (α,β,γ) solve (7.41).

7.1.3 Robust Large Population Games: Risk-Neutral Case

Consider a very large number of risk-neutral decision-makers and a malicious/disturbance term v. We assume that the disturbance term v takes values in a subset V of $\mathbb{R}$ and define the set of disturbance strategies $\mathcal{V}$ in a similar fashion as $\mathcal{U}$. There are as many minimizing decision-makers as adversaries [53, 55]. The pair of control action (u,v) of a generic decision-maker and a generic adversary does not affect the population mean field.

Next, we formulate a risk-neutral robust mean-field game as a minmax mean-field game as follows.

Let

$$L_2(u,v) = \frac{1}{2}\int_0^T \{q(t)x^2(t) + \bar{q}(t)(x(t)-m(t))^2 + r(t)u^2(t) - s(t)v^2(t)\}dt \\ +\frac{1}{2}\left[q(T)x^2(T) + \bar{q}(T)[x(T)-m(T)]^2\right] \tag{7.52}$$

be the cost functional of a generic decision-maker with strategy u under disturbance v when the mean-field process is m. The state dynamics of a generic decision-maker is $dx(t) = [ax + \bar{a}m(t) + bu(t) + cv(t)]dt + \sigma dB(t),\ x(0) \in \mathbb{R}$.

The best response of a generic decision-maker is the following risk-neutral linear-quadratic zero-sum differential game:

$$\begin{cases} \inf_{u(\cdot)\in\mathcal{U}} \sup_{v(\cdot)\in\mathcal{V}} E\left[L_2(u,v)\right], \\ \text{subject to} \\ dx(t) = [ax + \bar{a}m(t) + bu(t) + cv(t)]dt + \sigma dB(t), \\ x(0) \in \mathbb{R}, \end{cases} \tag{7.53}$$

where $q(t) \geq 0, \bar{q}(t) \geq 0, r(t) > 0$, and $a, \bar{a}, b, \sigma$ are real numbers and where $m(t)$ is the mean state trajectory generated by all decision-makers at robust equilibrium (if it exists).

Any $(\bar{u}(\cdot), \bar{v}(\cdot)) \in \mathcal{U} \times \mathcal{V}$ delivering the inf-sup in (7.53) is called a risk-neutral robust best response of a generic decision-maker to the mean-field process $(m(t))_t$ under worst-case disturbance $\bar{v}$. The corresponding state process, obtained from (7.53), will be denoted by $\bar{x}(\cdot) := x^{\bar{u},\bar{v}}[m](\cdot)$.

The robust mean-field equilibrium problem we are concerned with is characterization of the quadruple $(\bar{x}, \bar{u}, \bar{v}, m)$ solution of the problem (7.53), and the mean state generated by all the decision-makers coincides with m, i.e.,

$$m(t) = E[x^{\bar{u},\bar{v}}[m](t)],$$

which is a fixed-point equation.

A robust mean-field equilibrium is a quadruple $(\bar{x}, \bar{u}, \bar{v}, m)$ such that $\bar{u}$ is a minimizer in (7.53) under worst disturbance $\bar{v}$ and $E[\bar{x}] = m$. At a robust mean-field equilibrium, the best response to the mean field under worst-case disturbance should reproduce the mean field itself.

We use a direct method to obtain the robust best response of a decision-maker. Let $f(t,x) = \frac{1}{2}\beta(t)x^2(t) + \alpha(t)x(t) + \gamma(t)$. By applying Itô's formula, we have

$$\begin{aligned} & f(T, x(T)) - f(0, x(0)) \\ &= \int_0^T \{\frac{\dot{\beta}(t)}{2} + a\beta(t)\}x^2(t) + (\dot{\alpha}(t) + a\alpha(t) + \bar{a}\beta(t)m(t))x(t)dt \\ &+ \int_0^T (\dot{\gamma}(t) + \bar{a}\alpha(t)m(t) + \frac{\sigma^2}{2}\beta(t)) + (bu(t) + cv(t))(\beta(t)x(t) + \alpha(t))dt \\ &+ \int_0^T \sigma(\beta(t)x(t) + \alpha(t))dB(t). \end{aligned} \tag{7.54}$$

We now compute the difference $L_2(u,v) - \frac{1}{2}\beta(0)x^2(0) - \alpha(0)x(0) - \gamma(0)$.
We have

$$
\begin{aligned}
L_2(u,v) - \tfrac{1}{2}\beta(0)x^2(0) - \alpha(0)x(0) - \gamma(0) = &\int_0^T \{\tfrac{\dot{\beta}(t)}{2} + a\beta(t) + \tfrac{q(t)+\bar{q}(t)}{2}\}x^2(t)\,dt \\
&+ \int_0^T (\dot{\alpha}(t) + a\alpha(t) + \bar{a}\beta(t)m(t) - \bar{q}(t)m(t))x(t)dt \\
&+ \int_0^T (\dot{\gamma}(t) + \bar{a}\alpha(t)m(t) + \tfrac{\sigma^2}{2}\beta(t) + \tfrac{\bar{q}(t)}{2}m^2(t))dt \\
&+ \int_0^T [(bu(t) + cv(t))(\beta(t)x(t) + \alpha(t)) + \tfrac{r(t)}{2}u^2(t) - \tfrac{s(t)}{2}v^2(t)]\,dt \\
&+ \int_0^T \sigma(\beta(t)x(t) + \alpha(t))dB(t).
\end{aligned}
\tag{7.55}
$$

We now use the following relations:

$$
bu(\beta x + \alpha) + \frac{r}{2}u^2 = \frac{r}{2}\left(u + \frac{b}{r}(\beta x + \alpha)\right)^2 - \frac{b^2\beta^2}{2r}x^2 - \frac{b^2\alpha\beta}{r}x - \frac{b^2\alpha^2}{2r}
$$

and

$$
cv(\beta x + \alpha) - \frac{s}{2}v^2 = -\frac{s}{2}\left(v - \frac{c}{s}(\beta x + \alpha)\right)^2 + \frac{c^2\beta^2}{2s}x^2 - \frac{c^2\alpha\beta}{r}x - \frac{c^2\alpha^2}{2s}
$$

to obtain

$$
\begin{aligned}
&L_2(u,v) - \frac{1}{2}\beta(0)x^2(0) - \alpha(0)x(0) - \gamma(0) \\
&= \int_0^T \{\frac{\dot{\beta}(t)}{2} + a\beta(t) + \frac{q(t)+\bar{q}(t)}{2} + (-\frac{b^2}{2r(t)} + \frac{c^2}{2s(t)})\beta^2(t)\}x^2(t) \\
&+ \int_0^T (\dot{\alpha}(t) + a\alpha(t) + \bar{a}\beta(t)m(t) - \bar{q}(t)m(t) + (-\frac{b^2}{r(t)} + \frac{c^2}{s(t)})\alpha(t)\beta(t))x(t)dt \\
&+ \int (\dot{\gamma}(t) + \bar{a}\alpha(t)m(t) + \frac{\sigma^2}{2}\beta(t) + \frac{\bar{q}(t)}{2}m^2(t) + (-\frac{b^2}{2r(t)} + \frac{c^2}{2s(t)})\alpha^2(t))dt \\
&+ \int_0^T \frac{r(t)}{2}\left[u(t) + \frac{b}{r(t)}(\beta(t)x(t)+\alpha(t))\right]^2 - \frac{s(t)}{2}\left[v(t) - \frac{c}{s(t)}(\beta(t)x(t)+\alpha(t))\right]^2 dt \\
&+ \int_0^T \sigma(\beta(t)x(t) + \alpha(t))dB(t).
\end{aligned}
\tag{7.56}
$$

Therefore,

$$
E[L_2(u,v)] \geq \frac{1}{2}\beta(0)x^2(0) + \alpha(0)x(0) + \gamma(0), \quad (u,v) \in \mathscr{U} \times \mathscr{V}, \tag{7.57}
$$

if and only if the triple (α, β, γ) solves the following system of equations where we call the equation β a “robust” Riccati equation:

$$\begin{cases} \dot{\beta}(t) + 2a\beta(t) + (-\frac{b^2}{r(t)} + \frac{c^2}{s(t)})\beta^2(t) + q(t) + \bar{q}(t) = 0, \\ \beta(T) = q(T) + \bar{q}(T) \geq 0, \\ \dot{\alpha}(t) + a\alpha(t) + (\bar{a}\beta(t) - \bar{q}(t))m(t) + (-\frac{b^2}{r(t)} + \frac{c^2}{s(t)})\alpha(t)\beta(t) = 0, \\ \alpha(T) = -\bar{q}(T)m(T), \\ \dot{\gamma}(t) + \bar{a}\alpha(t)m(t) + \frac{\sigma^2}{2}\beta(t) + \frac{\bar{q}(t)}{2}m^2(t) + (-\frac{b^2}{2r(t)} + \frac{c^2}{2s(t)})\alpha^2(t) \\ \gamma(T) = \frac{\bar{q}(T)}{2}m^2(T), \end{cases} \tag{7.58}$$

with equality if and only if

$$\bar{u}(t) = -\frac{b}{r(t)}(\beta(t)x(t) + \alpha(t)), \quad \bar{v}(t) = \frac{c}{s(t)}(\beta(t)x(t) + \alpha(t)). \tag{7.59}$$

Since $b > 0, r > 0, s > 0$ by assumption, the robust Riccati equation in β in (7.58) has a unique positive solution if $\frac{b^2}{r} - \frac{c^2}{s} > 0$. Injecting β into the equation satisfied by α, we get a solution $\alpha[m]$ by direct integration. Thus, there exists a unique best-response strategy whenever $\frac{b^2}{r} > \frac{c^2}{s} > 0$.

Remark 7.1 Setting $c^2/s = \theta\sigma^2$ in the robust Riccati equation (7.58), we obtain the risk-sensitive Riccati equation (7.41).

Risk-Neutral Robust Mean-Field Equilibrium

If $(\bar{x}, \bar{u}, \bar{v}, m)$ is a robust mean-field equilibrium, then m solves the following fixed-point equation:

$$m = \Phi[m], \tag{7.60}$$

where

$$\Phi[m](t) := m_0 + \int_0^t (a + \bar{a})m(t') + (-\frac{b^2}{r(t')} + \frac{c^2}{s(t')})(\beta(t')m(t') + \alpha[m](t'))dt'.$$

Since the term $(-\frac{b^2}{r})$ is changed to $(-\frac{b^2}{r} + \frac{c^2}{s})$ compared to the previous analysis in (7.15) and

$$L_{\Phi} \leq T\left[g_3 + |-\frac{b^2}{r} + \frac{c^2}{s}|_T[\bar{q}(T) + \epsilon_3 e^{T|a+(-\frac{b^2}{r}+\frac{c^2}{s})\beta|_T}]\right],$$

where $g_3 = |a + \bar{a} + (-b^2/r + c^2/s)\beta|_T$, $\epsilon_3 = |-\frac{b^2}{r} + \frac{c^2}{s}|_T$.

For $\frac{b^2}{r} > \frac{c^2}{s}$, a sufficient condition for strict contraction is given by $L_{\Phi} < 1$, which holds when T is small enough.

Thus, we have proved the following.

Proposition 7.3 *Consider the robust linear-quadratic game associated with (7.53). Suppose that* $r > 0, s > 0$, *and* $\frac{b^2}{r} > \frac{c^2}{s} > 0$. *Then there exists a unique best-response strategy* $\bar{u} = -\frac{b}{r}(\beta x + \alpha)$, *and the worst-case disturbance is* $\bar{v} = \frac{c}{s}(\beta x + \alpha)$, *where* α *and* β *solve the system (7.58).*
In addition, if $T\left[g_3 + |-\frac{b^2}{r} + \frac{c^2}{s}|_T[\bar{q}(T) + \epsilon_3 e^{T|a+(-\frac{b^2}{r}+\frac{c^2}{s})\beta|_T}]\right] < 1$, *then there is a unique risk-neutral robust mean-field equilibrium.*

7.1.4 Linear Exponential-Quadratic Robust Large Population Games

We address the problem in Sect. 7.1.3 but now with the quadratic cost exponentiated.

The risk-sensitive robust best response of a decision-maker is a solution to the following minmax problem:

$$\begin{cases} \inf_{u(\cdot)\in\mathscr{U}} \sup_{v(\cdot)\in\mathscr{V}} Ee^{\theta L_2(u,v)}, \\ \text{subject to} \\ dx(t) = [ax(t) + \bar{a}m(t) + bu(t) + cv(t)]dt + \sigma dB(t), \\ x(0) \in \mathbb{R}. \end{cases} \tag{7.61}$$

Any $(\bar{u}(\cdot), \bar{v}(\cdot)) \in \mathscr{U} \times \mathscr{V}$ realizing the inf-sup in (7.61) is a risk-sensitive robust best response of a generic decision-maker to the mean-field process $(m(t))_t$ under worst-case disturbance v. The corresponding state process, generated under that pair, will be denoted by $\bar{x}(\cdot) := x^{\bar{u},\bar{v}}[m](\cdot)$.

The risk-sensitive robust mean-field equilibrium problem we are concerned with is characterization of the quadruple $(\bar{x}, \bar{u}, \bar{v}, m)$ solution of the problem (7.61) and establishment of the result that the mean of state generated by all the decision-makers agrees with m, i.e.,

$$m(t) = E[x^{\bar{u},\bar{v}}[m](t)],$$

which is a fixed-point equation. Completing with the term

$$-\frac{1}{2}\int_0^T \theta^2\sigma^2(\beta(t)x(t) + \alpha(t))^2 dt + \frac{1}{2}\theta\int_0^T \theta\sigma^2(\beta(t)x(t) + \alpha(t))^2 dt$$

in $L_2(u, v)$, we obtain

$$\theta[L_2(u, v) - \frac{1}{2}\beta(0)x^2(0) - \alpha(0)x(0) - \gamma(0)] \tag{7.62}$$
$$= \theta\int_0^T \{\frac{\dot{\beta}(t)}{2} + a\beta(t) + \frac{q(t) + \bar{q}(t)}{2} + (-\frac{b^2}{2r(t)} + \frac{c^2}{2s(t)} + \frac{\theta}{2}\sigma^2)\beta^2(t)\}x^2(t)dt$$

$$
\begin{aligned}
&+\theta\int(\dot{\alpha}(t)+a\alpha(t)+\bar{a}\beta(t)m(t)-\bar{q}(t)m(t)+(-\frac{b^2}{r(t)}+\frac{c^2}{s(t)}+\theta\sigma^2)\alpha(t)\beta(t))x(t)dt\\
&+\theta\int(\dot{\gamma}(t)+\bar{a}\alpha(t)m(t)+\frac{\sigma^2}{2}\beta(t)+\frac{\bar{q}(t)}{2}m^2(t)+(-\frac{b^2}{2r(t)}+\frac{c^2}{2s(t)}+\frac{\theta}{2}\sigma^2)\alpha^2(t))dt\\
&+\theta\int_0^T\frac{r(t)}{2}\left[u+\frac{b}{r(t)}(\beta(t)x(t)+\alpha(t))\right]^2-\frac{s(t)}{2}\left[v(t)-\frac{c}{s(t)}(\beta(t)x(t)+\alpha(t))\right]^2dt\\
&+\int_0^T\theta\sigma(\beta(t)x(t)+\alpha(t))dB(t)-\frac{1}{2}\int_0^T\theta^2\sigma^2(\beta(t)x(t)+\alpha(t))^2dt. \qquad (7.63)
\end{aligned}
$$

The processes of exponentiations followed by taking the expectation yield

$$
Ee^{\theta[L_2(u,v)]}\geq E\left[\mathscr{E}_T(x)\right]\exp\theta\left[\frac{1}{2}\beta(0)x^2(0)+\alpha(0)x(0)+\gamma(0)\right],\ (u,v)\in\mathscr{U}\times\mathscr{V},
$$

where

$$
\mathscr{E}_T(x):=\exp\left(\int_0^T\theta\sigma(\beta(t)x(t)+\alpha(t))dB(t)-\frac{1}{2}\int_0^T\theta^2\sigma^2(\beta(t)x(t)+\alpha(t))^2dt\right),
$$

if and only if

$$
\begin{cases}
\dot{\beta}(t)+2a\beta(t)+(-\frac{b^2}{r(t)}+\frac{c^2}{s(t)}+\theta\sigma^2)\beta^2(t)+q(t)+\bar{q}(t)=0,\\
\beta(T)=q(T)+\bar{q}(T)\geq 0,\\
\dot{\alpha}(t)+a\alpha(t)+(\bar{a}\beta(t)-\bar{q}(t))m(t)\\
\quad+(-\frac{b^2}{r(t)}+\frac{c^2}{s(t)}+\theta\sigma^2)\alpha(t)\beta(t)=0,\\
\alpha(T)=-\bar{q}(T)m(T),\\
\dot{\gamma}(t)+\bar{a}\alpha(t)m(t)+\frac{\sigma^2}{2}\beta(t)+\frac{\bar{q}(t)}{2}m^2(t)+(-\frac{b^2}{2r(t)}+\frac{c^2}{2s(t)}+\frac{\theta}{2}\sigma^2)\alpha^2(t),\\
\gamma(T)=\frac{\bar{q}(T)}{2}m^2(T),
\end{cases}
\qquad (7.64)
$$

with equality if and only if

$$
\bar{u}(t)=-\frac{b}{r(t)}(\beta(t)x(t)+\alpha(t)),\ \ \bar{v}(t)=\frac{c}{s(t)}(\beta(t)x(t)+\alpha(t)). \qquad (7.65)
$$

The risk-sensitive robust Riccati equation in β above has a unique positive solution $\beta(t)\geq 0$ if $\frac{b^2}{r}-\frac{c^2}{s}-\theta\sigma^2>0$, which is satisfied if θ and c are relatively small. In this case, in view of Lemma 7.1, $E[\mathscr{E}_T(x)]=1$, which yields

$$
Ee^{\theta[L(u,v)]}\geq e^{\theta\left[\frac{1}{2}\beta(0)x^2(0)+\alpha(0)x(0)+\gamma(0)\right]},\qquad (u,v)\in\mathscr{U}\times\mathscr{V},
$$

with equality if $(u,v)=(\bar{u},\bar{v})$, which is a unique best-response strategy to the mean field for $\frac{b^2}{r}>\frac{c^2}{s}+\theta\sigma^2>0$.

Note however that for large θ, or c, the solution β may blow up in finite time and the control $-\frac{b}{r}(\beta x+\alpha)$ would then become ill-defined.
Thus, we have proved the following.

Proposition 7.4 *Suppose that* $r>0, s>0, q>0, \bar{q}\geq 0$ *and* $\frac{b^2}{r}>\frac{c^2}{s}+\theta\sigma^2>0$, *then there exists a unique best-response strategy* $\bar{u}=-\frac{b}{r}(\beta x+\alpha)$, *and the worst-case disturbance is* $\bar{v}=\frac{c}{s}(\beta x+\alpha)$, *where* α *and* β *solve (7.64).*
In addition, if $T\left[g_4+|-\frac{b^2}{r}+\frac{c^2}{s}|_T[\bar{q}(T)+\epsilon_4 e^{T|a+(-\frac{b^2}{r}+\frac{c^2}{s}+\theta\sigma^2)\beta|_T}]\right]<1$,
where $g_4:=|a+(-\frac{b^2}{r}+\frac{c^2}{s}+\theta\sigma^2)\beta|_T$, $\epsilon_4=|\frac{b^2}{r}-\frac{c^2}{s}|_T$, *then there is a unique robust risk-sensitive mean-field equilibrium.*

Remark 7.2 If the mean-field term is a function of the control actions of all the decision-makers, written as $\xi(E[\bar{u}])$, instead of the mean state $\bar{m}$, a similar methodology applies, except that the fixed-point equation would now be different

$$E[\bar{u}]=-\frac{b}{r}\left\{\beta\bar{m}+\alpha(\xi(E[\bar{u}]))]\right\}.$$

This type of mean-field structure is observed in electricity distribution networks where the mean-field term that modulates the price of electricity is a function of aggregate demand and supply, where the demand is generated by the users (residential consumers, commercial, industrial, transportation). It is also relevant in wireless networking where the mean-field term is the interference $E[\bar{u}.|\bar{x}|^2]$ created by other users, where $\bar{u}\geq 0$ is the transmission power of a generic user.

7.1.5 Linear-Quadratic Large Population Games for Backward Equations

What we have presented so far in this section are forward mean-field-type games in measure-linear cost. In this subsection, we model dynamic situations where the state at a final time is somehow known or fixed in advance. Such situations can be modeled with backward dynamical systems. A backward stochastic differential equation (BSDE) is a stochastic differential equation (SDE) for which a random terminal condition on the state has been specified.

Why a Backward Model?
Our motivation for backward mean-field-type games comes from practical engineering applications:

- In road traffic engineering, the state of the traffic flow at terminal time (say, 3 am) is a known random variable and can be modeled with backward SDEs.
- In social networks, the information dynamics of an advertizement after a certain period of time is known. If we model the propagation of information (say, with

noisy logistic dynamics), then the state at the final time is a known random variable. This leads to a backward equation:

$$dx = x(1 - \frac{x}{C})dt + \sigma x dB, \quad x(T) = \xi$$

- In energy-efficient buildings, the air-conditioning (AC) control should set the indoor temperature toward a comfort temperature of a user. The temperature dynamics can be described by a backward equation where the temperature in building i is influenced by the outdoor temperature and the temperature of the neighboring and connecting buildings to i. The final temperature after a few minutes of testing should stabilize a desired comfort temperature, which is known because it is set by the user or the building manager. This leads to a backward SDE:

$$dT_i = \frac{1}{RC}[(T_{out} - T_i) + u]dt + \sigma dB, \quad T^i(T) = T_{comfort},$$

  where T_{out} is the outside air temperature, $T_{comfort}$ is the comfort temperature, $\sigma > 0$, R is the thermal resistance, and C is the capacity.
- Charging of electric vehicles over the smart grid is usually done in off-peak periods. The state of the charge of a battery can be described by a backward SDE where the final state is reached when the battery is fully charged (if known in advance). The SDE is $dx = \frac{I(x,u)}{Q_{\max}}dt + \sigma dB, \quad x(T) = 1$, where I is the current and $Q_{\max}$ is the charge capacity, $\sigma > 0$.

These backward SDE problems have been usually analyzed via constraints or penalization techniques with forward SDEs. However, the solutions provided by the penalization method is not the true solution of the original problem. Therefore, we will address the original problem in a more formal way here and derive optimality equations.

Why Is This a Challenging Problem in Terms of Decision-Making?

There are two technical challenges that arise when computing the best response under backward SDEs:

- Reverse time techniques do not provide an adapted control process. Since the future trajectory (before the terminal time) is unknown, the decision of a decision-maker at time t should be obtained from a measurable function of the current and past knowledge (it is called a behavioral strategy). With backward SDEs, when we reverse time by setting $t' = T - t$, we get $y(0) = x(T)$, which is not in the filtration generated by the state up to t.
- The best response at time t may not be $\mathcal{F}_t$-measurable and hence not admissible.

7.1.5.1 Backward Risk-Neutral Game

Consider a very large number of risk-neutral decision-makers. We formulate and solve a linear-quadratic mean-field game under backward stochastic differential equation (BSDE). In contrast to the earlier forward state model, unpredictability and non-admissibility issues arise in the optimization of systems described by BSDE. Although for the stochastic backward mean-field game problem the optimal strategy is no longer a feedback of the current state, it is indeed a linear state feedback of the entire past history of the state process. The optimal strategies of the decision-makers are given explicitly. We show that there is a unique best-response strategy to the mean field, and a simple sufficient condition of existence and uniqueness of mean-field equilibrium is provided. The best response of a decision-maker is the following risk-neutral linear-quadratic mean-field problem where the mean-field term is through the mean of all the states of all the decision-makers:

$$\begin{cases} \inf_{u(\cdot)\in\mathcal{U}} E[L(u,m)] \\ \text{subject to} \\ dx = -[ax + \bar{a}m + bu + cz]dt + z dB(t), \\ x(T) = x_T, \quad b \neq 0, \end{cases} \tag{7.66}$$

where z is a process associated with x to get a solution that is adapted to the natural filtration $\mathbb{F}$,

$$L(u,m) = \frac{1}{2}\Big[q(0)x^2(0) + \bar{q}(0)[x(0) - m(0)]^2 + \int_0^T \{qx^2 + \bar{q}(x-m)^2 + ru^2 + sz^2\}dt\Big],$$

$q(t) > 0, \bar{q}(t) > 0, r > 0, s \geq 0$ and $a, \bar{a}, b, c$ are real-valued and where $m(t)$ is the mean state trajectory created by all decision-makers at an equilibrium (if it exists).

Any $u_*(\cdot) \in \mathcal{U}$ satisfying the infimum in (7.66) is called a risk-neutral best-response strategy of a generic decision-maker to the mean-field term $(m(t))_t$. The corresponding state process, solution of (7.66), is denoted by $x_*(\cdot) := x^{u_*}[m](\cdot)$.

The risk-neutral mean-field equilibrium problem we are concerned with is characterization of the triplet (x_*, u_*, m_*) solution of the problem (7.5), and the mean state created by all the decision-makers coincides with m_*, i.e.,

$$m_*(t) = E[x^{u_*}[m_*](t)],$$

which is a fixed-point equation.

A mean-field equilibrium is a triplet (x_*, u_*, m_*) such that u_* minimizes the cost in (7.66) and $E[x_*] = m_*$.

Explicit Solution

We first provide an explicit representation of the state equation for each predictable control u.

Proposition 7.5 *Let the coefficients a, c be bounded over $[0, T]$, and the processes $\bar{a}m + bu$ and the terminal output x_T be square integrable. Then,*

$$dx = -[ax + \bar{a}m + bu + cz]dt + zdB, \tag{7.67}$$

$$x(T) = x_T \tag{7.68}$$

has a unique adapted solution x, and it is explicitly given by the following formula:

$$v(t)x(t) = E\left[v(T)x(T) + \int_0^T v(t')(\bar{a}m + bu)\, dt' |\, \mathscr{F}_t\right] \tag{7.69}$$

$$dv(t) = v(t)[adt + cdB(t)],\ v(0) = 1. \tag{7.70}$$

Proof Itô's formula for vx yields

$$d(vx) = [-\bar{a}vm - bvu]dt + [cvx + vz]dB.$$

It follows that

$$v(t)x(t) + \int_0^t v(t')(\bar{a}m + bu)\, dt' = v(0)x(0) + \int_0^t v[cx + z]dB(t'). \tag{7.71}$$

Since $v(0) = 1$, and under the assumptions above, $x(0) + \int_0^t v[cx + z]dB(s)$ defines a local martingale:

$$v(t)x(t) + \int_0^t v(s)(\bar{a}m + bu)\, ds = E[v(T)x(T) + \int_0^T v(s)(\bar{a}m + bu)\, ds \mid \mathscr{F}_t].$$

This completes the proof. ■

Determining the Best Response of a Decision-Maker

Proposition 7.6 *Each generic decision-maker has a unique best-response strategy given by*

$$u_*(t) = -\frac{b}{r}p(t), \tag{7.72}$$

where the process $p(t)$ is explicitly given by

$$p(t) = v(t)[p(0) + \int_0^t [\frac{qx(t') + \bar{q}(x(t') - m) - csz}{v(t')}]dt'$$
$$+ \int_0^t \frac{sz}{v(t')} dB(t')],$$

which is predictable and depends on the past history of the state $(x(s), z(s)),\ s \leq t$, *and* v *is the process defined by* $dv(t) = v(t)[adt + cdB(t)],\ v(0) = 1$.

Proof Let $H = -(ax + \bar{a}m + bu + cz)p - \frac{1}{2}[qx^2 + \bar{q}(x - m)^2 + ru^2 + sz^2]$. This function is strictly concave in (x, z, u) if $q + \bar{q}, r, s$ are positive. Therefore, the necessary condition for optimality is also sufficient. From the minimum principle of backward McKean–Vlasov SDEs, we have

$$dp = -H_x dt - H_z dB,$$

$$p(0) = -q(0)x(0) + \bar{q}(0)(x(0) - m(0)).$$

We compute $H_x = -ap - qx - \bar{q}(x - m),\ H_z = -cp - sz$. It follows that

$$dp = p[adt + cdB] + \{(qx + \bar{q}(x - m))dt + szdB\}.$$

The homogeneous part of the SDE is the forward equation satisfied by v. Let $p = \tilde{p}v$. By Itô's formula,

$$dp = vd\tilde{p} + \tilde{p}dv + d\langle \tilde{p}, v\rangle. \tag{7.73}$$

Thus,

$$d\tilde{p} = \frac{[qx + \bar{q}(x - m) - csz]dt + sz\, dB}{v}. \tag{7.74}$$

As $\tilde{p}(0) = \frac{p(0)}{v(0)} = p(0)$, we have

$$\tilde{p}(t) = p(0) + \int_0^t \frac{qx + \bar{q}(x - m) - csz}{v} dt' + \int_0^t \frac{sz}{v(t')} dB(t'),$$

and hence

$$p(t) = v(t)[p(0) + \int_0^t \frac{qx + \bar{q}(x - m) - csz}{v} dt'] + v(t)\int_0^t \frac{sz}{v(t')} dB(t').$$

We now show the optimality of the control $-\frac{b}{r}p$. We first note that when we perturb the control u, not only the process x is changed but also the process z changes in order to keep the new process adapted to the filtration. Let us compute the difference $L(u, m) - L(u_*, m)$:

$$
\begin{aligned}
&2(L(u,m) - L(u_*,m)) &(7.75)\\
&= 2q(0)x(0)[x(0) - x_*(0)] + 2\bar{q}(0)[x_*(0) - m(0)][x(0) - x_*(0)]\\
&\quad + \int_0^T -2bp(u + \frac{b}{r}p) + 2qx_*(x - x_*)dt\\
&\quad + \int_0^T 2\bar{q}(x_* - m)(x - x_*)dt\\
&\quad + \int_0^T 2sz_*(z - z_*)dt\\
&\quad + q(0)[x(0) - x_*(0)]^2 + \bar{q}(0)[x(0) - x_*(0)]^2\\
&\quad + \int r(u + \frac{b}{r}p)^2 + q[x - x_*]^2 + \bar{q}[x - x_*]^2 + s[z - z_*]^2 dt. &(7.76)
\end{aligned}
$$

Since $q(0), \bar{q}(0), r$ and s are positive,

$$
\begin{aligned}
&L(u,m) - L(u_*,m) &(7.77)\\
&\geq q(0)x(0)[x(0) - x_*(0)] + \bar{q}(0)[x_*(0) - m(0)][x(0) - x_*(0)]\\
&\quad + \int_0^T -bp(u + \frac{b}{r}p) + 2qx_*(x - x_*)dt\\
&\quad + \int_0^T [\bar{q}(x_* - m)(x - x_*) + sz_*(z - z_*)]dt.
\end{aligned}
$$

The first terms can be rewritten as

$$
\begin{aligned}
&q(0)x(0)[x(0) - x_*(0)] + \bar{q}(0)[x_*(0) - m(0)][x(0) - x_*(0)]\\
&= -p(0)[x(0) - x_*(0)].
\end{aligned}
$$

Using the differential formula, we get $-p(0)[x(0)-x_*(0)]+p(T)[x(T)-x_*(T)] = \int_0^T d(p(x - x_*))$. We know from the two terminal conditions that $x(T) = x_T = x_*(T)$. Hence,

$$p(T)[x(T) - x_*(T)] = 0$$

and one has

$$-\, p(0)[x(0) - x_*(0)] = \int_0^T d[p(x - x_*)].$$

Recall that process p solves

$$dp = [ap + qx_* + \bar{q}(x_* - m)]dt + (cp + sz_*)dB.$$

We compute $d(x - x_*)$:

$$d(x - x_*) = -[a(x - x_*) + b(u + \frac{b}{r}p) + c(z - z_*)]dt$$
$$+[z - z_*]dB. \tag{7.78}$$

We now apply Itô's formula to the product $p(x - x_*)$:

$$-p(0)[x(0) - x_*(0)] = \int_0^T d[p(x - x_*)] \tag{7.79}$$

$$= \int_0^T -bp(u + \frac{b}{r}p) + qx_*(x - x_*)dt$$
$$+ \int_0^T \bar{q}(x_* - m)(x - x_*)dt + \int_0^T sz_*(z - z_*)dt$$
$$+ \int_0^T [(x - x_*)(cp + sz_*) + p(z - z_*)]\, dB.$$

Taking the expectation, we conclude that

$$E[L(u, m) - L(u_*, m)] \geq 0,$$

with equality if and only $u_* = -\frac{b}{r}p$. Thus, there is a unique best response to m. ■

Note, in particular, that p is a linear functional of m.

Existence and Uniqueness of Mean-Field Equilibrium
The mean state trajectory when each decision-maker uses its best-response strategy solves the following backward ordinary differential equation:

$$\bar{x}(t) = \bar{x}(T) + \int_t^T [(a + \bar{a})\bar{x} - \frac{b^2}{r}\bar{p} + c\bar{z}]dt',$$

where

$$\bar{p}(t) = \bar{p}(0) + \int_0^t [a\bar{p} + q\bar{x}]dt',\ \bar{p}(0) = -q\bar{x}(0).$$

By equating m and $\bar{x}$, we arrive at a fixed-point equation. If (x, u, m) is a mean-field equilibrium, then m solves the following fixed-point equation:

$$m(t) = m(T) + \int_t^T [(a + \bar{a})m - \frac{b^2}{r}\bar{p}[m] + c\bar{z}]dt' =: \Phi[m](t),$$

which can be rewritten as $m = \Phi[m]$. The operator Φ maps square-integrable function over $[0, T]$ into itself. It follows that

$$\Phi[m](t) - \Phi[m'](t) = \int_t^T [(a+\bar{a})(m-m') - \frac{b^2}{r}(\bar{p}-\bar{p}')]dt'$$

and

$$(\bar{p}-\bar{p}')(t) = (\bar{p}-\bar{p}')(0) + \int_0^t [a(\bar{p}-\bar{p}') + q(m-m')]dt'.$$

By Grönwall inequality, $|p - p'| \leq \delta T q |m - m'|$ for a constant δ. Similarly, inequality holds for m :

$$|\Phi[m](t) - \Phi[m']| \leq T\left(|a+\bar{a}| + \frac{b^2}{r}\delta q T\right)|m-m'|.$$

The Lipschitz constant of Φ is less than $T\left(|a+\bar{a}| + \frac{b^2}{r}\delta q T\right)$, which can be strictly less than one when T is sufficiently small.

Therefore, the fixed-point equation becomes a strict contraction for T sufficiently small, and hence, there is a unique mean-field equilibrium when the length of the horizon is small enough. Note that the mean of process p can be written as $\bar{p} = \beta m + \alpha$, where β and α solve the following forward equations:

$$\dot{\beta} - \bar{a}\beta + \frac{b^2}{r}\beta^2 - q = 0, \quad \beta(0) = -q(0) \tag{7.80}$$

$$\dot{\alpha} + (-a + \frac{b^2}{r})\alpha - c\bar{z}\beta = 0, \quad \alpha(0) = 0. \tag{7.81}$$

The presence of the mean of process z suggests searching for stochastic Riccati-like equations.

The (backward) mean-field equilibrium is explicitly given by

$$\begin{cases} m(t) = m(T) + \int_t^T [(a+\bar{a} - \frac{b^2}{r}\beta)m - \frac{b^2}{r}\alpha + c\bar{z}]dt' \\ u_* = -\frac{b}{r}p, \\ p(t) = v(t)[p(0) + \int_0^t \frac{qx+\bar{q}(x-m)-csz}{v}dt'] + v(t)\int_0^t \frac{sz}{v(t')}dB(t'), \\ x(t) = x(T) + \int_t^T [ax + \bar{a}m - \frac{b^2}{r}p + cz]dt' - \int_t^T z dB(t'). \end{cases}$$

7.1.5.2 Backward Risk-Sensitive Game

We now focus on risk-sensitive games with exponential quadratic cost functionals and backward SDEs. We first establish the stochastic minimum principle for augmented state and use a logarithmic transformation to derive a risk-sensitive minimum principle. The mean-field best response is expressed as a function of the adjoint equation that solves a linear forward SDE. The best response of a generic decision-maker is obtained from

$$\begin{cases} \inf_{u(\cdot)\in\mathscr{U}} E[e^{\theta L(u,m)}] \\ \text{subject to} \\ dx = -[ax + \bar{a}m + bu + cz]dt + zdB(t), \\ x(T) = x_T. \end{cases} \tag{7.82}$$

where $\theta \neq 0$. Introduce the following backward integral equation:

$$y(t) = 0 - \int_T^t \{qx^2 + \bar{q}(x-m)^2 + ru^2 + sz^2\}dt' = 0 + \int_t^T \{qx^2 + \bar{q}(x-m)^2 + ru^2 + sz^2\}dt',$$

which has the associated BSDE:

$$dy = -\frac{1}{2}[qx^2 + \bar{q}(x-m)^2 + ru^2 + sz^2]dt, \quad y(T) = 0.$$

Then,

$$L(u, m) = \frac{1}{2}\left[q(0)x^2(0) + \bar{q}(0)[x(0) - m(0)]^2\right] + y(0),$$

and the risk-sensitive cost functional with the augmented state (x, y, z) has no running cost. The risk-sensitive Hamiltonian is

$$H^\theta := -p_1[ax + \bar{a}m + bu + cz] - \frac{p_2}{2}[qx^2 + \bar{q}(x-m)^2 + ru^2 + sz^2].$$

The stochastic maximum principle of backward SDE yields to the following forward adjoint equations:

$$\begin{cases} dp_1 = -H_x^\theta dt - H_z^\theta dB \\ p_1(0) = -\theta \{q(0)x(0) + \bar{q}(0)[x(0) - m(0)]\} e^{\theta L(\bar{u})}, \\ dp_2 = [z_2]dB \\ p_2(0) = -\theta e^{\theta L(\bar{u})}. \end{cases} \tag{7.83}$$

Proposition 7.7 *The process p_2 is a signed process and has the opposite sign of θ. By the martingale representation, the process p_2 remains negative over $(0, t)$.*

Proof From the initial relation $p_2(0) = -\theta e^{\theta L(\bar{u})}$, we observe $p_2(0)$ is negative and has opposite sign of θ. ■

Proposition 7.8 *Each generic decision-maker has a unique best-response strategy to m, and it is given by*

$$u_* = -\frac{b}{r}\frac{p_1}{p_2} = -\frac{b}{r}\tilde{p},$$

where $\tilde{p}$ solves the following forward SDE:

$$\begin{cases} d\tilde{p} = [a\tilde{p} + qx + \bar{q}(x-m)]dt + \tilde{q}(-\tilde{z}_2 dt + dB) \\ \tilde{p}(0) = q(0)x(0) + \bar{q}(0)[x(0) - m(0)], \\ \tilde{z}_2 = z_2/p_2 \end{cases} \tag{7.84}$$

Proof The case where $\theta = 0$ has been analyzed in the risk-neutral setup. Therefore, we focus on $\theta \neq 0$. Since $\theta \neq 0$, it follows from Proposition 7.7 that $p_2 \neq 0$. Thus, one can divide p_1 by p_2, and there is a unique maximizer to the risk-sensitive Hamiltonian and it is given by $u_* = -\frac{b}{r}\frac{p_1}{p_2}$. By strict concavity of H^θ in (x, u, z), it follows that u_* is the unique best-response strategy. We now transform the adjoint by setting $\tilde{p} = \frac{p_1}{p_2}$. Then, by Itô's formula, $\tilde{p}$ solves the following forward SDE:

$$\begin{cases} d\tilde{p} = -\tilde{H}_x dt + \tilde{q}(-\tilde{z}_2 dt + dB) \\ \tilde{p}(0) = q(0)x(0) + \bar{q}(0)[x(0) - m(0)], \end{cases} \tag{7.85}$$

where $\tilde{H}_x = H_x^\theta/p_2 = -a\tilde{p} - qx - \bar{q}(x-m)$, $\tilde{z}_2 = z_2/p_2$, $\tilde{q} = -\tilde{p}\tilde{z}_2 - \tilde{H}_z$. $\tilde{H}_z = -c\tilde{p} - sz$. This completes the proof. ■

The mean of risk-sensitive state trajectory when each decision-maker uses its best-response strategy solves the following backward integral equation:

$$\bar{x}(t) = \bar{x}(T) + \int_t^T ((a+\bar{a})\bar{x} - \frac{b^2}{r}E[\tilde{p}] + cE[z])dt',$$

where

$$E[\tilde{p}(t)] = E[\tilde{p}(0)] + \int_0^t (aE[\tilde{p}] + q\bar{x} - E[\tilde{q}\tilde{z}_2])dt', \ \bar{p}(0) = -q\bar{x}(0).$$

By equating m and $\bar{x}$, we obtain a fixed-point equation. It follows from the above analysis that for each fixed θ, there is a unique risk-sensitive equilibrium when T is sufficiently small.

7.1.6 *Linear-Quadratic Mean-Field-Type Global Optima*

In this subsection, we consider a single decision-maker with the following linear-quadratic optimization problem of mean-field type:

$$\begin{cases} \inf_{u(\cdot)\in\mathscr{U}} \frac{1}{2} E\left[q(T)x^2(T) + \bar{q}(T)[x(T) - E[x(T)]]^2 \right. \\ \quad \left. + \int_0^T q(t)x^2(t) + \bar{q}(t)(x(t) - E[x(t)])^2 + r(t)u^2(t)dt\right], \\ \text{subject to} \\ dx(t) = [ax(t) + \bar{a}E[x(t)] + bu(t)]dt + \sigma dB(t), \\ x(0) \in \mathbb{R}. \end{cases} \tag{7.86}$$

We use tools from mean-field-type optimization to find social optima of the game described in Sect. 7.1.1. Problem (7.86) is fundamentally different from the class of LQ mean-field-type games as studied of the previous subsections. In the mean-field game problem (7.5) of Sect. 7.1.1, the expected value of the state $E[x(t)]$ is replaced by the expected value of equilibrium state $m_*(t) = E[x_*(t)]$, which is generated by an infinite population of individual decision-makers. Therefore, in the mean-field game problem, the decision of a single individual does not influence the mean-field term $m_*(t) = E[x_*(t)]$.

The mean-field-type optimization (7.86) is totally different. There is one single decision-maker. The decision-maker's control action $u(t)$ influences the mean-field term $m^u(t) = E[x^u(t)]$. In (7.86), the term

$$\frac{q(T)}{2}x^2(T) + \frac{\bar{q}(T)}{2}(x(T) - E[x(T)])^2 + \frac{1}{2}\int_0^T qx^2 + \bar{q}(x - E[x])^2 + ru^2dt$$

should read

$$\frac{q(T)}{2}(x^u(T))^2 + \frac{\bar{q}(T)}{2}(x(T) - E[x^u(T)])^2 + \frac{1}{2}\int_0^T q(x^u)^2 + \bar{q}(x - E[x^u])^2 + ru^2dt,$$

which we have simplified for notational purpose.

Let

$$H(x, y, u, p, z) := (ax + \bar{a}y + bu)p + \frac{1}{2}[qx^2 + \bar{q}(x - y)^2 + ru^2] + \sigma z.$$

Then,

$$H_x = ap + qx + \bar{q}(x - y), \qquad H_y = \bar{a}p - \bar{q}(x - y).$$

The first-order adjoint equation is

$$\begin{aligned} dp &= -(H_x + E[H_y])dt + zdB \qquad (7.87)\\ &= -(ap + qx + \bar{q}(x - E[x]) + E[\bar{a}p])dt + zdB \\ p(T) &= q(T)x(T) + \bar{q}(T)(x(T) - E[x(T)]). \end{aligned}$$

The dynamics of the mean state is $\frac{d}{dt}E[x(t)] = (a + \bar{a})E[x(t)] + bE[u(t)]$.

We try a solution in the following form: $p(t) = \beta(t)x(t) + \eta(t)E[x(t)]$, with $\beta(T) = q(T) + \bar{q}(T), \quad \eta(T) = -\bar{q}(T)$.

$$d[\beta x + \eta E[x]] = (\dot{\beta} + a\beta - \frac{b^2}{r}\beta^2)xdt \tag{7.88}$$
$$+(\dot{\eta}E[x] + \eta(\frac{d}{dt}E[x]) + \bar{a}\beta E[x] - \frac{b^2}{r}\eta\beta E[x])dt + \sigma\beta dB.$$

The evolution of the mean states is given by

$$\frac{d}{dt}E[x] = (a+\bar{a})E[x] - \frac{b^2}{r}E[p] = \left[a + \bar{a} - \frac{b^2}{r}(\beta+\eta)\right]E[x],$$

$$d[\beta x + \eta E[x]] = (\dot{\beta} + a\beta - \frac{b^2}{r}\beta^2)xdt$$
$$+(\dot{\eta} + [a + \bar{a} - \frac{b^2}{r}(\beta+\eta)]\eta + \bar{a}\beta - \frac{b^2}{r}\eta\beta)E[x]dt + \sigma\beta dB.$$

On the other hand,

$$dp = -\{(a\beta + q + \bar{q})x + (a\eta - \bar{q} + \bar{a}(\beta+\eta))E[x]\}\, dt + zdB.$$

By identifying the coefficients, we obtain

$$\dot{\beta} + 2a\beta - \frac{b^2}{r}\beta^2 + q + \bar{q} = 0$$
$$\beta(T) = \bar{q}(T) + q(T),$$
$$\left\{\dot{\eta} + 2[(a+\bar{a}) - \frac{b^2}{r}\beta]\eta - \frac{b^2}{r}\eta^2 + 2\bar{a}\beta - \bar{q}\right\} E[x] = 0$$
$$\eta(T) = -q(T).$$

Let $y = \beta + \eta$. Then y solves the Riccati equation:

$$\dot{y} + 2(a+\bar{a})y - \frac{b^2}{r}y^2 + q = 0, \quad y(T) = q(T).$$

The optimal strategy can be expressed as a function of y :

$$u = -\frac{b}{r}(\beta x + \eta E[x]) = -\frac{b}{r}(\beta x + \eta x - \eta x + \eta E[x]) = -\frac{b}{r}[yx + \eta(E[x] - x)].$$

Note that these Riccati-like equations in η are different from the one obtained for the best-response strategy in (7.11) and (7.18). In contrast to the game case, the mean-field-type optimization problem (7.86) does not need restrictive condition for existence of a solution.

7.1.7 Notes

Papers [30, 31] consider differential games with a finite number of decision-makers with linear systems affected by white noise, running cost quadratic in the control and in the displacement of the state from a reference position, and with long-term-average integral cost functional. The author solves an associated system of Hamilton–Jacobi–Bellman and Kolmogorov–Fokker–Planck equations and derives explicit Nash equilibria in the form of linear feedbacks. The author further computes the limit as the number of decision-makers goes to infinity, assuming that they are almost identical and with suitable scalings of the parameters. This provides a quadratic-Gaussian solution to a system of two differential equations. The works in [60, 61] present a comprehensive study of a class of linear-quadratic games. The authors have adopted the adjoint equation approach to investigate the existence of equilibrium strategies. See [116, 117, 159, 179, 181, 195, 197, 200–203, 221–223, 238, 312, 325, 327, 353, 366] for more details on linear-quadratic games with measure-linear cost.

This section has taken the earlier lines of research a step further and lifted it to the level of linear-quadratic mean-field stochastic differential games where the solution concept is that of mean-field Nash equilibrium in state-and-mean-field feedback form. We have done this by providing a set of sufficient conditions for risk-sensitive mean-field Nash equilibrium to satisfy and by establishing connections with other robust mean-field stochastic differential games. A class of risk-sensitive mean-field-type games has been first introduced in [325, 327]. Robust mean-field-type games were introduced in [55]. Detailed investigations of risk-sensitive and robust mean-field-type games can be found in [26, 253, 287].

7.2 Linear-Quadratic Differential Mean-Field-Type Games

Mean-field-type games with linear-quadratic setup has the potential to provide approximate error control, variance reduction, consensus, and risk reduction problems. Due to the mean-field-type structure, the classical solution methods do not apply directly. This section identifies the challenges of mean-field-type games [102, 330, 331, 333] with variance term and provides insights into the methodologies used for finding mean-field-type equilibria.

In Sect. 7.1, we have discussed mean-field games in infinite-population regime with linear-quadratic structure. This section considers basic linear-quadratic mean-field-type games. The key difference between these two frameworks lies in the mean-field term. In the mean-field-type framework:

- decision of a single decision-maker may affect directly the mean-field term,
- the number of decision-makers is not necessarily large,
- e payoff and/or the control actions are decision-maker specific and do not require indistinguishability, interchangeability or symmetry (in distribution) between the decision-makers.

7.2.1 Variance-Aware Game Model

Definition 7.5 A mean-field-type game problem in which

- the state dynamics is linear in the states, expected value of states, control actions, expected values of the controls, and
- the instantaneous cost is quadratic in the states, expected value of states, control actions, expected value of the controls,

is called a *linear-quadratic mean-field-type game*.

Definition 7.6 (McKean–Vlasov Dynamics) A dynamical system is of McKean–Vlasov type if the transition kernel functional (drift, diffusion, jump coefficient functions) involves not only the states and the control actions but also the probability distribution of the state and/or control actions.

Note that the McKean–Vlasov dynamics differ from the Liouville dynamics (Definition 7.2), which are measure-dependent dynamics. As these measures are not necessarily the probability measures of the underlying state, the flow of these dynamics is different. In an infinite-population setting, the state dynamics of a generic deviating decision-maker is not a McKean-Vlasov dynamics but a Liouville dynamics as the infinite-population mean-field may differ from the probability law of the deviating decision-maker [102, 319, 333].

We examine linear-quadratic mean-field-type game problems with several risk-neutral decision-makers. Consider I risk-neutral decision-makers ($I \geq 2$) and $L_i(u_1, \ldots, u_I)$ to be the objective functional of decision-maker i, given by

$$\begin{aligned} L_i(u_1, \ldots, u_I) &= \tfrac{1}{2}q_i(T)x^2(T) + \tfrac{1}{2}\bar{q}_i(T)\,(E[x(T)])^2 \\ &+\tfrac{1}{2}\int_0^T \left\{q_i(t)x^2(t) + \bar{q}_i(t)(E[x(t)])^2 + r_i(t)u_i^2(t) + \bar{r}_i(t)(E[u_i(t)])^2\right\} dt. \end{aligned} \tag{7.89}$$

Then, the best response of decision-maker i to the control action process $u_{-i} := (u_1, \ldots, u_{i-1}, u_{i+1}, \ldots, u_I)$, solves the following risk-neutral linear-quadratic mean-field-type control problem:

$$\begin{cases} \inf_{u_i(\cdot)\in\mathscr{U}_i} E\left[L_i(u_1,\ldots,u_I)\right], \\ \text{subject to} \\ dx(t) = \\ \left\{a(t)x(t)+\bar{a}(t)E[x(t)]+\sum_{i\in\mathscr{I}} b_i(t)u_i(t)+\sum_{i\in\mathscr{I}}\bar{b}_i(t)E[u_i(t)]\right\}dt \\ +\sigma(t)dB(t)+\int_\Theta \mu(t,\theta)\tilde{N}(dt,d\theta), \\ x(0)\in\mathbb{R}, \end{cases} \tag{7.90}$$

where $q_i(t) > 0, q_i(t)+\bar{q}_i(t) > 0, r_i(t) \geq \delta > 0, r_i(t)+\bar{r}_i(t) \geq \delta > 0$ for some δ, and $a(t), \bar{a}(t), b_i(t), \sigma(t)$ are deterministic measurable real-valued functions and where $E[x(t)]$ is the expected value of the state created by all decision-makers under the control action profile $(u_1,\ldots,u_I) \in \prod_{j=1}^I \mathscr{U}_j$. The method below can handle time-varying coefficients. For simplicity, we impose integrability condition on these coefficient functions over the finite horizon $[0,T]$:

$$\int_0^T \left[|a(t)|+|\bar{a}(t)|+|b(t)|+|\bar{b}(t)|+\sigma^2(t)+\int_\Theta \mu^2(t,\theta)\nu(d\theta)\right]dt < +\infty. \tag{7.91}$$

Notice that the state dynamics of (7.90) is of the McKean-Vlasov type and differ from the dynamics in (7.1).

Let $\bar{x}(t) = E[x(t)], \bar{u}_i(t) = E[u_i(t)]$ denote the expected values of the state and control action of decision-maker i, respectively. We rewrite the expected objective functional and the state coefficients in terms of $x-\bar{x}$ and $\bar{x}$:

$$\begin{cases} E[L_i] = \frac{1}{2}[q_i(T)var(x(T))+[q_i(T)+\bar{q}_i(T)]\bar{x}^2(T) \\ \qquad +E\int_0^T q_i(t)var(x(t))+[q_i(t)+\bar{q}_i(t)]\bar{x}^2(t)dt \\ \qquad +E\int_0^T r_i(t)var(u_i(t))+(r_i(t)+\bar{r}_i(t))[\bar{u}_i]^2dt], \\ dx \quad = \{a(t)(x(t)-\bar{x}(t))+(a+\bar{a})\bar{x}(t) \\ \qquad +\sum_{i=1}^I b_i(t)(u_i(t)-\bar{u}_i(t))+\sum_{i=1}^I(b_i(t)+\bar{b}_i(t))\bar{u}_i(t)\Big\}dt \\ \qquad +\sigma(t)dB(t)+\int_\Theta \mu(t,\theta)\tilde{N}(dt,d\theta). \end{cases} \tag{7.92}$$

Note that the expected value of the first term in the integral in L_i can be seen as a weighted variance of the state.

Remark 7.2.1.1 The problem (7.90) is different from the infinite-population mean-field game problem introduced in 7.5. In fact, here the number of decision-makers is not necessarily large, and there is no symmetry in terms of cost functions, control actions, and/or parameters. In (7.90), the mean-field term is $E[x]$ and is influenced by all decision-makers. The fact that a single decision-maker can influence the mean-field term changes completely the optimality equations derived earlier in Sect. 7.1. As the variance is nonlinear in the probability distribution of the state, the expected costs in the problem (7.90) are nonlinear in the probability distribution of the state [102, 319, 333] .

Direct Method

The direct method consists of writing a generic structure of the anticipated optimal cost functional, with unknown deterministic functions to be identified. Inspired by the structure of the terminal cost function, we try a generic solution in a quadratic form. Let $f_i(t,x) = \frac{1}{2}\alpha_i(t)(x-\bar{x})^2 + \frac{1}{2}\beta_i(t)\bar{x}^2 + \gamma_i(t)\bar{x} + \delta_i(t)$, where $\alpha, \beta, \gamma, \delta$ are deterministic functions of time, such that

$$f_i(T, x(T)) = \frac{1}{2}\{q_i(T)E(x(T)-\bar{x}(T))^2 + [q_i(T)+\bar{q}_i(T)]\bar{x}^2(T)\}.$$

By identification of the processes at final time $t = T$, we obtain $\alpha_i(T) = q_i(T)$, $\beta_i(T) = q_i(T) + \bar{q}_i(T)$, $\gamma_i(T) = \delta_i(T) = 0$. Using Itô's formula for jump-diffusion processes to the $f_i(t,x)$, and taking expectation and re-arranging the terms lead to

$$\begin{aligned}
&E[f_i(T,x(T)) - f_i(0,x(0))] = \tfrac{1}{2}E\int_0^T \dot{\alpha}_i(x-\bar{x})^2 + \dot{\beta}_i\bar{x}^2 dt\\
&+\tfrac{1}{2}E\int_0^T 2\beta_i[(a+\bar{a})\bar{x}^2 + \textstyle\sum_{j=1}^I(b_j+\bar{b}_j)\bar{u}_j\bar{x}]dt\\
&+\tfrac{1}{2}E\int_0^T \left\{2a\alpha_i(x-\bar{x})^2 + 2\alpha_i\textstyle\sum_{j=1}^I b_j(u_j-\bar{u}_j)(x-\bar{x})\right\}dt\\
&+\tfrac{1}{2}\int_0^T[\sigma^2 + \int_\Theta \mu^2(t,\theta)\nu(d\theta)]\alpha_i dt\\
&+E\int_0^T \dot{\gamma}_i\bar{x} + \gamma_i[(a+\bar{a})\bar{x} + \textstyle\sum_{j=1}^I(b_j+\bar{b}_j)\bar{u}_j]dt + \int_0^T \dot{\delta}_i dt.
\end{aligned} \tag{7.93}$$

Therefore, we examine optimal strategies in state-and-mean-field feedback form.

7.2.1.1 Optimal Feedback Strategies

The information structure for feedback solution is as follows. The model and the objective functions are assumed to be common knowledge. We assume that the state is perfectly observed. We will show below that the mean-field term is computable (given the initial mean state and the knowledge of the coefficients in the linear state model). If the other decision-makers play their optimal state-and-mean-field feedback strategies, then the functions $\gamma_1, \ldots, \gamma_I$ are identically zero at any time. We compute again $E[L_i - f_i(0,x(0))]$ and complete the squares using the elements of $\{x-\bar{x},\ \bar{x}\}$:

$$\begin{aligned}
&E[L_i - f_i(0,x(0))]\\
&= \tfrac{1}{2}(q_i(T)-\alpha_i(T))E(x(T)-\bar{x}(T))^2\\
&\quad +\tfrac{1}{2}[q_i(T)+\bar{q}_i(T)-\beta_i(T)]\bar{x}^2(T)\\
&+\tfrac{1}{2}E\int_0^T[\dot{\alpha}_i + 2a\alpha_i - \tfrac{b_i^2}{r_i}\alpha_i^2 - 2\alpha_i\textstyle\sum_{j\neq i}\tfrac{b_j^2}{r_j}\alpha_j + q_i](x-\bar{x})^2\\
&+\tfrac{1}{2}E\int_0^T\Big\{\dot{\beta}_i + 2\beta_i(a+\bar{a}) - \beta_i^2\tfrac{(b_i+\bar{b}_i)^2}{r_i+\bar{r}_i}\\
&\qquad -2\beta_i\textstyle\sum_{j\neq i}\tfrac{(b_j+\bar{b}_j)^2}{r_j+\bar{r}_j}\beta_j + q_i + \bar{q}_i\Big\}\bar{x}^2 dt\\
&+\tfrac{1}{2}E\int_0^T r_i[u_i - \bar{u}_i + \tfrac{b_i}{r_i}\alpha_i(x-\bar{x})]^2 dt\\
&+\tfrac{1}{2}E\int_0^T(r_i+\bar{r}_i)[\bar{u}_i + \beta_i\tfrac{(b_i+\bar{b}_i)}{r_i+\bar{r}_i}\bar{x}]^2 dt\\
&+\tfrac{1}{2}E\int_0^T[\sigma^2 + \int_\Theta \mu^2(t,\theta)\nu(d\theta)]\,\alpha_i dt,
\end{aligned} \tag{7.94}$$

It follows that

$$\begin{cases} \inf_{u_i \in \mathscr{U}_i} E[L_i] \\ = \frac{1}{2}\alpha_i(0) var(x(0)) + \frac{1}{2}\beta_i(0)(E[x(0)])^2 + \frac{1}{2}\int_0^T [\sigma^2 + \int_\Theta \mu^2(t,\theta)\nu(d\theta)]\, \alpha_i dt, \\ u_i^* = -\frac{b_i}{r_i}\alpha_i(x - \bar{x}) - \beta_i \frac{(b_i+\bar{b}_i)}{r_i+\bar{r}_i}\bar{x}, \\ \dot{\alpha}_i + 2a\alpha_i - \frac{b_i^2}{r_i}\alpha_i^2 - 2\alpha_i \sum_{j\neq i} \frac{b_j^2}{r_j}\alpha_j + q_i = 0, \\ \alpha_i(T) = q_i(T), \\ \dot{\beta}_i + 2\beta_i(a+\bar{a}) - \beta_i^2 \frac{(b_i+\bar{b}_i)^2}{r_i+\bar{r}_i} - 2\beta_i \sum_{j\neq i} \frac{(b_j+\bar{b}_j)^2}{r_j+\bar{r}_j}\beta_j + q_i + \bar{q}_i = 0, \\ \beta_i(T) = q_i(T) + \bar{q}_i(T), \\ \bar{x}(t) = \bar{x}(0) e^{\int_0^t \{(a+\bar{a}) - \sum_{i=1}^I \beta_i \frac{(b_i+\bar{b}_i)^2}{r_i+\bar{r}_i}\} dt'}, \end{cases} \tag{7.95}$$

provides a mean-field-type Nash equilibrium in feedback strategies.

Feedback Solutions Differ from Open-Loop Solutions
These Riccati equations are different from those of open-loop control strategies as we will see below. The coefficient of the coupling terms $\beta_i\beta_j$, $\alpha_i\alpha_j$, is different, reflecting the coupling through the state and the mean state.

Notice that the optimal strategy is in state-and-mean-field feedback form, which is different than the standard LQG game solution. As $\bar{a}, \bar{b}, \bar{r}, \bar{q}$ vanish in (7.95), one gets the Nash equilibrium of the corresponding stochastic differential game in closed-loop strategies with $\alpha_i = \beta_i$, and u_i becomes mean-field-free. When the diffusion coefficient σ and the jump rate μ vanish, we obtain the noiseless deterministic game problem, and the optimal strategy solution will be given by the equation in β_i because $x - \bar{x} = 0$ in the deterministic case.

How to Feedback the Mean-Field Term $E[x(t)]$?
Here the mean-field term can be explicitly computed if the initial mean state x_0 is given and the model known. Indeed, we have

$$E[x(t)] = \bar{x}(t) = \bar{x}(0) e^{\int_0^t \{(a+\bar{a}) - \sum_{i=1}^I \beta_i \frac{(b_i+\bar{b}_i)^2}{r_i+\bar{r}_i}\} dt'},$$

and the semi-explicit equilibrium solution $((u_i^*)_{i\in\mathscr{I}}, \bar{x}(t))$ is provided by (7.95).

7.2.1.2 Cooperative Solutions

We present in this subsection the global optimum and Nash bargaining solution of the game. We first consider the fully cooperative scenario where all the decision-makers jointly optimize a single global objective $L_0 := \sum_i L_i$ given by

$$\begin{cases} \inf_{(u_1,\ldots,u_I)} E \sum_i L_i, \\ \\ dx = \{a(x-\bar{x}) + (a+\bar{a})\bar{x} \\ + \sum_{i=1}^I b_i(u_i - \bar{u}_i) + \sum_{i=1}^I (b_i + \bar{b}_i)\bar{u}_i \Big\} dt \\ + \sigma dB(t) + \int_\Theta \mu(t,\theta)\tilde{N}(dt, d\theta), \\ x(0) = x_0. \end{cases} \tag{7.96}$$

Following the same methodology as above with $q_0 = \sum_i q_i$, $\bar{q}_0 = \sum_i \bar{q}_i$ and $f_0(t,x) = \frac{1}{2}\alpha_0(t)(x-\bar{x})^2 + \frac{1}{2}\beta_0(t)\bar{x}^2$, we arrive at the following completion of squares:

$$\begin{cases} \inf_{(u_1,\ldots,u_I)} E \sum_i [L_i] = \frac{1}{2}\alpha_0(0) var(x(0)) + \frac{1}{2}\beta_0(0)[Ex(0)]^2 \\ + \frac{1}{2}\int_0^T [\sigma^2(t) + \int_\Theta \mu^2(t,\theta)\nu(d\theta)]\, \alpha_0(t) dt, \\ u_i^* = -\frac{b_i}{r_i}\alpha_0(x-\bar{x}) - \beta_0 \frac{(b_i+\bar{b}_i)}{r_i+\bar{r}_i}\bar{x}, \\ \dot{\alpha}_0 + 2a\alpha_0 - \alpha_0^2 \sum_i \frac{b_i^2}{r_i} + q_0 = 0, \\ \alpha_0(T) = q_0(T) = \sum_{i=1}^I q_i(T), \\ \dot{\beta}_0 + 2\beta_0(a+\bar{a}) - \beta_0^2 \sum_{i=1}^I \frac{(b_i+\bar{b}_i)^2}{r_i+\bar{r}_i} + q_0 + \bar{q}_0 = 0, \\ \beta_0(T) = \sum_{i=1}^I q_i(T) + \bar{q}_i(T). \\ \bar{x}(t) = \bar{x}(0) e^{\int_0^t \{(a+\bar{a}) - \beta_0 \sum_{i=1}^I \frac{(b_i+\bar{b}_i)^2}{r_i+\bar{r}_i}\} dt'}. \end{cases} \tag{7.97}$$

The functions α_0, β_0 are explicitly given by

$$\begin{cases} S := \sum_{i=1}^I \frac{b_i^2}{r_i}, \\ \alpha_0(t) = \frac{a}{S} + \sqrt{\frac{q_0}{S} + \frac{a^2}{S^2}} \left(-1 + \frac{q_0(T) - \frac{a}{S} + \sqrt{\frac{q_0}{S} + \frac{a^2}{S^2}}}{\Gamma} \right), \\ \Gamma := \frac{1}{2}(q_0(T) - \frac{a}{S} + \sqrt{\frac{q_0}{S} + \frac{a^2}{S^2}}) \\ -\frac{1}{2}(q_0(T) - \frac{a}{S} - \sqrt{\frac{q_0}{S} + \frac{a^2}{S^2}}) e^{-2(T-t)\sqrt{\frac{q_0}{S} + \frac{a^2}{S^2}}}, \\ \tilde{S} := \sum_{i=1}^I \frac{(b_i+\bar{b}_i)^2}{r_i+\bar{r}_i}, \\ \beta_0(t) = \frac{a+\bar{a}}{\tilde{S}} + \sqrt{\frac{q_0+\bar{q}_0}{\tilde{S}} + \frac{a^2}{\tilde{S}^2}} \left(-1 + \frac{q_0(T) + \bar{q}_0(T) - \frac{a+\bar{a}}{\tilde{S}} + \sqrt{\frac{q_0+\bar{q}_0}{\tilde{S}} + \frac{(a+\bar{a})^2}{\tilde{S}^2}}}{\tilde{\Gamma}} \right), \\ \tilde{\Gamma} := \frac{1}{2}(q_0(T) + \bar{q}_0(T) - \frac{a+\bar{a}}{\tilde{S}} + \sqrt{\frac{q_0+\bar{q}_0}{\tilde{S}} + \frac{(a+\bar{a})^2}{\tilde{S}^2}}) \\ -\frac{1}{2}(q_0(T) + \bar{q}_0(T) - \frac{a+\bar{a}}{\tilde{S}} - \sqrt{\frac{q_0+\bar{q}_0}{\tilde{S}} + \frac{(a+\bar{a})^2}{\tilde{S}^2}}) \times \\ e^{-2(T-t)\sqrt{\frac{q_0+\bar{q}_0}{\tilde{S}} + \frac{(a+\bar{a})^2}{\tilde{S}^2}}}. \end{cases} \tag{7.98}$$

The global optimum cost in the fully cooperative case is

$$L_0=\frac{1}{2}\alpha_0(0)var(x(0))+\frac{1}{2}\beta_0(0)(E[x(0)])^2+\frac{1}{2}\int_0^T[\sigma^2(t)+\int_\Theta\mu^2(t,\theta)\nu(d\theta)]\,\alpha_0(t)dt. \tag{7.99}$$

7.2.1.3 Nash Bargaining Solution

Mean-field-type bargaining theory deals with the situation in which decision-makers can realize—through cooperation—other better outcomes than the one that becomes effective when they do not cooperate. This noncooperative outcome is called the threat point $L^{NE}=(L_1^{NE},\ldots,L_I^{NE})$. The question is which outcome might the decision-makers possibly agree to. Let $\mathscr{V}$ be the set of feasible outcomes of the benefit of bargaining. We assume that, if the agents unanimously agree upon a point $V=(V_1,\ldots,V_I)\in\mathscr{V}$, they obtain V. Otherwise they obtain $L^{NE}=(L_1^{NE},\ldots,L_I^{NE})$. This presupposes that each decision-maker can enforce the threat point, when she does not agree with the proposal made. Another threat point could be a maximin point of the game. The outcome V the decision-makers will finally agree on is called the solution of the bargaining problem.

For a given set $\mathscr{V}$, the Nash bargaining solution selects the point at which the product of gains from d is maximal:

$$NBS(\mathscr{V},L^{NE})=\arg\max_{v\in\mathscr{V}}\{\prod_{i\in\mathscr{N}}[L_i^{NE}-v_i]\}.$$

Here we exploit the convexity of the functional $u\mapsto\langle w,L(u)\rangle:=\sum_{i=1}^{i}w_iL_i(u)$ for any given $w=(w_1,\ldots,w_I)\in\mathbb{R}^I_{++}$ such that $\sum_{i=1}^I w_i=1$ to reach any point in the Pareto frontier of the game. The maximization (in w) of the product $P:=\prod_{i=1}^I[L_i^{NE}-L_i(\hat{u}(w))]$ yields

$$\sum_{i=1}^{I}\frac{\partial}{\partial w_j}L_i(\hat{u}(w)).\{\prod_{j\neq i}[L_j^{NE}-L_j(\hat{u}(w))]\}=c$$

This is equivalent to

$$\sum_{i=1}^{I}\frac{P}{L_i^{NE}-L_i(\hat{u}(w))}\frac{\partial}{\partial w_j}L_i(\hat{u}(w))=c.$$

We set $y_i:=\frac{\frac{P}{L_i^{NE}-L_i(\hat{u}(w))}}{\sum_{i=1}^I\left(\frac{P}{L_i^{NE}-L_i(\hat{u}(w))}\right)}$. Then, it follows that $\sum_{i=1}^I y_i\frac{\partial}{\partial w_j}L_i(\hat{u}(w))=\frac{c}{\sum_{i=1}^I\left(\frac{P}{L_i^{NE}-L_i(\hat{u}(w))}\right)}$. Moreover $y_i\geq 0,\ \sum_{i=1}^I y_i=1.$

Assume that the matrix $(\frac{\partial}{\partial w_j} L_i(\hat{u}(w)))_{(i,j)\in\mathcal{N}^2}$ has at least rank $I-1$. Then the Nash bargaining solution is explicitly given by $V = (L_1(\hat{u}(w^*)), \ldots, L_I(\hat{u}(w^*)))$ with the weight

$$w_i^* = \frac{\prod_{j\neq i}[L_j^{NE} - L_j(\hat{u}(w^*))]}{\sum_{k=1}^{I}\prod_{j\neq k}[L_j^{NE} - L_j(\hat{u}(w^*))]} = y_i(w^*)$$

where the optimal bargaining strategy profile is $\hat{u}(w) \in \arg\min_u\{\langle w, L(u)\rangle\}$. It remains to compute the functional $\hat{u}(w) = (\hat{u}_1(w), \ldots, \hat{u}_I(w))$:

$$\begin{aligned}
&\inf_{(u_1,\ldots,u_I)} E\sum_i w_i L_i,\\
&dx = \{a(x-\bar{x}) + (a+\bar{a})\bar{x}\\
&+\sum_{i=1}^{I} b_i(u_i - \bar{u}_i) + \sum_{i=1}^{I}(b_i+\bar{b}_i)\bar{u}_i\Big\}\, dt\\
&+\sigma dB + \int_{\Theta}\mu(t,\theta)\tilde{N}(dt,d\theta),\\
&x(0) = x_0.
\end{aligned} \tag{7.100}$$

Following the same methodology as above with $q_0 = \sum_i w_i q_i$, $\bar{q}_0 = \sum_i w_i \bar{q}_i$, we obtain

$$\begin{aligned}
&\inf_{(u_1,\ldots,u_I)} E\sum_i w_i L_i =\\
&\tfrac{1}{2}\alpha_0(0)var(x(0)) + \tfrac{1}{2}\beta_0(0)[\bar{x}(0)]^2\\
&+\int_0^T[\sigma^2 + \int_{\Theta}\mu^2(t,\theta)\nu(d\theta)]\tfrac{\alpha_0}{2}dt,\\
&\hat{u}_i(w) = -\tfrac{b_i}{w_i r_i}\alpha_0(x-\bar{x}) - \tfrac{(b_i+\bar{b}_i)}{w_i(r_i+\bar{r}_i)}(\beta_0\bar{x}),\\
&\dot{\alpha}_0 + 2a\alpha_0 - \alpha_0^2\sum_i \tfrac{b_i^2}{w_i r_i} + \sum_i w_i q_i = 0,\\
&\alpha_0(T) = \sum_i w_i q_i(T),\\
&\dot{\beta}_0 + 2\beta_0(a+\bar{a}) - \beta_0^2\sum_i \tfrac{(b_i+\bar{b}_i)^2}{w_i(r_i+\bar{r}_i)} + \sum_i w_i(q_i+\bar{q}_i) = 0,\\
&\beta_0(T) = \sum_i w_i(q_i + \bar{q}_i(T)),\\
&\bar{x}(t) = \bar{x}(0)e^{[a+\bar{a}-\beta_0\sum_{i=1}^{I}\frac{(b_i+\bar{b}_i)^2}{w_i(r_i+\bar{r}_i)}]dt}.
\end{aligned} \tag{7.101}$$

7.2.2 Linear-Quadratic Robust Mean-Field-Type Games

We now consider a robust mean-field-type game with two decision-makers. Decision-maker 1 minimizes the cost functional with respect to u_1, and Decision-maker 2 maximizes it with respect to u_2. The minmax problem of mean-field type is given by

$\inf_{u_1} \sup_{u_2} E[L(u_1, u_2)],$

$$\left\{\begin{array}{l} dx = \{a(t)(x(t) - \bar{x}(t)) + (a(t) + \bar{a}(t))\bar{x}(t) \\ + \sum_{i=1}^{2} b_i(t)(u_i(t) - \bar{u}_i(t)) + \sum_{i=1}^{2}(b_i(t) + \bar{b}_i(t))\bar{u}_i(t)\Big\} dt \\ +\sigma(t)dB(t) + \int_{\Theta} \mu(t,\theta)\tilde{N}(t, d\theta), \\ x(0) = x_0 \in L^2(\mathbb{R}), \end{array}\right. \tag{7.102}$$

where the cost functional is

$$\begin{array}{l} L(u_1, u_2) = \frac{1}{2}[q(T)(x(T) - \bar{x}(T))^2 + [q(T) + \bar{q}(T)]\bar{x}^2(T) \\ \qquad + \int_0^T q(t)(x(t) - \bar{x}(t))^2 + [q(t) + \bar{q}(t)]\bar{x}^2(t)dt \\ + \int_0^T r_1(t)(u_1(t) - \bar{u}_1(t))^2 + (r_1(t) + \bar{r}_1(t))[\bar{u}_1(t)]^2 dt \\ + \int_0^T r_2(t)(u_2(t) - \bar{u}_2(t))^2 + (r_2(t) + \bar{r}_2(t))[\bar{u}_2(t)]^2 dt]. \end{array} \tag{7.103}$$

The risk-neutral robust mean-field-type equilibrium problem we are concerned with is characterization of the processes $(x^*, u_1^*, u_2^*, E[x^*])$ such that for the decision-maker, $\bar{u}_1^*$ is the minimizer and u_2^* is the maximizer of the best-response problem (7.102) and establishing the result that the expected value of the resulting common state created by all the decision-makers is $\bar{x}^*(t) = E[x^*(t)]$.

Below we solve Problem (7.102) for $r_1(t) > 0, \bar{r}_1(t) > 0, r_2(t) < 0, \ \bar{r}_2(t) < 0:$

$$\begin{array}{l} E[f(T, x(T)) - f(0, x(0))] \\ = \frac{1}{2}E\int_0^T \{\dot{\alpha} + 2a\alpha\}(x - \bar{x})^2 dt + \{\dot{\beta} + 2(a + \bar{a})\beta\}\bar{x}^2 dt \\ +\frac{1}{2}E\int_0^T 2\alpha \sum_{i=1}^{2} b_i(u_i - \bar{u}_i)(x - \bar{x})dt \\ +\frac{1}{2}E\int_0^T 2\beta\sum_{i=1}^{2}(b_i + \bar{b}_i)\bar{u}_i\bar{x}dt \\ +\frac{1}{2}\int_0^T [\sigma^2 + \int_{\Theta} \mu^2(t,\theta)\nu(d\theta)]\, \alpha dt \end{array} \tag{7.104}$$

$$\begin{array}{l} E[L - f(0, x(0))] \\ = \frac{1}{2}[(q(T) - \alpha(T))]E(x(T) - \bar{x}(T))^2 \\ +\frac{1}{2}[q(T) + \bar{q}(T) - \beta(T)]\bar{x}^2(T) \\ +\frac{1}{2}E\int_0^T \{\dot{\alpha} + 2a\alpha - (\frac{b_1^2}{r_1} + \frac{b_2^2}{r_2})\alpha^2 + q\}(x - \bar{x})^2 dt \\ +\frac{1}{2}E\int_0^T \Big\{\dot{\beta} + 2(a + \bar{a})\beta \\ \quad -(\frac{(b_1+\bar{b}_1)^2}{r_1+\bar{r}_1} + \frac{(b_2+\bar{b}_2)^2}{r_2+\bar{r}_2})\beta^2 + q + \bar{q}\Big\}\bar{x}^2 dt \\ +\frac{1}{2}E\int_0^T r_1[u_1 - \bar{u}_1 + \frac{b_1}{r_1}\alpha(x - \bar{x})]^2\, dt \\ +\frac{1}{2}E\int_0^T (r_1 + \bar{r}_1)[\bar{u}_1 + \frac{(b_1+\bar{b}_1)}{r_1+\bar{r}_1}\beta\bar{x}]^2\, dt \\ +\frac{1}{2}E\int_0^T r_2[u_2 - \bar{u}_2 + \frac{b_2}{r_2}\alpha(x - \bar{x})]^2 dt \\ +\frac{1}{2}E\int_0^T (r_2 + \bar{r}_2)[\bar{u}_2 + \frac{(b_2+\bar{b}_2)}{r_2+\bar{r}_2}\beta\bar{x}]^2 dt \\ +\frac{1}{2}\int_0^T [\sigma^2 + \int_{\Theta} \mu^2(t,\theta)\nu(d\theta)]\, \alpha dt, \end{array} \tag{7.105}$$

where we have used the following completion of squares:

$$
\begin{aligned}
& r_1(u_1-\bar{u}_1)^2+(r_1+\bar{r}_1)[\bar{u}_1]^2 \\
& +2\alpha b_1(u_1-\bar{u}_1)(x-\bar{x})+2\beta(b_1+\bar{b}_1)\bar{u}_1\bar{x} \\
& =r_1[u_1-\bar{u}_1+\tfrac{b_1}{r_1}\alpha(x-\bar{x})]^2-\tfrac{b_1^2}{r_1}\alpha^2(x-\bar{x})^2 \\
& +(r_1+\bar{r}_1)[\bar{u}_1+\tfrac{(b_1+\bar{b}_1)}{r_1+\bar{r}_1}\beta\bar{x}]^2-\tfrac{(b_1+\bar{b}_1)^2}{r_1+\bar{r}_1}\beta^2\bar{x}^2, \\
& r_2(u_2-\bar{u}_2)^2+(r_2+\bar{r}_2)[\bar{u}_2]^2 \\
& +2\alpha b_2(u_2-\bar{u}_2)(x-\bar{x})+2\beta(b_2+\bar{b}_2)\bar{u}_2\bar{x} \\
& =r_2[u_2-\bar{u}_2+\tfrac{b_2}{r_2}\alpha(x-\bar{x})]^2-\tfrac{b_2^2}{r_2}\alpha^2(x-\bar{x})^2 \\
& +(r_2+\bar{r}_2)[\bar{u}_2+\tfrac{(b_2+\bar{b}_2)}{r_2+\bar{r}_2}\beta\bar{x}]^2-\tfrac{(b_2+\bar{b}_2)^2}{r_2+\bar{r}_2}\beta^2\bar{x}^2.
\end{aligned}
\tag{7.106}
$$

It follows that the equilibrium solution is

$$
\begin{aligned}
& \inf_{u_1}\sup_{u_2} E[L(u_1,u_2)]=\tfrac{1}{2}\alpha(0)var(x(0))+\tfrac{1}{2}\beta(0)[E(x(0))]^2 \\
& +\tfrac{1}{2}\int_0^T[\sigma^2(t)+\int_\Theta \mu^2(t,\theta)\nu(d\theta)]\,\alpha(t)dt, \\
& u_1^*=-\tfrac{b_1}{r_1}\alpha(x-\bar{x})-\tfrac{(b_1+\bar{b}_1)}{r_1+\bar{r}_1}\beta\bar{x}, \\
& u_2^*=-\tfrac{b_2}{r_2}\alpha(x-\bar{x})-\tfrac{(b_2+\bar{b}_2)}{r_2+\bar{r}_2}\beta\bar{x}, \\
& \dot{\alpha}+2a\alpha-(\tfrac{b_1^2}{r_1}+\tfrac{b_2^2}{r_2})\alpha^2+q=0, \\
& \alpha(T)=q(T), \\
& \dot{\beta}+2(a+\bar{a})\beta-(\tfrac{(b_1+\bar{b}_1)^2}{r_1+\bar{r}_1}+\tfrac{(b_2+\bar{b}_2)^2}{r_2+\bar{r}_2})\beta^2+q+\bar{q}=0, \\
& \beta(T)=q(T)+\bar{q}(T), \\
& \bar{x}(t)=\bar{x}(0)e^{\int_0^t\{(a+\bar{a})-\beta\sum_{i=1}^2\frac{(b_i+\bar{b}_i)^2}{r_i+\bar{r}_i}\}dt'}.
\end{aligned}
\tag{7.107}
$$

When $r_1>0,\ \bar{r}_1\geq 0,\ r_2<0, \bar{r}_2\leq 0,\ \ S_2=\sum_{i=1}^2\frac{b_i^2}{r_i}>0$ and $\tilde{S}_2=\sum_{i=1}^2\frac{(b_i+\bar{b}_i)^2}{r_i+\bar{r}_i}>0$, are independent of time in $[0,T)$, then the functions α,β are explicitly given by

$$
\begin{aligned}
& \alpha(t)=\tfrac{a}{S_2}+\sqrt{\tfrac{q_0}{S_2}+\tfrac{a^2}{S_2^2}}\left(-1+\frac{q_0(T)-\frac{a}{S_2}+\sqrt{\frac{q_0}{S_2}+\frac{a^2}{S_2^2}}}{\Gamma}\right), \\
& \Gamma:=\tfrac{1}{2}(q_0(T)-\tfrac{a}{S_2}+\sqrt{\tfrac{q_0}{S_2}+\tfrac{a^2}{S_2^2}})-\tfrac{1}{2}(q_0(T)-\tfrac{a}{S_2}-\sqrt{\tfrac{q_0}{S_2}+\tfrac{a^2}{S_2^2}})e^{-2(T-t)\sqrt{\frac{q_0}{S_2}+\frac{a^2}{S_2^2}}}, \\
& \beta(t)=\tfrac{a+\bar{a}}{\tilde{S}_2}+\sqrt{\tfrac{q_0+\bar{q}_0}{\tilde{S}_2}+\tfrac{a^2}{\tilde{S}_2^2}}\left(-1+\frac{q_0(T)+\bar{q}_0(T)-\frac{a+\bar{a}}{\tilde{S}_2}+\sqrt{\frac{q_0+\bar{q}_0}{\tilde{S}_2}+\frac{(a+\bar{a})^2}{\tilde{S}_2^2}}}{\tilde{\Gamma}}\right), \\
& \tilde{\Gamma}:=\tfrac{1}{2}(q_0(T)+\bar{q}_0(T)-\tfrac{a+\bar{a}}{\tilde{S}_2}+\sqrt{\tfrac{q_0+\bar{q}_0}{\tilde{S}_2}+\tfrac{(a+\bar{a})^2}{\tilde{S}_2^2}}) \\
& -\tfrac{1}{2}(q_0(T)+\bar{q}_0(T)-\tfrac{a+\bar{a}}{\tilde{S}_2}-\sqrt{\tfrac{q_0+\bar{q}_0}{\tilde{S}_2}+\tfrac{(a+\bar{a})^2}{\tilde{S}_2^2}})e^{-2(T-t)\sqrt{\frac{q_0+\bar{q}_0}{\tilde{S}_2}+\frac{(a+\bar{a})^2}{\tilde{S}_2^2}}}.
\end{aligned}
\tag{7.108}
$$

Notice that under the conditions $r_1(t) > \delta,\ \bar{r}_1(t) > \delta,\ r_2(t) < -\delta, \bar{r}_2(t) < -\delta,\ \delta > 0,\ S_2(t) > 0,\ \tilde{S}_2(t) > 0$, the minmax solution is also a maximin solution: that is, (u_1^*, u_2^*) constitutes a saddle point for the zero-sum game satisfying It solves

$$E[L(u_1^*, u_2)] \leq E[L(u_1^*, u_2^*)] \leq E[L(u_1, u_2^*)],\ \forall (u_1, u_2) \in \mathcal{U}_1 \times \mathcal{U}_2. \tag{7.109}$$

The value of the game is

$$\begin{aligned} E[L(u_1^*, u_2^*)] = \tfrac{1}{2}\alpha(0)var(x(0)) + \tfrac{1}{2}\beta(0)[E[x(0)]]^2 \\ +\tfrac{1}{2}\int_0^T [\sigma^2 + \int_\Theta \mu^2(t,\theta)\nu(d\theta)]\,\alpha(t)dt. \end{aligned} \tag{7.110}$$

where $var(x(0))$ is the variance of $x(0)$.

Dynamic Programming Principle
We now verify the validity of the results of the direct method above using dynamic programming. Due to the presence of the expected value of the optimal state in the drift, the flow property, i.e., the equality $x^{t_0,x_0}(t) = x^{\tau, x^{t_0,x_0}(\tau)}(t)$, may not hold.

We consider the probability distribution $m(t, dx)$ of $x(t)$ as a state and reformulate the game problem as an infinite-dimensional problem with the state $m(t, dx)$. One can write the cost functional in terms of the probability distribution $m(t, dx)$. This leads to a dynamic programming principle in infinite dimensions. Below we use functional derivatives with respect to m. The Bellman equilibrium system (in infinite dimensions) is

$$\hat{V}_{i,t}(t,m) + \int_x H_i(x, m, \hat{V}_{i,m}, \hat{V}_{i,xm}, \hat{V}_{i,xxm})m(dx) = 0,$$

where the terminal equilibrium payoff functional at time T is

$$\hat{V}_i(T,m) = \tfrac{1}{2}q(T)\int_y (y-\bar{m})^2\, m(dy) + \tfrac{1}{2}[q(T)+\bar{q}(T)]\left[\int_y y\, m(dy)\right]^2, \tag{7.111}$$

and the integrand Hamiltonian is

$$\begin{aligned} H_i(x, m, \hat{V}_{i,m}, \hat{V}_{i,xm}, \hat{V}_{i,xmm}) = \inf_{u_i}\Big\{\tfrac{1}{2}q_i(x-\bar{x})^2 + \tfrac{1}{2}[q_i+\bar{q}_i]\bar{x}^2 \\ +\tfrac{1}{2}r_i(u_i-\bar{u}_i)^2 + \tfrac{1}{2}(r_i+\bar{r}_i)[\bar{u}_i]^2 + [a(x-\bar{x}) + (a+\bar{a})\bar{x}]\hat{V}_{i,xm} \\ + [\textstyle\sum_{i=1}^I b_i(u_i-\bar{u}_i) + \sum_{i=1}^I (b_i+\bar{b}_i)\bar{u}_i]\hat{V}_{i,xm}\Big\} \\ +\tfrac{\sigma^2}{2}\hat{V}_{i,xxm} + \int_\Theta [\hat{V}_{i,m}(t, x+\mu) - \hat{V}_{i,m} - \mu\hat{V}_{i,xm}]\nu(d\theta). \end{aligned} \tag{7.112}$$

Inspired from the structure of the final payoff $\hat{V}_i(T,m)$, we choose a guess functional in the following form:

$$\hat{V}_i(t,m) = \tfrac{\alpha_i}{2}\int_y (y-\bar{m})^2\, m(dy) + \tfrac{\beta_i}{2}\left[\int_y y\, m(dy)\right]^2 + \delta_i \int_y m(dy). \tag{7.113}$$

Note that the term $\int_y m(dy)$ plays a role when studying the variation of $\hat{V}_i$ in m.

Consider another measure $\tilde{m} \in L^2$ and differentiate $\hat{V}_i(t, m + \epsilon\tilde{m})$ with respect to scalar ϵ to obtain

$$\begin{array}{l} \frac{d}{d\epsilon}\hat{V}_i(t, m + \epsilon\tilde{m}) \\ = \frac{\alpha_i}{2}\int_y (y - \bar{m} - \epsilon\bar{\tilde{m}})^2\, \tilde{m}(dy) \\ + \frac{2\beta_i\bar{\tilde{m}}}{2}\left[\int_y y\, m(dy)\right] + \delta_i \int_y \tilde{m}(dy). \end{array} \tag{7.114}$$

We then deduce the following equalities:

$$\begin{array}{l} \hat{V}_{i,m}[t, \tilde{m}](x) \\ = \frac{\alpha_i}{2}\left[x - \int_y y\, m(dy)\right]^2 + \beta_i x\left[\int_y y\, m(dy)\right] + \delta_i, \\ \hat{V}_{i,xm}[t, m](x) = \\ \alpha_i\left[x - \int_y y\, m(dy)\right] + \beta_i\left[\int_y y\, m(dy)\right], \\ \hat{V}_{i,xxm}[t, m](x) = \alpha_i, \\ \hat{V}_{i,m}(t, x + \mu) - \hat{V}_{i,m} - \mu\hat{V}_{i,xm} = \frac{1}{2}\alpha_i\mu^2. \end{array} \tag{7.115}$$

The optimal strategies are then

$$\begin{array}{l} u_i^* = -\frac{b_i + \bar{b}_i}{r_i + \bar{r}_i}\beta_i\left[\int_y y\, m(t, dy)\right] - \frac{b_i}{r_i}\alpha_i\left[x - \int_y y\, m(t, dy)\right], \\ \bar{u}_i^* = -\frac{b_i + \bar{b}_i}{r_i + \bar{r}_i}\beta_i\left[\int_y y\, m(t, dy)\right], \end{array} \tag{7.116}$$

which are exactly the expressions of the optimal strategies obtained in (7.95):

$$\begin{array}{l} \tilde{H}_i(x, m, \hat{V}_{i,m}, \hat{V}_{i,xm}, \hat{V}_{i,xmm}) \\ = \frac{1}{2}q_i(x - \bar{x})^2 + \frac{1}{2}[q_i + \bar{q}_i]\bar{x}^2 \\ + a(x - \bar{x})(\hat{V}_{i,xm} - E\hat{V}_{i,xm}) + (a + \bar{a})\bar{x}E\hat{V}_{i,xm} \\ + \frac{\sigma^2}{2}\hat{V}_{i,xxm} + \int_\Theta [\hat{V}_{i,m}(t, x + \mu) - \hat{V}_{i,m} - \mu\hat{V}_{i,xm}]\nu(d\theta) \\ + \inf_{u_i}\left\{\frac{1}{2}r_i(u_i - \bar{u}_i)^2 + \frac{1}{2}(r_i + \bar{r}_i)[\bar{u}_i]^2\right. \\ + [\hat{V}_{i,xm} - E\hat{V}_{i,xm}]\sum_{j=1}^I b_j(u_j - \bar{u}_j) \\ \left. + [E\hat{V}_{i,xm}]\sum_{j=1}^I (b_j + \bar{b}_j)\bar{u}_j\right\} \\ = \frac{1}{2}\{2a\alpha_i - 2\alpha_i\sum_{j=1}^I \frac{b_j^2}{r_j}\alpha_j + \frac{b_i^2}{r_i}\alpha_i^2 + q_i\}(x - \bar{x})^2 \\ + \frac{1}{2}\{2(a + \bar{a})\beta_i - 2\beta_i\sum_{j=1}^I \frac{(b_j + \bar{b}_j)^2}{(r_j + \bar{r}_j)}\beta_j + \frac{(b_i + \bar{b}_i)^2}{r_i + \bar{r}_i}\beta_i^2 + q_i + \bar{q}_i\}\bar{x}^2 \\ + \frac{1}{2}\alpha_i[\sigma^2 + \int_\Theta \mu^2(t, \theta)\nu(d\theta)] \end{array} \tag{7.117}$$

Using the time derivative of $\hat{V}_i(t, m)$,

$$\hat{V}_{i,t} = \frac{\dot{\alpha}_i}{2} \int_y (y - \bar{m})^2 \, m(dy) + \frac{\dot{\beta}_i}{2} \left[\int_y y \, m(dy)\right]^2 + \dot{\delta}_i \tag{7.118}$$

and identifying the coefficients, we arrive at

$$\begin{aligned} &\dot{\alpha}_i + 2a\alpha_i - 2\alpha_i \sum_{j=1}^{I} \frac{b_j^2}{r_j}\alpha_j + \frac{b_i^2}{r_i}\alpha_i^2 + q_i = 0,\\ &\alpha_i(T) = q_i(T),\\ &\dot{\beta}_i + 2(a + \bar{a})\beta_i - 2\beta_i \sum_{j=1}^{I} \frac{(b_j+\bar{b}_j)^2}{(r_j+\bar{r}_j)}\beta_j\\ &\quad + \frac{(b_i+\bar{b}_i)^2}{r_i+\bar{r}_i}\beta_i^2 + q_i + \bar{q}_i = 0,\\ &\beta_i(T) = q_i(T) + \bar{q}_i(T),\\ &\dot{\delta}_i + \frac{1}{2}\alpha_i[\sigma^2 + \int_\Theta \mu^2(t,\theta)\nu(d\theta)] = 0,\\ &\delta_i(T) = 0. \end{aligned} \tag{7.119}$$

We retrieve the expressions in (7.95) confirming the validity of this approach.

Stochastic Maximum Principle
Here we address the problem of solving the risk-neutral linear-quadratic mean-field-type games using stochastic minimum principle (SMP). To do so, we first solve the best-response problem in (7.90).

The risk-neutral Hamiltonian is

$$H_i(t,x,y,u,p,z) := \left(ax + \bar{a}y + \sum_{j=1}^{I} b_j u_j\right) p_i + \frac{1}{2}[q_i x^2 + \bar{q}_i (x-y)^2 + r_i u_i^2] + \sigma z_i.$$

We have

$$H_{i,x} = ap_i + q_i x + \bar{q}_i(x - Ex), \qquad H_{i,u_i} = b_i p_i + r_i u_i, \quad H_{i,y} = \bar{a}p_i - \bar{q}_i(x - Ex).$$

Minimization of the Hamiltonian yields

$$\bar{u}_i(t) = -\frac{b_i}{r_i} p_i(t). \tag{7.120}$$

The associated state dynamics $\bar{x}$ solves then the SDE:

$$d\bar{x}(t) = \left(a\bar{x}(t) + \bar{a}E\bar{x}(t) - \sum_{i=1}^{I} \frac{b_i^2}{r_i} p_i(t)\right) dt + \sigma dB(t). \tag{7.121}$$

The best-response equation involves only the first-order adjoint equation because the diffusion coefficient in the state dynamics is independent of the control.

The first-order adjoint equation of the mean-field-type problem is

$$\begin{aligned} dp_i &= -[H_{i,x} + E[H_{i,y}]]dt + z_i dB & (7.122)\\ &= -[ap_i + q_i x + \bar{q}_i(x - \bar{x}) + \bar{a}E[p_i]]dt + z_i dB, & (7.123)\\ p_i(T) &= q_i(T)x(T) + \bar{q}_i(T)(x(T) - \bar{x}(T)). & (7.124)\end{aligned}$$

We try a solution of the form

$$p_i(t) := \beta_i(t)\bar{x}(t) + \alpha_i(t), \tag{7.125}$$

where $\beta_i(t),\ \alpha_i(t)$ are deterministic functions to be determined. From (7.124), it follows that

$$\beta_i(T) \ = q_i(T) + \bar{q}_i(T), \text{ and } \quad \alpha_i(T) = -\bar{q}_i(T)\bar{x}(T). \tag{7.126}$$

Applying Itô's formula to (7.125) yields

$$\begin{aligned} d[\beta_i x + \alpha_i] = \{\dot{\beta}_i + a\beta_i - \beta_i \sum_{j=1}^{I} \frac{b_j^2}{r_j}\beta_j\}xdt \\ +\{\dot{\alpha}_i + \bar{a}\beta_i Ex - \beta_i \sum_{j=1}^{I} \frac{b_j^2}{r_j}\alpha_j\}dt + \sigma\beta_i dB(t).\end{aligned}$$

Recomputing the first-order adjoint equation, we get

$$dp_i = -[(a\beta_i + q_i + \bar{q}_i)x + (a + \bar{a})\alpha_i + (\bar{a}\beta_i - \bar{q}_i)\bar{x}]dt + z_i dB(t). \tag{7.127}$$

Equating (7.127) and (7.127), we get

$$\begin{aligned} \dot{\beta}_i + 2a\beta_i - \beta_i \sum_{j=1}^{I} \frac{b_j^2}{r_j}\beta_j + q_i + \bar{q}_i = 0, && (7.128)\\ \beta_i(T) = q_i(T) + \bar{q}_i(T) && (7.129)\\ \dot{\alpha}_i + (a + \bar{a})\alpha_i + (2\bar{a}\beta_i - \bar{q}_i)Ex - \beta_i \sum_{j=1}^{I} \frac{b_j^2}{r_j}\alpha_j = 0, && (7.130)\\ \alpha_i(T) = -\bar{q}_i(T)\bar{x}(T) && (7.131)\\ \bar{u}_i \ = -\frac{b_i}{r_i}(\beta_i x + \alpha_i), && (7.132)\\ z_i = \sigma\beta_i. && (7.133)\end{aligned}$$

We refine the expression of the deterministic function $\alpha(t)$ by setting $\alpha_i(t) = \eta_i(t)E[x(t)]$ in (7.130). Then,

$$\dot{\alpha}_i = \{\dot{\eta}_i + \eta_i(a+\bar{a}) - \eta_i \sum_{j=1}^{I} \frac{b_j^2}{r_j}(\beta_j + \eta_j)\}\bar{x}. \tag{7.134}$$

Putting all this together with (7.130), we arrive at

$$\left\{ \dot{\eta}_i + 2(a+\bar{a})\eta_i - \eta_i \sum_{j=1}^{I} \frac{b_j^2}{r_j}(\beta_j+\eta_j) - \beta_i \sum_{j=1}^{I} \frac{b_j^2}{r_j}\eta_j + (2\bar{a}\beta_i - \bar{q}_i) \right\} \bar{x} = 0. \tag{7.135}$$

The equilibrium system is

$$\begin{cases} \dot{\beta}_i + 2a\beta_i - \beta_i \sum_{j=1}^{I} \frac{b_j^2}{r_j}\beta_j + q_i + \bar{q}_i = 0, \\ \beta_i(T) = q_i(T) + \bar{q}_i(T), \\ \dot{\eta}_i + 2(a+\bar{a})\eta_i - \eta_i(\sum_{j=1}^{I} \frac{b_j^2}{r_j}(\beta_j+\eta_j)) - \beta_i(\sum_{j=1}^{I} \frac{b_j^2}{r_j}\eta_j) + 2\bar{a}\beta_i - \bar{q}_i = 0, \\ \eta_i(T) = -\bar{q}_i(T), \\ u_i = -\frac{b_i}{r_i}(\beta_i x + \eta_i \bar{x}). \\ dx = \left(a - \sum_{j=1}^{I} \frac{b_j^2}{r_j}\beta_j\right) x dt + \left(\bar{a} - \sum_{j=1}^{I} \frac{b_j^2}{r_j}\eta_j\right) \bar{x} dt + \sigma dB(t), \\ \frac{d}{dt}\bar{x} = \left(a + \bar{a} - \sum_{j=1}^{I} \frac{b_j^2}{r_j}(\beta_j + \eta_j)\right) \bar{x}. \end{cases} \tag{7.136}$$

Let $y_i = \beta_i + \eta_i$. Then y solves the Riccati equation:

$$\dot{y}_i + 2(a+\bar{a})y_i + q_i - y_i \sum_{j=1}^{I} \frac{b_j^2}{r_j} y_j = 0.$$

The best-response strategy can be expressed as a function of y.

$$u_i = -\frac{b_i}{r_i}(\beta_i x + \eta_i \bar{x}) = -\frac{b_i}{r_i}(\beta_i x + \eta_i x - \eta_i x + \eta_i \bar{x}) = -\frac{b_i}{r_i}[y_i x + \eta_i(\bar{x} - x)].$$

We can also express the best response as

$$\begin{cases} \bar{u}_i = -\frac{b_i}{r_i}[\beta_i(x - \bar{x}) + y_i \bar{x}], \\ \dot{\beta}_i + 2a\beta_i - \beta_i \sum_{j=1}^{I} \frac{b_j^2}{r_j}\beta_j + q_i + \bar{q}_i = 0, \\ \beta_i(T) = q_i(T) + \bar{q}_i(T), \\ \dot{y}_i + 2(a+\bar{a})y_i - y_i \sum_{j=1}^{I} \frac{b_j^2}{r_j} y_j + q_i = 0, \\ y_i(T) = q_i(T). \end{cases} \tag{7.137}$$

in the basis $\{x - \bar{x}, \bar{x}\}$.

State and Mean-Field-Dependent Diffusion

The result of (7.136) is for state dynamics with constant diffusion. In this subsection, we focus on the case where the diffusion coefficient is both state and mean-field dependent. Consider the loss function

$$\begin{aligned}\tilde{L}_i(u_1,\ldots,u_I) = \tfrac{1}{2}[q_i(T)x^2(T) + \bar{q}_i(T)(\bar{x}(T))^2 \\ + \textstyle\int_0^T \{q_i(t)x^2(t) + \bar{q}_i(t)(E[x(t)])^2 + r_i(t)u_i^2(t)\}dt].\end{aligned} \tag{7.138}$$

The best response of decision-maker i to $u_{-i} := (u_1,\ldots,u_{i-1},u_{i+1},\ldots,u_I)$, solves the following risk-neutral linear-quadratic mean-field-type problem

$$\begin{cases}\inf_{u_i(\cdot)\in\mathscr{U}_i} E\left[\tilde{L}_i(u_1,\ldots,u_I)\right], \\ \text{subject to} \\ dx(t)=[ax(t)+\bar{a}E[x(t)]+\sum_{i=1}^I b_iu_i(t)]dt+(\sigma+cx(t)+\bar{c}E[x(t)])dB(t), \\ x(0)\in\mathbb{R},\end{cases} \tag{7.139}$$

where $q_i(t) \geq 0, \bar{q}_i(t) \geq 0, r_i(t) > 0$, and $a, \bar{a}, b_i, \sigma, c, \bar{c}$ are real numbers and where $E[x(t)]$ is the expected value of state trajectory created by all decision-makers under the control action profile $(u_1,\ldots,u_I)$.

The associated risk-neutral Hamiltonian for decision-maker i is

$$H_i(t,x,y,u,p,z) := (ax+\bar{a}y+\sum_{j=1}^I b_ju_j)p_i+\frac{1}{2}[q_ix^2+\bar{q}_iy^2+r_iu_i^2]+(\sigma+cx+\bar{c}y)z_i.$$

We have

$$H_{i,x} = ap_i + q_ix + cz_i, \qquad H_{i,u_i} = b_ip_i + r_iu_i, \quad H_{i,y} = \bar{a}p_i + \bar{q}_i\bar{x} + \bar{c}z_i.$$

Minimization of the Hamiltonian yields

$$\bar{u}_i(t) = -\frac{b_i}{r_i}p_i(t). \tag{7.140}$$

The associated state dynamics $\bar{x}$ solves then the SDE:

$$d\bar{x}(t) = \left(a\bar{x}(t) + \bar{a}E\bar{x}(t) - \sum_{i=1}^I \frac{b_i^2}{r_i}p_i(t)\right)dt + (\sigma + c\bar{x}(t) + \bar{c}E[\bar{x}(t)])dB(t). \tag{7.141}$$

The first-order adjoint equation yields

$$\begin{aligned} dp_i &= -[H_{i,x} + EH_{i,y}]dt + z_idB && (7.142) \\ &= -[ap_i + q_ix + cz_i + \bar{a}E[p_i] + \bar{q}_i\bar{x} + \bar{c}Ez_i]dt + z_idB && (7.143)\end{aligned}$$

$$p_i(T) = q_i(T)x(T) + \bar{q}_i(T)\bar{x}(T). \tag{7.144}$$

We try a solution of the following form:

$$p_i(t) := \beta_i(t)\bar{x}(t) + \eta_i(t)E[\bar{x}(t)], \tag{7.145}$$

where $\beta_i(t),\ \eta_i(t)$ are deterministic functions to be determined. From (7.144), it follows that

$$\beta_i(T) \ = q_i(T), \text{ and } \quad \eta_i(T) = \bar{q}_i(T). \tag{7.146}$$

Let $\alpha_i(t) = \eta_i(t)E[x(t)]$. Applying Itô's formula to (7.145) yields

$$\begin{aligned} d[\beta_i x + \alpha_i] = \{\dot{\beta}_i + a\beta_i - \beta_i \sum_{j=1}^{I} \frac{b_j^2}{r_j}\beta_j\}xdt \\ +\{\dot{\alpha}_i + \bar{a}\beta_i Ex - \beta_i \sum_{j=1}^{I} \frac{b_j^2}{r_j}\alpha_j\}dt + \beta_i(\sigma + cx + \bar{c}\bar{x})dB(t) \end{aligned}$$

and

$$\dot{\alpha}_i = \{\dot{\eta}_i + \eta_i(a+\bar{a}) - \eta_i \sum_{j=1}^{I} \frac{b_j^2}{r_j}(\beta_j + \eta_j)\}\bar{x}. \tag{7.147}$$

$$\begin{aligned} d[\beta_i x + \alpha_i] &= \{\dot{\beta}_i + a\beta_i - \beta_i \sum_{j=1}^{I} \frac{b_j^2}{r_j}\beta_j\}xdt + \beta_i(\sigma + cx + \bar{c}\bar{x})dB(t) \\ &+\{\dot{\eta}_i + \eta_i(a+\bar{a}) - \eta_i \sum_{j=1}^{I} \frac{b_j^2}{r_j}(\beta_j + \eta_j) + \bar{a}\beta_i - \beta_i \sum_{j=1}^{I} \frac{b_j^2}{r_j}\eta_j\}Exdt, \end{aligned} \tag{7.148}$$

and $z_i = \beta_i(\sigma + cx + \bar{c}\bar{x})$. Recomputing the first-order adjoint equation, we get

$$\begin{aligned} dp_i &= -[(a+c^2)\beta_i + q_i]xdt - (c+\bar{c})\beta_i\sigma dt + z_i dB \\ &\quad -[(a+\bar{a})\eta_i + \bar{q}_i + (\bar{a} + \bar{c}^2 + 2c\bar{c})\beta_i]\bar{x}dt. \end{aligned} \tag{7.149}$$

The equilibrium system is

$$\left\{\begin{array}{l} \dot{\beta}_i + (2a+c^2)\beta_i - \beta_i \sum_{j=1}^{I} \frac{b_j^2}{r_j}\beta_j + q_i = 0, \\ \beta_i(T) = q_i(T), \\ \left\{\dot{\eta}_i + 2(a+\bar{a})\eta_i - \eta_i \sum_{j=1}^{I} \frac{b_j^2}{r_j}(\beta_j+\eta_j) \; +(2\bar{a}+\bar{c}^2+2c\bar{c})\beta_i + \bar{q}_i - \beta_i \sum_{j=1}^{I} \frac{b_j^2}{r_j}\eta_j \right\} \bar{x} \\ \quad +(c+\bar{c})\beta_i\sigma = 0, \\ \eta_i(T) = \bar{q}_i(T), \\ u_i = -\frac{b_i}{r_i}(\beta_i x + \eta_i \bar{x}). \\ dx = \left(a - \sum_{j=1}^{I} \frac{b_j^2}{r_j}\beta_j\right) x dt + \left(\bar{a} - \sum_{j=1}^{I} \frac{b_j^2}{r_j}\eta_j\right) \bar{x} dt + (\sigma + cx + \bar{c}\bar{x}) dB(t), \\ \frac{d}{dt}\bar{x} = \left(a + \bar{a} - \sum_{j=1}^{I} \frac{b_j^2}{r_j}(\beta_j+\eta_j)\right)\bar{x}. \end{array}\right. \tag{7.150}$$

Remark 7.1 For $(c+\bar{c})\sigma \neq 0$, the system (7.151) is highly nonlinear in η_i and $\bar{x}$. To avoid this type of nonlinearity, we change our guess solution to counterbalance the term $(c+\bar{c})\beta_i\sigma$. This means that the adjoint process is $p_i(t) = \beta_i(t)x(t) + \eta_i(t)E[x(t)] + \gamma_i(t)$,

$$\left\{\begin{array}{l} \dot{\beta}_i + (2a+c^2)\beta_i - \beta_i \sum_{j=1}^{I} \frac{b_j^2}{r_j}\beta_j + q_i = 0, \\ \beta_i(T) = q_i(T) \\ \dot{\eta}_i + 2(a+\bar{a})\eta_i - \eta_i \sum_{j=1}^{I} \frac{b_j^2}{r_j}(\beta_j+\eta_j) + (2\bar{a}+\bar{c}^2+2c\bar{c})\beta_i + \bar{q}_i - \beta_i \sum_{j=1}^{I} \frac{b_j^2}{r_j}\eta_j = 0 \\ \eta_i(T) = \bar{q}_i(T), \\ \dot{\gamma}_i + (a+\bar{a})\gamma_i - \beta_i[\sum_j \frac{b_j^2}{r_j}\gamma_j] + (c+\bar{c})\beta_i\sigma = 0, \\ \gamma_i(T) = 0, \\ u_i = -\frac{b_i}{r_i}(\beta_i x + \eta_i \bar{x} + \gamma_i). \\ dx = \left(a - \sum_{j=1}^{I} \frac{b_j^2}{r_j}\beta_j\right) x dt + \left(\bar{a} - \sum_{j=1}^{I} \frac{b_j^2}{r_j}\eta_j\right) \bar{x} dt + (\sigma + cx + \bar{c}\bar{x}) dB(t), \\ \frac{d}{dt}\bar{x} = \left(a + \bar{a} - \sum_{j=1}^{I} \frac{b_j^2}{r_j}(\beta_j+\eta_j)\right)\bar{x}. \end{array}\right. \tag{7.151}$$

In particular, γ is not needed, if $\sigma = 0$, or, if $c + \bar{c} = 0$.

Example 7.1 (Network Security as a Public Good) Our smartphone provider tracks our interest and your location pattern and knows who are interacting with us. Our online and in-store purchasing patterns are recorded and reveal if we are unemployed, sick, or pregnant. Our e-mails and texts expose our intimate and casual friends. Several online search engine companies know what we are thinking by saving our private (historical) searches. Several social networks can determine our sexual orientation without us even revealing it. And these entities share some piece of information with each other or, even worse, have it stolen by cybercriminals in huge data breaches, andthere is a mean-field effect in the society. Everyone who is connected to the network is concerned with the security of the mass since its own security may depend on it.

Thomas Jefferson, the third President of the United States, described knowledge in the following way:

He who receives an idea from me, receives instruction himself without lessening mine; as he who lights his taper at mine, receives light without darkening me.

This clearly indicates that *knowledge* has a tendency toward the concept of a global public good.

Mimicking this argument, security can be seen as a public good. We consider here a simple dynamic model of private investment in network security where users have saturating payoffs and security investment costs. The novelty in the security-level dynamics is the presence of mean-field terms as the average security level. In this setting, we characterize the unique solution of the team problem, which we interpret as the case when the public good is provided collectively. Next, we investigate whether or not the public good will be provided if users privately invest.

The provision of acceptable security level is a key requirement of network security. Individual users can invest to computer security, hardware security, anti-spam, anti-virus, etc., up to a certain cost. Like classical public good examples such as stadium, national defense, clean air, knowledge, and communication network security can be seen as non-excludable and non-rivalrous up to some extent. While it is already possible to create exclusive more secure networks with VPN and https, the basic network security is nonexclusive. Network security is also non-rivalrous, even though encryption imposes some individual cost. Security of the network such as public internet, Wi-Fi (wireless local area networking with devices based on the IEEE 802.11) can therefore be seen as a public good provisioning.

We consider I decision-makers who can contribute by investing in security, whose level evolution is governed by

$$dx = (-ax - \bar{a}\bar{x} + \sum_{j=1}^{I} b_j u_j)dt + cxdB, \quad x(0) = x_0.$$

Each decision-maker i faces an investment cost given by $\rho_i u_i + \frac{r_i}{2}u_i^2, \ \rho_i \geq 0, r_i > 0$, and benefits from the common security of the network according to an individual payoff given by $q_i x(1 - \epsilon_i x), \ \epsilon_i > 0$. When the network security level is very large (positively), there is a saturation effect from the term $q_i x(1 - \epsilon_i x)$, which becomes negative. Payoff functional to be maximized is given by

$$- \, var(x(T)) + \int_0^T q_i(t)x(t)(1 - \epsilon_i(t)x(t)) - \rho_i(t)u_i(t) - \frac{r_i(t)}{2}u_i^2(t)dt.$$

The best response can be written in the following form:

$$\bar{u}_i^E(t) = -\frac{b_i}{r_i(t)}\beta_i(t)x(t) - \frac{b_i}{r_i(t)}\eta_{1i}(t)E[x(t)] - \frac{b_i}{r_i(t)}\eta_{2i}(t) - \frac{\rho_i(t)}{r_i(t)}, \tag{7.152}$$

where

$$\begin{cases} \dot{\beta}_i + (-2a + c^2)\beta_i - \beta_i \sum_{j=1}^I \frac{b_j^2}{r_j}\beta_j + 2q_i\epsilon_i = 0, \\ \beta_i(T) = 1, \\ \dot{\eta}_{1i} - 2(a+\bar{a})\eta_{1i} - 2\bar{a}\beta_i - \beta_i \sum_{j=1}^I \frac{b_j^2}{r_j}\eta_{1j} - \eta_{1i} \sum_{j=1}^I \frac{b_j^2}{r_j}(\beta_j + \eta_{1j}) = 0, \\ \eta_{1i}(T) = -1, \\ \dot{\eta}_{2i} - (a+\bar{a})\eta_{2i} - \beta_i \sum_{j=1}^I \frac{b_j}{r_j}(b_j\eta_{2j} + \rho_j) - \eta_{1i} \sum_{j=1}^I \frac{b_j}{r_j}(b_j\eta_{2j} + \rho_j) - q_i = 0, \\ \eta_{2i}(T) = 0. \end{cases}$$

The global optimum (GO) of the public good security game solves

$$\begin{cases} \bar{u}_i^{GO}(t) = -\frac{b_i}{r_i(t)}\beta(t)x(t) - \frac{b_i}{r_i(t)}\eta_1(t)E[x(t)] - \frac{b_i}{r_i(t)}\eta_2(t) - \frac{\rho_i(t)}{r_i(t)}, \\ \dot{\beta} + (-2a + c^2)\beta - \beta^2 \sum_i \frac{b_i^2}{r_i} + 2\sum_{i=1}^I q_i\epsilon_i = 0, \\ \beta(T) = I, \\ \dot{\eta}_1 - 2(a+\bar{a})\eta_1 - 2\bar{a}\beta - \beta\eta_1 \sum_i \frac{b_i^2}{r_i} - \eta_1(\beta + \eta_1)\sum_i \frac{b_i^2}{r_i} = 0, \\ \eta_1(T) = -I, \\ \dot{\eta}_2 - (a+\bar{a})\eta_2 - \beta \sum_i \frac{b_i}{r_i}(b_i\eta_2 + \rho_i) - \eta_1 \sum_i \frac{b_i}{r_i}(b_i\eta_2 + \rho_i) - \sum_{i=1}^I q_i = 0, \\ \eta_2(T) = 0. \end{cases}$$

The aims to look at the security level at equilibrium x^E versus the security level at global optimum x^{GO} over time. The contribution effect to the drift at equilibrium is

$$\sum_{i=1}^I b_i\bar{u}_i^E(t) = \sum_{i=1}^I -\frac{b_i^2}{r_i(t)}\beta_i(t)x^E(t) - \frac{b_i^2}{r_i(t)}\eta_{1i}(t)E[x^E(t)] - \frac{b_i^2}{r_i(t)}\eta_{2i}(t) - \frac{b_i\rho_i(t)}{r_i(t)},$$

and the contribution effect to the drift at global optimum is

$$\sum_{i=1}^I b_i\bar{u}_i^{GO} = \sum_{i=1}^I -\frac{b_i^2}{r_i(t)}\beta(t)x^{GO}(t) - \frac{b_i^2}{r_i(t)}\eta_1(t)E[x^{GO}(t)] - \frac{b_i^2}{r_i(t)}\eta_2(t) - \frac{b_i\rho_i(t)}{r_i(t)}.$$

Taking the expectation of the yields

$$\begin{cases} \sum_{i=1}^I b_i E[\bar{u}_i^{GO} - \bar{u}_i^E(t)] \\ = -\sum_{i=1}^I \frac{b_i^2}{r_i(t)}(\beta(t) + \eta_1(t) - \beta_i(t) - \eta_{1i})E[x^{GO}(t) - x^E(t)] \\ -\sum_{i=1}^I \frac{b_i^2}{r_i(t)}(\eta_2(t) - \eta_{2i}(t)) \end{cases}$$

Thus, the evolution of $y(t) = E[x^{GO}(t) - x^E(t)]$ can be obtained over $[0, T]$:

$$\begin{cases} \dot{y}(t) = -(a+\bar{a})y(t) + \sum_{i=1}^{I} b_i E[\bar{u}_i^{GO} - \bar{u}_i^{E}(t)] \\ = -[(a+\bar{a}) + \sum_{i=1}^{I} \frac{b_i^2}{r_i(t)}(\beta(t) + \eta_1(t) - \beta_i(t) - \eta_{1i})]y(t) \\ - \sum_{i=1}^{I} \frac{b_i^2}{r_i(t)}(\eta_2(t) - \eta_{2i}(t)), \\ y(0) = 0. \end{cases}$$

Let $\dot{w}_0(t) = [(a+\bar{a})+\sum_{i=1}^{I} \frac{b_i^2}{r_i(t)}(\beta(t)+\eta_1(t)-\beta_i(t)-\eta_{1i})]w_0(t)$ with $w_0(0) = 1$.

Then, the function y is explicitly given by $y(t) = -w_0^{-1}(t) \int_0^t w_0(t') \sum_{i=1}^{I} \frac{b_i^2}{r_i(t')} (\eta_2(t) - \eta_{2i}(t'))dt'$.

As $w_0^{-1}(t)$ is positive, the sign of y is completely determined by the opposite sign of $S(t) = \int_0^t w_0(t') \sum_{i=1}^{I} \frac{b_i^2}{r_i(t')}(\eta_2(t) - \eta_{2i}(t'))dt'$.

From the boundary condition of the coefficients, the term S is negative near T. This makes a higher security level at the global optimum than at equilibrium. The extension of this result to the entire time horizon remains an open issue.

Notes

In this section, we have introduced a basic linear-quadratic mean-field-type game with state-variance and action-variance cost. For more recent developments on mean-field-type games, see [14, 15, 15–17, 27–29, 33, 33, 33, 35–37, 39–43, 46–48, 50–52, 62, 99, 102, 124, 125, 127, 130, 132, 146, 157, 158, 166, 168, 185, 194, 245, 247, 281, 282, 293, 294, 296, 307, 308, 311, 320, 330–332, 334, 336–338, 350, 359, 367] and the references therein.

7.3 Linear-Quadratic Mean-Field-Type Games with Common Noise

In this section, we modify the state dynamics of Sect. 7.2 to include a common noise. We consider a multi-person, mean-field-type game that is formulated and solved, which is described by a linear jump-diffusion system of conditional mean-field type, where the conditioning is with respect to common noise and a quadratic cost functional involving the second moment, the square of the conditional expectation of the state, and the control actions of the decision-makers. We propose a simple method to solve the game problem with two filter terms involved in the state and in the objective functional. The optimal strategies for the decision-makers are shown to be in a state-and-mean-field feedback form. The equilibrium cost of the decision-makers is explicitly derived using a simple direct method. Moreover, the equilibrium cost is a weighted sum of the initial variance and an integral of a weighted variance of the diffusion and the jump process. Finally, the method is used to compute global optimum strategies as well as saddle-point strategies and Nash bargaining solution in state-and-mean-field feedback form.

Notation

Let $T > 0$ be a fixed length of the time horizon and $(\Omega, \mathscr{F}, \mathbb{F}^{B,N,B_o}, \mathbb{P})$ be a given filtered probability space on which a one-dimensional standard Brownian motion $B = \{B(t)\}_{t\geq 0}$ is given, $\tilde{N}(dt, d\theta) = N(dt, d\theta) - \nu(d\theta)dt$ is a compensated jump process with Lévy measure ν defined over Θ. $B_o = \{B_o(t)\}_{t\geq 0}$ is another one-dimensional standard Brownian motion that is observed by all decision-makers. The filtration $\mathbb{F} = \{\mathscr{F}_t^{B,N,B_o},\ 0 \leq t \leq T\}$ is the natural filtration generated by the unions of events from B, B_o or N augmented by $\mathbb{P}$-null sets of $\mathscr{F}$. The filtration $\mathscr{F}_t^{B_o}$ is the filtration generated by the common noise B_o up to t. The processes B, B_o and N are assumed to be independent. In practice, B is used to capture smaller disturbance and N is used for larger jumps of the system. B_o is an observed common noise [321]. Let $\hat{X}(t) = E[X(t)|\ \mathscr{F}_t^{B_o}]$ denote the conditional expectation of the random variable $X(t)$ with respect to the filtration $\mathscr{F}_t^{B_o}$. Note that $\hat{X}$ is a random process.

7.3.1 Games with Common Noise

We first extend Definition 7.6 to the conditional probability measure case.

Definition 7.3.1.1 (Conditional McKean-Vlasov) A dynamical system is of McKean–Vlasov type if the transition kernel functional (drift, diffusion, jump coefficient functions) involves not only the states and control actions but also the conditional probability measure of the state and/or control actions with respect to the available information and/or the observed process.

As indicated in Definition 7.6, conditional McKean–Vlasov dynamics differ from conditional Liouville dynamics.

Consider I risk-neutral decision-makers ($I \geq 2$) and $L_i(u_1, \ldots, u_I)$ to be the objective functional of decision-maker i, given by

$$\begin{aligned}
&L_i(u_1, \ldots, u_I) = \tfrac{1}{2}q_i(T)x^2(T) + \tfrac{1}{2}\bar{q}_i(T)[E(x(T)|\mathscr{F}_T^{B_o})]^2 \\
&+\tfrac{1}{2}\int_0^T \left\{q_i(t)x^2(t) + \bar{q}_i(t)(E[x(t)\ |\mathscr{F}_t^{B_o}\ ])^2 + r_i(t)u_i^2(t) + \bar{r}_i(t)[E(u_i(t)|\mathscr{F}_t^{B_o})]^2\right\} dt.
\end{aligned} \tag{7.153}$$

Then, the best response of decision-maker i to the process $u_{-i} := (u_1, \ldots, u_{i-1}, u_{i+1}, \ldots, u_I)$, solves the following risk-neutral linear-quadratic mean-field-type control problem (with common noise):

$$\begin{cases}
\inf_{u_i\in\mathscr{U}_i} E\left[L_i(u_1, \ldots, u_I)\right] \text{ subject to} \\
dx(t) = \left\{a(t)x(t) + \bar{a}(t)E[x(t)\mid \mathscr{F}_t^{B_o}]\right\} dt \\
+\left\{\sum_{i=1}^I b_i(t)u_i(t) + \sum_{i=1}^I \bar{b}_i(t)E[u_i(t)\mid \mathscr{F}_t^{B_o}]\right\} dt \\
+\sigma(t)dB(t) + \int_\Theta \mu(t,\theta)\tilde{N}(dt, d\theta) + \sigma_o(t)dB_o(t), \\
x(0) = x_0,
\end{cases} \tag{7.154}$$

where the coefficients $q_i(t), q_i(t) + \bar{q}_i(t), r_i(t), r_i(t) + \bar{r}_i(t) \geq \epsilon > 0$ and $a(t), \bar{a}(t), b_i(t), \sigma(t), \sigma_o(t)$ are deterministic real-valued functions and where $E[x(t)|\mathcal{F}_t^{B_o}]$ is the conditional expectation of the state created by all decision-makers under the control action profile $(u_1, \ldots, u_I) \in \prod_{j=1}^I \mathcal{U}_j$. The method below can handle time-varying coefficients as well.

Note that the state dynamics of (7.154) is of conditional McKean–Vlasov type (see Definition 7.3.1.1).

The risk-neutral mean-field-type Nash equilibrium problem we are concerned with is to find and characterize the processes $(x^*, u^*, E[x^* \mid \mathcal{F}^{B_o}], E[u^* \mid \mathcal{F}^{B_o}])$ such that for every decision-maker i, the strategy u_i^* is an optimizer of the best-response problem (7.154) and the conditional expected value of the resulting common state created by all the decision-makers is $E[x^* \mid \mathcal{F}_t^{B_o}]$. This means that an equilibrium is a fixed point of the best-response correspondence $BR = (BR_1, \ldots, BR_I)$, where $BR_i : \prod_{j\neq i} \mathcal{U}_j \rightarrow 2^{\mathcal{U}_i}$, is the best-response correspondence of decision-maker i. We rewrite the expected objective functional and the state coefficients in terms of $x - \hat{x}$ and $\hat{x}$:

$$
\begin{aligned}
E[L_i] = E[\hat{L}_i] &= \tfrac{1}{2}[q_i(T)E(x(T) - \hat{x}(T))^2 + [q_i(T) + \bar{q}_i(T)]E\hat{x}^2(T) \\
&\quad + E\int_0^T q_i(x - \hat{x})^2 + [q_i + \bar{q}_i]\hat{x}^2 + r_i(u_i - \hat{u}_i)^2 + (r_i + \bar{r}_i)[\hat{u}_i]^2 dt], \\
dx &= \left\{a(x - \hat{x}) + (a + \bar{a})\hat{x} + \textstyle\sum_{i=1}^I b_i(u_i - \hat{u}_i) + \sum_{i=1}^I (b_i + \bar{b}_i)\hat{u}_i\right\} dt \\
&\quad + \sigma dB + \int_\Theta \mu(t, \theta)\tilde{N}(dt, d\theta) + \sigma_o dB_o.
\end{aligned}
\tag{7.155}
$$

Note that the expected value of the first term in the integral in L_i can be seen as a weighted conditional variance. From the conditional McKean–Vlasov stochastic differential equation (SDE) of the state x, we derive the SDE satisfied by the filter $\hat{x}(t) = E[x(t) \mid \mathcal{F}_t^{B_o}]$. We take the conditioning of the state to obtain the filter:

$$
d\hat{x} = \{(a + \bar{a})\hat{x} + \textstyle\sum_{i=1}^I (b_i + \bar{b}_i)\hat{u}_i\}dt + \sigma_o dB_o, \tag{7.156}
$$

where we have used the fact that

$$
E[B(t) \mid \mathcal{F}_t^{B_o}] = 0, \;\; E[\tilde{N}(t) \mid \mathcal{F}_t^{B_o}] = 0. \tag{7.157}
$$

The direct method entails writing a generic structure of the cost functional, with unknown deterministic functions to be identified. Inspired from the structure of the terminal cost function, we try a generic solution in a quadratic form. Let $f_i(t, x) = \frac{1}{2}\alpha_i(t)(x - \hat{x})^2 + \frac{1}{2}\beta_i(t)\hat{x}^2 + \gamma_i(t)\hat{x} + \delta_i(t)$, where $\alpha, \beta, \gamma, \delta$ are random functions of time t, which are $\mathcal{F}_t^{B_o}$-measurable such that

$$
f_i(T, x(T)) = \frac{1}{2}\{q_i(T)E(x(T) - \hat{x}(T))^2 + [q_i(T) + \bar{q}_i(T)]\hat{x}^2(T)\},
$$

i.e., $\alpha_i(T) = q_i(T),\ \beta_i(T) = q_i(T) + \bar{q}_i(T),\ \gamma_i(T) = \delta_i(T) = 0$. Using Itô's formula for jump-diffusion processes, one obtains

$$\begin{array}{l}
f_i(T, x(T)) - f_i(0, x(0)) \\
= \frac{1}{2}\int_0^T \{d\alpha_i + 2a\alpha_i\}(x-\hat{x})^2 dt + \frac{1}{2}\int_0^T [\sigma^2 + \int_\Theta \mu^2(t,\theta)\nu(d\theta)]\alpha_i dt \\
+\frac{1}{2}\int_0^T 2\alpha_i \sum_{j=1}^I b_j(u_j - \hat{u}_j)(x-\hat{x})dt + \frac{1}{2}\int_0^T 2\alpha_i(x-\hat{x})(\sigma dB + \int_\Theta \mu(t,\theta)\tilde{N}(dt,d\theta)) \\
+\frac{1}{2}\int_0^T \{d\beta_i + 2(a+\bar{a})\beta_i\}\hat{x}^2 dt + \frac{1}{2}\int_0^T [\,\sigma_o^2\beta_i + 2\beta_{i,o}\sigma_o\hat{x}]dt \\
+\frac{1}{2}\int_0^T 2\beta_i \sum_{j=1}^I (b_j + \bar{b}_j)\hat{u}_j\hat{x}dt + \frac{1}{2}\int_0^T 2\beta_i\sigma_o\hat{x}dB_o \\
+\int_0^T \{\dot{\gamma}_i + (a+\bar{a})\gamma_i\}\hat{x}dt + \int_0^T \gamma_i \sum_{j=1}^I (b_j+\bar{b}_j)\hat{u}_j dt \\
+\int_0^T \gamma_{i,o}\sigma_o dt + \int_0^T d\delta_i + \int_0^T \gamma_i\sigma_o dB_o.
\end{array} \tag{7.158}$$

Feedback Strategies

The information structure for feedback solution is as follows. The model and the objective functions are assumed to be common knowledge. We assume that the state is of perfect observation and the mean-field term is computable. In addition, we assume that the common noise is of perfect observation. We compute again $\hat{L}_i - f_i(0, x(0))$ and complete the squares using the elements of $\{x - \hat{x},\ \hat{x}\}$:

$$\begin{array}{l}
\hat{L}_i - f_i(0, x(0)) = \frac{1}{2}(q_i(T) - \alpha_i(T))(x(T) - \hat{x}(T))^2 + \frac{1}{2}[q_i(T) + \bar{q}_i(T) - \beta_i(T)]\hat{x}^2(T) \\
+(0 - \gamma_i(T))\hat{x}(T) + (0 - \delta_i(T)) \\
+\frac{1}{2}\int_0^T \{d\alpha_i + 2a\alpha_i - \frac{b_i^2}{r_i}\alpha_i^2 - 2\alpha_i \sum_{j\neq i}\frac{b_j^2}{r_j}\alpha_j + q_i\}(x-\hat{x})^2 dt \\
+\frac{1}{2}\int_0^T \left\{d\beta_i + 2(a+\bar{a})\beta_i - \frac{(b_i+\bar{b}_i)^2}{r_i+\bar{r}_i}\beta_i^2 - 2\beta_i\sum_{j\neq i}\frac{(b_j+\bar{b}_j)^2}{r_j+\bar{r}_j}\beta_j + q_i + \bar{q}_i\right\}\hat{x}^2 dt \\
+\int_0^T \left\{d\gamma_i + (a+\bar{a})\gamma_i + \beta_{i,o}\sigma_o - \frac{(b_i+\bar{b}_i)^2}{r_i+\bar{r}_i}\beta_i\gamma_i\right. \\
\left. -\beta_i\sum_{j\neq i}\frac{(b_j+\bar{b}_j)^2}{r_j+\bar{r}_j}\gamma_j - \gamma_i\sum_{j\neq i}\frac{(b_j+\bar{b}_j)^2}{r_j+\bar{r}_j}\beta_j\right\}\hat{x}dt \\
+\int_0^T \left\{[\sigma^2 + \int_\Theta \mu^2(t,\theta)\nu(d\theta)]\frac{1}{2}\alpha_i + d\delta_i + \frac{1}{2}\sigma_o^2\beta_i + \gamma_{i,o}\sigma_o\right. \\
\left. -\gamma_i\sum_{j\neq i}\frac{(b_j+\bar{b}_j)^2}{r_j+\bar{r}_j}\gamma_j - \frac{(b_i+\bar{b}_i)^2}{r_i+\bar{r}_i}\frac{1}{2}\gamma_i^2\right\}dt \\
+\frac{1}{2}\int_0^T r_i[u_i - \hat{u}_i + \frac{b_i}{r_i}\alpha_i(x-\hat{x})]^2 dt + \frac{1}{2}\int_0^T (r_i+\bar{r}_i)[u_i - \hat{u}_i + \frac{b_i+\bar{b}_i}{r_i+\bar{r}_i}(\beta_i\hat{x} + \gamma_i)]^2 dt \\
+\frac{1}{2}\int_0^T 2\alpha_i(x-\hat{x})(\sigma dB + \int_\Theta \mu\tilde{N}(dt,d\theta)) + \frac{1}{2}\int_0^T 2\sigma_o(\beta_i\hat{x} + \gamma_i)dB_o.
\end{array} \tag{7.159}$$

The resulting system is then the following:

$$
\begin{aligned}
&\inf_{u_i\in\mathcal{U}_i} E[L_i] = \inf_{u_i\in\mathcal{U}_i} E[\hat{L}_i] = E\{\tfrac{1}{2}\alpha_i(0)(x(0)-\hat{x}(0))^2 + \tfrac{1}{2}\beta_i(0)[\hat{x}(0)]^2 \\
&\quad +\gamma_i(0)\hat{x}(0) + \delta_i(0)\} \\
&u_i^* = -\tfrac{b_i}{r_i}\alpha_i(x-\hat{x}) - \tfrac{(b_i+\bar{b}_i)}{r_i+\bar{r}_i}(\beta_i\hat{x}+\gamma_i), \\
&d\alpha_i = -\left\{2a\alpha_i - \tfrac{b_i^2}{r_i}\alpha_i^2 - 2\alpha_i\sum_{j\neq i}\tfrac{b_j^2}{r_j}\alpha_j + q_i\right\}dt, \\
&\alpha_i(T) = q_i(T), \\
&d\beta_i = -\left\{2\beta_i(a+\bar{a}) - \beta_i^2\tfrac{(b_i+\bar{b}_i)^2}{r_i+\bar{r}_i} - 2\beta_i\sum_{j\neq i}\tfrac{(b_j+\bar{b}_j)^2}{r_j+\bar{r}_j}\beta_j + q_i + \bar{q}_i\right\}dt, \\
&\beta_i(T) = q_i(T) + \bar{q}_i(T), \\
&d\gamma_i = -\left\{(a+\bar{a})\gamma_i + \beta_{i,o}\sigma_o - \tfrac{(b_i+\bar{b}_i)^2}{r_i+\bar{r}_i}\beta_i\gamma_i\right. \\
&\left. -\beta_i\sum_{j\neq i}\tfrac{(b_j+\bar{b}_j)^2}{r_j+\bar{r}_j}\gamma_j - \gamma_i\sum_{j\neq i}\tfrac{(b_j+\bar{b}_j)^2}{r_j+\bar{r}_j}\beta_j\right\}dt, \\
&\gamma_i(T) = 0, \\
&d\delta_i = -\left\{\tfrac{1}{2}\sigma_o^2\beta_i + \gamma_{i,o}\sigma_o - \gamma_i\sum_{j\neq i}\tfrac{(b_j+\bar{b}_j)^2}{r_j+\bar{r}_j}\gamma_j - \tfrac{(b_i+\bar{b}_i)^2}{r_i+\bar{r}_i}\tfrac{1}{2}\gamma_i^2\right. \\
&\left. + [\sigma^2 + \int_\Theta \mu^2(t,\theta)\nu(d\theta)]\tfrac{1}{2}\alpha_i\right\}dt, \\
&\delta_i(T) = 0, \\
&d\hat{x} = \{[a+\bar{a} - \textstyle\sum_{i=1}^I \tfrac{(b_i+\bar{b}_i)^2}{r_i+\bar{r}_i}\beta_i]\hat{x} - \textstyle\sum_{i=1}^I \tfrac{(b_i+\bar{b}_i)^2}{r_i+\bar{r}_i}\gamma_i\}dt + \sigma_o dB_o, \\
&\hat{x}(0) = \hat{x}_0,
\end{aligned}
\tag{7.160}
$$

(whenever it admits a solution) provides a conditional mean-field Nash equilibrium in feedback strategies that is conditioned on the common noise B_o.

These stochastic Riccati equations are different from those of open-loop control strategies. The coefficients of the coupling terms $\beta_i\beta_j$, $\alpha_i\alpha_j$, are different reflecting the coupling through the state and the mean state. Note that the optimal strategy is in state-and-conditional mean-field feedback form, which is different from the standard LQG game solution. When σ_o vanishes and no common noise is available, the processes γ and δ become deterministic functions of time. In the absence of common noise, the function γ_i is identically zero.

As $\bar{a}, \bar{b}, \bar{r}, \bar{q}, \sigma_o$ vanish in (7.160), one gets the Nash equilibrium of the corresponding stochastic differential game in closed-loop strategies with $\alpha_i = \beta_i$, and u_i becomes mean-field-free. When the diffusion coefficient σ and the jump rate μ vanish, one obtains the noiseless deterministic game problem, and the optimal strategy solution will then be given by the equation in β_i because $x - \hat{x} = 0$ in the deterministic case.

The problem and its solution are now succinctly presented in the following two boxes:

LQJD-MFTG Problem with Common Noise

$$
\begin{aligned}
&L_i := \tfrac{1}{2}q_i(T)x^2(T) + \tfrac{1}{2}\bar{q}_i(T)[E(x(T)|\mathscr{F}_T^{B_o})]^2+\\
&\tfrac{1}{2}\int_0^T \left\{q_i(t)x^2(t) + \bar{q}_i(t)(E[x(t)\,|\mathscr{F}_t^{B_o}\,])^2 + r_i(t)u_i^2(t) + \bar{r}_i(t)[E(u_i(t)|\mathscr{F}_t^{B_o})]^2\right\} dt\\
&\text{Decision-maker } i: \ \ \inf_{u_i} E[L_i]\\
&\text{subject to}\\
&dx(t) = \left\{a(t)x(t) + \bar{a}(t)E[x(t)|\mathscr{F}_t^{B_o}] + \textstyle\sum_{i=1}^I b_i(t)u_i(t) + \sum_{i=1}^I \bar{b}_i E[u_i(t)|\mathscr{F}_t^{B_o}]\right\} dt\\
&+\sigma(t)dB(t) + \textstyle\int_\Theta \mu(t,\theta)\tilde{N}(dt,d\theta) + \sigma_o(t)dB_o(t),\\
&x(0) \in \mathbb{R}.\\
&\text{Conditions:} q_i(T), q_i(T) + \bar{q}_i(T), q_i(t), q_i(t) + \bar{q}_i(t), r_i(t), r_i(t) + \bar{r}_i(t) \geq \epsilon > 0,
\end{aligned}
\tag{7.161}
$$

Solution to the LQJD-MFTG Problem with Common Noise (7.161)

$$
\begin{aligned}
&\inf_{u_i\in\mathscr{U}_i} E[L_i]=E\left\{\tfrac{1}{2}\alpha_i(0)(x(0)-\hat{x}(0))^2+\tfrac{1}{2}\beta_i(0)[\hat{x}(0)]^2+\gamma_i(0)\hat{x}(0)+\delta_i(0)\right\}\\
&u_i^* = -\tfrac{b_i}{r_i}\alpha_i(x-\hat{x}) - \tfrac{(b_i+\bar{b}_i)}{r_i+\bar{r}_j}(\beta_i\hat{x}+\gamma_i),\\
&d\hat{x} = \{[a+\bar{a} - \textstyle\sum_{i=1}^I \frac{(b_i+\bar{b}_i)^2}{r_i+\bar{r}_i}\beta_i]\hat{x} + \sum_{i=1}^I \frac{(b_i+\bar{b}_i)^2}{r_i+\bar{r}_i}\gamma_i\}dt+\sigma_o dB_o,\\
&\hat{x}(0) = \hat{x}_0,\\
&d\alpha_i = -\left\{2a\alpha_i - \tfrac{b_i^2}{r_i}\alpha_i^2 - 2\alpha_i\textstyle\sum_{j\neq i}\frac{b_j^2}{r_j}\alpha_j + q_i\right\}dt,\\
&\alpha_i(T) = q_i(T),\\
&d\beta_i = -\left\{2\beta_i(a+\bar{a}) - \beta_i^2\tfrac{(b_i+\bar{b}_i)^2}{r_i+\bar{r}_i}\right.\\
&\qquad \left.-2\beta_i\textstyle\sum_{j\neq i}\frac{(b_j+\bar{b}_j)^2}{r_j+\bar{r}_j}\beta_j + q_i + \bar{q}_i\right\}dt,\\
&\beta_i(T) = q_i(T) + \bar{q}_i(T),\\
&d\gamma_i = -\left\{(a+\bar{a})\gamma_i+\beta_{i,o}\sigma_o - \tfrac{(b_i+\bar{b}_i)^2}{r_i+\bar{r}_i}\beta_i\gamma_i\right.\\
&\left.-\beta_i\textstyle\sum_{j\neq i}\frac{(b_j+\bar{b}_j)^2}{r_j+\bar{r}_j}\gamma_j - \gamma_i\sum_{j\neq i}\frac{(b_j+\bar{b}_j)^2}{r_j+\bar{r}_j}\beta_j\right\}dt,\\
&\gamma_i(T) = 0,\\
&d\delta_i = -\left\{\tfrac{1}{2}\sigma_o^2\beta_i+\gamma_{i,o}\sigma_o - \gamma_i\textstyle\sum_{j\neq i}\frac{(b_j+\bar{b}_j)^2}{r_j+\bar{r}_j}\beta_j - \frac{(b_i+\bar{b}_i)^2}{r_i+\bar{r}_i}\frac{1}{2}\gamma_i^2\right.\\
&\left.+\ [\sigma^2+\textstyle\int_\Theta \mu^2(t,\theta)\nu(d\theta)]\frac{1}{2}\alpha_i\right\}dt,\\
&\delta_i(T) = 0
\end{aligned}
\tag{7.162}
$$

7.3.2 Fully Cooperative Game Problem with Common Noise

We now consider the fully cooperative scenario where all the decision-makers decide jointly to optimize a single global objective $L_0 := \sum_i L_i$ given by

$$\begin{array}{l} \inf_{(u_1,\ldots,u_I)} E\sum_i \hat{L}_i, \\ dx = \left\{a(x-\hat{x}) + (a+\bar{a})\hat{x} + \sum_{i=1}^I b_i(u_i-\hat{u}_i) + \sum_{i=1}^I (b_i+\bar{b}_i)\hat{u}_i\right\} dt \\ +\sigma dB + \int_\Theta \mu(t,\theta)\tilde{N}(dt,d\theta) + \sigma_o dB_o, \\ x(0) = x_0. \end{array} \tag{7.163}$$

Following the same methodology as above with $q_0 = \sum_i q_i$, $\bar{q}_0 = \sum_i \bar{q}_i$ and $f_0(t,x) = \frac{1}{2}\alpha_0(t)(x-\hat{x})^2 + \frac{1}{2}\beta_0(t)\hat{x}^2 + \gamma_0(t)\hat{x} + \delta_0(t)$, we compute the gap between $\sum_i \hat{L}_i$ and $f_0(0,x(0))$ as follows:

$$\begin{array}{l} \sum_i \hat{L}_i - f_0(0,x(0)) \\ = \frac{1}{2}(q_0(T)-\alpha_0(T))(x(T)-\hat{x}(T))^2 + \frac{1}{2}[q_0(T)+\bar{q}_0(T)-\beta_0(T)]\hat{x}^2(T) \\ +(0-\gamma_0(T))\hat{x}(T) + (0-\delta_0(T)) \\ \frac{1}{2}\int_0^T \{d\alpha_0 + 2a\alpha_0 + q_0\}(x-\hat{x})^2 dt + \frac{1}{2}\int_0^T \{d\beta_0 + 2(a+\bar{a})\beta_0 + q_0 + \bar{q}_0\}\hat{x}^2 dt \\ + \int_0^T \{d\gamma_0 + (a+\bar{a})\gamma_0 + \beta_{0,o}\sigma_o\}\hat{x}dt + \frac{1}{2}\int_0^T [\sigma^2 + \int_\Theta \mu^2(t,\theta)\nu(d\theta)]\alpha_0 dt \\ +\frac{1}{2}\int_0^T \sigma_o^2\beta_0 dt + \int_0^T \gamma_{0,o}\sigma_o dt + \int_0^T d\delta_0 \\ +\frac{1}{2}\int_0^T \sum_i r_i(u_i-\hat{u}_i)^2 + 2(u_i-\hat{u}_i)\alpha_0 b_i(x-\hat{x})dt \\ +\frac{1}{2}\int_0^T \sum_i (r_i+\bar{r}_i)\hat{u}_i^2 + 2\hat{u}_i(\beta_0\hat{x}+\gamma_0)(b_i+\bar{b}_i)dt \\ +\frac{1}{2}\int_0^T 2\alpha_0(x-\hat{x})(\sigma dB + \int_\Theta \mu\tilde{N}(dt,d\theta)) + \frac{1}{2}\int_0^T 2\sigma_o(\beta_0\hat{x}+\gamma_0)dB_o. \end{array} \tag{7.164}$$

We now use the following completion of squares:

$$\begin{array}{l} \sum_i r_i(u_i-\hat{u}_i)^2 + 2(u_i-\hat{u}_i)\alpha_0 b_i(x-\hat{x}) \\ = \sum_i r_i[u_i - \hat{u}_i + \alpha_0\frac{b_i}{r_i}(x-\hat{x})]^2 - [\alpha_0^2\sum_i \frac{b_i^2}{r_i}](x-\hat{x})^2, \\ \text{and} \\ \sum_i (r_i+\bar{r}_i)\hat{u}_i^2 + 2\hat{u}_i(\beta_0\hat{x}+\gamma_0)(b_i+\bar{b}_i) \\ = \sum_i (r_i+\bar{r}_i)[\hat{u}_i + \frac{(b_i+\bar{b}_i)}{r_i+\bar{r}_i}(\beta_0\hat{x}+\gamma_0)]^2 - \sum_i \frac{(b_i+\bar{b}_i)^2}{r_i+\bar{r}_i}(\beta_0^2\hat{x}^2 + 2\beta_0\gamma_0\hat{x} + \gamma_0^2). \end{array} \tag{7.165}$$

Thus,

$$
\begin{aligned}
&\textstyle\sum_i \hat{L}_i - f_0(0, x(0)) \\
&= \tfrac{1}{2}(q_0(T) - \alpha_0(T))(x(T) - \hat{x}(T))^2 + \tfrac{1}{2}[q_0(T) + \bar{q}_0(T) - \beta_0(T)]\hat{x}^2(T) \\
&+(0 - \gamma_0(T))\hat{x}(T) + (0 - \delta_0(T)) \\
&+\tfrac{1}{2}\textstyle\int_0^T \{d\alpha_0 + 2a\alpha_0 - \alpha_0^2 \sum_i \frac{b_i^2}{r_i} + q_0\}(x - \hat{x})^2 dt \\
&+\tfrac{1}{2}\textstyle\int_0^T \{d\beta_0 + 2(a + \bar{a})\beta_0 - \beta_0^2 \sum_i \frac{(b_i+\bar{b}_i)^2}{r_i+\bar{r}_i} + q_0 + \bar{q}_0\}\hat{x}^2 dt \\
&+\textstyle\int_0^T \{d\gamma_0 + (a + \bar{a})\gamma_0 + \beta_{0,o}\sigma_o - \beta_0\gamma_0 \sum_i \frac{(b_i+\bar{b}_i)^2}{r_i+\bar{r}_i}\}\hat{x} dt \\
&+\tfrac{1}{2}\textstyle\int_0^T [\sigma^2 + \int_\Theta \mu^2(t,\theta)\nu(d\theta)]\alpha_0 dt + \tfrac{1}{2}\int_0^T \sigma_o^2 \beta_0 dt \\
&+\textstyle\int_0^T d\delta_0 + \gamma_{0,o}\sigma_o - \tfrac{1}{2}\gamma_0^2 \sum_i \frac{(b_i+\bar{b}_i)^2}{r_i+\bar{r}_i} dt \\
&+\tfrac{1}{2}\textstyle\int_0^T \sum_i r_i[u_i - \hat{u}_i + \alpha_0 \frac{b_i}{r_i}(x - \hat{x})]^2 dt \\
&+\tfrac{1}{2}\textstyle\int_0^T (r_i + \bar{r}_i)[\hat{u}_i + \frac{(b_i+\bar{b}_i)}{r_i+\bar{r}_i}(\beta_0\hat{x} + \gamma_0)]^2 dt \\
&+\tfrac{1}{2}\textstyle\int_0^T 2\alpha_0(x - \hat{x})(\sigma dB + \int_\Theta \mu \tilde{N}(dt, d\theta)) + \int_0^T \sigma_o(\beta_0\hat{x} + \gamma_0) dB_o,
\end{aligned}
\tag{7.166}
$$

which implies that the globally optimal solution is

$$
\begin{aligned}
&\inf_{(u_1,\ldots,u_I)} E \textstyle\sum_i [L_i] = \\
&E\{\tfrac{1}{2}\alpha_0(0) var(x(0)) + \tfrac{1}{2}\beta_0(0)[\hat{x}(0)]^2 + \gamma_0(0)\hat{x}(0) + \delta_0(0)\}, \\
&u_i^* = -\tfrac{b_i}{r_i}\alpha_0(x - \hat{x}) - \tfrac{(b_i+\bar{b}_i)}{r_i+\bar{r}_i}(\beta_0\hat{x} + \gamma_0), \\
&d\alpha_0 = -\left\{2a\alpha_0 - \alpha_0^2 \textstyle\sum_i \frac{b_i^2}{r_i} + q_i\right\} dt, \\
&\alpha_0(T) = q_0(T), \\
&d\beta_0 = -\left\{2\beta_0(a + \bar{a}) - \beta_0^2 \textstyle\sum_i \frac{(b_i+\bar{b}_i)^2}{r_i+\bar{r}_i} + q_0 + \bar{q}_0\right\} dt, \\
&\beta_0(T) = q_0(T) + \bar{q}_0(T), \\
&d\gamma_0 = -\left\{(a + \bar{a})\gamma_0 + \beta_{0,o}\sigma_o - \beta_0\gamma_0 \textstyle\sum_i \frac{(b_i+\bar{b}_i)^2}{r_i+\bar{r}_i}\right\} dt, \\
&\gamma(T) = 0, \\
&d\delta_0 = -\left\{[\sigma^2 + \textstyle\int_\Theta \mu^2(t,\theta)\nu(d\theta)]\frac{\alpha_0}{2} + \sigma_o^2 \frac{\beta_0}{2} + \gamma_{0,o}\sigma_o - \frac{1}{2}\gamma_0^2 \sum_i \frac{(b_i+\bar{b}_i)^2}{r_i+\bar{r}_i}\right\} dt, \\
&d\hat{x} = \{[a + \bar{a} - \beta_0 \textstyle\sum_{i=1}^I \frac{(b_i+\bar{b}_i)^2}{r_i+\bar{r}_i}]\hat{x} - \gamma_0 \sum_{i=1}^I \frac{(b_i+\bar{b}_i)^2}{r_i+\bar{r}_i}\}dt + \sigma_o dB_o, \\
&\hat{x}(0) = \hat{x}_0.
\end{aligned}
\tag{7.167}
$$

The functions α_0, β_0 are explicitly given by

$$
\begin{array}{l}
S := \sum_i \frac{b_i^2}{r_i},\\
\alpha_0(t) = \frac{a}{S} + \sqrt{\frac{q_0}{S} + \frac{a^2}{S^2}}\left(-1 + \frac{q_0(T) - \frac{a}{S} + \sqrt{\frac{q_0}{S} + \frac{a^2}{S^2}}}{\Gamma}\right),\\
\Gamma := \frac{1}{2}(q_0(T) - \frac{a}{S} + \sqrt{\frac{q_0}{S} + \frac{a^2}{S^2}})\\
-\frac{1}{2}(q_0(T) - \frac{a}{S} - \sqrt{\frac{q_0}{S} + \frac{a^2}{S^2}})e^{-2(T-t)\sqrt{\frac{q_0}{S} + \frac{a^2}{S^2}}},\\
\tilde{S} := \sum_{i=1}^{I} \frac{(b_i+\bar{b}_i)^2}{r_i+\bar{r}_i},\\
\beta_0(t) = \frac{a+\bar{a}}{\tilde{S}}\\
+\sqrt{\frac{q_0+\bar{q}_0}{\tilde{S}} + \frac{a^2}{\tilde{S}^2}}\left(-1 + \frac{q_0(T)+\bar{q}_0(T) - \frac{a+\bar{a}}{\tilde{S}} + \sqrt{\frac{q_0+\bar{q}_0}{\tilde{S}} + \frac{(a+\bar{a})^2}{\tilde{S}^2}}}{\tilde{\Gamma}}\right),\\
\tilde{\Gamma} := \frac{1}{2}(q_0(T) + \bar{q}_0(T) - \frac{a+\bar{a}}{\tilde{S}} + \sqrt{\frac{q_0+\bar{q}_0}{\tilde{S}} + \frac{(a+\bar{a})^2}{\tilde{S}^2}})\\
-\frac{1}{2}(q_0(T) + \bar{q}_0(T) - \frac{a+\bar{a}}{\tilde{S}} - \sqrt{\frac{q_0+\bar{q}_0}{\tilde{S}} + \frac{(a+\bar{a})^2}{\tilde{S}^2}})\times\\
e^{-2(T-t)\sqrt{\frac{q_0+\bar{q}_0}{\tilde{S}} + \frac{(a+\bar{a})^2}{\tilde{S}^2}}}.
\end{array}
\tag{7.168}
$$

The global optimum cost in the fully cooperative case is

$$
E[L_0] = E\{\frac{1}{2}\alpha_0(0)var(x(0)) + \frac{1}{2}\beta_0(0)[\hat{x}(0)]^2 + \gamma_0(0)\hat{x}(0) + \delta_0(0)\},
$$

which is, as to be expected, less than the total cost at the Nash equilibrium, which was

$$
\begin{array}{l}
E\left\{\frac{1}{2}(x(0) - \hat{x}(0))^2[\sum_i \alpha_i(0)] + \frac{1}{2}[\hat{x}(0)]^2[\sum_i \beta_i(0)]\right.\\
\left.+\hat{x}(0)[\sum_i \gamma_i(0)] + [\sum_i \delta_i(0)]\right\}.
\end{array}
\tag{7.169}
$$

The problem and its solution are now succinctly presented in the following two boxes.

Full Cooperation Problem with Common Noise

$$
\begin{array}{l}
\text{Team}\ :\ \inf_{(u_1,\ldots,u_I)} E\sum_i L_i,\ \text{subject to:}\\
dx = \left\{a(x-\hat{x}) + (a+\bar{a})\hat{x} + \sum_{i=1}^{I} b_i(u_i - \hat{u}_i) + \sum_{i=1}^{I}(b_i+\bar{b}_i)\hat{u}_i\right\}dt\\
+\sigma dB + \int_{\Theta}\mu(t,\theta)\tilde{N}(dt,d\theta) + \sigma_o dB_o,\\
x(0) = x_0,
\end{array}
\tag{7.170}
$$

Solution to the full cooperation problem with common noise (7.170)

$$\begin{aligned}
&\inf_{(u_1,\ldots,u_I)} E\sum_i [L_i] = \\
&E\{\tfrac{1}{2}\alpha_0(0)var(x(0)) + \tfrac{1}{2}\beta_0(0)[\hat{x}(0)]^2 + \gamma_0(0)\hat{x}(0) + \delta_0(0)\}, \\
&u_i^* = -\tfrac{b_i}{r_i}\alpha_0(x-\hat{x}) - \tfrac{(b_i+\bar{b}_i)}{r_i+\bar{r}_i}(\beta_0\hat{x}+\gamma_0), \\
&d\hat{x} = \{[a+\bar{a} - \beta_0\textstyle\sum_{i=1}^I \tfrac{(b_i+\bar{b}_i)^2}{r_i+\bar{r}_i}]\hat{x} + \gamma_0\sum_{i=1}^I \tfrac{(b_i+\bar{b}_i)^2}{r_i+\bar{r}_i}\}dt + \sigma_o dB_o, \\
&\hat{x}(0) = \hat{x}_0. \\
&d\alpha_0 = -\left\{2a\alpha_0 - \alpha_0^2(\textstyle\sum_i \tfrac{b_i^2}{r_i}) + q_0\right\}dt, \\
&\alpha_0(T) = q_0(T), \\
&d\beta_0 = -\left\{2\beta_0(a+\bar{a}) - \beta_0^2\textstyle\sum_i \tfrac{(b_i+\bar{b}_i)^2}{r_i+\bar{r}_i} + q_0 + \bar{q}_0\right\}dt, \\
&\beta_0(T) = q_0(T) + \bar{q}_0(T), \\
&d\gamma_0 = -\left\{(a+\bar{a})\gamma_0 + \beta_{0,o}\sigma_o - \beta_0\gamma_0\textstyle\sum_i \tfrac{(b_i+\bar{b}_i)^2}{r_i+\bar{r}_i}\right\}dt, \\
&\gamma(T) = 0, \\
&d\delta_0 = -\left\{[\sigma^2 + \textstyle\int_\Theta \mu^2(t,\theta)\nu(d\theta)]\tfrac{\alpha_0}{2} + \sigma_o^2\tfrac{\beta_0}{2} + \gamma_{0,o}\sigma_o - \tfrac{1}{2}\gamma_0^2\sum_i \tfrac{(b_i+\bar{b}_i)^2}{r_i+\bar{r}_i}\right\}dt,
\end{aligned} \tag{7.171}$$

7.3.3 *Nash Bargaining Problem with Common Noise*

Mean-field-type bargaining theory deals with the situation in which players can realize—through cooperation—other better outcomes than the one that becomes effective when they do not cooperate. This noncooperative outcome based on Nash equilibrium is called the threat point $L^{NE} = (L_1^{NE}, \ldots, L_I^{NE})$. The question is which outcome might the players possibly agree to. Let $\mathscr{V}$ be the set of feasible outcomes of the benefit of bargaining. We assume that, if the agents unanimously agree on a point $V = (V_1, \ldots, V_I) \in \mathscr{V}$, they obtain v. Otherwise, they obtain $L^{NE} = (L_1^{NE}, \ldots, L_I^{NE})$. This presupposes that each player can enforce the threat point, when he does not agree with a proposal. The outcome V the players will finally agree on is called the solution of the bargaining problem. The Nash bargaining solution selects for a given set $\mathscr{V}$ of feasible payoffs the point at which the product of gains from L^{NE} is maximal:

$$NBS(\mathscr{V}, L^{NE}) = \arg\max_{v\in\mathscr{V}}\{\prod_{i\in\mathscr{I}}[L_i^{NE} - v_i]\}.$$

The difficulty is that $u \mapsto \prod_{i=1}^I [L_i^{NE} - L_i(u)]$ is not necessarily concave. Here we exploit the convexity of the functional $u \mapsto \langle w, L(u)\rangle$ for any given $w = (w_1, \ldots, w_I) \in \mathbb{R}_{++}^I$ such that $\sum_{i=1}^I w_i = 1$ to reach any point in the Pareto frontier of the game. The first-order condition of the maximization (in w) of the product $P := \prod_{i=1}^{I} [L_i^{NE} - L_i(\hat{u}(w))]$ yields

$$\sum_{i=1}^{I} \frac{\partial}{\partial w_j} L_i(\hat{u}(w)).\{\prod_{j\neq i} [L_j^{NE} - L_j(\hat{u}(w))]\} = c$$

This is equivalent to

$$\sum_{i=1}^{I} \frac{P}{L_i^{NE} - L_i(\hat{u}(w))} \frac{\partial}{\partial w_j} L_i(\hat{u}(w)) = c.$$

We set $y_i := \frac{\frac{P}{L_i^{NE} - L_i(\hat{u}(w))}}{\sum_{i=1}^{I}\left(\frac{P}{L_i^{NE} - L_i(\hat{u}(w))}\right)}$. Then, it follows that $\sum_{i=1}^{I} y_i \frac{\partial}{\partial w_j} L_i(\hat{u}(w)) = \frac{c}{\sum_{i=1}^{I}\left(\frac{P}{L_i^{NE} - L_i(\hat{u}(w))}\right)}$. Moreover, $y_i \geq 0,\ \sum_{i=1}^{I} y_i = 1.$

Assume that the matrix $(\frac{\partial}{\partial w_j} L_i(\hat{u}(w)))_{(i,j)\in \mathcal{I}^2}$ has at least rank $I-1$. Then the Nash bargaining solution is explicitly given by $V = (L_1(\hat{u}(w^*)), \ldots, L_I(\hat{u}(w^*)))$ with the weight

$$w_i^* = \frac{\prod_{j\neq i}[L_j^{NE} - L_j(\hat{u}(w^*))]}{\sum_{k=1}^{I}\prod_{j\neq k}[L_j^{NE} - L_j(\hat{u}(w^*))]} = y_i(w^*)$$

where the optimal bargaining strategy profile is $\hat{u}(w) \in \arg\min_u\{\langle w, L(u)\rangle\}$. It remains to compute the functional $\hat{u}(w) = (\hat{u}_1(w), \ldots, \hat{u}_I(w))$. To do so, we solve the weighted objective functional minimization problem:

$$\begin{aligned}
&\inf_{(u_1,\ldots,u_I)} E \sum_i w_i L_i,\\
&dx = \left\{a(x-\hat{x}) + (a+\bar{a})\hat{x} + \sum_{i=1}^{I} b_i(u_i - \hat{u}_i) + \sum_{i=1}^{I}(b_i + \bar{b}_i)\hat{u}_i\right\} dt\\
&+\sigma dB + \int_{\Theta} \mu(t,\theta)\tilde{N}(dt, d\theta) + \sigma_o dB_o,\\
&x(0) = x_0.
\end{aligned} \tag{7.172}$$

Following the same methodology as above, we obtain

$$\begin{array}{l}
\inf_{(u_1,\ldots,u_I)} E\sum_i w_i \hat{L}_i = \\
E\{\frac{1}{2}\alpha_0(0)var(x(0)) + \frac{1}{2}\beta_0(0)[\hat{x}(0)]^2 + \gamma_0(0)\hat{x}(0) + \delta_0(0)\}, \\
\hat{u}_i(w) = -\frac{b_i}{w_i r_i}\alpha_0(x-\hat{x}) - \frac{(b_i+\bar{b}_i)}{w_i(r_i+\bar{r}_i)}(\beta_0\hat{x}+\gamma_0), \\
d\alpha_0 = -\left\{2a\alpha_0 - \alpha_0^2\sum_i \frac{b_i^2}{w_i r_i} + \sum_i w_i q_i\right\}dt, \\
\alpha_0(T) = \sum_i w_i q_i(T), \\
d\beta_0 = -\left\{2\beta_0(a+\bar{a}) - \beta_0^2\sum_i \frac{(b_i+\bar{b}_i)^2}{w_i(r_i+\bar{r}_i)} + \sum_i w_i(q_i+\bar{q}_i)\right\}dt, \\
\beta_0(T) = \sum_i w_i(q_i+\bar{q}_i(T)), \\
d\gamma_0 = -\left\{(a+\bar{a})\gamma_0 + \beta_{0,o}\sigma_o - \beta_0\gamma_0\sum_i \frac{(b_i+\bar{b}_i)^2}{w_i(r_i+\bar{r}_i)}\right\}dt, \\
\gamma_0(T) = 0, \\
d\delta_0 = -\left\{[\sigma^2 + \int_\Theta \mu^2(t,\theta)\nu(d\theta)]\frac{\alpha_0}{2} + \sigma_o^2\frac{\beta_0}{2}\right. \\
\left. +\gamma_{0,o}\sigma_o - \frac{1}{2}\gamma_0^2\sum_i \frac{(b_i+\bar{b}_i)^2}{w_i(r_i+\bar{r}_i)}\right\}dt, \\
d\hat{x} = \{[a+\bar{a}-\beta_0\sum_{i=1}^I \frac{(b_i+\bar{b}_i)^2}{w_i(r_i+\bar{r}_i)}]\hat{x} - \gamma_0\sum_{i=1}^I \frac{(b_i+\bar{b}_i)^2}{w_i(r_i+\bar{r}_i)}\}dt, \\
\hat{x}(0) = \hat{x}_0.
\end{array} \tag{7.173}$$

The weight w satisfies

$$w_1(L_1^{NE} - L_1(u(w))) = \ldots = w_I(L_I^{NE} - L_I(u(w))).$$

We now summarize the problem and its solution in the following boxes.

Nash Bargaining Problem with Common Noise

$$NBS(\mathcal{V}, L^{NE}) = \arg\max_{v\in\mathcal{V}}\{\prod_{i\in\mathcal{I}}[L_i^{NE} - v_i]\} \tag{7.174}$$

Solution to the Nash Bargaining Problem with Common Noise (7.174)

$$\begin{array}{l}
\inf_{(u_1,\ldots,u_I)} E\sum_{i=1}^I w_i L_i = \\
E\{\frac{1}{2}\alpha_0(0)var(x(0)) + \frac{1}{2}\beta_0(0)[\hat{x}(0)]^2 + \gamma_0(0)\hat{x}(0) + \delta_0(0)\}, \\
\hat{u}(w) \in \arg\min_u \langle w, L(u)\rangle \\
\hat{u}_i(w) = -\frac{b_i}{w_i r_i}\alpha_0(x-\hat{x}) - \frac{(b_i+\bar{b}_i)}{w_i(r_i+\bar{r}_i)}(\beta_0\hat{x}+\gamma_0), \\
w_i^* = \frac{\prod_{j\neq i}[L_j^{NE} - L_j(\hat{u}(w^*))]}{\sum_{k=1}^I \prod_{j\neq k}[L_j^{NE} - L_j(\hat{u}(w^*))]}.
\end{array} \tag{7.175}$$

7.3.4 Robust Optimization Problem with Common Noise

We now consider a robust mean-field-type game with two decision-makers. Decision-maker 1 minimizes a cost function with respect to u_1, and decision-maker 2 maximizes the same cost function with respect to u_2. The inf-sup problem of mean-field type is given by

$$\begin{array}{l} \inf_{u_1}\sup_{u_2} E[L(u_1,u_2)],\\ dx = \Big\{a(t)(x(t)-\hat{x}(t)) + (a(t)+\bar{a}(t))\hat{x}(t)\\ +\sum_{i=1}^{2} b_i(t)(u_i(t)-\hat{u}_i(t)) + \sum_{i=1}^{2}(b_i(t)+\bar{b}_i(t))\hat{u}_i(t)\Big\}\, dt\\ +\sigma(t)dB(t) + \int_{\Theta}\mu(t,\theta)\tilde{N}(t,d\theta)+\sigma_o(t)dB_o(t),\\ x(0)=x_0 \in L^2(\mathbb{R}), \end{array} \tag{7.176}$$

where the objective functional is

$$\begin{array}{l} L(u_1,u_2) = \frac{1}{2}[q(T)(x(T)-\hat{x}(T))^2 + [q(T)+\bar{q}(T)]\hat{x}^2(T)\\ \qquad + \int_0^T q(t)(x(t)-\hat{x}(t))^2 + [q(t)+\bar{q}(t)]\hat{x}^2(t)dt\\ + \int_0^T r_1(t)(u_1(t)-\hat{u}_1(t))^2 + (r_1(t)+\bar{r}_1(t))[\hat{u}_1(t)]^2dt\\ + \int_0^T r_2(t)(u_2(t)-\hat{u}_2(t))^2 + (r_2(t)+\bar{r}_2(t))[\hat{u}_2(t)]^2dt] \end{array} \tag{7.177}$$

The risk-neutral robust mean-field-type equilibrium problem we are concerned with is characterization of the quadruple $(x^*, u_1^*, u_2^*, \hat{x}^*)$ such that u_1^* is the minimizer and u_2^* is the maximizer of the best-response problem (7.176) and the conditional expected value of the resulting common state created by both decision-makers is $\hat{x}^*(t) = E[x^*(t) \mid \mathcal{F}_t^{B_o}]$.

Below we solve Problem (7.176) for $r_1(t) > 0, \bar{r}_1(t) > 0, r_2(t) < 0, \bar{r}_2(t) < 0$. It follows that the equilibrium solution is

$$\begin{array}{l} \inf_{u_1}\sup_{u_2} E[L(u_1,u_2)] =\\ E\{\frac{1}{2}\alpha(0)var(x(0)) + \frac{1}{2}\beta(0)[\hat{x}(0)]^2 + \gamma(0)\hat{x}(0) + \delta(0)\},\\ u_1^* = -\frac{b_1}{r_1}\alpha(x-\hat{x}) - \frac{(b_1+\bar{b}_1)}{r_1+\bar{r}_1}(\beta\hat{x}+\gamma),\\ u_2^* = -\frac{b_2}{r_2}\alpha(x-\hat{x}) - \frac{(b_2+\bar{b}_2)}{r_2+\bar{r}_2}(\beta\hat{x}+\gamma),\\ d\alpha = -\left\{2a\alpha - \alpha^2\sum_{i=1}^{2}\frac{b_i^2}{r_i} + q\right\}dt,\\ \alpha(T) = q(T),\\ d\beta = -\left\{2\beta(a+\bar{a}) - \beta^2\sum_{i=1}^{2}\frac{(b_i+\bar{b}_i)^2}{r_i+\bar{r}_i} + q + \bar{q}\right\}dt,\\ \beta(T) = q(T)+\bar{q}(T),\\ d\gamma = -\left\{(a+\bar{a})\gamma + \beta_o\sigma_o - \beta\gamma\sum_{i=1}^{2}\frac{(b_i+\bar{b}_i)^2}{r_i+\bar{r}_i}\right\}dt,\\ \gamma(T) = 0,\\ d\delta = -\Big\{[\sigma^2 + \int_{\Theta}\mu^2(t,\theta)\nu(d\theta)]\frac{\alpha}{2} + \sigma_o^2\frac{\beta}{2}\\ +\gamma_o\sigma_o - \frac{1}{2}\gamma^2\sum_{i=1}^{2}\frac{(b_i+\bar{b}_i)^2}{r_i+\bar{r}_i}\Big\}dt,\\ d\hat{x} = \{[a+\bar{a}-\beta\sum_{i=1}^{2}\frac{(b_i+\bar{b}_i)^2}{r_i+\bar{r}_i}]\hat{x} - \gamma\sum_{i=1}^{2}\frac{(b_i+\bar{b}_i)^2}{r_i+\bar{r}_i}\}dt + \sigma_o dB_o,\\ \hat{x}(0) = \hat{x}_0. \end{array} \tag{7.178}$$

When $r_1 > 0,\ \bar{r}_1 \geq 0,\ r_2 < 0, \bar{r}_2 \leq 0,\quad S_2 = \sum_{i=1}^{2} \frac{b_i^2}{r_i} > 0$ and $\tilde{S}_2 = \sum_{i=1}^{2} \frac{(b_i+\bar{b}_i)^2}{r_i+\bar{r}_i} > 0$ are independent of time t, then the functions α, β are explicitly given by

$$
\begin{aligned}
\alpha(t) &= \tfrac{a}{S_2} + \sqrt{\tfrac{q}{S_2} + \tfrac{a^2}{S_2^2}} \left(-1 + \frac{q(T) - \frac{a}{S_2} + \sqrt{\frac{q}{S_2} + \frac{a^2}{S_2^2}}}{\Gamma} \right), \\
\Gamma &:= \tfrac{1}{2}\Big(q(T) - \tfrac{a}{S_2} + \sqrt{\tfrac{q}{S_2} + \tfrac{a^2}{S_2^2}}\Big) - \tfrac{1}{2}\Big(q(T) - \tfrac{a}{S_2} - \sqrt{\tfrac{q}{S_2} + \tfrac{a^2}{S_2^2}}\Big) e^{-2(T-t)\sqrt{\frac{q}{S_2} + \frac{a^2}{S_2^2}}}, \\
\beta(t) &= \tfrac{a+\bar{a}}{\tilde{S}_2} + \sqrt{\tfrac{q+\bar{q}}{\tilde{S}_2} + \tfrac{a^2}{\tilde{S}_2^2}} \left(-1 + \frac{q(T)+\bar{q}(T) - \frac{a+\bar{a}}{\tilde{S}_2} + \sqrt{\frac{q+\bar{q}}{\tilde{S}_2} + \frac{(a+\bar{a})^2}{\tilde{S}_2^2}}}{\tilde{\Gamma}} \right), \\
\tilde{\Gamma} &:= \tfrac{1}{2}\Big(q(T) + \bar{q}(T) - \tfrac{a+\bar{a}}{\tilde{S}_2} + \sqrt{\tfrac{q+\bar{q}}{\tilde{S}_2} + \tfrac{(a+\bar{a})^2}{\tilde{S}_2^2}}\Big) \\
&\quad -\tfrac{1}{2}\Big(q(T) + \bar{q}(T) - \tfrac{a+\bar{a}}{\tilde{S}_2} - \sqrt{\tfrac{q+\bar{q}}{\tilde{S}_2} + \tfrac{(a+\bar{a})^2}{\tilde{S}_2^2}}\Big) e^{-2(T-t)\sqrt{\frac{q+\bar{q}}{\tilde{S}_2} + \frac{(a+\bar{a})^2}{\tilde{S}_2^2}}}.
\end{aligned}
\tag{7.179}
$$

Notice that under the conditions $r_1(t) > 0,\ \bar{r}_1(t) > 0,\ r_2(t) < 0, \bar{r}_2(t) < 0,\ S_2(t) > 0, \tilde{S}_2(t) > 0$, the minimax solution is also a maximin solution: there is a saddle point and the saddle point is (u_1^*, u_2^*). It solves

$$
E[L(u_1^*, u_2)] \leq E[L(u_1^*, u_2^*)] \leq E[L(u_1, u_2^*)],\ \forall (u_1, u_2) \in \mathcal{U}_1 \times \mathcal{U}_2. \tag{7.180}
$$

The value of the game is $E[L(u_1^*, u_2^*)]$. The robust optimization problem (7.181) with common noise is summarized below.

Robust Optimization Problem with Common Noise

$$
\begin{aligned}
& L(u_1, u_2) = \tfrac{1}{2}[q(T)(x(T) - \hat{x}(T))^2 + [q(T) + \bar{q}(T)]\hat{x}^2(T) \\
& \qquad + \textstyle\int_0^T q(t)(x(t) - \hat{x}(t))^2 + [q(t) + \bar{q}(t)]\hat{x}^2(t)dt \\
& + \textstyle\int_0^T r_1(t)(u_1(t) - \hat{u}_1(t))^2 + (r_1(t) + \bar{r}_1(t))[\hat{u}_1(t)]^2 dt \\
& + \textstyle\int_0^T r_2(t)(u_2(t) - \hat{u}_2(t))^2 + (r_2(t) + \bar{r}_2(t))[\hat{u}_2(t)]^2 dt], \\
& \inf_{u_1} \sup_{u_2} E[L(u_1, u_2)] \text{ subject to} \\
& dx = \Big\{a(t)(x(t) - \hat{x}(t)) + (a(t) + \bar{a}(t))\hat{x}(t) \\
& + \textstyle\sum_{i=1}^{2} b_i(t)(u_i(t) - \hat{u}_i(t)) + \sum_{i=1}^{2}(b_i(t) + \bar{b}_i(t))\hat{u}_i(t)\Big\} dt \\
& + \sigma(t)dB(t) + \textstyle\int_{\Theta} \mu(t, \theta)\tilde{N}(t, d\theta) + \sigma_o(t)dB_o(t), \\
& x(0) = x_0 \in L^2(\mathbb{R}),
\end{aligned}
\tag{7.181}
$$

The solution to the robust optimization problem (7.181) with common noise is summarized below.

Solution to Robust Optimization Problem with Common Noise (7.181)

$$\begin{aligned}
&\inf_{u_1}\sup_{u_2} E[L(u_1,u_2)]=E\{\tfrac{1}{2}\alpha(0)var(x(0))+\tfrac{1}{2}\beta(0)[\hat{x}(0)]^2\\
&\qquad\qquad +\gamma(0)\hat{x}(0)+\delta(0)\},\\
&u_1^* = -\tfrac{b_1}{r_1}\alpha(x-\hat{x}) - \tfrac{(b_1+\bar{b}_1)}{r_1+\bar{r}_1}(\beta\hat{x}+\gamma),\\
&u_2^* = -\tfrac{b_2}{r_2}\alpha(x-\hat{x}) - \tfrac{(b_2+\bar{b}_2)}{r_2+\bar{r}_2}(\beta\hat{x}+\gamma),\\
&d\hat{x} = \{[a+\bar{a}-\beta\textstyle\sum_{i=1}^2 \tfrac{(b_i+\bar{b}_i)^2}{r_i+\bar{r}_i}]\hat{x}+\gamma\textstyle\sum_{i=1}^2\tfrac{(b_i+\bar{b}_i)^2}{r_i+\bar{r}_i}\}dt+\sigma_o dB_o,\\
&\hat{x}(0)=\hat{x}_0,\\
&d\alpha = -\left\{2a\alpha - \alpha^2\textstyle\sum_{i=1}^2\tfrac{b_i^2}{r_i}+q\right\}dt,\\
&\alpha(T)=q(T),\\
&d\beta = -\left\{2\beta(a+\bar{a})-\beta^2\textstyle\sum_{i=1}^2\tfrac{(b_i+\bar{b}_i)^2}{r_i+\bar{r}_i}+q+\bar{q}\right\}dt,\\
&\beta(T)=q(T)+\bar{q}(T),\\
&d\gamma = -\left\{(a+\bar{a})\gamma+\beta_o\sigma_o-\beta\gamma\textstyle\sum_{i=1}^2\tfrac{(b_i+\bar{b}_i)^2}{r_i+\bar{r}_i}\right\}dt,\\
&\gamma(T)=0,\\
&d\delta = -\left\{[\sigma^2+\textstyle\int_\Theta\mu^2(t,\theta)\nu(d\theta)]\tfrac{\alpha}{2}+\sigma_o^2\tfrac{\beta}{2}+\gamma_o\sigma_o-\tfrac{1}{2}\gamma^2\textstyle\sum_{i=1}^2\tfrac{(b_i+\bar{b}_i)^2}{r_i+\bar{r}_i}\right\}dt,
\end{aligned} \tag{7.182}$$

7.3.5 *Verification and Validation*

In this section, we verify the validity of our results above using the stochastic Bellman system. It is known that the conditional measure solves a stochastic Kolmogorov–Fokker–Planck equation given by

$$dm = \{-[mD]_x + \frac{1}{2}(\sigma^2 m)_{xx} + \frac{\sigma_o^2}{2}m_{xx} + J^*[m]\}dt - \sigma_o m_x dB_o,$$

where the drift term is $D := a(x-\hat{x}) + (a+\bar{a})\hat{x} + \sum_{i=1}^I b_i(u_i-\hat{u}_i) + \sum_{i=1}^I(b_i + \bar{b}_i)\hat{u}_i$, and the jump term is

$$J[\phi] := \int_\Theta [\phi(x+\mu) - \phi(x) - \mu\phi_x]\nu(d\theta),$$

J^* being the adjoint operator associated with J.

We write the objective functional in terms of m. Then we use the stochastic Hamilton–Jacobi-Bellman equation using Kunita's correction formula. We show below that Kunita's correction term yields $+\ \sigma_o^2 \int m_x \partial_x [\int_y m(dy) V_{i,xm}] dxdt$, which is zero in this case. The stochastic Bellman equilibrium system is

$$
\begin{aligned}
&-d\hat{V} + \Big\{ \hat{V}_{i,t}(t,m) + \textstyle\int_x H_i(x, m, \hat{V}_{i,m}, \hat{V}_{i,xm}, \hat{V}_{i,xxm}) m(dx) \\
&+\sigma_o \textstyle\int m_x \sigma_o \partial_x [\int_y m(dy) V_{i,xm}] dxdt + \frac{\sigma_o^2}{2} \int_{x,y} \hat{V}_{i,mm}(x,y) m_x(dx) m_x(dy) \Big\} dt \\
&+[\sigma_o \textstyle\int_x m(dx) V_{i,xm}] dB_o = 0,
\end{aligned}
\tag{7.183}
$$

where

$$
\hat{V}_i(T,m) = \frac{1}{2} q(T) \int_y (y-\bar{m})^2\, m(dy) + \frac{1}{2}[q(T)+\bar{q}(T)] \left[\int_y y\, m(dy) \right]^2,
$$

and the integrand Hamiltonian is

$$
\begin{aligned}
&H_i(x, m, \hat{V}_{i,m}, \hat{V}_{i,xm}, \hat{V}_{i,xmm}) = \inf_{u_i} \Big\{ \tfrac{1}{2} q_i (x-\hat{x})^2 + \tfrac{1}{2}[q_i + \bar{q}_i] \hat{x}^2 \\
&+\tfrac{1}{2} r_i (u_i - \hat{u}_i)^2 + \tfrac{1}{2}(r_i + \bar{r}_i)[\hat{u}_i]^2 + [a(x-\hat{x}) + (a+\bar{a})\hat{x}] \hat{V}_{i,xm} \\
&+ [\textstyle\sum_{i=1}^I b_i (u_i - \hat{u}_i) + \sum_{i=1}^I (b_i + \bar{b}_i) \hat{u}_i] \hat{V}_{i,xm} \Big\} \\
&+\tfrac{\sigma^2+\sigma_0^2}{2} \hat{V}_{i,xxm} + \textstyle\int_\Theta [\hat{V}_{i,m}(t, x+\mu) - \hat{V}_{i,m} - \mu \hat{V}_{i,xm}] \nu(d\theta)
\end{aligned}
\tag{7.184}
$$

Our guess functional is

$$
\hat{V}_i(t,m) = \frac{\alpha_i}{2} \int_y (y-\bar{m})^2\, m(dy) + \frac{\beta_i}{2} \left[\int_y y\, m(dy) \right]^2 + \gamma_i \int_y y\, m(dy) + \delta_i \int_y m(dy).
$$

$$
\begin{aligned}
&\tfrac{1}{2} r_i (u_i - \hat{u}_i)^2 + \tfrac{1}{2}(r_i + \bar{r}_i)[\hat{u}_i]^2 + [\textstyle\sum_{i=1}^I b_i (u_i - \hat{u}_i) + \sum_{j=1}^I (b_j + \bar{b}_j) \hat{u}_i] \hat{V}_{i,xm} \\
&= \tfrac{1}{2} r_i (u_i - \hat{u}_i)^2 + \tfrac{1}{2}(r_i + \bar{r}_i)[\hat{u}_i]^2 + \Big\{ \hat{V}_{i,xm} - E\hat{V}_{i,xm} | \mathscr{F}^{B_o} \Big\} \textstyle\sum_{j=1}^I b_j (u_j - \hat{u}_j) \\
&+[E\hat{V}_{i,xm} | \mathscr{F}^{B_o}] \textstyle\sum_{j=1}^I b_j (u_j - \hat{u}_j) + \Big\{ \hat{V}_{i,xm} - E\hat{V}_{i,xm} | \mathscr{F}^{B_o} \Big\} \sum_{j=1}^I (b_j + \bar{b}_j) \hat{u}_j \\
&+[E\hat{V}_{i,xm} | \mathscr{F}^{B_o}] \textstyle\sum_{j=1}^I (b_j + \bar{b}_j) \hat{u}_j,
\end{aligned}
\tag{7.185}
$$

where we have used the following decomposition:

$$
\hat{V}_{i,xm} = \hat{V}_{i,xm} - E[\hat{V}_{i,xm} | \mathscr{F}^{B_o}] + E[\hat{V}_{i,xm} | \mathscr{F}^{B_o}].
$$

Noting that the conditional expected value of the following term

$$
[E\hat{V}_{i,xm} | \mathscr{F}^{B_o}] \sum_{j=1}^I b_j (u_j - \hat{u}_j) + \Big\{ \hat{V}_{i,xm} - E\hat{V}_{i,xm} | \mathscr{F}^{B_o} \Big\} \sum_{j=1}^I (b_j + \bar{b}_j) \hat{u}_j
\tag{7.186}
$$

is zero, the problem boils down to the optimization of

$$\begin{aligned}&\tfrac{1}{2}r_i(u_i-\hat{u}_i)^2+\tfrac{1}{2}(r_i+\bar{r}_i)[\hat{u}_i]^2\\&+\left\{\hat{V}_{i,xm}-E\hat{V}_{i,xm}|\mathscr{F}^{B_o}\right\}\sum_{j=1}^{I}b_j(u_j-\hat{u}_j)+[E\hat{V}_{i,xm}|\mathscr{F}^{B_o}]\sum_{j=1}^{I}(b_j+\bar{b}_j)\hat{u}_j.\end{aligned}\tag{7.187}$$

The optimal strategies are then

$$\begin{aligned}u_i^*&=\hat{u}_i-\tfrac{b_i}{r_i}\left\{\hat{V}_{i,xm}-E\hat{V}_{i,xm}|\mathscr{F}^{B_o}\right\}\\&=-\tfrac{b_i+\bar{b}_i}{r_i+\bar{r}_i}(\beta_i\left[\int_y y\,m(t,dy)\right]+\gamma_i)-\tfrac{b_i}{r_i}\alpha_i\left[x-\int_y y\,m(t,dy)\right],\\\hat{u}_i^*&=-\tfrac{b_i+\bar{b}_i}{r_i+\bar{r}_i}E[\hat{V}_{i,xm}|\mathscr{F}^{B_o}]=-\tfrac{b_i+\bar{b}_i}{r_i+\bar{r}_i}(\beta_i\left[\int_y y\,m(t,dy)\right]+\gamma_i),\end{aligned}\tag{7.188}$$

which are exactly the expressions of the optimal strategies obtained in (7.160). Now note that

$$\begin{aligned}&\tilde{H}_i(x,m,\hat{V}_{i,m},\hat{V}_{i,xm},\hat{V}_{i,xxm})=\tfrac{1}{2}q_i(x-\hat{x})^2+\tfrac{1}{2}[q_i+\bar{q}_i]\hat{x}^2\\&+a(x-\hat{x})(\hat{V}_{i,xm}-E\hat{V}_{i,xm}|\mathscr{F}^{B_o})+(a+\bar{a})\hat{x}E\hat{V}_{i,xm}|\mathscr{F}^{B_o}\\&+\tfrac{(\sigma^2+\sigma_o^2)}{2}\hat{V}_{i,xxm}+\int_\Theta[\hat{V}_{i,m}(t,x+\mu)-\hat{V}_{i,m}-\mu\hat{V}_{i,xm}]\nu(d\theta)\\&+\inf_{u_i}\left\{\tfrac{1}{2}r_i(u_i-\hat{u}_i)^2+\tfrac{1}{2}(r_i+\bar{r}_i)[\hat{u}_i]^2\right.\\&\left.+[\hat{V}_{i,xm}-E\hat{V}_{i,xm}|\mathscr{F}^{B_o}]\sum_{j=1}^{I}b_j(u_j-\hat{u}_j)+[E\hat{V}_{i,xm}|\mathscr{F}^{B_o}]\sum_{j=1}^{I}(b_j+\bar{b}_j)\hat{u}_j\right\}\\&=\tfrac{1}{2}\{2a\alpha_i-2\alpha_i\sum_{j=1}^{I}\tfrac{b_j^2}{r_j}\alpha_j+\tfrac{b_i^2}{r_i}^2\alpha_i^2+q_i\}(x-\hat{x})^2\\&+\tfrac{1}{2}\left\{2(a+\bar{a})\beta_i-2\beta_i\sum_{j=1}^{I}\tfrac{(b_j+\bar{b}_j)^2}{(r_j+\bar{r}_j)}\beta_j+\tfrac{(b_i+\bar{b}_i)^2}{r_i+\bar{r}_i}\beta_i^2+q_i+\bar{q}_i\right\}\hat{x}^2\\&+\tfrac{1}{2}\left\{2(a+\bar{a})\gamma_i+2\gamma_i\beta_i\tfrac{(b_i+\bar{b}_i)^2}{(r_i+\bar{r}_i)}-2\beta_i\sum_j\tfrac{(b_j+\bar{b}_j)^2}{(r_j+\bar{r}_j)}\gamma_j-2\gamma_i\sum_j\tfrac{(b_j+\bar{b}_j)^2}{(r_j+\bar{r}_j)}\beta_j\right\}\hat{x}\\&+\tfrac{1}{2}\tfrac{(b_i+\bar{b}_i)^2}{(r_i+\bar{r}_i)}\gamma_i^2-\gamma_i\sum_j\tfrac{(b_j+\bar{b}_j)^2}{(r_j+\bar{r}_j)}\gamma_j+\tfrac{1}{2}\alpha_i[\sigma^2+\sigma_o^2+\int_\Theta\mu(t,\theta)\nu(d\theta)]\end{aligned}\tag{7.189}$$

$$\tfrac{\sigma_o^2}{2}\int_x\int_y\hat{V}_{i,mm}m_x(dx)m_x(dy)=\tfrac{\sigma_o^2}{2}\int_x\int_y\hat{V}_{i,xymm}m(dx)m(dy)=\tfrac{\sigma_o^2}{2}(\beta_i-\alpha_i),\tag{7.190}$$

and

$$\int_x\sigma_o\hat{V}_{xm}dB_om(dx)=\sigma_o\,dB_o(\beta_i\hat{x}+\gamma_i)=[\sigma_o\beta_i\,dB_o]\hat{x}+[\sigma_o\gamma_i\,dB_o].\tag{7.191}$$

Using the time derivative and the square terms in dt, we obtain

$$\begin{aligned}\hat{V}_{i,t}&=\tfrac{\dot{\alpha}_i}{2}\int_y(y-\bar{m})^2\,m(dy)+\tfrac{\dot{\beta}_i}{2}\left[\int_y y\,m(dy)\right]^2+\dot{\gamma}_i\int_y y\,m(dy)+\dot{\delta}_i\\&+\beta_{i,o}\sigma_o[\int_y y\,m(dy)]+\gamma_{i,o}\sigma_o,\end{aligned}\tag{7.192}$$

and the stochastic Bellman system:

$$
\begin{aligned}
&-d\hat{V}_i + \Big\{\hat{V}_{i,t}(t,m) + \int_x H_i(x,m,\hat{V}_{i,m},\hat{V}_{i,xm},\hat{V}_{i,xxm})m(t,dx) \\
&+\tfrac{\sigma_o^2}{2}\int_{x,y}\hat{V}_{i,mm}(x,y)m_x(dx)m_x(dy)\Big\}\,dt \\
&+\sigma_o\int_x m(t,dx)V_{i,xm}dB_o \\
&= \tfrac{1}{2}\{d\alpha_i + 2a\alpha_i - 2\alpha_i\textstyle\sum_{j=1}^{I}\frac{b_j^2}{r_j}\alpha_j + \frac{b_i^2}{r_i}\alpha_i^2 + q_i\}(x-\hat{x})^2 \\
&+\tfrac{1}{2}\Big\{d\beta_i + 2(a+\bar{a})\beta_i - 2\beta_i\textstyle\sum_{j=1}^{I}\frac{(b_j+\bar{b}_j)^2}{(r_j+\bar{r}_j)}\beta_j \\
&+\tfrac{(b_i+\bar{b}_i)^2}{r_i+\bar{r}_i}\beta_i^2 + q_i + \bar{q}_i\Big\}\hat{x}^2 \\
&+\tfrac{1}{2}\Big\{d\gamma_i + \beta_{i,o}\sigma_o + 2(a+\bar{a})\gamma_i + 2\gamma_i\beta_i\tfrac{(b_i+\bar{b}_i)^2}{(r_i+\bar{r}_i)} \\
&-2\beta_i\textstyle\sum_j\frac{(b_j+\bar{b}_j)^2}{(r_j+\bar{r}_j)}\gamma_j - 2\gamma_i\sum_j\frac{(b_j+\bar{b}_j)^2}{(r_j+\bar{r}_j)}\beta_j + 2[\sigma_o\beta_i\,dB_o]\Big\}\hat{x} \\
&+d\delta_i + \gamma_{i,o}\sigma_o + \tfrac{1}{2}\tfrac{(b_i+\bar{b}_i)^2}{(r_i+\bar{r}_i)}\gamma_i^2 - \gamma_i\textstyle\sum_j\frac{(b_j+\bar{b}_j)^2}{(r_j+\bar{r}_j)}\gamma_j \\
&+\tfrac{1}{2}\alpha_i[\sigma^2 + \textstyle\int_\Theta \mu(t,\theta)\nu(d\theta)] \\
&+\tfrac{\sigma_o^2}{2}\beta_i + [\sigma_o\gamma_i\,dB_o],
\end{aligned}
\tag{7.193}
$$

and by identifying the coefficients, we retrieve the expressions in (7.160), which confirms the validity of the feedback strategies obtained by the approach developed here [59, 121, 126, 129, 310, 318, 321, 339].

7.4 Other-Regarding Preferences

In this section, we formulate and solve mean-field-type games with heterogeneous empathy structure, which is described by a linear jump-diffusion system of conditional mean-field type where the conditioning is with respect to common noise and a quadratic empathetic cost functional involving the second moment, the square of the expected values of the state, and the control actions for the decision-makers. We discuss a direct method to solve the empathy game problem with two filter terms involving in the state dynamics and in the other-regarding objective functional. This solution approach does not require solving the master Bellman–Kolmogorov equations or backward–forward stochastic partial differential equations.

For empathy altruism, the optimal strategies for the decision-makers are shown to be in state-and-mean-field feedback form. The empathetic optimal strategies are given explicitly as the sum of the linear state-feedback strategy for the associated empathetic linear-quadratic game problem with random coefficients and a conditional mean-field feedback term. The empathetic equilibrium cost of the decision-makers are explicitly derived using a simple direct method. Moreover, the empathetic equilibrium cost is a weighted sum of the initial variance and an integral of a weighted variance of the diffusion and the jump process. Then, the method is used to compute global optimum strategies as well as saddle-point strategies and Nash bargaining solutions in state-and-mean-field feedback form. Empathy spite, however, may exhibit nonexistence of optimal admissible strategies.

We provide a basic game in which the optimal feedback strategy blows up in finite time when empathy spite is involved. It is also shown that, for this class of games, Nash bargaining solution can be obtained by designing empathy structure. Also the following conclusions are arrived at:

- Empathy structure can be inconsistent.
- Pure altruism can be inconsistent.
- Partial altruism may help in reducing payoff inequality.
- Empathy spite may lead to non-admissible optimal strategies.
- Empathy spite may lead to nonexistence of equilibria.

7.4.1 Empathetic Preferences

We consider $I \geq 2$ decision-makers in a strategic interaction. Let $\mathscr{I}$ be the set of decision-makers. Each decision-maker $i \in \mathscr{I}$ has a non-empty set of actions U_i. The classical non-empathetic objective function of decision-maker i is $L_i : \prod_{j=1}^{I} U_j \to \mathbb{R}$. Decision-makers have empathetic preferences on the joint strategy outcomes. The outcome of decision-maker i in the empathetic interaction by choosing action $u_i \in U_i$ combines her intrinsic preference for $u_i \in U_i$ with the intrinsic preference of i's neighbors (other than i herself), $\mathscr{I}_i$, where the weight given to the preference of any neighbor $j \in \mathscr{I}_i$ depends on the strength of the directed relationship between i and j. As a basic setup and for illustration purposes, this is modeled with the parameter λ_{ij}. By a self-regarding decision-maker, we refer to a decision-maker in the game who optimizes her own payoff (without empathy for others). A self-regarding decision-maker thus cares about the behavior that impacts her own payoff. This is scaled with a scalar λ_{ii}. The sign of λ_{ii} plays an important role as it determines if it is a maximization or minimization of own payoff. An other-regarding decision-maker considers not only her own payoff but also some of her network members' payoffs. Then, the other-regarding decision-maker will include these in her preferences and create an empathetic payoff. She is still acting to maximize her new empathetic payoff. Based on these basic empathy structures [40, 50, 127, 330, 346], we construct an empathetic payoff as a combination of payoffs through a real-valued matrix $\Lambda = (\lambda_{ij})_{i,j}$. Note that we include here not only positive value of λ_{ij} (referred to as partial altruism) but also negative value of λ_{ij} (referred to as partial spite). The one-hop empathy neighbors of i is determined by

$$\mathscr{I}_i = \{j \in \mathscr{I} \mid j \neq i,\ \lambda_{ij} \neq 0\}.$$

The instant empathetic payoff of i is

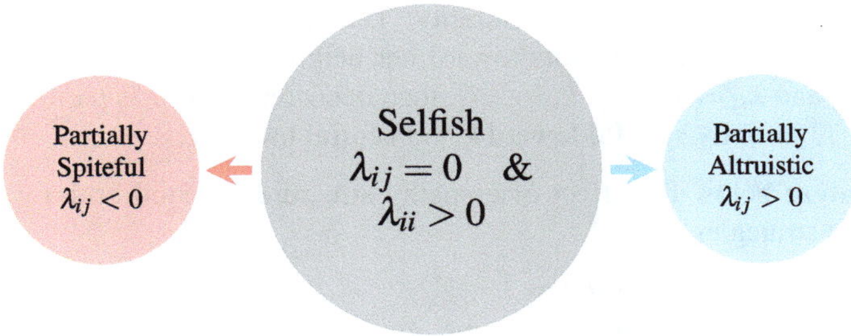

Fig. 7.1 Behavior of i toward j for different sign values of λ_{ij}

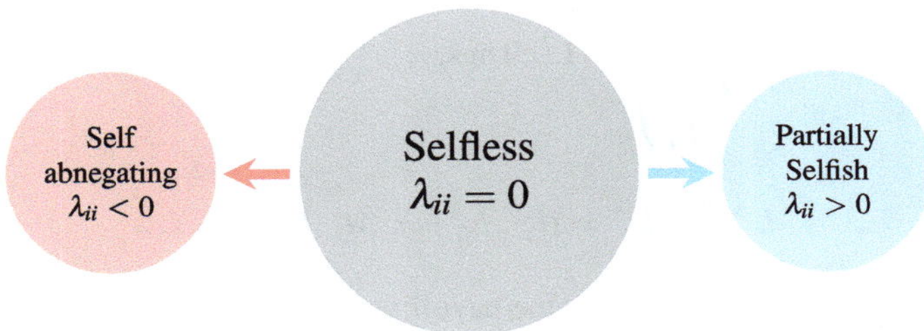

Fig. 7.2 Behavior of i toward herself for different sign values of λ_{ii}

$$L_i^A := \lambda_{ii} l_i + \sum_{j \in \mathscr{I}_i} \lambda_{ij} l_j,$$

where $L_i^I := L_i$ denotes the initial non-empathetic payoff of decision-maker i. Based on the material payoffs, a decision-maker can have empathy/malice for the others and selfishness/selflessness for herself (Fig. 7.1).

The empathetic payoff l_i^A can be interpreted as follows:

- Selfishness: Selfishness is being concerned for oneself or one's own payoff, regardless of others' payoffs. It is the lack of consideration for other decision-makers' payoffs. If $\lambda_{ij} = 0$, we say that i is neutrally empathetic toward j. Decision-maker i is self-regarding if $\lambda_{ii} \neq 0$, $\lambda_{ij} = 0$ for all $j \neq i$. If all the λ_{ij} are zeros for every (i, j), then each decision-maker focuses on her own-payoff positively or negatively depending on the sign of λ_{ii}. The case of $\lambda_{ii} > 0$ corresponds to partial selfishness. The case of $\lambda_{ii} < 0$ is seen as a self-abnegating behavior (see Fig. 7.2).
- Partial altruism: If $\lambda_{ij} > 0$, we say that i is partially altruistic toward j. If all the λ_{ij} are positive for every i, j, each decision-maker is considering the other decision-makers in her decision in a partially altruistic way.
- Partial spite: If $\lambda_{ij} < 0$, we say that i is partially spiteful toward j. If all the λ_{ij} are negative for every i, j each decision-maker is considering the other decision-makers in her decision in a partially spiteful way (Fig. 7.1).

- Mixed altruism–spitefulness–neutrality: The same decision-maker i may have different empathetic behaviors toward her neighbors. For example, if $\lambda_{ij} > 0$, $\lambda_{ik} < 0$, and $\lambda_{il} = 0$ for $j, k, l \in \mathcal{I}_i$, then decision-maker i is partially altruistic toward j, partially spiteful toward k and neutral toward l.

We analyze below the effect of empathy structure on equilibrium payoffs and equilibrium strategies.

Consistency of Empathy Structure

We examine the consistency of some empathy profiles at different levels of reasoning. The one-level game is the one obtained by applying the matrix Λ to the payoff vector $L = \begin{pmatrix} L_1 \\ L_2 \\ \vdots \\ L_I \end{pmatrix} = L^I$. Thus,

$$L^{\Lambda,1} = L^{\Lambda} = (l_1^{\Lambda}, L_2^{\Lambda}) = \Lambda l, \ \ L^{\Lambda,0} := L.$$

At the k-th level of empathy, the game payoff vector becomes

$$L^{\Lambda,k} = \Lambda L^{\Lambda,k-1} = \Lambda^k L.$$

Definition 7.7 The empathy structure Λ is consistent if the equilibrium structure of the k-th level game is unchanged for any $k \geq 1$.

It is not difficult to see that the identity matrix $\begin{pmatrix} 1 & 0 & \dots & 0 \\ 0 & 1 & \dots & 0 \\ \dots & \dots & \dots & \dots \\ 0 & 0 & 0 & 1 \end{pmatrix}$ is a consistent empathy structure.

Let $\rho > 0$, and $\Lambda_c = \frac{\rho}{I} \begin{pmatrix} 1 & \dots\dots & 1 \\ \dots\dots & & \dots \\ 1 & & 1 \end{pmatrix}$. Then for any $k \geq 1$, $\Lambda_c^k = \rho^{k-1} \Lambda_c$. It then turns out that the payoff vector at the k-th level empathetic game is

$$L^{\Lambda_c,k} = \Lambda_c^k L = \rho^{k-1} \Lambda_c L.$$

Since $\rho^{k-1} > 0$, the k-level empathy game is strategically equivalent to the 1-level empathy game. This means that Λ_c is consistent for $\rho > 0$, i.e., in the partially equally altruism case.

Let us now provide an inconsistent empathy profile. Let $\rho < 0$, and $\Lambda = \frac{\rho}{I} \begin{pmatrix} 1 & \dots\dots & 1 \\ \dots\dots & & \dots \\ 1 & & 1 \end{pmatrix}$. Then for any $k \geq 1$, $\Lambda^k = \rho^{k-1} \Lambda$. It then turns out that the payoff vector at the $(2k + 1)$-th level empathetic game is

$$l^{\Lambda,2k+1} = \Lambda^{2k+1} l = \rho^{2k} \Lambda l.$$

Since $\rho^{2k} > 0$, the $(2k+1)$-level empathy game is strategically equivalent to the 1-level empathy game. This means that the outcome is similar to the one from $\frac{\rho}{I}\begin{pmatrix} 1 & \ldots\ldots & 1 \\ \ldots\ldots & & \ldots \\ 1 & & 1 \end{pmatrix}$.

However, when $\rho^{2k+1} < 0$, the $2k$-level empathy game may not be strategically equivalent to the 1-level empathy game. This means that $\frac{\rho}{I}\begin{pmatrix} 1 & \ldots\ldots & 1 \\ \ldots\ldots & & \ldots \\ 1 & & 1 \end{pmatrix}$, is inconsistent for $\rho < 0$ and for $L = L^I$ nontrivially zero.

Consistent Empathy Structure

Consider an arbitrary number of decision-makers. If there exists a positive sequence $\epsilon_k > 0$ such that $\Lambda^k = \epsilon_k \Lambda$, then the empathy structure Λ is consistent.

Indeed, let $\epsilon_k > 0$ and the matrix Λ satisfy the relation $\Lambda^k = \epsilon_k \Lambda$. This means that the payoff vector of the k-th level game is $l^{\Lambda,k} = \epsilon_k \Lambda l$. Since $\epsilon_k > 0$, the k-th level empathy game is strategically equivalent to the 1-level empathy game for any $k \geq 1$.

Infinite Hierarchy

We examine the limit of the matrix Λ^k as k goes to infinity. As we have seen above, Λ^k may not converge in general. If the maximum modulus of the eigenvalues satisfies $\rho(\Lambda) < 1$, then Λ^k converges to $\begin{pmatrix} 0 & 0 & 0 & 0 \\ \ldots & \ldots & \ldots & \ldots \\ 0 & 0 & 0 & 0 \end{pmatrix}$ as k goes to infinity. Thus, the infinite hierarchy empathy game becomes a trivial one in this case.

If $\lim_k \Lambda^k = \Lambda_\infty$ and the infinite hierarchy of empathy game payoff vector is $\Lambda_\infty l$, the equilibrium structure of the game $\Lambda_\infty l$ may be different from the equilibrium structure of the finite hierarchy of empathy with payoff vector $\Lambda^k l$.

Infinitely Consistent Empathy Profiles

Let $I = 2$. The only empathy profiles in a generic empathetic mean-field-type game, which are infinitely consistent, are I and and $\begin{pmatrix} \lambda_{11} & \frac{\lambda_{11}(1-\lambda_{11})}{\lambda_{21}} \\ \lambda_{21} & 1-\lambda_{11} \end{pmatrix}$ for $(\lambda_{11}, \lambda_{21}) \in \mathbb{R}^2$.

The proof goes as follows. If the empathy structure Λ is a diagonalizable matrix but has a single eigenvalue, it must be I. If Λ is diagonalizable and has two eigenvalues, they must be 1 and 0. There is a basis of matrix P such that $\Lambda = PDP^{-1}$ with $D = \begin{pmatrix} 1 & 0 \\ 0 & 0 \end{pmatrix}$. Thus,

$$\Lambda^2 = PD^2P^{-1} = PDP^{-1} = \Lambda.$$

Since the eigenvalues are

$$\frac{tr(\Lambda)}{2} \pm \sqrt{\frac{tr(\Lambda)^2}{4} - \det\Lambda},$$

it implies that $\mathrm{tr}\Lambda = 1$ and $\det\Lambda = 0$. Thus, given λ_{11} and λ_{21}, row 1 must be a multiple of row 2 by $\frac{\lambda_{11}}{\lambda_{21}}$.

Pure Altruism Leads to an Inconsistent Empathy Profile

From the previous analysis, we deduce that pure altruism with profiles $\begin{pmatrix} 0 & \lambda_{12} > 0 \\ \lambda_{21} > 0 & 0 \end{pmatrix}$ for $(\lambda_{11}, \lambda_{21}) \in \mathbb{R}^2_{++}$ are inconsistent. For arbitrary number of decision-makers, the square matrix $\begin{pmatrix} 0 & 1 & \ldots\ldots & 1 \\ 1 & 0 & \ldots & 1 \\ \ldots & \ldots & \ldots & \ldots \\ 1 & 1 & \ldots & 1 \; 0 \end{pmatrix}^2$ has positive diagonal elements and hence different equilibrium structures as for $\Lambda = \begin{pmatrix} 0 & 1 & \ldots\ldots & 1 \\ 1 & 0 & \ldots & 1 \\ \ldots & \ldots & \ldots & \ldots \\ 1 & 1 & \ldots & 1 \; 0 \end{pmatrix}$. The two-level empathy game does not the same equilibrium set as the one-level empathy game. Thus, pure altruism is inconsistent.

7.4.2 *Empathetic Game with Common Noise*

The empathetic objective functional of i is $L_i^{\Lambda}(u_1, \ldots, u_I)$ given by

$$\begin{aligned} & L_i^{\Lambda}(u_1, \ldots, u_I) = \tfrac{1}{2}(\lambda_{ii} q_i(T) + \textstyle\sum_{j\in\mathcal{I}_i} \lambda_{ij} q_j(T)) x^2(T) \\ & + \tfrac{1}{2}(\lambda_{ii} \bar{q}_i(T) + \textstyle\sum_{j\in\mathcal{I}_i} \lambda_{ij} \bar{q}_j(T))[E(x(T)|\mathcal{F}_T^{B_o})]^2 \\ & + \tfrac{1}{2}\int_0^T \Big\{ (\lambda_{ii} q_i(t) + \textstyle\sum_{j\in\mathcal{I}_i} \lambda_{ij} q_j(t)) x^2(t) \\ & \quad + (\lambda_{ii} \bar{q}_i(t) + \textstyle\sum_{j\in\mathcal{I}_i} \lambda_{ij} \bar{q}_j(t))(E[x(t) \,|\mathcal{F}_t^{B_o}\,])^2 \\ & \quad + \lambda_{ii} r_i(t) u_i^2(t) + \textstyle\sum_{j\in\mathcal{I}_i} \lambda_{ij} r_j(t) u_j^2(t) \\ & \quad + \lambda_{ii} \bar{r}_i(t)[E(u_i(t)|\mathcal{F}_t^{B_o})]^2 + \textstyle\sum_{j\in\mathcal{I}_i} \lambda_{ij} \bar{r}_j(t)[E(u_j(t)|\mathcal{F}_t^{B_o})]^2 \Big\}\, dt. \end{aligned} \tag{7.194}$$

Then, the empathetic best response of decision-maker i to the process

$$(u_{-i}, E[x \,|\mathcal{F}^{B_o}\,], E[u_{-i} \,|\mathcal{F}^{B_o}\,])$$

$$:= (u_1, \ldots, u_{i-1}, u_{i+1}, \ldots, u_I, E[x \,|\mathcal{F}^{B_o}\,], E[u_{-i} \,|\mathcal{F}^{B_o}\,]),$$

solves the following risk-neutral conditional mean-field-type control problem

$$\begin{cases} \inf_{u_i\in\mathscr{U}_i} E\left[L_i^{\Lambda}(u_1,\ldots,u_I)\right], \\ \text{subject to} \\ dx(t) = \Big\{a(t)x(t) + \bar{a}(t)E[x(t)\mid \mathscr{F}_t^{B_o}] \\ +\sum_{i=1}^I b_i(t)u_i(t) + \sum_{i=1}^I \bar{b}_i E[u_i(t)\mid \mathscr{F}_t^{B_o}]\Big\}\, dt \\ +\sigma(t)dB(t) + \int_\Theta \mu(t,\theta)\tilde{N}(dt,d\theta) + \sigma_o(t)dB_o(t), \\ x(0)\in\mathbb{R}, \end{cases} \tag{7.195}$$

where $q_i(t) > 0, q_i(t)+\bar{q}_i(t) > 0, r_i(t) > 0, r_i(t)+\bar{r}_i(t) > 0$ and $a(t), \bar{a}(t), b_i(t), \sigma(t), \sigma_o(t)$ are real-valued functions and where $E[x(t)]$ is the expected value of the state created by all decision-makers under the control action profile $(u_1,\ldots,u_I) \in \prod_{j=1}^I \mathscr{U}_j$. The method below can handle time-varying coefficients.

We rewrite the expected empathetic objective functional and the state coefficients in terms of $x-\hat{x}$ and $\hat{x}$:

$$\begin{array}{l} EL_i^{\Lambda}(u_1,\ldots,u_I) = E\hat{L}_i^{\Lambda}(u_1,\ldots,u_I), \\ \hat{L}_i^{\Lambda}(u_1,\ldots,u_I) = \frac{1}{2}(\lambda_{ii}q_i(T) + \sum_{j\in\mathscr{I}_i}\lambda_{ij}q_j(T))(x(T)-\hat{x}(T))^2 \\ +\frac{1}{2}(\lambda_{ii}(q_i(T)+\bar{q}_i(T)) + \sum_{j\in\mathscr{I}_i}\lambda_{ij}(q_j(T)+\bar{q}_j(T)))\hat{x}^2(T) \\ +\frac{1}{2}\int_0^T \Big\{(\lambda_{ii}q_i + \sum_{j\in\mathscr{I}_i}\lambda_{ij}q_j)(x-\hat{x})^2 \\ \quad +(\lambda_{ii}(q_i+\bar{q}_i) + \sum_{j\in\mathscr{I}_i}\lambda_{ij}(q_j+q_j))\hat{x}^2 \\ \quad +\lambda_{ii}r_i(u_i-\hat{u}_i)^2 + \sum_{j\in\mathscr{I}_i}\lambda_{ij}r_j(u_j-\hat{u}_j)^2 \\ \quad +\lambda_{ii}(r_i+\bar{r}_i)\hat{u}_i^2 + \sum_{j\in\mathscr{I}_i}\lambda_{ij}(r_j+\bar{r}_j)\hat{u}_j^2\Big\}\, dt, \end{array} \tag{7.196}$$

$$\begin{array}{l} \inf_{u_i} E\hat{L}_i^{\Lambda}(u_1,\ldots,u_I), \\ dx = \Big\{a(x-\hat{x}) + (a+\bar{a})\hat{x} + \sum_{i=1}^I b_i(u_i-\hat{u}_i) + \sum_{i=1}^I (b_i+\bar{b}_i)\hat{u}_i\Big\}\, dt \\ \quad +\sigma dB + \int_\Theta \mu(t,\theta)\tilde{N}(dt,d\theta) + \sigma_o dB_o, \\ x(0) = x_0,\ \ \hat{x}(0) = \hat{x}_0. \end{array} \tag{7.197}$$

From the conditional McKean–Vlasov stochastic differential equation, we derive the SDE satisfied by the filter $\hat{x}(t) = E[x(t)\mid \mathscr{F}_t^{B_o}]$. We take the conditioning of the state to obtain the following equation:

$$d\hat{x} = \{(a+\bar{a})\hat{x} + \sum_{i=1}^I (b_i+\bar{b}_i)\hat{u}_i\}dt + \sigma_o dB_o,\ \hat{x}(0) = \hat{x}_0, \tag{7.198}$$

where we have used the fact that

$$E[B(t)\mid \mathscr{F}_t^{B_o}] = 0,\ \ E[\tilde{N}(t,.)\mid \mathscr{F}_t^{B_o}] = 0. \tag{7.199}$$

The direct method consists of writing a generic structure of the cost functional, with unknown deterministic functions to be identified. Inspired from the structure of the terminal cost function, we try a generic solution in a quadratic form. Let $f_i(t,x) = \frac{1}{2}\alpha_i(t)(x-\hat{x})^2 + \frac{1}{2}\beta_i(t)\hat{x}^2 + \gamma_i(t)\hat{x} + \delta_i(t)$, where $\alpha, \beta, \gamma, \delta$ are random functions of time t, which are $\mathscr{F}_t^{B_o}$-measurable such that

$$f_i(T,x(T)) = \frac{1}{2}\Big\{(\lambda_{ii}q_i(T) + \sum_{j\in\mathscr{I}_i}\lambda_{ij}q_j(T))(x(T)-\hat{x}(T))^2 \\ +[\lambda_{ii}(q_i(T)+\bar{q}_i(T)) + \sum_{j\in\mathscr{I}_i}\lambda_{ij}(q_j(T)+\bar{q}_j(T))]\hat{x}^2(T)\Big\}, \tag{7.200}$$

By identification, one obtains

$$\begin{aligned} \alpha_i(T) &= \lambda_{ii}q_i(T) + \sum_{j\in\mathscr{I}_i}\lambda_{ij}q_j(T), \\ \beta_i(T) &= \lambda_{ii}(q_i(T)+\bar{q}_i(T)) + \sum_{j\in\mathscr{I}_i}\lambda_{ij}(q_j(T)+\bar{q}_j(T)), \\ \gamma_i(T) &= \delta_i(T) = 0. \end{aligned} \tag{7.201}$$

Using Itô's formula for jump-diffusion processes, we compute the gap between $\hat{L}_i^{\Lambda}$ and $f_i(0,x(0))$ as

$$\begin{aligned} &\hat{L}_i^{\Lambda} - f_i(0,x(0)) \\ &= \frac{1}{2}(\lambda_{ii}q_i(T) + \sum_{j\in\mathscr{I}_i}\lambda_{ij}q_j(T) - \alpha_i(T))(x(T)-\hat{x}(T))^2 \\ &+\frac{1}{2}[\lambda_{ii}(q_i(T)+\bar{q}_i(T)) + \sum_{j\in\mathscr{I}_i}\lambda_{ij}(q_j(T)+\bar{q}_j(T)) - \beta_i(T)]\hat{x}^2(T) \\ &+(0-\gamma_i(T))\hat{x}(T) + (0-\delta_i(T)) \\ &+\frac{1}{2}\int_0^T\{d\alpha_i + (2a\alpha_i + \lambda_{ii}q_i + \sum_{j\in\mathscr{I}_i}\lambda_{ij}q_j)dt\}(x-\hat{x})^2 \\ &+\frac{1}{2}\int_0^T\{d\beta_i + (2(a+\bar{a})\beta_i + \lambda_{ii}(q_i+\bar{q}_i) + \sum_{j\in\mathscr{I}_i}\lambda_{ij}(q_j+\bar{q}_j))dt\}\hat{x}^2 \\ &+\int_0^T\{d\gamma_i + ((a+\bar{a})\gamma_i + \beta_{i,o}\sigma_o)dt\}\hat{x} \\ &+\frac{1}{2}\int_0^T[\sigma^2 + \int_\Theta \mu^2(t,\theta)\nu(d\theta)]\alpha_i dt \\ &+\frac{1}{2}\int_0^T \sigma_o^2\beta_i dt \\ &+\int_0^T \gamma_{i,o}\sigma_o dt + \int_0^T d\delta_i \\ &+\frac{1}{2}\int_0^T \lambda_{ii}r_i(u_i-\hat{u}_i)^2 + \sum_{j\in\mathscr{I}_i}\lambda_{ij}r_j(u_j-\hat{u}_j)^2 + 2\alpha_i\sum_{j=1}^I b_j(u_j-\hat{u}_j)(x-\hat{x})dt \\ &+\frac{1}{2}\int_0^T \lambda_{ii}(r_i+\bar{r}_i)\hat{u}_i^2 + \sum_{j\in\mathscr{I}_i}\lambda_{ij}(r_j+\bar{r}_j)\hat{u}_j^2 + 2\beta_i\sum_{j=1}^I(b_j+\bar{b}_j)\hat{u}_j\hat{x}dt \\ &+\int_0^T \gamma_i\sum_{j=1}^I(b_j+\bar{b}_j)\hat{u}_j dt \\ &+\frac{1}{2}\int_0^T 2\alpha_i(x-\hat{x})(\sigma dB + \int_\Theta \mu\tilde{N}(dt,d\theta)) + \frac{1}{2}\int_0^T 2\sigma_o(\beta_i\hat{x}+\gamma_i)dB_o. \end{aligned} \tag{7.202}$$

Empathetic Feedback Strategies

The information structure for feedback solution is as follows. The model and the objective functions are assumed to be common knowledge. We assume that the state is of perfect observation and the mean-field term is computable. In addition, we assume that the common noise is of perfect observation. We compute again $\hat{L}_i^{\Lambda} - f_i(0,x(0))$ and complete the squares using the elements of $\{x-\hat{x},\ \hat{x}\}$:

$$
\begin{aligned}
&\hat{L}_i^{\Lambda} - f_i(0, x(0)) \\
&= \tfrac{1}{2}(\lambda_{ii}q_i(T) + \textstyle\sum_{j\in\mathcal{I}_i}\lambda_{ij}q_j(T) - \alpha_i(T))(x(T) - \hat{x}(T))^2 \\
&+\tfrac{1}{2}[\lambda_{ii}(q_i(T) + \bar{q}_i(T)) + \textstyle\sum_{j\in\mathcal{I}_i}\lambda_{ij}(q_j(T) + \bar{q}_j(T)) - \beta_i(T)]\hat{x}^2(T) \\
&+(0 - \gamma_i(T))\hat{x}(T) + (0 - \delta_i(T)) \\
&+\tfrac{1}{2}\int_0^T \Big\{ d\alpha_i + (2a\alpha_i - \tfrac{b_i^2}{\lambda_{ii}r_i}\alpha_i^2 - 2\alpha_i \textstyle\sum_{j\neq i}\tfrac{b_j^2}{\lambda_{jj}r_j}\alpha_j)dt \\
&+(\lambda_{ii}q_i + \textstyle\sum_{j\in\mathcal{I}_i}\lambda_{ij}q_j + \sum_{j\in\mathcal{I}_i}\tfrac{b_j^2\lambda_{ij}}{\lambda_{jj}^2 r_j}\alpha_j^2)dt \Big\} (x - \hat{x})^2 \\
&+\tfrac{1}{2}\int_0^T \Big\{ d\beta_i + (2(a+\bar{a})\beta_i - \tfrac{(b_i+\bar{b}_i)^2}{\lambda_{ii}(r_i+\bar{r}_i)}\beta_i^2 - 2\beta_i \textstyle\sum_{j\neq i}\tfrac{(b_j+\bar{b}_j)^2}{\lambda_{jj}(r_j+\bar{r}_j)}\beta_j)dt \\
&+(\lambda_{ii}(q_i+\bar{q}_i) + \textstyle\sum_{j\in\mathcal{I}_i}\lambda_{ij}(q_j+\bar{q}_j) + \sum_{j\in\mathcal{I}_i}\lambda_{ij}\tfrac{(b_j+\bar{b}_j)^2}{\lambda_{jj}^2(r_j+\bar{r}_j)}\beta_j^2)dt \Big\} \hat{x}^2 \\
&+\int_0^T \Big\{ d\gamma_i + ((a+\bar{a})\gamma_i + \beta_{i,o}\sigma_o - \tfrac{(b_i+\bar{b}_i)^2}{\lambda_{ii}(r_i+\bar{r}_i)}\beta_i\gamma_i)dt \\
&+(-\beta_i \textstyle\sum_{j\neq i}\tfrac{(b_j+\bar{b}_j)^2}{\lambda_{jj}(r_j+\bar{r}_j)}\gamma_j - \gamma_i \sum_{j\neq i}\tfrac{(b_j+\bar{b}_j)^2}{\lambda_{jj}(r_j+\bar{r}_j)}\beta_j)dt \\
&+\textstyle\sum_{j\in\mathcal{I}_i}\lambda_{ij}\tfrac{(b_j+\bar{b}_j)^2}{\lambda_{jj}^2(r_j+\bar{r}_j)}\beta_j\gamma_j dt \Big\} \hat{x} \\
&+\int_0^T \Big\{ d\delta_i + ([\sigma^2 + \textstyle\int_\Theta \mu^2(t,\theta)\nu(d\theta)]\tfrac{1}{2}\alpha_i + + \tfrac{1}{2}\sigma_o^2\beta_i + \gamma_{i,o}\sigma_o)dt \\
&-\gamma_i \textstyle\sum_{j\neq i}\tfrac{(b_j+\bar{b}_j)^2}{\lambda_{jj}(r_j+\bar{r}_j)}\gamma_j - \tfrac{(b_i+\bar{b}_i)^2}{\lambda_{ii}(r_i+\bar{r}_i)}\tfrac{1}{2}\gamma_i^2 + \tfrac{1}{2}\sum_{j\in\mathcal{I}_i}\lambda_{ij}\tfrac{(b_j+\bar{b}_j)^2}{\lambda_{jj}^2(r_j+\bar{r}_j)}\gamma_j^2 \Big\} dt \\
&+\tfrac{1}{2}\int_0^T \lambda_{ii}r_i[u_i - \hat{u}_i + \tfrac{b_i}{\lambda_{ii}r_i}\alpha_i(x-\hat{x})]^2 + \lambda_{ii}(r_i+\bar{r}_i)[\hat{u}_i + \tfrac{b_i+\bar{b}_i}{\lambda_{ii}(r_i+\bar{r}_i)}(\beta_i\hat{x}+\gamma_i)]^2 dt \\
&+\tfrac{1}{2}\int_0^T 2\alpha_i(x-\hat{x})(\sigma dB + \textstyle\int_\Theta \mu\tilde{N}(dt,d\theta)) + \tfrac{1}{2}\int_0^T 2\sigma_o(\beta_i\hat{x}+\gamma_i)dB_o
\end{aligned}
\tag{7.203}
$$

It follows that the following system

$$
\begin{aligned}
&\inf_{u_i\in\mathcal{U}_i} E[\hat{L}_i] = E\{\tfrac{1}{2}\alpha_i(0)(x(0)-\hat{x}(0))^2 + \tfrac{1}{2}\beta_i(0)[\hat{x}(0)]^2 + \gamma_i(0)\hat{x}(0) + \delta_i(0)\} \\
&u_i^* = -\tfrac{b_i}{\lambda_{ii}r_i}\alpha_i(x-\hat{x}) - \tfrac{(b_i+\bar{b}_i)}{\lambda_{ii}(r_i+\bar{r}_i)}(\beta_i\hat{x}+\gamma_i), \\
&d\alpha_i = -\Big\{ 2a\alpha_i - \tfrac{b_i^2}{\lambda_{ii}r_i}\alpha_i^2 - 2\alpha_i\textstyle\sum_{j\neq i}\tfrac{b_j^2}{\lambda_{jj}r_j}\alpha_j + \lambda_{ii}q_i + \sum_{j\in\mathcal{I}_i}\lambda_{ij}q_j \\
&+\textstyle\sum_{j\in\mathcal{I}_i}\tfrac{b_j^2\lambda_{ij}}{\lambda_{jj}^2 r_j}\alpha_j^2 \Big\} dt, \\
&\alpha_i(T) = \lambda_{ii}q_i(T) + \textstyle\sum_{j\in\mathcal{I}_i}\lambda_{ij}q_j(T), \\
&d\beta_i = -\Big\{ 2(a+\bar{a})\beta_i - \tfrac{(b_i+\bar{b}_i)^2}{\lambda_{ii}(r_i+\bar{r}_i)}\beta_i^2 - 2\beta_i\textstyle\sum_{j\neq i}\tfrac{(b_j+\bar{b}_j)^2}{\lambda_{jj}(r_j+\bar{r}_j)}\beta_j + \lambda_{ii}(q_i+\bar{q}_i) \\
&+\textstyle\sum_{j\in\mathcal{I}_i}\lambda_{ij}(q_j+\bar{q}_j) + \sum_{j\in\mathcal{I}_i}\lambda_{ij}\tfrac{(b_j+\bar{b}_j)^2}{\lambda_{jj}^2(r_j+\bar{r}_j)}\beta_j^2 \Big\} dt, \\
&\beta_i(T) = \lambda_{ii}(q_i(T)+\bar{q}_i(T)) + \textstyle\sum_{j\in\mathcal{I}_i}\lambda_{ij}(q_j(T)+\bar{q}_j(T)), \\
&d\gamma_i = -\Big\{ (a+\bar{a})\gamma_i + \beta_{i,o}\sigma_o - \tfrac{(b_i+\bar{b}_i)^2}{\lambda_{ii}(r_i+\bar{r}_i)}\beta_i\gamma_i \\
&-\beta_i\textstyle\sum_{j\neq i}\tfrac{(b_j+\bar{b}_j)^2}{\lambda_{jj}(r_j+\bar{r}_j)}\gamma_j - \gamma_i\sum_{j\neq i}\tfrac{(b_j+\bar{b}_j)^2}{\lambda_{jj}(r_j+\bar{r}_j)}\beta_j + \sum_{j\in\mathcal{I}_i}\lambda_{ij}\tfrac{(b_j+\bar{b}_j)^2}{\lambda_{jj}^2(r_j+\bar{r}_j)}\beta_j\gamma_j \Big\} dt, \\
&\gamma_i(T) = 0, \\
&d\delta_i = -\Big\{ [\sigma^2 + \textstyle\int_\Theta \mu^2(t,\theta)\nu(d\theta)]\tfrac{1}{2}\alpha_i + \tfrac{1}{2}\sigma_o^2\beta_i + \gamma_{i,o}\sigma_o \\
&-\gamma_i\textstyle\sum_{j\neq i}\tfrac{(b_j+\bar{b}_j)^2}{\lambda_{jj}(r_j+\bar{r}_j)}\gamma_j - \tfrac{(b_i+\bar{b}_i)^2}{\lambda_{ii}(r_i+\bar{r}_i)}\tfrac{1}{2}\gamma_i^2 + \tfrac{1}{2}\sum_{j\in\mathcal{I}_i}\lambda_{ij}\tfrac{(b_j+\bar{b}_j)^2}{\lambda_{jj}^2(r_j+\bar{r}_j)}\gamma_j^2 \Big\} dt, \\
&\delta_i(T) = 0, \\
&d\hat{x} = \{[a+\bar{a} - \textstyle\sum_{i=1}^I \tfrac{(b_i+\bar{b}_i)^2}{\lambda_{ii}(r_i+\bar{r}_i)}\beta_i]\hat{x} - \sum_{i=1}^I \tfrac{(b_i+\bar{b}_i)^2}{\lambda_{ii}(r_i+\bar{r}_i)}\gamma_i\}dt + \sigma_o dB_o, \\
&\hat{x}(0) = \hat{x}_0,
\end{aligned}
\tag{7.204}
$$

(whenever it has a solution) provides a conditional mean-field empathetic Nash equilibrium in feedback strategies that is conditioned on the common noise B_o, and this holds for empathy structures such that $\lambda_{ii}r_i > 0,\ \lambda_{ii}(r_i+\bar{r}_i) > 0,\ i \in \mathcal{I}$.

These empathetic stochastic Riccati equations (7.204) are different from those of open-loop control strategies.

The underlying game formulation and the feedback solutions are now collectively presented in the following two boxes.

Empathetic Game Problem with Common Noise

$$
\begin{array}{l}
EL_i^{\Lambda}(u_1,\ldots,u_I) = E\hat{L}_i^{\Lambda}(u_1,\ldots,u_I)\\
= \frac{1}{2}(\lambda_{ii}q_i(T) + \sum_{j\in\mathscr{I}_i}\lambda_{ij}q_j(T))E(x(T)-\hat{x}(T))^2\\
+\frac{1}{2}(\lambda_{ii}(q_i(T)+\bar{q}_i(T)) + \sum_{j\in\mathscr{I}_i}\lambda_{ij}(q_j(T)+\bar{q}_j(T)))E\hat{x}^2(T)\\
+\frac{1}{2}E\int_0^T\Big\{(\lambda_{ii}q_i + \sum_{j\in\mathscr{I}_i}\lambda_{ij}q_j)(x-\hat{x})^2 + (\lambda_{ii}(q_i+\bar{q}_i) + \sum_{j\in\mathscr{I}_i}\lambda_{ij}(q_j+\bar{q}_j))\hat{x}^2\\
+\lambda_{ii}r_i(u_i-\hat{u}_i)^2 + \sum_{j\in\mathscr{I}_i}\lambda_{ij}r_j(u_j-\hat{u}_j)^2 + \lambda_{ii}(r_i+\bar{r}_i)\hat{u}_i^2\\
\quad + \sum_{j\in\mathscr{I}_i}\lambda_{ij}(r_j+\bar{r}_j)\hat{u}_j^2\Big\}\,dt,\\
\text{subject to}\\
dx = \Big\{a(x-\hat{x}) + (a+\bar{a})\hat{x} + \sum_{i=1}^I b_i(u_i-\hat{u}_i) + \sum_{i=1}^I(b_i+\bar{b}_i)\hat{u}_i\Big\}\,dt\\
\quad +\sigma dB + \int_\Theta \mu(t,\theta)\tilde{N}(dt,d\theta)+\sigma_o dB_o.
\end{array}
\tag{7.205}
$$

Solution to the Empathetic Game Problem with Common Noise (7.205)

$$
\begin{array}{l}
\inf_{u_i\in\mathscr{U}_i} E[L_i^{\Lambda}]\\
= E\{\frac{1}{2}\alpha_i(0)(x(0)-\hat{x}(0))^2 + \frac{1}{2}\beta_i(0)[\hat{x}(0)]^2 + \gamma_i(0)\hat{x}(0) + \delta_i(0)\}\\
u_i^* = -\frac{b_i}{\lambda_{ii}r_i}\alpha_i(x-\hat{x}) - \frac{(b_i+\bar{b}_i)}{\lambda_{ii}(r_i+\bar{r}_i)}(\beta_i\hat{x}+\gamma_i),\\
d\hat{x} = \{[a+\bar{a} - \sum_{i=1}^I \frac{(b_i+\bar{b}_i)^2}{\lambda_{ii}(r_i+\bar{r}_i)}\beta_i]\hat{x} - \sum_{i=1}^I \frac{(b_i+\bar{b}_i)^2}{\lambda_{ii}(r_i+\bar{r}_i)}\gamma_i\}dt+\sigma_o dB_o,\\
\hat{x}(0) = \hat{x}_0,\\
d\alpha_i = -\Big\{2a\alpha_i - \frac{b_i^2}{\lambda_{ii}r_i}\alpha_i^2 - 2\alpha_i\sum_{j\neq i}\frac{b_j^2}{\lambda_{jj}r_j}\alpha_j + \lambda_{ii}q_i + \sum_{j\in\mathscr{I}_i}\lambda_{ij}q_j\\
+\sum_{j\in\mathscr{I}_i}\frac{b_j^2\lambda_{ij}}{\lambda_{jj}^2 r_j}\alpha_j^2\Big\}\,dt,\\
\alpha_i(T) = \lambda_{ii}q_i(T) + \sum_{j\in\mathscr{I}_i}\lambda_{ij}q_j(T),\\
d\beta_i = -\Big\{2(a+\bar{a})\beta_i - \frac{(b_i+\bar{b}_i)^2}{\lambda_{ii}(r_i+\bar{r}_i)}\beta_i^2 - 2\beta_i\sum_{j\neq i}\frac{(b_j+\bar{b}_j)^2}{\lambda_{jj}(r_j+\bar{r}_j)}\beta_j\\
+\lambda_{ii}(q_i+\bar{q}_i) + \sum_{j\in\mathscr{I}_i}\lambda_{ij}(q_j+\bar{q}_j)+\sum_{j\in\mathscr{I}_i}\lambda_{ij}\frac{(b_j+\bar{b}_j)^2}{\lambda_{jj}^2(r_j+\bar{r}_j)}\beta_j^2\Big\}\,dt,\\
\beta_i(T) = \lambda_{ii}(q_i(T)+\bar{q}_i(T)) + \sum_{j\in\mathscr{I}_i}\lambda_{ij}(q_j(T)+\bar{q}_j(T)),\\
d\gamma_i = -\Big\{(a+\bar{a})\gamma_i + \beta_{i,o}\sigma_o - \frac{(b_i+\bar{b}_i)^2}{\lambda_{ii}(r_i+\bar{r}_i)}\beta_i\gamma_i\\
-\beta_i\sum_{j\neq i}\frac{(b_j+\bar{b}_j)^2}{\lambda_{jj}(r_j+\bar{r}_j)}\gamma_j - \gamma_i\sum_{j\neq i}\frac{(b_j+\bar{b}_j)^2}{\lambda_{jj}(r_j+\bar{r}_j)}\beta_j + \sum_{j\in\mathscr{I}_i}\lambda_{ij}\frac{(b_j+\bar{b}_j)^2}{\lambda_{jj}^2(r_j+\bar{r}_j)}\beta_j\gamma_j\Big\}\,dt,\\
\gamma_i(T) = 0,\\
d\delta_i = -\Big\{[\sigma^2 + \int_\Theta \mu^2(t,\theta)\nu(d\theta)]\frac{1}{2}\alpha_i+\frac{1}{2}\sigma_o^2\beta_i + \gamma_{i,o}\sigma_o\\
-\gamma_i\sum_{j\neq i}\frac{(b_j+\bar{b}_j)^2}{\lambda_{jj}(r_j+\bar{r}_j)}\gamma_j - \frac{(b_i+\bar{b}_i)^2}{\lambda_{ii}(r_i+\bar{r}_i)}\frac{1}{2}\gamma_i^2 + \frac{1}{2}\sum_{j\in\mathscr{I}_i}\lambda_{ij}\frac{(b_j+\bar{b}_j)^2}{\lambda_{jj}^2(r_j+\bar{r}_j)}\gamma_j^2\Big\}\,dt,\\
\delta_i(T) = 0,
\end{array}
\tag{7.206}
$$

7.4.2.1 Verification of the Empathetic Optimal Strategies

We verify the validity of the expression of empathetic strategies above using the stochastic Bellman system. The conditional measure solves a stochastic Kolmogorov–Fokker–Planck equation given by

$$dm = \{-[mD]_x + \frac{(\sigma^2+\sigma_o^2)}{2} m_{xx} + J^*[m]\}dt - \sigma_o m_x dB_o,$$

where the drift term is $D := a(x-\hat{x}) + (a+\bar{a})\hat{x} + \sum_{i=1}^{I} b_i(u_i - \hat{u}_i) + \sum_{i=1}^{I}(b_i + \bar{b}_i)\hat{u}_i$, and the jump term is

$$J[\phi] := \int_{\Theta} [\phi(x+\mu) - \phi(x) - \mu\phi_x]\nu(d\theta),$$

J^* being the adjoint operator associated with the jump operator J. We first write the objective functional in terms of m. Then, we seek for a solution to the stochastic Hamilton–Jacobi–Bellman equation using Kunita's correction formula. We show below that Kunita's correction term is $\sigma_o^2 \int_x m_x \partial_x [\int_y m(dy) V_{i,xm}] dxdt$, which is zero in this case. The stochastic Bellman equilibrium system is

$$\begin{aligned}
&-d\hat{V} + \Big\{ \hat{V}_{i,t}(t,m) + \textstyle\int_x H_i(x,m,\hat{V}_{i,m},\hat{V}_{i,xm},\hat{V}_{i,xxm}) m(dx) \\
&+\sigma_o \textstyle\int m_x \sigma_o \partial_x [\int_y m(dy) V_{i,xm}] dx \\
&+\tfrac{\sigma_o^2}{2} \textstyle\int_{x,y} \hat{V}_{i,mm}(x,y) m_x(dx) m_x(dy) \Big\} dt + [\sigma_o \int_x m(dx) V_{i,xm}] dB_o = 0,
\end{aligned} \tag{7.207}$$

where

$$\begin{aligned}
&\hat{V}_i(T,m) = \tfrac{1}{2}[\lambda_{ii} q_i(T) + \textstyle\sum_{j\in\mathscr{I}_i} \lambda_{ij} q_j(T)] \int_y (y-\bar{m})^2\, m(dy) \\
&+\tfrac{1}{2}[\lambda_{ii}(q_i(T)+\bar{q}_i(T)) + \textstyle\sum_{j\in\mathscr{I}_i} \lambda_{ij}(q_j(T)+\bar{q}_j(T))] \left[\int_y y\, m(dy)\right]^2,
\end{aligned} \tag{7.208}$$

and the integrand Hamiltonian is

$$\begin{aligned}
&H_i(x,m,\hat{V}_{i,m},\hat{V}_{i,xm},\hat{V}_{i,xmm}) \\
&= \inf_{u_i} \Big\{ \tfrac{1}{2}[\lambda_{ii} q_i + \textstyle\sum_{j\in\mathscr{I}_i} \lambda_{ij} q_j](x-\hat{x})^2 + \tfrac{1}{2}[\lambda_{ii}(q_i+\bar{q}_i) + \sum_{j\in\mathscr{I}_i} \lambda_{ij}(q_j+\bar{q}_j)]\hat{x}^2 \\
&+\tfrac{1}{2}\lambda_{ii} r_i (u_i - \hat{u}_i)^2 + \tfrac{1}{2} \textstyle\sum_{j\in\mathscr{I}_i} \lambda_{ij} r_j (u_j - \hat{u}_j)^2 \\
&+\tfrac{1}{2}\lambda_{ii}(r_i + \bar{r}_i)[\hat{u}_i]^2 + \tfrac{1}{2} \textstyle\sum_{j\in\mathscr{I}_i} \lambda_{ij}(r_j + \bar{r}_j)[\hat{u}_j]^2 \\
&+[a(x-\hat{x}) + (a+\bar{a})\hat{x}]\hat{V}_{i,xm} + [\textstyle\sum_{i=1}^{I} b_i(u_i - \hat{u}_i) + \sum_{i=1}^{I}(b_i + \bar{b}_i)\hat{u}_i]\hat{V}_{i,xm} \Big\} \\
&+\tfrac{\sigma^2+\sigma_0^2}{2} \hat{V}_{i,xxm} + \textstyle\int_{\Theta} [\hat{V}_{i,m}(t,x+\mu) - \hat{V}_{i,m} - \mu \hat{V}_{i,xm}] \nu(d\theta)
\end{aligned} \tag{7.209}$$

Inspired by the terminal payoff structure, we choose a guess functional in the following form:

$$\hat{V}_i(t,m)=\frac{\alpha_i}{2}\int_y (y-\bar{m})^2\, m(dy)+\frac{\beta_i}{2}\left[\int_y y\, m(dy)\right]^2+\gamma_i\int_y y\, m(dy)+\delta_i\int_y m(dy). \tag{7.210}$$

The optimal strategies then become

$$\begin{aligned}
u_i^* &= \hat{u}_i - \frac{b_i}{\lambda_{ii}r_i}\left\{\hat{V}_{i,xm} - E\hat{V}_{i,xm}|\mathscr{F}^{B_o}\right\}\\
&= \hat{u}_i - \frac{b_i}{\lambda_{ii}r_i}\alpha_i\left[x-\int_y y\, m(t,dy)\right]\\
&= -\frac{b_i+\bar{b}_i}{\lambda_{ii}(r_i+\bar{r}_i)}(\beta_i\left[\int_y y\, m(t,dy)\right]+\gamma_i) - \frac{b_i}{\lambda_{ii}r_i}\alpha_i\left[x-\int_y y\, m(t,dy)\right],\\
\hat{u}_i^* &= -\frac{b_i+\bar{b}_i}{\lambda_{ii}(r_i+\bar{r}_i)}E[\hat{V}_{i,xm}|\mathscr{F}^{B_o}] = -\frac{b_i+\bar{b}_i}{\lambda_{ii}(r_i+\bar{r}_i)}(\beta_i\left[\int_y y\, m(t,dy)\right]+\gamma_i),
\end{aligned} \tag{7.211}$$

which are exactly the expressions of the optimal strategies obtained in (7.204).

$$\begin{aligned}
&\tilde{H}_i(x,m,\hat{V}_{i,m},\hat{V}_{i,xm},\hat{V}_{i,xxm})\\
&= \frac{1}{2}\Big\{2a\alpha_i - 2\alpha_i\sum_{j=1}^{I}\frac{b_j^2}{\lambda_{jj}r_j}\alpha_j + \frac{b_i^2}{\lambda_{ii}r_i}\alpha_i^2 + \lambda_{ii}q_i + \sum_{j\in\mathscr{I}_i}\lambda_{ij}q_j\\
&+\sum_{j\in\mathscr{I}_i}\lambda_{ij}\frac{b_j^2}{\lambda_{jj}^2 r_j}\alpha_j^2\Big\}(x-\hat{x})^2\\
&+\frac{1}{2}\Big\{2(a+\bar{a})\beta_i - 2\beta_i\sum_{j=1}^{I}\frac{(b_j+\bar{b}_j)^2}{\lambda_{jj}(r_j+\bar{r}_j)}\beta_j\\
&+\frac{(b_i+\bar{b}_i)^2}{\lambda_{ii}(r_i+\bar{r}_i)}\beta_i^2 + \lambda_{ii}(q_i+\bar{q}_i) + \sum_{j\in\mathscr{I}_i}\lambda_{ij}(q_j+\bar{q}_j) + \sum_{j\in\mathscr{I}_i}\lambda_{ij}\frac{(b_j+\bar{b}_j)^2}{\lambda_{jj}^2(r_j+\bar{r}_j)}\beta_j^2\Big\}\hat{x}^2\\
&+\frac{1}{2}\Big\{2(a+\bar{a})\gamma_i + 2\gamma_i\beta_i\frac{(b_i+\bar{b}_i)^2}{\lambda_{ii}(r_i+\bar{r}_i)} - 2\beta_i\sum_j\frac{(b_j+\bar{b}_j)^2}{\lambda_{jj}(r_j+\bar{r}_j)}\gamma_j - 2\gamma_i\sum_j\frac{(b_j+\bar{b}_j)^2}{\lambda_{jj}(r_j+\bar{r}_j)}\beta_j\\
&+2\sum_{j\in\mathscr{I}_i}\lambda_{ij}\frac{(b_j+\bar{b}_j)^2}{\lambda_{jj}^2(r_j+\bar{r}_j)}\beta_j\gamma_j\Big\}\hat{x} + \frac{1}{2}\frac{(b_i+\bar{b}_i)^2}{(r_i+\bar{r}_i)}\gamma_i^2 - \gamma_i\sum_j\frac{(b_j+\bar{b}_j)^2}{(r_j+\bar{r}_j)}\gamma_j\\
&+\frac{1}{2}\sum_{j\in\mathscr{I}_i}\lambda_{ij}\frac{(b_j+\bar{b}_j)^2}{\lambda_{jj}^2(r_j+\bar{r}_j)}\gamma_j^2 + \frac{1}{2}\alpha_i[\sigma^2+\sigma_o^2+\int_\Theta\mu(t,\theta)\nu(d\theta)]
\end{aligned} \tag{7.212}$$

$$\frac{\sigma_o^2}{2}\int_x\int_y \hat{V}_{i,mm}m_x(dx)m_x(dy) = \frac{\sigma_o^2}{2}\int_x\int_y \hat{V}_{i,xymm}m(dx)m(dy) = \frac{\sigma_o^2}{2}(\beta_i-\alpha_i), \tag{7.213}$$

$$\int_x \sigma_o\hat{V}_{xm}dB_o m(dx) = \sigma_o\, dB_o(\beta_i\hat{x}+\gamma_i) = [\sigma_o\beta_i\, dB_o]\hat{x} + [\sigma_o\gamma_i\, dB_o] \tag{7.214}$$

Using the time derivative and the square terms in dt, we obtain

$$\begin{aligned}
\hat{V}_{i,t} &= \frac{\dot{\alpha}_i}{2}\int_y (y-\bar{m})^2\, m(dy)\\
&+\frac{\dot{\beta}_i}{2}\left[\int_y y\, m(dy)\right]^2 + \dot{\gamma}_i\int_y y\, m(dy) + \dot{\delta}_i + \beta_{i,o}\sigma_o[\int_y y\, m(dy)] + \gamma_{i,o}\sigma_o,
\end{aligned} \tag{7.215}$$

$$
\begin{aligned}
&-d\hat{V} + \Big\{\hat{V}_{i,t}(t,m) + \textstyle\int_x H_i(x,m,\hat{V}_i,\hat{V}_{i,xm},\hat{V}_{i,xxm})m(t,dx)\\
&+\tfrac{\sigma_o^2}{2}\textstyle\int_{x,y}\hat{V}_{i,mm}(x,y)m_x(dx)m_x(dy)\Big\}\,dt + \sigma_o\textstyle\int_x m(t,dx)V_{i,xm}dB_o\\
&= \tfrac{1}{2}\Big\{d\alpha_i + 2a\alpha_i - 2\alpha_i\textstyle\sum_{j=1}^I \frac{b_j^2}{\lambda_{jj}r_j}\alpha_j + \frac{b_i^2}{\lambda_{ii}r_i}\alpha_i^2\\
&+\lambda_{ii}q_i + \textstyle\sum_{j\in\mathscr{I}_i}\lambda_{ij}q_j + \sum_{j\in\mathscr{I}_i}\lambda_{ij}\frac{b_j^2}{\lambda_{jj}^2 r_j}\alpha_j^2\Big\}(x-\hat{x})^2\\
&+\tfrac{1}{2}\Big\{d\beta_i + 2(a+\bar{a})\beta_i - 2\beta_i\textstyle\sum_{j=1}^I\frac{(b_j+\bar{b}_j)^2}{\lambda_{jj}(r_j+\bar{r}_j)}\beta_j\\
&+\tfrac{(b_i+\bar{b}_i)^2}{\lambda_{ii}(r_i+\bar{r}_i)}\beta_i^2 + \lambda_{ii}(q_i+\bar{q}_i) + \textstyle\sum_{j\in\mathscr{I}_i}\lambda_{ij}(q_j+\bar{q}_j) + \sum_{j\in\mathscr{I}_i}\lambda_{ij}\frac{(b_j+\bar{b}_j)^2}{\lambda_{jj}^2(r_j+\bar{r}_j)}\beta_j^2\Big\}\hat{x}^2\\
&+\tfrac{1}{2}\Big\{d\gamma_i + \beta_{i,o}\sigma_o + 2(a+\bar{a})\gamma_i + 2\gamma_i\beta_i\tfrac{(b_i+\bar{b}_i)^2}{\lambda_{ii}(r_i+\bar{r}_i)}\\
&-2\beta_i\textstyle\sum_j\frac{(b_j+\bar{b}_j)^2}{\lambda_{jj}(r_j+\bar{r}_j)}\gamma_j - 2\gamma_i\sum_j\frac{(b_j+\bar{b}_j)^2}{\lambda_{jj}(r_j+\bar{r}_j)}\beta_j + 2\sum_{j\in\mathscr{I}_i}\lambda_{ij}\frac{(b_j+\bar{b}_j)^2}{\lambda_{jj}^2(r_j+\bar{r}_j)}\beta_j\gamma_j + [\sigma_o\beta_i\,dB_o]\Big\}\hat{x}\\
&+d\delta_i + \gamma_{i,o}\sigma_o + \tfrac{1}{2}\tfrac{(b_i+\bar{b}_i)^2}{\lambda_{ii}(r_i+\bar{r}_i)}\gamma_i^2 - \gamma_i\textstyle\sum_j\frac{(b_j+\bar{b}_j)^2}{\lambda_{jj}(r_j+\bar{r}_j)}\gamma_j + \frac{1}{2}\sum_{j\in\mathscr{I}_i}\lambda_{ij}\frac{(b_j+\bar{b}_j)^2}{\lambda_{jj}^2(r_j+\bar{r}_j)}\gamma_j^2\\
&+\tfrac{1}{2}\alpha_i[\sigma^2 + \textstyle\int_\Theta \mu(t,\theta)\nu(d\theta)] + \frac{\sigma_o^2}{2}\beta_i + [\sigma_o\gamma_i\,dB_o],
\end{aligned}
\tag{7.216}
$$

and by identifying the coefficients, we retrieve the expressions in (7.204), which confirms the validity of the feedback strategies obtained by the approach developed here.

7.4.3 *Empathy Spite and Nonexistence of Empathetic Equilibria*

In this section, we the role of empathy spite in the existence of equilibria. We first observe that if $\lambda_{ii}r_i < 0$, the equilibrium payoff can be infinite. In order to understand such behavior, we consider a particular case with two decision-makers 1 and 2 , $\lambda_{11} = 1, \lambda_{12} = -1,\ \lambda_{22} = 1, \lambda_{21} = -1$. Let $r_i = 1, \bar{r}_i = 0,\ a = \bar{a} = 0,\ b = (\sqrt{2}, -1),\ \bar{b} = (0,0), q_1(t) = q_2(t) = 0,\ t < T$. We analyze the problem as function of $q_1(T) - \bar{q}_2(T)$. The problem is mean-field dependent because of the terminal payoff.

Below we show that empathy spite may lead to nonexistence of empathetic equilibria in feedback strategies. We also show that for some range of parameters, there is an empathetic equilibrium in feedback strategies:

$$
\begin{aligned}
&EL_1^A(u_1,u_2) = \tfrac{1}{2}(q_1(T)-q_2(T))E(x(T)-\hat{x}(T))^2\\
&+\tfrac{1}{2}(q_1(T)+\bar{q}_1(T)-q_2(T)-\bar{q}_2(T))E\hat{x}^2(T)\\
&+\tfrac{1}{2}E\textstyle\int_0^T\{+(u_1-\hat{u}_1)^2-(u_2-\hat{u}_2)^2+\hat{u}_1^2-\hat{u}_2^2\}\,dt,\\
&EL_2^A(u_1,u_2) = \tfrac{1}{2}(q_2(T)-q_1(T))E(x(T)-\hat{x}(T))^2\\
&+\tfrac{1}{2}(q_2(T)+\bar{q}_2(T)-q_1(T)-\bar{q}_1(T))E\hat{x}^2(T)\\
&+\tfrac{1}{2}E\textstyle\int_0^T\{+(u_2-\hat{u}_2)^2-(u_1-\hat{u}_1)^2+\hat{u}_2^2-\hat{u}_1^2\}\,dt,\\
&dx=\Big\{\textstyle\sum_{i=1}^I b_i(u_i-\hat{u}_i)+\sum_{i=1}^I b_i\hat{u}_i\Big\}dt+\sigma dB+\int_\Theta \mu(t,\theta)\tilde{N}(dt,d\theta)+\sigma_o dB_o.
\end{aligned}
\tag{7.217}
$$

The solution is given by

$$
\begin{array}{l}
\inf_{u_i\in\mathcal{U}_i} E[L_i] = \inf_{u_i\in\mathcal{U}_i} E[\hat{L}_i]\\
= E\{\frac{1}{2}\alpha_i(0)(x(0)-\hat{x}(0))^2 + \frac{1}{2}\beta_i(0)[\hat{x}(0)]^2 + \gamma_i(0)\hat{x}(0) + \delta_i(0)\}\\
u_1^* = -\sqrt{2}\alpha_1(x-\hat{x}) - \sqrt{2}(\beta_1\hat{x}+\gamma_1),\\
u_2^* = -\alpha_2(x-\hat{x}) - (\beta_2\hat{x}+\gamma_2),\\
d\alpha_1 = -\left\{-2\alpha_1^2 - 2\alpha_1\alpha_2 - \alpha_2^2\right\}dt,\\
\alpha_1(T) = q_1(T) - q_2(T),\\
d\alpha_2 = -\left\{-\alpha_2^2 - 4\alpha_2\alpha_1 - 2\alpha_1^2\right\}dt,\\
\alpha_2(T) = -(q_1(T) - q_2(T)) = -\alpha_1(T),\\
d\beta_1 = -\left\{-2\beta_1^2 - 2\beta_1\beta_2 - \beta_2^2\right\}dt,\\
\beta_1(T) = (q_1(T)+\bar{q}_1(T)) - (q_2(T)+\bar{q}_2(T)),\\
d\beta_2 = -\left\{-\beta_2^2 - 4\beta_2\beta_1 - 2\beta_1^2\right\}dt,\\
\beta_2(T) = (q_2(T)+\bar{q}_2(T)) - (q_1(T)+\bar{q}_1(T)) = -\beta_1(T),\\
d\gamma_1 = -\left\{\beta_{1,o}\sigma_o - 2\beta_1\gamma_1 - \beta_1\gamma_2 - \gamma_1\beta_2 - \beta_2\gamma_2\right\}dt,\\
\gamma_1(T) = 0,\\
d\gamma_2 = -\left\{\beta_{2,o}\sigma_o - \beta_2\gamma_2 - 2\beta_2\gamma_1 - 2\gamma_2\beta_1 - 2\beta_1\gamma_1\right\}dt,\\
\gamma_2(T) = 0,\\
d\delta_1 = -\left\{[\sigma^2 + \int_\Theta \mu^2(t,\theta)\nu(d\theta)]\frac{1}{2}\alpha_1 + \frac{1}{2}\sigma_o^2\beta_1 + \gamma_{1,o}\sigma_o - \gamma_1\gamma_2 - \gamma_1^2 - \frac{1}{2}\gamma_2^2\right\}dt,\\
\delta_1(T)=0,\\
d\delta_2 = -\left\{[\sigma^2 + \int_\Theta \mu^2(t,\theta)\nu(d\theta)]\frac{1}{2}\alpha_2 + \frac{1}{2}\sigma_o^2\beta_2 + \gamma_{2,o}\sigma_o - 2\gamma_2\gamma_1 - \frac{1}{2}\gamma_2^2 - \gamma_1^2\right\}dt,\\
\delta_2(T) = 0,\\
d\hat{x} = \{-[2\beta_1+\beta_2]\hat{x} - [2\gamma_1+\gamma_2]\}dt + \sigma_o dB_o,\ \hat{x}(0)=\hat{x}_0.
\end{array}
\tag{7.218}
$$

Assume that $q_1(T) \neq q_2(T)$, and consider the stochastic Riccati system:

$$
\begin{array}{l}
d\alpha_1 = -\left\{-2\alpha_1^2 - 2\alpha_1\alpha_2 - \alpha_2^2\right\}dt,\\
\alpha_1(T) = q_1(T) - q_2(T),\\
d\alpha_2 = -\left\{-\alpha_2^2 - 4\alpha_2\alpha_1 - 2\alpha_1^2\right\}dt,\\
\alpha_2(T) = -(q_1(T) - q_2(T)) = -\alpha_1(T),
\end{array}
\tag{7.219}
$$

Taking the sum of the two processes $\alpha = \alpha_1 + \alpha_2$ yields

$$
\begin{array}{l}
d(\alpha_1+\alpha_2) = -\left\{-4\alpha_1^2 - 6\alpha_1\alpha_2 - 2\alpha_2^2\right\}dt,\\
\alpha_1(T) + \alpha_2(T) = 0,
\end{array}
\tag{7.220}
$$

which is equivalent to

$$
\begin{array}{l}
d(\alpha_1+\alpha_2) = (\alpha_1+\alpha_2)\{4\alpha_1 + 2\alpha_2\}dt,\\
\alpha_1(T) + \alpha_2(T) = 0.
\end{array}
\tag{7.221}
$$

The process $\alpha = \alpha_1 + \alpha_2$ is $\mathcal{F}^{B_o}$-measurable and is identically zero at the terminal time $T,\ \alpha_1(T) + \alpha_2(T) = 0$. It then follows that α is identically zero for all time $t \in [0, T]$ and thus $\alpha_2(t) = -\alpha_1(t)$ for $t \in [0, T]$:

$$d\alpha_1 = \alpha_1^2 dt, \quad \alpha_1(T) = q_1(T) - q_2(T). \tag{7.222}$$

Since $q_1(T) \neq q_2(T)$, it follows that

$$\alpha_1(t) = \frac{1}{T - t + \frac{1}{q_1(T)-q_2(T)}}$$

does not blow up within $[0, T]$ if $q_1(T) - q_2(T) > 0$. Thus, if $q_1(T) - q_2(T) > 0$, there is an empathetic mean-field Nash equilibrium in feedback strategies for every horizon $[0, T],\ T > 0$.

Two cases are of interest:

- If $q_1(T) - q_2(T) < 0$, α_i does not exist on $[0, T]$ if $T + \frac{1}{q_1(T)-q_2(T)} > 0$. It has a finite time escape at $t = T + \frac{1}{q_1(T)-q_2(T)} < T$ if $T + \frac{1}{q_1(T)-q_2(T)} \geq t_0 = 0$. So the empathetic conditional mean-field-type game has no linear feedback Nash equilibrium in that case for a chosen horizon length T satisfying $T + \frac{1}{q_1(T)-q_2(T)} > 0$. The stochastic Riccati has a local solution for a small time window $[T - \epsilon, T]$ but not a global solution on $[0, T]$. We conclude that the use of empathy structure (such as partial spite) may lead to nonexistence of Nash equilibria even in finite horizon MFT-LQJD games. Note that α_1 blows up (singularity formation), and it leads to a non-admissible strategy on $[0, T]$.
- For $q_1(T) = q_2(T)$, however, the system becomes admissible and zero is a solution.

7.4.4 Linear-Quadratic Mean-Field-Type Games with Regime Switching

One way of modeling structural changes in dynamics as well as in the payoffs is through regime switching. For example, an electricity demand distribution may change states from two-peak per day mode to a linear mode as the season moves from a period of hot/cool to a spring/fall with comfortable temperatures. Regime switching enables a very powerful dynamical system of mean-field-type by incorporating both of these periods.

Section 7.4.4.1 presents a risk-neutral game problem with regime switching. Section 7.4.4.2 focuses on team problems with regime switching.

7.4.4.1 Equilibrium Strategies Under Regime Switching

In this subsection, we seek for equilibrium strategies and costs under regime switching. Let $\tilde{s}(t)$ be a regime that switches between a finite number of elements in $\tilde{S}$, with switching rate $\tilde{q}_{\tilde{s}k}$, and $L_i(u_1, \dots, u_I)$ be the objective functional of decision-maker i, given by

$$
\begin{aligned}
&L_i(u_1, \dots, u_I) = \tfrac{1}{2}q_i(T, \tilde{s}(T))x^2(T) + \tfrac{1}{2}\bar{q}_i(T, \tilde{s}(T))[E(x(T)|\mathscr{F}_T^{B_o})]^2 \\
&+\tfrac{1}{2}\int_0^T \Big\{q_i(t, \tilde{s}(t))x^2(t) + \bar{q}_i(t, \tilde{s}(t))(E[x(t)\mid\mathscr{F}_t^{B_o}])^2 \\
&\qquad +r_i(t, \tilde{s}(t))u_i^2(t) + \bar{r}_i(t, \tilde{s}(t))[E(u_i(t)|\mathscr{F}_t^{B_o})]^2\Big\}\, dt.
\end{aligned}
\tag{7.223}
$$

Then, the best response of decision-maker i to the process $(u_{-i}, E[x|\mathscr{F}^{B_o}], E[u_{-i}|\mathscr{F}^{B_o}])$ solves the following risk-neutral linear-quadratic mean-field-type control problem.

Game Problem with Regime Switching

$$
\begin{aligned}
&\inf_{u_i(\cdot)\in\mathscr{U}_i} E\left[L_i(u_1, \dots, u_I)\right], \\
&\text{subject to} \\
&dx(t) = \Big\{a(t, \tilde{s}(t))x(t) + \bar{a}(t, \tilde{s}(t))E[x(t)\mid\mathscr{F}_t^{B_o}] \\
&+\textstyle\sum_{i=1}^I b_i(t, \tilde{s}(t))u_i(t) + \sum_{i=1}^I \bar{b}_i(t, \tilde{s}(t))E[u_i(t)\mid\mathscr{F}_t^{B_o}]\Big\}\, dt \\
&+\sigma(t, \tilde{s}(t))dB(t) + \int_\Theta \mu(t, \theta, \tilde{s}(t))\tilde{N}(dt, d\theta) + \kappa(t, .).\tilde{M}(dt) \\
&+\sigma_o(t, \tilde{s}(t))dB_o(t), \\
&x(0) \in \mathbb{R},\ \tilde{s}(0) \in \tilde{S}
\end{aligned}
\tag{7.224}
$$

where $q_i(t, \tilde{s}(t)) > 0, q_i(t, \tilde{s}(t)) + \bar{q}_i(t, \tilde{s}(t)) > 0, r_i(t, \tilde{s}(t)) > 0, r_i(t, \tilde{s}(t)) + \bar{r}_i(t, \tilde{s}(t)) > 0$ are $\mathscr{F}^{B_o,\tilde{s}}$-measurable and $a(t, \tilde{s}(t)), \bar{a}(t, \tilde{s}(t)), b_i(t, \tilde{s}(t)), \sigma(t, \tilde{s}(t))$, $\sigma_o(t, \tilde{s}(t))$ are measurable real-valued functions and where $E[x(t)|\mathscr{F}_t^{B_o}]$ is the conditional expectation of the state created by all decision-makers under the control action profile $(u_1, \dots, u_I) \in \prod_{j\neq i}\mathscr{U}_j$. $\tilde{M}(t), \tilde{N}(t), B(t), B_o(t)$ are mutually independent.

Solution to the Game Problem with Regime Switching (7.224)

$$
\begin{aligned}
&\inf_{u_i\in\mathcal{U}_i} E[L_i]\\
&=E\{\tfrac{1}{2}\alpha_i(0,\tilde{s}(0))(x(0)-\hat{x}(0))^2+\tfrac{1}{2}\beta_i(0,\tilde{s}(0))\hat{x}(0)^2+\gamma_i(0,\tilde{s}(0))\hat{x}+\delta_i(0,\tilde{s}(0))\},\\
&u_i^* = -\tfrac{b_i}{r_i}\alpha_i(x-\hat{x}) - \tfrac{(b_i+\bar{b}_i)}{r_i+\bar{r}_j}(\beta_i\hat{x}+\gamma_i),\\
&d\hat{x} = \{[a+\bar{a}-\textstyle\sum_{i=1}^I \tfrac{(b_i+\bar{b}_i)^2}{r_i+\bar{r}_i}\beta_i]\hat{x} - \sum_{i=1}^I \tfrac{(b_i+\bar{b}_i)^2}{r_i+\bar{r}_i}\gamma_i\}dt + \sigma_o dB_o,\\
&\hat{x}(0)=\hat{x}_0,\\
&d\alpha_i = -\left\{2a\alpha_i - \tfrac{b_i^2}{r_i}\alpha_i^2 - 2\alpha_i\textstyle\sum_{j\neq i}\tfrac{b_j^2}{r_j}\alpha_j + q_i + \sum_{k\neq\tilde{s}}[\alpha_i(t,k)-\alpha_i(t,\tilde{s})]\tilde{q}_{\tilde{s}k}\right\}dt,\\
&\alpha_i(T,\tilde{s}) = q_i(T,\tilde{s}),\\
&d\beta_i = -\Big\{2(a+\bar{a})\beta_i - \tfrac{(b_i+\bar{b}_i)^2}{r_i+\bar{r}_i}\beta_i^2 - 2\beta_i\textstyle\sum_{j\neq i}\tfrac{(b_j+\bar{b}_j)^2}{r_j+\bar{r}_j}\beta_j + q_i + \bar{q}_i\\
&\qquad + \textstyle\sum_{k\neq\tilde{s}}[\beta_i(t,k)-\beta_i(t,\tilde{s})]\tilde{q}_{\tilde{s}k}\Big\}dt,\\
&\beta_i(T,\tilde{s}) = q_i(T,\tilde{s}) + \bar{q}_i(T,\tilde{s}),\\
&d\gamma_i = -\Big\{(a+\bar{a})\gamma_i - \tfrac{(b_i+\bar{b}_i)^2}{r_i+\bar{r}_i}\beta_i\gamma_i - \beta_i\textstyle\sum_{j\neq i}\tfrac{(b_j+\bar{b}_j)^2}{r_j+\bar{r}_j}\gamma_j\\
&\qquad -\gamma_i\textstyle\sum_{j\neq i}\tfrac{(b_j+\bar{b}_j)^2}{r_j+\bar{r}_j}\beta_j + \sum_{k\neq\tilde{s}}[\gamma_i(t,k)-\gamma_i(t,\tilde{s})]\tilde{q}_{\tilde{s}k}\Big\}dt,\\
&\gamma_i(T,\tilde{s}) = 0,\\
&d\delta_i = -\Big\{\tfrac{1}{2}\alpha_i[\sigma^2 + \textstyle\int_\Theta \mu^2(t_-,.,\theta)\nu(d\theta)] + \tfrac{1}{2}\sigma_o^2\beta_i\\
&\qquad +\tfrac{1}{2}\textstyle\sum_{k\neq\tilde{s}}\alpha_i(t,k)\kappa^2(t_-,k)\tilde{q}_{\tilde{s}k} + \sum_{k\neq\tilde{s}}[\delta_i(t,k)-\delta_i(t,\tilde{s}]\tilde{q}_{\tilde{s}k}\\
&\qquad -\tfrac{(b_i+\bar{b}_i)^2}{r_i+\bar{r}_i}\tfrac{1}{2}\gamma_i^2 - \gamma_i\textstyle\sum_{j\neq i}\tfrac{(b_j+\bar{b}_j)^2}{r_j+\bar{r}_j}\gamma_j\Big\}dt,\\
&\delta_i(T,\tilde{s}) = 0,
\end{aligned}
\tag{7.225}
$$

7.4.4.2 Mean-Field-Type Weighted Global Optima

This subsection presents global optimum strategies and costs under regime switching. Let $w_1, \ldots, w_I$ be positive real numbers.

Weighted Global Optimization Problem with Regime Switching

$$
\begin{aligned}
&\inf_{(u_1,\ldots,u_I)} E\textstyle\sum_i w_i L_i,\\
&dx = \left\{a(x-\hat{x}) + (a+\bar{a})\hat{x} + \textstyle\sum_{i=1}^I b_i(u_i-\hat{u}_i) + \sum_{i=1}^I (b_i+\bar{b}_i)\hat{u}_i\right\}dt\\
&+\sigma dB + \textstyle\int_\Theta \mu(t,\theta)\tilde{N}(dt,d\theta) + \sigma_o dB_o + \kappa(t).\tilde{M}(dt),\\
&x(0) = x_0,
\end{aligned}
\tag{7.226}
$$

The weighted global optimization (7.226) of mean-field-type is solved explicitly below.

Solution to the Weighted Global Optimization Problem with Regime Switching (7.226)

$$
\begin{aligned}
&\inf_{(u_1,\ldots,u_I)} E\sum_i w_i L_i = E\left\{\tfrac{1}{2}\alpha_0(0)var(x(0)) + \tfrac{1}{2}\beta_0(0)[\hat{x}(0)]^2 \right.\\
&\qquad\qquad \left. + \gamma_0(0)\hat{x}(0) + \delta_0(0)\right\},\\
&\hat{u}_i(w) = -\frac{b_i}{w_i r_i}\alpha_0(x-\hat{x}) - \frac{(b_i+\bar{b}_i)}{w_i(r_i+\bar{r}_i)}(\beta_0\hat{x}+\gamma_0),\\
&d\hat{x} = \{[a+\bar{a}-\beta_0\textstyle\sum_{i=1}^{I}\frac{(b_i+\bar{b}_i)^2}{w_i(r_i+\bar{r}_i)}]\hat{x} - \gamma_0\sum_{i=1}^{I}\frac{(b_i+\bar{b}_i)^2}{w_i(r_i+\bar{r}_i)}\}dt + \sigma_o dB_o,\\
&\hat{x}(0) = \hat{x}_0,\\
&d\alpha_0 = -\left\{2a\alpha_0 - \alpha_0^2\textstyle\sum_i \frac{b_i^2}{w_i r_i} + \sum_i w_i q_i + \sum_{k\neq\tilde{s}}[\alpha_0(t,k)-\alpha_0(t,\tilde{s})]\tilde{q}_{\tilde{s}k}\right\}dt,\\
&\alpha_0(T,\tilde{s}) = \textstyle\sum_i w_i q_i(T,\tilde{s}),\\
&d\beta_0 = -\left\{2\beta_0(a+\bar{a}) - \beta_0^2\textstyle\sum_i \frac{(b_i+\bar{b}_i)^2}{w_i(r_i+\bar{r}_i)} + \sum_i w_i(q_i+\bar{q}_i)\right.\\
&\qquad \left. + \textstyle\sum_{k\neq\tilde{s}}[\beta_0(t,k)-\beta_0(t,\tilde{s})]\tilde{q}_{\tilde{s}k}\right\}dt,\\
&\beta_0(T,\tilde{s}) = \textstyle\sum_i w_i(q_i+\bar{q}_i(T,\tilde{s})),\\
&d\gamma_0 = -\left\{(a+\bar{a})\gamma_0 + \beta_{0,o}\sigma_o - \beta_0\gamma_0\textstyle\sum_i \frac{(b_i+\bar{b}_i)^2}{w_i(r_i+\bar{r}_i)} + \sum_{k\neq\tilde{s}}[\gamma_0(t,k)-\gamma_0(t,\tilde{s})]\tilde{q}_{\tilde{s}k}\right\}dt,\\
&\gamma_0(T,\tilde{s}) = 0,\\
&d\delta_0 = -\left\{[\sigma^2 + \textstyle\int_\Theta \mu^2(t,\theta)\nu(d\theta)]\frac{\alpha_0}{2} + \sigma_o^2\frac{\beta_0}{2} + \gamma_{0,o}\sigma_o - \frac{1}{2}\gamma_0^2\sum_i \frac{(b_i+\bar{b}_i)^2}{w_i(r_i+\bar{r}_i)}\right.\\
&\qquad \left. + \tfrac{1}{2}\textstyle\sum_{k\neq\tilde{s}}\alpha_0(t,k)\kappa^2(t_-,k)\tilde{q}_{\tilde{s}k} + \sum_{k\neq\tilde{s}}[\delta_0(t,k)-\delta_0(t,\tilde{s})]\tilde{q}_{\tilde{s}k}\right\}dt,\\
&\delta(T,\tilde{s}) = 0.
\end{aligned}
\tag{7.227}
$$

7.5 Price of Simplicity of Linear-Quadratic Mean-Field Games

This section presents the price of simplicity of mean-field games through basic examples with linear state dynamics and variance-aware cost functions. We show that the game solution obtained using the mean-field game approach (which freezes the mean-field term) may lead to poor performance compared to the original game problem. The price of simplicity, which is the gap between the solution of the original problem and the solution achieved by the mean-field game problem, is unbounded. This suggests a careful design of the decomposition of the original problem instead of freezing the mean-field terms. To illustrate and confirm this statement, we present a simple example that exhibits this behavior. In contrast, we show that the price of simplicity of the mean-field-type game approach is zero. See also Chapter 15 of Volume II of this book.

7.5.1 Variance-Aware Strategic Games

This subsection introduces a description of the problem, which is solved in the follow-up subsubsections by using different game approaches. Let us consider I decision-makers with associated state dynamics as follows:

$$
\begin{aligned}
&x(0) = x_0 \neq 0,\\
&dx(t) = \left(b_1 x(t) + \bar{b}_1 \bar{x}(t) + \sum_{i\in\mathscr{I}} b_{2i} u_i(t) + \sum_{j\in\mathscr{I}} \bar{b}_{2j} \bar{u}_j(t)\right) dt \\
&+x(t)\left(\sigma dB(t) + \int_{\theta\in\Theta} \mu(\theta)\tilde{N}(dt, d\theta)\right),
\end{aligned} \tag{7.228}
$$

where $\mathscr{I} := \{1, 2, \ldots, I\}$, $I \geq 2, \bar{x} = E[x]$ and $\bar{u}_i = E[u_i]$. Clearly, there is a coupling in both the system dynamics throughout the term $\sum_{j=1}^{I} \bar{b}_{2j}\bar{u}_j$ and the cost function throughout the term $\sum_{j\neq i} \varepsilon_{ij}\bar{u}_i\bar{u}_j$, with $i \in \mathscr{I}$. In (7.228), B is a standard Brownian motion representing uncertainty or noise, and N denotes a jump process with Lévy measure $\nu(d\theta)$ defined over Θ. It is assumed that ν is a Radon measure over the jump space Θ. $\tilde{N}(dt, d\theta) = N(dt, d\theta) - \nu(d\theta)dt$, is a compensated jump process. The objective of decision-maker $i \in \mathscr{I}$ is to minimize the expected long-term cost $E[L_i]$ where

$$
\begin{aligned}
&L_i(x_0, u) = \tfrac{1}{2} q_i x^2(T) + \tfrac{1}{2} \bar{q}_i \bar{x}^2(T)\\
&+\tfrac{1}{2}\int_0^T \left(q_i x^2(t) + \bar{q}_i \bar{x}^2(t) + r_i u_i^2(t) + \bar{r}_i \bar{u}_i^2(t) + 2\sum_{j\neq i} \varepsilon_{ij}\bar{u}_i(t)\bar{u}_j(t) \right) dt,
\end{aligned} \tag{7.229}
$$

with $q_i, \bar{q}_i, r_i,\ r_i + \bar{r}_i \geq \epsilon > 0$. Because of the structure of the cost L_i, we consider control strategies u_i that are progressively measurable and square integrable over $[0, T]$. We denote by $\mathscr{U}_i$ the set of strategies for decision-maker i. The goal is to solve the following problem (P) :

$$
(P) \begin{cases} i \in \mathscr{I}, \\ \inf_{u_i\in\mathscr{U}_i} E L_i(x_0, u), \\ \text{subject to (7.228).} \end{cases} \tag{7.230}
$$

We first present an approximation methodology for problem (P) called mean-field game approach in Sect. 7.5.1.1. Second, we utilize a mean-field-type game approach in Sect. 7.5.1.2. We show that the mean-field-type game approach provides an exact method for problem (P). Thus, the gap between the mean-field game solution and the mean-field-type game solution are presented as the *price of simplicity* of mean-field games in Sect. 7.5.1.3.

7.5.1.1 Multi-population Mean-Field Game Approach

In the multi-population mean-field game approach, there are I populations. In each population, there is a large number of virtual agents approximating the mean-field terms. The expected value term $[E(u_i(t))]^2$, for simplicity, is approximated to a mean population control denoted by $[y_i(t)]^2$, which is a frozen quantity in the sense that the control action of a generic single virtual agent does not influence the term y_i. Thus, the average control of population j is approximated by

$$y_j(t) = \lim_{K_j \to +\infty} \frac{1}{K_j} \sum_{k=1}^{K_j} u_{jk}(t) \approx E[u_j(t)].$$

Such a method has been widely used in different works in statistical physics, quantum mechanics, and thermodynamics. The mean-field individual cost over the horizon $[0, T]$ of virtual agent k in population $i \in \mathscr{I}$ is

$$\begin{aligned}
&L_{\text{mfg},ik}(x_i(0), u_{ik}, m, y) = \tfrac{1}{2}q_i x^2(T) + \tfrac{1}{2}\bar{q}_i m^2(T) \\
&+ \int_0^T \left(\tfrac{1}{2}q_i x^2(t) + \tfrac{1}{2}\bar{q}_i m^2(t) + \tfrac{1}{2}r_i u_{ik}^2(t) + \tfrac{1}{2}\bar{r}_i y_i^2(t) + \textstyle\sum_{j\neq i} \varepsilon_{ij} y_i(t) y_j(t) \right) dt,
\end{aligned} \tag{7.231}$$

The state of the virtual agent k from population i is given by

$$\begin{aligned}
&x_{ik}(0) = x_{i,0}, \\
&dx_{ik}(t) = \left(b_1 x_{ik}(t) + \bar{b}_1 m(t) + b_{2i} u_{ik}(t) + \sum_{j\neq i} b_{2j} u_j(t) + \sum_{j=1}^{I} \bar{b}_{2j} y_j(t) \right) dt \\
&+ x_{ik}(t) \left(\sigma dB_{ik}(t) + \int_{\theta\in\Theta} \mu(\theta) \tilde{N}_{ik}(dt, d\theta) \right).
\end{aligned} \tag{7.232}$$

where B_{ik} and $\tilde{N}_{ik}$ are independent across decision-makers and populations. This is now a multi-population mean-field game problem as the decision-makers are coupled through the mean-field term $m(t)$ and $y_i(t)$.

Given m, y, the multi-population mean-field best-response problem of a generic decision-maker k in population i is to solve $\inf_{u_{ik}\in\mathscr{U}_i} EL_{ik}(x_0, u_{ik}, m, y)$, subject to (7.232). The mean-field Nash equilibrium problem of the multi-population mean-field game is a situation in which every decision-maker k in population i is playing its best-response strategy u_{ik}^* and the resulting limiting arithmetic average state/control of all decision-makers coincides with m^*, y^* by consistency.

Proposition 7.5.1.2 presents the solution of the mean-field Nash equilibrium of multi-population mean-field game problem.

Proposition 7.5.1.2 *The mean-field equilibrium strategy is semi-explicitly given by*

$$\begin{cases} u^*_{\mathrm{mfg},ik}(t) = -\frac{b_{2i}}{r_i}\left(\alpha_i(t)x(t) + \gamma_i(t)m(t)\right), \\ \textit{where } i \in \{1,2,\ldots,I\}, \\ m(t) = m(0)\exp\left(\int_0^t g(s)ds\right), \\ g(t) = b_1 + \bar{b}_1 - \sum_{j=1}^{I} \frac{b_{2j}(b_{2j}+\bar{b}_{2j})}{r_j}\left(\alpha_j(t)+\gamma_j(t)\right), \\ y_i(t) = -\frac{b_{2i}(\alpha_i(t)+\gamma_i(t))}{r_i}m(t), \end{cases} \tag{7.233}$$

The mean-field equilibrium cost of k in population i is semi-explicitly given by

$$L^*_{\mathrm{mfg},ik} = E\left(\frac{1}{2}\alpha_i(0)x_0^2 + \frac{1}{2}\beta_i(0)m^2(0) + \gamma_i(0)m(0)x_0\right), \tag{7.234}$$

where $\alpha_i(t)$, $\gamma_i(t)$, and $\beta_i(t)$ solve the following differential equations:

$$\begin{aligned}
&\dot{\alpha}_i(t) = -q_i - 2\alpha_i(t)b_1 - \alpha_i(t)c + 2\alpha_i(t)\sum_{j\neq i}\tfrac{b_{2j}^2}{r_i}\alpha_j(t) + \tfrac{b_{2i}^2}{r_i}\alpha_i^2(t),\\
&\dot{\gamma}_i(t) = -\gamma_i(t)g(t) - \bar{b}_1\alpha_i(t) - \gamma_i(t)b_1 + \alpha_i(t)\gamma_i(t)\tfrac{b_{2i}}{r_i} + \alpha_i(t)\sum_{j=1}^{I}\tfrac{b_{2j}\bar{b}_{2j}}{r_j}(\alpha_j(t)+\gamma_j(t))\\
&+\alpha_i(t)\sum_{j\neq i}\tfrac{b_{2j}^2}{r_j}\gamma_j(t) + \gamma_i(t)\sum_{j\neq i}\tfrac{b_{2j}^2}{r_j}\alpha_j(t),\\
&\dot{\beta}_i(t) = -2\tfrac{b_{2i}(\alpha_i(t)+\gamma_i(t))}{r_i}\sum_{j\neq i}\varepsilon_{ij}\tfrac{b_{2j}(\alpha_j(t)+\gamma_j(t))}{r_j}\\
&+2\gamma_i(t)\left(\sum_{j\neq i}\tfrac{b_{2j}^2}{r_j}\gamma_j(t) + \sum_{j=1}^{I}\tfrac{b_{2j}\bar{b}_{2j}}{r_j}(\alpha_j(t)+\gamma_j(t))\right)\\
&-\bar{q}_i - 2\beta_i(t)g(t) - 2\bar{b}_1\gamma_i(t) + \tfrac{b_{2i}^2}{r_i}\gamma_i^2(t) - \bar{r}_i\tfrac{b_{2i}^2(\alpha_i(t)+\gamma_i(t))^2}{r_i^2},
\end{aligned} \tag{7.235}$$

with the boundary conditions $\alpha_i(T) = q_i$, $\gamma_i(T) = 0$, and $\beta_i(T) = \bar{q}_i$, and $c = \left(\sigma^2 + \int_{\theta\in\Theta}\mu^2(\theta)\nu(d\theta)\right)$.

Notice that the mean-field equilibrium strategies are functions of $x(t)$ and $m(t)$.

Proof Consider the following candidate guess functional:

$$f_i(t,x) = \frac{1}{2}\alpha_i(t)x^2(t) + \frac{1}{2}\beta_i(t)m^2(t) + \gamma_i(t)m(t)x(t). \tag{7.236}$$

Applying Itô's formula yields

$$\begin{aligned}
f_i(T,x(T)) - f_i(0,x_0) =&\int_0^T\Big(\tfrac{1}{2}\dot{\alpha}_i(t)x^2(t) + \tfrac{1}{2}\dot{\beta}_i(t)m^2(t)\\
&+\beta_i(t)m(t)\dot{m}(t) + [\dot{\gamma}_i(t)m(t) + \gamma_i(t)\dot{m}(t)]x(t)\\
&+(\alpha_i(t)x(t)+\gamma_i(t)m(t))\Big(b_1x(t) + \bar{b}_1m(t)\\
&+b_{2i}u_i(t) + \sum_{j\neq i}b_{2j}u_j(t) + \sum_{j=1}^{n}\bar{b}_{2j}y_j(t)\Big)\Big)dt\\
&+\int_0^T\tfrac{1}{2}\alpha_i(t)\left(\sigma^2 + \int_{\theta\in\Theta}\mu^2(\theta)\nu(d\theta)\right)x^2(t)dt + \int_0^T\epsilon_i dt,
\end{aligned} \tag{7.237}$$

where $E[\epsilon_i] = 0$. Then

$$
\begin{aligned}
&E[f_i(T, x(T)) - f_i(0, x_0)] = E\int_0^T \Bigg(\tfrac{1}{2}\dot{\alpha}_i(t) + \alpha_i(t)b_1 \\
&+\tfrac{1}{2}\alpha_i(t)\left(\sigma^2 + \int_{\theta\in\Theta} \mu^2(\theta)\nu(d\theta)\right) \Bigg)x^2(t)dt \\
&+E\int_0^T \Bigg(\tfrac{1}{2}\dot{\beta}_i(t)m^2(t) + \beta_i(t)m(t)\dot{m}(t) + \bar{b}_1\gamma_i(t)m^2(t) \Bigg)dt \\
&+E\int_0^T \Bigg(\dot{\gamma}_i(t)m(t) + \gamma_i(t)\dot{m}(t) + \bar{b}_1\alpha_i(t)m(t) + \gamma_i(t)m(t)b_1 \Bigg)x(t)dt \\
&+E\int_0^T \Bigg((\alpha_i(t)x(t) + \gamma_i(t)m(t))b_{2i}u_i(t) \\
&+(\alpha_i(t)x(t) + \gamma_i(t)m(t))\left(\sum_{j\neq i} b_{2j}u_j(t) + \sum_{j=1}^n \bar{b}_{2j}y_j(t)\right) \Bigg)dt.
\end{aligned}
\tag{7.238}
$$

Completion of squares leads to the following equations:

$$
\begin{aligned}
&\tfrac{1}{2}r_i u_{ik}^2(t) + b_{2i}\left(\alpha_i(t)x(t) + \gamma_i(t)m(t)\right)u_{ik}(t) = \tfrac{1}{2}r_i\left(u_{ik}(t) + \frac{b_{2i}(\alpha_i(t)x(t)+\gamma_i(t)m(t))}{r_i}\right)^2 \\
&-\frac{b_{2i}^2}{2r_i}\left(\alpha_i^2(t)x^2(t) + 2\gamma_i(t)\alpha_i(t)x(t)m(t) + \gamma_i^2(t)m^2(t)\right),\ i \in \mathscr{I}.
\end{aligned}
\tag{7.239}
$$

Putting all this together, we arrive that

$$
\begin{aligned}
&E[L_i(T, x(T)) - f_i(0, x_0)] = \tfrac{1}{2}\left[q_i(T) - \alpha_i(T)\right]x_i(T)^2 \\
&+\tfrac{1}{2}\left[\bar{q}_i(T) - \beta_i(T)\right]m(T)^2 + \left[0 - \gamma_i(T)\right]x(T)m(T) \\
&+E\int_0^T \tfrac{1}{2}\Bigg(\dot{\alpha}_i(t) + q_i + 2\alpha_i(t)b_1 \\
&+\alpha_i(t)\left(\sigma^2 + \int_{\theta\in\Theta}\mu^2(\theta)\nu(d\theta)\right) - 2\alpha_i(t)\sum_{j\neq i}\frac{b_{2j}^2}{r_i}\alpha_j(t) - \frac{b_{2i}^2}{r_i}\alpha_i^2(t)\Bigg)x^2(t)dt \\
&+E\int_0^T \left(\dot{\gamma}_i(t)m(t) + \gamma_i(t)\dot{m}(t) + \bar{b}_1\alpha_i(t)m(t) + \gamma_i(t)m(t)b_1 - \alpha_i(t)\gamma_i(t)m(t)\frac{b_{2i}}{r_i}\right)x(t)dt \\
&+E\int_0^T \Bigg(-\alpha_i(t)\sum_{j=1}^I \frac{b_{2j}\bar{b}_{2j}}{r_j}(\alpha_j(t) + \gamma_j(t)) \\
&-\alpha_i(t)\sum_{j\neq i}\frac{b_{2j}^2}{r_j}\gamma_j(t) - \gamma_i(t)\sum_{j\neq i}\frac{b_{2j}^2}{r_j}\alpha_j(t)\Bigg)x(t)m(t)dt \\
&+E\int_0^T \Bigg(\sum_{j\neq i}\varepsilon_{ij}\frac{b_{2i}(\alpha_i(t)+\gamma_i(t))}{r_i}\frac{b_{2j}(\alpha_j(t)+\gamma_j(t))}{r_j}m^2(t) \\
&-\gamma_i(t)\left(\sum_{j\neq i}\frac{b_{2j}^2}{r_j}\gamma_j(t) + \sum_{j=1}^I\frac{b_{2j}\bar{b}_{2j}}{r_j}(\alpha_j(t) + \gamma_j(t))\right)m^2(t) \\
&+\tfrac{1}{2}\bar{q}_i m^2(t) + \tfrac{1}{2}\dot{\beta}_i(t)m^2(t) + \beta_i(t)m(t)\dot{m}(t) + \bar{b}_1\gamma_i(t)m^2(t) \\
&-\frac{b_{2i}^2}{2r_i}\gamma_i^2(t)m^2(t) + \tfrac{1}{2}\bar{r}_i\frac{b_{2i}^2(\alpha_i(t)+\gamma_i(t))^2}{r_i^2}m^2\Bigg)dt \\
&+E\int_0^T \tfrac{1}{2}r_i\left(u_{ik}(t) + \frac{b_{2i}(\alpha_i(t)x(t)+\gamma_i(t)m(t))}{r_i}\right)^2 dt.
\end{aligned}
\tag{7.240}
$$

The result then follows by minimizing this expression. ■

7.5.1.2 Mean-Field-Type Game Approach

In this subsection, we solve game problem (P) semi-explicitly. As initially stated in (P), the mean-field terms $E(x), E(u_i)$ are not frozen. A change in the control action u_i affects the state x, and therefore it modifies its expected value. Proposition 7.5.1.3 presents the solution of the game problem (P) in (7.230) by using a mean-field-type game approach.

A mean-field-type equilibrium is a strategy profile that solves problem (P), i.e., $(u_1^*, \ldots, u_n^*) \in \prod_{j=1}^{I} \mathcal{U}_j$ such that

$$L^*_{\text{mftg},i} = E[L_i(x_0, u^*)] = \inf_{u_i \in \mathcal{U}_i} E[L_i(x_0, u_1^*, \ldots, u_{i-1}^*, u_i, u_{i+1}^*, \ldots, u_I^*)]. \tag{7.241}$$

Proposition 7.5.1.3 *Consider the mean-field-type game problem (P) in* (7.230) *with an arbitrary number of decision-makers. The mean-field-type optimal strategy is semi-explicitly given by*

$$\begin{cases} u^*_{\text{mftg},i}(t) = -\dfrac{\chi_i(t) b_{2i}}{r_i}(x(t) - E[x(t)]) - \bar{\eta}_i(t) E(x(t)), \\ \bar{\eta}(t) = A^{-1} v(t), \\ a_{ii} = 1, \;\; a_{ij} = \dfrac{\bar{\varepsilon}_{ij}}{r_i + \bar{r}_i}, \\ v_i(t) = \dfrac{\bar{\chi}_i(t)(b_{2i} + \bar{b}_{2i})}{r_i + \bar{r}_i}, \\ \dot{\chi}_i(t) = -q_i - 2b_1 \chi_i(t) - c\chi_i(t) \\ \qquad + \dfrac{\chi_i^2(t) b_{2i}^2}{r_i} + 2\chi_i(t) \displaystyle\sum_{j \neq i} b_{2j} \dfrac{\chi_j(t) b_{2j}}{r_j}, \\ \dot{\bar{\chi}}_i(t) = -(q_i + \bar{q}_i) - 2(b_1 + \bar{b}_1)\bar{\chi}_i(t) - c\chi_i(t) \\ \qquad + (r_i + \bar{r}_i)\bar{\eta}_i^2(t) + 2\bar{\chi}_i(t) \displaystyle\sum_{j \neq i} (b_{2j} + \bar{b}_{2j}) \bar{\eta}_j(t), \end{cases} \tag{7.242}$$

and with boundary conditions $\chi_i(T) = q_i$ *and* $\bar{\chi}_i(T) = q_i + \bar{q}_i$. *The mean-field-type equilibrium cost is semi-explicitly given by* $L^*_{\text{mftg},i} = \frac{\chi_i(0)}{2} E(x_0 - E[x_0])^2 + \frac{\bar{\chi}_i(0)}{2}(E[x_0])^2$.

Notice that, different from the mean-field game solution, the mean-field-type game solution requires the state x, and its expected values $E(x)$ feedback.

Proof Consider the following candidate guess functional:

$$\tilde{f}_i(t, x(t)) = \frac{1}{2}\chi_i(t)(x(t) - E[x(t)])^2 + \frac{1}{2}\bar{\chi}_i(t) E[x(t)]^2. \tag{7.243}$$

We apply Itô's formula to arrive at

$$
\begin{aligned}
&E[L_i - \tilde{f}_i(0)] = \tfrac{1}{2}\left(q_i(T) - \chi_i(T)\right)(x(T) - E[x(T)])^2 \\
&+\tfrac{1}{2}\left(q_i + \bar{q}_i - \bar{\chi}_i(T)\right) E[x(T)]^2 \\
&+E\textstyle\int_0^T \tfrac{1}{2}q_i\,(x(t) - E[x(t)])^2 + \tfrac{1}{2}(q_i + \bar{q}_i)E[x(t)]^2 \\
&+\tfrac{1}{2}r_i\,(u_i(t) - E[u_i(t)])^2 + \tfrac{1}{2}(r_i + \bar{r}_i)E[u_i(t)]^2 dt \\
&+E\textstyle\int_0^T \sum_{j\neq i} \bar{\epsilon}_{ij} E[u_i(t)]E[u_j(t)]dt \\
&+E\textstyle\int_0^T \Big(\tfrac{1}{2}\dot{\chi}_i(t)(x(t) - E[x(t)])^2 + \tfrac{1}{2}\dot{\bar{\chi}}_i(t)E[x(t)] \\
&+\chi_i(t)(x(t) - E[x(t)])\cdot\Big(b_1(x(t) - E[x(t)]) + \textstyle\sum_{j=1}^I b_{2j}(u_j(t) - E[u_j(t)])\Big) \\
&+\bar{\chi}_i(t)E[x(t)]\Big((b_1 + \bar{b}_1)E[x(t)] + \textstyle\sum_{j=1}^I (b_{2j} + \bar{b}_{2j})E[u_j(t)]\Big) \\
&+\tfrac{1}{2}\chi_i(t)\left[(x(t) - E[x(t)])^2 c + E[x(t)]^2 c\right]\Big)dt,
\end{aligned}
\tag{7.244}
$$

where $c = \sigma^2 + \int_{\theta\in\Theta}\mu^2(\theta)\nu(d\theta)$. After performing completion of squares, we obtain

$$
\begin{aligned}
&E[L_i - \tilde{f}_i(0)] = \tfrac{1}{2}\left(q_i(T) - \chi_i(T)\right)(x(T) - E[x(T)])^2 + \tfrac{1}{2}\left(q_i + \bar{q}_i - \bar{\chi}_i(T)\right) E[x(T)]^2 \\
&+E\textstyle\int_0^T \left(\dot{\chi}_i(t) + q_i + 2b_1\chi_i(t) + c\chi_i(t)\right)\tfrac{1}{2}(x(t) - E[x(t)])^2 dt \\
&+E\textstyle\int_0^T \Big(\dot{\bar{\chi}}_i(t) + (q_i + \bar{q}_i) + 2(b_1 + \bar{b}_1)\bar{\chi}_i(t) + c\chi_i(t)\Big)\tfrac{1}{2}E[x(t)]^2 dt \\
&+E\textstyle\int_0^T \tfrac{1}{2}r_i\Big(u_i(t) - E[u_i(t)] + \tfrac{\chi_i(t)b_{2i}}{r_i}(x(t) - E[x(t)])\Big)^2 \\
&-\Big(\tfrac{\chi_i^2(t)b_{2i}^2}{r_i} + 2\chi_i(t)\textstyle\sum_{j\neq i} b_{2j}\tfrac{\chi_j(t)b_{2j}}{r_j}\Big)\tfrac{1}{2}(x(t) - E[x(t)])^2 dt \\
&+E\textstyle\int_0^T \tfrac{1}{2}(r_i + \bar{r}_i)\left(E[u_i(t)] + \bar{\eta}_i(t)E[x(t)]\right)^2 \\
&-\Big((r_i + \bar{r}_i)\bar{\eta}_i^2 + 2\bar{\chi}_i(t)\textstyle\sum_{j\neq i}(b_{2j} + \bar{b}_{2j})\bar{\eta}_j(t)\Big)\tfrac{1}{2}E[x(t)]^2 dt
\end{aligned}
\tag{7.245}
$$

with parameters $a_{ii} = 1$, $a_{ij} = \frac{\bar{\epsilon}_{ij}}{r_i+\bar{r}_i}$, $v_i(t) = \frac{\bar{\chi}_i(t)(b_{2i}+\bar{b}_{2i})}{r_i+\bar{r}_i}$, $\bar{\eta}_i(t) = \left(A^{-1}v(t)\right)_i$, and $E[u(t)] = -\left(A^{-1}v\right)E[x(t)]$, thus completing the proof. ■

7.5.1.3 Price of Simplicity

In the above, we have computed the mean-field equilibrium $u^*_{\text{mfg},i}(t)$ and the mean-field-type equilibrium $u^*_{\text{mftg},i}(t)$, for every i.

Definition 7.5.1.4 The price of simplicity of the mean-field game approach is the (total) cost gap between the mean-field game equilibrium cost and the total cost of the original problem (P): $\sum_i [L^*_{\text{mfg,i}} - L^*_{\text{mftg,i}}]$.

The mean-field-type game approach (Sect. 7.5.1.2) provides the exact equilibrium solution as it solves the original problem (P). Therefore, the price of simplicity of the mean-field-type game approach is zero. However, the price of simplicity of the mean-field game approach (Sect. 7.5.1.1) can be different than zero as illustrated below.

Proposition 7.5.1.5 *Let the initial distribution of state to be a Dirac measure concentrated at* x_0. *There exists a parameter* $c_0 > 0$ *such that for*

$$c = \left(\sigma^2 + \int_{\theta\in\Theta} \mu^2(\theta)\nu(d\theta)\right) \geq c_0,$$

the price of simplicity $L^*_{\mathrm{mfg}} - L^*_{\mathrm{mftg}}$, *is positive.*

Proof The optimal mean-field cost is given by

$$\begin{aligned} L^*_{\mathrm{mfg},ik} &= E\left[\tfrac{1}{2}\alpha_i(0)x_0^2 + \tfrac{1}{2}\beta_i(0)m^2(0) + \gamma_i(0)m(0)x_0\right] \\ &= E\Bigg[\tfrac{1}{2}\alpha_i(0)(x_0 - E[x_0])^2 + \tfrac{1}{2}(\beta_i(0)+\alpha_i(0))m^2(0) + \gamma_i(0)m(0)x_0\Bigg]. \end{aligned} \tag{7.246}$$

Let $x_0 = m(0)$; then

$$\begin{aligned} L^*_{\mathrm{mfg}} &= \tfrac{1}{2}\left(\sum_i (\beta_i(0) + \alpha_i(0) + 2\gamma_i(0))\right)x_0^2, \\ L^*_{\mathrm{mftg}} &= \tfrac{1}{2}\left(\sum_i \bar{\chi}_i(0)\right)x_0^2, \end{aligned} \tag{7.247}$$

and the price of simplicity is

$$L^*_{\mathrm{mfg}} - L^*_{\mathrm{mftg}} = \frac{\left(\sum\limits_i (\beta_i(0) + \alpha_i(0) + 2\gamma_i(0) - \bar{\chi}_i(0))\right)}{2} x_0^2, \tag{7.248}$$

Let $h_i(t) = \alpha_i(t) + \beta_i(t) + 2\gamma_i(t)$. Taking that the derivative $\dot{h}_i(t) - \dot{\bar{\chi}}_i(t)$ yields the evaluation of $\beta_i(t) + \alpha_i(t) + 2\gamma_i(t) - \bar{\chi}_i(t)$.

The gap $L^*_{\mathrm{mfg}} - L^*_{\mathrm{mftg}}$ can then be rewritten as $(\beta_i(t) + 2\gamma_i(t))c + \omega(t)$, where $\omega(t)$ is independent of $c = \left(\sigma^2 + \int_{\theta\in\Theta}\mu^2(\theta)\nu(d\theta)\right)$. ■

7.5.2 *Price of Simplicity: Explicit Gap*

Proposition 7.5.2.1 provides a refined result of the price of simplicity, i.e., the gap between both equilibrium costs given by $L^*_{\mathrm{mfg}} - L^*_{\mathrm{mftg}}$ by choosing specific parameters. See also Chapter 15 of Volume II of this book.

Proposition 7.5.2.1 *Let* $I = 1$, *for* $t = T$, $q_i(T) = q$, $\bar{q}_i(T) = \bar{q}$, *and for* $t < T$, $q_i(t) = 0$, $\bar{q}_i(t) = 0$, $b_1 = 0$, $b_2 = 1$, $\bar{\epsilon}_{ij} = 0$. *The mean-field game equilibrium cost is*

$$\begin{cases} L^*_{\text{mfg}} = \left(\bar{q}\exp\left(-2\int_0^T \alpha(s)\, ds\right) + \alpha(0)\right) x^2(0), \\ \alpha(t) = c\left(1 - \dfrac{1}{1 + \dfrac{q}{c-q}\exp(c(T-t))}\right), \\ m(t) = m(0)\exp\left(-\int_0^t \alpha(s)\, ds\right), \end{cases} \tag{7.249}$$

The mean-field-type equilibrium cost is

$$\begin{cases} L^*_{\text{mftg}} = \bar{\chi}(0)x_0^2, \\ \dot{\bar{\chi}}(t) + c\alpha(t) - \bar{\chi}^2(t) = 0, \\ \bar{\chi}(T) = q + \bar{q}, \\ E(x_i(t)) = [Ex(0)]\exp\left(-\int_0^T \bar{\chi}(s)\, ds\right). \end{cases} \tag{7.250}$$

The gap between the mean-field equilibrium cost and the mean-field-type equilibrium cost is positive and given by

$$L^*_{\text{mfg}} - L^*_{\text{mftg}} = \bar{q}x(0)^2\exp\left(-\int_0^T \alpha(t)dt\right)\left(-\exp\left(-\int_0^T \bar{\chi}(t)dt\right) + \exp\left(-\int_0^T \alpha(t)dt\right)\right). \tag{7.251}$$

The inefficiency gap $L^*_{\text{mfg}} - L^*_{\text{mftg}}$ grows without bound with $\bar{q}$ and x_0^2.

Let us consider the following decoupled system:

$$dx_i(t) = u_i(t)dt + x_i(t)\left(\sigma dB_i + \int_{\theta\in\Theta} \mu(\theta)\tilde{N}_i(dt, d\theta)\right), \tag{7.252}$$

with $i \in \{1, 2, \ldots, I\}$, and where B_i is standard Brownian motion representing individual uncertainty or noise and $\tilde{N}_i(dt, d\theta) = N_i(dt, d\theta) - \nu(d\theta)dt$ is a compensated process where N_i is a jump process with Lévy measure $\nu(d\theta)$ defined over Θ. It is assumed that ν is a Radon measure over the jump space Θ. In the mean-field game approach, the expected-value term $[E(x_i(T))]^2$, for simplicity, is approximated to a mean population state denoted by $[m(T)]^2$, which is a frozen quantity. The objective of decision-maker i is to minimize the arithmetic average global cost

$$L = \lim \sup_{I\to+\infty} \frac{1}{I}\sum_{i=1}^{I} L_i(x_i(0), u_i),$$

where L_i is the individual long-term cost given by

$$L_i(x_i(0), u_i) = qx_i^2(T) + \bar{q}\,[m_{I-1}(T)]^2 + \int_0^T u_i^2(t)dt, \tag{7.253}$$

with $i \in \{1, 2, \dots\}$ and $m_{I-1}(T) = \frac{1}{I-1}\sum_{j\neq i} x_j(t)$. Therefore, the mean-field individual costs in the limiting regime become

$$\begin{aligned} L^*_{\mathrm{mfg},i} &= \min\nolimits_{u_i\in\mathcal{U}_i} L_{\mathrm{mfg},i}(x_i(0), u_i),\\ &= \bar{q}[m(T)]^2 + \min\nolimits_{u_i\in\mathcal{U}_i} E\left[qx_i^2(T) + \textstyle\int_0^T u_i^2(t)dt\right], \end{aligned} \tag{7.254}$$

which is LQ problem under the jump-diffusion process. Consider the best-response problem of the mean-field game:

$$\begin{cases} L^*_{\mathrm{mfg},i} = \bar{q}[m(T)]^2 + \inf_{u_i\in\mathcal{U}_i} E\left[qx_i^2(T) + \displaystyle\int_0^T u_i^2(t)dt\right],\\ dx_i(t) = u_i(t)dt + x_i(t)\left(\sigma dB_i + \displaystyle\int_{\theta\in\Theta} \mu(\theta)\tilde{N}_i(dt, d\theta)\right)\\ x_i(0) = x_0. \end{cases} \tag{7.255}$$

Then, the mean-field equilibrium strategy is explicitly given by

$$\begin{cases} u^*_{\mathrm{mfg},i}(t) = -\alpha(t)x_i(t), \quad i \in \{1, 2, \dots, I\}\\ \alpha(t) = c\left[1 - \dfrac{1}{1 + \frac{q}{c-q}e^{c(T-t)}}\right],\\ m(t) = m(0)e^{-\int_0^t \alpha(s)\, ds},\\ c = \sigma^2 + \displaystyle\int_{\theta\in\Theta} \mu^2(\theta)\nu(d\theta). \end{cases} \tag{7.256}$$

The mean-field equilibrium global cost is explicitly given by

$$\begin{cases} L^*_{\mathrm{mfg}} = \bar{q}[m(T)]^2 + \alpha(0)x_0^2 = \left[\bar{q}e^{-2\int_0^t \alpha(s)\, ds} + \alpha(0)\right]x_0^2,\\ \alpha(0) = c\left[1 - \dfrac{1}{1 + \frac{q}{c-q}e^{cT}}\right]. \end{cases} \tag{7.257}$$

We provide a simple proof for these statements by using a direct method. It is based on the Itô's formula for jump-diffusion processes applied to the functional $\alpha(t)x^2(t)$. First, note that

$$\begin{cases} \alpha(T)x^2(T) - \alpha(0)x^2(0) \\ = \int_0^T \dot{\alpha}(t)x^2(t) + 2\alpha(t)x(t)u(t)dt \\ + \int_0^T \alpha x^2(t)[\sigma^2 + \int_\Theta \mu^2(\theta)\nu(d\theta)]dt \\ + \int_0^T 2\alpha(t)\sigma x^2(t)dB(t) \\ + \int_0^T \int_\Theta \alpha(t)x^2(t)[(1+\mu(\theta))^2 - 1]\tilde{N}(dt, d\theta), \end{cases} \tag{7.258}$$

It then follows that

$$\begin{cases} qx_i^2(T) + \int_0^T u_i^2(t)dt - \alpha(0)x^2(0) \\ = [q - \alpha(T)]x_i^2(T) + \\ \int_0^T \{\dot{\alpha}(t) + \alpha[\sigma^2 + \int_\Theta \mu^2(\theta)\nu(d\theta)] - \alpha^2(t)\}x^2(t)dt \\ + \int_0^T [u_i(t) + \alpha(t)x(t)]^2 dt \\ + \int_0^T 2\alpha(t)\sigma x^2(t)dB(t) \\ + \int_0^T \int_\Theta \alpha(t)x^2(t)[(1+\mu(\theta))^2 - 1]\tilde{N}(dt, d\theta) \end{cases} \tag{7.259}$$

By identification, we obtain

$$\begin{cases} \alpha(T) = q, \\ \dot{\alpha}(t) + \alpha(t)[\sigma^2 + \int_\Theta \mu^2(\theta)\nu(d\theta)] - \alpha^2(t) = 0, \\ u_i(t) = -\alpha(t)x_i(t), \\ \dot{m}(t) = -\alpha m(t), m(0) = x_0, \\ E[x_i^2(T) + \int_0^T u_i^2(t)] = \alpha(0)x^2(0). \end{cases} \tag{7.260}$$

By direct computation,

$$\alpha(0) = c\left[1 - \frac{1}{1 + \frac{q}{c-q}e^{cT}}\right],$$

with $c = [\sigma^2 + \int_\Theta \mu^2(\theta)\nu(d\theta)]$. This completes the proof.

The state trajectory at the mean-field game equilibrium satisfies

$$x_i(t) = \left(x_0 e^{\int_0^t -\alpha(t)dt}\right) e^{\sigma B_i(t) + \int_0^t \int_{\theta\in\Theta} \mu(\theta)\tilde{N}_i(dt, d\theta)}.$$

We now employ the mean-field-type method, i.e., the mean-field term Ex_i is not frozen. A change in the control action u_i affects the state x_i, and therefore it affects its expected value. Consider the best-response problem of the mean-field-type game. There is no interdependence between decision-makers in this trivial case:

$$\begin{cases} i \in \{1, 2, \ldots, I\}, \\ L^*_{\text{mftg}} = \inf_{u_i \in \mathcal{U}_i} E\left[\bar{q}[Ex_i(T)]^2 + qx_i^2(T) + \int_0^T u_i^2(t)dt\right] \\ dx_i(t) = u_i(t)dt + x_i(t)\left(\sigma dB_i + \int_{\theta\in\Theta} \mu(\theta)\tilde{N}_i(dt, d\theta)\right). \end{cases} \tag{7.261}$$

The mean-field-type optimal strategy is explicitly given by

$$\begin{cases} u^*_{\text{mftg},i}(t) = -\alpha(t)\left[x_i(t) - Ex_i(t)\right] - \beta(t)Ex_i(t), \\ \quad i \in \{1, 2, \ldots, I\}, \\ \dot{\beta}(t) + c\alpha(t) - \beta^2(t) = 0, \quad \beta(T) = q + \bar{q}, \\ Ex_i(t) = x_0 e^{-\int_0^t \beta(s)\, ds}. \end{cases} \tag{7.262}$$

The mean-field-type global optimal cost is explicitly given by

$$L^*_{\text{mftg}} = \beta(0)x_0^2. \tag{7.263}$$

To prove these statements, we use a direct method for the jump-diffusion process. We write the cost functional as

$$\begin{cases} L_{\text{mftg},i} = qE(x_i(T) - Ex_i(T))^2 + (q + \bar{q})[Ex_i(T)]^2 \\ +E\displaystyle\int_0^T (u_i - Eu_i(t))^2 + (Eu_i(t))^2\, dt \\ = q\text{var}(x_i(T)) + (q + \bar{q})[Ex_i(T)]^2 + \displaystyle\int_0^T \text{var}(u_i(t)) + (Eu_i(t))^2\, dt \end{cases} \tag{7.264}$$

The minimization problem of i involves a variance reduction problem. Following the completion of squares technique and using the stochastic process $\alpha(t)(x(t) - Ex_i(t))^2 + \beta(t)[Ex_i(t)]^2$, we arrive at the desired result.

The state trajectory at the mean-field-type game solution is then

$$\begin{aligned} dx_i(t) &= (\alpha(t) - \beta(t))x_0 e^{-\int_0^t \beta(s)\, ds} \\ &\quad + x_i(t)\left\{-\alpha(t)dt + \left(\sigma dB_i + \int_{\theta\in\Theta} \mu(\theta)\tilde{N}_i(dt, d\theta)\right)\right\}. \end{aligned} \tag{7.265}$$

7.5.3 Price of Simplicity Is Unbounded

This subsection examines the price of simplicity of the mean-field game approach. The gap between the mean-field equilibrium cost and global optimum cost is

$$L^*_{\text{mfg}} - L^*_{\text{mftg}} = \bar{q}x_0^2 e^{-\int_0^T \alpha(t)dt}\left[-e^{-\int_0^T \beta(t)dt} + e^{-\int_0^T \alpha(t)dt}\right], \tag{7.266}$$

which is positive. Moreover, the inefficiency gap $L^*_{\text{mfg}} - L^*_{\text{mftg}}$ grows without bound with $\bar{q}$ and x_0^2.

The proof follows as a byproduct of the previous subsection calculations. Introduce $\epsilon(t) = \alpha(t) - \beta(t)$. Taking the difference between the ordinary differential equation satisfied by α and the one satisfied by β, one obtains that

$$\dot{\epsilon}(t) = \epsilon(t)(\alpha(t) + \beta(t)), \ \ \epsilon(T) = -\bar{q} < 0.$$

Integrating the previous equation over $[t, T]$ yields

$$\epsilon(t) = -\bar{q}e^{-\int_t^T (\alpha(s)+\beta(s))ds}.$$

Thus,

$$L^*_{\text{mfg}} - L^*_{\text{mftg}} = \bar{q}x_0^2 e^{-\int_0^T [\alpha(t)+\beta(t)]dt}\left[e^{\bar{q}\int_0^T e^{-\int_t^T (\alpha(s)+\beta(s))ds}dt} - 1\right] \geq 0. \tag{7.267}$$

This shows that the approach that consists of freezing the mean-field term by assuming that a single decision-maker has a negligible effect on the mean-field term may lead to inefficient solution when compared with global optimization. In order words, the mean-field game approach to this problem is extremely simple, but it leads to a high level of price of simplicity, which is unbounded. As the mean-field term here is the expected value of its own state, a single decision-maker does affect the mean-field term by just changing its control action.

7.5.3.1 An Illustrative Example

We set $\mu = 0$, $\sigma = \sqrt{2}$, $x_0 = 1$, $q = 1$, $T = 40$. We then have $c = 2$, and the function α is

$$\alpha(t) = 2[1 - \frac{1}{1 + e^{2(40-t)}}].$$

The mean-field equilibrium cost (obtained by freezing the mean-field term) is

$$L^*_{\text{mfg}} = \bar{q}e^{-2\int_0^{40} \alpha(s)\,ds} + 2[1 - \frac{1}{1 + e^{80}}].$$

Figure 7.3 plots the variation of the total mean-field equilibrium cost L^*_{mfg} as a function of the coefficient $\bar{q}$. We observe that the mean-field equilibrium cost increases very rapidly with $\bar{q}$.

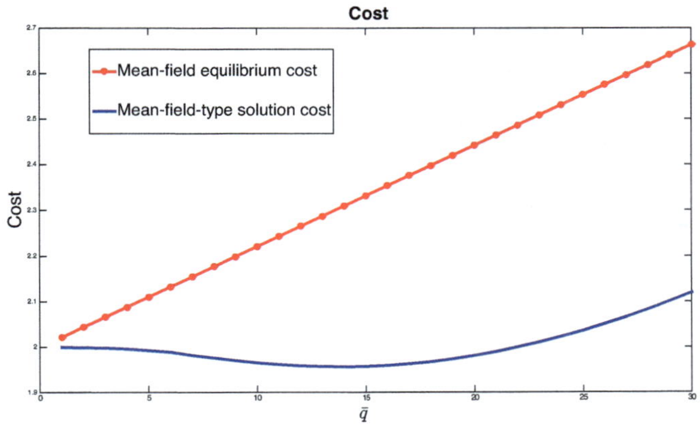

Fig. 7.3 Mean-field equilibrium cost versus mean-field-type solution cost

The mean-field-type solution (which does not freeze the mean-field term) cost is

$$L^*_{\mathrm{mftg}} = \beta(0),$$

with

$$\dot{\beta}(t) - \beta^2(t) = -4[1 - \frac{1}{1 + e^{2(T-t)}}]. \ \beta(1) = 1 + \bar{q},$$

Figure 7.3 plots also the variation of the total mean-field-type solution cost L^*_{mftg} as a function of the coefficient $\bar{q}$. We observe that the cost remains surprisingly small when the length of the horizon T is large enough.

We then plot the gap between the two solution approaches. Figure 7.4 represents the gap $L^*_{\mathrm{mfg}} - L^*_{\mathrm{mftg}}$. As we can see, when $\bar{q} = 0$, the problem becomes mean-field-free, and both approaches coincide with the standard LQ (with jump-diffusion) and hence provides identical costs. However, when the problem is mean-field dependent, i.e., $\bar{q} > 0$, the two approaches lead to different performances. The mean-field game approach becomes inefficient when $\bar{q}$ increases. The gap grows without bound, i.e., the price of simplicity is unbounded. Figure 7.5 represents the ratio $L^*_{\mathrm{mfg}}/L^*_{\mathrm{mftg}}$, which is known as the price of anarchy. The price of anarchy of the mean-field game approach is higher than one, showing the inefficiency of mean-field game solution.

Table 7.1 summarizes the key results from this example.

We evaluated the *price of simplicity* in mean-field game approaches, finding that freezing the mean-field term may result in poor performance compared to the original game, with an unbounded gap between the two solutions, highlighting the importance of a thoughtful decomposition of the original game problem. Similarly, we have illustrated through a simple example that simplifying optimization problems by means of freezing mean-field terms may lead to a sub-optimal solution. We have quantified the price of simplicity of the mean-field game approach and

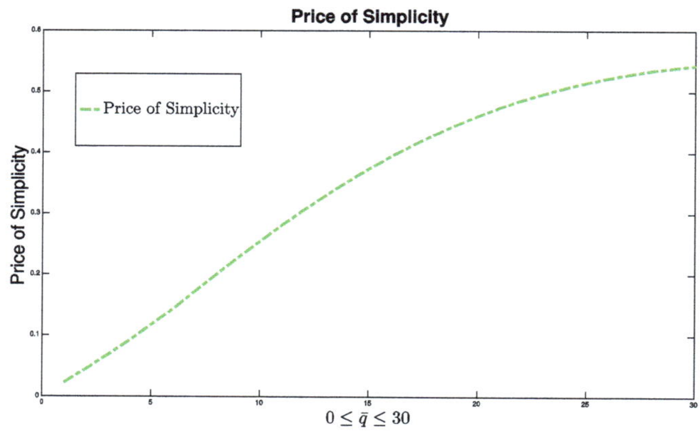

Fig. 7.4 The price of simplicity of the mean-field game approach is unbounded

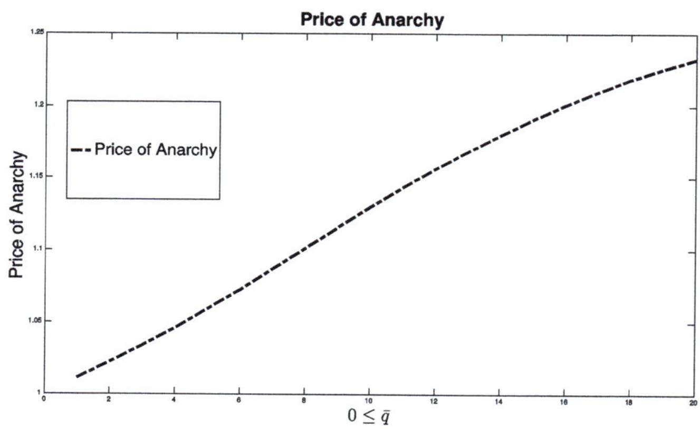

Fig. 7.5 The price of anarchy of the mean-field game approach is higher than one

Table 7.1 MFG solution versus MFTG solution in the single population case. MFG stands for the mean-field game approach. MFTG stands for the mean-field-type game approach

MFG	MFTG
The mean-field term is frozen	The mean-field term is not frozen
Leads to a poor performance	Provides a better performance
Price of simplicity is unbounded	No performance loss
Higher expected value state (cost)	Lower expected value state
Less stable	More stable

have shown that it can be inefficient. On the other hand, we have shown that the solution obtained by the mean-field-type game approach (which does not freeze the mean-field term) is globally optimal in this specific example [33, 37, 39, 41–44, 102, 157, 296, 332, 334, 350].

7.6 Linear-Quadratic Differential Mean-Field-Type Games

This section extends the problem examined in Sect. 7.1 to multiple decision-makers. We consider a linear-quadratic mean-field-type game problem under a jump-diffusion-regime switching state dynamics that include mean-field terms that are:

- the conditional expectation of the state with respect to the filtration of the regime switching, in the drift, diffusion, and jump coefficient functions,
- the conditional expectation of the control actions, in the drift, diffusion, and jump coefficient functions.

The state process is in one dimension. The instantaneous cost function includes:

- the conditional variance of the state,
- the square of the conditional expected value of the state,
- the conditional variance of the control-action process,
- the square of the conditional expected value of the control-action process,
- covariance between the state and the control,
- a product term between the conditional expected value of the control action and the conditional expected value of the state.

We follow a direct method to solve the mean-field-type game in a semi-explicit way. The method starts by identifying a partial guess functional where the coefficients are regime switching dependent. Then, it uses Itô's formula for jump-diffusion-regime switching processes, followed by a completion of squares for both control and mean control. The completion of squares technique used in Sect. 7.1 cannot be directly used due to extra dependence from other decision-makers. A careful matrix inversion is used to derive the interdependence between coefficients. Finally, the processes are identified using an orthogonal decomposition technique, and a system of ordinary differential equations is derived in a semi-explicit way.

We introduce the following notations. Let $[0, T],\ T > 0$ be a fixed time horizon and $(\Omega, \mathcal{F}, \mathbb{F}^{B_o,N_o,B,N,s}, \mathbb{P})$ be a given filtered probability space.The filtration $\mathbb{F} = \{\mathcal{F}_t^{B,N,s},\ 0 \le t \le T\}$ is the natural filtration of the union $\{B, N, s\}$ augmented by $\mathbb{P}$-null sets of $\mathcal{F}$. In practice, B is used to capture smaller disturbance and N is used for larger jumps in the system.

Let $k \ge 1$. $L^k([0, T] \times \mathcal{S}; \mathbb{R})$ be the set of measurable functions $f : [0, T] \times \mathcal{S} \to \mathbb{R}$ such that $\int_0^T |f(t, s)|^k dt < \infty$, $\mathcal{L}^k_{\mathbb{F}}([0, T] \times \mathcal{S}; \mathbb{R})$ be the set of $\mathbb{F}$-adapted $\mathbb{R}$-valued processes $X(\cdot)$ such that $E\left[\int_0^T |X(t)|^k dt\right] < \infty$. The stochastic quantity $\bar{X}(t) = E[X(t)|\ \mathcal{F}^s_{t-}]$ denotes the conditional expectation of the random variable $X(t)$ with respect to the filtration $\mathcal{F}^s_t$. Note that $\bar{X}$ is a random process.

An admissible control strategy u of the decision-maker is an $\mathbb{F}$-adapted and square-integrable process with values in a non-empty subset $U_i = \mathbb{R}$. We denote the set of all admissible controls by $\mathcal{U}_i$:

$$\mathcal{U}_i = \{u_i(\cdot) \in \mathcal{L}^2_{\mathbb{F}}([0, T] \times \mathcal{S}; \mathbb{R});\ \ u_i(.) \in U_i\ a.e.t \in [0, T],\ \ \mathbb{P} - a.s.\}$$

7.6.1 Jump-Diffusion-Regime Switching Process

For the sake of simplicity, we consider the one-dimensional case. Consider the following state dynamics with jump-diffusion and regime switching:

$$\begin{aligned} & x(0) = x_0, \quad s(0) = s_0, \\ & dx = bdt + \sigma dB + \int_\Theta \mu(\cdot, \theta)\tilde{N}(dt, d\theta), \end{aligned} \tag{7.268}$$

where

$$\begin{aligned} & b = b_0 + b_1 x + \bar{b}_1 \bar{x} + \sum_{j\in\mathcal{I}} b_{j2} u_j + \sum_{j\in\mathcal{I}} \bar{b}_{j2}\bar{u}_j, \\ & \sigma = \sigma_0 + \sigma_1 x + \bar{\sigma}_1 \bar{x} + \sum_{j\in\mathcal{I}} \sigma_{j2} u_j + \sum_{j\in\mathcal{I}} \bar{\sigma}_{j2}\bar{u}_j, \\ & \mu(\cdot, \theta) = \mu_0 + \mu_1 x + \bar{\mu}_1 \bar{x} + \sum_{j\in\mathcal{I}} \mu_{j2} u_j + \sum_{j\in\mathcal{I}} \bar{\mu}_{j2}\bar{u}_j, \end{aligned} \tag{7.269}$$

with $b_k(t,s); \sigma_k(t,s); \bar{b}_k(t,s); \bar{\sigma}_k(t,s) \in \mathbb{R}$ and $s \in \mathcal{S}$, where $\mathcal{S}$ is the set of switching regimes. Moreover, $\mu_k(t,s,\theta) \in \mathbb{R}$, $\theta \in \Theta$ where Θ is the set of jump sizes, for all $k \in \{0,1,2\}$. The processes B is a standard one-dimensional Brownian motions, $\tilde{N}(dt, d\theta) = N(dt, d\theta) - \nu(d\theta)dt$, is a compensated jump process with Radon measure ν over Θ. The switching process s switches between elements in $\mathcal{S}$ with transition rate $Q = (\tilde{q}_{ss'}, \ (s,s') \in \mathcal{S}^2)$ such that

$$s' \neq s, \ \tilde{q}_{ss'} > 0, \ \ \tilde{q}_{ss} = -\sum_{s'\neq s} \tilde{q}_{ss'}.$$

It is assumed that the processes $x_0, s(t), B(t), N(t,.)$ are mutually independent.

The term $\bar{x} = E[x|\mathcal{F}^s]$ denotes the expected value of x given $\mathcal{F}^s$, and $\bar{u} = E[u|\mathcal{F}^s]$. These are called conditional state mean-field and conditional control-action mean-field terms. The state dynamics is therefore of conditional McKean–Vlasov type as the conditional probability measure of the state appears in the drift, diffusion, and jump coefficients via $(\bar{x}, \ \bar{u})$. We denote a solution (if any) to (7.268) by $x := x^{u,x_0,s_0}$.

7.6.2 Variance and State-Action Covariance Cost

We associate a cost functional of mean-field type for decision-maker i as

$$\begin{aligned} & L_i(x,s,u) = \tfrac{1}{2} q_{iT} x^2(T) + \tfrac{1}{2}\bar{q}_{iT}\bar{x}^2(T) + \epsilon_{i3}(T)\bar{x}(T) \\ & + \tfrac{1}{2}\int_0^T \left(q_i x^2 + \bar{q}_i \bar{x}^2 + r_i u_i^2 + \bar{r}_i \bar{u}_i^2 \right) dt + \int_0^T \left(\epsilon_{i1} x u_i + \bar{\epsilon}_{i1}\bar{x}\bar{u}_i + \epsilon_{i2}\bar{u}_i + \epsilon_{i3}\bar{x} \right) dt \end{aligned} \tag{7.270}$$

where $q_{iT} = q_i(T,)q_i = q_i(t,s)$, $\bar{q}_{iT} = q_i(T,s)$, $\bar{q}_i = q_i(t,s)$, $r_i = r_i(t,s(t)) > 0$, $\bar{r}_i = \bar{r}_i(t,s)$, $r_i + \bar{r}_i > 0$, $\epsilon_{ik} = \epsilon_{ik}(t,s(t)) \in \mathbb{R}$.

Mean-Field-Type Game Problem with Jump-Diffusion-Regime Switching

For every decision-maker i:

Given $(u_j,\ j \in \mathscr{I}\backslash\{i\})$, the best-response problem of the decision-maker i is

$$\begin{cases} \inf_{u_i \in \mathscr{U}_i} E\left[L_i(x,s,u) | x(0) = x_0, s(0) = s_0\right] \\ \text{such that (7.268).} \end{cases} \tag{7.271}$$

We consider the following assumption:

$$\begin{cases} q_{iT}, q_i > 0,\ q_{iT} + \bar{q}_{iT} > 0, q_i + \bar{q}_i > 0,\ r_i > 0, r_i + \bar{r}_i > 0, \\ E[x_0^2] < +\infty,\ |\mathscr{S}| < +\infty, \\ \int_0^T \int_{\Theta} \phi_{ik}^2 \nu(d\theta) < +\infty, \text{ with } \phi_{ik} \in \{\mu_{ik}, \bar{\mu}_{ik}\}, \\ b_k, \sigma_k^2, \bar{b}_k, \bar{\sigma}_k^2 \in L^1([0,T] \times \mathscr{S}, \mathbb{R}) \end{cases} \tag{7.272}$$

The integrability condition above yields

$$\begin{array}{l} j \in \mathscr{I}, \\ \int_0^T [|b_0(t,s)| + |b_1(t,s)| + |\bar{b}_1(t,s)| + |b_{j2}(t,s)| + |\bar{b}_{j2}(t,s)|]dt \\ + \int_0^T [\sigma_0^2(t,s) + \sigma_1^2(t,s) + \bar{\sigma}_1^2(t,s) + \sigma_{j2}^2(t,s) + \bar{\sigma}_{j2}^2(t,s)]dt \\ + \int_0^T \int_{\Theta} [\mu_0^2(t,s,\theta) + \mu_1^2(t,s,\theta) + \bar{\mu}_1^2(t,s,\theta)]\nu(d\theta) \\ + \int_0^T \int_{\Theta} [\mu_{j2}^2(t,s,\theta) + \bar{\mu}_{j2}^2(t,s,\theta)]\nu(d\theta)dt < +\infty. \end{array} \tag{7.273}$$

Remark 7.2 Under Assumption (7.272), the state dynamics has solution for each $u \in \mathscr{U}$. Note that we do not impose boundedness conditions on the coefficients or Lipschitz conditions on the cost. Under Assumption (7.272), Problem (7.271) is well defined.

We now state the main result of this section.

Solution to Mean-Field-Type Game Problem (7.271)

Let Assumption (7.272) hold. Then, the equilibrium strategy and the equilibrium cost are given by:

$$\begin{array}{l} u_i^* = \bar{u}_i^* - \frac{\tilde{\tau}_i}{c_i}(x - \bar{x}) \\ \bar{u}_i^* = -\frac{\bar{\tau}_i}{\bar{c}_i} = -\frac{\bar{\lambda}_{i2}\bar{x}}{\bar{c}_i} - \frac{\bar{\lambda}_{i3}}{\bar{c}_i} \\ L_i^*(x,s,u^*) = \frac{1}{2}E[\alpha_i(0,s(0))(x_0 - \bar{x}_0)^2] \\ + \frac{1}{2}E[\beta_i(0,s(0))\bar{x}_0^2] + E[\gamma_i(0,s(0))\bar{x}_0] + E[\delta_i(0,s(0))], \end{array} \tag{7.274}$$

Coefficients of the Solution to Problem (7.271)

$$\begin{aligned}
&\dot{\alpha}_i = -2\alpha_i b_1 - \alpha_i \left(\sigma_1^2 + \int_\Theta \mu_1^2 \nu(d\theta)\right) - q_i - \sum_{s'\in\mathcal{S}}[\alpha_i(.,s') - \alpha_i(.,s)]\tilde{q}_{ss'}\\
&-\alpha_i\left(\sum_{j\neq i}\sigma_{j2}\frac{\tilde{\tau}_j}{c_j}\right)^2 - \int_\Theta \alpha_i\left(\sum_{j\neq i}^n \mu_{j2}\frac{\tilde{\tau}_j}{c_j}\right)^2 \nu(d\theta)\\
&+2\alpha_i\sum_{j\neq i}\left(\sigma_1\sigma_{j2} + \int_\Theta \mu_1\mu_{j2}\nu(d\theta)\right)\frac{\tilde{\tau}_j}{c_j} + 2\alpha_i\sum_{j\neq i} b_{j2}\frac{\tilde{\tau}_j}{c_j} + \frac{\tilde{\tau}_i^2}{c_i},\\
&\alpha_i(T,s) = q_i(T,s),\\
&\dot{\beta}_i = -2\beta_i(b_1+\bar{b}_1) - \alpha_i\left((\sigma_1+\bar{\sigma}_1)^2 + \int_\Theta(\mu_1+\bar{\mu}_1)^2\nu(d\theta)\right)\\
&-(q_i+\bar{q}_i) - \sum_{s'\in\mathcal{S}}[\beta_i(.,s') - \beta_i(.,s)]\tilde{q}_{ss'}\\
&-\alpha_i\left(\sum_{j\neq i}(\sigma_{j2}+\bar{\sigma}_{j2})\frac{\bar{\lambda}_{j2}}{\bar{c}_j}\right)^2 - \alpha_i\int_\Theta\left(\sum_{j\neq i}(\mu_{j2}+\bar{\mu}_{j2})\frac{\bar{\lambda}_{j2}}{\bar{c}_j}\right)^2\nu(d\theta) + \frac{\bar{\lambda}_{i2}^2}{\bar{c}_i}\\
&+2\sum_{j\neq i}\alpha_i\left((\sigma_1+\bar{\sigma}_1)(\sigma_{j2}+\bar{\sigma}_{j2}) + \int_\Theta(\mu_1+\bar{\mu}_1)(\mu_{j2}+\bar{\mu}_{j2})\nu(d\theta)\right)\frac{\bar{\lambda}_{j2}}{\bar{c}_j}\\
&+2\sum_{j\neq i}\beta_i(b_{j2}+\bar{b}_{j2})\frac{\bar{\lambda}_{j2}}{\bar{c}_j},\\
&\beta_i(T,s) = q_i(T,s) + \bar{q}_i(T,s),\\
&\dot{\gamma}_i = -\beta_i b_0 - \gamma_i(b_1+\bar{b}_1) - \alpha_i\left(\sigma_0(\sigma_1+\bar{\sigma}_1) + \int_\Theta \mu_0(\mu_1+\bar{\mu}_1)\nu(d\theta)\right)\\
&-\epsilon_{i3} - \sum_{s'\in\mathcal{S}}[\gamma_i(.,s') - \gamma_i(.,s)]\tilde{q}_{ss'}\\
&-\alpha_i\left(\sum_{j\neq i}(\sigma_{j2}+\bar{\sigma}_{j2})\frac{\bar{\lambda}_{j2}}{\bar{c}_j}\right)\left(\sum_{j\neq i}(\sigma_{j2}+\bar{\sigma}_{j2})\frac{\bar{\lambda}_{j3}}{\bar{c}_j}\right)\\
&-\alpha_i\int_\Theta\left(\sum_{j\neq i}(\mu_{j2}+\bar{\mu}_{j2})\frac{\bar{\lambda}_{j2}}{\bar{c}_j}\right)\left(\sum_{j\neq i}(\mu_{j2}+\bar{\mu}_{j2})\frac{\bar{\lambda}_{j3}}{\bar{c}_j}\right)\nu(d\theta)\\
&+\sum_{j\neq i}\left[\alpha_i\left(\sigma_0(\sigma_{j2}+\bar{\sigma}_{j2}) + \int_\Theta \mu_0(\mu_{j2}+\bar{\mu}_{j2})\nu(d\theta)\right) + \gamma_i(b_{j2}+\bar{b}_{j2})\right]\frac{\bar{\lambda}_{j2}}{\bar{c}_j}\\
&+\sum_{j\neq i}\left[\alpha_i\left((\sigma_1+\bar{\sigma}_1)(\sigma_{j2}+\bar{\sigma}_{j2}) + \int_\Theta(\mu_1+\bar{\mu}_1)(\mu_{j2}+\bar{\mu}_{j2})\nu(d\theta)\right)\right]\frac{\bar{\lambda}_{j3}}{\bar{c}_j}\\
&+\sum_{j\neq i}\beta_i(b_{j2}+\bar{b}_{j2})\frac{\bar{\lambda}_{j3}}{\bar{c}_j} + \frac{\bar{\lambda}_{i2}\bar{\lambda}_{i3}}{\bar{c}_i},\\
&\gamma_i(T,s) = \epsilon_{i3}(T,s),\\
&\dot{\delta}_i = -\gamma_i b_0 - \frac{\alpha_i}{2}\left(\sigma_0^2 + \int_\Theta \mu_0^2\nu(d\theta)\right) - \sum_{s'\in\mathcal{S}}[\delta_i(.,s') - \delta_i(.,s)]\tilde{q}_{ss'}\\
&-\alpha_i\frac{1}{2}\left(\sum_{j\neq i}(\sigma_{j2}+\bar{\sigma}_{j2})\frac{\bar{\lambda}_{j3}}{\bar{c}_j}\right)^2 - \alpha_i\int_\Theta\frac{1}{2}\left(\sum_{j\neq i}(\mu_{j2}+\bar{\mu}_{j2})\frac{\bar{\lambda}_{j3}}{\bar{c}_j}\right)^2\nu(d\theta)\\
&+\sum_{j\neq i}\left[\alpha_i\left(\sigma_0(\sigma_{j2}+\bar{\sigma}_{j2}) + \int_\Theta \mu_0(\mu_{j2}+\bar{\mu}_{j2})\nu(d\theta)\right)\right]\frac{\bar{\lambda}_{j3}}{\bar{c}_j}\\
&+\sum_{j\neq i}\gamma_i(b_{j2}+\bar{b}_{j2})\frac{\bar{\lambda}_{j3}}{\bar{c}_j} + \frac{\bar{\lambda}_{i3}^2}{2\bar{c}_i},\\
&\delta_i(T,s) = 0,
\end{aligned} \tag{7.275}$$

whenever these ordinary differential equations have a solution that does not blow up in $[0, T]$.

Relationship between the coefficients

$$
\begin{aligned}
&A\tilde{\tau} = v,\\
&\bar{A}\bar{\tau} = \bar{v},\\
&A_{ii} = \bar{A}_{ii} = 1,\\
&A_{ij} = \frac{\alpha_i}{c_j}\{\sigma_{i2}\sigma_{j2} + \int_\Theta \mu_{i2}\,\mu_{j2}\nu(d\theta)\},\ \forall j \neq i\\
&v_i = \left[\epsilon_{i1} + \alpha_i\left\{b_{i2} + \sigma_1\sigma_{i2} + \int_\Theta \mu_1\mu_{i2}\nu(d\theta)\right\}\right],\\
&\bar{A}_{ij} = \frac{\alpha_i}{\bar{c}_j}\left[(\sigma_{i2}+\bar{\sigma}_{i2})(\sigma_{j2}+\bar{\sigma}_{j2}) + \int_\Theta(\mu_{i2}+\bar{\mu}_{i2})(\mu_{j2}+\bar{\mu}_{j2})\nu(d\theta)\right],\ \forall j \neq i\\
&\bar{v}_i = \epsilon_{i2} + \alpha_i\left(\sigma_0(\sigma_{i2}+\bar{\sigma}_{i2}) + \int_\Theta \mu_0(\mu_{i2}+\bar{\mu}_{i2})\nu(d\theta)\right) + \gamma_i(b_{i2}+\bar{b}_{i2})\\
&+\left[\epsilon_{i1} + \bar{\epsilon}_{i1} + \alpha_i\left((\sigma_1+\bar{\sigma}_1)(\sigma_{i2}+\bar{\sigma}_{i2}) + \int_\Theta(\mu_1+\bar{\mu}_1)(\mu_{i2}+\bar{\mu}_{i2})\nu(d\theta)\right)\right]\bar{x}\\
&+\beta_i(b_{i2}+\bar{b}_{i2})\bar{x},\\
&\bar{A}\bar{\tau} = \bar{A}\left(\bar{\lambda}_2\bar{x} + \bar{\lambda}_3\right) = \bar{v},\\
&c_i = r_i + \alpha_i\sigma_{i2}^2 + \int_\Theta \alpha_i\mu_{i2}^2\nu(d\theta),\\
&\bar{c}_i = (r_i + \bar{r}_i) + \alpha_i\left[(\sigma_{i2}+\bar{\sigma}_{i2})^2 + \int_\Theta(\mu_{i2}+\bar{\mu}_{i2})^2\nu(d\theta)\right],
\end{aligned}
\tag{7.276}
$$

whenever the matrices A and $\bar{A}$ are invertible.

Notice the following:

- The matrices A and $\bar{A}$ become the identity matrix for problems where the control and the expected value of the control do not appear in the diffusion and jump coefficient functions.
- The structure of the differential equations changes from linear, to quadratic and fraction in α. As it is expected for game problems, the coefficients $\alpha_i(t, s)$ are interdependent not only across the switching regime but also coupled between the decision-makers $\{\alpha_j(t, s)\}_{j \neq i}$.

7.6.3 Construction of a Semi-explicit Equilibrium Strategy

In this subsection, we develop a proof of optimality using a direct method decomposed in several steps.

7.6.3.1 Reformulation of the Game Problem

First, consider the following problem.

Equivalent reformulation of Problem (7.271)

$$\inf_{u_i\in\mathscr{U}_i} E\left[\hat{L}_i(x,s,u)|x(0)=x_0, s(0)=s_0\right], \tag{7.277}$$

where

$$\begin{aligned}
\hat{L}_i(x,s,u) &= \tfrac{1}{2}q_{iT}(x_T-\bar{x}_T)^2+\tfrac{1}{2}(q_{iT}+\bar{q}_{iT})\bar{x}_T^2+\epsilon_{i3}(T)\bar{x}(T)\\
&+\tfrac{1}{2}\int_0^T\left(q_i(x-\bar{x})^2+(q_i+\bar{q}_i)\bar{x}^2+r_i(u_i-\bar{u}_i)^2+(r_i+\bar{r}_i)\bar{u}_i^2\right)dt\\
&+\int_0^T\left(\epsilon_{i1}(x-\bar{x})(u_i-\bar{u}_i)+(\epsilon_{i1}+\bar{\epsilon}_{i1})\bar{x}\bar{u}_i+\epsilon_{i2}\bar{u}_i+\epsilon_{i3}\bar{x}\right)dt,
\end{aligned} \tag{7.278}$$

subject to

$$\begin{aligned}
dx &= \left(b_0+b_1(x-\bar{x})+(b_1+\bar{b}_1)\bar{x}+\textstyle\sum_{j\in\mathscr{I}}b_{j2}(u_j-\bar{u}_j)+\sum_{j\in\mathscr{I}}(b_{j2}+\bar{b}_{j2})\bar{u}_j\right)dt\\
&+\left(\sigma_0+\sigma_1(x-\bar{x})+(\sigma_1+\bar{\sigma}_1)\bar{x}+\textstyle\sum_{j\in\mathscr{I}}\sigma_{j2}(u_j-\bar{u}_j)+\sum_{j\in\mathscr{I}}(\sigma_{j2}+\bar{\sigma}_{j2})\bar{u}_j\right)dB\\
&+\int_\Theta\left(\mu_0+\mu_1(x-\bar{x})+(\mu_1+\bar{\mu}_1)\bar{x}+\textstyle\sum_{j\in\mathscr{I}}\mu_{j2}(u_j-\bar{u}_j)\right)\tilde{N}(dt,d\theta)\\
&+\int_\Theta\textstyle\sum_{j\in\mathscr{I}}(\mu_{j2}+\bar{\mu}_{j2})\bar{u}_j\tilde{N}(dt,d\theta),\\
x(0)&=x_0,\ s(0)=s_0.
\end{aligned} \tag{7.279}$$

Proposition 7.6.3.2 *Problem (7.271) is exactly Problem (7.277).*

Proof Use the decompositions $x=x-\bar{x}+\bar{x}$ and $u_j=u_j-\bar{u}_j+\bar{u}_j$ noticing that both $E[\bar{x}(x-\bar{x})]=0$ and $E[\bar{u}_j(u_j-\bar{u}_j)]=0$, i.e., $x-\bar{x}\perp\bar{x}$ and $u_j-\bar{u}_j\perp\bar{u}_j$. ■

Remark 7.3 Notice that $E[\hat{L}_i(x,s,u)|x(0)=x_0, s(0)=s_0]$ can be rewritten as

$$\begin{aligned}
E[\hat{L}_i(x,s,u)|x(0)&=x_0, s(0)=s_0]=\tfrac{1}{2}q_{iT}\,\mathrm{var}(x_T)+\tfrac{1}{2}(q_{iT}+\bar{q}_{iT})\bar{x}_T^2+\epsilon_{i3}(T)\bar{x}(T)\\
&+\tfrac{1}{2}\int_0^T\left(q_i\,\mathrm{var}(x)+(q_i+\bar{q}_i)\bar{x}^2+r_i\,\mathrm{var}(u_i)+(r_i+\bar{r}_i)\bar{u}_i^2\right)dt\\
&+\int_0^T\left(\epsilon_{i1}\,\mathrm{cov}(x,u_i)+(\epsilon_{i1}+\bar{\epsilon}_{i1})\bar{x}\bar{u}_i+\epsilon_{i2}\bar{u}_i+\epsilon_{i3}\bar{x}\right)dt,
\end{aligned} \tag{7.280}$$

where $\mathrm{cov}(x,u_j)=E(x-\bar{x})(u_j-\bar{u}_j)$ is the conditional covariance and $\mathrm{var}[y]=E[y-E[y]]^2$ is the variance of y, which clearly indicates a weighted conditional variance reduction problem depending on switching regime s.

7.6.3.2 Semi-explicit Solution by Means of Direct Method

We construct a semi-explicit solution to Problems (7.271) and (7.277) via a direct method.

7.6.3.3 Itô's Formula

Let $f_i(t,x,s) = \frac{1}{2}\alpha_i(x-\bar{x})^2 + \frac{1}{2}\beta_i\bar{x}^2 + \gamma_i\bar{x} + \delta_i$, where $\alpha_i = \alpha_i(t,s)$, $\beta_i = \beta_i(t,s)$, $\gamma_i = \gamma_i(t,s)$, $\delta_i = \delta_i(t,s)$, and $x = x^{u,x_0,s_0}(t)$ solution to (7.279).

Taking the expectation of the Itô's formula applied to f_i yields

$$
\begin{aligned}
&d(E[f_i(t,x,s)])\\
&=E\Bigg[\tfrac{1}{2}\left\{\dot{\alpha}_i+2\alpha_i b_1+\alpha_i\left(\sigma_1^2+\textstyle\int_\Theta \mu_1^2\nu(d\theta)\right)+\sum_{s'\in\mathcal{S}}[\alpha_i(.,s')-\alpha_i(.,s)]\tilde{q}_{ss'}\right\}(x-\bar{x})^2\\
&+\tfrac{1}{2}\left\{\dot{\beta}_i+2\beta_i(b_1+\bar{b}_1)+\alpha_i\left((\sigma_1+\bar{\sigma}_1)^2+\textstyle\int_\Theta(\mu_1+\bar{\mu}_1)^2\nu(d\theta)\right)\right.\\
&\left.+\sum_{s'\in\mathcal{S}}[\beta_i(.,s')-\beta_i(.,s)]\tilde{q}_{ss'}\right\}\bar{x}^2\\
&+\left\{\dot{\gamma}_i+\beta_i b_0+\gamma_i(b_1+\bar{b}_1)+\alpha_i\left(\sigma_0(\sigma_1+\bar{\sigma}_1)+\textstyle\int_\Theta \mu_0(\mu_1+\bar{\mu}_1)\nu(d\theta)\right)\right.\\
&\left.+\sum_{s'\in\mathcal{S}}[\gamma_i(.,s')-\gamma_i(.,s)]\tilde{q}_{ss'}\right\}\bar{x}\\
&+\left(\dot{\delta}_i+\gamma_i b_0+\tfrac{\alpha_i}{2}\left(\sigma_0^2+\textstyle\int_\Theta \mu_0^2\nu(d\theta)\right)+\sum_{s'\in\mathcal{S}}[\delta_i(.,s')-\delta_i(.,s)]\tilde{q}_{ss'}\right)\\
&+\tfrac{\alpha_i}{2}\{\textstyle\sum_{j\in\mathcal{I}}\sigma_{j2}(u_j-\bar{u}_j)\}^2+\int_\Theta \frac{\alpha_i}{2}\{\sum_{j\in\mathcal{I}}\mu_{j2}(u_j-\bar{u}_j)\}^2\nu(d\theta)\\
&+\alpha_i\textstyle\sum_{j\in\mathcal{I}}\left\{\sigma_1\sigma_{j2}+\int_\Theta \mu_1\mu_{j2}\nu(d\theta)\right\}(x-\bar{x})(u_j-\bar{u}_j)\\
&+\alpha_i\textstyle\sum_{j\in\mathcal{I}} b_{j2}(x-\bar{x})(u_j-\bar{u}_j)\\
&+\tfrac{\alpha_i}{2}\left\{\textstyle\sum_{j\in\mathcal{I}}(\sigma_{j2}+\bar{\sigma}_{j2})\bar{u}_j\right\}^2+\int_\Theta \frac{\alpha_i}{2}\left\{\sum_{j\in\mathcal{I}}(\mu_{j2}+\bar{\mu}_{j2})\bar{u}_j\right\}^2\nu(d\theta)\\
&+\alpha_i\textstyle\sum_{j\in\mathcal{I}}\left\{\sigma_0(\sigma_{j2}+\bar{\sigma}_{j2})+\int_\Theta \mu_0(\mu_{j2}+\bar{\mu}_{j2})\nu(d\theta)\right\}\bar{u}_j\\
&+\gamma_i\textstyle\sum_{j\in\mathcal{I}}(b_{j2}+\bar{b}_{j2})\bar{u}_j\\
&+\alpha_i\textstyle\sum_{j\in\mathcal{I}}\left\{(\sigma_1+\bar{\sigma}_1)(\sigma_{j2}+\bar{\sigma}_{j2})+\int_\Theta(\mu_1+\bar{\mu}_1)(\mu_{j2}+\bar{\mu}_{j2})\nu(d\theta)\right\}\bar{x}\bar{u}_j\\
&+\beta_i\textstyle\sum_{j\in\mathcal{I}}(b_{j2}+\bar{b}_{j2})\bar{x}\bar{u}_j\Bigg]dt.
\end{aligned}
\tag{7.281}
$$

7.6.3.4 Guess Functional Identification

We compute $E[\hat{L}_i(x,s,u) - f_i(0,x_0,s_0)]$, i.e.,

$$
\begin{aligned}
&E[\hat{L}_i(x,s,u)|x(0)=x_0,s(0)=s_0]-E[f(0,x_0,s_0)|s]\\
&=\tfrac{1}{2}E[(q_{iT}-\alpha_i(T))(x_T-\bar{x}_T)^2|s]\\
&+\tfrac{1}{2}E[(q_{iT}+\bar{q}_{iT}-\beta_i(T))\bar{x}_T^2|s]+(\epsilon_{i3}(T)-\gamma_{iT})\bar{x}(T)\\
&+E\int_0^T\Bigg[\tfrac{1}{2}\left\{\dot{\alpha}_i+2\alpha_i b_1+\alpha_i\left(\sigma_1^2+\int_\Theta\mu_1^2\nu(d\theta)\right)+q_i+\sum_{s'\in\mathcal{S}}[\alpha_i(.,s')-\alpha_i(.,s)]\tilde{q}_{ss'}\right\}\\
&(x-\bar{x})^2\\
&+\tfrac{1}{2}\left\{\dot{\beta}_i+2\beta_i(b_1+\bar{b}_1)+\alpha_i\left((\sigma_1+\bar{\sigma}_1)^2+\int_\Theta(\mu_1+\bar{\mu}_1)^2\nu(d\theta)\right)+q_i+\bar{q}_i\right.\\
&\left.+\sum_{s'\in\mathcal{S}}[\beta_i(.,s')-\beta_i(.,s)]\tilde{q}_{ss'}\right\}\bar{x}^2\\
&+\left\{\dot{\gamma}_i+\beta_i b_0+\gamma_i(b_1+\bar{b}_1)+\alpha_i\left(\sigma_0(\sigma_1+\bar{\sigma}_1)+\int_\Theta\mu_0(\mu_1+\bar{\mu}_1)\nu(d\theta)\right)+\epsilon_{i3}\right.\\
&\left.+\sum_{s'\in\mathcal{S}}[\gamma_i(.,s')-\gamma_i(.,s)]\tilde{q}_{ss'}\right\}\bar{x}\\
&+\left(\dot{\delta}_i+\gamma_i b_0+\tfrac{\alpha_i}{2}\left(\sigma_0^2+\int_\Theta\mu_0^2\nu(d\theta)\right)+\sum_{s'\in\mathcal{S}}[\delta_i(.,s')-\delta_i(.,s)]\tilde{q}_{ss'}\right)\\
&+\tfrac{1}{2}r_i(u_i-\bar{u}_i)^2+\tfrac{\alpha_i}{2}\{\textstyle\sum_{j\in\mathcal{I}}\sigma_{j2}(u_j-\bar{u}_j\}^2+\int_\Theta\tfrac{\alpha_i}{2}\{\sum_{j\in\mathcal{I}}\mu_{j2}(u_j-\bar{u}_j)\}^2\nu(d\theta)\\
&+\epsilon_{i1}(x-\bar{x})(u_i-\bar{u}_i)+\alpha_i\textstyle\sum_{j\in\mathcal{I}}\left\{\sigma_1\sigma_{j2}+\int_\Theta\mu_1\mu_{j2}\nu(d\theta)\right\}(x-\bar{x})(u_j-\bar{u}_j)\\
&+\alpha_i\textstyle\sum_{j\in\mathcal{I}}b_{j2}(x-\bar{x})(u_j-\bar{u}_j)\\
&+\tfrac{1}{2}(r_i+\bar{r}_i)\bar{u}_i^2+\tfrac{\alpha_i}{2}\{\textstyle\sum_{j\in\mathcal{I}}(\sigma_{j2}+\bar{\sigma}_{j2})\bar{u}_j\}^2+\int_\Theta\tfrac{\alpha_i}{2}\{\sum_{j\in\mathcal{I}}(\mu_{j2}+\bar{\mu}_{j2})\bar{u}_j\}^2\nu(d\theta)\\
&+\epsilon_{i2}\bar{u}_i+\alpha_i\textstyle\sum_{j\in\mathcal{I}}\left\{\sigma_0(\sigma_{j2}+\bar{\sigma}_{j2})+\int_\Theta\mu_0(\mu_{j2}+\bar{\mu}_{j2})\nu(d\theta)\right\}\bar{u}_j\\
&+\gamma_i\textstyle\sum_{j\in\mathcal{I}}(b_{j2}+\bar{b}_{j2})\bar{u}_j+(\epsilon_{i1}+\bar{\epsilon}_{i1})\bar{x}\bar{u}_i\\
&+\alpha_i\textstyle\sum_{j\in\mathcal{I}}\left\{(\sigma_1+\bar{\sigma}_1)(\sigma_{j2}+\bar{\sigma}_{j2})+\int_\Theta(\mu_1+\bar{\mu}_1)(\mu_{j2}+\bar{\mu}_{j2})\nu(d\theta)\right\}\bar{x}\bar{u}_j\\
&+\beta_i\textstyle\sum_{j\in\mathcal{I}}(b_{j2}+\bar{b}_{j2})\bar{x}\bar{u}_j\Bigg]dt.
\end{aligned}
\tag{7.282}
$$

7.6.3.5 Completion of Squares for i-th Decision-Maker

We use the following equalities. We start with terms in $(u_i-\bar{u}_i)$.

$$
\begin{aligned}
&\tfrac{1}{2}r_i(u_i-\bar{u}_i)^2+\tfrac{\alpha_i}{2}\{\textstyle\sum_{j\in\mathcal{I}}\sigma_{j2}(u_j-\bar{u}_j)\}^2+\int_\Theta\tfrac{\alpha_i}{2}\{\sum_{j\in\mathcal{I}}\mu_{j2}(u_j-\bar{u}_j)\}^2\nu(d\theta)\\
&+\epsilon_{i1}(x-\bar{x})(u_i-\bar{u}_i)+\alpha_i\textstyle\sum_{j\in\mathcal{I}}\left\{\sigma_1\sigma_{j2}+\int_\Theta\mu_1\mu_{j2}\nu(d\theta)\right\}(x-\bar{x})(u_j-\bar{u}_j)\\
&+\alpha_i\textstyle\sum_{j\in\mathcal{I}}b_{j2}(x-\bar{x})(u_j-\bar{u}_j)\\
&=\tfrac{1}{2}c_i(u_i-\bar{u}_i+\tfrac{\tau_i}{c_i})^2-\tfrac{\tau_i^2}{2c_i}+\tfrac{\alpha_i}{2}\{\textstyle\sum_{j\neq i}\sigma_{j2}(u_j-\bar{u}_j)\}^2\\
&+\int_\Theta\tfrac{\alpha_i}{2}\{\textstyle\sum_{j\neq i}^n\mu_{j2}(u_j-\bar{u}_j)\}^2\nu(d\theta)\\
&+\alpha_i\textstyle\sum_{j\neq i}\left\{\sigma_1\sigma_{j2}+\int_\Theta\mu_1\mu_{j2}\nu(d\theta)\right\}(x-\bar{x})(u_j-\bar{u}_j)\\
&+\alpha_i\textstyle\sum_{j\neq i}b_{j2}(x-\bar{x})(u_j-\bar{u}_j)\\
&=\tfrac{1}{2}c_i\left(u_i-\bar{u}_i+\tfrac{\tilde{\tau}_i}{c_i}(x-\bar{x})\right)^2\\
&+(x-\bar{x})^2\Bigg(\tfrac{\alpha_i}{2}\left\{\textstyle\sum_{j\neq i}\sigma_{j2}\tfrac{\tilde{\tau}_j}{c_j}\right\}^2+\int_\Theta\tfrac{\alpha_i}{2}\left\{\sum_{j\neq i}^n\mu_{j2}\tfrac{\tilde{\tau}_j}{c_j}\right\}^2\nu(d\theta)\\
&-\alpha_i\textstyle\sum_{j\neq i}\left\{\sigma_1\sigma_{j2}+\int_\Theta\mu_1\mu_{j2}\nu(d\theta)\right\}\tfrac{\tilde{\tau}_j}{c_j}-\alpha_i\sum_{j\neq i}b_{j2}\tfrac{\tilde{\tau}_j}{c_j}-\tfrac{\tilde{\tau}_i^2}{2c_i}\Bigg)
\end{aligned}
\tag{7.283}
$$

where

$$
\begin{array}{l}
c_i := r_i + \alpha_i \sigma_{i2}^2 + \int_\Theta \alpha_i \mu_{i2}^2 \nu(d\theta), \\
\\
\tau_i := \alpha_i \sigma_{i2} \sum_{j\neq i} \sigma_{j2}(u_j - \bar{u}_j) + \int_\Theta \alpha_i \mu_{i2}\{\sum_{j\neq i}^n \mu_{j2}(u_j - \bar{u}_j)\}\nu(d\theta) \\
+\left[\epsilon_{i1} + \alpha_i \left\{b_{i2} + \sigma_1\sigma_{i2} + \int_\Theta \mu_1\mu_{i2}\nu(d\theta)\right\}\right](x - \bar{x}).
\end{array}
\tag{7.284}
$$

The vectorial function τ is obtained by solving the linear system of functions and is given by $\tau = [A^{-1}v](x - \bar{x})$, where

$$
\begin{array}{l}
A_{ii} = 1, \\
A_{ij} = \frac{\alpha_i}{c_j}\{\sigma_{i2}\sigma_{j2} + \int_\Theta \mu_{i2}\, \mu_{j2}\nu(d\theta)\}, \ \forall j \neq i \\
v_i = \left[\epsilon_{i1} + \alpha_i \left\{b_{i2} + \sigma_1\sigma_{i2} + \int_\Theta \mu_1\mu_{i2}\nu(d\theta)\right\}\right],
\end{array}
\tag{7.285}
$$

whenever the matrix A is invertible. Thus, τ_i can be written as $\tau_i = \tilde{\tau}_i(x - \bar{x})$ with $\tilde{\tau}_i := [A^{-1}v]_i$, whenever the matrix A is invertible.

Next we examine the completion of squares for the terms $\bar{u}_i$.

$$
\begin{array}{l}
\frac{1}{2}(r_i + \bar{r}_i)\bar{u}_i^2 + \alpha_i \left[\frac{1}{2}(\sigma_{i2} + \bar{\sigma}_{i2})^2\bar{u}_i^2 + \frac{1}{2}\left(\sum_{j\neq i}(\sigma_{j2} + \bar{\sigma}_{j2})\bar{u}_j \right)^2 \right. \\
\left. +(\sigma_{i2} + \bar{\sigma}_{i2})\bar{u}_i \sum_{j\neq i} (\sigma_{j2} + \bar{\sigma}_{j2})\bar{u}_j \right] \\
+\alpha_i \int_\Theta \left[\frac{1}{2}(\mu_{i2} + \bar{\mu}_{i2})^2\bar{u}_i^2 + \frac{1}{2}\left(\sum_{j\neq i}(\mu_{j2} + \bar{\mu}_{j2})\bar{u}_j \right)^2 \right. \\
\left. +(\mu_{i2} + \bar{\mu}_{i2})\bar{u}_i \sum_{j\neq i} (\mu_{j2} + \bar{\mu}_{j2})\bar{u}_j \right] \nu(d\theta) \\
+\left[\epsilon_{i2} + \alpha_i \left(\sigma_0(\sigma_{i2} + \bar{\sigma}_{i2}) + \int_\Theta \mu_0(\mu_{i2} + \bar{\mu}_{i2})\nu(d\theta) \right) + \gamma_i(b_{i2} + \bar{b}_{i2}) \right] \bar{u}_i \\
+\sum_{j\neq i} \left[\alpha_i \left(\sigma_0(\sigma_{j2} + \bar{\sigma}_{j2}) + \int_\Theta \mu_0(\mu_{j2} + \bar{\mu}_{j2})\nu(d\theta) \right) + \gamma_i(b_{j2} + \bar{b}_{j2}) \right] \bar{u}_j \\
+\left[\epsilon_{i1} + \bar{\epsilon}_{i1} + \alpha_i \left((\sigma_1 + \bar{\sigma}_1)(\sigma_{i2} + \bar{\sigma}_{i2}) + \int_\Theta (\mu_1 + \bar{\mu}_1)(\mu_{i2} + \bar{\mu}_{i2})\nu(d\theta) \right) \right. \\
\left. +\beta_i(b_{i2} + \bar{b}_{i2}) \right] \bar{x}\bar{u}_i \\
+\sum_{j\neq i} \left[\alpha_i \left((\sigma_1 + \bar{\sigma}_1)(\sigma_{j2} + \bar{\sigma}_{j2}) + \int_\Theta (\mu_1 + \bar{\mu}_1)(\mu_{j2} + \bar{\mu}_{j2})\nu(d\theta) \right) \right. \\
\left. +\beta_i(b_{j2} + \bar{b}_{j2}) \right] \bar{x}\bar{u}_j \\
= \frac{\bar{c}_i}{2}\left(\bar{u}_i + \frac{\bar{\lambda}_{i2}\bar{x} + \bar{\lambda}_{i3}}{\bar{c}_i} \right)^2
\end{array}
$$

$$
\begin{aligned}
&+\Bigg\{\alpha_i \frac{1}{2}\Bigg(\sum_{j\neq i}(\sigma_{j2}+\bar{\sigma}_{j2})\frac{\bar{\lambda}_{j2}}{\bar{c}_j}\Bigg)^2+\alpha_i\int_{\Theta}\frac{1}{2}\Bigg(\sum_{j\neq i}(\mu_{j2}+\bar{\mu}_{j2})\frac{\bar{\lambda}_{j2}}{\bar{c}_j}\Bigg)^2\nu(d\theta)-\frac{\bar{\lambda}_{i2}^2}{2\bar{c}_i}\\
&\quad-\sum_{j\neq i}\Bigg[\alpha_i\Bigg((\sigma_1+\bar{\sigma}_1)(\sigma_{j2}+\bar{\sigma}_{j2})+\int_{\Theta}(\mu_1+\bar{\mu}_1)(\mu_{j2}+\bar{\mu}_{j2})\nu(d\theta)\Bigg)\\
&+\beta_i(b_{j2}+\bar{b}_{j2})\Bigg]\frac{\bar{\lambda}_{j2}}{\bar{c}_j}\Bigg\}\bar{x}^2\\
&+\Bigg\{\alpha_i\Bigg(\sum_{j\neq i}(\sigma_{j2}+\bar{\sigma}_{j2})\frac{\bar{\lambda}_{j2}}{\bar{c}_j}\Bigg)\Bigg(\sum_{j\neq i}(\sigma_{j2}+\bar{\sigma}_{j2})\frac{\bar{\lambda}_{j3}}{\bar{c}_j}\Bigg)\\
&+\alpha_i\int_{\Theta}\Bigg(\sum_{j\neq i}(\mu_{j2}+\bar{\mu}_{j2})\frac{\bar{\lambda}_{j2}}{\bar{c}_j}\Bigg)\Bigg(\sum_{j\neq i}(\mu_{j2}+\bar{\mu}_{j2})\frac{\bar{\lambda}_{j3}}{\bar{c}_j}\Bigg)\nu(d\theta)\\
&\quad-\sum_{j\neq i}\Bigg[\alpha_i\Bigg(\sigma_0(\sigma_{j2}+\bar{\sigma}_{j2})+\int_{\Theta}\mu_0(\mu_{j2}+\bar{\mu}_{j2})\nu(d\theta)\Bigg)+\gamma_i(b_{j2}+\bar{b}_{j2})\Bigg]\frac{\bar{\lambda}_{j2}}{\bar{c}_j}\\
&\quad-\sum_{j\neq i}\Bigg[\alpha_i\Bigg((\sigma_1+\bar{\sigma}_1)(\sigma_{j2}+\bar{\sigma}_{j2})+\int_{\Theta}(\mu_1+\bar{\mu}_1)(\mu_{j2}+\bar{\mu}_{j2})\nu(d\theta)\Bigg)\\
&+\beta_i(b_{j2}+\bar{b}_{j2})\Bigg]\frac{\bar{\lambda}_{j3}}{\bar{c}_j}-\frac{\bar{\lambda}_{i2}\bar{\lambda}_{i3}}{\bar{c}_i}\Bigg\}\bar{x}\\
&+\alpha_i\frac{1}{2}\Bigg(\sum_{j\neq i}(\sigma_{j2}+\bar{\sigma}_{j2})\frac{\bar{\lambda}_{j3}}{\bar{c}_j}\Bigg)^2+\alpha_i\int_{\Theta}\frac{1}{2}\Bigg(\sum_{j\neq i}(\mu_{j2}+\bar{\mu}_{j2})\frac{\bar{\lambda}_{j3}}{\bar{c}_j}\Bigg)^2\nu(d\theta)\\
&-\sum_{j\neq i}\Bigg[\alpha_i\Bigg(\sigma_0(\sigma_{j2}+\bar{\sigma}_{j2})+\int_{\Theta}\mu_0(\mu_{j2}+\bar{\mu}_{j2})\nu(d\theta)\Bigg)+\gamma_i(b_{j2}+\bar{b}_{j2})\Bigg]\frac{\bar{\lambda}_{j3}}{\bar{c}_j}-\frac{\bar{\lambda}_{i3}^2}{2\bar{c}_i},
\end{aligned}
\tag{7.286}
$$

where

$$
\begin{aligned}
\bar{c}_i &= (r_i+\bar{r}_i)+\alpha_i\Bigg[(\sigma_{i2}+\bar{\sigma}_{i2})^2+\int_{\Theta}(\mu_{i2}+\bar{\mu}_{i2})^2\nu(d\theta)\Bigg]\\
\bar{\tau}_i &= \alpha_i\Bigg[(\sigma_{i2}+\bar{\sigma}_{i2})\sum_{j\neq i}(\sigma_{j2}+\bar{\sigma}_{j2})\bar{u}_j\Bigg]\\
&+\alpha_i\int_{\Theta}\Bigg[(\mu_{i2}+\bar{\mu}_{i2})\sum_{j\neq i}(\mu_{j2}+\bar{\mu}_{j2})\bar{u}_j\Bigg]\nu(d\theta)\\
&+\epsilon_{i2}+\alpha_i\Bigg(\sigma_0(\sigma_{i2}+\bar{\sigma}_{i2})+\int_{\Theta}\mu_0(\mu_{i2}+\bar{\mu}_{i2})\nu(d\theta)\Bigg)+\gamma_i(b_{i2}+\bar{b}_{i2})\\
&+\Bigg[\epsilon_{i1}+\bar{\epsilon}_{i1}+\alpha_i\Bigg((\sigma_1+\bar{\sigma}_1)(\sigma_{i2}+\bar{\sigma}_{i2})+\int_{\Theta}(\mu_1+\bar{\mu}_1)(\mu_{i2}+\bar{\mu}_{i2})\nu(d\theta)\Bigg)\\
&+\beta_i(b_{i2}+\bar{b}_{i2})\Bigg]\bar{x}.
\end{aligned}
\tag{7.287}
$$

Therefore,

$$
\begin{aligned}
&\bar{\tau}_i = -\alpha_i\left[(\sigma_{i2}+\bar{\sigma}_{i2})\sum_{j\neq i}(\sigma_{j2}+\bar{\sigma}_{j2})\frac{\bar{\tau}_j}{\bar{c}_j}\right]\\
&-\alpha_i\int_\Theta\left[(\mu_{i2}+\bar{\mu}_{i2})\sum_{j\neq i}(\mu_{j2}+\bar{\mu}_{j2})\frac{\bar{\tau}_j}{\bar{c}_j}\right]\nu(d\theta)\\
&+\epsilon_{i2}+\alpha_i\left(\sigma_0(\sigma_{i2}+\bar{\sigma}_{i2})+\int_\Theta\mu_0(\mu_{i2}+\bar{\mu}_{i2})\nu(d\theta)\right)+\gamma_i(b_{i2}+\bar{b}_{i2})\\
&+\left[\epsilon_{i1}+\bar{\epsilon}_{i1}+\alpha_i\left((\sigma_1+\bar{\sigma}_1)(\sigma_{i2}+\bar{\sigma}_{i2})+\int_\Theta(\mu_1+\bar{\mu}_1)(\mu_{i2}+\bar{\mu}_{i2})\nu(d\theta)\right)\right.\\
&\left.+\beta_i(b_{i2}+\bar{b}_{i2})\right]\bar{x}.
\end{aligned}
\tag{7.288}
$$

The vectorial function $\bar{\tau}$ is obtained by solving the linear system of functions and is given by $\bar{\tau}=[\bar{A}^{-1}\bar{v}]$, where

$$
\begin{aligned}
&\bar{A}_{ii}=1,\\
&\bar{A}_{ij}=\frac{\alpha_i}{\bar{c}_j}\left[(\sigma_{i2}+\bar{\sigma}_{i2})(\sigma_{j2}+\bar{\sigma}_{j2})+\int_\Theta(\mu_{i2}+\bar{\mu}_{i2})(\mu_{j2}+\bar{\mu}_{j2})\nu(d\theta)\right],\\
&\bar{v}_i=\epsilon_{i2}+\alpha_i\left(\sigma_0(\sigma_{i2}+\bar{\sigma}_{i2})+\int_\Theta\mu_0(\mu_{i2}+\bar{\mu}_{i2})\nu(d\theta)\right)+\gamma_i(b_{i2}+\bar{b}_{i2})\\
&+\left[\epsilon_{i1}+\bar{\epsilon}_{i1}+\alpha_i\left((\sigma_1+\bar{\sigma}_1)(\sigma_{i2}+\bar{\sigma}_{i2})+\int_\Theta(\mu_1+\bar{\mu}_1)(\mu_{i2}+\bar{\mu}_{i2})\nu(d\theta)\right)\right.\\
&\left.+\beta_i(b_{i2}+\bar{b}_{i2})\right]\bar{x},,
\end{aligned}
\tag{7.289}
$$

Hence, $\bar{\tau}_i$ is expressed as

$$
\begin{aligned}
&\bar{\tau}=\bar{A}^{-1}\bar{v}=\bar{\lambda}_2\bar{x}+\bar{\lambda}_3,\\
&\bar{\tau}_i=\bar{\lambda}_{i2}\bar{x}+\bar{\lambda}_{i3},\\
&\frac{\bar{\tau}_i^2}{2\bar{c}_i}=\frac{\bar{\lambda}_{i2}^2}{2\bar{c}_i}\bar{x}^2+\frac{\bar{\lambda}_{i2}\bar{\lambda}_{i3}}{\bar{c}_i}\bar{x}+\frac{\bar{\lambda}_{i3}^2}{2\bar{c}_i}.
\end{aligned}
\tag{7.290}
$$

7.6.3.6 Matching

$$
\begin{aligned}
&E[\hat{L}_i(x,s,u)|x(0)=x_0,s(0)=s_0]-E[f(0,x_0,s_0)|s]\\
&=\frac{1}{2}E[(q_{iT}-\alpha_i(T))(x_T-\bar{x}_T)^2|s]\\
&+\frac{1}{2}E[(q_{iT}+\bar{q}_{iT}-\beta_i(T))\bar{x}_T^2|s]+(\epsilon_{i3}(T)-\gamma_{iT})\bar{x}(T)\\
&+Z_2+Z_1,
\end{aligned}
\tag{7.291}
$$

$$
\begin{aligned}
Z_2 = E\int_0^T \Bigg[\frac{1}{2}\Bigg\{ & \dot{\alpha}_i + 2\alpha_i b_1 + \alpha_i\left(\sigma_1^2 + \int_\Theta \mu_1^2 \nu(d\theta)\right) + q_i \\
& + \sum_{s'\in\mathcal{S}}[\alpha_i(.,s') - \alpha_i(.,s)]\tilde{q}_{ss'} \\
& + \alpha_i\Bigg(\sum_{j\neq i}\sigma_{j2}\frac{\tilde{\tau}_j}{c_j}\Bigg)^2 + \int_\Theta \alpha_i\Bigg(\sum_{j\neq i}^n \mu_{j2}\frac{\tilde{\tau}_j}{c_j}\Bigg)^2\nu(d\theta) \\
& -2\alpha_i\sum_{j\neq i}\Bigg(\sigma_1\sigma_{j2} + \int_\Theta \mu_1\mu_{j2}\nu(d\theta)\Bigg)\frac{\tilde{\tau}_j}{c_j} - 2\alpha_i\sum_{j\neq i} b_{j2}\frac{\tilde{\tau}_j}{c_j} - \frac{\tilde{\tau}_i^2}{c_i}\Bigg\}(x-\bar{x})^2 \\
& + \frac{1}{2}\Bigg\{\dot{\beta}_i + 2\beta_i(b_1+\bar{b}_1) + \alpha_i\left((\sigma_1+\bar{\sigma}_1)^2 + \int_\Theta(\mu_1+\bar{\mu}_1)^2\nu(d\theta)\right) + q_i + \bar{q}_i \\
& + \sum_{s'\in\mathcal{S}}[\beta_i(.,s') - \beta_i(.,s)]\tilde{q}_{ss'} \\
& + \alpha_i\Bigg(\sum_{j\neq i}(\sigma_{j2}+\bar{\sigma}_{j2})\frac{\bar{\lambda}_{j2}}{\bar{c}_j}\Bigg)^2 + \alpha_i\int_\Theta\Bigg(\sum_{j\neq i}(\mu_{j2}+\bar{\mu}_{j2})\frac{\bar{\lambda}_{j2}}{\bar{c}_j}\Bigg)^2\nu(d\theta) - \frac{\bar{\lambda}_{i2}^2}{\bar{c}_i} \\
& -2\sum_{j\neq i}\Bigg[\alpha_i\Bigg((\sigma_1+\bar{\sigma}_1)(\sigma_{j2}+\bar{\sigma}_{j2}) + \int_\Theta(\mu_1+\bar{\mu}_1)(\mu_{j2}+\bar{\mu}_{j2})\nu(d\theta)\Bigg) \\
& + \beta_i(b_{j2}+\bar{b}_{j2})\Bigg]\frac{\bar{\lambda}_{j2}}{\bar{c}_j}\Bigg\}\bar{x}^2\Bigg]dt,
\end{aligned}
\tag{7.292}
$$

$$
\begin{aligned}
Z_1 = E\int_0^T \Bigg[+ \Bigg\{ & \dot{\gamma}_i + \beta_i b_0 + \gamma_i(b_1+\bar{b}_1) + \alpha_i\left(\sigma_0(\sigma_1+\bar{\sigma}_1) + \int_\Theta \mu_0(\mu_1+\bar{\mu}_1)\nu(d\theta)\right) \\
& + \epsilon_{i3} + \sum_{s'\in\mathcal{S}}[\gamma_i(.,s') - \gamma_i(.,s)]\tilde{q}_{ss'} \\
& + \alpha_i\Bigg(\sum_{j\neq i}(\sigma_{j2}+\bar{\sigma}_{j2})\frac{\bar{\lambda}_{j2}}{\bar{c}_j}\Bigg)\Bigg(\sum_{j\neq i}(\sigma_{j2}+\bar{\sigma}_{j2})\frac{\bar{\lambda}_{j3}}{\bar{c}_j}\Bigg) \\
& + \alpha_i\int_\Theta\Bigg(\sum_{j\neq i}(\mu_{j2}+\bar{\mu}_{j2})\frac{\bar{\lambda}_{j2}}{\bar{c}_j}\Bigg)\Bigg(\sum_{j\neq i}(\mu_{j2}+\bar{\mu}_{j2})\frac{\bar{\lambda}_{j3}}{\bar{c}_j}\Bigg)\nu(d\theta) \\
& - \sum_{j\neq i}\Bigg[\alpha_i\Bigg(\sigma_0(\sigma_{j2}+\bar{\sigma}_{j2}) + \int_\Theta \mu_0(\mu_{j2}+\bar{\mu}_{j2})\nu(d\theta)\Bigg) + \gamma_i(b_{j2}+\bar{b}_{j2})\Bigg]\frac{\bar{\lambda}_{j2}}{\bar{c}_j} \\
& - \sum_{j\neq i}\Bigg[\alpha_i\Bigg((\sigma_1+\bar{\sigma}_1)(\sigma_{j2}+\bar{\sigma}_{j2}) + \int_\Theta(\mu_1+\bar{\mu}_1)(\mu_{j2}+\bar{\mu}_{j2})\nu(d\theta)\Bigg) \\
& + \beta_i(b_{j2}+\bar{b}_{j2})\Bigg]\frac{\bar{\lambda}_{j3}}{\bar{c}_j} - \frac{\bar{\lambda}_{i2}\bar{\lambda}_{i3}}{\bar{c}_i}\Bigg\}\bar{x} \\
& + \Bigg\{\dot{\delta}_i + \gamma_i b_0 + \frac{\alpha_i}{2}\left(\sigma_0^2 + \int_\Theta \mu_0^2\nu(d\theta)\right) + \sum_{s'\in\mathcal{S}}[\delta_i(.,s') - \delta_i(.,s)]\tilde{q}_{ss'} \\
& + \alpha_i\frac{1}{2}\Bigg(\sum_{j\neq i}(\sigma_{j2}+\bar{\sigma}_{j2})\frac{\bar{\lambda}_{j3}}{\bar{c}_j}\Bigg)^2 \\
& + \alpha_i\int_\Theta\frac{1}{2}\Bigg(\sum_{j\neq i}(\mu_{j2}+\bar{\mu}_{j2})\frac{\bar{\lambda}_{j3}}{\bar{c}_j}\Bigg)^2\nu(d\theta) \\
& - \sum_{j\neq i}\Bigg[\alpha_i\Bigg(\sigma_0(\sigma_{j2}+\bar{\sigma}_{j2}) + \int_\Theta \mu_0(\mu_{j2}+\bar{\mu}_{j2})\nu(d\theta)\Bigg) + \gamma_i(b_{j2}+\bar{b}_{j2})\Bigg]\frac{\bar{\lambda}_{j3}}{\bar{c}_j} - \frac{\bar{\lambda}_{i3}^2}{2\bar{c}_i}\Bigg\} \\
& + \frac{1}{2}c_i\Bigg(u_i - \bar{u}_i + \frac{\tilde{\tau}_i}{c_i}(x-\bar{x})\Bigg)^2 + \frac{\bar{c}_i}{2}\Bigg(\bar{u}_i + \frac{\bar{\lambda}_{i2}\bar{x} + \bar{\lambda}_{i3}}{\bar{c}_i}\Bigg)^2\Bigg]dt,
\end{aligned}
\tag{7.293}
$$

where the minimization of the latest expression leads to the results in the statement provided that c_i and $\bar{c}_i$ are positive.

To recap, in this section, we have introduced a linear-quadratic mean-field-type game [102, 319, 333] under a jump-diffusion-regime switching state dynamics that includes mean-field terms which are:

- the conditional expectation of the state with respect to the filtration of the regime switching, in the drift, diffusion, and jump coefficient functions,
- the conditional expectation of the control actions, in the drift, diffusion, and jump coefficient functions.

We have taken the state process to be one-dimensional. The instantaneous cost function included:

- the conditional variance of the state,
- the square of the conditional expected value of the state,
- the conditional variance of the control-action process,
- the square of the conditional expected value of the control-action process,
- covariance between the state and the control,
- a product term between the conditional expected value of the control action and the conditional expected value of the state.

We will examine the extension that includes more cross terms such as $(u_i - \bar{u}_i)(u_j - \bar{u}_j)$ and $\bar{u}_i\bar{u}_j$ in Chap. 11.

For more developments on mean-field-type games, see [15, 15–17, 27–29, 33, 33, 33, 35–37, 39–43, 46–48, 50–52, 62, 99, 102, 124, 125, 127, 132, 146, 157, 158, 166, 168, 185, 194, 245, 247, 281, 282, 293, 294, 296, 307, 308, 311, 320, 330–332, 334, 336–338, 350, 359, 367] and the references therein.

7.7 Linear-Quadratic Conditional Mean-Field-Type Games

This section extends the problem examined in Sect. 7.6 to noisy cost measurement. We consider a generic linear-quadratic mean-field-type game problem under a jump-diffusion-regime switching state dynamics that include common noise and mean-field terms that are:

- the conditional expectation of the state with respect to the filtration of the regime switching, in the drift, diffusion, and jump coefficient functions,
- the conditional expectation of the control actions, in the drift, diffusion, and jump coefficient functions.

The state process is in one dimension. The common noise is of two types: a Brownian motion and a Poisson random process. The instant cost function includes:

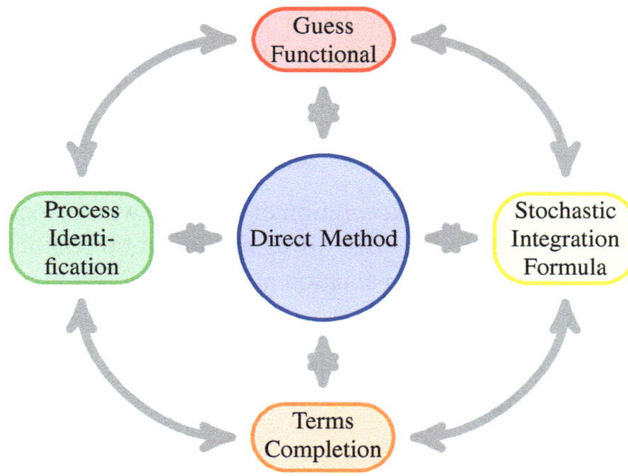

Fig. 7.6 Direct method and its key steps

- the conditional variance of the state,
- the square of the conditional expected value of the state,
- the conditional variance of the control-action process,
- the square of the conditional expected value of the control-action process,
- a cross-term such as covariance between the state and the control,
- a product term between the conditional expected value of the control action and the conditional expected value of the state,
- a noisy cost measurement, which is a jump-diffusion-regime switching and common noise dependent.

We follow a direct method to solve the game problem in a semi-explicit way. The method starts by identifying a partial guess functional where the coefficients are regime switching dependent. Then, it uses Itô's formula for jump-diffusion-regime switching processes, followed by a completion of squares for both control and mean control. The completion of squares technique used in the method developed in Sect. 7.6 cannot be directly used here due to extra dependence from other decision-makers. A careful matrix inversion is used to derive the interdependence between coefficients. Finally, the processes are identified using an orthogonal decomposition technique, and a system of stochastic differential equations is derived in a semi-explicit way. The procedure is summarized in Fig. 7.6.

We introduce the following notations. Let $[0, T]$, $T > 0$ be a fixed time horizon and $(\Omega, \mathcal{F}, \mathbb{F}^{B_o,N_o,B,N,s}, \mathbb{P})$ be a given filtered probability space.The filtration $\mathbb{F} = \{\mathcal{F}_t^{B_o,N_o,B,N,s},\ 0 \leq t \leq T\}$ is the natural filtration of the union $\{B, N, B_o, N_o, s\}$ augmented by $\mathbb{P}$-null sets of $\mathcal{F}$. In practice, B is used to capture smaller disturbance and N is used for larger jumps of the system. Let $k \geq 1$. $L^k([0, T] \times \mathcal{S}; \mathbb{R})$ be the set of measurable functions $f : [0, T] \times \mathcal{S} \to \mathbb{R}$ such that $\int_0^T |f(t, s)|^k dt < \infty$, $\mathcal{L}_{\mathbb{F}}^k([0, T] \times \mathcal{S}; \mathbb{R})$ be the set of $\mathbb{F}$-adapted $\mathbb{R}$-valued processes $X(\cdot)$ such that $E\left[\int_0^T |X(t)|^k dt\right] < \infty$. The stochastic quantity $\hat{X}(t) = E[X(t)|\ \mathcal{F}_t^{B_o,N_o,s}]$ denotes the conditional expectation of the random variable $X(t)$ with respect to the filtration $\mathscr{F}_t^{B_o,N_o,s}$. Note that $\hat{x}$ is a random process. There are $I \geq 2$ decision-makers. An admissible control strategy u_i of decision-maker i is an $\mathbb{F}$-adapted and square-integrable process with values in a non-empty subset $U_i = \mathbb{R}$. We denote the set of all admissible controls by $\mathscr{U}_i$:

$$\mathscr{U}_i = \{u_i(\cdot) \in \mathscr{L}^2_{\mathbb{F}}([0,T] \times \mathscr{S}; \mathbb{R});\ u_i(.) \in U_i\ a.e.t \in [0,T],\ \mathbb{P} - a.s.\}$$

7.7.1 Games Under Jump-Diffusion-Regime Switching Common Noise

For the sake of simplicity, we consider the one-dimensional case. Consider the following state dynamics with jump-diffusion and regime switching:

$$\begin{aligned}
&x(0) = x_0, \quad s(0) = s_0,\\
&dx = bdt + \sigma dB + \textstyle\int_\Theta \mu(\cdot,\theta)\tilde{N}(dt,d\theta) + \sigma_o dB_o + \int_\Theta \mu_o(\cdot,\theta)\tilde{N}_o(dt,d\theta),
\end{aligned} \tag{7.294}$$

where

$$\begin{aligned}
&b = b_0 + b_1 x + \bar{b}_1\hat{x} + \textstyle\sum_{j=1}^I b_{j2}u_j + \sum_{j=1}^I \bar{b}_{j2}\hat{u}_j,\\
&\sigma = \sigma_0 + \sigma_1 x + \bar{\sigma}_1\hat{x} + \textstyle\sum_{j=1}^I \sigma_{j2}u_j + \sum_{j=1}^I \bar{\sigma}_{j2}\hat{u}_j,\\
&\mu(\cdot,\theta) = \mu_0 + \mu_1 x + \bar{\mu}_1\hat{x} + \textstyle\sum_{j=1}^I \mu_{j2}u_j + \sum_{j=1}^I \bar{\mu}_{j2}\hat{u}_j,\\
&\sigma_o = \sigma_{o0} + \sigma_{o1} x + \bar{\sigma}_{o1}\hat{x} + \textstyle\sum_{j=1}^I \sigma_{oj2}u_j + \sum_{j=1}^I \bar{\sigma}_{oj2}\hat{u}_j,\\
&\mu_o(\cdot,\theta) = \mu_{o0} + \mu_{o1} x + \bar{\mu}_{o1}\hat{x} + \textstyle\sum_{j=1}^I \mu_{oj2}u_j + \sum_{j=1}^I \bar{\mu}_{oj2}\hat{u}_j,
\end{aligned} \tag{7.295}$$

with $b_k(t,s); \sigma_k(t,s); \bar{b}_k(t,s); \bar{\sigma}_k(t,s), \sigma_o(t,s) \in \mathbb{R}$, and $s \in \mathscr{S}$, where $\mathscr{S}$ is a finite set of switching regimes. Moreover, $\mu_k(t,s,\theta), \mu_{ok}(t,s,\theta) \in \mathbb{R}$, $\theta \in \Theta$ where Θ is the set of jump sizes for all $k \in \{0,1,2\}$:

$$\begin{aligned}
&b_k, \sigma_k, \bar{b}_k, \bar{\sigma}_k : [0,T] \times \mathscr{S} \to \mathbb{R},\ k \in \{0,1\}\\
&b_{j2}, \sigma_{j2}, \bar{b}_{j2}, \bar{\sigma}_{j2} : [0,T] \times \mathscr{S} \to \mathbb{R},\ j \in \mathscr{I},\\
&\mu_k, \mu_{ok}, \bar{\mu}_k, \bar{\mu}_{ok} : [0,T] \times \mathscr{S} \times \Theta \to \mathbb{R},\ k \in \{0,1\}\\
&\mu_{j2}, \mu_{oj2}, \bar{\mu}_{j2}, \bar{\mu}_{oj2} : [0,T] \times \mathscr{S} \times \Theta \to \mathbb{R},\ j \in \mathscr{I},
\end{aligned} \tag{7.296}$$

The stochastic differential equation (7.294) consists of

- three different processes:
  - The switching process s switches between different elements in the finite set $\mathscr{S}$ with transition rate $Q = (\tilde{q}_{ss'},\ (s,s') \in \mathscr{S}^2)$ such that

$$s' \neq s,\ \tilde{q}_{ss'} > 0,\ \ \tilde{q}_{ss} = -\sum_{s' \neq s} \tilde{q}_{ss'}.$$

- The regime switching dependent jump-diffusion process $\sigma dB + \int_{\Theta} \mu(\cdot,\theta)\tilde{N}(dt,d\theta)$, where the process B is a standard one-dimensional Brownian motion, $\tilde{N}(dt,d\theta) = N(dt,d\theta) - \nu(d\theta)dt$ is a compensated jump process with Radon measure ν over Θ.
- Common noise which is also a regime switching dependent jump-diffusion process $\sigma_o dB_o + \int_{\Theta} \mu_o(\cdot,\theta)\tilde{N}_o(dt,d\theta)$, where the process B_o is a standard one-dimensional Brownian motion, $\tilde{N}_o(dt,d\theta) = N_o(dt,d\theta) - \nu_o(d\theta)dt$ is a compensated jump process with Radon measure ν_o over Θ. The common noise B_o is used to capture smaller fluctuations. The Poisson random measure N_o as an observable common noise allows us to capture larger jumps (big changes) in the realization of the common noise.

  It is assumed that the processes $s(t), B(t), N(t,.), B_o(t), N_o(t,.), x_0$ are mutually independent, and independent of $B_o(t), N_o(t,.)$ which are two types of common noises.

- The term $\hat{x} = E[x|\mathscr{F}^{B_o,N_0,s}]$ denotes the conditional expectation of x given $\mathscr{F}_t^{B_o,N_0,s}$, and $\hat{u} = E[u|\mathscr{F}^{B_o,N_0,s}]$. These are called conditional state mean-field and conditional control-action mean-field terms.

The state dynamics is therefore of conditional McKean–Vlasov type as the conditional probability measure of the state appears in the drift, diffusion, and jump coefficients via $(\hat{x},\ \hat{u})$. We denote a solution (if any) to (7.294) by $x := x^{u,x_0,s_0}$.

7.7.1.1 Noisy Cost Functional per Decision-Maker

We associate a cost functional of mean-field type of decision-maker i as

$$\begin{aligned}
&L_i(x,s,u) = \tfrac{1}{2}q_{iT}x^2(T) + \tfrac{1}{2}\bar{q}_{iT}\hat{x}^2(T) + \epsilon_{i3}(T)x(T) + \bar{\epsilon}_{i3}(T)\hat{x}(T)\\
&+\tfrac{1}{2}\int_0^T \left(q_i x^2 + \bar{q}_i\hat{x}^2 + r_i u_i^2 + \bar{r}_i\hat{u}_i^2\right)dt\\
&+\int_0^T \left(\epsilon_{i1}xu_i + \bar{\epsilon}_{i1}\hat{x}\hat{u}_i + \epsilon_{i2}u_i + \bar{\epsilon}_{i2}\hat{u}_i + \epsilon_{i3}x + \bar{\epsilon}_{i3}\hat{x}\right)dt\\
&+\int_0^T \left(\xi_{i0,B_o} + \xi_{i1,B_o}\hat{x} + \tfrac{1}{2}\xi_{i2,B_o}x^2 + \tfrac{1}{2}\bar{\xi}_{i2,B_o}\hat{x}^2\right)dB_o\\
&+\int_0^T\int_{\Theta} \left(\xi_{i0,N_o} + \xi_{i1,N_o}\hat{x} + \tfrac{1}{2}\xi_{i2,N_o}x^2 + \tfrac{1}{2}\bar{\xi}_{i2,N_o}\hat{x}^2\right)\tilde{N}_o(dt,d\theta),
\end{aligned} \tag{7.297}$$

where $q_{iT} = q_i(T,s(T)),\ q_i = q_i(t,s(t)),\ \bar{q}_{iT} = \bar{q}_i(T,s(T)),\ \bar{q}_i = q_i(t,s(t))$; and $r_i = r_i(t,s(t)),\ \bar{r}_i = \bar{r}_i(t,s(t)) \in \mathbb{R},\ \epsilon_{ik} = \epsilon_{ik}(t,s(t)),\ \bar{\epsilon}_{ik} = \bar{\epsilon}_{ik}(t,s(t)) \in \mathbb{R}$. $\xi_{ik,B_o}(t,s(t)), \bar{\xi}_{ik,B_o}(t,s(t)) \in \mathbb{R},\ \xi_{ik,N_o}(t,s(t)), \bar{\xi}_{ik,N_o}(t,s(t),\theta) \in \mathbb{R}$.

$$
\begin{aligned}
&q_{iT}, \bar{q}_{iT}, \epsilon_{i3T}, \epsilon_{i3T} : \mathscr{S} \to \mathbb{R}, \\
&q_i, \bar{q}_i, r_i, \bar{r}_i, \epsilon_{ik}, \epsilon_{ik} : [0, T] \times \mathscr{S} \to \mathbb{R}, \; k \in \{1, 2, 3\} \\
&\xi_{ik,B_o}, \bar{\xi}_{i2,B_o} : [0, T] \times \mathscr{S} \to \mathbb{R}, \; k \in \{0, 1, 2\}, \\
&\xi_{ik,N_o}, \bar{\xi}_{i2,N_o} : [0, T] \times \mathscr{S} \times \Theta \to \mathbb{R}, \; k \in \{0, 1, 2\}.
\end{aligned} \tag{7.298}
$$

All the coefficient functions in the cost are random and $\mathscr{F}^{B_o,N_o}$-measurable. The last term in (7.427) given by

$$
(\xi_{i0,B_o} + \xi_{i1,B_o}\hat{x} + \xi_{i2,B_o}x^2 + \bar{\xi}_{i2,B_o}\hat{x}^2)dB_o
$$

$$
+ \int_{\Theta} (\xi_{i0,N_o} + \xi_{i1,N_o}\hat{x} + \xi_{i2,N_o}x^2 + \bar{\xi}_{i2,N_o}\hat{x}^2)\tilde{N}_o(dt, d\theta),
$$

denotes a noisy cost measurement of decision-maker i.

7.7.1.2 Equilibrium Solution Concept

We look for Nash equilibria in state-and-conditional mean-field-type feedback strategies. We start with best-response strategies.

Mean-Field-Type Best-Response Problem with Jump-Diffusion-Regime Switching and Common Noise
Given the strategies of the other decision-makers $u_{-i} := (u_j \in \mathscr{U}_j, \; j \neq i)$, the best-response problem of the decision-maker is

$$
\begin{cases}
\inf_{u_i \in \mathscr{U}_i} E\,[L_i(x, s, u) \mid x(0) = x_0, s(0) = s_0] \\
\text{such that (7.294).}
\end{cases} \tag{7.299}
$$

Mean-Field-Type Game Problem Under Common Noise
The mean-field-type Nash equilibrium is a strategy profile $u^* \in \prod_j \mathscr{U}_j$, such that for every decision-maker i, u_i^* solves (7.299) given the strategy profile from others' $u^*_{-i} = (u^*_j \in \mathscr{U}_j, \; j \neq i)$.

$$
\begin{cases}
i \in \mathscr{I} := \{1, 2, \ldots, n\}, \\
E\left[L_i(x^*, s, u^*) | x(0) = x_0, s(0) = s_0\right] \\
= \inf_{u_i \in \mathscr{U}_i} E\left[L_i(x, s, u_i, u^*_{-i}) \; | x(0) = x_0, s(0) = s_0\right], \\
\text{such that } x = x^{u_i, u^*_{-i}} \text{ solves (7.294) under } (u_i, u^*_{-i}), \\
\text{such that } x^* = x^{u^*} \text{ solves (7.294) under } u^*.
\end{cases} \tag{7.300}
$$

We consider the following assumption:

$$\begin{cases} q_{iT} > 0, q_{iT} + \bar{q}_{iT} > 0, q_i > 0,\ q_i + \bar{q}_i > 0, \\ r_i > 0, r_i + \bar{r}_i > 0, \\ E[x_0^2] < +\infty,\ |\mathscr{S}| < +\infty, \\ L^2_\nu([0,T] \times \mathscr{S} \times \Theta) : \displaystyle\int_0^T \int_\Theta \phi^2 \nu(d\theta) < +\infty, \text{ with } \phi \in \{\mu_{ik}, \bar{\mu}_{ik}\}, \\ L^2_{\nu_o}([0,T] \times \mathscr{S} \times \Theta) : \displaystyle\int_0^T \int_\Theta \phi^2 \nu_o(d\theta) < +\infty, \text{ with } \phi \in \{\mu_{oik}, \bar{\mu}_{oik}\}, \\ b_k, \bar{b}_k, \sigma_k^2, \bar{\sigma}_k^2, \sigma_{ok}^2, \bar{\sigma}_{ok}^2 \in L^1([0,T] \times \mathscr{S}, \mathbb{R}) \end{cases} \tag{7.301}$$

The integrability condition above yields

$$\begin{array}{l} s \in \mathscr{S}, \\ \int_0^T [|b_0(t,s)| + |b_1(t,s)| + |\bar{b}_1(t,s)| + |b_{j2}(t,s)| + |\bar{b}_{j2}(t,s)|]dt \\ + \int_0^T [\sigma_0^2(t,s) + \sigma_1^2(t,s) + \bar{\sigma}_1^2(t,s) + \sigma_{j2}^2(t,s) + \bar{\sigma}_{j2}^2(t,s)]dt \\ + \int_0^T \int_\Theta [\mu_0^2(t,s,\theta) + \mu_1^2(t,s,\theta) + \bar{\mu}_1^2(t,s,\theta) + \mu_{j2}^2(t,s,\theta) + \bar{\mu}_{j2}^2(t,s,\theta)]\nu(d\theta)dt \\ + \int_0^T \int_\Theta \big[\mu_{o0}^2(t,s,\theta) + \mu_{o1}^2(t,s,\theta) + \bar{\mu}_{o1}^2(t,s,\theta) \\ \quad + \mu_{oj2}^2(t,s,\theta) + \bar{\mu}_{oj2}^2(t,s,\theta)\big] \nu_o(d\theta)dt < +\infty. \end{array} \tag{7.302}$$

Remark 7.4 Under Assumption (7.301), the state dynamics has a solution for each strategy profile $u = (u_i)_{i \in \mathscr{I}} \in \mathscr{U}$. Note that this assumption does not impose boundedness conditions on the coefficients or Lipschitz conditions on the cost over the horizon $[0, T]$. Under Assumption (7.301), Problem (7.299) is well-defined.

We now state the main result of this section.

Solution to Mean-Field-Type Game Problem (7.299)

Let Assumption (7.301) hold. Then, the equilibrium strategy and the equilibrium cost are given by

$$\begin{array}{l} u_i^* = \hat{u}_i^* - \frac{\tilde{\tau}_i}{c_i}(x - \hat{x}), \\ \hat{u}_i^* = -\frac{\bar{\tau}_{i1}\hat{x}}{\bar{c}_i} - \frac{\bar{\tau}_{i0}}{\bar{c}_i} \\ L_i^*(x, s, u^*) = \frac{1}{2}E[\alpha_i(0, s(0))(x_0 - \hat{x}_0)^2] \\ + \frac{1}{2}E[\beta_i(0, s(0))\hat{x}_0^2] + E[\gamma_i(0, s(0))\hat{x}_0] + E[\delta_i(0, s(0))], \end{array} \tag{7.303}$$

Coefficient Processes $\alpha_i, \alpha_{i,B_o}, \alpha_{i,N_o}$ of the Solution to Problem (7.299)

$$
\begin{aligned}
d\alpha_i =& -\{2b_1\alpha_i + \alpha_i[\sigma_1^2+\sigma_{o1}^2+\int_\Theta \mu_1^2\nu(d\theta)+\int_\Theta \mu_{o1}^2\nu_o(d\theta)]\\
&+q_i - \frac{\bar{\tau}_i^2}{c_i} + \sum_{s'\neq s}\tilde{q}_{ss'}[\alpha_i(t,s')-\alpha_i(t,s)] + 2\alpha_{i,B_o}\sigma_{o1} + 2\int_\Theta \alpha_{i,N_o}\mu_{o1}\nu_o(d\theta)\\
&+\alpha_i[\sum_{j\neq i}\sigma_{j2}\frac{\bar{\tau}_j}{c_j}]^2 + \alpha_i\int_\Theta[\sum_{j\neq i}\mu_{j2}\frac{\bar{\tau}_j}{c_j}]^2\nu(d\theta)\\
&+\alpha_i[\sum_{j\neq i}\sigma_{oj2}\frac{\bar{\tau}_j}{c_j}]^2 + \alpha_i\int_\Theta[\sum_{j\neq i}\mu_{oj2}\frac{\bar{\tau}_j}{c_j}]^2\nu_o(d\theta)\\
&-2\alpha_i\sum_{j\neq i}\{b_{j2}+\sigma_1\sigma_{j2}+\int_\Theta \mu_1\mu_{j2}\nu(d\theta)\}\frac{\bar{\tau}_j}{c_j}\\
&-2\alpha_i\sum_{j\neq i}\{\sigma_{o1}\sigma_{oj2}+\int_\Theta \mu_{o1}\mu_{oj2}\nu_o(d\theta)\}\frac{\bar{\tau}_j}{c_j}\\
&-2\sum_{j\neq i}\{\alpha_{i,B_o}\sigma_{oj2}+\int_\Theta \alpha_{i,N_o}\mu_{oj2}\nu_o(d\theta)\}\frac{\bar{\tau}_j}{c_j}\Big\}dt\\
&-\left(\xi_{i2,B_o}+\alpha_i[2\sigma_{o1}-2\sum_j\sigma_{oj2}\frac{\bar{\tau}_j}{c_j}]\right)dB_o\\
&-\int_\Theta\left(\xi_{i2,N_o}+\alpha_i[2\mu_{o1}-2\sum_j\mu_{oj2}\frac{\bar{\tau}_j}{c_j}]\right)\tilde{N}_o(dt,d\theta),\\
\alpha_i(T,s) =& \, q_i(T,s),\\
\alpha_{i,B_o} =& -\left(\xi_{i2,B_o}+\alpha_i[2\sigma_{o1}-2\sum_j\sigma_{oj2}\frac{\bar{\tau}_j}{c_j}]\right),\\
\alpha_{i,N_o} =& -\left(\xi_{i2,N_o}+\alpha_i[2\mu_{o1}-2\sum_j\mu_{oj2}\frac{\bar{\tau}_j}{c_j}]\right).
\end{aligned}
\tag{7.304}
$$

whenever these stochastic differential equations have a solution that is $\mathscr{F}^{B_o,N_o}$-adapted and does not blow up in $[0, T]$, and

Coefficient processes $\beta_i, \beta_{i,B_o}, \beta_{i,N_o}$ of the Solution to Problem (7.299)

$$
\begin{aligned}
d\beta_i =& -\{2(b_1+\bar{b}_1)\beta_i + \beta_i[(\sigma_{o1}+\bar{\sigma}_{o1})^2+\int_\Theta(\mu_{o1}+\bar{\mu}_{o1})^2\nu_o(d\theta)]\\
&+(q_i+\bar{q}_i) - \frac{\bar{\tau}_{i1}^2}{c_i} + \sum_{s'\neq s}\tilde{q}_{ss'}[\beta_i(t,s')-\beta_i(t,s)]\\
&+\alpha_i[(\sigma_1+\bar{\sigma}_1)^2+\int_\Theta(\mu_1+\bar{\mu}_1)^2\nu(d\theta)]\\
&+2\beta_{i,B_o}(\sigma_{o1}+\bar{\sigma}_{o1}) + 2\int_\Theta \beta_{i,N_o}(\mu_{o1}+\bar{\mu}_{o1})\nu_o(d\theta)\\
&+\alpha_i[\sum_{j\neq i}(\sigma_{j2}+\bar{\sigma}_{j2})\frac{\bar{\tau}_{j1}}{c_j}]^2 + \alpha_i\int_\Theta[\sum_{j\neq i}(\mu_{j2}+\bar{\mu}_{j2})\frac{\bar{\tau}_{j1}}{c_j}]^2\nu(d\theta)\\
&+\beta_i[\sum_{j\neq i}(\sigma_{oj2}+\bar{\sigma}_{oj2})\frac{\bar{\tau}_{j1}}{c_j}]^2 + \beta_i\int_\Theta[\sum_{j\neq i}(\mu_{oj2}+\bar{\mu}_{oj2})\frac{\bar{\tau}_{j1}}{c_j}]^2\nu_o(d\theta)\\
&-2\alpha_i\sum_{j\neq i}\{(\sigma_1+\bar{\sigma}_1)(\sigma_{j2}+\bar{\sigma}_{j2})+\int_\Theta(\mu_1+\bar{\mu}_1)(\mu_{j2}+\bar{\mu}_{j2})\nu(d\theta)\}\frac{\bar{\tau}_{j1}}{c_j}\\
&-2\beta_i\sum_{j\neq i}\{(\sigma_{o1}+\bar{\sigma}_{o1})(\sigma_{oj2}+\bar{\sigma}_{oj2})+\int_\Theta(\mu_{o1}+\bar{\mu}_{o1})(\mu_{oj2}+\bar{\mu}_{oj2})\nu_o(d\theta)\}\frac{\bar{\tau}_{j1}}{c_j}\\
&-2\beta_i\sum_{j\neq i}(b_{j2}+\bar{b}_{j2})\frac{\bar{\tau}_{j1}}{c_j}\\
&-2\sum_{j\neq i}\{\beta_{i,B_o}(\sigma_{oj2}+\bar{\sigma}_{oj2})+\int_\Theta \beta_{i,N_o}(\mu_{oj2}+\bar{\mu}_{oj2})\}\frac{\bar{\tau}_{j1}}{c_j}\Big\}dt\\
&-\left(\xi_{i2,B_o}+\bar{\xi}_{i2,B_o}+\beta_i[2(\sigma_{o1}+\bar{\sigma}_{o1})-2\sum_j(\sigma_{oj2}+\bar{\sigma}_{oj2})\frac{\bar{\tau}_{j1}}{c_j}]\right)dB_o\\
&-\int_\Theta\left(\xi_{i2,N_o}+\bar{\xi}_{i2,N_o}+\beta_i[2(\mu_{o1}+\bar{\mu}_{o1})-2\sum_j(\mu_{oj2}+\bar{\mu}_{oj2})\frac{\bar{\tau}_{j1}}{c_j}]\right)\tilde{N}_o(dt,d\theta),\\
\beta_i(T,s) =& \, q_i(T,s)+\bar{q}_i(T,s),\\
\beta_{i,B_o} =& -\left(\xi_{i2,B_o}+\bar{\xi}_{i2,B_o}+\beta_i[2(\sigma_{o1}+\bar{\sigma}_{o1})-2\sum_j(\sigma_{oj2}+\bar{\sigma}_{oj2})\frac{\bar{\tau}_{j1}}{c_j}]\right),\\
\beta_{i,N_o} =& -\left(\xi_{i2,N_o}+\bar{\xi}_{i2,N_o}+\beta_i[2(\mu_{o1}+\bar{\mu}_{o1})-2\sum_j(\mu_{oj2}+\bar{\mu}_{oj2})\frac{\bar{\tau}_{j1}}{c_j}]\right),
\end{aligned}
\tag{7.305}
$$

whenever these stochastic differential equations have a solution that is $\mathscr{F}^{B_o,N_o}$-adapted and does not blow up in $[0, T]$.

Coefficient Processes $\gamma_i, \gamma_{i,B_o}, \gamma_{i,N_o}$ of the Solution to Problem (7.299)

$$
\begin{aligned}
d\gamma_i = &-\Big\{(b_1+\bar{b}_1)\gamma_i + [(\epsilon_{i3}+\bar{\epsilon}_{i3}) - \tfrac{\bar{\tau}_{i1}\bar{\tau}_{i0}}{\bar{c}_i}]dt + \textstyle\sum_{s'\neq s}\tilde{q}_{ss'}[\gamma_i(t,s')-\gamma_i(t,s)] \\
&+\alpha_i[\sigma_0(\sigma_1+\bar{\sigma}_1) + \textstyle\int_\Theta \mu_0(\mu_1+\bar{\mu}_1)\nu(d\theta)] \\
&+\beta_i[b_0 + \sigma_{o0}(\sigma_{o1}+\bar{\sigma}_{o1}) + \textstyle\int_\Theta \mu_{o0}(\mu_{o1}+\bar{\mu}_{o1})\nu_o(d\theta)] \\
&+\beta_{i,B_o}\sigma_{o0}dt + \textstyle\int_\Theta \beta_{i,N_o}\mu_{o0}\nu_o(d\theta) \\
&+\gamma_{i,B_o}(\sigma_{o1}+\bar{\sigma}_{o1})dt + \textstyle\int_\Theta \gamma_{i,N_o}(\mu_{o1}+\bar{\mu}_{o1})\nu_o(d\theta) \\
&+\alpha_i[\textstyle\sum_{j\neq i}(\sigma_{j2}+\bar{\sigma}_{j2})\frac{\bar{\tau}_{j0}}{\bar{c}_j}][\sum_{k\neq i}(\sigma_{k2}+\bar{\sigma}_{k2})\frac{\bar{\tau}_{k1}}{\bar{c}_k}] \\
&+\alpha_i \textstyle\int_\Theta[\sum_{j\neq i}(\mu_{j2}+\bar{\mu}_{j2})\frac{\bar{\tau}_{j0}}{\bar{c}_j}][\sum_{k\neq i}(\mu_{k2}+\bar{\mu}_{k2})\frac{\bar{\tau}_{k1}}{\bar{c}_k}]\nu(d\theta) \\
&+\beta_i[\textstyle\sum_{j\neq i}(\sigma_{oj2}+\bar{\sigma}_{oj2})\frac{\bar{\tau}_{j0}}{\bar{c}_j}][\sum_{j\neq i}(\sigma_{oj2}+\bar{\sigma}_{oj2})\frac{\bar{\tau}_{j1}}{\bar{c}_j}] \\
&+\beta_i \textstyle\int_\Theta[\sum_{j\neq i}(\mu_{oj2}+\bar{\mu}_{oj2})\frac{\bar{\tau}_{j0}}{\bar{c}_j}][\sum_{j\neq i}(\mu_{oj2}+\bar{\mu}_{oj2})\frac{\bar{\tau}_{j1}}{\bar{c}_j}]\nu_o(d\theta) \\
&-\alpha_i \textstyle\sum_{j\neq i}\{(\sigma_1+\bar{\sigma}_1)(\sigma_{j2}+\bar{\sigma}_{j2}) + \int_\Theta(\mu_1+\bar{\mu}_1)(\mu_{j2}+\bar{\mu}_{j2})\nu(d\theta)\}\frac{\bar{\tau}_{j0}}{\bar{c}_j} \\
&-\beta_i \textstyle\sum_{j\neq i}\{(\sigma_{o1}+\bar{\sigma}_{o1})(\sigma_{oj2}+\bar{\sigma}_{oj2}) + \int_\Theta(\mu_{o1}+\bar{\mu}_{o1})(\mu_{oj2}+\bar{\mu}_{oj2})\nu_o(d\theta)\}\frac{\bar{\tau}_{j0}}{\bar{c}_j} \\
&-\beta_i \textstyle\sum_{j\neq i}(b_{j2}+\bar{b}_{j2})\frac{\bar{\tau}_{j0}}{\bar{c}_j} \\
&-\textstyle\sum_{j\neq i}\{\beta_{i,B_o}(\sigma_{oj2}+\bar{\sigma}_{oj2}) + \int_\Theta \beta_{i,N_o}(\mu_{oj2}+\bar{\mu}_{oj2})\}\frac{\bar{\tau}_{j0}}{\bar{c}_j} \\
&-\alpha_i \textstyle\sum_{j\neq i}\{\sigma_0(\sigma_{j2}+\bar{\sigma}_{j2}) + \int_\Theta \mu_0(\mu_{j2}+\bar{\mu}_{j2})\nu(d\theta)\}\frac{\bar{\tau}_{j1}}{\bar{c}_j} \\
&-\beta_i \textstyle\sum_{j\neq i}\{\sigma_{o0}(\sigma_{oj2}+\bar{\sigma}_{oj2}) + \int_\Theta \mu_{o0}(\mu_{oj2}+\bar{\mu}_{oj2})\nu_o(d\theta)\}\frac{\bar{\tau}_{j1}}{\bar{c}_j} \\
&-\textstyle\sum_{j\neq i}\{\gamma_{i,B_o}(\sigma_{oj2}+\bar{\sigma}_{oj2}) + \int_\Theta \gamma_{i,N_o}(t,\theta)(\mu_{oj2}+\bar{\mu}_{oj2})\nu_o(d\theta)\}\frac{\bar{\tau}_{j1}}{\bar{c}_j} \\
&-\gamma_i \textstyle\sum_{j\neq i}(b_{j2}+\bar{b}_{j2})\frac{\bar{\tau}_{j1}}{\bar{c}_j}\ \}dt \\
&-\Big[\xi_{i1,B_o} + \beta_i(\sigma_{o0} - \textstyle\sum_j(\sigma_{oj2}+\bar{\sigma}_{oj2})\frac{\bar{\tau}_{j0}}{\bar{c}_j}) \\
&\qquad +\gamma_i((\sigma_{o1}+\bar{\sigma}_{o1}) - \textstyle\sum_j(\sigma_{oj2}+\bar{\sigma}_{oj2})\frac{\bar{\tau}_{j1}}{\bar{c}_j})\Big]dB_o \\
&-\textstyle\int_\Theta\Big[\xi_{i1,N_o} + \beta_i(\mu_{o0} - \sum_j(\mu_{oj2}+\bar{\mu}_{oj2})\frac{\bar{\tau}_{j0}}{\bar{c}_j}) \\
&\qquad +\gamma_i((\mu_{o1}+\bar{\mu}_{o1}) - \textstyle\sum_j(\mu_{oj2}+\bar{\mu}_{oj2})\frac{\bar{\tau}_{j1}}{\bar{c}_j})\Big]\tilde{N}_o(dt,d\theta), \\
\gamma_i(T,s) = &\ \epsilon_{i3}(T,s) + \bar{\epsilon}_{i3}(T,s), \quad s\in\mathcal{S} \\
\gamma_{i,B_o} = &-\Big[\xi_{i1,B_o} + \beta_i(\sigma_{o0} - \textstyle\sum_j(\sigma_{oj2}+\bar{\sigma}_{oj2})\frac{\bar{\tau}_{j0}}{\bar{c}_j}) \\
&\qquad +\gamma_i((\sigma_{o1}+\bar{\sigma}_{o1}) - \textstyle\sum_j(\sigma_{oj2}+\bar{\sigma}_{oj2})\frac{\bar{\tau}_{j1}}{\bar{c}_j})\Big], \\
\gamma_{i,N_o} = &-\Big[\xi_{i1,N_o} + \beta_i(\mu_{o0} - \textstyle\sum_j(\mu_{oj2}+\bar{\mu}_{oj2})\frac{\bar{\tau}_{j0}}{\bar{c}_j}) \\
&\qquad +\gamma_i((\mu_{o1}+\bar{\mu}_{o1}) - \textstyle\sum_j(\mu_{oj2}+\bar{\mu}_{oj2})\frac{\bar{\tau}_{j1}}{\bar{c}_j})\Big],
\end{aligned}
\tag{7.306}
$$

whenever the system of stochastic differential equations has a solution that is $\mathcal{F}^{B_o,N_o}$-adapted for each regime.

Coefficient Processes δ_i, δ_{i,B_o}, δ_{i,N_o} of the Solution to Problem (7.299)

$$
\begin{aligned}
d\delta_i =& -\Big\{\tfrac{\alpha_i}{2}(\sigma_0^2+\int_\Theta \mu_0^2\nu(d\theta))+\tfrac{\beta_i}{2}(\sigma_{o0}^2+\int_\Theta \mu_{o0}^2\nu_o(d\theta))\\
&+b_0\gamma_i dt+\gamma_{i,B_o}\sigma_{o0}dt+\int_\Theta \gamma_{i,N_o}(t,\theta)\mu_{o0}\nu_o(d\theta)+\sum_{s'\neq s}\tilde{q}_{ss'}[\delta_i(t,s')-\delta_i(t,s)]\\
&+\tfrac{\alpha_i}{2}[\sum_{j\neq i}(\sigma_{j2}+\bar{\sigma}_{j2})\tfrac{\bar{\tau}_{j0}}{\bar{c}_j}]^2+\int_\Theta \tfrac{\alpha_i}{2}[\sum_{j\neq i}(\mu_{j2}+\bar{\mu}_{j2})\tfrac{\bar{\tau}_{j0}}{\bar{c}_j}]^2\nu(d\theta)\\
&+\tfrac{\beta_i}{2}[\sum_{j\neq i}(\sigma_{oj2}+\bar{\sigma}_{oj2})\tfrac{\bar{\tau}_{j0}}{\bar{c}_j}]^2+\tfrac{\beta_i}{2}\int_\Theta[\sum_{j\neq i}(\mu_{oj2}+\bar{\mu}_{oj2})\tfrac{\bar{\tau}_{j0}}{\bar{c}_j}]^2\nu_o(d\theta)\\
&-\tfrac{\bar{\tau}_{i0}^2}{2\bar{c}_i}-\alpha_i\sum_{j\neq i}\{\sigma_0(\sigma_{j2}+\bar{\sigma}_{j2})+\int_\Theta \mu_0(\mu_{j2}+\bar{\mu}_{j2})\nu(d\theta)\}\tfrac{\bar{\tau}_{j0}}{\bar{c}_j}\\
&-\beta_i\sum_{j\neq i}\{\sigma_{o0}(\sigma_{oj2}+\bar{\sigma}_{oj2})+\int_\Theta \mu_{o0}(\mu_{oj2}+\bar{\mu}_{oj2})\nu_o(d\theta)\}\tfrac{\bar{\tau}_{j0}}{\bar{c}_j}\\
&-\sum_{j\neq i}\{\gamma_{i,B_o}(\sigma_{oj2}+\bar{\sigma}_{oj2})+\int_\Theta \gamma_{i,N_o}(t,\theta)(\mu_{oj2}+\bar{\mu}_{oj2})\nu_o(d\theta)\}\tfrac{\bar{\tau}_{j0}}{\bar{c}_j}\\
&-\gamma_i\sum_{j\neq i}(b_{j2}+\bar{b}_{j2})\tfrac{\bar{\tau}_{j0}}{\bar{c}_j}\ \Big\}dt\\
&-\Big(\xi_{i0,B_o}+\gamma_i[\sigma_{o0}-\sum_j(\sigma_{oj2}+\bar{\sigma}_{oj2})\tfrac{\bar{\tau}_{j0}}{\bar{c}_j}]\Big)dB_o\\
&-\int_\Theta\Big(\xi_{i0,N_o}+\gamma_i[\mu_{o0}-\sum_j(\mu_{oj2}+\bar{\mu}_{oj2})\tfrac{\bar{\tau}_{j0}}{\bar{c}_j}]\Big)\tilde{N}_o(dt,d\theta),\\
\delta_i&(T,s)=0,\\
\delta_{i,B_o}&=-\Big(\xi_{i0,B_o}+\gamma_i[\sigma_{o0}-\sum_j(\sigma_{oj2}+\bar{\sigma}_{oj2})\tfrac{\bar{\tau}_{j0}}{\bar{c}_j}]\Big),\\
\delta_{i,N_o}&=-\Big(\xi_{i0,N_o}+\gamma_i[\mu_{o0}-\sum_j(\mu_{oj2}+\bar{\mu}_{oj2})\tfrac{\bar{\tau}_{j0}}{\bar{c}_j}]\Big),
\end{aligned}
\tag{7.307}
$$

whenever the system of stochastic differential equations has a solution that is $\mathcal{F}^{B_o,N_o}$-adapted.

Relationship Between the coefficients

$$
\begin{aligned}
&A\tilde{\tau}=v,\\
&\bar{A}\bar{\tau}_k=\bar{v}_k,\ k\in\{0,1\}\\
&A_{ii}=\bar{A}_{ii}=1,\\
&A_{ij}=\tfrac{\alpha_i}{c_j}[(\sigma_{i2}\sigma_{j2}+\sigma_{oi2}\sigma_{oj2})+\int_\Theta \mu_{i2}\mu_{j2}\nu(d\theta)+\mu_{oi2}\mu_{oj2}\nu_o(d\theta)],\ j\neq i,\\
&v_i=\epsilon_{i1}+\alpha_i\{b_{i2}+\sigma_1\sigma_{i2}+\int_\Theta \mu_1\mu_{i2}\nu(d\theta)\}\\
&+\alpha_i\ \{\sigma_{o1}\sigma_{oi2}+\int_\Theta \mu_{o1}\mu_{oi2}\}\nu_o(d\theta)+\{\alpha_{i,B_o}\sigma_{oi2}+\int_\Theta \alpha_{i,N_o}\mu_{oi2}\nu_o(d\theta)\}\\
&\bar{A}_{ij}=\tfrac{\alpha_i}{\bar{c}_j}\{(\sigma_{i2}+\bar{\sigma}_{i2})(\sigma_{j2}+\bar{\sigma}_{j2})+\int_\Theta(\mu_{i2}+\bar{\mu}_{i2})(\mu_{j2}+\bar{\mu}_{j2})\nu(d\theta)\}\\
&+\tfrac{\beta_i}{\bar{c}_j}\{(\sigma_{oi2}+\bar{\sigma}_{oi2})(\sigma_{oj2}+\bar{\sigma}_{oj2})+\int_\Theta(\mu_{oi2}+\bar{\mu}_{oi2})(\mu_{oj2}+\bar{\mu}_{oj2})\nu_o(d\theta)\},\\
&\bar{v}_{i1}=(\epsilon_{i1}+\bar{\epsilon}_{i1})+\alpha_i[(\sigma_1+\bar{\sigma}_1)(\sigma_{i2}+\bar{\sigma}_{i2})+\int_\Theta(\mu_1+\bar{\mu}_1)(\mu_{i2}+\bar{\mu}_{i2})\nu(d\theta)]\\
&+\beta_i[(b_{i2}+\bar{b}_{i2})+(\sigma_{o1}+\bar{\sigma}_{o1})(\sigma_{oi2}+\bar{\sigma}_{oi2})+\int_\Theta(\mu_{o1}+\bar{\mu}_{o1})(\mu_{oi2}+\bar{\mu}_{oi2})\nu_o(d\theta)]\\
&+\beta_{i,B_o}(\sigma_{oi2}+\bar{\sigma}_{oi2})+\int_\Theta \beta_{i,N_o}(\mu_{oi2}+\bar{\mu}_{oi2})\\
&\bar{v}_{i0}=+(\epsilon_{i2}+\bar{\epsilon}_{i2})+\alpha_i[\sigma_0(\sigma_{i2}+\bar{\sigma}_{i2})+\int_\Theta \mu_0(\mu_{i2}+\bar{\mu}_{i2})\nu(d\theta)]\\
&+\beta_i[\sigma_{o0}(\sigma_{oi2}+\bar{\sigma}_{oi2})+\int_\Theta \mu_{o0}(\mu_{oi2}+\bar{\mu}_{oi2})\nu_o(d\theta)]\\
&+\gamma_i(b_{i2}+\bar{b}_{i2})+\gamma_{i,B_o}(\sigma_{oi2}+\bar{\sigma}_{oi2})+\int_\Theta \gamma_{i,N_o}(t,\theta)(\mu_{oi2}+\bar{\mu}_{oi2})\nu_o(d\theta)\\
&c_i=r_i+\alpha_i\left(\sigma_{i2}^2+\int_\Theta \mu_{i2}^2\nu(d\theta)+\sigma_{oi2}^2+\int_\Theta \mu_{oi2}^2\nu_o(d\theta)\ \right),\\
&\bar{c}_i=(r_i+\bar{r}_i)+\alpha_i[(\sigma_{i2}+\bar{\sigma}_{i2})^2+\int_\Theta(\mu_{i2}+\bar{\mu}_{i2})^2\nu(d\theta)]\\
&+\beta_i[(\sigma_{oi2}+\bar{\sigma}_{oi2})^2+\int_\Theta(\mu_{oi2}+\bar{\mu}_{oi2})^2\nu_o(d\theta)],
\end{aligned}
\tag{7.308}
$$

whenever the matrices A and $\bar{A}$ are invertible.

Notice the following:

- The matrices A and $\bar{A}$ become the I-dimensional identity matrix for problems in which the control strategy and the expected value of the control strategy do not appear in the diffusion and jump coefficient functions. Thus, the invertibility conditions are systematically satisfied for such systems.
- The structure of the differential equations changes from linear to quadratic and fraction in α. As it is expected, for game problems, the coefficients $\alpha_i(t,s)$ are interdependent not only across the switching regime but also coupled between the decision-makers $\{\alpha_j(t,s)\}_{j\neq i}$. Furthermore, these coefficients are regularized by the common noises B_o, N_o.
- The optimal strategies are interrelated with each other via the coefficients $\tau, \alpha, \beta, \gamma$. The vector optimal strategy is given by

$$u^* = -\tilde{A}^{-1}\tilde{v}(x-\hat{x}) - \tilde{\bar{A}}^{-1}\tilde{\bar{v}}_1\hat{x} - \tilde{\bar{A}}^{-1}\tilde{\bar{v}}_0,$$

where the matrices and vectors are given by

$$\tilde{A}_{ij} = \frac{A_{ij}c_j}{c_i},\ \tilde{A}_{ii} = 1,\ \tilde{v}_i = \frac{v_i}{c_i},\ \tilde{\bar{A}}_{ij} = \frac{\bar{A}_{ij}\bar{c}_j}{\bar{c}_i},\ \tilde{\bar{A}}_{ii} = 1,\ \tilde{\bar{v}}_{ik} = \frac{\bar{v}_{ik}}{\bar{c}_i}.$$

7.7.2 *Construction of a Semi-explicit Equilibrium Strategy*

We develop a proof of optimality using a direct method decomposed into several steps.

7.7.2.1 Reformulation of the Game Problem

First, consider the following problem:

Equivalent reformulation of Problem (7.299)

$$\inf_{u_i\in\mathcal{U}_i} E\left[\hat{L}_i(x,s,u) \mid x(0)=x_0, s(0)=s_0\right], \tag{7.309}$$

where

(continued)

$$\hat{L}_i(x,s,u) = \frac{1}{2}q_{iT}(x(T)-\hat{x}(T))^2 + \frac{1}{2}(q_{iT}+\bar{q}_{iT})\hat{x}^2(T) + (\epsilon_{i3}(T)+\bar{\epsilon}_{i3}(T))\hat{x}(T)$$
$$+\frac{1}{2}\int_0^T \left(q_i(x-\hat{x})^2 + (q_i+\bar{q}_i)\hat{x}^2 + r_i(u_i-\hat{u}_i)^2 + (r_i+\bar{r}_i)\hat{u}_i^2\right)dt$$
$$+\int_0^T \left(\epsilon_{i1}(x-\hat{x})(u_i-\hat{u}_i) + (\epsilon_{i1}+\bar{\epsilon}_{i1})\hat{x}\hat{u}_i + (\epsilon_{i2}+\bar{\epsilon}_{i2})\hat{u}_i + (\epsilon_{i3}+\bar{\epsilon}_{i3})\hat{x}\right)dt$$
$$+\int_0^T \left(\xi_{i0,B_o} + \xi_{i1,B_o}\hat{x} + \frac{1}{2}\xi_{i2,B_o}(x-\hat{x})^2 + \frac{1}{2}(\xi_{i2,B_o}+\bar{\xi}_{i2,B_o})\hat{x}^2\right)dB_o$$
$$+\int_0^T\int_\Theta \left(\xi_{i0,N_o} + \xi_{i1,N_o}\hat{x} + \frac{1}{2}\xi_{i2,N_o}(x-\hat{x})^2 + \frac{1}{2}(\xi_{i2,N_o}+\bar{\xi}_{i2,N_o})\hat{x}^2\right)\tilde{N}_o(dt,d\theta), \quad (7.310)$$

subject to

$$dx = \left(b_0 + b_1(x-\hat{x}) + (b_1+\bar{b}_1)\hat{x} + \sum_{j=1}^I b_{j2}(u_j-\hat{u}_j) + \sum_{j=1}^I (b_{j2}+\bar{b}_{j2})\hat{u}_j\right)dt$$
$$+\left(\sigma_0 + \sigma_1(x-\hat{x}) + (\sigma_1+\bar{\sigma}_1)\hat{x} + \sum_{j=1}^I \sigma_{j2}(u_j-\hat{u}_j) + \sum_{j=1}^I(\sigma_{j2}+\bar{\sigma}_{j2})\hat{u}_j\right)dB$$
$$+\int_\Theta \{\mu_0 + \mu_1(x-\hat{x}) + (\mu_1+\bar{\mu}_1)\hat{x}$$
$$+\sum_{j=1}^I \mu_{j2}(u_j-\hat{u}_j) + (\mu_{j2}+\bar{\mu}_{j2})\hat{u}_j\}\,\tilde{N}(dt,d\theta)$$
$$+\{\sigma_{o0} + \sigma_{o1}(x-\hat{x}) + (\sigma_{o1}+\bar{\sigma}_{o1})\hat{x}$$
$$+\sum_{j=1}^I \sigma_{oj2}(u_j-\hat{u}_j) + \sum_{j=1}^I(\sigma_{oj2}+\bar{\sigma}_{oj2})\hat{u}_j\}\,dB_o$$
$$+\int_\Theta \{\mu_{o0} + \mu_{o1}(x-\hat{x}) + (\mu_{o1}+\bar{\mu}_{o1})\hat{x}$$
$$+\sum_{j=1}^I \mu_{oj2}(u_j-\hat{u}_j) + (\mu_{oj2}+\bar{\mu}_{oj2})\hat{u}_j\}\,\tilde{N}_o(dt,d\theta)$$
$$x(0) = x_0,\ s(0) = s_0. \quad (7.311)$$

Problem (7.299) is exactly Problem (7.309). To prove it, we use the decompositions $x = x - \hat{x} + \hat{x}$ and $u_j = u_j - \hat{u}_j + \hat{u}_j$ noticing that both $E[\hat{x}(x - \hat{x}) \mid \mathscr{F}^{B_o,N_o,s}] = 0$ and $E[\hat{u}_j(u_j - \hat{u}_j) \mid \mathscr{F}^{B_o,N_o,s}] = 0$, i.e., $x - \hat{x} \perp \hat{x}$ and $u_j - \hat{u}_j \perp \hat{u}_j$ conditioned on the filtration $\mathscr{F}^{B_o,N_o,s}$.

7.7.2.2 Semi-explicit Representation

We construct a semi-explicit solution to Problems (7.299) and (7.309) via a direct method. Let $f_i(t,x,s) = \frac{1}{2}\alpha_i(x-\hat{x})^2 + \frac{1}{2}\beta_i\hat{x}^2 + \gamma_i\hat{x} + \delta_i$, where $\alpha_i = \alpha_i(t,s)$, $\beta_i = \beta_i(t,s)$, $\gamma_i = \gamma_i(t,s)$, $\delta_i = \delta_i(t,s)$, are stochastic processes adapted to the filtration $\mathscr{F}^{B_o,N_o,s}$, and $x = x^{u,x_0,s_0}$ is a solution to (7.311) starting from x_0, s_0 at time 0 under the strategy profile u. We compute the stochastic integro-differential equation satisfied by $\hat{x}$. It is given by

$$d\hat{x} = \left(b_0 + (b_1+\bar{b}_1)\hat{x} + \sum_{j=1}^I(b_{j2}+\bar{b}_{j2})\hat{u}_j\right)dt$$
$$+\left(\sigma_{o0} + (\sigma_{o1}+\bar{\sigma}_{o1})\hat{x} + \sum_{j=1}^I(\sigma_{oj2}+\bar{\sigma}_{oj2})\hat{u}_j\right)dB_o \quad (7.312)$$
$$+\int_\Theta\left(\mu_{o0} + (\mu_{o1}+\bar{\mu}_{o1})\hat{x} + \sum_{j=1}^I(\mu_{oj2}+\bar{\mu}_{oj2})\hat{u}_j\right)\tilde{N}_o(dt,d\theta),$$
$$\hat{x}(0) = \hat{x}_0,\ s(0) = s_0.$$

Using Eq. (7.312), it follows that the process $x - \hat{x}$ solves

$$
\begin{aligned}
& d(x-\hat{x}) = \left(b_1(x-\hat{x}) + \sum_{j=1}^{I} b_{j2}(u_j - \hat{u}_j)\right) dt \\
& + \left(\sigma_0 + \sigma_1(x-\hat{x}) + (\sigma_1+\bar{\sigma}_1)\hat{x} + \sum_{j=1}^{I} \sigma_{j2}(u_j - \hat{u}_j) + \sum_{j=1}^{I}(\sigma_{j2}+\bar{\sigma}_{j2})\hat{u}_j\right) dB \\
& + \int_\Theta \left(\mu_0 + \mu_1(x-\hat{x}) + (\mu_1+\bar{\mu}_1)\hat{x} + \sum_{j=1}^{I} \mu_{j2}(u_j - \hat{u}_j)\right) \tilde{N}(dt, d\theta) \\
& + \int_\Theta \sum_{j=1}^{I} (\mu_{j2} + \bar{\mu}_{j2})\hat{u}_j \tilde{N}(dt, d\theta) \\
& + \left(\sigma_{o1}(x-\hat{x}) + \sum_{j=1}^{I} \sigma_{oj2}(u_j - \hat{u}_j)\right) dB_o \\
& + \int_\Theta \left(\mu_{o1}(x-\hat{x}) + \sum_{j=1}^{I} \mu_{oj2}(u_j - \hat{u}_j)\right) \tilde{N}_o(dt, d\theta), \\
& x(0) = x_0, \; s(0) = s_0.
\end{aligned}
\tag{7.313}
$$

7.7.2.3 Guess Functional Identification

We compute difference between guess and actual cost functional $\hat{L}_i(x, s, u) - f_i(0, x_0, s_0) \mid \mathscr{F}^{B_o, N_o, s}$, which is

$$
\begin{aligned}
& \hat{L}_i(x, s, u) - f_i(0, x_0, s_0) \\
& = \tfrac{1}{2}(q_{iT} - \alpha_i(T))(x_T - \hat{x}_T)^2|s] + \tfrac{1}{2}(q_{iT} + \bar{q}_{iT} - \beta_i(T))\hat{x}_T^2 + (\epsilon_{i3}(T) + \bar{\epsilon}_{i3}(T) - \gamma_{iT})\hat{x}(T) \\
& + \int_0^T \Bigg[\Big\{d\alpha_i + 2b_1\alpha_i dt + \alpha_i[\sigma_1^2 + \sigma_{o1}^2 + \int_\Theta \mu_1^2 \nu(d\theta) + \int_\Theta \mu_{o1}^2 \nu_o(d\theta)]dt \\
& + q_i dt + \sum_{s' \neq s} \tilde{q}_{ss'}[\alpha_i(t, s') - \alpha_i(t, s)]dt + 2\alpha_{i,B_o}\sigma_{o1}dt + 2\int_\Theta \alpha_{i,N_o}\mu_{o1}\nu_o(d\theta)dt \\
& + \{\xi_{i2,B_o} + 2\sigma_{o1}\alpha_i\}dB_o + \int_\Theta \{\xi_{i2,N_o} + 2\mu_{o1}\alpha_i\}\tilde{N}_o(dt, d\theta)\Big\} \tfrac{1}{2}(x-\hat{x})^2 \\
& + \Big\{d\beta_i + 2(b_1 + \bar{b}_1)\beta_i dt + \beta_i[(\sigma_{o1} + \bar{\sigma}_{o1})^2 dt + \int_\Theta (\mu_{o1} + \bar{\mu}_{o1})^2 \nu_o(d\theta)]dt \\
& + (q_i + \bar{q}_i)dt + \sum_{s' \neq s} \tilde{q}_{ss'}[\beta_i(t, s') - \beta_i(t, s)]dt \\
& + \alpha_i[(\sigma_1 + \bar{\sigma}_1)^2 + \int_\Theta (\mu_1 + \bar{\mu}_1)^2 \nu(d\theta)]dt \\
& + 2\beta_{i,B_o}(\sigma_{o1} + \bar{\sigma}_{o1})dt + 2\int_\Theta \beta_{i,N_o}(\mu_{o1} + \bar{\mu}_{o1})\nu_o(d\theta)dt \\
& \{\xi_{i2,B_o} + \bar{\xi}_{i2,B_o} + 2(\sigma_{o1} + \bar{\sigma}_{o1})\beta_i\}dB_o \\
& + \int_\Theta \{\xi_{i2,N_o} + \bar{\xi}_{i2,N_o} + 2(\mu_{o1} + \bar{\mu}_{o1})\beta_i\}\tilde{N}_o(dt, d\theta)\Big\} \tfrac{1}{2}\hat{x}^2 dt \\
& + \Big\{d\gamma_i + (b_1 + \bar{b}_1)\gamma_i dt + (\epsilon_{i3} + \bar{\epsilon}_{i3})dt + \sum_{s' \neq s} \tilde{q}_{ss'}[\gamma_i(t, s') - \gamma_i(t, s)]dt \\
& + \alpha_i[\sigma_0(\sigma_1 + \bar{\sigma}_1) + \int_\Theta \mu_0(\mu_1 + \bar{\mu}_1)\nu(d\theta)]dt \\
& + \beta_i[b_0 + \sigma_{o0}(\sigma_{o1} + \bar{\sigma}_{o1}) + \int_\Theta \mu_{o0}(\mu_{o1} + \bar{\mu}_{o1})\nu_o(d\theta)]dt + \beta_{i,B_o}\sigma_{o0}dt \\
& + \int_\Theta \beta_{i,N_o}\mu_{o0}\nu_o(d\theta)dt \\
& + \gamma_{i,B_o}(\sigma_{o1} + \bar{\sigma}_{o1})dt + \int_\Theta \gamma_{i,N_o}(\mu_{o1} + \bar{\mu}_{o1})\nu_o(d\theta)dt \\
& + \{\xi_{i1,B_o} + \sigma_{o0}\beta_i + (\sigma_{o1} + \bar{\sigma}_{o1})\gamma_i\}dB_o \\
& + \int_\Theta \{\xi_{i1,N_o} + \mu_{o0}\beta_i + (\mu_{o1} + \bar{\mu}_{o1})\gamma_i\}\tilde{N}_o(dt, d\theta)\Big\} \hat{x} \Bigg] + Z_0,
\end{aligned}
\tag{7.314}
$$

$$
\begin{aligned}
Z_0 = \Bigg[& d\delta_i + \tfrac{\alpha_i}{2}(\sigma_0^2 + \textstyle\int_\Theta \mu_0^2 \nu(d\theta))dt \\
& + \tfrac{\beta_i}{2}\{\sigma_{o0}^2 + \textstyle\int_\Theta \mu_{o0}^2 \nu_o(d\theta)\}dt \\
& + b_0\gamma_i dt + \gamma_{i,B_o}\sigma_{o0}dt + \textstyle\int_\Theta \gamma_{i,N_o}(t,\theta)\mu_{o0}\nu_o(d\theta)dt \\
& + \textstyle\sum_{s'\neq s} \tilde{q}_{ss'}[\delta_i(t,s') - \delta_i(t,s)]dt \\
& + \{\xi_{i0,B_o} + \sigma_{o0}\gamma_i\}dB_o + \textstyle\int_\Theta \{\xi_{i0,N_o} + \mu_{o0}\gamma_i\}\tilde{N}_o(dt,d\theta) \\
& + \tfrac{1}{2}r_i(u_i-\hat{u}_i)^2 + \tfrac{\alpha_i}{2}[\textstyle\sum_{j=1}^I \sigma_{j2}(u_j-\hat{u}_j)]^2 + \int_\Theta \tfrac{\alpha_i}{2}[\sum_{j=1}^I \mu_{j2}(u_j-\hat{u}_j)]^2\nu(d\theta)dt \\
& + \epsilon_{i1}(x-\hat{x})(u_i-\hat{u}_i) + \textstyle\sum_{j=1}^I b_{j2}\alpha_i(x-\hat{x})(u_j-\hat{u}_j)dt \\
& + \alpha_i \textstyle\sum_{j=1}^I \{\sigma_1\sigma_{j2} + \int_\Theta \mu_1\mu_{j2}\nu(d\theta)\}(x-\hat{x})(u_j-\hat{u}_j)\, dt \\
& + \tfrac{\alpha_i}{2}[\textstyle\sum_{j=1}^I \sigma_{oj2}(u_j-\hat{u}_j)]^2 + \tfrac{\alpha_i}{2}\int_\Theta [\sum_{j=1}^I \mu_{oj2}(u_j-\hat{u}_j)]^2\nu_o(d\theta)\, dt \\
& + \alpha_i \textstyle\sum_{j=1}^I \{\sigma_{o1}\sigma_{oj2} + \int_\Theta \mu_{o1}\mu_{oj2}\}(x-\hat{x})(u_j-\hat{u}_j)\nu_o(d\theta)\, dt \\
& + \textstyle\sum_{j=1}^I \{\alpha_{i,B_o}\sigma_{oj2} + \int_\Theta \alpha_{i,N_o}\mu_{oj2}\nu_o(d\theta)\}(x-\hat{x})(u_j-\hat{u}_j)dt \\
& + \textstyle\sum_{j=1}^I \sigma_{oj2}\alpha_i(x-\hat{x})(u_j-\hat{u}_j)dB_o + \int_\Theta \sum_{j=1}^I \mu_{oj2}\alpha_i(x-\hat{x})(u_j-\hat{u}_j)\tilde{N}_o(dt,d\theta) \\
& + \tfrac{1}{2}(r_i+\bar{r}_i)\hat{u}_i^2 + \tfrac{\alpha_i}{2}[\textstyle\sum_{j=1}^I (\sigma_{j2}+\bar{\sigma}_{j2})\hat{u}_j]^2 + \int_\Theta \tfrac{\alpha_i}{2}[\sum_{j=1}^I (\mu_{j2}+\bar{\mu}_{j2})\hat{u}_j]^2\nu(d\theta)dt \\
& + \tfrac{\beta_i}{2}[\textstyle\sum_{j=1}^I (\sigma_{oj2}+\bar{\sigma}_{oj2})\hat{u}_j]^2 dt + \tfrac{\beta_i}{2}\int_\Theta [\sum_{j=1}^I (\mu_{oj2}+\bar{\mu}_{oj2})\hat{u}_j]^2\nu_o(d\theta)dt \\
& + (\epsilon_{i1}+\bar{\epsilon}_{i1})\hat{x}\hat{u}_i dt \\
& + \alpha_i \textstyle\sum_{j=1}^I \{(\sigma_1+\bar{\sigma}_1)(\sigma_{j2}+\bar{\sigma}_{j2}) + \int_\Theta (\mu_1+\bar{\mu}_1)(\mu_{j2}+\bar{\mu}_{j2})\nu(d\theta)\}\hat{x}\hat{u}_j dt \\
& + \beta_i \textstyle\sum_{j=1}^I \{(b_{j2}+\bar{b}_{j2}) + (\sigma_{o1}+\bar{\sigma}_{o1})(\sigma_{oj2}+\bar{\sigma}_{oj2}) \\
& + \textstyle\int_\Theta (\mu_{o1}+\bar{\mu}_{o1})(\mu_{oj2}+\bar{\mu}_{oj2})\nu_o(d\theta)\}\hat{u}_j\hat{x}dt \\
& + \textstyle\sum_{j=1}^I \{\beta_{i,B_o}(\sigma_{oj2}+\bar{\sigma}_{oj2}) + \int_\Theta \beta_{i,N_o}(\mu_{oj2}+\bar{\mu}_{oj2})\}\hat{x}\hat{u}_j dt \\
& + \textstyle\sum_{j=1}^I (\sigma_{oj2}+\bar{\sigma}_{oj2})\beta_i\hat{x}\hat{u}_j dB_o + \int_\Theta \sum_{j=1}^I (\mu_{oj2}+\bar{\mu}_{oj2})\beta_i\hat{x}\hat{u}_j\tilde{N}_o(dt,d\theta) \\
& + (\epsilon_{i2}+\bar{\epsilon}_{i2})\hat{u}_i + \alpha_i \textstyle\sum_{j=1}^I \sigma_0(\sigma_{j2}+\bar{\sigma}_{j2})\hat{u}_j + \alpha_i \sum_{j=1}^I \int_\Theta \mu_0(\mu_{j2}+\bar{\mu}_{j2})\hat{u}_j\nu(d\theta)dt \\
& + \beta_i \textstyle\sum_{j=1}^I \{\sigma_{o0}(\sigma_{oj2}+\bar{\sigma}_{oj2}) + \int_\Theta \mu_{o0}(\mu_{oj2}+\bar{\mu}_{oj2})\nu_o(d\theta)\}\hat{u}_j \\
& + \gamma_i \textstyle\sum_{j=1}^I (b_{j2}+\bar{b}_{j2})\hat{u}_j dt \\
& + \textstyle\sum_{j=1}^I \{\gamma_{i,B_o}(\sigma_{oj2}+\bar{\sigma}_{oj2}) + \int_\Theta \gamma_{i,N_o}(t,\theta)(\mu_{oj2}+\bar{\mu}_{oj2})\nu_o(d\theta)\}\hat{u}_j dt \\
& + \textstyle\sum_{j=1}^I (\sigma_{oj2}+\bar{\sigma}_{oj2})\gamma_i\hat{u}_j dB_o + \sum_{j=1}^I \int_\Theta (\mu_{oj2}+\bar{\mu}_{oj2})\gamma_i\hat{u}_j\tilde{N}_o(dt,d\theta) + \tilde{\tilde{\epsilon}} \Bigg].
\end{aligned}
\tag{7.315}
$$

7.7.2.4 Completion of Squares for i-th Decision-Maker

We use the following equalities. We start with terms in $(u_i - \hat{u}_i)$:

$$
\begin{aligned}
&\tfrac{1}{2}c_i(u_i-\hat{u}_i+\tfrac{\tilde{\tau}_i}{c_i}(x-\hat{x}))^2\\
&+\Big[-\tfrac{\tilde{\tau}_i^2}{c_i}+\alpha_i[\textstyle\sum_{j\neq i}\sigma_{j2}\tfrac{\tilde{\tau}_j}{c_j}]^2\\
&+\textstyle\int_\Theta \alpha_i[\sum_{j\neq i}\mu_{j2}\tfrac{\tilde{\tau}_j}{c_j}]^2\nu(d\theta)\\
&-2\alpha_i\textstyle\sum_{j\neq i}\{b_{j2}+\sigma_1\sigma_{j2}+\int_\Theta \mu_1\mu_{j2}\nu(d\theta)\}\tfrac{\tilde{\tau}_j}{c_j}\\
&+\alpha_i[\textstyle\sum_{j\neq i}\sigma_{oj2}\tfrac{\tilde{\tau}_j}{c_j}]^2\\
&+\alpha_i\textstyle\int_\Theta[\sum_{j\neq i}\mu_{oj2}\tfrac{\tilde{\tau}_j}{c_j}]^2\nu_o(d\theta)\\
&-2\alpha_i\ \textstyle\sum_{j\neq i}\{\sigma_{o1}\sigma_{oj2}+\int_\Theta \mu_{o1}\mu_{oj2}\}\tfrac{\tilde{\tau}_j}{c_j}\nu_o(d\theta)\\
&-2\textstyle\sum_{j\neq i}\{\alpha_{i,B_o}\sigma_{oj2}+\int_\Theta \alpha_{i,N_o}\mu_{oj2}\nu_o(d\theta)\}\tfrac{\tilde{\tau}_j}{c_j}\\
&-2\textstyle\sum_j\sigma_{oj2}\alpha_i\tfrac{\tilde{\tau}_j}{c_j}dB_o-2\sum_j\int_\Theta \mu_{oj2}\alpha_i\tfrac{\tilde{\tau}_j}{c_j}\tilde{N}_o(dt,d\theta)\Big]\tfrac{1}{2}(x-\hat{x})^2
\end{aligned}
\tag{7.316}
$$

where

$$
\begin{aligned}
&c_i=\{r_i+\alpha_i\left(\sigma_{i2}^2+\textstyle\int_\Theta \mu_{i2}^2\nu(d\theta)+\sigma_{oi2}^2+\int_\Theta \mu_{oi2}^2\nu_o(d\theta)\ \right)\},\\
&A\tilde{\tau}=v,\\
&A_{ii}=1,\ A_{ij}=\tfrac{\alpha_i}{c_j}[(\sigma_{i2}\sigma_{j2}+\sigma_{oi2}\sigma_{oj2})+\textstyle\int_\Theta(\mu_{i2}\mu_{j2}+\mu_{oi2}\mu_{oj2})\nu(d\theta)],\ j\neq i,\\
&v_i=+\epsilon_{i1}+\alpha_i\{b_{i2}+\sigma_1\sigma_{i2}+\textstyle\int_\Theta \mu_1\mu_{i2}\nu(d\theta)\}\\
&+\alpha_i\ \{\sigma_{o1}\sigma_{oi2}+\textstyle\int_\Theta \mu_{o1}\mu_{oi2}\}\nu_o(d\theta)+\{\alpha_{i,B_o}\sigma_{oi2}+\int_\Theta \alpha_{i,N_o}\mu_{oi2}\nu_o(d\theta)\},
\end{aligned}
\tag{7.317}
$$

whenever the matrix A is invertible.

Next we examine the completion of squares for the terms in $\hat{u}_i$. By expanding the square terms and reorganizing, we arrive at

$$
\begin{aligned}
&\tfrac{\bar{c}_i}{2}\left(\hat{u}_i+\tfrac{\bar{\tau}_{i1}\hat{x}+\bar{\tau}_{i0}}{\bar{c}_i}\right)^2-\tfrac{\bar{\tau}_{i1}^2}{2\bar{c}_i}\hat{x}^2-\tfrac{2\bar{\tau}_{i1}\bar{\tau}_{i0}}{2\bar{c}_i}\hat{x}-\tfrac{\bar{\tau}_{i0}^2}{2\bar{c}_i}\\
&+\tfrac{\alpha_i}{2}[\textstyle\sum_{j\neq i}(\sigma_{j2}+\bar{\sigma}_{j2})\hat{u}_j]^2+\int_\Theta \tfrac{\alpha_i}{2}[\sum_{j\neq i}(\mu_{j2}+\bar{\mu}_{j2})\hat{u}_j]^2\nu(d\theta)\\
&+\tfrac{\beta_i}{2}[\textstyle\sum_{j\neq i}(\sigma_{oj2}+\bar{\sigma}_{oj2})\hat{u}_j]^2\\
&+\tfrac{\beta_i}{2}\textstyle\int_\Theta[\sum_{j\neq i}(\mu_{oj2}+\bar{\mu}_{oj2})\hat{u}_j]^2\nu_o(d\theta)\\
&+\alpha_i\textstyle\sum_{j\neq i}\{(\sigma_1+\bar{\sigma}_1)(\sigma_{j2}+\bar{\sigma}_{j2})+\int_\Theta(\mu_1+\bar{\mu}_1)(\mu_{j2}+\bar{\mu}_{j2})\nu(d\theta)\}\hat{x}\hat{u}_j\\
&+\beta_i\textstyle\sum_{j\neq i}\{(b_{j2}+\bar{b}_{j2})+(\sigma_{o1}+\bar{\sigma}_{o1})(\sigma_{oj2}+\bar{\sigma}_{oj2})\\
&+\textstyle\int_\Theta(\mu_{o1}+\bar{\mu}_{o1})(\mu_{oj2}+\bar{\mu}_{oj2})\nu_o(d\theta)\}\hat{x}\hat{u}_j\\
&+\textstyle\sum_{j\neq i}\{\beta_{i,B_o}(\sigma_{oj2}+\bar{\sigma}_{oj2})+\int_\Theta\ \beta_{i,N_o}(\mu_{oj2}+\bar{\mu}_{oj2})\}\hat{x}\hat{u}_j\\
&+\textstyle\sum_{j\neq i}(\sigma_{oj2}+\bar{\sigma}_{oj2})\beta_i\hat{x}\hat{u}_jdB_o\\
&+\textstyle\int_\Theta\sum_{j\neq i}(\mu_{oj2}+\bar{\mu}_{oj2})\beta_i\hat{x}\hat{u}_j\ \tilde{N}_o(dt,d\theta)\\
&+\alpha_i\textstyle\sum_{j\neq i}\{\sigma_0(\sigma_{j2}+\bar{\sigma}_{j2})+\int_\Theta \mu_0(\mu_{j2}+\bar{\mu}_{j2})\nu(d\theta)\}\hat{u}_j\\
&+\beta_i\textstyle\sum_{j\neq i}\{\sigma_{o0}(\sigma_{oj2}+\bar{\sigma}_{oj2})+\int_\Theta \mu_{o0}(\mu_{oj2}+\bar{\mu}_{oj2})\nu_o(d\theta)\}\hat{u}_j\\
&+\textstyle\sum_{j\neq i}\{\gamma_{i,B_o}(\sigma_{oj2}+\bar{\sigma}_{oj2})+\int_\Theta \gamma_{i,N_o}(t,\theta)(\mu_{oj2}+\bar{\mu}_{oj2})\nu_o(d\theta)\}\hat{u}_jdt\\
&+\gamma_i\textstyle\sum_{j\neq i}(b_{j2}+\bar{b}_{j2})\hat{u}_j\\
&+\gamma_i\textstyle\sum_{j\neq i}(\sigma_{oj2}+\bar{\sigma}_{oj2})\hat{u}_jdB_o+\gamma_i\sum_{j\neq i}\int_\Theta(\mu_{oj2}+\bar{\mu}_{oj2})\hat{u}_j\tilde{N}_o(dt,d\theta).
\end{aligned}
\tag{7.318}
$$

It follows that the completion of squares in $\hat{u}_i$ yields

$$
\begin{aligned}
&\frac{\bar{c}_i}{2}\left(\hat{u}_i+\frac{\bar{\tau}_{i1}\hat{x}+\bar{\tau}_{i0}}{\bar{c}_i}\right)^2+\Big[-\frac{\bar{\tau}_{i1}^2}{\bar{c}_i}+\alpha_i[\textstyle\sum_{j\neq i}(\sigma_{j2}+\bar{\sigma}_{j2})\frac{\bar{\tau}_{j1}}{\bar{c}_j}]^2\\
&+\alpha_i\int_\Theta[\textstyle\sum_{j\neq i}(\mu_{j2}+\bar{\mu}_{j2})\frac{\bar{\tau}_{j1}}{\bar{c}_j}]^2\nu(d\theta)\\
&+\beta_i[\textstyle\sum_{j\neq i}(\sigma_{oj2}+\bar{\sigma}_{oj2})\frac{\bar{\tau}_{j1}}{\bar{c}_j}]^2+\beta_i\int_\Theta[\sum_{j\neq i}(\mu_{oj2}+\bar{\mu}_{oj2})\frac{\bar{\tau}_{j1}}{\bar{c}_j}]^2\nu_o(d\theta)\\
&-2\alpha_i\textstyle\sum_{j\neq i}\{(\sigma_1+\bar{\sigma}_1)(\sigma_{j2}+\bar{\sigma}_{j2})+\int_\Theta(\mu_1+\bar{\mu}_1)(\mu_{j2}+\bar{\mu}_{j2})\nu(d\theta)\}\frac{\bar{\tau}_{j1}}{\bar{c}_j}\\
&-2\beta_i\textstyle\sum_{j\neq i}\{(b_{j2}+\bar{b}_{j2})+(\sigma_{o1}+\bar{\sigma}_{o1})(\sigma_{oj2}+\bar{\sigma}_{oj2})\\
&+\int_\Theta(\mu_{o1}+\bar{\mu}_{o1})(\mu_{oj2}+\bar{\mu}_{oj2})\nu_o(d\theta)\}\frac{\bar{\tau}_{j1}}{\bar{c}_j}\\
&-2\textstyle\sum_{j\neq i}\{\beta_{i,B_o}(\sigma_{oj2}+\bar{\sigma}_{oj2})+\int_\Theta\beta_{i,N_o}(\mu_{oj2}+\bar{\mu}_{oj2})\}\frac{\bar{\tau}_{j1}}{\bar{c}_j}\\
&-2\textstyle\sum_j(\sigma_{oj2}+\bar{\sigma}_{oj2})\beta_i\frac{\bar{\tau}_{j1}}{\bar{c}_j}dB_o-2\int_\Theta\sum_j(\mu_{oj2}+\bar{\mu}_{oj2})\beta_i\frac{\bar{\tau}_{j1}}{\bar{c}_j}\tilde{N}_o(dt,d\theta)\Big]\frac{1}{2}\hat{x}^2\\
&+\Big[-\frac{2\bar{\tau}_{i1}\bar{\tau}_{i0}}{2\bar{c}_i}+\alpha_i[\textstyle\sum_{j\neq i}(\sigma_{j2}+\bar{\sigma}_{j2})\frac{\bar{\tau}_{j0}}{\bar{c}_j}][\sum_{k\neq i}(\sigma_{k2}+\bar{\sigma}_{k2})\frac{\bar{\tau}_{k1}}{\bar{c}_k}]\\
&+\alpha_i\int_\Theta[\textstyle\sum_{j\neq i}(\mu_{j2}+\bar{\mu}_{j2})\frac{\bar{\tau}_{j0}}{\bar{c}_j}][\sum_{k\neq i}(\mu_{k2}+\bar{\mu}_{k2})\frac{\bar{\tau}_{k1}}{\bar{c}_k}]\nu(d\theta)\\
&+\beta_i[\textstyle\sum_{j\neq i}(\sigma_{oj2}+\bar{\sigma}_{oj2})\frac{\bar{\tau}_{j0}}{\bar{c}_j}][\sum_{j\neq i}(\sigma_{oj2}+\bar{\sigma}_{oj2})\frac{\bar{\tau}_{j1}}{\bar{c}_j}]\\
&+\beta_i\int_\Theta[\textstyle\sum_{j\neq i}(\mu_{oj2}+\bar{\mu}_{oj2})\frac{\bar{\tau}_{j0}}{\bar{c}_j}][\sum_{j\neq i}(\mu_{oj2}+\bar{\mu}_{oj2})\frac{\bar{\tau}_{j1}}{\bar{c}_j}]\nu_o(d\theta)\\
&-\alpha_i\textstyle\sum_{j\neq i}\{(\sigma_1+\bar{\sigma}_1)(\sigma_{j2}+\bar{\sigma}_{j2})+\int_\Theta(\mu_1+\bar{\mu}_1)(\mu_{j2}+\bar{\mu}_{j2})\nu(d\theta)\}\frac{\bar{\tau}_{j0}}{\bar{c}_j}\\
&-\beta_i\textstyle\sum_{j\neq i}\{(b_{j2}+\bar{b}_{j2})+(\sigma_{o1}+\bar{\sigma}_{o1})(\sigma_{oj2}+\bar{\sigma}_{oj2})\\
&+\int_\Theta(\mu_{o1}+\bar{\mu}_{o1})(\mu_{oj2}+\bar{\mu}_{oj2})\nu_o(d\theta)\}\frac{\bar{\tau}_{j0}}{\bar{c}_j}\\
&-\textstyle\sum_{j\neq i}\{\beta_{i,B_o}(\sigma_{oj2}+\bar{\sigma}_{oj2})+\int_\Theta\beta_{i,N_o}(\mu_{oj2}+\bar{\mu}_{oj2})\}\frac{\bar{\tau}_{j0}}{\bar{c}_j}\\
&-\alpha_i\textstyle\sum_{j\neq i}\{\sigma_0(\sigma_{j2}+\bar{\sigma}_{j2})+\int_\Theta\mu_0(\mu_{j2}+\bar{\mu}_{j2})\nu(d\theta)\}\frac{\bar{\tau}_{j1}}{\bar{c}_j}\\
&-\beta_i\textstyle\sum_{j\neq i}\{\sigma_{o0}(\sigma_{oj2}+\bar{\sigma}_{oj2})+\int_\Theta\mu_{o0}(\mu_{oj2}+\bar{\mu}_{oj2})\nu_o(d\theta)\}\frac{\bar{\tau}_{j1}}{\bar{c}_j}\\
&-\textstyle\sum_{j\neq i}\{\gamma_{i,B_o}(\sigma_{oj2}+\bar{\sigma}_{oj2})+\int_\Theta\gamma_{i,N_o}(t,\theta)(\mu_{oj2}+\bar{\mu}_{oj2})\nu_o(d\theta)\}\frac{\bar{\tau}_{j1}}{\bar{c}_j}\\
&-\gamma_i\textstyle\sum_{j\neq i}(b_{j2}+\bar{b}_{j2})\frac{\bar{\tau}_{j1}}{\bar{c}_j}\\
&-\beta_i\textstyle\sum_j(\sigma_{oj2}+\bar{\sigma}_{oj2})\frac{\bar{\tau}_{j0}}{\bar{c}_j}dB_o-\gamma_i\sum_j(\sigma_{oj2}+\bar{\sigma}_{oj2})\frac{\bar{\tau}_{j1}}{\bar{c}_j}dB_o\\
&-\beta_i\int_\Theta\textstyle\sum_j(\mu_{oj2}+\bar{\mu}_{oj2})\frac{\bar{\tau}_{j0}}{\bar{c}_j}\tilde{N}_o(dt,d\theta)\\
&-\gamma_i\textstyle\sum_j\int_\Theta(\mu_{oj2}+\bar{\mu}_{oj2})\frac{\bar{\tau}_{j1}}{\bar{c}_j}\tilde{N}_o(dt,d\theta)\Big]\hat{x}\\
&-\frac{\bar{\tau}_{i0}^2}{2\bar{c}_i}+\frac{\alpha_i}{2}[\textstyle\sum_{j\neq i}(\sigma_{j2}+\bar{\sigma}_{j2})\frac{\bar{\tau}_{j0}}{\bar{c}_j}]^2+\int_\Theta\frac{\alpha_i}{2}[\sum_{j\neq i}(\mu_{j2}+\bar{\mu}_{j2})\frac{\bar{\tau}_{j0}}{\bar{c}_j}]^2\nu(d\theta)\\
&+\frac{\beta_i}{2}[\textstyle\sum_{j\neq i}(\sigma_{oj2}+\bar{\sigma}_{oj2})\frac{\bar{\tau}_{j0}}{\bar{c}_j}]^2\\
&+\frac{\beta_i}{2}\int_\Theta[\textstyle\sum_{j\neq i}(\mu_{oj2}+\bar{\mu}_{oj2})\frac{\bar{\tau}_{j0}}{\bar{c}_j}]^2\nu_o(d\theta)\\
&-\alpha_i\textstyle\sum_{j\neq i}\{\sigma_0(\sigma_{j2}+\bar{\sigma}_{j2})+\int_\Theta\mu_0(\mu_{j2}+\bar{\mu}_{j2})\nu(d\theta)\}\frac{\bar{\tau}_{j0}}{\bar{c}_j}\\
&-\beta_i\textstyle\sum_{j\neq i}\{\sigma_{o0}(\sigma_{oj2}+\bar{\sigma}_{oj2})+\int_\Theta\mu_{o0}(\mu_{oj2}+\bar{\mu}_{oj2})\nu_o(d\theta)\}\frac{\bar{\tau}_{j0}}{\bar{c}_j}\\
&-\textstyle\sum_{j\neq i}\{\gamma_{i,B_o}(\sigma_{oj2}+\bar{\sigma}_{oj2})+\int_\Theta\gamma_{i,N_o}(t,\theta)(\mu_{oj2}+\bar{\mu}_{oj2})\nu_o(d\theta)\}\frac{\bar{\tau}_{j0}}{\bar{c}_j}\\
&-\gamma_i\textstyle\sum_{j\neq i}(b_{j2}+\bar{b}_{j2})\frac{\bar{\tau}_{j0}}{\bar{c}_j}\\
&-\gamma_i\textstyle\sum_j(\sigma_{oj2}+\bar{\sigma}_{oj2})\frac{\bar{\tau}_{j0}}{\bar{c}_j}dB_o-\gamma_i\sum_j\int_\Theta(\mu_{oj2}+\bar{\mu}_{oj2})\frac{\bar{\tau}_{j0}}{\bar{c}_j}\tilde{N}_o(dt,d\theta),
\end{aligned}
\tag{7.319}
$$

where $\bar{\tau}_{i0}, \bar{\tau}_{i1}, \bar{c}_i$ are given by

$$
\begin{aligned}
&\bar{A}\bar{\tau}_k = \bar{v}_k, \quad k \in \{0, 1\},\\
&\bar{A}_{ii} = 1, \; \bar{A}_{ij} = \tfrac{\alpha_i}{\bar{c}_j}\{(\sigma_{i2} + \bar{\sigma}_{i2})(\sigma_{j2} + \bar{\sigma}_{j2}) + \textstyle\int_\Theta (\mu_{i2} + \bar{\mu}_{i2})(\mu_{j2} + \bar{\mu}_{j2})\nu(d\theta)\}\\
&+\tfrac{\beta_i}{\bar{c}_j}\{(\sigma_{oi2} + \bar{\sigma}_{oi2})(\sigma_{oj2} + \bar{\sigma}_{oj2}) + \textstyle\int_\Theta (\mu_{oi2} + \bar{\mu}_{oi2})(\mu_{oj2} + \bar{\mu}_{oj2})\nu_o(d\theta)\},\\
&\bar{c}_i = (r_i + \bar{r}_i) + \alpha_i[(\sigma_{i2} + \bar{\sigma}_{i2})^2 + \textstyle\int_\Theta (\mu_{i2} + \bar{\mu}_{i2})^2\nu(d\theta)]\\
&+\beta_i[(\sigma_{oi2} + \bar{\sigma}_{oi2})^2 + \textstyle\int_\Theta (\mu_{oi2} + \bar{\mu}_{oi2})^2\nu_o(d\theta)],\\
&\bar{v}_{i1} = (\epsilon_{i1} + \bar{\epsilon}_{i1}) + \alpha_i[(\sigma_1 + \bar{\sigma}_1)(\sigma_{i2} + \bar{\sigma}_{i2}) + \textstyle\int_\Theta (\mu_1 + \bar{\mu}_1)(\mu_{i2} + \bar{\mu}_{i2})\nu(d\theta)]\\
&+\beta_i[(b_{i2}+\bar{b}_{i2})+(\sigma_{o1}+\bar{\sigma}_{o1})(\sigma_{oi2}+\bar{\sigma}_{oi2}) + \textstyle\int_\Theta (\mu_{o1} + \bar{\mu}_{o1})(\mu_{oi2} + \bar{\mu}_{oi2})\nu_o(d\theta)]\\
&+\beta_{i,B_o}(\sigma_{oi2} + \bar{\sigma}_{oi2}) + \textstyle\int_\Theta \; \beta_{i,N_o}(\mu_{oi2} + \bar{\mu}_{oi2}),\\
&\bar{v}_{i0} = +(\epsilon_{i2} + \bar{\epsilon}_{i2}) + \alpha_i[\sigma_0(\sigma_{i2} + \bar{\sigma}_{i2}) + \textstyle\int_\Theta \mu_0(\mu_{i2} + \bar{\mu}_{i2})\nu(d\theta)]\\
&+\beta_i[\sigma_{o0}(\sigma_{oi2} + \bar{\sigma}_{oi2}) + \textstyle\int_\Theta \mu_{o0}(\mu_{oi2} + \bar{\mu}_{oi2})\nu_o(d\theta)]\\
&+\gamma_i(b_{i2} + \bar{b}_{i2}) + \gamma_{i,B_o}(\sigma_{oi2} + \bar{\sigma}_{oi2}) + \textstyle\int_\Theta \gamma_{i,N_o}(t,\theta)(\mu_{oi2} + \bar{\mu}_{oi2})\nu_o(d\theta)
\end{aligned}
\tag{7.320}
$$

7.7.2.5 Matching the Processes

We now replace the expressions of (7.316) and (7.319) by their corresponding values in (7.314) to obtain the following equality:

$$
\begin{aligned}
&\hat{L}_i(x,s,u) - f_i(0,x_0,s_0)\\
&= \tfrac{1}{2}(q_{iT} - \alpha_i(T))(x(T) - \hat{x}(T))^2] + \tfrac{1}{2}(q_{iT} + \bar{q}_{iT} - \beta_i(T))\hat{x}^2(T)\\
&+(\epsilon_{i3}(T) + \bar{\epsilon}_{i3}(T) - \gamma_{iT})\hat{x}(T)\\
&+\int_0^T \Bigg[\Big\{d\alpha_i + 2b_1\alpha_i dt + \alpha_i[\sigma_1^2 + \sigma_{o1}^2 + \int_\Theta \mu_1^2\nu(d\theta) + \int_\Theta \mu_{o1}^2\nu_o(d\theta)]dt\\
&+[q_i - \tfrac{\tilde{\tau}_i^2}{c_i}]dt + \sum_{s'\neq s}\tilde{q}_{ss'}[\alpha_i(t,s') - \alpha_i(t,s)]dt\\
&+2\alpha_{i,B_o}\sigma_{o1}dt + 2\int_\Theta \alpha_{i,N_o}\mu_{o1}\nu_o(d\theta)dt\\
&+\alpha_i[\sum_{j\neq i}\sigma_{j2}\tfrac{\tilde{\tau}_j}{c_j}]^2 + \int_\Theta \alpha_i[\sum_{j\neq i}\mu_{j2}\tfrac{\tilde{\tau}_j}{c_j}]^2\nu(d\theta) + \alpha_i[\sum_{j\neq i}\sigma_{oj2}\tfrac{\tilde{\tau}_j}{c_j}]^2\\
&+\alpha_i\int_\Theta[\sum_{j\neq i}\mu_{oj2}\tfrac{\tilde{\tau}_j}{c_j}]^2\nu_o(d\theta)\, dt\\
&-2\alpha_i\sum_{j\neq i}\{b_{j2} + \sigma_1\sigma_{j2} + \int_\Theta \mu_1\mu_{j2}\nu(d\theta)\}\tfrac{\tilde{\tau}_j}{c_j}\\
&-2\alpha_i\ \sum_{j\neq i}\{\sigma_{o1}\sigma_{oj2} + \int_\Theta \mu_{o1}\mu_{oj2}\nu_o(d\theta)\}\tfrac{\tilde{\tau}_j}{c_j}\ dt\\
&-2\sum_{j\neq i}\{\alpha_{i,B_o}\sigma_{oj2} + \int_\Theta \alpha_{i,N_o}\mu_{oj2}\nu_o(d\theta)\}\tfrac{\tilde{\tau}_j}{c_j}dt\\
&+\{\xi_{i2,B_o} + \alpha_i[2\sigma_{o1} - 2\sum_j\sigma_{oj2}\tfrac{\tilde{\tau}_j}{c_j}]\}dB_o\\
&+\int_\Theta\{\xi_{i2,N_o} + \alpha_i[2\mu_{o1} - 2\sum_j\mu_{oj2}\tfrac{\tilde{\tau}_j}{c_j}]\}\tilde{N}_o(dt,d\theta)\Big\}\,\tfrac{1}{2}(x-\hat{x})^2\\
&+\Big\{d\beta_i + 2(b_1+\bar{b}_1)\beta_i dt + \beta_i[(\sigma_{o1}+\bar{\sigma}_{o1})^2dt + \int_\Theta(\mu_{o1}+\bar{\mu}_{o1})^2\nu_o(d\theta)]dt\\
&+[(q_i+\bar{q}_i) - \tfrac{\bar{\tau}_{i1}^2}{\bar{c}_i}]dt + \sum_{s'\neq s}\tilde{q}_{ss'}[\beta_i(t,s') - \beta_i(t,s)]dt\\
&+\alpha_i[(\sigma_1+\bar{\sigma}_1)^2 + \int_\Theta(\mu_1+\bar{\mu}_1)^2\nu(d\theta)]dt\\
&+2\beta_{i,B_o}(\sigma_{o1}+\bar{\sigma}_{o1})dt + 2\int_\Theta \beta_{i,N_o}(\mu_{o1}+\bar{\mu}_{o1})\nu_o(d\theta)dt\\
&+\alpha_i[\sum_{j\neq i}(\sigma_{j2}+\bar{\sigma}_{j2})\tfrac{\bar{\tau}_{j1}}{\bar{c}_j}]^2dt + \alpha_i\int_\Theta[\sum_{j\neq i}(\mu_{j2}+\bar{\mu}_{j2})\tfrac{\bar{\tau}_{j1}}{\bar{c}_j}]^2\nu(d\theta)dt\\
&+\beta_i[\sum_{j\neq i}(\sigma_{oj2}+\bar{\sigma}_{oj2})\tfrac{\bar{\tau}_{j1}}{\bar{c}_j}]^2dt + \beta_i\int_\Theta[\sum_{j\neq i}(\mu_{oj2}+\bar{\mu}_{oj2})\tfrac{\bar{\tau}_{j1}}{\bar{c}_j}]^2\nu_o(d\theta)dt\\
&-2\alpha_i\sum_{j\neq i}\{(\sigma_1+\bar{\sigma}_1)(\sigma_{j2}+\bar{\sigma}_{j2}) + \int_\Theta(\mu_1+\bar{\mu}_1)(\mu_{j2}+\bar{\mu}_{j2})\nu(d\theta)\}\tfrac{\bar{\tau}_{j1}}{\bar{c}_j}dt\\
&-2\beta_i\sum_{j\neq i}\{(b_{j2}+\bar{b}_{j2}) + (\sigma_{o1}+\bar{\sigma}_{o1})(\sigma_{oj2}+\bar{\sigma}_{oj2})\\
&+\int_\Theta(\mu_{o1}+\bar{\mu}_{o1})(\mu_{oj2}+\bar{\mu}_{oj2})\nu_o(d\theta)\}\tfrac{\bar{\tau}_{j1}}{\bar{c}_j}\\
&-2\sum_{j\neq i}\{\beta_{i,B_o}(\sigma_{oj2}+\bar{\sigma}_{oj2}) + \int_\Theta \beta_{i,N_o}(\mu_{oj2}+\bar{\mu}_{oj2})\}\tfrac{\bar{\tau}_{j1}}{\bar{c}_j}dt\\
&+\{\xi_{i2,B_o} + \bar{\xi}_{i2,B_o} + \beta_i[2(\sigma_{o1}+\bar{\sigma}_{o1}) - 2\sum_j(\sigma_{oj2}+\bar{\sigma}_{oj2})\tfrac{\bar{\tau}_{j1}}{\bar{c}_j}]\}dB_o\\
&+\int_\Theta\{\xi_{i2,N_o}+\bar{\xi}_{i2,N_o}+\beta_i[2(\mu_{o1}+\bar{\mu}_{o1})-2\sum_j(\mu_{oj2}+\bar{\mu}_{oj2})\tfrac{\bar{\tau}_{j1}}{\bar{c}_j}]\}\tilde{N}_o(dt,d\theta)\Big\}\,\tfrac{1}{2}\hat{x}^2\Bigg]\\
&+Z_1,
\end{aligned}
\tag{7.321}
$$

$$
\begin{aligned}
Z_1 = \Bigg[\Big\{ & d\gamma_i + (b_1+\bar{b}_1)\gamma_i dt + [(\epsilon_{i3}+\bar{\epsilon}_{i3}) - \tfrac{\bar{\tau}_{i1}\bar{\tau}_{i0}}{\bar{c}_i}]dt \\
&+ \textstyle\sum_{s'\neq s} \tilde{q}_{ss'}[\gamma_i(t,s') - \gamma_i(t,s)]dt + \alpha_i[\sigma_0(\sigma_1+\bar{\sigma}_1) + \int_\Theta \mu_0(\mu_1+\bar{\mu}_1)\nu(d\theta)]dt \\
&+\beta_i[b_0 + \sigma_{o0}(\sigma_{o1}+\bar{\sigma}_{o1}) + \textstyle\int_\Theta \mu_{o0}(\mu_{o1}+\bar{\mu}_{o1})\nu_o(d\theta)]dt \\
&+\beta_{i,B_o}\sigma_{o0}dt + \textstyle\int_\Theta \beta_{i,N_o}\mu_{o0}\nu_o(d\theta)dt \\
&+\gamma_{i,B_o}(\sigma_{o1}+\bar{\sigma}_{o1})dt + \textstyle\int_\Theta \gamma_{i,N_o}(\mu_{o1}+\bar{\mu}_{o1})\nu_o(d\theta)dt \\
&+\alpha_i[\textstyle\sum_{j\neq i}(\sigma_{j2}+\bar{\sigma}_{j2})\frac{\bar{\tau}_{j0}}{\bar{c}_j}][\sum_{k\neq i}(\sigma_{k2}+\bar{\sigma}_{k2})\frac{\bar{\tau}_{k1}}{\bar{c}_k}] \\
&+\alpha_i \textstyle\int_\Theta [\sum_{j\neq i}(\mu_{j2}+\bar{\mu}_{j2})\frac{\bar{\tau}_{j0}}{\bar{c}_j}][\sum_{k\neq i}(\mu_{k2}+\bar{\mu}_{k2})\frac{\bar{\tau}_{k1}}{\bar{c}_k}]\nu(d\theta) \\
&+\beta_i[\textstyle\sum_{j\neq i}(\sigma_{oj2}+\bar{\sigma}_{oj2})\frac{\bar{\tau}_{j0}}{\bar{c}_j}][\sum_{j\neq i}(\sigma_{oj2}+\bar{\sigma}_{oj2})\frac{\bar{\tau}_{j1}}{\bar{c}_j}] \\
&+\beta_i \textstyle\int_\Theta [\sum_{j\neq i}(\mu_{oj2}+\bar{\mu}_{oj2})\frac{\bar{\tau}_{j0}}{\bar{c}_j}][\sum_{j\neq i}(\mu_{oj2}+\bar{\mu}_{oj2})\frac{\bar{\tau}_{j1}}{\bar{c}_j}]\nu_o(d\theta) \\
&-\alpha_i \textstyle\sum_{j\neq i}\{(\sigma_1+\bar{\sigma}_1)(\sigma_{j2}+\bar{\sigma}_{j2}) + \int_\Theta (\mu_1+\bar{\mu}_1)(\mu_{j2}+\bar{\mu}_{j2})\nu(d\theta)\}\frac{\bar{\tau}_{j0}}{\bar{c}_j} \\
&-\beta_i \textstyle\sum_{j\neq i}\{(b_{j2}+\bar{b}_{j2}) + (\sigma_{o1}+\bar{\sigma}_{o1})(\sigma_{oj2}+\bar{\sigma}_{oj2}) \\
&+ \textstyle\int_\Theta (\mu_{o1}+\bar{\mu}_{o1})(\mu_{oj2}+\bar{\mu}_{oj2})\nu_o(d\theta)\}\frac{\bar{\tau}_{j0}}{\bar{c}_j} \\
&- \textstyle\sum_{j\neq i}\{\beta_{i,B_o}(\sigma_{oj2}+\bar{\sigma}_{oj2}) + \int_\Theta \beta_{i,N_o}(\mu_{oj2}+\bar{\mu}_{oj2})\}\frac{\bar{\tau}_{j0}}{\bar{c}_j}dt \\
&-\alpha_i \textstyle\sum_{j\neq i}\{\sigma_0(\sigma_{j2}+\bar{\sigma}_{j2}) + \int_\Theta \mu_0(\mu_{j2}+\bar{\mu}_{j2})\nu(d\theta)\}\frac{\bar{\tau}_{j1}}{\bar{c}_j}dt \\
&-\beta_i \textstyle\sum_{j\neq i}\{\sigma_{o0}(\sigma_{oj2}+\bar{\sigma}_{oj2}) + \int_\Theta \mu_{o0}(\mu_{oj2}+\bar{\mu}_{oj2})\nu_o(d\theta)\}\frac{\bar{\tau}_{j1}}{\bar{c}_j}dt \\
&- \textstyle\sum_{j\neq i}\{\gamma_{i,B_o}(\sigma_{oj2}+\bar{\sigma}_{oj2}) + \int_\Theta \gamma_{i,N_o}(t,\theta)(\mu_{oj2}+\bar{\mu}_{oj2})\nu_o(d\theta)\}\frac{\bar{\tau}_{j1}}{\bar{c}_j}dt \\
&-\gamma_i \textstyle\sum_{j\neq i}(b_{j2}+\bar{b}_{j2})\frac{\bar{\tau}_{j1}}{\bar{c}_j}dt \\
&+\Big[\xi_{i1,B_o} + \beta_i(\sigma_{o0} - \textstyle\sum_j(\sigma_{oj2}+\bar{\sigma}_{oj2})\frac{\bar{\tau}_{j0}}{\bar{c}_j}) + \gamma_i((\sigma_{o1}+\bar{\sigma}_{o1}) - \sum_j(\sigma_{oj2}+\bar{\sigma}_{oj2})\frac{\bar{\tau}_{j1}}{\bar{c}_j})\Big]dB_o \\
&+\textstyle\int_\Theta \Big[\xi_{i1,N_o} + \beta_i(\mu_{o0} - \sum_j(\mu_{oj2}+\bar{\mu}_{oj2})\frac{\bar{\tau}_{j0}}{\bar{c}_j}) + \gamma_i((\mu_{o1}+\bar{\mu}_{o1}) \\
&- \textstyle\sum_j(\mu_{oj2}+\bar{\mu}_{oj2})\frac{\bar{\tau}_{j1}}{\bar{c}_j})\Big]\tilde{N}_o(dt,d\theta)\ \Big\}\hat{x} \\
&+d\delta_i + \tfrac{\alpha_i}{2}(\sigma_0^2 + \textstyle\int_\Theta \mu_0^2\nu(d\theta))dt + \frac{\beta_i}{2}[\sigma_{o0}^2 + \int_\Theta \mu_{o0}^2\nu_o(d\theta)]dt \\
&+b_0\gamma_i dt + \gamma_{i,B_o}\sigma_{o0}dt + \textstyle\int_\Theta \gamma_{i,N_o}(t,\theta)\mu_{o0}\nu_o(d\theta)dt + \sum_{s'\neq s}\tilde{q}_{ss'}[\delta_i(t,s') - \delta_i(t,s)]dt \\
&+\tfrac{\alpha_i}{2}[\textstyle\sum_{j\neq i}(\sigma_{j2}+\bar{\sigma}_{j2})\frac{\bar{\tau}_{j0}}{\bar{c}_j}]^2dt + \int_\Theta \frac{\alpha_i}{2}[\sum_{j\neq i}(\mu_{j2}+\bar{\mu}_{j2})\frac{\bar{\tau}_{j0}}{\bar{c}_j}]^2\nu(d\theta)dt \\
&+\tfrac{\beta_i}{2}[\textstyle\sum_{j\neq i}(\sigma_{oj2}+\bar{\sigma}_{oj2})\frac{\bar{\tau}_{j0}}{\bar{c}_j}]^2dt + \frac{\beta_i}{2}\int_\Theta [\sum_{j\neq i}(\mu_{oj2}+\bar{\mu}_{oj2})\frac{\bar{\tau}_{j0}}{\bar{c}_j}]^2\nu_o(d\theta) \\
&-\tfrac{\bar{\tau}_{i0}^2}{2\bar{c}_i}dt - \alpha_i \textstyle\sum_{j\neq i}\{\sigma_0(\sigma_{j2}+\bar{\sigma}_{j2}) + \int_\Theta \mu_0(\mu_{j2}+\bar{\mu}_{j2})\nu(d\theta)\}\frac{\bar{\tau}_{j0}}{\bar{c}_j}dt \\
&-\beta_i \textstyle\sum_{j\neq i}\{\sigma_{o0}(\sigma_{oj2}+\bar{\sigma}_{oj2}) + \int_\Theta \mu_{o0}(\mu_{oj2}+\bar{\mu}_{oj2})\nu_o(d\theta)\}\frac{\bar{\tau}_{j0}}{\bar{c}_j}dt \\
&- \textstyle\sum_{j\neq i}\{\gamma_{i,B_o}(\sigma_{oj2}+\bar{\sigma}_{oj2}) + \int_\Theta \gamma_{i,N_o}(t,\theta)(\mu_{oj2}+\bar{\mu}_{oj2})\nu_o(d\theta)\}\frac{\bar{\tau}_{j0}}{\bar{c}_j}dt \\
&-\gamma_i \textstyle\sum_{j\neq i}(b_{j2}+\bar{b}_{j2})\frac{\bar{\tau}_{j0}}{\bar{c}_j}dt \\
&+\{\xi_{i0,B_o} + \gamma_i[\sigma_{o0} - \textstyle\sum_j(\sigma_{oj2}+\bar{\sigma}_{oj2})\frac{\bar{\tau}_{j0}}{\bar{c}_j}]\}dB_o \\
&+\textstyle\int_\Theta \{\xi_{i0,N_o} + \gamma_i[\mu_{o0} - \sum_j(\mu_{oj2}+\bar{\mu}_{oj2})\frac{\bar{\tau}_{j0}}{\bar{c}_j}]\}\tilde{N}_o(dt,d\theta) \\
&+\tfrac{1}{2}c_i(u_i - \hat{u}_i + \tfrac{\bar{\tau}_i}{c_i}(x-\hat{x}))^2dt + \tfrac{\bar{c}_i}{2}(\hat{u}_i + \tfrac{\bar{\tau}_{i1}\hat{x}+\bar{\tau}_{i0}}{\bar{c}_i})^2dt + d\tilde{\bar{\epsilon}}\Bigg].
\end{aligned}
\tag{7.322}
$$

The minimization of the last expression leads to the desired results provided that c_i and $\bar{c}_i$ are positive. This completes the proof. ■

7.8 Adversarial Linear-Quadratic Conditional Mean-Field-Type Games

This section extends the problem examined in Sect. 7.7 in the context adversarial games. We consider a generic adversarial linear-quadratic mean-field-type game problem under a jump-diffusion-regime switching state dynamics that include common noise and mean-field terms that are:

- the conditional expectation of the state with respect to the filtration of the regime switching, in the drift, diffusion, and jump coefficient functions,
- the conditional expectation of the control actions, in the drift, diffusion, and jump coefficient functions.

The state process is one-dimensional. The common noise is of two types: a Brownian motion and a Poisson random process. The stage-wise cost function includes:

- the conditional variance of the state,
- the square of the conditional expected value of the state,
- the conditional variance of the control-action process,
- the square of the conditional expected value of the control-action process,
- a cross-term such as covariance between the state and the control,
- a product term between the conditional expected value of the control action and the conditional expected value of the state
- a noisy cost measurement which is a jump-diffusion-regime switching and common noise dependent.

7.8.1 *Adversarial Game*

For the sake of simplicity, we consider the one-dimensional case. There are 2 decision-makers, one attacker and one defender. The set of decision-makers is denoted by $\mathcal{I} = \{1, 2\}$. Decision-maker i, the defender, chooses a strategy $u_i \in \mathcal{U}_i$, to minimize its performance functional defined below. Decision-maker $j \neq i$ is the attacker and chooses a control strategy $u_j \in \mathcal{U}_j$ to create maximum damage. The information structure of the adversarial problem entails perfect state observation and common noise observation. The state dynamics is driven by a jump-diffusion and regime switching and is given by

$$
\begin{array}{l}
x(0) = x_0, \quad s(0) = s_0, \\
dx = bdt + \sigma dB + \int_\Theta \mu(\cdot, \theta)\tilde{N}(dt, d\theta) + \sigma_o dB_o + \int_\Theta \mu_o(\cdot, \theta)\tilde{N}_o(dt, d\theta),
\end{array}
\tag{7.323}
$$

where

$$
\begin{cases}
b = b_0 + b_1 x + \bar{b}_1 \hat{x} + \sum_{k\in\mathscr{I}} b_{k2} u_k + \sum_{k\in\mathscr{I}} \bar{b}_{k2} \hat{u}_k, \\
\sigma = \sigma_0 + \sigma_1 x + \bar{\sigma}_1 \hat{x} + \sum_{k\in\mathscr{I}} \sigma_{k2} u_k + \sum_{k\in\mathscr{I}} \bar{\sigma}_{k2} \hat{u}_k, \\
\mu(\cdot, \theta) = \mu_0 + \mu_1 x + \bar{\mu}_1 \hat{x} + \sum_{k\in\mathscr{I}} \mu_{k2} u_k + \sum_{k\in\mathscr{I}} \bar{\mu}_{k2} \hat{u}_k, \\
\sigma_o = \sigma_{o0} + \sigma_{o1} x + \bar{\sigma}_{o1} \hat{x} + \sum_{k\in\mathscr{I}} \sigma_{ok2} u_k + \sum_{k\in\mathscr{I}} \bar{\sigma}_{ok2} \hat{u}_k, \\
\mu_o(\cdot, \theta) = \mu_{o0} + \mu_{o1} x + \bar{\mu}_{o1} \hat{x} + \sum_{k\in\mathscr{I}} \mu_{ok2} u_k + \sum_{k\in\mathscr{I}} \bar{\mu}_{ok2} \hat{u}_k,
\end{cases}
\tag{7.324}
$$

with $b_{k'}(t, s); \sigma_{k'}(t, s); \bar{b}_{k'}(t, s); \bar{\sigma}_{k'}(t, s), \sigma_o(t, s) \in \mathbb{R}$, and $s \in \mathscr{S}$, where $\mathscr{S}$ is a finite set of switching regimes. Moreover, $\mu_{k'}(t, s, \theta), \mu_o(t, s, \theta) \in \mathbb{R}, \theta \in \Theta$ where Θ is the set of jump sizes for all $k' \in \{0, 1, 2\}$.

$$
\begin{array}{l}
b_{k'}, \sigma_{k'}, \bar{b}_{k'}, \bar{\sigma}_{k'} : [0, T] \times \mathscr{S} \to \mathbb{R}, \ k' \in \{0, 1\}, \\
b_{k2}, \sigma_{k2}, \bar{b}_{k2}, \bar{\sigma}_{k2} : [0, T] \times \mathscr{S} \to \mathbb{R}, \ k \in \mathscr{I}, \\
\mu_{k'}, \mu_{ok'}, \bar{\mu}_{k'}, \bar{\mu}_{ok'} : [0, T] \times \mathscr{S} \times \Theta \to \mathbb{R}, \ k' \in \{0, 1\}, \\
\mu_{k2}, \mu_{ok2}, \bar{\mu}_{k2}, \bar{\mu}_{ok2} : [0, T] \times \mathscr{S} \times \Theta \to \mathbb{R}, \ k \in \mathscr{I},
\end{array}
\tag{7.325}
$$

The stochastic differential equation (7.323) consists of the following:

- The switching process switches between different elements in the finite set $\mathscr{S}$ with transition rate $Q = (\tilde{q}_{ss'}, \ (s, s') \in \mathscr{S}^2)$ such that

$$
s' \neq s, \ \tilde{q}_{ss'} > 0, \ \tilde{q}_{ss} = -\sum_{s' \neq s} \tilde{q}_{ss'}.
$$

- The regime switching-dependent jump-diffusion process $\sigma dB + \int_\Theta \mu(\cdot, \theta) \tilde{N}(dt, d\theta)$, where the process B is a standard one-dimensional Brownian motion, $\tilde{N}(dt, d\theta) = N(dt, d\theta) - \nu(d\theta)dt$ is a compensated jump process with Radon measure ν over Θ.
- The common noise, which is also a regime switching-dependent jump-diffusion process $\sigma_o dB_o + \int_\Theta \mu_o(\cdot, \theta)\tilde{N}_o(dt, d\theta)$, where the process B_o is a standard one-dimensional Brownian motion, $\tilde{N}_o(dt, d\theta) = N_o(dt, d\theta) - \nu_o(d\theta)dt$ is a compensated jump process with Radon measure ν_o over Θ.
- The processes $s(t)$, $B(t)$, $N(t, .)$, $B_o(t)$, $N_o(t, .)$ are mutually independent. In particular, the Poisson random measure N_o as an observable common noise allows us to capture larger jumps (big changes) in the realization of the common noise.
- The term $\hat{x} = E[x|\mathscr{F}_t^{B_o, N_0, s}]$ denotes the conditional expectation of x given $\mathscr{F}_t^{B_o, N_0, s}$, and $\hat{u} = E[u|\mathscr{F}_t^{B_o, N_0, s}]$. These are called conditional state mean-field and conditional control-action mean-field terms.

The state dynamics (7.323) is therefore of conditional McKean–Vlasov type as the conditional probability measure of the state appears in the drift, diffusion, and jump coefficients via $(\hat{x},\ \hat{u})$. Henceforth, we denote a solution (if any) to (7.323) by $x := x^{u,x_0,s_0}$.

Next we address the question of controlling it through a performance criterion for each decision-maker.

7.8.1.1 Noisy Cost Functional per Decision-Maker

We associate a cost functional of conditional mean-field type of decision-maker i (minimizer) as

$$\begin{aligned}
&L(x,s,u)\\
&= \tfrac{1}{2}q_T x^2(T) + \tfrac{1}{2}\bar{q}_T \hat{x}^2(T) + \epsilon_{3T} x(T) + \bar{\epsilon}_{3T}\hat{x}(T)\\
&+ \int_0^T (\tfrac{1}{2}q x^2 + \tfrac{1}{2}\bar{q}\hat{x}^2 + \epsilon_3 x + \bar{\epsilon}_3 \hat{x})dt\\
&+ \int_0^T (\tfrac{1}{2}r_i u_i^2 + \tfrac{1}{2}\bar{r}_i \hat{u}_i^2 + \epsilon_{i1} x u_i + \bar{\epsilon}_{i1}\hat{x}\hat{u}_i + \epsilon_{i2} u_i + \bar{\epsilon}_{i2}\hat{u}_i)dt\\
&+ \int_0^T (\tfrac{1}{2}r_j u_j^2 + \tfrac{1}{2}\bar{r}_j \hat{u}_j^2 + \epsilon_{j1} x u_j + \bar{\epsilon}_{j1}\hat{x}\hat{u}_j + \epsilon_{j2} u_j + \bar{\epsilon}_{j2}\hat{u}_j)dt\\
&+ \int_0^T (\xi_{0,B_o} + \xi_{1,B_o}\hat{x} + \tfrac{1}{2}\xi_{2,B_o} x^2 + \tfrac{1}{2}\bar{\xi}_{2,B_o}\hat{x}^2)dB_o\\
&+ \int_0^T \int_\Theta (\xi_{0,N_o} + \xi_{1,N_o}\hat{x} + \tfrac{1}{2}\xi_{2,N_o} x^2 + \tfrac{1}{2}\bar{\xi}_{2,N_o}\hat{x}^2)\tilde{N}_o(dt,d\theta),
\end{aligned} \tag{7.326}$$

where

$$\begin{aligned}
&q_T, \bar{q}_T, \epsilon_{3T}, \bar{\epsilon}_{3T} : \mathscr{S} \to \mathbb{R},\\
&q, \bar{q}, r_k, \bar{r}_k, \epsilon_{kk'}, \bar{\epsilon}_{kk'} : [0,T]\times\mathscr{S} \to \mathbb{R},\ k' \in \{1,2,3\},\\
&\xi_{k',B_o}, \bar{\xi}_{k',B_o} : [0,T]\times\mathscr{S} \to \mathbb{R},\ k' \in \{0,1,2\},\\
&\xi_{k',N_o}, \bar{\xi}_{k',N_o} : [0,T]\times\mathscr{S}\times\Theta \to \mathbb{R},\ k' \in \{0,1,2\}.
\end{aligned} \tag{7.327}$$

All these random coefficient functions are $\mathscr{F}^{B_o,N_o}$-measurable for each regime s. The last term of (7.326) given by

$$\begin{aligned}
&(\xi_{0,B_o} + \xi_{1,B_o}\hat{x} + \xi_{2,B_o} x^2 + \bar{\xi}_{2,B_o}\hat{x}^2)dB_o\\
&+ \int_\Theta (\xi_{0,N_o} + \xi_{1,N_o}\hat{x} + \xi_{2,N_o} x^2 + \bar{\xi}_{2,N_o}\hat{x}^2)\tilde{N}_o(dt,d\theta),
\end{aligned} \tag{7.328}$$

denotes a noisy cost measurement of adversaries. In practice, measurement and observations are usually done with specific evaluation criteria and possibly with machines. Here the measurement is considered as imperfect, and the measurement noise is captured as the latter term depending on the realization of the common noise.

7.8.1.2 Equilibrium Solution Concept

Given $u_{-i} = (u_k \in \mathcal{U}_k,\ k \neq i)$ of the attacker, the best-response problem of the defender is

$$\begin{cases} \inf_{u_i \in \mathcal{U}_i} E\left[L(x, s, u) \mid x(0) = x_0, s(0) = s_0\right], \\ \text{such that (7.323).} \end{cases} \tag{7.329}$$

Given $u_{-j} = (u_k \in \mathcal{U}_k,\ k \neq j)$ of the defender, the best-response problem of the attacker is

$$\begin{cases} \sup_{u_j \in \mathcal{U}_j} E\left[L(x, s, u) \mid x(0) = x_0, s(0) = s_0\right], \\ \text{such that (7.323).} \end{cases} \tag{7.330}$$

A risk-neutral minmax mean-field-type equilibrium is a strategy profile $u^* \in \prod_{k \in \mathcal{I}} \mathcal{U}_k$, such that

$$\begin{cases} E\left[L(x, s, u^*) | x(0) = x_0, s(0) = s_0\right] \\ = \inf_{u_i \in \mathcal{U}_i} \sup_{u_j \in \mathcal{U}_j} E\left[L(x, s, u_i, u_j) | x(0) = x_0, s(0) = s_0\right], \\ \text{such that (7.323).} \end{cases} \tag{7.331}$$

A risk-neutral maxmin mean-field-type equilibrium is a strategy profile $u^* \in \prod_{k \in \mathcal{I}} \mathcal{U}_k$, such that

$$\begin{cases} E\left[L(x, s, u^*) | x(0) = x_0, s(0) = s_0\right] \\ = \sup_{u_j \in \mathcal{U}_j} \inf_{u_i \in \mathcal{U}_i} E\left[L(x, s, u_i, u_j) | x(0) = x_0, s(0) = s_0\right], \\ \text{such that (7.323).} \end{cases} \tag{7.332}$$

A saddle point (u_i^*, u_j^*) solves the inequalities:

$$\begin{aligned} &E[L(x, s, u_i^*, u_j)] \leq E[L(x, s, u^*)] \leq E[L(x, s, u_i, u_j^*)], \\ &\text{for every pair } (u_i, u_j) \in \mathcal{U}_i \times \mathcal{U}_j. \end{aligned} \tag{7.333}$$

We are interested in finding mean-field-type robust Nash equilibria (saddle point) when the information structure of each decision-maker is $\{t, x, B_o, N_o, s\}$. Based on the information available, the decision-maker can construct the process $\hat{x}$. Note that expectation of $\hat{x}$ is the same as the expectation of the original process x. We show below that the optimal strategy involves two processes $(x - \hat{x})$ and $\hat{x}$ among the other coefficients that are regime and common noise dependent. Here the observation process is the vector (x, B_o, N_o, s), and the structure of the game model is assumed to be known to both decision-makers.

We invoke the following assumption:

$$\begin{cases} q_T, q > 0, q_T + \bar{q}_T > 0, q + \bar{q} > 0, \\ r_i > 0, r_i + \bar{r}_i > 0,\ r_j < 0, r_j + \bar{r}_j < 0, \\ E[x_0^2] < +\infty, \int_0^T \int_\Theta \phi^2 \nu(d\theta) < +\infty, \text{with } \phi \in \{\mu_{kk'}, \bar{\mu}_{kk'}\}, \\ \int_0^T \int_\Theta \phi^2 \nu_o(d\theta) < +\infty, \text{with } \phi \in \{\mu_{okk'}, \bar{\mu}_{okk'}\} \\ |\mathcal{S}| < +\infty, b_{k'}, \sigma_{k'}^2, \bar{b}_{k'}, \bar{\sigma}_k^2 \in L^1([0,T] \times \mathcal{S}, \mathbb{R}), \end{cases} \tag{7.334}$$

Let Assumption (7.334) hold. Then, generically, the equilibrium strategy and the equilibrium cost, associated with the mean-field-type game problem (7.333), are given by

$$\begin{aligned}
&u_k^* = \hat{u}_k^* - \tfrac{\tilde{\tau}_k}{c_k}(x - \hat{x}), \\
&\hat{u}_k^* = -\tfrac{\bar{\tau}_{k1}}{\bar{c}_k}\hat{x} - \tfrac{\bar{\tau}_{k0}}{\bar{c}_k}, \\
&(u_i - \hat{u}_i) = -[A^{-1}v]_i(x - \hat{x}) = -\tfrac{1}{1 - a_{ij}a_{ji}}(v_i - a_{ij}v_j)(x - \hat{x}), \\
&\hat{u}_i = -[\bar{A}^{-1}v_1]_i\hat{x} - [\bar{A}^{-1}v_0]_i = -\tfrac{1}{1 - \bar{a}_{ij}\bar{a}_{ji}}\left([v_{i1} - \bar{a}_{ij}v_{j1}]\hat{x} + [v_{i0} - \bar{a}_{ij}v_{j0}]\right), \\
&L * (x, s, u^*) = E[\tfrac{1}{2}\alpha(0, s(0))(x_0 - \hat{x}_0)^2 + \tfrac{1}{2}\beta(0, s(0))\hat{x}_0^2 + \gamma(0, s(0))\hat{x}_0 + \delta(0, s(0))],
\end{aligned}$$

The coefficient processes $\alpha, \alpha_{B_o}, \alpha_{N_o}$ of the solution to Problem (7.333) are given by the following backward stochastic differential system:

$$\begin{aligned}
&d\alpha = -\Big\{2b_1\alpha + \alpha[\sigma_1^2 + \sigma_{o1}^2 + \textstyle\int_\Theta \mu_1^2 \nu(d\theta) + \int_\Theta \mu_{o1}^2 \nu_o(d\theta)] \\
&+q - (\tfrac{\tilde{v}_i^2}{c_i} + \tfrac{\tilde{\tau}_j^2}{c_j}) + \textstyle\sum_{s' \neq s} \tilde{q}_{ss'}[\alpha(t, s') - \alpha(t, s)] \\
&+2\alpha_{B_o}\sigma_{o1} + 2\textstyle\int_\Theta \alpha_{N_o}\mu_{o1}\nu_o(d\theta)\Big\}dt \\
&-\{\xi_{2,B_o} + 2\sigma_{o1}\alpha - 2\alpha[\sigma_{oi2}\tfrac{\tilde{\tau}_i}{c_i} + \sigma_{oj2}\tfrac{\tilde{\tau}_j}{c_j}]\}dB_o \\
&-\textstyle\int_\Theta \{\xi_{2,N_o} + 2\mu_{o1}\alpha - 2\alpha[\mu_{oi2}\tfrac{\tilde{\tau}_i}{c_i} + \mu_{oj2}\tfrac{\tilde{\tau}_j}{c_j}]\}\tilde{N}_o(dt, d\theta), \\
&\alpha(T, s) = q(T, s), \\
&\alpha_{B_o} = -\left(\xi_{2,B_o} + \alpha[2\sigma_{o1} - 2\textstyle\sum_j \sigma_{oj2}\tfrac{\tilde{\tau}_j}{c_j}]\right), \\
&\alpha_{N_o} = -\left(\xi_{2,N_o} + \alpha[2\mu_{o1} - 2\textstyle\sum_j \mu_{oj2}\tfrac{\tilde{\tau}_j}{c_j}]\right),
\end{aligned}$$

whenever these stochastic differential equations have a solution that is $\mathcal{F}^{B_o,N_o}$-adapted and does not blow up in $[0, T]$, and $\hat{\phi} = \phi + \bar{\phi}$, with $\phi \in \{q_T, \epsilon_{3T}, q, r_i, \epsilon_{ik}, \xi_{2,B_o}, \xi_{2,N_o}, b_k, \sigma, \sigma_o, \mu, \mu_o\}$, the coefficient processes β, β_{B_o}, β_{N_o} of the solution to Problem (7.333) are given by

$$
\begin{aligned}
d\beta = & -\Big\{2\hat{b}_1\beta + \beta[\hat{\sigma}_{o1}^2 + \int_\Theta \hat{\mu}_{o1}^2 \nu_o(d\theta)] \\
& + \hat{q} - (\tfrac{\bar{v}_{j1}^2}{\bar{c}_i} + \tfrac{\bar{\tau}_{j1}^2}{\bar{c}_j}) + \sum_{s'\neq s} \tilde{q}_{ss'}[\beta(t,s') - \beta(t,s)] \\
& + \alpha[\hat{\sigma}_1^2 + \int_\Theta \hat{\mu}_1^2 \nu(d\theta)] + 2\beta_{B_o}\hat{\sigma}_{o1} + 2\int_\Theta \beta_{N_o}\hat{\mu}_{o1}\nu_o(d\theta)\Big\}dt \\
& -\{\hat{\xi}_{i2,B_o} + 2\hat{\sigma}_{o1}\beta - 2\beta\sum_j \hat{\sigma}_{oj2}\tfrac{\bar{\tau}_{j1}}{c_j}\}dB_o \\
& - \int_\Theta \{\hat{\xi}_{2,N_o} + 2\hat{\mu}_{o1}\beta - 2\beta\sum_j \hat{\mu}_{oj2}\tfrac{\bar{\tau}_{j1}}{c_j}\}\tilde{N}_o(dt, d\theta), \\
\beta(T,s) = & \ \hat{q}(T,s), \\
\beta_{B_o} = & -\left(\hat{\xi}_{2,B_o} + \beta[2\hat{\sigma}_{o1} - 2\sum_k \hat{\sigma}_{ok2}\tfrac{\bar{\tau}_{k1}}{\bar{c}_k}]\right), \\
\beta_{N_o} = & -\left(\hat{\xi}_{2,N_o} + \beta[2\hat{\mu}_{o1} - 2\sum_k \hat{\mu}_{ok2}\tfrac{\bar{\tau}_{k1}}{\bar{c}_k}]\right),
\end{aligned}
$$

whenever these stochastic differential equations have a solution that is $\mathscr{F}^{B_o,N_o}$-adapted and does not blow up in $[0, T]$, and the coefficient processes $\gamma, \gamma_{B_o}, \gamma_{N_o}$ of the solution to Problem (7.333) are given by

$$
\begin{aligned}
d\gamma = & -\Big\{\hat{b}_1\gamma + \hat{\epsilon}_3 - (\tfrac{\bar{v}_{i1}\bar{v}_{i0}}{\bar{c}_i} + \tfrac{\bar{\tau}_{j1}\bar{\tau}_{j0}}{\bar{c}_j}) + \sum_{s'\neq s} \tilde{q}_{ss'}[\gamma(t,s') - \gamma(t,s)] \\
& + \alpha[\sigma_0\hat{\sigma}_1 + \int_\Theta \mu_0\hat{\mu}_1\nu(d\theta)] + \beta[b_0 + \sigma_{o0}\hat{\sigma}_{o1} + \int_\Theta \mu_{o0}\hat{\mu}_{o1}\nu_o(d\theta)] \\
& + \beta_{B_o}\sigma_{o0} + \int_\Theta \beta_{N_o}\mu_{o0}\nu_o(d\theta) + \gamma_{B_o}\hat{\sigma}_{o1} + \int_\Theta \gamma_{N_o}\hat{\mu}_{o1}\nu_o(d\theta)\Big\}dt \\
& - \Big\{\xi_{1,B_o} + \beta[\sigma_{o0} - \sum_j \hat{\sigma}_{oj2}\tfrac{\bar{\tau}_{j0}}{c_j}] + \gamma[\hat{\sigma}_{o1} - \sum_j \hat{\sigma}_{oj2}\tfrac{\bar{\tau}_{j1}}{c_j}]\Big\}\, dB_o \\
& - \int_\Theta \Big\{\xi_{1,N_o} + \beta[\mu_{o0} - \sum_j \hat{\mu}_{oj2}\tfrac{\bar{\tau}_{j0}}{c_j}]\gamma[\hat{\mu}_{o1} - \sum_j \hat{\mu}_{oj2}\tfrac{\bar{\tau}_{j1}}{c_j}]\Big\}\, \tilde{N}_o(dt, d\theta), \\
\gamma(T,s) = & \ \hat{\epsilon}_{3T}, \quad s \in \mathscr{S} \\
\gamma_{B_o} = & -\Big[\xi_{1,B_o} + \beta(\sigma_{o0} - \sum_j \hat{\sigma}_{oj2}\tfrac{\bar{\tau}_{j0}}{\bar{c}_j}) + \gamma(\hat{\sigma}_{o1} - \sum_j \hat{\sigma}_{oj2}\tfrac{\bar{\tau}_{j1}}{\bar{c}_j})\Big], \\
\gamma_{N_o} = & -\Big[\xi_{1,N_o} + \beta(\mu_{o0} - \sum_j \hat{\mu}_{oj2}\tfrac{\bar{\tau}_{j0}}{\bar{c}_j}) + \gamma(\hat{\mu}_{o1} - \sum_j \hat{\mu}_{oj2}\tfrac{\bar{\tau}_{j1}}{\bar{c}_j})\Big],
\end{aligned}
$$

whenever the system of stochastic differential equations has a solution that is $\mathscr{F}^{B_o,N_O}$-adapted, the coefficient processes $\delta, \delta_{i,B_o}, \delta_{i,N_o}$ of the solution to Problem (7.333) are given by

$$
\begin{aligned}
d\delta = & -\Big\{b_0\gamma - (\tfrac{\bar{v}_{i0}^2}{2\bar{c}_i} + \tfrac{\bar{\tau}_{j0}^2}{2\bar{c}_j}) + \tfrac{\alpha}{2}(\sigma_0^2 + \int_\Theta \mu_0^2\nu(d\theta)) + \tfrac{\beta}{2}(\sigma_{o0}^2 + \int_\Theta \mu_{o0}^2\nu_o(d\theta)) \\
& + \gamma_{B_o}\sigma_{o0} + \int_\Theta \gamma_{N_o}(t,\theta)\mu_{o0}\nu_o(d\theta) + \sum_{s'\neq s} \tilde{q}_{ss'}[\delta(t,s') - \delta(t,s)]\Big\}dt \\
& -\{\xi_{0,B_o} + \gamma[\sigma_{o0} - \sum_j \hat{\sigma}_{oj2}\tfrac{\bar{\tau}_{j0}}{c_j}]\}dB_o \\
& - \int_\Theta \{\xi_{0,N_o} + \gamma[\mu_{o0} - \sum_j \hat{\mu}_{oj2}\tfrac{\bar{\tau}_{j0}}{c_j}]\}\tilde{N}_o(dt, d\theta), \\
\delta(T,s) = & \ 0, \\
\delta_{i,B_o} = & -\left(\xi_{0,B_o} + \gamma[\sigma_{o0} - \sum_j \hat{\sigma}_{oj2}\tfrac{\bar{\tau}_{j0}}{\bar{c}_j}]\right), \\
\delta_{i,N_o} = & -\left(\xi_{0,N_o} + \gamma[\mu_{o0} - \sum_j \hat{\mu}_{oj2}\tfrac{\bar{\tau}_{j0}}{\bar{c}_j}]\right),
\end{aligned}
$$

whenever the system of backward stochastic differential equations has a solution that is $\mathscr{F}^{B_o,N_o}$-adapted. The relationship between the coefficients is through $\tilde{\tau}, \bar{\tau}$ given by

$$\begin{array}{l}
c_i = \{r_i + \alpha\left(\sigma_{i2}^2 + \int_\Theta \mu_{i2}^2\nu(d\theta) + \sigma_{oi2}^2 + \int_\Theta \mu_{oi2}^2\nu_o(d\theta)\right)\},\\
c_j = \{r_j + \alpha\left(\sigma_{j2}^2 + \int_\Theta \mu_{j2}^2\nu(d\theta) + \sigma_{oj2}^2 + \int_\Theta \mu_{oj2}^2\nu_o(d\theta)\right)\},\\
\tilde{A}_{ii} = 1 = \tilde{A}_{jj},\\
j \neq i,\ \tilde{A}_{ij} = \frac{\alpha}{c_i}[(\sigma_{i2}\sigma_{j2} + \sigma_{oi2}\sigma_{oj2}) + \int_\Theta(\mu_{i2}\mu_{j2} + \mu_{oi2}\mu_{oj2})\nu(d\theta)],\\
v_i = \epsilon_{i1} + \alpha\{b_{i2} + \sigma_1\sigma_{i2} + \int_\Theta \mu_1\mu_{i2}\nu(d\theta)\}\\
+\alpha\ \{\sigma_{o1}\sigma_{oi2} + \int_\Theta \mu_{o1}\mu_{oi2}\}\nu_o(d\theta)\\
+\{\alpha_{B_o}\sigma_{oi2} + \int_\Theta \alpha_{N_o}\mu_{oi2}\nu_o(d\theta)\},\\
v_j = \epsilon_{j1} + \alpha\{b_{j2} + \sigma_1\sigma_{j2} + \int_\Theta \mu_1\mu_{j2}\nu(d\theta)\}\\
+\alpha\ \{\sigma_{o1}\sigma_{oj2} + \int_\Theta \mu_{o1}\mu_{oj2}\}\nu_o(d\theta) + \{\alpha_{B_o}\sigma_{oj2} + \int_\Theta \alpha_{N_o}\mu_{oj2}\nu_o(d\theta)\},\\
\tilde{A}(u - \hat{u}) + \tilde{v}(x - \hat{x}) = 0,\\
\tilde{v}_i = \frac{v_i}{c_i},\\
\frac{\tau_i}{c_i} = [\tilde{A}^{-1}v]_i,
\end{array} \tag{7.335}$$

$$\begin{array}{l}
\tilde{\bar{A}}_{ii}=1=\tilde{\bar{A}}_{jj},\ \tilde{\bar{A}}_{ij}=\frac{\alpha}{\bar{c}_i}\{\hat{\sigma}_{i2}\hat{\sigma}_{j2}+\int_\Theta \hat{\mu}_{i2}\hat{\mu}_{j2}\nu(d\theta)\}+\frac{\beta}{\bar{c}_j}\{\hat{\sigma}_{oi2}\hat{\sigma}_{oj2}+\int_\Theta \hat{\mu}_{oi2}\hat{\mu}_{oj2}\nu_o(d\theta)\},\\
\bar{c}_i = \hat{r}_i + \alpha[\hat{\sigma}_{i2}^2 + \int_\Theta \hat{\mu}_{i2}^2\nu(d\theta)] + \beta[\hat{\sigma}_{oi2}^2 + \int_\Theta \hat{\mu}_{oi2}^2\nu_o(d\theta)],\\
\bar{c}_i = \hat{r}_j + \alpha[\hat{\sigma}_{j2}^2 + \int_\Theta \hat{\mu}_{j2}^2\nu(d\theta)] + \beta[\hat{\sigma}_{oj2}^2 + \int_\Theta \hat{\mu}_{oj2}^2\nu_o(d\theta)],\\
\bar{v}_{i1} = \hat{\epsilon}_{i1} + \alpha[\hat{\sigma}_1\hat{\sigma}_{i2} + \int_\Theta \hat{\mu}_1\hat{\mu}_{i2}\nu(d\theta)]\\
+\beta[\hat{b}_{i2} + \hat{\sigma}_{o1}\hat{\sigma}_{oi2} + \int_\Theta \hat{\mu}_{o1}\hat{\mu}_{oi2}\nu_o(d\theta)] + \beta_{B_o}\hat{\sigma}_{oi2} + \int_\Theta \beta_{N_o}\hat{\mu}_{oi2},\\
\bar{v}_{i0} = +\hat{\epsilon}_{i2} + \alpha[\sigma_0\hat{\sigma}_{i2} + \int_\Theta \mu_0\hat{\mu}_{i2}\nu(d\theta)]\\
+\beta[\sigma_{o0}\hat{\sigma}_{oi2} + \int_\Theta \mu_{o0}\hat{\mu}_{oi2}\nu_o(d\theta)] + \gamma\hat{b}_{i2} + \gamma_{B_o}\hat{\sigma}_{oi2} + \int_\Theta \gamma_{N_o}(t,\theta)\hat{\mu}_{oi2}\nu_o(d\theta),\\
\frac{\tilde{\tau}_{jk}}{c_j} = [\tilde{\bar{A}}^{-1}\tilde{\bar{v}}_k]_j,\ \bar{A}\hat{u} + \tilde{\bar{v}}_1\hat{x} + \tilde{\bar{v}}_0 = 0.
\end{array} \tag{7.336}$$

We note the following:

- The matrices A and $\bar{A}$ become the identity matrix (and hence invertible) for problems in which the control and the expected value of the control do not appear in the diffusion and jump coefficient functions.
- The structure of the differential equations changes from linear to quadratic and fractional in α. As it is expected for game problems, the coefficients $\alpha(t, s)$ are interdependent not only across the switching regime. Furthermore, these coefficients are regularized by the common noises B_o, N_o.
- The saddle-point strategy is in state-and-conditional mean-field-type feedback form and driven by common noise.
- The solution above is for essential-inf sup as it is in conditional expectation. When taking the expectation in E_{B_o}, the terms in ξ vanish.

7.8.2 Construction of a Semi-explicit Saddle-Point Strategy

In this subsection, we develop a proof of optimality using a direct method that consists of several steps.

7.8.2.1 Reformulation of the Minmax Game Problem

First, consider the following minmax problem:

$$\inf_{u_i\in\mathcal{U}_i} \sup_{u_j\in\mathcal{U}_j} E\left[\hat{L}(x,s,u) \mid x(0)=x_0, s(0)=s_0\right], \tag{7.337}$$

subject to

$$\begin{aligned}
dx = &\Big\{b_0 + b_1(x-\hat{x}) + \hat{b}_1\hat{x} \\
&+\textstyle\sum_{j\in\mathcal{I}} b_{j2}(u_j-\hat{u}_j) + \sum_{j\in\mathcal{I}} \hat{b}_{j2}\hat{u}_j\Big\}dt \\
&+\Big\{\sigma_0 + \sigma_1(x-\hat{x}) + \hat{\sigma}_1\hat{x} \\
&+\textstyle\sum_{j\in\mathcal{I}} \sigma_{j2}(u_j-\hat{u}_j) + \sum_{j\in\mathcal{I}} \hat{\sigma}_{j2}\hat{u}_j\Big\}dB \\
&+\textstyle\int_\Theta \Big\{\mu_0 + \mu_1(x-\hat{x}) + \hat{\mu}_1\hat{x} \\
&+\textstyle\sum_{j\in\mathcal{I}} \mu_{j2}(u_j-\hat{u}_j) + \sum_{j\in\mathcal{I}} \hat{\mu}_{j2}\hat{u}_j\Big\}\tilde{N}(dt,d\theta) \\
&+\Big\{\sigma_{o0} + \sigma_{o1}(x-\hat{x}) + \hat{\sigma}_{o1}\hat{x} \\
&+\textstyle\sum_{j\in\mathcal{I}} \sigma_{oj2}(u_j-\hat{u}_j) + \sum_{j\in\mathcal{I}} \hat{\sigma}_{oj2}\hat{u}_j\Big\}dB_o \\
&+\textstyle\int_\Theta \Big\{\mu_{o0} + \mu_{o1}(x-\hat{x}) + \hat{\mu}_{o1}\hat{x} \\
&+\textstyle\sum_{j\in\mathcal{I}} \mu_{oj2}(u_j-\hat{u}_j) + \sum_{j\in\mathcal{I}} \hat{\mu}_{oj2}\hat{u}_j\Big\}\tilde{N}_o(dt,d\theta), \\
&x(0)=x_0,\ s(0)=s_0,
\end{aligned} \tag{7.338}$$

where

$$\begin{aligned}
&\hat{L}(x,s,u) \\
&= \tfrac{1}{2}q_T(x(T)-\hat{x}(T))^2 + \tfrac{1}{2}\hat{q}_T\hat{x}^2(T) + \hat{\epsilon}_{3T}\hat{x}(T) \\
&+\textstyle\int_0^T (\tfrac{1}{2}q(x-\hat{x})^2 + \tfrac{1}{2}\hat{q}\hat{x}^2 + \hat{\epsilon}_3\hat{x})dt \\
&+\textstyle\int_0^T (\tfrac{1}{2}r_i(u_i-\hat{u}_i)^2 + \tfrac{1}{2}\hat{r}_i\hat{u}_i^2 + \epsilon_{i1}(x-\hat{x})(u_i-\hat{u}_i) + \hat{\epsilon}_{i1}\hat{x}\hat{u}_i + \hat{\epsilon}_{i2}\hat{u}_i)dt \\
&+\textstyle\int_0^T (\tfrac{1}{2}r_j(u_j-\hat{u}_j)^2 + \tfrac{1}{2}\hat{r}_j\hat{u}_j^2 + \epsilon_{j1}(x-\hat{x})(u_j-\hat{u}_j) + \hat{\epsilon}_{j1}\hat{x}\hat{u}_j + \hat{\epsilon}_{j2}\hat{u}_j)dt \\
&+\textstyle\int_0^T (\xi_{0,B_o} + \xi_{1,B_o}\hat{x} + \tfrac{1}{2}\xi_{2,B_o}(x-\hat{x})^2 + \tfrac{1}{2}\hat{\xi}_{2,B_o}\hat{x}^2)dB_o \\
&+\textstyle\int_0^T\int_\Theta (\xi_{0,N_o} + \xi_{1,N_o}\hat{x} + \tfrac{1}{2}\xi_{2,N_o}(x-\hat{x})^2 + \tfrac{1}{2}\hat{\xi}_{2,N_o}\hat{x}^2)\tilde{N}_o(dt,d\theta),
\end{aligned} \tag{7.339}$$

and

$$\hat{\phi} = \phi + \bar{\phi},$$
$$\phi \in \{q_T, \epsilon_{3T}, q, r_i, \epsilon_{ik}, \xi_{2,B_o}, \xi_{2,N_o}, b_k, \sigma, \sigma_o, \mu, \mu_o\}.$$

We note that Problem (7.333) is exactly the same as Problem (7.337). To see it, use the decompositions $x = x - \hat{x} + \hat{x}$ and $u_j = u_j - \hat{u}_j + \hat{u}_j$ noticing that both $E[\hat{x}(x - \hat{x}) \mid \mathscr{F}^{B_o,N_o,s}] = 0$ and $E[\hat{u}_j(u_j - \hat{u}_j) \mid \mathscr{F}^{B_o,N_o,s}] = 0$, i.e., $x - \hat{x} \perp \hat{x}$ and $u_j - \hat{u}_j \perp \hat{u}_j$ conditioned on the filtration $\mathscr{F}^{B_o,N_o,s}$.

7.8.2.2 Semi-explicit Solution by Means of the Direct Method

We construct a semi-explicit solution to Problems (7.333) and (7.337) via a direct method. To do so, we consider the following candidate guess cost functional:

$$f(t, x, s) = \frac{1}{2}\alpha(x - \hat{x})^2 + \frac{1}{2}\beta\hat{x}^2 + \gamma\hat{x} + \delta,$$

where $\alpha = \alpha(t, s)$, $\beta = \beta(t, s)$, $\gamma = \gamma(t, s)$, and $\delta = \delta(t, s)$, are stochastic processes adapted to the filtration $\mathscr{F}^{B_o,N_o,s}$, and $x = x^{u,x_0,s_0}$ is a solution to (7.338) starting from x_0, s_0 at time 0 under the strategy profile u. The terms $(x - \hat{x})\hat{x}$ and $(x - \hat{x})$ are not included in the guess cost functional f because their conditional expectations do not contribute to the cost.

7.8.2.3 Itô's Formula

We aim to develop an Itô's formula for the guess functional. We compute the stochastic integro-differential equation satisfied by $\hat{x}$. It is given by

$$\begin{aligned}
d\hat{x} &= \left(b_0 + \hat{b}_1\hat{x} + \textstyle\sum_{j\in\mathscr{I}} \hat{b}_{j2}\hat{u}_j\right) dt \\
&+ \left(\sigma_{o0} + \hat{\sigma}_{o1}\hat{x} + \textstyle\sum_{j\in\mathscr{I}} \hat{\sigma}_{oj2}\hat{u}_j\right) dB_o \\
&+ \textstyle\int_\Theta \left(\mu_{o0} + \hat{\mu}_{o1}\hat{x} + \sum_{j\in\mathscr{I}} \hat{\mu}_{oj2}\hat{u}_j\right) \tilde{N}_o(dt, d\theta), \\
\hat{x}(0) &= \hat{x}_0,\ s(0) = s_0,
\end{aligned} \tag{7.340}$$

where we have used the fact that $E[\tilde{N}|\mathscr{F}_{t-}^{B_o,N_o,s}] = 0,\ E[B|\mathscr{F}_{t-}^{B_o,N_o,s}] = 0$. It is clear that $\hat{x}$ is a jump-diffusion-regime switching process driven by B_o, N_o, s.

$$
\begin{aligned}
df_i = &\Big\{d\alpha + 2b_1\alpha dt + \alpha[\sigma_1^2 + \sigma_{o1}^2 + \textstyle\int_\Theta \mu_1^2\nu(d\theta) \\
&+ \textstyle\int_\Theta \mu_{o1}^2\nu_o(d\theta)] + \sum_{s'\neq s}\tilde{q}_{ss'}[\alpha(t,s') - \alpha(t,s)] \\
&+ 2\alpha_{B_o}\sigma_{o1} + 2\textstyle\int_\Theta \alpha_{N_o}\mu_{o1}\nu_o(d\theta) \\
&+ 2\sigma_{o1}\alpha dB_o + 2\textstyle\int_\Theta \mu_{o1}\alpha\tilde{N}_o(dt,d\theta)\Big\}\frac{1}{2}(x-\hat{x})^2 \\
&+ \Big\{d\beta + 2\hat{b}_1\beta dt + \hat{\sigma}_{o1}^2\beta + \textstyle\int_\Theta \hat{\mu}_{o1}^2\nu_o(d\theta)\beta dt \\
&+ \textstyle\sum_{s'\neq s}\tilde{q}_{ss'}[\beta(t,s') - \beta(t,s)]dt \\
&+ \alpha[\hat{\sigma}_1^2 + \textstyle\int_\Theta \hat{\mu}_1^2\nu(d\theta)]dt \\
&+ 2\beta_{B_o}\hat{\sigma}_{o1} + 2\textstyle\int_\Theta \beta_{N_o}\hat{\mu}_{o1}\nu_o(d\theta)dt \\
&+ 2(\sigma_{o1} + \bar{\sigma}_{o1})\beta dB_o + 2\textstyle\int_\Theta \hat{\mu}_{o1}\beta\tilde{N}_o(dt,d\theta)\Big\}\frac{1}{2}\hat{x}^2 dt \\
&+ \Big\{d\gamma + \hat{b}_1\gamma + \textstyle\sum_{s'\neq s}\tilde{q}_{ss'}[\gamma(t,s') - \gamma(t,s)] \\
&+ \alpha[\sigma_0\hat{\sigma}_1 + \textstyle\int_\Theta \mu_0\hat{\mu}_1\nu(d\theta)] \\
&+ \beta[b_0 + \sigma_{o0}\hat{\sigma}_{o1} + \textstyle\int_\Theta \mu_{o0}\hat{\mu}_{o1}\nu_o(d\theta)] + \beta_{B_o}\sigma_{o0} \\
&+ \textstyle\int_\Theta \beta_{N_o}\mu_{o0}\nu_o(d\theta) + \gamma_{B_o}\hat{\sigma}_{o1} + \int_\Theta \gamma_{N_o}\hat{\mu}_{o1}\nu_o(d\theta)dt \\
&+ \{\sigma_{o0}\beta + \hat{\sigma}_{o1}\gamma\}dB_o + \textstyle\int_\Theta\{\mu_{o0}\beta + \hat{\mu}_{o1}\gamma\}\tilde{N}_o(dt,d\theta)\Big\}\hat{x} \\
&+ d\delta + \frac{\alpha}{2}(\sigma_0^2 + \textstyle\int_\Theta \mu_0^2\nu(d\theta))dt \\
&+ \frac{\beta}{2}\{\sigma_{o0}^2 + \textstyle\int_\Theta \mu_{o0}^2\nu_o(d\theta)\}dt \\
&+ b_0\gamma dt + \gamma_{B_o}\sigma_{o0}dt + \textstyle\int_\Theta \gamma_{N_o}(t,\theta)\mu_{o0}\nu_o(d\theta)dt \\
&+ \textstyle\sum_{s'\neq s}\tilde{q}_{ss'}[\delta(t,s') - \delta(t,s)]dt \\
&+ \sigma_{o0}\gamma dB_o + \textstyle\int_\Theta \mu_{o0}\gamma\tilde{N}_o(dt,d\theta) \\
&+ \frac{\alpha}{2}[\textstyle\sum_{j=1}^2 \sigma_{j2}(u_j - \hat{u}_j)]^2 \\
&+ \textstyle\int_\Theta \frac{\alpha}{2}[\sum_{j=1}^2 \mu_{j2}(u_j - \hat{u}_j)]^2\nu(d\theta) \\
&+ \textstyle\sum_{j=1}^2 b_{j2}\alpha(x-\hat{x})(u_j - \hat{u}_j)dt \\
&+ \alpha\textstyle\sum_{j=1}^2\{\sigma_1\sigma_{j2} + \int_\Theta \mu_1\mu_{j2}\nu(d\theta)\}(x-\hat{x})(u_j - \hat{u}_j)
\end{aligned}
$$

$$
\begin{aligned}
&+ \frac{\alpha}{2}[\textstyle\sum_{j=1}^2 \sigma_{oj2}(u_j - \hat{u}_j)]^2 \\
&+ \frac{\alpha}{2}\textstyle\int_\Theta[\sum_{j=1}^2 \mu_{oj2}(u_j - \hat{u}_j)]^2\nu_o(d\theta) \\
&+ \alpha\textstyle\sum_{j=1}^2\{\sigma_{o1}\sigma_{oj2} + \int_\Theta \mu_{o1}\mu_{oj2}\}(x-\hat{x})(u_j - \hat{u}_j)\nu_o(d\theta) \\
&+ \textstyle\sum_{j=1}^2\{\alpha_{B_o}\sigma_{oj2} \\
&+ \textstyle\int_\Theta \alpha_{N_o}\mu_{oj2}\nu_o(d\theta)\}(x-\hat{x})(u_j - \hat{u}_j)dt \\
&+ \textstyle\sum_{j=1}^2 \sigma_{oj2}\alpha(x-\hat{x})(u_j - \hat{u}_j)dB_o \\
&+ \textstyle\int_\Theta\sum_{j=1}^2 \mu_{oj2}\alpha(x-\hat{x})(u_j - \hat{u}_j)\tilde{N}_o(dt,d\theta) \\
&+ \frac{\alpha}{2}[\textstyle\sum_{j=1}^2 \hat{\sigma}_{j2}\hat{u}_j]^2 + \int_\Theta \frac{\alpha}{2}[\sum_{j=1}^2 \hat{\mu}_{j2}\hat{u}_j]^2\nu(d\theta) \\
&+ \frac{\beta}{2}[\textstyle\sum_{j=1}^2 \hat{\sigma}_{oj2}\hat{u}_j]^2 dt + \frac{\beta}{2}\int_\Theta[\sum_{j=1}^2 \hat{\mu}_{oj2}\hat{u}_j]^2\nu_o(d\theta)dt \\
&+ \alpha\textstyle\sum_{j=1}^2\{\hat{\sigma}_1\hat{\sigma}_{j2} + \int_\Theta \hat{\mu}_1\hat{\mu}_{j2}\nu(d\theta)\}\hat{x}\hat{u}_j \\
&+ \beta\textstyle\sum_{j=1}^2\{\hat{b}_{j2} + \hat{\sigma}_{o1}\hat{\sigma}_{oj2} + \int_\Theta \hat{\mu}_{o1}\hat{\mu}_{oj2}\nu_o(d\theta)\}\hat{u}_j\hat{x} \\
&+ \textstyle\sum_{j=1}^2\{\beta_{B_o}\hat{\sigma}_{oj2} + \int_\Theta \beta_{N_o}\hat{\mu}_{oj2}\}\hat{x}\hat{u}_j dt \\
&+ \textstyle\sum_{j=1}^2 \hat{\sigma}_{oj2}\beta\hat{x}\hat{u}_j dB_o + \int_\Theta\sum_{j=1}^2 \hat{\mu}_{oj2}\beta\hat{x}\hat{u}_j\ \tilde{N}_o(dt,d\theta) \\
&+ \alpha\textstyle\sum_{j=1}^2 \sigma_0\hat{\sigma}_{j2}\hat{u}_j + \alpha\sum_{j=1}^2\int_\Theta \mu_0\hat{\mu}_{j2}\hat{u}_j\nu(d\theta) \\
&+ \beta\textstyle\sum_{j=1}^2\{\sigma_{o0}\hat{\sigma}_{oj2} + \int_\Theta \mu_{o0}\hat{\mu}_{oj2}\nu_o(d\theta)\}\hat{u}_j \\
&+ \textstyle\sum_{j=1}^2 \hat{b}_{j2}\gamma\hat{u}_j dt \\
&+ \textstyle\sum_{j=1}^2\{\gamma_{B_o}\hat{\sigma}_{oj2} + \int_\Theta \gamma_{N_o}(t,\theta)\hat{\mu}_{oj2}\nu_o(d\theta)\}\hat{u}_j dt \\
&+ \textstyle\sum_{j=1}^2 \hat{\sigma}_{oj2}\gamma\hat{u}_j dB_o + \sum_{j=1}^2\int_\Theta \hat{\mu}_{oj2}\gamma\hat{u}_j\tilde{N}_o(dt,d\theta) + \tilde{\tilde{\epsilon}}.
\end{aligned}
$$

7.8.2.4 Guess Functional Identification

We compute difference between the cost functional and candidate guess one $\hat{L}(x, s, u) - f(0, x_0, s_0) \mid \mathcal{F}^{B_o,N_o,s}$, which is

$$
\begin{aligned}
&\hat{L}(x, s, u) - f(0, x_0, s_0) \\
&= \tfrac{1}{2}(q_T - \alpha(T))(x(T) - \hat{x}(T))^2] \\
&+\tfrac{1}{2}(\hat{q}_T - \beta(T))\hat{x}_T^2 + (\hat{\epsilon}_{3T} - \gamma_T)\hat{x}(T) \\
&+\int_0^T \Big[\{d\alpha + 2b_1\alpha dt \\
&+\alpha[\sigma_1^2 + \sigma_{o1}^2 + \textstyle\int_\Theta \mu_1^2\nu(d\theta) + \int_\Theta \mu_{o1}^2\nu_o(d\theta)]dt \\
&+q dt + \textstyle\sum_{s'\neq s} \tilde{q}_{ss'}[\alpha(t, s') - \alpha(t, s)]dt \\
&+2\alpha_{B_o}\sigma_{o1}dt + 2\textstyle\int_\Theta \alpha_{N_o}\mu_{o1}\nu_o(d\theta)dt \\
&+\{\xi_{2,B_o} + 2\sigma_{o1}\alpha\}dB_o \\
&+\textstyle\int_\Theta\{\xi_{2,N_o} + 2\mu_{o1}\alpha\}\tilde{N}_o(dt, d\theta)\Big\}\tfrac{1}{2}(x - \hat{x})^2 \\
&+\Big\{d\beta + 2\hat{b}_1\beta dt + \beta[\hat{\sigma}_{o1}^2 dt + \textstyle\int_\Theta \hat{\mu}_{o1}^2\nu_o(d\theta)]dt \\
&+\hat{q}dt + \textstyle\sum_{s'\neq s} \tilde{q}_{ss'}[\beta(t, s') - \beta(t, s)]dt \\
&+\alpha[\hat{\sigma}_1^2 + \textstyle\int_\Theta \hat{\mu}_1^2\nu(d\theta)]dt \\
&+2\beta_{B_o}\hat{\sigma}_{o1}dt + 2\textstyle\int_\Theta \beta_{N_o}\hat{\mu}_{o1}\nu_o(d\theta)dt \\
&\{\hat{\xi}_{i2,B_o} + 2(\sigma_{o1} + \bar{\sigma}_{o1})\beta\}dB_o \\
&+\textstyle\int_\Theta\{\hat{\xi}_{2,N_o} + 2\hat{\mu}_{o1}\beta\}\tilde{N}_o(dt, d\theta)\Big\}\tfrac{1}{2}\hat{x}^2 dt \\
&+\Big\{d\gamma + \hat{b}_1\gamma dt + \hat{\epsilon}_3 dt \\
&+\textstyle\sum_{s'\neq s} \tilde{q}_{ss'}[\gamma(t, s') - \gamma(t, s)]dt \\
&+\alpha[\sigma_0\hat{\sigma}_1 + \textstyle\int_\Theta \mu_0\hat{\mu}_1\nu(d\theta)]dt \\
&+\beta[b_0 + \sigma_{o0}\hat{\sigma}_{o1} + \textstyle\int_\Theta \mu_{o0}\hat{\mu}_{o1}\nu_o(d\theta)]dt \\
&+\beta_{B_o}\sigma_{o0}dt + \textstyle\int_\Theta \beta_{N_o}\mu_{o0}\nu_o(d\theta)dt \\
&+\gamma_{B_o}\hat{\sigma}_{o1}dt + \textstyle\int_\Theta \gamma_{N_o}\hat{\mu}_{o1}\nu_o(d\theta)dt \\
&+\{\xi_{1,B_o} + \sigma_{o0}\beta + \hat{\sigma}_{o1}\gamma\}dB_o \\
&+\textstyle\int_\Theta\{\xi_{1,N_o} + \mu_{o0}\beta + \hat{\mu}_{o1}\gamma\}\tilde{N}_o(dt, d\theta)\Big\}\hat{x} \\
&+d\delta + \tfrac{\alpha}{2}(\sigma_0^2 + \textstyle\int_\Theta \mu_0^2\nu(d\theta))dt \\
&+\tfrac{\beta}{2}\{\sigma_{o0}^2 + \textstyle\int_\Theta \mu_{o0}^2\nu_o(d\theta)\}dt \\
&+b_0\gamma dt + \gamma_{B_o}\sigma_{o0}dt + \textstyle\int_\Theta \gamma_{N_o}(t, \theta)\mu_{o0}\nu_o(d\theta)dt \\
&+\textstyle\sum_{s'\neq s} \tilde{q}_{ss'}[\delta(t, s') - \delta(t, s)]dt + \{\xi_{0,B_o} + \sigma_{o0}\gamma\}dB_o \\
&+\textstyle\int_\Theta\{\xi_{0,N_o} + \mu_{o0}\gamma\}\tilde{N}_o(dt, d\theta)
\end{aligned}
\tag{7.341}
$$

$$
\begin{aligned}
&+\tfrac{1}{2}r_i(u_i-\hat{u}_i)^2+\tfrac{\alpha}{2}[\textstyle\sum_{j=1}^2\sigma_{j2}(u_j-\hat{u}_j)]^2\\
&+\int_\Theta \tfrac{\alpha}{2}[\textstyle\sum_{j=1}^2\mu_{j2}(u_j-\hat{u}_j)]^2\nu(d\theta)dt\\
&+\epsilon_{i1}(x-\hat{x})(u_i-\hat{u}_i)+\textstyle\sum_{j=1}^2 b_{j2}\alpha(x-\hat{x})(u_j-\hat{u}_j)dt\\
&+\alpha\textstyle\sum_{j=1}^2\{\sigma_1\sigma_{j2}+\int_\Theta\mu_1\mu_{j2}\nu(d\theta)\}(x-\hat{x})(u_j-\hat{u}_j)\,dt\\
&+\tfrac{\alpha}{2}[\textstyle\sum_{j=1}^2\sigma_{oj2}(u_j-\hat{u}_j)]^2\\
&+\tfrac{\alpha}{2}\int_\Theta[\textstyle\sum_{j=1}^2\mu_{oj2}(u_j-\hat{u}_j)]^2\nu_o(d\theta)\,dt\\
&+\alpha\ \textstyle\sum_{j=1}^2\{\sigma_{o1}\sigma_{oj2}\\
&+\int_\Theta\mu_{o1}\mu_{oj2}\}(x-\hat{x})(u_j-\hat{u}_j)\nu_o(d\theta)\,dt\\
&+\ \textstyle\sum_{j=1}^2\{\alpha_{B_o}\sigma_{oj2}\\
&+\int_\Theta\alpha_{N_o}\mu_{oj2}\nu_o(d\theta)\}(x-\hat{x})(u_j-\hat{u}_j)dt\\
&+\textstyle\sum_{j=1}^2\sigma_{oj2}\alpha(x-\hat{x})(u_j-\hat{u}_j)dB_o\\
&+\int_\Theta\textstyle\sum_{j=1}^2\mu_{oj2}\alpha(x-\hat{x})(u_j-\hat{u}_j)\tilde{N}_o(dt,d\theta)\\
&+\tfrac{1}{2}\hat{r}_i\hat{u}_i^2+\tfrac{\alpha}{2}[\textstyle\sum_{j=1}^2\hat{\sigma}_{j2}\hat{u}_j]^2+\int_\Theta\tfrac{\alpha}{2}[\textstyle\sum_{j=1}^2\hat{\mu}_{j2}\hat{u}_j]^2\nu(d\theta)dt\\
&+\tfrac{\beta}{2}[\textstyle\sum_{j=1}^2\hat{\sigma}_{oj2}\hat{u}_j]^2dt+\tfrac{\beta}{2}\int_\Theta[\textstyle\sum_{j=1}^2\hat{\mu}_{oj2}\hat{u}_j]^2\nu_o(d\theta)dt\\
&+\hat{\epsilon}_{i1}\hat{x}\hat{u}_i+\alpha\textstyle\sum_{j=1}^2\{\hat{\sigma}_1\hat{\sigma}_{j2}+\int_\Theta\hat{\mu}_1\hat{\mu}_{j2}\nu(d\theta)\}\hat{x}\hat{u}_jdt\\
&+\beta\textstyle\sum_{j=1}^2\{\hat{b}_{j2}+\hat{\sigma}_{o1}\hat{\sigma}_{oj2}+\int_\Theta\hat{\mu}_{o1}\hat{\mu}_{oj2}\nu_o(d\theta)\}\hat{u}_j\hat{x}dt\\
&+\textstyle\sum_{j=1}^2\{\beta_{B_o}\hat{\sigma}_{oj2}+\int_\Theta\ \beta_{N_o}\hat{\mu}_{oj2}\}\hat{x}\hat{u}_jdt\\
&+\textstyle\sum_{j=1}^2\hat{\sigma}_{oj2}\beta\hat{x}\hat{u}_jdB_o+\int_\Theta\textstyle\sum_{j=1}^2\hat{\mu}_{oj2}\beta\hat{x}\hat{u}_j\ \tilde{N}_o(dt,d\theta)\\
&+\hat{\epsilon}_{i2}\hat{u}_i+\alpha\textstyle\sum_{j=1}^2\sigma_0\hat{\sigma}_{j2}\hat{u}_j+\alpha\textstyle\sum_{j=1}^2\int_\Theta\mu_0\hat{\mu}_{j2}\hat{u}_j\nu(d\theta)dt\\
&+\beta\textstyle\sum_{j=1}^2\{\sigma_{o0}\hat{\sigma}_{oj2}+\int_\Theta\mu_{o0}\hat{\mu}_{oj2}\nu_o(d\theta)\}\hat{u}_j\\
&+\gamma\textstyle\sum_{j=1}^2\hat{b}_{j2}\hat{u}_jdt\\
&+\textstyle\sum_{j=1}^2\{\gamma_{B_o}\hat{\sigma}_{oj2}+\int_\Theta\gamma_{N_o}(t,\theta)\hat{\mu}_{oj2}\nu_o(d\theta)\}\hat{u}_jdt\\
&+\textstyle\sum_{j=1}^2\hat{\sigma}_{oj2}\gamma\hat{u}_jdB_o+\textstyle\sum_{j=1}^2\int_\Theta\hat{\mu}_{oj2}\gamma\hat{u}_j\tilde{N}_o(dt,d\theta)+\tilde{\tilde{\epsilon}}\\
&+\tfrac{1}{2}r_k(u_k-\hat{u}_k)^2+\tfrac{1}{2}\hat{r}_k\hat{u}_k^2\\
&+\epsilon_{k1}(x-\hat{x})(u_k-\hat{u}_k)+\hat{\epsilon}_{k1}\hat{x}\hat{u}_k+\hat{\epsilon}_{k2}\hat{u}_k\Big].
\end{aligned}
$$

7.8.2.5 Completion of Squares for Attacker/Defender

We start with the completion of the terms in $(u_i-\hat{u}_i)$. Toward to end, we use the following equalities:

$$
\begin{aligned}
&\tfrac{1}{2}\left\{r_i + \alpha\left(\sigma_{i2}^2 + \int_\Theta \mu_{i2}^2\nu(d\theta)\right.\right.\\
&\left.\left.+\sigma_{oi2}^2 + \int_\Theta \mu_{oi2}^2\nu_o(d\theta)\right)\right\}(u_i-\hat{u}_i)^2\\
&+\tfrac{\alpha}{2}2\sigma_{i2}\sigma_{j2}(u_j-\hat{u}_j)(u_i-\hat{u}_i)\\
&+\int_\Theta \tfrac{\alpha}{2}2\mu_{i2}\mu_{j2}(u_j-\hat{u}_j)\nu(d\theta)(u_i-\hat{u}_i)\\
&+\tfrac{\alpha}{2}2\sigma_{oi2}\sigma_{oj2}(u_j-\hat{u}_j)(u_i-\hat{u}_i)\\
&+\tfrac{\alpha}{2}2\int_\Theta \mu_{oi2}\mu_{oj2}(u_j-\hat{u}_j)\nu_o(d\theta)\,(u_i-\hat{u}_i)\\
&+\{\epsilon_{i1}+b_{i2}\alpha\}(x-\hat{x})(u_i-\hat{u}_i)\\
&+\alpha\{\sigma_1\sigma_{i2}+\int_\Theta \mu_1\mu_{i2}\nu(d\theta)\}(x-\hat{x})(u_i-\hat{u}_i)\\
&+\alpha\,\{\sigma_{o1}\sigma_{oi2}+\int_\Theta \mu_{o1}\mu_{oi2}\nu_o(d\theta)\}(x-\hat{x})(u_i-\hat{u}_i)\\
&+\{\alpha_{B_o}\sigma_{oi2}+\int_\Theta \alpha_{N_o}\mu_{oi2}\nu_o(d\theta)\}(x-\hat{x})(u_i-\hat{u}_i)\\
&+[\sigma_{oi2}\alpha(x-\hat{x})dB_o+\int_\Theta \mu_{oi2}\alpha(x-\hat{x})\tilde{N}_o(dt,d\theta)](u_i-\hat{u}_i)\\
&+\tfrac{1}{2}(r_j+\alpha(\sigma_{j2}^2+\int_\Theta \mu_{j2}^2\nu(d\theta)))(u_j-\hat{u}_j)^2\\
&+\tfrac{\alpha}{2}(\sigma_{oj2}^2+\int_\Theta \mu_{oj2}^2\nu_o(d\theta))(u_j-\hat{u}_j)^2\\
&+(\epsilon_{j1}+b_{j2}\alpha)(x-\hat{x})(u_j-\hat{u}_j)\\
&+\alpha\{\sigma_1\sigma_{j2}+\int_\Theta \mu_1\mu_{j2}\nu(d\theta)\}(x-\hat{x})(u_j-\hat{u}_j)\\
&+\alpha\,\{\sigma_{o1}\sigma_{oj2}+\int_\Theta \mu_{o1}\mu_{oj2}\}(x-\hat{x})(u_j-\hat{u}_j)\nu_o(d\theta)\\
&+\{\alpha_{B_o}\sigma_{oj2}+\int_\Theta \alpha_{N_o}\mu_{oj2}\nu_o(d\theta)\}(x-\hat{x})(u_j-\hat{u}_j)\\
&+\sigma_{oj2}\alpha(x-\hat{x})(u_j-\hat{u}_j)dB_o\\
&+\int_\Theta \mu_{oj2}\alpha(x-\hat{x})(u_j-\hat{u}_j)\tilde{N}_o(dt,d\theta)\\
&=\tfrac{1}{2}c_i(u_i-\hat{u}_i+\tfrac{\tilde{\tau}_i}{c_i}(x-\hat{x}))^2+\tfrac{1}{2}\tilde{c}_j(u_j-\hat{u}_j+\tfrac{\tilde{\tau}_j}{c_j}(x-\hat{x}))^2\\
&-(\tfrac{v_i^2}{c_i}+\tfrac{\tilde{\tau}_j^2}{c_j})(x-\hat{x})^2\\
&-\alpha[\sigma_{oi2}\tfrac{\tau_i}{c_i}+\sigma_{oj2}\tfrac{\tau_j}{c_j}]dB_o(x-\hat{x})^2\\
&-\alpha\int_\Theta[\mu_{oi2}\tfrac{\tau_i}{c_i}+\mu_{oj2}\tfrac{\tau_j}{c_j}]\tilde{N}_o(dt,d\theta)(x-\hat{x})^2,
\end{aligned}
\tag{7.342}
$$

where

$$
\begin{aligned}
&c_i=\{r_i+\alpha\left(\sigma_{i2}^2+\int_\Theta \mu_{i2}^2\nu(d\theta)+\sigma_{oi2}^2+\int_\Theta \mu_{oi2}^2\nu_o(d\theta)\right)\},\\
&c_j=\{r_j+\alpha\left(\sigma_{j2}^2+\int_\Theta \mu_{j2}^2\nu(d\theta)+\sigma_{oj2}^2+\int_\Theta \mu_{oj2}^2\nu_o(d\theta)\right)\},\\
&\tilde{A}_{ii}=1=\tilde{A}_{jj},\\
&j\neq i,\ \tilde{A}_{ij}=\tfrac{\alpha}{c_i}[(\sigma_{i2}\sigma_{j2}+\sigma_{oi2}\sigma_{oj2})\\
&+\int_\Theta(\mu_{i2}\mu_{j2}+\mu_{oi2}\mu_{oj2})\nu(d\theta)],\\
&v_i=+\epsilon_{i1}+\alpha\{b_{i2}+\sigma_1\sigma_{i2}+\int_\Theta \mu_1\mu_{i2}\nu(d\theta)\}\\
&+\alpha\,\{\sigma_{o1}\sigma_{oi2}+\int_\Theta \mu_{o1}\mu_{oi2}\}\nu_o(d\theta)\\
&+\{\alpha_{B_o}\sigma_{oi2}+\int_\Theta \alpha_{N_o}\mu_{oi2}\nu_o(d\theta)\},\\
&v_j=\epsilon_{j1}+\alpha\{b_{j2}+\sigma_1\sigma_{j2}+\int_\Theta \mu_1\mu_{j2}\nu(d\theta)\}\\
&+\alpha\,\{\sigma_{o1}\sigma_{oj2}+\int_\Theta \mu_{o1}\mu_{oj2}\}\nu_o(d\theta)\\
&+\{\alpha_{B_o}\sigma_{oj2}+\int_\Theta \alpha_{N_o}\mu_{oj2}\nu_o(d\theta)\},\\
&\tilde{A}u+\tilde{v}(x-\hat{x})=0,\\
&\tilde{v}_i=\tfrac{v_i}{c_i},\\
&\tfrac{\tau_i}{c_i}=[\tilde{A}^{-1}v]_i,
\end{aligned}
\tag{7.343}
$$

Next we examine the completion of squares for the terms in $\hat{u}_i$. By expanding the square terms and reorganizing, we arrive at

$$
\begin{aligned}
&\tfrac{1}{2}\left\{\hat{r}_i+\alpha[\hat{\sigma}_{i2}^2+\int_\Theta \hat{\mu}_{i2}^2\nu(d\theta)]\right.\\
&\left.+\beta[\hat{\sigma}_{oi2}^2+\int_\Theta \hat{\mu}_{oi2}^2\nu_o(d\theta)]\right\}\hat{u}_i^2\\
&+\alpha\hat{\sigma}_{i2}\hat{\sigma}_{j2}\hat{u}_j\hat{u}_i\\
&+\int_\Theta \alpha\hat{\mu}_{i2}\hat{\mu}_{j2}\hat{u}_j\nu(d\theta)u_i\\
&+\beta\hat{\sigma}_{oi2}\hat{\sigma}_{oj2}\hat{u}_j\hat{u}_i\\
&+\beta\int_\Theta \hat{\mu}_{oi2}\hat{\mu}_{oj2}\hat{u}_j\nu_o(d\theta)\hat{u}_i\\
&+\hat{\epsilon}_{i1}\hat{x}\hat{u}_i+\alpha\{\hat{\sigma}_1\hat{\sigma}_{i2}+\int_\Theta \hat{\mu}_1\hat{\mu}_{i2}\nu(d\theta)\}\hat{x}\hat{u}_i\\
&+\beta\{\hat{b}_{i2}+\hat{\sigma}_{o1}\hat{\sigma}_{oi2}+\int_\Theta \hat{\mu}_{o1}\hat{\mu}_{oi2}\nu_o(d\theta)\}\hat{x}\hat{u}_i\\
&+\{\beta_{B_o}\hat{\sigma}_{oi2}+\int_\Theta\ \beta_{N_o}\hat{\mu}_{oi2}\}\hat{x}\hat{u}_i\\
&+\hat{\sigma}_{oi2}\beta\hat{x}\hat{u}_i dB_o+\int_\Theta \hat{\mu}_{oi2}\beta\hat{x}\hat{u}_i\ \tilde{N}_o(dt,d\theta)\\
&+\hat{\epsilon}_{i2}\hat{u}_i+\alpha\{\sigma_0\hat{\sigma}_{i2}+\int_\Theta \mu_0\hat{\mu}_{i2}\nu(d\theta)\}\hat{u}_i\\
&+\beta\{\sigma_{o0}\hat{\sigma}_{oi2}+\int_\Theta \mu_{o0}\hat{\mu}_{oi2}\nu_o(d\theta)\}\hat{u}_i\\
&+\{\gamma_{B_o}\hat{\sigma}_{oi2}+\int_\Theta \gamma_{N_o}(t,\theta)\hat{\mu}_{oi2}\nu_o(d\theta)\}\hat{u}_i dt\\
&+\gamma[\hat{b}_{i2}+\hat{\sigma}_{oi2}\, dB_o+\int_\Theta \hat{\mu}_{oi2}\tilde{N}_o(dt,d\theta)]\hat{u}_i\\
&+\tfrac{\alpha}{2}[\hat{\sigma}_{j2}\hat{u}_j]^2+\int_\Theta \tfrac{\alpha}{2}[\hat{\mu}_{j2}\hat{u}_j]^2\nu(d\theta)\\
&+\tfrac{\beta}{2}[\hat{\sigma}_{oj2}\hat{u}_j]^2\\
&+\tfrac{\beta}{2}\int_\Theta[\hat{\mu}_{oj2}\hat{u}_j]^2\nu_o(d\theta)\\
&+\alpha\{\hat{\sigma}_1\hat{\sigma}_{j2}+\int_\Theta \hat{\mu}_1\hat{\mu}_{j2}\nu(d\theta)\}\hat{x}\hat{u}_j\\
&+\beta\{\hat{b}_{j2}+\hat{\sigma}_{o1}\hat{\sigma}_{oj2}+\int_\Theta \hat{\mu}_{o1}\hat{\mu}_{oj2}\nu_o(d\theta)\}\hat{x}\hat{u}_j\\
&+\{\beta_{B_o}\hat{\sigma}_{oj2}+\int_\Theta\ \beta_{N_o}\hat{\mu}_{oj2}\}\hat{x}\hat{u}_j\\
&+\hat{\sigma}_{oj2}\beta\hat{x}\hat{u}_j dB_o\\
&+\int_\Theta \hat{\mu}_{oj2}\beta\hat{x}\hat{u}_j\ \tilde{N}_o(dt,d\theta)\\
&+\alpha\{\sigma_0\hat{\sigma}_{j2}+\int_\Theta \mu_0\hat{\mu}_{j2}\nu(d\theta)\}\hat{u}_j\\
&+\beta\{\sigma_{o0}\hat{\sigma}_{oj2}+\int_\Theta \mu_{o0}\hat{\mu}_{oj2}\nu_o(d\theta)\}\hat{u}_j\\
&+\{\gamma_{B_o}\hat{\sigma}_{oj2}+\int_\Theta \gamma_{N_o}(t,\theta)\hat{\mu}_{oj2}\nu_o(d\theta)\}\hat{u}_j dt\\
&+\gamma\hat{b}_{j2}\hat{u}_j\\
&+\gamma\hat{\sigma}_{oj2}\hat{u}_j dB_o+\gamma\int_\Theta \hat{\mu}_{oj2}\hat{u}_j\tilde{N}_o(dt,d\theta)\\
&+\tfrac{1}{2}\hat{r}_k\hat{u}_k^2\\
&+\hat{\epsilon}_{k1}\hat{x}\hat{u}_k+\hat{\epsilon}_{k2}\hat{u}_k=\frac{\bar{c}_i}{2}\left(\hat{u}_i+\frac{\bar{\tau}_{i1}\hat{x}+\bar{\tau}_{i0}}{\bar{c}_i}\right)^2-\frac{\bar{c}_i}{2}\tilde{A}_{ij}^2\hat{u}_j^2-\bar{c}_i\tilde{A}_{ij}\hat{u}_j\left(\frac{\bar{\tau}_{i1}\hat{x}+\bar{\tau}_{i0}}{\bar{c}_i}\right)\\
&-\frac{\bar{v}_{i1}^2}{2\bar{c}_i}\hat{x}^2-\frac{2\bar{v}_{i1}\bar{v}_{i0}}{2\bar{c}_i}\hat{x}-\frac{\bar{v}_{i0}^2}{2\bar{c}_i}\\
&+\tfrac{\alpha}{2}[\hat{\sigma}_{j2}\hat{u}_j]^2+\int_\Theta \tfrac{\alpha}{2}[\hat{\mu}_{j2}\hat{u}_j]^2\nu(d\theta)\\
&+\tfrac{\beta}{2}[\hat{\sigma}_{oj2}\hat{u}_j]^2+\tfrac{\beta}{2}\int_\Theta[\hat{\mu}_{oj2}\hat{u}_j]^2\nu_o(d\theta)\\
&+\alpha\{\hat{\sigma}_1\hat{\sigma}_{j2}+\int_\Theta \hat{\mu}_1\hat{\mu}_{j2}\nu(d\theta)\}\hat{x}\hat{u}_j\\
&+\beta\{\hat{b}_{j2}+\hat{\sigma}_{o1}\hat{\sigma}_{oj2}+\int_\Theta \hat{\mu}_{o1}\hat{\mu}_{oj2}\nu_o(d\theta)\}\hat{x}\hat{u}_j\\
&+\{\beta_{B_o}\hat{\sigma}_{oj2}+\int_\Theta\ \beta_{N_o}\hat{\mu}_{oj2}\}\hat{x}\hat{u}_j\\
&+\hat{\sigma}_{oj2}\beta\hat{x}\hat{u}_j dB_o\\
&+\int_\Theta \hat{\mu}_{oj2}\beta\hat{x}\hat{u}_j\ \tilde{N}_o(dt,d\theta)\\
&+\alpha\{\sigma_0\hat{\sigma}_{j2}+\int_\Theta \mu_0\hat{\mu}_{j2}\nu(d\theta)\}\hat{u}_j\\
&+\beta\{\sigma_{o0}\hat{\sigma}_{oj2}+\int_\Theta \mu_{o0}\hat{\mu}_{oj2}\nu_o(d\theta)\}\hat{u}_j\\
&+\{\gamma_{B_o}\hat{\sigma}_{oj2}+\int_\Theta \gamma_{N_o}(t,\theta)\hat{\mu}_{oj2}\nu_o(d\theta)\}\hat{u}_j dt\\
&+\gamma\hat{b}_{j2}\hat{u}_j+\gamma\hat{\sigma}_{oj2}\hat{u}_j dB_o\\
&+\gamma\int_\Theta \hat{\mu}_{oj2}\hat{u}_j\tilde{N}_o(dt,d\theta)\\
&+\tfrac{1}{2}\hat{r}_k\hat{u}_k^2+\hat{\epsilon}_{k1}\hat{x}\hat{u}_k+\hat{\epsilon}_{k2}\hat{u}_k,
\end{aligned}
$$

which means that

$$
\begin{aligned}
&\tfrac{\bar{c}_i}{2}\left(\hat{u}_i + \tfrac{\bar{\tau}_{i1}\hat{x}+\bar{\tau}_{i0}}{\bar{c}_i}\right)^2\\
&-\tfrac{\bar{c}_i}{2}\tilde{A}_{ij}^2\hat{u}_j^2 - \bar{c}_i\tilde{A}_{ij}\hat{u}_j(\tfrac{\bar{\tau}_{i1}\hat{x}+\bar{\tau}_{i0}}{\bar{c}_i})\\
&-\tfrac{\bar{v}_{i1}^2}{2\bar{c}_i}\hat{x}^2 - \tfrac{2\bar{v}_{i1}\bar{v}_{i0}}{2\bar{c}_i}\hat{x} - \tfrac{\bar{v}_{i0}^2}{2\bar{c}_i}\\
&+\tfrac{\alpha}{2}[\hat{\sigma}_{j2}\hat{u}_j]^2 + \int_\Theta \tfrac{\alpha}{2}[\hat{\mu}_{j2}\hat{u}_j]^2\nu(d\theta)\\
&+\tfrac{\beta}{2}[\hat{\sigma}_{oj2}\hat{u}_j]^2 + \tfrac{\beta}{2}\int_\Theta[\hat{\mu}_{oj2}\hat{u}_j]^2\nu_o(d\theta)\\
&+\alpha\{\hat{\sigma}_1\hat{\sigma}_{j2} + \int_\Theta \hat{\mu}_1\hat{\mu}_{j2}\nu(d\theta)\}\hat{x}\hat{u}_j\\
&+\beta\{\hat{b}_{j2} + \hat{\sigma}_{o1}\hat{\sigma}_{oj2} + \int_\Theta \hat{\mu}_{o1}\hat{\mu}_{oj2}\nu_o(d\theta)\}\hat{x}\hat{u}_j\\
&+\{\beta_{B_o}\hat{\sigma}_{oj2} + \int_\Theta \beta_{N_o}\hat{\mu}_{oj2}\}\hat{x}\hat{u}_j\\
&+\hat{\sigma}_{oj2}\beta\hat{x}\hat{u}_j dB_o\\
&+\int_\Theta \hat{\mu}_{oj2}\beta\hat{x}\hat{u}_j\ \tilde{N}_o(dt,d\theta)\\
&+\alpha\{\sigma_0\hat{\sigma}_{j2} + \int_\Theta \mu_0\hat{\mu}_{j2}\nu(d\theta)\}\hat{u}_j\\
&+\beta\{\sigma_{o0}\hat{\sigma}_{oj2} + \int_\Theta \mu_{o0}\hat{\mu}_{oj2}\nu_o(d\theta)\}\hat{u}_j\\
&+\{\gamma_{B_o}\hat{\sigma}_{oj2} + \int_\Theta \gamma_{N_o}(t,\theta)\hat{\mu}_{oj2}\nu_o(d\theta)\}\hat{u}_j dt\\
&+\gamma\hat{b}_{j2}\hat{u}_j + \gamma\hat{\sigma}_{oj2}\hat{u}_j dB_o\\
&+\gamma\int_\Theta \hat{\mu}_{oj2}\hat{u}_j\tilde{N}_o(dt,d\theta)\\
&+\tfrac{1}{2}\hat{r}_k\hat{u}_k^2 + \hat{\epsilon}_{k1}\hat{x}\hat{u}_k + \hat{\epsilon}_{k2}\hat{u}_k\\
&= \tfrac{\bar{c}_i}{2}\left(\hat{u}_i + \tfrac{\bar{\tau}_{i1}\hat{x}+\bar{\tau}_{i0}}{\bar{c}_i}\right)^2 + \tfrac{\bar{\bar{c}}_j}{2}\left(\hat{u}_j + \tfrac{\bar{\tau}_{j1}\hat{x}+\bar{\tau}_{j0}}{\bar{c}_j}\right)^2\\
&-(\tfrac{\bar{v}_{i1}^2}{\bar{c}_i} + \tfrac{\bar{\tau}_{j1}^2}{\bar{c}_j})\tfrac{1}{2}\hat{x}^2\\
&-(\tfrac{\bar{v}_{i1}\bar{v}_{i0}}{\bar{c}_i} + \tfrac{\bar{\tau}_{j1}\bar{\tau}_{j0}}{\bar{c}_j})\hat{x}\\
&-(\tfrac{\bar{v}_{i0}^2}{2\bar{c}_i} + \tfrac{\bar{\tau}_{j0}^2}{2\bar{c}_j})\\
&+\textstyle\sum_j \hat{\sigma}_{oj2}\beta\hat{x}\hat{u}_j dB_o\\
&+\int_\Theta\textstyle\sum_j \hat{\mu}_{oj2}\beta\hat{x}\hat{u}_j\ \tilde{N}_o(dt,d\theta)\\
&+\gamma\textstyle\sum_j \hat{\sigma}_{oj2}\hat{u}_j dB_o + \gamma\sum_j\int_\Theta \hat{\mu}_{oj2}\hat{u}_j\tilde{N}_o(dt,d\theta)\\
&= \tfrac{\bar{c}_i}{2}\left(\hat{u}_i + \tfrac{\bar{\tau}_{i1}\hat{x}+\bar{\tau}_{i0}}{\bar{c}_i}\right)^2 + \tfrac{\bar{\bar{c}}_j}{2}\left(\hat{u}_j + \tfrac{\bar{\tau}_{j1}\hat{x}+\bar{\tau}_{j0}}{\bar{c}_j}\right)^2 - (\tfrac{\bar{\tau}_{i1}^2}{\bar{c}_i} + \tfrac{\bar{\tau}_{j1}^2}{\bar{c}_j})\tfrac{1}{2}\hat{x}^2\\
&-2\beta\textstyle\sum_j \hat{\sigma}_{oj2}\tfrac{\bar{\tau}_{j1}}{c_j}\tfrac{1}{2}\hat{x}^2 dB_o\\
&-2\beta\int_\Theta\textstyle\sum_j \hat{\mu}_{oj2}\tfrac{\bar{\tau}_{j1}}{c_j}\tfrac{1}{2}\hat{x}^2\ \tilde{N}_o(dt,d\theta)\\
&-(\tfrac{\bar{v}_{i1}\bar{v}_{i0}}{\bar{c}_i} + \tfrac{\bar{\tau}_{j1}\bar{\tau}_{j0}}{\bar{c}_j})\hat{x} - \textstyle\sum_j \hat{\sigma}_{oj2}\beta\tfrac{\bar{\tau}_{jo}}{c_j}\hat{x}dB_o\\
&-\gamma\textstyle\sum_j \hat{\sigma}_{oj2}\tfrac{\bar{\tau}_{j1}}{c_j}\hat{x}dB_o - \int_\Theta\sum_j \hat{\mu}_{oj2}\beta\tfrac{\bar{\tau}_{j0}}{c_j}\hat{x}\ \tilde{N}_o(dt,d\theta)\\
&-\gamma\textstyle\sum_j\int_\Theta \hat{\mu}_{oj2}\tfrac{\bar{\tau}_{j1}}{c_j}\hat{x}\tilde{N}_o(dt,d\theta)\\
&-(\tfrac{\bar{v}_{i0}^2}{2\bar{c}_i} + \tfrac{\bar{\tau}_{j0}^2}{2\bar{c}_j})\\
&-\gamma\textstyle\sum_j \hat{\sigma}_{oj2}\tfrac{\bar{\tau}_{j0}}{c_j}dB_o - \gamma\sum_j\int_\Theta \hat{\mu}_{oj2}\tfrac{\bar{\tau}_{j0}}{c_j}\tilde{N}_o(dt,d\theta),
\end{aligned}
\tag{7.344}
$$

where $\bar{\tau}_{i0}, \bar{\tau}_{i1}, \bar{c}_i$ are given by

$$
\begin{aligned}
&\tilde{\tilde{A}}_{ii} = 1 = \tilde{\tilde{A}}_{jj},\ \tilde{\tilde{A}}_{ij} = \frac{\alpha}{\bar{c}_i}\{\hat{\sigma}_{i2}\hat{\sigma}_{j2} + \int_\Theta \hat{\mu}_{i2}\hat{\mu}_{j2}\nu(d\theta)\} \\
&+\frac{\beta}{\bar{c}_i}\{\hat{\sigma}_{oi2}\hat{\sigma}_{oj2} + \int_\Theta \hat{\mu}_{oi2}\hat{\mu}_{oj2}\nu_o(d\theta)\}, \\
&\bar{c}_i = \hat{r}_i + \alpha[\hat{\sigma}_{i2}^2 + \int_\Theta \hat{\mu}_{i2}^2\nu(d\theta)] \\
&+\beta[\hat{\sigma}_{oi2}^2 + \int_\Theta \hat{\mu}_{oi2}^2\nu_o(d\theta)], \\
&\bar{c}_j = \hat{r}_j + \alpha[\hat{\sigma}_{j2}^2 + \int_\Theta \hat{\mu}_{j2}^2\nu(d\theta)] \\
&+\beta[\hat{\sigma}_{oj2}^2 + \int_\Theta \hat{\mu}_{oj2}^2\nu_o(d\theta)], \\
&\bar{v}_{i1} = \hat{\epsilon}_{i1} + \alpha[\hat{\sigma}_1\hat{\sigma}_{i2} + \int_\Theta \hat{\mu}_1\hat{\mu}_{i2}\nu(d\theta)] \\
&+\beta[\hat{b}_{i2} + \hat{\sigma}_{o1}\hat{\sigma}_{oi2} + \int_\Theta \hat{\mu}_{o1}\hat{\mu}_{oi2}\nu_o(d\theta)] \\
&+\beta_{B_o}\hat{\sigma}_{oi2} + \int_\Theta \beta_{N_o}\hat{\mu}_{oi2}, \\
&\bar{v}_{i0} = +\hat{\epsilon}_{i2} + \alpha[\sigma_0\hat{\sigma}_{i2} + \int_\Theta \mu_0\hat{\mu}_{i2}\nu(d\theta)] \\
&+\beta[\sigma_{o0}\hat{\sigma}_{oi2} + \int_\Theta \mu_{o0}\hat{\mu}_{oi2}\nu_o(d\theta)] \\
&+\gamma\hat{b}_{i2} + \gamma_{B_o}\hat{\sigma}_{oi2} + \int_\Theta \gamma_{N_o}(t,\theta)\hat{\mu}_{oi2}\nu_o(d\theta), \\
&\tilde{\bar{c}}_j = \bar{c}_j\det(\tilde{A}), \\
&\tilde{\bar{v}}_{ik} = \frac{\bar{v}_{ik}}{c_i}, \\
&\frac{\bar{\tau}_{jk}}{c_j} = [\tilde{A}^{-1}v_k]_j,
\end{aligned}
\tag{7.345}
$$

7.8.2.6 Identification Process

We now replace the expressions of (7.342) by their corresponding values in (7.341) to obtain the following equivalent expression:

$$
\begin{aligned}
&\hat{L}(x,s,u) - f(0,x_0,s_0)\\
&= \tfrac{1}{2}(q_T - \alpha(T))(x(T) - \hat{x}(T))^2]\\
&+\tfrac{1}{2}(\hat{q}_T - \beta(T))\hat{x}_T^2 + (\hat{\epsilon}_{3T} - \gamma_T)\hat{x}(T)\\
&+\int_0^T \Big[\ \{d\alpha + 2b_1\alpha dt\\
&+\alpha[\sigma_1^2 + \sigma_{o1}^2 + \textstyle\int_\Theta \mu_1^2\nu(d\theta) + \int_\Theta \mu_{o1}^2\nu_o(d\theta)]dt\\
&+qdt - (\tfrac{\bar{v}_i^2}{c_i} + \tfrac{\bar{\tau}_j^2}{c_j})dt + \textstyle\sum_{s'\neq s}\tilde{q}_{ss'}[\alpha(t,s') - \alpha(t,s)]dt\\
&+2\alpha_{B_o}\sigma_{o1}dt + 2\textstyle\int_\Theta \alpha_{N_o}\mu_{o1}\nu_o(d\theta)dt\\
&+\{\xi_{2,B_o} + 2\sigma_{o1}\alpha - 2\alpha[\sigma_{oi2}\tfrac{\bar{\tau}_i}{c_i} + \sigma_{oj2}\tfrac{\bar{\tau}_j}{c_j}]\}dB_o\\
&+\textstyle\int_\Theta\{\xi_{2,N_o} + 2\mu_{o1}\alpha - 2\alpha[\mu_{oi2}\tfrac{\bar{\tau}_i}{c_i} + \mu_{oj2}\tfrac{\bar{\tau}_j}{c_j}]\}\tilde{N}_o(dt,d\theta)\\
&\}\ \tfrac{1}{2}(x-\hat{x})^2\\
&+\Big\{d\beta + 2\hat{b}_1\beta dt + \beta[\hat{\sigma}_{o1}^2 dt + \textstyle\int_\Theta \hat{\mu}_{o1}^2\nu_o(d\theta)]dt\\
&+\hat{q}dt - (\tfrac{\bar{v}_{i1}^2}{\bar{c}_i} + \tfrac{\bar{\tau}_{j1}^2}{\bar{c}_j})dt + \textstyle\sum_{s'\neq s}\tilde{q}_{ss'}[\beta(t,s') - \beta(t,s)]dt\\
&+\alpha[\hat{\sigma}_1^2 + \textstyle\int_\Theta \hat{\mu}_1^2\nu(d\theta)]dt\\
&+2\beta_{B_o}\hat{\sigma}_{o1}dt + 2\textstyle\int_\Theta \beta_{N_o}\hat{\mu}_{o1}\nu_o(d\theta)dt\\
&\{\hat{\xi}_{i2,B_o} + 2\hat{\sigma}_{o1}\beta - 2\beta\textstyle\sum_j \hat{\sigma}_{oj2}\tfrac{\bar{\tau}_{j1}}{c_j}\}dB_o\\
&+\textstyle\int_\Theta\{\hat{\xi}_{2,N_o} + 2\hat{\mu}_{o1}\beta - 2\beta\sum_j \hat{\mu}_{oj2}\tfrac{\bar{\tau}_{j1}}{c_j}\}\tilde{N}_o(dt,d\theta)\Big\}\ \tfrac{1}{2}\hat{x}^2 dt\\
&+\Big\{d\gamma + \hat{b}_1\gamma dt + \hat{\epsilon}_3 dt - (\tfrac{\bar{v}_{i1}\bar{v}_{i0}}{\bar{c}_i} + \tfrac{\bar{\tau}_{j1}\bar{\tau}_{j0}}{\bar{c}_j})dt\\
&+\textstyle\sum_{s'\neq s}\tilde{q}_{ss'}[\gamma(t,s') - \gamma(t,s)]dt\\
&+\alpha[\sigma_0\hat{\sigma}_1 + \textstyle\int_\Theta \mu_0\hat{\mu}_1\nu(d\theta)]dt\\
&+\beta[b_0 + \sigma_{o0}\hat{\sigma}_{o1} + \textstyle\int_\Theta \mu_{o0}\hat{\mu}_{o1}\nu_o(d\theta)]dt\\
&+\beta_{B_o}\sigma_{o0}dt + \textstyle\int_\Theta \beta_{N_o}\mu_{o0}\nu_o(d\theta)dt\\
&+\gamma_{B_o}\hat{\sigma}_{o1}dt + \textstyle\int_\Theta \gamma_{N_o}\hat{\mu}_{o1}\nu_o(d\theta)dt\\
&+\{\xi_{1,B_o} + \beta[\sigma_{o0} - \textstyle\sum_j \hat{\sigma}_{oj2}\tfrac{\bar{\tau}_{j0}}{c_j}] + \gamma[\hat{\sigma}_{o1} - \sum_j \hat{\sigma}_{oj2}\tfrac{\bar{\tau}_{j1}}{c_j}]\}dB_o\\
&+\textstyle\int_\Theta\Big\{\xi_{1,N_o} + \beta[\mu_{o0} - \sum_j \hat{\mu}_{oj2}\tfrac{\bar{\tau}_{j0}}{c_j}]\\
&+\gamma[\hat{\mu}_{o1} - \textstyle\sum_j \hat{\mu}_{oj2}\tfrac{\bar{\tau}_{j1}}{c_j}]\Big\}\ \tilde{N}_o(dt,d\theta)\Big\}\ \hat{x}\\
&+d\delta - (\tfrac{\bar{v}_{i0}^2}{2\bar{c}_i} + \tfrac{\bar{\tau}_{j0}^2}{2\bar{c}_j})dt + \tfrac{\alpha}{2}(\sigma_0^2 + \textstyle\int_\Theta \mu_0^2\nu(d\theta))dt\\
&+\tfrac{\beta}{2}(\sigma_{o0}^2 + \textstyle\int_\Theta \mu_{o0}^2\nu_o(d\theta))dt\\
&+b_0\gamma dt + \gamma_{B_o}\sigma_{o0}dt + \textstyle\int_\Theta \gamma_{N_o}(t,\theta)\mu_{o0}\nu_o(d\theta)dt\\
&+\textstyle\sum_{s'\neq s}\tilde{q}_{ss'}[\delta(t,s') - \delta(t,s)]dt\\
&+\{\xi_{0,B_o} + \gamma[\sigma_{o0} - \textstyle\sum_j \hat{\sigma}_{oj2}\tfrac{\bar{\tau}_{j0}}{c_j}]\}dB_o\\
&+\textstyle\int_\Theta\{\xi_{0,N_o} + \gamma[\mu_{o0} - \sum_j \hat{\mu}_{oj2}\tfrac{\bar{\tau}_{j0}}{c_j}]\}\tilde{N}_o(dt,d\theta)\\
&+\tfrac{1}{2}c_i(u_i - \hat{u}_i + \tfrac{\bar{\tau}_i}{c_i}(x-\hat{x}))^2 + \tfrac{\bar{c}_i}{2}\Big(\hat{u}_i + \tfrac{\bar{\tau}_{i1}\hat{x} + \bar{\tau}_{i0}}{\bar{c}_i}\Big)^2\\
&+\tfrac{1}{2}\tilde{c}_j(u_j - \hat{u}_j + \tfrac{\bar{\tau}_j}{c_j}(x-\hat{x}))^2 + \tfrac{\bar{\tilde{c}}_j}{2}\Big(\hat{u}_j + \tfrac{\bar{\tau}_{j1}\hat{x} + \bar{\tau}_{j0}}{\bar{c}_j}\Big)^2\Big],
\end{aligned}
\tag{7.346}
$$

The minimization/maximization (minimax) of the above expression (u_i, u_j) establishes the desired results provided that c_i and $\bar{c}_i$ are positive and $\tilde{c}_j$ and $\tilde{\bar{c}}_j$ are negative.

The inf sup of the following term

$$\begin{aligned}&+\frac{1}{2}c_i(u_i-\hat{u}_i+\frac{\tilde{\tau}_i}{c_i}(x-\hat{x}))^2+\frac{\bar{c}_i}{2}\left(\hat{u}_i+\frac{\bar{\tau}_{i1}\hat{x}+\bar{\tau}_{i0}}{\bar{c}_i}\right)^2\\&+\frac{1}{2}\tilde{c}_j(u_j-\hat{u}_j+\frac{\tilde{\tau}_j}{c_j}(x-\hat{x}))^2+\frac{\tilde{\bar{c}}_j}{2}\left(\hat{u}_j+\frac{\bar{\tau}_{j1}\hat{x}+\bar{\tau}_{j0}}{\bar{c}_j}\right)^2\end{aligned} \tag{7.347}$$

is achieved whenever

$$c_i>0, \bar{c}_i>0, \tilde{c}_j<0, \tilde{\bar{c}}_j<0,$$

thanks to strict convex–concave property. Moreover, $\inf\sup = \sup\inf$, thus providing a saddle-point solution. This completes the proof.

7.8.3 Direct Method Versus Dynamic Programming

We now develop a dynamic programming method to find the expressions for optimal strategies in state-mean-field feedback form. For the sake of simplicity, we examine only the case where the common noises are state-and-control independent. The stochastic process is given by

$$\begin{aligned}x(t)&=x(0)+\int_0^t bdt+\int_0^t\sigma dB+\int_0^t\int_\Theta \mu\tilde{N}(.,d\theta)\\&+\int_0^t\sigma_o dB_o+\int_0^t\int_\Theta\mu_o\tilde{N}_o(.,d\theta)\\&=x_{nc}+x_c,\end{aligned} \tag{7.348}$$

with $x_c=x(0)+\int_0^t\sigma_o dB_o+\int_0^t\int_\Theta\mu_o\tilde{N}_o(.,d\theta)$, and $x_{nc}(t)=\int_0^t bdt+\int_0^t\sigma dB+\int_0^t\int_\Theta\mu\tilde{N}(.,d\theta)$. $\sigma_o=\sigma_o(t)$, $\mu_o=\mu_o(t,\theta)$.

The probability measure of x is m, defined in precise terms as

$$\begin{aligned}m(t,\mathscr{X},s)&=\mathbb{P}(x(t)\in\mathscr{X},s)=\mathbb{P}(x_{nc}(t)+x_c(t)\in\mathscr{X},s)\\&=\mathbb{P}(x_{nc}(t)\in\mathscr{X}-x_c(t),s)=m_{nc}(t,\mathscr{X}-x_c(t),s).\end{aligned} \tag{7.349}$$

Combining integration by parts and Itô's formula for jump-diffusion processes, it is easily be shown that the probability distribution function of x_{nc} solves (in the distribution sense) the Fokker–Planck–Kolmogorov equation given by

$$\begin{aligned}&m_{nc,t}+(m_{nc}b)_x-\frac{1}{2}(m_{nc}\sigma^2)_{xx}+J^*[m_{nc}]+S^*[m_{nc}]=0,\\&m_{nc}(0,.)=m(0,.),\end{aligned} \tag{7.350}$$

where $m(t,x,s) = m_{nc}(t,y,s)$, with $y = x - x_c(t)$, which means that the process y solves the stochastic differential equation:

$$dy = -\sigma_o dB_o - \int_{\Theta} \mu_o \tilde{N}_o(.,d\theta),$$

where J is the linear jump operator

$$J[\phi] := \int_{\Theta}[\phi(x+\mu) - \phi(x) - \mu\phi_x]\nu(d\theta), \tag{7.351}$$

and S is the linear switching operator given by

$$S[\phi] := \sum_{s'\neq s}[\phi(.,s') - \phi(.,s)]\tilde{q}_{ss'}, \tag{7.352}$$

where ϕ is a test function, J^* (respectively S^*) is the adjoint operator of J (respectively to S) satisfying

$$\langle J[\phi], m\rangle = \langle \phi, J^*[m]\rangle. \tag{7.353}$$

Next we apply Itô's formula to $m_{nc}(t,y,s)$ to derive the following equalities:

$$\begin{array}{l} dm = -\Big\{(mb)_x - \frac{1}{2}(m\sigma^2)_{xx} - \frac{1}{2}\sigma_o^2 m_{xx} \\ +J^*[m] - J_o[m] + S^*[m]\} \, dt \\ -\sigma_o m_x dB_o + \int_{\Theta}[m(x-\mu_o) - m(x)]\tilde{N}_o(.,d\theta) \\ =: \tilde{b}dt + \tilde{\sigma}dB_o + \int_{\Theta}\tilde{\mu}\tilde{N}_o(.,d\theta), \end{array} \tag{7.354}$$

It follows that m solves the stochastic Fokker–Planck–Kolmogorov equation:

$$\begin{array}{l} dm = -\Big\{(mb)_x - \frac{1}{2}(m\sigma^2)_{xx} - \frac{1}{2}\sigma_o^2 m_{xx} \\ +J^*[m] - J_o[m] + S^*[m]\} \, dt \\ -\sigma_o m_x dB_o + \int_{\Theta}[m(x-\mu_o,s) - m(x,s)]\tilde{N}_o(.,d\theta) \\ =: \tilde{b}dt + \tilde{\sigma}dB_o + \int_{\Theta}\tilde{\mu}\tilde{N}_o(.,d\theta), \end{array} \tag{7.355}$$

which is a stochastic partial integro-differential equation involving $m, m_x, m_{xx}, \int m$, and $\sum m$. In the case of differentiability with respect to the measure m, we use the concept of Gâteaux differentiability. If $F : L^2(\mathbb{R}) \to \mathbb{R}$, then

$$\lim_{\epsilon\to 0} \frac{d}{d\epsilon} F(m+\epsilon\tilde{m}) = \int_{\xi\in\mathbb{R}} \frac{\partial F(m)}{\partial m}(\xi)\tilde{m}(d\xi),$$

with the functional $\xi \in \mathbb{R} \mapsto \frac{\partial F}{\partial m}(m)(\xi) \in L^2(\mathbb{R})$. Note that in the smooth case, one can connect the derivative with respect to x of the Gâteaux derivative b_m to the notion of functional derivative with respect to measure m and also to the Wasserstein gradient.

Introduce the functional:

$$
\begin{aligned}
O_i &:= \langle l_i, m\rangle + \langle \tilde{b}, \hat{V}_{i,m}\rangle + \frac{1}{2}\int_x \int_{x'} \hat{V}_{i,mm}(x,x')\sigma_o(x,.)m_x(dx)\sigma_o(x',.)m_x(dx') \\
&+ \int_\Theta [\hat{V}_i(m+\tilde{\mu}) - \hat{V}_i(m) - \langle \tilde{\mu}, \hat{V}_{i,m}\rangle]\nu_o(d\theta)
\end{aligned}
\tag{7.356}
$$

where l_i is the instant noisy cost functional:

$$
\begin{aligned}
l_i &= \frac{1}{2}\left(q_i(x-\hat{x})^2 + \hat{q}_i\hat{x}^2 + r_i(u_i-\hat{u}_i)^2 + \hat{r}_i\hat{u}_i^2 \right)dt \\
&+\left(\epsilon_{i1}(x-\hat{x})(u_i-\hat{u}_i) + \hat{\epsilon}_{i1}\hat{x}\hat{u}_i + \hat{\epsilon}_{i2}\hat{u}_i + (\epsilon_{i3}+\bar{\epsilon}_{i3})\hat{x} \right)dt \\
&+\left(\xi_{i0,B_o} + \xi_{i1,B_o}\hat{x} + \frac{1}{2}\xi_{i2,B_o}(x-\hat{x})^2 + \frac{1}{2}\hat{\xi}_{i2,B_o}\hat{x}^2 \right)dB_o \\
&+\int_\Theta \left(\xi_{i0,N_o} + \xi_{i1,N_o}\hat{x} + \frac{1}{2}\xi_{i2,N_o}(x-\hat{x})^2 + \frac{1}{2}\hat{\xi}_{i2,N_o}\hat{x}^2 \right)\tilde{N}_o(dt,d\theta),
\end{aligned}
\tag{7.357}
$$

and introduce the Hamiltonian functional:

$$
\hat{H}_i = \inf_{u_i} O_i(x,m,u,s).
$$

If there exists $\hat{V}$ such that

$$
\left\{
\begin{array}{l}
i \in \mathcal{I}, \\
\hat{V}_{i,t}(t,m,s) + \hat{H}_i(m,s,\hat{V}_i,\hat{V}_{i,m},\hat{V}_{i,xm},\hat{V}_{i,xxm},\hat{V}_{i,mm}) = 0, \\
\hat{V}_i(T,m,s) = \int m(T,dx,s)h_i(x,m(T,.),s), \\
m(t,.,.) = \mathbb{P}_{x(t)|\mathcal{F}_{t-}^{B_o,N_o}}, \text{ solves } (7.355),
\end{array}
\right.
\tag{7.358}
$$

then $\hat{V}_i(0,m_0,s_0)$ is an equilibrium payoff and the optimal strategy u_i of i minimizes $O_i(x,m,u,s)$.

We provide an outline of the proof. In order to employ the dynamic programming principle, we look for an augmented state such that one has a Markovian system. The pair $(m,\tilde{s})$ represents the driving force of the objective functions. By rewriting the payoff functional $E[L_i|\mathcal{F}_T^{N_o,B_o,s}]$ as a function of the measures $m(t,.)$ and the strategies $u_i(.)$, we obtain a classical control setup but in infinite dimensions. Hence, we only need the evolution of the measures $m(t,.)$, $s(.)$, which is given by (7.355). Then, in a vectorial sense $(m(t,dx,s))_{s\in\mathcal{S}}$, one obtains a stochastic differential game problem with m as a state (in infinite dimensions). The best response of i is

$$
\hat{V}_i(0,m_0,s_0) = \left\{
\begin{array}{l}
\inf_{u_i\in\mathcal{U}_i} E[L_i(m_0,u,s)], \\
m(t,.) \text{ solves } (7.355).
\end{array}
\right.
$$

Applying the classical DPP to the measure-valued control problem yields

$$\hat{V}_{i,t} + \inf_{u_i} O_i = 0.$$

We apply the theory of distributions to establish the following equalities from the drift part of dm:

$$\begin{aligned}
&\langle \tilde{b}, \hat{V}_{i,m} \rangle \\
&= \int_x \tilde{b}(m)\hat{V}_{i,m}\, dx \\
&= \int_x \left\{ b\hat{V}_{i,xm} + \frac{\sigma^2}{2}\hat{V}_{i,xxm} + J[\hat{V}_{i,m}] + S[\hat{V}_{i,m}] + \frac{1}{2}(\sigma_o^2 \hat{V}_{i,m})_{xx} - J_o^*[\hat{V}_{i,m}] \right\} m(t, dx, \tilde{s}),
\end{aligned} \tag{7.359}$$

which is obtained by integration by part (in the sense of distribution). The second-order (diffusion) term in B_o is

$$\frac{1}{2}\int_x \int_{x'} \sigma_o(x) m_x(dx) \sigma_o(x') m_x(dx') \hat{V}_{i,mm}(x, x'), \tag{7.360}$$

and the second-order (jump) term in N_o is

$$\int [\hat{V}_i(t, m + \tilde{\mu}, s) - \hat{V}_i(t, m, s) - \int \tilde{\mu}(dx)\hat{V}_{i,m}(t, m, s)(x)]\nu_o(d\theta).$$

This completes the proof of the dynamic programming system.

From the dynamic programming system in $\hat{V}_i$, one can obtain the PIDE satisfied by $\hat{V}_{i,m}$, which is the master system under jump-diffusion-regime switching common noise. We now establish t the equilibrium system satisfied by $\hat{V}$. The terminal cost of decision-maker i is

$$\begin{aligned}
&\hat{V}_i(T, m, s) = \frac{1}{2}\int q_i(T, s)[x - \int y m(dy)]^2 m(dx) \\
&+ \frac{1}{2}(q_i(T, s) + \bar{q}_i(T, s))[\int y m(dy)]^2 + \int [\epsilon_{i3}(T, s) + \bar{\epsilon}_{i3}(T, s)] x m(dx),
\end{aligned} \tag{7.361}$$

Next we compute the variation of loss function in the direction of $\tilde{m}$. We start with the terminal loss functional:

$$\begin{aligned}
&\hat{V}_i(T, m + \eta\tilde{m}, s) = \frac{1}{2}\int_x q_i(T, s)[x - \int y m(dy, s) - \eta \int y\tilde{m}(dy)]^2 m(dx) \\
&+ \eta\frac{1}{2}\int_x q_i(T, s)[x - \int y m(dy) - \eta \int y\tilde{m}(dy)]^2 \tilde{m}(dx) \\
&+ \frac{1}{2}(q_i(T, s) + \bar{q}_i(T, s))[\int y m(dy) + \eta \int y\tilde{m}(dy)]^2 \\
&+ \int [\epsilon_{i3}(T, s) + \bar{\epsilon}_{i3}(T, s)] x m(dx) + \eta \int \epsilon_{i3}(T, s) y\tilde{m}(dy),
\end{aligned} \tag{7.362}$$

$$\begin{aligned}
&\hat{V}_{i,m}(T, m, s)(x') = \frac{1}{2}q_i(T, s)(x' - \int y m(dy))^2 + (q_i(T, s) + \bar{q}_i(T, s))[\int y m(dy)]x' \\
&+ (\epsilon_{i3} + \bar{\epsilon}_{i3})(T, s)x', \\
&\hat{V}_{i,xm}(T, m, s)(x) = q_i(T, s)(x' - \int y m(dy)) + (q_i + \bar{q}_i)(T, s)\int y m(dy) + \epsilon_{i3}(T, s), \\
&\hat{V}_{i,xxm}(T, m, s)(x) = q_i(T, s).
\end{aligned} \tag{7.363}$$

The second-order derivative is

$$\hat{V}_{i,mm}(T,m,s)(x',x'')=-q_i(T,s)(x'-\int ym(dy))x''+(q_i(T,s)+\bar{q}_i(T,s))x''x'. \tag{7.364}$$

Similarly, the derivatives of the guess cost functional yields

$$\begin{aligned}
&\hat{V}_i(t,m,s) = \tfrac{1}{2}\alpha_i \int_x (x-\int ym(dy))^2 m(dx) + \tfrac{1}{2}\beta_i(\int ym(dy))^2 \\
&+\gamma_i \int ym(dy) + \delta_i \int m(dy), \\
&\hat{V}_{i,m}(t,m,s)(x) = \tfrac{1}{2}\alpha_i(x-\int ym(dy))^2 + \beta_i x(\int ym(dy)) + \gamma_i x + \delta_i, \\
&\hat{V}_{i,xm}(t,m,s)(x) = \alpha_i(x-\int ym(dy)) + \beta_i \int ym(dy) + \gamma_i, \\
&E[\hat{V}_{i,xm}(t,m,s)(x)|\mathscr{F}^{B_o,N_o}] = \beta_i \int ym(dy) + \gamma_i, \\
&\hat{V}_{i,xxm}(t,m,s)(x) = \alpha_i, \\
&\hat{V}_{i,mm}(t,m,s)(x,y) = -\alpha_i y(x-\int ym(dy)) + \beta_i xy, \\
&\hat{V}_{i,xymm}(t,m,s)(x,y) = \beta_i - \alpha_i.
\end{aligned} \tag{7.365}$$

The random coefficients $\alpha, \beta, \gamma, \delta$ are progressively $\mathscr{F}^{B_o,N_o}_{t-}$-measurable. We now compute the second-order jump term:

$$\begin{aligned}
&\hat{V}_i(t,m+\tilde{\mu},s)=\hat{V}_i(t,m,s)+\int \tilde{\mu}(dx)\hat{V}_{i,m}(t,m,s)(x)+\tfrac{1}{2}\int_{x,y}[(\beta_i-\alpha_i)xy]\tilde{\mu}(dx)\tilde{\mu}(dy) \\
&+\tfrac{1}{2}\alpha_i(\int y\tilde{\mu}(dy))^2 \int \tilde{\mu}(dx)+\alpha_i[\int ym(dy)][\int x\tilde{\mu}(dx)] \int \tilde{\mu}(dy).
\end{aligned} \tag{7.366}$$

It follows that

$$\begin{aligned}
&\hat{V}_i(t,m+\tilde{\mu},s) - \hat{V}_i(t,m,s) - \int \tilde{\mu}(dx)\hat{V}_{i,m}(t,m,s)(x) \\
&= \tfrac{1}{2}(\beta_i-\alpha_i)[\int x\tilde{\mu}(dx)]^2,
\end{aligned} \tag{7.367}$$

whenever $\int \tilde{\mu}(dx) = 0$.

The jump term in x is

$$\begin{aligned}
J[\hat{V}_{i,m}] &= \hat{V}_{i,m}(x+\mu) - \hat{V}_{i,m}(x) - \mu\hat{V}_{i,xm}(x) \\
&= \tfrac{1}{2}\alpha_i\mu^2 = \tfrac{1}{2}\hat{V}_{i,xxm}(x)\mu^2.
\end{aligned} \tag{7.368}$$

$$\langle J_o^*[\hat{V}_{i,m}], m\rangle = \langle J_o[m], \hat{V}_{i,m}\rangle = \int [\tfrac{1}{2}\int \alpha_i\mu^2\nu_o(d\theta)]m(dx) \tag{7.369}$$

The adjoint operator of switching is

$$S^*[\phi] := \textstyle\sum_{s'} q_{s's}\phi(.,s') - \phi(.,s)]\tilde{q}_{ss'}. \tag{7.370}$$

Note that $S^*[m]$ is the drift of the mean-field dynamics widely used in evolutionary game dynamics for the evolution of mass action. It is also called master equation in chemical reaction networks [326, 358, and the references therein].

7.8.3.1 Determination of the Optimal Strategy

Let

$$\begin{aligned}
O_i &= \int\{l_i + b\hat{V}_{i,xm} + \tfrac{\sigma^2}{2}\hat{V}_{i,xxm} + J[\hat{V}_{i,m}] + S[\hat{V}_{i,m}] + \tfrac{1}{2}(\sigma_o^2\hat{V}_{i,m})_{xx}\}m(dx) \\
&-\langle J_o[m], \hat{V}_{i,m}\rangle\rangle \\
&+\tfrac{1}{2}\int_x\int_{x'} \hat{V}_{i,mm}(x,x')\sigma_o(x,.)m_x(dx)\sigma_o(x',.)m_x(dx') \\
&+\tfrac{1}{2}\int_\Theta\int_{x,y} xy\hat{V}_{i,xymm}(t,m,s)(x,y)\tilde{\mu}(dx)\tilde{\mu}(dy)\nu_o(d\theta)
\end{aligned} \tag{7.371}$$

The optimization of O_i in u_i leads to the optimization of the following term:

$$\int\left\{l_i + b\hat{V}_{i,xm} + \tfrac{\sigma^2}{2}\hat{V}_{i,xxm} + J[\hat{V}_{i,m}]\right\}m(dx). \tag{7.372}$$

This is equivalent to the minimization (in u_i) of the conditional expectation of the following integrand:

$$l_i + b\hat{V}_{i,xm} + \tfrac{\sigma^2}{2}\hat{V}_{i,xxm} + J[\hat{V}_{i,m}]. \tag{7.373}$$

We use the following decomposition:

$$X = \left(X - E[X|\mathscr{F}^{B_o,N_o,s}]\right) + E[X|\mathscr{F}^{B_o,N_o,s}] \tag{7.374}$$

when X is $\hat{V}_{i,xm}$.

The minimization yields

$$\begin{aligned}
&\tfrac{1}{2}r_i(u_i-\hat{u}_i)^2+\epsilon_{i1}(x-\hat{x})(u_i-\hat{u}_i)+[\textstyle\sum_j b_{2j}(u_j-\hat{u}_j)]\left(\hat{V}_{i,xm}-E[\hat{V}_{i,xm}|\mathscr{F}^{B_o,N_o,s}]\right) \\
&+\tfrac{\alpha_i}{2}[\textstyle\sum_{j=1}^n \sigma_{j2}(u_j-\hat{u}_j)]^2 + \int_\Theta \tfrac{\alpha_i}{2}[\sum_{j=1}^n \mu_{j2}(u_j-\hat{u}_j)]^2\nu(d\theta) \\
&+\alpha_i\textstyle\sum_{j=1}^n\{\sigma_1\sigma_{j2} + \int_\Theta \mu_1\mu_{j2}\nu(d\theta)\}(x-\hat{x})(u_j-\hat{u}_j) \\
&+\tfrac{1}{2}\hat{r}_i\hat{u}_i^2 + \tfrac{\alpha_i}{2}[\textstyle\sum_{j=1}^n \hat{\sigma}_{j2}\hat{u}_j]^2 + \int_\Theta \tfrac{\alpha_i}{2}[\sum_{j=1}^n \hat{\mu}_{j2}\hat{u}_j]^2\nu(d\theta) \\
&+\hat{\epsilon}_{i1}\hat{x}\hat{u}_i + [\textstyle\sum_j \hat{b}_{2j}\hat{u}_j)]E[\hat{V}_{i,xm}|\mathscr{F}^{B_o,N_o,s}] \\
&+\alpha_i\textstyle\sum_{j=1}^n\{\hat{\sigma}_1\hat{\sigma}_{j2} + \int_\Theta \hat{\mu}_1\hat{\mu}_{j2}\nu(d\theta)\}\hat{x}\hat{u}_j \\
&+\hat{\epsilon}_{i2}\hat{u}_i + \alpha_i\textstyle\sum_{j=1}^n \sigma_0\hat{\sigma}_{j2}\hat{u}_j + \alpha_i\sum_{j=1}^n\int_\Theta \mu_0\hat{\mu}_{j2}\hat{u}_j\nu(d\theta).
\end{aligned} \tag{7.375}$$

This means that

$$\begin{aligned}
&c_i(u_i-\hat{u}_i) + \alpha_i\textstyle\sum_{j\neq i}[\sigma_{i2}\sigma_{j2} + \int_\Theta \mu_{i2}\mu_{j2}\nu(d\theta)](u_j-\hat{u}_j) \\
&+\left\{\epsilon_{i1} + \alpha_i[b_{2i} + \sigma_1\sigma_{i2} + \textstyle\int_\Theta \mu_1\mu_{i2}\nu(d\theta)]\right\}(x-\hat{x}) \\
&= 0, \\
&c_i = r_i + \alpha_i[\sigma_{i2}^2 + \textstyle\int_\Theta \mu_{i2}^2\nu(d\theta)],
\end{aligned} \tag{7.376}$$

which means that $A(u-\hat{u}) + v(x-\hat{x}) = 0$ with

$$\begin{aligned} &a_{ii} = 1, \\ &j \neq i,\ a_{ij} = \frac{\alpha_i}{c_i}[\sigma_{i2}\sigma_{j2} + \int_\Theta \mu_{i2}\mu_{j2}\nu(d\theta)], \\ &v_i = \frac{1}{c_i}\left\{\epsilon_{i1} + \alpha_i[b_{2i} + \sigma_1\sigma_{i2} + \int_\Theta \mu_1\mu_{i2}\nu(d\theta)]\right\} \end{aligned} \tag{7.377}$$

Hence,

$$(u - \bar{u}) = -A^{-1}v(x - \hat{x}).$$

Similarly,

$$\begin{aligned} &\bar{c}_i\hat{u}_i + \alpha_i \sum_{j\neq i}[\hat{\sigma}_{i2}\hat{\sigma}_{j2} + \int_\Theta \hat{\mu}_{i2}\hat{\mu}_{j2}\nu(d\theta)]\hat{u}_j \\ &+[\hat{\epsilon}_{i1} + \beta_i\hat{b}_{2i}]\hat{x} + \alpha_i\{\hat{\sigma}_1\hat{\sigma}_{i2} + \int_\Theta \hat{\mu}_1\hat{\mu}_{i2}\nu(d\theta)\}\hat{x} \\ &\hat{\epsilon}_{i2} + \alpha_i[\sigma_0\hat{\sigma}_{i2} + \int_\Theta \mu_0\hat{\mu}_{i2}\nu(d\theta)] \\ &+\gamma_i(b_{2i} + \bar{b}_{2i}) = 0, \\ &\bar{c}_i = \hat{r}_i + \alpha_i[\sigma_{i2}^2 + \int_\Theta \mu_{i2}^2\nu(d\theta)], \end{aligned} \tag{7.378}$$

which means that $\bar{A}\hat{u} + \bar{v}_1\hat{x} + \bar{v}_0 = 0$ with

$$\begin{aligned} &\bar{a}_{ii} = 1, \\ &j \neq i,\ \bar{a}_{ij} = \frac{\alpha_i}{\bar{c}_i}[\hat{\sigma}_{i2}\hat{\sigma}_{j2} + \int_\Theta \hat{\mu}_{i2}\hat{\mu}_{j2}\nu(d\theta)], \\ &\bar{v}_{i1} = \frac{1}{\bar{c}_i}[\hat{\epsilon}_{i1} + \beta_i\hat{b}_{2i} + \alpha_i\{\hat{\sigma}_1\hat{\sigma}_{i2} + \int_\Theta \hat{\mu}_1\hat{\mu}_{i2}\nu(d\theta)\}], \\ &\bar{v}_{i0} = \frac{1}{\bar{c}_i}\{\hat{\epsilon}_{i2} + \alpha_i[\sigma_0\hat{\sigma}_{i2} + \int_\Theta \mu_0\hat{\mu}_{i2}\nu(d\theta)] + \gamma_i\hat{b}_{2i}\} \end{aligned} \tag{7.379}$$

$$\hat{u} = -\bar{A}^{-1}\bar{v}_1\hat{x} - \bar{A}^{-1}\bar{v}_0.$$

The optimal strategy profile is then

$$u = -A^{-1}v(x - \hat{x}) - \bar{A}^{-1}\bar{v}_1\hat{x} - \bar{A}^{-1}\bar{v}_0.$$

7.8.4 Adversarial Case

We now consider the zero-sum case with performance functional:

$$\begin{aligned} &\hat{L}(x, s, u) \\ &= \frac{1}{2}q_T(x(T) - \hat{x}(T))^2 + \frac{1}{2}\hat{q}_T\hat{x}^2(T) + \hat{\epsilon}_{3T}\hat{x}(T) \\ &+ \int_0^T (\frac{1}{2}q(x - \hat{x})^2 + \frac{1}{2}\hat{q}\hat{x}^2 + \hat{\epsilon}_3\hat{x})dt \\ &+ \int_0^T (\frac{1}{2}r_i(u_i - \hat{u}_i)^2 + \frac{1}{2}\hat{r}_i\hat{u}_i^2 + \epsilon_{i1}(x - \hat{x})(u_i - \hat{u}_i) + \hat{\epsilon}_{i1}\hat{x}\hat{u}_i + \hat{\epsilon}_{i2}\hat{u}_i)dt \\ &+ \int_0^T (\frac{1}{2}r_j(u_j - \hat{u}_j)^2 + \frac{1}{2}\hat{r}_j\hat{u}_j^2 + \epsilon_{j1}(x - \hat{x})(u_j - \hat{u}_j) + \hat{\epsilon}_{j1}\hat{x}\hat{u}_j + \hat{\epsilon}_{j2}\hat{u}_j)dt \\ &+ \int_0^T (\xi_{0,B_o} + \xi_{1,B_o}\hat{x} + \frac{1}{2}\xi_{2,B_o}(x - \hat{x})^2 + \frac{1}{2}\hat{\xi}_{2,B_o}\hat{x}^2)dB_o \\ &+ \int_0^T \int_\Theta (\xi_{0,N_o} + \xi_{1,N_o}\hat{x} + \frac{1}{2}\xi_{2,N_o}(x - \hat{x})^2 + \frac{1}{2}\hat{\xi}_{2,N_o}\hat{x}^2)\tilde{N}_o(dt, d\theta), \end{aligned} \tag{7.380}$$

where l is the running cost and h is the terminal cost. Then, we have

$$
\begin{aligned}
&O := \int\{l + b\hat{V}_{xm} + \tfrac{\sigma^2}{2}\hat{V}_{xxm} + J[\hat{V}_m] + S[\hat{V}_m] + \tfrac{1}{2}(\sigma_o^2\hat{V}_m)_{xx}\}m(dx)\\
&-\langle J_o[m], \hat{V}_m\rangle\rangle\\
&+\tfrac{1}{2}\int_x\int_{x'}\hat{V}_{mm}(x,x')\sigma_o(x,.)m_x(dx)\sigma_o(x',.)m_x(dx')\\
&+\int_\Theta[\hat{V}(m+\tilde{\mu}) - \hat{V}(m) - \langle\tilde{\mu}, \hat{V}_m\rangle]\nu_o(d\theta),\\
&\hat{H} = \inf_{u_i}\sup_{u_j} O = \sup_{u_j}\inf_{u_i} O,
\end{aligned}
\tag{7.381}
$$

$$
\begin{cases}
\hat{V}_t(t,m,s) + \hat{H}(m,s,\hat{V},\hat{V}_m,\hat{V}_{xm},\hat{V}_{xxm},\hat{V}_{mm}) = 0,\\
\hat{V}(T,m,s) = \int m(T,dx,s)h(x,m(T,.),s),\\
m(t,.,.) = \mathbb{P}_{x(t)|\mathscr{F}_t^{B_o,N_o,s}},
\end{cases}
\tag{7.382}
$$

where $\hat{V}(0, m_0, s_0)$ is an equilibrium payoff and the saddle-point strategies u_i, u_j are the saddle points of the functional O.

The system (7.382) can be seen an extension of the Hamilton–Jacobi–Isaacs to the mean-field-type games context. The condition $\inf_{u_i}\sup_{u_j} O = \sup_{u_j}\sup_{u_i} O$, is fullfiled when $c_i, \bar{c}_i$ are positive and $\tilde{c}_j, \tilde{\bar{c}}_j$ are negative. The minmax (resp. maxmin) yields

$$
\begin{aligned}
&\tfrac{1}{2}r_i(u_i-\hat{u}_i)^2 + \tfrac{1}{2}\hat{r}_i\hat{u}_i^2 + \epsilon_{i1}(x-\hat{x})(u_i-\hat{u}_i) + \hat{\epsilon}_{i1}\hat{x}\hat{u}_i + \hat{\epsilon}_{i2}\hat{u}_i\\
&+\tfrac{1}{2}r_j(u_j-\hat{u}_j)^2 + \tfrac{1}{2}\hat{r}_j\hat{u}_j^2 + \epsilon_{j1}(x-\hat{x})(u_j-\hat{u}_j) + \hat{\epsilon}_{j1}\hat{x}\hat{u}_j + \hat{\epsilon}_{j2}\hat{u}_j\\
&+[\textstyle\sum_j b_{2j}(u_j-\hat{u}_j)]\left(\hat{V}_{i,xm} - E[\hat{V}_{i,xm}|\mathscr{F}^{B_o,N_o,s}]\right)\\
&+\tfrac{\alpha_i}{2}[\textstyle\sum_{j=1}^n\sigma_{j2}(u_j-\hat{u}_j)]^2 + \int_\Theta\tfrac{\alpha_i}{2}[\sum_{j=1}^n\mu_{j2}(u_j-\hat{u}_j)]^2\nu(d\theta)\\
&+\alpha_i\textstyle\sum_{j=1}^n\{\sigma_1\sigma_{j2} + \int_\Theta\mu_1\mu_{j2}\nu(d\theta)\}(x-\hat{x})(u_j-\hat{u}_j)\\
&+[\textstyle\sum_j\hat{b}_{2j}\hat{u}_j]E[\hat{V}_{i,xm}|\mathscr{F}^{B_o,N_o,s}]\\
&+\tfrac{\alpha_i}{2}[\textstyle\sum_{j=1}^n\hat{\sigma}_{j2}\hat{u}_j]^2 + \int_\Theta\tfrac{\alpha_i}{2}[\sum_{j=1}^n\hat{\mu}_{j2}\hat{u}_j]^2\nu(d\theta)\\
&+\alpha_i\textstyle\sum_{j=1}^n\sigma_0\hat{\sigma}_{j2}\hat{u}_j + \alpha_i\sum_{j=1}^n\int_\Theta\mu_0\hat{\mu}_{j2}\hat{u}_j\nu(d\theta)\\
&+\alpha_i\textstyle\sum_{j=1}^n\{\hat{\sigma}_1\hat{\sigma}_{j2} + \int_\Theta\hat{\mu}_1\hat{\mu}_{j2}\nu(d\theta)\}\hat{x}\hat{u}_j
\end{aligned}
\tag{7.383}
$$

The minimization in $(u_i - \hat{u}_i, \hat{u}_i)$ provides

$$
\begin{aligned}
&\{r_i+\alpha[\sigma_{i2}^2+\textstyle\int_\Theta\mu_{i2}^2\nu(d\theta)]\}(u_i-\hat{u}_i)+\alpha\sum_{j\neq i}[\sigma_{i2}\sigma_{j2}+\int_\Theta\mu_{i2}\mu_{j2}\nu(d\theta)](u_j-\hat{u}_j)\\
&\{\epsilon_{i1}+\alpha[b_{i2}+\sigma_1\sigma_{i2}+\textstyle\int_\Theta\mu_1\mu_{i2}\nu(d\theta)]\}(x-\hat{x}) = 0,\\
&\{\hat{r}_i+\alpha[\hat{\sigma}_{i2}^2+\textstyle\int_\Theta\hat{\mu}_{i2}^2\nu(d\theta)]\}\hat{u}_i + \sum_{j\neq i}\alpha\{\hat{\sigma}_{i2}\hat{\sigma}_{j2}+\int_\Theta\hat{\mu}_{i2}\hat{\mu}_{j2}\nu(d\theta)\}\hat{u}_j\\
&+\{\hat{\epsilon}_{i1}+\hat{b}_{2i}\beta+\alpha[\hat{\sigma}_1\hat{\sigma}_{i2}+\textstyle\int_\Theta\hat{\mu}_1\hat{\mu}_{i2}\nu(d\theta)]\}\hat{x}\\
&+\hat{\epsilon}_{i2}+\hat{b}_{2i}\gamma+\alpha[\sigma_0\hat{\sigma}_{i2}+\textstyle\int_\Theta\mu_0\hat{\mu}_{i2}\nu(d\theta)] = 0,
\end{aligned}
\tag{7.384}
$$

Putting all this together, one arrives at

$$
\begin{aligned}
& A(u-\hat{u})+v(x-\hat{x})=0, \\
& \bar{A}\hat{u}+\bar{v}_1\hat{x}+\bar{v}_0=0, \\
& a_{ii}=1, \\
& a_{ij}=\frac{\alpha[\sigma_{i2}\sigma_{j2}+\int_\Theta \mu_{i2}\mu_{j2}\nu(d\theta)]}{r_i+\alpha[\hat{\sigma}_{i2}^2+\int_\Theta \hat{\mu}_{i2}^2\nu(d\theta)]} \\
& A^{-1}=\frac{1}{1-a_{ij}a_{ji}}\begin{pmatrix} 1 & -a_{ij} \\ -a_{ji} & 1 \end{pmatrix}, \\
& v_i=\epsilon_{i1}+\alpha[b_{i2}+\sigma_1\sigma_{i2}+\textstyle\int_\Theta \mu_1\mu_{i2}\nu(d\theta)], \\
& \bar{a}_{ii}=1, \\
& \bar{a}_{ij}=\frac{\alpha[\hat{\sigma}_{i2}\hat{\sigma}_{j2}+\int_\Theta \hat{\mu}_{i2}\hat{\mu}_{j2}\nu(d\theta)]}{\hat{r}_i+\alpha[\hat{\sigma}_{i2}^2+\int_\Theta \hat{\mu}_{i2}^2\nu(d\theta)]}, \\
& \bar{v}_{i1}=\frac{\{\hat{\epsilon}_{i1}+\hat{b}_{2i}\beta+\alpha[\hat{\sigma}_1\hat{\sigma}_{i2}+\int_\Theta \hat{\mu}_1\hat{\mu}_{i2}\nu(d\theta)]\}}{\hat{r}_i+\alpha[\hat{\sigma}_{i2}^2+\int_\Theta \hat{\mu}_{i2}^2\nu(d\theta)]}, \\
& \bar{v}_{i0}=\frac{\hat{\epsilon}_{i2}+\hat{b}_{2i}\gamma+\alpha[\sigma_0\hat{\sigma}_{i2}+\int_\Theta \mu_0\hat{\mu}_{i2}\nu(d\theta)]}{\hat{r}_i+\alpha[\hat{\sigma}_{i2}^2+\int_\Theta \hat{\mu}_{i2}^2\nu(d\theta)]}, \\
& \bar{A}^{-1}=\frac{1}{1-\bar{a}_{ij}\bar{a}_{ji}}\begin{pmatrix} 1 & -\bar{a}_{ij} \\ -\bar{a}_{ji} & 1 \end{pmatrix}, \\
& (u_i-\hat{u}_i)=-[A^{-1}v]_i(x-\hat{x})=-\frac{1}{1-a_{ij}a_{ji}}(v_i-a_{ij}v_j)(x-\hat{x}), \\
& \hat{u}_i=-[\bar{A}^{-1}v_1]_i\hat{x}-[\bar{A}^{-1}v_0]_i=-\frac{1}{1-a_{ij}a_{ji}}\left([v_{i1}-\bar{a}_{ij}v_{j1}]\hat{x}+[v_{i0}-\bar{a}_{ij}v_{j0}]\right)
\end{aligned}
\tag{7.385}
$$

We retrieve the same expressions as in the case of the direct method, which confirms the validity of the direct method.

To recap, in this section, we have shown that a mean-field-type saddle-point equilibrium can be determined in a semi-explicit way for the adversarial linear-quadratic game problem with jump-diffusion process and regime switching, where all drift, diffusion, and jump terms are state, control input, conditional expected value state, conditional expected value control, and switching dependent. In addition, the common noise includes not only a diffusion but also jump process, which is crucial in many application areas. The presented method is fairly straightforward. It does not require partial integral–differential equations with switching terms or backward–forward stochastic maximum principle theory.

7.9 Risk-Sensitive Linear-Quadratic Conditional Mean-Field-Type Games

This section focuses on a special class of risk-sensitive linear-quadratic mean-field-type game problems under a diffusion state dynamics with common noise that include mean-field terms that are:

- the conditional expectation of the state with respect to the filtration of the common noise, in the drift functions,
- the conditional expectation of the control actions, in the drift functions.

The state process is one-dimensional. The instantaneous cost function includes:

- the conditional variance of the state,
- the square of the conditional expected value of the state,
- the conditional variance of the control-action process,
- the square of the conditional expected value of the control-action process,
- a cross-term such as covariance between the state and the control,
- a product term between the conditional expected value of the control action and the conditional expected value of the state,
- a noisy measurement of the cost where the measurement noise is also state and conditional mean-field-dependent.

We follow a direct method to solve the risk-sensitive mean-field-type game problem in a semi-explicit way. The method starts by identifying a partial guess functional where the coefficients are regime switching dependent. Then, it uses Itô's formula for jump-diffusion-regime switching processes, followed by a completion of squares for both control and conditional mean control. The completion of squares technique used in expected utility approach cannot be directly used due to extra dependence from other decision-makers. Finally, the processes are identified using an orthogonal decomposition technique, and a system of stochastic differential equations is derived in a semi-explicit way.

We introduce the following notations. Let $[0, T],\ T > 0$ be a fixed time horizon and $(\Omega, \mathscr{F}, \mathbb{F}^{B,B_o}, \mathbb{P})$ be a given filtered probability space.The filtration $\mathbb{F} = \{\mathscr{F}_t^{B,B_o},\ 0 \leq t \leq T\}$ is the natural filtration of the union $\{B, B_o, s\}$ augmented by $\mathbb{P}$-null sets of $\mathscr{F}$. In practice, B is used to capture smaller disturbance.

Let $k \geq 1$. $L^k([0, T]; \mathbb{R})$ is the set of functions $f : [0, T] \to \mathbb{R}$ such that $\int_0^T |f(t)|^k dt < \infty$. $\mathscr{L}_{\mathbb{F}}^k([0, T]; \mathbb{R})$ is the set of $\mathbb{F}$-adapted $\mathbb{R}$-valued processes $X(\cdot)$ such that $E\left[\int_0^T |X(t)|^k dt\right] < \infty$. The stochastic quantity $\hat{X} = E[X(t)|\ \mathscr{F}_t^{B_o}]$ denotes the conditional expectation of the random variable $X(t)$ with respect to the filtration $\mathscr{F}_t^{B_o}$. Note that $\hat{X}$ is a random process. $expL_{\mathbb{F}}^2([0, T]; \mathbb{R})$ is the set of $\mathbb{F}$-adapted $\mathbb{R}$-valued processes $X(\cdot)$ such that $E\left[e^{x_0^2+\int_0^T x^2(t)dt}\right] < \infty$.

An admissible control strategy u of the decision-maker is an $\mathbb{F}$-adapted and square-integrable process with values in a non-empty subset $U_i = \mathbb{R}$. We denote the set of all admissible controls by $\mathscr{U}_i$:

$$\mathscr{U}_i = \{u_i(\cdot) \in expL_{\mathbb{F}}^2([0, T]; \mathbb{R});\ \ u_i(.) \in U_i\ a.e.t \in [0, T],\ \mathbb{P} - a.s.\}$$

7.9.1 Covariance and Noisy Cost

Consider the following state dynamics with jump-diffusion and regime switching:

$$\begin{array}{l} x(0) = x_0, \\ dx = bdt + \sigma dB + \int_{\Theta} \mu(\cdot,\theta)\tilde{N}(dt,d\theta) + \sigma_o dB_o, \end{array} \tag{7.386}$$

where

$$\begin{array}{l} b = b_0 + b_1(x-\hat{x}) + \hat{b}_1\hat{x} + \sum_{j=1}^{I} b_{j2}(u_j - \hat{u}_j) + \sum_{j=1}^{I} \hat{b}_{j2}\hat{u}_j, \\ \sigma = \sigma_0, \sigma_o, \end{array} \tag{7.387}$$

with $b_k(t);\ \sigma_k(t);\ \hat{b}_k(t);\ \hat{\sigma}_k(t) \in \mathbb{R}$. The process B is a standard one-dimensional Brownian motion. The process B_o is a standard one-dimensional Brownian motion. It is assumed that the processes $B(t)$ are independent of $B_o(t)$, which is a common noise. The term $\hat{x} = E[x|\mathscr{F}^{B_o}]$ denotes the conditional expectation of x given $\mathscr{F}_t^{B_o,s}$, and $\hat{u} = E[u|\mathscr{F}^{B_o}]$. These are called conditional state mean-field and conditional control-action mean-field terms. The state dynamics is therefore of conditional McKean–Vlasov type as the conditional probability measure of the state appears in the drift, diffusion, and jump coefficients via $(\hat{x},\ \hat{u})$. Below we denote a solution (if any) to (7.386) by $x := x^{u,x_0}$.

We associate a cost functional of conditional mean-field type of decision-maker i as

$$\begin{array}{l} L_i(x,u) = \frac{1}{2}q_{iT}(x-\hat{x})^2(T) + \frac{1}{2}\hat{q}_{iT}\hat{x}^2(T) + \epsilon_{i3}(T)\hat{x}(T) \\ +\frac{1}{2}\int_0^T \left(q_i(x-\hat{x})^2 + \hat{q}_i\hat{x}^2 + r_i(u_i-\hat{u}_i)^2 + \hat{r}_i\hat{u}_i^2\right)dt \\ +\int_0^T \left(\epsilon_{i1}(x-\hat{x})(u_i-\hat{u}_i) + \hat{\epsilon}_{i1}\hat{x}\hat{u}_i + \epsilon_{i2}\hat{u}_i + \epsilon_{i3}\hat{x}\right)dt \\ +\int_0^T (\xi_{i0,B_o} + \xi_{i1,B_o}\hat{x} + \frac{1}{2}\xi_{i2,B_o}(x-\hat{x})^2 + \frac{1}{2}\hat{\xi}_{i2,B_o}\hat{x}^2)dB_o, \end{array} \tag{7.388}$$

where $q_{iT} = q_i(T)$, $q_i = q_i(t)$, $\hat{q}_{iT} = q_i(T)$, $\hat{q}_i = q_i(t)$; and $r_i = r_i(t)$, $\hat{r}_i = \hat{r}_i(t)$, $\epsilon_{ik} = \epsilon_{ik}(t)$ are $\mathscr{F}^{B_o}$-measurable.

The last term

$$(\xi_{i0,B_o} + \xi_{i1,B_o}\hat{x} + \xi_{i2,B_o}(x-\hat{x})^2 + \hat{\xi}_{i2,B_o}\hat{x}^2)dB_o$$

denotes a noisy cost measurement of decision-maker i.

The best-response problem of the risk-sensitive decision-maker i (with risk-sensitive parameter $\eta_i \neq 0$) given $u_{-i} = (u_j \in \mathscr{U}_j,\ j \neq i)$ is

Risk-Sensitive Mean-Field-Type Best-Response Problem with Jump-Diffusion

$$\left\{ \begin{array}{l} \inf_{u_i \in \mathscr{U}_i} \frac{1}{\eta_i} \log\ E\left[e^{\eta_i L_i(x,s,u)}|x(0) = x_0\right] \\ \text{such that (7.386)} \end{array} \right. \tag{7.389}$$

The risk-sensitive mean-field-type equilibrium is a strategy profile $u^* \in \prod_{j\in\mathcal{I}} \mathcal{U}_j$, such that for every decision-maker $i \in \mathcal{I}$, u_i^* solves (7.389) given $u_{-i}^* = (u_j^* \in \mathcal{U}_j,\ j \neq i)$

Risk-Sensitive Mean-Field-Type Equilibrium Problem

$$\begin{cases} i \in \mathcal{I} := \{1, 2, \ldots, I\}, \\ \frac{1}{\eta_i} \log\ E\left[e^{\eta_i L_i(x,u^*)} | x(0) = x_0\right] \\ = \inf_{u_i\in\mathcal{U}_i} \frac{1}{\eta_i} \log\ E\left[e^{\eta_i L_i(x,u_i,u_{-i}^*)} | x(0) = x_0\right], \\ \text{such that (7.386)} \end{cases} \tag{7.390}$$

We consider the following assumption:

$$\begin{cases} q_{iT}, q_{iT} + \bar{q}_{iT} q_i > 0, q_i > 0,\ q_i + \bar{q}_i > 0, r_i > 0, r_i + \bar{r}_i > 0, \\ E[e^{\eta_i x_0^2}] < +\infty, \\ \int_0^T \int_\Theta \phi_{ik}^2 \nu(d\theta) < +\infty, \text{ with } \phi_{ik} \in \{\mu_{ik}, \hat{\mu}_{ik}\}, \\ b_k, \sigma_k^2, \hat{b}_k, \hat{\sigma}_k^2 \in L^1([0, T], \mathbb{R}) \end{cases} \tag{7.391}$$

The integrability condition above yields

$$\int_0^T [|b_0(t)| + |b_1(t)| + |\hat{b}_1(t)| + |b_{j2}(t)| + |\hat{b}_{j2}(t)| + \sigma_0^2(t)]dt < +\infty. \tag{7.392}$$

Remark 7.5 Under Assumption (7.391), the state dynamics has a solution for each $u \in \mathcal{U}$. Note that we do not impose boundedness conditions on the coefficients or Lipschitz conditions on the cost. Under Assumption (7.391), Problem (7.390) is well defined.

We now state the main result of this section.

Solution to Risk-Sensitive Mean-Field-Type Game Problem (7.390)
Let Assumption 7.391 hold. Then, the Nash equilibrium strategy and the equilibrium cost are given by

$$\begin{aligned} & u_i^* = \hat{u}_i^* - \frac{(\epsilon_{i1}+\alpha_i b_{i2})}{r_i}(x - \hat{x}), \\ & \hat{u}_i^* = -\frac{1}{\hat{r}_i}[(\hat{\epsilon}_{i1} + \beta_i \hat{b}_{i2})\hat{x} + (\epsilon_{i2} + \gamma_i \hat{b}_{i2})], \\ & \frac{1}{\eta_i} \log E[e^{\eta_i L_i^*(x^*,u^*)}] = \\ & \frac{1}{\eta_i} \log E e^{\eta_i [\frac{1}{2}\alpha_i(0)(x_0-\hat{x}_0)^2 + \frac{1}{2}\beta_i(0)\hat{x}_0^2 + \gamma_i(0)\hat{x}_0 + \delta_i(0)]}, \end{aligned} \tag{7.393}$$

Coefficients of the Solution to Problem (7.390)

$$
\begin{aligned}
& d\alpha_i = -\{2b_1\alpha_i + q_i \\
& -\frac{(\epsilon_{i1}+\alpha_i b_{i2})^2}{r_i} + \eta_i\alpha_i^2\sigma^2 - 2\alpha_i \sum_{j\neq i} b_{j2}\frac{(\epsilon_{j1}+\alpha_j b_{j2})}{r_j}\Big\} dt - \xi_{i,2,B_o} dB_o, \\
& \alpha_i(T) = q_i(T), \\
& d\beta_i = -\Big\{2\hat{b}_1\beta_i + \hat{q}_i \\
& -\frac{1}{\hat{r}_i}(\hat{\epsilon}_{i1} + \beta_i\hat{b}_{i2})^2 - 2\beta_i \sum_{j\neq i} \hat{b}_{j2}\frac{1}{\hat{r}_j}(\hat{\epsilon}_{j1} + \beta_j\hat{b}_{j2})\Big\} dt - \hat{\xi}_{i2,B_o} dB_o, \\
& \beta_i(T) = \hat{q}_i(T), \\
& d\gamma_i = -\Big\{\beta_i b_0 + \gamma_i\hat{b}_1 + \epsilon_{i3} \\
& -\frac{1}{\hat{r}_i}(\hat{\epsilon}_{i1} + \beta_i\hat{b}_{i2})(\epsilon_{i2} + \gamma_i\hat{b}_{i2}) \\
& -\beta_i \sum_{j\neq i} \frac{1}{\hat{r}_j}(\epsilon_{j2} + \gamma_j\hat{b}_{j2})\hat{b}_{j2} - \gamma_i \sum_{j\neq i} \frac{1}{\hat{r}_j}(\hat{\epsilon}_{j1} + \beta_j\hat{b}_{j2})\hat{b}_{j2} + \sigma_o\beta_{i,o,B_o}\Big\} dt \\
& -(\sigma_o\beta_i + \xi_{i1,B_o}) dB_o, \\
& \gamma_i(T) = \epsilon_{i3}(T), \\
& d\delta_i = -\{\gamma_i b_0 \\
& -\frac{1}{2\hat{r}_i}(\epsilon_{i2} + \gamma_i\hat{b}_{i2})^2 - \gamma_i \sum_{j\neq i} \frac{1}{\hat{r}_j}(\epsilon_{j2} + \gamma_j\hat{b}_{j2})\hat{b}_{j2} + \frac{\alpha_i}{2}\sigma^2 + \frac{\beta_i}{2}\sigma_o^2 \\
& +\sigma_o\gamma_{i,o,B_o}\Big\} dt - (\sigma_o\gamma_i + \xi_{i0,B_o}) dB_o, \\
& \delta_i(T) = 0,
\end{aligned}
\tag{7.394}
$$

whenever these stochastic differential equations have a solution that does not blow up in $[0, T]$.

We note the following:

- As all the risk-sensitivity indices $\{\eta_i\}_i$ go to zero, one retrieves the risk-neutral solution under jump-diffusion common noise.
- The method presented here does not include costs of the form $\frac{1}{\eta_i} \log E e^{\eta_i [q_i x^2(T) + \epsilon_i \hat{x}(T)]}$. Indeed, the last expression can be rewritten as

$$
\frac{1}{\eta_i} \log E e^{\eta_i [q_i (x(T)-\hat{x}(T))^2 + q_i\hat{x}^2(T) + 2q_i(x(T)-\hat{x}(T))\hat{x}(T) + \epsilon\hat{x}(T)]},
$$

  which involves another coefficient process associated with $(x(T) - \hat{x}(T))\hat{x}(T)$. This means that the cross-term $(x(T) - \hat{x}(T))\hat{x}(T)$ needs to be considered in the guess functional.
- When the jump-diffusion coefficients depend on the conditional mean-field terms, the change of measure leads to a higher-order polynomial. For example,

$$
e^{\int_0^T \sigma_1\hat{x}(x-\hat{x})dB - \frac{1}{2}\int_0^T \sigma_1^2\hat{x}^2(x-\hat{x})^2 dt},
$$

  will require a matching with a polynomial of degree four. Hence, one has to lead with exponential non-quadratic terms.

7.9.2 Semi-explicit Equilibrium Strategy

We develop a proof of optimality using a direct method decomposed in several steps.

7.9.2.1 Semi-explicit Solution by Means of Direct Method

We construct a semi-explicit solution to Problems (7.390) via a direct method. Let $f_i(t,x) = \frac{1}{2}\alpha_i(x-\hat{x})^2 + \frac{1}{2}\beta_i\hat{x}^2 + \gamma_i\hat{x} + \delta_i$, where the processes $\alpha_i = \alpha_i(t)$, $\beta_i = \beta_i(t)$, $\gamma_i = \gamma_i(t)$, $\delta_i = \delta_i(t)$, are $\mathscr{F}_t^{N_o,B_o}$ measurable, and $x = x^{u,x_0}(t)$ is the solution to (7.386). Furthermore, the processes $(x-\hat{x})$ and $\hat{x}$ solve the following stochastic differential equations of conditional mean-field type:

$$
\begin{aligned}
&\hat{x}(0) = \hat{x}_0,\\
&d\hat{x} = [b_0 + \hat{b}_1\hat{x} + \textstyle\sum_{j=1}^I \hat{b}_{j2}\hat{u}_j]dt + \sigma_o dB_o,\\
&x(0) = x_0,\\
&d(x-\hat{x}) = [b_1(x-\hat{x}) + \textstyle\sum_{j=1}^I b_{j2}(u_j - \hat{u}_j)]dt + \sigma dB,
\end{aligned}
\tag{7.395}
$$

The two processes $(x-\hat{x})$ and $\hat{x}$ satisfy $E[(x-\hat{x})\hat{x}|\mathscr{F}^{B_o}] = 0$. However, $E[e^{(x-\hat{x})\hat{x}}|\mathscr{F}^{B_o,s}]$ creates an offset dependent on the strategies employed by the decision-makers:

$$
\begin{aligned}
&\eta_i[\hat{L}_i(x,u) - f_i(0,x_0)]\\
&= \tfrac{\eta_i}{2}(q_{iT} - \alpha_i(T))(x_T - \hat{x}_T)^2 + \tfrac{\eta_i}{2}(\hat{q}_{iT} - \beta_i(T))\hat{x}_T^2 + \eta_i(\epsilon_{i3}(T) - \gamma_{iT})\hat{x}(T)\\
&+\eta_i(0 - \delta_{iT})\\
&+\eta_i \int_0^T \Bigg[\tfrac{1}{2}\{d\alpha_i + 2b_1\alpha_i dt + q_i dt\\
&+(-\tfrac{(\epsilon_{i1}+\alpha_i b_{i2})^2}{r_i} + \eta_i\alpha_i^2\sigma^2 - 2\alpha_i \textstyle\sum_{j\neq i} b_{j2}\tfrac{(\epsilon_{j1}+\alpha_j b_{j2})}{r_j})dt + \xi_{i2,B_o}dB_o\Big\}(x-\hat{x})^2\\
&+\tfrac{1}{2}\Big\{d\beta_i + 2\hat{b}_1\beta_i dt + \hat{q}_i dt\\
&-\tfrac{1}{\hat{r}_i}(\hat{\epsilon}_{i1} + \beta_i\hat{b}_{i2})^2 dt - 2\beta_i \textstyle\sum_{j\neq i} \hat{b}_{j2}\tfrac{1}{\hat{r}_j}(\hat{\epsilon}_{j1} + \beta_j\hat{b}_{j2})dt + \hat{\xi}_{i2,B_o}dB_o\Big\}\hat{x}^2\\
&+\Big\{d\gamma_i + \beta_i b_0 dt + \gamma_i\hat{b}_1 dt + \epsilon_{i3}dt\\
&-\tfrac{1}{\hat{r}_i}(\hat{\epsilon}_{i1} + \beta_i\hat{b}_{i2})(\epsilon_{i2} + \gamma_i\hat{b}_{i2})dt\\
&-\beta_i \textstyle\sum_{j\neq i}\tfrac{1}{\hat{r}_j}(\epsilon_{j2} + \gamma_j\hat{b}_{j2})\hat{b}_{j2}dt - \gamma_i \textstyle\sum_{j\neq i}\hat{b}_{j2}\tfrac{1}{\hat{r}_j}(\hat{\epsilon}_{j1} + \beta_j\hat{b}_{j2})\,dt\\
&+\sigma_o\beta_{i,o,B_o}dt + \sigma_o\beta_i dB_o + \xi_{i1,B_o}dB_o\Big\}\hat{x}\\
&+d\delta_i + \gamma_i b_0 dt\\
&-\tfrac{1}{2\hat{r}_i}(\epsilon_{i2} + \gamma_i\hat{b}_{i2})^2 dt - \gamma_i \textstyle\sum_{j\neq i}\tfrac{1}{\hat{r}_j}(\epsilon_{j2} + \gamma_j\hat{b}_{j2})\hat{b}_{j2}dt\\
&+\tfrac{\alpha_i}{2}\sigma^2 dt + \tfrac{\beta_i}{2}\sigma_o^2 dt + \sigma_o\gamma_{i,o,B_o}dt + \sigma_o\gamma_i dB_o + \xi_{i0,B_o}dB_o\\
&+\tfrac{1}{2}r_i\{u_i - \hat{u}_i + \tfrac{(\epsilon_{i1}+\alpha_i b_{i2})}{r_i}(x-\hat{x})\}^2\,dt\\
&+\tfrac{1}{2}\hat{r}_i\{\hat{u}_i + \tfrac{1}{\hat{r}_i}[(\hat{\epsilon}_{i1} + \beta_i\hat{b}_{i2})\hat{x} + (\epsilon_{i2} + \gamma_i\hat{b}_{i2})]\}^2\,dt\Bigg]\\
&+\eta_i \int_0^T \alpha_i\sigma(x-\hat{x})dB - \tfrac{1}{2}\int_0^T \eta_i^2\alpha_i^2\sigma^2(x-\hat{x})^2 dt.
\end{aligned}
\tag{7.396}
$$

7.9.2.2 Matching the Processes

The minimization of the latest expression leads to the desired results provided that r_i, $\hat{r}_i$ and the coefficient of $-\alpha_i^2$, i.e., $\frac{b_{i2}^2}{r_i} - \eta_i\sigma^2$, are positive. In particular, this limits the range of the risk-sensitivity index of decision-maker i :

$$\eta_i < \frac{b_{i2}^2}{r_i\sigma^2}.$$

Otherwise, there may be a finite time escape (blow-up) within $[0, T)$ when the risk-sensitivity index is higher in value.

7.9.2.3 Connection with the Risk-Neutral Case

When the decision-makers are risk-neutral, i.e., they belong to $\mathcal{I}_0 = \{i \mid \eta_i \to 0\}$, one retrieves the risk-neutral mean-field-type games with conditional variance, conditional covariance under jump-diffusion common noise, and noisy cost measurement:

$$\begin{cases} i \in \mathcal{I} := \{1, 2 \ldots, I\}, \\ E\left[L_i(x, u^*)|x(0) = x_0\right] = \inf_{u_i \in \mathcal{U}_i} E\left[L_i(x, u_i, u^*_{-i})|x(0) = x_0\right], \\ \text{such that (7.386)} \end{cases} \tag{7.397}$$

The risk-neutral solution is given in a semi-explicit way with the coefficients solving a stochastic Riccati equation:

$$\begin{array}{l} i \in \mathcal{I}_0, \\ d\alpha_i = -\left\{2b_1\alpha_i + q_i - \frac{(\epsilon_{i1}+\alpha_i b_{i2})^2}{r_i} - 2\alpha_i \sum_{j\neq i} b_{j2}\frac{(\epsilon_{j1}+\alpha_j b_{j2})}{r_j}\right\} dt - \xi_{i,2,B_o} dB_o, \\ i \in \mathcal{I}_0, \alpha_i(T) = q_i(T), \end{array} \tag{7.398}$$

7.9.3 *Different Attitudes Toward the Risk*

A risk-averse decision-maker (with cost functional) is a decision-maker who prefers higher cost with known risks rather than lower cost with unknown risks. In other words, among various control strategies leading to the same cost value with different levels of risks, this decision-maker always prefers the one with the lowest risk.

We consider a finite population of decision-makers classified as follows:

- risk-neutral decision-makers: $\mathcal{I}_0 = \{i \mid \eta_i \to 0\}$
- risk-seeking decision-makers $\mathcal{I}_+ = \{i \mid \eta_i < 0\}$,

- risk-averse decision-makers $\mathscr{I}_- = \{i \mid \eta_i > 0\}$.

This leads to a mixture of stochastic Riccati systems described by

$$\begin{array}{l} i \in \mathscr{I}_0, \\ d\alpha_i = -\left\{2b_1\alpha_i + q_i - \frac{(\epsilon_{i1}+\alpha_i b_{i2})^2}{r_i} - 2\alpha_i \sum_{j\neq i} b_{j2}\frac{(\epsilon_{j1}+\alpha_j b_{j2})}{r_j}\right\} dt - \xi_{i2,B_o} dB_o, \\ i \in \mathscr{I}, \alpha_i(T) = q_i(T), \\ \\ i \in \mathscr{I}_+ \cup \mathscr{I}_-, \\ d\alpha_i = -\left\{2b_1\alpha_i + q_i - \frac{(\epsilon_{i1}+\alpha_i b_{i2})^2}{r_i} + \eta_i\alpha_i^2\sigma^2 - 2\alpha_i \sum_{j\neq i} b_{j2}\frac{(\epsilon_{j1}+\alpha_j b_{j2})}{r_j}\right\} dt \\ -\xi_{i2,B_o} dB_o, \\ i \in \mathscr{I}, \alpha_i(T) = q_i(T), \end{array} \tag{7.399}$$

The risk-neutral coefficient α_i, $i \in \mathscr{I}_0$ can be obtained from the risk-sensitive coefficient α_i, $i \in \mathscr{I}_+ \cup \mathscr{I}_-$, as the risk-sensitivity index η_i goes to zero.

In this section, we have seen that when the orthogonal of the processes is well designed, the risk-sensitive mean-field-type games can be solved. However, a small variation of the cost function may create an offset in the Girsanov transform, which leads to higher-order polynomials in the state [14, 37, 62, 123, 345]. The specific class of cost structure and methodology developed here extend to fractional Brownian motions and Gauss–Volterra processes [37, 344]. The (square) matrix version of the model can be found in Sect. 7.10.

7.10 Matrix-Valued Mean-Field-Type Games

In the study of time-dependent mean-field-type games, we have thus far considered the scalar case. The matrix case gives us a second representation. A special feature of the matrix-valued case is noncommutativity and nonsymmetric structures. This section extends the problem examined in Sect. 7.4 to multidimensional state in matrix form. We consider a matrix-valued linear-quadratic mean-field-type game problem under a jump-diffusion-regime switching state dynamics that include mean-field terms that are:

- the conditional expectation of the state with respect to the filtration of the regime switching, in the drift, diffusion, and jump coefficient functions,
- the conditional expectation of the control actions, in the drift, diffusion, and jump coefficient functions.

The state process is in dimension $(d \times d)$. The instant cost function includes:

- the conditional variance of the state,
- the square of the conditional expected value of the state,
- the conditional variance of the control-action process,

- the square of the conditional expected value of the control-action process,
- covariance between the state and the control,
- a product term between the conditional expected value of the control action and the conditional expected value of the state.

The key methodology used in this section is a direct method to solve the mean-field-type game problem in a semi-explicit way. The method starts by identifying a partial guess functional where the coefficient are regime switching dependent. Then, it uses Itô's formula for jump-diffusion-regime switching processes, followed by a completion of squares for both control and mean control. A careful matrix inversion is used to derive the interdependence between coefficients. Finally, the processes are identified and a system of ordinary differential equations is derived.

We introduce the following notations. Let $[0, T]$, $T > 0$ be a fixed time horizon and $(\Omega, \mathcal{F}, \mathbb{F}^{B,N,s}, \mathbb{P})$ be a given filtered probability space.The filtration $\mathbb{F} = \{\mathcal{F}_t^{B,N,s},\ 0 \leq t \leq T\}$ is the natural filtration of the union $\{B, N, s\}$ augmented by $\mathbb{P}$-null sets of $\mathcal{F}$. In practice, B is used to capture smaller disturbance and N is used for larger jumps of the system.

An admissible control strategy U_i of decision-maker i is an $\mathbb{F}$-adapted and square-integrable process with values in a $\mathbb{U}_i = \mathbb{R}^{d\times d}$, $d \geq 1$. We denote the set of all such admissible controls by $\mathcal{U}_i$.

7.10.1 Matrix-Valued Risk-Neutral Problems

A matrix-valued system is a system in which the output value is a matrix. In the differential setup, it should provide a matrix at any time within the horizon.

We consider $I \geq 1$ decision-makers over the time horizon $[0, T]$, $T > 0$. Each decision-maker $i \in \mathcal{I} := \{1, \dots, I\}$ chooses a matrix-valued strategy U_i over the horizon $[0, T]$. The state satisfies the following matrix-valued linear jump-diffusion-regime switching system of mean-field type:

$$\begin{aligned} dX &= [B_1(X - \hat{X}) + (B_1 + \bar{B}_1)\hat{X} + \textstyle\sum_{j\in\mathcal{I}} B_{2j}(U_j - \hat{U}_j) \\ &\quad + \textstyle\sum_{j=1}^{I} (B_{2j} + \bar{B}_{2j})\hat{U}_j]dt \\ &\quad + S_0 dB + \textstyle\int_\theta M_0(\cdot, \theta)\tilde{N}(dt, d\theta), \\ X(0) &= X_0, \end{aligned} \tag{7.400}$$

where $X(t)$, $\hat{X}(t) = E[X(t)|\mathcal{F}_t^s]$, $B_1(t, s(t))$, $\bar{B}_1(t, s(t)) \in \mathbb{R}^{d\times d}$, $B_{2j}(t, s(t))$, $\bar{B}_{2j}(t, s(t))$, $S_0(t, s(t))$, $M_0(t, s(t)) \in \mathbb{R}^{d\times d}$, B is a $\mathbb{R}^{d\times d}$ matrix-valued Brownian motion, $U_i \in \mathcal{U}_i$, s is a regime switching process with transition rates $\tilde{q}_{ss'}$ satisfying

$$\tilde{q}_{ss'} > 0,\ \tilde{q}_{ss} = -\sum_{s' \neq s} \tilde{q}_{ss'}.$$

$N(t,.)$ is a $\mathbb{R}^{d\times d}$ matrix-valued Poisson random process with compensated process $\tilde{N}(dt,d\theta) = N(dt,d\theta) - \nu(d\theta)dt$, ν is a matrix of Radon measure over the set of jump sizes Θ. $\mathscr{F}_t^s$ is the filtration generated by the regime switching process s. By abuse of notation we suppress the use of $s(t-)$ and $X(t-)$ for the left value of switching $s(t)$ and the jump $X(t)$, respectively.

To the state system (7.400), we associate the cost functional of decision-maker i:

$$\begin{array}{l} L_i(X,s,U) = \langle Q_i(T,s(T))(X(T)-\hat{X}(T)), X(T)-\hat{X}(T)\rangle \\ +\langle (Q_i(T,s(T)) + \bar{Q}_i(T,s(T)))\hat{X}(T), \hat{X}(T)\rangle \\ +\int_0^T \langle Q_i(X-\hat{X}), X-\hat{X}\rangle + \langle (Q_i+\bar{Q}_i)\hat{X}, \hat{X}\rangle dt \\ +\int_0^T \langle R_i(U_i-\hat{U}_i), U_i-\hat{U}_i\rangle + \langle (R_i+\bar{R}_i)\hat{U}_i, \hat{U}_i\rangle dt, \end{array} \tag{7.401}$$

where $\langle M, N\rangle = \text{trace}(M^*N) = \text{trace}(N^*M)$, M^* being the adjoint operator of M (transposition), the coefficients $Q_i := Q_i(t,s(t))$, $R_i := R_i(t,s(t))$, $\bar{Q}_i := \bar{Q}_i(t,s(t))$, $\bar{R}_i := \bar{R}_i(t,s(t))$ are possibly time and regime-switching dependent with values in $\mathbb{R}^{d\times d}$.

7.10.1.1 Risk-Neutral Nash Equilibria

We provide basic definitions of the risk-neutral problems and their solution concepts.

Definition 7.8 (Mean-Field-Type Risk-Neutral Nash Equilibrium) Given $(U_j,\ j\neq i)$, a risk-neutral best-response strategy of decision-maker i is a strategy that solves $\inf_{U_i\in\mathscr{U}_i} E[L_i]$ subject to (7.400). The set of risk-neutral best responses of i is denoted by $\text{rnBR}_i(U_{-i})$.

A mean-field-type risk-neutral Nash equilibrium is a strategy profile $(U_j^{\text{rn}},\ j\in\mathscr{I})$, of all decision-makers such that for every decision-maker i,

$$U_i^{\text{rn}} \in \text{rnBR}_i(U_{-i}^{\text{rn}}). \tag{7.402}$$

Theorem 7.1 *Assume that $Q_i, R_i, Q_i+\bar{Q}_i, R_i+\bar{R}_i$ are symmetric positive definite. Then the matrix-valued mean-field-type (risk-neutral) Nash equilibrium strategy (7.402) and the (risk-neutral) equilibrium cost are given by*

$$\begin{cases} U_i^{\text{rn}} = -R_i^{-1}B_{2i}^*P_i(X-\hat{X}) - (R_i+\bar{R}_i)^{-1}(B_{2i}+\bar{B}_{2i})^*\bar{P}_i\hat{X}, \\ L_i^{\text{rn}}(X,s,U^{\text{rn}}) = E\langle P_i(0,s(0))(X_0-\hat{X}_0), X_0-\hat{X}_0\rangle \\ +E\langle \bar{P}_i(0,s(0))\hat{X}_0, \hat{X}_0\rangle + E[\delta_i(0,s(0))], \\ i\in\mathscr{I}, \end{cases}$$

where P_i, $\bar{P}_i$, and δ_i solve the following differential equations:

$$\begin{cases} \dot{P}_i + Q_i + B_1^* P_i + P_i B_1 + \sum_{s' \neq s} (P_i(t,s') - P_i(t,s))\tilde{q}_{ss'} \\ -P_i B_{2i} R_i^{-1} B_{2i}^* P_i - \sum_{j \neq i} P_j B_{2j} R_j^{-1*} B_{2j}^* P_i - \sum_{j \neq i} P_i B_{2j} R_j^{-1*} B_{2j}^* P_j = 0, \\ P_i(T,s) = Q_i(T,s), \\ \dot{\bar{P}}_i + Q_i + \bar{Q}_i + \bar{P}_i^*(B_1 + \bar{B}_1)^* + (B_1 + \bar{B}_1)^* \bar{P}_i \\ + \sum_{s' \neq s} (\bar{P}_i(t,s') - \bar{P}_i(t,s))\tilde{q}_{ss'} \\ -\bar{P}_i (B_{2i} + \bar{B}_{2i})(R_i + \bar{R}_i)^{-1} (B_{2i} + \bar{B}_{2i})^* \bar{P}_i \\ - \sum_{j \neq i} \bar{P}_j (B_{2j} + \bar{B}_{2j})(R_j + \bar{R}_j)^{-1*} (B_{2j} + \bar{B}_{2j})^* \bar{P}_i \\ - \sum_{j \neq i} \bar{P}_i (B_{2j} + \bar{B}_{2j})(R_j + \bar{R}_j)^{-1*} (B_{2j} + \bar{B}_{2j})^* \bar{P}_j = 0, \\ \bar{P}_i(T,s) = Q_i(T,s) + \bar{Q}_i(T,s), \\ \dot{\delta}_i + \langle P_i S_0, S_0 \rangle + \int_\Theta \langle P_i M_0, M_0 \nu(d\theta) \rangle + \sum_{s' \neq s} (\delta_i(t,s') - \delta_i(t,s))\tilde{q}_{ss'} = 0, \\ \delta_i(T,s) = 0, \\ s \in \mathcal{S}, \end{cases} \tag{7.403}$$

whenever these matrix-valued differential equations have a unique solution that does not blow up within $[0, T]$.

A proof of Theorem 7.1 can be obtained from the direct method.

Generically, the optimal strategy of decision-maker i is in state-and-mean-field feedback form. It is also dependent on the switching regime:

$$\begin{aligned} U_i^{\mathrm{rn}} = & -\sum_{s \in \mathcal{S}} R_i^{-1}(t,s) B_{2i}^*(t,s) P_i(t,s)(X(t) - \hat{X}(t)) \mathbb{1}_{\{s(t)=s\}} \\ & -\sum_{s \in \mathcal{S}} (R_i(t,s) + \bar{R}_i(t,s))^{-1} (B_{2i}(t,s) + \bar{B}_{2i}(t,s))^* \bar{P}_i(t,s) \hat{X}(t) \mathbb{1}_{\{s(t)=s\}}. \end{aligned} \tag{7.404}$$

Note that if P is a solution, then P^* is also a solution when the coefficients of the cost functions are symmetric positive definite. Therefore, $P_i^*(t,s) = P_i(t,s)$, $(t,s) \in [0,T] \times \mathcal{S}$.

From the matrix state system (7.400), the conditional expected matrix $\hat{X}(t) := E[X(t)|\mathcal{F}_t^s]$, where $\mathcal{F}^s$ is the natural filtration of the regime switching process s up to t, solves the following system:

$$\begin{aligned} d\hat{X} &= [(B_1 + \bar{B}_1)\hat{X} + \textstyle\sum_{j=1}^I (B_{2j} + \bar{B}_{2j})\hat{U}_j]dt, \\ \hat{X}(0) &= E[X_0]. \end{aligned}$$

which means that

$$d\hat{X}=[(B_1+\bar{B}_1)-\frac{1}{2}\sum_{j=1}^{I}(B_{2j}+\bar{B}_{2j})(R_j+\bar{R}_j)^{-1}(B_{2j}+\bar{B}_{2j})^*(\bar{P}_j^*+\bar{P}_j)]\hat{X}dt,$$
$$\hat{X}(0)=E[X_0],$$

which will be used for feedback in the optimal strategy.

7.10.1.2 Risk-Neutral Full Cooperation

Definition 7.9 (Mean-Field-Type Risk-Neutral Full Cooperation) A mean-field-type risk-neutral fully cooperative solution is a strategy profile $(U_j^{\mathrm{rn,g}},\ j\in\mathscr{I})$, of all decision-makers such that

$$E[L_0(X,s,U^{rn,g})]=\inf{}_{(U_1,\dots,U_I)}E[L_0(X,s,U],\tag{7.405}$$

where $L_0:=\sum_{j\in\mathscr{I}}L_j$ is the social (global) cost.

Solution to the Risk-Neutral Fully Cooperative Game (7.405)
Assume that $Q_0:=\sum_{j\in\mathscr{I}}Q_j$, $Q_0+\bar{Q}_0:=\sum_{j\in\mathscr{I}}[Q_j+\bar{Q}_j]$, R_i, $R_i+\bar{R}_i$ are symmetric positive definite. The fully cooperative solution of the problem $\inf{}_{(U_1,\dots,U_I)}E[\sum_{j\in\mathscr{I}}L_j]$ is given by

$$\begin{cases} i\in\mathscr{I},\\ U_i^{\mathrm{rn,g}}-\hat{U}_i^{\mathrm{rn,g}}=-R_i^{-1}B_{2i}^*P_0(X-\hat{X}),\\ \hat{U}_i^{\mathrm{rn,g}}=-(R_i+\bar{R}_i)^{-1}(B_{2i}+\bar{B}_{2i})^*\bar{P}_0\hat{X},\\ L_0^{\mathrm{rn,g}}(X,s,U^{\mathrm{rn,g}})\\ =E\langle P_0(0,s(0))(X_0-\hat{X}_0),X_0-\hat{X}_0\rangle+E\langle\bar{P}_0(0,s(0))\hat{X}_0,\hat{X}_0\rangle+E[\delta_0(0,s(0))], \end{cases}$$

where P_0, $\bar{P}_0$, and δ_0 solve the following differential equations:

$$\begin{cases} \dot{P}_0+Q_0+P_0B_1+B_1^*P_0+\sum_{s'\neq s}(P_0(t,s')-P_0(t,s))\tilde{q}_{ss'}\\ -P_0\Big[\sum_{i\in\mathscr{I}}B_{2i}R_i^{-1}B_{2i}^*\Big]P_0=0,\\ P_0(T,s)=Q_0(T,s),\\ \dot{\bar{P}}_0+Q_0+\bar{Q}_0+\bar{P}_0(B_1+\bar{B}_1)+(B_1+\bar{B}_1)^*\bar{P}_0\\ +\sum_{s'\neq s}(\bar{P}_0(t,s')-\bar{P}_0(t,s))\tilde{q}_{ss'}\\ -\bar{P}_0\Big[\sum_{i\in\mathscr{I}}(B_{2i}+\bar{B}_{2i})(R_i+\bar{R}_i)^{-1}(B_{2i}+\bar{B}_{2i})^*\Big]\bar{P}_0=0,\\ \bar{P}_0(T,s)=Q_0(T,s)+\bar{Q}_0(T,s),\\ \dot{\delta}_0+\langle P_0S_0,S_0\rangle+\int_{\Theta}\langle P_0M_0,M_0\nu(d\theta)\rangle\\ +\sum_{s'\neq s}(\delta_0(t,s')-\delta_0(t,s))\tilde{q}_{ss'}=0,\\ \delta_0(T,s)=0,\\ \forall s\in\mathscr{S}. \end{cases}\tag{7.406}$$

The proof is immediate by considering one single team and with a choice vector of matrices $U = (U_i)_{i\in\mathcal{I}}$ from Theorem 7.1.

7.10.1.3 Risk-Neutral Adversarial Interaction

Definition 7.10 (Mean-Field-Type Risk-Neutral Saddle Point Solution) The set of decision-makers is divided into two teams: a team of defenders and a team of attackers. The defenders set is $I_+ := \{i \in \mathcal{I} \mid R_i \succ 0, (R_i + \bar{R}_i) \succ 0\}$ and the attackers set is $I_- := \{j \in \mathcal{I} \mid -R_j \succ 0,\ -(R_j + \bar{R}_j) \succ 0\}$. A mean-field-type risk-neutral saddle point is a strategy profile $(U_j^{\mathrm{ad}},\ j \in I_+)$, of the team of defenders and $(U_j^{\mathrm{ad}},\ j \in I_-)$ of the team of attackers such that

$$L^{\mathrm{ad}}(X, s, (U_i^{\mathrm{ad}})_{i\in I_+}, (U_j)_{j\in I_-}) \leq L^{\mathrm{ad}}(X, s, U^{\mathrm{ad}}) \leq L^{\mathrm{ad}}(X, s, (U_i)_{i\in I_+}, (U_j^{\mathrm{ad}})_{j\in I_-}) \tag{7.407}$$

and $L^{\mathrm{ad}}(X, s, U^{\mathrm{ad}})$ is the value of the adversarial team (risk-neutral) game, where the cost functional is

$$\begin{aligned}
&L^{\mathrm{ad}}(X, s, U)\\
&:= \langle Q(T, s(T))(X(T) - \hat{X}(T)), X(T) - \hat{X}(T)\rangle\\
&+\langle (Q(T, s(T)) + \bar{Q}(T, s(T)))\hat{X}(T), \hat{X}(T)\rangle\\
&+\int_0^T \langle Q(X - \hat{X}), X - \hat{X}\rangle + \langle (Q + \bar{Q})\hat{X}, \hat{X}\rangle dt\\
&+\int_0^T \sum_{i\in\mathcal{I}} \langle R_i(U_i - \hat{U}_i), U_i - \hat{U}_i\rangle + \sum_{i\in\mathcal{I}} \langle (R_i + \bar{R}_i)\hat{U}_i, \hat{U}_i\rangle dt.
\end{aligned}$$

Team of Defenders vs Team of Attackers (7.407)
Assume that Q, $Q+\bar{Q}$, R_i, $R_i+\bar{R}_i$ are symmetric positive definite for $i \in \mathcal{I}_+$ and $-R_j$, $-(R_j+\bar{R}_j)$ are symmetric positive definite for $j \in \mathcal{I}_-$. We assume that $\mathcal{I} := \mathcal{I}_+ \cup \mathcal{I}_-$. The adversarial game problem of the team attackers $\mathcal{I}_-$ and the team of defenders $\mathcal{I}_+$ has a saddle, and it is given by

$$\begin{cases}
U_i^{\mathrm{ad}} - \hat{U}_i^{\mathrm{ad}} = -R_i^{-1} B_{2i}^* P^{\mathrm{ad}}(X - \hat{X}),\\
\hat{U}_i^{\mathrm{ad}} = -(R_i + \bar{R}_i)^{-1}(B_{2i} + \bar{B}_{2i})^* \bar{P}^{\mathrm{ad}}\hat{X},\ i \in \mathcal{I}_+,\\
U_j^{\mathrm{ad}} - \hat{V}_j^{\mathrm{ad}} = -R_j^{-1} B_{2j}^* P^{\mathrm{ad}}(X - \hat{X}),\ j \in \mathcal{I}_-,\\
\bar{U}_j^{\mathrm{ad}} = -(R_j + \bar{R}_j)^{-1}(B_{2j} + \bar{B}_{2j})^* \bar{P}^{\mathrm{ad}}\hat{X},\ j \in \mathcal{I}_-,\\
L^{\mathrm{ad}}(X, s, U^{\mathrm{ad}}) = E\langle P^{\mathrm{ad}}(0, s(0))(X_0 - \hat{X}_0), X_0 - \hat{X}_0\rangle\\
+E\langle \bar{P}^{\mathrm{ad}}(0, s(0))\hat{X}_0, \hat{X}_0\rangle + E[\delta^{\mathrm{ad}}(0, s(0))],
\end{cases}$$

where P^{ad}, $\bar{P}^{\mathrm{ad}}$, and δ^{ad} solve the following differential equations:

(continued)

$$
\begin{cases}
\dot{P} + Q + PB_1 + B_1^* P + \sum_{s' \neq s}(P(t,s') - P(t,s))\tilde{q}_{ss'} \\
-P\Big[\sum_{i \in \mathcal{I}_+} B_{2i} R_i^{-1} B_{2i}^* + \sum_{j \in \mathcal{I}_-} B_{2j} R_j^{-1} B_{2j}^*\Big] P = 0, \\
P(T,s) = Q(T,s), \\
\dot{\bar{P}} + Q + \bar{Q} + \bar{P}(B_1 + \bar{B}_1) + (B_1 + \bar{B}_1)^* \bar{P} \\
+\sum_{s' \neq s}(\bar{P}(t,s') - \bar{P}(t,s))\tilde{q}_{ss'} \\
-\bar{P}\Big[\sum_{i \in \mathcal{I}_+} (B_{2i} + \bar{B}_{2i})(R_i + \bar{R}_i)^{-1}(B_{2i} + \bar{B}_{2i})^* \\
+\sum_{j \in \mathcal{I}_-} (B_{2j} + \bar{B}_{2j})(R_j + \bar{R}_j)^{-1}(B_{2j} + \bar{B}_{2j})^*\Big]\bar{P} = 0, \\
\bar{P}(T,s) = Q(T,s) + \bar{Q}(T,s), \\
\dot{\delta} + \langle PS_0, S_0 \rangle + \int_\Theta \langle PM_0, M_0 \nu(d\theta) \rangle \\
+\sum_{s' \neq s}(\delta(t,s') - \delta(t,s))\tilde{q}_{ss'} = 0, \\
\delta(T,s) = 0, \\
\forall s \in \mathcal{S}.
\end{cases}
\tag{7.408}
$$

The proof is immediate by considering two adversarial teams and with choice vector of matrices $U_+ = (U_i)_{i \in \mathcal{I}_+}$ and $U_- = (U_i)_{i \in \mathcal{I}_-}$, respectively.

Note that the Riccati equations in (7.408) have a unique positive definite solution P^{ad} if in addition

$$
\Big[\sum_{i \in \mathcal{I}_+} B_{2i} R_i^{-1} B_{2i}^* + \sum_{j \in \mathcal{I}_-} B_{2j} R_j^{-1} B_{2j}^*\Big] \succ 0,
$$

which does not blow up within $[0, T]$. Similarly, there is unique positive definite solution $\bar{P}^{\text{ad}}$ if in addition

$$
\begin{aligned}
&\Big[\sum_{i \in \mathcal{I}_+} (B_{2i} + \bar{B}_{2i})(R_i + \bar{R}_i)^{-1}(B_{2i} + \bar{B}_{2i})^* \\
&+\sum_{j \in \mathcal{I}_-} (B_{2j} + \bar{B}_{2j})(R_j + \bar{R}_j)^{-1}(B_{2j} + \bar{B}_{2j})^*\Big] \succ 0, \; t \in [0, T].
\end{aligned}
$$

7.10.2 Matrix-Valued Risk-Sensitive Problems

We provide basic definitions of risk-sensitive problems and their solution concepts.

7.10.2.1 Risk-Sensitive Nash Equilibria

Definition 7.11 (Mean-Field-Type Risk-Sensitive Equilibrium) Given $(U_j,\ j \neq i)$, a risk-sensitive best-response strategy of decision-maker i is a strategy that solves

$$\inf_{U_i} \frac{1}{\lambda_i} \log \left(E[e^{\lambda_i L_i}] \right), \tag{7.409}$$

$$\text{subject to } (7.400). \tag{7.410}$$

The set of risk-sensitive best responses of i is denoted by $\text{rsBR}_i(U_{-i})$.

A mean-field-type risk-sensitive Nash equilibrium is a strategy profile $(U_j^{\text{rs}},\ j \in \mathcal{I})$, of all decision-makers such that for every decision-maker i,

$$U_i^{\text{rs}} \in \text{rsBR}_i(U_{-i}^{\text{rs}}). \tag{7.411}$$

When $M_0 \neq 0$, the exponential martingale of compensated Poisson random process times a linear process yields to an exponential of non-quadratic terms:

$$E\exp[\int_0^T \int_\Theta \langle (P_i^* + P_i)(X - \hat{X}), M_0 \tilde{N}(dt, d\theta) \rangle],$$

whose quadratic covariation has an exponential non-quadratic term. Therefore, we consider only the risk-sensitive case when M_0 vanishes (no Poisson jump) and there is a single regime $\mathcal{S} = \{s_0\}$. When $\lambda_i > 0$, decision-maker i is risk-averse and when $\lambda_i < 0$ is risk-seeking. As λ_i goes to zero, decision-maker i becomes a risk-neutral decision-maker. The best-response problem of decision-maker i is well posed only for $\lambda_i \leq \bar{\lambda}_i$ where $\bar{\lambda}_i$ will be determined from the solution region of the differential system derived below.

Risk-Sensitive NonZero-Sum Game (7.411)
Assume that $Q_i, R_i, Q_i + \bar{Q}_i, R_i + \bar{R}_i$ are symmetric positive definite. There is no jump $M_0 = 0$ and one single regime $\mathcal{S} = \{s_0\}$. Then the matrix-valued mean-field-type (risk-sensitive) Nash equilibrium strategy and the (risk-sensitive) equilibrium costs are given by

$$\begin{cases} U_i^{\text{rs}} - \hat{U}_i^{\text{rs}} = -R_i^{-1} B_{2i}^* P_i^{\text{rs}}(X - \hat{X}) - (R_i + \bar{R}_i)^{-1}(B_{2i} + \bar{B}_{2i})^* \bar{P}_i \hat{X}, \\ \hat{U}_i^{\text{rs}} = -(R_i + \bar{R}_i)^{-1}(B_{2i} + \bar{B}_{2i})^* \bar{P}_i \hat{X}, \\ L_i^{\text{rs}}(X, s, U^{\text{rs}}) \\ = \frac{1}{\lambda_i} \log E\exp\{\lambda_i[\langle P_i^{\text{rs}}(0, s(0))(X_0 - \hat{X}_0), X_0 - \hat{X}_0 \rangle \\ +\langle \bar{P}_i(0, s(0))\hat{X}_0, \hat{X}_0 \rangle + \delta_i^{\text{rs}}(0, s(0))]\}, \\ i \in \mathcal{I}, \end{cases} \tag{7.412}$$

where P_i^{rs}, $\bar{P}_i^{\text{rs}} = \bar{P}_i$, and δ_i^{rs} solve the following differential equations:

(continued)

$$\begin{cases} \dot{P}_i + Q_i + P_i B_1^* + B_1 P_i \\ -P_i[B_{2i} R_i^{-1} B_{2i}^* - 2\lambda_i S_0 S_0^*] P_i \\ -\sum_{j\neq i} P_j B_{2j} R_j^{-1*} B_{2j}^* P_i \\ -\sum_{j\neq i} P_i B_{2j} R_j^{-1*} B_{2j}^* P_j = 0, \\ P_i(T,s) = Q_i(T,s), \\ \dot{\bar{P}}_i + Q_i + \bar{Q}_i + \bar{P}_i(B_1 + \bar{B}_1)^* + (B_1 + \bar{B}_1)\bar{P}_i \\ -\bar{P}_i[(B_{2i} + \bar{B}_{2i})(R_i + \bar{R}_i)^{-1}(B_{2i} + \bar{B}_{2i})^*]\bar{P}_i \\ -\sum_{j\neq i} \bar{P}_j(B_{2j} + \bar{B}_{2j})(R_j + \bar{R}_j)^{-1*}(B_{2j} + \bar{B}_{2j})^* \bar{P}_i \\ -\sum_{j\neq i} \bar{P}_i(B_{2j} + \bar{B}_{2j})(R_j + \bar{R}_j)^{-1*}(B_{2j} + \bar{B}_{2j})^* \bar{P}_j = 0, \\ \bar{P}_i(T) = Q_i(T) + \bar{Q}_i(T), \\ \dot{\delta}_i + \langle P_i S_0, S_0 \rangle, \\ \delta_i(T) = 0 \end{cases} \tag{7.413}$$

whenever these differential system of equations have a unique solution that does not blow up in $[0, T]$.

A proof can be obtained by modifying the above steps. The martingale term in Itô's formula is $2\lambda_i \int_0^T \langle S_0^* P_i(X - \hat{X}), dB \rangle$. By adding and removing the term

$$2\lambda_i^2 \int_0^T \langle P_i S_0 S_0^* P_i(X - \hat{X}), (X - \hat{X}) \rangle dt,$$

to Itô's formula, we reorganize the exponential quadratic terms. Using that relation

$$E\left[e^{2\lambda_i \int_0^T \langle S_0^* P_i(X-\hat{X}), dB\rangle - 2\lambda_i^2 \int_0^T \langle P_i S_0 S_0^* P_i(X-\hat{X}),(X-\hat{X})\rangle dt}\right] = 1,$$

and matching the term in P_i, we arrive at the announced result. This completes the proof.

As all λ_i vanish, the matrix-valued differential system (7.413) becomes the risk-neutral system (7.419) above, and the risk-sensitive optimal strategy coincides with the risk-neutral one. A bound for $\bar{\lambda}_i$ can be obtained from the positivity condition of the matrices:

$$[B_{2i} R_i^{-1} B_{2i}^* - 2\lambda_i S_0 S_0^*] \succ 0. \tag{7.414}$$

Note that $\bar{P}_i^{\text{rs}} = \bar{P}_i$ because this coefficient is associated with the term $\langle \hat{X}, \hat{X} \rangle$ which is independent of the Brownian motion.

Next we consider a mixture of risk-neutral, risk-seeking, and risk-averse decision-maker optimality coefficients that are obtained by solving the following system:

$$\begin{cases} i \in \mathcal{I}_0 : \\ \dot{P}_i + Q_i + P_i B_1^* + B_1 P_i \\ -P_i B_{2i} R_i^{-1} B_{2i}^* P_i \\ -\sum_{j\neq i} P_j B_{2j} R_j^{-1*} B_{2j}^* P_i - \sum_{j\neq i} P_i B_{2j} R_j^{-1*} B_{2j}^* P_j = 0, \end{cases} \tag{7.415}$$

$$\begin{cases} i \in \mathcal{I}_+ : \\ \dot{P}_i + Q_i + P_i B_1^* + B_1 P_i \\ -P_i [B_{2i} R_i^{-1} B_{2i}^* - 2\lambda_i S_0 S_0^*] P_i \\ -\sum_{j\neq i} P_j B_{2j} R_j^{-1*} B_{2j}^* P_i - \sum_{j\neq i} P_i B_{2j} R_j^{-1*} B_{2j}^* P_j = 0, \end{cases} \tag{7.416}$$

$$\begin{cases} i \in \mathcal{I}_- : \\ \dot{P}_i + Q_i + P_i B_1^* + B_1 P_i \\ -P_i [B_{2i} R_i^{-1} B_{2i}^* + 2(-\lambda_i) S_0 S_0^*] P_i \\ -\sum_{j\neq i} P_j B_{2j} R_j^{-1*} B_{2j}^* P_i - \sum_{j\neq i} P_i B_{2j} R_j^{-1*} B_{2j}^* P_j = 0, \end{cases} \tag{7.417}$$

$$P_i(T) = Q_i(T), i \in \mathcal{I}, \tag{7.418}$$

7.10.2.2 Risk-Sensitive Full Cooperation

Definition 7.12 (Mean-Field-Type Risk-Sensitive Full Cooperation) A mean-field-type risk-sensitive fully cooperative solution is a strategy profile $(U_j^{\text{rs,g}},\ j \in \mathcal{I})$, of all decision-makers such that

$$\inf{}_{(U_1,\dots,U_I)} \frac{1}{\lambda} \log \left[E e^{\lambda L_0(X,s,U)} \right] = \frac{1}{\lambda} \log \left[E e^{\lambda L_0(X,s,U^{rs,g})} \right],$$

Risk-Sensitive Fully Cooperative Game
Assume that $Q_0 := \sum_{j\in\mathcal{I}} Q_j$, $Q_0 + \bar{Q}_0 := \sum_{j\in\mathcal{I}} [Q_j, +\bar{Q}_j]$, R_i, $R_i + \bar{R}_i$ are symmetric positive definite. The risk-sensitive fully cooperative solution of the problem $\inf_{(U_1,\dots,U_I)} \frac{1}{\lambda} \log E[e^{\lambda \sum_{j\in\mathcal{I}} L_j}]$ is given by

$$\begin{cases} U_i^{\text{rs,g}} - \hat{U}_i^{\text{rs,g}} = -R_i^{-1} B_{2i}^* P_0^{\text{rs,g}} (X - \hat{X}), \\ \hat{U}_i^{\text{rs,g}} = -(R_i + \bar{R}_i)^{-1} (B_{2i} + \bar{B}_{2i})^* \bar{P}_0 \hat{X}, \\ L_0^{\text{rs,g}}(X, s, U^{\text{rs,g}}) \\ = \frac{1}{\lambda} \log E\exp\{\lambda[\langle P_0^{\text{rs,g}}(0, s(0))(X_0 - \hat{X}_0), X_0 - \hat{X}_0\rangle \\ +\langle \bar{P}_0(0, s(0))\hat{X}_0, \hat{X}_0\rangle + \delta_0^{\text{rs,g}}(0, s(0))]\}, \\ i \in \{1, \dots, I\}, \end{cases}$$

where $P_0^{\text{rs,g}}$, $\bar{P}_0^{\text{rs,g}} = \bar{P}_0$, and $\delta_0^{\text{rs,g}}$ solve the following differential equations:

(continued)

$$
\begin{cases}
\dot{P}_0 + Q_0 + P_0 B_1 + B_1^* P_0 \\
-P_0\Big[\sum_{i\in\mathcal{I}} B_{2i} R_i^{-1} B_{2i}^* - 2\lambda S_0 S_0^*\Big] P_0 = 0, \\
P_0(T,s) = Q_0(T,s), \\
\dot{\bar{P}}_0 + Q_0 + \bar{Q}_0 + \bar{P}_0(B_1 + \bar{B}_1) + (B_1 + \bar{B}_1)^* \bar{P}_0 \\
\\
-\bar{P}_0\Big[\sum_{i\in\mathcal{I}} (B_{2i} + \bar{B}_{2i})(R_i + \bar{R}_i)^{-1}(B_{2i} + \bar{B}_{2i})^*\Big]\bar{P}_0 = 0, \\
\bar{P}_0(T,s) = Q_0(T,s) + \bar{Q}_0(T,s), \\
\dot{\delta}_0 + \langle P_0 S_0, S_0\rangle = 0, \\
\delta_0(T,s) = 0,
\end{cases}
\tag{7.419}
$$

Notice that the bound $\bar{\lambda}$ is obtained from the positivity condition $\sum_{i\in\mathcal{I}} B_{2i} R_i^{-1} B_{2i}^* - 2\lambda S_0 S_0^* \succ 0$. Hence, the risk condition is relaxed, thanks to the full cooperation in comparison with the noncooperative risk consideration, i.e.,

$$
\begin{aligned}
\bar{\lambda} &= \sup\left\{\lambda \,\Big|\, \begin{matrix}\text{full cooperation risk-sensitive} \\ \text{solution is well-posed}\end{matrix}\right\}, \\
\bar{\lambda}_i &= \sup\left\{\lambda_i \,\Big|\, \begin{matrix}\text{noncooperative risk-sensitive} \\ \text{solution is well-posed}\end{matrix}\right\},
\end{aligned}
\tag{7.420}
$$

where

$$
B_{2i} R_i^{-1} B_{2i}^* \succ 2\lambda_i S_0 S_0, \quad \text{(from (7.452))}.
$$

By taking the sum over i, we arrive at $\sum_{i\in\mathcal{I}} B_{2i} R_i^{-1} B_{2i}^* \succ 2\sum_{i\in\mathcal{I}} \lambda_i S_0 S_0$; hence,

$$
\sum_{i\in\mathcal{I}} \bar{\lambda}_i \le \bar{\lambda}.
$$

We conclude that a full cooperation among the decision-makers increases the well-posed domain due to sharing of risk.

7.10.2.3 Risk-Sensitive Adversarial Interaction

Definition 7.13 (Mean-Field-Type Risk-Sensitive Saddle Point Solution) The set of decision-makers is divided into two teams: a team of defenders and a team of attackers. The defenders set is $I_+ := \{i \in \mathcal{I} \mid R_i \succ 0, (R_i + \bar{R}_i) \succ 0\}$ and the attackers set is $I_- := \{j \in \mathcal{I} \mid -R_j \succ 0,\ -(R_j + \bar{R}_j) \succ 0\}$. A mean-field-type risk-sensitive saddle point is a strategy profile $(U_j^{\mathrm{ad}},\ j \in I_+)$, of the team of defenders and $(U_j^{\text{ad}},\ j \in I_-)$ of the team of attackers such that

$$\begin{aligned}
&\tfrac{1}{\lambda}\log\left[Ee^{\lambda L^{\text{ad}}(X,s,(U_i^{\text{ad}})_{i\in I_+},(U_j)_{j\in I_-})}\right]\\
&\leq \tfrac{1}{\lambda}\log\left[Ee^{\lambda L^{\text{ad}}(X,s,U^{\text{ad}})}\right]\\
&\leq \tfrac{1}{\lambda}\log\left[Ee^{\lambda L^{\text{ad}}(X,s,(U_i)_{i\in I_+},(U_j^{\text{ad}})_{j\in I_-})}\right]
\end{aligned}$$

Adversarial Risk-Sensitive Game

Assume that Q, $Q+\bar{Q}$, R_i, $R_i+\bar{R}_i$ are symmetric positive definite for $i \in \mathcal{I}_+$ and $-R_j, -(R_j+\bar{R}_j)$ are symmetric positive definite for $j \in \mathcal{I}_-$. We assume that $\mathcal{I} := \mathcal{I}_+ \cup \mathcal{I}_-$. The adversarial risk-sensitive game problem of the team of attackers $\mathcal{I}_-$ and the team of defenders $\mathcal{I}_+$ has a saddle point, and it is given by

$$\begin{cases}
U_i^{\text{rs,ad}}=-R_i^{-1}B_{2i}^*P^{rs,ad}(X-\hat{X})-(R_i+\bar{R}_i)^{-1}(B_{2i}+\bar{B}_{2i})^*\bar{P}^{rs,ad}\hat{X},\ i\in\mathcal{I}_+,\\
U_j^{\text{rs,ad}}=-R_j^{-1}B_{2j}^*P^{rs,ad}(X-\hat{X})-(R_j+\bar{R}_j)^{-1}(B_{2j}+\bar{B}_{2j})^*\bar{P}^{rs,ad}\hat{X},\ j\in\mathcal{I}_-,\\
L^{\text{ad}}(X,s,U^{\text{ad}}) = E\langle P^{\text{rs,ad}}(0,s(0))(X_0-\hat{X}_0), X_0-\hat{X}_0\rangle\\
+E\langle \bar{P}^{\text{ad}}(0,s(0))\hat{X}_0,\hat{X}_0\rangle + E[\delta^{\text{rs,ad}}(0,s(0))],
\end{cases}$$

where $P^{\text{rs,ad}}, \bar{P}^{\text{rs,ad}} = \bar{P}^{\text{ad}}$, and $\delta^{rs,ad}$ solve the following differential equations:

$$\begin{cases}
\dot{P}+Q+PB_1+B_1^*P\\
-P\left[\sum_{i\in\mathcal{I}_+}B_{2i}R_i^{-1}B_{2i}^*+\sum_{j\in\mathcal{I}_-}B_{2j}R_j^{-1}B_{2j}^*-2\lambda S_0S_0^*\right]P=0,\\
P(T,s)=Q(T,s),\\
\dot{\bar{P}}+Q+\bar{Q}+\bar{P}(B_1+\bar{B}_1)+(B_1+\bar{B}_1)^*\bar{P}\\
-\bar{P}\left[\sum_{i\in\mathcal{I}_+}(B_{2i}+\bar{B}_{2i})(R_i+\bar{R}_i)^{-1}(B_{2i}+\bar{B}_{2i})^*\right.\\
\left.+\sum_{j\in\mathcal{I}_-}(B_{2j}+\bar{B}_{2j})(R_j+\bar{R}_j)^{-1}(B_{2j}+\bar{B}_{2j})^*-2\lambda S_0S_0^*\right]\bar{P}=0,\\
\bar{P}(T,s)=Q(T,s)+\bar{Q}(T,s),\\
\dot{\delta}+\langle PS_0,S_0\rangle=0, \delta(T,s)=0,\\
\forall s\in\mathcal{S}.
\end{cases} \tag{7.421}$$

Notice that these risk-sensitive adversarial Riccati equations have positive definite solution $P^{rs,ad}$ if in addition

$$\left[\sum_{i\in\mathcal{I}_+}B_{2i}R_i^{-1}B_{2i}^*+\sum_{j\in\mathcal{I}_-}B_{2j}R_j^{-1}B_{2j}^*-2\lambda S_0S_0^*\right]\succ 0,$$

which does not blow up within $[0, T]$.

7.10.3 Adversarial Versus Risk-Sensitive Problems

This section discusses some relationship between the solutions obtained from a risk-neutral adversarial game problem and the risk-sensitive game problem.

7.10.3.1 Mean-Field-Free Control

Consider one risk-averse decision-maker who aims to solve the following risk-sensitive control problem with $\lambda > 0$:

$$\begin{array}{l} \inf_{U^{\text{rs}}} L^{rs}(X^{\text{rs}}, s, U^{\text{rs}}) \\ \text{subject to} \\ dX^{\text{rs}} = [B_1 X^{\text{rs}} + B_2 U^{\text{rs}}]dt + S_0 dB, \\ X(0) = X_0,\ s(t) = s_0 \end{array}$$

where

$$\begin{array}{l} L^{rs}(X^{\text{rs}}, s, U^{\text{rs}}) = \frac{1}{\lambda} \log \left[E e^{\lambda L(X^{\text{rs}}, s, U^{\text{rs}})} \right], \\ L(X^{\text{rs}}, s, U^{\text{rs}}) = \langle Q(T, s(T)) X^{\text{rs}}(T), X^{\text{rs}}(T) \rangle + \int_0^T \langle Q X^{\text{rs}}, X^{\text{rs}} \rangle dt \\ \qquad + \int_0^T \langle R U^{\text{rs}}, U^{\text{rs}} \rangle dt. \end{array}$$

Its solution is given by

$$\begin{array}{l} U^{\text{rs}} = -R^{-1} B_2^* P^{\text{rs}} X^{\text{rs}}, \\ L^{\text{rs}}(X^{\text{rs}}, s, U^{\text{rs}}) = \frac{1}{\lambda} \log E\exp\{\lambda [\langle P_0^{\text{rs}}(0, s(0)) X_0^{\text{rs}}, X_0^{\text{rs}} \rangle + \delta_0^{\text{rs}}(0, s(0))]\}, \end{array}$$

Now consider the following adversarial risk-neutral control problem:

$$\begin{array}{l} \inf_{U^{\text{ad}}} \sup_{V^{\text{ad}}} E[L^{\text{ad}}(U^{\text{ad}}, V^{\text{ad}})], \\ \text{subject to} \\ dX^{\text{ad}} = \left[B_1 X^{\text{ad}} + B_2 U^{\text{ad}} + \left[\sqrt{2\lambda} S_0 [-\tilde{R}]^{\frac{1}{2}} \right] V^{\text{ad}} \right] dt + S_0 dB, \end{array}$$

with

$$L^{\text{ad}}(X^{\text{ad}}, s, U^{\text{ad}}) = L(X^{\text{ad}}, s, U^{\text{ad}}) + \int_0^T \langle \tilde{R} V^{\text{ad}}, V^{\text{ad}} \rangle dt,$$

where $\tilde{R} \prec 0$. Its explicit solution is given by

$$
\begin{aligned}
U^{\text{ad}} &= -R^{-1}B_2^* P^{\text{ad}} X^{\text{ad}},\\
V^{\text{ad}} &= \sqrt{2\lambda}[-\tilde{R}]^{-\frac{1}{2}} S_0^* P^{\text{ad}} X^{\text{ad}},\\
L^{\text{ad}}(X^{\text{ad}}, s, U^{\text{ad}}) &= E\langle P^{\text{ad}}(0, s(0)) X_0^{\text{ad}}, X_0^{\text{ad}}\rangle + E[\delta^{\text{ad}}(0, s(0))],
\end{aligned}
$$

the coefficients P^{ad} and P^{rs} solve the same differential system which is given by

$$
\begin{aligned}
&\dot{P} + Q + PB_1 + B_1^* P - P\Big[B_2 R^{-1} B_2^* - 2\lambda S_0 S_0^*\Big]P = 0,\\
&P(T, s) = Q(T, s),
\end{aligned}
$$

Notice that the evolution of the system states X^{rs} and X^{ad} may not have the same probability law even if they start at the same initial random variable X_0. When X_0 is a deterministic matrix and $\mathscr{S}$ is singleton, the two optimal costs L^{ad} and L^{rs} coincide. In general, these costs are different.

By taking the expectation of the state, it follows that

$$
\begin{aligned}
d\hat{X}^{\text{rs}} &= [B_1 - B_2 R^{-1} B_2^* P]\hat{X}^{\text{rs}} dt,\\
d\hat{X}^{\text{ad}} &= [B_1 - (B_2 R^{-1} B_2^* - 2\lambda S_0 S_0^*)P]\hat{X}^{\text{ad}} dt.
\end{aligned}
\tag{7.422}
$$

Hence,

$$
\hat{X}^{\text{ad}}(t) = \hat{X}^{\text{rs}}(t) e^{2\lambda \int_0^t S_0 S_0^* P(t')dt'}.
$$

We conclude that the first moments of the two problems are different for $\lambda S_0 S_0^* \neq 0$ and $EX_0 \neq 0$.

7.10.3.2 Mean-Field-Free Team

The observation of Sect. 7.10.3.1 can be extended to team problems. Consider the following risk-sensitive team problem with team members in $\mathscr{I}_+$:

$$
\begin{aligned}
&\inf_{U^{\text{rs}}} \tfrac{1}{\lambda} \log\left[E e^{\lambda L_{\text{team}}(X^{\text{rs}}, s, U^{\text{rs}})}\right],\\
&\text{subject to}\\
&dX^{\text{rs}} = [B_1 X^{\text{rs}} + \textstyle\sum_{i\in\mathscr{I}_+} B_{2i} U_i^{\text{rs}}]dt + S_0 dB,
\end{aligned}
$$

where

$$
\begin{aligned}
L_{\text{team}}(X^{\text{rs}}, s, U^{\text{rs}}) &= \langle Q(T, s(T)) X^{\text{rs}}(T), X^{\text{rs}}(T)\rangle\\
&+ \textstyle\int_0^T \langle QX^{\text{rs}}, X^{\text{rs}}\rangle dt + \int_0^T \sum_{i\in\mathscr{I}_+} \langle R_i U_i^{\text{rs}}, U_i^{\text{rs}}\rangle dt,
\end{aligned}
$$

Its semi-explicit solution is given by

$$U_i^{\mathrm{rs}} = -R_i^{-1} B_{2i}^* P^{\mathrm{rs}} X^{\mathrm{rs}},$$
$$L_{\mathrm{team}}^{\mathrm{rs}}(X^{\mathrm{rs}}, s, U^{\mathrm{rs}}) = \frac{1}{\lambda} \log E\exp\{\lambda[\langle P_0^{\mathrm{rs}}(0, s(0)) X_0^{\mathrm{rs}}, X_0^{\mathrm{rs}}\rangle + \delta_0^{\mathrm{rs}}(0, s(0))]\}.$$

We now design the following adversarial risk-neutral game problem between two teams in I_+ and I_-, respectively:

$$\inf_{U^{\mathrm{ad}}} \sup_{V^{\mathrm{ad}}} E[L_{\mathrm{team}}^{\mathrm{ad}}(U, V)],$$
subject to
$$dX^{\mathrm{ad}} = \Big[B_1 X^{\mathrm{ad}} + \textstyle\sum_{i\in\mathscr{I}_+} B_{2i} U^{\mathrm{ad}} + |\mathscr{I}_-|^{-1} \sum_{j\in\mathscr{I}_-} \Big[\sqrt{2\lambda} S_0 [-\tilde{R}_j]^{\frac{1}{2}}\Big] V_j^{\mathrm{ad}}\Big] dt + S_0 dB,$$

with

$$L_{\mathrm{team}}^{\mathrm{ad}}(X^{\mathrm{ad}}, s, U^{\mathrm{ad}}) = \langle Q(T, s(T)) X^{\mathrm{ad}}(T), X^{\mathrm{ad}}(T)\rangle$$
$$+ \int_0^T \langle Q X^{\mathrm{ad}}, X^{\mathrm{ad}}\rangle + \textstyle\sum_{i\in\mathscr{I}_+} \langle R_i U_i^{\mathrm{ad}}, U_i^{\mathrm{ad}}\rangle + \sum_{j\in\mathscr{I}_-} \langle \tilde{R}_j V_j^{\mathrm{ad}}, V_j^{\mathrm{ad}}\rangle dt,$$

with $\tilde{R}_j \prec 0,\ j \in \mathscr{I}_-$. Its semi-explicit solution is given by

$$U_i^{\mathrm{ad}} = -R_i^{-1} B_{2i}^* P^{\mathrm{ad}} X^{\mathrm{ad}},$$
$$V_j^{\mathrm{ad}} = \sqrt{2\lambda} [-\tilde{R}]_j^{\frac{1}{2}} S_0^* P^{\mathrm{ad}} X^{\mathrm{ad}},$$
$$L_{\mathrm{team}}^{\mathrm{ad}}(X^{\mathrm{ad}}, s, U^{\mathrm{ad}}) = E\langle P^{\mathrm{ad}}(0, s(0)) X_0^{\mathrm{ad}}, X_0^{\mathrm{ad}}\rangle + E[\delta^{\mathrm{ad}}(0, s(0))].$$

The coefficients P^{ad} and P^{rs} solve the same differential equation, which is given by

$$\dot{P} + Q + P B_1 + B_1^* P - P\Big[\textstyle\sum_{i\in\mathscr{I}_+} B_{2i} R_i^{-1} B_{2i}^* - 2\lambda S_0 S_0^*\Big] P = 0,$$
$$P(T, s) = Q(T, s),$$

The risk-sensitive optimal team control U^{rs} and the risk-neutral adversarial team optimal strategy U^{rs} have a similar structure. However, the probability law of the system states X^{rs} and X^{ad} may be different even when they start from the same initial random variable X_0.

This means that, in a problem where the first moments areinvolved in the cost, the means of the mean-field trajectory of risk-sensitive and robust design are not necessarily equal. Below we show that one can modify the robust control law such that the first moments of these designs are equal. It will be achieved by means mean-field-type robust control.

7.10.3.3 Mean-Field-Type Team Problem

We now examine possible relationship between risk-sensitive team (full cooperation) and robust team game in the mean-field-dependent case. Consider the following risk-sensitive mean-field-type team problem, involving the teams $\mathscr{I}_+$:

$$\begin{aligned}
&\inf_{U^{\text{rs}}} \tfrac{1}{\lambda}\log\left[Ee^{\lambda L_{\text{team}}(X,s,U^{\text{rs}})}\right],\\
&\text{subject to}\\
&dX=[B_1(X-\hat{X})+(B_1+\bar{B}_1)\hat{X}+\textstyle\sum_{i\in\mathscr{I}_+}B_{2i}(U_i-\hat{U}_i)\\
&+\textstyle\sum_{i\in\mathscr{I}_+}(B_{2i}+\bar{B}_{2i})\hat{U}_i]dt+S_0dB,
\end{aligned}$$

where

$$\begin{aligned}
&L_{\text{team}}(X,s,(U_i)_{i\in\mathscr{I}_+})\\
&=\langle Q(T,s(T))(X(T)-\hat{X}(T)),X(T)-\hat{X}(T)\rangle+\langle (Q(T,s(T)+\bar{Q}(T,s(T))\hat{X}(T),\hat{X}(T)\rangle\\
&+\textstyle\int_0^T\langle Q(X-\hat{X}),X-\hat{X}\rangle+\langle (Q+\bar{Q})\hat{X},\hat{X}\rangle dt\\
&+\textstyle\int_0^T\sum_{i\in\mathscr{I}_+}\langle R_i(U_i-\hat{U}_i),U_i-\hat{U}_i\rangle+\sum_{i\in\mathscr{I}_+}\langle (R_i+\bar{R}_i)\hat{U}_i,\hat{U}_i\rangle dt.
\end{aligned}$$

Its semi-explicit solution is given by

$$\begin{cases}
U_i^{\text{rs}}=-R_i^{-1}B_{2i}^*P^{\text{rs}}(X^{\text{rs}}-\hat{X}^{\text{rs}})-(R_i+\bar{R}_i)^{-1}(B_{2i}+\bar{B}_{2i})^*\bar{P}^{\text{rs}}\hat{X}^{\text{rs}},\\
L_{\text{team}}^{\text{rs}}(X^{\text{rs}},s,U^{\text{rs}})=\frac{1}{\lambda}\log E\exp\{\lambda[\langle P^{\text{rs}}(0,s(0))(X_0^{\text{rs}}-\hat{X}_0^{\text{rs}}),X_0^{\text{rs}}-\hat{X}_0^{\text{rs}}\rangle\\
+\langle \bar{P}^{\text{rs}}(0,s(0))\hat{X}_0^{\text{rs}},\hat{X}_0^{\text{rs}}\rangle+\delta_0^{\text{rs}}(0,s(0))]\},\\
i\in\mathscr{I}_+.
\end{cases}$$

We design the following adversarial risk-neutral mean-field-type game problem with two teams in $\mathscr{I}_+$ and $\mathscr{I}_-$:

$$\begin{aligned}
&\inf_{U^{\text{ad}}}\sup_{V^{\text{ad}}}E[L_{\text{team}}^{\text{ad}}(U,V)],\\
&\text{subject to}\\
&dX^{\text{ad}}=[B_1(X^{\text{ad}}-\hat{X}^{\text{ad}})+(B_1+\bar{B}_1)\hat{X}^{\text{ad}}\\
&+\textstyle\sum_{i\in\mathscr{I}_+}B_{2i}(U_i^{\text{ad}}-\hat{U}_i^{\text{ad}})+\sum_{i\in\mathscr{I}_+}(B_{2i}+\bar{B}_{2i})\hat{U}_i^{\text{ad}}\\
&+|\mathscr{I}_-|^{-1}\textstyle\sum_{j\in\mathscr{I}_-}\sqrt{2\lambda}S_0[-\tilde{R}_j]^{\frac{1}{2}}(V_j^{\text{ad}}-\hat{V}_j^{\text{ad}})]dt+S_0dB,
\end{aligned}$$

with

$$\begin{aligned}
&L_{\text{team}}^{\text{ad}}(X^{\text{ad}},s,U^{\text{ad}})=L_{\text{team}}(X^{\text{ad}},s,U^{\text{ad}})\\
&+\textstyle\int_0^T\sum_{j\in\mathscr{I}_-}\langle \tilde{R}_j(V_j^{\text{ad}}-\hat{V}_j^{\text{ad}}),V_j^{\text{ad}}-\hat{V}_j^{\text{ad}}\rangle dt,
\end{aligned}$$

with $\tilde{R}_j \prec 0$. Its semi-explicit solution is given by

$$\begin{cases}
U_i^{\text{ad}}-\hat{U}_i^{\text{ad}}=-R_i^{-1}B_{2i}^*P^{\text{ad}}(X^{\text{ad}}-\hat{X}^{\text{ad}}),\\
\hat{U}_i^{\text{ad}}=-(R_i+\bar{R}_i)^{-1}(B_{2i}+\bar{B}_{2i})^*\bar{P}^{\text{ad}}\hat{X}^{\text{ad}},\ i\in\mathscr{I}_+,\\
V_j^{\text{ad}}-\hat{V}_j^{\text{ad}}=\sqrt{2\lambda}[-\tilde{R}_j]^{-\frac{1}{2}}S_0^*P^{\text{ad}}(X^{\text{ad}}-\hat{X}^{\text{ad}}),\ j\in\mathscr{I}_-,\\
\hat{V}_j^{\text{ad}}=0,\\
L^{\text{ad}}(X^{\text{ad}},s,U^{\text{ad}})\\
=E\langle P^{\text{ad}}(0,s(0))(X_0^{\text{ad}}-\hat{X}_0^{\text{ad}}),X_0^{\text{ad}}-\hat{X}_0^{\text{ad}}\rangle\\
+E\langle \bar{P}^{\text{ad}}(0,s(0))\hat{X}_0^{\text{ad}},\hat{X}_0^{\text{ad}}\rangle+E[\delta^{\text{ad}}(0,s(0))],
\end{cases}$$

Thus, it is obtained that P^{rs} and P^{ad} solve the same equation:

$$\begin{aligned}&\dot{P} + Q + PB_1 + B_1^* P\\ &-P\Big[\sum_{i\in\mathscr{I}_+} B_{2i}R_i^{-1}B_{2i}^* - 2\lambda S_0 S_0^*\Big]P = 0,\end{aligned}$$

where λ is such that $[\sum_{i\in\mathscr{I}_+} B_{2i}R_i^{-1}B_{2i}^* - 2\lambda S_0 S_0^*\Big] \succ 0$, $\bar{P}^{\text{rs}}$ and $\bar{P}^{\text{ad}}$ solve the same Riccati equations:

$$\begin{aligned}&\dot{\bar{P}} + Q + \bar{Q} + \bar{P}(B_1 + \bar{B}_1) + (B_1 + \bar{B}_1)^*\bar{P}\\ &-\bar{P}\Big[\sum_{i\in\mathscr{I}_+}(B_{2i} + \bar{B}_{2i})(R_i + \bar{R}_i)^{-1}(B_{2i} + \bar{B}_{2i})^*\Big]\bar{P} = 0.\end{aligned}$$

with positive define matrix $[\sum_{i\in\mathscr{I}_+}(B_{2i} + \bar{B}_{2i})(R_i + \bar{R}_i)^{-1}(B_{2i} + \bar{B}_{2i})^*\Big] \succ 0$. In contrast to the mean-free cases above, the mean-field-type robust design provides has an interesting feature: the expectation of the states of the two problems coincide: $\hat{X}^{\text{rs}} = \hat{X}^{\text{ad}}$, and solve the following matrix-valued differential system:

$$\begin{aligned}&d\hat{X} = [(B_1 + \bar{B}_1) - \sum_{i\in\mathscr{I}_+}(B_{2i} + \bar{B}_{2i})(R_i + \bar{R}_i)^{-1}(B_{2i} + \bar{B}_{2i})^*\bar{P}]\hat{X}dt,\\ &\hat{X}(0) = E[X_0^{\text{rs}}] = E[X_0^{\text{ad}}].\end{aligned}$$

This means that when facing a risk-sensitive problem, the "true" team decision-maker can design a robust control law of mean-field type such that

$$V_j^{\text{ad}} = \sqrt{2\lambda}[-\tilde{R}_j]^{-\frac{1}{2}} S_0^* P^{\text{ad}}(X^{\text{ad}} - \hat{X}^{\text{ad}}),\ j \in \mathscr{I}_-.$$

Moreover, this robust control law of mean-field type preserves the mean trajectory of the states because $EV_j^{\text{ad}} = 0$.

7.10.3.4 Mean-Field-Type Game Problem

We now discuss the relationship between robustness and risk sensitivity in the context of games. Consider the risk-sensitive mean-field-type game problem given by

$$\begin{aligned}&\inf_{U_i\in\mathscr{U}_i}\tfrac{1}{\lambda_i}\log\left[Ee^{\lambda_i L_i(X,s,U)}\right],\\ &\text{subject to}\\ &dX = [B_1(X - \hat{X}) + (B_1 + \bar{B}_1)\hat{X} + \textstyle\sum_{i=1}^I B_{2i}(U_i - \hat{U}_i)\\ &+\textstyle\sum_{i=1}^I(B_{2i} + \bar{B}_{2i})\hat{U}_i]dt + S_0 dB,\end{aligned}$$

where $\lambda_i > 0$, $L_i(X, s, U)$ is the cost presented in (7.401). Hence, by Theorem 7.412, it follows that the solution is given by

$$U_i = -R_i^{-1} B_{2i}^* P_i^{rs}(X - \hat{X}) - (R_i + \bar{R}_i)^{-1}(B_{2i} + \bar{B}_{2i})^* \bar{P}_i^{rs} \hat{X},$$

We design the following robust mean-field-type game problem in which each "true" decision-maker i of the game above can also treat the risk as a *fictitious adversary* ik who picks a robust control strategy V_{ik}.

$$\begin{array}{l} \inf_{U_i \in \mathscr{U}_i} \left[\sup_{V_{ik}} L_i^{\mathrm{ad}}(X, s, U, V) \right] \\ \text{subject to} \\ dX = [B_1(X - \hat{X}) + (B_1 + \bar{B}_1)\hat{X} + \sum_{i=1}^{I} B_{2i}(U_i - \hat{U}_i) + \sum_{i=1}^{I} (B_{2i} + \bar{B}_{2i})\hat{U}_i]dt \\ + S_0 dB + \sum_{i \in \mathscr{I}} \sqrt{2\lambda_i} S_0 [-\tilde{R}_{ik}]^{\frac{1}{2}} (V_{ik} - \hat{V}_{ik}) dt \end{array}$$

where $-\tilde{R}_{ik} \succ 0$, $L_i(X, s, U)$ is the cost presented in (7.401), the robust cost functional

$$\begin{array}{l} L_i^{\mathrm{ad}}(X, s, U, V) = L_i(X, s, U) \\ + \int_0^T \sum_{i \in \mathscr{I}} \langle \tilde{R}_{ik}(V_{ik} - \hat{V}_{ik}), V_{ik} - \hat{V}_{ik} \rangle + \sum_{i \in \mathscr{I}} \langle (\tilde{R}_{ik} + \bar{\tilde{R}}_{ik}) \hat{V}_{ik}, \hat{V}_{ik} \rangle dt, \end{array}$$

where $-(\tilde{R}_{ik} + \bar{\tilde{R}}_{ik}) \succ 0$. By Theorem 7.412, it follows that the solution is given by

$$\begin{cases} U_i^{\mathrm{ad}} = -R_i^{-1} B_{2i}^* P_i^{\mathrm{ad}}(X^{\mathrm{ad}} - \hat{X}^{\mathrm{ad}}) - (R_i + \bar{R}_i)^{-1}(B_{2i} + \bar{B}_{2i})^* \bar{P}_i^{\mathrm{ad}} \hat{X}^{\mathrm{ad}}, \\ V_{ik}^{\mathrm{ad}} = -\tilde{R}_{ik}^{-1} B_{2ik}^* P_i^{\mathrm{ad}}(X^{\mathrm{ad}} - \hat{X}^{\mathrm{ad}}), \\ \hat{V}_{ik}^{\mathrm{ad}} = 0, \end{cases}$$

The coefficients P_i^{rs} and P_i^{ad} solve the corresponding differential equation presented next:

$$\begin{cases} \dot{P}_i + Q_i + P_i B_1^* + B_1 P_i - P_i [B_{2i} R_i^{-1} B_{2i}^* - 2\lambda_i S_0 S_0^*] P_i \\ - \sum_{j \neq i} P_j B_{2j} R_j^{-1*} B_{2j}^* P_i - \sum_{j \neq i} P_i B_{2j} R_j^{-1*} B_{2j}^* P_j = 0, \\ P_i(T, s) = Q_i(T, s), \\ \dot{\bar{P}}_i + Q_i + \bar{Q}_i + \bar{P}_i(B_1 + \bar{B}_1)^* + (B_1 + \bar{B}_1)\bar{P}_i \\ - \bar{P}_i[(B_{2i} + \bar{B}_{2i})(R_i + \bar{R}_i)^{-1}(B_{2i} + \bar{B}_{2i})^*]\bar{P}_i \\ - \sum_{j \neq i} \bar{P}_j (B_{2j} + \bar{B}_{2j})(R_j + \bar{R}_j)^{-1*}(B_{2j} + \bar{B}_{2j})^* \bar{P}_i \\ - \sum_{j \neq i} \bar{P}_i (B_{2j} + \bar{B}_{2j})(R_j + \bar{R}_j)^{-1*}(B_{2j} + \bar{B}_{2j})^* \bar{P}_j = 0, \\ \bar{P}_i(T, s) = Q_i(T, s) + \bar{Q}_i(T, s). \end{cases}$$

Moreover, the expected matrix value of the states $\hat{X}^{rs}$ and $\hat{X}^{\mathrm{ad}}$ coincides. This result is very interesting because it says that for this class of risk-sensitive mean-field-type game problem, one can design a risk-neutral robust mean-field-type control strategy that preserves the mean trajectory in addition to sharing similar equilibrium control law structures. The optimal control laws of the fictitious adversaries V_{ik}^{ad} satisfies

$EV_{ik}^{\rm ad} = \hat{V}_{ik}^{\rm ad} = 0$. By choosing the following specific design

$$\begin{aligned} S_0 &= \sqrt{\mu} D, \\ -\tilde{R}_i &= \gamma_i^2 \mathbb{I}_{d\times d} \succ 0, \\ \gamma_i \sqrt{2\lambda_i \mu} &= 1, \end{aligned} \tag{7.423}$$

the robust mean-field-type game yields

$$\begin{aligned} & L_i^{\rm ad}(X, s, U, V) = \\ & L_i(X, s, U) - \gamma_i^2 \int_0^T \langle (V_{ik} - \hat{V}_{ik}), V_{ik} - \hat{V}_{ik} \rangle dt, \\ & dX = [B_1(X - \hat{X}) + (B_1 + \bar{B}_1)\hat{X} + \sum_{i=1}^I B_{2i}(U_i - \hat{U}_i) + \sum_{i=1}^I (B_{2i} + \bar{B}_{2i})\hat{U}_i] dt \\ & + \sqrt{\mu} D dB + \sum_{i\in\mathscr{I}} D(V_{ik} - \hat{V}_{ik}) dt, \end{aligned} \tag{7.424}$$

Thanks to the mean-field term $(V_{ik} - \hat{V}_{ik})$, the resulting means of mean-field terms of the two problems coincide. This new result is due to variance reduction design $\langle (V_{ik} - \hat{V}_{ik}), V_{ik} - \hat{V}_{ik} \rangle$. The fictitious adversary control law can be seen as a noise (zero mean): $EV_{ik}^{\rm ad} = 0$ by setting

$$d\tilde{B} = \sqrt{\mu} D dB + \sum_{i\in\mathscr{I}} D(V_{ik} - \hat{V}_{ik}) dt,$$

Table 7.2 summarizes the main relationship between the two class of game problems.

Table 7.2 Relationship between the game models. RS, risk-sensitive; RN, risk-neutral; DM, decision-maker

RS mean-field-type dependence	Risk-neutral designs
1 RS decision-maker	✓ RN mean-field-type robust control design
	✓ Same mean state as the original RS problem
	✓ Designed a fictitious RN adversary
	With zero mean contribution to the state
1 RS team (full cooperation)	✓ RN mean-field-type robust control design
	✓ Same mean state as the original RS problem
	✓ Designed 1 fictitious RN adversary
	With zero mean contribution to the state
I-RS noncooperative DMs	✓ RN mean-field-type robust control design
	✓ Same mean state as the original RS problem
	✓ Designed I-fictitious RN adversaries
	With zero mean contribution to the state

7.10.4 State-Control and Mean-Field-Dependent Noise

Consider the following state dynamics with jump-diffusion and regime switching:

$$\begin{aligned}
&X(0) = X_0,\\
&s(0) = s_0,\\
&dX = [B_0 + B_1 X + \bar{B}_1 \hat{X} + \textstyle\sum_{j\in\mathscr{I}} B_{j2} U_j + \sum_{j\in\mathscr{I}} \bar{B}_{j2} \hat{U}_j] dt\\
&+[S_0 + S_1 X + \bar{S}_1 \hat{X} + \textstyle\sum_{j\in\mathscr{I}} S_{j2} U_j + \sum_{j\in\mathscr{I}} \bar{S}_{j2} \hat{U}_j] dB\\
&+ \textstyle\int_{\Theta} [M_0 + M_1 X + \bar{M}_1 \hat{X} + \sum_{j\in\mathscr{I}} M_{j2} U_j + \sum_{j\in\mathscr{I}} \bar{M}_{j2} \hat{U}_j] \tilde{N}(dt, d\theta),
\end{aligned} \tag{7.425}$$

where

$$\begin{aligned}
&B_k, S_k, \bar{B}_k, \bar{S}_k : [0,T] \times \mathscr{S} \to \mathbb{R}^{d\times d},\ k \in \{0,1\}\\
&B_{j2}, S_{j2}, \bar{B}_{j2}, \bar{S}_{j2} : [0,T] \times \mathscr{S} \to \mathbb{R}^{d\times d},\ j \in \mathscr{I},\\
&M_k, \bar{M}_k : [0,T] \times \mathscr{S} \times \Theta \to \mathbb{R}^{d\times d},\ k \in \{0,1\}\\
&M_{j2}, \bar{M}_{j2} : [0,T] \times \mathscr{S} \times \Theta \to \mathbb{R}^{d\times d},\ j \in \mathscr{I},
\end{aligned} \tag{7.426}$$

with $B_k(t,s)$; $S_k(t,s)$; $\bar{B}_k(t,s)$; $\bar{S}_k(t,s) \in \mathbb{R}^{d\times d}$ and $s \in \mathscr{S}$, where $\mathscr{S}$ is the set of switching regimes. Moreover, $M_k(t,s,\theta) \in \mathbb{R}^{d\times d}$, $\theta \in \Theta$ where Θ is the set of jump sizes, for all $k \in \{0,1,2\}$. The process B is $(d \times d)$-dimensional Brownian motions, $\tilde{N}(dt, d\theta) = N(dt, d\theta) - \nu(d\theta)dt$, is a $(d \times d)$-dimensional compensated jump processes with matrix Radon measures ν over Θ. The switching process s switches between element in $\mathscr{S}$ with transition rate $\tilde{Q} = (\tilde{q}_{ss'},\ (s,s') \in \mathscr{S}^2)$ such that

$$s' \neq s,\ \tilde{q}_{ss'} > 0,\ \tilde{q}_{ss} = -\sum_{s'\neq s} \tilde{q}_{ss'}.$$

It is assumed that the processes X_0, s, B, N are mutually independent.

The term $\hat{X} = E[X|\mathscr{F}^s]$ denotes the expected (matrix-)value of X given the filtration $\mathscr{F}^s$, and $\hat{U} = E[U|\mathscr{F}^s]$. These are called conditional state mean-field and conditional control-action mean-field terms. The state dynamics is therefore of conditional McKean–Vlasov type as the conditional probability measure of the state appears in the drift, diffusion, and jump coefficients via $(\hat{X},\ \hat{U})$. We denote a solution (if any) to (7.425) by $X := X^{U,X_0,s_0}$, as used in the development below.

To the state dynamics (7.425) we associate a cost functional of mean-field type of decision-maker i as

$$\begin{aligned}
&L_i(X,s,U) = \tfrac{1}{2}\langle Q_{iT} X(T), X(T)\rangle + \tfrac{1}{2}\langle \bar{Q}_{iT} \hat{X}(T), \hat{X}(T)\rangle\\
&+\langle E_{i3}(T), X(T)\rangle + \langle \bar{E}_{i3}(T), \hat{X}(T)\rangle\\
&+\tfrac{1}{2}\int_0^T \Big(\langle Q_i X, X\rangle + \langle \bar{Q}_i \hat{X}, \hat{X}\rangle + \langle R_i U_i, U_i\rangle + \langle \bar{R}_i \hat{U}_i, \hat{U}_i\rangle \Big) dt\\
&+\int_0^T \Big(\langle E_{i1} X, U_i\rangle + \langle \bar{E}_{i1} \hat{X}, \hat{U}_i\rangle + \langle E_{i2}, U_i\rangle + \langle \bar{E}_{i2}, \hat{U}_i\rangle + \langle E_{i3}, X\rangle + \langle \bar{E}_{i3}, \hat{X}\rangle \Big) dt,
\end{aligned} \tag{7.427}$$

where $Q_{iT} = Q_i(T, s(T))$, $Q_i = Q_i(t, s(t))$, $\bar{Q}_{iT} = \bar{Q}_i(T, s(T))$, $\bar{Q}_i = Q_i(t, s(t))$; and $R_i = R_i(t, s(t))$, $\bar{R}_i = \bar{R}_i(t, s(t))$, $E_{ik} = E_{ik}(t, s(t))$, $\bar{E}_{ik} = \bar{E}_{ik}(t, s(t))$, are $d \times d$ matrices.

$$\begin{aligned}
&Q_{iT}, \bar{Q}_{iT}, E_{i3T}, \bar{E}_{i3T} : \mathscr{S} \to \mathbb{R}^{d\times d},\ i \in \mathscr{I},\\
&Q_i, \bar{Q}_i, R_i, \bar{R}_i, E_{ik}, \bar{E}_{ik} : [0, T] \times \mathscr{S} \to \mathbb{R}^{d\times d},\ k \in \{1, 2, 3\},\ i \in \mathscr{I}
\end{aligned} \tag{7.428}$$

Matrix-Valued Mean-Field-Type Equilibrium Problem

$$\begin{cases} i \in \mathscr{I},\\ \inf_{U_i\in\mathscr{U}_i} E\left[L_i(x, s, U)|X(0) = X_0, s(0) = s_0\right] \\ \text{such that (7.425).}\end{cases} \tag{7.429}$$

We invoke the following assumption:

$$\begin{cases} Q_{iT}, Q_i \text{ symmetric positive definite},\\ Q_{iT} + \bar{Q}_{iT},\ Q_i + \bar{Q}_i \text{ symmetric positive definite},\\ R_i,\ R_i + \bar{R}_i \text{ symmetric positive definite},\\ E[\langle X_0, X_0\rangle] < +\infty,\ |\mathscr{S}| < +\infty,\\ \int_0^T \int_\Theta \langle \phi, \phi\nu(d\theta)\rangle < +\infty \text{ with } \phi \in \{M_{ik}, \bar{M}_{ik}\},\\ B_k, S_k^* S_k, \bar{B}_k, \bar{S}_k^* \bar{S}_k \in L^1([0, T] \times \mathscr{S}, \mathbb{R}^{d\times d}) \end{cases} \tag{7.430}$$

We provide a sufficient condition for Nash equilibrium:

$$\begin{cases}
\hat{U} = \bar{A}^{-1}[\bar{N}_1\hat{X} + \bar{N}_0],\\
U - \hat{U} = -A^{-1}N(X - \hat{X}),\\
A_{ij} = C_i^{-1} S_{2i}^* \frac{1}{2}(P_i^* + P_i)S_{2j} + \int_\Theta C_i^{-1} M_{2i}^* \frac{1}{2}(P_i^* + P_i)\nu(d\theta)M_{2j},\\
A_{ii} = \bar{A}_{ii} = \mathbb{1},\\
\bar{A}_{ij} = \bar{C}_i^{-1}(S_{2i} + \bar{S}_{2i})^* \frac{1}{2}(P_i^* + P_i)(S_{2j} + \bar{S}_{2j})\\
+ \int_\Theta \bar{C}_i^{-1}(M_{2i} + \bar{M}_{2i})^* \frac{1}{2}(P_i^* + P_i)\nu(d\theta)(M_{2j} + \bar{M}_{2j}),\\
\bar{N}_{i1} = \bar{C}_i^{-1}[(B_{2i} + \bar{B}_{2i})^* \frac{1}{2}(\bar{P}_i^* + \bar{P}_i) + (S_{2i} + \bar{S}_{2i})^* \frac{1}{2}(P_i^* + P_i)(S_1 + \bar{S}_1)\\
+ \int_\Theta (M_{2i} + \bar{M}_{2i})^* \frac{1}{2}(P_i^* + P_i)\nu(d\theta)(M_1 + \bar{M}_1)],\\
\bar{N}_{i0} = \bar{C}_i^{-1}[(B_{2i} + \bar{B}_{2i})^* \Gamma_i + (E_{i2} + \bar{E}_{i2}) + (S_{2i} + \bar{S}_{2i})^* \frac{1}{2}(P_i^* + P_i)S_0\\
+ \int_\Theta (M_{2i} + \bar{M}_{2i})^* \frac{1}{2}(P_i^* + P_i)\nu(d\theta)M_0],\\
N_i = C_i^{-1}[B_{2i}^* \frac{1}{2}(P_i^* + P_i) + S_{2i}^* \frac{1}{2}(P_i^* + P_i)S_1 + M_{2i}^* \frac{1}{2}(P_i^* + P_i)\nu(d\theta)M_1],\\
\bar{C}_i = (R_i + \bar{R}_i) + (S_{2i} + \bar{S}_{2i})^* \frac{1}{2}(P_i^* + P_i)(S_{2i} + \bar{S}_{2i})\\
+ \int_\Theta (M_{2i} + \bar{M}_{2i})^* \frac{1}{2}(P_i^* + P_i)\nu(d\theta)(M_{2i} + \bar{M}_{2i}),\\
C_i = R_i + S_{2i}^* \frac{1}{2}(P_i^* + P_i)S_{2i} + \int_\Theta M_{2i}^* \frac{1}{2}(P_i^* + P_i)\nu(d\theta)M_{2i},
\end{cases} \tag{7.431}$$

$$
\begin{cases}
\delta_i(T,s) = 0, \\
0 = \dot{\delta}_i + \langle \Gamma_i, B_0 \rangle + \sum_{s' \neq s} (\delta_i(t,s') - \delta_i(t,s)) \tilde{q}_{ss'} \\
+\frac{1}{2} \langle \frac{1}{2}(P_i^* + P_i) S_0, S_0 \rangle + \frac{1}{2} \int_{\Theta} \langle \frac{1}{2}(P_i^* + P_i) M_0, M_0 \nu(d\theta) \rangle \\
-\frac{1}{2} \langle \bar{C}_i \tilde{\bar{N}}_{i0}, \tilde{\bar{N}}_{i0} \rangle \\
+\frac{1}{2} \langle \frac{1}{2}(P_i^* + P_i) \sum_{j \neq i} (S_{2j} + \bar{S}_{2j}) \tilde{\bar{N}}_{j0}, \sum_{j \neq i} (S_{2j} + \bar{S}_{2j}) \tilde{\bar{N}}_{j0} \rangle \\
+\frac{1}{2} \int_{\Theta} \langle \frac{1}{2}(P_i^* + P_i) \sum_{j \neq i} (M_{2j} + \bar{M}_{2j}) \tilde{\bar{N}}_{j0}, \sum_{j \neq i} (M_{2j} + \bar{M}_{2j}) \tilde{\bar{N}}_{j0} \nu(d\theta) \rangle \\
-\langle \Gamma_i, \sum_{j \neq i} (B_{2j} + \bar{B}_{2j}) \tilde{\bar{N}}_{j0} \rangle \\
-\frac{1}{2} \langle \frac{1}{2}(P_i^* + P_i) \sum_{j \neq i} (S_{2j} + \bar{S}_{2j}) \tilde{\bar{N}}_{j0}, S_0 \rangle \\
-\frac{1}{2} \int_{\Theta} \langle \frac{1}{2}(P_i^* + P_i) \sum_{j \neq i} (M_{2j} + \bar{M}_{2j}) \tilde{\bar{N}}_{j0}, M_0 \nu(d\theta) \rangle \\
-\frac{1}{2} \langle \frac{1}{2}(P_i^* + P_i) S_0, \ \sum_{j \neq i} (S_{2j} + \bar{S}_{2j}) \tilde{\bar{N}}_{j0} \rangle \\
-\frac{1}{2} \int_{\Theta} \langle \frac{1}{2}(P_i^* + P_i) M_0, \ \sum_{j \neq i} (M_{2j} + \bar{M}_{2j}) \tilde{\bar{N}}_{j0} \nu(d\theta) \rangle,
\end{cases}
\tag{7.432}
$$

$$
\begin{cases}
\Gamma_i(T,s) = (E_{i3}(T,s) + \bar{E}_{i3}(T,s)), \\
0 = \dot{\Gamma}_i + (B_1 + \bar{B}_1)^* \Gamma_i + (E_{i3} + \bar{E}_{i3}) + \frac{(\bar{P}_i^* + \bar{P}_i)}{2} B_0 + \sum_{s' \neq s} (\Gamma_i(t,s') - \Gamma_i(t,s)) \tilde{q}_{ss'}] \\
+\frac{1}{2}[(S_1 + \bar{S}_1)^* \frac{1}{2}(P_i^* + P_i) S_0 + S_0^* \frac{1}{2}(P_i^* + P_i)(S_1 + \bar{S}_1)] \\
+\frac{1}{2} \int_{\Theta} [(M_1 + \bar{M}_1)^* \frac{1}{2}(P_i^* + P_i) M_0 + M_0^* (P_i^* + P_i)(M_1 + \bar{M}_1)] \nu(d\theta) \\
-\frac{1}{2}[\tilde{\bar{N}}_{i1}^* \bar{C}_i \tilde{\bar{N}}_{i0} + \tilde{\bar{N}}_{i0}^* \bar{C}_i \tilde{\bar{N}}_{i1}] + \frac{1}{2}[\sum_{j \neq i} (S_{2j} + \bar{S}_{2j}) \tilde{\bar{N}}_{j0}]^* \frac{1}{2}(P_i^* + P_i) \sum_{j \neq i} (S_{2j} + \bar{S}_{2j}) \tilde{\bar{N}}_{j1} \\
+\frac{1}{2}[\sum_{j \neq i} (S_{2j} + \bar{S}_{2j}) \tilde{\bar{N}}_{j1}]^* \frac{1}{2}(P_i^* + P_i) \sum_{j \neq i} (S_{2j} + \bar{S}_{2j}) \tilde{\bar{N}}_{j0} \rangle \\
+\frac{1}{2} \int_{\Theta} [\sum_{j \neq i} (M_{2j} + \bar{M}_{2j}) \tilde{\bar{N}}_{j0}]^* \frac{1}{2}(P_i^* + P_i) \sum_{j \neq i} (M_{2j} + \bar{M}_{2j}) \tilde{\bar{N}}_{j1} \nu(d\theta) \\
+\frac{1}{2} \int_{\Theta} [\sum_{j \neq i} (M_{2j} + \bar{M}_{2j}) \tilde{\bar{N}}_{j1}]^* \frac{1}{2}(P_i^* + P_i) \sum_{j \neq i} (M_{2j} + \bar{M}_{2j}) \tilde{\bar{N}}_{j0} \nu(d\theta) \\
- \sum_{j \neq i} (B_{2j} + \bar{B}_{2j}) \tilde{\bar{N}}_{j0}]^* \frac{1}{2}(\bar{P}_i^* + \bar{P}_i) \\
- \sum_{j \neq i} (B_{2j} + \bar{B}_{2j}) \tilde{\bar{N}}_{j1}]^* \Gamma_i \\
-\frac{1}{2}(S_1 + \bar{S}_1)^* \frac{1}{2}(P_i^* + P_i) \sum_{j \neq i} (S_{2j} + \bar{S}_{2j}) \tilde{\bar{N}}_{j0} \\
-\frac{1}{2}[\sum_{j \neq i} (S_{2j} + \bar{S}_{2j}) \tilde{\bar{N}}_{j0}]^* \frac{1}{2}(P_i^* + P_i)(S_1 + \bar{S}_1) \\
-\frac{1}{2} S_0^* \frac{1}{2}(P_i^* + P_i) \sum_{j \neq i} (S_{2j} + \bar{S}_{2j}) \tilde{\bar{N}}_{j1} - \frac{1}{2}[\sum_{j \neq i} (S_{2j} + \bar{S}_{2j}) \tilde{\bar{N}}_{j1}]^* \frac{1}{2}(P_i^* + P_i) S_0 \\
-\frac{1}{2} \int_{\Theta} M_0^* \frac{1}{2}(P_i^* + P_i) \sum_{j \neq i} (M_{2j} + \bar{M}_{2j}) \tilde{\bar{N}}_{j1} \nu(d\theta) \\
-\frac{1}{2} \int_{\Theta} [\sum_{j \neq i} (M_{2j} + \bar{M}_{2j}) \tilde{\bar{N}}_{j1}]^* \frac{1}{2}(P_i^* + P_i) M_0 \nu(d\theta) \\
-\frac{1}{2} \int_{\Theta} (M_1 + \bar{M}_1)^* \frac{1}{2}(P_i^* + P_i) \sum_{j \neq i} (M_{2j} + \bar{M}_{2j}) \tilde{\bar{N}}_{j0} \nu(d\theta) \\
-\frac{1}{2} \int_{\Theta} [\sum_{j \neq i} (M_{2j} + \bar{M}_{2j}) \tilde{\bar{N}}_{j0}]^* \frac{1}{2}(P_i^* + P_i)(M_1 + \bar{M}_1) \nu(d\theta)
\end{cases}
\tag{7.433}
$$

$$\begin{cases} \bar{P}_i(T,s) = Q_i(T,s) + \bar{Q}_i(T,s), \\ 0 = \dot{\bar{P}}_i + (Q_i + \bar{Q}_i) + \bar{P}_i(B_1 + \bar{B}_1) + (B_1 + \bar{B}_1)^* \bar{P}_i + \sum_{s' \neq s}(\bar{P}_i(t,s') - \bar{P}_i(t,s))\tilde{q}_{ss'}] \\ +(S_1 + \bar{S}_1)^* \frac{1}{2}(P_i^* + P_i)(S_1 + \bar{S}_1) + \int_\Theta (M_1 + \bar{M}_1)^* \frac{1}{2}(P_i^* + P_i)(M_1 + \bar{M}_1)\nu(d\theta) \\ -\tilde{\bar{N}}_{i1}^* \bar{C}_i \tilde{\bar{N}}_{i1} \\ +[\sum_{j\neq i}(S_{2j} + \bar{S}_{2j})\tilde{\bar{N}}_{j1}]^* \frac{1}{2}(P_i^* + P_i)\sum_{j\neq i}(S_{2j} + \bar{S}_{2j})\tilde{\bar{N}}_{j1} \\ + \int_\Theta [\sum_{j\neq i}(M_{2j} + \bar{M}_{2j})\tilde{\bar{N}}_{j1}]^* \frac{1}{2}(P_i^* + P_i)\sum_{j\neq i}(M_{2j} + \bar{M}_{2j})\tilde{\bar{N}}_{j1}\nu(d\theta) \\ -2[\sum_{j\neq i}(B_{2j} + \bar{B}_{2j})\tilde{\bar{N}}_{j1}]^* \frac{1}{2}(\bar{P}_i^* + \bar{P}_i) \\ -(S_1 + \bar{S}_1)^* \frac{1}{2}(P_i^* + P_i)\sum_{j\neq i}(S_{2j} + \bar{S}_{2j})\tilde{\bar{N}}_{j1} \\ -[\sum_{j\neq i}(S_{2j} + \bar{S}_{2j})\tilde{\bar{N}}_{j1}]^* \frac{1}{2}(P_i^* + P_i)[(S_1 + \bar{S}_1) \\ - \int_\Theta (M_1 + \bar{M}_1)^* \frac{1}{2}(P_i^* + P_i)\sum_{j\neq i}(M_{2j} + \bar{M}_{2j})\tilde{\bar{N}}_{j1}\nu(d\theta) \\ - \int_\Theta [\sum_{j\neq i}(M_{2j} + \bar{M}_{2j})\tilde{\bar{N}}_{j1}]^* \frac{1}{2}(P_i^* + P_i)[(M_1 + \bar{M}_1)\nu(d\theta), \end{cases} \tag{7.434}$$

$$\begin{cases} P_i(T,s) = Q_i(T,s), \\ 0 = \dot{P}_i + Q_i + P_i B_1 + B_1^* P_i + \sum_{s'\neq s}(P_i(t,s') - P_i(t,s))\tilde{q}_{ss'} \\ +S_1^* \frac{1}{2}(P_i^* + P_i)S_1 + \int_\Theta \langle M_1^* \frac{1}{2}(P_i^* + P_i)M_1\nu(d\theta) \\ -\tilde{N}_i^* C_i \tilde{N}_i \\ +[\sum_{j\neq i} S_{2j}\tilde{N}_j]^* \frac{1}{2}(P_i^* + P_i)\sum_{j\neq i} S_{2j}\tilde{N}_j + \int_\Theta [\sum_{j\neq i} M_{2j}\tilde{N}_j]^* \\ \frac{1}{2}(P_i^* + P_i)\sum_{j\neq i} M_{2j}\tilde{N}_j\nu(d\theta) \\ -2[\sum_{j\neq i} B_{2j}\tilde{N}_j]^* \frac{1}{2}(P_i^* + P_i) \\ -S_1^* \frac{1}{2}(P_i^* + P_i)\sum_{j\neq i} S_{2j}\tilde{N}_j - [\sum_{j\neq i} S_{2j}\tilde{N}_j]^* \frac{1}{2}(P_i^* + P_i)S_1 \\ - \int_\Theta \langle M_1^* \frac{1}{2}(P_i^* + P_i)\sum_{j\neq i} M_{2j}\tilde{N}_j\nu(d\theta) \\ - \int_\Theta [\sum_{j\neq i} M_{2j}\tilde{N}_j]^* \frac{1}{2}(P_i^* + P_i)M_1\nu(d\theta) \end{cases} \tag{7.435}$$

To prove the mean-field-type Nash equilibrium statement above, we consider a candidate guess functional for the equilibrium cost in the following form:

$$F_i = \frac{1}{2}\langle P_i(X - \hat{X}), (X - \hat{X})\rangle + \frac{1}{2}\langle \bar{P}_i\hat{X}, \hat{X}\rangle + \langle \Gamma_i, \hat{X}\rangle + \delta_i,$$

where the coefficients $P_i(t,s)$, $\bar{P}_i(t,s)$, $\Gamma_i(t,s)$, $\delta_i(t,s)$ are deterministic functions to be determined. We then apply the direct method. ■

7.10.5 Additive Common Noises

7.10.5.1 Risk-Neutral Matrix-Valued Games Under Common Noise

We modify the state dynamics to include an additional common noise, which is a jump-diffusion process $S_{o0}dB_o + \int_\theta M_{o0}(.,\theta)\tilde{N}_o(dt,d\theta)$:

$$
\begin{array}{l}
dX=[B_1(X-\hat{X})+(B_1+\bar{B}_1)\hat{X}+\sum_{j=1}^n B_{2j}(U_j-\hat{U}_j)+\sum_{j=1}^n(B_{2j}+\bar{B}_{2j})\hat{U}_j]dt\\
+S_0dB+\int_\theta M_0(\cdot,\theta)\tilde{N}(dt,d\theta)+S_{o0}dB_o+\int_\theta M_{o0}(.,\theta)\tilde{N}_o(dt,d\theta),\\
X(0)=X_0,\\
s(0)=s_0,
\end{array}
\tag{7.436}
$$

and the cost functional is

$$
\begin{array}{l}
L_i(X,s,U)=\langle Q_i(T,s(T))(X(T)-\hat{X}(T)),X(T)-\hat{X}(T)\rangle\\
+\langle (Q_i(T,s(T))+\bar{Q}_i(T,s(T)))\hat{X}(T),\hat{X}(T)\rangle\\
+\int_0^T\langle Q_i(X-\hat{X}),X-\hat{X}\rangle+\langle (Q_i+\bar{Q}_i)\hat{X},\hat{X}\rangle dt\\
+\int_0^T\langle R_i(U_i-\hat{U}_i),U_i-\hat{U}_i\rangle+\langle (R_i+\bar{R}_i)\hat{U}_i,\hat{U}_i\rangle dt,
\end{array}
\tag{7.437}
$$

Theorem 7.2 *Consider the risk-neutral equilibrium problem:*

$$
\begin{array}{l}
i\in\mathscr{I},\\
inf_{U_i\in\mathscr{U}_i}EL_i(X,s,U),\\
subject\ to\quad (7.436).
\end{array}
\tag{7.438}
$$

Assume that Q_i, R_i, $Q_i+\bar{Q}_i$, $R_i+\bar{R}_i$ are symmetric positive definite. Then the matrix-valued mean-field-type (risk-neutral) Nash equilibrium strategy and the (risk-neutral) equilibrium cost are given by

$$
\begin{cases}
U_i^{\rm rn}=-R_i^{-1}B_{2i}^*P_i(X-\hat{X})-(R_i+\bar{R}_i)^{-1}(B_{2i}+\bar{B}_{2i})^*[\bar{P}_i\hat{X}+\Gamma_i]\\
L_i^{\rm rn}(X,s,U^{\rm rn})=E\langle P_i(0,s(0))(X_0-\hat{X}_0),X_0-\hat{X}_0\rangle\\
+E[\langle \bar{P}_i(0,s(0))\hat{X}_0,\hat{X}_0\rangle+2\langle \Gamma_i(0,s(0)),\hat{X}_0\rangle+\delta_i(0,s(0))],\\
i\in\mathscr{I},
\end{cases}
\tag{7.439}
$$

where $(P_i, P_{i,B_o}, P_{i,N_o})$, $(\bar{P}_i, \bar{P}_{i,B_o}, \bar{P}_{i,N_o})$, $(\Gamma_i, \Gamma_{i,B_o}, \Gamma_{i,N_o})$, $(\delta_i, \delta_{i,B_o}, \delta_{i,N_o})$ are adapted to the filtration $\mathscr{F}^{s,B_o,N_o}$ and solve the following differential equations:

$$\begin{cases} -dP_i = \Big\{Q_i + B_1^* P_i + P_i B_1 + \sum_{s'\neq s}(P_i(t,s') - P_i(t,s))\tilde{q}_{ss'} \\ -P_i B_{2i} R_i^{-1} B_{2i}^* P_i - \sum_{j\neq i}[P_i B_{2j} R_j^{-1} B_{2j}^* P_j + P_j B_{2j} R_j^{-1} B_{2j}^* P_i] \,\}dt, \\ P_i(T,s) = Q_i(T,s), \\ -d\bar{P}_i = \{Q_i + \bar{Q}_i + ((B_1+\bar{B}_1)^*\bar{P}_i + \bar{P}_i(B_1+\bar{B}_1)) \\ +\sum_{s'\neq s}(\bar{P}_i(t,s') - \bar{P}_i(t,s))\tilde{q}_{ss'} \\ -\bar{P}_i[(B_{2i}+\bar{B}_{2i})(R_i+\bar{R}_i)^{-1}(B_{2i}+\bar{B}_{2i})^*]\bar{P}_i \\ -\sum_{j\neq i}\bar{P}_j(B_{2j}+\bar{B}_{2j})(R_j+\bar{R}_j)^{-1*}(B_{2j}+\bar{B}_{2j})^*\bar{P}_i \\ -\sum_{j\neq i}\bar{P}_i(B_{2j}+\bar{B}_{2j})(R_j+\bar{R}_j)^{-1*}(B_{2j}+\bar{B}_{2j})^*\bar{P}_j \,\}dt, \\ \bar{P}_i(T,s) = Q_i(T,s) + \bar{Q}_i(T,s), \\ -d\Gamma_i = \{\bar{P}_{i,B_o} S_{o0} + \int_\theta \bar{P}_{i,N_o} M_{o0}(.,\theta)\nu_o(d\theta) \\ +(B_1+\bar{B}_1)^*\Gamma_i + \sum_{s'\neq s}(\Gamma_i(t,s') - \Gamma_i(t,s))\tilde{q}_{ss'} \\ -\bar{P}_i(B_{2i}+\bar{B}_{2i})(R_i+\bar{R}_i)^{-1}(B_{2i}+\bar{B}_{2i})^*\Gamma_i \\ -\sum_{j\neq i}\bar{P}_i(B_{2j}+\bar{B}_{2j})(R_j+\bar{R}_j)^{-1}(B_{2j}+\bar{B}_{2j})^*\Gamma_j \\ -\sum_{j\neq i}\bar{P}_j(B_{2j}+\bar{B}_{2j})(R_j+\bar{R}_j)^{-1}(B_{2j}+\bar{B}_{2j})^*\Gamma_i \,\}dt, \\ \Gamma_i(T,s) = 0, \\ -d\Delta_i = \{S_0^* P_i S_0 + \int_\Theta M_0^* P_i M_0 \nu(\theta) \\ +S_{o0}^*\bar{P}_i S_{o0} + \int_\theta M_{o0}^*\bar{P}_i M_{o0}\nu_o(d\theta) \\ +\sum_{s'\neq s}(\Delta_i(t,s') - \Delta_i(t,s))\tilde{q}_{ss'} \\ +S_{o0}^*\Gamma_{i,B_o} + \int_\theta M_{o0}^*\nu_o(d\theta)\Gamma_{i,N_o} \\ +\Gamma_{i,B_o}^* S_{o0} + \int_\Theta \Gamma_{i,N_o}^* M_{o0}\nu_o(d\theta) \\ -\Gamma_i^*\sum_{i\in\mathscr{I}}(B_{2i}+\bar{B}_{2i})(R_i+\bar{R}_i)^{-1}(B_{2i}+\bar{B}_{2i})^*\Gamma_i \\ -\sum_{j\neq i}\Gamma_i^*(B_{2j}+\bar{B}_{2j})(R_j+\bar{R}_j)^{-1}(B_{2j}+\bar{B}_{2j})^*\Gamma_j \,\}dt \\ \delta_i(T,s) = trace(\Delta_i(T,s)) = 0, \\ \forall s \in \mathscr{S}, \end{cases} \tag{7.440}$$

whenever these matrix-valued stochastic differential equations have a unique solution that does not blow up within $[0, T]$.

Note that the stochastic Riccati system in $(P_i, P_{i,B_o}, P_{i,N_o})$ can be rewritten in a generic form as follows:

$$\begin{cases} i \in \mathscr{I}, \\ -dP_i \\ = \Big\{Q_i + [B_1 - \sum_{j\in\mathscr{I}} B_{2j}R_j^{-1}B_{2j}^*P_j]^*P_i + P_i[B_1 - \sum_{j\in\mathscr{I}} B_{2j}R_j^{-1}B_{2j}^*P_j] \\ +P_i B_{2i} R_i^{-1} B_{2i}^* P_i + \sum_{s'\neq s}(P_i(t,s') - P_i(t,s))\tilde{q}_{ss'} \,\}dt, \\ P_i(T,s) = Q_i(T,s), \end{cases} \tag{7.441}$$

A proof of Theorem 7.2 can be obtained from the direct method. From the state system (7.436), the conditional expected matrix $\hat{X}(t) := E[X(t)|\mathscr{F}_t^{s,B_o,N_o}]$, where $\mathscr{F}_t^{s,B_o,N_o}$ is the natural filtration of the process s, B_o, N_o up to t, solves the following matrix-valued stochastic differential system:

$$\begin{array}{l} d\hat{X} = [(B_1 + \bar{B}_1)\hat{X} + \sum_{j=1}^n (B_{2j} + \bar{B}_{2j})\hat{U}_j]dt \\ +S_{o0}dB_o + \int_\theta M_{o0}(.,\theta)\tilde{N}_o(dt, d\theta), \\ \hat{X}(0) = E[X_0]. \end{array} \tag{7.442}$$

which means that

$$\begin{array}{l} d\hat{X} \\ = [(B_1 + \bar{B}_1) - \sum_{j=1}^n (B_{2j} + \bar{B}_{2j})(R_j + \bar{R}_j)^{-1}(B_{2j} + \bar{B}_{2j})^* \bar{P}_j]\hat{X}dt \\ - \sum_{j=1}^n (B_{2j} + \bar{B}_{2j})(R_j + \bar{R}_j)^{-1}(B_{2j} + \bar{B}_{2j})^* \Gamma_j dt \\ +S_{o0}dB_o + \int_\theta M_{o0}(.,\theta)\tilde{N}_o(dt, d\theta), \\ \hat{X}(0) = E[X_0]. \end{array} \tag{7.443}$$

which will be used for feedback in the optimal strategy. In the absence of the common noise ($S_{o0} = 0, M_{o0} = 0$) the process Γ_i is identically equal to zero. The processes (P_{i,B_o}, P_{i,N_o}), $(\bar{P}_{i,B_o}, \bar{P}_{i,N_o})$, are identically zero.

Next, we provide a semi-explicit solution to the full cooperation case.

Corollary 7.1 *Assume that* $Q_0 := \sum_{j\in\mathscr{I}} Q_j$, $Q_0 + \bar{Q}_0 := \sum_{j\in\mathscr{I}}[Q_j + \bar{Q}_j]$, R_i, $R_i + \bar{R}_i$ *are symmetric positive definite. The fully cooperative solution of the problem* $\inf_{(U_1,\ldots,U_n)} E[\sum_{j\in\mathscr{I}} L_j]$ *is given by*

$$\begin{cases} i \in \mathscr{I} \\ U_i^{\mathrm{rn,g}} = -R_i^{-1}B_{2i}^* P_0(X - \hat{X}) \\ -(R_i + \bar{R}_i)^{-1}(B_{2i} + \bar{B}_{2i})^*(\bar{P}_0\hat{X} + \Gamma_0), \\ L^{\mathrm{rn,g}}(X, s, U^{\mathrm{rn,g}}) \\ = E\langle P_0(0, s(0))(X_0 - \hat{X}_0), X_0 - \hat{X}_0\rangle \\ +E[\langle \bar{P}_0(0, s(0))\hat{X}_0, \hat{X}_0\rangle + 2\langle \Gamma_0(0, s(0)), \hat{X}_0\rangle + \delta_0(0, s(0))], \end{cases} \tag{7.444}$$

where P_0, $\bar{P}_0$, Γ_0, *and* δ_0 *solve the following differential equations:*

$$
\begin{cases}
-d P_0 \\
= Q_0 + B_1^* P_0 + P_0 B_1 + \sum_{s'\neq s}(P_0(t,s') - P_0(t,s))\tilde{q}_{ss'}dt \\
-P_0[\sum_{i\in\mathcal{I}} B_{2i} R_i^{-1} B_{2i}^*] P_0 dt \\
-P_{B_o} dB_o - \int_\theta P_{N_o}\tilde{N}_o(dt,d\theta), \\
P_0(T,s) = Q_0(T,s), \\
-d\bar{P}_0 = Q_0 + \bar{Q}_0 + (B_1+\bar{B}_1)^*\bar{P}_0 + \bar{P}_0(B_1+\bar{B}_1)dt \\
+\sum_{s'\neq s}(\bar{P}_0(t,s') - \bar{P}_0(t,s))\tilde{q}_{ss'}dt \\
-\bar{P}_0[\sum_{i\in\mathcal{I}}(B_{2i}+\bar{B}_{2i})(R_i+\bar{R}_i)^{-1}(B_{2i}+\bar{B}_{2i})^*]\bar{P}_0 dt, \\
\bar{P}_0(T,s) = Q_0(T,s) + \bar{Q}_0(T,s), \\
-d\Gamma_0 = \bar{P}_{0,B_o} S_{o0} + \int_\theta \bar{P}_{0,N_o} M_{o0}(.,\theta)\nu_o(d\theta) \\
+(B_1+\bar{B}_1)^*\Gamma_0 + \sum_{s'\neq s}(\Gamma_0(t,s') - \Gamma_0(t,s))\tilde{q}_{ss'}]dt \\
-\bar{P}_0\sum_{i\in\mathcal{I}}(B_{2i}+\bar{B}_{2i})(R_i+\bar{R}_i)^{-1}(B_{2i}+\bar{B}_{2i})^*\Gamma_0 dt, \\
\Gamma_0(T,s) = 0, \\
-d\Delta_0 = +[S_0^* P_0 S_0 + \int_\Theta M_0^* P_0 M_0 \nu(\theta)]dt \\
+[S_{o0}^*\bar{P}_0 S_{o0} + \int_\theta M_{o0}^*\bar{P}_0 M_{o0}\nu_o(d\theta)]dt \\
+[\sum_{s'\neq s}(\Delta_0(t,s') - \Delta_0(t,s))\tilde{q}_{ss'}]dt \\
+[S_{o0}^*\Gamma_{0,B_o} + \int_\theta M_{o0}^*\nu_o(d\theta)\Gamma_{0,N_o}]dt \\
+[\Gamma_{0,B_o}^* S_{o0} + \int_\Theta \Gamma_{0,N_o}^* M_{o0}\nu_o(d\theta)]dt \\
-\Gamma_0^*\sum_{i\in\mathcal{I}}(B_{2i}+\bar{B}_{2i})(R_i+\bar{R}_i)^{-1}(B_{2i}+\bar{B}_{2i})^*\Gamma_0 dt, \\
\delta_0(T,s) = trace(\Delta_0(T,s)) = 0, \\
s\in\mathcal{S}.
\end{cases}
\tag{7.445}
$$

Notice that the deterministic part of these Riccati equations (7.445) has positive solution P_0, $\bar{P}_0$ and there is no blow up in $[0,T]$.

The proof of the result Corollary 7.1 is immediate from Theorem 7.2 by considering one single team (full cooperation) and with a choice vector of matrices $U = (U_i)_{i\in\mathcal{I}}$.

Corollary 7.2 *Assume that Q, $Q+\bar{Q}$, R_i, $R_i+\bar{R}_i$ are symmetric positive definite for $i\in\mathcal{I}_+$ and $-R_j$, $-(R_j+\bar{R}_j)$ are symmetric positive definite for $j\in\mathcal{I}_-$. We assume that $\mathcal{I}_+\cup\mathcal{I}_- = \mathcal{I}$. The adversarial game problem of the team of attackers $\mathcal{I}_-$ and the team of defenders $\mathcal{I}_+$ has a saddle, and it is given by*

$$
\begin{cases}
U_i^{\mathrm{ad}} = -R_i^{-1} B_{2i}^* P_0^{ad}(X-\hat{X}) \\
-(R_i+\bar{R}_i)^{-1}(B_{2i}+\bar{B}_{2i})^*(\bar{P}_0^{ad}\hat{X} + \Gamma_0^{ad}),\ i\in\mathcal{I}_+, \\
U_j^{\mathrm{ad}} = -R_j^{-1} B_{2i}^* P_0^{ad}(X-\hat{X}) \\
-(R_j+\bar{R}_j)^{-1}(B_{2j}+\bar{B}_{2j})^*(\bar{P}_0^{ad}\hat{X} + \Gamma_0^{ad}),\ j\in\mathcal{I}_-, \\
L^{\mathrm{ad}}(X,s,U^{\mathrm{ad}}) = E\langle P^{ad}(0,s(0))(X_0-\hat{X}_0), X_0-\hat{X}_0\rangle \\
+E[\langle\bar{P}^{ad}(0,s(0))\hat{X}_0,\hat{X}_0\rangle] \\
+E[2\langle\Gamma^{ad}(0,s(0)),\hat{X}_0\rangle + \delta^{ad}(0,s(0))], \\
i\in\mathcal{I},
\end{cases}
\tag{7.446}
$$

where P^{ad}, $\bar{P}^{ad}$, Γ^{ad}, *and* δ^{ad} *solve the following differential equations:*

$$\begin{cases} -dP \\ = Qdt + (B_1^* P + PB_1)dt + \sum_{s'\neq s}(P(t,s') - P(t,s))\tilde{q}_{ss'}dt \\ -P[\sum_{i\in\mathcal{I}_+} B_{2i}R_i^{-1}B_{2i}^*]Pdt - P[\sum_{j\in\mathcal{I}_-} B_{2j}R_j^{-1}B_{2j}^*]Pdt, \\ P(T,s) = Q(T,s), \\ -d\bar{P} = ((B_1+\bar{B}_1)^*\bar{P} + \bar{P}(B_1+\bar{B}_1))dt \\ \sum_{s'\neq s}(\bar{P}(t,s') - \bar{P}(t,s))\tilde{q}_{ss'}dt \\ +(Q+\bar{Q})dt \\ -\bar{P}[\sum_{i\in\mathcal{I}_+}(B_{2i}+\bar{B}_{2i})(R_i+\bar{R}_i)^{-1}(B_{2i}+\bar{B}_{2i})^*]\bar{P}dt \\ -\bar{P}[\sum_{j\in\mathcal{I}_-}(B_{2j}+\bar{B}_{2j})(R_j+\bar{R}_j)^{-1}(B_{2j}+\bar{B}_{2j})^*]\bar{P}dt, \\ \bar{P}(T,s) = Q(T,s) + \bar{Q}(T,s), \\ -d\Gamma = \bar{P}_{B_o}S_{o0} + \int_\theta \bar{P}_{N_o}M_{o0}(.,\theta)\nu_o(d\theta)dt \\ +(B_1+\bar{B}_1)^*\Gamma + \sum_{s'\neq s}(\Gamma(t,s') - \Gamma(t,s))\tilde{q}_{ss'}]dt \\ -\bar{P}[\sum_{i\in\mathcal{I}_+}(B_{2i}+\bar{B}_{2i})(R_i+\bar{R}_i)^{-1}(B_{2i}+\bar{B}_{2i})^*]\Gamma dt \\ -\bar{P}[\sum_{j\in\mathcal{I}_-}(B_{2j}+\bar{B}_{2j})(R_j+\bar{R}_j)^{-1}(B_{2j}+\bar{B}_{2j})^*]\Gamma dt, \\ \Gamma(T,s) = 0, \\ -d\Delta = +[S_0^* P S_0 + \int_\Theta M_0^* P M_0 \nu(\theta)]dt \\ +[S_{o0}^* \bar{P} S_{o0} + \int_\theta M_{o0}^* \bar{P} M_{o0}\nu_o(d\theta)]dt \\ +[\sum_{s'\neq s}(\Delta(t,s') - \Delta_0(t,s))\tilde{q}_{ss'}]dt \\ +[S_{o0}^*\Gamma_{B_o} + \int_\theta M_{o0}^*\nu_o(d\theta)\Gamma_{N_o}]dt \\ +[\Gamma_{B_o}^* S_{o0} + \int_\Theta \Gamma_{N_o}^* M_{o0}\nu_o(d\theta)]dt \\ -\Gamma^*[\sum_{i\in\mathcal{I}_+}(B_{2i}+\bar{B}_{2i})(R_i+\bar{R}_i)^{-1}(B_{2i}+\bar{B}_{2i})^*]\Gamma dt \\ -\Gamma^*[\sum_{j\in\mathcal{I}_-}(B_{2j}+\bar{B}_{2j})(R_j+\bar{R}_j)^{-1}(B_{2j}+\bar{B}_{2j})^*]\Gamma dt, \\ \delta(T,s) = trace(\Delta(T,s)) = 0, \end{cases} \tag{7.447}$$

The proof of Corollary 7.2 is immediate from Theorem 7.2 by considering two adversarial teams and with choice vector of matrices $U_+ = (U_i)_{i\in\mathcal{I}_+}$ and $U_- = (U_i)_{i\in\mathcal{I}_-}$, respectively.

Notice that the deterministic part of the Riccati equations in (7.447) has a unique positive definite solution P^{ad} if in addition

$$\left[\sum_{i\in\mathcal{I}_+} B_{2i}R_i^{-1}B_{2i}^* + \sum_{j\in\mathcal{I}_-} B_{2j}R_j^{-1}B_{2j}^*\right] \succ 0, \tag{7.448}$$

which does not blow up within $[0,T]$, and positive solution $\bar{P}^{\text{ad}}$ if in addition

$$\begin{aligned} &\Big[\sum_{i\in\mathcal{I}_+}(B_{2i}+\bar{B}_{2i})(R_i+\bar{R}_i)^{-1}(B_{2i}+\bar{B}_{2i})^* \\ &+\sum_{j\in\mathcal{I}_-}(B_{2j}+\bar{B}_{2j})(R_j+\bar{R}_j)^{-1}(B_{2j}+\bar{B}_{2j})^*\Big] \succ 0, \end{aligned} \tag{7.449}$$

within $[0, T]$. Next, we study the risk-sensitive case and point out some facts regarding the comparison of its solution with respect to the risk-neutral case as the risk-sensitivity index vanishes.

7.10.5.2 Risk-Sensitive Matrix-Valued Games Under Common Noise

This subsection extends Sect. 7.10.2 to include additive common noises that are observed by all decision-makers.

Theorem 7.3 *Assume that $Q_i, R_i, Q_i + \bar{Q}_i, R_i + \bar{R}_i$ are symmetric positive definite and $M_0 = 0$. Then the matrix-valued mean-field-type (risk-sensitive) Nash equilibrium strategy and the (risk-sensitive) equilibrium cost are given by*

$$\begin{cases} U_i^{\mathrm{rs}} = -R_i^{-1} B_{2i}^* P_i^{rs}(X - \hat{X}) - (R_i + \bar{R}_i)^{-1}(B_{2i} + \bar{B}_{2i})^*(\bar{P}_i^{rs}\hat{X} + \Gamma_i^{rs}), \\ L_i^{\mathrm{rs}}(X, s, U^{\mathrm{rs}}) = \frac{1}{\lambda_i} \log E\exp\{\lambda_i[\langle P_i^{\mathrm{rs}}(0, s(0))(X_0 - \hat{X}_0), X_0 - \hat{X}_0\rangle \\ +\langle \bar{P}_i^{rs}(0, s(0))\hat{X}_0, \hat{X}_0\rangle + 2\langle \Gamma_i^{rs}(0, s(0)), \hat{X}_0\rangle + \delta_i^{\mathrm{rs}}(0, s(0))]\}, \\ i \in \mathscr{I}, \end{cases} \tag{7.450}$$

where $(P_i^{\mathrm{rs}}, P_{i,B_o}^{\mathrm{rs}}, P_{i,N_o}^{\mathrm{rs}}), (\bar{P}_i^{\mathrm{rs}}, \bar{P}_{i,B_o}^{\mathrm{rs}}, \bar{P}_{i,N_o}^{\mathrm{rs}}) = (\bar{P}_i, \bar{P}_{i,B_o}, \bar{P}_{i,N_o}), (\Gamma_i, \Gamma_{i,B_o}, \Gamma_{i,N_o})$, $(\delta_i^{\mathrm{rs}}, \delta_{i,B_o}^{\mathrm{rs}}, \delta_{i,N_o}^{\mathrm{rs}})$ are adapted to the filtration $\mathscr{F}^{s,B_o,N_o}$ and solve the following differential equations:

$$
\begin{cases}
dP_i = \Big\{Q_i + (B_1^* P_i + P_i B_1) + \sum_{s'\neq s}(P_i(t,s') - P_i(t,s))\tilde{q}_{ss'} \\
-P_i[B_{2i} R_i^{-1} B_{2i}^* - 2\lambda_i S_0 S_0^*]P_i \\
-\sum_{j\neq i}[P_i B_{2j} R_j^{-1} B_{2j}^* P_j + P_j B_{2j} R_j^{-1} B_{2j}^* P_i]\,\}dt \\
P_i(T,s) = Q_i(T,s), \\
-d\bar{P}_i = \big\{(Q_i + \bar{Q}_i) + ((B_1 + \bar{B}_1)^* \bar{P}_i + \bar{P}_i(B_1 + \bar{B}_1)) \\
\sum_{s'\neq s}(\bar{P}_i(t,s') - \bar{P}_i(t,s))\tilde{q}_{ss'} \\
-\bar{P}_i[(B_{2i} + \bar{B}_{2i})(R_i + \bar{R}_i)^{-1}(B_{2i} + \bar{B}_{2i})^*]\bar{P}_i \\
-\sum_{j\neq i} \bar{P}_j(B_{2j} + \bar{B}_{2j})(R_j + \bar{R}_j)^{-1*}(B_{2j} + \bar{B}_{2j})^* \bar{P}_i \\
-\sum_{j\neq i} \bar{P}_i(B_{2j} + \bar{B}_{2j})(R_j + \bar{R}_j)^{-1*}(B_{2j} + \bar{B}_{2j})^* \bar{P}_j\ \} \\
\bar{P}_i(T,s) = Q_i(T,s) + \bar{Q}_i(T,s), \\
-d\Gamma_i = \bar{P}_{i,B_o} S_{o0} + \int_\theta \bar{P}_{i,N_o} M_{o0}(.,\theta)\nu_o(d\theta) \\
+(B_1 + \bar{B}_1)^* \Gamma_i + \sum_{s'\neq s}(\Gamma_i(t,s') - \Gamma_i(t,s))\tilde{q}_{ss'}]dt \\
-\bar{P}_i(B_{2i} + \bar{B}_{2i})(R_i + \bar{R}_i)^{-1}(B_{2i} + \bar{B}_{2i})^* \Gamma_i \\
-\sum_{j\neq i} \bar{P}_i(B_{2j} + \bar{B}_{2j})(R_j + \bar{R}_j)^{-1}(B_{2j} + \bar{B}_{2j})^* \Gamma_j \\
-\sum_{j\neq i} \bar{P}_j(B_{2j} + \bar{B}_{2j})(R_j + \bar{R}_j)^{-1}(B_{2j} + \bar{B}_{2j})^* \Gamma_i \\
\Gamma_i(T,s) = 0, \\
-d\Delta_i = +[S_0^* P_i S_0 + \int_\Theta M_0^* P_i M_0 \nu(\theta)]dt \\
+[S_{o0}^* \bar{P}_i S_{o0} + \int_\theta M_{o0}^* \bar{P}_i M_{o0}\nu_o(d\theta)]dt \\
+[\sum_{s'\neq s}(\Delta_i(t,s') - \Delta_i(t,s))\tilde{q}_{ss'}]dt \\
+[S_{o0}^* \Gamma_{i,B_o} + \int_\theta M_{o0}^* \nu_o(d\theta)\Gamma_{i,N_o}]dt \\
+[\Gamma_{i,B_o}^* S_{o0} + \int_\Theta \Gamma_{i,N_o}^* M_{o0}\nu_o(d\theta)]dt \\
-\Gamma_i^*(B_{2i} + \bar{B}_{2i})(R_i + \bar{R}_i)^{-1}(B_{2i} + \bar{B}_{2i})^* \Gamma_i dt \\
-\sum_{j\neq i} \Gamma_i^*(B_{2j} + \bar{B}_{2j})(R_j + \bar{R}_j)^{-1}(B_{2j} + \bar{B}_{2j})^* \Gamma_j dt \\
\delta_i(T,s) = trace(\Delta_i(T,s)) = 0, \\
\forall s \in \mathcal{S},
\end{cases}
\tag{7.451}
$$

whenever these differential system of equations have a unique solution that does not blow up in $[0, T]$.

Note that the P_i equation in (7.467) is modified and the risk-sensitivity appears in (7.451) through the term $P_i^{rs} 2\lambda_i S_0 S_0^* P_i^{rs}$.

As all λ_i vanish, the matrix-valued differential system (7.451) becomes the risk-neutral system (7.467) above, and the risk-sensitive optimal strategy coincides with the risk-neutral one. A bound for $\bar{\lambda}_i$ can be obtained from the positivity condition of the matrices:

$$
[B_{2i} R_i^{-1} B_{2i}^* - 2\lambda_i S_0 S_0^*] \succ 0. \tag{7.452}
$$

Note that $\bar{P}_i^{\text{rs}} = \bar{P}_i$ because this coefficient is associated with the term $\langle \hat{X}, \hat{X} \rangle$ which depends on B_o, N_o but is independent of the Brownian motion B.

Risk-Averse—Risk-Neutral—Risk-Seeking Decision-Makers

Consider a finite population of decision-makers $\mathscr{I} := \{1, \dots I\}$ is classified as follows:

- risk-neutral decision-makers: $\mathscr{I}_0 = \{i \in \mathscr{I} \mid \lambda_i \to 0\}$,
- risk-averse decision-makers: $\mathscr{I}_+ = \{i \in \mathscr{I} \mid \lambda_i > 0\}$,
- risk-seeking decision-makers: $\mathscr{I}_- = \{i \in \mathscr{I} \mid \lambda_i < 0\}$.

Corollary 7.3 *A mixture of risk-neutral, risk-seeking and risk-averse are obtained solving the following system:*

$$\begin{cases} i \in \mathscr{I}_0 : \\ -dP_i = \Big\{Q_i + B_1^* P_i + P_i B_1 + \sum_{s' \neq s}(P_i(t,s') - P_i(t,s))\tilde{q}_{ss'} \\ -P_i[B_{2i} R_i^{-1} B_{2i}^*]P_i \\ -\sum_{j \neq i} P_j B_{2j} R_j^{-1*} B_{2j}^* P_i - \sum_{j \neq i} P_i B_{2j} R_j^{-1*} B_{2j}^* P_j \ \}dt, \end{cases} \tag{7.453}$$

$$\begin{cases} i \in \mathscr{I}_+ : \\ -dP_i = \Big\{Q_i + B_1^* P_i + P_i B_1 + \sum_{s' \neq s}(P_i(t,s') - P_i(t,s))\tilde{q}_{ss'} \\ -P_i[B_{2i} R_i^{-1} B_{2i}^* - 2\lambda_i S_0 S_0^*]P_i \\ -\sum_{j \neq i} P_j B_{2j} R_j^{-1*} B_{2j}^* P_i - \sum_{j \neq i} P_i B_{2j} R_j^{-1*} B_{2j}^* P_j \ \}dt \end{cases} \tag{7.454}$$

$$\begin{cases} i \in \mathscr{I}_- : \\ -dP_i = \Big\{Q_i + B_1^* P_i + P_i B_1 + \sum_{s' \neq s}(P_i(t,s') - P_i(t,s))\tilde{q}_{ss'} \\ -P_i[B_{2i} R_i^{-1} B_{2i}^* + 2(-\lambda_i) S_0 S_0^*]P_i \\ -\sum_{j \neq i} P_j B_{2j} R_j^{-1*} B_{2j}^* P_i - \sum_{j \neq i} P_i B_{2j} R_j^{-1*} B_{2j}^* P_j \ \}dt \end{cases} \tag{7.455}$$

$$P_i(T,s) = Q_i(T,s), i \in \mathscr{I}, \tag{7.456}$$

$$\forall s \in \mathscr{S}. \tag{7.457}$$

Corollary 7.4 *Assume that* $Q_0 := \sum_{j \in \mathscr{I}} Q_j$, $Q_0 + \bar{Q}_0 := \sum_{j \in \mathscr{I}}[Q_j, +\bar{Q}_j]$, R_i, $R_i + \bar{R}_i$ *are symmetric positive definite. The risk-sensitive fully cooperative solution of the problem* $\inf_{(U_1,\dots,U_n)} \frac{1}{\lambda} \log E[e^{\lambda \sum_{j \in \mathscr{I}} L_j}]$ *is given by*

$$\begin{cases} U_i^{\mathrm{rs,g}} - \hat{U}_i^{\mathrm{rs,g}} = -R_i^{-1} B_{2i}^* P_0^{\mathrm{rs,g}}(X - \hat{X}), \\ \hat{U}_i^{\mathrm{rs,g}} = -(R_i + \bar{R}_i)^{-1}(B_{2i} + \bar{B}_{2i})^*(\bar{P}_0 \hat{X} + \Gamma_0), \\ L_0^{\mathrm{rs,g}}(X, s, U^{\mathrm{rs,g}}) = \frac{1}{\lambda} \log E\exp\{\lambda[\langle P_0^{\mathrm{rs,g}}(0, s(0))(X_0 - \hat{X}_0), X_0 - \hat{X}_0\rangle \\ +\langle \bar{P}_0(0, s(0))\hat{X}_0, \hat{X}_0\rangle + 2\langle \Gamma_0(0, s(0)), \hat{X}_0\rangle + \delta_0^{\mathrm{rs,g}}(0, s(0))]\}, \\ i \in \mathscr{I}, \end{cases} \tag{7.458}$$

where $P_0^{\mathrm{rs,g}}$, $\bar{P}_0^{\mathrm{rs,g}} = \bar{P}_0$, Γ_0, *and* $\delta_0^{\mathrm{rs,g}}$ *solve the following differential equations:*

$$
\begin{cases}
-dP_0 \\
= \Big\{ Q_0 + (B_1^* P_0 + P_0 B_1) + \sum_{s' \neq s} (P_0(t,s') - P_0(t,s))\tilde{q}_{ss'} \\
-P_0[\sum_{i\in\mathcal{I}} B_{2i} R_i^{-1} B_{2i}^* - 2\lambda S_0 S_0^*] P_0 \Big\} dt \\
-P_{B_o} dB_o - \int_\theta P_{N_o} \tilde{N}_o(dt, d\theta), \\
P_0(T,s) = Q_0(T,s), \\
-d\bar{P}_0 = \big\{ Q_0 + \bar{Q}_0 + (B_1 + \bar{B}_1)^* \bar{P}_0 + \bar{P}_0 (B_1 + \bar{B}_1) \\
\sum_{s' \neq s} (\bar{P}_0(t,s') - \bar{P}_0(t,s))\tilde{q}_{ss'} dt \\
-\bar{P}_0[\sum_{i\in\mathcal{I}} (B_{2i} + \bar{B}_{2i})(R_i + \bar{R}_i)^{-1}(B_{2i} + \bar{B}_{2i})^*]\bar{P}_0 \big\} dt, \\
\bar{P}_0(T,s) = Q_0(T,s) + \bar{Q}_0(T,s), \\
-d\Gamma_0 = \big\{ \bar{P}_{0,B_o} S_{o0} + \int_\theta \bar{P}_{0,N_o} M_{o0}(.,\theta)\nu_o(d\theta) \\
+(B_1 + \bar{B}_1)^* \Gamma_0 + \sum_{s' \neq s} (\Gamma_0(t,s') - \Gamma_0(t,s))\tilde{q}_{ss'}] \\
-\bar{P}_0[\sum_{i\in\mathcal{I}} (B_{2i} + \bar{B}_{2i})(R_i + \bar{R}_i)^{-1}(B_{2i} + \bar{B}_{2i})^*]\Gamma_0 \big\} dt, \\
\Gamma_0(T,s) = 0, \\
-d\Delta_0 = +[S_0^* P_0 S_0 + \int_\Theta M_0^* P_0 M_0 \nu(\theta)] dt \\
+[S_{o0}^* \bar{P}_0 S_{o0} + \int_\theta M_{o0}^* \bar{P}_0 M_{o0} \nu_o(d\theta)] dt \\
+[\sum_{s' \neq s} (\Delta_0(t,s') - \Delta_0(t,s))\tilde{q}_{ss'}] dt \\
+[S_{o0}^* \Gamma_{0,B_o} + \int_\theta M_{o0}^* \nu_o(d\theta) \Gamma_{0,N_o}] dt \\
+[\Gamma_{0,B_o}^* S_{o0} + \int_\Theta \Gamma_{0,N_o}^* M_{o0} \nu_o(d\theta)] dt \\
-\Gamma_0^* \sum_{i\in\mathcal{I}} (B_{2i} + \bar{B}_{2i})(R_i + \bar{R}_i)^{-1}(B_{2i} + \bar{B}_{2i})^* \Gamma_0 dt, \\
\delta_0(T,s) = trace(\Delta_0) = 0, \\
\forall s \in \mathcal{S}.
\end{cases}
\tag{7.459}
$$

Corollary 7.5 *Assume that Q, $Q + \bar{Q}$, R_i, $R_i + \bar{R}_i$ are symmetric positive definite for $i \in \mathcal{I}_+$ and $-R_j$, $-(R_j + \bar{R}_j)$ are symmetric positive definite for $j \in \mathcal{I}_-$. We assume that $\mathcal{I}_+ \cup \mathcal{I}_- = \mathcal{I}$. The adversarial risk-sensitive game problem between the team of attackers $\mathcal{I}_-$ and the team of defenders $\mathcal{I}_+$ has a saddle (matrix) point, and it is given by*

$$
\begin{cases}
U_i^{\mathrm{rs,ad}} = -R_i^{-1} B_{2i}^* P_0^{rs,ad} (X - \hat{X}) \\
-(R_i + \bar{R}_i)^{-1} (B_{2i} + \bar{B}_{2i})^* (\bar{P}^{ad} \hat{X} + \Gamma^{ad}), \ i \in \mathcal{I}_+, \\
U_j^{\mathrm{rs,ad}} = -R_j^{-1} B_{2i}^* P^{rs,ad} (X - \hat{X}) \\
-(R_j + \bar{R}_j)^{-1} (B_{2j} + \bar{B}_{2j})^* (\bar{P}^{ad} \hat{X} + \Gamma^{ad}), \ j \in \mathcal{I}_-, \\
L^{\mathrm{ad}}(X, s, U^{\mathrm{rs,ad}}) = E\langle P^{\mathrm{rs,ad}}(0, s(0))(X_0 - \hat{X}_0), X_0 - \hat{X}_0 \rangle \\
+E\langle \bar{P}^{\mathrm{ad}}(0, s(0)) \hat{X}_0, \hat{X}_0 \rangle \\
+2E\langle \Gamma^{\mathrm{ad}}(0, s(0), \hat{X}_0 \rangle + E[\delta^{\mathrm{rs,ad}}(0, s(0))],
\end{cases}
\tag{7.460}
$$

where $P^{\mathrm{rs,ad}}$, $\bar{P}^{\mathrm{rs,ad}} = \bar{P}^{\mathrm{ad}}$, Γ^{ad}, and $\delta^{\mathrm{rs,ad}}$ solve the following differential equations:

$$\begin{cases} -dP \\ = (B_1^* P + P B_1)dt + \sum_{s' \neq s} (P(t,s') - P(t,s))\tilde{q}_{ss'}dt \\ +Qdt \\ -P[\sum_{i \in \mathcal{I}_+} B_{2i} R_i^{-1} B_{2i}^* - 2\lambda S_0 S_0^*] P dt - P[\sum_{j \in \mathcal{I}_-} B_{2j} R_j^{-1} B_{2j}^*] P dt \\ P(T,s) = Q(T,s), \\ -d\bar{P} = ((B_1 + \bar{B}_1)^* \bar{P} + \bar{P}(B_1 + \bar{B}_1))dt \\ \sum_{s' \neq s} (\bar{P}(t,s') - \bar{P}(t,s))\tilde{q}_{ss'}dt \\ +(Q + \bar{Q})dt \\ -\bar{P}[\sum_{i \in \mathcal{I}_+} (B_{2i} + \bar{B}_{2i})(R_i + \bar{R}_i)^{-1}(B_{2i} + \bar{B}_{2i})^*] \bar{P} dt \\ -\bar{P}[\sum_{j \in \mathcal{I}_-} (B_{2j} + \bar{B}_{2j})(R_j + \bar{R}_j)^{-1}(B_{2j} + \bar{B}_{2j})^*] \bar{P} dt \\ \bar{P}(T,s) = Q(T,s) + \bar{Q}(T,s), \\ -d\Gamma = \bar{P}_{B_o} S_{o0} + \int_\theta \bar{P}_{N_o} M_{o0}(.,\theta)\nu_o(d\theta) \\ +(B_1 + \bar{B}_1)^* \Gamma + \sum_{s' \neq s} (\Gamma(t,s') - \Gamma(t,s))\tilde{q}_{ss'}]dt \\ -\bar{P}[\sum_{i \in \mathcal{I}_+} (B_{2i} + \bar{B}_{2i})(R_i + \bar{R}_i)^{-1}(B_{2i} + \bar{B}_{2i})^*] \Gamma dt \\ -\bar{P}[\sum_{j \in \mathcal{I}_-} (B_{2j} + \bar{B}_{2j})(R_j + \bar{R}_j)^{-1}(B_{2j} + \bar{B}_{2j})^*] \Gamma dt, \\ \Gamma(T,s) = 0, \\ -d\Delta = +[S_0^* P S_0 + \int_\Theta M_0^* P M_0 \nu(\theta)]dt \\ +[S_{o0}^* \bar{P} S_{o0} + \int_\theta M_{o0}^* \bar{P} M_{o0} \nu_o(d\theta)]dt \\ +[\sum_{s' \neq s} (\Delta(t,s') - \Delta_0(t,s))\tilde{q}_{ss'}]dt \\ +[S_{o0}^* \Gamma_{B_o} + \int_\theta M_{o0}^* \nu_o(d\theta) \Gamma_{N_o}]dt \\ +[\Gamma_{B_o}^* S_{o0} + \int_\Theta \Gamma_{N_o}^* M_{o0} \nu_o(d\theta)]dt \\ -\Gamma^*[\sum_{i \in \mathcal{I}_+} (B_{2i} + \bar{B}_{2i})(R_i + \bar{R}_i)^{-1}(B_{2i} + \bar{B}_{2i})^*] \Gamma dt \\ -\Gamma^*[\sum_{j \in \mathcal{I}_-} (B_{2j} + \bar{B}_{2j})(R_j + \bar{R}_j)^{-1}(B_{2j} + \bar{B}_{2j})^*] \Gamma dt, \\ \delta(T,s) = trace(\Delta(T,s)) = 0, \\ s \in \mathcal{S}. \end{cases} \tag{7.461}$$

Notice that the risk-sensitive adversarial Riccati equation has a positive definite solution $P^{\mathrm{rs,ad}}$ if in addition

$$\left[\sum_{i \in \mathcal{I}_+} B_{2i} R_i^{-1} B_{2i}^* + \sum_{j \in \mathcal{I}_-} B_{2j} R_j^{-1} B_{2j}^* - 2\lambda S_0 S_0^* \right] \succ 0, \tag{7.462}$$

does not blow up within $[0, T]$. Following Sect. 7.10.3.4, the risk-sensitive adversarial problem can also be seen as a risk-neutral game between three teams: one team of regular decisions-makers, one team of adversaries, and one team of "virtual" adversaries. The team of virtual adversaries will replace the riskiness part of the problem.

7.10.5.3 Coopetition and Other-Regarding Preferences

We now examine other-regarding preferences and coopetition between decision-makers. Each decision-maker i has empathy matrix Λ_{ij} toward decision-maker j. This enables an empathetic payoff:

$$L_i^{\Lambda} = \text{trace}[\textstyle\sum_{j\in\mathcal{I}} \Lambda_{ij} H_j], \tag{7.463}$$

If $\Lambda_{ij} = \lambda_{ij}\mathbb{I}_{n\times n}$, one obtains the scalar empathetic payoff:

$$L_i^{\Lambda} = \lambda_{ii} L_i + \textstyle\sum_{j\neq i} \lambda_{ij} L_j, \tag{7.464}$$

which can be seen as a mixture of selfish, selfless, altruism, spite, and self-abnegation depending on the sign of λ_{ij} :

$$\begin{aligned}
&H_i^{\Lambda}(X, s, U)\\
&= (X(T) - \hat{X}(T))^*[\textstyle\sum_{j\in\mathcal{I}} \lambda_{ij} Q_j(T, s(T))](X(T) - \hat{X}(T))\\
&+\hat{X}^*(T)[\textstyle\sum_{j\in\mathcal{I}} \lambda_{ij}(Q_j(T, s(T)) + \bar{Q}_j(T, s(T)))]\hat{X}(T)\\
&+\textstyle\int_0^T (X - \hat{X})^*[\sum_{j\in\mathcal{I}} \lambda_{ij} Q_j](X - \hat{X}) + \hat{X}^*[\sum_{j\in\mathcal{I}} \lambda_{ij}(Q_j + \bar{Q}_j)]\hat{X}dt\\
&+\textstyle\int_0^T \sum_{j\in\mathcal{I}} (U_j - \hat{U}_j)^* \lambda_{ij} R_j (U_j - \hat{U}_j) + \sum_{j\in\mathcal{I}} \hat{U}_j^* \lambda_{ij}(R_j + \bar{R}_j)\hat{U}_j dt\\
&= (X(T) - \hat{X}(T))^*[Q_i^{\Lambda}(T, s(T))](X(T) - \hat{X}(T))\\
&+\hat{X}^*(T)[Q_i^{\Lambda}(T, s(T)) + \bar{Q}_i^{\Lambda}(T, s(T))]\hat{X}(T)\\
&+\textstyle\int_0^T (X - \hat{X})^*[Q_i^{\Lambda}](X - \hat{X}) + \hat{X}^*[Q_i^{\Lambda} + \bar{Q}_i^{\Lambda}]\hat{X}dt\\
&+\textstyle\int_0^T (U_i - \hat{U}_i)^* \lambda_{ii} R_i (U_i - \hat{U}_i) + \hat{U}_i^* \lambda_{ii}(R_i + \bar{R}_i)\hat{U}_i dt\\
&+\textstyle\int_0^T \sum_{j\neq i} (U_j - \hat{U}_j)^* \lambda_{ij} R_j (U_j - \hat{U}_j) + \sum_{j\neq i} \hat{U}_j^* \lambda_{ij}(R_j + \bar{R}_j)\hat{U}_j dt,
\end{aligned} \tag{7.465}$$

where $\Phi_i^{\Lambda} = \sum_{j\in\mathcal{I}} \lambda_{ij}\Phi_j,\ \ \Phi \in \{Q, \bar{Q}, H\}$.

Theorem 7.4 *Assume that* $Q_i^{\Lambda}, \lambda_{ii}R_i, (Q_i^{\Lambda} + \bar{Q}_i^{\Lambda}), \lambda_{ii}(R_i + \bar{R}_i)$ *are symmetric positive definite. Then, the matrix-valued mean-field-type (risk-neutral) Nash equilibrium strategy and the (risk-neutral) equilibrium cost are given by*

$$\begin{cases}
U_i^{\text{rn}} = -\frac{1}{\lambda_{ii}} R_i^{-1} B_{2i}^* P_i (X - \hat{X}) - \frac{1}{\lambda_{ii}}(R_i + \bar{R}_i)^{-1}(B_{2i} + \bar{B}_{2i})^*[\bar{P}_i\hat{X} + \Gamma_i]\\
L_i^{\text{rn}}(X, s, U^{\text{rn}}) = E\langle P_i(0, s(0))(X_0 - \hat{X}_0), X_0 - \hat{X}_0\rangle\\
+E[\langle \bar{P}_i(0, s(0))\hat{X}_0, \hat{X}_0\rangle + 2\langle \Gamma_i(0, s(0)), \hat{X}_0\rangle + \delta_i(0, s(0))],\\
i \in \mathcal{I},
\end{cases} \tag{7.466}$$

where $(P_i, P_{i,B_o}, P_{i,N_o})$, $(\bar{P}_i, \bar{P}_{i,B_o}, \bar{P}_{i,N_o})$, $(\Gamma_i, \Gamma_{i,B_o}, \Gamma_{i,N_o})$, $(\delta_i, \delta_{i,B_o}, \delta_{i,N_o})$ *are adapted to the filtration* $\mathbb{F}^{s,B_o,N_o}$ *and solve the following differential equations:*

$$
\left\{
\begin{array}{l}
-dP_i = \Big\{ Q_i^{\Lambda} + B_1^* P_i + P_i B_1 + \sum_{s'\neq s}(P_i(t,s') - P_i(t,s))\tilde{q}_{ss'} \\
-\frac{1}{\lambda_{ii}} P_i B_{2i} R_i^{-1} B_{2i}^* P_i \\
-\sum_{j\neq i} \frac{1}{\lambda_{jj}} P_i B_{2j} R_j^{-1} B_{2j}^* P_j - \sum_{j\neq i} \frac{1}{\lambda_{jj}} P_j B_{2j} R_j^{-1} B_{2j}^* P_i \\
+[\sum_{j\neq i} \frac{\lambda_{ij}}{\lambda_{jj}^2} P_j B_{2j} R_j^{-1} B_{2j}^* P_j] \Big\} dt \\
P_i(T,s) = Q_i^{\Lambda}(T,s), \\
-d\bar{P}_i = \{ (Q_i^{\Lambda} + \bar{Q}_i^{\Lambda}) + (B_1+\bar{B}_1)^* \bar{P}_i + \bar{P}_i (B_1 + \bar{B}_1) \\
+\sum_{s'\neq s}(\bar{P}_i(t,s') - \bar{P}_i(t,s))\tilde{q}_{ss'} \\
-\frac{1}{\lambda_{ii}} \bar{P}_i [(B_{2i}+\bar{B}_{2i})(R_i+\bar{R}_i)^{-1}(B_{2i}+\bar{B}_{2i})^*]\bar{P}_i \\
-\sum_{j\neq i} \frac{1}{\lambda_{jj}} \bar{P}_j (B_{2j}+\bar{B}_{2j})(R_j+\bar{R}_j)^{-1*}(B_{2j}+\bar{B}_{2j})^* \bar{P}_i \\
-\sum_{j\neq i} \frac{1}{\lambda_{jj}} \bar{P}_i (B_{2j}+\bar{B}_{2j})(R_j+\bar{R}_j)^{-1*}(B_{2j}+\bar{B}_{2j})^* \bar{P}_j \\
+[\sum_{j\neq i} \frac{\lambda_{ij}}{\lambda_{jj}^2} \bar{P}_j (B_{2j}+\bar{B}_{2j})(R_j+\bar{R}_j)^{-1}(B_{2j}+\bar{B}_{2j})^* \bar{P}_j] \} dt \\
\bar{P}_i(T,s) = Q_i^{\Lambda}(T,s) + \bar{Q}_i^{\Lambda}(T,s), \\
-d\Gamma_i = \{ \bar{P}_{i,B_o} S_{o0} + \int_{\theta} \bar{P}_{i,N_o} M_{o0}(.,\theta)\nu_o(d\theta) \\
+(B_1+\bar{B}_1)^* \Gamma_i + \sum_{s'\neq s}(\Gamma_i(t,s') - \Gamma_i(t,s))\tilde{q}_{ss'} \\
-\frac{1}{\lambda_{ii}} \bar{P}_i (B_{2i}+\bar{B}_{2i})(R_i+\bar{R}_i)^{-1}(B_{2i}+\bar{B}_{2i})^* \Gamma_i \\
-\sum_{j\neq i} \frac{1}{\lambda_{jj}} \bar{P}_i (B_{2j}+\bar{B}_{2j})(R_j+\bar{R}_j)^{-1}(B_{2j}+\bar{B}_{2j})^* \Gamma_j \\
-\sum_{j\neq i} \frac{1}{\lambda_{jj}} \bar{P}_j (B_{2j}+\bar{B}_{2j})(R_j+\bar{R}_j)^{-1}(B_{2j}+\bar{B}_{2j})^* \Gamma_i \\
+[\sum_{j\neq i} \frac{\lambda_{ij}}{\lambda_{jj}^2} \bar{P}_j (B_{2j}+\bar{B}_{2j})(R_j+\bar{R}_j)^{-1}(B_{2j}+\bar{B}_{2j})^* \Gamma_j] \} dt \\
\Gamma_i(T,s) = 0, \\
-d\Delta_i = \{ + S_0^* P_i S_0 + \int_{\Theta} M_0^* P_i M_0 \nu(\theta) \\
+ S_{o0}^* \bar{P}_i S_{o0} + \int_{\theta} M_{o0}^* \bar{P}_i M_{o0} \nu_o(d\theta) \\
+\sum_{s'\neq s}(\Delta_i(t,s') - \Delta_i(t,s))\tilde{q}_{ss'} \\
+S_{o0}^* \Gamma_{i,B_o} + \int_{\theta} M_{o0}^* \nu_o(d\theta) \Gamma_{i,N_o} \\
+\Gamma_{i,B_o}^* S_{o0} + \int_{\Theta} \Gamma_{i,N_o}^* M_{o0} \nu_o(d\theta) \\
-\frac{1}{\lambda_{ii}} \Gamma_i^* (B_{2i}+\bar{B}_{2i})(R_i+\bar{R}_i)^{-1}(B_{2i}+\bar{B}_{2i})^* \Gamma_i \\
-\sum_{j\neq i} \frac{1}{\lambda_{jj}} \Gamma_i^* (B_{2j}+\bar{B}_{2j})(R_j+\bar{R}_j)^{-1}(B_{2j}+\bar{B}_{2j})^* \Gamma_j \\
+[\sum_{j\neq i} \frac{\lambda_{ij}}{\lambda_{jj}^2} \Gamma_j (B_{2j}+\bar{B}_{2j})(R_j+\bar{R}_j)^{-1}(B_{2j}+\bar{B}_{2j})^* \Gamma_j] \} dt \\
\delta_i(T,s) = trace(\Delta_i(T,s)) = 0, \\
\forall s \in \mathcal{S},
\end{array}
\right.
\tag{7.467}
$$

whenever these matrix-valued differential equations have a unique solution that does not blow up within $[0, T]$.

The following results established in Sect. 7.4 in the time-independent game case hold also in the time-dependent matrix-valued case:

- Pure altruism can be inconsistent in the sense that the equilibrium set may change when the level of empathy hierarchy changes. For example, empathy pure altruism at the first level of empathy is obtained for $\lambda_{11} = 0, \lambda_{12} = 1, \lambda_{21} =$

$1, \lambda_{22} = 0$. The second level of empathy is when one empathetizes of the first level of empathetic payoffs. This corresponds to a new empathy structure $\lambda_{11}^{(2)} = 1, \lambda_{12}^{(2)} = 0, \lambda_{21}^{(2)} = 0, \lambda_{22} = 1$, i.e., pure selfishness. Clearly, this operation does not preserve the set of equilibria.
- Empathy spite may lead to non-admissible optimal strategies. One can choose an empathy spite $\lambda_{ij} < 0$ such that $Q_i^{\Lambda} \prec 0$. In this case, there can be an escape of P_i within $[0, T]$. Thus, empathy spite may lead to nonexistence of equilibria.

Note that the state dynamics (7.400) can be modified to accommodate matrices of compatible dimensions: $(d_1, d_2) = (d_1, k)(k, d_2)$, and $(d_1, d_2) = (d_1, l)(l, d_2)$. However, the results are subject to the existence of a positive solution to nonsymmetric Riccati systems within the horizon of the interaction. In the control-dependent jump-diffusion case, the resulting optimality system is not necessarily a Riccati one because it is non-quadratic in $P_i, \bar{P}_i$. It is of fractional order with the terms C_i^{-1}, and $\bar{C}_i^{-1}$. Matrix-valued MFTGs are investigated in [43, 51, 320, 335].

Chapter 8
Mean-Field-Type Games with Jump and Regime Switching

In this chapter, we present a generic class of mean-field-type games in which the payoffs and the state dynamics depend not only on the state-action profile of the decision-makers but also on a measure of the state-action pair. The state dynamics is a measure-dependent process with jump-diffusion and regime switching. We derive novel equilibrium systems to be solved. Two solution approaches are presented: (i) dynamic programming principle and (ii) stochastic maximum principle. The relationship between dual function and adjoint processes is provided. It is shown that the extension to the risk-sensitive case generates a nonlinearity to the adjoint process, and it involves three other processes associated with the diffusion, jump, and regime switching, respectively.

In Chap. 3, we examined mean-field-type games with discrete state spaces and established master adjoint systems [333, 338]. Here, we introduce risk-neutral MFTGs with jump-diffusion and regime switching:

- The framework is presented in three different steps: (i) The probability distribution (with all switching regimes) is identified as an augmented state. (ii) A dynamic programming principle in infinite dimensions is established, followed by a derivation of a sufficient optimality system for best-response strategies which are in state-and-mean-field feedback form. (iii) Following that, a dual function is introduced, and first-order adjoint process and second-order adjoint process candidates are established. We develop a stochastic maximum principle for finding open-loop best-response strategies. Under weak derivative conditions, we establish relationships between first- and second-order adjoint processes of the maximum principle and the dynamic programming principle where the state process is governed by a continuous-time Markov regime-switching jump-diffusion model of mean-field type. The methodology is illustrated in a simple non-quadratic example where a semi-explicit solution and an approximated solution are provided as solutions to a coupled system of ordinary integrodifferential equations, where the regime switching makes a coupling between solutions.

T. Başar et al., *Mean-Field-Type Game Theory I*, Static & Dynamic Game Theory: Foundations & Applications, https://doi.org/10.1007/978-3-032-07027-2_8

- A relationship between the dynamic programming principle and the stochastic maximum principle is established (see Proposition 8.4).

Then, risk-sensitive MFTGs with jump-diffusion and regime switching are examined.

- The above method is extended to risk-sensitive MFTGs using a logarithmic transformation. In contrast to the risk-neutral case, three additional adjoint processes are involved in the diffusion, Poisson jump, and switching. A particular structure of the linear-exponentiated quadratic case with terminal payoff that is mean-field-dependent is discussed.
- A relationship between risk-sensitive dynamic programming and risk-sensitive maximum principle is established (see Propositions 8.7 and 8.8) by introducing three additional adjoint processes associated respectively with the diffusion, jump, and regime-switching processes. The correspondences are in the weak-derivative sense.

It is worth mentioning that for the risk-sensitive MFTG problems under jump-diffusion and regime-switching models, there is an additional risk-sensitive parameter and a set of jump martingales related to the Markov chain of mean-field type that appear in the adjoint processes system, which are coupled between regimes.

The rest of the chapter is structured as follows. Section 8.1 presents a generic risk-neutral model and focuses on solution methods. Section 8.2 presents some extensions to risk-sensitive MFTGs with drift, jump, diffusion, and regime switching. Following that, we present master adjoint systems (MASSs) with jumps.

Table 8.1 summarizes the notation used in this chapter.

8.1 Risk-Neutral Mean-Field-Type Games with Jumps and Regime Switching

We start with a game model with risk-neutral decision-makers. A class of MFTGs in continuous time is described by

- The set of decision-makers $\mathscr{I} = \{1, 2, \ldots, I\}$, where $I \in \mathbb{N}$.
- The horizon of decision-making is the interval $[0, T]$, with $T > 0$.
- There is a non-empty state space $\mathscr{S}$. The set of probability measures on $\mathscr{S}$ is denoted by $\Delta(\mathscr{S})$. The jump set is denoted by Θ, and there is a finite set $\tilde{S}$ of switching regimes.
- For each decision-maker i, a non-empty control action set A_i is available. The set A_i is not necessarily convex. The set of all control actions of all the decision-makers is $A = \prod_{i\in\mathscr{I}} A_i$.
- An instant payoff of decision-maker i is $r_i : \mathscr{S} \times \Delta(\mathscr{S}) \times A \times \tilde{\mathscr{S}} \to \mathbb{R}$.
- The state evolution is described by a stochastic differential equation of mean-field type with drift, jumps, jump-diffusion, and regime switching:

Table 8.1 Table of notation

$\mathcal{I}$	$\triangleq$	Set of decision-makers
T	$\triangleq$	Length of the horizon
$[0, T]$	$\triangleq$	Horizon of the MFTG
t	$\triangleq$	Time index
$\mathcal{S}$	$\triangleq$	State space
$\tilde{\mathcal{S}}$	$\triangleq$	Switching regime space
Θ	$\triangleq$	Lévy jump space
$s(t)$	$\triangleq$	State at time t
$\tilde{s}(t)$	$\triangleq$	Switching regime at time t
$\Delta(\mathcal{S})$	$\triangleq$	Set of probability measures on $\mathcal{S}$
$m(t, .)$	$\triangleq$	Probability measure of the state at time t
A_i	$\triangleq$	Control action set of decision-maker $i \in \mathcal{I}$
$a_i(.)$	$\triangleq$	Strategy of decision-maker $i \in \mathcal{I}$
$a = (a_i)_{i\in\mathcal{I}}$	$\triangleq$	Strategy profile
$b(t, s, m, a, \tilde{s})$	$\triangleq$	Drift coefficient functional
$\sigma(t, s, m, a, \tilde{s})$	$\triangleq$	Diffusion coefficient functional
$\mu(t, s, m, a, \tilde{s}, \theta)$	$\triangleq$	Jump rate coefficient functional
$\delta(t, s, m, a, \tilde{s})$	$\triangleq$	Switching rate coefficient functional
$r_i(t, s, m, a, \tilde{s})$	$\triangleq$	Instant payoff of decision-maker $i \in \mathcal{I}$
$g_i(s, m, \tilde{s})$	$\triangleq$	Terminal payoff of decision-maker $i \in \mathcal{I}$
$\mathcal{R}_{i,T}(m_0, a, \tilde{s}_0)$	$\triangleq$	Cumulative payoff of i
$\hat{V}_i(t, m, \tilde{s})$	$\triangleq$	Equilibrium payoff of $i \in \mathcal{I}$
$(\bar{p}_i, \bar{q}_i, \bar{r}_i, \bar{s}_i)$	$\triangleq$	First-order adjoint process of $i \in \mathcal{I}$
$(\bar{P}_i, \bar{Q}_i, \bar{R}_i, \bar{S}_i)$	$\triangleq$	Second-order adjoint process of $i \in \mathcal{I}$
$\hat{v}_i(t, s, \tilde{s})$	$\triangleq$	Dual function of $i \in \mathcal{I}$
$\bar{H}_i$	$\triangleq$	$r_i + b\bar{p}_i + \sigma\bar{q}_i + \int \mu\bar{r}_i(., \theta)\nu(d\theta) + \sum_{k\in\tilde{S}} \delta(., k)\bar{s}_i(., k)q_k$
$\bar{H}_{i,s}$	$\triangleq$	$r_{i,s} + b_s\bar{p}_i + \sigma_s\bar{q}_i + \int_\Theta \mu_s\bar{r}(., \theta)\nu(d\theta) + \sum_k \delta_s(.)\bar{s}(., ., k)q_k$
$\bar{H}_{i,ss}$	$\triangleq$	$r_{i,ss} + b_{ss}\bar{p}_i + \sigma_{ss}\bar{q}_i + \int_\Theta \mu_{ss}\bar{r}(., \theta, \tilde{s})\nu(d\theta) + \sum_k \delta_{ss}\bar{s}(., k)q_k$

$$s(t)=s_0+\int_0^t b dt'+\int_0^t \sigma dB(t')+\int_0^t\int_\Theta \mu\tilde{N}(dt', d\theta, \tilde{s})+\int_0^t \delta.\tilde{M}(dt'),\ \ t>0,$$

where $\delta.\tilde{M}(dt) := \sum_{k\in\tilde{S}} \delta_k\tilde{M}(dt, k)$, $s(0) = s_0 \in \mathcal{S}$, $\tilde{s}(0) = \tilde{s}_0 \in \tilde{\mathcal{S}}$, $b, \sigma, \delta :$ $\mathcal{S}\times\Delta(\mathcal{S})\times A\times\tilde{\mathcal{S}} \to \mathbb{R}$, b is the drift coefficient functional, σ is the diffusion coefficient functional, and B is a one-dimensional standard Brownian motion. $\mu : \mathcal{S}\times\Delta(\mathcal{S})\times A\times\tilde{\mathcal{S}}\times\Theta \to \mathbb{R}$ is the jump rate, $\delta(.)$ is the switching rate of regimes of $\tilde{s}(t)$ with generator $(q_{\tilde{s}\tilde{s}'})$, and intensity $q_{\tilde{s}\tilde{s}'}\ \mathbb{1}_{\{\tilde{s}(t_-)=\tilde{s}\}}$ and the associated compensated martingale to $\tilde{s}$ is

$$\tilde{M}(t, \tilde{s}, \tilde{s}') := \sum_{t'\in[0,t)} \mathbb{1}_{\{\tilde{s}(t'_-)=\tilde{s}\}}\mathbb{1}_{\{\tilde{s}(t')=\tilde{s}'\}} - \int_0^t q_{\tilde{s}\tilde{s}'}(t')\mathbb{1}_{\{\tilde{s}(t'_-)=\tilde{s}\}}dt'.$$

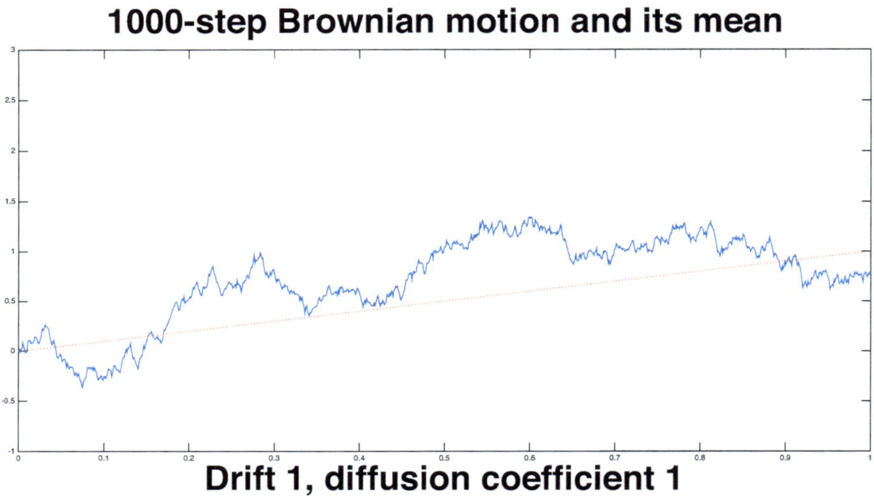

Fig. 8.1 $B(t)$ Brownian motion sample path. The Brownian motion will be used for capturing smaller noise changes in the state

The process N is a jump process with Lévy measure $\nu(d\theta, \tilde{s})$ defined over Θ and $\tilde{N}(dt, d\theta, \tilde{s}) = N(dt, d\theta, \tilde{s}) - \nu(d\theta)dt$, for each switching regime $\tilde{s} \in \mathcal{S}$. The above processes $\tilde{s}, N, B$ are $\mathbb{F}$-adapted and all defined on a given probability space $(\Omega, \mathbb{P}, \mathbb{F})$. Figures 8.1 and 8.2 respectively depict sample paths of Brownian motion and Poisson jump process. When the two processes are combined, one may capture a realistic behavior of block-wise seasonal events as illustrated in Fig. 8.3.

We consider functions $r_i, b, \sigma, \delta(s, m, a, \tilde{s}), \mu(s, m, a, \tilde{s}, \theta)$ where the arguments are $s \in \mathbb{R}$; m is a probability measure on $\mathbb{R}$, but we will remain mostly in the regular case (with respect to Lebesgue measure), in which m represents the probability density, assumed to be in $L^2(\mathbb{R})$, and $a \in \prod_{i \in \mathcal{I}} A_i$ is a control action profile of all decision-makers. For each regime $\tilde{s} \in \tilde{S}$, the functions b, σ, δ are scalar. All these functions are smooth. In the case of differentiability with respect to the measure m, we use the concept of Gâteaux differentiability. If $F : L^2(\mathbb{R}) \to \mathbb{R}$, then

$$\lim_{\varepsilon \to 0} \frac{d}{d\varepsilon} F(m + \varepsilon \tilde{m}) = \int_{\xi \in \mathbb{R}} \frac{\partial F(m)}{\partial m}(\xi)\tilde{m}(\xi)d\xi,$$

with the functional $\xi \in \mathbb{R} \mapsto \frac{\partial F}{\partial m}(m)(\xi) \in L^2(\mathbb{R})$. Note that in the smooth case, one can connect the derivative with respect to s of the Gâteaux derivative b_m to the notion of functional derivative with respect to measure and also to the Wasserstein gradient.

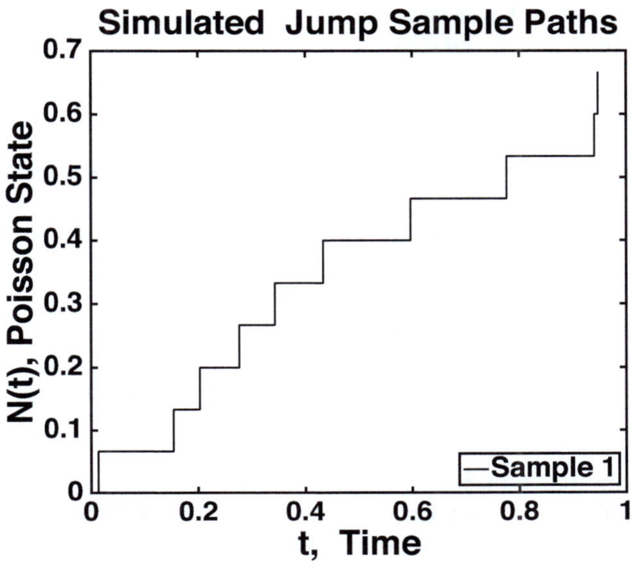

Fig. 8.2 A sample path of a Poisson jump process $N(t, \tilde{s})$. The Poisson process will be used for larger jumps

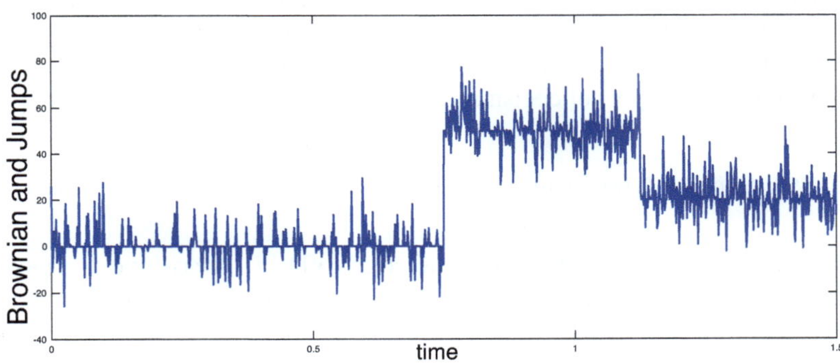

Fig. 8.3 A sample path of a jump-diffusion process

How Is the MFTG Played?

Given an initial regime $\tilde{s}_0$ and initial state s_0 which is drawn from the initial distribution m_0, the game $\mathcal{G}(s_0, \tilde{s}_0)$ proceeds as follows. At each instant, each decision-maker observes the current regime, state, and mean-field (perfect monitoring, perfect state observation), chooses a control action according to their strategy (defined below), and observes/measures their payoff.

For this game, we adopt the following notions: An admissible control of decision-maker $i \in \mathcal{I}$ is progressively a measurable process with respect to the filtration $\mathcal{F}_t$, taking values in A_i. The set of admissible controls of decision-maker i is $\mathscr{A}_i$. We identify two processes a_i and $\tilde{a}_i$ if

$$\mathbb{P}(a_i = \tilde{a}_i, \ \ a.e. \ \ on \ [0, T]) = 1.$$

A strategy for decision-maker i starting at time 0 is a Borel-measurable map:

$$a_i \, : \, [0, T] \times C([0, T], \mathscr{X}) \times \prod_{j \neq i} L^2([0, T], A_j) \to A_i,$$

for which there exists $\varepsilon > 0$ such that for any

$$(t', f_1, f_2) \in (0, T] \times \{C([0, T], \mathscr{X}) \times \prod_{j \neq i} L^2([0, T], A_j)\}^2,$$

if $f_1 = f_2$ on $[0, t']$, then $a_i(., f_1) = a_i(., f_2)$ on $[0, t' + \varepsilon]$. To each strategy profile, one can associate a control process profile. This will allow us to work with both open-loop and feedback forms of strategies by considering $a_i(., s^{s_0,a}, m)$. For open-loop strategies, the information structure is limited to $\{t\}$, with a_i being a measurable function of time. The stochastic maximum principle can be used as a methodology for finding optimal open-loop strategies. The information structure for feedback strategies is s and its distribution. The dual adjoint functions, which are obtained from the Bellman functions, can be used for finding feedback strategies where the feedback is a state-and-mean-field feedback.

The cumulative payoff of decision-maker i is

$$\mathscr{R}_{i,T}(s_0, a, \tilde{s}_0) := g_i(s(T), m(T, .), \tilde{s}(T)) + \int_0^T r_i(s(t), m(t, .), a(t), \tilde{s}(t)) \, dt,$$

where $g_i(s(T), m(T, .), \tilde{s}(T))$ is the terminal payoff of i. The risk-neutral payoff of i is $E[\mathscr{R}_{i,T}(s_0, a, \tilde{s}_0)]$. The risk-neutral MFTG $\mathscr{G}_{0,T}(m_0, \tilde{s}_0)$ is the strategic-form game $(\mathscr{I}, (\mathscr{A}_i, E\mathscr{R}_{i,T})_{i \in \mathscr{I}})$.

8.1.1 Solution Concept

We now introduce the key problems addressed in this chapter and some solution concepts.

8.1.2 Best-Response Correspondence

Given $a_{-i} := (a_1, \ldots, a_{i-1}, a_{i+1}, \ldots)$, the best-response problem associated with decision-maker i is

$$V_i(0, m_0, \tilde{s}_0) = \begin{cases} \sup_{a_i \in \mathscr{A}_i} E[\mathscr{R}_{i,T}(s_0, a, \tilde{s}_0)] \\ s(t) = s^{s_0, a, \tilde{s}_0}(t), \\ s(0) = s_0 \sim m_0, \ \tilde{s}(0) = \tilde{s}_0, \\ m(t, .) = m^{m_0, a, \tilde{s}_0}(t, .) = \mathbb{P}_{s^{s_0, a, \tilde{s}_0}(t)}. \end{cases} \tag{8.1}$$

The strategies a_i of i that achieve the above best-response value $V_i(0, s_0, \tilde{s}_0)$ are called best-response strategies of i to $a_{-i} := (a_j)_{j \neq i}$. The set of such strategies is denoted by $BR_i(a_{-i})$. We are interested in characterizing the best-response strategies of every decision-maker.

8.1.3 Mean-Field-Type Equilibria

A Nash equilibrium point is a strategy profile a such that for every decision-maker i, $a_i \in BR_i(a_{-i})$, and the resulting mean-field is $m(t, .) = \mathbb{P}_{s^{s_0, a, \tilde{s}_0}(t)}$. We are interested in characterizing equilibrium strategies of every decision-maker. As in classical games, the structure of the information available to the decision-makers is very important in establishing best-response strategies. We will investigate both open-loop strategies and state-and-mean-field feedback strategies.

8.1.4 Well-Posedness of the State Equation

As the state dynamics is a stochastic differential equation of mean-field type, the existence and uniqueness of a solution to the state equation are not always guaranteed. Below, we provide sufficient conditions for the existence and uniqueness of a solution.

Existence
If the coefficient functions b, σ, μ, δ are continuously differentiable with respect to s, m for each switching regime $\tilde{s}$ and the Gâteaux derivatives are bounded, then for each strategy profile $(a_i)_{i \in \mathscr{I}} \in \prod_{i \in \mathscr{I}} \mathscr{A}_i$, there is a unique solution to the state dynamics which will be denoted by $s(t) := s^{s_0, a, \tilde{s}_0}(t)$.

The proof of the existence result follows immediately by choosing a strategy profile $(a_i)_{i\in\mathcal{I}} \in \prod_{i\in\mathcal{I}} \mathcal{A}_i$, as coefficients are assumed bounded.

The next lemma provides Itô's formula for a jump-diffusion process with regime switching. We omit the proof and refer to [373].

Lemma 8.1 *Choose a test function f and set $y(t) := f(t, s(t), \tilde{s}(t))$. Then*

Jump-Diffusion Regime-Switching Integration Formula

$$\begin{aligned}
&f(T, s(T), \tilde{s}(T)) - f(0, s(0), \tilde{s}(0))\\
&= \int_0^T \{f_t + b.f_s + \tfrac{\sigma^2}{2} f_{ss} + J[f] + \tilde{S}[f]\}dt\\
&+ \int_0^T \sigma f_s dB\\
&+ \int_0^T \int_\Theta \{f(t, s(t_-) + \mu(t_-, ., \theta), \tilde{s}(t_-)) - f(t, s(t_-), \tilde{s}(t_-))\}\tilde{N}(dt, d\theta, \tilde{s}(t)),\\
&+ \int_0^T \sum_k [f(t, s(t_-) + \delta(.), k) - f(t, s(t_-), \tilde{s})]\tilde{M}(dt, k),
\end{aligned}$$

where the scripts t, s of f, f_t, f_s denote the derivatives with respect to t, s, f_{ss} denotes the second derivative with respect s, the jump operator is

$$J[f] := \int_\Theta \{f(t, s(t_-) + \mu(t_-, ., \theta)) - f(t, s(t_-)) - f_s.\mu(t, ., \theta)\}\nu(d\theta),$$

the switching operator is

$$\tilde{S}[f] := \sum_k [f(t, s(t_-) + \delta(t, .), k) - f(t, s(t_-), \tilde{s}) - f_s.\delta(t, .)]q_{\tilde{s}k},$$

and the martingale term of the switching process $\tilde{s}$ is

$$\tilde{M}(t, \tilde{s}) = \sum_j \sum_{t'\in[0,t)} \mathbb{1}_{\{\tilde{s}(t')=j\}} \mathbb{1}_{\{\tilde{s}(t'_-)=\tilde{s}\}} - \int_0^t q_{\tilde{s}j}(.)\mathbb{1}_{\{\tilde{s}(t'_-)=\tilde{s}\}},$$

and $q_{\tilde{s}} = \sum_{\tilde{s}'\neq\tilde{s}} \int_0^t q_{\tilde{s}\tilde{s}'}(t')\mathbb{1}_{\{\tilde{s}(t'_-)=\tilde{s}\}}dt'$ be the measure associated to $\tilde{s}$.

Let O be a linear operator. We define O^* the adjoint of the measure-dependent operator as

$$\langle O[f], m\rangle = \langle f, O^*[m]\rangle.$$

Lemma 8.2 *The probability distribution of the state solves (in the weak sense) as follows:*

Fokker-Planck-Kolmogorov Equation of Mean-Field Type

$$\begin{aligned} & m_t = -(bm)_s + \tfrac{1}{2}(\sigma^2 m)_{ss} + J^*[m] + \tilde{S}^*[m], \\ & m(0,.) = m_0(.), \end{aligned} \tag{8.2}$$

where J^* is the adjoint operator of J and $\tilde{S}^*$ is the adjoint of $\tilde{S}$.

Proof For the proof, we use a standard expected value approach in the sense of distribution using integration by parts. Let f be a test function over $[0, T] \times S \times \tilde{S}$. On one hand, one has

$$\begin{aligned} & E[f(t, s(t), \tilde{s}(t)) - f(0, s(0), \tilde{s}(0))|.] \\ & = \int_0^t Ef_t(t', s(t'), \tilde{s}(t'))dt' \\ & + \int_0^t \langle f(t', s(t'), \tilde{s}(t')), -m_t - (bm)_s + \tfrac{1}{2}(\sigma^2 m)_{ss} + J^*[m] + \tilde{S}^*[m] \rangle dt'ds. \end{aligned}$$

On the other hand, one gets

$$\begin{aligned} & E[f(t, s(t), \tilde{s}(t)) - f(0, s(0), \tilde{s}(0))|.] \\ & = \int_0^t Ef_t(t', s(t'), \tilde{s}(t'))dt' + \int_0^t \langle f(t', s, .), m_t(t', .) \rangle ds dt'. \end{aligned}$$

By identification, we obtain

$$m_t = -(bm)_s + \frac{1}{2}(\sigma^2 m)_{ss} + J^*[m] + \tilde{S}^*[m],$$

and this completes the proof. ■

Equation (8.2) together with the switching regime information will be used to construct a state of the game.

Solution Approaches
We now present two different solution approaches for solving MFTG problems. We also provide some relationships between the two methods.

8.1.5 Dynamic Programming Principle

In contrast to classical differential games or classical mean-field games in which one can directly use the standard dynamic programming principle (DPP) on $s(t)$, here, the presence of the term $m(t, .)$ in the functions $r_i, g_i, b, \sigma, \mu, \delta$ may create a time inconsistency, and $s(t)$ is not the appropriate state for DPP. The next proposition identifies an appropriate state space under which DPP can be obtained and a Hamilton-Jacobi-Bellman (HJB) equation can be derived.

Proposition 8.1 *Introduce the functional*

$$\hat{H}_i(t, s, m, a, \tilde{s}, \hat{V}_{sm}, \hat{V}_{ssm}, \hat{V}_m) := r_i + b\hat{V}_{i,sm} + \frac{\sigma^2}{2}\hat{V}_{i,ssm} + J[\hat{V}_{i,m}] + \tilde{S}[\hat{V}_{i,m}].$$

Hamilton-Jacobi-Bellman System with Drift-Jump-Difffusion and Regime Switching

If there exists $\hat{V}$ *such that*

$$\begin{cases} i \in \mathscr{I}, \\ \hat{V}_{i,t} + \int_s \left\{\sup_{a_i \in \mathscr{A}_i} \hat{H}_i(s, m, a, \tilde{s})\right\} m(t, ds, \tilde{s}) = 0, \\ \hat{V}_i(T, m, \tilde{s}) = \int_s m(T, ds, \tilde{s}) g_i(s, m(T, .), \tilde{s}), \\ m(t, ., .) = \mathbb{P}_{s(t)}, \text{ solves (8.2)}, \end{cases} \tag{8.3}$$

then $\hat{V}_i(0, m_0, \tilde{s}_0)$ *is an equilibrium payoff, and the optimal strategy* a_i *of* i *maximizes* $r_i + b\hat{V}_{i,sm} + \frac{\sigma^2}{2}\hat{V}_{i,ssm} + J[\hat{V}_{i,m}] + \tilde{S}[\hat{V}_{i,m}]$.

Proof In order to use a dynamic programming principle, we look for an augmented state so that one has a Markovian system. The pair $(m, \tilde{s})$ represents the driving force of the objective functions. By rewriting the payoff functional $E\mathscr{R}_{i,T}$ as a function of the measures $m(t, .)$ and the strategies $a_i(.)$, we obtain a classical dynamic game setup, but in the space of measures. Hence, we only need the evolution of the measure $m(t, .)$ which is given by (8.2). Then, in a vectorial sense $(m(t, ds, \tilde{s}))_{\tilde{s} \in \tilde{\mathscr{S}}}$, one obtains a deterministic differential game with m as state (in infinite dimensions). The best response of decision-maker i is

$$\hat{V}_i(0, m_0, \tilde{s}_0) = \begin{cases} \sup_{a_i \in \mathscr{A}_i} E[\mathscr{R}_{i,T}(m_0, a, \tilde{s})] \\ m(t, .) \text{ solves (8.2)}, \\ m_0. \end{cases}$$

Applying the classical DPP to the deterministic problem (with respect to the measure m) yields

$$\hat{V}_{i,t} + \sup_{a_i \in \mathscr{A}_i} [\int_s r_i m(t, ds, \tilde{s}) + \langle \tilde{b}, \hat{V}_{i,m} \rangle] = 0,$$

where $\tilde{b}(m)(t, s, \tilde{s}) = -(bm)_s + \frac{1}{2}(\sigma^2 m)_{ss} + J^*[m] + \tilde{S}^*[m]$. We apply the theory of distributions to establish the following equalities:

$$\langle \tilde{b}, \hat{V}_{i,m} \rangle = \int_s \tilde{b}(m) \hat{V}_{i,m}\, ds, \tag{8.4}$$

$$= \int_s -(bm)_s \hat{V}_{i,m} + \frac{1}{2}(\sigma^2 m)_{ss} \hat{V}_{i,m} + J^*[m]\hat{V}_{i,m} + \tilde{S}^*[m]\hat{V}_{i,m} ds,$$

$$= \int_s \left\{ b\hat{V}_{i,sm} + \frac{\sigma^2}{2} \hat{V}_{i,ssm} + J[\hat{V}_{i,m}] + \tilde{S}[\hat{V}_{i,m}] \right\} m(t, ds, \tilde{s}), \tag{8.5}$$

which is obtained by integration by parts (in the sense of distribution). This completes the proof. ■

Remark 8.1 The optimal strategy of decision-maker i maximizing the term

$$r_i + b\hat{V}_{i,sm} + \frac{\sigma^2}{2} \hat{V}_{i,ssm} + J[\hat{V}_{i,m}] + \tilde{S}[\hat{V}_{i,m}],$$

is generically a functional of $s, m, \tilde{s}$ which is called a state-and-mean-field-type feedback strategy $a_i^*(t, s, \tilde{s}, m, \hat{V}_m, \hat{V}_{sm}, \hat{V}_{ssm})$. The presence of the term $V_{i,m}$ is due to the jump term and the switching whose appearance as integral of $V_{i,m}$ in the dual Bellman system.

Proposition 8.1 provides a sufficient condition for equilibria in state-and-mean-field feedback form, i.e., $a_i(t, s, m, \tilde{s})$. However, the existence of a classical solution to the infinite-dimensional HJB equation is not guaranteed, and the value function may not be differentiable as in the classical case. A weaker notion of solution, such as viscosity solution using weak sub/super-differential set, will be introduced in the next subsection.

Remark 8.2 In classical stochastic control problems (with mean-field-free data), it is well known that optimizing subject to state dynamics is equivalent to optimizing subject to the Fokker-Planck-Kolmogorov equation. By contrast, in mean-field-type optimization, these two approaches are not equivalent, as we will show in the next subsection.

8.1.6 Dual Functions

Since the DPP equilibrium system (8.3) is in infinite dimensions, the solvability of such a partial-integrodifferential equation poses some technical difficulties. However, when the value function $\hat{V}$ is weakly differentiable with respect to (s, m) then a finite-dimensional quantity can be obtained as it is in the dual space.

Definition 8.1 The function $\hat{v}^*(t, s, \tilde{s}) = \hat{V}_m(t, s, \tilde{s})$ is the dual function associated with $\hat{V}$.

Proposition 8.2 *If $\hat{V}_m, \hat{V}_{sm}, \hat{V}_{ssm}$ exist and are continuous, then the equilibrium dual function $\hat{v}^* = \hat{V}_m$ solves the following system of partial-integrodifferential equations:*

$$\begin{cases} i \in \mathscr{I}, \\ \hat{v}^*_{i,t} + H_i + EH_{i,m} = 0, \\ \hat{v}^*_i(T, s, \tilde{s}) = g_i(s, m(T, .), \tilde{s}) + Eg_{i,m}, \\ m_t = -(bm)_s + \frac{1}{2}(\sigma^2 m)_{ss} + J^*[m] + \tilde{S}^*[m], \\ m(0, .) = m_0(.), \end{cases} \tag{8.6}$$

where $H_i(t, s, \tilde{s}, m, \hat{v}^*_i, \hat{v}^*_{i,s}, \hat{v}^*_{i,ss}) = \sup_{a_i} [r_i + b\hat{v}^*_{i,s} + \frac{\sigma^2}{2}\hat{v}^*_{i,ss} + J[\hat{v}^*_i] + \tilde{S}[\hat{v}^*_i]]$. The best-response strategy in state-and-mean-field feedback strategies is $a^*_i(s, \tilde{s}, m, \hat{v}^*_i, \hat{v}^*_{i,s}, \hat{v}^*_{i,ss})$ maximizing H_i.

Proof As $\hat{V}_m, \hat{V}_{sm}, \hat{V}_{ssm}$ exist and are continuous, we examine the HJB system (8.3) in the direction $\tilde{m} = m$ to obtain the following system:

$$\begin{cases} i \in \mathscr{I}, \\ (\hat{V}_{i,t})_m + \sup_{a_i} \hat{H}_i + \int_{s'} [\sup_{a_i} \hat{H}_i]_m(t, s', m, .)(s) m(t, ds', .) = 0, \\ \hat{V}_{i,m}(T, s, \tilde{s}) = g_i(s, m(T, .), \tilde{s}) \\ + \int_{s'} m(T, ds', \tilde{s}) g_{i,m}(s', m(T, .), \tilde{s}), \\ m(t, .) = \mathbb{P}_{s(t)}. \end{cases} \tag{8.7}$$

Noting that $(\hat{V}_{i,t})_m = \hat{v}^*_{i,t}$, $\hat{V}_{i,sm} = \hat{v}^*_{i,s}$ and $\hat{V}_{i,ssm} = \hat{v}^*_{i,ss}$, the latter system (8.7) becomes the announced one. This completes the proof. ∎

The dual function $\hat{v}^*_i(0, s_0, \tilde{s}_0)$ is not the equilibrium payoff of the decision-maker i. The dual function plays an important role in establishing a stochastic maximum principle, as we will see in the next subsection.

Corollary 8.1 (Mean-Field-Free Equations with Jump and Regime Switching) *If $r_i, g_i, b, \sigma, \mu, \delta$ are all independent of m, then $\hat{V}_i(t, m, \tilde{s}) = \int_s v_i(t, s, \tilde{s})\, m(t, ds, \tilde{s}) = E_{m(t,.)} v_i(t, s(t), \tilde{s})$, where the function $(v_i(t, s, \tilde{s}))_i$ solves the mean-field-free Bellman system with jump and switching regime:*

$$\begin{cases} i \in \mathscr{I}, \\ v_{i,t} + H_i = 0, \\ v_i(T, s, \tilde{s}) = g_i(s, \tilde{s}), \end{cases} \tag{8.8}$$

where $v_i = \hat{V}_{i,m}$ is mean-field free and

$$H_i = \sup_{a_i} \left\{ r_i + bv_{i,s} + \frac{\sigma^2}{2} v_{i,ss} + J[v_i] + \tilde{S}[v_i] \right\}.$$

Moreover v_i is the equilibrium payoff of i and solves

$$v_i(0, s_0, \tilde{s}_0) = \begin{cases} \sup_{a_i \in \mathcal{A}_i} E[\mathcal{R}_{i,T}(s_0, a, \tilde{s}_0)] \\ s(t) = s^{s_0, a, \tilde{s}_0}(t), \\ s(0) = s_0, \tilde{s}(0) = \tilde{s}_0. \end{cases} \tag{8.9}$$

The dual function $\hat{v}^*(t, s, \tilde{s}) = \hat{V}_m(t, s, \tilde{s})$ coincides with the function $v(t, s, \tilde{s})$ in the mean-field-free case. However, the two functions do not need to coincide in the mean-field-dependent setting.

8.1.7 Stochastic Maximum Principle

We present first-order and second-order adjoint processes that are useful in establishing necessary conditions for mean-field-type equilibria. The advantage of the adjoint processes is that they solve a class of linear backward SDEs of mean-field type. The following result provides existence and uniqueness conditions.

Lemma 8.3 *Consider the following mean-field backward SDE:*

$$\begin{aligned} \bar{p}(t) &= \bar{p}(T) + \int_t^T \tilde{E}[\hat{f}(t', \tilde{\bar{p}}(t'), \tilde{\bar{q}}(t'), \tilde{\bar{r}}(t'), \tilde{\bar{s}}(t')), \bar{p}(t'), \bar{q}(t'), \bar{r}(t'), \bar{s}(t'))]\, dt' \\ &- \int_t^T \bar{q}(t')dB(t') - \int_t^T \int_\Theta \bar{r}(t', \theta)\tilde{N}(dt', d\theta) - \int_t^T \bar{s}(t')\tilde{M}(dt'), \end{aligned} \tag{8.10}$$

where $\bar{p}(T)$ is a progressively measurable, square integrable random variable. Let $\hat{f}(t, ., ., ., .)$ be Lipschitz for all time $t \in [0, T]$ and $t \mapsto \hat{f}(t, 0, 0, 0, 0)$ be the square integrable over $[0, T]$. Then, the mean-field backward SDE has a unique adapted solution satisfying

$$E\left[\sup_{t\in[0,T]}|\bar{p}(t)|^2 + \int_0^T |\bar{q}(t)|^2 + \int_\Theta |\bar{r}(t, \theta)|^2 \nu(d\theta) + \sum_k |\bar{s}(t, k)|^2 q_k dt\right] < \infty. \tag{8.11}$$

We introduce the following functional

$$\bar{H}_i := r_i + b\hat{p}_i + \sigma\hat{q}_i + \int_\Theta \mu\hat{r}_i(., \theta)\nu(d\theta) + \sum_k \delta\hat{s}_i(., k)q_k.$$

Consider the first-order adjoint processes $(\bar{p}_i, \bar{q}_i, \bar{r}_i, \bar{s}_i)_{i\in\mathcal{I}}$

$$\begin{cases} d\bar{p}_i = -[\bar{H}_{i,s} + \tilde{E}H_{i,sm}]dt + \bar{q}_i dB + \int_\Theta \bar{r}_i \tilde{N}(dt, d\theta) + \bar{s}_i.d\tilde{M}(dt), \\ \bar{p}_i(T) = g_{i,s}(T) + \tilde{E}\{g_{i,sm}(T)\}, \\ i \in \mathcal{I}. \end{cases} \tag{8.12}$$

Lemma 8.4 *Let the functions $b, \sigma, \mu, \delta, r_i, g_i$ be continuously differentiable with respect to (s, m) and all their first-order derivatives with respect to (s, m) be continuous in (s, m, a) and be bounded. Then, the first-order adjoint system is a linear SIDE with almost surely bounded coefficient functions. Further, there is a unique $\mathscr{F}$-adapted solution such that*

$$E\left[\sup_{t\in[0,T]}|\bar{p}_i(t)|^2 + \int_0^T \{|\bar{q}_i(t)|^2 + \int_\Theta |\bar{r}_i(t,\theta)|^2\nu(d\theta) + |\bar{s}_i|^2\}dt\right] < +\infty.$$

Proof By choosing

$$\hat{f}(t, \tilde{\bar{p}}(t), \tilde{\bar{q}}(t), \tilde{\bar{r}}(t), \tilde{\bar{s}}(t), \bar{p}(t), \bar{q}(t), \bar{r}(t), \bar{s}(t)) =$$

$$\alpha_0(t, .) + \alpha_1(t, .)\tilde{\bar{p}}(t) + \alpha_2(t, .)\tilde{\bar{q}}(t) + \langle\alpha_3(t, .), \tilde{\bar{r}}(t)\rangle + \langle\alpha_4(t, .).\tilde{\bar{s}}(t)\rangle$$

$$+ \alpha_5(t, .)\bar{p}(t) + \alpha_6(t, .)\bar{q}(t) + \int_\Theta \alpha_7(t, \theta)\bar{r}(t, \theta)\nu(d\theta) + \sum_k \alpha_8(t, k).\bar{s}(t, k)q_k,$$

where $\alpha_j(t, .)$, $j \in \{1, 2, \ldots, 8\}$ are measurable bounded coefficient functions, one gets a backward equation in the form of the adjoint equations. Under the assumptions imposed on $b, \sigma, \mu, \delta, r_i, g_i$, the coefficients $\{\alpha_i\}_i$ fulfill the requirement, and the result follows as a direct application of Lemma 8.3. ■

These strong smoothness conditions on $b, \sigma, \mu, \delta, r_i, g_i$ can be considerably weakened using representations of weak sub/super-differential sets.

Note that for linear-quadratic MFTGs, the payoff function $r_i = -s^2 - \int y^2 m(t, dy) - a_i^2 - \int_{a_i'} (a_i')^2 \mathbb{P}_{a_i}(da_i')$ does not satisfy the above assumptions. In that case, one can relax the boundedness assumption and replace it by L^2 estimates if the second moment is finite.

Consider the second-order adjoint processes $(\bar{P}_i, \bar{Q}_i, \bar{R}_i, \bar{S}_i)_{i\in\mathscr{I}}$

$$\begin{cases} d\bar{P}_i = -\{\bar{H}_{i,ss} + \tilde{E}H_{i,ssm}\}dt \\ -\{2b_s + \sigma_s^2 + \int_\Theta \mu_s^2\nu(d\theta) + \sum_j \delta_s^2 q_j\}\bar{P}_i dt \\ -2\sigma_s\bar{Q}_i dt - \int_\Theta (2\mu_s + \mu_s^2)\bar{R}_i(t, \theta, .)\nu(d\theta)dt \\ -\sum_k (2\delta_s + \delta_s^2)\bar{S}_i(t, .)q_k dt, \\ +\bar{Q}_i dB + \int_\Theta \bar{R}_i(t, \theta)\tilde{N}(dt, d\theta) + \bar{S}_i.\tilde{M}(dt) \\ \bar{P}_i(T) = g_{i,ss}(T) + \tilde{E}\{g_{i,ssm}(T)\}, \\ i \in \mathscr{I}. \end{cases} \tag{8.13}$$

Lemma 8.5 *Let the functions $b, \sigma, \mu, \delta, r_i, g_i$ be twice continuously differentiable with respect to (s, m) and all their derivatives up to the second order with respect* *to* (s, m) *be continuous in* (s, m, a) *and be bounded. Then, the second-order adjoint system (8.13) is a linear backward SDE with almost surely bounded coefficient functionals. Further, there is a unique* $\mathscr{F}$*-adapted solution such that*

$$E\Big[\sup_{t\in[0,T]}|\bar{P}_i(t)|^2 + \int_0^T |\bar{Q}_i(t)|^2 + \int_\Theta |\bar{R}(t,\theta)|^2\nu(d\theta) \\ + \sum_k q_k|\bar{S}_i(t,k)|^2 dt\Big] < +\infty.$$

Proof The result readily follows from Lemma 8.3. ■

Proposition 8.3 (Stochastic Maximum Principle)

If (s, a) is a pair of equilibrium state and equilibrium strategy, then there is a vector of $\mathbb{F}$-adapted processes $(\bar{p}_i, \bar{q}_i, \bar{r}_i, \bar{s}_i, \bar{P}_i, \bar{Q}_i, \bar{R}_i, \bar{S}_i)_{i\in\mathscr{I}}$ such that

$$\begin{aligned} &\bar{H}_i(s,m,a,\tilde{s}) - \bar{H}_i(s,m,a_i',a_{-i},\tilde{s}) \\ &+\tfrac{1}{2}\bar{P}_i(\sigma(s,m,a_i',a_{-i},\tilde{s}) - \sigma(s,m,a,\tilde{s}))^2 \\ &+\tfrac{1}{2}\int_\Theta[\bar{P}_i + \bar{R}_i(t,\theta)](\mu(s,m,a_i',a_{-i},.,\theta) - \mu(s,m,a,.,\theta))^2\nu(d\theta) \\ &+\tfrac{1}{2}\sum_k[\bar{P}_i + \bar{S}_i(t,k)](\delta(s,m,a_i',a_{-i},.) - \delta(s,m,a,.))^2 q_k \geq 0, \end{aligned} \tag{8.14}$$

for all $a_i' \in \mathscr{A}_i$, almost every time $t \in [0,T]$ and $\mathbb{P}$-almost surely.

In particular, if σ, μ, δ *are independent of* a_i *then*

$$\sup_{a_i'\in\mathscr{A}_i} \bar{H}_i(s,m,a_i',a_{-i},\tilde{s}) \leq \bar{H}_i(s,m,a,\tilde{s}).$$

8.1.8 Relationships Between DPP and SMP

In order to provide weaker conditions, we introduce super- and sub-differentials of the viscosity solution $\hat{V}$. Define the first-order super-differential of the dual function $\hat{V}_m$ as

$$D_{t,w}^{1,+}\hat{V}_m(t,w) = \Big\{(d_1,d_2) \in \mathbb{R}\times T(\mathscr{S}) \mid \\ \limsup_{t',w'\to t,w} \frac{\hat{V}_m(t',w') - \hat{V}_m(t,w) - d_1(t'-t) - d_2(w'-w)}{|t'-t| + \big| w'-w \big|} \leq 0\Big\}. \tag{8.15}$$

Similarly, the first-order sub-differential of the function $\hat{V}_m$ is

$$D^{1,-}_{t,w}\hat{V}_m(t,w) = \left\{\begin{array}{l}(d_1,d_2)\in\mathbb{R}\times T(\mathscr{S}) \mid \\ \liminf_{t',w'\to t,w} \frac{\hat{V}_m(t',w')-\hat{V}_m(t,w)-d_1(t'-t)-d_2(w'-w)}{|t'-t|+\left|w'-w\right|} \geq 0\end{array}\right\}. \tag{8.16}$$

Using these weak derivatives, one has a relationship between the stochastic maximum principle and the dynamic programming principle in terms of inclusion: $\bar{p}(t) \in D^{1,+}_s \hat{V}_m(t,w)$. In particular, if $\hat{V}$ is Gâteaux differentiable with respect to m and s, then

$$\begin{cases} \bar{p}_i(t) = \hat{V}_{i,sm}(s(t), m(t), \tilde{s}(t)) = d_2, \\ \bar{q}_i(t) = \sigma \hat{V}_{i,ssm}(s(t), m(t), \tilde{s}(t)). \end{cases}$$

Proposition 8.4 (Process Identification)

If the first- and second-order weak derivatives with respect to s of $\hat{V}_m$ exist, then the process $\hat{p}_i(t) = \hat{V}_{i,sm}(s(t), \tilde{s}(t))$ solves the backward SDE:

$$\begin{array}{l} d\hat{p}_i = -[\bar{H}_{i,s} + \tilde{E}H_{i,sm}]dt + \hat{q}_i dB + \int_\Theta \hat{r}_i(t,\theta)\tilde{N}(dt,d\theta) + \hat{s}_i.\tilde{M}(dt), \\ \bar{H}_{i,s} := r_{i,s} + b_s\hat{p}_i + \sigma_s\hat{q}_i + \int_\Theta \mu_s\hat{r}_i(.,\theta)\nu(d\theta) + \sum_k \delta_s\hat{s}_i(.,k)q_k, \\ \hat{p}_i = \hat{v}^*_{i,s} = \hat{V}^*_{i,sm} = \bar{p}_i, \\ \hat{q}_i = \sigma\hat{v}^*_{i,ss} = \sigma\hat{V}^*_{i,ssm} = \bar{q}_i, \\ \hat{r}_i(.,\theta) = \hat{v}^*_{i,s}(s+\mu,.) - \hat{v}^*_{i,s} = \hat{V}^*_{i,sm}(s+\mu,.) - \hat{V}^*_{i,sm} = \bar{r}_i, \\ \hat{s}(.,k) = \hat{v}^*_{i,s}(s+\delta,k) - \hat{v}^*_{i,s} = \hat{V}^*_{i,sm}(s(t_-)+\delta,k) - \hat{V}^*_{i,sm} = \bar{s}_i(.,k). \end{array} \tag{8.17}$$

The results from Proposition 8.4 are very important since they establish connections between DPP and SMP approaches.

Proof We provide a proof in the smooth case. Let $\hat{v}^*_{i,s}(t,s,\tilde{s}) = \hat{V}_{i,sm}(t,s,\tilde{s})$ be continuously differentiable in the first argument and twice continuously differentiable in the second argument for each switching regime $\tilde{s}$. Itô's formula applied to $\hat{p}_i(t) = \hat{v}^*_{i,s}(t, s(t), \tilde{s}(t))$ yields

$$\begin{cases} d\hat{p}_i = d\hat{v}^*_{i,s} \\ = [\hat{v}^*_{i,ts} + b.\hat{v}^*_{i,ss} + \frac{\sigma^2}{2}\hat{v}^*_{i,sss} + J[\hat{v}^*_{i,s}] + \tilde{S}[\hat{v}^*_{i,s}]]dt \\ +\sigma\hat{v}^*_{i,ss}dB \\ + \int_\Theta\{\hat{v}^*_{i,s}(s+\mu(t_-,.,\theta),\tilde{s}) - \hat{v}^*_{i,s}(s,\tilde{s})\}\tilde{N}(dt,d\theta,\tilde{s}), \\ + \sum_k[\hat{v}^*_{i,s}(s+\delta,k) - \hat{v}^*_{i,s}(s,\tilde{s})]\tilde{M}(dt,k), \end{cases} \tag{8.18}$$

We identify the coefficient processes $\hat{q}_i = \sigma \hat{v}^*_{i,ss} = \sigma \hat{V}_{i,ssm}$. It remains to find the drift coefficient of $\hat{p}_i$. Set $-\alpha := \hat{v}^*_{i,ts} + b.\hat{v}^*_{i,ss} + \frac{\sigma^2}{2}\hat{v}^*_{i,sss} + J[\hat{v}^*_{i,s}] + \tilde{S}[\hat{v}^*_{i,s}]$. By differentiating with respect to s the system (8.6) of $\hat{v}^*_i$, we obtain

$$\hat{v}^*_{i,st} + H_{i,s}(t, s, \tilde{s}, m, \hat{v}^*_i, \hat{v}^*_{i,s}, \hat{v}^*_{i,ss}) + EH_{i,sm} = 0.$$

By expanding the terms in $H_{i,s}$, one gets

$$\begin{cases} H_i = \sup_{a_i} [r_i + b\hat{v}^*_{i,s} + \frac{\sigma^2}{2}\hat{v}^*_{i,ss} + J[\hat{v}^*_i] + \tilde{S}[\hat{v}^*_i]], \\ \\ H_{i,s} = r_{i,s} + b_s(\hat{v}^*_{i,s}) + b\hat{v}^*_{i,ss} + \frac{\sigma^2}{2}\hat{v}^*_{i,sss} + \sigma_s(\sigma\hat{v}^*_{i,ss}) \\ +(J[\hat{v}^*_i] + \tilde{S}[\hat{v}^*_i]])_s. \end{cases}$$

Expanding the derivatives of the jump and the switch terms, we arrive at

$$\begin{cases} H_{i,s} = r_{i,s} + b_s(\hat{v}^*_{i,s}) + b\hat{v}^*_{i,ss} + \frac{\sigma^2}{2}\hat{v}^*_{i,sss} + \sigma_s(\sigma\hat{v}^*_{i,ss}) \\ +J[\hat{v}^*_{i,s}] + \int_\Theta \mu_s[\hat{v}^*_{i,s}(s+\mu, .) - \hat{v}^*_{i,s}]\nu(d\theta) \\ +\tilde{S}[\hat{v}^*_{i,s}] + \sum_k q_k\delta_s[\hat{v}^*_{i,s}(s+\delta, k) - \hat{v}^*_{i,s}]. \end{cases}$$

Summing together, one obtains the following equality:

$$\begin{cases} 0 = \hat{v}^*_{i,st} + \partial_s H_i + EH_{i,sm} \\ = \hat{v}^*_{i,st} + EH_{i,sm} \\ +r_{i,s} + b_s(\hat{v}^*_{i,s}) + b\hat{v}^*_{i,ss} + \frac{\sigma^2}{2}\hat{v}^*_{i,sss} + \sigma_s(\sigma\hat{v}^*_{i,ss}) \\ +J[\hat{v}^*_{i,s}] + \int_\Theta \mu_s[\hat{v}^*_{i,s}(s+\mu, .) - \hat{v}^*_{i,s}]\nu(d\theta) \\ +\tilde{S}[\hat{v}^*_{i,s}] + \sum_k q_k\delta_s[\hat{v}^*_{i,s}(s+\delta, k) - \hat{v}^*_{i,s}]. \end{cases}$$

We deduce that

$$\begin{cases} \hat{v}^*_{i,ts} + b.\hat{v}^*_{i,ss} + \frac{\sigma^2}{2}\hat{v}^*_{i,sss} + J[\hat{v}^*_{i,s}] + \tilde{S}[\hat{v}^*_{i,s}] \\ = -\{r_{i,s} + b_s(\hat{v}^*_{i,s}) + \sigma_s(\sigma\hat{v}^*_{i,ss}) \\ + \int_\Theta \mu_s[\hat{v}^*_{i,s}(s+\mu, .) - \hat{v}^*_{i,s}]\nu(d\theta) \\ + \sum_k \delta_s[\hat{v}^*_{i,s}(s+\delta, k) - \hat{v}^*_{i,s}]q_k + EH_{i,sm}\}. \end{cases} \tag{8.19}$$

Thus, the process $\hat{p}_i$ solves

$$\begin{cases} d\hat{p}_i = d\hat{v}^*_{i,s} \\ = -\{r_{i,s} + b_s(\hat{v}^*_{i,s}) + \sigma_s(\sigma\hat{v}^*_{i,ss})\}dt \\ - \int_\Theta \mu_s[\hat{v}^*_{i,s}(s+\mu,.) - \hat{v}^*_{i,s}]\nu(d\theta)dt \\ - \sum_k \delta_s[\hat{v}^*_{i,s}(s(t_-)+\delta,k) - \hat{v}^*_{i,s}]q_k dt \\ -EH_{i,sm}dt \\ +\sigma\hat{v}^*_{i,ss}dB \\ + \int_\Theta\{\hat{v}^*_{i,s}(s+\mu,\tilde{s}) - \hat{v}^*_{i,s}(s,\tilde{s})\}\tilde{N}(dt,d\theta) \\ + \sum_k[\hat{v}^*_{i,s}(s+\delta,k) - \hat{v}^*_{i,s}(s,\tilde{s})]\tilde{M}(dt,k). \end{cases} \tag{8.20}$$

The latter stochastic differential equation can be rewritten as

$$\begin{array}{l} d\hat{p}_i = -[\bar{H}_{i,s} + \tilde{E}H_{i,sm}]dt + \hat{q}_i dB \\ + \int_\Theta \hat{r}_i(t,\theta)\tilde{N}(dt,d\theta) + \hat{s}_i.\tilde{M}(dt) \end{array},$$

where

$$\bar{H}_{i,s} = r_{i,s} + b_s\hat{p}_i + \sigma_s\hat{q}_i + \int_\Theta \mu_s\hat{r}(.,\theta)\nu(d\theta) + \sum_k \delta_s\hat{s}(.,k)q_k,$$

$$\begin{array}{l} \hat{p}_i = \hat{v}^*_{i,s}, \\ \hat{q}_i = \sigma\hat{v}^*_{i,ss}, \\ \hat{r}(.,\theta) = [\hat{v}^*_{i,s}(s+\mu,.) - \hat{v}^*_{i,s}] \\ \hat{s}(.,k,.) = [\hat{v}^*_{i,s}(s(t_-)+\delta,k) - \hat{v}^*_{i,s}]\mathbb{1}_{\{\tilde{s}\to k\}}. \end{array}$$

The processes $(\bar{p}_i, \bar{q}_i, \bar{r}_i, \bar{s}_i)_i$ and $(\hat{p}_i = \hat{v}^*_{i,s}, \hat{q}_i = \sigma\hat{v}^*_{i,ss}, \hat{r}_i, \hat{s}_i)_i$ solve the same backward SIDE system with the same terminal condition. Under the assumptions above, these two processes are almost surely identical. This completes the proof. ■

Second-Order Adjoint Process

Since A_i is not necessarily convex, the first-order adjoint process is not enough to describe an optimality system. We use a second-order adjoint system to derive a necessary condition for optimality. Define the second-order super-differential as $D^{1,2,+}_{t,w}\hat{V}_m(t,w) = \{(d_1,d_2,d_3) \in \mathbb{R}^3 \mid$

$$\limsup_{t',w'\to t,w} \frac{\hat{V}_m(t',w') - V_m(t,w) - d_1(t'-t) - d_2(w'-w) - \frac{1}{2}(w'-w)'d_3(w'-w)}{|t'-t| + \left|w'-w\right|^2} \leq 0.\}$$

Similarly, the second-order sub-differential is $D^{1,2,-}_{t,w}\hat{V}_m(t,w) = \{(d_1,d_2,d_3) \in \mathbb{R}^3 \mid$

$$\liminf_{t',w'\to t,w} \frac{\hat{V}_m(t',w') - V_m(t,w) - d_1(t'-t) - d_2(w'-w) - \frac{1}{2}(w'-w)'d_3(w'-w)}{|t'-t| + \left|w'-w\right|^2} \geq 0.\}$$

$\hat{V}_i$ is a viscosity solution if and only if $\hat{V}_i(T,m,.) = \int_s g_i m(T,ds,.)$ for any measure m and for all $(t,m) \in [0,T) \times \Delta(\mathcal{S})$

$$\begin{cases} d_1 + \int \sup_{a_i(.)\in\mathscr{A}_i} H_i(t,s,\tilde{s},m,a,d_2,d_3) \leq 0, \ \forall\, d \in D^{1,2,+}_{t,w}\hat{V}_{i,m}(t,w), \\ d_1 + \int \sup_{a_i(.)\in\mathscr{A}_i} H_i \geq 0, \ \forall\, d \in D^{1,2,-}_{t,w}\hat{V}_{i,m}(t,w). \end{cases}$$

In particular, if $\hat{V}_{i,m} \in C^{1,3}$, then

$$\begin{cases} \bar{P}_i(t) = \hat{V}_{i,ssm}(t,s(t)) = d_3, \\ \bar{Q}_i(t) = \sigma \hat{V}_{i,sssm}(t,s(t)). \end{cases}$$

Proposition 8.5 (Identification of Second-Order Adjoint Processes)

If the first- and second-order weak derivatives with respect to s of $\hat{V}_m$ exist, then the process $\hat{P}_i(t) = \hat{V}_{i,ssm}(t,s(t),\tilde{s}(t))$ solves the backward SIDE:

$$\begin{array}{l} d\hat{P}_i = -\{\bar{H}_{i,ss} + \tilde{E}H_{i,ssm}\}dt \\ -\{2b_s + \sigma_s^2 + \int_\Theta \mu_s^2 \nu(d\theta) + \sum_k \delta_s^2 \hat{v}^*_{i,ss}(s,.)q_k\}\hat{P}_i dt \\ -2\sigma_s \hat{Q}_i dt - \int_\Theta (2\mu_s + \mu_s^2)\hat{R}_i(t,\theta,.)\nu(d\theta)dt \\ -\sum_k (2\delta_s + \delta_s^2)\hat{S}_i(t,k)q_k dt \\ +\hat{Q}_i dB + \int_\Theta \hat{R}_i(t,\theta)\tilde{N}(dt,d\theta,\tilde{s}) + \sum_k \hat{S}_i \tilde{M}(dt,k), \\ \hat{P}_i(T) = g_{i,ss}(s(T),m(T)) + \int_s g_{i,ssm} m(T,ds), \end{array},$$

where

$$\begin{array}{l} \bar{H}_{i,ss} := r_{i,ss} + b_{ss}\hat{P}_i + \sigma_{ss}\hat{Q}_i \\ + \int_\Theta \mu_{ss}\hat{R}_i(.,\theta,\tilde{s})\nu(d\theta) + \sum_k \delta_{ss}\hat{S}_i(.,k)q_k, \end{array}$$

and the relationships with dynamic programming are

$$\begin{array}{l} \hat{P}_i = \hat{v}^*_{i,ss} = \hat{V}^*_{i,ssm} = \bar{P}_i, \\ \hat{Q}_i = \sigma \hat{v}^*_{i,sss} = \sigma \hat{V}^*_{i,sssm} = \bar{Q}_i, \\ \hat{R}_i(.,\theta) = [\hat{v}^*_{i,ss}(s+\mu,.) - \hat{v}^*_{i,ss}] = [\hat{V}^*_{i,ssm}(s+\mu,.) - \hat{V}^*_{i,ssm}] = \bar{R}_i, \\ \hat{S}(.,) = [\hat{v}^*_{i,ss}(s+\delta,k) - \hat{v}^*_{i,ss}] = [\hat{V}^*_{i,ssm}(s+\delta,k) - \hat{V}^*_{i,ssm}] = \bar{S}_i. \end{array}$$

Proof We compute explicitly the second-order terms. Let $\tilde{g}_i(t,s) = \hat{V}_{i,ssm}(t,s) = \hat{v}^*_{i,ss}(t,s,.)$. Applying Itô's formula to $\hat{v}^*_{i,ss}(t,s,\tilde{s}). = \tilde{g}_i(t,s,\tilde{s})$ yields

$$\begin{cases} d\hat{P}_i = d\hat{v}^*_{i,ss} \\ = [\hat{v}^*_{i,tss} + b.\hat{v}^*_{i,sss} + \frac{\sigma^2}{2}\hat{v}^*_{i,ssss} + J[\hat{v}^*_{i,ss}] + \tilde{S}[\hat{v}^*_{i,ss}]]dt \\ +\sigma \hat{v}^*_{i,sss} dB \\ + \int_\Theta \{\hat{v}^*_{i,ss}(s+\mu,\tilde{s}) - \hat{v}^*_{i,ss}(s,\tilde{s})\}\tilde{N}(dt,d\theta,\tilde{s}) \\ + \sum_k [\hat{v}^*_{i,ss}(s+\delta,k) - \hat{v}^*_{i,ss}(s,\tilde{s})]\tilde{M}(dt,k). \end{cases} \tag{8.21}$$

We identify the coefficient processes $\hat{Q}_i = \sigma(\tilde{g}_i)_s = \sigma \hat{v}^*_{i,sss} = \sigma V_{i,sssm} = \bar{Q}_i$. It remains to find the drift coefficient of $\hat{P}_i$:

$$\begin{cases} \hat{P}_i = \hat{v}^*_{i,ss}(t,s) = \hat{V}_{i,ssm}(t,s), \\ (\tilde{g}_i)_t = \hat{v}^*_{i,tss}, \\ b(\tilde{g}_i)_s = bv^*_{i,sss}, \\ \frac{\sigma^2}{2}(\tilde{g}_i)_{ss} = \frac{\sigma^2}{2}\hat{v}^*_{i,ssss}. \end{cases}$$

The drift term is

$$\hat{v}^*_{i,tss} + b\hat{v}^*_{i,sss} + \frac{\sigma^2}{2}\hat{v}^*_{i,ssss} + J[\hat{v}_{i,ss}] + \tilde{S}[\hat{v}^*_{i,ss}]. \tag{8.22}$$

In order to identify the latter term, one can differentiate twice with respect to s Eq. (8.6) that is satisfied by the dual function $\hat{v}^*_i$:

$$\begin{cases} i \in \mathcal{I}, \\ 0 = \hat{v}^*_{i,sst} + r_{i,ss} + (b\hat{v}^*_{i,s})_{ss} + (\frac{\sigma^2}{2}\hat{v}^*_{i,ss})_{ss} \\ +(J[\hat{v}_i] + \tilde{S}[\hat{v}^*_i])_{ss} + \tilde{E}H_{i,ssm}, \\ \hat{v}^*_{i,ss}(T,m) = g_{i,ss}(s, m(T,.)) + \tilde{E}g_{i,ssm}, \\ m(t,.) = \mathbb{P}_{s(t)},\ m(0,.) = m_0(.). \end{cases} \tag{8.23}$$

$$\begin{cases} (bv^*_{i,s})_{ss} = (b_s v^*_{i,s} + bv^*_{i,ss})_s \\ = b_{ss}v^*_{i,s} + 2b_s v^*_{i,ss} + bv^*_{i,sss}, \\ (\frac{\sigma^2}{2}v^*_{i,ss})_{ss} = [\sigma_s\sigma v^*_{i,ss} + \frac{\sigma^2}{2}v^*_{i,sss}]_s \\ = (\sigma_{ss}\sigma + \sigma_s^2)v^*_{i,ss} + 2\sigma_s\sigma v^*_{i,sss} + \frac{\sigma^2}{2}v^*_{i,ssss}, \\ \\ (J[\hat{v}_i])_{ss} = \left(J[\hat{v}^*_{i,s}] + \int_\Theta \mu_s[\hat{v}^*_{i,s}(s+\mu,.) - \hat{v}^*_{i,s}]\nu(d\theta)\right)_s \\ = J[\hat{v}^*_{i,ss}] + 2\int_\Theta \mu_s[\hat{v}^*_{i,ss}(s+\mu,.) - \hat{v}^*_{i,ss}]\nu(d\theta) \\ + \int_\Theta \mu_{ss}[\hat{v}^*_{i,s}(s+\mu,.) - \hat{v}^*_{i,s}]\nu(d\theta) \\ + \int_\Theta \mu_s^2\hat{v}^*_{i,ss}(s+\mu,.)\nu(d\theta), \\ \\ (\tilde{S}[\hat{v}^*_i])_{ss} = \tilde{S}[\hat{v}^*_{i,ss}] + 2\sum_k \delta_s[\hat{v}^*_{i,ss}(s+\delta,.) - \hat{v}^*_{i,ss}]q_k \\ + \sum_k \delta_{ss}[\hat{v}^*_{i,s}(s+\delta,.) - \hat{v}^*_{i,s}]q_k + \sum_k \delta_s^2\hat{v}^*_{i,ss}(s+\delta,.)q_k. \end{cases}$$

We deduce that

$$
\begin{cases}
-\{v^*_{i,tss} + bv^*_{i,sss} + \frac{\sigma^2}{2} v^*_{i,ssss} + J[\hat{v}_{i,ss}] + \tilde{S}[\hat{v}^*_{i,ss}]\} \\
= r_{i,ss} + b_{ss} v^*_{i,s} + \sigma_{ss}(\sigma v^*_{i,ss}) \\
+ \int_\Theta \mu_{ss}[\hat{v}^*_{i,s}(s+\mu,.) - \hat{v}^*_{i,s}]\nu(d\theta) \\
+ \sum_k \delta_{ss}[\hat{v}^*_{i,s}(s+\delta,k) - \hat{v}^*_{i,s}]q_k \\
+(2b_s + \sigma_s^2)v^*_{i,ss} + 2\sigma_s(\sigma v^*_{i,sss}) + \tilde{E}H_{i,ssm} \\
+2\int_\Theta \mu_s[\hat{v}^*_{i,ss}(s+\mu,.) - \hat{v}^*_{i,ss}]\nu(d\theta) \\
+ \int_\Theta \mu_s^2[\hat{v}^*_{i,ss}(s+\mu,.) - \hat{v}^*_{i,ss}]\nu(d\theta) \\
+ \int_\Theta \mu_s^2 \hat{v}^*_{i,ss}\nu(d\theta) \\
+2\sum_k \delta_s[\hat{v}^*_{i,ss}(s+\delta,k) - \hat{v}^*_{i,ss}]q_k \\
+ \sum_k \delta_s^2[\hat{v}^*_{i,ss}(s+\delta,k) - \hat{v}^*_{i,ss}(s,.)]q_k \\
+ \sum_k \delta_s^2 \hat{v}^*_{i,ss}(s,.)q_k \\
= \bar{H}_{i,ss} + \tilde{E}H_{i,ssm} \\
+\{2b_s + \sigma_s^2 + \int_\Theta \mu_s^2\nu(d\theta) + \sum_k \delta_s^2 q_k\}\hat{P}_i + 2\sigma_s\hat{Q}_i \\
+ \int_\Theta (2\mu_s + \mu_s^2)\hat{R}_i(t,\theta,.)\nu(d\theta) \\
+ \sum_k (2\delta_s + \delta_s^2)\hat{S}_i(t,k)q_k.
\end{cases}
$$

Thus, the SIDE becomes

$$
\begin{cases}
d\hat{P}_i = d\hat{v}^*_{i,ss} \\
= -\{\bar{H}_{i,ss} + \tilde{E}H_{i,ssm}\}dt \\
-\{2b_s + \sigma_s^2 + \int_\Theta \mu_s^2\nu(d\theta) + \sum_k \delta_s^2 q_k\}\hat{P}_i dt - 2\sigma_s\hat{Q}_i dt \\
- \int_\Theta (2\mu_s + \mu_s^2)\hat{R}_i(t,\theta,.)\nu(d\theta)dt \\
- \sum_k (2\delta_s + \delta_s^2)\hat{S}_i(t,\theta,.)q_k dt \\
+\hat{Q}_i dB \\
+ \int_\Theta \hat{R}_i(t,.,\theta,.)\tilde{N}(dt,d\theta,\tilde{s}), \\
+ \sum_k \hat{S}_i(t,k)\tilde{M}(dt,k),
\end{cases}
\tag{8.24}
$$

where

$$
\hat{R}_i(t,.,\theta,.) := \{\hat{v}^*_{i,ss}(s+\mu,\tilde{s}) - \hat{v}^*_{i,ss}(s,\tilde{s})\}
$$

$$
\hat{S}_i := [\hat{v}^*_{i,ss}(s+\delta,k) - \hat{v}^*_{i,ss}(s,\tilde{s})].
$$

This completes the proof. ■

8.2 Risk-Sensitive Mean-Field-Type Games with Jumps and Regime Switching

Each decision-maker has a risk-sensitivity index $\tau_i \in \mathbb{R}$. When τ_i vanishes, one gets the risk-neutral setup discussed above. In what follows, we assume $\tau_i \neq 0$ for all $i \in \mathcal{I}$.

Consider the risk-sensitive best response of decision-maker i as

$$\left\{\begin{array}{l} \sup_{a_i \in \mathscr{A}_i} \frac{1}{\tau_i} \log\left(E e^{\tau_i \mathscr{R}_{i,T}}\right) \text{ subject to} \\ s(t) = s_0 + \int_0^t b dt' + \int_0^t \sigma dB(t') \\ + \int_0^t \int_\Theta \mu \tilde{N}(dt', d\theta) + \int_0^t \delta \tilde{M}(dt'),\ t > 0 \\ s(t) = s^{s_0, a, \tilde{s}_0}(t),\ s(0) = s_0 \sim m_0, \tilde{s}(0) = \tilde{s}_0 \\ m(t, .) = m^{m_0, a, \tilde{s}_0}(t, .) = \mathbb{P}_{s^{s_0, a, \tilde{s}_0}(t)}. \end{array}\right. \tag{8.25}$$

Introduce the following augmented state (s, z) such that $z_i(0) = 0$ and $dz_i(t) = r_i(t, s(t), m(t), a(t), \tilde{s}(t))\, dt$. Then

$$e^{\tau_i \mathscr{R}_{i,T}} = e^{\tau_i z_i(T) + \tau_i g_i(s(T), m(T), \tilde{s}(T))}.$$

Thus, we obtain the following mean-field-type optimization problem for decision-maker i:

$$\left\{\begin{array}{l} \sup_{a_i} E e^{\tau_i [z_i(T) + g_i(s(T), m(T), \tilde{s}(T))]}, \\ z_i(t) = \int_0^t r_i\, dt', \\ z_i(0) = 0, \\ s(t) = s_0 + \int_0^t b dt' + \int_0^t \sigma dB(t') \\ + \int_0^t \int_\Theta \mu \tilde{N}(dt', d\theta) + \int_0^t \delta \tilde{M}(dt'), \\ s(0) = s_0. \end{array}\right. \tag{8.26}$$

Let us introduce $\tilde{m}(t, dsdz, .) = \mathbb{P}_{s(t), z(t)}$, the probability measure associated with the joint process $(s(t), z(t), .)$. Generically, the measure $\tilde{m}$ solves the Fokker-Planck-Kolmogorov forward equation in the distributional (weak) sense. The term $E e^{\tau_i \mathscr{R}_{i,T}}$ can be rewritten in a deterministic manner as

$$\int \tilde{m}(T, dsdz)\, e^{\tau_i (z_i + g_i(s, \int_{\tilde{z}} \tilde{m}(T, ., d\tilde{z})))}.$$

This is a terminal cost in the sense that it is evaluated only at $\tilde{m}(T, .)$. $\tilde{m}$ is an infinite-dimensional quantity that serves as a state in the problem. The advantage now is that $\tilde{m}(.)$ is a deterministic quantity. Since there is no running cost, one can write directly the HJB equation using classical calculus of variations for

$$\hat{V}_i^\tau(t, \tilde{m}, \tilde{s}) = \sup_{a_i \in \mathscr{A}_i} \int \tilde{m}(T, dsdz)\, e^{\tau_i (z_i + g_i(s, \int_{\tilde{z}} \tilde{m}(T, ., d\tilde{z})))},$$

starting from $\tilde{m}(t,.) = \tilde{m}$ at time t : The risk-sensitive best value is $V_i^{\tau}(t,\tilde{m}) := \frac{1}{\tau_i}\log \hat{V}_i^{\tau}(t,\tilde{m})$:

$$\tilde{m}_t = -\partial_s[b\tilde{m}] - (f\tilde{m})_z + \frac{1}{2}(\sigma'\sigma\tilde{m})_{ss} + J^*[\tilde{m}] + \tilde{S}^*[\tilde{m}], \tag{8.27}$$

with initial distribution $\tilde{m}(0, dsdz) = m_0(ds)\delta_0(dz)$ and σ' is the transpose of σ.

The risk-sensitive HJB system yields

$$\begin{aligned} 0 &= \hat{V}_{i,t}^{\tau}(t,\tilde{m},\tilde{s}) + \int H_i^{\tau}\tilde{m}(t,dsdz), \\ &\hat{V}_i^{\tau}(T,\tilde{m},\tilde{s}) = \int \tilde{m}(T,dsdz)\, e^{\tau_i(z_i+g_i(s,\int_{\tilde{z}}\tilde{m}(T,.,d\tilde{z})))}, \\ &i \in \mathcal{I}. \end{aligned} \tag{8.28}$$

where $H_i^{\tau}(t,s,z,\tilde{m},\hat{V}_{i,\tilde{m}},\hat{V}_{i,z\tilde{m}},\hat{V}_{i,ss\tilde{m}}) = \sup_{a_i} b\partial_s\hat{V}_{i,\tilde{m}} + \langle r, \partial_z\hat{V}_{i,\tilde{m}}\rangle + \frac{\sigma^2}{2}\partial_{ss}\hat{V}_{i,\tilde{m}} + J[\hat{V}_{i,\tilde{m}}] + \tilde{S}[\hat{V}_{i,\tilde{m}}]$. Note that

$$\begin{aligned} &\hat{V}_i^{\tau}(T,\tilde{m},\tilde{s}) \\ &= \int \tilde{m}(T,dsdz)\, e^{\tau_i(z_i+g_i(s,\int_{\tilde{z}}\tilde{m}(T,.,d\tilde{z},\tilde{s})))} \qquad (8.29) \\ &= \int_{s,z_i} e^{\tau_i(z_i+g_i(s,\int_{\tilde{z}_i}\int_{\tilde{z}_{-i}}\tilde{m}(T,.,d\tilde{z})))} \int_{z_{-i}} \tilde{m}(T,dsdz,\tilde{s}) \\ &= \int_{s,z_i} e^{\tau_i(z_i+g_i(s,\int_{\tilde{z}_i}\tilde{m}_i(T,.,d\tilde{z}_i)))}\tilde{m}_i(T,dsdz_i,\tilde{s}) \qquad (8.30) \\ &= \hat{V}_i^{\tau}(T,\tilde{m}_i,\tilde{s}), \qquad (8.31) \end{aligned}$$

where $\tilde{m}_i(t,dsdz_i,\tilde{s}) = \int_{z_{-i}}\tilde{m}(t,dsdz_idz_{-i},\tilde{s})$ is the marginal of $\tilde{m}$ with respect to (s,z_i) and $\hat{V}_i^{\tau}(t,\tilde{m},\tilde{s}) = \hat{V}_i^{\tau}(t,\tilde{m}_i,\tilde{s})$. The above calculation shows that the value function $\frac{1}{\tau_i}\log\hat{V}_i^{\tau}$ depends only on the measure $\tilde{m}_i$.

Proposition 8.6 (Risk-Sensitive HJB of Mean-Field Type)

If there exists a function $\tilde{V}$ with

$$\tilde{V}_{i,t},\ \tilde{V}_{i,s\tilde{m}_i},\ \tilde{V}_{i,ss\tilde{m}_i},\ \tilde{V}_{i,zz\tilde{m}_i},$$

(continued)

satisfying

$$
\begin{aligned}
0 &= \tilde V_{i,t} + \int H_i^{\tau}\tilde m_i(t, dsdz_i, .),\\
\tilde V_i(T,\tilde m_i,\tilde s) &= \int \tilde m_i(T, dsdz_i, \tilde s)\, e^{\tau_i (z_i + g_i(s,.,\int_{\tilde z}\tilde m_i(T,.,d\tilde z,\tilde s)))},\\
\tilde m_i(t, dsdz_i,\tilde s) &= \int_{z_{-i}} \tilde m(t, dsdz_i dz_{-i},\tilde s),\\
\tilde m_{i,t} &= -\partial_s[b\tilde m_i] - (r_i\tilde m_i)_{z_i} + \frac{1}{2}(\sigma^2\tilde m_i)_{ss} + J^*[\tilde m_i] + \tilde S^*[\tilde m_i],\\
\tilde m_i(0, dsdz_i,\tilde s) &= m_0(ds,\tilde s)\delta_0(dz_i),\\
i &\in \mathscr{I},
\end{aligned}
\tag{8.32}
$$

where the integrand Hamiltonian is

$$
\begin{aligned}
&H_i^{\tau}(t, s, \tilde s, z, \tilde m_i, \tilde V_{i,\tilde m_i}, \tilde V_{i,s\tilde m_i}, \tilde V_{i,z_i\tilde m_i}, \tilde V_{i,ss\tilde m_i})\\
&= \sup_{a_i}[b\tilde V_{i,s\tilde m_i} + r_i\tilde V_{i,z_i\tilde m_i} + \tfrac{\sigma^2}{2}\tilde V_{i,ss\tilde m_i} + J[\tilde V_{i,m}] + \tilde S[\tilde V_{i,m}]\\
&= \tilde V_{i,z_i\tilde m_i}\sup_{a_i}[r_i + b\frac{\tilde V_{i,s\tilde m_i}}{\tilde V_{i,z_i\tilde m_i}} + \frac{\sigma^2}{2}\frac{\tilde V_{i,ss\tilde m_i}}{\tilde V_{i,z_i\tilde m_i}} + \frac{J[\tilde V_{i,\tilde m_i}]+\tilde S[\tilde V_{i,\tilde m_i}]}{\tilde V_{i,z_i\tilde m_i}}]
\end{aligned}
$$

then $\frac{1}{\tau_i}\log\tilde V_i(0,\tilde m,\tilde s)$ is an equilibrium payoff for decision-maker i, and the best-response strategy a_i minimizes H_i^{τ} given the other decision-makers' strategies a_{-i}.

Proof Apply the recipe of Proposition 8.1 with the augmented state (s, z_i) and the measure $\tilde m_i$. ■

Let $\tilde v_i^*(t,s,\tilde s,z_i) := \tilde V_{i,\tilde m_i}(\tilde m_i)(t,s,\tilde s,z_i)$. It follows that $\tilde v_i^*(t,s,\tilde s,z_i)$ solves the following PIDE system:

$$
0 = \tilde v^*_{i,t} + \tilde v^*_{i,z_i} H_i(t, s, z_i, \tilde m_i, \frac{\tilde v^*_{i,s}}{\tilde v^*_{i,z_i}}, \frac{\tilde v^*_{i,ss}}{\tilde v^*_{i,z_i}}) + \int_{s',z_i'} \tilde v^*_{i,\tilde z_i} H_{i,\tilde m_i}\tilde m_i(t, ds', dz_i'). \tag{8.33}
$$

The terminal condition for this PIDE is

$$
\tilde v_i^*(T,s,\tilde s,z_i) = e^{\tau_i(z_i+g_i(s,m(T),\tilde s))}
$$

$$
+\tau_i\int g_{i,m}(s', m(T),\tilde s)(s)e^{\tau_i(z_i'+g_i(s',m(T)))}\,\tilde m_i(T, ds',\tilde s, dz_i').
$$

From the equality $\tilde{V}^{\tau}_{i,\tilde{m}}(t,\tilde{m}) = \tau_i V^{\tau}_{i,\tilde{m}}(t,\tilde{m},\tilde{s})\tilde{V}^{\tau}_i(t,\tilde{m},\tilde{s})$, we deduce that the function $v_i^*(t,s,\tilde{s},z_i) := V^{\tau}_{i,\tilde{m}_i}(\tilde{m}_i)(t,s,\tilde{s},z_i) = \frac{\tilde{V}^{\tau}_{i,\tilde{m}}}{\tau_i \tilde{V}^{\tau}_i} = \frac{\tilde{v}_i^*(t,s,\tilde{s},z_i)}{\tau_i \tilde{V}^{\tau}_i}$ is the dual function associated with the risk-sensitive best-response value of decision-maker i.

Note that, as in the risk-neutral case, in general, the risk-sensitive dual function $v^*(t,s,\tilde{s},z_i)$ is different from the risk-sensitive payoff function $V^{\tau}(t,\tilde{m}_i,\tilde{s})$ or $V^{\tau}(t,\tilde{m},\tilde{s})$.

Lemma 8.6 *For $\tau_i \neq 0$, the process η_i has a constant sign. It satisfies $\eta(t,\tilde{s}) \neq 0$ almost everywhere and has the same sign as θ_i.*

Lemma 8.7 *The process $\eta_i := \tilde{V}_{i,z_i\tilde{m}_i}$ solves the following SIDE:*

$$\begin{cases} d\eta_i = d\tilde{v}^*_{i,z} \\ = \sigma \hat{v}^*_{i,sz} dB \\ + \int_{\Theta}\{\tilde{v}^*_{i,z}(s+\mu,\tilde{s}) - \tilde{v}^*_{i,z}(s,\tilde{s})\}\tilde{N}(dt,d\theta,\tilde{s}), \\ + \sum_k[\tilde{v}^*_{i,z}(s+\delta,k) - \tilde{v}^*_{i,z}(s,\tilde{s})]\tilde{M}(dt,k), \end{cases} \tag{8.34}$$

and the function $\eta_i(t,s,\tilde{s},z) = \tilde{v}^*_{i,z}(t,s,\tilde{s},z) = \tilde{V}_{i,z\tilde{m}_i}(t,s,\tilde{s},z)$ solves the backward PIDE

$$0 = \eta_{i,t} + r_i.\eta_{i,z} + b.\eta_{i,s} + \frac{\sigma^2}{2}\eta_{i,ss} + J[\eta_i] + \tilde{S}[\eta_i],$$

with the terminal condition

$$\eta_i(T,s,z,\tilde{s}) = \tau_i e^{\tau_i(z_i+g_i(s,m(T,.),\tilde{s}))}.$$

Proof The risk-sensitive Hamiltonian is $H_i^{\tau} = \eta_i H_i$ with

$$H_i := \sup_{a_i}\left[r_i + b\frac{\tilde{V}_{i,s\tilde{m}_i}}{\eta_i} + \frac{\sigma^2}{2}\frac{\tilde{V}_{i,ss\tilde{m}_i}}{\eta_i} + \frac{J[\tilde{V}_{i,\tilde{m}_i}] + \tilde{S}[\tilde{V}_{i,\tilde{m}_i}]}{\eta_i}\right].$$

The dual function solves the PIDE:

$$0 = \tilde{v}^*_{i,t} + H_i^{\tau} + \int_{s',z'} H^{\tau}_{i,\tilde{m}_i}\tilde{m}_i(t,ds',\tilde{s},dz_i').$$

Differentiating it with respect to z, we arrive at

$$\begin{aligned} &0 = \tilde{v}^*_{i,zt} + r_i.\tilde{v}^*_{i,zz} + b.\tilde{v}^*_{i,zs} + \tfrac{\sigma^2}{2}\tilde{v}^*_{i,zss} + (J[\tilde{v}_i] + \tilde{S}[\tilde{v}_i^*])_z, \\ &\tilde{v}^*_{i,z}(T,.) = \eta_i(T) = \tau_i e^{\tau_i(z_i+g_i(s,m(T)))}. \end{aligned} \tag{8.35}$$

Since μ and δ are independent of z, the derivatives of

$$J[\tilde{v}_i] := \int_{\Theta} \{\tilde{v}_i(., s+\mu, s, z) - \tilde{v}_i(., s, z) - \tilde{v}_{i,s}.\mu\}\nu(d\theta),$$

with respect to z yields

$$(J[\tilde{v}_i])_z := \int_{\Theta} \{\tilde{v}_{i,z}(., s+\mu, s, z) - \tilde{v}_{i,z}(., s, z) - \tilde{v}_{i,zs}.\mu\}\nu(d\theta) = J[\tilde{v}_{iz}],$$

and the switching term is

$$\tilde{S}[\tilde{v}_i] := \sum_k q_k[\tilde{v}_i(t, s+\delta, z, k) - \tilde{v}_i(t, s, z, \tilde{s}), \tilde{s}) - \tilde{v}_{i,s}.\delta],$$

and its z-derivative yields

$$(\tilde{S}[\tilde{v}_i])_z := \sum_k q_k[\tilde{v}_{i,z}(t, s+\delta, z, k) - \tilde{v}_{i,z}(t, s, z, \tilde{s}), \tilde{s}) - \tilde{v}_{i,zs}.\delta] = \tilde{S}[\tilde{v}_{i,z}].$$

It follows from (8.35) that the functional $\tilde{v}^*_{i,z}$ solves

$$\begin{array}{l} 0 = \tilde{v}^*_{i,zt} + r_i.\tilde{v}^*_{i,zz} + b.\tilde{v}^*_{i,zs} + \frac{\sigma^2}{2}\tilde{v}^*_{i,zss} + J[\tilde{v}^*_{iz}] + \tilde{S}[\tilde{v}^*_{i,z}], \\ \tilde{v}^*_{i,z}(T, .) = \eta_i(T) = \tau_i e^{\tau_i(z_i + g_i(s, m(T)))}. \end{array} \tag{8.36}$$

On the other hand, one can use Itô's formula to get a stochastic evolution of the process $\eta(t)$:

$$\begin{cases} d\tilde{v}^*_{i,z} \\ = [\tilde{v}^*_{i,tz} + r_i.\tilde{v}^*_{i,zz} + b.\tilde{v}^*_{i,sz} + \frac{\sigma^2}{2}\hat{v}^*_{i,ssz} + J[\hat{v}^*_{i,z}] + \tilde{S}[\hat{v}^*_{i,z}]]dt \\ +\sigma\hat{v}^*_{i,sz}dB \\ + \int_{\Theta}\{\tilde{v}^*_{i,z}(s+\mu, z, \tilde{s}) - \tilde{v}^*_{i,z}(s, z, \tilde{s})\}\tilde{N}(dt, d\theta, \tilde{s}), \\ + \sum_k[\tilde{v}^*_{i,z}(s+\delta, k) - \tilde{v}^*_{i,z}(s, \tilde{s})]\tilde{M}(dt, k). \end{cases} \tag{8.37}$$

The process η has a mixture of martingale representation, and its evolution is given by

$$\begin{cases} d\tilde{v}^*_{i,z} \\ = +\sigma\hat{v}^*_{i,sz}dB \\ + \int_{\Theta}\{\tilde{v}^*_{i,z}(s+\mu, z, \tilde{s}) - \tilde{v}^*_{i,z}(s, z, \tilde{s})\}\tilde{N}(dt, d\theta, \tilde{s}), \\ + \sum_k[\tilde{v}^*_{i,z}(s+\delta, z, k) - \tilde{v}^*_{i,z}(s, z, \tilde{s})]\tilde{M}(dt, k). \end{cases} \tag{8.38}$$

This completes the proof. ■

Proposition 8.7 (Risk-Sensitive First-Order Adjoint Processes)

For $\tau_i \neq 0$, the risk-sensitive first-order adjoint process $(p_i^\tau, q_i^\tau, \eta_i, l_i, \omega_i^1, \omega_i^2)_{i\in\mathcal{I}}$ solves

$$\begin{cases} dp_i^\tau = -\{\bar{H}_{i,s}^\tau + \frac{1}{\eta_i}E[\eta_i \bar{H}_{i,s\tilde{m}_i}^\tau]\}dt \\ -q_i^\tau l_i dt - \int_\Theta r_i^\tau \omega_i^1 \nu(d\theta)dt - \sum_k s_i^\tau \omega_i^2 q_k dt \\ +q_i^\tau dB + \int_\Theta r_i^\tau . \tilde{N}(dt, d\theta, .) + s_i^\tau . \tilde{M}(dt), \\ p_i^\tau(T) = g_{i,s} + \frac{1}{\eta_i(T)}E\{\eta_i(T) g_{i,sm}\}, \\ d\eta_i = \eta_i l_i dB \\ +\eta_i \left[\int \omega_i^1(t,\theta,.)\tilde{N}(dt,d\theta,\tilde{s}) + \sum_k \omega_i^2(t,k)\tilde{M}(dt,k)\right], \\ \eta_i(T) = \tau_i e^{\tau_i[z_i(T)+g_i(s(T),m(T,.),\tilde{s}(T))}, \end{cases} \tag{8.39}$$

where the risk-sensitive Hamiltonian is $\eta_i \bar{H}_i^\tau$ with

$$\bar{H}_i^\tau = r_i + bp_i^\tau + \sigma(q_i^\tau + l_i p_i^\tau)$$

$$+ \int_\Theta \mu[r_i^\tau + \omega_i^1 p_i^\tau]\nu(d\theta)dt + \sum_k \delta[s_i^\tau + \omega_i^2 p_i^\tau]q_k,$$

$$\bar{H}_{i,s}^\tau = r_{i,s} + bp_i^\tau + \sigma_s(q_i^\tau + l_i p_i^\tau)$$

$$+ \int \mu_s[r_i^\tau + \omega_i^1 p_i^\tau]\nu(d\theta)dt + \sum_k \delta_s[s_i^\tau + \omega_i^2 p_i^\tau]q_k.$$

Proposition 8.8 (Process Identification)

Under the assumptions above, the risk-sensitive processes are given by

(continued)

$$\begin{cases} p_i^\tau = \frac{\tilde{v}_{i,s}^*}{\tilde{v}_{i,z}^*} = \frac{\tilde{v}_{i,s}^*}{\eta_i}, \\ q_i^\tau = \frac{1}{\eta_i}[\sigma \tilde{v}_{i,ss}^* - \sigma p_i^\tau \hat{v}_{i,sz}^*], \\ r_i^\tau = \frac{1}{\eta_i}\{\tilde{v}_{i,s}^*(s+\mu, \tilde{s}(t_-)) - \tilde{v}_{i,s}^*(s,\tilde{s})\} - \frac{1}{\eta_i} p_i^\tau \{\tilde{v}_{i,z}^*(s+\mu,\tilde{s}) - \tilde{v}_{i,z}^*(s,\tilde{s})\}, \\ s_i^\tau(t) = \frac{1}{\eta_i}[\tilde{v}_{i,s}^*(s+\delta,k) - \tilde{v}_{i,s}^*(s,\tilde{s})] - \frac{1}{\eta_i} p_i^\tau [\tilde{v}_{i,z}^*(s+\delta,k) - \tilde{v}_{i,z}^*(s,\tilde{s})], \\ \eta_i = \tilde{v}_{i,z}^* = \tilde{V}_{i,z\tilde{m}_i}, \\ l_i = \sigma \frac{\hat{v}_{i,sz}^*}{\tilde{v}_{i,z}^*} = \sigma \frac{\hat{v}_{i,sz}^*}{\eta_i}, \\ \omega_i^1 := \frac{1}{\eta_i}(\tilde{v}_{i,z}^*(s+\mu,.) - \tilde{v}_{i,z}^*), \\ \omega_i^2 := \frac{1}{\eta_i}(\tilde{v}_{i,z}^*(s+\delta,k) - \tilde{v}_{i,z}^*). \end{cases} \tag{8.40}$$

Proof We start by recalling the integration by parts formula for jump-diffusion processes. Let

$$\begin{cases} dx_k = b_k dt + \sigma_k dB + \int \mu_k \tilde{N}(dt, d\theta, \tilde{s}) + \delta_k.\tilde{M}(dt), \\ k \in \{1, 2\}. \end{cases} \tag{8.41}$$

Then

$$d[x_1 x_2] = x_1 dx_2 + x_2 dx_1 + [\sigma_1\sigma_2 + \int \mu_1\mu_2 \nu(d\theta) + \sum_k \delta_1.\delta_2.q_{\tilde{s}k}]dt.$$

Applying the latter integration rule to the process $[\eta p]$, we obtain $d[\eta p] = \eta dp + p d\eta + x_3 dt$ where $x_3 = \langle \eta, p\rangle_t$:

$$\Big\{ dp = \frac{1}{\eta}\{d[\eta p] - p d\eta - x_3 dt\} = \frac{1}{\eta}\{d[\tilde{v}_{i,s}^*] - p d\eta - x_3 dt\}$$

$$\begin{cases} d\hat{v}_{i,s}^* = [\tilde{v}_{i,ts}^* + r_i.\tilde{v}_{i,zs}^* + b.\tilde{v}_{i,ss}^* + \frac{\sigma^2}{2}\tilde{v}_{i,sss}^* + J[\tilde{v}_{i,s}] + \tilde{S}[\tilde{v}_{i,s}^*]]dt + \sigma \tilde{v}_{i,ss}^* dB \\ + \int_\Theta \{\tilde{v}_{i,s}^*(s+\mu,\tilde{s}) - \tilde{v}_{i,s}^*(s,\tilde{s})\}\tilde{N}(dt, d\theta, \tilde{s}) + \sum_k [\tilde{v}_{i,s}^*(s+\delta,k) \\ \quad -\tilde{v}_{i,s}^*(s,\tilde{s})]\tilde{M}(dt,k), \end{cases} \tag{8.42}$$

$$\begin{cases} d\eta_i = d\tilde{v}_{i,z}^* = \sigma \hat{v}_{i,sz}^* dB + \int_\Theta \{\tilde{v}_{i,z}^*(s+\mu,\tilde{s}) - \tilde{v}_{i,z}^*(s,\tilde{s})\}\tilde{N}(dt,d\theta,\tilde{s}) \\ + \sum_k [\tilde{v}_{i,z}^*(s+\delta,k) - \tilde{v}_{i,z}^*(s,\tilde{s})]\tilde{M}(dt,k). \end{cases} \tag{8.43}$$

Thus

$$\begin{cases} \frac{x_3}{\eta_i} = \frac{1}{\eta_i}[\sigma \tilde{v}_{i,ss}^* - \sigma p_i^\tau \hat{v}_{i,sz}^*].\frac{\sigma \hat{v}_{i,sz}^*}{\eta_i} \\ + \int \frac{1}{\eta_i^2}[\{\tilde{v}_{i,s}^*(s+\mu,\tilde{s}) - \tilde{v}_{i,s}^*(s,\tilde{s})\} \\ -p_i^\tau \{\tilde{v}_{i,z}^*(s+\mu,\tilde{s}) - \tilde{v}_{i,z}^*(s,\tilde{s})\}][\{\tilde{v}_{i,z}^*(s+\mu,\tilde{s}) - \tilde{v}_{i,z}^*(s,\tilde{s})\}\nu(d\theta) \\ + \sum_k \frac{1}{\eta_i^2}[\tilde{v}_{i,s}^*(s+\delta,k) - \tilde{v}_{i,s}^*(s,\tilde{s}) \\ -p_i^\tau [\tilde{v}_{i,z}^*(s+\delta,k) - \tilde{v}_{i,z}^*(s,\tilde{s})][\{\tilde{v}_{i,z}^*(s+\delta,k) - \tilde{v}_{i,z}^*(s,\tilde{s})\}q_k. \end{cases} \tag{8.44}$$

$$\begin{cases} \eta_i dp_i^\tau = -x_3 dt \\ +[\tilde{v}_{i,ts}^* + r_i.\tilde{v}_{i,zs}^* + b.\tilde{v}_{i,ss}^* + \frac{\sigma^2}{2}\tilde{v}_{i,sss}^* + J[\tilde{v}_{i,s}^*] + \tilde{S}[\tilde{v}_{i,s}^*]]dt \\ +[\sigma\tilde{v}_{i,ss}^* - \sigma p_i^\tau \hat{v}_{i,sz}^*]dB \\ + \int_\Theta \{\tilde{v}_{i,s}^*(s+\mu,\tilde{s}) - \tilde{v}_{i,s}^*(s,\tilde{s})\}\tilde{N}(dt,d\theta) \\ - \int_\Theta p_i^\tau \{\tilde{v}_{i,z}^*(s+\mu,\tilde{s}) - \tilde{v}_{i,z}^*(s,\tilde{s})\}\tilde{N}(dt,d\theta,\tilde{s}) \\ + \sum_k [\tilde{v}_{i,s}^*(s+\delta,k) - \tilde{v}_{i,s}^*(s,\tilde{s})]\tilde{M}(dt,k) \\ - \sum_k p_i^\tau [\tilde{v}_{i,z}^*(s+\delta,k) - \tilde{v}_{i,z}^*(s,\tilde{s})]\tilde{M}(dt,k). \end{cases} \tag{8.45}$$

Similar calculation as in (8.19) yields

$$\begin{cases} \{\tilde{v}_{i,ts}^* + r_i.\tilde{v}_{i,sz}^* + b.\tilde{v}_{i,ss}^* + \frac{\sigma^2}{2}\tilde{v}_{i,sss}^* + J[\tilde{v}_{i,s}^*] + \tilde{S}[\tilde{v}_{i,s}^*]\} \\ = -\{r_{i,s}(\hat{v}_{i,z}^*) + b_s(\hat{v}_{i,s}^*) + \sigma_s(\sigma\hat{v}_{i,ss}^*) \\ \quad + \int_\Theta \mu_s[\hat{v}_{i,s}^*(s+\mu,.) - \hat{v}_{i,s}^*]\nu(d\theta) \\ + \sum_k \delta_s[\hat{v}_{i,s}^*(s(t_-)+\delta,k) - \hat{v}_{i,s}^*]q_k + EH_{i,s\tilde{m}_i}^\tau\}. \end{cases} \tag{8.46}$$

Thus

$$\begin{cases} \eta_i dp_i^\tau = -x_3 dt \\ -\{r_{i,s}(\hat{v}_{i,z}^*) + b_s(\hat{v}_{i,s}^*) + \sigma_s(\sigma\hat{v}_{i,sss}^*) \\ \quad + \int_\Theta \mu_s[\hat{v}_{i,s}^*(s+\mu,.) - \hat{v}_{i,s}^*]\nu(d\theta) \\ + \sum_k \delta_s[\hat{v}_{i,s}^*(s(t_-)+\delta,k) - \hat{v}_{i,s}^*]q_k + EH_{i,s\tilde{m}_i}^\tau\}dt \\ +[\sigma\tilde{v}_{i,ss}^* - \sigma p_i^\tau \hat{v}_{i,sz}^*]dB \\ + \int_\Theta \{\tilde{v}_{i,s}^*(s+\mu,\tilde{s}) - \tilde{v}_{i,s}^*(s,\tilde{s})\}\tilde{N}(dt,d\theta,\tilde{s}) \\ - \int_\Theta p_i^\tau \{\tilde{v}_{i,z}^*(s+\mu,\tilde{s}) - \tilde{v}_{i,z}^*(s,\tilde{s})\}\tilde{N}(dt,d\theta,\tilde{s}) \\ + \sum_k [\tilde{v}_{i,s}^*(s+\delta,k) - \tilde{v}_{i,s}^*(s,\tilde{s})]\tilde{M}(dt,k) \\ - \sum_k p_i^\tau [\tilde{v}_{i,z}^*(s+\delta,k) - \tilde{v}_{i,z}^*(s,\tilde{s})]\tilde{M}(dt,k). \end{cases} \tag{8.47}$$

We need to make sure that η_i does not vanish. This is examined in Lemma 8.6 which states that η_i is always positive: $\eta_i(t) > 0$. We can therefore divide by $\eta_i(t) = \tilde{v}_{i,z}^*(s(t), z(t), \tilde{s}(t))$.

Thus

$$\begin{cases} dp_i^\tau = -\frac{x_3}{\eta_i} dt \\ -\frac{1}{\eta_i}\{r_{i,s}(\hat{v}_{i,z}^*) + b_s(\hat{v}_{i,s}^*) + \sigma_s(\sigma\hat{v}_{i,ss}^*) \\ +\frac{1}{\eta_i} \int_\Theta \mu_s[\hat{v}_{i,s}^*(s+\mu,.) - \hat{v}_{i,s}^*]\nu(d\theta) \\ +\frac{1}{\eta_i} \sum_k \delta_s[\hat{v}_{i,s}^*(s+\delta,k) - \hat{v}_{i,s}^*]q_k + EH_{i,s\tilde{m}_i}\}dt \\ +\frac{1}{\eta_i}[\sigma\tilde{v}_{i,ss}^* - \sigma p_i^\tau \hat{v}_{i,sz}^*]dB \\ +\frac{1}{\eta_i} \int_\Theta \{\tilde{v}_{i,s}^*(s+\mu,\tilde{s}) - \tilde{v}_{i,s}^*(s,\tilde{s})\}\tilde{N}(dt,d\theta,\tilde{s}) \\ -\frac{1}{\eta_i} \int_\Theta p_i^\tau \{\tilde{v}_{i,z}^*(s+\mu,\tilde{s}) - \tilde{v}_{i,z}^*(s,\tilde{s})\}\tilde{N}(dt,d\theta,\tilde{s}) \\ +\frac{1}{\eta_i} \sum_k [\tilde{v}_{i,s}^*(s+\delta,k) - \tilde{v}_{i,s}^*(s,\tilde{s})]\tilde{M}(dt,k) \\ -\frac{1}{\eta_i} \sum_k p_i^\tau [\tilde{v}_{i,z}^*(s+\delta,k) - \tilde{v}_{i,z}^*(s,\tilde{s})]\tilde{M}(dt,k). \end{cases} \tag{8.48}$$

Set

$$
\begin{array}{l}
p_i^\tau = \frac{\tilde{v}^*_{i,s}}{\tilde{v}^*_{i,z}} = \frac{\tilde{v}^*_{i,s}}{\eta_i},\\
q_i^\tau = \frac{1}{\eta_i}[\sigma \tilde{v}^*_{i,ss} - \sigma p_i^\tau \hat{v}^*_{i,sz}],\\
r_i^\tau = \frac{1}{\eta_i}\{\tilde{v}^*_{i,s}(s+\mu,\tilde{s}) - \tilde{v}^*_{i,s}(s,\tilde{s})\}\\
-\frac{1}{\eta_i} p_i^\tau \{\tilde{v}^*_{i,z}(s+\mu,\tilde{s}) - \tilde{v}^*_{i,z}(s,\tilde{s})\},\\
s_i^\tau = \frac{1}{\eta_i}[\tilde{v}^*_{i,s}(s+\delta,k) - \tilde{v}^*_{i,s}(s,\tilde{s})]\\
-\frac{1}{\eta_i} p_i^\tau [\tilde{v}^*_{i,z}(s+\delta,k) - \tilde{v}^*_{i,z}(s,\tilde{s})],\\
l_i^\tau := (\sigma \frac{\hat{v}^*_{i,sz}}{\eta_i}).
\end{array}
$$

It follows that

$$
q_i^\tau + p_i^\tau(\sigma \frac{\hat{v}^*_{i,sz}}{\eta_i}) = \sigma \frac{\tilde{v}^*_{i,ss}}{\eta_i} = q_i^\tau + p_i^\tau l_i,
$$

$$
-\frac{1}{\eta_i}[\sigma \tilde{v}^*_{i,ss} - \sigma p_i^\tau \hat{v}^*_{i,sz}].\frac{\sigma \hat{v}^*_{i,sz}}{\eta_i} = -q_i^\tau l_i = -[\sigma \frac{\tilde{v}^*_{i,ss}}{\eta_i} l_i - p_i^\tau |l_i|^2].
$$

$$
\begin{cases}
-\frac{x_3}{\eta_i} = -q_i^\tau l_i - \int r_i^\tau \frac{1}{\eta_i}\{\tilde{v}^*_{i,z}(s(t_-)+\mu(t_-,.,\theta),\tilde{s}(t_-))\\
\quad -\tilde{v}^*_{i,z}(s(t_-),\tilde{s}(t_-))\}\nu(d\theta)\\
-\sum_k s_i^\tau \frac{1}{\eta_i}\{\tilde{v}^*_{i,z}(s(t_-)+\delta,k) - \tilde{v}^*_{i,z}(s(t_-),\tilde{s})\}q_k.
\end{cases}
\tag{8.49}
$$

Substituting the above notation, we obtain

$$
\begin{cases}
dp_i^\tau = q_i^\tau[-l_i dt + dB]\\
+\int_\Theta r_i^\tau.\tilde{N}(dt,d\theta,\tilde{s}) + s_i^\tau.\tilde{M}(dt)\\
-\{r_{i,s} + b_s p_i^\tau + \sigma_s(q_i^\tau + p_i^\tau l_i) + \frac{1}{\eta_i} E H^\tau_{i,s\tilde{m}_i}\}dt\\
-\int r_i^\tau(t,.)\frac{1}{\eta_i}\{\tilde{v}^*_{i,z}(s(t_-)+\mu(t_-,.,\theta),\tilde{s}(t_-))\\
\quad -\tilde{v}^*_{i,z}(s(t_-),\tilde{s}(t_-))\}\tilde{m}(d\theta)dt\\
-\sum_k s_i^\tau(t)\frac{1}{\eta_i}\{\tilde{v}^*_{i,z}(s(t_-)+\delta,k) - \tilde{v}^*_{i,z}(s(t_-),\tilde{s})\}q_{\tilde{s}k}\\
-\int_\Theta \mu_s \frac{1}{\eta_i}[\hat{v}^*_{i,s}(s+\mu,.) - \hat{v}^*_{i,s}]\tilde{m}(d\theta)dt\\
-\sum_k \delta_s \frac{1}{\eta_i}[\hat{v}^*_{i,s}(s(t_-)+\delta,k) - \hat{v}^*_{i,s}]q_{\tilde{s}k}dt.
\end{cases}
\tag{8.50}
$$

We compute the extra terms:

$$
\begin{cases}
\frac{1}{\eta_i}[\tilde{v}^*_{i,s}(s+\mu,.) - \tilde{v}^*_{i,s}]\\
= r_i^\tau + p_i^\tau \frac{1}{\eta_i}(\tilde{v}^*_{i,z}(s+\mu,.) - \tilde{v}^*_{i,z})\\
= r_i^\tau + p_i^\tau \omega_i^1,\\
\frac{1}{\eta_i}[\tilde{v}^*_{i,s}(s(t_-)+\delta,k) - \tilde{v}^*_{i,s}]\\
= s_i^\tau + p_i^\tau \frac{1}{\eta_i}(\tilde{v}^*_{i,z}(s+\delta,k) - \tilde{v}^*_{i,z}) = s_i^\tau + p_i^\tau \omega_i^2,
\end{cases}
\tag{8.51}
$$

$$\omega_i^1 := \frac{1}{\eta_i}(\tilde{v}_{i,z}^*(s+\mu,.) - \tilde{v}_{i,z}^*),$$

$$\omega_i^2 := \frac{1}{\eta_i}(\tilde{v}_{i,z}^*(s+\delta,k) - \tilde{v}_{i,z}^*).$$

$$\begin{cases} dp_i^\tau = q_i^\tau[-l_i dt + dB] \\ + \int_\Theta r_i^\tau.\tilde{N}(dt,d\theta,\tilde{s}) + s_i^\tau.\tilde{M}(dt) \\ -\{r_{i,s} + b_s p_i^\tau + \sigma_s(q_i^\tau + p_i^\tau l_i) + \frac{1}{\eta_i} E H_{i,s\tilde{m}_i}^\tau\}dt \\ - \int r_i^\tau(t,.)\omega_i^1 \nu(d\theta)dt \\ - \sum_k s_i^\tau \omega_i^2 q_k - \int_\Theta \mu_s(r_i^\tau + p_i^\tau \omega_i^1)\nu(d\theta)dt \\ - \sum_k \delta_s[s_i^\tau + p_i^\tau \omega_i^2]q_{\tilde{s}k}dt. \end{cases} \tag{8.52}$$

Introducing the risk-sensitive non-optimized Hamiltonian is $\eta_i \bar{H}_i$ with

$$\bar{H}_i^\tau = r_i + bp_i^\tau + \sigma(q_i^\tau + l_i p_i^\tau)$$

$$+ \int \mu[r_i^\tau + \omega_i^1 p_i^\tau]\tilde{m}(d\theta)dt + \sum_k \delta[s_i^\tau + \omega_i^2 p_i^\tau]q_k,$$

and

$$\bar{H}_{i,s}^\tau = r_{i,s} + b_s p_i^\tau + \sigma_s(q_i^\tau + l_i p_i^\tau) + \int \mu_s[r_i^\tau + \omega_i^1 p_i^\tau]\tilde{m}(d\theta)dt + \sum_k \delta_s[s_i^\tau + \omega_i^2 p_i^\tau]q_k,$$

the risk-sensitive first-order adjoint system becomes

$$\begin{cases} dp_i^\tau = -\{\bar{H}_{i,s}^\tau + \frac{1}{\eta_i} E[\eta_i \bar{H}_{i,s\tilde{m}_i}^\tau]\}dt \\ -q_i^\tau l_i dt \\ - \int_\Theta r_i^\tau \omega_i^1 \tilde{m}(d\theta)dt \\ - \sum_k s_i^\tau \omega_i^2 q_k dt \\ +q_i^\tau dB + \int_\Theta r_i^\tau.\tilde{N}(dt,d\theta,\tilde{s}) + s_i^\tau.\tilde{M}(dt). \end{cases} \tag{8.53}$$

■

Remark 8.3 In the mean-field-free case, we obtain

$$
\begin{cases}
\tilde{v}^*_{i,z} = \tau_i v_i, \\
p_i^\tau = \frac{v_{i,s}}{\tau_i v_i} \\
q_i^\tau = \frac{1}{\tau_i v_i}[\sigma \tilde{v}_{i,ss} - \tau_i \sigma p_i^\tau v_{i,s}] \\
r_i^\tau = \frac{1}{\tau_i v_i}\{v_{i,s}(s+\mu, \tilde{s}) - v_{i,s}(s, \tilde{s})\} \\
-\frac{1}{v_i} p_i^\tau \{v_i(s+\mu, \tilde{s}) - v_i(s, \tilde{s})\}, \\
s_i^\tau(t) = \frac{1}{\tau_i v_i}[v_{i,s}(s+\delta, k) - v_{i,s}(s, \tilde{s})] \\
-\frac{1}{v_i} p_i^\tau [v_i(s+\delta, k) - v_i(s, \tilde{s})], \\
\eta_i = \tau_i v_i, \\
l_i = \sigma v_{i,s} = \sigma p_i, \\
\omega_i^1 := \frac{1}{v_i}(v_i(s+\mu, .) - v_i), \\
\omega_i^2 := \frac{1}{v_i}(v_i(s+\delta, k) - v_i).
\end{cases}
\tag{8.54}
$$

Example 8.1 Consider two decision-makers (investors) who would like to decide how much to invest in traditional energy sources and renewable energy sources. In terms of portfolio, the traditional energy sources are considered to be less risky with respect to external volatility and weather conditions. The renewable energy investment is more risky but allows for meeting growing energy demand and reducing waste/recycling concerns but also enabling sustainable development and growth with significant socioeconomic, environmental, and health benefits. We consider the following data: The control action spaces are $\mathscr{A}_1 = \mathbb{R}_+, \ \mathscr{A}_2 = [0, 1]$. The drift function is $b := -a_1 + (1 - a_2)\kappa_1 s + a_2 \kappa_2 s$, which is state- and action-dependent with a regime-switching-dependent capital $\kappa_j = \kappa_j(t, \tilde{s})$ for both options $j \in \{1, 2\}$. The diffusion function is $\sigma := a_2 s \tilde{\sigma}(t, \tilde{s})$, the jump rate is $\mu := a_2 s \tilde{\mu}(t, \theta, \tilde{s})$, and the switching is between a finite number of regimes with $\delta = 0$. The instantaneous reward functional is $r := e^{-\beta t} \frac{a_1(t)^{\alpha(\tilde{s})}}{\alpha(\tilde{s})}$, where $\beta > 0$ and the terminal reward is

$$
g := e^{-\beta T} \frac{s(T)^{\alpha(\tilde{s})}}{\alpha(\tilde{s})} + e^{-\beta T} u^{\varepsilon}[\frac{1}{\alpha(\tilde{s})} E s(T)^{\alpha(\tilde{s})}].
$$

with $\alpha(\tilde{s}) \in (0, 1)$, and the function u^{ε} is assumed to be a nondecreasing real-valued function which vanishes with ε. Note that the terminal reward is mean-field-dependent.

We first look for the team solution, when both decision-makers play cooperatively by selecting jointly an action profile $a = (a_1, a_2)$. We start looking for an equilibrium in feedback strategies. We apply the DPP above with a new state as m. The non-optimized Hamiltonian is

$$
\begin{aligned}
&\mathscr{H}(t, s, \tilde{s}, m, a, \hat{V}_m, \hat{V}_{sm}, \hat{V}_{ssm}) \\
&= e^{-\beta t}\frac{a_1^{\alpha(\tilde{s})}}{\alpha(\tilde{s})} + [-a_1 + (1-a_2)\kappa_1 s + a_2\kappa_2 s]\hat{V}_{sm}(t, s, \tilde{s}) \\
&+\tfrac{1}{2}[a_2 s(t)\tilde{\sigma}(t, \tilde{s})]^2\hat{V}_{ssm}(t, s, \tilde{s}) \\
&+\textstyle\int_{\mathbb{R}}[\hat{V}_m(t, s + a_2 s\tilde{\mu}(t, \theta, \tilde{s}), \tilde{s}) - \hat{V}_m(t, s, \tilde{s})]\nu(d\theta) \\
&-\textstyle\int_{\mathbb{R}} a_2 s(t)\tilde{\mu}(t, \theta, \tilde{s})\hat{V}_{sm}(t, s, \tilde{s})\nu(d\theta) \\
&+\textstyle\sum_{k\neq\tilde{s}} q_{\tilde{s}k}[\hat{V}_m(t, s, k) - \hat{V}_m(t, s, \tilde{s})].
\end{aligned}
\tag{8.55}
$$

Applying DPP yields the following system:

$$
\hat{V}_t(t, m, \tilde{s}) + \int_s m(t, ds, \tilde{s}) \sup_{a_1, a_2} \mathscr{H}(t, s, \tilde{s}, m, a, \hat{V}_m, \hat{V}_{sm}, \hat{V}_{ssm}) = 0,
$$

with the terminal condition

$$
\begin{aligned}
\hat{V}(T, m, \tilde{s}) &= e^{-\beta T}\int_y \frac{y^{\alpha(\tilde{s})})}{\alpha(\tilde{s})} m(T, dy, \tilde{s}) \\
&+ e^{-\beta T} u^{\varepsilon}[\frac{1}{\alpha(\tilde{s})}\int_{\mathbb{R}_+} y^{\alpha(\tilde{s})} m(T, dy, \tilde{s})].
\end{aligned}
$$

Note that this is a coupled system between the switching regimes. Next, we determine the best-response strategies. Assuming that $\alpha(\tilde{s}) \in (0, 1)$, the function $a_1 \mapsto e^{-\beta t}\frac{a_1^{\alpha(\tilde{s})}}{\alpha(\tilde{s})} - a_1\hat{V}_{sm}(t, s, \tilde{s})$ is concave in $\mathbb{R}_{++}$. The structure of the optimal strategy of decision-maker 1 is

$$
a_1^* := [e^{+\beta t}\hat{V}_{sm}(t, s, \tilde{s})]^{\frac{1}{\alpha(\tilde{s})-1}} = [e^{+\beta t}\hat{v}_s^*(t, s, \tilde{s})]^{\frac{1}{\alpha(\tilde{s})-1}},
$$

provided that $\hat{V}_{sm}(t, s, \tilde{s})$ is positive. By substituting the expression of a_1^*, we look for decision-maker 2's optimal strategy by maximizing

$$
\begin{aligned}
&a_2 \mapsto [(1-a_2)\kappa_1 s + a_2\kappa_2 s]\hat{V}_{sm}(t, s, \tilde{s}) \\
&+\tfrac{1}{2}[a_2 s(t)\tilde{\sigma}(t, \tilde{s})]^2\hat{V}_{ssm}(t, s, \tilde{s}) \\
&+\textstyle\int_{\mathbb{R}}[\hat{V}_m(t, s + a_2 s\tilde{\mu}(t, \theta, \tilde{s}), \tilde{s}) - \hat{V}_m(t, s, \tilde{s})]\nu(d\theta) \\
&-\textstyle\int_{\mathbb{R}} a_2 s(t)\tilde{\mu}(t, \theta, \tilde{s})\hat{V}_{sm}(t, s, \tilde{s})\nu(d\theta),
\end{aligned}
\tag{8.56}
$$

which involves the functional $\hat{V}_m(t, s + a_2 s\tilde{\mu})$ which depends on a_2. We need to check both first- and second-order optimality:

$$
\begin{aligned}
0 &= (-\kappa_1 + \kappa_2)s\hat{V}_{sm} + a_2[s\tilde{\sigma}(t, \tilde{s})]^2\hat{V}_{ssm} \\
&+\textstyle\int_{\mathbb{R}} s(t)\tilde{\mu}(t, \theta, \tilde{s})[\hat{V}_{sm}(t, s + a_2 s\tilde{\mu}(t, \theta, \tilde{s}), \tilde{s}) - \hat{V}_{sm}(t, s, \tilde{s})]\nu(d\theta).
\end{aligned}
\tag{8.57}
$$

The second-order derivative is negative if

$$[s\tilde{\sigma}(t,\tilde{s})]^2\hat{V}_{ssm} \; + \int_{\mathbb{R}}(s\tilde{\mu})^2\hat{V}_{ssm}(t,s+a_2s\tilde{\mu}(t,\theta,\tilde{s}),\tilde{s})\nu(d\theta) \leq 0. \tag{8.58}$$

We try a solution in the form of the terminal payoff:

$$\hat{V}(t,m) = \xi_1(t)\int_{y\in\mathscr{S}}\frac{y^{\alpha(\tilde{s})}}{\alpha(\tilde{s})}m(t,dy) + \xi_2(t)u^{\varepsilon}[\frac{1}{\alpha(\tilde{s})}\int y^{\alpha(\tilde{s})}m(t,dy)] + o(\varepsilon),$$

$$\begin{aligned}\hat{V}_m(t,s) &= \xi_1(t)\frac{s^{\alpha(\tilde{s})}}{\alpha(\tilde{s})} + \xi_2(t)\frac{s^{\alpha(\tilde{s})}}{\alpha(\tilde{s})}u^{\varepsilon}_s[\frac{1}{\alpha(\tilde{s})}\int y^{\alpha(\tilde{s})}m(t,dy)]\\ &= \frac{s^{\alpha(\tilde{s})}}{\alpha(\tilde{s})}\left[\xi_1(t)+\xi_2(t)u^{\varepsilon}_s(\bar{m}_\alpha)\right],\end{aligned}$$

with $\bar{m}_\alpha := \frac{1}{\alpha(\tilde{s})}\int y^{\alpha(\tilde{s})}m(t,dy)$.

$$\hat{V}_{sm}(t,s) = \left[\xi_1(t)+\xi_2(t)u^{\varepsilon}_s(\bar{m}_\alpha)\right]s^{\alpha(\tilde{s})-1},$$

$$\hat{V}_{ssm}(t,s) = \left[\xi_1(t)+\xi_2(t)u^{\varepsilon}_s(\bar{m}_\alpha)\right][\alpha(\tilde{s})-1]s^{\alpha(\tilde{s})-2} \leq 0.$$

Since $[\alpha(\tilde{s})-1] \leq 0$, $\hat{V}_{ssm}(t,s)$ is negative if $\xi_1(t) and \xi_2(t)$ are positive (which will be proved below):

$$\begin{aligned}&[\hat{V}_{sm}(t,s+a_2s\tilde{\mu}(t,\theta,\tilde{s}),\tilde{s}) - \hat{V}_{sm}(t,s,\tilde{s})]\\ &= \left[\xi_1(t)+\xi_2(t)u^{\varepsilon}_s(\bar{m}_\alpha)\right]s^{\alpha(\tilde{s})-1}[(1+a_2\tilde{\mu})^{\alpha(\tilde{s})-1}-1].\end{aligned} \tag{8.59}$$

The optimal strategy for 1 is

$$a_1^* = s\left(e^{\beta t}[\xi_1+\xi_2u^{\varepsilon}_s(\bar{m}_\alpha)]\right)^{\frac{1}{\alpha(\tilde{s})-1}}.$$

The optimal strategy of decision-maker 2 solves

$$\begin{aligned}0 &= (-\kappa_1+\kappa_2) + a_2\tilde{\sigma}(t,\tilde{s})^2[\alpha(\tilde{s})-1]\\ &+ \int_{\mathbb{R}}\tilde{\mu}(t,\theta,\tilde{s})[(1+a_2\tilde{\mu})^{\alpha(\tilde{s})-1}-1]\nu(d\theta).\end{aligned} \tag{8.60}$$

We take the projection of the root of the latter equation into [0, 1] to obtain the structure of the optimal strategy of decision-maker 2. It remains to determine the functions $\xi_1(t)$ and $\xi_2(t)$.

Let

$$\Omega := [\kappa_1 - (\kappa_1-\kappa_2)a_2 + \frac{1}{2}[a_2\tilde{\sigma}(t,\tilde{s})]^2(\alpha(\tilde{s})-1)]\alpha$$

$$+\int_{\mathbb{R}}\{(1+a_2\tilde{\mu})^{\alpha}-1-a_2\tilde{\mu}\alpha\}\nu(d\theta).$$

The time derivative of the functional $\hat{V}(t,m,\tilde{s})$ is

$$\begin{aligned}&\hat{V}_t(t,m,\tilde{s})=\xi_{1,t}\int_y \frac{y^{\alpha(\tilde{s})}}{\alpha(\tilde{s})}m(dy,\tilde{s})+\xi_{2,t}u^{\varepsilon}[\bar{m}_{\alpha}]\\&=\int_y m(dy,\tilde{s})(\xi_{1,t}\frac{y^{\alpha(\tilde{s})}}{\alpha(\tilde{s})}+\xi_{2,t}u^{\varepsilon}[\bar{m}_{\alpha}])\\&=\int_y m(dy,\tilde{s})\frac{y^{\alpha(\tilde{s})}}{\alpha(\tilde{s})}(\xi_{1,t}+\xi_{2,t}\frac{u^{\varepsilon}[\bar{m}_{\alpha}]}{\bar{m}_{\alpha}}).\end{aligned}\tag{8.61}$$

The Bellman equation reduces to

$$\begin{aligned}&[\xi_{1,t}+\xi_{2,t}\frac{u^{\varepsilon}[\bar{m}_{\alpha}]}{\bar{m}_{\alpha}}]\frac{s^{\alpha(\tilde{s})}}{\alpha(\tilde{s})}\\&+e^{\frac{\beta}{\alpha-1}t}(1-\alpha).[\xi_1+\xi_2u_s^{\varepsilon}(\bar{m}_{\alpha})]^{1+\frac{1}{\alpha(\tilde{s})-1}}\frac{s^{\alpha(\tilde{s})}}{\alpha(\tilde{s})}\\&+\Omega.[\xi_1+\xi_2u_s^{\varepsilon}(\bar{m}_{\alpha})]\frac{s^{\alpha(\tilde{s})}}{\alpha(\tilde{s})}\\&+\frac{s^{\alpha(\tilde{s})}}{\alpha(\tilde{s})}\sum_{k\neq\tilde{s}}q_{\tilde{s}k}\{\xi_1(t,k)+\xi_2(t,k)u_s^{\varepsilon}(\bar{m}_{\alpha(\tilde{s})})\}\\&-\frac{s^{\alpha(\tilde{s})}}{\alpha(\tilde{s})}\sum_{k\neq\tilde{s}}q_{\tilde{s}k}\{\xi_1(t,\tilde{s})+\xi_2(t,\tilde{s})u_s^{\varepsilon}(\bar{m}_{\alpha(\tilde{s})})\}.\end{aligned}\tag{8.62}$$

We use the following approximation

$$[\xi_1+\xi_2u_s^{\varepsilon}(\bar{m}_{\alpha})]^{\frac{\alpha(\tilde{s})}{\alpha(\tilde{s})-1}}=$$

$$\xi_1^{\frac{\alpha(\tilde{s})}{\alpha(\tilde{s})-1}}+\xi_2u_s^{\varepsilon}(\bar{m}_{\alpha})\frac{\alpha(\tilde{s})}{\alpha(\tilde{s})-1}\xi_1^{\frac{\alpha(\tilde{s})}{\alpha(\tilde{s})-1}-1}+\varepsilon.$$

Hence

$$\begin{aligned}&e^{\frac{\beta}{\alpha(\tilde{s})-1}t}(1-\alpha(\tilde{s})).[\xi_1+\xi_2u_s(\bar{m}_{\alpha})]^{1+\frac{1}{\alpha(\tilde{s})-1}}\frac{s^{\alpha(\tilde{s})}}{\alpha(\tilde{s})}\\&=e^{\frac{\beta}{\alpha(\tilde{s})-1}t}(1-\alpha(\tilde{s})).\xi_1^{\frac{\alpha(\tilde{s})}{\alpha(\tilde{s})-1}}\frac{s^{\alpha(\tilde{s})}}{\alpha(\tilde{s})}\\&+e^{\frac{\beta}{\alpha(\tilde{s})-1}t}(1-\alpha(\tilde{s})).\xi_2u_s^{\varepsilon}(\bar{m}_{\alpha})\frac{\alpha(\tilde{s})}{\alpha(\tilde{s})-1}\xi_1^{\frac{\alpha(\tilde{s})}{\alpha(\tilde{s})-1}-1}\frac{s^{\alpha(\tilde{s})}}{\alpha(\tilde{s})}\\&+\varepsilon e^{\frac{\beta}{\alpha(\tilde{s})-1}t}(1-\alpha(\tilde{s})).\frac{s^{\alpha(\tilde{s})}}{\alpha(\tilde{s})}.\end{aligned}\tag{8.63}$$

It follows that as ε vanishes, ξ_1 solves the ordinary differential system:

$$\begin{aligned}&\xi_{1,t}+e^{\frac{\beta}{\alpha(\tilde{s})-1}t}(1-\alpha(\tilde{s})).\xi_1^{1+\frac{1}{\alpha(\tilde{s})-1}}+\Omega.\xi_1+\sum_{k\neq\tilde{s}}q_{\tilde{s}k}[\xi_1(t,k)-\xi_1(t,\tilde{s})]=0,\\&\xi_1(T)=e^{-\beta T}.\end{aligned}\tag{8.64}$$

Lemma 8.8 *The system (8.64) has a unique solution $\xi_1(t,k)$, and the solution is nonnegative.*

Lemma 8.9 *As ε goes to zero, the problem becomes mean-field free and is explicitly solvable. It admits a unique closed-loop solution given by*

$$\hat{V}(t, m, \tilde{s}) = \int_s \xi_1(t, \tilde{s}) \frac{s^{\alpha(\tilde{s})}}{\alpha(\tilde{s})} m(t, ds, \tilde{s}).$$

The dual function is given by $\hat{v}^*(t, s, \tilde{s}) = \xi_1(t, \tilde{s}) \frac{s^{\alpha(\tilde{s})}}{\alpha(\tilde{s})}$, and it corresponds to the equilibrium payoff function of the mean-field-free problem. The optimal strategy of decision-maker 1 in state feedback form is given by

$$a_1^* = s \left(e^{\beta t} \xi_1\right)^{\frac{1}{\alpha(\tilde{s})-1}}. \tag{8.65}$$

Decision-maker 2's optimal strategy depends only on the switching regime and is a root of

$$0=(-\kappa_1+\kappa_2)+a_2\tilde{\sigma}(t,\tilde{s})^2[\alpha(\tilde{s}) - 1]+ \int_{\mathbb{R}} \tilde{\mu}(t, \theta, \tilde{s})[(1 + a_2\tilde{\mu})^{\alpha(\tilde{s})-1} - 1]\nu(d\theta), \tag{8.66}$$

where $\xi_1(t, k)$ is the unique solution to the system (8.64).

For the mean-field-dependent case, however, we need to determine ξ_2. Given $\xi_1(t, k)$, the function ξ_2 can be approximated by solving the ordinary differential system:

$$\begin{aligned} &\xi_{2,t} \frac{u^{\varepsilon}[\bar{m}_\alpha]}{\bar{m}_\alpha u_s^{\varepsilon}(\bar{m}_\alpha)} + \xi_2[e^{\frac{\beta}{\alpha-1}t}\alpha(\tilde{s})\xi_1^{\frac{1}{\alpha(\tilde{s})-1}}] + \frac{\varepsilon}{u_s^{\varepsilon}(\bar{m}_\alpha)} e^{\frac{\beta}{\alpha-1}t}(1-\alpha) \\ &+\Omega.\xi_2 + \textstyle\sum_{k\neq\tilde{s}} q_{\tilde{s}k}\{\xi_2(t, k) - \xi_2(t, \tilde{s})\}. \end{aligned} \tag{8.67}$$

The approximated function solves

$$\begin{aligned} &\xi_{2,t} \frac{u^{\varepsilon}[\bar{m}_\alpha]}{\bar{m}_\alpha u_s^{\varepsilon}(\bar{m}_\alpha)} + \xi_2[e^{\frac{\beta}{\alpha-1}t}\alpha(\tilde{s})\xi_1^{\frac{1}{\alpha(\tilde{s})-1}} + \Omega] \\ &+ \textstyle\sum_{k\neq\tilde{s}} q_{\tilde{s}k}\{\xi_2(t, k) - \xi_2(t, \tilde{s})\}, \\ &\xi_2(T, .) = e^{-\beta T}, \\ &\frac{d}{dt}\bar{m}_\alpha = \bar{m}_\alpha\{-\alpha e^{\frac{\beta}{\alpha-1}t}(\xi_1 + \xi_2 u_s^{\varepsilon})^{\frac{1}{\alpha-1}} + \Omega\} \\ &+ \textstyle\sum_{k\neq\tilde{s}} q_{\tilde{s}k}\{\xi_2(t, k) - \xi_2(t, \tilde{s})\}. \end{aligned} \tag{8.68}$$

Example 8.2 We consider a risk-sensitive linear-quadratic problem in the one-dimensional case:

$$\begin{cases} \sup_{a_i(\cdot)\in\mathscr{A}_i} E e^{\tau_i\left[-\frac{1}{2}s^2(T)-\beta_i(\tilde{s})E[s(T)]+z(T)\right]}, \\ \text{subject to} \\ ds(t) = (b_0 s(t) + b_1 a_1(t) + b_2 a_2(t))dt + \sigma dB(t) + \int \theta \tilde{N}(dt, d\theta), \\ dz(t) = -[\frac{1}{2}a_1^2(t) + \frac{1}{2}a_2^2(t)]dt, \\ s(0) = s_0, \quad z(0) = 0, \end{cases} \tag{8.69}$$

with $b(s,m,a) := b_0 s + b_1 a_1 + b_2 a_2,\ \ r_i(s,m,a) := -\frac{1}{2}a_1^2 - \frac{1}{2}a_2^2,\ \ g_i(s,m,\tilde{s}) := -\frac{1}{2}s^2 - \beta_i \int y m(dy),\ \ \sigma(s) := \sigma, \mu(\theta) = \theta, \delta = 0, q_{ss'} > 0$ if $s \neq s'$. Note that $r_i, b, \sigma, \mu, \delta$ are independent of m. However, g_i depends on m when $\beta_i(\tilde{s}) \neq 0$. The corresponding Hamiltonian

$$H_i(s, \tilde{s}, \tilde{m}, \hat{V}_{i,\tilde{m}}, \hat{V}_{s\tilde{m}}, \hat{V}_{i,z\tilde{m}}, \hat{V}_{i,ss\tilde{m}}) \tag{8.70}$$

$$= \max_{a_i}[(b_0 s + b_1 a_1 + b_2 a_2)\hat{V}_{i,s\tilde{m}} - [\frac{1}{2}a_1^2 + \frac{1}{2}a_2^2]\hat{V}_{i,z\tilde{m}}]$$

$$+\frac{1}{2}\sigma^2 \hat{V}_{i,ss\tilde{m}} + \int_\Theta [\hat{V}_{i,\tilde{m}}(t, s+\theta, \tilde{s}) - \hat{V}_{i,\tilde{m}}(t, s, \tilde{s}) - \theta \hat{V}_{i,s\tilde{m}}]\nu(d\theta) \tag{8.71}$$

$$+\sum_{k\neq\tilde{s}}[\hat{V}_{i,\tilde{m}}(t, s, k) - \hat{V}_{i,\tilde{m}}(t, s, \tilde{s})]q_{\tilde{s},k}. \tag{8.72}$$

The risk-sensitive Bellman system writes

$$\hat{V}_{i,t}(t, \tilde{m}, \tilde{s}) + H_i(s, \tilde{s}, \tilde{m}, \hat{V}_{i,\tilde{m}}, \hat{V}_{s\tilde{m}}, \hat{V}_{i,z\tilde{m}}, \hat{V}_{i,ss\tilde{m}}) = 0, \tag{8.73}$$

$$\hat{V}_{i,t}(T, \tilde{m}, \tilde{s}) = \int_{s,z} e^{\tau_i\left[-\frac{1}{2}s^2 - \beta_i(\tilde{s})[\int_{s'} s' m(T,ds')] + z\right]}\tilde{m}(T, dsdz, \tilde{s}), \tag{8.74}$$

where the best-response strategy is

$$\bar{a}_i = b_i \frac{\hat{V}_{i,s\tilde{m}}}{\hat{V}_{i,s\tilde{m}}} = b_i \frac{\hat{v}^*_{i,s}}{\hat{v}^*_{i,z}}, \tag{8.75}$$

noting that, by Lemma 8.6, $\hat{v}^*_{i,z} \neq 0$ for any time and state. The optimal state $(s(t), z(t))$ solves

$$\begin{cases} ds(t) = (b_0 s(t) + b_1^2 \frac{\hat{v}^*_{1,s}}{\hat{v}^*_{1,z}} + b_2^2 \frac{\hat{v}^*_{2,s}}{\hat{v}^*_{2,z}})dt + \sigma dB(t) + \int \theta \tilde{N}(dt, d\theta), \\ dz(t) = -\sum_i \frac{1}{2}(b_i \frac{\hat{v}^*_{i,s}}{\hat{v}^*_{i,z}})^2 dt, \\ s(0) = s_0, \quad z(0) = 0. \end{cases} \tag{8.76}$$

Its associated infinitesimal generator is

$$\mathscr{A}\psi(s, z, \tilde{s}) = \frac{\sigma^2}{2}\psi_{ss}(s, z, \tilde{s}) + \left(b_0 s + \sum_i b_i^2 \frac{\hat{v}^*_{i,s}}{\hat{v}^*_{i,z}}\right)\psi_s(s, z, \tilde{s}) \tag{8.77}$$

$$-\frac{1}{2}[\sum_i (b_i \frac{\hat{v}^*_{i,s}}{\hat{v}^*_{i,z}})^2]\frac{1}{2}\psi_z(s, z, \tilde{s})$$

$$+\int_\Theta [\psi(s+\theta, z, \tilde{s}) - \psi(s, z, \tilde{s}) - \theta \psi_s]\nu(d\theta, \tilde{s})$$

$$+\sum_{k\neq\tilde{s}}[\psi(s,z,k)-\psi(s,z,\tilde{s})]q_{\tilde{s},k}. \tag{8.78}$$

The dual function $(\hat{v}_i^*)$ is the solution of

$$\hat{v}_{i,t}^*(t,s,z,\tilde{s})+\mathscr{A}\hat{v}_i^*(t,s,z,\tilde{s})=0, \tag{8.79}$$

with terminal value

$$\begin{array}{l}\hat{v}_i^*(T,s,z,\tilde{s})=e^{\tau_i(z+g(s,m(T,\tilde{s})))}\\ -\tau_i\beta_i(\tilde{s})s\int_{\zeta,y}e^{\tau_i(\zeta+g(y,m(T,\tilde{s})))}\tilde{m}(T,d\zeta,dy,\tilde{s}).\end{array} \tag{8.80}$$

The best-response strategies are $b_i\frac{\hat{v}_{i,s}^*}{\hat{v}_{i,z}^*}=b_ip_i^\tau$. By Proposition 8.7, the process $p_i^\tau:=\frac{\hat{v}_{i,s}^*}{\hat{v}_{i,z}^*}$ satisfies

$$\begin{cases}dp_i^\tau(t)=-\{b_0p_i^\tau(t)+l_i(t)q_i^\tau(t)\}dt\\ -\int_\Theta r_i^\tau\omega_i^1\nu(d\theta)dt-\sum_k s_i^\tau\omega_i^2q_kdt\\ +q_i^\tau dB+\int_\Theta r_i^\tau.\tilde{N}(dt,d\theta,.),\\ p_i^\tau(T)=-s(T)-\beta_i(\tilde{s})\frac{\eta_i(0)}{\eta_i(T)},\\ d\eta_i=\eta_i\left[l_idB+\int\omega_i^1\tilde{N}(dt,d\theta,\tilde{s})+\omega_i^2.\tilde{M}(dt)\right],\\ \eta_i(T)=\tau_ie^{\tau_i[z_i(T)+g_i(s(T),m(T,\tilde{s}),\tilde{s}(T))}.\end{cases} \tag{8.81}$$

The process (p_i^τ,q_i^τ,l_i) has an explicit solution in terms of $\eta_i(0)$ and deterministic functions π_i,ω_i and is given by $p_i^\tau(t,\tilde{s})=\pi_i(t,\tilde{s})s(t)+\omega_i(t,\tilde{s})\frac{\eta_i(0)}{\eta_i(t)}$, $q_i^\tau=\sigma\pi_i-l_i(t)\omega_i\eta_i(0)\eta_i^{-1}(t)$, where π_i,ω_i solve the Riccati equations.

The expected value $y(t)=E[s(t)]$ of the optimal state solves the ODE

$$\begin{cases}\dot{y}(t)=b_0y(t)+\sum_i b_i^2E[p_i^\tau],\\ y(0)=E[s(0)],\end{cases} \tag{8.82}$$

which is explicit in terms of $\eta(0)$ to be determined. The equilibrium payoff of the problem is $v_i^\tau:=\frac{\hat{v}_{i,z}(0,s_0,0)}{\tau_i}=\frac{\eta_i(0)}{\tau_i}$ which is again explicit except the dependence on $\eta_i(0)$ to be determined.

Approximation of the Risk-Sensitive Value

We now assume that $\beta_i\neq 0$ is small. Earlier in this section, we provided an explicit solution of the value function in terms of $\frac{\eta_i(0)}{\tau_i}$. However, $\eta_i(0)$ needs to be determined. When $\beta_i=0$, this can be explicitly computed from $\pi_i(0),\omega_i(0)$. The optimal adjoint function is $\hat{v}_{i,0}^*(s,z,t)=e^{\tau_i(z-\frac{1}{2}\Pi_i(t)s^2+\rho_i(t))}$, where Π_i,ρ_i are solutions of a risk-sensitive Riccati equation in the mean-field-free case. Now, when $\beta_i\neq 0$, the problem becomes mean-field-dependent, and we aim to approximate $\hat{v}_{i,\beta}^*(T):=e^{-\tau_i\beta_iy(T)}E\left[e^{\tau_i[z(T)-\frac{1}{2}s^2(T)]}\right]$. For β_i small enough, we test the guess (ansatz) function $\hat{v}^*_{i,\beta}(s,z,t) = e^{\tau_i\beta_i y(T)}\hat{v}^*_{i,0}(s,z,t)$, where $y(T) = y_\beta(T) \sim y_0(T) = \bar{s}_0 e^{\int_0^T [b_0+\sum_i b_i^2 \Pi_i(t)]dt}$,

$$E[e^{\tau_i[z(T)-\frac{1}{2}s^2(T)]}] \sim \int_s\int_z u_0(s,z,0)\tilde{m}(s,z,0)dxdz,$$

which is expressed as $\int_s u_0(s,0,0)m_0(s)ds = \int_s e^{-\frac{1}{2}\Pi_i(0)s^2+\rho_i(0)}m_0(s)ds$.

Example 8.3 (Semi-Explicit Solution)

$$\begin{array}{l}\inf_{u_i} E[L_i] \text{ subject to}\\ dx = [b_1(x-E[x]) + \sum_j b_{2j}(u_j - E[u_j]) + \bar{b}_1 E[x] + \sum_j \bar{b}_{2j}E[u_j]]dt\\ +(x-E[x])[\sigma dB + \int_\Theta \mu d\tilde{N}],\\ \mathbb{P}(\tilde{s}(t+\varepsilon)=\tilde{s}'|\tilde{s},u) = \int_t^{t+\varepsilon}\tilde{q}_{\tilde{s}\tilde{s}'}dt' + o(\varepsilon),\ \tilde{s}'\neq\tilde{s}\\ x(0)=x_0,\ \tilde{s}(0)=\tilde{s}_0,\end{array} \tag{8.83}$$

where the cost functional is

$$\begin{array}{l}L_i = q_{iT}\frac{(x_T-E[x_T])^{2k_i}}{2k_i} + \bar{q}_{iT}\frac{(E[x_T])^{2\bar{k}_i}}{2\bar{k}_i}\\ +\int_0^T q_i\frac{(x-E[x])^{2k_i}}{2k_i} + r_i\frac{(u_i-E[u_i])^{2k_i}}{2k_i} + \bar{q}_i\frac{(E[x])^{2\bar{k}_i}}{2\bar{k}_i} + \bar{r}_i\frac{(E[u_i])^{2\bar{k}_i}}{2\bar{k}_i}dt,\end{array} \tag{8.84}$$

where $k_i \geq 1, \bar{k}_i \geq 1$ are natural numbers and the coefficients are time- and switching-dependent.

Assume $q_i > 0, \bar{q}_i > 0, r_i, \bar{r}_i > 0,\ Ex_0^{2k_i} < +\infty$. The Nash equilibrium of the MFTG (8.83) is given by

$$\begin{array}{l}u_i^* = -\sum_{\tilde{s}\in\mathscr{S}}\left[(\frac{b_{2i}\alpha_i}{r_i})^{\frac{1}{2k_i-1}}(x-E[x]) - (\frac{\bar{b}_{2i}\bar{\alpha}_i}{\bar{r}_i})^{\frac{1}{2\bar{k}_i-1}}E[s]\right]\mathbb{1}_{\{\tilde{s}(t)=\tilde{s}\}},\\ EL_i = E[\alpha_i(0,\tilde{s}_0)\frac{(x_0-E[x_0])^{2k_i}}{2k_i} + \bar{\alpha}_i(0,\tilde{s}_0)\frac{(E[x_0])^{2\bar{k}_i}}{2\bar{k}_i}],\end{array} \tag{8.85}$$

whenever the following system of ordinary differential equations admits a positive solution $[0,T]$.

$$\begin{array}{l}\dot{\alpha}_i + q_i + 2k_i\alpha_i b_1 + \alpha_i k_i(2k_i-1)\sigma^2 + \alpha_i\int_\theta[(1+\mu)^{2k_i} - 1 - 2k_i\mu]\nu(d\theta)\\ +(\sum_{\tilde{s}'}[\alpha_i(t,\tilde{s}') - \alpha_i(t,\tilde{s})]\tilde{q}_{\tilde{s}\tilde{s}'})\\ -(2k_i-1)r_i(\frac{b_{2i}\alpha_i}{r_i})^{\frac{2k_i}{2k_i-1}} - 2k_i\alpha_i\sum_{j\neq i} b_{2j}(\frac{b_{2j}\alpha_j}{r_j})^{\frac{1}{2k_j-1}} = 0,\\ \alpha_i(T,\tilde{s}) = q_i(T,\tilde{s}),\\ \dot{\bar{\alpha}}_i + \bar{q}_i + 2\bar{k}_i\bar{\alpha}_i\bar{b}_1 + (\sum_{\tilde{s}'}[\bar{\alpha}_i(t,\tilde{s}') - \bar{\alpha}_i(t,\tilde{s})]\tilde{q}_{\tilde{s}\tilde{s}'})\\ -(2\bar{k}_i-1)\bar{r}_i(\frac{\bar{b}_{2i}\bar{\alpha}_i}{\bar{r}_i})^{\frac{2\bar{k}_i}{2\bar{k}_i-1}} - 2\bar{k}_i\bar{\alpha}_i[\sum_{j\neq i}\bar{b}_{2j}(\frac{\bar{b}_{2j}\bar{\alpha}_j}{\bar{r}_j})^{\frac{1}{2\bar{k}_j-1}}] = 0,\\ \bar{\alpha}_i(T,\tilde{s}) = \bar{q}_i(T,\tilde{s}),\ \tilde{s}\in\tilde{\mathscr{S}}.\end{array} \tag{8.86}$$

8.3 MASS: Master Adjoint System

In this section, the decision-makers are minimizers, and loss functions are considered. Consider $I \geq 2$ decision-makers with distribution-dependent quantities of interest interacting over a finite time horizon (t_0, t_1), $t_0 < t_1$. A strategy a_i of decision-maker $i \in \mathscr{I} = \{1, \ldots, I\}$ is an A_i-valued measurable function, where A_i is a non-empty convex set. Let $a = (a_i)_{i \in \mathscr{I}}$ be the action profile of all decision-makers. The collective state s of all decision-makers, which can be a common and/or collection of individual states, is in a finite-dimensional Euclidean space $\mathscr{S} := \mathbb{R}^d$, $d \geq 1$. Let $\mathscr{M}_2(\mathbb{R}^d)$ be the set of Borel measures on $\mathbb{R}^d$ with finite second moments. The running cost function of decision-maker $i \in \mathscr{I}$ is l_i, and the terminal cost function is h_i :

$$\begin{aligned} l_i &: D \to \mathbb{R}, \\ h_i &: \mathscr{S} \times \mathscr{M}_2(\mathscr{S}) \to \mathbb{R}, \end{aligned} \tag{8.87}$$

where $D := (t_0, t_1) \times \mathscr{S} \times \prod_i A_i \times \mathscr{M}_2(\mathscr{S}) \times \mathscr{M}_2(\prod_i A_i)$. The measure μ of the state is given by a Kolmogorov equation with coefficients

$$\begin{aligned} b, \tilde{b}, \tilde{d} &: D \to \mathbb{R}^d, \\ \sigma &: D \to \mathbb{R}^{d \times k}, \\ \gamma &: D \times \Theta \to \mathbb{R}^d, \end{aligned} \tag{8.88}$$

where b represents the drift, $\tilde{b}$ represents the inflow into the state, $\tilde{d}$ represents the outflow from state, σ represents the diffusion, and γ represents a jump coefficient with jump measure ν over Θ. Let μ be a Borel measure of the state on $\mathbb{R}^d$. The data of the MFTG is given by

$$\mathscr{G} = \{[t_0, t_1], \mathscr{I}, (A_i, l_i, h_i)_{i \in \mathscr{I}}, b, \sigma, \gamma, \tilde{b}, \tilde{d}, \mu_0\}.$$

The game is played as follows. Given an initial state distribution μ_0 at time t_0, the game $\mathscr{G}$ proceeds as follows. At each instant $t \in (t_0, t_1)$, each decision-maker observes the state (perfect monitoring, perfect state observation), chooses a control action according to their strategy, and observes/measures their cost. We restrict ourselves to the class of admissible strategies in state-and-mean-field-type feedback form. The risk-neutral mean-field-type Nash equilibrium problem is then introduced as

$$\begin{cases} \inf_{a_i} \int h_i(s, \mu(t_1))\mu(t_1, ds) + \int_{t_0}^{t_1} \int l_i(t, s, a, \mu, \tilde{\mu})\mu(t, ds)dt, \\ \text{subject to} \\ \mu_t = -\text{div}_s(b\mu) + \frac{1}{2} trace[(\sigma'\sigma\mu)_{ss}] \\ + \int_{\Theta} [\mu(s-\gamma) - \mu(s) + \langle \mu, \gamma \rangle_s] \nu(d\theta) + (\tilde{b} - \tilde{d})\mu, \\ \mu(t_0, ds) = \mu_0(ds), \\ i \in \mathscr{I}. \end{cases} \tag{8.89}$$

8.3.1 Bellman Systems with Finitely Many Decision-Makers

We introduce the equilibrium cost (if any) as $\hat{V}_i : (t_0, t_1) \times \mathcal{M}_2(\mathcal{S}) \to \mathbb{R}$ given by

$$\begin{cases} \hat{V}_i(t,\mu) := \inf_{a_i} \int h_i(s, \hat{\mu}(t_1))\hat{\mu}(t_1, ds) + \int_t^{t_1} \int l_i(t,s,a,\hat{\mu},\tilde{\mu})\hat{\mu}(t,ds)dt, \\ \text{subject to} \\ \hat{\mu}_t = -\text{div}_s(b\hat{\mu}) + \frac{1}{2}trace[(\sigma'\sigma\hat{\mu})_{ss}] \\ + \int_\Theta [\hat{\mu}(s-\gamma) - \hat{\mu}(s) + \langle \hat{\mu}, \gamma\rangle_s]\nu(d\theta) + (\tilde{b} - \tilde{d})\hat{\mu}, \\ \hat{\mu}(t, ds) = \mu(ds), \\ i \in \mathcal{I}, \end{cases} \tag{8.90}$$

starting from $t \in (t_0, t_1)$ with the measure μ.

A sufficient condition for a mean-field-type Nash equilibrium is obtained by the HJB system given below:

$$\begin{cases} \hat{V}_{i,t}(t,\mu) + H_i(t,\mu,\hat{V}_\mu,\hat{V}_{s\mu},\hat{V}_{ss\mu}) = 0, & \text{on } (t_0,t_1) \times \mathcal{M}_2(\mathcal{S}) \\ \hat{V}_i(t_1,\mu) = \int h_i(s,\mu)\mu(ds), & \text{on } \{t_1\} \times \mathcal{M}_2(\mathcal{S}) \\ i \in \mathcal{I}, \end{cases} \tag{8.91}$$

where the Hamiltonian H_i is given by

$$H_i(t,\mu,\hat{V}_\mu,\hat{V}_{s\mu},\hat{V}_{ss\mu}) = \inf_{a_i \in A_i} \int \Big\{ l_i + \langle b, \hat{V}_{i,s\mu}\rangle + \frac{1}{2}trace[\sigma'\sigma\hat{V}_{i,ss\mu}]$$
$$+ \int_\Theta [\hat{V}_{i,\mu}(t_-, s+\gamma) - \hat{V}_{i,\mu} - \langle \gamma, \hat{V}_{i,s\mu}\rangle]\nu(d\theta) + \hat{V}_{i,\mu}(t,s,\mu)(\tilde{b} - \tilde{d}) \Big\} \mu(ds).$$

The system (8.91) is a system of partial-integrodifferential equations (PIDEs) in the space of measures. It provides not only a mean-field-type Nash equilibrium $(a_i^*)_i$ solution to (8.89) but also the associated equilibrium cost $\hat{V}_i(t_0, \mu_0)$ for decision-maker i. One of the main challenges of the Nash MFTG problem is to understand (8.91). In particular, the existence of a solution to the infinite-dimensional HJB $(\hat{V}_i)_{i \in \mathcal{I}}$ is an open issue in the general setting.

8.3.2 MASS with Finitely Many Decision-Makers

Unnormalized MASS for Jumps

The unnormalized master adjoint system [339, 340] associated with the equilibrium problem (8.89) is given by

(continued)

$$\begin{cases} \hat{U}_{i,t}(t,s,\mu) + \partial_\mu[H_i(t,\mu,\hat{U},\hat{U}_s,\hat{U}_{ss})](s) = 0, & \text{on } (t_0,t_1)\times\mathcal{S}\times\mathcal{M}_2(\mathcal{S}) \\ \hat{U}_i(t_1,s,\mu) = h_i(s,\mu) + \int h_{i,\mu}(s,s',\mu)\mu(ds'), & \text{on } \{t_1\}\times\mathcal{S}\times\mathcal{M}_2(\mathcal{S}), \\ i \in \mathcal{I}, \end{cases} \tag{8.92}$$

where $\partial_\mu[f(p,\mu)](s) := \frac{\delta f}{\delta\mu}(p,s,\mu) := f_\mu(p,s,\mu)$ is the Gâteaux differential of f and the Hamiltonian is given by

$$H_i(t,\mu,\hat{U},\hat{U}_s,\hat{U}_{ss}) = \inf_{a_i\in A_i}\int \Big\{ l_i + \langle b, \hat{U}_{i,s}\rangle + \tfrac{1}{2}trace[\sigma'\sigma\hat{U}_{i,ss}] + \int_\Theta [\hat{U}_i(t_-,s+\gamma,\mu) - \hat{U}_i - \langle\gamma,\hat{U}_{i,s}\rangle]\nu(d\theta) + \hat{U}_i(t,s,\mu)(\tilde{b}-\tilde{d})\Big\}\,\mu(ds).$$

Again, one of the challenges in (8.92) is the well posedness of MASS $\hat{U}$.

We provide an example where MASS admits a solution even with a non-convex and non-Lipschitz Hamiltonian.

Example 8.4 (Distributed Variance Reduction) Consider the distributed variance reduction problem with finitely many atomic agents over a Euclidean space, with the following data:

$$\begin{cases} l_i = \frac{1}{2}r_i\Big|a_i(t,s,\mu) - \int a_i(t,y,\mu)\mu(t,dy)\Big|^2 + \frac{1}{2}\bar{r}_i\Big|\int a_i(t,y,\mu)\mu(t,dy)\Big|^2 \\ +\frac{1}{2}q_i\Big|s - \int y\mu(t,dy)\Big|^2 + \frac{1}{2}\bar{q}_i\Big|\int y\mu(t,dy)\Big|^2, \\ b = \sum_{j\in\mathcal{I}} a_j(t,s,\mu), \\ h_i = q_i\int\Big|s - \int y\mu(t,dy)\Big|^2\mu(ds) + \bar{q}_i\Big|\int y\mu(t,dy)\Big|^2, \\ \sigma,\gamma \ \text{state-action-mean-field independent}, \\ \tilde{b} = \tilde{d}, \\ i\in\mathcal{I}, \end{cases}$$

where $\Big|.\Big| = \Big|.\Big|_2$ is the 2-norm, $q_i, \bar{q}_i, r_i, \bar{r}_i > 0$, which is given by

$$a_i^*(t,s,\mu) = -\frac{1}{r_i}(\hat{V}_{i,s\mu} - \int \hat{V}_{i,y\mu}\mu(dy)) - \frac{1}{\bar{r}_i}\int \hat{V}_{i,y\mu}\mu(dy).$$

The MFTG equilibrium strategy is rewritten as

$$a_i^*(t,s,\mu) = -\frac{1}{r_i}(\hat{U}_{i,s} - \int \hat{U}_{i,y}\mu(dy)) - \frac{1}{\bar{r}_i}\int \hat{U}_{i,y}\mu(dy).$$

The MFTG equilibrium cost system is given by the following HJB system:

$$\begin{aligned}
&\hat{V}_{i,t} + \tfrac{1}{2}q_i \int \Big|s\Big|^2 \mu(ds) + \tfrac{1}{2}(\bar{q}_i - q_i)\Big|\int y\mu(dy)\Big|^2 \\
&-\tfrac{1}{2r_i}\int \Big|\hat{V}_{i,s\mu}\Big|^2 \mu(ds) - (\tfrac{1}{2\bar{r}_i} - \tfrac{1}{2r_i})\Big|\int \hat{V}_{i,s\mu}\mu(ds)\Big|^2 \\
&-\int \langle \textstyle\sum_{j\neq i} \tfrac{1}{r_j}\hat{V}_{j,s\mu}, \hat{V}_{i,s\mu}\rangle \mu(ds) \\
&-\langle \textstyle\sum_{j\neq i}(\tfrac{1}{\bar{r}_j} - \tfrac{1}{r_j})\int \hat{V}_{j,s\mu}\mu(ds), \Big(\int \hat{V}_{i,y\mu}\mu(dy)\Big)\rangle \\
&+\tfrac{1}{2}\int trace[\sigma'\sigma \hat{V}_{i,ss\mu}]\mu(ds) \\
&+\int\int_\Theta [\hat{V}_{i,\mu}(t_-, s+\gamma) - \hat{V}_{i,\mu} - \langle \gamma, \hat{V}_{i,s\mu}\rangle]\nu(d\theta)\mu(ds) \ = 0, \\
&\hat{V}_i(t_1,\mu) = q_i \int \Big|s - \int y\mu(dy)\Big|^2 \mu(ds) + \bar{q}_i\Big|\int y\mu(dy)\Big|^2.
\end{aligned}$$

We observe that the HJB system exhibits a nonlinearity in the measure μ via the terms $\Big|\int \hat{V}_{i,s\mu}\mu(ds)\Big|^2$ and $\langle\int \hat{V}_{j,s\mu}\mu(ds), \Big(\int \hat{V}_{i,y\mu}\mu(dy)\Big)\rangle$. The Hamiltonian of i is non-convex in μ and non-Lipschitz in μ. The system $(\hat{V}_i)_{i\in\mathcal{I}}$ is semi-explicitly solvable, and it is given by

$$\hat{V}_i(t,\mu) = \alpha_i(t)\int \Big|s - \int y\mu(dy)\Big|^2 \mu(ds) + \bar{\alpha}_i(t)\Big|\int y\mu(dy)\Big|^2 + \delta_i(t)\int \mu(dy),$$

where $(\alpha_i, \bar{\alpha}_i, \delta_i)_{i\in\mathcal{I}}$ solve coupled ordinary differential systems provided in Chap. 7, Sect. 7.6.

The master adjoint systems offer the benefit of being a time, state, and mean-field functionality, thereby yielding equilibrium strategies. However, a notable drawback of master adjoint systems is their inability to furnish the equilibrium costs incurred by decision-makers. To determine these costs, a reevaluation of the equilibrium strategies derived from MASS is necessary. It is important to note that, in general, the MASS does not serve as the system of value functions.

8.4 A Class of Coalitional Mean-Field-Type Games

In the standard formulation of a game, a player's payoff function depends on the states and actions of all the players. Yet real-world applications suggest considering also a functional of the probability measure of states and actions of all the players. We consider cooperative MFTGs in which the state dynamics and the payoffs depend not only on the state and action profiles but also on a functional of their probability measure. We establish a stochastic maximum principle (SMP) and provide a time-dependent payoff allocation procedure for coalitions. The allocated payoff considers not only a fairness property but also the cost of forming the coalition. In order to understand the sustainability of cooperation in the long term, time consistency and subgame perfectness solution concept equations are established. We then establish a risk-sensitive SMP for an exponential-integral payoff functional of mean-field type.

Modeling the joint interactive behavior of players in a finite or infinite dynamical system has been one of the major challenges. We examine the possibility for players to cooperate in such a scenario. One of the fundamental elements in the theory of cooperative games is the formulation of the optimal behavior for the players. Player behavior (control action and imputations) satisfying specific optimality behaviors then constitutes a solution of the game. In other words, a candidate solution concept for a dynamic cooperative game is a set of optimality principles, such as dynamic bargaining solution and payoff allocation procedure.

Altruism and cooperation are fascinating research areas. There is a general belief that the players are better off when they all work cooperatively. However, we often observe strange behaviors that are far from cooperation. Thus, if cooperation is the answer, what is the question, and how can we explain the strange behaviors observed (further detailed below)?

Let us consider a simple example with two players. Assume that if they work together (jointly), they will be able to get $v(\{12\})$. Player 1 gets $v(\{1\})$ if he or she works alone and Player 2 gets $v(\{2\})$. From these three numbers, it is not clear why these players should work together.

Depending on the relative ordering of these three values, it is not clear that cooperation is necessarily the most favorable arrangement for the players. We now make this argument more concrete in the following.

Let us introduce a cost of forming a coalition, $c(\{12\}) \geq 0$ which is the cost incurred when both players pool their efforts (it includes information exchange cost and coalition creation cost). While this cost is often neglected in the literature, it may be important in many settings.

Thus, a necessary condition for possible cooperation between the players is

$$v(\{12\}) - c(\{12\}) \geq v(\{1\}) + v(\{2\}).$$

Then, the next question is as follows: What will be their payoff if they cooperate? To answer this question, we need to know how to share the outcome of the cooperation. It is clear that allocating the equal share $\frac{1}{2}[v(\{12\}) - c(\{12\})]$ to each player is not necessarily appropriate since it can be less than $\max(v(\{1\}), v(\{2\}))$. Thus, the allocation has to be done in a more clever way.

Cooperative game-theoretic solutions such as Bargaining, Core, Shapley, and Nucleolus have been introduced to deal with such problems. When stochasticity and time-dependency are involved, these solution concepts require a careful adaptation, and a new necessary condition needs to be established depending on whether the players are risk-seeking or risk-averse. In addition, if the payoff functions and the state dynamics are of mean-field type, the optimality equations need to be established (see Results 8.1 and 8.2).

Specifically, we consider cooperative MFTGs. The general cooperative game of mean-field type of interest can be formulated as a collective reaction to the evolutive states, actions, and distribution of states and actions of the players. The fact that the payoff functional (grand coalition value) may be nonlinear with respect to the conditional expectation makes the cooperative game problem time inconsistent in the sense that Bellman's optimality principle, based on applying the law of iterated conditional expectations on the payoff functional, does not hold. A way to solve this cooperative game problem is to derive an extended version of the dynamic programming principle and an extended Hamilton-Jacobi-Bellman equation, which turn out to be very complex.

We first study cooperative MFTGs in a risk-neutral setup. We establish a payoff allocation procedure for a class of cooperative MFTGs where the distribution enters not only through states, but also through the control actions. The second part of the chapter addresses the risk-sensitive aspect of cooperative interaction. Risk sensitivity is an important aspect for the sustainability of cooperation.

8.4.1 Cooperative MFTGs

Let $T > 0$, $t \in \mathcal{T} = [0, T]$, B be a standard Brownian motion on $\mathcal{T}$ defined on a given filtered probability space $(\Omega, \mathcal{F}, \mathbb{P}, \{\mathcal{F}_t\}_{t\geq 0})$. We consider a cooperativeMFTG given by $\mathcal{G} = (\mathcal{N}, \mathcal{T}, c(., \mathcal{T}), v(., \mathcal{T}))$, where $\mathcal{I} = \{1, 2 \dots, I\}$ is the set of players, $n \geq 1$, $v(., \mathcal{T}) : 2^{\mathcal{N}} \to \mathbb{R}$ is the value associated to a coalition, and $c(., \mathcal{T}) : 2^{\mathcal{N}} \to \mathbb{R}$ is a cost of coalition making. The grand coalition problem $v = v(\mathcal{N}, \mathcal{T})$ is given by

$$\begin{cases} \sup_{(a_1,\dots,a_I)} E\left[g(x(T), E\xi(x(T))) + \int_{t\in\mathcal{T}} r(t, x(t), m(t), a(t))\, dt\right], \\ \text{subject to} \\ dx(t) = b(t, x(t), m(t), a(t))dt + \sigma(t, x(t), m(t), a(t))dB(t) \\ x(0) = x_0 \in \mathcal{X} \subseteq \mathbb{R}^d, \\ m(t) = E\phi(t, x(t), a(t)) \in \mathbb{R}^{d_\phi},\ t \in \mathcal{T}, \end{cases} \tag{8.93}$$

where $\mathcal{I}$ is the set of players (assumed to be finite), $t \in \mathcal{T}$; $x(t) \in \mathcal{X} \subseteq \mathbb{R}^d$ is a d-dimensional state vector, $d \geq 1$; $a_i(t) \in \mathcal{A}_i$, is the control of the player i at time t with $\mathcal{A} = \prod_{i\in\mathcal{I}} \mathcal{A}_i \subseteq \mathbb{R}^k$; the control action set $\mathcal{A}_i$ is non-empty; B is a standard Brownian motion in $\mathbb{R}^l$; $m(t) = E\phi(t, x(t), a(t)) \in \mathbb{R}^{d_\phi}$ is the aggregative term which uses the probability distribution of $(x(t), a(t))$ at time t; and d_ϕ is the dimension of ϕ. The aggregate function is $\phi : \mathcal{T} \times \mathcal{X} \times \mathcal{A} \to \mathbb{R}^{d_\phi}$.

Specification of the Payoffs

The instantaneous payoff function is $r = \sum_{i\in\mathcal{I}} r_i : \mathcal{T} \times \mathcal{X} \times \mathbb{R}^{d_\phi} \times \mathcal{A} \to \mathbb{R}$, and the terminal aggregate function is $\xi : \mathcal{X} \to \mathbb{R}^{d_\xi}$. $g = \sum_i g_i : \mathcal{X} \times \mathbb{R}^{d_\xi} \to \mathbb{R}$ is the terminal payoff.

Specification of the State Dynamics
The drift coefficient function is $b: \mathscr{T} \times \mathscr{X} \times \mathbb{R}^{d_\phi} \times \mathscr{A} \to \mathbb{R}^d$, and the diffusion coefficient function is $\sigma: \mathscr{T} \times \mathscr{X} \times \mathbb{R}^{d_\phi} \times \mathscr{A} \to \mathbb{R}^{d \times l}$.

Definition 8.2 An admissible cooperative control is an $\mathscr{F}$-adapted and square integrable process with values in a non-empty subset $\mathscr{A}$. Denote by $\mathbb{A}$ the class of admissible controls. Any admissible control that solves (8.93) is called an optimal cooperative strategy.

Our objective is to find or characterize (through a stochastic maximum principle) an optimal cooperative strategy $a^* \in \arg\max_a v(.)$, $a \in \mathbb{A}$.

On the Well Posedness of the Problem
We will make the following assumptions:

Hypothesis H1
The functions b, σ, r, g are twice continuously differentiable with the respect to (x, m). Moreover, b, σ, r, g, ϕ and all their first- and second-order derivatives with respect to (x, m) are continuous in (x, m, a) and bounded.

Lemma 8.10 (Existence) *Under H1, for each control $a(.) \in \mathscr{A}$, the state dynamics admits a unique strong solution, $x(t) := x^a(t)$.*

The proof is classical, and a constructive proof can be obtained using the Banach-Picard iteration argument.

Next, we focus on how the cooperation outcome can be allocated between the players in a dynamic setting.

Payoff Allocation
Noncooperative equilibrium plays the role of a benchmark in a cooperative game, i.e., it dictates what players could secure for themselves if there is no agreement, i.e., $v(\{i\}, \mathscr{T}) =$

$$\begin{cases} \sup_{a_i} E\left[g_i(x(T), E\xi(x(T))) + \int_{t\in\mathscr{T}} r_i(t, x(t), m(t), a(t))\, dt \right], \\ \text{subject to} \\ dx(t) = b(t, x(t), m(t), a(t))dt + \sigma(t, x(t), m(t), a(t))dB(t), \\ x(0) = x_0 \in \mathscr{X} \subseteq \mathbb{R}^d, \\ m(t) = E\phi(t, x(t), a(t)) \in \mathbb{R}^{d_\phi} \\ t \in \mathscr{T}. \end{cases} \quad (8.94)$$

Let $R_{c,i}$ be the cooperative payoff of player i, i.e.

$$R_{c,i} = E\left[g_i(x_c^*(T), E\xi(x_c^*(T))) + \int_{t\in\mathscr{T}} r_i(t, x_c^*(t), m_c^*(t), a_c^*(t))\, dt \right].$$

One has $\sum_{i\in\mathscr{I}} R_{c,i} = v(\mathscr{I},\mathscr{T})$. However, one needs to find a better way to share the payoff $v(\mathscr{I},\mathscr{T}) - c(\mathscr{I},\mathscr{T})$. This motivates the introduction of the notion of imputation, i.e., a vector profile $\gamma_1,\ldots,\gamma_I$ such that $\sum_{i\in\mathscr{I}'}\gamma_i \geq v(\mathscr{I}',\mathscr{T}) - c(\mathscr{I}',\mathscr{T})$ for any $\mathscr{I}'\subset\mathscr{I}$, and $\sum_{i\in\mathscr{I}}\gamma_i = v(\mathscr{I},\mathscr{T}) - c(\mathscr{I},\mathscr{T})$.

By virtue of mean-field-type joint optimization, the sum of individual payoffs under cooperation is greater or equal to its noncooperative counterpart, i.e.

$$\sum_i R_{c,i} \geq \sum_i v(\{i\},\mathscr{T}).$$

Thus, the dividend of cooperation, not including the coalition-making cost, is $DC = \sum_i R_{c,i} - \sum_i v(\{i\},\mathscr{T}) \geq 0$.

Therefore, the dividend of cooperation with the inclusion of coalition-making cost to be distributed among the players is $DC - c(\mathscr{I},\mathscr{T})$.

As a first consequence, it is clear that if the coalition-making cost is too high (compared to the game coalition value), then there is no reason for the players to form a coalition.

Therefore, for cooperation, we require the positivity of $DC - c(\mathscr{I},\mathscr{T})$.

Employing a cooperative game approach yields individual payoffs for the entire interval $\mathscr{T}$. The selected imputation has, by definition, the property that each player's payoff in the cooperative game is higher or equal to what they would get in a noncooperative game played on the same time interval.

Let $\gamma_i(s) = \gamma(\{i\},[s,T])$ be the cooperative payoff-to-go after side payment for player i at position $[s,T], 0 < s < T$ of the game. This is the amount that player i will actually pocket. One way to share the payoff is to use a dynamical Shapley value.

8.4.1.1 Dynamical Shapley Value

The allocated payoff to player i is

$$\gamma_i = \sum_{\mathscr{I}'\subset\mathscr{I}, i\notin\mathscr{I}'} \frac{|\mathscr{I}'|!(|\mathscr{I}|-|\mathscr{I}'|-1)!}{|\mathscr{I}|!} \\ [(v-c)(\mathscr{I}'\cup\{i\},[s,T]) - (v-c)(\mathscr{I}',[s,T])]. \tag{8.95}$$

For the two-player case, the payoffs of the players are

$$\gamma_1 = v(\{1\}) + \frac{1}{2}\left[v(\{12\}) - c(\{12\}) - v(\{1\}) - v(\{2\})\right], \tag{8.96}$$

$$\gamma_2 = v(\{2\}) + \frac{1}{2}\left[v(\{12\}) - c(\{12\}) - v(\{1\}) - v(\{2\})\right]. \tag{8.97}$$

Definition 8.3 A cooperative solution is time-consistent if, at any time, $s \in [0, T]$ the cooperative solution payoff-to-go $\gamma(\{i\}, [s, T]) \geq v(\{i\}, [s, T])$ where the deviating payoff $v(\{i\}, [s, T])$ is computed along the state trajectory of the cooperative one $x_c^*(t)$.

A stronger notion of time consistency is that the cooperative payoff-to-go dominates the noncooperative payoff-to-go for any state $x(s)$, $s \in \mathcal{T}$. This is called subgame-consistent solution.

8.4.1.2 Dynamical Proportional Allocation

Though one of the most commonly used allocation principles is the dynamical Shapley value, in the case when players may be asymmetric in their powers and sizes of payoffs, equal imputation of cooperative gains may not be totally agreeable to asymmetric players. To overcome this, one can suggest the allocation principle in which the players' shares of the gain from cooperation are proportional to the relative sizes of their expected deviating payoffs. Thus, a proportional time-consistent solution is given by

$$\begin{aligned}\gamma_i(s) &= \frac{v(\{i\}, [s, T])}{\sum_i v(\{i\}, [s, T])} \sum_i \gamma(\{i\}, [s, T]) \\ &= \frac{v(\{i\}, [s, T])}{\sum_i v(\{i\}, [s, T])} [v(\mathcal{I}, [s, T]) - c(\mathcal{I}, [s, T])]. \qquad (8.98)\end{aligned}$$

Assuming these quantities are positive, one has $\gamma(\{i\}, [s, T]) \geq v(\{i\}, [s, T]), \forall s$, (individual rationality) and $\sum_i \gamma(\{i\}, [s, T]) = v(\mathcal{I}, [s, T]) - c(\mathcal{I}, [s, T])$ (efficiency at any time).

Then, one needs to find the appropriate instantaneous distribution of the payoff γ_i, i.e., find functions $\hat{\gamma}_i$ such that

$$\gamma_i(0) = \hat{\gamma}_i(T) + \int_0^T \hat{\gamma}_i(s)\, ds.$$

Next, we establish necessary conditions for optimality under H1. These conditions may not be sufficient in general. However, if in addition convexity holds, these conditions become sufficient.

Stochastic Maximum Principle

Let us introduce the non-optimized Hamiltonian $H(t, x, m, a, p, q) = b'p + trace(\sigma' q) + r$. This function introduces two adjoint processes p an q. The adjoint process p is associated with the drift b, and q is associated with the diffusion coefficient σ.

First-Order Adjoint Equation
Let

$$\alpha_\phi(t) = r_x + (b_x)'p + \sum_{i=1}^{l}(\sigma_x^i)'q_i$$
$$+(\partial\phi)'\left(E[r_m] + E[(b_m)'p] + E[\sum_{i=1}^{l}(\sigma_m^i)'q_i]\right), \quad (8.99)$$

then $\alpha_\phi(t) \in \mathbb{R}^d$, and the first-order adjoint equation is given by

$$dp = -\alpha_\phi(t)dt + q(t)dB(t), \quad p(T) = g_x(T) + (\partial\xi)'E[g_m(T)]. \quad (8.100)$$

Lemma 8.11 (Existence) *Under assumption H1, the mean-field-type first-order adjoint equation admits a unique $\mathcal{F}$-adapted solution (p, q) such that $E[\sup_{t\in\mathcal{T}}|p(t)|^2 + \int_0^T |q(t)|^2\, dt] < +\infty$.*

Proof The proof follows from Lemma 8.4. ∎

Second-Order Adjoint Equation
Let

$$\beta_\phi = [P'b_x + b_x'P + \sum_{i=1}^{l}(\sigma_x^i)'P\sigma_x^i + \sum_{i=1}^{l} Q_i\sigma_x^i + (\sigma_x^i)'Q_i + H_{xx}]$$
$$+\left(E[r_m'] + E[P'b_m] + E[\sum_{i=1}^{l} Q_i\sigma_m^i + (\sigma_m^i)'Q_i]\right)\partial^2\phi. \quad (8.101)$$

Following Chap. 4, the second-order adjoint equation is given by

$$dP = -\beta_\phi dt + QdB(t), \quad P(T) = g_{xx} + E[g_m']\partial^2\xi. \quad (8.102)$$

Lemma 8.12 (Existence) *Under assumption H1, the mean-field second-order adjoint equation admits a unique $\mathcal{F}$-adapted solution (P, Q) such that $E[\sup_{t\in\mathcal{T}}|P(t)|^2 + \int_0^T |Q(t)|^2\, dt] < +\infty$.*

The proof follows from Lemma 8.5.

8.4.1.3 A Necessary Condition Under H1

Theorem 8.1 *Let H1 hold. If (x^*, a^*) is an optimal solution of the cooperative MFTG, then there exist two pairs of processes (p, q), (P, Q) that satisfy the first-order and second-order adjoint equations, such that*

$$H(t,x^*,m^*,a^*,p,q) - H(t,x^*,m^*,a,p,q) +$$
$$\frac{1}{2}P(t)\left(\sigma(t,x^*,m^*,a^*) - \sigma(t,x^*,m^*,a)\right)^2 \geq 0, \tag{8.103}$$

for all $a(.) \in \mathscr{A}$, almost every t and $\mathbb{P}$-almost surely where

$$\begin{cases} dp(t) = -\alpha_\phi dt + qdB(t), \ p(T) = g_x(T) + (\partial\xi)'E[g_m(T)], \\ dP = -\beta_\phi dt + QdB(t), \ P(T) = g_{xx} + E[g_m']\partial^2\xi. \end{cases}$$

The proof of Theorem 8.1 uses the analogue of the single-player global optimization of mean-field type. The Pontryagin approach via spike variation is established using the same steps as in Sect. 4.5. Therefore, we do not repeat the proof here. The inequality involving the Hamiltonian function in Theorem 8.1 may be achieved at several control laws. In that case, to any optimal control law, we associate and properly define a specific state trajectory. Note that even if the optimal control law is unique, the optimal state trajectory may not be unique (discontinuous) depending on the starting point. This leads to the so-called Skiba point. However, Theorem 8.1 does not say anything about the cooperative game $\mathscr{G}$. For cooperative MFTG, one needs to check the stability of the coalitions and the allocation procedure, which is missing in the global optimization of the mean-field type. The cooperative strategy obtained in Theorem 8.1 may not be time-consistent, i.e., for $t' > t_0$

$$a^*(t_0, x_0, .) \neq \arg\max_a v(\mathscr{I}, x', [t', T]).$$

In order to characterize the payoffs of the deviating players $v(\{i\}, \mathscr{T})$, one needs to understand the noncooperative game solution, whose necessary condition under H1 is given by the following system:

If (x_i^*, a_i^*) is an equilibrium solution of the noncooperativeMFTG, then there are $2I$ pairs of processes $(p_i, q_i), (P_i, Q_i), \ i \in \mathscr{I}$ that satisfy the first-order and second-order system of adjoint equations, such that

$$H_i(t,x^*,m^*,a^*,p_i,q_i) - H_i(t,x^*,m^*,a,p_i,q_i)$$
$$+\frac{1}{2}P_i(t)\left(\sigma(t,x^*,m^*,a^*) - \sigma(t,x^*,m^*,a)\right)^2 \geq 0, \tag{8.104}$$

for all $a_i(.) \in \mathscr{A}_i$, almost every t and $\mathbb{P}$-almost surely where

$$\begin{cases} dp_i(t) = -\alpha_{i,\phi}dt + q_i dB(t), \ p_i(T) = g_{i,x}(T) + (\partial\xi)'E[g_{i,m}(T)], \\ \alpha_{i,\phi} = r_{i,x} + (b_x)'p_i + (\sigma_x)'q_i + (\partial\phi)'\left(E[r_{i,m}] + E[(b_m)'p_i] + E[(\sigma_m)'q_i]\right) \\ dP_i = -\beta_{i,\phi}dt + Q_i dB(t), \ P_i(T) = g_{i,xx} + E[g_{i,m}']\partial^2\xi \\ \beta_{i,\phi} = [P_i'b_x + b_x'P_i + \sigma_x'P_i\sigma_x + Q_i\sigma_x + \sigma_x'Q_i' + H_{i,xx}] \\ \quad + \left(E[r_{i,m}'] + E[P_i'b_m] + E[\sigma_m'Q_i]'\right)\partial^2\phi. \end{cases}$$

In the next subsection, we focus on time-consistent and subgame-consistent solutions.

8.4.1.4 Aggregative Term as the Vector of Mean States and Mean Actions

We consider a particular case where $\phi(t, x(t), a(t)) = (x(t), a(t))$, i.e., the aggregative term is exactly the vector of means of states and means of actions. Subsequently, the payoff has the form of $r(t, x(t), Ex(t), Ea(t), a(t))$, with a similar structure for the coefficient functions b and σ. This means that $d_\phi = 2$ and $\alpha_\phi(t)$ become $r_x + (b_x)'p + (\sigma_x)'q + E[r_m] + E[(b_m)'p] + E[(\sigma_m)'q]$.

The following result follows from Theorem 8.1

Corollary 8.2 *Let H1 hold. If (x^*, a^*) is an optimal solution of the cooperative MFTG, then there are two pairs of processes $(p, q), (P, Q)$ that satisfy the first-order and second-order adjoint equations, such that*

$$H(t, x^*, m^*, a^*, p, q) - H(t, x^*, m^*, a, p, q)$$
$$+\frac{1}{2}P(t)\left(\sigma(t, x^*, m^*, a^*) - \sigma(t, x^*, m^*, a)\right)^2 \geq 0, \tag{8.105}$$

for all $a(.) \in \mathscr{A}$, almost every t and $\mathbb{P}$-almost surely where

$$\begin{cases} dp = -(b_x p + E[b_m p] + \sigma_x q + E[\sigma_m q] + r_x + E[r_m])dt + q dB(t), \\ p(T) = g_x(T) + E[g_m(T)], \\ dP = -[2b_x P + \sigma_x^2 P + 2\sigma_x Q + H_{xx}]dt + Q dB(t), \\ P(T) = g_{xx}. \end{cases}$$

Example 8.5 (Variance Reduction) Let

$-r(t, x(t), Ex(t), Ea(t), a(t)) = \langle Qx, x\rangle + \langle \hat{Q}Ex, Ex\rangle + \langle Ra, a\rangle + \langle \hat{R}Ea, Ea\rangle$.
$-g(T, x(T), Ex(T)) = \langle Gx(T), x(T)\rangle + \langle \hat{G}Ex(T), Ex(T)\rangle$.
$b(t, x(t), Ex(t), Ea(t), a(t)) = Ax + \hat{A}Ex + Ba + \hat{B}Ea$.

$\sigma(t, x(t), Ex(t), Ea(t), a(t)) = A_1 x + \hat{A}_1 Ex + B_1 a + \hat{B}_1 Ea$, where $Q, \hat{Q}, R, \hat{R}, A, \hat{A}, A_1, \hat{A}_1, B_1, \hat{B}_1$ are deterministic matrix-valued functions chosen appropriately with the dimension of the control process $a(t)$.

If $R, Q, G\ \hat{R}, \hat{Q}, \hat{G}$ are symmetric, then our necessary condition above becomes

$$\begin{cases} 2Ra + 2\hat{R}Ea + B'p + \hat{B}'Ep + B_1'q + \hat{B}_1'Eq = 0 \\ dp = -[2Qx + A'p + A_1'q + 2\hat{Q}Ex + \hat{A}'Ep + \hat{A}_1'Eq]dt + q dB(t) \\ \frac{1}{2}p(T) = Gx + \hat{G}E[x]. \end{cases}$$

The optimal cooperative strategy under this procedure may depend on the initial state x_0.

Mean of Actions as Aggregative Term
For the particular case where $\phi(t, x(t), a(t)) = a(t)$, the aggregative term is exactly the mean of actions. The payoff has the form $r(t, x(t), Ea(t), a(t))$, with a similar structure for the coefficients b and σ. Therein, the actions are the electricity demand of the users and the price of a function of aggregate demand and aggregate supply from the producers of electricity.

Mean of States as Aggregative Term
For the particular case where $\phi(t, x(t), a(t)) = x(t)$, the aggregative is exactly the mean of states. The payoff has the form $r(t, x(t), Ex(t), a(t))$, with a similar structure for the coefficients b and σ.

8.4.2 Toward Credible Cooperation

For dynamic games, an additional and stringent condition on the solutions is required: The specific optimality principle must dictate optimality at any instant of time throughout the game duration along the optimal state trajectory. This condition is known as dynamic stability or time consistency.

In the context of MFTGs, the notion of time consistency is crucial since the initial distribution of states and starting time naturally influence the Kolmogorov forward equation. A cooperative solution is subgame-consistent if an extension of the cooperative strategy to a situation with a later starting time and to any possible state brought about by the prior optimal behavior of the players remains optimal. Subgame consistency is a stronger notion of time consistency. In the presence of stochastic elements, subgame consistency is required in a credible cooperative solution. In the field of cooperative MFTGs, little research has been published to date on subgame-consistent solutions.

If the set $\mathscr{A}$ is not necessarily convex, but a general separable complete metric space (Polish space), then Pontryagin's approach suggests the following perturbation method called spike variation. The approach is well adapted to subgame perfection in games. Fix $(s, x) \in [0, T] \times \mathscr{X}$, and define the control law a_ε as the spike variation of $\hat{a}$ over the set $[s, s+\varepsilon],\ \varepsilon > 0$, i.e.

$$a_\varepsilon(t) = a(t)\mathbb{1}_{[s,s+\varepsilon]}(t) + \hat{a}(t)\mathbb{1}_{\mathscr{T}\setminus[s,s+\varepsilon]}(t),$$

where a is an arbitrary admissible control and $\mathbb{1}_{[s,s+\varepsilon]}$ is the indicator function over the set $[s, s+\varepsilon]$.

Definition 8.4 Let R be an objective to be maximized in v. We say that $\hat{a}$ is a subgame perfect cooperative strategy under spike variation if for any t_0, x_0, a

$$\lim_{\varepsilon\to 0} \frac{1}{\varepsilon}\left[R([t_0, T], x_0, \hat{a}) - R([t_0, T], x_0, a_\varepsilon)\right] \geq 0.$$

Note that a subgame perfect cooperative strategy under spike variation is, in particular, a time-consistent solution.

The key difference here is that the solution that we are looking for should not depend on the initial data (when and where we started).

Let H^{t_0,x_0} be the Hamiltonian function associated with the random variable x that starts from x_0 at $t_0 \in \mathcal{T}$. $H^{t_0,x_0}(t,x,m,a,p,q) = bp^{t_0,x_0} + \sigma q^{t_0,x_0} + r^{t_0,x_0}$, where the notation r^{t_0,x_0} is obtained from r when the aggregate term is $m = E_{t_0,x_0}\,\phi(t,x,a)$.

Let

$$
\begin{aligned}
\alpha_\phi^{t_0,x_0}(t) = r_x^{t_0,x_0} &+ (b_x)' p^{t_0,x_0} + (\sigma_x)' q^{t_0,x_0} \\
&+ (\partial\phi)' \left(E[r_m^{t_0,x_0}] + E[(b_m)' p^{t_0,x_0}] + E[(\sigma_m)' q^{t_0,x_0}]\right).
\end{aligned}
$$

Then, the first-order adjoint equation is given by

$$
dp^{t_0,x_0} = -\alpha_\phi^{t_0,x_0}(t)dt + q^{t_0,x_0} dB(t), \tag{8.106}
$$

$$
p^{t_0,x_0}(T) = g_x^{t_0,x_0}(T) + (\partial\xi)' E_{t_0,x_0}[g_m^{t_0,x_0}(T)]. \tag{8.107}
$$

Under assumption H1, the mean-field-type first-order adjoint equation admits a unique $\mathcal{F}$-adapted solution $(p^{t_0,x_0}, q^{t_0,x_0})$ such that $E[\sup_{t\in[t_0,T]} |p^{t_0,x_0}(t)|^2 + \int_{t_0}^T |q^{t_0,x_0}(t)|^2\, dt] < +\infty$.

Let

$$
\begin{aligned}
\beta_\phi^{t_0,x_0} &= P^{t_0,x_0} b_x + b_x' P'^{t_0,x_0} + P^{t_0,x_0} \sigma_x^2 \\
&\quad + Q^{t_0,x_0} \sigma_x^{t_0,x_0} + \sigma_x' Q'^{t_0,x_0} + H_{xx}^{t_0,x_0} \\
&\quad + \left(E[r_m'^{t_0,x_0}] + E[P'^{t_0,x_0} b_m] + E[\sigma_m' Q^{t_0,x_0}]'\right) \partial^2\phi.
\end{aligned}
$$

The second-order adjoint equation is given by

$$
dP^{t_0,x_0} = -\beta_\phi^{t_0,x_0} dt + Q^{t_0,x_0} dB(t), \quad P^{t_0,x_0}(T) = g_{xx}^{t_0,x_0} + E_{t_0,x_0}[g_m'^{t_0,x_0}]\partial^2\xi. \tag{8.108}
$$

Under assumption H1, the mean-field second-order adjoint equation admits a unique $\mathcal{F}$-adapted solution $(P^{t_0,x_0}, Q^{t_0,x_0})$ such that $E[\sup_{t\in[t_0,T]} |P^{t_0,x_0}(t)|^2 + \int_{t_0}^T |Q^{t_0,x_0}(t)|^2\, dt] < +\infty$.

Theorem 8.2 *Let H1 hold. If $(\hat{x}, \hat{a})$ is an optimal solution of the cooperative MFTG, then there are two pairs of processes $(p, q), (P, Q)$ that satisfy the first-order and second-order adjoint equations, such that*

$$
H^{t,\hat{x}(t)}(t, \hat{x}, \hat{m}, \hat{a}, p^{t,\hat{x}(t)}, q^{t,\hat{x}(t)}) - H^{t,\hat{x}(t)}(t, \hat{x}, \hat{m}, a, p^{t,\hat{x}(t)}, q^{t,\hat{x}(t)})
$$

$$+\frac{1}{2}P^{t,\hat{x}(t)}(t)\left(\sigma(t,\hat{x},\hat{m},\hat{a})-\sigma(t,\hat{x},\hat{m},a)\right)^2\geq 0,$$

for all $a(.)\in\mathscr{A}$, almost every t and $\mathbb{P}$-almost surely.

The proof of Theorem 8.2 follows similar steps as that of Theorem 8.1 conditioning on $t_0=t$, $x_0=x$. Note that now the cooperative solution $\hat{a}$ does not depend on the initial data x_0 and can therefore be used for any $t\in[0,T]$ in any subgame.

Example 8.6 (Mean-Variance Approach) In the line of the variance reduction Example 8.5 above, we consider a terminal payoff g as $Ex(T)-\bar{\lambda}\,var(x(T))$, $\bar{\lambda}>0$ and $Q,\hat{Q},R,\hat{R}$ are zeros, which is obviously of mean-field type. The state dynamics is control-dependent and is obtained by setting $\hat{A},\hat{B},\hat{A}_1,\hat{B}_1$ to zero. By Theorem 8.2, one has

$$(*)\left\{\begin{array}{c} B_1'q+B'p=0\\ dp=-Apdt+qdB\\ p(T)=\mathbb{1}-\lambda[x(T)-E_{t_0,x_0}x(T)],\ \lambda=2\bar{\lambda}. \end{array}\right.$$

The final form of $p(T)$ suggests looking for a solution in the following form:

$$p^{t_0,x_0}(t)=\eta_1(t)-\eta_2(t)[x(t)-E_{t_0,x_0}x(t)],$$

where the terminal values are $\eta_1(T)=\mathbb{1}$ and $\eta_2(T)=\lambda I$. By conditioning on t_0,x_0, it is clear that

$$p^{t_0,x_0}(t_0)=\eta_1(t_0),$$

which is independent of x_0. Using system (*), one deduces $q^{t_0,x_0}(t_0)=-(B_1')^{-1}B'\eta_1(t_0)$. By Itô's formula applied to p, and arranging the coefficient functions, one gets by identification

$$(\dot{\eta}_1+A\eta_1)-(\dot{\eta}_2+\eta_2A+A\eta_2)[x(t)-E_{t_0,x_0}x(t)]=0,$$

and $q=-\eta_2B_1\hat{a}$.

$$\left\{\begin{array}{l}\dot{\eta}_1+A\eta_1=0,\ \eta_1(T)=\mathbb{1}\\ \dot{\eta}_2+\eta_2A+A\eta_2=0,\ \eta_2(T)=\lambda I.\end{array}\right.$$

The optimal cooperative control action is

$$\hat{a}(t)=-(\eta_2B_1)^{-1}q=(\eta_2B_1)^{-1}(B_1')^{-1}B'\eta_1(t),$$

which is time-consistent and time-independent of which state the system starts at time t_0 and is a subgame perfect optimal cooperative strategy.

In the unidimensional case

$$(**)\begin{cases} \eta_1(t) = e^{A(T-t)} \\ \eta_2(T) = \lambda e^{2A(T-t)} \\ p^{t,x}(t) = e^{A(T-t)} \\ q^{t,x}(t) = -(B_1')^{-1} B' e^{-A(T-t)} \\ \hat{a}(t) = \frac{1}{\lambda} B_1^{-2} B' e^{A(T-t)}. \end{cases}$$

Note that, now, the optimal control action $\hat{a}(t)$ is independent of the initial data x_0.

8.5 Risk-Sensitive Cooperative Mean-Field-Type Games

We consider the following risk-sensitive cooperative MFTG problem:

$$\mathscr{P}_{rs} = \begin{cases} \sup_{a(.)} \frac{1}{\theta} \log \left[E e^{\theta [g(x(T), E\xi(x(T))) + \int_0^T r(t,x(t),m(t),a(t))]} \right], \\ \text{subject to} \\ dx(t) = b(t, x(t), m(t), a(t))dt + \sigma(t, x(t), m(t), a(t))dB(t), \\ x(0) = x_0 \in \mathscr{X} \subseteq \mathbb{R}^d, \end{cases} \tag{8.109}$$

where $\theta \in \mathbb{R}$ is the risk-sensitivity index.

Note that the above payoff functional $\rho(R) = \frac{1}{\theta} \log \left\{ E[e^{\theta R}] \right\}$ returns a real number or $\pm\infty$ and satisfies the following properties:

(P1) Translation invariance: $\rho(R + c) = \rho(R) + c$ for any random variable R with well-defined moment generating function and $c \in \mathbb{R}$.
(P2) Convexity: $\rho(\lambda R_1 + (1-\lambda)R_2) \le \lambda\rho(R_1) + (1-\lambda)\rho(R_2)$ for all R_1, R_2 and $\lambda \in (0, 1)$.
(P3) Monotonicity: If R_1, R_2 have well-defined moment generating functions and $R_1 \le R_2$, then $\rho(R_1) \le \rho(R_2)$.
(P4) It is additive on independent risks.
(P5) It is normalized, $\rho(0) = 0$.

Moreover, one has a variational representation using the Donsker-Varadhan variational formula.

We note that following Chap. 1, the function ρ is not homogeneous. Thus, it is not a coherent risk measure.

Lemma 8.13 (Donsker-Varadhan Variational Formula) *For any R bounded Borel measurable*

$$\log[Ee^R] = \sup_{\mathbb{Q}<<\mathbb{P}} [E_{\mathbb{Q}} R - d_{KL}(\mathbb{Q}, \mathbb{P})].$$

where $d_{KL}(\mathbb{Q}, \mathbb{P}) := \int \frac{d\mathbb{Q}}{d\mathbb{P}} \ln[\frac{d\mathbb{Q}}{d\mathbb{P}}] d\mathbb{P}$ *is the relative entropy [also called the Kullback-Leibler divergence] of* $\mathbb{Q}$ *with respect to* $\mathbb{P}$.

The main difference between the risk-sensitive problem and usual risk-neutral problems $[\theta \to 0]$ is the exponential-of-integral-type payoff functional (8.109). In this subsection, we first reformulate problem (8.110) with an augmented state and terminal payoff function. An intermediate stochastic maximum principle is then obtained by applying SMP for loss functions without running cost. Then, we transform the intermediate first- and second-order adjoint processes to a more simple form.

Risk-Sensitive SMP: Augmented States
Let $R_\theta = E[e^{\theta R}]$. The risk-sensitive cooperative game problem consists of optimizing $\frac{1}{\theta} \log(R_\theta)$ which is equivalent to optimizing R_θ. This means that the stochastic optimal mean-field-type control problem $(\mathscr{P}_{rs})$ is equivalent to

$$\mathscr{P}_{rs}(x, y) = \begin{cases} \sup_{a(.)} Ee^{\theta[g(x(T), E\xi(x(T))) + y(T)]}, \\ \text{subject to} \\ dx(t) = b(t, x(t), m(t), a(t))dt \\ \qquad + \sigma(t, x(t), m(t), a(t))dB(t), \\ dy(t) = r(t, x(t), m(t), a(t))dt, \\ x(0) = x_0 \in \mathscr{X} \subseteq \mathbb{R}^d, \ y(0) = 0. \end{cases} \tag{8.110}$$

We apply the SMP of Theorem 8.1 with the augmented state $z = (x, y)$ to derive the risk-sensitive first-order adjoint equation.

First-Order Adjoint Equation: Risk-Sensitive Case
Let

$$\alpha_{rs,\phi}(t) = (b_z)' p + (\sigma_z)' q + (\partial\phi)' \left(E[(b_m)' p] + E[(\sigma_m)' q] \right) \tag{8.111}$$

$$= \begin{pmatrix} b_x & 0 \\ r_x & 0 \end{pmatrix}' p + \begin{pmatrix} \sigma_x & 0 \\ 0 & 0 \end{pmatrix}' q + \tag{8.112}$$

$$+(\partial\phi)' E \left[\begin{pmatrix} b_m & 0 \\ r_m & 0 \end{pmatrix}' p + \begin{pmatrix} \sigma_m & 0 \\ 0 & 0 \end{pmatrix}' q \right]. \tag{8.113}$$

Then, $\alpha_\phi(t) \in \mathbb{R}^{d+1}$, and the risk-sensitive first-order adjoint equation is given by

$$dp = -\alpha_{rs,\phi} dt + q dB(t), \tag{8.114}$$

$$p'(T) = \vec{p}(T) = \theta v_T \binom{g_x}{1} + \theta(\partial\xi)' E \left[v_T \binom{g_m(T)}{0} \right]. \tag{8.115}$$

Let H_θ be

$$H_\theta(t,x,m,a,p,q)=\langle (\begin{smallmatrix} b\\ r\end{smallmatrix}),\vec{p}\rangle+\langle (\begin{smallmatrix} \sigma\\ 0\end{smallmatrix}),\vec{q}\rangle.$$

The risk-sensitive second-order adjoint equation is

$$dP=-\beta_{rs,\phi}dt+QdB(t),$$

where

$$\beta_{rs,\phi}=\left\{P\begin{pmatrix} b_x & 0\\ r_x & 0\end{pmatrix}+\begin{pmatrix} b_x & 0\\ r_x & 0\end{pmatrix}'P'+\begin{pmatrix} \sigma_x & 0\\ 0 & 0\end{pmatrix}P\begin{pmatrix} \sigma_x & 0\\ 0 & 0\end{pmatrix}'\right. \tag{8.116}$$

$$\left.+Q\begin{pmatrix} \sigma_x & 0\\ 0 & 0\end{pmatrix}+\begin{pmatrix} \sigma_x & 0\\ 0 & 0\end{pmatrix}'Q'+\begin{pmatrix} H_{\theta,xx} & 0\\ 0 & 0\end{pmatrix}\right\} \tag{8.117}$$

$$+\left(E[P'\begin{pmatrix} b_m & 0\\ r_m & 0\end{pmatrix}]+E[\begin{pmatrix} \sigma_m & 0\\ 0 & 0\end{pmatrix}'Q]'\right)\partial^2\phi. \tag{8.118}$$

with $P(T)=\theta v_T\begin{pmatrix} \theta g_x^2+g_{xx} & \theta g_x\\ \theta g_x & \theta\end{pmatrix}+\theta(\partial^2\xi)'E\left[v_T(\begin{smallmatrix} g_m(T)\\ 0\end{smallmatrix})\right].$

The risk-sensitive SMP with augmented state is stated as follows.

Corollary 8.3 *Let H1 hold. If (x^*,a^*) is an optimal solution of the risk-sensitive cooperative mean-field-type game problem, then there are two pairs of processes (p,q), (P,Q) that satisfy the first-order and second-order risk-sensitive adjoint equations, such that*

$$H_\theta(t,x^*,m^*,a^*,p,q)-H_\theta(t,x^*,m^*,a,p,q)$$

$$+\frac{1}{2}\left(\begin{smallmatrix} \sigma(t,x^*,m^*,a^*)-\sigma(t,x^*,m^*,a)\\ 0\end{smallmatrix}\right)'P(t)\left(\begin{smallmatrix} \sigma(t,x^*,m^*,a^*)-\sigma(t,x^*,m^*,a)\\ 0\end{smallmatrix}\right)\geq 0, \tag{8.119}$$

for all $a\in\mathbb{A}$, almost every t and $\mathbb{P}$-almost surely.

In particular, if $\sigma(t,x,m,a):=\sigma(t,x,m)$ is independent of the action a, then $H_\theta(t,x^*,m^*,a^*,p,q)=\max_a H_\theta(t,x^*,m^*,a,p,q)$.

Transformation of the First-Order Adjoint Process

Although the above result is a good SMP for the risk-sensitive mean-field-type control, the fact that the second component y remains in the adjoint variables makes it difficult to apply. To solve this problem, we need to transform the adjoint processes. The key idea behind such transformation is to eliminate the processes p_2 and q_2.

Since the original problem is $\frac{1}{\theta}\log R_\theta$, it is reasonable to use a logarithmic transformation. The value function is, in general, not smooth. A simple example exhibiting this non-smooth behavior is given by $r(x, a) = xa,\ b(x, a) = -x + a$, and $\mathscr{A} = [-1, 1]$. In this case, the value function is discontinuous at zero, hence non-differentiable. Since Itô's formula is not directly applicable in the non-smooth value function case, we cannot use it here.

Therefore, we need to establish a SMP that applies to more general classes of functions. The idea is to do a change of functions via a generic function $v(t, x(t))$ which is an L^2-martingale, and v is not necessarily related to the adjoint of the control problem.

Noting that $dp_2 = q_2 dB$ and $p_2(T) = -\theta v_T$, an explicit expression for the process p_2 is given by

$$p_2(t) = E[-\theta v_T \mid \mathscr{F}_t].$$

A way to eliminate the pair (p_2, q_2) from the SMP above is to transform it into the form $(\tilde{p}_2, 0)$. Since the process $\tilde{p}_2$ is a martingale, this would mean that $\tilde{p}_2(t) = \tilde{p}_2(T)$ is a constant.

At the terminal time T, one has $\frac{p_2(T)}{\theta v_T} = -1$. Thus, we set $\tilde{p}_2(T) := \frac{p_2(T)}{\theta v_T} = -1$. This suggests also that at any time t, $\tilde{p}_2(t) = -1$.

A natural transformation is then to choose $p_2(t) = \tilde{p}_2(t)\theta v(t) = -\theta v(t)$, with $v(T) = v_T$. Since p_2 is an L^2-martingale, we ask v to be an L^2-martingale. Thus, $v(t) = E[v_T \mid \mathscr{F}_t]$. By the martingale representation theorem, one can also write

$$dv = z dB,\ v(T) = v_T, \tag{8.120}$$

for a certain process z in L^2. The process z can be seen as a Malliavin derivative.

Since $\phi_T > 0$, $v > 0$, assuming that $|h| + |b| \leq c$, we deduce that

$$e^{-\theta[1+Tc]} \leq v(T) \leq e^{\theta[1+Tc]},$$

and

$$\forall t < T,\ e^{-\theta[1+Tc]} \leq v(t) \leq e^{\theta[1+Tc]}.$$

Let $\tilde{p} = \frac{1}{\theta v}\vec{p}$, and assume that $\tilde{p}$ has the following form:

$$d\tilde{p} = -\tilde{\alpha} dt + \tilde{q} dB(t).$$

We apply Itô's formula to $\vec{p} = \theta v \tilde{p}$, $d\vec{p} = \theta v d\tilde{p} + \theta \tilde{p} dv + \theta . d\langle v, \tilde{p}\rangle_t$. Using the expression of dv in (8.120), one arrives at $d\vec{p} = \theta v d\tilde{p} + \theta . z\tilde{q} dt + \theta \tilde{p} z dB(t)$. Thus

$$d\tilde{p} = \frac{d\vec{p}}{\theta v} - \tilde{q}\frac{z}{v} dt - \tilde{p}\frac{z}{v} dB(t).$$

Substituting the expression of $d\vec{p}$, we find the diffusion function $\tilde{q}$ as

$$\tilde{q} = \frac{\vec{q}}{\theta v} - \tilde{p}\frac{z}{v}, \tag{8.121}$$

and the drift function of the process $\tilde{p}$ is

$$- \tilde{\alpha} = -\frac{\alpha}{\theta v} - \tilde{q}\frac{z}{v} = -\begin{pmatrix} b_x & 0 \\ r_x & 0 \end{pmatrix} [\frac{1}{\theta v}\vec{p}] - \frac{1}{\theta v}(\partial\phi)' E[\begin{pmatrix} b_m & 0 \\ r_m & 0 \end{pmatrix} \vec{p}]$$
$$-\begin{pmatrix} \sigma_x & 0 \\ 0 & 0 \end{pmatrix} [\frac{1}{\theta v}\vec{q}] - \frac{1}{\theta v}(\partial\phi)' E[\begin{pmatrix} \sigma_m & 0 \\ 0 & 0 \end{pmatrix} \vec{q}] - \tilde{q}\frac{z}{v}. \tag{8.122}$$

Now, use the relations

$$\begin{cases} \vec{p} = \theta v \tilde{p} \\ \vec{q} = \theta v \tilde{q} + \theta \tilde{p} z, \end{cases}$$

to arrive at

$$\tilde{\alpha}(t) = \begin{pmatrix} b_x & 0 \\ r_x & 0 \end{pmatrix} \tilde{p} + \frac{1}{v}(\partial\phi)' E[\begin{pmatrix} v b_m & 0 \\ v f_m & 0 \end{pmatrix} \tilde{p}]$$
$$+ \begin{pmatrix} \sigma_x & 0 \\ 0 & 0 \end{pmatrix} \left\{\tilde{q} + \tilde{p}\frac{z}{v}\right\} + \frac{1}{v}(\partial\phi)' E[\begin{pmatrix} \sigma_m & 0 \\ 0 & 0 \end{pmatrix} \{v\tilde{q} + \tilde{p}z\}] + \tilde{q}\frac{z}{v}.$$

$\vec{p}(T) = \theta e^{\theta(g(x(T),E\xi x(T))+y(T))}\binom{g_x}{1} + \theta(\partial\xi)' E\left[e^{\theta(g(x(T),E\xi x(T))+y(T))}\binom{g_m}{0}\right] = \theta v(T)\tilde{p}(T)$.

By choosing $v(T) = e^{\theta(g(x(T),\bar{x}(T))+y(T))}$, one gets

$$\tilde{p}(T) = -\binom{g_x + \frac{1}{v(T)}(\partial\xi)' E[v(T) g_m]}{1}.$$

It is easily verified that $d\tilde{p}_2(t) = -\tilde{q}_2\frac{z}{v}dt + \tilde{q}_2 dB = \tilde{q}_2[-\frac{z}{v}dt + dB],\ \tilde{p}_2(T) = -1$.

Since $z \in L^2$ and $e^{-\theta[1+Tc]} < \frac{1}{v} < e^{\theta[1+Tc]}$ almost surely, it is clear that $\frac{z}{v} \in L^2$. We are now in a position to use the Girsanov theorem, which states that

$$d\tilde{p}_2 = \tilde{q}_2 dB^{\mathbb{Q}},\ \mathbb{Q} - almost\ surely,$$

where $\frac{d\mathbb{Q}}{d\mathbb{P}}|_{\mathcal{F}_t} = e^{\int_0^t -\frac{z}{v}dB - \frac{z^2}{2v^2}ds}$.

Hence

$$\tilde{\alpha}_1 = [b_x' + \sigma_x'\frac{z}{v}]\tilde{p}_1 + [\sigma_x' + \frac{z}{v}]\tilde{q}_1 + r_x$$
$$+\frac{1}{v}(\partial\phi)' E\left(v[b_m' + \sigma_m'\frac{z}{v}]\tilde{p}_1 + v\sigma_m'\tilde{q}_1 + v r_m\right), \tag{8.123}$$

and the main risk-sensitive first-order adjoint equation becomes

$$d\tilde{p}_1 = -\tilde{\alpha}_1 dt + \tilde{q}_1 dB(t), \quad \tilde{p}_1(T) = g_x + \frac{1}{v(T)}(\partial\xi)' E[v(T) g_m].$$

Remark 8.4 Note that the risk-sensitive first-order adjoint process dynamics are nonlinear anymore. It exhibits a quadratic coupling between the process z and the processes $\tilde{p}_1$ and $\tilde{q}_1$.

The risk-sensitive Hamiltonian becomes $H_\theta(t, x, m, a, p, q) = [\theta v]\tilde{H}_\theta(t, x, m, a, \tilde{p}_1, \tilde{q}_1)$ where

$$\tilde{H}_\theta(t, x, m, a, \tilde{p}_1, \tilde{q}_1) = b\frac{p_1}{\theta v} + f\frac{p_2}{\theta v} + trace[\sigma\frac{q_1}{\theta v}]$$
$$= b\tilde{p}_1 + r + \sigma[\tilde{q}_1 + \tilde{p}_1\frac{z}{v}] = b\tilde{p}_1 + r + \sigma\tilde{q}_1 + \sigma\tilde{p}_1\frac{z}{v}.$$

Transformation of the Second-Order Adjoint Process
For the second-order adjoint equation, let $\tilde{P}(t) = \frac{P}{\theta v} + \theta\tilde{p}\tilde{p}$. Then, the matrix $P(t)$ is $\theta v\tilde{P} - \theta^2 v\tilde{p}\tilde{p}'$, and the matrix term becomes

$$\frac{1}{2}[\theta v]\begin{pmatrix}\sigma(t, x^*, m^*, a^*) - \sigma(t, x^*, m^*, a)\\ 0\end{pmatrix}\left[\tilde{P} - \theta\tilde{p}\tilde{p}'\right]$$
$$\begin{pmatrix}\sigma(t, x^*, m^*, a^*) - \sigma(t, x^*, m^*, a),\\ 0\end{pmatrix}$$

which proves the following result.

Theorem 8.3 (Risk-Sensitive SMP) *Let H1 hold. If (x^*, a^*) is an optimal solution of the risk-sensitive cooperative game problem, then there are two pairs of processes $(\tilde{p}, \tilde{q})$, $(\tilde{P}, \tilde{Q})$ that satisfy the first-order and second-order risk-sensitive adjoint equations, such that*

$$\tilde{H}_\theta(t, x^*, m^*, a^*, \tilde{p}_1, \tilde{q}_1) - \tilde{H}_\theta(t, x^*, m^*, a, \tilde{p}_1, \tilde{q}_1)$$
$$+\frac{1}{2}\begin{pmatrix}\sigma(., a^*) - \sigma(., a)\\ 0\end{pmatrix}\left[\tilde{P} - \theta\tilde{p}_1\tilde{p}_1'\right]\begin{pmatrix}\sigma(., a^*) - \sigma(., a)\\ 0\end{pmatrix} \geq 0, \quad (8.124)$$

for all $a(.) \in \mathbb{A}$, almost every t and $\mathbb{P}$-almost surely.

In particular, if $\sigma(t, x, m, a) := \sigma(t, x, m)$ is independent of a, then

$$\tilde{H}_\theta(t, x^*, m^*, a^*, \tilde{p}_1, \tilde{q}_1) = \max_a \tilde{H}_\theta(t, x^*, m^*, a, \tilde{p}_1, \tilde{q}_1).$$

Remark 8.5 In particular, for a regular function v that is related to the adjoint process, one obtains the standard MP result if $\sigma_m = 0 = g_m = f_m = r_m$, i.e, the mean-field-free case.

8.6 Notes

In this chapter, two solution approaches to mean-field-type games were presented [102, 330, 331, 333]. The state-of-the-art of stochastic maximum principle includes measure-dependent drifts, diffusions, jumps, and regime switching. The state-of-the-art of the dynamic programming principle works on infinite dimensions, in the space of measures. We have extended the dynamic programming setting by including jump and regime switching in the context of MFTGs. By introducing dual functions which are weak Gâteaux-derivatives with respect to the measure, several important relationships between the stochastic maximum principle and the dynamic programming principle were established. This has allowed us to determine both open-loop strategies and feedback best-response strategies. We have presented master adjoint systems with finitely many decision-makers [62, 338, 340]. Mean-field learning algorithms can be found in [344–346].

Chapter 9
MASS: Master Adjoint Systems with Atomic and Nonatomic Decision-Makers

The term atomic/nonatomic has been widely used in the literature to refer to the measure associated with the set of all decision-makers. If that measure has no atom, the game is sometimes referred to as a nonatomic game. If the index of a single decision-maker is of nonzero measure, that decision-maker is an atomic one. Here, we use a slightly different definition, but the idea is similar: An atomic decision-maker has a non-negligible effect on the population mean-field of states and or actions.

This chapter presents unified interaction problems with distribution-dependent quantities of interest in which a finite number of atomic decision-makers and multiple classes of nonatomic decision-makers are involved. Different types of mean-field terms influence the payoffs and the state dynamics:

- an *individual mean-field term* which may be affected by the decision of the corresponding generic individual, and
- a *population* mean-field term which is generated by the entire population of decision-makers.

In this setting, a single decision-maker has a non-negligible effect on its individual mean-field term but has a negligible influence on the population mean-field term. This particular combined and unified structure changes the classical optimality equations. We characterize mean-field-type Nash equilibria in state-and-mean-field-type feedback strategies using a system of partial integrodifferential equations in the space of measures of both state and control actions. The features are displayed in Table 9.1.

The rest of the chapter is organized as follows. In the next section, we formulate the problem, define a solution concept, and derive a sufficient condition for optimality in state-and-mean-field-type feedback strategies. We then provide a basic example with a polynomial cost. A specific class of games under partial observation is also discussed.

T. Başar et al., *Mean-Field-Type Game Theory I*, Static & Dynamic Game Theory: Foundations & Applications, https://doi.org/10.1007/978-3-032-07027-2_9

Table 9.1 Difference between infinite vs. finite population MFTG

Properties	Infinite population MFTG	MFTG
Model focus	Strategic interactions influenced by population state-mf, action-mf individual state-mf, and action-mf	Broader range, including aggregate effects of state-mf, action-mf of all decision-makers
Equilibrium concept	Mean-field-type equilibria in individual state, individual-mf, population state-mf, and population action-mf behavioral strategies	Cournot, Nash, Stackelberg, Berge, and Saddle point of mean-field type
Population size	Typically infinite	Flexible, accommodates finite and large populations
Population structure	One or multiple infinite populations	Flexible; each decision-maker has its distinguished features; accommodates individuals, teams, and adversaries; finite and large populations
Equilibrium principles	Relies on PDEs, backward-forward SDE	Direct method, DPP, SMP, and Wiener chaos expansion
Applications	Economics, finance, traffic management, biology, etc.	Risk engineering, blockchain, energy, crowd, transportation, food, water, risk AI, social networks, marriage, psychology, economics, finance, network routing, social dynamics, etc.
Congestion effects	Coming from the crowd	Coming from individual decisions
Asymptotic indistinguishability per class	Considered	Flexibly treated, allowing for diverse modeling with atomic, nonatomic, or a mixture of both
Discrete time noises	Fairly general	Fairly general
Continuous time noises	Brownian motion, Poisson process, regime switching, fractional Brownian motion, Gauss-Volterra, and Rosenblatt	Brownian motion, Poisson process, regime switching, fractional Brownian motion, Gauss-Volterra, Rosenblatt, anda mixture of these
Model complexity	Involves solving infinite dimensional sPIDE in the space of measures	Involves PIDEs and backward-forward SIDEs

9.1 The Interaction Model

In this section, we set up the notations and specify the interaction model. We start with the data of arbitrarily many atomic decision-makers. Then, a description of several classes of nonatomic decision-makers is followed. Then, an equilibrium problem is formulated. Below, we consider a probability space $(\Omega, \mathscr{F}, \mathbb{P})$, two time instants $t_0 < t_1$. Define the filtrations $\mathscr{F}_t^{do} := \mathscr{F}_t^{B_{do}, N_{do} x_{do}(t_0)}$ generated by union of events in $\{B_{do}(t'), N_{do}(t'), x_{do}(t_0), t' \leq t\}$ up to time t where B_{do} is a d-dimensional Brownian motion, $N_{do}(dt, d\theta)$ is a jump process with the set of jump sizes $\Theta = \mathbb{R}_+ \backslash \{0\}$, $\tilde{N}_{do}(dt, d\theta) = N_{do}(dt, d\theta) - \nu_{do}(d\theta)dt$ is the compensated jump, and ν_{do} is a Radon measure over Θ. Similarly, $\mathscr{F}_t^{do,c}$ generated by union of events in $\{B_{do}(t'), N_{do}(t'), x_{do}(t_0), B_c(t'), N_c(t'), x_c(t_0), t' \leq t\}$ up to time t. For a given random variable X, its probability measure is denoted by

$$\mathbb{P}_X(\Sigma) := \mathbb{P}(X \in \Sigma) = \mathbb{P}[X^{-1}(\Sigma)] = [\mathbb{P} \circ X^{-1}](\Sigma),$$

for every Σ a $\mathbb{P}_X$-measurable subset of Ω. Given two measures μ_1 and μ_2, the product measure is defined by $[\mu_1 \otimes \mu_2](dx_1, dx_2) = \mu_1(dx_1)\mu_2(dx_2)$.

- $\mathscr{I}$ be set of atomic decision-makers,
- x_{do} the state of all atomic decision-makers,
- μ_{do} be the state-mean-field of all atomic decision-maker,
- a_i denotes the control action of atomic decision-maker i,
- $\mathbb{P}_{a_i}$ action-mean-field of atomic decision-maker i,
- $\mathscr{C}$ the set of classes of nonatomic decision-makers,
- x_c be the state of a generic nonatomic decision-maker in class c,
- m_c be the population state-mean-field of all nonatomic decision-maker in class c,
- $\tilde{m}_c$ be the population action-mean-field of all nonatomic decision-maker in class c,
- μ_c be the individual state-mean-field of a generic nonatomic decision-maker in class c,
- a_c the individual action of a generic nonatomic decision-maker in class c,
- $\mathbb{P}_{a_c}$ the individual action-mean-field of a generic nonatomic decision-maker in class c.

9.1.1 Data of the Atomic Decision-Makers

There are $I \geq 2$ number of atomic decision-makers interacting within the time horizon $[t_0, t_1]$, $t_0 < t_1$. The set of atomic decision-makers is denoted by $\mathscr{I} = \{1, 2, \ldots, I\}$. Decision-maker $i \in \mathscr{I}$ has a control action $a_i \in A_i = \mathbb{R}^d$, $d \geq$

1. The (intrinsic) state process x_{do} of all atomic decision-makers is driven by a multidimensional drift-jump-diffusion process of mean-field type, that is $\mathcal{X}_{do} = \mathbb{R}^d$-valued and is given by

$$\begin{cases} dx_{do} = b_{do}dt + \sigma_{do}dB_{do} + \int_{\Theta} \gamma_{do}(.,\theta)\tilde{N}_{do}(dt, d\theta), \text{ on } (t_0, t_1) \\ x_{do}(t_0) \sim \mu_{do}(t_0, .) \in \mathscr{P}(\mathcal{X}_{do}), \end{cases} \tag{9.1}$$

where

$$\begin{array}{l} (t, x_{do}, (a_j)_{j\in\mathscr{I}}, \mu_{do}, (\mathbb{P}_{a_j})_{j\in\mathscr{I}}, m, \tilde{m}) \\ \in D_{do} := [t_0, t_1] \times \mathcal{X}_{do} \times \prod_{i\in\mathscr{I}} A_i \times \mathscr{P}(\mathcal{X}_{do}) \times \prod_{i\in\mathscr{I}} \mathscr{P}(A_i) \\ \quad \times \prod_{c\in\mathscr{C}}[\mathscr{P}(\mathcal{X}_c) \times \mathscr{P}(A_c)], \\ \text{Drift coefficient: } b_{do} : D_{do} \to \mathcal{X}_{do}, \\ \text{Diffusion coefficient: matrix } \sigma_{do} : D_{do} \to \mathbb{R}^{d\times d}, \\ \text{Jump rate coefficient: } \gamma_{do} : D_{do} \times \Theta \to \mathbb{R}^d. \end{array}$$

where $\mathscr{P}(\mathcal{X}_{do})$ denotes the set of Borel probability measures on $\mathcal{X}_{do}$. We assume that the random processes $x_{do}(t_0)$, B_{do} and N_{do} are mutually independent. A solution to (9.1) will be denoted by $x_{do}(t) := x_{do}^{x_{do}(t_0),\mu_{do}(t_0,.),a}(t)$. The (long-term) performance functional of the atomic decision-maker i is given by

$$L_i((a_j)_{j\in\mathscr{I}}, m, \tilde{m}) =$$

$$h_i(x_{do}(t_1), \mu_{do}(t_1), m(t_1)) + \int_{t_0}^{t_1} l_i(t, x_{do}, (a_j)_{j\in\mathscr{I}}, \mu_{do}, (\mathbb{P}_{a_j})_{j\in\mathscr{I}}, m, \tilde{m})dt,$$

where $\mu_{do}(t, dy) = \mathbb{P}_{x_{do}(t)}(dy)$ is the probability measure of the state $x_{do}(t)$ at time t and

$$\begin{array}{l} l_i : D_{do} \to \mathbb{R}, \\ h_i : \mathcal{X}_{do} \times \mathscr{P}(\mathcal{X}_{do}) \times \prod_{c\in\mathscr{C}} \mathscr{P}(\mathcal{X}_c) \to \mathbb{R}, \end{array}$$

where l_i is the running cost and h_i is the terminal cost at time t_1.

The expected value of the performance functional of the atomic decision-maker i can be written as

$$\begin{array}{l} \hat{L}_i((a_j)_{j\in\mathscr{I}}) := EL_i((a_j)_{j\in\mathscr{I}}, m) \\ = \int h_i(x_{do}, \mu_{do}(t_1, .), m(t_1)).\mu_{do}(t_1, dx_{do}) \\ + \int_{t_0}^{t_1} \int l_i(t, x_{do}, (a_j)_{j\in\mathscr{I}}, \mu_{do}, (\mathbb{P}_{a_j})_{j\in\mathscr{I}}, m, \tilde{m})\mu_{do}(t, dx_{do})dt. \end{array}$$

The expected performance functional $\hat{L}_i$ of i is completely determined by the conditional probability distribution $\mu_{do} \otimes m = \mathbb{P}_{x_{do}} \otimes m$ of the process (x_{do}, m). Since m is conditioned on x_{do}, the dynamics of the state x_{do} is of the conditional McKean-Vlasov type, also called *quenched dynamics*.

9.1.2 Data of the Nonatomic Decision-Makers

There are $C \geq 1$ classes of nonatomic decision-makers denoted by $\mathscr{C} := \{I+1, I+2, \ldots, I+C\}$. A representative decision-maker in class $c \in \mathscr{C}$ has a control action $a_c \in A_c = \mathbb{R}^{d_c}$. The class-specific generic state x_c is driven by a multidimensional drift-jump-diffusion process of mean-field type, that is, $\mathscr{X}_c = \mathbb{R}^{d_c}$−valued and given by

$$dx_c = b_c dt + \sigma_c dB_c + \int_\Theta \gamma_c(., \theta)\tilde{N}_c(dt, d\theta), \quad x_c(t_0) \sim \mu_c(t_0, .), \tag{9.2}$$

where

$$\begin{array}{l} D_c := [t_0, t_1] \times \mathscr{X}_{do} \times \mathscr{X}_c \times \prod_{i\in\mathscr{I}} A_i \times \mathscr{A}_c \times \mathscr{P}(\mathscr{X}_{do}) \times \mathscr{P}(\mathscr{X}_c) \\ \times \prod_{i\in\mathscr{I}} \mathscr{P}(A_i) \times \mathscr{P}(A_c) \times \prod_{c'\in\mathscr{C}}[\mathscr{P}(\mathscr{X}_{c'}) \times \mathscr{P}(A_{c'})], \\ b_c : \ D_c \rightarrow \mathscr{X}_c, \\ \sigma_c : \ D_c \rightarrow \mathbb{R}^{d_c \times d_c}, \\ \gamma_c : D_c \times \Theta \rightarrow \mathbb{R}^{d_c}, \\ \nu_c \text{ is a Radon measure over } \Theta. \end{array}$$

The processes $B_c, N_c, x_c(t_0)$ are assumed to be mutually independent and independent of each of $B_{do}, N_{do}, x_{do}(t_0)$. The performance functional a representative decision-maker in class $c \in \mathscr{C}$ is

$$\begin{array}{l} L_c(a_{do}, a_c, m, \tilde{m}) = h_c(x_{do}(t_1), x_c(t_1), \mu_{do,c}(t_1), m(t_1)) \\ + \int_{t_0}^{t_1} l_c(t, x_{do}, x_c, (a_i)_{i\in\mathscr{I}}, a_c, \mu_{do,c}, (\mathbb{P}_{a_i})_{i\in\mathscr{I}}, \mathbb{P}_{a_c}, m, \tilde{m})dt, \end{array} \tag{9.3}$$

where $\mu_c(t, .) = \mathbb{P}_{x_c(t)}$ is the probability measure of the generic individual state $x_c(t)$.

Note that the expected value of the performance functional L_c can be rewritten as

$$\begin{array}{l} \hat{L}_c((a_i)_{i\in\mathscr{I}}, a_c, m, \tilde{m}) := EL_c(a_{do}, a_c, m, \tilde{m}) \\ = \int h_c(x_{do}, x_c, \mu_{do,c}(t_1, .), m(t_1)) \, \mu_{do,c}(t_1, dx_{do}dx_c) \\ + \int_{t_0}^{t_1} \int l_c(t, x_{do}, x_c, (a_i)_{i\in\mathscr{I}}, a_c, \mu_{do,c}, (\mathbb{P}_{a_i})_{i\in\mathscr{I}}, \mathbb{P}_{a_c}, m, \tilde{m})\mu_{do,c}(t, dx_{do}dx_c)dt. \end{array} \tag{9.4}$$

The expected performance functional $\hat{L}_c$ of c is completely determined by the probability distribution $\mu_{do,c} \otimes m$ of the process (x_{do}, x_c, m).

Key Features of the Proposed MFTG

Two types of mean-field terms are involved in each of the two categories. For atomic decision-makers, the two mean-field terms are defined as follows:

- $\mu_{do} = \mathbb{P}_{x_{do}}$ is the probability measure of x_{do}, called individual mean-field of the state of the atomic decision-makers.
- $\mathbb{P}_{a_i}$ is the individual mean-field of the action a_i of the atomic decision-maker i.

For nonatomic decision-makers, the two mean-field terms are defined as follows:

- m_c is the population mean-field of states in class c, i.e., the probability distribution of states generated by all nonatomic decision-makers of class c.
- $\tilde{m}_c$ is the distribution of actions picked by all nonatomic decision-makers in class c in equilibrium. This is another population mean-field of actions of class c. The population mean-field of actions of nonatomic decision-makers is $\tilde{m} = (\tilde{m}_c)_{c\in\mathscr{C}}$.

A deviating nonatomic decision-maker also has two individual mean-field terms defined as follows:

- $\mu_c = \mathbb{P}_{x_c}$ is the probability measure of x_c and is called individual mean-field of the state of a generic deviating nonatomic decision-makers.
- $\mathbb{P}_{a_c}$ is the individual mean-field of the action a_c of a deviating nonatomic decision-maker in class c.

Some consequences of the model are in order:

- From the model above on $b_c, \sigma_c, \gamma_c, l_c, h_c$, the state process x_c of a generic representative nonatomic decision-maker in class c is correlated with the state x_{do} of the atomic decision-makers.
- What makes the atomic decision-maker in $\mathscr{I}$ a dominating decision-maker is that its coefficients depend on $(x_{do}, \mu_{do}, (\mathbb{P}_{a_i})_{i\in\mathscr{I}}, m, \tilde{m})$ but is not affected by $x_c, a_c, \mathbb{P}_{a_c}$ of a single generic representative nonatomic decision-maker in class c. Each atomic decision-maker $i \in \mathscr{I}$ can affect significantly the population mean-field terms $(m, \tilde{m})$ via $(x_{do}, \mu_{do}, \mathbb{P}_{a_i})$. A change in x_{do} will change m_c for each $c \in \mathscr{C}$ making a conditional probability measure of the process $x_c|x_{do}$, where the conditioning is with respect to the state x_{do} of all atomic decision-makers. The individual dynamics of each of the nonatomic decision-makers influences the common process $(x_{do}, a_i, i \in \mathscr{I})$. The population mean-field terms $m, \tilde{m}$ are random measures.
- What makes a nonatomic decision-maker in class $c \in \mathscr{C}$ a dominated decision-maker is that its coefficients depend on

$$(x_{do}, x_c, (a_i)_{i\in\mathscr{I}}, a_c, \mu_{do}, \mu_c, (\mathbb{P}_{a_i})_{i\in\mathscr{I}}, \mathbb{P}_{a_c}, m, \tilde{m}),$$

  and hence is significantly affected by the data $x_{do}, \mu_{do}, (\mathbb{P}_{a_i})_{i\in\mathscr{I}}$ of the atomic decision-makers. A representative nonatomic decision-maker in class $c \in \mathscr{C}$ has a negligible effect on the population mean-field terms $(m, \tilde{m})$. For each nonatomic decision-maker in class $c \in \mathscr{C}$, there is an interaction with the class c via $(m_c, \tilde{m}_c)$, but also there is an interaction across other classes via $(m_{c'}, \tilde{m}_{c'})_{c'\in\mathscr{C}}$ and also an interaction with the atomic decision-maker.

Key Differences

- What makes this model <u>different</u> from multi-population mean-field games? The MFTG model introduced in this chapter differs from multi-population mean-field games in many ways. In multi-population mean-field games, the functions $l_c, h_c, b_c, \sigma_c, \gamma_c$ are *independent* of $\mu_c, \mathbb{P}_{a_c}$. The classical multi-population MFG models do not include, for example, the variance of the individual state and covariance between state and action of a representative decision-maker.

- What makes this model <u>different</u> from mean-field games with multiple major players and multi-class minor players?

  In the classical mean-field games with multiple major players and multi-class minor players, $l_{do}, h_{do}, b_{do}, \sigma_{do}, \gamma_{do}$ are *independent* of $\mu_{do}, (\mathbb{P}_{a_i})_{i\in\mathcal{I}}$, and $l_c, h_c, b_c, \sigma_c, \gamma_c$ are *independent* of $\mu_c, \mathbb{P}_{a_c}$. These major/dominating-minor models do not include, for example, the variance (or higher order moments) of the individual state of a representative decision-maker. A typical cost function of a representative agent, which is not covered in the prior (extended) MFG models, is given by

$$c(x_{do}, m, \tilde{m}) \sum_{i\in\mathcal{I}} \left(a_i - \int a_i d\mu_{do}\right)^6$$

$$+ q(x_{do}, m, \tilde{m}) \left(a_c - \int a_c d\mu_{do}\right)^4 + \left(a_c - \int a_c d\mu_{do}\right)\left(x_c - \int x_c d\mu_{do}\right).$$

9.1.3 Equilibrium Problem

Consider the following information structure:

- Perfect state measurement, i.e., all the decision-makers observe the state x_{do}.
- Perfect knowledge of the model, i.e., the model above is known by all decision-makers.
- Each *nonatomic* decision-maker in class c observes its own state x_c.
- In addition, each decision-maker is assumed to have a computational capability such as being able to compute an aggregative term of m from the model.

Let $\mathcal{A}_i$ be the set of control strategies of atomic decision-maker i that are A_i-valued and $\mathcal{F}^{do}$-progressively measurable with respect to the filtration generated by the union of events in $\{B, N, x_{do}(t_0)\}$. Let $\mathcal{A}_c$ be the set of control strategies of a nonatomic decision-maker in class c that are $\mathcal{F}^{do,c}$-progressively measurable. We restrict our attention to state-and-mean-field-type feedback strategies. This allow us to write $\tilde{m}$ as function of m and $\mathbb{P}_{a_c}$ as function of μ_c. The instantaneous costs l_c, l_i are then written as functions of m, μ_c, and by abuse of notation, we use the same functions.

9.1.3.1 Evolution of Probability Measures

Each atomic decision-maker will need the joint process (x_{do}, m) in its cost functional. The evolution of the joint process is given by

$$d\begin{pmatrix} x_{do} \\ m \end{pmatrix} = d\begin{pmatrix} b \\ O^*_{\mathscr{C}}[m] \end{pmatrix} dt + \begin{pmatrix} \sigma_{do} & 0 \\ 0 & 0 \end{pmatrix} d\begin{pmatrix} B_{do} \\ 0 \end{pmatrix} + \int_{\Theta} \begin{pmatrix} \gamma_{do} & 0 \\ 0 & 0 \end{pmatrix} \begin{pmatrix} \tilde{N}_{do}(dt, d\theta) \\ 0. \end{pmatrix} \tag{9.5}$$

Proposition 9.1 *The infinitesimal generator of the $(x_{do}(t), m(t))$ is given by*

$$\begin{aligned} &O_{do,m}[f_{\mu_{do}}] = f_{x_{do}\mu_{do}} b_{do} + \tfrac{1}{2}\langle \sigma_{do}, \sigma_{do} f_{x_{do}x_{do}\mu_{do}} \rangle \\ &+ \textstyle\int_{\Theta} \{ f_{\mu_{do}}(t, x_{do}(t_-) + \gamma_{do}(t_-, ., \theta), m) - f_{\mu_{do}}(t, x_{do}(t_-), m) \\ &- \langle f_{x_{do}\mu_{do}}, \gamma_{do}(t, ., \theta) \rangle \} \nu_{do}(d\theta) \\ &+ \textstyle\int O_{\mathscr{C}}[f_{m\mu_{do}}] m(dy), \end{aligned} \tag{9.6}$$

and

$$\begin{aligned} &O_{\mathscr{C}}[f_{m\mu_{do}} | x_{do}, a] := \textstyle\sum_c \langle \bar{b}_c, f_{y_c m \mu_{do}} \rangle + \tfrac{1}{2} \textstyle\sum_c \langle \bar{\sigma}_c, \bar{\sigma}_c f_{y_c y_c m \mu_{do}} \rangle \\ &+ \textstyle\int_{\Theta} \{ f_{m\mu_{do}}(t, y(t_-) + \bar{\gamma}(t_-, ., \theta), m) - f_{m\mu_{do}}(t, y(t_-), m) \\ &- \textstyle\sum_{c\in\mathscr{C}} \langle f_{y_c m \mu_{do}}, \bar{\gamma}_c(t, ., \theta) \rangle \} \nu_{\mathscr{C}}(d\theta), \end{aligned} \tag{9.7}$$

where O^* is the adjoint operator of O, ν is the jump intensity measure, $\langle x, y \rangle = trace[x^* y]$ where x^* is the transpose of x, f is a generic real-valued test function of $\mu_{do} \otimes m$, and $\bar{\phi}$, $\phi \in \{b, \sigma, \gamma\}$ denotes the evaluation at the optimal strategies of other nonatomic decision-makers $(a^*_c, \mathbb{P}_{a^*_c})$, i.e.

$$\bar{\phi} = \phi(t, x_{do}, x^*_c, (a_i)_{i\in\mathscr{I}}, a^*_c, \mu_{do,c}, (\mathbb{P}_{a_i})_{i\in\mathscr{I}}, \mathbb{P}_{a^*_c}, m, \tilde{m}).$$

Proof Let g be a test function on $\mathscr{X}_{do} \times \mathscr{P}(\prod_{c\in\mathscr{C}} \mathscr{X}_c)$. By applying functional Itô's formula for jump-diffusion processes, we obtain

$$\begin{aligned} &g(x_{do}(t_1), m(t_1)) \\ &= g(x_{do}(t_0), m(t_0)) + \textstyle\int_{t_0}^{t_1} [g_{x_{do}} b_{do} + \int O_{\mathscr{C}}[g_m] m(ds) + \tfrac{1}{2}\langle \sigma_{do}, \sigma_{do} g_{x_{do}x_{do}} \rangle] dt \\ &+ \textstyle\int_{t_0}^{t_1} \int [g(x_{do} + \gamma_{do}, m) - g(x_{do}, m) - \langle \gamma_{do}, g_{x_{do}} \rangle] \nu_{do}(d\theta)] dt \\ &+ \textstyle\int_{t_0}^{t_1} \langle g_{x_{do}}, \sigma_{do} dB_{do} \rangle + \int_{t_0}^{t_1} \int [g(x_{do} + \gamma_{do}, m) - g(x_{do}, m)] \tilde{N}(dt, d\theta). \end{aligned} \tag{9.8}$$

Taking the expected value, we arrive at the announced result.

Remark 9.1 Note that the multi-class population mean-fields $m = (m_c)_{c\in\mathscr{C}}$ and $\tilde{m} = (\tilde{m}_c)_{c\in\mathscr{C}}$ are random processes as they are conditioned on $(x_{do}, (a_j)_j, \mu_{do}, (\mathbb{P}_{a_j})_j)$. Equation (9.6) can be modified to include correlations between $B_{do}, B_c, c \in \mathscr{C}$ by adding

$$\sum_{c\in\mathscr{C}} trace[\sigma_{do}^{*}\sigma_c f_{x_{do}x_c\mu_{do,c}}] + \frac{1}{2}\sum_{(c,c')\in\mathscr{C}^2,\ c\neq c'} trace[\sigma_c^{*}\sigma_{c'} f_{x_c s_{c'}\mu_{do,c}}].$$

Similarly, the conditional probability law of the process x_c is given by

$$d\mu_c = O_c^{*}[\mu_c|x_{do}, (a_i)_{i\in\mathscr{I}}]\, dt, \tag{9.9}$$

where

$$\begin{aligned} &O_c[f|x_{do}, a] := \langle b_c, f_{x_c}\rangle + \tfrac{1}{2}\langle \sigma_c, \sigma_c f_{x_c x_c}\rangle \\ &+ \int_{\Theta}\{f(t, x_c(t_-) + \gamma_c(t_-, ., \theta)) - f(t, x_c(t_-)) - \langle f_{x_c}, \gamma_c(t, ., \theta)\rangle\}\nu_c(d\theta). \end{aligned} \tag{9.10}$$

9.1.3.2 Best-Response Problem of Atomic Decision-Maker

Given the multi-class population mean-field $(m, \tilde{m})$ and the strategies $(a_j)_{j\in\mathscr{I}\backslash\{i\}}$ of the other atomic decision-makers than i, the atomic decision-maker $i \in \mathscr{I}$ solves a best-response problem of mean-field type given by

$$\begin{aligned} &\inf_{a_i\in\mathscr{A}_i} \hat{L}_i((a_j)_{j\in\mathscr{I}}, \mu_{do}\otimes m) \text{ subject to} \\ &d\mu_{do} = O_{do}^{*}[\mu_{do}]\, dt, \\ &dm = O_{\mathscr{C}}^{*}[m|s_{do}, (a_i)_{i\in\mathscr{I}}]\, dt, \\ &\mu_{do}(t_0, .)\otimes m(t_0, .). \end{aligned} \tag{9.11}$$

The atomic decision-makers are influenced by the collective motion of the multi-class population mean-fields $m = (m_c)_{c\in\mathscr{C}}$ and $\tilde{m} = (\tilde{m}_c)_{c\in\mathscr{C}}$.

9.1.3.3 Best-Response Problem of Nonatomic Decision-Maker

Given $((a_j, \mathbb{P}_{a_j})_{j\in\mathscr{I}}, m, \tilde{m})$, a representative decision-maker of class $c \in \mathscr{C}$ solves the best-response problem of mean-field type given by

$$\begin{aligned} &\inf_{a_c} \hat{L}_c((a_i)_{i\in\mathscr{I}}, a_c, , \mu_{do,c}\otimes m) \text{ subject to} \\ &d\mu_{do,c} = O_{do,c}^{*}[\mu_{do,c}]\, dt, \\ &dm = O_{\mathscr{C}}^{*}[m|s_{do}, (a_i)_{i\in\mathscr{I}}]\, dt, \\ &\mu_{do}(t_0, .)\otimes \mu_c(t_0, .)\otimes m(t_0, .). \end{aligned} \tag{9.12}$$

Following Proposition 9.1, the infinitesimal generator of the process $(x_{do}(t), x_c(t), m(t))$ is given by

$$
\begin{aligned}
&O_{do,c,m}[f_{\mu_{do,c}}] = f_{x_{do}\mu_{do,c}} b_{do} + \tfrac{1}{2}\langle \sigma_{do}, \sigma_{do} f_{x_{do}x_{do}\mu_{do,c}}\rangle + f_{x_c\mu_{do,c}} b_c \\
&+\tfrac{1}{2}\langle \sigma_c, \sigma_c f_{x_c x_c \mu_{do,c}}\rangle \\
&+\int_\Theta \left\{ f_{\mu_{do,c}}(t, x_{do}(t_-) + \gamma_{do}(t_-, \theta_{do}), x_c(t_-) + \gamma_c(t_-, ., \theta_c), .) \right. \\
&- f_{\mu_{do,c}}(t, x_{do}(t_-), x_c(t_-), .) \\
&\left. -\langle f_{x_{do}\mu_{do,c}}, \gamma_{do}(t, ., \theta_{do})\rangle - \langle f_{x_c\mu_{do,c}}, \gamma_c(t, ., \theta_c)\rangle \right\} \nu_{do}(d\theta_{do})\nu_c(d\theta_c) \\
&+\int O_{\mathscr{C}}[f_{m\mu_{do,c}}] m(dy).
\end{aligned}
\tag{9.13}
$$

9.1.4 Mean-Field-Type Nash Equilibrium Problem

A mean-field-type Nash equilibrium is a strategy profile in which no decision-maker has an incentive to change her strategy unilaterally. This means that no atomic decision-maker can lower her expected cost by changing her choice alone, and one nonatomic decision-maker can improve her performance by unilateral deviation. At a Nash equilibrium, each decision-maker is playing its best response to the others' strategies and/or reacting to the mean-field. Therefore, the mean-field-type Nash equilibrium problem combines the best-response problems (9.11) (9.12) together with a consistency check $\mu_c = m_c$ for every class $c \in \mathscr{C}$:

$$
\text{(MF-NE)} \begin{cases}
\text{Find } (a_i^*)_i, (a_c^*)_c, \ \mu_{do}, (\mu_c)_c, (\mathbb{P}_{a_j^*})_j, m, \tilde{m}, (\hat{V}_j)_j, (\hat{V}_c)_c \text{ such that} \\
\forall\, i \in \mathscr{I}, \ (a_i^*, \mathbb{P}_{a_i^*}) \text{ is a best response strategy to } a_{-i}^*, (\mathbb{P}_{a_j^*})_{j\neq i}, m, \tilde{m}, \\
\qquad \hat{V}_i \text{ is an equilibrium cost in Problem (9.11)} \\
\forall c \in \mathscr{C}, \ (a_c^*, \mathbb{P}_{a_c^*}) \text{ is a best response strategy to } (a_i^*)_{i\in\mathscr{I}}, (\mathbb{P}_{a_i^*})_{i\in\mathscr{I}}, m, \tilde{m}, \\
\qquad \hat{V}_c \text{ is an equilibrium cost in Problem (9.12)}, \\
\mathbb{P}_{a_c^*} = \tilde{m}_c, \\
\mu_c^* = m_c.
\end{cases}
\tag{9.14}
$$

The main difficulty with the problem (MF-NE) is that it involves a fixed point at different levels: an inner fixed point within each class, an outer fixed point with other classes, and a fixed point with the atomic decision-makers:

$$
(a_i, a_c)_{i,c} \to (\mathbb{P}_{a_i}, \mathbb{P}_{a_c})_{i,c} \to (x_{do}, x_c, \mu_{do}, \mu_c, m, \tilde{m}) \to (\hat{V}_i, \hat{V}_c)_{i,c} \to (a_i, a_c)_{i,c}
$$

Next, we would like to derive sufficient conditions for mean-field-type Nash equilibria in state-and-mean-field type feedback strategies.

9.1.5 Equilibrium System

In order to apply the dynamic programming principle in infinite dimensions to the MFTG problem above, we introduce the integrand Hamiltonians. The integrand Hamiltonian of atomic decision-maker i is

$$
\begin{aligned}
\hat{H}_i &= \hat{H}_i(t, x_{do}, \mu_{do}, m, \tilde{m}, (\hat{V}_{j,\mu_{do}}, \hat{V}_{j,x_{do}\mu_{do}}, \hat{V}_{j,x_{do}x_{do}\mu_{do}})_j, \\
&(\hat{V}_{j,m\mu_{do}}, \hat{V}_{j,x_c m\mu_{do}}, \hat{V}_{j,x_c x_c m\mu_{do}})_{j,c}) \\
&:= \inf_{a_i} \left\{ l_i + O_{do,m}[\hat{V}_{i,\mu_{do}}] \right\} \\
&= \inf_{a_i} \Big\{ l_i + \hat{V}_{i,s_{do}\mu_{do}} b_{do} + \tfrac{1}{2}\langle \sigma_{do}, \sigma_{do} \hat{V}_{i,s_{do}s_{do}\mu_{do}} \rangle \\
&+ \textstyle\int_\Theta \Big\{ \hat{V}_{i,\mu_{do}}(t, s_{do}(t_-) + \gamma_{do}(t_-, ., \theta), m) \\
&- \hat{V}_{i,\mu_{do}}(t, s_{do}(t_-), m) - \langle \hat{V}_{i,s_{do}\mu_{do}}, \gamma_{do}(t, ., \theta) \rangle \Big\} \nu_{do}(d\theta) \\
&+ \textstyle\int [\sum_c \langle \bar{b}_c, \hat{V}_{i,y_c m\mu_{do}} \rangle + \tfrac{1}{2} \sum_c \langle \bar{\sigma}_c, \bar{\sigma}_c \hat{V}_{i,y_c y_c m\mu_{do}} \rangle] m(dy) \\
&+ \textstyle\int \int_\Theta \Big\{ \hat{V}_{i,m\mu_{do}}(t, s_{do}, \mu_{do}, y(t_-) + \bar{\gamma}(t_-, ., \theta), m) \\
&- \hat{V}_{i,m\mu_{do}}(t, s_{do}, \mu_{do}, y(t_-), m) - \textstyle\sum_{c\in\mathscr{C}} \langle \hat{V}_{i,y_c m\mu_{do}}, \bar{\gamma}_c(t, ., \theta) \rangle \Big\} \nu_{\mathscr{C}}(d\theta) m(dy) \Big\},
\end{aligned} \tag{9.15}
$$

and $\hat{V}_i$ is a function of $(t, \mu_{do} \otimes m)$. The process m appears as part of the (augmented) state in the Hamiltonian of atomic decision-maker i because m may depend heavily on x_{do}, μ_{do} via $l_c, h_c, b_c, \sigma_c, \gamma_c,\ c \in \mathscr{C}$. An optimizer of the integrand Hamiltonian $\hat{H}_i$ (if any) is in state-and-mean-field feedback form and is generically a function of $(t, x_{do}, m, \tilde{m}, \mu_{do})$.

The integrand Hamiltonian of a representative nonatomic decision-maker in class c is given by

$$
\begin{aligned}
\hat{H}_c &= \hat{H}_c(t, x_{do}, x_c, \mu_{do,c}, m, \tilde{m}, \hat{V}_{c,\mu_{do,c}}, \hat{V}_{c,x_c\mu_{do,c}}, \hat{V}_{c,x_c x_c\mu_{do,c}}, Z) \\
&:= \inf_{a_c} \left\{ l_c + O_{do,c,m}[\hat{V}_{c,\mu_{do,c}}] \right\} \\
&= \inf_{a_c} \Big\{ l_c + \hat{V}_{c,x_{do}\mu_{do,c}} b_{do} + \tfrac{1}{2}\langle \sigma_{do}, \sigma_{do} \hat{V}_{c,s_{do}s_{do}\mu_{do,c}} \rangle \\
&+ \hat{V}_{c,s_c\mu_{do,c}} b_c + \tfrac{1}{2}\langle \sigma_c, \sigma_c \hat{V}_{c,s_c s_c\mu_{do,c}} \rangle \\
&+ \textstyle\int_\Theta \Big\{ \hat{V}_{c,\mu_{do,c}}(t, s_{do}(t_-) + \gamma_{do}(t_-, \theta_{do}), s_c(t_-) + \gamma_c(t_-, ., \theta_c), .) \\
&- \hat{V}_{c,\mu_{do,c}}(t, s_{do}(t_-), s_c(t_-), .) \\
&- \langle \hat{V}_{c,s_{do}\mu_{do,c}}, \gamma_{do}(t, ., \theta_{do}) \rangle - \langle \hat{V}_{c,s_c\mu_{do,c}}, \gamma_c(t, ., \theta_c) \} \nu_{do}(d\theta_{do}) \nu_c(d\theta_c) \\
&\} \\
&+ \textstyle\int [\sum_{c'} \langle \bar{b}_{c'}, \hat{V}_{c,y_{c'} m\mu_{do,c}} \rangle + \tfrac{1}{2} \sum_{c'} \langle \bar{\sigma}_{c'}, \bar{\sigma}_{c'} \hat{V}_{c,y_{c'} y_{c'} m\mu_{do,c}} \rangle] m(dy) \\
&+ \textstyle\int \int_\Theta \Big\{ \hat{V}_{c,m\mu_{do,c}}(t, s_{do}, s_c, \mu_{do,c}, y(t_-) + \bar{\gamma}(t_-, ., \theta), m) \\
&- \hat{V}_{c,m\mu_{do,c}}(t, s_{do}, s_c, \mu_{do,c}, y(t_-), m) \\
&- \textstyle\sum_{c\in\mathscr{C}} \langle \hat{V}_{c,y_{c'} m\mu_{do,c}}, \bar{\gamma}_{c'}(t, ., \theta) \rangle \} \nu_{\mathscr{C}}(d\theta) \Big\} m(dy),
\end{aligned} \tag{9.16}
$$

where $Z = (\hat{V}_{j,\mu_{do}}, \hat{V}_{j,x_{do}\mu}, \hat{V}_{j,x_{do}x_{do}\mu_{do}})_j, (\hat{V}_{j,m\mu_{do}}, \hat{V}_{j,s_{c'}m\mu_{do}}, \hat{V}_{j,s_{c'}s_{c'}m\mu_{do}})_{j,c'},$ $\hat{V}_{c,x_{do}\mu_{do,c}}, \hat{V}_{c,x_{do}x_{do}\mu_{do,c}}, \hat{V}_{c,x_{do}\mu_{do,c}}, \hat{V}_{c,x_{do}x_{do}\mu_{do,c}}, \hat{V}_{c,m\mu_{do,c}}, \hat{V}_{c,sm\mu_{do,c}},$ $\hat{V}_{c,ssm\mu_{do,c}})$.

An optimizer of the integrand Hamiltonian $\hat{H}_c$ (if any) is in state-and-mean-field feedback form and is generically a function of $(t, x_{do}, x_c, m, \mu_{do,c})$. The process m appears as part of the (augmented) state in the Hamiltonian of a nonatomic decision-maker in class c because x_{do} alone does not satisfy the Markov property.

9.1.5.1 Bellman System: Equilibrium Cost and Strategies

Let $\hat{V}_i: (t_0, t_1) \times \mathscr{P}(\mathscr{X}_{do}) \times \prod_c \mathscr{P}(\mathscr{X}_c) \to \mathbb{R}$ defined by

$$\hat{V}_i(t, \mu_{do} \otimes m) = \inf_{a_i(.)\in\mathscr{A}_i} \left[\int_t^{t_1} \int l_i \mu_{do}(t', dx_{do}) dt' + \int h_i \mu_{do}(t_1, dx_{do}) \right],$$

subject the (deterministic) dynamics of μ_{do}.

The dynamic programming principle provides the following result.

Proposition 9.2 *If there exists $\hat{V}$ such that*

$$\begin{cases} \hat{V}_{i,t}(t, \mu_{do} \otimes m) + \int \hat{H}_i \mu_{do}(t, dx_{do}) = 0, \text{ on } (t_0, t_1) \times \mathscr{P}(\mathscr{X}_{do}) \times \prod_c \mathscr{P}(\mathscr{X}_c) \\ \hat{V}_i(t_1, \mu_{do} \otimes m) = \int h_i(x_{do}, \mu_{do} \otimes m)\mu_{do}(dx_{do}), \\ \\ \hat{V}_{c,t}(t, \mu_{do,c} \otimes m) + \int \hat{H}_c \mu_{do,c}(t, dx_{do}dx_c) = 0, \text{ on } (t_0, t_1) \times \mathscr{P}(\mathscr{X}_{do} \times \mathscr{X}_c) \times \prod_{c'} \mathscr{P}(\mathscr{X}_{c'}), \\ \hat{V}_c(t_1, \mu_{do,c} \otimes m) = \int h_c(x_{do}, x_c, \mu_{do,c}, m)\mu_{do,c}(dx_{do}dx_c), \\ \\ \mu_c(t, .) = m_c(t, .), \\ \mathbb{P}_{a_c^*} = \tilde{m}_c, \\ i \in \mathscr{I}, \ c \in \mathscr{C}, \end{cases} \tag{9.17}$$

then

- *$\hat{V}_c(t_0, \mu_{do,c}(t_0, .) \times m(t_0, .))$ is an equilibrium cost of a generic nonatomic decision-maker in class c, and $\hat{V}_i(t_0, \mu_{do,c}(t_0, .) \otimes m(t_0, .))$ is an equilibrium cost of the atomic decision-maker $i \in \mathscr{I}$.*
- *The optimal strategy a_c^* minimizes the expression in $\hat{H}_c$. Generically, a_c^* is a function of $(t, x_{do}, x_c, m, \mu_{do,c})$ which is in a state-and-mean-field feedback form. The optimal strategy a_i^* minimizes $\hat{H}_i$. Generically, a_i^* is a function of (t, x_{do}, m, μ_{do}) which is in a state-and-mean-field feedback form.*

Proof By the integration formula, one has

$$\hat{V}_i(t_1, \mu_{do} \otimes m(t_1)) - \hat{V}_i(0, \mu_{do} \otimes m(t_0, .)) = \int_{t_0}^{t_1} [\hat{V}_{i,t} + \int O_{do,m}[\hat{V}_{i,\mu_{do}}]\mu_{do}(t, dx_{do})]dt. \tag{9.18}$$

We apply a verification method. In order to do so, we compute the difference between the expected cost functional $\hat{L}_i$ and the candidate guess functional $\hat{V}_i(0, \mu_{do} \otimes m(t_0, .))$:

$$\begin{aligned} &\hat{L}_i - \hat{V}_i(0, \mu_{do}(t_0, .) \otimes m) = \int h_i \mu_{do}(t_1, dx_{do}) - \hat{V}_i(0, \mu_{do} \otimes m(t_1, .)) \\ &+ \int_{t_0}^{t_1} \hat{V}_{i,t} + \int \{l_i + \int O_{do,m}[\hat{V}_{i,\mu_{do}}]\}\mu_{do}(t, dx_{do}). \end{aligned} \tag{9.19}$$

We do a similar computation with $\hat{V}_c$. Then, the announced result follows by taking the minimization in a_i (respectively in a_c) and by matching the functionals with $\hat{V}$. This completes the proof.

A solution (if any) to the infinite-dimensional PIDE (9.17) provides the equilibrium values and the equilibrium strategies of all the decision-makers. System (9.17) is nonlinear in (μ_{do}, μ_c) as long as

- $\hat{H}_i$ or h_i depend on $\mu_{do}(t, dx_{do})$ and/or $(\mathbb{P}_{a_j})_j$,
- $\hat{H}_c$ or h_c depend on μ_c or $\mathbb{P}_{a_c}$.

Hence, MFTG problems are nonlinear in $(\mu_{do}, \mu_c, \mu_{do,c})$ in general. In addition, the system involves three integrals. One integral is with respect to the measure (μ_{do}, μ_c). The second integral is with respect to the population measure m. And the third integral is with respect to the jump measure $\nu(\theta)$ the known functional $\hat{V}_{\mu_{do}}, \hat{V}_{\mu_{do,c}}$. These integrals lead to a PIDE system instead of a PDE system.

9.1.5.2 MASS: Master Adjoint System for Equilibrium Strategies

The master adjoint system (MASS) can be seen as an adjoint system obtained from the Gâteaux derivative (when it exists) of $(\hat{V}_i, \hat{V}_c)$ with respect to μ_{do} and μ_{doc}, respectively. Let $\hat{U}_i(t, x_{do}, \mu_{do} \otimes m) := \hat{V}_{i,\mu_{do}}(t, x_{do}, \mu_{do} \otimes m)$ and $\hat{U}_c(t, x_{do}, x_c, m) := \hat{V}_{c,\mu_{do,c}}(t, x_{do}, x_c, \mu_{do} \otimes m)$. Using (9.17), we obtain

$$
\begin{cases}
\hat{U}_{i,t}(t, x_{do}, \mu_{do} \otimes m) + \hat{H}_i + \int [\partial_{\mu_{do}} \hat{H}_i] \mu_{do}(t, ds'_{do}) = 0, \\
\text{on } (t_0, t_1) \times \mathcal{X}_{do} \times \mathcal{P}(\mathcal{X}_{do}) \times \prod_c \mathcal{P}(\mathcal{X}_c) \\
\hat{U}_i(t_1, x_{do}, \mu_{do} \otimes m) = h_i(x_{do}, \mu_{do}, m) + \int h_{i,\mu_{do}}(x_{do}, \mu_{do}) \mu_{do}(ds'_{do}), \\
\\
\hat{U}_{c,t}(t, x_{do}, x_c, \mu_{do,c} \otimes m) + \hat{H}_c + \int [\partial_{\mu_{do,c}} \hat{H}_c] \mu_{do,c}(t, ds'_{do} ds'_c) = 0, \\
\text{on } (t_0, t_1) \times \mathcal{X}_{do} \times \mathcal{X}_c \times \mathcal{P}(\mathcal{X}_{do} \times \mathcal{X}_c) \times \prod_{c'} \mathcal{P}(\mathcal{X}_{c'}), \\
\hat{U}_c(t_1, x_{do}, x_c, \mu_{do,c} \otimes m) = h_c(x_{do}, x_c, \mu_{do}, \mu_c, m) \\
\qquad + \int h_{c,\mu_{do,c}}(x_{do}, x_c, \mu_{do}, \mu_c, m) \mu_{do,c}(ds'_{do} ds'_c), \\
\\
\mu_c(t, .) = m_c(t, .), \\
\mathbb{P}_{a_c^*} = \tilde{m}_c, \\
i \in \mathcal{I}, \ c \in \mathcal{C},
\end{cases}
\tag{9.20}
$$

where the integrand Hamiltonians are given by

$$
\begin{cases}
\hat{H}_i = \inf_{a_i} \left\{ l_i + O_{do,m}[\hat{U}_i] \right\}, \\
\hat{H}_c = \inf_{a_c} \left\{ l_c + O_{do,c,m}[\hat{U}_c]. \right\}
\end{cases}
\tag{9.21}
$$

Example 9.1 (When the Nonatomic Decision-Makers Have Finite State Space) We assume that the nonatomic decision-makers in class $c \in \mathcal{C}$ have a finite state space $\mathscr{X}_c$. The transition rate of the individual state $x_c(t)$ of a nonatomic decision-maker in class c is

$$q_{c,x_c s'_c}(t, x_{do}, (a_i)_{i\in\mathscr{I}}, a_c, \mu_{do}, \mu_c, (\mathbb{P}_{a_i})_{i\in\mathscr{I}}, \mathbb{P}_{a_c}, m, \tilde{m}),$$

where

$$\begin{array}{l} x_c \neq s'_c,\ q_{c,x_c s'_c} > 0, \\ q_{c,x_c x_c} = -\sum_{s'_c\neq x_c} q_{c,x_c s'_c}. \end{array} \tag{9.22}$$

Here, the rate also depends on the atomic decision-makers making $\mu_c, m, \tilde{m}$ random.

The evolution of the random measure μ_c is given by

$$d\mu_c(t, x_c) = \left\{\sum_{s'_c\in\mathscr{X}_c} \mu_c(t, s'_c) q_{c,s'_c x_c} - \sum_{s'_c\in\mathscr{X}_c} \mu_c(t, x_c) q_{c,x_c s'_c}\right\} dt, \tag{9.23}$$

and the evolution of the stochastic process m is given by

$$dm_c(t, x_c) = \left\{\sum_{s'_c\in\mathscr{X}_c} m_c(t, s'_c) q^*_{c,s'_c x_c} - \sum_{s'_c\in\mathscr{X}_c} m_c(t, x_c) q^*_{c,x_c s'_c}\right\} dt, \tag{9.24}$$

where q^* is obtained from q by evaluating it at $(a^*_c)_c$. The decision-maker in class c influences significantly the distribution of its own-state μ_c but not the population distribution m_c. The Bellman system reduces to the following system:

$$\begin{array}{l} \dot{V}_{i,t}(t, \mu_{do}\otimes m) + \int \hat{H}_i \mu_{do}(dx_{do}) = 0, \\ \hat{V}_i(t_1, \mu_{do}\otimes m) = \int h_i(x_{do}, \mu_{do}, m)\mu_{do}(dx_{do}), \\ \dot{V}_{c,t}(t, \mu_{do,c}\otimes m) + \int \sum_{x_c} \hat{H}_c[\mu_{do}(dx_{do})\otimes \mu_c(x_c)] = 0, \\ \hat{V}_i(t_1, \mu_{do,c}\otimes m) = \int \sum_{x_c} h_c(x_{do}, x_c, \mu_{do}, \mu_c, m)[\mu_{do}(dx_{do})\otimes \mu_c(x_c)], \end{array} \tag{9.25}$$

where the integrand Hamiltonians are

$$\begin{array}{l} \hat{H}_i \\ = \inf_{a_i} \Big\{ l_i + \hat{V}_{i,x_{do}\mu_{do}} b_{do} + \frac{1}{2}\langle \sigma_{do}, \sigma_{do} \hat{V}_{i,x_{do}x_{do}\mu_{do}}\rangle \\ + \int_{\Theta} \{ \hat{V}_{i,\mu_{do}}(t, x_{do}(t_-) + \gamma_{do}(t_-, ., \theta), m) - \hat{V}_{i,\mu_{do}}(t, x_{do}(t_-), m) \\ -\langle \hat{V}_{i,x_{do}\mu_{do}}, \gamma_{do}(t, ., \theta)\rangle \} \nu_{do}(d\theta) \\ + \sum_c \sum_{x_c, x'_c} [\hat{V}_{i,m_c(.,s'_c)\mu_{do}} - \hat{V}_{i,m_c(.,x_c)\mu_{do}}] q_{c,x_c x'_c} m_c(t, x_c) \Big\}, \\ \\ \hat{H}_c = \inf_{a_c} \Big\{ l_c + \hat{V}_{c,x_{do}\mu_{do,c}} b_{do} + \frac{1}{2}\langle \sigma_{do}, \sigma_{do} \hat{V}_{c,x_{do}x_{do}\mu_{do,c}}\rangle \\ + \int_{\Theta} \{ \hat{V}_{c,\mu_{do,c}}(t, x_{do} + \gamma_{do}, x_c, m) - \hat{V}_{c,\mu_{do,c}}(t, x_{do}, x_c, m) \\ -\langle \hat{V}_{c,x_{do}\rho}, \gamma_{do}(t, ., \theta)\rangle \rangle \} \nu_{do}(d\theta) \\ + \sum_{s'_c} [\hat{V}_{c,\mu_c(.,s'_c)\mu_{do,c}} - \hat{V}_{c,\mu_c(.,x_c)\mu_{do,c}}] q_{c,x_c s'_c} \Big\} \\ + \sum_{c'\in\mathscr{C}} \sum_{s_{c'}, s'_{c'}} [\hat{V}_{c',m_{c'}(.,s'_{c'})\mu_{do,c}} - \hat{V}_{c',m_{c'}(.,s_{c'})\mu_{do,c}}] q^*_{c',s_{c'}s'_{c'}} m_{c'}(t, s_{c'}). \end{array} \tag{9.26}$$

The master adjoint system is given by

$$
\begin{aligned}
&\hat{U}_{i,t} + \hat{H}_i + \int \hat{H}_{i,\mu_{do}}[x_{do}, \mu_{do}, m](s'_{do})\mu_{do}(ds'_{do}) = 0,\\
&\hat{U}_i(t_1, x_{do}, \mu_{do} \otimes m) = h_i(x_{do}, \mu_{do}, m) + \int h_{i,\mu_{do}}[x_{do}, \mu_{do}, m](s'_{do})\mu_{do}(ds'_{do}),\\
&\\
&\hat{U}_{c,t}(t, x_{do}, x_c, \mu_{do,c} \otimes m) + \hat{H}_c\\
&+\int \sum_{s'_c} \hat{H}_{c,\mu_c(x_c)\mu_{do}}[x_{do}, x_c, \mu_{do}, \mu_c, m](s'_{do}, s'_c)[\mu_{do}(ds'_{do}) \otimes \mu_c(s'_c)] = 0,\\
&\hat{U}_c(t_1, \mu_{do,c} \otimes m) = h_c(x_{do}, x_c, \mu_{do}, \mu_c, m)\\
&+\int \sum_{s'_c} h_{c,\mu_c(x_c)\mu_{do}}[x_{do}, x_c, \mu_{do}, \mu_c, m](s'_{do}, s'_c)[\mu_{do}(ds'_{do}) \otimes \mu_c(s'_c)],\\
&i \in \mathscr{I},\ c \in \mathscr{C}, x_c \in \mathscr{X}_c.
\end{aligned}
\tag{9.27}
$$

Next, we provide a basic example with a polynomial cost for which existence will be established by a direct computation and verification.

Example 9.2 (Polynomial Cost)

We investigate a mean-field-type game with the following data. The data of the atomic decision-makers are given by

$$
\begin{aligned}
&0 \leq t_0 < t_1,\\
&l_i(t, s, a, \mu, m) = q_i \tfrac{(s-\bar{s})^{2k_i}}{2k_i} + r_i \tfrac{(a_i-\bar{a}_i)^{2k_i}}{2k_i}\\
&+c_i(s-\bar{s})^{2k_i-1}(a_i - \bar{a}_i) + \sum_{j\in\mathscr{I}\setminus\{i\}} \epsilon_{ij}(s-\bar{s})^{2(k_i-1)}(a_i - \bar{a}_i)(a_j - \bar{a}_j)\\
&+\bar{q}_i \tfrac{\bar{s}^{2\bar{k}_i}}{2\bar{k}_i} + \bar{r}_i \tfrac{\bar{a}_i^{2\bar{k}_i}}{2\bar{k}_i} + \bar{c}_i \bar{s}^{2\bar{k}_i-1}\bar{a}_i + \sum_{j\neq i} \bar{\epsilon}_{ij}\bar{s}^{2(\bar{k}_i-1)}\bar{a}_i\bar{a}_j,\\
&h_i(s, \mu, m) = q_{i,t_1} \tfrac{(s(t_1)-\bar{s}(t_1))^{2k_i}}{2k_i} + \bar{q}_{i,t_1} \tfrac{[\bar{s}(t_1)]^{2\bar{k}_i}}{2\bar{k}_i},\\
&b(t, s, a, \mu, m) = b_1(s-\bar{s}) + \bar{b}_1\bar{s} + \sum_{j\in\mathscr{I}} \left[b_{2j}(a_j - \bar{a}_j) + \bar{b}_{2j}\bar{a}_j\right],\\
&\sigma(t, s, a, \mu, m) = (s-\bar{s})\tilde{\sigma},\\
&\gamma(t, s, a, \mu, m, \theta) = (s-\bar{s})\tilde{\gamma}(., \theta),\\
&\bar{s}(t) = \int y\mu(t, dy),\\
&\bar{a}_i(t) = \int a_i(t, y, \mu, m)\mu(t, dy),\ i \in \mathscr{I},
\end{aligned}
\tag{9.28}
$$

where $k_i \geq 1$, $\bar{k}_i \geq 1$ are natural numbers and the coefficients are time-dependent. The coefficient functions $q_i, r_i, \bar{q}_i, \bar{r}_i$, are nonnegative functions and $\int_{t_0}^{t_1}\int_\Theta \tilde{\gamma}^2(t, ., \theta)\nu(d\theta)dt < +\infty$. Assume that $q_i, r_i, c_i, \epsilon_{ij}, \bar{q}_i, \bar{r}_i, \bar{c}_i, \bar{\epsilon}_{ij}, b_1, b_{2j}, \bar{b}_1, \bar{b}_{2j}, \tilde{\sigma}, \tilde{\gamma}$ are functions of $(t, \lambda_1(t) = \bar{m}(t) = \int ym(t, dy), \lambda_2(t) = \int \langle y, y\rangle m(t, dy))$. Hence, the nonatomic decision-makers enter collectively into the above through the pair (λ_1, λ_2) which corresponds to the population mean-field term. Interestingly, the gradient (with respect to the measure) of the polynomial cost l_i considered here does not satisfy the linear growth condition.

The data of the nonatomic decision-makers are given by

$$
\begin{aligned}
&2 \le C < +\infty,\\
&l_c(t, s, x_c, a, a_c, \mu, \mu_c, m) = -\tilde{p}_c(\bar{s}, \bar{m}, -\textstyle\sum_{i\in\mathscr{I}} \bar{a}_i)a_c + \frac{3}{2}(a_c - \bar{a}_c)^2 + \frac{1}{2}(\bar{a}_c)^2,\\
&h_c(s, x_c, \mu, m) = \tilde{h}_{1c}(\bar{m}_c(t_1))x_c^2(t_1) + \tilde{h}_{2c}(\bar{m}_c(t_1))(\bar{x}_c(t_1))^2,\\
&b_c(t, s, x_c, a, \mu, m) = -a_c,\\
&\sigma_c(t, s, x_c, a, \mu, m) = \sqrt{2},\ \gamma_c = 0,
\end{aligned}
\tag{9.29}
$$

with $\tilde{p}_c, \tilde{h}_c$ being real-valued positive functions. There is an inter-class interaction through the terms $(\bar{m}_c)_{c\in\mathscr{C}}$, and there is an interaction with the atomic decision-makers via $\bar{s}$. The optimal strategies of a generic nonatomic decision-maker in class c are

$$
a_c = \tilde{p}_c + \frac{1}{3}\hat{V}_{c,x_c\mu_{do,c}} + \frac{2}{3}\int \hat{V}_{c,x_c\mu_{do,c}}\mu_{do,c}(dx_{do}dx_c dm).
$$

In this particular case, a solution to the Bellman system can be obtained in a semi-explicit way, subject to a system of backward ordinary differential equations.

Example 9.3 (Legendre-Fenchel Transform) For illustration purposes of the mean-field-type Nash equilibrium system (9.17), let us consider one atomic and a single class of nonatomic decision-makers, i.e., $I = 1, C = 1$. In order to use the special case of the Legendre-Fenchel transforms between instantaneous costs and Hamiltonians, we choose $b_c = a_c, \sigma_c = \sqrt{2}\mathbb{I}, \gamma_c = 0$ and $b_{do} = a_i, \sigma_{do} = \sqrt{2}\mathbb{I}$. We compute the integrand Hamiltonians and derive the Bellman system of PIDEs:

$$
\begin{cases}
-\hat{V}_{i,t}(t, \mu_{do} \otimes m) + \int\{H_i(x_{do}, \mu_{do}, m, \tilde{m}, \hat{V}_{i,x_{do}\mu_{do}}) - trace[\hat{V}_{i,x_{do}x_{do}\mu_{do}}]\}\mu_{do}(dx_{do})\\
+\int\{\langle H_{c,p_c}, \hat{V}_{i,y_c m\mu_{do}}\rangle - trace[\hat{V}_{i,y_c y_c m\mu_{do}}]\}m_c(dy_c)\mu_{do}(dx_{do}) = 0,\\
\hat{V}_i(t_1, \mu_{do} \otimes m) = \int h_i(x_{do}, \mu_{do}, m)\mu_{do}(dx_{do}),\\
\\
-\hat{V}_{c,t}(t, \mu_{do,c} \otimes m)\\
+\int\{H_c(t, x_{do}, x_c, \mu_{do}, m, \tilde{m}, \hat{V}_{c,x_c\mu_{do,c}}) - trace[\hat{V}_{c,x_c x_c\mu_{do,c}}]\}\mu_{do,c}(dx_{do}dx_c)\\
+\int\{\langle H_{i,p_i}, \hat{V}_{c,x_{do}\mu_{do,c}}\rangle - trace[\hat{V}_{c,x_{do}x_{do}\mu_{do,c}}]\}\mu_{do,c}(dx_{do}dx_c)\\
+\int\{\langle H_{c,p_c}, \hat{V}_{c,y_c m\mu_{do,c}}\rangle - trace[\hat{V}_{c,y_c y_c m\mu_{do,c}}]\}m_c(dy_c)\mu_{do,c}(dx_{do}dx_c) = 0,\\
\hat{V}_c(t_1, \mu_{do,c} \otimes m) = \int h_c(x_{do}, x_c, \mu_c, m)\mu_{do,c}(dx_{do}dx_c),\\
\\
i \in \mathscr{I},\ c \in \mathscr{C}\\
\\
a_i^* = -H_{i,p_i}.
\end{cases}
\tag{9.30}
$$

The Measure-Linear Case When the functionals l_i, h_i, l_c, h_c are independent of the individual mean-field terms $\mu_{do}, \mathbb{P}_{a_i}, \mu_c, \mathbb{P}_{a_c}$ the Hamiltonian becomes linear in the measure μ_{do}, and in this case, we retrieve a mean-field game model with a major and a large population of minor decision-makers. We also retrieve the master system of mean-field games:

$$\left\{\begin{array}{l}-\hat{U}_{i,t}(t, x_{do}, m) + \ H_i(x_{do}, m, \tilde{m}, \hat{U}_{i,x_{do}}) - trace[\hat{U}_{i,x_{do}x_{do}}] \\ + \int\{\langle H_{c,p_c}, \hat{U}_{i,y_c m}\rangle - trace[\hat{U}_{i,y_c y_c m}]\}m_c(dy_c) = 0, \\ \hat{U}_i(t_1, x_{do}, m) = h_i(x_{do}, m), \\ \\ -\hat{U}_{c,t}(t, x_{do}, x_c, m) + H_c(t, x_{do}, x_c, m, \tilde{m}, \hat{U}_{c,x_c}) - \Delta_{x_c}(\hat{U}_c) \\ +\{\langle H_{i,p_i}, \hat{U}_{c,x_{do}}\rangle - \Delta_{x_{do}}(\hat{U}_c)\} \\ + \int\{\langle H_{c,p_c}, \hat{U}_{c,y_c m}\rangle - trace[\hat{U}_{c,y_c y_c m}]\}m_c(dy_c) = 0, \\ \hat{U}_c(t_1, x_{do}, x_c, m) = h_c(x_{do}, x_c, m).\end{array}\right. \tag{9.31}$$

Note that the master system in ($\hat{U}_i = \hat{V}_{i,\rho}, \hat{U}_c = \hat{V}_{c,\rho}$) is not the equilibrium value function. It provides the Nash equilibrium strategies a_i^*, a_c^*. We instead work with the primitive functions $(\hat{V}_i, \hat{V}_c)$ in (9.17) to get both equilibrium values and equilibrium strategies directly.

9.2 Maximum Principle Under Partial Observation

We now consider a particular class of mean-field-type games under partial observation. We use a spike variation technique to establish optimality equations under suitable conditions. We provide the existence of the best response to the mean-field in both pure and (behavioral) mixed strategies. Sufficient conditions for the existence of mean-field equilibria are provided. Let $(\Omega, \mathscr{F}, \mathbb{F}, \mathbb{P})$ be a given filtered probability space on which a standard Brownian motion $W = \{W(t)\}_{t\geq 0}$ is defined. Let $\mathscr{F}_t^W$ be the $\mathbb{P}$-completed natural filtrations generated by W up to t. Set $\mathscr{F}^W := \{\mathscr{F}_t^W, \ t_0 \leq t \leq t_1\}$ and $\mathbb{F} := \{\mathscr{F}_t, \ t_0 \leq t \leq t_1\}$, where $\mathscr{F}_t = \mathscr{F}_t^W \vee \sigma(x_0)$. Consider an infinite number of decision-makers. An admissible control a_i of decision-maker i is an $\mathscr{F}^{W_i}$-adapted process with values in a non-empty, closed, and bounded subset (not necessarily convex) A_i of $\mathbb{R}^d$ and satisfies $E[\int_0^T |a_i(t)|^2 dt] < \infty$. Those are nonanticipative measurable functionals of the Brownian motions.

In the dynamic problem, each decision-maker has a different information structure (decentralized information). Let $\mathscr{A}_i$ be the set of admissible strategies of i (with decentralized partial information), $\mathscr{G}_{i,t} \subset \mathscr{F}_{i,t}$:

$$\mathscr{A}_i := \{a_i \in L^2_{\mathscr{G}_{i,T}}([0,T], \mathbb{R}^d), \ a_i(t,.) \in A_i \ \mathbb{P} - a.s\}.$$

Given a strategy $a_i \in \mathscr{A}_i$ and m a (population) mean-field term generated by other decision-makers, we consider the signal-observation $x_i^{a_i}$ which satisfies the following stochastic differential equation of mean-field type to which we associate a best response to mean-field:

$$\begin{cases} \inf_{a_i(\cdot)\in\mathscr{A}_i} J(a_i, m) \\ \\ \text{subject to} \\ dx_i^{a_i}(t) = b(t, x_i^{a_i}(t), Ex_i^{a_i}(t), a_i(t), m(t))dt + \sigma(t, x_i^{a_i}(t), Ex_i^{a_i}(t), a_i(t), m(t))dW_{i,t}, \\ x_i^{a_i}(0) \sim \mathscr{L}(X_0), \\ m(t) = \text{mean-field of other players}, \end{cases} \tag{9.32}$$

where

$$b(t, x, y, a, m) : \; [t_0, t_1] \times \mathbb{R}^n \times \mathbb{R}^n \times A \times \Lambda \longrightarrow \mathbb{R}, \tag{9.33}$$

$$\sigma(t, x, y, a, m) : \; [t_0, t_1] \times \mathbb{R} \times \mathbb{R} \times A \times \Lambda \longrightarrow \mathbb{R}. \tag{9.34}$$

$$\begin{aligned} J(a_i, m) &= E[h(x_i^{a_i}(t_1), Ex_i^{a_i}(t_1), m(t_1))] \\ &\quad + E\left[\int_{t_0}^{t_1} l(t, x_i^{a_i}(t), Ex_i^{a_i}(t), a_i(t), m(t))dt\right], \end{aligned}$$

h is the terminal cost and l is the running cost.

Definition 9.1 (Best Response to Mean-Field) Given m, any $\bar{a}_i(\cdot) \in \mathscr{A}_i$ which satisfies

$$J(\bar{a}_i(\cdot), m) = \inf_{a_i(\cdot)\in\mathscr{A}_i} J(a_i(\cdot), m), \tag{9.35}$$

is called a *pure* best-response strategy to m.

Example 9.4 (Two Different Mean-Field Terms) The action a_i of i has a big influence on its own-state x_i and hence on the local mean-field $E[x_i]$. However, a_i does not influence m (the mean-field generated by the entire population). The cost is

$$l = var(x_i) + \int_j [var(x_j) - var(x_i)]\epsilon(var(x_j) - var(x_i))\, dj,$$

where ϵ is given by $\epsilon(y) = \alpha\mathbb{1}_{\{y>0\}} - \beta\mathbb{1}_{\{y<0\}}$ and $var(x_i) = E\{(x_i - E[x_i])^2\}$. Here, the population mean-field m is $\int_j [var(x_j) - var(x_i)]\epsilon(var(x_j) - var(x_i))\, dj$.

Example 9.5 Consider a large population of consumers, some producers, and prosumers. Each controller aims to control the room temperature, which is described by a controlled SDE. The control action a_i of controller i depends on the price of electricity p (demand, supply, location), which is designed as a function of demand= $\int_I consumption(i)m_1(di)$, supply= $\int_J supply(j)m_2(dj)$, and the geographical location. The payoff of a consumer is a sort of tradeoff between comfort temperature and electricity cost $a_i p$(demand, supply, location). There is also a cost for driving the temperature to the comfort zone temperature: $(T_i - T_{i,comfort})^2 + var(T_i)$. Clearly, there are several different mean-field terms: Demand is a mean-field generated by the population consumers and prosumers. The supply is a mean-field generated by the population *regular* producers and prosumers. Finally, $var(T_i)$ is a mean-field term generated by the state T_i of controller i.

9.2.1 Well Posedness of the Individual State Dynamics

We study here the existence of a solution to the state dynamics.

Proposition 9.3 (See [18], Proposition 1.2.) *Suppose that the functions b, σ, l, h are twice continuously differentiable with respect to (x, y, m). Moreover, assume that b, σ, l, h and all their derivatives up to the second order with respect to (x, y, m) are continuous in (x, y, a, m) and bounded. Then, for each admissible strategy profile $a_i \in \mathscr{A}_i$, the stochastic differential equation (SDE) in problem (9.32) admits a unique strong solution $x_i^{a_i} := x_i^{a_i,m}$.*

9.2.1.1 Conditional Stochastic Maximum Principle

Note that the presence of the mean-field terms (the expectations $E[x_i^{a_i}(s)]$, $t_0 \leq t \leq t_1$), in the loss function l, may cause time inconsistency, in which case the classical Bellman principle is no longer valid. The extension of the Bellman principle to this situation leads to a higher-dimensional partial differential equation with random coefficients, and this motivates the use of the stochastic maximum principle (SMP) approach instead of trying extensions of the dynamic programming principle (DPP).

For convenience, we will use the following notation throughout the subsection. For $\phi \in \{b, \sigma, l, h\}$, we define

$$\begin{cases} \delta\phi(t) = \phi(t, \bar{x}_i(t), E[\bar{x}_i(t)], a_i(t), m(t)) \\ -\phi(t, \bar{x}_i(t), E[\bar{x}_i(t)], \bar{a}_i(t), m(t)); \\ \phi_x(t) = \frac{\partial\phi}{\partial x}(t, \bar{x}_i(t), E[\bar{x}_i(t)], \bar{a}_i(t), m(t)), \\ \phi_{xx}(t) = \frac{\partial^2\phi}{\partial x^2}(t, \bar{x}_i(t), E[\bar{x}_i(t)], \bar{a}_i(t), m(t)); \\ \phi_y(t) = \frac{\partial\phi}{\partial y}(t, \bar{x}_i(t), E[\bar{x}_i(t)], \bar{a}_i(t), m(t)). \end{cases} \tag{9.36}$$

where a_i is an admissible pure strategy from $\mathscr{A}_i$.

We define the Hamiltonian $H(t, X, a, m, p, q)$ associated with random variable $X \in L^1(\Omega, \mathscr{F}, \mathbb{P})$ as follows, for $(p, q) \in \mathbb{R} \times \mathbb{R}$:

$$H := b(t, X, E[X], a, m)p + \sigma(t, X, E[X], a, m)q - l(t, X, E[X], a, m). \tag{9.37}$$

Moreover, we let

$$\begin{aligned}\delta H(t) &:= p(t)\delta b(t) + q(t)\delta\sigma(t) - \delta l(t),\\ H_k(t) &:= b_k(t)p + \sigma_k(t)q - l_k(t),\end{aligned} \tag{9.38}$$

for $k \in \{x, y, xx\}$.

Given m, we apply the SMP for risk-neutral mean-field type control from [78], Theorem 2.1, to the state dynamics x to derive the first-order adjoint equation.

Under the assumptions of Proposition 9.3, there exists a unique $\mathbb{F}$-adapted pair of processes (p, q), which solves the backward SDE:

$$\begin{aligned} p(t) &= -h_x(T) - E[h_y(T)] \\ &\quad + \int_t^T \{H_x(s) + E[H_y(s)]\}ds - \int_t^T q(s)dW_s, \end{aligned} \tag{9.39}$$

such that $E\left[\sup_{t\in[0,T]} |p(t)|^2 + \int_0^T |q(t)|^2 dt\right] < +\infty$. However, these processes (p, q) may not be adapted to decentralized information $\mathcal{G}_{i,t}$. This is why their conditioning will appear in the maximum principle below.

Again by Buckdahn et al. [78], Theorem 2.1, there exists a unique $\mathbb{F}$-adapted pair of processes (P, Q), which solves the second-order adjoint equation:

$$P(t) = -h_{xx}(T) + \tag{9.40}$$

$$\int_t^T \{2b_x(s)P(s) + \sigma_x^2 P(s) + 2\sigma_x(s)Q(s) + H_{xx}(s)\}ds$$

$$-\int_t^T Q(s)dW_s, \tag{9.41}$$

such that $E\left[\sup_{t\in[0,T]} |P(t)|^2 + \int_0^T |Q(t)|^2 dt\right] < +\infty$.

Proposition 9.4 *Let the assumptions of Proposition 9.3 hold. Let m be a given mean-field term. If $(\bar{x}_i, \bar{a}_i)$ is a best response, then there are two pairs of $\mathbb{F}$-adapted processes (p, q) and (P, Q) that satisfy (9.39) and (9.40), respectively, such that*

$$i \in \mathcal{I} : \left[\delta H(t) + \tfrac{1}{2}\delta\sigma(t)' P(t) d\sigma(t) \mid \mathcal{G}_{i,t}\right] \le 0, \tag{9.42}$$

for all $a_i \in \mathcal{A}_i$, almost every t and $\mathbb{P}$-almost surely, where

$$\delta H(t) := H(t, \bar{x}(t), a_i, m(t), p(t), q(t)) - H(t, \bar{x}(t), \bar{a}_i(t), m(t), p(t), q(t)).$$

We now discuss the existence of a pure best-response strategy.

Suppose $\{(b, \sigma'\sigma, l) \ |a_i \in A_i\}$ is a convex set. Then, each player i in problem (9.32) has at least one best-response $\bar{a}_i$ in pure strategies, where $\bar{a}_i$ is A_i-valued progressively $\mathcal{G}_i$-measurable strategy.

Concave Hamiltonian Case

If A_i is convex, the Hamiltonian is jointly concave in (x, Ex, a_i), and the final cost function h is convex in (x, Ex), then the SMP becomes sufficient conditions for optimality, i.e., the necessary conditions of optimality become sufficient conditions of global optimality. Moreover, the optimal strategies are given by the maximizers of the Hamiltonian (no need for the second-order process P). The SMP of the best-response problem is simply given by the variational inequality $\forall a_i \in A_i$:

$$\langle E[\partial_{\bar{a}_i} H \mid \mathscr{G}_{i,t}], a_i - \bar{a}_i\rangle = \sum_k \langle E[\partial_{\bar{a}_{ik}} H \mid \mathscr{G}_{i,t}], a_{ik} - \bar{a}_{ik}\rangle \geq 0.$$

9.2.2 *Mixed Extension*

A best-response strategy may fail to exist in pure strategies if the set of action spaces is not convex. Therefore, we propose a mixed extension of the problem. In fact, for player i, we may wonder, if there is no best response to mean-field which belongs to $\mathscr{A}_i$, is it still possible to determine the minimal value of the performance functional J? Ideally, we would like to have

$$\inf_{a_i \in \mathscr{A}_i} J(a_i, m) = \inf_{\mu_i \in \mathscr{M}_i} \tilde{J}(\mu_i, m) = \tilde{J}(\bar{\mu}_i, m), \tag{9.43}$$

where

$$\bar{\mu}_i \in \arg\min_{\mu_i \in \mathscr{M}_i} \tilde{J}(\mu_i, m),$$

for some larger set $\mathscr{M}_i$ of strategies that "contains" $\mathscr{A}_i$ and has a richer topological structure that allows for the existence of a minimum $\bar{\mu}_i$ of the performance function J. In other words, if the set $\mathscr{A}_i$ is too small to contain a minimizer of J, would it be possible to "embed" it into a larger class of controls $\mathscr{M}_i$, with a richer topological structure, for which the control problem becomes solvable.

This "embedding" is known as relaxed control in the control theory literature. It is also known as mixed extension in game theory. Let $\mathscr{M}_i$ be the set of behavioral *mixed strategies* or relaxed controls. In this setting, $\mathscr{A}_i$ is the set of *pure* strategies or strict controls.

The mixed extension of best-response problem (9.32) is the following:

Let $\mathscr{M}_i$ be the set of admissible behavioral mixed strategies of i (with decentralized partial information), $\mathscr{G}_{i,t} \subset \mathscr{F}_{i,t}$, $\mu_i(t, .) \in \mathscr{P}(A_i)$, which is the set of probability measures over A_i:

$$\begin{cases} \inf_{\mu(\cdot)\in\mathcal{M}_i} \tilde{J}(\mu_i, m), \\ \\ \text{subject to} \\ dx_i^{\mu_i}(t) = \int_{A_i} b(t, x_i^{\mu_i}(t), E[x_i^{\mu_i}(t)], a_i, m(t))\mu_{i,t}(da_i)dt \\ +\int_{A_i} \sigma\left(t, x_i^{\mu_i}(t), E[x^{\mu_i}(t)], a_i(t), m(t)\right) M_i(da_i, dt), \\ x_i^{\mu_i}(0) \sim \mathcal{L}(X_0), \end{cases} \tag{9.44}$$

where $M_i(da_i, dt)$ is an **orthogonal** martingale measure whose intensity is the relaxed behavioral mixed strategy (relaxed control) $\mu_{it}(da_i)dt$ and

$$\tilde{J}(\mu_i, m) = E[h(x_i^{\mu_i}(T), Ex_i^{\mu_i}(T), m(T))] \tag{9.45}$$

$$+ E\left[\int_0^T \int_{A_i} l(t, x_i^{\mu_i}(t), Ex_i^{\mu_i}(t), a_i, m(t))\mu_{it}(da_i)dt\right].$$

Definition 9.2 (Mixed Best Response) Any $\bar{\mu}_i(\cdot) \in \mathcal{M}_i$ which satisfies

$$\tilde{J}(\bar{\mu}_i(\cdot), m) = \inf_{\mu_i(\cdot)\in\mathcal{M}_i} \tilde{J}(\mu_i(\cdot), m), \tag{9.46}$$

is called a behavioral *mixed* best-response strategy to the mean-field m.

Remark 9.2 Following the mixed extension procedure with orthogonal measure, the best-response value remains unchanged:

$$\inf_{\mu_i(\cdot)\in\mathcal{M}_i} \tilde{J}(\mu_i(\cdot), m) = \inf_{a_i(\cdot)\in\mathcal{A}_i} J(a_i(\cdot), m).$$

We derive a maximum principle for weakly relaxed state dynamics:

$$\begin{aligned} &\mathcal{H}^{(x_i(\cdot),\mu_i(\cdot))}(t, x_i, a_i, m) = \\ &H(t, x_i, E[x_i], a_i, m, p(t), q(t) - P(t).[\textstyle\int_{u_i\in A_i} \mu_{it}(du_i)\sigma(t, x_i, E[x_i], u_i, m)]) \\ &+\tfrac{1}{2}\sigma'(t, x_i, E[x_i], a_i, m)P(t)\sigma(t, x_i, E[x_i], a_i, m), \end{aligned} \tag{9.47}$$

where $(t, x_i, a_i) \in [0, T] \times \mathbb{R}^d \times A$, $(p(t), q(t))$, and $(P(t), Q(t))$ are solutions of the adjoint equations associated with (x_i, a_i, m).

If the mixed pair $(\bar{\mu}_i, \bar{x}_i)$ denotes a best response to mean-field m, then for $a.e\ t \in [0, T]$, $\mathbb{P}-$ a.s.

$$E\left[\mathcal{H}^{(\bar{x}_i(t),\bar{\mu}_i(t))}(t, \bar{x}_i(t), \bar{\mu}_i(t), m(t)) \mid \mathcal{G}_{i,t}\right]$$

$$= \max_{\upsilon_i\in\mathcal{P}(A_i)} E\left[\mathcal{H}^{(\bar{x}_i(t),\bar{\mu}_i(t))}(t, x_i(t), \upsilon_i, m(t)) \mid \mathcal{G}_{i,t}\right].$$

9.2.3 Mean-Field-Type Equilibrium

We now analyze the equilibrium of the MFTG.

Definition 9.3 (Mean-Field-Type Equilibrium) Any $(\bar{a}_i(\cdot) \in \mathscr{A}_i, i \in \{1, 2, \ldots\}, m)$ such that for every player i

$$J(\bar{a}_i(\cdot), m) = \inf_{a_i(\cdot)\in\mathscr{A}_i} \tilde{J}(a_i, m), \tag{9.48}$$

$$m(t) = \text{generated by } (\bar{x}_j(t))_j, \tag{9.49}$$

is called a behavioral *mean-field-type equilibrium* strategy of the MFTG under decentralized information.

Example 9.6 (Computation and Structure of Equilibria) Let $\sigma(t, x_i, E[x_i], a_i, m) = \hat{\sigma}(t, x_i, E[x_i], m)$. If there is a best-response strategy, it maximizes

$$E\left[-l(x_i, E[x_i], a_i, m) + \langle b(x_i, E[x_i], a_i, m), p\rangle \mid \mathscr{G}_{i,t}\right].$$

If $l = l_1(x_i(t), E[x_i(t)], m(t)) + \langle R(t, m(t))a_i, a_i\rangle$ and $b = A(t, m(t))x_i + B(t, m(t))a_i + c(t, E[x_i(t)], m(t))$ with $R = R'$ positive-definite matrix, then

$$E\left[-2R(t, m(t))a_i + B(t, m(t))p(t) \mid \mathscr{G}_{i,t}\right] = 0.$$

If R, B are deterministic matrix functions, then $\bar{a}_i = \frac{1}{2}R^{-1}BE[p \mid \mathscr{G}_{i,t}]$. This means that the best-response strategy is a function of own filter $\hat{p}(t) = E[p(t) \mid \mathscr{G}_{i,t}]$ and the population mean-field $m(t)$. In the quadratic cost case, the equilibrium strategy has the following form:

$$\alpha_1(t)E[x_i(t) \mid \mathscr{G}_{i,t}] + \alpha_2(t)E[x_i(t)] + \alpha_3(t)m(t) + \gamma(t),$$

where α_i are deterministic coefficients.

We now focus on the mixed extension case:
A pair $(\bar{\mu}_i, m)$ such that for every player i

$$\tilde{J}(\bar{\mu}_i(\cdot), m) = \inf_{\mu_i(\cdot)\in\mathscr{M}_i} \tilde{J}(\mu_i, m), \tag{9.50}$$

and m is the mean-field term generated by the equilibrium $(\bar{x}_j, \bar{\mu}_j)_j$ and is called a behavioral *mixed equilibrium* strategy of the MFTG under decentralized information.

$$\tilde{p}(t) = -h_x(T) - E[h_y(T)] + \int_t^T \{\tilde{H}_x(s) + E[\tilde{H}_y(s)]\}ds - \int_t^T \tilde{q}(s)dW_s, \tag{9.51}$$

such that $E\left[\sup_{t\in[0,T]} |\tilde{p}(t)|^2 + \int_0^T |\tilde{q}(t)|^2 dt\right] < +\infty$.

Proposition 9.5 (Mixed SMP) *Let the assumptions of Proposition 9.3 hold. If* $(\bar{x}_i, \bar{\mu}_i, m)$ *is a behavioral mixed equilibrium, then there is a pair of* $\mathbb{F}$*-adapted processes* $(\tilde{p}(t), \tilde{q}(t))$ *that satisfies (9.51) such that*

$$i \in \mathscr{I} : E\left[E_{\mu_i}\delta\tilde{H}(t) + \tfrac{1}{2}PE_{\mu_i}\{(\delta\tilde{\sigma}(t))^2\}|\, \mathscr{G}_{i,t}\right] \leq 0, \tag{9.52}$$

for all $\mu_i(.) \in \mathscr{P}(A_i)$, *almost every* t *and* $\mathbb{P}-$*almost surely, where*

$$\delta\tilde{H}(t) := H(t, \bar{x}_i(t), a_i, m(t), \tilde{p}(t), \tilde{q}(t)) - E_{\bar{\mu}_i}H(t, \bar{x}_i(t), \bar{a}_i(t), m(t), \tilde{p}(t), \tilde{q}(t)),$$

$$\delta\tilde{\sigma}(t) = \sigma(t, \bar{x}_i(t), E[\bar{x}_i(t)], a_i(t), m(t)) - E_{\bar{\mu}}\left[\sigma(t, \bar{x}_i(t), E[\bar{x}_i(t)], \bar{a}_i(t), m(t))\right].$$

Proof The proof is similar to that in Chap. 4.

Proposition 9.6 (Existence of a Best Response) *Let the assumptions of Proposition 9.3 hold. Then, the mean-field-type best-response problem (9.32) has at least one solution in behavioral mixed strategies. Equivalently, problem (9.44) has at least one best response for a given* m.

Note that, with this type of relaxation, the set $\mathscr{P}(A_i)$ is convex; however, one still needs the second-order matrix P.

Proposition 9.7 (Existence of Equilibria) *Let the assumptions of Proposition 9.3 hold. Then, there exists an equilibrium.*

Proof With behavioral mixed strategies, we have a compact set, and the objective functional $J(\mu_i, m)$ is linear in μ_i. We use the following steps: $m \overset{BR}{\longrightarrow} BR(m) := \{\mu_i^*, \mid J(\mu_i^*, m) = \inf_{\mu_i \in \mathscr{M}_i} J(\mu_i, m)\} \overset{\xi}{\to} \{(\mathscr{L}(x^{\mu^*}))\, |\mu_i^* \in BR\}$. Thus, at a mean-field equilibrium, the mean-field m satisfies the fixed-point inclusion $m \in \xi(m)$. We show the existence of an equilibrium we would like to check the sufficient conditions for Kakutani-Fan-Glicksberg's fixed point theorem. To do so, we need to show compactness of $\{(\mu_i, m) \in \mathscr{M}_i \times \mathscr{P}(\mathbb{R}), \int |a|^2\mu_i(., da) < +\infty$, lower hemicontinuity of ξ, and convex set-valued $\xi(m)$.

Relative compactness $\xi(\mathscr{P}(\mathbb{R}))$: Under assumption of Proposition 9.3, the range of ξ, i.e., $\xi(\mathscr{P}(\mathbb{R}))$, is relatively compact. Since $\mathscr{P}(A_i)$ is non-empty, convex, and compact, $\mathscr{M}_i$, equipped with the weak convergence, provides a relatively compact set. J is continuous, and $BR()$ is hemicontinuous. The existence follows from fixed-point theory.

9.3 Summary of the Chapter

In this chapter, we have presented MFTGs with both atomic and multi-class nonatomic decision-makers. We proposed a PIDE approach in infinite-dimensional spaces and provided sufficient conditions for mean-field-type Nash equilibria in state-and-mean-field-type feedback strategies using a Hamilton-Jacobi-Bellman system in infinite dimensions. As a corollary, master adjoint systems (MASSs) are derived. In general, the master adjoint systems differ from the master equations used [10, 63, 67, 68, 85–87, 89, 93–95, 113, 114, 161–163, 171, 180, 182, 206, 240, 254, 255, 277, 354].

9.4 Notes

9.4.1 Without Individual Mean-Field Terms

Several numerical schemes for solving mean-field games in the infinite population regime have been proposed in the literature. The computational complexity or learning parameters are functions of the numerical scheme parameters, such as step size and number of discretized time-space grid. The global error depends significantly on the choice scheme. The work in [80] introduces a high-order scheme for mean-field games using a computational approach. An optimization formulation of mean-field games is explored in the context of control and optimization [183]. The work [172] discusses mean-field games applied to controlling coherent structures in nonlinear fluid systems. The work in [274] explores imitation in mean-field games, providing insights into neural information processing systems. [372] addresses learning regularized monotone graphon mean-field games within the framework of neural information processing systems. [244] investigates mean-field game systems under displacement monotonicity. [227] examined discrete-time indefinite linear-quadratic mean-field games and control in the finite-population case. A class of mean-field games with optimal stopping and its applications can be found in [198]. The work [214] establishes Lipschitz stability estimates and uniqueness for the retrospective analysis of mean-field games, [148] discusses linear-quadratic mean-field games in Hilbert spaces, and [215] focuses on the mean-field games system with lateral Cauchy data via Carleman estimates. In the context of learning in games, [12] introduces maximum causal entropy inverse reinforcement learning for mean-field games, [118] explores extended mean-field games with multidimensional singular controls and nonlinear jump impact, [234] addresses well posedness for mean-field games with finite state and action space, [189] investigates mean-field games with common Poisson noise using a maximum principle approach, [196] proposes unsupervised solution operator learning for mean-field games via sampling-invariant parametrizations, [199] discusses that model-based RL for mean-field games is not statistically harder than single-agent RL, and [291] explores Bayesian learning in mean-field games.

In the context of inverse game theory, the work [370] introduces a bilevel optimization method for inverse mean-field games. The work [115] discusses intrinsic regularization by noise for 1D mean-field games, [262] explores time periodic solutions of first-order mean-field games from the perspective of the Mather theory,

[232] provides a convergence result for a local planning problem for mean-field games and a rigorous proof of a Freidlin-Ventchel-type large deviation principle for the $1 + 1$ KPZ equation, [225] discusses transition-informed reinforcement learning for large-scale Stackelberg mean-field games, [292] explores the existence and computation of stationary solutions for congestion-type mean-field games via bifurcation theory and forward-forward problems, [90] discusses Newton schemes for mean-field games, [106] introduces a Hessian-dependent functional with free boundaries and applications to mean-field games, [368] investigates when mean-field reinforcement learning is tractable and relevant, [266] analyzes and provides numerical approximations of stationary second-order mean-field game partial differential inclusions, and [275] discusses reconstructing a state-independent cost function in a mean-field game model. [261] explores finite and mean-field games for optimal investment with HARA utility function and the presence of risk-seeking agents; [228] uses a mean-field game approach to relative investment-consumption games with habit formation; [91] introduces a Lagrange-Galerkin scheme for first-order mean-field game systems; [135] explores energy transition under scenario uncertainty using a mean-field game of stopping with common noise; [66] discusses approximation and perturbations of stable solutions to a stationary mean-field game system; [280] discusses Markovian-switching systems, including backward and forward-backward stochastic differential equations, mean-field interactions, and nonzero-sum differential games; [233] addresses the inverse boundary problem for a mean-field game system with a probability density constraint; [120] explores multiple equilibria in mean-field game models for large oligopolies with strategic complementarities; [207] analyzes the convergence rate of the equilibrium measure for the hybrid LQG mean-field games; [204] presents a mean-field game for a forward-backward stochastic system with partial observation and common noise; [219] introduces mirror descent-ascent for the mean-field min-max problem; and [72] discusses mean-field SDEs driven by G-Brownian motion.

In the context of weak solutions of games, [374] explores viscosity solutions for HJB equations on the process space, with an application to mean-field control with common noise, [64] introduces a mean-field game between informed traders and a broker, [235] proposes a design for a finite-time H_∞ controller for stochastic mean-field systems, [276] discusses optimal strategies for large-scale pursuers against one evader using a mean-field game-based hierarchical control approach, [105] explores nonlinear graphon mean-field systems, [375] presents a deep learning algorithm for computing mean-field control problems using forward-backward score dynamics, [230] discusses mean-field-reflected BSDEs driven by a marked point process. [306] introduces an accurate numerical scheme for mean-field forward and backward SDEs with jumps, [208] establishes a stochastic maximum principle for a time-changed mean-field game, [152] explores kernel-based statistical learning theory in the mean-field limit, [369] proposes a fast proximal gradient method for dynamic mean-field planning, [226] presents sequential propagation of chaos for mean-field BSDE systems, [101] provides remarks on linear-quadratic closed-loop games with many players, [216] discusses sampling from the mean-field stationary distribution, and [236] establishes a Laplace principle for large population games with control interaction.

The work in [186] addresses learning high-dimensional McKean-Vlasov forward-backward stochastic differential equations with general distribution dependence; [11] explores the generalization error of graph neural networks in the mean-field regime; [371] discusses Wasserstein proximal operators describing score-based generative models and resolving memorization; [365] analyzes the strong convergence of Euler-Maruyama schemes for doubly perturbed McKean-Vlasov stochastic differential equations; [288] studies conditional McKean-Vlasov SDEs with jumps and Markovian regime switching, addressing well posedness, propagation of chaos, and the averaging principle; and [107] explores the master Bellman equation in the Wasserstein space, establishing the uniqueness of viscosity solutions.

9.4.2 With Individual Mean-Field Terms

In 2014, the work in [102] used the term *mean-field-type games* to distinguish the equilibrium systems of games in which the individual mean-field terms are present and which differ from the equilibrium equations obtained in mean-field games where the individual mean-field terms are not considered.

Since then, several works have considered MFTGs. The work [331] examines uncertainty quantification in mean-field-type teams and games. MFTGs with risk-seeking, risk-averse, and risk-neutral decision-makers are discussed in [43, 330]. Equilibrium principles of MFTGs are provided in [132, 333].

The work in [338, 340] examines master adjoint systems in MFTGs. [40, 41, 43, 50, 51, 294] examine co-opetitive linear-quadratic MFTGs, and [33, 124, 125, 127, 311] examine price dynamics in MFTGs under common noise. The work in [282] studies how much users' psychology matters in MFTGs, and [47, 132, 168, 336] examines MFTG with both individual and population mean-field terms.

Stackelberg MFTGs with non-quadratic cost are discussed in [146, 157, 158]. Berge solutions and mutual support are discussed in [350]. MFTGs with jump and regime switching are examined in [52, 62, 185].

The work in [37] studies fractional MFTGs with non-quadratic costs. MATLAB numerical implementation can be found in [48]. The work in [334] examines how important payoff measurement noise is.

Chapter 10
Semi-Explicit Solutions in Non-Quadratic Mean-Field-Type Games

In this chapter, we are dealing with state models where all quantities are scalars. In Chap. 7, we have studied multi-person MFTGs that are described by a linear jump-diffusion system of conditional mean-field type and a quadratic cost functional involving the second moment of the states and control actions, the square of the conditional expectation of the state, and the control actions of the decision-makers.

This chapter examines MFTGs by means of a direct method. We provide various solvable examples beyond the classical linear-quadratic games. It includes quadratic-quadratic games and games with power, logarithmic, sine square, and hyperbolic sine square payoffs. Nonlinear state dynamics such as log state, control-dependent regime switching, quadratic state, cotangent state, and hyperbolic cotangent state are considered. We identify equilibrium strategies and equilibrium payoffs in state-and-conditional mean-field-type feedback form. It is shown that a simple direct method can be used to solve broader classes of non-quadratic MFTGs under jump-diffusion-regime-switching Gauss–Volterra processes, which include fractional Brownian motion and multi-fractional Brownian motion. We provide semi-explicit solutions to the fully cooperative, noncooperative, nonzero-sum, and adversarial games. In many cases involving fractional Brownian motions and Gauss–Volterra, we restrict the analysis to linear and mean-field-type feedback strategies due to memory dependence.

10.1 Direct Method Beyond LQ-MFTG

The direct method is not limited to the linear-quadratic case and can be extended to a class of linear-exponentiated quadratic MFTGs (LEQ-MFTG), minimax linear-quadratic MFTGs (minimax LQ-MFTG), and minimax LEQ-MFTG. Below, we present several examples to illustrate how the direct method addresses nonlinear

T. Başar et al., *Mean-Field-Type Game Theory I*, Static & Dynamic Game Theory: Foundations & Applications, https://doi.org/10.1007/978-3-032-07027-2_10

and/or non-quadratic MFTGs. The examples below go beyond LQ-MFTG, LEQ-MFTG, and minimax LQ problems.

We provide semi-explicit solutions to mean-field-free games presented in Table 10.1 and to MFTGs presented in Table 10.2. Several noises are examined: Brownian motion B, regime switching s, jump process N, and Gauss–Volterra process B_{gv} which is introduced below.

The Gauss–Volterra noise processes are obtained from the integral of a Brownian motion with a suitable kernel function. It includes fractional Brownian motion and multi-fractional Brownian motion.

The rest of this chapter is structured as follows. In Sect. 10.2, state-and-mean-field-type feedback equilibria and equilibrium cost of some nonlinear, non-quadratic stochastic differential games are presented in a semi-explicit way. In Sect. 10.3, we formulate and solve various MFTGs with a non-quadratic quantity of interest and provide semi-explicit solutions using a direct method. Section 10.4 presents semi-explicit solutions to some non-quadratic MFTGs driven by the Gauss–Volterra processes. Time-fractional MFTGs are presented in Sect. 10.5. Numerical examples are presented in Sect. 10.6.

10.2 Some Solvable Mean-Field-Free Games

We start with mean-field-free settings where logarithm, logarithm square, Legendre-Fenchel duality payoff, and power payoffs are presented. The cost functions are not necessarily quadratic, and the state dynamics are not necessarily linear.

10.2.1 Logarithmic Scale

Consider a set of decision-makers $\mathcal{I} = \{1, \ldots, I\}$, $I \geq 2$ interacting in the following nonlinear, non-quadratic mean-field-free game:

$$\begin{cases} \inf_{u_i} E[L_i(x, u)], \\ \text{subject to} \\ dx = \left(b_1 x \ln(x) + \sum_{j\in\mathcal{I}} b_{2j} x u_j\right) dt + x[\sigma dB + \int \mu d\tilde{N}], \\ x(0) \triangleq x_0 >> e, \\ \mathbb{P}(s(t+\epsilon) = s'|s, u) = \int_t^{t+\epsilon} \tilde{q}_{ss'} dt' + o(\epsilon), \ s' \neq s \\ s(0) = s_0, \\ \tilde{q}_{ss'} > 0, s \neq s', \\ \tilde{q}_{ss} := -\sum_{s'\neq s} \tilde{q}_{ss'}, \end{cases} \tag{10.1}$$

Table 10.1 Semi-explicit solutions of some mean-field-free games. Note that the process B_{gv} considered in the multiplicative noise case does not take into account x in the memory but considers the deterministic coefficients that are a function of time

Problem	State	Cost	Noise
Proposition 10.1	Drift: $x \ln x + x \sum_{j\in\mathcal{I}} b_{2j} u_j$ $\mathcal{I} = \{1, \ldots, I\}$	$q_i \ln(x) + r_i u_i^{2k}$,	Brownian: $x dB$ Jump: $\int_\Theta x \mu d\tilde{N}$ Switching: $\tilde{q}_{ss'}$
Proposition 10.2	Drift: $b_1 x \ln(x) + \sum_{j\in\mathcal{I}} b_{2j} x u_j$ $\mathcal{I} = \{1, \ldots, I\}$	$q_i \ln^2(x) + r_i u_i^2$	Brownian: $x\sqrt{\ln(x)}$ Switching: $\tilde{q}_{ss'}$
Proposition 10.3	Drift: $\frac{b_1 l_1(x)}{l_1'(x)} + \frac{h(x)\sum_j b_{2j} u_j}{l_1'(x)}$ $\mathcal{I} = \{1, \ldots, I\}$	$q_i l_1(x) + \sum_{j\in\mathcal{I}} r_{ij} l_1(u_j)$	Brownian: $\sqrt{\frac{\sigma_1^2 + \sigma_2^2 l_1(x)}{l_1''(x)}}$ Switching: $\tilde{q}_{ss'}$
Proposition 10.4	Drift: $2b\sqrt{x}$ $+x\Big[\sigma^2 + \sigma_o^2 + \sigma_{cogv}^2 + \sigma_{o,cogv}^2$ $+\int_\Theta \mu^2 \nu(d\theta) + \int_\Theta \mu_o^2 \nu_o(d\theta)\Big]$ $\mathcal{I} = \{1, \ldots, I\}$	$q_i \frac{x^k}{k} + r_i \frac{u_i^{2k}}{k}$	Brownian: $2x\sigma dB$ Common Noise: $2x\sigma_o dB_o$ Jump: $2x \int_\Theta \mu d\tilde{N}$ Common Jump: $2x \int_\Theta \mu_o d\tilde{N}_o$ Switching: $\tilde{q}_{ss'}$ Common G-V: $2x\sigma_{o,gv} dB_{o,gv}$ Gauss–Volterra: $2x\sigma_{gv} dB_{gv}$

Table 10.2 Semi-explicit solutions of some MFTGs presented in this chapter. Note that the process B_{gv} considered in the multiplicative noise case does not take into account x in the memory but considers the coefficients that are a function of time

Proposition 10.5	Drift: 0 $s(t) \in \mathscr{S}$ $\mathcal{I} = \{1, \ldots, I\}$	$r_i(u_i - \bar{u}_i)^2 + \bar{r}_i \bar{u}_i^2 + \epsilon_i \bar{u}_i$	Switching: $\tilde{q}_{ss'}(u)$ $= \sum_j b_{2jss'}(u_j - \bar{u}_j)^2 + \bar{b}_{2jss'}\bar{u}_j^2$ $+ \sum_j b_{1jss'}(u_j - \bar{u}_j) + \bar{b}_{1jss'}\bar{u}_j$ $+ \sum_j \bar{b}_{ojss'}$,
Proposition 10.6	Drift: $\sum_{j\in\mathcal{I}}[q_j(u_j - \bar{u}_j)^2 + \bar{q}_j \bar{u}_j^2$ $+\epsilon_{1j}(u_j - \bar{u}_j) + \bar{\epsilon}_{1j}\bar{u}_j]$ $\mathcal{I} = \{1, \ldots, I\}$	$r_i(u_i - \bar{u}_i)^2 + \bar{r}_i \bar{u}_i^2 + \bar{\epsilon}_{2i}\bar{u}_i$	Brownian: σdB Jump: $\int_\Theta \mu d\tilde{N}$
Proposition 10.7	Drift: $\frac{1}{2}cot(\frac{x-\bar{x}}{2}) + \sum_j b_{2j}(u_j - \bar{u}_j)$ $+\frac{1}{2}cot(\frac{\bar{x}}{2}) + \sum_j \bar{b}_{2j}\bar{u}_j$ $\mathcal{I} = \{1, \ldots, I\}$	$((u_i - \bar{u}_i)^2 - q_i)\cos^2(\frac{x-\bar{x}}{4}) + q_i$ $+(\bar{u}_i^2 - \bar{q}_i)\cos^2(\frac{\bar{x}}{4}) + \bar{q}_i$	Brownian: σdB Switching: $\tilde{q}_{ss'}$
Proposition 10.8	Drift: $\frac{1}{2}coth(\frac{x-\bar{x}}{2}) + \sum_j b_{2j}(u_j - \bar{u}_j)$ $+\frac{1}{2}coth(\frac{\bar{x}}{2}) + \sum_j \bar{b}_{2j}\bar{u}_j$ $\mathcal{I} = \{1, \ldots, I\}$	$((u_i - \bar{u}_i)^2 + q_i)\cosh^2(\frac{x-\bar{x}}{4}) - q_i$ $+(\bar{u}_i^2 + \bar{q}_i)\cosh^2(\frac{\bar{x}}{4}) - \bar{q}_i$	Brownian: σdB Switching: $\tilde{q}_{ss'}$
Proposition 10.9	Drift: $-(u_1 - \bar{u}_1) + b_1(x - \bar{x})$ $+b_2\epsilon(x - \bar{x})(u_2 - \bar{u}_2)$ $-\bar{u}_1 + \bar{b}_{11}\bar{x} + \bar{b}_{12}\bar{y} + \bar{b}_{13}\bar{z} + \bar{b}_2\bar{u}_2\bar{x}$ $\mathcal{I} = \{1, \ldots, I\}$	$-q\ var(x) - r_1\ var(u_1)$ $+\bar{r}_1 \frac{\bar{u}_1^\rho}{\rho}$	Brownian: $\sigma(x - \bar{x})(u_2 - \bar{u}_2)dB$ Common noise: $\bar{\sigma}\bar{x}\bar{u}_2 dB_o$ Switching: $\tilde{q}_{ss'}$
Proposition 10.10	Drift: $b_1(x - \bar{x}) + \sum_j b_{2j}(u_j - \bar{u}_j)$ $+\bar{b}_1\bar{x} + \sum_j \bar{b}_{2j}\bar{u}_j$ $\mathcal{I} = \{1, \ldots, I\}$	$q_i \frac{(x-\bar{x})^{2k_i}}{2k_i} + r_i \frac{(u_i-\bar{u}_i)^{2k_i}}{2k_i}$ $+\bar{q}_i \frac{\bar{x}^{2\bar{k}_i}}{2\bar{k}_i} + \bar{r}_i \frac{\bar{u}_i^{2\bar{k}_i}}{2\bar{k}_i}$	Brownian: $(x - \bar{x})\sigma dB$ Jump: $(x - \bar{x})\int_\Theta \mu d\tilde{N}$ Switching: $\tilde{q}_{ss'}$ Gauss–Volterra: $(x - \bar{x})\sigma_{gv} dB_{gv}$
Proposition 10.11	Global optimum of Proposition 10.10		
Proposition 10.12	Adversarial solution of Proposition 10.10		
Proposition 10.13	Time-fractional solution of Proposition 10.10		

where

$$L_i(x,u) = -q_i(T,s(T))\ln(x(T)) + \int_0^T \left(-q_i\ln(x) + r_i u_i^{2k}\right) dt, \tag{10.2}$$

with $k \geq 1$ as a natural number. The state dynamics in (10.1) involves a jump-diffusion process (B, N) and regime-switching process s. The drift function in (10.1) has some nonlinearities. It involves the functions $x\ln x$ and $xu_j,\ j \in \mathcal{I}$, which are nonlinear. The payoff functional in (10.1) of decision-maker i involves the function $-\ln x$ which is non-quadratic and polynomial $u_i^{2k}, k \geq 1$, which is non-quadratic for $k \neq 1$. Thus, we are dealing with a non-LQ game. The next proposition provides a sufficient condition for Nash equilibria in state-feedback strategies. The explicit expression of the equilibrium and equilibrium cost is subject to the solvability of ordinary differential equations instead of partial differential equations. We refer to these expressions as semi-explicit solutions whenever the ordinary differential system has a global solution within the interval of interaction of the game.

Proposition 10.1 *Assume that*

$$\begin{aligned} &x_0 >> e,\ q_i(t,s) \geq 0, r_i(t,s) > \delta > 0,\\ &\int_{\theta\in\Theta}[\ln(1+\mu(\theta)) - \mu(\theta)]\nu(d\theta) < \infty. \end{aligned} \tag{10.3}$$

Then, a mean-field-free Nash equilibrium and corresponding equilibrium cost are given by

$$\begin{aligned} &u_i^* = \sum_{s\in\mathcal{S}}[\tfrac{1}{2k}\tfrac{\alpha_i(t,s)b_{2i}(t,s)}{r_i(t,s)}]^{\frac{1}{2k-1}}\mathbb{1}_{\{s(t)=s\}},\\ &E[L_i(x,u^*)] = E[-\alpha_i(0,s_0)\ln(x_0) + \delta_i(0,s_0)], \end{aligned} \tag{10.4}$$

where α_i and δ_i solve the following differential equations:

$$\begin{aligned} &\dot{\alpha}_i + q_i + \alpha_i b_1 + \sum_{s'}[\alpha_i(t,s') - \alpha_i(t,s)]\tilde{q}_{ss'} = 0,\\ &\alpha_i(T,s) = q_i(T,s),\\ &\dot{\delta}_i + \alpha_i[\tfrac{\sigma^2}{2} - \int_{\theta\in\Theta}[\ln(1+\mu(\theta)) - \mu(\theta)]\nu(d\theta)]\\ &+\sum_{s'}[\delta_i(t,s') - \delta_i(t,s)]\tilde{q}_{ss'} - (2k-1)r_i(\tfrac{1}{2kr_i}b_{2i}\alpha_i)^{\frac{2k}{2k-1}}\\ &-\alpha_i\sum_{j\neq i} b_{2j}[\tfrac{1}{2k}\tfrac{\alpha_j b_{2j}}{r_j}]^{\frac{1}{2k-1}} = 0,\\ &\delta_i(T,s) = 0,\ i \in \mathcal{I}. \end{aligned} \tag{10.5}$$

.

Proof Consider the following guess functional: $f_i(t,x,s) = -\alpha_i\ln(x) + \delta_i$.

By applying Itô's formula for jump-diffusion-regime-switching processes, the gap between the cost and the guess functional $E[L_i(x,u) - f_i(0,x_0)]$ can be computed, and it is given by

$$
\begin{aligned}
&E[L_i(x,u) - f_i(0,x_0,s_0)]\\
&= E\left(-q_i(T,s(T)) + \alpha_i(T,s(T))\right)\ln(x(T)) + \delta_i(T,s(T))\\
&E\int_0^T -\{\dot{\alpha}_i + q_i + \alpha_i b_1 + \sum_{s'}[\alpha_i(t,s') - \alpha_i(t,s)]\tilde{q}_{ss'}\}ln(x)dt\\
&+\int_0^T \dot{\delta}_i + \frac{\sigma^2}{2}\alpha_i + \sum_{s'}[\delta_i(t,s') - \delta_i(t,s)]\tilde{q}_{ss'}dt\\
&+\int_0^T -(2k-1)r_i(\frac{1}{2kr_i}b_{2i}\alpha_i)^{\frac{2k}{2k-1}} - \alpha_i\sum_{j\neq i} b_{2j}[\frac{1}{2k}\frac{\alpha_j b_{2j}}{r_j}]^{\frac{1}{2k-1}}\\
&-\int_0^T \alpha_i\int_{\theta\in\Theta}[\ln(1+\mu(\theta)) - \mu(\theta)]\nu(d\theta)dt\\
&+E\int_0^T[-b_{2i}\alpha_i u_i + r_i u_i^{2k} + (2k-1)r_i(\frac{1}{2kr_i}b_{2i}\alpha_i)^{\frac{2k}{2k-1}}]dt.
\end{aligned}
\tag{10.6}
$$

. Noting that

$$
[-b_{2i}\alpha_i u_i + r_i u_i^{2k} + (2k-1)r_i(\frac{1}{2kr_i}b_{2i}\alpha_i)^{\frac{2k}{2k-1}}] \geq 0,
$$

with equality iff $u_i = u_i^* := [\frac{1}{2k}\frac{\alpha_i b_{2i}}{r_i}]^{\frac{1}{2k-1}}$, the desired result follows. ■

Remark 10.1 For $k = 1$, the system reduces to the following ordinary differential equations:

$$
\begin{aligned}
&u_i^* = \sum_{s\in\mathcal{S}}[\frac{1}{2}\frac{\alpha_i b_{2i}}{r_i}]\mathbb{1}_{\{s(t)=s\}},\\
&E[L_i(x,u^*)] = E[-\alpha_i(0)\ln(x_0)] + \delta_i(0)],\\
&\dot{\alpha}_i + q_i + \alpha_i b_1 + \sum_{s'}[\alpha_i(t,s') - \alpha_i(t,s)]\tilde{q}_{ss'} = 0,\\
&\dot{\delta}_i + \frac{\sigma^2}{2}\alpha_i + \sum_{s'}[\delta_i(t,s') - \delta_i(t,s)]\tilde{q}_{ss'}\\
&-\frac{1}{4r_i}b_{2i}^2\alpha_i^2 - \frac{1}{2}\alpha_i\sum_{j\neq i} b_{2j}^2\frac{\alpha_j}{r_j}\\
&-\alpha_i\int_{\theta\in\Theta}[\ln(1+\mu(\theta)) - \mu(\theta)]\nu(d\theta) = 0.
\end{aligned}
\tag{10.7}
$$

10.2.2 Logarithm Square

Consider the nonlinear, non-quadratic mean-field-free game formulated as

$$
\begin{cases}
L_i(x,u) = q_i(T)\ln^2(x(T)) + \int_0^T \left(q_i\ln^2(x) + r_i u_i^2\right)dt,\\
\inf_{u_i} E[L_i(x,u)],\\
\text{subject to}\\
dx = \left(b_1 x\ln(x) + \sum_{j\in\mathcal{I}} b_{2j}xu_j\right)dt,\\
x(0) \triangleq x_0 >> e,\\
\mathbb{P}(s(t+\epsilon) = s'|s,u) = \int_t^{t+\epsilon}\tilde{q}_{ss'}dt' + o(\epsilon),\ s' \neq s.
\end{cases}
\tag{10.8}
$$

The state dynamics in (10.8) involve a jump-diffusion process (B, N) and regime-switching process s. The drift function in (10.8) has some nonlinearities.

It involves the functions $x\ln x$ and $xu_j,\ j\in\mathcal{I}$ which are nonlinear. The payoff functional in (10.8) of decision-maker i involves the function $-\ln x$ which is non-quadratic and polynomial $u_i^{2k}, k\geq 1$ which is non-quadratic for $k\neq 1$. Thus, we are dealing with a non-LQ game.

The next proposition provides a sufficient condition for Nash equilibria in state-feedback strategies. The explicit expression of the equilibrium and equilibrium cost is subject to the solvability of ordinary differential equations instead of partial differential equations. We refer to these expressions as semi-explicit solutions whenever the ordinary differential system has a global solution within the interval of interaction of the game.

Proposition 10.2 *Assume that* $x_0 >> e,\ q_i(t,s)\geq 0, r_i(t,s)>\delta>0$ *and* $\int_{\theta\in\Theta}[\ln(1+\mu(\theta))-\mu(\theta)]\nu(d\theta)<\infty$. *A nonlinear, non-quadratic mean-field-free Nash equilibrium and corresponding equilibrium cost are given by*

$$\begin{aligned} u_i^* &= -\sum_{s\in\mathcal{S}}\mathbb{1}_{\{s(t)=s\}}[\tfrac{\alpha_i b_{2i}}{r_i}\ln(x)],\\ E[L_i(x,u^*)] &= E[\alpha_i(0)\ln^2(x_0)], \end{aligned} \tag{10.9}$$

where α_i *solves the following system of coupled differential equations:*

$$\begin{aligned} &i\in\mathcal{I},\\ &\dot{\alpha}_i=-q_i-2b_1\alpha_i+2\alpha_i\sum_{j\in\mathcal{I}\setminus\{i\}}\alpha_j\frac{b_{2j}^2}{r_j}\\ &+\alpha_i^2\frac{b_{2i}^2}{r_i}+\sum_{s'\in\mathcal{S}}[\alpha_i(t,s')-\alpha_i(t,s)]\tilde{q}_{ss'},\\ &\alpha_i(T,s)=q_i(T,s). \end{aligned} \tag{10.10}$$

Proof Consider the guess functional: $f_i(t,x)=\alpha_i\ln^2(x)$. Applying Itô's formula yields

$$\begin{aligned} &f_i(T,x(T))-f_i(0,x_0)=\int_0^T\dot{\alpha}_i\ln^2(x)dt\\ &+\int_0^T 2\alpha_i\ln(x)\left(b_1\ln(x)+\sum_{j\in\mathcal{I}}b_{2j}u_j\right)dt\\ &+\int_0^T\sum_{s'\in\mathcal{S}}[\alpha_i(t,s')-\alpha_i(t,s)]\tilde{q}_{ss'}\ln^2(x)dt. \end{aligned} \tag{10.11}$$

By the stochastic integration formula, the gap between the cost and the guess functional $E[L_i(x,u)-f_i(0,x_0)]$ is given by

$$\begin{aligned} &E[L_i(x,u)-f_i(0,x_0)]=E\left(q_i(T)-\alpha_i(T)\right)\ln^2(x(T))\\ &+E\int_0^T q_i\ln^2(x)dt+E\int_0^T\dot{\alpha}_i\ln^2(x)dt\\ &+E\int_0^T\left(2\alpha_i b_1\ln^2(x)+2\alpha_i\ln(x)\sum_{j\in\mathcal{I}\setminus\{i\}}b_{2j}u_j\right)dt\\ &+E\int_0^T r_i\left(u_i^2+2\alpha_i\frac{\ln(x)b_{2i}}{r_i}u_i\right)dt\\ &+E\int_0^T\sum_{s'\in\mathcal{S}}[\alpha_i(t,s')-\alpha_i(t,s)]\tilde{q}_{ss'}\ln^2(x)dt. \end{aligned} \tag{10.12}$$

By performing completion, one obtains

$$\left(u_i + \alpha_i \frac{\ln(x) b_{2i}}{r_i}\right)^2 - \alpha_i^2 \frac{\ln^2(x) b_{2i}^2}{r_i^2} = u_i^2 + 2\alpha_i \frac{\ln(x) b_{2i}}{r_i} u_i, \tag{10.13}$$

leading to

$$\begin{aligned}
&E[L_i(x,u) - f_i(0,x_0)] = E\,(q_i(T) - \alpha_i(T)) \ln^2(x(T)) \\
&+E \int_0^T q_i \ln^2(x) dt + E \int_0^T \dot{\alpha}_i \ln^2(x) dt \\
&+E \int_0^T \left(2\alpha_i b_1 \ln^2(x) - 2\alpha_i \ln^2(x) \sum_{j \in \mathcal{I}\setminus\{i\}} \alpha_j \frac{b_{2j}^2}{r_j}\right) dt \\
&E \int_0^T r_i \left(u_i + \alpha_i \frac{\ln(x) b_{2i}}{r_i}\right)^2 dt \\
&-E \int_0^T \alpha_i^2 \frac{\ln^2(x) b_{2i}^2}{r_i} dt \\
&+E \int_0^T \sum_{s' \in \mathcal{S}} [\alpha_i(t,s') - \alpha_i(t,s)] \tilde{q}_{ss'} \ln^2(x) dt.
\end{aligned} \tag{10.14}$$

Finally, the desired result follows from the minimization of the terms. This completes the proof. ■

10.2.3 Legendre-Fenchel Cost

We consider convex running loss functions l_1 and l_2 :

$$\begin{cases}
L_i = q_{iT} l_1(x_T) + \int_0^T q_i l_1(x) + \sum_{j \in \mathcal{I}} r_{ij} l_2(u_j) dt, \\
\inf_{u_i} E[L_i], \\
\text{subject to} \\
\mathbb{P}(s(t+\epsilon) = s'|s) = \int_t^{t+\epsilon} \tilde{q}_{ss'} dt' + o(\epsilon), \ s' \neq s \\
dx = [\frac{b_1 l_1(x)}{l_1'(x)} + \frac{h(x) \sum_j b_{2j} u_j}{l_1'(x)}] dt + \sqrt{\frac{\sigma_1^2 + \sigma_2^2 l_1(x)}{l_1''(x)}} dB, \\
x(0) = x_0, \ s(0) = s_0,
\end{cases} \tag{10.15}$$

where $\tilde{q}_{ss'} > 0, s \neq s'$ and $\tilde{q}_{ss} := -\sum_{s' \neq s} \tilde{q}_{ss'}$. Recall that the Legendre-Fenchel transform of l is

$$-l^*(x) = \inf_u \{l(u) - xu\}.$$

Proposition 10.3 *Assume that $l_1, l_2, l_2'', r_{ij}, q_i$ are positive. Then, the game (10.15) has a solution:*

$$\begin{aligned}
&u_i^* = \sum_{s \in \mathcal{S}} \mathbb{1}_{\{s(t)=s\}} (l_2^*)'(-\frac{b_{2i}\alpha_i}{r_{ii}} h(x)), \\
&E[L_i(x,u)] = E[\alpha_i(0,s_0) l(x_0) + \delta_i(0,s_0)],
\end{aligned} \tag{10.16}$$

with

$$\begin{array}{l}\dot{\alpha}_i + q_i + \alpha_i(b_1 + \frac{\sigma_2^2}{2}) + \sum_{s'}(\alpha_i(t,s') - \alpha_i(t,s))\tilde{q}_{ss'} \\ -\eta_{ii} + \sum_{j\neq i}\eta_{ij} + \alpha_i b_{2j}\mu_j = 0, \\ \dot{\delta}_i + \frac{\sigma_1^2}{2} + \sum_{s'}(\delta_i(t,s') - \delta_i(t,s))\tilde{q}_{ss'} = 0,\end{array} \tag{10.17}$$

where

$$\begin{array}{l}r_{ii}l_2^*(-\frac{b_{2i}\alpha_i}{r_{ii}}h(x)) = \eta_{ii}l_1(x), \\ r_{ij}l_2(u_j^*) = \eta_{ij}l_1(x), \\ h(x)(l_2^*)'[-\frac{\alpha_j b_{2j}}{r_{jj}}h(x)] = \gamma_j l_1(x).\end{array} \tag{10.18}$$

These conditions are fulfilled by choosing for example

$$l_2(y) = yh(y),\ l_1 = \kappa l_2, h(x) = \frac{x^{2k-1}}{2k}.$$

Proof Step 1: Note that the structure of the problem is mainly driven by the evolution of the function l_1.

Step 2: Inspired by the structure of the problem, we propose a candidate guess functional in the form of l_1 with deterministic coefficients α_i, δ_i. Let $f_i(t,x,s) = \alpha_i(t,s)l_1(x) + \delta_i(t,s)$.

Step 3: We apply a stochastic integration formula for diffusion-regime switching to obtain the difference between the cost and the guess functional as

$$\begin{array}{l}E[L_i(x,s,u) - f_i(0,x_0,s_0)] \\ = E(q_i(T,s(T)) - \alpha_i(T,s(T)))l_1(x(T)) + (0 - \delta_i(T,s(T))) \\ +E\int_0^T[\dot{\alpha}_i + q_i + \alpha_i(b_1 + \frac{\sigma_2^2}{2}) \\ +\sum_{s'}(\alpha_i(t,s') - \alpha_i(t,s))\tilde{q}_{ss'}]l(x) \\ +\dot{\delta}_i + \frac{\sigma_1^2}{2} + \sum_{s'}(\delta_i(t,s') - \delta_i(t,s))\tilde{q}_{ss'} \\ -r_{ii}l_2^*(-\frac{b_{2i}\alpha_i}{r_{ii}}h(x)) + \sum_{j\neq i}\{r_{ij}l_2(u_j^*) + \alpha_i b_{2j}h(x)u_j^*\} \\ +\{r_{ii}[l_2(u_i) + \frac{\alpha_i b_{2i}h(x)}{r_{ii}}u_i] + r_{ii}l_2^*(-\frac{\alpha_i b_{2i}h(x)}{r_{ii}})\}dt.\end{array} \tag{10.19}$$

Step 4: Observing that

$$\{[l_2(u_i) + \frac{\alpha_i b_{2i}h(x)}{r_{ii}}u_i] + l_2^*(-\frac{\alpha_i b_{2i}h(x)}{r_{ii}})\} \geq 0,$$

with equality if, and only if, $u_i = u_i^*$, the one-shot optimization provides $u_j^* = (l_2^*)'[-\frac{\alpha_j b_{2j}}{r_{jj}}h(x)]$.

Step 5: By the identification of processes, the desired result follows. This completes the proof. ■

10.2.4 Geometric Gauss–Volterra Game

Consider the following geometric Gauss–Volterra game with unobserved processes B_o, $B_{o,gv}$, N_o which are assumed to be independent:

$$\begin{cases} i \in \mathcal{I} = \{1, \ldots, I\},\ I \geq 1, \\ L_i(x,u) = q_{iT}\frac{x_T^k}{k} + \int_0^T \left(q_i\frac{x^k}{k} + r_i\frac{u_i^{2k}}{k}\right)dt, \\ \inf_{u_i} EL_i(x,u), \\ \text{subject to} \\ dx = \left(2b\sqrt{x} + x\left[\sigma^2 + \sigma_o^2 + \sigma_{cogv}^2 + \sigma_{o,cogv}^2 + \int_\Theta \mu^2\nu(d\theta) + \int_\Theta \mu_o^2\nu_o(d\theta)\right]\right)dt \\ +2x\left[\sigma dB + \sigma_o dB_o + \int_\Theta \mu d\tilde{N} + \int_\Theta \mu_o d\tilde{N}_o + \sigma_{gv}dB_{gv} + \sigma_{o,gv}dB_{o,gv}\right], \\ x(0) = x_0 > 0, \\ \mathbb{P}(s(t+\epsilon) = s'|s) = \int_t^{t+\epsilon} \tilde{q}_{ss'}dt' + o(\epsilon),\ s' \neq s, \\ \tilde{q}_{ss'} > 0,\ s \neq s', \\ \tilde{q}_{ss} := -\sum_{s'\neq s} \tilde{q}_{ss'}, \end{cases} \tag{10.20}$$

where $b = b_1\sqrt{x} + \sum_{j\in\mathcal{I}} b_{2j}u_j$, and σ, σ_o, σ_{gv}, $\sigma_{o,gv}$, μ, and μ_o are real-valued and regime-switching-dependent functions $s(t)$ with $\tilde{q}_{ss'} > 0$, $s \neq s'$ and $\tilde{q}_{ss} := -\sum_{s'\neq s} \tilde{q}_{ss'}$. Note that the process B_{gv} considered in the multiplicative noise case does not take into account x in the memory but considers the coefficients that are a function of time.

Proposition 10.4 *Assume that $q_i \geq 0, r_i > \delta > 0, x_0 > 0, k \geq 1$. The mean-field-free equilibrium for the geometric Gauss–Volterra game in (10.20) is given by*

$$\begin{aligned} u_i^* &= -\sum_{s\in\mathcal{S}} \mathbb{1}_{\{s(t)=s\}} \left(\frac{\alpha_i b_{2i}}{r_i}\right)^{\frac{1}{2k-1}} \sqrt{x}, \\ EL_i(x,u^*) &= E\alpha_i(0,s_0)\frac{x_0^k}{k}, \end{aligned} \tag{10.21}$$

where α_i solves the following differential equation:

$$\begin{aligned} & i \in \mathcal{I}, \\ & \dot{\alpha}_i + q_i + k\alpha_i([\sigma^2 + \sigma_o^2 + \sigma_{gv}^2 + \sigma_{o,gv}^2] \\ & + \int_\Theta \mu^2\nu(d\theta) + \int_\Theta \mu_o^2\nu_o(d\theta)) \\ & +2kb_1\alpha_i + 2k(k-1)\alpha_i(\sigma^2 + \sigma_o^2 + \sigma_{gv}^2 + \sigma_{o,gv}^2 \\ & + \int_\Theta \mu^2\nu(d\theta) + \int_\Theta \mu_o^2\nu_o(d\theta)) \\ & - \left(\frac{\alpha_i b_{2i}}{r_i}\right)^{\frac{2k}{2k-1}} r_i(k - \tfrac{1}{2}) \\ & -2\alpha_i k \sum_{j\neq i} b_{2j} \left(\frac{\alpha_j b_{2j}}{r_j}\right)^{\frac{1}{2k-1}} + \sum_{s'}(\alpha_i(t,s') - \alpha_i(t,s))\tilde{q}_{ss'} = 0, \end{aligned} \tag{10.22}$$

Proof This proof is developed following a direct method.

Step 1: Note that the problem is a mean-field-free game problem driven by $\sqrt{x}$.

Step 2: Based on the structure of the problem, we propose the following guess functional:

$$f_i(t,x,s) = \alpha_i(t,s)\frac{x^k}{k}. \tag{10.23}$$

Step 3: We apply the stochastic integration formula for jump-diffusion-regime-switching Gauss–Volterra and common noises to compute the difference between the costs and the guess functionals, i.e.

$$\begin{aligned}
&E[L_i(x,u) - f_i(0,s_0)] = E(q_{i,T} - \alpha_i(T,s(T)))\frac{x^k(T)}{k}\\
&+E\int_0^T (q_i + \dot{\alpha}_i)\frac{x^k}{k} + k\alpha_i\frac{x^k}{k}([\sigma^2 + \sigma_o^2 + \sigma_{cogv}^2 + \sigma_{o,cogv}^2]\\
&+\int_\Theta \mu^2 d\nu(\theta) + \int_\Theta \mu_o^2 d\nu_o(\theta))\\
&+2kb_1\alpha_i\frac{x^k}{k} + 2k(k-1)\alpha_i\frac{x^k}{k}(\sigma^2 + \sigma_o^2 + \sigma_{cogv}^2 + \sigma_{o,cogv}^2\\
&+\int_\Theta \mu^2 d\nu(\theta) + \int_\Theta \mu_o^2 d\nu_o(\theta))\\
&+[r_i\frac{u_i^{2k}}{k} + 2\alpha_i x^{k-\frac{1}{2}} b_{2i}u_i] + 2\alpha_i x^{k-\frac{1}{2}}\sum_{j\neq i} b_{2j}u_j\\
&+\frac{x^k}{k}\sum_{s'}(\alpha_i(t,s') - \alpha_i(t,s))\tilde{q}_{ss'}dt.
\end{aligned} \tag{10.24}$$

Step 4: We perform completion:

$$u_i^* = \left(-\frac{\alpha_i b_{2i}}{r_i}\right)^{\frac{1}{2k-1}}\sqrt{x}. \tag{10.25}$$

Step 5: We perform the process of matching terms after replacing back the optimal control inputs in the gap $E[L_i(x,u) - f_i(0,s_0)]$, i.e.

$$\begin{aligned}
&E[L_i(x,u) - f_i(0,s_0)] = E(q_{i,T} - \alpha_i(T,s(T)))\frac{x^k(T)}{k}\\
&+E\int_0^T (q_i + \dot{\alpha}_i)\frac{x^k}{k} + k\alpha_i\frac{x^k}{k}([\sigma^2 + \sigma_o^2 + \sigma_{cogv}^2 + \sigma_{o,cogv}^2]\\
&+\int_\Theta \mu^2 d\nu(\theta) + \int_\Theta \mu_o^2 d\nu_o(\theta))\\
&+2kb_1\alpha_i\frac{x^k}{k}\\
&+2k(k-1)\alpha_i\frac{x^k}{k}(\sigma^2 + \sigma_o^2 + \sigma_{cogv}^2 + \sigma_{o,cogv}^2 + \int_\Theta \mu^2 d\nu(\theta) + \int_\Theta \mu_o^2 d\nu_o(\theta))\\
&+[r_i\frac{u_i^{2k}}{k} + 2\alpha_i x^{k-\frac{1}{2}} b_{2i}u_i + \left(-\frac{\alpha_i b_{2i}}{r_i}\right)^{\frac{2k}{2k-1}} r_i(1-\frac{1}{2k})x^k]\\
&+\left[-k\left(-\frac{\alpha_i b_{2i}}{r_i}\right)^{\frac{2k}{2k-1}} r_i(k-\frac{1}{2}) + 2\alpha_i k\sum_{j\neq i} b_{2j}\left(-\frac{\alpha_j b_{2j}}{r_j}\right)^{\frac{1}{2k-1}}\right]\frac{x^k}{k}\\
&+\frac{x^k}{k}\sum_{s'}(\alpha_i(t,s') - \alpha_i(t,s))\tilde{q}_{ss'}dt.
\end{aligned} \tag{10.26}$$

Finally, the desired result is arrived at minimizing terms, which completes the proof. ■

When $s, B_o, B_{o,gv}, N_o$ are noises observed by all decision-makers (observed common noises), the ordinary differential system in α becomes a differential system driven by extra coefficients of the union of events with B_o, $B_{o,gv}$, N_o :

$$
\begin{aligned}
&d\alpha_i + (2kb_1\alpha_i + q_i)dt + k\alpha_i(\sigma^2 + \sigma_o^2 + \sigma_{gv}^2 + \sigma_{o,gv}^2 + \int_\Theta \mu^2\nu(d\theta) \\
&+ \int_\Theta \mu_o^2\nu_o(d\theta))dt \\
&+2k(k-1)\alpha_i(\sigma^2 + \sigma_o^2 + \sigma_{gv}^2 + \sigma_{o,gv}^2 + \int_\Theta \mu^2\nu(d\theta) + \int_\Theta \mu_o^2\nu_o(d\theta))dt \\
&- \left(\tfrac{\alpha_i b_{2i}}{r_i}\right)^{\frac{2k}{2k-1}} r_i(k-\tfrac{1}{2})dt - 2\alpha_i k \sum_{j\neq i} b_{2j}\left(\tfrac{\alpha_j b_{2j}}{r_j}\right)^{\frac{1}{2k-1}} dt \\
&+ \sum_{s'}(\alpha_i(t,s') - \alpha_i(t,s))\tilde{q}_{ss'}dt \\
&+2k\Big[\alpha_{i,B_o}\sigma_o + \int_\Theta \mu_o\alpha_{i,N_o}\nu_o(d\theta) + cov[\sigma_{o,gv}dB_{gv}, \alpha_{i,B_{gv}}dB_{gv}]\Big]dt \\
&+2k\alpha_i\Big[\sigma dB + \sigma_o dB_o + \int_\Theta \mu d\tilde{N} + \int_\Theta \mu_o d\tilde{N}_o + \sigma_{gv}dB_{gv} + \sigma_{o,gv}dB_{o,gv}\Big] = 0,
\end{aligned}
\tag{10.27}
$$

with the terminal condition $q_i(T,s) \geq 0$ being $\mathcal{F}^{B_o,B_{o,gv},N_o}$-measurable random coefficient.

10.3 Some Solvable Mean-Field-Type Games

10.3.1 Control-Dependent Switching MFTG

In most continuous time MFTG models with regime switching considered in the literature, it is assumed that the switching rate $\tilde{q}_{ss'}$ is control-independent. In this subsection, we provide an example with control-dependent switching rate $\tilde{q}_{ss'}(u)$ in which the MFTG problem can be solved semi-explicitly.

$$
\begin{cases}
i \in \mathcal{I}, \\
\inf_{u_i} E[L_i], \\
\text{subject to} \\
\mathbb{P}(s(t+\epsilon) = s'|s,u) = \int_t^{t+\epsilon} \tilde{q}_{ss'}(u)dt' + o(\epsilon),\ s' \neq s \\
\tilde{q}_{ss'}(u) = \sum_j b_{2jss'}(u_j - \bar{u}_j)^2 + \bar{b}_{2jss'}\bar{u}_j^2 + b_{1jss'}(u_j - \bar{u}_j) + \bar{b}_{1jss'}\bar{u}_j + \bar{b}_{ojss'},
\end{cases}
\tag{10.28}
$$

where $L_i = q_{iT} + \int_0^T r_i(u_i - \bar{u}_i)^2 + \bar{r}_i\bar{u}_i^2 + \epsilon_i\bar{u}_i dt$, $b_{kjss'} > 0$ for $s' \neq s$ and $Eb_{1jss'} = 0$.

Proposition 10.5 *Assume that $r_i, \bar{r}_i > \delta > 0$. The equilibrium strategy is*

$$
u_i^* = -\frac{1}{2}\cdot\frac{\sum_{s'\in\mathcal{S}} V_i(t,s')b_{1iss'}}{r_i + \sum_{s'\in\mathcal{S}} V_i(t,s')b_{2iss'}} - \frac{1}{2}\cdot\frac{\epsilon_i + \sum_{s'\in\mathcal{S}} V_i(t,s')\bar{b}_{1iss'}}{\bar{r}_i + \sum_{s'\in\mathcal{S}} V_i(t,s')\bar{b}_{2iss'}},
\tag{10.29}
$$

and the equilibrium cost is $V_i(t,s)$, which satisfies the following ordinary differential system:

$$
\begin{aligned}
&\dot{V}_i + \sum_{s'\in\mathcal{S}} V_i(t,s')\bar{b}_{ojss'} \\
&-\frac{1}{4}\frac{(\sum_{s'\in\mathcal{S}} V_i(t,s')b_{1iss'})^2}{[r_i+\sum_{s'\in\mathcal{S}} V_i(t,s')b_{2iss'}]} - \frac{1}{4}\frac{(\epsilon_i+\sum_{s'\in\mathcal{S}} V_i(t,s')\bar{b}_{1iss'})^2}{[\bar{r}_i+\sum_{s'\in\mathcal{S}} V_i(t,s')\bar{b}_{2iss'}]} \\
&+\frac{1}{4}\sum_{j\neq i}[\sum_{s'\in\mathcal{S}} V_i(t,s')b_{2jss'}](\frac{\sum_{s'\in\mathcal{S}} V_j(t,s')b_{1jss'}}{r_j+\sum_{s'\in\mathcal{S}} V_j(t,s')b_{2jss'}})^2 \\
&+\frac{1}{4}\sum_{j\neq i}[\sum_{s'\in\mathcal{S}} V_i(t,s')\bar{b}_{2jss'}](\frac{\epsilon_j+\sum_{s'\in\mathcal{S}} V_j(t,s')\bar{b}_{1jss'}}{\bar{r}_j+\sum_{s'\in\mathcal{S}} V_j(t,s')\bar{b}_{2jss'}})^2 \\
&-\frac{1}{2}\sum_{j\neq i}[\sum_{s'\in\mathcal{S}} V_i(t,s')b_{1jss'}](\frac{\sum_{s'\in\mathcal{S}} V_j(t,s')b_{1jss'}}{r_j+\sum_{s'\in\mathcal{S}} V_j(t,s')b_{2jss'}}) \\
&-\frac{1}{2}\sum_{j\neq i}[\sum_{s'\in\mathcal{S}} V_i(t,s')\bar{b}_{1jss'}](\frac{\epsilon_j+\sum_{s'\in\mathcal{S}} V_j(t,s')\bar{b}_{1jss'}}{\bar{r}_j+\sum_{s'\in\mathcal{S}} V_j(t,s')\bar{b}_{2jss'}}) = 0, \\
&V_i(T,s) = q_i(T,s),\ s\in\mathcal{S}.
\end{aligned}
\tag{10.30}
$$

Proof Step 1: We observe that the structure of the problem does not have a drift and is driven by regime switching.

Step 2: Based on Step 1, we propose a guess in the following form: $V_i(t,s)$.

Step 3: We use the stochastic integration formula for regime switching to compute the difference between the cost and the guess functional as

$$
\begin{aligned}
&E[L_i(s,u) - V_i(0,s_0)] = E[q_{iT} - V_i(T,s(T))] \\
&+E\int_0^T \dot{V}_i + \sum_{s'\in\mathcal{S}} V_i(t,s')\bar{b}_{ojss'} \\
&-\frac{1}{4}\frac{(\sum_{s'\in\mathcal{S}} V_i(t,s')b_{1iss'})^2}{r_i+\sum_{s'\in\mathcal{S}} V_i(t,s')b_{2iss'}} \\
&+\frac{1}{4}\sum_{j\neq i}[\sum_{s'\in\mathcal{S}} V_i(t,s')b_{2jss'}](\frac{\sum_{s'\in\mathcal{S}} V_j(t,s')b_{1jss'}}{r_j+\sum_{s'\in\mathcal{S}} V_j(t,s')b_{2jss'}})^2 \\
&-\frac{1}{2}\sum_{j\neq i}[\sum_{s'\in\mathcal{S}} V_i(t,s')b_{1jss'}](\frac{\sum_{s'\in\mathcal{S}} V_j(t,s')b_{1jss'}}{r_j+\sum_{s'\in\mathcal{S}} V_j(t,s')b_{2jss'}}) \\
&-\frac{1}{4}\frac{(\epsilon_i+\sum_{s'\in\mathcal{S}} V_i(t,s')\bar{b}_{1iss'})^2}{[\bar{r}_i+\sum_{s'\in\mathcal{S}} V_i(t,s')\bar{b}_{2iss'}]}, \\
&+\frac{1}{4}\sum_{j\neq i}[\sum_{s'\in\mathcal{S}} V_i(t,s')\bar{b}_{2jss'}](\frac{\epsilon_j+\sum_{s'\in\mathcal{S}} V_j(t,s')\bar{b}_{1jss'}}{\bar{r}_j+\sum_{s'\in\mathcal{S}} V_j(t,s')\bar{b}_{2jss'}})^2 \\
&-\frac{1}{2}\sum_{j\neq i}[\sum_{s'\in\mathcal{S}} V_i(t,s')\bar{b}_{1jss'}](\frac{\epsilon_j+\sum_{s'\in\mathcal{S}} V_j(t,s')\bar{b}_{1jss'}}{\bar{r}_j+\sum_{s'\in\mathcal{S}} V_j(t,s')\bar{b}_{2jss'}}) \\
&+[r_i+\sum_{s'\in\mathcal{S}} V_i(t,s')b_{2iss'}]\{u_i - \bar{u}_i + \frac{1}{2}\frac{\sum_{s'\in\mathcal{S}} V_i(t,s')b_{1iss'}}{r_i+\sum_{s'\in\mathcal{S}} V_i(t,s')b_{2iss'}}\}^2 dt \\
&+[\bar{r}_i+\sum_{s'\in\mathcal{S}} V_i(t,s')\bar{b}_{2iss'}]\{\bar{u}_i + \frac{1}{2}\frac{\epsilon_i+\sum_{s'\in\mathcal{S}} V_i(t,s')\bar{b}_{1iss'}}{\bar{r}_i+\sum_{s'\in\mathcal{S}} V_i(t,s')\bar{b}_{2iss'}}\}^2 dt.
\end{aligned}
\tag{10.31}
$$

Step 4: Assuming that $[r_i + \sum_{s'\in\mathcal{S}} V_i(t,s')b_{2iss'}] > 0$, and $[\bar{r}_i + \sum_{s'\in\mathcal{S}} V_i(t,s')\bar{b}_{2iss'}] > 0$, the completion of terms leads to a one-shot optimization of a strictly convex and coercive function.

Step 5: Minimization and matching of various terms lead to the desired result. ■

10.3.2 Quadratic-Quadratic MFTG

We examine here a class of quadratic-quadratic MFTG (QQ-MFTG) problem where the state is nonlinear in u, for which we derive a semi-explicit solution. The problem formulation is

$$\begin{cases} i \in \mathcal{I}, \\ L_i(x,u) = q_{iT}x_T + \int_0^T r_i(u_i-\bar{u}_i)^2 + \bar{r}_i\bar{u}_i^2 + \bar{\epsilon}_{2i}\bar{u}_i dt \\ \inf_{u_i} \ E[L_i(x,u)], \\ \text{subject to} \\ dx = \sum_{j\in\mathcal{I}}[q_j(u_j-\bar{u}_j)^2 + \bar{q}_j\bar{u}_j^2 + \epsilon_{1j}(u_j-\bar{u}_j) + \bar{\epsilon}_{1j}\bar{u}_j]dt \\ +\sigma dB + \int \mu d\tilde{N}, \\ x(0) = x_0, \\ \mathbb{P}(s(t+\epsilon)=s'|s) = \int_t^{t+\epsilon} \tilde{q}_{ss'}dt' + o(\epsilon), \ s' \neq s, \\ \tilde{q}_{ss'} > 0, \ s \neq s', \\ \tilde{q}_{ss} := -\sum_{s'\neq s}\tilde{q}_{ss'}, \end{cases} \tag{10.32}$$

where the coefficients are regime-switching-dependent with switching rate matrix $\tilde{Q} = (\tilde{q}_{ss'}, (s,s') \in \mathscr{S}^2)$.

Proposition 10.6 *Assume that $r_i > \delta > 0, q_i \geq 0, \ r_i > \delta > 0, \bar{r}_i > 0, \bar{q}_i \geq 0, \bar{r}_i > \delta > 0, \ E[\epsilon_{1i}] = 0, \ \forall i \in \mathcal{I}$. The QQ-MFTG problem (10.32) admits a unique solution, which is given by*

$$\begin{aligned} &u_i^* - \bar{u}_i^* = -\tfrac{\alpha_i}{2(r_i+\alpha_i q_i)}\epsilon_{1i}, \\ &\bar{u}_i^* = -\tfrac{[\bar{\epsilon}_{2i}+\alpha_i\bar{\epsilon}_{1i}]}{2[\bar{r}_i+\alpha_i\bar{q}_i]}, \\ &L_i^* = E[\alpha_i(0,s(0))x_0] + E\int_0^T \{ \\ &\alpha_i\sum_{j\neq i} q_j\left(\tfrac{\alpha_j}{2(r_j+\alpha_j q_j)}\epsilon_{1j}\right)^2 + \alpha_i\sum_{j\neq i}\bar{q}_j\left(\tfrac{[\bar{\epsilon}_{2j}+\alpha_j\bar{\epsilon}_{1j}]}{2[\bar{r}_j+\alpha_j\bar{q}_j]}\right)^2 \\ &-\left(\alpha_i\sum_{j\neq i}\epsilon_{1j}\left(\tfrac{\alpha_j}{2(r_j+\alpha_j q_j)}\epsilon_{1j}\right) + \alpha_i\sum_{j\neq i}\bar{\epsilon}_{1j}\left(\tfrac{[\bar{\epsilon}_{2j}+\alpha_j\bar{\epsilon}_{1j}]}{2[\bar{r}_j+\alpha_j\bar{q}_j]}\right)\right) \\ &-\tfrac{\alpha_i^2}{4(r_i+\alpha_i q_i)}\epsilon_{1i}^2 - \tfrac{[\bar{\epsilon}_{2i}+\alpha_i\bar{\epsilon}_{1i}]^2}{4[\bar{r}_i+\alpha_i\bar{q}_i]}\ \}dt. \end{aligned} \tag{10.33}$$

Note that the semi-explicit solution is in fact an explicit solution. Letting $\vec{\alpha}_i(t) = [\alpha_i(t,s)]_{s\in\mathscr{S}}$ and $\vec{q}_i(T) = [q_i(T,s)]_{s\in\mathscr{S}}$, we have that $\vec{\alpha}_i(t)$ is explicitly given by

$$\vec{\alpha}_i(t) = \vec{q}_i(T)e^{-\int_t^T \tilde{Q}dt'}. \tag{10.34}$$

In particular, $\vec{\alpha}_i(0) = \vec{q}_i(T)e^{-\int_0^T \tilde{Q}dt'}$.

Proof Let us consider the following guess functional: $V_i = \alpha_i x$. Then

$$\begin{aligned} &\partial_t V_i = \dot{\alpha}_i x, \ \partial_x V_i = \alpha_i, \\ &V_i(t,x+\mu) - V_i(t,x) - \partial_x V_i \mu = 0. \end{aligned} \tag{10.35}$$

Itô's formula yields

$$
\begin{array}{l}
V_i(T) - V_i(0) = \int_0^T \left(\dot{\alpha}_i x + \alpha_i \sum_{j\in\mathcal{I}} q_j(u_j - \bar{u}_j)^2\right.\\
+\alpha_i \sum_{j\in\mathcal{I}} \bar{q}_j\bar{u}_j^2 + \alpha_i \sum_{j\in\mathcal{I}} \epsilon_{1j}(u_j - \bar{u}_j)\\
+\alpha_i \sum_{j\in\mathcal{I}} \bar{\epsilon}_{1j}\bar{u}_j + \sum_{s'\in\mathcal{S}} x \left.\left(\alpha_i(.,s') - \alpha_i(.,s)\right)\tilde{q}_{ss'}\right) dt\\
+\int_0^T \alpha_i\sigma dB + \int_\Theta \alpha_i\mu\tilde{N}(dt,d\theta).
\end{array}
\tag{10.36}
$$

Thus, the difference $E[L_i - V_i(0)]$ is given by

$$
\begin{array}{l}
E[L_i - V_i(0)] = E\left[q_i(T)x(T) - V_i(T)\right]\\
+E\int_0^T r_i(u_i - \bar{u}_i)^2 + \bar{r}_i\bar{u}_i^2 + \bar{\epsilon}_{2i}\bar{u}_i\\
+\dot{\alpha}_i x + \alpha_i \sum_{j\in\mathcal{I}} q_j(u_j - \bar{u}_j)^2\\
+\alpha_i \sum_{j\in\mathcal{I}} \bar{q}_j\bar{u}_j^2 + \alpha_i \sum_{j\in\mathcal{I}} \epsilon_{1j}(u_j - \bar{u}_j)\\
+\alpha_i \sum_{j\in\mathcal{I}} \bar{\epsilon}_{1j}\bar{u}_j + x\sum_{s'\in\mathcal{S}} \left(\alpha_i(.,s') - \alpha_i(.,s)\right)\tilde{q}_{ss'}\, dt.
\end{array}
\tag{10.37}
$$

Performing the completion of squares yields

$$
\begin{array}{l}
\left((u_i - \bar{u}_i)^2 + \frac{\alpha_i}{r_i+\alpha_i q_i}\epsilon_{1i}(u_i - \bar{u}_i)\right) = \left(u_i - \bar{u}_i + \frac{\alpha_i}{2(r_i+\alpha_i q_i)}\epsilon_{1i}\right)^2 - \frac{\alpha_i^2}{4(r_i+\alpha_i q_i)^2}\epsilon_{1i}^2,\\
\left(\bar{u}_i^2 + \frac{[\bar{\epsilon}_{2i}+\alpha_i\bar{\epsilon}_{1i}]}{[\bar{r}_i+\alpha_i\bar{q}_i]}\bar{u}_i\right) = \left(\bar{u}_i + \frac{[\bar{\epsilon}_{2i}+\alpha_i\bar{\epsilon}_{1i}]}{2[\bar{r}_i+\alpha_i\bar{q}_i]}\right)^2 - \frac{[\bar{\epsilon}_{2i}+\alpha_i\bar{\epsilon}_{1i}]^2}{4[\bar{r}_i+\alpha_i\bar{q}_i]^2},
\end{array}
\tag{10.38}
$$

$$
\begin{array}{l}
E[L_i - V_i(0)] = E\left[q_i(T)x(T) - V_i(T)\right]\\
+E\int_0^T \dot{\alpha}_i x + \alpha_i \sum_{j\neq i} q_j(u_j - \bar{u}_j)^2 + \alpha_i \sum_{j\neq i} \bar{q}_j\bar{u}_j^2\\
+\alpha_i \sum_{j\neq i} \epsilon_{1j}(u_j - \bar{u}_j) + \alpha_i \sum_{j\neq i} \bar{\epsilon}_{1j}\bar{u}_j\\
+(r_i + \alpha_i q_i)\left(u_i - \bar{u}_i + \frac{\alpha_i}{2(r_i+\alpha_i q_i)}\epsilon_{1i}\right)^2\\
-\frac{\alpha_i^2}{4(r_i+\alpha_i q_i)}\epsilon_{1i}^2 + [\bar{r}_i + \alpha_i\bar{q}_i]\left(\bar{u}_i + \frac{[\bar{\epsilon}_{2i}+\alpha_i\bar{\epsilon}_{1i}]}{2[\bar{r}_i+\alpha_i\bar{q}_i]}\right)^2\\
-\frac{[\bar{\epsilon}_{2i}+\alpha_i\bar{\epsilon}_{1i}]^2}{4[\bar{r}_i+\alpha_i\bar{q}_i]} + \sum_{s'\in\mathcal{S}} x\left(\alpha_i(.,s') - \alpha_i(.,s)\right)\tilde{q}_{ss'}.
\end{array}
\tag{10.39}
$$

Minimization of various terms yields

$$
\begin{array}{l}
\dot{\alpha}_i(t,s) = -\sum_{s'\in\mathcal{S}} \left(\alpha_i(.,s') - \alpha_i(.,s)\right)\tilde{q}_{ss'},\\
\alpha_i(T,s) = q_i(T,s),
\end{array}
\tag{10.40}
$$

which completes the proof. ■

Remark 10.2 Notice that using the result presented in Proposition 10.6, the following quadratic-exponential-quadratic MFTG (QEQ-MFTG) can be solved explicitly:

$$\begin{cases} L_i(x,u) = q_{iT}x_T + \int_0^T r_i(u_i-\bar{u}_i)^2 + \bar{r}_i\bar{u}_i^2 + \bar{\epsilon}_{2i}\bar{u}_i dt, \\ \inf_{u_i} E\exp\left[\lambda_i L_i(x,u)\right], \\ \text{subject to} \\ dx = \sum_{j\in\mathcal{I}}[q_j(u_j-\bar{u}_j)^2 + \bar{q}_j\bar{u}_j^2 + \epsilon_{1j}(u_j-\bar{u}_j) + \bar{\epsilon}_{1j}\bar{u}_j]dt \\ +\sigma dB, \\ x(0) = x_0. \end{cases} \tag{10.41}$$

10.3.3 Quadratic State and Power Utility

This subsection examines a class of MFTGs with power payoffs and a nonlinear state. The model is inspired by the modern portfolio optimization under a shared asset platform by several decision-makers. The state $x(t)$ is the total amount of money. Decision-maker i can decide to consume certain amount u_{1i} and reallocate the remaining between less-risky assets $(1-u_{2i})\kappa_2 x$ and more risky assets $u_{2i}\kappa_2 x + \bar{u}_{2i}x[\sigma dB + \int \mu d\tilde{N}]$. The coefficients κ_1, κ_2 depend on time t and on the switching regime $s(t)$ which takes values in $\mathcal{S}$. The set S is non-empty and finite. We have modified the model to include mean-field terms: a function of the expected value of the state and a function of the expected value of the control action:

$$\begin{cases} R_i(x,u) = -q_{iT}\dfrac{(x_T-\bar{x}_T)^{2k_i}}{2k_i} + \bar{q}_{iT}\dfrac{\bar{x}_T^{\bar{\rho}_i}}{\bar{\rho}_i} \\ +\int_0^T -q_i\dfrac{(x-\bar{x})^{2k_i}}{2k_i} + \bar{q}_i\dfrac{\bar{x}^{\bar{\rho}_i}}{\bar{\rho}_i} - r_i\dfrac{(u_{1i}-\bar{u}_{1i})^{2k_i}}{2k_i} + \bar{r}_i\dfrac{\bar{u}_{1i}^{\bar{\rho}_i}}{\bar{\rho}_i}dt, \\ \sup_{u_i} E[R_i(x,u)], \\ \text{subject to} \\ dx = \sum_{i\in\mathcal{I}}[-(u_{1i}-\bar{u}_{1i}) + (1-(u_{2i}-\bar{u}_{2i}))\kappa_1(x-\bar{x})] \\ +\sum_{i\in\mathcal{I}}(u_{2i}-\bar{u}_{2i})\kappa_2(x-\bar{x})dt \\ +\sum_{i\in\mathcal{I}}(u_{2i}-\bar{u}_{2i})(x-\bar{x})[\sigma dB + \int \mu d\tilde{N}] \\ +\sum_{i\in\mathcal{I}}[-\bar{u}_{1i} + (1-\bar{u}_{2i})\bar{\kappa}_1\bar{x} + \bar{u}_{2i}\bar{\kappa}_2\bar{x}]dt \\ +\sum_{i\in\mathcal{I}}\bar{u}_{2i}\bar{x}[\bar{\sigma}dB_o + \int \bar{\mu}d\tilde{N}_o], \\ x(0) = x_0, \end{cases} \tag{10.42}$$

where the coefficients are regime-switching-dependent and $k_i \geq 1, \bar{\rho}_i \in (0,1)$. The coefficients $q_i, \bar{q}_i, r_i, \bar{r}_i$ are positive. Note that the state dynamics (10.42) is not linear in (x,u).

Following the same method as in the problem (10.42), a semi-explicit solution can be obtained.

Note that a similar method can be used to derive a semi-explicit solution to the following game problem in which decision-makers are minimizers and $k_i(t,s) > 1,\ \bar{\rho}_i(t,s) > 1$.

$$\begin{cases} i \in \mathcal{I}, \\ L_i(x,u) = q_{iT}\frac{(x_T-\bar{x}_T)^{2k_i}}{2k_i} + \bar{q}_{iT}\frac{\bar{x}_T^{\bar{\rho}_i}}{\bar{\rho}_i} \\ + \int_0^T q_i\frac{(x-\bar{x})^{2k_i}}{2k_i} + \bar{q}_i\frac{\bar{x}^{\bar{\rho}_i}}{\bar{\rho}_i} + r_i\frac{(u_{1i}-\bar{u}_{1i})^{2k_i}}{2k_i} + \bar{r}_i\frac{\bar{u}_{1i}^{\bar{\rho}_i}}{\bar{\rho}_i}dt, \\ \inf_{u_i} E[L_i(x,u)], \\ \text{subject to} \\ dx = \sum_i (u_{1i}-\bar{u}_{1i}) + (1-(u_{2i}-\bar{u}_{2i}))k_1(x-\bar{x}) \\ + \sum_i (u_{2i}-\bar{u}_{2i})k_2(x-\bar{x})dt \\ + \sum_i (u_{2i}-\bar{u}_{2i})(x-\bar{x})[\sigma dB + \int \mu d\tilde{N}] \\ + \sum_i [-\bar{u}_{1i} + (1-\bar{u}_{2i})\bar{k}_1\bar{x} + \bar{u}_{2i}\bar{k}_2\bar{x}]dt \\ + \sum_i \bar{u}_{2i}\bar{x}[\bar{\sigma} dB_o + \int \bar{\mu} d\tilde{N}_o], \\ x(0) = x_0. \end{cases} \tag{10.43}$$

This can be easily extended to include multi-type power utilities in the following form:

$$\sum_{k=1}^K q_{ik}\frac{(x-\bar{x})^{\rho_{ik}}}{\rho_{ik}} + \bar{q}_{ik}\frac{\bar{x}^{\bar{\rho}_{ik}}}{\bar{\rho}_{ik}} + \sum_{k=1}^K r_{ik}\frac{(u_{1i}-\bar{u}_{1i})^{\rho_{ik}}}{\rho_{ik}} + \bar{r}_{ik}\frac{\bar{u}_{1i}^{\bar{\rho}_{ik}}}{\bar{\rho}_{ik}}. \tag{10.44}$$

10.3.4 Nonlinear State and Log Utility

We consider a logarithm of the Cobb-Douglas utility given by

$$\begin{cases} i \in \mathcal{I}, \\ R_i(x,u) = -q_{iT}\frac{(x_T-\bar{x}_T)^{2k_i}}{2k_i} + \log[(\bar{x}_T)^{\bar{q}_{iT}}] \\ + \int_0^T -q_i\frac{(x-\bar{x})^{2k_i}}{2k_i} - r_i\frac{(u_{1i}-\bar{u}_{1i})^{2k_i}}{2k_i} + \log[\bar{x}^{\bar{q}_i}\bar{u}_{1i}^{\bar{r}_i}]dt \\ \sup_{u_i} E[R_i(x,u)], \\ \text{subject to} \\ dx = \sum_i[-(u_{1i}-\bar{u}_{1i}) + (1-(u_{2i}-\bar{u}_{2i}))\kappa_1(x-\bar{x})] \\ + \sum_i (u_{2i}-\bar{u}_{2i})\kappa_2(x-\bar{x})dt \\ + \sum_i (u_{2i}-\bar{u}_{2i})(x-\bar{x})[\sigma dB + \int \mu d\tilde{N}] \\ + \sum_i [-\bar{u}_{1i} + (1-\bar{u}_{2i})\bar{\kappa}_1\bar{x} + \bar{u}_{2i}\bar{\kappa}_2\bar{x}]dt \\ + \sum_i \bar{u}_{2i}\bar{x}[\bar{\sigma} dB_o + \int \bar{\mu} d\tilde{N}_o], \\ x(0) = x_0,\ s(0) = s_0, \\ \mathbb{P}(s(t+\epsilon) = s'|s,u) = \int_t^{t+\epsilon} \tilde{q}_{ss'}dt' + o(\epsilon),\ s' \neq s, \end{cases} \tag{10.45}$$

where the coefficients are regime-switching-dependent.

Note that the state dynamics (10.45) is not linear in (x, u).

Following the same method as in (10.42), the problem (10.45) can be solved explicitly.

10.3.5 Cotangent Drift

In this subsection, we examine MFTGs with cotangent drift. Using trigonometric relations, a semi-explicit equilibrium solution is derived:

$$\begin{array}{l} i \in \mathcal{I}, \\ L_i(x,u) = \\ \int_0^T ((u_i-\bar{u}_i)^2 - q_i)\cos^2(\frac{x-\bar{x}}{4}) + q_i + (\bar{u}_i^2 - \bar{q}_i)\cos^2(\frac{\bar{x}}{4}) + \bar{q}_i dt, \\ \inf_{u_i} E[L_i(x,u)], \\ \text{subject to} \\ dx = [\frac{1}{2}cot(\frac{x-\bar{x}}{2}) + \sum_j b_{2j}(u_j-\bar{u}_j) + \frac{1}{2}cot(\frac{\bar{x}}{2}) + \sum_j \bar{b}_{2j}\bar{u}_j]dt \\ +\sigma dB, \\ x(0) = x_0, \end{array} \tag{10.46}$$

where the coefficients are regime-switching-dependent and $\cot(\theta) = \frac{1}{\tan(\theta)} = -cot(-\theta)$.

Proposition 10.7 *Assume that $q_i > 0, \bar{q}_i > 0, i \in \mathcal{I}$. The MFTG problem with cotangent drift (10.46) has a unique equilibrium solution, which is given by*

$$\begin{array}{l} i \in \mathcal{I}, \\ u_i^* - \bar{u}_i^* = -\sum_{s\in\mathcal{S}} \mathbb{1}_{\{s(t)=s\}} b_{2i}\frac{\alpha_i}{4} tan(\frac{x-\bar{x}}{4}) \\ \bar{u}^* = -\sum_{s\in\mathcal{S}} \mathbb{1}_{\{s(t)=s\}} \frac{\bar{\alpha}_i}{4}\bar{b}_{2i} tan(\frac{\bar{x}}{4}) \\ E[L_i(x,u^*)] = \\ E[\alpha_i(0,s_0)\sin^2(\frac{x_0-\bar{x}_0}{4}) + \bar{\alpha}_i(0,s_0)\sin^2(\frac{\bar{x}_0}{4}) + \delta_i(0,s_0)], \end{array} \tag{10.47}$$

whenever the following system

$$\begin{array}{l} i \in \mathcal{I}, \\ \dot{\alpha}_i + q_i - (2+\sigma^2)\frac{\alpha_i}{8} - \frac{\alpha_i^2}{16} b_{2i}^2 \\ +\sum_{s'}[\alpha_i(t,s') - \alpha_i(t,s)]\tilde{q}_{ss'} - \frac{\alpha_i}{8}\sum_{j\neq i}\alpha_j b_j^2 = 0, \\ \alpha_i(T,s) = 0, \\ \dot{\bar{\alpha}}_i + \bar{q}_i - \frac{\bar{\alpha}_i}{4} - \frac{\bar{\alpha}_i^2}{16}\bar{b}_{2i}^2 \\ +\sum_{s'}[\bar{\alpha}_i(t,s') - \bar{\alpha}_i(t,s)]\tilde{q}_{ss'} - \frac{\bar{\alpha}_i}{8}\sum_{j\neq i}\bar{b}_{2j}^2\bar{\alpha}_j = 0, \\ \bar{\alpha}_i(T,s) = 0, \\ \dot{\delta}_i + \frac{\alpha_i}{16}(2+\sigma^2) + \frac{\bar{\alpha}_i}{8} + \sum_{s'}[\delta_i(t,s') - \delta_i(t,s)]\tilde{q}_{ss'} = 0, \\ \delta_i(T,s) = 0, \end{array} \tag{10.48}$$

has unique solution with positive $\alpha_i, \bar{\alpha}_i$ *which do not blow up within* $[0, T]$. *The mean-field term* $\bar{x}$ *solves*

$$\begin{aligned} d\bar{x} &= [\tfrac{1}{2}cot(\tfrac{\bar{x}}{2}) - \textstyle\sum_j \bar{b}_{2j}^2 \frac{\bar{\alpha}_j}{4} \tan(\tfrac{\bar{x}}{4})]dt, \\ \bar{x}(0) &= \bar{x}_0, \end{aligned} \quad (10.49)$$

which has a unique solution for $\bar{x}_0 \in (0, \pi)$.

Proof We prove the result using a direct method.

Step 1: We observe that the MFTG is driven by functionals of $x - \bar{x}$ and $\bar{x}$ which are conditionally orthogonal processes.

Step 2: Given the structure of the problem, we propose the following guess functional:

$$f_i = \alpha_i(t, s)\sin^2(\frac{x - \bar{x}}{4}) + \bar{\alpha}_i(t, s)\sin^2(\frac{\bar{x}}{4}) + \delta_i(t, s).$$

Step 3: We apply Brownian with regime switching to obtain the difference between the cost functional and the guess functional as

$$\begin{aligned} & E[L_i - f_i(0)] = -Ef_i(T) \\ & +E\textstyle\int_0^T \cos^2(\frac{x-\bar{x}}{4})\left(u - \bar{u} + b_{2i}\frac{\alpha_i}{4}tan(\frac{x-\bar{x}}{4})\right)^2 dt \\ & +E\textstyle\int_0^T \sin^2(\frac{x-\bar{x}}{4})\Big\{\dot{\alpha}_i + q_i - (2+\sigma^2)\frac{\alpha_i}{8} - \frac{\alpha_i^2}{16}b_{2i}^2 \\ & + \textstyle\sum_{s'}[\alpha_i(t, s') - \alpha_i(t, s)]\tilde{q}_{ss'} - \frac{\alpha_i}{8}\sum_{j\neq i}\alpha_j b_j^2\Big\}dt \\ & +\dot{\delta}_i + \frac{\alpha_i}{16}(2+\sigma^2) + \textstyle\sum_{s'}[\delta_i(t, s') - \delta_i(t, s)]\tilde{q}_{ss'} \\ & +\Big\{\dot{\bar{\alpha}}_i + \bar{q}_i - \frac{\bar{\alpha}_i}{4} - \frac{\bar{\alpha}_i^2}{16}\bar{b}_{2i}^2 \\ & + \textstyle\sum_{s'}[\bar{\alpha}_i(t, s') - \bar{\alpha}_i(t, s)]\tilde{q}_{ss'} - \frac{\bar{\alpha}_i}{8}\sum_{j\neq i}\bar{b}_{2j}^2\bar{\alpha}_j\ \Big\}\sin^2(\frac{\bar{x}}{4})dt \\ & +\cos^2(\tfrac{\bar{x}}{4})[\bar{u}_i + \tfrac{\bar{\alpha}_i}{4}\bar{b}_{2i}tan(\tfrac{\bar{x}}{4})]^2 + \tfrac{\bar{\alpha}_i}{8}dt. \end{aligned} \quad (10.50)$$

Step 4: Completion of various terms leads to a strictly convex one-shot optimization with a coercive function $\cos^2(\frac{\bar{x}}{4})[\bar{u}_i + \frac{\bar{\alpha}_i}{4}\bar{b}_{2i}tan(\frac{\bar{x}}{4})]^2$ whenever $\cos^2(\frac{\bar{x}}{4}) > 0$.

Step 5: By the process of matching terms, one obtains the desired result. This completes the proof. ■

10.3.6 Hyperbolic coTangent Drift

Problem (10.46) can be modified to handle the hyperbolic cotangent drift case as specified below. The functions cos, sin, tan, *and* cot are replaced by cosh, sinh, tanh, *and* coth, respectively:

$$
\begin{aligned}
& i \in \mathcal{I}, \\
& L_i(x,u) = \int_0^T ((u_i - \bar{u}_i)^2 + q_i)\cosh^2(\tfrac{x-\bar{x}}{4}) - q_i \\
& +(\bar{u}_i^2 + \bar{q}_i)\cosh^2(\tfrac{\bar{x}}{4}) - \bar{q}_i dt, \\
& \inf_{u_i} E[L_i(x,u)], \\
& \text{subject to} \\
& dx = [\tfrac{1}{2}coth(\tfrac{x-\bar{x}}{2}) + \textstyle\sum_j b_{2j}(u_j - \bar{u}_j) \\
& +\tfrac{1}{2}coth(\tfrac{\bar{x}}{2}) + \textstyle\sum_j \bar{b}_{2j}\bar{u}_j]dt + \sigma dB, \\
& x(0) = x_0,
\end{aligned}
\tag{10.51}
$$

where the coefficients are regime-switching-dependent and $\coth(\theta) = \frac{e^{\theta}+e^{\theta}}{e^{\theta}-e^{-\theta}} = -\coth(-\theta)$.

Proposition 10.8 *For the MFTG formulated above by (10.51), the equilibrium strategies and the equilibrium costs are given by*

$$
\begin{aligned}
& u_i^* - \bar{u}_i^* = -\textstyle\sum_{s\in\mathcal{S}} \mathbb{1}_{\{s(t)=s\}} b_{2i}\frac{\alpha_i}{4}\tanh(\frac{x-\bar{x}}{4}), \\
& \bar{u}_i^* = -\textstyle\sum_{s\in\mathcal{S}} \mathbb{1}_{\{s(t)=s\}} \frac{\bar{\alpha}_i}{4}\bar{b}_{2i}\tanh(\frac{\bar{x}}{4}), \\
& E[L_i(x,u^*)] = E[\alpha_i(0,s_0)\sinh^2(\tfrac{x_0-\bar{x}_0}{4})] + E[\bar{\alpha}_i(0,s_0)\sinh^2(\tfrac{\bar{x}_0}{4}) + \delta_i(0,s_0)],
\end{aligned}
\tag{10.52}
$$

whenever the following system

$$
\begin{aligned}
& i \in \mathcal{I}, \\
& \dot{\alpha}_i + q_i + (2+\sigma^2)\tfrac{\alpha_i}{8} - \tfrac{\alpha_i^2}{16}b_{2i}^2 + \textstyle\sum_{s'}[\alpha_i(t,s') - \alpha_i(t,s)]\tilde{q}_{ss'} - \tfrac{\alpha_i}{8}\textstyle\sum_{j\neq i}\alpha_j b_j^2 = 0, \\
& \alpha_i(T,s) = 0, \\
& \dot{\bar{\alpha}}_i + \bar{q}_i + \tfrac{\bar{\alpha}_i}{4} - \tfrac{\bar{\alpha}_i^2}{16}\bar{b}_{2i}^2 + \textstyle\sum_{s'}[\bar{\alpha}_i(t,s') - \bar{\alpha}_i(t,s)]\tilde{q}_{ss'} - \tfrac{\bar{\alpha}_i}{8}\textstyle\sum_{j\neq i}\bar{b}_{2j}^2\bar{\alpha}_j = 0, \\
& \bar{\alpha}_i(T,s) = 0, \\
& \dot{\delta}_i + \tfrac{\alpha_i}{16}(2+\sigma^2) + \tfrac{\bar{\alpha}_i}{8} + \textstyle\sum_{s'}[\delta_i(t,s') - \delta_i(t,s)]\tilde{q}_{ss'} = 0 \\
& \delta_i(T,s) = 0,
\end{aligned}
\tag{10.53}
$$

has unique solution with positive $\alpha_i, \bar{\alpha}_i$ which do not blow up within $[0, T]$.

The mean-field term $\bar{x}$ solves

$$
\begin{aligned}
& d\bar{x} = [\tfrac{1}{2}coth(\tfrac{\bar{x}}{2}) - \textstyle\sum_j \bar{b}_{2j}^2\frac{\bar{\alpha}_j}{4}\tanh(\frac{\bar{x}}{4})]dt, \\
& \bar{x}(0) = \bar{x}_0 \neq 0,
\end{aligned}
\tag{10.54}
$$

which has a unique global solution within $[0, T]$.

The system in (10.53) shares some similarities with the system in (10.48) of problem (10.46), but still two systems are different. In particular, the sign of the terms $(2+\sigma^2)\frac{\alpha_i}{8}$, and $\frac{\bar{\alpha}_i}{4}$ have changed.

Proof We prove the statement on the hyperbolic game using a direct method. Let

$$
\begin{array}{l}
i \in \mathcal{I}, \\
f_i(t,x,s) = \alpha_i(t,s(t))\sinh^2(\frac{x(t)-\bar{x}(t)}{4}) \\
+\bar{\alpha}_i(t,s(t))\sinh^2(\frac{\bar{x}(t)}{4}) + \delta_i(t,s(t)),
\end{array}
$$

be a guess functional combining hyperbolic functions.

$$
\begin{array}{l}
E[L_i - f_i(0)] = -Ef_i(T) \\
+E\int_0^T \cosh^2(\frac{x-\bar{x}}{4})\left(u-\bar{u}+b_{2i}\frac{\alpha_i}{4}\tanh(\frac{x-\bar{x}}{4})\right)^2 dt \\
+E\int_0^T \sinh^2(\frac{x-\bar{x}}{4})\left\{\dot{\alpha}_i + q_i + (2+\sigma^2)\frac{\alpha_i}{8} - \frac{\alpha_i^2}{16}b_{2i}^2\right. \\
\left. +\sum_{s'}[\alpha_i(t,s')-\alpha_i(t,s)]\tilde{q}_{ss'} - \frac{\alpha_i}{8}\sum_{j\neq i}\alpha_j b_j^2\right\} dt \\
+\dot{\delta}_i + \frac{\alpha_i}{16}(2+\sigma^2) + \frac{\bar{\alpha}_i}{8} + \sum_{s'}[\delta_i(t,s')-\delta_i(t,s)]\tilde{q}_{ss'} \\
+\left\{\dot{\bar{\alpha}}_i + \bar{q}_i + \frac{\bar{\alpha}_i}{4} - \frac{\bar{\alpha}_i^2}{16}\bar{b}_{2i}^2\right. \\
\left. +\sum_{s'}[\bar{\alpha}_i(t,s')-\bar{\alpha}_i(t,s)]\tilde{q}_{ss'} - \frac{\bar{\alpha}_i}{8}\sum_{j\neq i}\bar{b}_{2j}^2\bar{\alpha}_j\right\}\sinh^2(\frac{\bar{x}}{4})dt \\
+\cosh^2(\frac{\bar{x}}{4})[\bar{u}_i + \frac{\bar{\alpha}_i}{4}\bar{b}_{2i}\tanh(\frac{\bar{x}}{4})]^2 dt.
\end{array}
\tag{10.55}
$$

Through the matching of terms, one obtains the desired result. This completes the proof. ■

10.3.7 A Delayed and Trend-Based MFTG

We present a cooperative MFTG with basic state dynamics $x(t)$, regime switching $s(t)$, a trend $y(t) := \int_{-\tau}^{0} e^{\lambda t'}x(t+t')dt'$ on the time window $[t-\tau, t]$, and delayed state $z(t) = x(t-\tau)$. This class of games plays an important role in real-world applications as the effects of actions are not instantaneous in general. It may take a certain time delay. This leads to delayed and trend-based stochastic differential equations of mean-field type:

$$
\begin{array}{l}
i \in \mathcal{I}, \\
R(x,u) = -q_T var(x_T) + \frac{(\bar{x}_T+\bar{\eta}\bar{y}_T)^\rho}{\rho} \\
\int_0^T -q\ var(x) - r_1\ var(u_1) + \bar{r}_1\frac{\bar{u}_1^\rho}{\rho}dt, \\
\sup_{(u_1,u_2)} E[R(x,u)], \\
\text{subject to} \\
dx = [-(u_1-\bar{u}_1) + b_1(x-\bar{x}) + b_2\epsilon(x-\bar{x})(u_2-\bar{u}_2)]dt \\
+[-\bar{u}_1 + \bar{b}_{11}\bar{x} + \bar{b}_{12}\bar{y} + \bar{b}_{13}\bar{z} + \bar{b}_2\bar{u}_2\bar{x}]dt \\
+\sigma(x-\bar{x})(u_2-\bar{u}_2)dB + \bar{\sigma}\bar{x}\bar{u}_2 dB_o, \\
x(t') = x_0(t'),\ t' \in (-\delta, 0],\ s(0) = s_0,
\end{array}
\tag{10.56}
$$

where $var(X)$ denotes the variance of the random variable X and $\bar{X}(t) = E[X(t)|\ \mathcal{F}^{s,B_o}]$ is the conditional expectation with respect to the common noises s, B_o.

Lemma 10.1 *The conditional expected trend $\bar{y}$ satisfies the following stochastic differential equation:*

$$d\bar{y} = [\bar{x} - \lambda\bar{y} - e^{-\lambda\tau}\bar{z}]dt.$$

Proof Let

$$dy = dt\int_{-\tau}^{0} e^{\lambda s}[\frac{d}{dt}x(t+s)]ds = [x(t) - e^{-\lambda\tau}z(t) - \lambda y(t)]dt.$$

Taking the conditional expectations, one obtains

$$d\bar{y} = [\bar{x} - \lambda\bar{y} - e^{-\lambda\tau}\bar{z}]dt.$$

This completes the proof. ■

Proposition 10.9 *The equilibrium strategies and the equilibrium payoff of the delayed MFTG (10.56) are given by*

$$\begin{array}{l} u_1^* - \bar{u}_1^* = \sum_{s\in\mathscr{S}} \mathbb{1}_{\{s(t)=s\}}\frac{\alpha}{r_1}(x-\bar{x}),\\ \bar{u}_1^* = \sum_{s\in\mathscr{S}} \mathbb{1}_{\{s(t)=s\}}(\bar{x}+\bar{\eta}\bar{y})(\frac{\beta}{\bar{r}_1})^{\frac{1}{\rho-1}},\\ u_2^* - \bar{u}_2^* = -\sum_{s\in\mathscr{S}} \mathbb{1}_{\{s(t)=s\}}\frac{b_2\epsilon}{\sigma^2},\\ \bar{u}_2^* = -\sum_{s\in\mathscr{S}} \mathbb{1}_{\{s(t)=s\}}\frac{(\bar{x}+\bar{\eta}\bar{y})(\beta\bar{b}_2+\beta_{B_o}\bar{\sigma})}{2(\rho-1)\beta\bar{\sigma}^2\bar{x}},\\ E[R(x,u^*)] = E[-\alpha(0,s_0)var(x) + \beta(0,s_0)\frac{(\bar{x}_0+\bar{\eta}\bar{y}_0)^{\rho}}{\rho}], \end{array} \tag{10.57}$$

whenever the following system

$$\begin{array}{l} d\alpha + (2b_1\alpha + q + \sum_{s'}[\alpha(t,s') - \alpha(t,s)]\tilde{q}_{ss'} - \frac{\alpha^2}{r_1} - b_2^2\epsilon\alpha)dt = 0\\ \alpha(t,s) = q_T,\\ d\beta - \frac{\rho}{\rho-1}\frac{(\beta\bar{b}_2+\beta_{B_o}\bar{\sigma})^2}{4\beta\bar{\sigma}^2}dt + (\beta\rho(b_{11}+b_{13}e^{\lambda\tau}) + \beta^{\frac{\rho}{\rho-1}}(\rho-1)(\bar{r}_1)^{-\frac{1}{\rho-1}})dt\\ +\sum_{s'}[\beta(t,s') - \beta(t,s)]\tilde{q}_{ss'}dt = 0,\\ \beta(T,s) = 1,\\ \bar{\eta} := b_{13}e^{\lambda\tau}, \end{array} \tag{10.58}$$

has a unique solution $\alpha, \beta, \beta_{B_o}$.

Note that the system in α has a positive solution provided that if $q \geq 0, q_T \geq 0,\ r_1 > 0$. With a single regime (no switching) $\mathscr{S} = \{s_0\}$, the equation in β yields

$$
\begin{aligned}
&\dot{\beta} + \beta[-\tfrac{\rho}{\rho-1}\tfrac{\bar{b}_2^2}{4\bar{\sigma}^2} + \rho(b_{11} + b_{13}e^{\lambda\tau})] + \beta^{\frac{\rho}{\rho-1}}(\rho-1)(\bar{r}_1)^{-\frac{1}{\rho-1}} \\
&+ \textstyle\sum_{s'}[\beta(t,s') - \beta(t,s)]\tilde{q}_{ss'} = 0, \\
&\beta(T,s) = 1.
\end{aligned} \tag{10.59}
$$

This is completely solvable with the explicit solution given by

$$
\beta(t,s) = E[((1-\frac{c}{\omega})e^{\frac{-\omega}{\rho-1}(T-t)} + \frac{c}{\omega})^{1-\rho} \mid s(t) = s],
$$

where

$$
c := (1-\rho)\bar{r}_1^{\frac{1}{1-\rho}}, \ \omega := \frac{\rho}{1-\rho}\frac{\bar{b}_2^2}{4\bar{\sigma}^2} + \rho(b_{11} + b_{13}e^{\lambda\tau}) > 0.
$$

Proof Let $f(x) = -\alpha\, var(x) + \beta\frac{(\bar{x}+\bar{\eta}\bar{y})^\rho}{\rho}$ be a guess functional:

$$
\begin{aligned}
&E[R - f(0)] = E - (q_T - \alpha_T)var(x_T) + (1-\beta_T)\tfrac{(\bar{x}+\bar{\eta}\bar{y})^\rho}{\rho} \\
&+E\int_0^T - \{d\alpha + (2b_1\alpha + q)dt \\
&+(\textstyle\sum_{s'}[\alpha(t,s') - \alpha(t,s)]\tilde{q}_{ss'} - \tfrac{\alpha^2}{r_1} - b_2^2\epsilon\alpha)dt\Big\}(x-\bar{x})^2dt \\
&-r_1[u_1 - \bar{u}_1 - \tfrac{\alpha}{r}(x-\bar{x})]^2dt \\
&-\alpha\sigma^2[u_2 - \bar{u}_2 + \tfrac{b_2\epsilon}{\sigma^2}]^2(x-\bar{x})^2dt \\
&+\{d\beta - \tfrac{\rho}{\rho-1}\tfrac{(\beta\bar{b}_2+\beta_{B_o}\bar{\sigma})^2}{4\beta\bar{\sigma}^2}dt - \tfrac{\rho}{\rho-1}\tfrac{(\beta\bar{b}_2+\beta_{B_o}\bar{\sigma})}{2\bar{\sigma}}dB_o \\
&+\beta\rho(b_{11} + b_{13}e^{\lambda\tau})dt + \beta^{\frac{\rho}{\rho-1}}(\rho-1)(\bar{r}_1)^{-\frac{1}{\rho-1}}dt \\
&+\textstyle\sum_{s'}[\beta(t,s') - \beta(t,s)]\tilde{q}_{ss'}dt\}\tfrac{(\bar{x}+\bar{\eta}\bar{y})^\rho}{\rho} \\
&+(\rho-1)\beta\bar{\sigma}^2\bar{x}^2(\bar{x}+\bar{\eta}\bar{y})^{\rho-2}[\bar{u}_2 + \tfrac{(\bar{x}+\bar{\eta}\bar{y})(\beta\bar{b}_2+\beta_{B_o}\bar{\sigma})}{2(\rho-1)\beta\bar{\sigma}^2\bar{x}}]^2dt \\
&+[-(\bar{x}+\bar{\eta}\bar{y})^{\rho-1}\beta\bar{u}_1 + \tfrac{\bar{r}_1\bar{u}_1^\rho}{\rho} - \beta^{\frac{\rho}{\rho-1}}(\bar{x}+\bar{\eta}\bar{y})^\rho(1-\tfrac{1}{\rho})(\bar{r}_1)^{-\frac{1}{\rho-1}}]dt,
\end{aligned} \tag{10.60}
$$

with the following careful matching $\eta = \bar{b}_{13}e^{\lambda\tau}$, $\bar{b}_{12} = \bar{b}_{13}e^{\lambda\tau}(b_{11} + \lambda + \bar{b}_{13}e^{\lambda\tau})$. The joint optimization over (u_1, u_2) together with the mean-field terms $(\bar{u}_1, \bar{u}_2)$ yields the desired result provided that $\rho < 1$. This completes the proof. ■

10.3.8 Mean-Field of MFTG

In this subsection, we examine a class of mean-fields of MFTGs.

In view of the delayed MFTG (10.56), we have modified the coefficient function $\bar{r}_1$ to be $\bar{r}_1(m)$, where m is the conditional total consumption of the large population. Then, m is obtained as

$$m = \int \bar{u}_1(t, \bar{x}, \bar{y})\mu^m(t, d\bar{x}, d\bar{y}),$$

where $\mu^m(t, d\bar{x}, d\bar{y})$ is the conditional distribution of all players' states and trends in the large population under m, which reduces to the fixed-point problem $m = (\bar{x}(m) + \bar{\eta}\bar{y}(m))(\frac{\beta(m)}{\bar{r}_1(m)})^{\frac{1}{\rho-1}}$.

Now consider the following modified Cournot-Ross game with I producers and a large population of potential consumers:

$$\begin{aligned}
&i \in \mathcal{I},\\
&R_i(x, u) = \bar{R} - \tfrac{(x_T - \bar{x}_T)^{2k}}{2k} + \tfrac{(\bar{x}_T)^{2k}}{2k}\\
&+ \int_0^T -\bar{r}_i \tfrac{(u_i - \bar{u}_i)^{2k}}{2k} + \bar{x}^{2k-1}\bar{u}_i - \bar{r}_i \tfrac{\bar{u}_i^{2k}}{2k} dt,\\
&\sup_{u_i} E[R_i(x, u)],\\
&\text{subject to}\\
&dx = [D(m) - S]dt + \sigma(x - \bar{x})dB,\\
&x(0) = x_0, \ s(0) = s_0.
\end{aligned} \tag{10.61}$$

Let $D(m)$ be the demand generated by a large population of consumers. Given a demand $D(m)$, each macro-player i has a certain utility of mean-field type. The payoff function in the Cournot game $\bar{x}\bar{u}_i - \bar{r}_i \frac{\bar{u}_i^2}{2}$ is modified to be $\bar{x}^{2k-1}\bar{u}_i - \bar{r}_i \frac{\bar{u}_i^{2k}}{2k}$ and some extra mean-field-dependent terms.

By means of a direct method, one can fully characterize the mean-field equilibrium of (10.61). It is given by the following set of equations:

$$\begin{aligned}
&i \in \mathcal{I},\\
&E[R_i(x, u^*)] = E[\bar{R} - \alpha_i(0, s_0)\tfrac{(x_0 - \bar{x}_0)^{2k}}{2k} - \bar{\alpha}_i(0, s_0)\tfrac{(\bar{x}_0)^{2k}}{2k}],\\
&u_i^* = (\tfrac{\alpha_i}{r_i})^{\frac{1}{2k-1}}(x - \bar{x}) + (\tfrac{\bar{\alpha}_i + 1}{\bar{r}_i})^{\frac{1}{2k-1}}\bar{x},\\
&D(m) = [\textstyle\sum_i (\tfrac{\bar{\alpha}_i + 1}{\bar{r}_i})^{\frac{1}{2k-1}}]\bar{x}.
\end{aligned} \tag{10.62}$$

where $\alpha_i, \bar{\alpha}_i$ solve a system of ordinary differential equations.

10.4 MFTG Beyond Brownian Motions and Poisson

In this subsection, a class of MFTGs with a state-dependent Gauss–Volterra noise is formulated and solved with a polynomial and mean-field-dependent payoff for an arbitrary number of players and a finite time horizon. The control strategies are linear state and mean-field feedbacks. A mean-field-type Nash equilibrium is verified for the game, and the optimal strategies are obtained using a direct method that does not require solving nonlinear partial integrodifferential equations or forward-backward stochastic differential equations.

10.4.1 Gauss–Volterra Mean-Field-Type Game

The Gauss–Volterra processes are singular integrals of a standard Brownian motion and include (i) fractional Brownian motions, (ii) Liouville fractional Brownian motions, and (iii) multi-fractional Brownian motions. The difficulty of finding a semi-explicit solution is significantly increased if the noise process, and thereby the state process, is driven by non-Markov processes or non-martingales.

Let B_{gv} be a Gauss–Volterra process with mean zero and covariance:

$$c(t,t') = E[B_{gv}(t)B_{gv}(t')] = \int_0^{\min(t,t')} K(t,t'')K(t',t'')dt''.$$

The variance of the process $\int_0^t \sigma_{gv}(t',s)dB_{gv}(t')$ is given by

$$\sigma_{cogv}^2 = \frac{d}{dt}\left[\int_0^t \left\{K(t'_+,t')\sigma_{gv}(t') + \int_{t'}^t \sigma_{gv}(t'')K(t',t'')dt''\right\}^2 dt'\right].$$

Definition 10.1 $B_{gv}(t) = \int_0^t K(t,t')dB(t')$

Following similar steps, a linear-state-and-mean-field-type equilibrium solution can be obtained for the class of noncooperative MFTGs with non-quadratic cost and state driven by the additive Gauss–Volterra processes:

$$\begin{array}{l} i \in \mathcal{I}, \\ L_i(x,u) = q_{iT}\frac{(x_T-\bar{x}_T)^2}{2} + \bar{q}_{iT}\frac{\bar{x}_T^{2\bar{k}_i}}{2\bar{k}_i} \\ +\int_0^T q_i\frac{(x-\bar{x})^2}{2} + r_i\frac{(u_i-\bar{u}_i)^2}{2} + \bar{q}_i\frac{\bar{x}^{2\bar{k}_i}}{2\bar{k}_i} + \bar{r}_i\frac{\bar{u}_i^{2\bar{k}_i}}{2\bar{k}_i}dt, \\ \inf_{u_i} E[L_i(x,u)], \\ \text{subject to} \\ x(t) = x_0 + \int_0^t [b_1(x-\bar{x}) + \sum_j b_{2j}(u_j-\bar{u}_j) + \bar{b}_1\bar{x} + \sum_j \bar{b}_{2j}\bar{u}_j]dt' \\ +\int_0^t \sigma_{gv}(t')dB_{gv}(t'), \end{array} \quad (10.63)$$

where $\bar{k}_i \geq 1$ are natural numbers and the coefficients are time. The process B_{gv} considered here is the true Gauss–Volterra process as given in Definition 10.1.

10.4.2 Noncooperative MFTG Under Modulated Gauss–Volterra Processes

This section examines a class of noncooperative MFTGs with non-quadratic cost and state driven by the modulated Gauss–Volterra processes:

$$
\begin{array}{l}
i \in \mathcal{I}, \\
L_i(x,u) = q_{iT}\frac{(x_T-\bar{x}_T)^{2k_i}}{2k_i} + \bar{q}_{iT}\frac{\bar{x}_T^{2\bar{k}_i}}{2\bar{k}_i} \\
+\int_0^T q_i\frac{(x-\bar{x})^{2k_i}}{2k_i} + r_i\frac{(u_i-\bar{u}_i)^{2k_i}}{2k_i} + \bar{q}_i\frac{\bar{x}^{2\bar{k}_i}}{2\bar{k}_i} + \bar{r}_i\frac{\bar{u}_i^{2\bar{k}_i}}{2\bar{k}_i}dt, \\
\inf_{u_i} E[L_i(x,u)], \\
\text{subject to} \\
dx = [b_1(x-\bar{x}) + \sum_j b_{2j}(u_j-\bar{u}_j) + \bar{b}_1\bar{x} + \sum_j \bar{b}_{2j}\bar{u}_j]dt \\
+(x-\bar{x})[\sigma dB + \int_{\Theta}\mu d\tilde{N} + \sigma_{gv}dB_{gv}], \\
\\
\mathbb{P}(s(t+\epsilon)=s'|s,u) = \int_t^{t+\epsilon}\tilde{q}_{ss'}dt' + o(\epsilon),\ s'\neq s \\
x(0)=x_0,\ s(0)=s_0,
\end{array}
\tag{10.64}
$$

where $k_i \geq 1, \bar{k}_i \geq 1$ are natural numbers and the coefficients are time- and switching-dependent. Note that the process B_{gv} considered in the multiplicative noise case does not take into account x in the memory but considers the deterministic coefficients that are a function of time.

Remark 10.3 The cost functional is clearly non-quadratic for $k_i > 1$ or $\bar{k}_i > 1$. As a special case, for $k_i = 1$, we have the variance reduction game (10.64) under the Gauss–Volterra process.

Proposition 10.10 *Assume $q_i > 0, \bar{q}_i > 0, r_i, \bar{r}_i > 0$. The mean-field-type Nash equilibrium of the mean-field-type game (10.64) under the Gauss–Volterra process is given by*

$$
\begin{array}{l}
i \in \mathcal{I}, \\
u_i^* = -\sum_{s\in\mathcal{S}} \mathbb{1}_{\{s(t)=s\}}[\left(\frac{b_{2i}\alpha_i}{r_i}\right)^{\frac{1}{2k_i-1}}(x-\bar{x}) + \left(\frac{\bar{b}_{2i}\bar{\alpha}_i}{\bar{r}_i}\right)^{\frac{1}{2\bar{k}_i-1}}\bar{x}], \\
EL_i = E[\alpha_i(0,s_0)\frac{(x_0-\bar{x}_0)^{2k_i}}{2k_i} + \bar{\alpha}_i(0,s_0)\frac{\bar{x}_0^{2\bar{k}_i}}{2\bar{k}_i}],
\end{array}
\tag{10.65}
$$

whenever the following system of ordinary differential equations admits a positive solution which does not have a singularity within $[0, T]$:

$$
\begin{aligned}
&i \in \mathcal{I},\\
&\dot{\alpha}_i + q_i + 2k_i\alpha_i b_1 + \alpha_i k_i(2k_i - 1)(\sigma^2 + \sigma^2_{cogv})\\
&+\alpha_i \int_\theta [(1+\mu)^{2k_i} - 1 - 2k_i\mu]\nu(d\theta)\\
&+(\sum_{s'}[\alpha_i(t,s') - \alpha_i(t,s)]\tilde{q}_{ss'}) - (2k_i - 1)r_i(-\tfrac{b_{2i}\alpha_i}{r_i})^{\frac{2k_i}{2k_i-1}}\\
&+2k_i\alpha_i[\sum_{j\neq i} b_{2j}(-\tfrac{b_{2j}\alpha_j}{r_j})^{\frac{1}{2k_j-1}} = 0, \qquad (10.66)\\
&\alpha_i(T,s) = q_i(T,s),\\
&\dot{\bar{\alpha}}_i + \bar{q}_i + 2\bar{k}_i\bar{\alpha}_i\bar{b}_1 + (\sum_{s'}[\bar{\alpha}_i(t,s') - \bar{\alpha}_i(t,s)]\tilde{q}_{ss'})\\
&-(2\bar{k}_i - 1)\bar{r}_i(-\tfrac{\bar{b}_{2i}\bar{\alpha}_i}{\bar{r}_i})^{\frac{2\bar{k}_i}{2\bar{k}_i-1}} + 2\bar{k}_i\bar{\alpha}_i[\sum_{j\neq i}\bar{b}_{2j}(-\tfrac{\bar{b}_{2j}\bar{\alpha}_j}{\bar{r}_j})^{\frac{1}{2\bar{k}_j-1}}] = 0.\\
&\bar{\alpha}_i(T,s) = \bar{q}_i(T,s),
\end{aligned}
$$

Proof Consider the guess functional $f_i = \alpha_i\frac{(x-\bar{x})^{2k_i}}{2k_i} + \bar{\alpha}_i\frac{\bar{x}^{2\bar{k}_i}}{2\bar{k}_i}$:

$$
\begin{aligned}
&E[L_i - f_i(0)] = E(q_{iT} - \alpha_i(T,s(T)))\tfrac{(x_T-\bar{x}_T)^{2k_i}}{2k_i}\\
&+(\bar{q}_{iT} - \bar{\alpha}_i(T,s(T)))\tfrac{\bar{x}_T^{2\bar{k}_i}}{2\bar{k}_i}\\
&+E\int_0^T q_i\tfrac{(x-\bar{x})^{2k_i}}{2k_i} + \dot{\alpha}_i\tfrac{(x-\bar{x})^{2k_i}}{2k_i} + \alpha_i b_1(x-\bar{x})^{2k_i}\\
&+r_i\tfrac{(u_i-\bar{u}_i)^{2\bar{k}_i}}{2\bar{k}_i} + [\sum_j \alpha_i(x-\bar{x})^{2k_i-1}b_{2j}(u_j - \bar{u}_j)]\\
&+\alpha_i k_i(2k_i-1)(\sigma^2+\sigma^2_{cogv})\tfrac{(x-\bar{x})^{2k_i}}{2k_i}\\
&+\alpha_i\int_\theta[(1+\mu)^{2k_i} - 1 - \mu 2k_i]\nu(d\theta)\tfrac{(x-\bar{x})^{2k_i}}{2k_i} \qquad (10.67)\\
&+(\sum_{s'}[\alpha_i(t,s') - \alpha_i(t,s)]\tilde{q}_{ss'})\tfrac{(x-\bar{x})^{2k_i}}{2k_i}\\
&+\bar{q}_i\tfrac{\bar{x}^{2\bar{k}_i}}{2\bar{k}_i} + +\dot{\bar{\alpha}}_i\tfrac{\bar{x}^{2\bar{k}_i}}{2\bar{k}_i} + \bar{\alpha}_i\bar{x}^{2\bar{k}_i}\bar{b}_1\\
&+\bar{r}_i\tfrac{\bar{u}_i^{2\bar{k}_i}}{2\bar{k}_i} + \bar{\alpha}_i\bar{x}^{2\bar{k}_i-1}[\sum_j \bar{b}_{2j}\bar{u}_j]\\
&+(\sum_{s'}[\bar{\alpha}_i(t,s') - \bar{\alpha}_i(t,s)]\tilde{q}_{ss'})\tfrac{\bar{x}^{2\bar{k}_i}}{2\bar{k}_i}.
\end{aligned}
$$

We complete the following term:

$$
\begin{aligned}
&r_i\tfrac{(u_i-\bar{u}_i)^{2\bar{k}}}{2k_i} + \alpha_i(x-\bar{x})^{2k_i-1}b_{2i}(u_i - \bar{u}_i)\\
&= r_i\tfrac{(u_i-\bar{u}_i)^{2\bar{k}}}{2k_i} + \alpha_i(x-\bar{x})^{2k_i-1}b_{2i}(u_i - \bar{u}_i)\\
&= r_i\tfrac{(u_i-\bar{u}_i)^{2\bar{k}_i}}{2k_i} + \alpha_i(x-\bar{x})^{2k_i-1}b_{2i}(u_i - \bar{u}_i)\\
&+(2k_i - 1)r_i(-\tfrac{b_{2i}\alpha_i}{r_i})^{\frac{2k_i}{2k_i-1}}\tfrac{(x-\bar{x})^{2k_i}}{2k_i} - (2k_i-1)r_i(-\tfrac{b_{2i}\alpha_i}{r_i})^{\frac{2k_i}{2k_i-1}}\tfrac{(x-\bar{x})^{2k_i}}{2k_i}. \qquad (10.68)
\end{aligned}
$$

A similar completion can be done for the terms in $\bar{u}_i$. Thus

$$
\begin{aligned}
&E[L_i - f_i(0)] = E(q_{iT} - \alpha_i(T))\tfrac{(x_T-\bar{x}_T)^{2k_i}}{2k_i} + (\bar{q}_{iT} - \bar{\alpha}_i(T))\tfrac{\bar{x}_T^{2\bar{k}_i}}{2\bar{k}_i}\\
&+E\int_0^T \Big\{[\dot{\alpha}_i + q_i + 2k_i\alpha_i b_1 + \alpha_i k_i(2k_i-1)(\sigma^2+\sigma^2_{cogv})\\
&+\alpha_i\int_\theta[(1+\mu)^{2k_i} - 1 - \mu 2k_i]\nu(d\theta)\\
&+\textstyle\sum_{s'}[\alpha_i(t,s') - \alpha_i(t,s)]\tilde{q}_{ss'}\\
&-(2k_i-1)r_i(-\tfrac{b_{2i}\alpha_i}{r_i})^{\frac{2k_i}{2k_i-1}} + 2k_i\alpha_i\textstyle\sum_{j\neq i} b_{2j}(-\tfrac{b_{2j}\alpha_j}{r_j})^{\frac{1}{2k_j-1}}\ \}\ \tfrac{(x-\bar{x})^{2k_i}}{2k_i}\\
&+\big\{\dot{\bar{\alpha}}_i + \bar{q}_i + 2\bar{k}_i\bar{\alpha}_i\bar{b}_1 + (\textstyle\sum_{s'}[\bar{\alpha}_i(t,s') - \bar{\alpha}_i(t,s)]\tilde{q}_{ss'})\\
&-(2\bar{k}_i-1)\bar{r}_i(-\tfrac{\bar{b}_{2i}\bar{\alpha}_i}{\bar{r}_i})^{\frac{2\bar{k}_i}{2\bar{k}_i-1}} + 2\bar{k}_i\bar{\alpha}_i[\textstyle\sum_{j\neq i}\bar{b}_{2j}(-\tfrac{\bar{b}_{2j}\bar{\alpha}_j}{\bar{r}_j})^{\frac{1}{2\bar{k}_j-1}}]\ \}\ \tfrac{\bar{x}^{2\bar{k}_i}}{2\bar{k}_i}\\
&+r_i[\tfrac{(u_i-\bar{u}_i)^{2k_i}}{2k_i} + \alpha_i(x-\bar{x})^{2k_i-1}b_{2i}(u_i-\bar{u}_i)]\\
&+(2k_i-1)r_i(-\tfrac{b_{2i}\alpha_i}{r_i})^{\frac{2k_i}{2k_i-1}}\tfrac{(x-\bar{x})^{2k_i}}{2k_i}\\
&+\bar{r}_i[\tfrac{\bar{u}_i^{2\bar{k}_i}}{2\bar{k}_i} + \bar{\alpha}_i\bar{x}^{2\bar{k}_i-1}b_{2i}\bar{u}_i] + (2\bar{k}_i-1)\bar{r}_i(-\tfrac{\bar{b}_{2i}\bar{\alpha}_i}{\bar{r}_i})^{\frac{2\bar{k}}{2\bar{k}_i-1}}\tfrac{\bar{x}^{2\bar{k}_i}}{2\bar{k}_i}.
\end{aligned}
\tag{10.69}
$$

We now note that for $k_i > \frac{1}{2}, r_i > 0$, one has

$$
\begin{aligned}
&r_i\tfrac{(u_i-\bar{u}_i)^{2\bar{k}_i}}{2\bar{k}_i} + \alpha_i(x-\bar{x})^{2k_i-1}b_{2i}(u_i-\bar{u}_i) + (2k_i-1)r_i(-\tfrac{b_{2i}\alpha_i}{r_i})^{\frac{2k_i}{2k_i-1}}\tfrac{(x-\bar{x})^{2k_i}}{2k_i} \geq 0,\\
&\bar{r}_i\tfrac{\bar{u}_i^{2\bar{k}}}{2\bar{k}_i} + \bar{\alpha}_i\bar{x}^{2\bar{k}_i-1}b_{2i}\bar{u}_i + (2\bar{k}_i-1)\bar{r}_i(-\tfrac{\bar{b}_{2i}\bar{\alpha}_i}{\bar{r}_i})^{\frac{2\bar{k}_i}{2\bar{k}_i-1}}\tfrac{\bar{x}^{2\bar{k}_i}}{2\bar{k}_i} \geq 0,
\end{aligned}
\tag{10.70}
$$

with equalities in (10.85) if, and only if

$$
u_i - \bar{u}_i = (-\frac{b_{2i}\alpha_i}{r_i})^{\frac{1}{2k_i-1}}(x-\bar{x}),\ \ \bar{u}_i = (-\frac{\bar{b}_{2i}\bar{\alpha}_i}{\bar{r}_i})^{\frac{1}{2\bar{k}_i-1}}\bar{x}.
$$

Through matching, one obtains the desired result. This completes the proof. ■

At equilibrium, the mean-field-type term $\bar{x}$ in Proposition 10.10 solves

$$
\begin{aligned}
&d\bar{x} = [\bar{b}_1 + \textstyle\sum_j \bar{b}_{2j}\Big(-\tfrac{\bar{b}_{2j}\bar{\alpha}_j}{\bar{r}_j}\Big)^{\frac{1}{2\bar{k}_j-1}}]\bar{x}dt,\\
&\bar{x}(0) = \bar{x}_0,
\end{aligned}
\tag{10.71}
$$

which admits a unique solution within $[0, T]$ subject to the integrability of the regime-switching -dependent coefficient $[\bar{b}_1 + \sum_j \bar{b}_{2j}\Big(-\frac{\bar{b}_{2j}\bar{\alpha}_j}{\bar{r}_j}\Big)^{\frac{1}{2\bar{k}_j-1}}]$ over $[0, T]$.

10.4.3 Fully Cooperative MFTG Under Gauss–Volterra Noise

In this subsection, we choose $k_i = k,\ \bar{k}_i = \bar{k}$ and assume that the I decision-makers are fully cooperative. They jointly (and cooperatively) solve the following problem:

$$\begin{array}{l} \inf_{(u_1,\ldots,u_I)} E[\sum_i L_i(x,u)], \\ \text{subject to} \\ dx = [b_1(x-\bar{x}) + \sum_j b_{2j}(u_j - \bar{u}_j) + \bar{b}_1\bar{x} + \sum_j \bar{b}_{2j}\bar{u}_j]dt \\ +(x-\bar{x})[\sigma dB + \int_\Theta \mu d\tilde{N} + \sigma_{gv} dB_{gv}], \\ \mathbb{P}(s(t+\epsilon) = s'|s,u) = \int_t^{t+\epsilon} \tilde{q}_{ss'}dt' + o(\epsilon),\ s' \neq s \\ x(0) = x_0,\ s(0) = s_0. \end{array} \tag{10.72}$$

Proposition 10.11 *The global optimum of the fully cooperative mean-field-type game (10.72) under the Gauss–Volterra process is given by*

$$\begin{array}{l} i \in \mathcal{I}, \\ u_i^* = -\sum_{s\in\mathcal{S}} \mathbb{1}_{\{s(t)=s\}}[\left(\frac{b_{2i}\alpha_0}{r_i}\right)^{\frac{1}{2k-1}}(x-\bar{x}) + \left(\frac{\bar{b}_{2i}\bar{\alpha}_0}{\bar{r}_i}\right)^{\frac{1}{2\bar{k}-1}}\bar{x}], \\ EL_i = E[\alpha_0(0,s_0)\frac{(x_0-\bar{x}_0)^{2k}}{2k} + \bar{\alpha}_0(0,s_0)\frac{\bar{x}_0^{2\bar{k}}}{2\bar{k}}], \end{array} \tag{10.73}$$

whenever the following system of ordinary differential equations admits a positive solution which does not blow up within $[0, T]$:

$$\begin{array}{l} [\dot{\alpha}_0 + (\sum_i q_i) + 2k\alpha_0 b_1 + \alpha_0 k(2k-1)(\sigma^2 + \sigma_{cogv}^2) \\ +\alpha_0 \int_\theta [(1+\mu)^{2k} - 1 - 2k\mu]\nu(d\theta) \\ +\sum_{s'}[\alpha_0(t,s') - \alpha_0(t,s)]\tilde{q}_{ss'} \\ -(2k-1)\sum_i r_i(\frac{b_{2i}\alpha_0}{r_i})^{\frac{2k}{2k-1}} = 0, \\ \alpha_0(T,s) = \sum_i q_i(T,s), \\ \dot{\bar{\alpha}}_0 + (\sum_i \bar{q}_i) + 2\bar{k}\bar{\alpha}_0\bar{b}_1 + (\sum_{s'}[\bar{\alpha}_0(t,s') - \bar{\alpha}_0(t,s)]\tilde{q}_{ss'}) \\ -(2\bar{k}-1)\sum_i \bar{r}_i(\frac{\bar{b}_{2i}\bar{\alpha}_0}{\bar{r}_i})^{\frac{2\bar{k}}{2\bar{k}-1}} = 0, \\ \bar{\alpha}_0(T,s) = \sum_i \bar{q}_i(T,s). \end{array} \tag{10.74}$$

Proof The proof follows form steps similar to those in Proposition 10.10. This completes the proof. ■

Remark 10.4 A sufficient condition for the existence and uniqueness of the global optimum of mean-field type is obtained under the following conditions.

If $r_i > \delta > 0$, $\sum_i q_i > 0$, and all coefficients are continuous, then there is no escape of α_0 within $[T-\epsilon, T]$. If in addition the coefficient functions $b_1, \sigma^2, \sigma_{cogv}^2, \int_\theta[(1+\mu)^{2k} - 1 - 2k\mu]\nu(d\theta)$, $r_i(\frac{b_{2i}}{r_i})^{\frac{2k}{2k-1}}$, and $(\frac{b_{2i}}{r_i})^{\frac{1}{2k-1}}$ are all integrable within $[0, T]$, then there is no escape of α_0 within the entire interval

$[0, T]$. Similar reasoning works for $\bar{\alpha}_0$ when $\bar{r}_i > \delta > 0, i \in \mathcal{I}, \sum_j \bar{q}_j > 0$, and the coefficient functions $\bar{b}_1, \bar{r}_i(\frac{\bar{b}_{2i}}{\bar{r}_i})^{\frac{2\bar{k}}{2\bar{k}-1}}, (\frac{\bar{b}_{2i}}{\bar{r}_i})^{\frac{1}{2\bar{k}-1}}$ are integrable within $[0, T]$.

At the global optimum, the mean-field term $\bar{x}$ in Proposition 10.11 solves

$$\begin{aligned} d\bar{x} &= \left[\bar{b}_1 + \sum_j \bar{b}_{2j}\left(-\frac{\bar{b}_{2j}\bar{\alpha}_0}{\bar{r}_j}\right)^{\frac{1}{2\bar{k}-1}}\right]\bar{x}dt, \\ \bar{x}(0) &= \bar{x}_0, \end{aligned} \tag{10.75}$$

which admits a unique solution within $[0, T]$ subject to the integrability of the regime-switching -dependent coefficient $\bar{b}_1 + \sum_j \bar{b}_{2j}\left(-\frac{\bar{b}_{2j}\bar{\alpha}_0}{\bar{r}_j}\right)^{\frac{1}{2\bar{k}-1}}$ over $[0, T]$.

10.4.4 Adversarial MFTG Under Gauss–Volterra Noise

In this subsection, we choose $k_i = k \geq 1$, $\bar{k}_i = \bar{k} \geq 1$ and assume that the I decision-makers are divided into two teams: $\mathcal{I}_+ = \{i \in \mathcal{I} \mid r_i > 0, \bar{r}_i > 0\} and \mathcal{I}_- = \{i \in \mathcal{I} \mid r_i < 0, \bar{r}_i < 0\}, \mathcal{I} = \mathcal{I}_+ \cup \mathcal{I}_-$. The decision-makers in team $\mathcal{I}_+$ minimize the functional $EL_{ad}(x, u)$ over $(u_i)_{i\in\mathcal{I}_+}$. The decision-makers in team $\mathcal{I}_-$ maximize $EL_{ad}(x, u)$ over $(u_j)_{j\in\mathcal{I}_-}$. This leads to a minimax game problem:

$$\begin{aligned} &L_{ad}(x, u) = q_T\frac{(x_T-\bar{x}_T)^{2k}}{2k} + \bar{q}_T\frac{\bar{x}_T^{2\bar{k}}}{2\bar{k}} \\ &+\int_0^T q\frac{(x-\bar{x})^{2k}}{2k} + \sum_i r_i\frac{(u_i-\bar{u}_i)^{2k}}{2k} + \bar{q}\frac{\bar{x}^{2k}}{2\bar{k}} + \sum_i \bar{r}_i\frac{\bar{u}_i^{2\bar{k}}}{2\bar{k}}dt, \\ &\inf_{(u_i)_{i\in\mathcal{I}_+}} \sup_{(u_j)_{j\in\mathcal{I}_-}} E[L_{ad}(x, u)], \\ &\text{subject to} \\ &dx = [b_1(x-\bar{x}) + \sum_j b_{2j}(u_j - \bar{u}_j) + \bar{b}_1\bar{x} + \sum_j \bar{b}_{2j}\bar{u}_j]dt \\ &+(x-\bar{x})[\sigma dB + \int_\Theta \mu d\tilde{N} + \sigma_{gv}dB_{gv}], \\ &\mathbb{P}(s(t+\epsilon) = s'|s, u) = \int_t^{t+\epsilon} \tilde{q}_{ss'}dt' + o(\epsilon), \; s' \neq s \\ &x(0) = x_0, \; s(0) = s_0. \end{aligned} \tag{10.76}$$

A mean-field-type risk-neutral saddle point is a strategy profile $(u_j^*, \; j \in I_+)$, of the team of defenders and $(u_j^*, \; j \in I_-)$ of the team of attackers such that

$$\begin{aligned} &EL_{\text{ad}}(x, s, (u_i^*)_{i\in I_+}, (u_j)_{j\in I_-}) \leq EL_{\text{ad}}(X, s, u^*) \\ &\leq EL_{\text{ad}}(X, s, (u_i)_{i\in I_+}, (u_j^*)_{j\in I_-}), \; \forall \, (u_i)_{i\in I_+}, (u_j)_{j\in I_-}. \end{aligned} \tag{10.77}$$

Proposition 10.12 *With $\mathcal{I}_+ \cup \mathcal{I}_- = \mathcal{I}$ and $q > 0, \bar{q} > 0$, the saddle point solution of the adversarial mean-field-type game (10.76) under the Gauss–Volterra process is given by*

$$
\begin{aligned}
& i \in \mathcal{I}_+, \\
& u_i^* = -\sum_{s\in\mathscr{S}} 1\!\!1_{\{s(t)=s\}}[\left(\frac{b_{2i}\alpha_{ad}}{r_i}\right)^{\frac{1}{2k-1}}(x-\bar{x}) + \left(\frac{\bar{b}_{2i}\bar{\alpha}_{ad}}{\bar{r}_i}\right)^{\frac{1}{2\bar{k}-1}}\bar{x}], \\
& j \in \mathcal{I}_-, \\
& u_j^* = -\sum_{s\in\mathscr{S}} 1\!\!1_{\{s(t)=s\}}[\left(\frac{b_{2j}\alpha_{ad}}{r_j}\right)^{\frac{1}{2k-1}}(x-\bar{x}) - \left(\frac{\bar{b}_{2j}\bar{\alpha}_{ad}}{\bar{r}_j}\right)^{\frac{1}{2\bar{k}-1}}\bar{x}], \\
& EL_{ad} = E[\alpha_{ad}(0,s_0)\frac{(x_0-\bar{x}_0)^{2k}}{2k} + \bar{\alpha}_{ad}(0,s_0)\frac{\bar{x}_0^{2\bar{k}}}{2\bar{k}}],
\end{aligned}
\tag{10.78}
$$

whenever the following system of ordinary differential equations admits a positive solution which does not blow up within the horizon $[0, T]$.

$$
\begin{aligned}
& 0 = \dot{\alpha}_{ad} + q + 2k\alpha_{ad}b_1 + \alpha_{ad}k(2k-1)(\sigma^2+\sigma^2_{cogv}) \\
& +\alpha_{ad}\int_\theta[(1+\mu)^{2k} - 1 - 2k\mu]\nu(d\theta) \\
& +\sum_{s'}[\alpha_{ad}(t,s') - \alpha_{ad}(t,s)]\tilde{q}_{ss'} \\
& -(2k-1)\alpha_{ad}^{\frac{2k}{2k-1}}[\sum_{i\in\mathcal{I}_+} r_i(-\frac{b_{2i}}{r_i})^{\frac{2k}{2k-1}} + \sum_{j\in\mathcal{I}_-} r_j(-\frac{b_{2j}}{r_j})^{\frac{2k}{2k-1}}], \\
& \alpha_{ad}(T,s) = q(T,s), \\
& 0 = \dot{\bar{\alpha}}_{ad} + \bar{q} + 2\bar{k}\bar{\alpha}_{ad}\bar{b}_1 + (\sum_{s'}[\bar{\alpha}_{ad}(t,s') - \bar{\alpha}_{ad}(t,s)]\tilde{q}_{ss'}) \\
& -(2\bar{k}-1)\bar{\alpha}_{ad}^{\frac{2\bar{k}}{2\bar{k}-1}}[\sum_{i\in\mathcal{I}_+} \bar{r}_i(-\frac{\bar{b}_{2i}}{\bar{r}_i})^{\frac{2\bar{k}}{2\bar{k}-1}} + \sum_{j\in\mathcal{I}_-} \bar{r}_j(-\frac{\bar{b}_{2j}}{\bar{r}_j})^{\frac{2\bar{k}}{2\bar{k}-1}}], \\
& \bar{\alpha}_{ad}(T,s) = \bar{q}(T,s).
\end{aligned}
\tag{10.79}
$$

Note that here the minmax and maxmin solutions are the same, and $(u_i^*)_{i\in\mathcal{I}_+}$, $(u_j^*)_{j\in\mathcal{I}_-}$ *is a saddle point. Thus, the adversarial MFTG has a value* $EL_{ad}(x, (u_i)_{i\in\mathcal{I}_+}, (u_j)_{j\in\mathcal{I}_-})$.

The proof follows from the direct method by exploiting the strict convex-concave and coercivity properties of the cost functional. ∎

Remark 10.5 A sufficient condition for the existence and uniqueness of the saddle point of mean-field type is

$$
2k-1>0,\ \sum_{i\in\mathcal{I}_+} r_i(-\frac{b_{2i}}{r_i})^{\frac{2k}{2k-1}} + \sum_{j\in\mathcal{I}_-} r_j(-\frac{b_{2j}}{r_j})^{\frac{2k}{2k-1}} > 0,
$$

$$
2\bar{k}-1>0,\ \sum_{i\in\mathcal{I}_+} \bar{r}_i(-\frac{\bar{b}_{2i}}{\bar{r}_i})^{\frac{2\bar{k}}{2\bar{k}-1}} + \sum_{j\in\mathcal{I}_-} \bar{r}_j(-\frac{\bar{b}_{2j}}{\bar{r}_j})^{\frac{2\bar{k}}{2\bar{k}-1}} > 0,
$$

$q > 0, \bar{q} > 0.$

10.5 Time-Fractional Mean-Field-Type Games

In this section, we examine MFTG, where the evolution of the state is given by a time-fractional or time-scaled differential equation.

Let $\alpha \in (0, 1]$. The α-fractional integral of f is defined as

$$[I^\alpha f](t) = \int_0^t s^{\alpha-1} f(s)ds,$$

whenever the integral exists. For $\alpha = 1$, $I^\alpha = I^1$, the 1-fractional integral becomes a standard integral.

Let $\alpha \in (0, 1]$. The α-fractional derivative of f is defined as

$$[d^\alpha f](t) = \lim_{\epsilon \to 0} \frac{f(te^{\epsilon t^{-\alpha}}) - f(t)}{\epsilon}.$$

If f is differentiable, then

$$[d^\alpha f](t) = t^{1-\alpha} df(t).$$

Note that under this definition of α-fractional derivative, one has $d^\alpha[\frac{t^\alpha}{\alpha}] = 1$.

Further, it can be seen that a function could be α-differential at a point but not differentiable in the classical sense. As a case in point, the function $f(t) = 5t^{\frac{1}{5}}$ is everywhere $\frac{1}{5}$-differentiable and has the $\frac{1}{5}$-derivative of 1 for $t > 0$, while it is not 1-differentiable at $t = 0$. Another example is the exponential:

$$d^\alpha[e^{\frac{t^\alpha}{\alpha}}] = e^{\frac{t^\alpha}{\alpha}}.$$

We consider a class of MFTGs with non-quadratic cost and with state driven by the Gauss–Volterra processes and time-fractional derivative:

$$\begin{aligned}
&i \in \mathcal{I},\\
&L_i^\alpha(x, u) = q_{iT}\tfrac{(x_T-\bar{x}_T)^{2k_i}}{2k_i} + \bar{q}_{iT}\tfrac{\bar{x}_T^{2\bar{k}_i}}{2\bar{k}_i}\\
&+I^\alpha[q_i\tfrac{(x-\bar{x})^{2k_i}}{2k_i} + r_i\tfrac{(u_i-\bar{u}_i)^{2k_i}}{2k_i} + \bar{q}_i\tfrac{\bar{x}^{2\bar{k}_i}}{2\bar{k}_i} + \bar{r}_i\tfrac{\bar{u}_i^{2\bar{k}_i}}{2\bar{k}_i}],\\
&\inf_{u_i} E[L_i^\alpha(x, u)],\\
&\text{subject to}\\
&d^\alpha x = [b_1(x-\bar{x}) + \textstyle\sum_j b_{2j}(u_j-\bar{u}_j) + \bar{b}_1\bar{x} + \sum_j \bar{b}_{2j}\bar{u}_j]dt\\
&+(x-\bar{x})[\sigma dB + \textstyle\int_\Theta \gamma d\tilde{N} + \sigma_{gv}dB_{gv}],\\
&\mathbb{P}(s(t+\delta) = s'|s, u) = \textstyle\int_t^{t+\delta}(t')^{\alpha-1}\tilde{q}_{ss'}dt' + o(\delta), \quad s' \neq s,\\
&x(0) = x_0, \quad s(0) = s_0,
\end{aligned} \tag{10.80}$$

where $k_i \geq 1, \bar{k}_i \geq 1, \alpha \in (0, 1)$ are natural numbers and the coefficients are time- and switching-dependent. The (fractional) cost functional is clearly non-quadratic in (x, u) for $k_i > 1$ or $\bar{k}_i > 1$. Note that the process B_{gv} considered in the multiplicative noise case does not take into account x in the memory but considers the deterministic coefficients that are a function of time.

Proposition 10.13 *Assume $q_i > 0, \bar{q}_i > 0, r_i, \bar{r}_i > 0$. The mean-field Nash equilibrium of the time-fractional MFTG (10.80) under the Gauss–Volterra process is given by*

$$
\begin{aligned}
& i \in \mathcal{I}, \\
& u_i^* = -\sum_{s\in\mathcal{S}} \mathbb{1}_{\{s(t)=s\}}\Big[\left(\frac{b_{2i}\lambda_i}{r_i}\right)^{\frac{1}{2k_i-1}}(x-\bar{x}) + \left(\frac{\bar{b}_{2i}\bar{\lambda}_i}{\bar{r}_i}\right)^{\frac{1}{2\bar{k}_i-1}}\bar{x}\Big], \\
& EL_i^{\alpha} = E[\lambda_i(0, s_0)\frac{(x_0-\bar{x}_0)^{2k_i}}{2k_i} + \bar{\lambda}_i(0, s_0)\frac{\bar{x}_0^{2\bar{k}_i}}{2\bar{k}_i}],
\end{aligned}
\tag{10.81}
$$

whenever the following system of ordinary differential equations admits a positive solution which does not have a singularity within $[0, T]$.

$$
\begin{aligned}
& i \in \mathcal{I}, \\
& d^{\alpha}\lambda_i + q_i + 2k_i\lambda_i b_1 + \lambda_i k_i(2k_i-1)(t^{\alpha-1}\sigma^2 + t^{1-\alpha}\tilde{\sigma}^2_{cogv}) \\
& +\lambda_i \int_{\theta}\Big[\frac{(1+\gamma t^{\alpha-1})^{2k_i}-1}{t^{\alpha-1}} - 2k_i\gamma\Big]\nu(d\theta) \\
& +\Big(\sum_{s'}[\lambda_i(t, s') - \lambda_i(t, s)]\tilde{q}_{ss'}\Big) \\
& -(2k_i-1)r_i\Big(-\frac{b_{2i}\lambda_i}{r_i}\Big)^{\frac{2k_i}{2k_i-1}} \\
& +2k_i\lambda_i\Big[\sum_{j\neq i} b_{2j}\Big(-\frac{b_{2j}\lambda_j}{r_j}\Big)^{\frac{1}{2k_j-1}} = 0, \\
& \lambda_i(T, s) = q_i(T, s), \\
& d^{\alpha}\bar{\lambda}_i + \bar{q}_i + 2\bar{k}_i\bar{\lambda}_i\bar{b}_1 + \Big(\sum_{s'}[\bar{\lambda}_i(t, s') - \bar{\lambda}_i(t, s)]\tilde{q}_{ss'}\Big) \\
& -(2\bar{k}_i-1)\bar{r}_i\Big(-\frac{\bar{b}_{2i}\bar{\lambda}_i}{\bar{r}_i}\Big)^{\frac{2\bar{k}_i}{2\bar{k}_i-1}} \\
& +2\bar{k}_i\bar{\lambda}_i\Big[\sum_{j\neq i} \bar{b}_{2j}\Big(-\frac{\bar{b}_{2j}\bar{\lambda}_j}{\bar{r}_j}\Big)^{\frac{1}{2\bar{k}_j-1}}\Big] = 0, \\
& \bar{\lambda}_i(T, s) = \bar{q}_i(T, s), \\
& \tilde{\sigma}^2_{cogv} = \frac{d}{dt}\left[\int_0^t \left\{K(t'_+, t')\tilde{\sigma}_{gv}(t') + \int_{t'}^t \tilde{\sigma}_{gv}(t'')K(t', t'')dt''\right\}^2 dt'\right], \\
& \tilde{\sigma}(t) = \sigma t^{\alpha-1}.
\end{aligned}
\tag{10.82}
$$

Proof Consider the guess functional $f_i = \lambda_i\frac{(x-\bar{x})^{2k_i}}{2k_i} + \bar{\lambda}_i\frac{\bar{x}^{2\bar{k}_i}}{2\bar{k}_i}$:

$$
\begin{aligned}
&E[L_i^{\alpha} - f_i(0)] = E(q_{iT} - \lambda_i(T, s(T)))\tfrac{(x_T-\bar{x}_T)^{2k_i}}{2k_i} \\
&+(\bar{q}_{iT} - \bar{\lambda}_i(T, s(T)))\tfrac{\bar{x}_T^{2\bar{k}_i}}{2\bar{k}_i} \\
&+E\int_0^T t^{\alpha-1} q_i \tfrac{(x-\bar{x})^{2k_i}}{2k_i} + \dot{\lambda}_i \tfrac{(x-\bar{x})^{2k_i}}{2k_i} + \lambda_i b_1 t^{\alpha-1}(x-\bar{x})^{2k_i} \\
&+r_i t^{\alpha-1}\tfrac{(u_i-\bar{u}_i)^{2\bar{k}_i}}{2\bar{k}_i} + [\textstyle\sum_j \lambda_i (x-\bar{x})^{2k_i-1} b_{2j} t^{\alpha-1}(u_j - \bar{u}_j)] \\
&+\lambda_i k_i(2k_i-1)(t^{2\alpha-2}\sigma^2 + \tilde{\sigma}^2_{cogv})\tfrac{(x-\bar{x})^{2k_i}}{2k_i} \\
&+\lambda_i \int_{\theta}[(1+\gamma t^{\alpha-1})^{2k_i} - 1 - \gamma t^{\alpha-1} 2k_i]\nu(d\theta)\tfrac{(x-\bar{x})^{2k_i}}{2k_i} \\
&+(\textstyle\sum_{s'}[\lambda_i(t,s') - \lambda_i(t,s)]\tilde{q}_{ss'}t^{\alpha-1})\tfrac{(x-\bar{x})^{2k_i}}{2k_i} \\
&+\bar{q}_i t^{\alpha-1}\tfrac{\bar{x}^{2\bar{k}_i}}{2\bar{k}_i} + \dot{\bar{\lambda}}_i \tfrac{\bar{x}^{2\bar{k}_i}}{2\bar{k}_i} + \bar{\lambda}_i \bar{x}^{2\bar{k}_i}\bar{b}_1 t^{\alpha-1} \\
&+\bar{r}_i t^{\alpha-1}\tfrac{\bar{u}_i^{2\bar{k}_i}}{2\bar{k}_i} + \bar{\lambda}_i \bar{x}^{2\bar{k}_i-1}[\textstyle\sum_j \bar{b}_{2j} t^{\alpha-1}\bar{u}_j] \\
&+(\textstyle\sum_{s'}[\bar{\lambda}_i(t,s') - \bar{\lambda}_i(t,s)]\tilde{q}_{ss'}t^{\alpha-1})\tfrac{\bar{x}^{2\bar{k}_i}}{2\bar{k}_i}.
\end{aligned}
\tag{10.83}
$$

We complete the following term:

$$
\begin{aligned}
&r_i \tfrac{(u_i-\bar{u}_i)^{2\bar{k}}}{2k_i} + \lambda_i (x-\bar{x})^{2k_i-1} b_{2i}(u_i - \bar{u}_i) \\
&= r_i[\tfrac{(u_i-\bar{u}_i)^{2\bar{k}}}{2k_i} + \lambda_i (x-\bar{x})^{2k_i-1} b_{2i}(u_i - \bar{u}_i)] \\
&= r_i[\tfrac{(u_i-\bar{u}_i)^{2k_i}}{2k_i} + \lambda_i (x-\bar{x})^{2k_i-1} b_{2i}(u_i - \bar{u}_i)] \\
&+(2k_i-1)r_i(-\tfrac{b_{2i}\lambda_i}{r_i})^{\frac{2k_i}{2k_i-1}}\tfrac{(x-\bar{x})^{2k_i}}{2k_i} \\
&-(2k_i-1)r_i(-\tfrac{b_{2i}\lambda_i}{r_i})^{\frac{2k_i}{2k_i-1}}\tfrac{(x-\bar{x})^{2k_i}}{2k_i}.
\end{aligned}
\tag{10.84}
$$

A similar completion can be done for the terms in $\bar{u}_i$. We then note that for $k_i > \frac{1}{2}, r_i > 0$, one has

$$
\begin{aligned}
&r_i[\tfrac{(u_i-\bar{u}_i)^{2\bar{k}_i}}{2\bar{k}_i} + \lambda_i (x-\bar{x})^{2k_i-1} b_{2i}(u_i - \bar{u}_i)] \\
&+(2k_i-1)r_i(-\tfrac{b_{2i}\lambda_i}{r_i})^{\frac{2k_i}{2k_i-1}}\tfrac{(x-\bar{x})^{2k_i}}{2k_i} \geq 0, \\
&\bar{r}_i[\tfrac{\bar{u}_i^{2\bar{k}}}{2\bar{k}_i} + \bar{\lambda}_i \bar{x}^{2\bar{k}_i-1} b_{2i}\bar{u}_i] \\
&+(2\bar{k}_i-1)\bar{r}_i(-\tfrac{\bar{b}_{2i}\bar{\lambda}_i}{\bar{r}_i})^{\frac{2\bar{k}_i}{2\bar{k}_i-1}}\tfrac{\bar{x}^{2\bar{k}_i}}{2\bar{k}_i} \geq 0,
\end{aligned}
\tag{10.85}
$$

with equalities in (10.85) iff

$$
u_i - \bar{u}_i = (-\frac{b_{2i}\lambda_i}{r_i})^{\frac{1}{2k_i-1}}(x-\bar{x}), \quad \bar{u}_i = (-\frac{\bar{b}_{2i}\bar{\lambda}_i}{\bar{r}_i})^{\frac{1}{2\bar{k}_i-1}}\bar{x}.
$$

Via proper matching of the terms, one arrives at the desired result. This completes the proof. ■

10.6 Example

In this section, we present an illustration of problem (10.64) by choosing the Gauss–Volterra process with the following kernel:

$$K(t,t')=K_H(t,t')=c_H(t-t')^{H-\frac{1}{2}}+c_H(\tfrac{1}{2}-H)\int_{t'}^{t}(z-t')^{H-\frac{3}{2}}(1-(\tfrac{t'}{z})^{\frac{1}{2}-H})dz, \tag{10.86}$$

where $H \in (0,1)$, $c_H = \sqrt{\frac{2H\Gamma(\frac{3}{2}-H)}{\Gamma(\frac{1}{2}+H)\Gamma(2-2H)}}$ is a normalizing constant, and Γ is the gamma function:

$$\Gamma(z) = \int_0^{+\infty} e^{-t}t^{z-1}dt, \quad \text{Re}(z) > 0.$$

The Gauss–Volterra process with kernel K_H is a fractional Brownian motion with the Hurst parameter H. The parameters of the numerical setting are given in Table 10.3.

It is important to note that under this setting, the problem (10.64) is not Markov, and the cost is not quadratic. From (10.10), we know that, in state-and-mean-field-type feedback strategies, the mean-field Nash equilibrium of the mean-field-type game (10.64) is given by

Table 10.3 Parameters used in the numerical example

Numerical setting	Value
Kernel K	$K = K_{0.8}$
T	1
Switching $\mathscr{S}$	$\{s_*, s^*\},\ s_* \neq s^*$
I	2019
$(k_i, \bar{k}_i)$	(2, 2) if $i \leq 2018$
	(4, 2) if $i = 2019$
$q_i, \bar{q}_i, r_i, \bar{r}_i$	1 if $i \leq 2018$
	100 if $i = 2019$
$b_1, b_{2i}, \bar{b}_{2i}, \sigma, \sigma_{gv}$	1 if $i \leq 2018$
$b_{2i}, \bar{b}_{2i}$	10 if $i = 2019$
$\mu(t, \theta, s)$	θ
Θ	$\mathbb{R}_+^*$
$\nu(d\theta)$	$ce^{-5\lvert\theta\rvert}d\theta$
$\sigma_{gv}(t, s^*)$	1
$\sigma_{gv}(t, s_*)$	10^{-2}
$\tilde{q}_{s^*s_*}$	0.7
$\tilde{q}_{s_*s^*}$	0.4

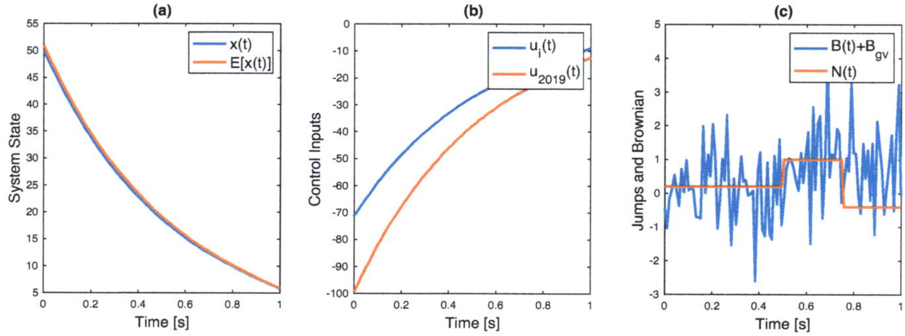

Fig. 10.1 Sample path of the optimal state, optimal strategy under Gauss–Volterra, jump, and diffusion. Initial state value is 50

$$
\begin{aligned}
u_i^* &= -\alpha_i^{\frac{1}{3}}(x-\bar{x}) - \bar{\alpha}_i^{\frac{1}{3}}\bar{x}, \ \ i \leq 2018, \\
u_i^* &= -(\tfrac{\alpha_i}{10})^{\frac{1}{7}}(x-\bar{x}) - (\tfrac{\bar{\alpha}_i}{10})^{\frac{1}{3}}\bar{x}, \ i = 2019.
\end{aligned}
\tag{10.87}
$$

Figure 10.1 plots (a) a sample path of the optimal state trajectory starting from $x_0 = 50$, (b) the optimal strategies of all decision-makers $i \leq 2018$ and 2019, and (c) sample noises. As expected, the state is moving toward zero when the optimal strategies are employed.

10.7 Notes

In this chapter, we have shown that a mean-field-type equilibrium can be determined in a semi-explicit way in state-and-mean-field feedback strategies for a broader class of nonlinear, non-quadratic game problems with nonlinearly distribution-dependent payoffs where the state dynamics are driven by conditional expected values of states, controls, Brownian motions, Gauss–Volterra processes, jump, and regime switching. The method does not require a sophisticated nonelementary extension to backward-forward systems.

Chapter 11
Stackelberg Mean-Field-Type Games

This chapter deals with bi-level hierarchical mean-field-type game problems. The decision-makers act in a sequential order with informational asymmetry. The state dynamics are driven by jump-diffusion-regime switching and Gauss-Volterra processes, which include fractional Brownian motion and multi-fractional Brownian motion. Throughout the entire chapter, we restrict the strategies of the decision-makers to the class of linear-state-and-mean-field feedback strategy set.

11.1 Introduction

One of the central goals in game theory is to understand the difficulties that parties have in reaching a solution when decision-makers act in a sequential order with different information structures. Informational differences provide an appealing explanation for the solution concept. We distinguish between the first mover, called the primary decision-maker (the leader), and the second mover, called the secondary decision-maker (the follower). The idea is that the follower can observe the policy or strategy of the leader and subsequently act.

This chapter introduces hierarchical mean-field-type games under Gauss-Volterra, jump-diffusion, and regime-switching processes. State-and-conditional mean-field feedback Stackelberg solutions are discussed. It provides

- semi-explicit Stackelberg solutions for the risk-neutral linear-quadratic case with a single leader and a single follower,
- semi-explicit Stackelberg risk-sensitive solutions for risk-seeking and risk-averse linear-quadratic case with a single leader and a single follower,
- semi-explicit Stackelberg solutions for the state-dependent noise case with a single leader and a single follower,
- semi-explicit Stackelberg solutions for the risk-neutral case with a single leader, multiple followers, and non-quadratic costs.

T. Başar et al., *Mean-Field-Type Game Theory I*, Static & Dynamic Game Theory: Foundations & Applications, https://doi.org/10.1007/978-3-032-07027-2_11

The direct method for the Stackelberg solution entails two stages:

- The first stage consists of a direct method for a mean-field-type control problem of the follower given the policy (strategy) of the leader. It consists of five elementary steps. The first step starts by setting the mean-field terms of the problem. The second step consists of identifying a partial guess functional for each decision-maker, where the coefficient functionals are possibly random and regime-switching-dependent. The third step uses the partial integration formula. The fourth step uses a completion of terms in a one-shot optimization for both control actions and the conditional expected value of the control actions of all decision-makers. The fifth and last step uses an algebraic basis of linearly independent processes to match the coefficients. The matching leads to a (possibly stochastic) differential system of equations, providing a semi-explicit representation of the best-response solution of the follower (as a function of the leader control action).
- The second stage uses again the direct method to solve the problem of the leader's mean-field-type control problem by inserting the best response of the follower in the cost functions as well as in the state dynamics.
- Return to the first stage to provide semi-explicit solutions for both.

The rest of this chapter is structured as follows. Section 11.2 examines a single leader, single follower problem in both risk-sensitive and risk-neutral settings. The state dynamics are extended to include state-dependent noise of the Gauss-Volterra type with jumps. Section 11.3 examines a non-quadratic cost setting with multiple followers.

11.2 Single Leader-Single Follower

We present and examine three classes of Stackelberg game problems of mean-field type. We start with two decision-makers: 0 and 1. Decision-maker 0 (leader) moves first. Decision-maker 0 announces its strategy u_0 which is restricted to a linear-state-and-mean-field feedback strategy set $\mathscr{U}_0$. Then, Decision-maker 1 (follower) reacts to 0's strategy u_0 by choosing a strategy $u_1 \in \mathscr{U}_1$. Each decision-maker $i \in \{0, 1\}$ has a cost functional $L_i(x, u)$ to be optimized.

To solve this hierarchical decision-making problem, we use a backward induction method. We start by solving the problem of decision-maker 1 given decision-maker 0's strategy. In the risk-neutral setting, decision-maker 1 solves

$$\inf_{u_1 \in \mathscr{U}_1} E[L_1(x, u_0, u_1)|u_0(.)].$$

This leads to a set-valued map called the reaction set or best-response set of decision-maker 1:

$$\text{rnBR}_1(u_0) = \arg \min_{v_1 \in \mathcal{U}_1} E[L_1(x, u_0, v_1)].$$

When the set $\text{rnBR}_1(u_0)$ has two or more elements, there is an ambiguity for decision-maker 0 on the possible response of decision-maker 1. In this case, several choices are possible: Pessimistic, probabilistic, and optimistic Stackelberg viewpoints have been proposed. Here, we provide sufficient conditions under which the reaction set $\text{rnBR}_1(u_0)$ of decision-maker 1 is a singleton set. Then, decision-maker 0, who is the leader, solves a mean-field-type control problem under jump-diffusion-regime switching to determine its best strategy. The leader 0 solves the following mean-field-type control problem:

$$\inf_{u_0 \in \mathcal{U}_0} \{E[L_0(x, u_0, u_1)], \ u_1 \in \text{rnBR}_1(u_0) \cap \mathcal{U}_1\}.$$

11.2.1 Stackelberg MFTG: Risk-Neutral LQ Case

We now make the description above precise by providing basic definitions of the risk-neutral hierarchical problems and associated solution concepts. The cost functionals are

$$\begin{aligned}
&i, j \in \mathcal{I} = \{0, 1\}, \\
&L_i(x, u) = q_{iT} \tfrac{(x_T - \bar{x}_T)^2}{2} + \bar{q}_{iT} \tfrac{\bar{x}_T^2}{2} \\
&+ \int_0^T q_i \tfrac{(x-\bar{x})^2}{2} + r_i \tfrac{(u_i - \bar{u}_i)^2}{2} + \varepsilon_{ij}(u_i - \bar{u}_i)(u_j - \bar{u}_j) + \bar{q}_i \tfrac{\bar{x}^2}{2} + \bar{r}_i \tfrac{\bar{u}_i^2}{2} + \bar{\varepsilon}_{ij} \bar{u}_i \bar{u}_j dt.
\end{aligned} \tag{11.1}$$

Definition 11.1 (Mean-Field-Type Risk-Neutral Response to the Leader) Given $u_0 \in \mathcal{U}_0$, a risk-neutral best-response strategy of the follower is a strategy that solves $\inf_{\mathcal{U}_1} EL_1(x, u)$, with L_1 defined in (11.1). The set of risk-neutral best responses of 1 is denoted by $\text{rnBR}_1(u_0)$.

Definition 11.2 (Mean-Field-Type Risk-Neutral Stackelberg Solution) A mean-field-type risk-neutral Stackelberg solution strategy $u_0^{\text{rn}}, u_1^{\text{rn}}$ such that for the leader

$$u_0^{\text{rn}} \in \arg \min_{u_0 \in \mathcal{U}_0} \{E[L_0(x, u_0, u_1^{\text{rn}})] : \ u_1^{\text{rn}} \in \text{rnBR}_1(u_0)\},$$

and, for the follower, $u_1^{\text{rn}} \in \text{rnBR}_1(u_0^{\text{rn}})$.

The mean-field-type risk-neutral Stackelberg problem is given by

$$
\begin{array}{ll}
\inf_{u_0\in\mathscr{U}_0} E[L_0(x,u)], & \\
\text{subject to} & \\
dx = \left\{b_1(x-\bar{x}) + \sum_{l\in\mathscr{I}} b_{2l}(u_l-\bar{u}_l) + \bar{b}_1\bar{x} + \sum_{l\in\mathscr{I}} \bar{b}_{2l}\bar{u}_l\right\} dt & \\
+[\sigma dB + \int_\Theta \mu d\tilde{N} + \sigma_{gv} dB_{gv}], & (11.2)\\
\mathbb{P}(s(t+\delta)=s'|s,u) = \int_t^{t+\delta} \tilde{q}_{ss'}dt' + o(\delta),\ s'\neq s, & \\
x(0)=x_0,\ s(0)=s_0, & \\
u_1 \in \text{rnBR}_1(u_0) = \arg\min_{v_1\in\mathscr{U}_1} E[L_1(x,u_0,v_1)], &
\end{array}
$$

with $\delta > 0,\ s_0 \in \mathscr{S}$.

The next proposition provides a Stackelberg solution in a semi-explicit form.

Proposition 11.1 *Consider risk-neutral problem (11.2) with a single leader and a single follower. We impose the following conditions:*

$$
\begin{array}{l}
r_0>0,\ r_1>0,\ q_0>0,\ q_1>0\\
\bar{r}_0>0,\ \bar{r}_1>0,\ \bar{q}_0\geq 0,\ \bar{q}_1\geq 0\\
\frac{1}{2}r_0-\frac{\varepsilon_{01}\varepsilon_{10}}{r_1}>\delta>0,\\
\frac{1}{2}\bar{r}_0-\frac{\bar{\varepsilon}_{01}\bar{\varepsilon}_{10}}{\bar{r}_1}>\delta>0,
\end{array}
\qquad (11.3)
$$

for some $\delta > 0$.

Assume that the strategies of the decision-makers are restricted to the class of linear-state-and-mean-field feedback strategy set.

Then, the optimal linear (state-and-mean-field feedback) Stackelberg strategies and costs are

$$
\begin{array}{ll}
u_0 = -\frac{1}{2}[\frac{1}{2}r_0-\frac{\varepsilon_{01}\varepsilon_{10}}{r_1}]^{-1}\{-\frac{\varepsilon_{01}b_{21}}{r_1}\alpha_1+\alpha_0(b_{20}-\frac{b_{21}\varepsilon_{10}}{r_1})\}(x-\bar{x}) & \\
-\frac{1}{2}[\frac{1}{2}\bar{r}_0-\frac{\bar{\varepsilon}_{01}\bar{\varepsilon}_{10}}{\bar{r}_1}]^{-1}\{-\frac{\bar{\varepsilon}_{01}\bar{b}_{21}}{\bar{r}_1}\beta_1+\beta_0(\bar{b}_{20}-\frac{\bar{b}_{21}\bar{\varepsilon}_{10}}{\bar{r}_1})\}\bar{x}, & \\
u_1 = (\frac{\varepsilon_{10}}{r_1}\frac{1}{2}\frac{(-\frac{\varepsilon_{01}b_{21}}{r_1}\alpha_1+\alpha_0(b_{20}-\frac{b_{21}\varepsilon_{10}}{r_1}))}{(\frac{1}{2}r_0-\frac{\varepsilon_{01}\varepsilon_{10}}{r_1})}-\frac{b_{21}}{r_1}\alpha_1)(x-\bar{x}) & \\
+[\frac{\bar{\varepsilon}_{10}}{\bar{r}_1}\frac{1}{2}\frac{\{-\frac{\bar{\varepsilon}_{01}\bar{b}_{21}}{\bar{r}_1}\beta_1+\beta_0(\bar{b}_{20}-\frac{\bar{b}_{21}\bar{\varepsilon}_{10}}{\bar{r}_1})\}}{\frac{1}{2}\bar{r}_0-\frac{\bar{\varepsilon}_{01}\bar{\varepsilon}_{10}}{\bar{r}_1}}-\beta_1\frac{\bar{b}_{21}}{\bar{r}_1}]\bar{x}, & (11.4)\\
E[L_0(x,u^{ss})] = E\alpha_0(0,s_0))\frac{1}{2}(x_0-\bar{x}_0)^2+\beta_0(0,s_0))\frac{1}{2}\bar{x}_0^2 & \\
+\frac{1}{2}E\int_0^T[\sigma^2+\int_\Theta\mu^2\nu(d\theta)+\sigma_{cogv}^2]\alpha_0 dt, & \\
E[L_1(x,u^{ss})] = E\alpha_1(0,s_0))\frac{1}{2}(x_0-\bar{x}_0)^2+\beta_1(0,s_0))\frac{1}{2}\bar{x}_0^2 & \\
+\frac{1}{2}E\int_0^T[\sigma^2+\int_\Theta\mu^2\nu(d\theta)+\sigma_{cogv}^2]\alpha_1 dt, &
\end{array}
$$

where the coefficients are given by

$$\begin{cases} 0=\dot{\alpha}_0+2\alpha_0(b_1-\frac{b_{21}^2}{r_1}\alpha_1)+q_0+\sum_{s'}[\alpha_0(t,s')-\alpha_0(t,s)]\tilde{q}_{ss'} \\ -\frac{1}{2}[\frac{1}{2}r_0-\frac{\varepsilon_{01}\varepsilon_{10}}{r_1}]^{-1}\{-\frac{\varepsilon_{01}b_{21}}{r_1}\alpha_1+\alpha_0(b_{20}-\frac{b_{21}\varepsilon_{10}}{r_1})\}^2, \\ \alpha_0(T,s)=q_0(T,s),\ s\in\mathscr{S}, \\ 0=\dot{\beta}_0+2\beta_0(\bar{b}_1-\frac{\bar{b}_{21}^2}{\bar{r}_1}\beta_1)+\bar{q}_0+\sum_{s'}[\beta_0(t,s')-\beta_0(t,s)]\tilde{q}_{ss'} \\ -\frac{1}{2}[\frac{1}{2}\bar{r}_0-\frac{\bar{\varepsilon}_{01}\bar{\varepsilon}_{10}}{\bar{r}_1}]^{-1}\{-\frac{\bar{\varepsilon}_{01}\bar{b}_{21}}{\bar{r}_1}\beta_1+\beta_0(\bar{b}_{20}-\frac{\bar{b}_{21}\bar{\varepsilon}_{10}}{\bar{r}_1})\}^2, \\ \beta_0(T,s)=\bar{q}_0(T,s),\ s\in\mathscr{S}, \\ 0=\dot{\alpha}_1+q_1+2\alpha_1b_1+\sum_{s'}[\alpha_1(t,s')-\alpha_1(t,s)]\tilde{q}_{ss'} \\ -\frac{1}{r_1}[b_{21}\alpha_1-\frac{\varepsilon_{10}}{2}[\frac{1}{2}r_0-\frac{\varepsilon_{01}\varepsilon_{10}}{r_1}]^{-1}\{-\frac{\varepsilon_{01}b_{21}}{r_1}\alpha_1+\alpha_0(b_{20}-\frac{b_{21}\varepsilon_{10}}{r_1})\}]^2 \\ -\alpha_1b_{20}[\frac{1}{2}r_0-\frac{\varepsilon_{01}\varepsilon_{10}}{r_1}]^{-1}\{-\frac{\varepsilon_{01}b_{21}}{r_1}\alpha_1+\alpha_0(b_{20}-\frac{b_{21}\varepsilon_{10}}{r_1})\}, \\ \alpha_1(T,s)=q_1(T,s),\ s\in\mathscr{S}, \\ 0=[\dot{\beta}_1+\bar{q}_1+2\beta_1\bar{b}_1+\sum_{s'}[\beta_1(t,s')-\beta_1(t,s)]\tilde{q}_{ss'}] \\ -\frac{1}{\bar{r}_1}[\beta_1\bar{b}_{21}-\frac{\bar{\varepsilon}_{10}}{2}[\frac{1}{2}\bar{r}_0-\frac{\bar{\varepsilon}_{01}\bar{\varepsilon}_{10}}{\bar{r}_1}]^{-1}\{-\frac{\bar{\varepsilon}_{01}\bar{b}_{21}}{\bar{r}_1}\beta_1+\beta_0(\bar{b}_{20}-\frac{\bar{b}_{21}\bar{\varepsilon}_{10}}{\bar{r}_1})\}]^2 \\ -\beta_1\bar{b}_{20}[\frac{1}{2}\bar{r}_0-\frac{\bar{\varepsilon}_{01}\bar{\varepsilon}_{10}}{\bar{r}_1}]^{-1}\{-\frac{\bar{\varepsilon}_{01}\bar{b}_{21}}{\bar{r}_1}\beta_1+\beta_0(\bar{b}_{20}-\frac{\bar{b}_{21}\bar{\varepsilon}_{10}}{\bar{r}_1})\}, \\ \beta_1(T,s)=\bar{q}_1(T,s),\ s\in\mathscr{S}, \end{cases} \tag{11.5}$$

whenever these Riccati systems admit a positive solution α, β which does not have singularity within $[0, T]$.

Proof We prove the statement in two stages.

Step 1 of stage 1: We start by identifying the mean-field terms.

Step 2: Based on the structure of the problem, we propose the following guess functional f_1

Step 3: We apply Itô's formula to compute the difference between the guess and cost functional of the follower:

$$\begin{aligned} &E[L_1-f_1(0)]= \\ &E(q_{1T}-\alpha_1(T,.))\tfrac{1}{2}(x_T-\bar{x}_T)^2+(\bar{q}_{1T}-\beta_1(T,.))\tfrac{1}{2}\bar{x}_T^2 \\ &+E\int_0^T[\dot{\alpha}_1+q_1+2\alpha_1b_1+\sum_{s'}[\alpha_1(t,s')-\alpha_1(t,s)]\tilde{q}_{ss'}]\tfrac{(x-\bar{x})^2}{2} \\ &+r_1\tfrac{(u_1-\bar{u}_1)^2}{2}+\{\varepsilon_{10}(u_0-\bar{u}_0)+\alpha_1b_{21}(x-\bar{x})\}(u_1-\bar{u}_1) \\ &+\tfrac{[\varepsilon_{10}(u_0-\bar{u}_0)+\alpha_1b_{21}(x-\bar{x})]^2}{2r_1} \\ &-\tfrac{[\varepsilon_{10}(u_0-\bar{u}_0)+\alpha_1b_{21}(x-\bar{x})]^2}{2r_1}+\alpha_1b_{20}(u_0-\bar{u}_0)(x-\bar{x}) \\ &+[\dot{\beta}_1+\bar{q}_1+2\beta_1\bar{b}_1+\sum_{s'}[\beta_1(t,s')-\beta_1(t,s)]\tilde{q}_{ss'}]\tfrac{1}{2}\bar{x}^2 \\ &+\bar{r}_1\tfrac{\bar{u}_1^2}{2}+\{\bar{\varepsilon}_{10}\bar{u}_0+\beta_1\bar{b}_{21}\bar{x}\}\bar{u}_1+\tfrac{[\bar{\varepsilon}_{10}\bar{u}_0+\beta_1\bar{b}_{21}\bar{x}]^2}{2\bar{r}_1} \\ &-\tfrac{[\bar{\varepsilon}_{10}\bar{u}_0+\beta_1\bar{b}_{21}\bar{x}]^2}{2\bar{r}_1}+\beta_1\bar{b}_{20}\bar{u}_0\bar{x} \\ &+\tfrac{1}{2}[\sigma^2+\int_\Theta\mu^2\nu(d\theta)+\sigma_{cogv}^2]\alpha_1dt. \end{aligned} \tag{11.6}$$

It can be readily seen that generically, there is a unique response strategy of the follower, and it is given by

$$\text{rnBR}_1(u_0) := \left\{ -\frac{\varepsilon_{10}(u_0 - \bar{u}_0) + \alpha_1 b_{21}(x - \bar{x})}{r_1} - \frac{\bar{\varepsilon}_{10}\bar{u}_0 + \beta_1 \bar{b}_{21}\bar{x}}{\bar{r}_1} \right\},$$

where $r_1, \bar{r}_1$ are assumed to be positive.

The reaction of 1, $\text{rnBR}_1(u_0)$ is a singleton set, and it is given by

$$u_1 = -\frac{\varepsilon_{10}(u_0 - \bar{u}_0) + \alpha_1 b_{21}(x - \bar{x})}{r_1} - \frac{\bar{\varepsilon}_{10}\bar{u}_0 + \beta_1 \bar{b}_{21}\bar{x}}{\bar{r}_1}.$$

We now move to the second stage of the direct method:

$$\begin{aligned}
&L_0(x, u) = q_{0T}\tfrac{(x_T - \bar{x}_T)^2}{2} + \bar{q}_{0T}\tfrac{\bar{x}_T^2}{2} \\
&+ \int_0^T q_0 \tfrac{(x-\bar{x})^2}{2} + [\tfrac{1}{2}r_0 - \tfrac{\varepsilon_{01}\varepsilon_{10}}{r_1}](u_0 - \bar{u}_0)^2 - \tfrac{\varepsilon_{01}b_{21}}{r_1}\alpha_1(x - \bar{x})(u_0 - \bar{u}_0) \\
&+ \bar{q}_0\tfrac{\bar{x}^2}{2} + [\tfrac{1}{2}\bar{r}_0 - \tfrac{\bar{\varepsilon}_{01}\bar{\varepsilon}_{10}}{\bar{r}_1}]\bar{u}_0^2 - \tfrac{\bar{\varepsilon}_{01}\bar{b}_{21}}{\bar{r}_1}\beta_1\bar{u}_0\bar{x}dt, \\
&dx = (b_1 - \tfrac{b_{21}^2}{r_1}\alpha_1)(x - \bar{x}) + (b_{20} - \tfrac{b_{21}\varepsilon_{10}}{r_1})(u_0 - \bar{u}_0) \\
&+ (\bar{b}_1 - \tfrac{\bar{b}_{21}^2}{\bar{r}_1}\beta_1)\bar{x} + (\bar{b}_{20} - \tfrac{\bar{b}_{21}\bar{\varepsilon}_{10}}{\bar{r}_1})\bar{u}_0 dt \\
&+ [\sigma dB + \int_\Theta \mu d\tilde{N} + \sigma_{gv} dB_{gv}].
\end{aligned} \tag{11.7}$$

The difference between the cost and the guess functional of the leader (decision-maker 0) is

$$\begin{aligned}
&E[L_0 - f_0(0)] = \\
&E(q_{0T} - \alpha_0(T, .))\tfrac{1}{2}(x_T - \bar{x}_T)^2 + (\bar{q}_{0T} - \beta_0(T, .))\tfrac{1}{2}\bar{x}_T^2 \\
&+ E\int_0^T + [\dot{\alpha}_0 + 2\alpha_0(b_1 - \tfrac{b_{21}^2}{r_1}\alpha_1) + q_0 + \textstyle\sum_{s'}[\alpha_0(t, s') - \alpha_0(t, s)]\tilde{q}_{ss'}]\tfrac{(x-\bar{x})^2}{2} \\
&+ [\tfrac{1}{2}r_0 - \tfrac{\varepsilon_{01}\varepsilon_{10}}{r_1}](u_0 - \bar{u}_0)^2 + \{-\tfrac{\varepsilon_{01}b_{21}}{r_1}\alpha_1 + \alpha_0(b_{20} - \tfrac{b_{21}\varepsilon_{10}}{r_1})\}(x - \bar{x})(u_0 - \bar{u}_0) \\
&+ \tfrac{1}{2}[\tfrac{1}{2}r_0 - \tfrac{\varepsilon_{01}\varepsilon_{10}}{r_1}]^{-1}\{-\tfrac{\varepsilon_{01}b_{21}}{r_1}\alpha_1 + \alpha_0(b_{20} - \tfrac{b_{21}\varepsilon_{10}}{r_1})\}^2\tfrac{(x-\bar{x})^2}{2} \\
&- \tfrac{1}{2}[\tfrac{1}{2}r_0 - \tfrac{\varepsilon_{01}\varepsilon_{10}}{r_1}]^{-1}\{-\tfrac{\varepsilon_{01}b_{21}}{r_1}\alpha_1 + \alpha_0(b_{20} - \tfrac{b_{21}\varepsilon_{10}}{r_1})\}^2\tfrac{(x-\bar{x})^2}{2} \\
&+ [\dot{\beta}_0 + 2\beta_0(\bar{b}_1 - \tfrac{\bar{b}_{21}^2}{\bar{r}_1}\beta_1) + \bar{q}_0 + \textstyle\sum_{s'}[\beta_0(t, s') - \beta_0(t, s)]\tilde{q}_{ss'}]\tfrac{\bar{x}^2}{2} \\
&+ [\tfrac{1}{2}\bar{r}_0 - \tfrac{\bar{\varepsilon}_{01}\bar{\varepsilon}_{10}}{\bar{r}_1}]\bar{u}_0^2 \\
&+ \{-\tfrac{\bar{\varepsilon}_{01}\bar{b}_{21}}{\bar{r}_1}\beta_1 + \beta_0(\bar{b}_{20} - \tfrac{\bar{b}_{21}\bar{\varepsilon}_{10}}{\bar{r}_1})\}\bar{u}_0\bar{x} \\
&+ \tfrac{1}{2}[\tfrac{1}{2}\bar{r}_0 - \tfrac{\bar{\varepsilon}_{01}\bar{\varepsilon}_{10}}{\bar{r}_1}]^{-1}\{-\tfrac{\bar{\varepsilon}_{01}\bar{b}_{21}}{\bar{r}_L}\beta_1 + \beta_0(\bar{b}_{20} - \tfrac{\bar{b}_{21}\bar{\varepsilon}_{10}}{\bar{r}_1})\}^2\tfrac{\bar{x}^2}{2} \\
&- \tfrac{1}{2}[\tfrac{1}{2}\bar{r}_0 - \tfrac{\bar{\varepsilon}_{01}\bar{\varepsilon}_{10}}{\bar{r}_1}]^{-1}\{-\tfrac{\bar{\varepsilon}_{01}\bar{b}_{21}}{\bar{r}_1}\beta_1 + \beta_0(\bar{b}_{20} - \tfrac{\bar{b}_{21}\bar{\varepsilon}_{10}}{\bar{r}_1})\}^2\tfrac{\bar{x}^2}{2} \\
&+ \tfrac{1}{2}[\sigma^2 + \int_\Theta \mu^2\nu(d\theta) + \sigma_{cogv}^2]\alpha_0 dt.
\end{aligned} \tag{11.8}$$

By minimization and matching the terms, the Stackelberg optimal state-and-mean-field feedback strategies are

$$
\begin{aligned}
u_0^{ss} &= -\tfrac{1}{2}[\tfrac{1}{2}r_0 - \tfrac{\varepsilon_{01}\varepsilon_{10}}{r_1}]^{-1}\{-\tfrac{\varepsilon_{01}b_{21}}{r_1}\alpha_1 + \alpha_0(b_{20} - \tfrac{b_{21}\varepsilon_{10}}{r_1})\}(x-\bar{x})\\
&-\tfrac{1}{2}[\tfrac{1}{2}\bar{r}_0 - \tfrac{\bar{\varepsilon}_{01}\bar{\varepsilon}_{10}}{\bar{r}_1}]^{-1}\{-\tfrac{\bar{\varepsilon}_{01}\bar{b}_{21}}{\bar{r}_1}\beta_1 + \beta_0(\bar{b}_{20} - \tfrac{\bar{b}_{21}\bar{\varepsilon}_{10}}{\bar{r}_1})\}\bar{x},\\
u_1^{ss} &= -\tfrac{\varepsilon_{10}(u_0-\bar{u}_0)+\alpha_1 b_{21}(x-\bar{x})}{r_1} - \tfrac{\bar{\varepsilon}_{10}\bar{u}_0+\beta_1\bar{b}_{21}\bar{x}}{\bar{r}_1}\\
&= [\tfrac{\varepsilon_{10}}{r_1}\tfrac{1}{2}[\tfrac{1}{2}r_0 - \tfrac{\varepsilon_{01}\varepsilon_{10}}{r_1}]^{-1}\{-\tfrac{\varepsilon_{01}b_{21}}{r_1}\alpha_1 + \alpha_0(b_{20} - \tfrac{b_{21}\varepsilon_{10}}{r_1})\} - \tfrac{b_{21}}{r_1}\alpha_1](x-\bar{x})\\
&+[\tfrac{\bar{\varepsilon}_{10}}{\bar{r}_1}\tfrac{1}{2}[\tfrac{1}{2}\bar{r}_0 - \tfrac{\bar{\varepsilon}_{01}\bar{\varepsilon}_{10}}{\bar{r}_1}]^{-1}\{-\tfrac{\bar{\varepsilon}_{01}\bar{b}_{21}}{\bar{r}_1}\beta_1 + \beta_0(\bar{b}_{20} - \tfrac{\bar{b}_{21}\bar{\varepsilon}_{10}}{\bar{r}_1})\} - \beta_1\tfrac{\bar{b}_{21}}{\bar{r}_1}]\bar{x},\\
E[L_0(x,u^{ss})] &= E\alpha_0(0,s_0))\tfrac{1}{2}(x_0-\bar{x}_0)^2 + \beta_0(0,s_0))\tfrac{1}{2}\bar{x}_0^2\\
&+\tfrac{1}{2}E\int_0^T[\sigma^2 + \int_\Theta \mu^2\nu(d\theta) + \sigma_{cogv}^2]\alpha_0 dt\\
E[L_1(x,u^{ss})] &= E\alpha_1(0,s_0))\tfrac{1}{2}(x_0-\bar{x}_0)^2 + \beta_1(0,s_0))\tfrac{1}{2}\bar{x}_0^2\\
&+\tfrac{1}{2}E\int_0^T[\sigma^2 + \int_\Theta \mu^2\nu(d\theta) + \sigma_{cogv}^2]\alpha_1 dt.
\end{aligned}
\tag{11.9}
$$

11.2.1.1 Effect of the Cross Terms

The cross terms $\varepsilon_{ij}u_iu_j$ and $\varepsilon_{ij}\bar{u}_i\bar{u}_j$ significantly affect the structure of the Stackelberg solution in feedback strategies. We show that there is no (admissible) feedback Nash equilibrium when the determinant of one of the matrices:

$$
\begin{pmatrix} r_0 & \varepsilon_{01}\\ \varepsilon_{10} & r_1\end{pmatrix}, \quad \begin{pmatrix} \bar{r}_0 & \bar{\varepsilon}_{01}\\ \bar{\varepsilon}_{10} & \bar{r}_1\end{pmatrix},
$$

is zero: $r_1r_0 = \varepsilon_{10}\varepsilon_{01}$ or $\bar{r}_1\bar{r}_0 = \bar{\varepsilon}_{10}\bar{\varepsilon}_{01}$. To see this, the running cost functions can be written as

$$
\begin{aligned}
\text{Decision-maker 1: } & \tfrac{1}{2}\bar{r}_1\bar{u}_1^2 + [\bar{\varepsilon}_{10}\bar{u}_0 + \beta_1\bar{b}_{21}\bar{x}]\bar{u}_1\\
&+\tfrac{1}{2}r_1(u_1-\bar{u}_1)^2 + [\varepsilon_{10}(u_0-\bar{u}_0) + \alpha_1 b_{21}(x-\bar{x})](u_1-\bar{u}_1),\\
\text{Decision-maker 0: } & \tfrac{1}{2}\bar{r}_0\bar{u}_0^2 + [\bar{\varepsilon}_{01}\bar{u}_1 + \beta_0\bar{b}_{20}\bar{x}]\bar{u}_0\\
&+\tfrac{1}{2}r_0(u_0-\bar{u}_0)^2 + [\varepsilon_{01}(u_1-\bar{u}_1) + \alpha_0 b_{20}(x-\bar{x})](u_0-\bar{u}_0).
\end{aligned}
\tag{11.10}
$$

The minimization will involve the matrices:

$$
\begin{pmatrix} \bar{r}_1 & \bar{\varepsilon}_{10}\\ \bar{\varepsilon}_{01} & \bar{r}_0\end{pmatrix}\begin{pmatrix}\bar{u}_1\\ \bar{u}_0\end{pmatrix} = -\begin{pmatrix}\beta_1\bar{b}_{21}\\ \beta_0\bar{b}_{20}\end{pmatrix}\bar{x}, \tag{11.11}
$$

$$
\begin{pmatrix} r_1 & \varepsilon_{10}\\ \varepsilon_{01} & r_0\end{pmatrix}\begin{pmatrix}u_1-\bar{u}_1\\ u_0-\bar{u}_0\end{pmatrix} = -\begin{pmatrix}\alpha_1 b_{21}\\ \alpha_0 b_{20}\end{pmatrix}(x-\bar{x}). \tag{11.12}
$$

Sufficient conditions for linear-state-and-mean-field-type feedback Nash equilibrium are

$$\begin{array}{l} r_0(t,s) > \delta > 0,\ r_1(t,s) > \delta > 0,\ q_0(t,s) \geq 0,\ q_1(t,s) \geq 0 \\ \bar{r}_0(t,s) > \delta > 0,\ \bar{r}_1(t,s) > \delta > 0,\ \bar{q}_0(t,s) \geq 0,\ \bar{q}_1(t,s) \geq 0, \\ r_0(t,s)r_1(t,s) - \varepsilon_{01}(t,s)\varepsilon_{10}(t,s) > \delta > 0, \\ \bar{r}_0(t,s)\bar{r}_1(t,s) - \bar{\varepsilon}_{01}(t,s)\bar{\varepsilon}_{10}(t,s) > \delta > 0. \end{array} \tag{11.13}$$

However, the linear-state-and-mean-field-type feedback Stackelberg solution conditions are slightly different:

$$\begin{array}{l} r_0(t,s) > \delta > 0,\ r_1(t,s) > \delta > 0,\ q_0(t,s) \geq 0,\ q_1(t,s) \geq 0 \\ \bar{r}_0(t,s) > \delta > 0,\ \bar{r}_1(t,s) > \delta > 0,\ \bar{q}_0(t,s) \geq 0,\ \bar{q}_1(t,s) \geq 0 \\ r_0(t,s)r_1(t,s) - 2\varepsilon_{01}(t,s)\varepsilon_{10}(t,s) > \delta > 0, \\ \bar{r}_0(t,s)\bar{r}_1(t,s) - 2\bar{\varepsilon}_{01}(t,s)\bar{\varepsilon}_{10}(t,s) > \delta > 0, \end{array} \tag{11.14}$$

when $\varepsilon_{01}\varepsilon_{10} = 0 = \bar{\varepsilon}_{01}\bar{\varepsilon}_{10}$ the two conditions (11.13) and (11.14) coincide.

11.2.2 Risk-Sensitive Stackelberg Solution

We provide basic definitions of the risk-sensitive hierarchical problems and their solution concepts.

Definition 11.3 (Mean-Field-Type Risk-Sensitive Response to the Leader) Given $u_0 \in \mathscr{U}_0$, a risk-sensitive best-response strategy of the follower is a strategy that solves $\inf_{\mathscr{U}_1} \frac{1}{\lambda_1} \log E[e^{\lambda_1 L_1(x,u)}]$. The set of risk-sensitive best responses of the follower is denoted by $\text{rsBR}_1(u_0)$.

Definition 11.4 (Mean-Field-Type Risk-Sensitive Stackelberg Solution) A mean-field-type risk-sensitive Stackelberg solution strategy $u_0^{\text{rs}}, u_1^{\text{rs}}$ such that for the leader

$$u_0^{\text{rs}} \in \arg\min_{u_0 \in \mathscr{U}_0} \left\{ \frac{1}{\lambda_0} \log E[e^{\lambda_0 L_0(x,u_0,u_1^{\text{rs}})}] :\ u_1^{\text{rs}} \in \text{rsBR}_1(u_0) \right\},$$

and, for the follower, $u_1^{\text{rs}} \in \text{rnBR}_1(u_0^{\text{rs}})$.

The mean-field-type risk-sensitive Stackelberg problem is formulated as

$$\begin{array}{l} \inf_{u_0 \in \mathscr{U}_0} \frac{1}{\lambda_0} \log E[e^{\lambda_0 L_0(x,u)}], \\ \text{subject to} \\ dx = \left\{ b_1(x - \bar{x}) + \sum_{l \in \mathscr{I}} b_{2l}(u_l - \bar{u}_l) + \bar{b}_1 \bar{x} + \sum_{l \in \mathscr{I}} \bar{b}_{2l} \bar{u}_l \right\} dt \\ +[\sigma dB + \sigma_{gv} dB_{gv}], \\ u_1 \in \text{rnBR}_1^{rs}(u_0) := \arg\min_{v_1 \in \mathscr{U}_1} \frac{1}{\lambda_1} \log E[e^{\lambda_1 L_1(x,u_0,v_1)}], \end{array} \tag{11.15}$$

where only a single regime is considered.

In the risk-sensitive case, the noise term in the cost functional will be completed by

$$\begin{aligned}
&\int_0^T \lambda_0\alpha_0(x-\bar{x})[\sigma dB + \sigma_{gv} dB_{gv}] \\
&= \lambda_0 \int_0^T \alpha_0(x-\bar{x})[\sigma dB + \sigma_{gv} dB_{gv}] - \tfrac{1}{2}\int_0^T \lambda_0^2 (x-\bar{x})^2[\sigma^2\alpha_0^2 + \tilde{\sigma}^2_{0,cogv}]dt \\
&+\lambda_0 \int_0^T \lambda_0[\sigma^2\alpha_0^2 + \tilde{\sigma}^2_{0,cogv}].\tfrac{1}{2}(x-\bar{x})^2 dt,
\end{aligned} \tag{11.16}$$

where

$$\tilde{\sigma}^2_{0,cogv} = \frac{d}{dt}\left[\int_0^t \left\{ K(t'_+, t')\alpha_0\sigma_{gv}(t') + \int_{t'}^t \alpha_0\sigma_{gv}(t'')K(t',t'')dt'' \right\}^2 dt' \right]. \tag{11.17}$$

Hence, the α system of Eq. (11.5) will be modified as follows:

$$\begin{aligned}
&0 = \dot{\alpha}_0 + 2\alpha_0(b_1 - \tfrac{b_{21}^2}{r_1}\alpha_1) + q_0 \\
&-\tfrac{1}{2}[\tfrac{1}{2}r_0 - \tfrac{\varepsilon_{01}\varepsilon_{10}}{r_1}]^{-1}\{-\tfrac{\varepsilon_{01}b_{21}}{r_1}\alpha_1 + \alpha_0(b_{20} - \tfrac{b_{21}\varepsilon_{10}}{r_1})\}^2 \\
&+\lambda_0[\sigma^2\alpha_0^2 + \tilde{\sigma}^2_{0,cogv}], \\
&0 = \dot{\alpha}_1 + q_1 + 2\alpha_1 b_1 \\
&-\tfrac{1}{r_1}[b_{21}\alpha_1 - \tfrac{\varepsilon_{10}}{2}[\tfrac{1}{2}r_0 - \tfrac{\varepsilon_{01}\varepsilon_{10}}{r_1}]^{-1}\{-\tfrac{\varepsilon_{01}b_{21}}{r_1}\alpha_1 + \alpha_0(b_{20} - \tfrac{b_{21}\varepsilon_{10}}{r_1})\}]^2 \\
&-\alpha_1 b_{20}[\tfrac{1}{2}r_0 - \tfrac{\varepsilon_{01}\varepsilon_{10}}{r_1}]^{-1}\{-\tfrac{\varepsilon_{01}b_{21}}{r_1}\alpha_1 + \alpha_0(b_{20} - \tfrac{b_{21}\varepsilon_{10}}{r_1})\} \\
&+\lambda_1[\sigma^2\alpha_1^2 + \tilde{\sigma}^2_{1,cogv}].
\end{aligned} \tag{11.18}$$

The risk-sensitive equilibrium state x will be different since the coefficients α_0, α_1 are different. As the risk-sensitive index vector (λ_0, λ_1) vanishes, we retrieve the risk-neutral Stackelberg solution above.

11.2.3 Stackelberg MFTG: LQ with State-Dependent Noise

We consider a state-dependent noise in the diffusion, jump, and Gauss-Volterra processes:

$$\begin{aligned}
&L_i(x,u) = q_{iT}\frac{(x_T-\bar{x}_T)^2}{2} + \bar{q}_{iT}\frac{\bar{x}_T^2}{2} \\
&+\int_0^T q_i\frac{(x-\bar{x})^2}{2} + r_i\frac{(u_i-\bar{u}_i)^2}{2} + \varepsilon_{ij}(u_i-\bar{u}_i)(u_j-\bar{u}_j) + \bar{q}_i\frac{\bar{x}^2}{2} + \bar{r}_i\frac{\bar{u}_i^2}{2} + \bar{\varepsilon}_{ij}\bar{u}_i\bar{u}_j dt, \\
&\inf_{u_0\in\mathcal{U}_0} E[L_0(x,u)], \\
&\text{subject to} \\
&dx = [b_1(x-\bar{x}) + \textstyle\sum_{l\in\mathcal{I}} b_{2l}(u_l-\bar{u}_l) + \bar{b}_1\bar{x} + \sum_{l\in\mathcal{I}} \bar{b}_{2l}\bar{u}_l]dt \\
&+(x-\bar{x})[\sigma dB + \textstyle\int_\Theta \mu d\tilde{N} + \sigma_{gv}dB_{gv}], \\
&\mathbb{P}(s(t+\delta) = s'|s,u) = \textstyle\int_t^{t+\delta} \tilde{q}_{ss'}dt' + o(\delta),\ s' \neq s, \\
&x(0) = x_0,\ s(0) = s_0, \\
&u_1 \in \text{rnBR}_1(u_0) = \arg\min_{v_1\in\mathcal{U}_1} L_1(x,u_0,v_1).
\end{aligned} \tag{11.19}$$

Following the same methodology as earlier, we obtain the optimal linear Stackelberg strategies and costs:

$$\begin{aligned}
u_0^{ss} &= -\frac{1}{2}[\frac{1}{2}r_0 - \frac{\varepsilon_{01}\varepsilon_{10}}{r_1}]^{-1}\{-\frac{\varepsilon_{01}b_{21}}{r_1}\alpha_1 + \alpha_0(b_{20} - \frac{b_{21}\varepsilon_{10}}{r_1})\}(x-\bar{x}) \\
&-\frac{1}{2}[\frac{1}{2}\bar{r}_0 - \frac{\bar{\varepsilon}_{01}\bar{\varepsilon}_{10}}{\bar{r}_1}]^{-1}\{-\frac{\bar{\varepsilon}_{01}\bar{b}_{21}}{\bar{r}_1}\beta_1 + \beta_0(\bar{b}_{20} - \frac{\bar{b}_{21}\bar{\varepsilon}_{10}}{\bar{r}_1})\}\bar{x}, \\
u_1^{ss} &= (\frac{\varepsilon_{10}}{r_1}\frac{1}{2}\frac{(-\frac{\varepsilon_{01}b_{21}}{r_1}\alpha_1+\alpha_0(b_{20}-\frac{b_{21}\varepsilon_{10}}{r_1}))}{(\frac{1}{2}r_0-\frac{\varepsilon_{01}\varepsilon_{10}}{r_1})} - \frac{b_{21}}{r_1}\alpha_1)(x-\bar{x}) \\
&+[\frac{\bar{\varepsilon}_{10}}{\bar{r}_1}\frac{1}{2}\frac{\{-\frac{\bar{\varepsilon}_{01}\bar{b}_{21}}{\bar{r}_1}\beta_1+\beta_0(\bar{b}_{20}-\frac{\bar{b}_{21}\bar{\varepsilon}_{10}}{\bar{r}_1})\}}{\frac{1}{2}\bar{r}_0-\frac{\bar{\varepsilon}_{01}\bar{\varepsilon}_{10}}{\bar{r}_1}} - \beta_1\frac{\bar{b}_{21}}{\bar{r}_1}]\bar{x}, \\
E[L_0(x,u^{ss})] &= E\alpha_0(0,s_0))\frac{1}{2}(x_0-\bar{x}_0)^2 + \beta_0(0,s_0))\frac{1}{2}\bar{x}_0^2, \\
E[L_1(x,u^{ss})] &= E\alpha_1(0,s_0))\frac{1}{2}(x_0-\bar{x}_0)^2 + \beta_1(0,s_0))\frac{1}{2}\bar{x}_0^2,
\end{aligned} \tag{11.20}$$

where the coefficients are solutions of the system:

$$\begin{aligned}
0 &= \dot{\alpha}_0 + 2\alpha_0(b_1 - \frac{b_{21}^2}{r_1}\alpha_1) + q_0 + \sum_{s'}[\alpha_0(t,s') - \alpha_0(t,s)]\tilde{q}_{ss'} \\
&-\frac{1}{2}[\frac{1}{2}r_0 - \frac{\varepsilon_{01}\varepsilon_{10}}{r_1}]^{-1}\{-\frac{\varepsilon_{01}b_{21}}{r_1}\alpha_1 + \alpha_0(b_{20} - \frac{b_{21}\varepsilon_{10}}{r_1})\}^2 \\
&+[\sigma^2 + \int_\Theta \mu^2\nu(d\theta) + \sigma_{cogv}^2]\alpha_0, \\
0 &= \dot{\beta}_0 + 2\beta_0(\bar{b}_1 - \frac{\bar{b}_{21}^2}{\bar{r}_1}\beta_1) + \bar{q}_0 + \sum_{s'}[\beta_0(t,s') - \beta_0(t,s)]\tilde{q}_{ss'} \\
&-\frac{1}{2}[\frac{1}{2}\bar{r}_0 - \frac{\bar{\varepsilon}_{01}\bar{\varepsilon}_{10}}{\bar{r}_1}]^{-1}\{-\frac{\bar{\varepsilon}_{01}\bar{b}_{21}}{\bar{r}_1}\beta_1 + \beta_0(\bar{b}_{20} - \frac{\bar{b}_{21}\bar{\varepsilon}_{10}}{\bar{r}_1})\}^2, \\
0 &= \dot{\alpha}_1 + q_1 + 2\alpha_1 b_1 + \sum_{s'}[\alpha_1(t,s') - \alpha_1(t,s)]\tilde{q}_{ss'} \\
&-\frac{1}{r_1}[b_{21}\alpha_1 - \frac{\varepsilon_{10}}{2}[\frac{1}{2}r_0 - \frac{\varepsilon_{01}\varepsilon_{10}}{r_1}]^{-1}\{-\frac{\varepsilon_{01}b_{21}}{r_1}\alpha_1 + \alpha_0(b_{20} - \frac{b_{21}\varepsilon_{10}}{r_1})\}]^2 \\
&-\alpha_1 b_{20}[\frac{1}{2}r_0 - \frac{\varepsilon_{01}\varepsilon_{10}}{r_1}]^{-1}\{-\frac{\varepsilon_{01}b_{21}}{r_1}\alpha_1 + \alpha_0(b_{20} - \frac{b_{21}\varepsilon_{10}}{r_1})\} \\
&+[\sigma^2 + \int_\Theta \mu^2\nu(d\theta) + \sigma_{cogv}^2]\alpha_1, \\
0 &= [\dot{\beta}_1 + \bar{q}_1 + 2\beta_1\bar{b}_1 + \sum_{s'}[\beta_1(t,s') - \beta_1(t,s)]\tilde{q}_{ss'}] \\
&-\frac{1}{\bar{r}_1}[\beta_1\bar{b}_{21} - \frac{\bar{\varepsilon}_{10}}{2}[\frac{1}{2}\bar{r}_0 - \frac{\bar{\varepsilon}_{01}\bar{\varepsilon}_{10}}{\bar{r}_1}]^{-1}\{-\frac{\bar{\varepsilon}_{01}\bar{b}_{21}}{\bar{r}_1}\beta_1 + \beta_0(\bar{b}_{20} - \frac{\bar{b}_{21}\bar{\varepsilon}_{10}}{\bar{r}_1})\}]^2 \\
&-\beta_1\bar{b}_{20}[\frac{1}{2}\bar{r}_0 - \frac{\bar{\varepsilon}_{01}\bar{\varepsilon}_{10}}{\bar{r}_1}]^{-1}\{-\frac{\bar{\varepsilon}_{01}\bar{b}_{21}}{\bar{r}_1}\beta_1 + \beta_0(\bar{b}_{20} - \frac{\bar{b}_{21}\bar{\varepsilon}_{10}}{\bar{r}_1})\}.
\end{aligned} \tag{11.21}$$

Note that the structures of the coefficients and the costs in (11.21) have changed compared with (11.5). Thus, the state-dependent noise solution differs from the state-independent noise Stackelberg solution of MFTG.

11.3 Single Leader-Multiple Followers with Non-Quadratic Costs

This section examines a class of mean-field-type games with one leader and multiple followers. The running costs are not necessarily quadratic, and the state dynamics are driven by Gauss-Volterra, jump, and regime-switching processes. The cost functionals are

$$
\begin{array}{l}
L_0(x,u) = q_{0T}\frac{(x_T-\bar{x}_T)^{2k_0}}{2k_0} + \bar{q}_{0T}\frac{\bar{x}_T^{2\bar{k}_0}}{2\bar{k}_0}\\
+\int_0^T q_0\frac{(x-\bar{x})^{2k_0}}{2k_i} + r_0\frac{(u_0-\bar{u}_0)^{2k_0}}{2k_0} + \sum_{j\neq 0}\varepsilon_{0j}(x-\bar{x})^{2(k_0-1)}(u_0-\bar{u}_0)(u_j-\bar{u}_j)\\
+\bar{q}_0\frac{\bar{x}^{2\bar{k}_0}}{2\bar{k}_0} + \bar{r}_0\frac{\bar{u}_0^{2\bar{k}_0}}{2\bar{k}_0} + \sum_{j\neq 0}\bar{\varepsilon}_{0j}\bar{x}^{2(\bar{k}_0-1)}\bar{u}_0\bar{u}_j dt,\\
\text{for } i\in\mathscr{I}\backslash\{0\},\\
L_i(x,u) = q_{iT}\frac{(x_T-\bar{x}_T)^{2k_i}}{2k_i} + \bar{q}_{iT}\frac{\bar{x}_T^{2\bar{k}_i}}{2\bar{k}_i}\\
+\int_0^T q_i\frac{(x-\bar{x})^{2k_i}}{2k_i} + r_i\frac{(u_i-\bar{u}_i)^{2k_i}}{2k_i} + \varepsilon_{i0}(x-\bar{x})^{2(k_i-1)}(u_i-\bar{u}_i)(u_0-\bar{u}_0)\\
+\bar{q}_i\frac{\bar{x}^{2\bar{k}_i}}{2\bar{k}_i} + \bar{r}_i\frac{\bar{u}_i^{2\bar{k}_i}}{2\bar{k}_i} + \bar{\varepsilon}_{i0}\bar{x}^{2(\bar{k}_i-1)}\bar{u}_i\bar{u}_0 dt,
\end{array}
\tag{11.22}
$$

where $k_i \geq 1$, $\bar{k}_i \geq 1$ are natural numbers and the coefficients are time- and switching-dependent.

Definition 11.5 (Mean-Field-Type Risk-Neutral Response to the Leader) Given $u_0 \in \mathscr{U}_0$, and $(u_j)_{j\in\mathscr{I}\backslash\{0,i\}}$, a risk-neutral best-response strategy of follower i is a strategy that solves $\inf_{\mathscr{U}_i} EL_i$. The set of risk-neutral best responses of follower i is denoted by $\mathrm{rnBR}_i(u_0; (u_j)_{j\in\mathscr{I}\backslash\{0,i\}})$.

Definition 11.6 (Mean-Field-Type Risk-Neutral Nash Equilibrium Between Followers) A mean-field-type risk-neutral Nash equilibrium between followers given the leader strategy u_0 is a strategy profile $(u_i^{\mathrm{rn}},\ i\in\mathscr{I}\backslash\{0\})$ of all followers such that for each decision-maker $i\in\mathscr{I}\backslash\{0\}$, $u_i^{\mathrm{rn}} \in \mathrm{rnBR}_i(u_0; (u_j^{\mathrm{rn}})_{j\in\mathscr{I}\backslash\{0,i\}})$.

Definition 11.7 (Mean-Field-Type Risk-Neutral Stackelberg Solution) A mean-field-type risk-neutral Stackelberg solution between the leader and multiple followers is a strategy u_0, $(u_i^{\mathrm{rn}},\ i\in\mathscr{I}\backslash\{0\})$, of all decision-makers such that for the leader

$$
\begin{array}{l}
u_0^{\mathrm{rn}} \in \arg\min_{u_0\in\mathscr{U}_0}\{EL_0(x,u_0,(u_i^{\mathrm{rn}})_{i\in\mathscr{I}\backslash\{0\}}) : u_i^{\mathrm{rn}} \in \mathrm{rnBR}_i(u_0; (u_j^{\mathrm{rn}})_{j\in\mathscr{I}\backslash\{0,i\}}),\\
i\in\mathscr{I}\backslash\{0\},
\end{array}
$$

and for every follower $i\in\mathscr{I}\backslash\{0\}$, $u_i^{\mathrm{rn}} \in \mathrm{rnBR}_i(u_0; (u_j^{\mathrm{rn}})_{j\in\mathscr{I}\backslash\{0,i\}})$.

$$
\begin{array}{l}
\inf_{u_0} E[L_0(x,u)],\\
\text{subject to}\\
dx = [b_1(x-\bar{x}) + \sum_{j\in\mathscr{I}} b_{2j}(u_j-\bar{u}_j) + \bar{b}_1\bar{x} + \sum_j \bar{b}_{2j}\bar{u}_j]dt\\
+(x-\bar{x})[\sigma dB + \int_\Theta \mu d\tilde{N} + \sigma_{gv} dB_{gv}],\\
\mathbb{P}(s(t+\delta)=s'|s,u) = \int_t^{t+\delta}\tilde{q}_{ss'}dt' + o(\delta),\ s'\neq s\\
x(0)=x_0,\ s(0)=s_0,\\
u_i \in \mathrm{rnBR}_i(.|u_0) = \arg\min_{v_i\in\mathscr{U}_i} L_i(x,u_0,u_1,\ldots,u_{i-1},v_i,u_{i+1},\ldots,u_I),
\end{array}
\tag{11.23}
$$

where L_0 and L_i are as given by (11.22).

The next proposition presents a Stackelberg solution with multiple followers.

Proposition 11.2 *Consider problem (11.23) with a single leader and multiple followers. The Stackelberg strategy and the cost of the leader are*

$$
\begin{aligned}
&u_0^{rn} = -\eta_0(x-\bar{x}) - \bar{\eta}_0\bar{x},\\
&EL_0 := E\alpha_0(0,.)\tfrac{(x_0-\bar{x}_0)^{2k_0}}{2k_0} + \beta_0(0,.)\tfrac{\bar{x}_0^{2\bar{k}_0}}{2\bar{k}_0},\\
&d\bar{x} = \left[\bar{b}_1 - \bar{b}_{20}\bar{\eta}_0 + \textstyle\sum_{j\neq 0}\bar{b}_{2j}\left(-\tfrac{[-\bar{\varepsilon}_{j0}\bar{\eta}_0+\beta_j\bar{b}_{2j}]}{\bar{r}_j}\right)^{\frac{1}{2\bar{k}_j-1}}\right]\bar{x}dt\\
&0 = [\dot{\alpha}_0 + 2k_0\alpha_0 b_1 + q_0 + 2k_0\Omega_0 + \textstyle\sum_{s'\neq s}(\alpha_0(t,s')-\alpha_0(t,s))\tilde{q}_{ss'}]\\
&+(2k_0-1)k_0(\sigma^2+\sigma^2_{cogv})\alpha_0. + \textstyle\int_\Theta \nu(d\theta)\{(1+\mu)^{2k_0}-1-2k_0\mu\}\alpha_0,\\
&0 = [\dot{\beta}_0 + 2\bar{k}_0\beta_0\bar{b}_1 + \bar{q}_0 + 2\bar{k}_0\bar{\Omega}_0 + \textstyle\sum_{s'\neq s}[\beta_0(t,s')-\beta_0(t,s)]\tilde{q}_{ss'})]\tfrac{\bar{x}^{2\bar{k}_0}}{2\bar{k}_0},\\
&\Omega_0 := r_0\tfrac{\eta_0^{2k_0}}{2k_0} - \textstyle\sum_{j\neq 0}\varepsilon_{0j}\left(-\tfrac{[-\varepsilon_{j0}\eta_0+\alpha_j b_{2j}]}{r_j}\right)^{\frac{1}{2k_j-1}}\eta_0\\
&+\alpha_0\left[-b_{20}\eta_0 + \textstyle\sum_{j\neq 0}b_{2j}\left(-\tfrac{[-\varepsilon_{j0}\eta_0+\alpha_j b_{2j}]}{r_j}\right)^{\frac{1}{2k_j-1}}\right],\\
&r_0\eta_0^{2k_0-1} = \textstyle\sum_{j\neq 0}\varepsilon_{0j}\left(-\tfrac{[-\varepsilon_{j0}\eta_0+\alpha_j b_{2j}]}{r_j}\right)^{\frac{1}{2k_j-1}}\\
&+\textstyle\sum_{j\neq 0}(\tfrac{\varepsilon_{j0}}{r_j}\tfrac{1}{2k_j-1})\varepsilon_{0j}\left(-\tfrac{[-\varepsilon_{j0}\eta_0+\alpha_j b_{2j}]}{r_j}\right)^{\frac{1}{2k_j-1}-1}\\
&+\alpha_0\left[b_{20} - \textstyle\sum_{j\neq 0}(\tfrac{\varepsilon_{j0}}{r_j}\tfrac{1}{2k_j-1})b_{2j}\left(-\tfrac{[-\varepsilon_{j0}\eta_0+\alpha_j b_{2j}]}{r_j}\right)^{\frac{1}{2k_j-1}-1}\right],\\
&\bar{\Omega}_0 := \bar{r}_0\tfrac{\bar{\eta}_0^{2\bar{k}_0}}{2\bar{k}_0} - \textstyle\sum_{j\neq 0}\bar{\varepsilon}_{0j}\left(-\tfrac{[-\bar{\varepsilon}_{j0}\bar{\eta}_0+\beta_j\bar{b}_{2j}]}{\bar{r}_j}\right)^{\frac{1}{2\bar{k}_j-1}}\bar{\eta}_0\\
&+\beta_0\left[-\bar{b}_{20}\bar{\eta}_0 + \textstyle\sum_{j\neq 0}\bar{b}_{2j}\left(-\tfrac{[-\bar{\varepsilon}_{j0}\bar{\eta}_0+\beta_j\bar{b}_{2j}]}{\bar{r}_j}\right)^{\frac{1}{2\bar{k}_j-1}}\right],\\
&\bar{r}_0\bar{\eta}_0^{2\bar{k}_0-1} = \textstyle\sum_{j\neq 0}\bar{\varepsilon}_{0j}\left(-\tfrac{[-\bar{\varepsilon}_{j0}\bar{\eta}_0+\beta_j\bar{b}_{2j}]}{\bar{r}_j}\right)^{\frac{1}{2\bar{k}_j-1}}.\\
&+\textstyle\sum_{j\neq 0}\left(\tfrac{\bar{\varepsilon}_{j0}}{\bar{r}_j}\tfrac{1}{2\bar{k}_j-1}\right)\bar{\varepsilon}_{0j}\left(-\tfrac{[-\bar{\varepsilon}_{j0}\bar{\eta}_0+\beta_j\bar{b}_{2j}]}{\bar{r}_j}\right)^{\frac{1}{2\bar{k}_j-1}-1}\\
&+\beta_0\left[\bar{b}_{20} - \textstyle\sum_{j\neq 0}\left(\tfrac{\bar{\varepsilon}_{j0}}{\bar{r}_j}\tfrac{1}{2\bar{k}_j-1}\right)\bar{b}_{2j}\left(-\tfrac{[-\bar{\varepsilon}_{j0}\bar{\eta}_0+\beta_j b_{2j}]}{\bar{r}_j}\right)^{\frac{1}{2\bar{k}_j-1}-1}\right].
\end{aligned}
\tag{11.24}
$$

The Stackelberg strategy and the cost of the followers are

$$\begin{aligned}
u_i &:= \left(-\tfrac{[-\varepsilon_{i0}\eta_0+\alpha_i b_{2i}]}{r_i}\right)^{\frac{1}{2k_i-1}}(x-\bar{x}) + \left(-\tfrac{[-\bar{\varepsilon}_{i0}\bar{\eta}_0+\beta_i \bar{b}_{2i}]}{\bar{r}_i}\right)^{\frac{1}{2\bar{k}_i-1}}\bar{x},\\
L_i &:= E\alpha_i(0,.)\tfrac{(x_0-\bar{x}_0)^{2k_i}}{2k_i} + \beta_i(0,.)\tfrac{\bar{x}_0^{2\bar{k}_i}}{2\bar{k}_i},\\
0 &= (\dot{\alpha}_i + 2k_i\alpha_i b_1 + q_i + 2k_i\Omega_i + \textstyle\sum_{s'\neq s}[\alpha_i(t,s') - \alpha_i(t,s)]\tilde{q}_{ss'})\\
&+(2k_i-1)k_i(\sigma^2+\sigma^2_{cogv})\alpha_i + \textstyle\int_\Theta \nu(d\theta)\{(1+\mu)^{2k_i} - 1 - 2k_i\mu\}\alpha_i,\\
0 &= \dot{\beta}_i + 2\bar{k}_i\beta_i\bar{b}_1 + \bar{q}_i + 2\bar{k}_i\bar{\Omega}_i + \textstyle\sum_{s'\neq s}[\beta_i(t,s') - \beta_i(t,s)]\tilde{q}_{ss'},\\
\Omega_i &:= -r_i\tfrac{2k_i-1}{2k_i}\left\{(-\tfrac{[-\varepsilon_{i0}\eta_0+\alpha_i b_{2i}]}{r_i})^{\frac{2k_i}{2k_i-1}}\right\} - \alpha_i b_{20}\eta_0,\\
\bar{\Omega}_i &:= -\bar{r}_i\tfrac{2\bar{k}_i-1}{2\bar{k}_i}\left\{(-\tfrac{[-\bar{\varepsilon}_{i0}\bar{\eta}_0+\beta_i \bar{b}_{2i}]}{\bar{r}_i})^{\frac{2\bar{k}_i}{2\bar{k}_i-1}}\right\} - \beta_i\bar{b}_{20}\bar{\eta}_0.
\end{aligned} \tag{11.25}$$

Proof We start from the set of followers' responses given the choice u_0 of the leader. Let $f_i = \alpha_i\frac{(x-\bar{x})^{2k_i}}{2k_i} + \beta_i\frac{\bar{x}^{2\bar{k}_i}}{2\bar{k}_i}$ be a guess functional for decision-maker i. The expected value of the difference $L_i - f_i(0)$ can be rewritten as follows:

$$\begin{aligned}
&E[L_i - f_i(0)]\\
&= E[q_{iT} - \alpha_i(T,.)]\tfrac{(x_T-\bar{x}_T)^{2k_i}}{2k_i} + [\bar{q}_{iT} - \beta_i(t,.)]\tfrac{\bar{x}_T^{2\bar{k}_i}}{2\bar{k}_i}\\
&+E\textstyle\int_0^T(\dot{\alpha}_i + q_i + \sum_{s'\neq s}[\alpha_i(t,s') - \alpha_i(t,s)]\tilde{q}_{ss'})\frac{1}{2k_i}(x-\bar{x})^{2k_i}\\
&+2k_i\alpha_i b_1.\tfrac{1}{2k_i}(x-\bar{x})^{2k_i}\\
&+(2k_i-1)k_i(\sigma^2+\sigma^2_{cogv})\alpha_i.\tfrac{1}{2k_i}(x-\bar{x})^{2k_i}\\
&+\textstyle\int_\Theta \nu(d\theta)\{(1+\mu)^{2k_i} - 1 - 2k_i\mu\}\alpha_i.\frac{1}{2k_i}(x-\bar{x})^{2k_i}\\
&r_i\tfrac{(u_i-\bar{u}_i)^{2k_i}}{2k_i} + \varepsilon_{ij}(x-\bar{x})^{2(k_i-1)}(u_i-\bar{u}_i)(u_j-\bar{u}_j)\\
&+\alpha_i\textstyle\sum_{j\in\mathcal{I}} b_{2j}(u_j-\bar{u}_j)(x-\bar{x})^{2k_i-1}\\
&+(\dot{\beta}_i + \bar{q}_i + \textstyle\sum_{s'\neq s}[\beta_i(t,s') - \beta_i(t,s)]\tilde{q}_{ss'}).\frac{1}{2\bar{k}_i}\bar{x}^{2\bar{k}_i}\\
&+2\bar{k}_i\beta_i\bar{b}_1.\tfrac{1}{2\bar{k}_i}\bar{x}^{2\bar{k}_i}\\
&+\bar{r}_i\tfrac{\bar{u}_i^{2\bar{k}_i}}{2\bar{k}_i} + \bar{\varepsilon}_{ij}\bar{x}^{2(\bar{k}_i-1)}\bar{u}_i\bar{u}_j + \beta_i\textstyle\sum_j \bar{b}_{2j}\bar{u}_j\bar{x}^{2\bar{k}_i-1}.
\end{aligned} \tag{11.26}$$

We now complete the following terms:

$$\begin{aligned}
&r_i\tfrac{(u_i-\bar{u}_i)^{2k_i}}{2k_i}\\
&+\varepsilon_{i0}(x-\bar{x})^{2(k_i-1)}(u_i-\bar{u}_i)(u_0-\bar{u}_0)\\
&+\alpha_i\textstyle\sum_{l\in\mathcal{I}} b_{2l}(u_l-\bar{u}_l)(x-\bar{x})^{2k_i-1}\\
&= r_i\left\{\tfrac{(u_i-\bar{u}_i)^{2k_i}}{2k_i} + \tfrac{[\varepsilon_{i0}(u_0-\bar{u}_0)+\alpha_i b_{2i}(x-\bar{x})]}{r_i}(x-\bar{x})^{2(k_i-1)}(u_i-\bar{u}_i)\right\}\\
&+r_i\tfrac{2k_i-1}{2k_i}\left\{(-\tfrac{[\varepsilon_{i0}(u_0-\bar{u}_0)+\alpha_i b_{2i}(x-\bar{x})]}{r_i}(x-\bar{x})^{2(k_i-1)})^{\frac{2k_i}{2k_i-1}}\right\}\\
&-r_i\tfrac{2k_i-1}{2k_i}\left\{(-\tfrac{[\varepsilon_{i0}(u_0-\bar{u}_0)+\alpha_i b_{2i}(x-\bar{x})]}{r_i}(x-\bar{x})^{2(k_i-1)})^{\frac{2k_i}{2k_i-1}}\right\}\\
&+\alpha_i b_{20}(u_0-\bar{u}_0)(x-\bar{x})^{2k_i-1}\\
&+\alpha_i\textstyle\sum_{l\neq\{i,0\}} b_{2l}(u_l-\bar{u}_l)(x-\bar{x})^{2k_i-1}.
\end{aligned} \tag{11.27}$$

It can be readily seen that generically, there is a unique best-response strategy of the followers, and it is given by

$$\mathrm{rnBR}_i(u_0) := \left\{ \left(-\frac{[\varepsilon_{i0}(u_0-\bar{u}_0)(x-\bar{x})^{-1} + \alpha_i b_{2i}}{r_i} \right)^{\frac{1}{2k_i-1}} (x-\bar{x}) \right.$$
$$\left. + \left(-\frac{[\bar{\varepsilon}_{i0}\bar{u}_0\bar{x}^{-1} + \beta_i \bar{b}_{2i}]}{\bar{r}_i} \right)^{\frac{1}{2\bar{k}_i-1}} \bar{x} \right\}, \tag{11.28}$$

where $r_i, \bar{r}_i$ are assumed to be positive.

Note that $\mathrm{rnBR}_i(u_0)$ is a singleton set, and the follower i's difference between the guess and the performance functional can be written as

$$\begin{aligned}
& E[L_i - f_i(0)] \\
& = E[q_{iT} - \alpha_i(T,.)]\tfrac{(x_T-\bar{x}_T)^{2k_i}}{2k_i} + [\bar{q}_{iT} - \beta_i(t,.)]\tfrac{\bar{x}_T^{2\bar{k}_i}}{2\bar{k}_i} \\
& + E\int_0^T (\dot{\alpha}_i + q_i + \textstyle\sum_{s'\neq s}[\alpha_i(t,s') - \alpha_i(t,s)]\tilde{q}_{ss'})\tfrac{1}{2k_i}(x-\bar{x})^{2k_i} \\
& + 2k_i\alpha_i b_1.\tfrac{1}{2k_i}(x-\bar{x})^{2k_i} \\
& + (2k_i-1)k_i(\sigma^2+\sigma_{gv}^2)\alpha_i.\tfrac{1}{2k_i}(x-\bar{x})^{2k_i} \\
& + \int_\Theta \nu(d\theta)\{(1+\mu)^{2k_i} - 1 - 2k_i\mu\}\alpha_i.\tfrac{1}{2k_i}(x-\bar{x})^{2k_i} \\
& - r_i\tfrac{2k_i-1}{2k_i}\left\{ \left(-\tfrac{[\varepsilon_{i0}(u_0-\bar{u}_0)+\alpha_i b_{2i}(x-\bar{x})]}{r_i}(x-\bar{x})^{2(k_i-1)} \right)^{\frac{2k_i}{2k_i-1}} \right\} \\
& + \alpha_i b_{20}(u_0-\bar{u}_0)(x-\bar{x})^{2k_i-1} \\
& + r_i\left\{ \tfrac{(u_i-\bar{u}_i)^{2k_i}}{2k_i} + \tfrac{[\varepsilon_{i0}(u_0-\bar{u}_0)+\alpha_i b_{2i}(x-\bar{x})]}{r_i}(x-\bar{x})^{2(k_i-1)}(u_i-\bar{u}_i) \right\} \\
& + r_i\tfrac{2k_i-1}{2k_i}\left\{ \left(-\tfrac{[\varepsilon_{i0}(u_0-\bar{u}_0)+\alpha_i b_{2i}(x-\bar{x})]}{r_i}(x-\bar{x})^{2(k_i-1)} \right)^{\frac{2k_i}{2k_i-1}} \right\} \\
& + (\dot{\beta}_i + \bar{q}_i + \textstyle\sum_{s'\neq s}[\beta_i(t,s') - \beta_i(t,s)]\tilde{q}_{ss'}).\tfrac{1}{2\bar{k}_i}\bar{x}^{2\bar{k}_i} \\
& + 2\bar{k}_i\beta_i\bar{b}_1.\tfrac{1}{2\bar{k}_i}\bar{x}^{2\bar{k}_i} \\
& - \bar{r}_i\tfrac{2\bar{k}_i-1}{2\bar{k}_i}\left\{ \left(-\tfrac{[\bar{\varepsilon}_{i0}\bar{u}_0+\beta_i\bar{b}_{2i}\bar{x}]}{\bar{r}_i}\bar{x}^{2(\bar{k}_i-1)} \right)^{\frac{2\bar{k}_i}{2\bar{k}_i-1}} \right\} \\
& + \beta_i\bar{b}_{20}\bar{u}_0\bar{x}^{2\bar{k}_i-1} \\
& + \bar{r}_i\left\{ \tfrac{\bar{u}_i^{2\bar{k}_i}}{2\bar{k}_i} + \tfrac{[\bar{\varepsilon}_{i0}\bar{u}_0+\beta_i\bar{b}_{2i}\bar{x}]}{\bar{r}_i}\bar{x}^{2(\bar{k}_i-1)}\bar{u}_i \right\} \\
& + \bar{r}_i\tfrac{2\bar{k}_i-1}{2\bar{k}_i}\left\{ \left(-\tfrac{[\bar{\varepsilon}_{i0}\bar{u}_0+\beta_i\bar{b}_{2i}\bar{x}]}{\bar{r}_i}\bar{x}^{2(\bar{k}_i-1)} \right)^{\frac{2\bar{k}_i}{2\bar{k}_i-1}} . \right\}
\end{aligned} \tag{11.29}$$

We now examine the leader's strategy and cost. By replacing the response of all followers to u_0, the leader problem (11.22) becomes

$$
\begin{aligned}
&L_0 = q_{0T}\frac{(x_T-\bar{x}_T)^{2k_0}}{2k_0} + \bar{q}_{0T}\frac{\bar{x}_T^{2\bar{k}_0}}{2\bar{k}_0}\\
&+\int_0^T q_0\frac{(x-\bar{x})^{2k_0}}{2k_0} + r_0\frac{(u_0-\bar{u}_0)^{2k_0}}{2k_0}\\
&+\sum_{j\neq 0}\varepsilon_{0j}(x-\bar{x})^{2k_0-1}\left(-\frac{[\varepsilon_{j0}(u_0-\bar{u}_0)(x-\bar{x})^{-1}+\alpha_j b_{2j}]}{r_j}\right)^{\frac{1}{2k_j-1}}(u_0-\bar{u}_0)\\
&+\bar{q}_0\frac{\bar{x}^{2\bar{k}_0}}{2\bar{k}_0} + \bar{r}_0\frac{\bar{u}_0^{2\bar{k}_0}}{2\bar{k}_0}\\
&+\sum_{j\neq 0}\bar{\varepsilon}_{0j}\bar{x}^{2\bar{k}_0-1}\bar{u}_0\left(-\frac{[\bar{\varepsilon}_{j0}\bar{u}_0\bar{x}^{-1}+\beta_j\bar{b}_{2j}]}{\bar{r}_j}\right)^{\frac{1}{2\bar{k}_j-1}}dt,\\
&\inf_{u_0} E[L_0(x,u)],\\
&\text{subject to}\\
&dx = b_1(x-\bar{x}) + b_{20}(u_0-\bar{u}_0)\\
&+\sum_{j\neq 0} b_{2j}\left(-\frac{[\varepsilon_{j0}(u_0-\bar{u}_0)(x-\bar{x})^{-1}+\alpha_j b_{2j}]}{r_j}\right)^{\frac{1}{2k_j-1}}(x-\bar{x})\\
&+\bar{b}_1\bar{x} + \bar{b}_{20}\bar{u}_0 + \sum_{j\neq 0}\bar{b}_{2j}\left(-\frac{[\bar{\varepsilon}_{j0}\bar{u}_0\bar{x}^{-1}+\beta_j\bar{b}_{2j}]}{\bar{r}_j}\right)^{\frac{1}{2\bar{k}_j-1}}\bar{x}dt\\
&+(x-\bar{x})[\sigma dB + \int_\Theta \mu d\tilde{N} + \sigma_{gv}dB_{gv}],\\
&\\
&\mathbb{P}(s(t+\delta)=s'|s,u) = \int_t^{t+\delta}\tilde{q}_{ss'}dt' + o(\varepsilon),\ s'\neq s\\
&x(0)=x_0,\ s(0)=s_0.
\end{aligned}
\tag{11.30}
$$

Problem (11.30) is a mean-field-type control problem with nonlinear state dynamics and non-quadratic cost functions.

Below, we compute the difference between the guess and cost functional of the leader. We have

$$
\begin{aligned}
&E[L_0 - f_0(0)]\\
&= E[q_{0T}-\alpha_0(T,.)]\frac{(x_T-\bar{x}_T)^{2k_0}}{2k_0} + [\bar{q}_{0T}-\beta_0(t,.)]\frac{\bar{x}_T^{2\bar{k}_0}}{2\bar{k}_0}\\
&+E\int_0^T[\dot{\alpha}_0 + 2k_0\alpha_0 b_1 + q_0 + \sum_{s'\neq s}(\alpha_0(t,s')-\alpha_0(t,s))\tilde{q}_{ss'}].\frac{(x-\bar{x})^{2k_0}}{2k_0}\\
&+(2k_0-1)k_0(\sigma^2+\sigma_{cogv}^2)\alpha_0.\frac{1}{2k_0}(x-\bar{x})^{2k_0}\\
&+\int_\Theta \nu(d\theta)\{(1+\mu)^{2k_0}-1-2k_0\mu\}\alpha_0.\frac{1}{2k_0}(x-\bar{x})^{2k_0}\\
&+r_0\frac{(u_0-\bar{u}_0)^{2k_0}}{2k_0} + \sum_{j\neq 0}\varepsilon_{0j}\left(-\frac{[\varepsilon_{j0}(u_0-\bar{u}_0)(x-\bar{x})^{-1}+\alpha_j b_{2j}]}{r_j}\right)^{\frac{1}{2k_j-1}}.\\
&(x-\bar{x})^{2k_0-1}(u_0-\bar{u}_0)\\
&+\alpha_0(x-\bar{x})^{2k_0-1}\Bigg[b_{20}(u_0-\bar{u}_0)\\
&+\sum_{j\neq 0} b_{2j}\left(-\frac{[\varepsilon_{j0}(u_0-\bar{u}_0)(x-\bar{x})^{-1}+\alpha_j b_{2j}]}{r_j}\right)^{\frac{1}{2k_j-1}}(x-\bar{x})\Bigg]\\
&+[\dot{\beta}_0 + 2\bar{k}_0\beta_0\bar{b}_1 + \bar{q}_0 + \sum_{s'\neq s}[\beta_0(t,s')-\beta_0(t,s)]\tilde{q}_{ss'})]\frac{\bar{x}^{2\bar{k}_0}}{2\bar{k}_0}\\
&+\bar{r}_0\frac{\bar{u}_0^{2\bar{k}_0}}{2\bar{k}_0} + \sum_{j\neq 0}\bar{\varepsilon}_{0j}\left(-\frac{[\bar{\varepsilon}_{j0}\bar{u}_0\bar{x}^{-1}+\beta_j\bar{b}_{2j}]}{\bar{r}_j}\right)^{\frac{1}{2\bar{k}_j-1}}\bar{x}^{2\bar{k}_0-1}\bar{u}_0\\
&+\beta_0\bar{x}^{2\bar{k}_0-1}\left[\bar{b}_{20}\bar{u}_0 + \sum_{j\neq 0}\bar{b}_{2j}\left(-\frac{[\bar{\varepsilon}_{j0}\bar{u}_0\bar{x}^{-1}+\beta_j\bar{b}_{2j}]}{\bar{r}_j}\right)^{\frac{1}{2\bar{k}_j-1}}\bar{x}\right]dt.
\end{aligned}
\tag{11.31}
$$

We complete the term of the leader as follows:

$$
\begin{aligned}
&r_0 \tfrac{(u_0-\bar{u}_0)^{2k_0}}{2k_0} + \textstyle\sum_{j\neq 0} \varepsilon_{0j} \left(- \tfrac{[\varepsilon_{j0}(u_0-\bar{u}_0)(x-\bar{x})^{-1}+\alpha_j b_{2j}]}{r_j} \right)^{\frac{1}{2k_j-1}} .(x-\bar{x})^{2k_0-1}(u_0-\bar{u}_0)\\
&+\alpha_0 (x-\bar{x})^{2k_0-1} \Bigg[b_{20}(u_0-\bar{u}_0)\\
&+\textstyle\sum_{j\neq 0} b_{2j} \left(- \tfrac{[\varepsilon_{j0}(u_0-\bar{u}_0)(x-\bar{x})^{-1}+\alpha_j b_{2j}]}{r_j} \right)^{\frac{1}{2k_j-1}} (x-\bar{x}) \Bigg]\\
&= r_0 \tfrac{(u_0-\bar{u}_0)^{2k_0}}{2k_0} + \textstyle\sum_{j\neq 0} \varepsilon_{0j} \left(- \tfrac{[\varepsilon_{j0}(u_0-\bar{u}_0)(x-\bar{x})^{-1}+\alpha_j b_{2j}]}{r_j} \right)^{\frac{1}{2k_j-1}} .\\
&(x-\bar{x})^{2k_0-1}(u_0-\bar{u}_0)\\
&+\alpha_0 (x-\bar{x})^{2k_0-1} \Bigg[b_{20}(u_0-\bar{u}_0)\\
&+\textstyle\sum_{j\neq 0} b_{2j} \left(- \tfrac{[\varepsilon_{j0}(u_0-\bar{u}_0)(x-\bar{x})^{-1}+\alpha_j b_{2j}]}{r_j} \right)^{\frac{1}{2k_j-1}} (x-\bar{x}) \Bigg]\\
&-\Omega_0 (x-\bar{x})^{2k_0} + \Omega_0 (x-\bar{x})^{2k_0},
\end{aligned}
\tag{11.32}
$$

where

$$
\begin{aligned}
&\Omega_0 := r_0 \tfrac{\eta_0^{2k_0}}{2k_0} - \textstyle\sum_{j\neq 0} \varepsilon_{0j} \left(- \tfrac{[-\varepsilon_{j0}\eta_0+\alpha_j b_{2j}]}{r_j} \right)^{\frac{1}{2k_j-1}} \eta_0\\
&+\alpha_0 \left[-b_{20}\eta_0 + \textstyle\sum_{j\neq 0} b_{2j} \left(- \tfrac{[-\varepsilon_{j0}\eta_0+\alpha_j b_{2j}]}{r_j} \right)^{\frac{1}{2k_j-1}} \right]\\
&r_0 \eta_0^{2k_0-1} = \textstyle\sum_{j\neq 0} \varepsilon_{0j} \left(- \tfrac{[-\varepsilon_{j0}\eta_0+\alpha_j b_{2j}]}{r_j} \right)^{\frac{1}{2k_j-1}} .\\
&+\textstyle\sum_{j\neq 0} \left(\tfrac{\varepsilon_{j0}}{r_j} \tfrac{1}{2k_j-1} \right) \varepsilon_{0j} \left(- \tfrac{[-\varepsilon_{j0}\eta_0+\alpha_j b_{2j}]}{r_j} \right)^{\frac{1}{2k_j-1}-1}\\
&+\alpha_0 \left[b_{20} - \textstyle\sum_{j\neq 0} \left(\tfrac{\varepsilon_{j0}}{r_j} \tfrac{1}{2k_j-1} \right) b_{2j} \left(- \tfrac{[-\varepsilon_{j0}\eta_0+\alpha_j b_{2j}]}{r_j} \right)^{\frac{1}{2k_j-1}-1} \right],\\
&\bar{\Omega}_0 := \bar{r}_0 \tfrac{\bar{\eta}_0^{2\bar{k}_0}}{2\bar{k}_0} - \textstyle\sum_{j\neq 0} \bar{\varepsilon}_{0j} \left(- \tfrac{[-\bar{\varepsilon}_{j0}\bar{\eta}_0+\beta_j \bar{b}_{2j}]}{\bar{r}_j} \right)^{\frac{1}{2\bar{k}_j-1}} \bar{\eta}_0\\
&+\beta_0 \left[-\bar{b}_{20}\bar{\eta}_0 + \textstyle\sum_{j\neq 0} \bar{b}_{2j} \left(- \tfrac{[-\bar{\varepsilon}_{j0}\bar{\eta}_0+\beta_j \bar{b}_{2j}]}{\bar{r}_j} \right)^{\frac{1}{2\bar{k}_j-1}} \right]\\
&\bar{r}_0 \bar{\eta}_0^{2\bar{k}_0-1} = \textstyle\sum_{j\neq 0} \bar{\varepsilon}_{0j} \left(- \tfrac{[-\bar{\varepsilon}_{j0}\bar{\eta}_0+\beta_j \bar{b}_{2j}]}{\bar{r}_j} \right)^{\frac{1}{2\bar{k}_j-1}} .\\
&+\textstyle\sum_{j\neq 0} \left(\tfrac{\bar{\varepsilon}_{j0}}{\bar{r}_j} \tfrac{1}{2\bar{k}_j-1} \right) \bar{\varepsilon}_{0j} \left(- \tfrac{[-\bar{\varepsilon}_{j0}\bar{\eta}_0+\beta_j \bar{b}_{2j}]}{\bar{r}_j} \right)^{\frac{1}{2\bar{k}_j-1}-1}\\
&+\beta_0 \left[\bar{b}_{20} - \textstyle\sum_{j\neq 0} \left(\tfrac{\bar{\varepsilon}_{j0}}{\bar{r}_j} \tfrac{1}{2\bar{k}_j-1} \right) \bar{b}_{2j} \left(- \tfrac{[-\bar{\varepsilon}_{j0}\bar{\eta}_0+\beta_j b_{2j}]}{\bar{r}_j} \right)^{\frac{1}{2\bar{k}_j-1}-1} \right].
\end{aligned}
\tag{11.33}
$$

It follows that

$$
\begin{aligned}
&E[L_0 - f_0(0)] \\
&= E[q_{0T} - \alpha_0(T,.)]\tfrac{(x_T-\bar{x}_T)^{2k_0}}{2k_0} + [\bar{q}_{0T} - \beta_0(t,.)]\tfrac{\bar{x}_T^{2\bar{k}_0}}{2\bar{k}_0} \\
&+E\int_0^T [\dot{\alpha}_0 + 2k_0\alpha_0 b_1 + q_0 + 2k_0\Omega_0 + \textstyle\sum_{s'\neq s}(\alpha_0(t,s') - \alpha_0(t,s))\tilde{q}_{ss'}].\tfrac{(x-\bar{x})^{2k_0}}{2k_0} \\
&+(2k_0-1)k_0(\sigma^2 + \sigma_{cogv}^2)\alpha_0.\tfrac{1}{2k_0}(x-\bar{x})^{2k_0} \\
&+\int_\Theta \nu(d\theta)\{(1+\mu)^{2k_0} - 1 - 2k_0\mu\}\alpha_0.\tfrac{1}{2k_0}(x-\bar{x})^{2k_0} \\
&r_0\tfrac{(u_0-\bar{u}_0)^{2k_0}}{2k_0} + \textstyle\sum_{j\neq 0}\varepsilon_{0j}\left(-\tfrac{[\varepsilon_{j0}(u_0-\bar{u}_0)(x-\bar{x})^{-1}+\alpha_j b_{2j}]}{r_j}\right)^{\frac{1}{2k_j-1}}.(x-\bar{x})^{2k_0-1}(u_0-\bar{u}_0) \\
&+\alpha_0(x-\bar{x})^{2k_0-1}\Bigg[b_{20}(u_0-\bar{u}_0) \\
&+\textstyle\sum_{j\neq 0} b_{2j}\left(-\tfrac{[\varepsilon_{j0}(u_0-\bar{u}_0)(x-\bar{x})^{-1}+\alpha_j b_{2j}]}{r_j}\right)^{\frac{1}{2k_j-1}}(x-\bar{x})\Bigg] - \Omega_0(x-\bar{x})^{2k_0} \\
&+[\dot{\beta}_0 + 2\bar{k}_0\beta_0\bar{b}_1 + \bar{q}_0 + 2\bar{k}_0\bar{\Omega}_0 + \textstyle\sum_{s'\neq s}[\beta_0(t,s') - \beta_0(t,s)]\tilde{q}_{ss'})]\tfrac{\bar{x}^{2\bar{k}_0}}{2\bar{k}_0} \\
&+\bar{r}_0\tfrac{\bar{u}_0^{2\bar{k}_0}}{2\bar{k}_0} + \textstyle\sum_{j\neq 0}\bar{\varepsilon}_{0j}\left(-\tfrac{[\bar{\varepsilon}_{j0}\bar{u}_0\bar{x}^{-1}+\beta_j\bar{b}_{2j}]}{\bar{r}_j}\right)^{\frac{1}{2\bar{k}_j-1}}\bar{x}^{2\bar{k}_0-1}\bar{u}_0 \\
&+\beta_0\bar{x}^{2\bar{k}_0-1}[\bar{b}_{20}\bar{u}_0 + \textstyle\sum_{j\neq 0}\bar{b}_{2j}\left(-\tfrac{[\bar{\varepsilon}_{j0}\bar{u}_0\bar{x}^{-1}+\beta_j\bar{b}_{2j}]}{\bar{r}_j}\right)^{\frac{1}{2\bar{k}_j-1}}\bar{x}] - \bar{\Omega}_0\bar{x}^{2\bar{k}_0}dt.
\end{aligned}
\tag{11.34}
$$

The conditional mean-field term $\bar{x}$ is given by

$$
d\bar{x} = [\bar{b}_1 - \bar{b}_{20}\bar{\eta}_0 + \textstyle\sum_{j\neq 0}\bar{b}_{2j}\left(-\tfrac{[-\bar{\varepsilon}_{j0}\bar{\eta}_0+\beta_j\bar{b}_{2j}]}{\bar{r}_j}\right)^{\frac{1}{2\bar{k}_j-1}}]\bar{x}dt. \tag{11.35}
$$

The decision-makers can therefore filter $\bar{x}$ using (11.35) and use it for feedback. The leader Stackelberg strategy is

$$
u_0 = -\eta_0(x-\bar{x}) - \bar{\eta}_0\bar{x}.
$$

Reorganizing the followers' term yields

$$
\begin{aligned}
&E[L_i - f_i(0)]\\
&= E[q_{iT} - \alpha_i(T,.)]\tfrac{(x_T-\bar{x}_T)^{2k_i}}{2k_i} + [\bar{q}_{iT} - \beta_i(t,.)]\tfrac{\bar{x}_T^{2\bar{k}_i}}{2\bar{k}_i}\\
&+E\int_0^T (\dot{\alpha}_i + 2k_i\alpha_i b_1 + q_i + 2k_i\Omega_i + \textstyle\sum_{s'\neq s}[\alpha_i(t,s') - \alpha_i(t,s)]\tilde{q}_{ss'})\tfrac{1}{2k_i}(x-\bar{x})^{2k_i}\\
&+(2k_i-1)k_i(\sigma^2 + \sigma_{cogv}^2)\alpha_i.\tfrac{1}{2k_i}(x-\bar{x})^{2k_i}\\
&+\int_\Theta \nu(d\theta)\{(1+\mu)^{2k_i} - 1 - 2k_i\mu\}\alpha_i.\tfrac{1}{2k_i}(x-\bar{x})^{2k_i}\\
&+r_i\left\{\tfrac{(u_i-\bar{u}_i)^{2k_i}}{2k_i} + \tfrac{[\varepsilon_{i0}(u_0-\bar{u}_0)+\alpha_i b_{2i}(x-\bar{x})]}{r_i}(x-\bar{x})^{2(k_i-1)}(u_i-\bar{u}_i)\right\}\\
&+r_i\tfrac{2k_i-1}{2k_i}\left\{(-\tfrac{[\varepsilon_{i0}(u_0-\bar{u}_0)+\alpha_i b_{2i}(x-\bar{x})]}{r_i}(x-\bar{x})^{2(k_i-1)})^{\frac{2k_i}{2k_i-1}}\right\}\\
&+(\dot{\beta}_i + 2\bar{k}_i\beta_i\bar{b}_1 + \bar{q}_i + 2\bar{k}_i\bar{\Omega}_i + \textstyle\sum_{s'\neq s}[\beta_i(t,s') - \beta_i(t,s)]\tilde{q}_{ss'}).\tfrac{1}{2\bar{k}_i}\bar{x}^{2\bar{k}_i}\\
&+\bar{r}_i\left\{\tfrac{\bar{u}_i^{2\bar{k}_i}}{2\bar{k}_i} + \tfrac{[\bar{\varepsilon}_{i0}\bar{u}_0+\beta_i\bar{b}_{2i}\bar{x}]}{\bar{r}_i}\bar{x}^{2(\bar{k}_i-1)}\bar{u}_i\right\}\\
&+\bar{r}_i\tfrac{2\bar{k}_i-1}{2\bar{k}_i}\left\{(-\tfrac{[\bar{\varepsilon}_{i0}\bar{u}_0+\beta_i\bar{b}_{2i}\bar{x}]}{\bar{r}_i}\bar{x}^{2(\bar{k}_i-1)})^{\frac{2\bar{k}_i}{2\bar{k}_i-1}}\right\},
\end{aligned}
\tag{11.36}
$$

$$
\begin{aligned}
\Omega_i &:= -r_i\frac{2k_i-1}{2k_i}\left\{\left(-\frac{[-\varepsilon_{i0}\eta_0+\alpha_i b_{2i}]}{r_i}\right)^{\frac{2k_i}{2k_i-1}}\right\} - \alpha_i b_{20}\eta_0,\\
\bar{\Omega}_i &:= -\bar{r}_i\frac{2\bar{k}_i-1}{2\bar{k}_i}\left\{\left(-\frac{[-\bar{\varepsilon}_{i0}\bar{\eta}_0+\beta_i\bar{b}_{2i}]}{\bar{r}_i}\right)^{\frac{2\bar{k}_i}{2\bar{k}_i-1}}\right\} - \beta_i\bar{b}_{20}\bar{\eta}_0,\\
u_i &:= \left(-\frac{[-\varepsilon_{i0}\eta_0+\alpha_i b_{2i}]}{r_i}\right)^{\frac{1}{2k_i-1}}(x-\bar{x}) + \left(-\frac{[-\bar{\varepsilon}_{i0}\bar{\eta}_0+\beta_i\bar{b}_{2i}]}{\bar{r}_i}\right)^{\frac{1}{2\bar{k}_i-1}}\bar{x}.
\end{aligned}
\tag{11.37}
$$

This completes the proof.

11.3.1 Numerical Illustration

We present some numerical illustrations of problem (11.23) by choosing multiple followers under the Gauss-Volterra process with the kernel:

$$
\begin{aligned}
K(t,t') = K_H(t,t') &= c_H(t-t')^{H-\frac{1}{2}}\\
&+ c_H\left(\frac{1}{2} - H\right)\int_{t'}^{t}(z-t')^{H-\frac{3}{2}}\left(1-\left(\frac{t'}{z}\right)^{\frac{1}{2}-H}\right)dz,
\end{aligned}
\tag{11.38}
$$

and $H \in (0,1)$, $c_H = \sqrt{\frac{2H\Gamma(\frac{3}{2}-H)}{\Gamma(\frac{1}{2}+H)\Gamma(2-2H)}}$, is a normalizing constant, where Γ is the gamma function:

Table 11.1 Parameters used in the numerical example

Numerical setting	Value
Kernel K	$K = K_{0.8}$
T	1
Switching $\mathscr{S}$	$\{s_*, s^*\},\ s_* \neq s^*$
I	2019
$(k_i, \bar{k}_i)$	$(2, 2)$ if $1 \leq i \leq 2019$ $(8, 4)$ if $i = 0$
$q_i, \bar{q}_i, r_i, \bar{r}_i$	1 if $1 \leq i \leq 2019$ 200 if $i = 0$
$\varepsilon_{ij}, \bar{\varepsilon}_{ij}$	10^{-3}
$b_1, b_{2i}, \bar{b}_{2i}, \sigma, \sigma_{gv}$	1 if $1 \leq i \leq 2019$
$b_{2i}, \bar{b}_{2i}$	20 if $i = 0$
$\mu(t, \theta, s)$	θ
Θ	$\mathbb{R}_+^*$
$\nu(d\theta)$	$ce^{-5\|\theta\|}d\theta$
$\sigma_{gv}(t, s^*)$	1
$\sigma_{gv}(t, s_*)$	10^{-2}
$\tilde{q}_{s^* s_*}$	0.7
$\tilde{q}_{s_* s^*}$	0.4

$$\Gamma(z) = \int_0^{+\infty} e^{-t} t^{z-1} dt, \quad \mathrm{Re}(z) > 0.$$

The Gauss-Volterra process with kernel K_H is a fractional Brownian motion with the Hurst parameter H. The parameters of the numerical setting are displayed in Table 11.1.

It is important to notice that under this setting, the problem (11.23) is not Markov, and the cost is not quadratic. From (11.2), we know that the Stackelberg solution of the mean-field-type game (11.23) is given by

$$\begin{aligned}
&u_i^{ss} := \left(-\tfrac{[-\varepsilon_{i0}\eta_0+\alpha_i b_{2i}]}{r_i}\right)^{\frac{1}{3}} (x - \bar{x}) + \left(-\tfrac{[-\bar{\varepsilon}_{i0}\bar{\eta}_0+\beta_i \bar{b}_{2i}]}{\bar{r}_i}\right)^{\frac{1}{3}} \bar{x}, \quad 1 \leq i \leq 2019\\
&u_0^{ss} = -\eta_0(x - \bar{x}) - \bar{\eta}_0 \bar{x},\\
&r_0\eta_0^{15} = \textstyle\sum_{j\neq 0} \varepsilon_{0j} \left(-\tfrac{[-\varepsilon_{j0}\eta_0+\alpha_j b_{2j}]}{r_j}\right)^{\frac{1}{3}} + \sum_{j\neq 0} \left(\tfrac{\varepsilon_{j0}}{3r_j}\right) \varepsilon_{0j} \left(-\tfrac{[-\varepsilon_{j0}\eta_0+\alpha_j b_{2j}]}{r_j}\right)^{-\frac{2}{3}}\\
&+\alpha_0 \left[b_{20} - \textstyle\sum_{j\neq 0} \left(\tfrac{\varepsilon_{j0}}{3r_j}\right) b_{2j} \left(-\tfrac{[-\varepsilon_{j0}\eta_0+\alpha_j b_{2j}]}{r_j}\right)^{-\frac{2}{3}}\right],\\
&\bar{r}_0\bar{\eta}_0^{7} = \textstyle\sum_{j\neq 0} \bar{\varepsilon}_{0j} \left(-\tfrac{[-\bar{\varepsilon}_{j0}\bar{\eta}_0+\beta_j \bar{b}_{2j}]}{\bar{r}_j}\right)^{\frac{1}{3}} + \sum_{j\neq 0} \left(\tfrac{\bar{\varepsilon}_{j0}}{3\bar{r}_j}\right) \bar{\varepsilon}_{0j} \left(-\tfrac{[-\bar{\varepsilon}_{j0}\bar{\eta}_0+\beta_j \bar{b}_{2j}]}{\bar{r}_j}\right)^{-\frac{2}{3}}\\
&+\beta_0 \left[\bar{b}_{20} - \textstyle\sum_{j\neq 0} \left(\tfrac{\bar{\varepsilon}_{j0}}{3\bar{r}_j}\right) \bar{b}_{2j} \left(-\tfrac{[-\bar{\varepsilon}_{j0}\bar{\eta}_0+\beta_j b_{2j}]}{\bar{r}_j}\right)^{-\frac{2}{3}}\right].
\end{aligned} \tag{11.39}$$

Figure 11.1 plots (a) a sample path of the Stackelberg state trajectory starting from $x_0 = 100$, (b) the optimal strategies of all decision-makers $1 \leq i \leq 2019$

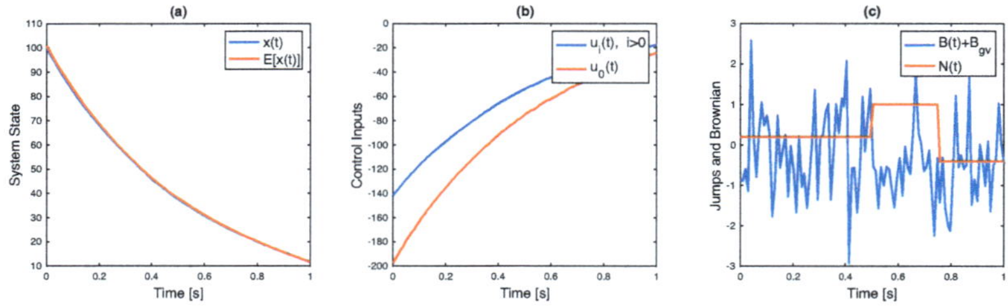

Fig. 11.1 Sample paths of the Stackelberg state and strategy of the leader and the followers under Gauss-Volterra, jump, and diffusion. Initial state value is 100

and 0, and (c) sample noises. As expected, the state is moving toward zero when the Stackelberg strategies are employed.

11.4 Hierarchical Mean-Field-Type Games

In this section, we use a dynamic programming principle approach in the space of measures.

11.4.1 One Layer of Hierarchy

11.4.1.1 The Setup

There are $I \geq 2$ decision-makers interacting within the horizon $[0, T],\ \ 0 < T$. The set of decision-makers is denoted by $\mathscr{I} = \{1, 2, \ldots, I\}$. Decision-maker $i \in \mathscr{I}$ has a control action $u_i \in U_i = \mathbb{R}$. The state x is driven by a drift-jump-diffusion process of mean-field type, evolving as

$$dx = bdt + \sigma dB + \int_{\Theta} \mu(., \theta)\tilde{N}(dt, d\theta), \quad x(0) \sim m(0, .),$$

where

Drift: b,
Diffusion coefficient: σ, Brownian motion B,
Set of jump size Θ,
Jump: $N(dt, d\theta)$,
Compensated jump:$\tilde{N}(dt, d\theta) = N(dt, d\theta) - \nu(d\theta)dt$,
$b, \sigma, \mu(., \theta), :\ [0, T] \times \mathbb{R} \times \prod_{j=1}^{I} U_j \times \mathscr{P}(\mathbb{R}) \to \mathbb{R}.$

where $\mathscr{P}(\mathbb{R})$ denotes the set of probability measures on $\mathbb{R}$. We assume that $x(0)$, B, and N are mutually independent.

The performance functional of decision-maker i is

$$L_i(u, m_0) = h_i(x(T), m(T)) + \int_0^T l_i(t, x, u, m)dt,$$

where $m(t, dy) = \mathbb{P}_{x(t)}(dy)$ is the probability measure of the state $x(t)$ at time t, and

$$\begin{aligned} &l_i : [0, T] \times \mathbb{R} \times \prod_{j=1}^{I} U_j \times \mathscr{P}(\mathbb{R}) \to \mathbb{R}, \\ &h_i : \mathbb{R} \times \mathscr{P}(\mathbb{R}) \to \mathbb{R}, \end{aligned}$$

We make the following information structure:

- perfect state measurement, i.e., all the decision-makers observe the state x,
- perfect knowledge of the model, i.e., the model above is known by all decision-makers.

In addition, each decision-maker is assumed to have a computability capability such that they are able to compute an aggregative term of m from the model, i.e., $\bar{x}(t) = \int y m(t, dy)$. Let $\mathscr{U}_i$ be the set of control strategies of decision-maker i that are progressively measurable with respect to the filtration generated by the unions of events in $\{B, N\}$.

11.4.1.2 Games with Polynomial Cost

We will investigate the MFTG with the following data:

$$\begin{aligned} &l_i(t, x, u, m) = q_i \tfrac{(x-\bar{x})^{2k_i}}{2k_i} + r_i \tfrac{(u_i-\bar{u}_i)^{2k_i}}{2k_i} + c_i (x - \bar{x})^{2k_i-1}(u_i - \bar{u}_i) \\ &+ \textstyle\sum_{j\in\mathscr{I}\setminus\{i\}} \varepsilon_{ij} (x - \bar{x})^{2(k_i-1)}(u_i - \bar{u}_i)(u_j - \bar{u}_j) \\ &+ \bar{q}_i \tfrac{\bar{x}^{2\bar{k}_i}}{2\bar{k}_i} + \bar{r}_i \tfrac{\bar{u}_i^{2\bar{k}_i}}{2\bar{k}_i} + \bar{c}_i \bar{x}^{2\bar{k}_i-1}\bar{u}_i + \textstyle\sum_{j\neq i} \bar{\varepsilon}_{ij} \bar{x}^{2(\bar{k}_i-1)}\bar{u}_i\bar{u}_j, \\ &h_i(x, m) = q_{iT} \tfrac{(x_T-\bar{x}_T)^{2k_i}}{2k_i} + \bar{q}_{iT} \tfrac{\bar{x}_T^{2\bar{k}_i}}{2\bar{k}_i}, \\ &b(t, x, u, m) = b_1(x - \bar{x}) + \bar{b}_1\bar{x} \\ &+ \textstyle\sum_{j\in\mathscr{I}} \left[b_{2j}(u_j - \bar{u}_j) + \bar{b}_{2j}\bar{u}_j\right], \\ &\sigma(t, x, u, m) = (x - \bar{x})\tilde{\sigma}, \\ &\mu(t, x, u, m, \theta) = (x - \bar{x})\tilde{\mu}(., \theta), \\ &\bar{x}(t) = \textstyle\int y m(t, dy), \\ &\bar{u}(t) = \textstyle\int u(t, y, m)m(t, dy), \end{aligned}$$

where $k_i \geq 1$, $\bar{k}_i \geq 1$ are natural numbers and the coefficients are time-dependent. Further, the coefficient functions $q_i, r_i, \bar{q}_i, \bar{r}_i$ are nonnegative functions.

Nash Mean-Field-Type Equilibrium
The risk-neutral MFTG is given by

$$(\mathcal{I}, U_i, \mathcal{U}_i, E[L_i])_{i\in\mathcal{I}}.$$

A risk-neutral mean-field-type Nash equilibrium is a solution of the following fixed-point problem:

$$\begin{aligned}&i \in \mathcal{I},\\ &E[L_i(u^*, m_0)] = \inf_{u_i\in\mathcal{U}_i} E[L_i(u_1^*, \ldots, u_{i-1}^*, u_i, u_{i+1}^*, \ldots, u_I^*, m_0)].\end{aligned} \tag{11.40}$$

Let $\hat{V}_i(t, m)$ be the optimal cost-to-go from m at time $t \in (0, T)$ given the strategies of the others:

$$\hat{V}_i(t, m) = \inf_{u_i} E[h_i(x(T), m(T)) + \int_t^T l_i(t, x, u, m)dt' | m(t) = m]. \tag{11.41}$$

Let $\hat{V}_{i,m}$ be the Gâteaux derivative of $\hat{V}_i(t, .)$ with respect to the measure m. Introduce the integrand Hamiltonian as

$$\begin{aligned}&H_i(x, m, (\hat{V}_{j,m}, \hat{V}_{j,xm}, \hat{V}_{j,xxm})_j) =\\ &\inf_{u_i\in U_i} \left\{ l_i + b\, \hat{V}_{i,xm} + \frac{\sigma^2}{2}\hat{V}_{i,xxm} + \int_\Theta [\hat{V}_{i,m}(t_-, x+\mu) - \hat{V}_{i,m} - \mu \hat{V}_{i,xm}]\nu(d\theta) \right\}.\end{aligned} \tag{11.42}$$

Denote the jump operator J as

$$J[\phi_i] := \int_\Theta [\phi_{i,m}(t_-, x+\mu) - \phi_{i,m} - \mu\phi_{i,xm}]\nu(d\theta),$$

and J^* be the adjoint operator of J:

$$\langle J[\phi], m\rangle = \langle \phi, J^*[m]\rangle.$$

A sufficient condition for a risk-neutral Nash equilibrium system is given by the following (backward-forward) partial integrodifferential system:

$$\begin{aligned}&i \in \mathcal{I},\\ &0 = \hat{V}_{i,t}(t, m) + \int_x H_i(x, m, (\hat{V}_{j,m}, \hat{V}_{j,xm}, \hat{V}_{j,xxm})_{j\in\mathcal{I}})m(t, dx),\\ &\hat{V}_i(T, m) = \int m(dy)h_i(y, m),\\ &m_t = -(mb)_x + \tfrac{1}{2}(m\sigma^2)_{xx} + J^*[m],\\ &m(0) = m_0.\end{aligned}$$

The system (11.43) is an infinite PIDE system in m, and it provides the Nash equilibrium values of the MFTG. Note that the PIDE system satisfied by $\{\hat{V}_{i,m}\}_{i\in\mathcal{I}}$ is not the equilibrium value $\{\hat{V}_i\}_{i\in\mathcal{I}}$ in (11.43). We use the system (11.43) to find a risk-neutral mean-field-type Nash equilibrium.

Proposition 11.3 *A risk-neutral mean-field-type Nash equilibrium is given in a semi-explicit form as follows:*

$$
\begin{aligned}
&u_i^{ne} = -\eta_i\left(x - \int ym(dy)\right) - \bar{\eta}_i \int ym(dy),\\
&0 = -r_i\eta_i^{2k_i-1} - \sum_{j\neq i}\varepsilon_{ij}\eta_j + b_{2i}\alpha_i + c_i,\\
&0 = -\bar{r}_i\bar{\eta}_i^{2\bar{k}_i-1} - \sum_{j\neq i}\bar{\varepsilon}_{ij}\bar{\eta}_j + \bar{b}_{2i}\bar{\alpha}_i + \bar{c}_i,\\
&\hat{V}_i(t,m) = \alpha_i \int_x \frac{(x-\int ym(dy))^{2k_i}}{2k_i} m(dx) + \bar{\alpha}_i \frac{(\int ym(dy))^{2\bar{k}_i}}{2\bar{k}_i},\\
&0 = \dot{\alpha}_i + q_i + r_i\eta_i^{2k_i} - 2k_ic_i\eta_i + 2k_i\sum_{j\neq i}\varepsilon_{ij}\eta_i\eta_j\\
&+2k_i\alpha_i[b_1 - \sum_{j\in\mathcal{I}} b_{2j}\eta_j] + 2k_i(2k_i-1)\alpha_i\tfrac{1}{2}\tilde{\sigma}^2\\
&+\alpha_i\int_\Theta[(1+\tilde{\mu})^{2k_i} - 1 - 2k_i\tilde{\mu}]\nu(d\theta),\\
&\alpha_i(T) = q_{iT},\\
&0 = \dot{\bar{\alpha}}_i + \bar{q}_i + \bar{r}_i\bar{\eta}_i^{2\bar{k}_i} - 2\bar{k}_i\bar{c}_i\bar{\eta}_i + 2\bar{k}_i\sum_{j\neq i}\bar{\varepsilon}_{ij}\bar{\eta}_i\bar{\eta}_j\\
&+2\bar{k}_i\bar{\alpha}_i[\bar{b}_1 - \sum_j \bar{b}_{2j}\bar{\eta}_j],\\
&\bar{\alpha}_i(T) = \bar{q}_i(T),
\end{aligned}
$$

for all $i \in \mathcal{I}$ *with*

$$
\int ym(t,dy) = \left[\int ym(0,dy)\right] e^{\int_0^t\left[\bar{b}_1 - \sum_j \bar{b}_{2j}\bar{\eta}_j\right]dt'}, \tag{11.43}
$$

whenever the above coefficient system admits a solution that does not have a singularity within $[0, T]$.

Proof of Proposition 11.3 Assuming perfect state observation and perfect knowledge of the model, we aim to solve the PIDE system (11.43): We start with the following guess functional of decision-maker i as

$$
\hat{V}_i(t,m) = \alpha_i(t) \int_x \frac{(x-\int ym(dy))^{2k_i}}{2k_i} m(dx) + \bar{\alpha}_i(t)\frac{(\int ym(dy))^{2\bar{k}_i}}{2\bar{k}_i}, \tag{11.44}
$$

where the coefficient functions $\alpha_i, \bar{\alpha}_i$ need to be determined. We compute the key terms $\hat{V}_{i,m}(t,m)$, $\hat{V}_{i,xm}(t,m)$, and $\hat{V}_{i,xxm}(t,m)$:

$$
\begin{aligned}
&\hat{V}_{i,m}(t,m) = -\alpha_i x \int \left(y - \int zm(dz)\right)^{2k_i-1} m(dy)\\
&+\alpha_i \frac{(x-\int ym(dy))^{2k_i}}{2k_i}\\
&+\bar{\alpha}_i x \left(\int ym(dy)\right)^{2\bar{k}_i-1},\\
&\hat{V}_{i,xm}(t,m) = -\alpha_i \int \left(y - \int zm(dz)\right)^{2k_i-1} m(dy)\\
&+\alpha_i \left(x - \int ym(dy)\right)^{2k_i-1}\\
&+\bar{\alpha}_i \left(\int ym(dy)\right)^{2\bar{k}_i-1},\\
&\hat{V}_{i,xxm}(t,m) = (2k_i-1)\alpha_i\left(x - \int ym(dy)\right)^{2(k_i-1)},\\
&\hat{V}_{i,m}(t,m)(x+\mu) - \hat{V}_{i,m}(t,m)(x) - \mu\hat{V}_{i,xm}(t,m)(x)\\
&= \alpha_i \frac{(x-\int ym(dy))^{2k_i}}{2k_i}\left[(1+\tilde{\mu})^{2k_i} - 1 - 2k_i\tilde{\mu}\right] + \tilde{\varepsilon},
\end{aligned}
$$

with $\int \tilde{\varepsilon} m(dy) = 0$. The integrand Hamiltonian is strictly convex in $(u_i - \bar{u}_i, \bar{u}_i)$, and the optimal control strategy is the unique minimizer of

$$
\begin{aligned}
& r_i \tfrac{(u_i-\bar{u}_i)^{2k_i}}{2k_i} + c_i (x-\bar{x})^{2k_i-1}(u_i-\bar{u}_i) \\
& + \textstyle\sum_{j\neq i} \varepsilon_{ij}(x-\bar{x})^{2(k_i-1)}(u_i-\bar{u}_i)(u_j-\bar{u}_j) \\
& + \left[\hat{V}_{i,xm}(t,m) - \int \hat{V}_{i,xm}(t,m)(x)m(dx)\right] \textstyle\sum_{j\in\mathcal{I}} b_{2j}(u_j-\bar{u}_j) \\
& + \bar{r}_i \tfrac{\bar{u}_i^{2\bar{k}_i}}{2\bar{k}_i} + \bar{c}_i \bar{x}^{2\bar{k}_i-1}\bar{u}_i + \textstyle\sum_{j\neq i} \bar{\varepsilon}_{ij}\bar{x}^{2(\bar{k}_i-1)}\bar{u}_i\bar{u}_j \\
& + \left[\int \hat{V}_{i,xm}(t,m)(x)m(dx)\right] \textstyle\sum_j \bar{b}_{2j}\bar{u}_j .
\end{aligned}
$$

By strict convexity and by orthogonality between $(u_i - \bar{u}_i)$ and $\bar{u}_i$, the f condition system below follows:

$$
\begin{aligned}
& i \in \mathcal{I}, \\
& 0 = r_i(u_i-\bar{u}_i)^{2k_i-1} + c_i(x-\bar{x})^{2k_i-1} + \textstyle\sum_{j\neq i} \varepsilon_{ij}(x-\bar{x})^{2(k_i-1)}(u_j-\bar{u}_j) \\
& + \left[\hat{V}_{i,xm}(t,m) - \int \hat{V}_{i,xm}(t,m)(x)m(dx)\right] b_{2i}, \\
& 0 = \bar{r}_i\bar{u}_i^{2\bar{k}_i-1} + \bar{c}_i\bar{x}^{2\bar{k}_i-1} + \textstyle\sum_{j\neq i} \bar{\varepsilon}_{ij}\bar{x}^{2(\bar{k}_i-1)}\bar{u}_j + \left[\int \hat{V}_{i,xm}(t,m)(x)m(dx)\right] \bar{b}_{2i}.
\end{aligned}
$$

The linear-state-and-mean-field-type feedback strategy $u_i = -\eta_i(x - \int ym(dy)) - \bar{\eta}_i \int ym(dy),\ i \in \mathcal{I}$ solves the system if the coefficients satisfy

$$
\begin{aligned}
& i \in \mathcal{I}, \\
& 0 = -r_i\eta_i^{2k_i-1} - \textstyle\sum_{j\neq i} \varepsilon_{ij}\eta_j + b_{2i}\alpha_i + c_i, \\
& 0 = -\bar{r}_i\bar{\eta}_i^{2\bar{k}_i-1} - \textstyle\sum_{j\neq i} \bar{\varepsilon}_{ij}\bar{\eta}_j + \bar{b}_{2i}\bar{\alpha}_i + \bar{c}_i.
\end{aligned}
$$

The integrand Hamiltonian of i becomes

$$
\begin{aligned}
H_i = & \left[q_i + r_i\eta_i^{2k_i} - 2k_ic_i\eta_i + 2k_i \textstyle\sum_{j\neq i} \varepsilon_{ij}\eta_i\eta_j\right] \tfrac{(x-\int ym(dy))^{2k_i}}{2k_i} \\
& + 2k_i\alpha_i \left[b_1 - \textstyle\sum_{j\in\mathcal{I}} b_{2j}\eta_j\right] \tfrac{(x-\int ym(dy))^{2k_i}}{2k_i} \\
& + 2k_i(2k_i-1)\alpha_i \tfrac{1}{2}\tilde{\sigma}^2 \tfrac{(x-\int ym(dy))^{2k_i}}{2k_i} \\
& + \alpha_i \int_\Theta [(1+\tilde{\mu})^{2k_i} - 1 - 2k_i\tilde{\mu}]\nu(d\theta) \tfrac{(x-\int ym(dy))^{2k_i}}{2k_i} \\
& + \left[\bar{q}_i + \bar{r}_i\bar{\eta}_i^{2\bar{k}_i} - 2\bar{k}_i\bar{c}_i\bar{\eta}_i + 2\bar{k}_i \textstyle\sum_{j\neq i} \bar{\varepsilon}_{ij}\bar{\eta}_i\bar{\eta}_j\right] \tfrac{(\int ym(dy))^{2\bar{k}_i}}{2\bar{k}_i} \\
& + 2\bar{k}_i\bar{\alpha}_i \left[\bar{b}_1 - \textstyle\sum_j \bar{b}_{2j}\bar{\eta}_j\right] \tfrac{(\int ym(dy))^{2\bar{k}_i-1}}{2\bar{k}_i} + \tilde{\varepsilon}_2.
\end{aligned}
$$

Through matching, the coefficients α_i solve the following ordinary differential equation:

$$
\begin{aligned}
&0 = \dot{\alpha}_i + q_i + r_i \eta_i^{2k_i} - 2k_i c_i \eta_i + 2k_i \sum_{j \neq i} \varepsilon_{ij} \eta_i \eta_j \\
&+2k_i \alpha_i [b_1 - \sum_{j \in \mathcal{I}} b_{2j} \eta_j] + 2k_i (2k_i - 1) \alpha_i \tfrac{1}{2} \tilde{\sigma}^2 \\
&+\alpha_i \int_{\Theta} [(1 + \tilde{\mu})^{2k_i} - 1 - 2k_i \tilde{\mu}] \nu(d\theta), \\
&\alpha_i(T) = q_{iT}, \\
&0 = \dot{\bar{\alpha}}_i + \bar{q}_i + \bar{r}_i \bar{\eta}_i^{2\bar{k}_i} - 2\bar{k}_i \bar{c}_i \bar{\eta}_i + 2\bar{k}_i \sum_{j \neq i} \bar{\varepsilon}_{ij} \bar{\eta}_i \bar{\eta}_j \\
&+2\bar{k}_i \bar{\alpha}_i [\bar{b}_1 - \sum_j \bar{b}_{2j} \bar{\eta}_j], \\
&\bar{\alpha}_i(T) = \bar{q}_{iT}.
\end{aligned}
$$

The aggregate mean-field term $\int ym(t, dy)$ can be derived in a semi-explicit form by taking the expected value of the state dynamics. It follows that

$$
\int ym(t, dy) = \left[\int ym(0, dy) \right] e^{\int_0^t \left[\bar{b}_1 - \sum_j \bar{b}_{2j} \bar{\eta}_j \right] dt'},
$$

which completes the proof. ■

The uniqueness of the coefficient system (11.43) in η requires a strong condition:

$$
0 = -r_i \eta_i^{2k_i - 1} - \sum_{j \neq i} \varepsilon_{ij} \eta_j + b_{2i} \alpha_i + c_i, \ i \in \mathcal{I}. \tag{11.45}
$$

This is a system of polynomials of odd degrees. The system (11.45) in η belongs to $\mathbb{R}[\eta_1, \ldots, \eta_I]$ the ring of polynomials with real coefficients in I variables $\eta_1, \ldots, \eta_I$.

- For arbitrary number of decision-makers $I \geq 3$ and $k_i = k = 1$, the system in η becomes linear and has a unique solution if and only if the determinant of the matrix M is nonzero, with $M_{ii} = r_i$ and $M_{ij} = \varepsilon_{ij}, \ i \neq j$. When the determinant is zero, the resulting control strategies become non-admissible, and the costs become infinite.
- For $k_i = k = 2$ and $I = 2$, the system in η becomes a cubic polynomial given by

$$
\begin{aligned}
r_1 \eta_1^3 + \varepsilon_{12} \eta_2 - b_{21} \alpha_1 - c_1 = 0, \\
r_2 \eta_2^3 + \varepsilon_{21} \eta_1 - b_{22} \alpha_2 - c_2 = 0.
\end{aligned} \tag{11.46}
$$

For $\varepsilon_{12} = 0$, then there is a unique solution given by

$$
\eta_1 = \left(\frac{b_{21} \alpha_1 + c_1}{r_1} \right)^{\frac{1}{3}}, \ \eta_2 = \left(\frac{-\varepsilon_{21} \eta_1 + b_{22} \alpha_2 + c_2}{r_2} \right)^{\frac{1}{3}}.
$$

For $\varepsilon_{12} \neq 0$, we derive from the first equation that $\eta_2 = \frac{-r_1 \eta_1^3 + b_{21} \alpha_1 + c_1}{\varepsilon_{12}}$. By substituting it to the second equation, we arrive at

$$r_2\left(\frac{-r_1\eta_1^3+b_{21}\alpha_1+c_1}{\varepsilon_{12}}\right)^3+\varepsilon_{21}\eta_1-b_{21}\alpha_1-c_1=0.$$

The latter equation is a polynomial. It has a unique real root in η_1 if its derivative has a constant sign. Its derivative is

$$\varepsilon_{21}-9\frac{r_1r_2}{\varepsilon_{12}}\eta_1^2\left(\frac{-r_1\eta_1^3+b_{21}\alpha_1+c_1}{\varepsilon_{12}}\right)^2.$$

It has a constant sign if ε_{21} and $\frac{r_1r_2}{\varepsilon_{12}}$ have opposite signs. If r_1, r_2 are positive, then the condition is reduced to

$$\varepsilon_{21}\varepsilon_{12}\leq 0.$$

- For $I=2$ and arbitrary k_i, k_j, a sufficient condition is that ε_{ji} and $(2k_i-1)(2k_j-1)\frac{r_ir_j}{\varepsilon_{ij}}$ have opposite signs. In particular, if $k_i\geq 1, k_j\geq 2,\ r_i>0, r_j>0$, then the condition reduces to

$$\varepsilon_{ij}\varepsilon_{ji}\leq 0.$$

The same reasoning applies to the system in $\bar{\eta}$ and has a unique solution if

$$\bar{\varepsilon}_{ij}\bar{\varepsilon}_{ji}\leq 0.$$

11.4.2 Two Layers of Hierarchy Interaction

This section presents a class of Stackelberg MFTGs with two layers of hierarchy. As in all Stackelberg games, the decision-makers act in sequential order, with the leaders announcing their strategies first and then the followers responding to these within the structure of MFTG. The state dynamics is driven by jump-diffusion processes, and the cost function is non-quadratic and has a polynomial structure. The structures of Stackelberg strategies and costs of the leaders and followers are given in a semi-explicit form in state-and-mean-field-type feedback form. A sufficient condition is provided using an infinite-dimensional partial integrodifferential system. We also identify a specific range of parameters for which the Nash equilibrium coincides with the hierarchical solution.

Numerical illustrations show that, in the symmetric case, having only one leader is suboptimal for the social cost. Having too many leaders is also suboptimal for the social cost.

In the next section, we investigate the bi-level hierarchical interaction case with multiple leaders and multiple followers.

We consider the description in (11.40) in a bi-level hierarchical game with two or more leaders, i.e., $|\mathcal{I}_L| \geq 2$, and two or more followers, i.e., $|\mathcal{I}_F| \geq 2$. We restrict our attention to the admissible strategies which are Lipschitz in the state x and $\bar{x}$. Given the strategies of the leaders $(u_i)_{i\in\mathcal{I}_L} \in \prod_{i\in\mathcal{I}_L} \mathcal{U}_i$ and the other followers $(u_{j'})_{j'\in\mathcal{I}_F\backslash\{j\}}$, a risk-neutral best-response strategy of follower j is a strategy that solves $\inf_{\mathcal{U}_j} EL_j$. The set of risk-neutral best responses of j is denoted by $\mathrm{rnBR}_j((u_i)_{i\in\mathcal{I}_L}, (u_{j'})_{j'\in\mathcal{I}_F\backslash\{j\}})$.

A mean-field-type risk-neutral Nash equilibrium between the followers given the leaders' strategies $(u_i)_{i\in\mathcal{I}_L} \in \prod_{i\in\mathcal{I}_L} \mathcal{U}_i$, is a strategy profile $(u_j,\ j \in \mathcal{I}_F)$, of all followers such that for every decision-maker $j \in \mathcal{I}_F$

$$u_j \in \mathrm{rnBR}_j((u_i)_{i\in\mathcal{I}_L}; (u_{j'})_{j'\in\mathcal{I}_F\backslash\{j\}}).$$

The followers solve the following Nash game given the strategies of the leaders $(u_i)_{i\in\mathcal{I}_L}$:

$$\begin{array}{l}
j \in \mathcal{I}_F,\\
0 = \hat{V}_{j,t}(t,m)+\\
\int_x H_j^r(x,m,(\hat{V}_{j',m},\hat{V}_{j',xm},\hat{V}_{j',xxm})_{j'\in\mathcal{I}_F}|(u_i)_{i\in\mathcal{I}_L})m(t,dx),\\
\hat{V}_j(T,m) = \int m(T,dy)h_j(y,m(T)),\\
H_j^r = \inf_{u_j\in U_j}\left\{l_j + b\,\hat{V}_{j,xm} + \frac{\sigma^2}{2}\hat{V}_{j,xxm} + J[\hat{V}_{j,m}]|(u_i)_{i\in\mathcal{I}_L}\right\}\\
m_t = -(mb)_x + \frac{1}{2}(m\sigma^2)_{xx} + J^*[m],\ m(0) = m_0.
\end{array}$$

Taking this into account, the leaders solve another Nash game, leading to the following PIDE system:

$$\begin{array}{l}
i \in \mathcal{I}_L:\\
0 = \hat{V}_{i,t}(t,m)+\\
\int_x H_i^r(x,m,(\hat{V}_{i',m},\hat{V}_{i',xm},\hat{V}_{i,xxm})_{i'\in\mathcal{I}_L\cup\mathcal{I}_F})m(t,dx),\\
\hat{V}_i(T,m) = \int m(T,dy)h_i(y,m(T)),\\
H_i^r = \inf_{u_i\in U_i}\left\{l_i + b\,\hat{V}_{i,xm} + \frac{\sigma^2}{2}\hat{V}_{i,xxm} + J[\hat{V}_{i,m}] \mid \{u_j^*(.,(u_i)_{i\in\mathcal{I}_L})\}_{j\in\mathcal{I}_F}\right\}\\
m_t = -(mb)_x + \frac{1}{2}(m\sigma^2)_{xx} + J^*[m],\ m(0) = m_0.
\end{array}$$

A minimizer of the integrand Hamiltonian H_i^r, denoted by

$$u_i^{ss} = u_i^{ss}(t,x,m,(\hat{V}_{i',m},\hat{V}_{i',xm},\hat{V}_{i',xxm})_{i'\in\mathcal{I}_L\cup\mathcal{I}_F}),$$

provides a candidate Stackelberg strategy for leader i. Note that u_i^{ss} depends on the strategy of the other leaders and of the followers. This can be seen through the following term:

$$(\hat{V}_{i',m}, \hat{V}_{i',xm}, \hat{V}_{i',xxm})_{i\in\mathscr{I}_L\cup\mathscr{I}_F}.$$

A mean-field-type risk-neutral Stackelberg solution between multiple leaders and multiple followers is a strategy $(u_i^{ss})_{i\in\mathscr{I}_L}, (u_j^{ss})_{j\in\mathscr{I}_F}$, satisfying for the leaders

$$\begin{aligned}
&i \in \mathscr{I}_L,\\
&u_i^{ss} \in \arg\min_{u_i\in\mathscr{U}_i}\Big\{EL_i(x, u_i, (u_{i'}^{ss})_{i\in\mathscr{I}_L\backslash\{i\}}, (u_j^{ss})_{j\in\mathscr{I}_F}) :\\
&u_j^{ss} \in \text{rnBR}_j((u_i^{ss})_{i\in\mathscr{I}_L}; (u_{j'}^{ss})_{j'\in\mathscr{I}_F\backslash\{j\}},\ j \in \mathscr{I}_F\Big\},
\end{aligned} \tag{11.47}$$

and for every follower

$$j \in \mathscr{I}_F,\ u_j^{ss} \in \text{rnBR}_j((u_i^{ss})_{i\in\mathscr{I}_L}; (u_{j'}^{ss})_{j'\in\mathscr{I}_F\backslash\{j\}}).$$

Proposition 11.4 *The risk-neutral mean-field-type Stackelberg solution with multiple leaders and multiple followers is given in a semi-explicit form as follows:*

$$\begin{aligned}
&u_j^{ss} = -\eta_j\left(x - \textstyle\int y m(dy)\right) - \bar{\eta}_j \int y m(dy),\ j \in \mathscr{I}_F,\\
&j \in \mathscr{I}_F,\\
&0 = -r_j\eta_j^{2k_j-1} - \textstyle\sum_{j'\in\mathscr{I}_F\backslash\{j\}} \varepsilon_{jj'}\eta_{j'} - \sum_{i\in\mathscr{I}_L} \varepsilon_{ji}\eta_i + b_{2j}\alpha_j + c_j,\\
&0 = -\bar{r}_j\bar{\eta}_j^{2\bar{k}_j-1} - \textstyle\sum_{j'\in\mathscr{I}_F\backslash\{j\}} \bar{\varepsilon}_{jj'}\bar{\eta}_{j'} - \sum_{i\in\mathscr{I}_L} \bar{\varepsilon}_{ji}\bar{\eta}_i\\
&+\bar{b}_{2j}\bar{\alpha}_j + \bar{c}_j,
\end{aligned}$$

$$\begin{aligned}
&i \in \mathscr{I}_L,\\
&0 = -r_i\eta_i^{2k_i-1} - \textstyle\sum_{i'\in\mathscr{I}_L\backslash\{i\}} \varepsilon_{ii'}\eta_{i'} - \sum_{j\in\mathscr{I}_F} \varepsilon_{ij}\eta_j + b_{2i}\alpha_i\\
&+\textstyle\sum_{j\in\mathscr{I}_F} \varepsilon_{ij}\eta_i \frac{\varepsilon_{ji}}{(2k_j-1)r_j\eta_j^{2k_j-2}}\\
&-\textstyle\sum_{j\in\mathscr{I}_F} b_{2j} \frac{\varepsilon_{ji}}{(2k_j-1)r_j\eta_j^{2k_j-2}}\alpha_i + c_i,\\
&0 = -\bar{r}_i\bar{\eta}_i^{2\bar{k}_i-1} - \textstyle\sum_{i'\in\mathscr{I}_L\backslash\{i\}} \bar{\varepsilon}_{ii'}\bar{\eta}_{i'} - \sum_{j\in\mathscr{I}_F} \bar{\varepsilon}_{ij}\bar{\eta}_j + \bar{b}_{2i}\bar{\alpha}_i\\
&+\textstyle\sum_{j\in\mathscr{I}_F} \bar{\varepsilon}_{ij}\bar{\eta}_i \frac{\bar{\varepsilon}_{ji}}{(2\bar{k}_j-1)\bar{r}_j\bar{\eta}_j^{2\bar{k}_j-2}}\\
&-\textstyle\sum_{j\in\mathscr{I}_F} \bar{b}_{2j} \frac{\bar{\varepsilon}_{ji}}{(2\bar{k}_j-1)\bar{r}_j\bar{\eta}_j^{2\bar{k}_j-2}}\bar{\alpha}_i + \bar{c}_i,
\end{aligned}$$

and

$$
\begin{aligned}
&\hat{V}_i(0,m) = \alpha_i(0)\int_x \frac{(x-\int y m_0(dy))^{2k_i}}{2k_i} m_0(dx) + \bar{\alpha}_i(0)\frac{(\int y m_0(dy))^{2\bar{k}_i}}{2\bar{k}_i},\\
&0 = \dot{\alpha}_i + q_i + r_i\eta_i^{2k_i} - 2k_i c_i \eta_i + 2k_i \textstyle\sum_{i'\in\mathcal{I}_L\setminus\{i\}} \varepsilon_{ii'}\eta_i\eta_{i'}\\
&+2k_i \textstyle\sum_{j\in\mathcal{I}_F} \varepsilon_{ij}\eta_i\eta_j\\
&+2k_i\left[b_1 - \textstyle\sum_{i'\in\mathcal{I}_L} b_{2i'}\eta_{i'} - \sum_{j\in\mathcal{I}_F} b_{2j}\eta_j\right]\alpha_i\\
&+2k_i(2k_i-1)\alpha_i\tfrac{1}{2}\tilde{\sigma}^2\\
&+\alpha_i\int_\Theta[(1+\tilde{\mu})^{2k_i} - 1 - 2k_i\tilde{\mu}]\nu(d\theta),\\
&\alpha_i(T) = q_{iT},\\
&0 = \dot{\bar{\alpha}}_i + \bar{q}_i + \bar{r}_i\bar{\eta}_i^{2\bar{k}_i} - 2\bar{k}_i\bar{c}_i\bar{\eta}_i\\
&+2\bar{k}_i \textstyle\sum_{i'\in\mathcal{I}_L\setminus\{i\}} \bar{\varepsilon}_{ii'}\bar{\eta}_i\bar{\eta}_{i'} + 2\bar{k}_i\sum_{j\in\mathcal{I}_F}\bar{\varepsilon}_{ij}\bar{\eta}_i\bar{\eta}_j\\
&+2\bar{k}_i\{\bar{b}_1 - \textstyle\sum_{i'\in\mathcal{I}_L}\bar{b}_{2i'}\bar{\eta}_{i'} - \sum_{j\in\mathcal{I}_F}\bar{b}_{2j}\bar{\eta}_j\}\bar{\alpha}_i,\\
&\bar{\alpha}_i(T) = \bar{q}_{iT},
\end{aligned}
$$

with

$$
\int y m(t,dy) = \left[\int y m(0,dy)\right] e^{\int_0^t [\bar{b}_1 - \sum_j \bar{b}_{2j}\bar{\eta}_j]dt'}, \tag{11.48}
$$

whenever the above coefficient system admits a unique solution.

Proof of Proposition 11.4 For the data in (11.40), the integrand Hamiltonian H_j^r, $j\in\mathcal{I}_F$ has a unique minimizer, denoted by

$$
u_j^* = u_j^*(t,x,m,(\hat{V}_{j',m},\hat{V}_{j',xm},\hat{V}_{j',xxm})_{j'\in\mathcal{I}_F},(u_i)_{i\in\mathcal{I}_L}),
$$

which provides the reaction strategies of the follower agents. Following (11.40) with leaders in $\mathcal{I}_L$ and followers in $\mathcal{I}_F$, the first-order optimality condition yields

$$
\begin{aligned}
&j\in\mathcal{I}_F,\\
&0 = r_j(u_j-\bar{u}_j)^{2k_j-1} + c_i(x-\bar{x})^{2k_i-1}\\
&+\textstyle\sum_{j'\in\mathcal{I}_F\setminus\{j\}}\varepsilon_{jj'}(x-\bar{x})^{2(k_j-1)}(u_{j'}-\bar{u}_{j'})\\
&+\textstyle\sum_{i\in\mathcal{I}_L}\varepsilon_{ji}(x-\bar{x})^{2(k_j-1)}(u_i-\bar{u}_i)\\
&+\left[\hat{V}_{j,xm}(t,m) - \int \hat{V}_{j,xm}(t,m)(x)m(dx)\right]b_{2j},\\
&0 = \bar{r}_j\bar{u}_j^{2\bar{k}_j-1} + \bar{c}_i\bar{x}^{2\bar{k}_i-1} + \textstyle\sum_{j'\in\mathcal{I}_F\setminus\{j\}}\bar{\varepsilon}_{jj'}\bar{x}^{2(\bar{k}_j-1)}\bar{u}_{j'}\\
&+\textstyle\sum_{i\in\mathcal{I}_L}\bar{\varepsilon}_{ji}\bar{x}^{2(\bar{k}_j-1)}\bar{u}_i + \left[\int \hat{V}_{j,xm}(t,m)(x)m(dx)\right]\bar{b}_{2j},
\end{aligned}
$$

and

$$
\begin{aligned}
&j\in\mathcal{I}_F,\\
&\textstyle\sum_{i\in\mathcal{I}_L}\varepsilon_{ji}\eta_i = -r_j\eta_j^{2k_j-1}\sum_{j'\in\mathcal{I}_F\setminus\{j\}}\varepsilon_{jj'}\eta_{j'} + b_{2j}\alpha_j + c_j,\\
&\textstyle\sum_{i\in\mathcal{I}_L}\bar{\varepsilon}_{ji}\bar{\eta}_i = -\bar{r}_j\eta_j^{2\bar{k}_j-1}\sum_{j'\in\mathcal{I}_F\setminus\{j\}}\bar{\varepsilon}_{jj'}\bar{\eta}_{j'} + \bar{b}_{2j}\bar{\alpha}_j + \bar{c}_j,
\end{aligned}
$$

which provides $\{\eta_j, \bar{\eta}_j\}_{j\in\mathscr{I}_F}$ as functions of $\{\eta_i, \bar{\eta}_i\}_{i\in\mathscr{I}_L}$ and $\alpha, \bar{\alpha}$. Following (11.40) with leaders in $\mathscr{I}_L$ and followers in $\mathscr{I}_F$, the leaders' integrand Hamiltonian can be rewritten as follows:

$$
\begin{aligned}
H_i^r &= \inf_{u_i\in U_i}\{l_i + b\,\hat{V}_{i,xm}\} + \tfrac{\sigma^2}{2}\hat{V}_{i,xxm} + J[\hat{V}_{i,m}],\\
&= \inf_{u_i\in U_i} q_i\tfrac{(x-\bar{x})^{2k_i}}{2k_i} + r_i\tfrac{(u_i-\bar{u}_i)^{2k_i}}{2k_i} + c_i(x-\bar{x})^{2k_i-1}(u_i-\bar{u}_i)\\
&+\textstyle\sum_{i'\in\mathscr{I}_L\backslash\{i\}}\varepsilon_{ii'}(x-\bar{x})^{2(k_i-1)}(u_i-\bar{u}_i)(u_{i'}-\bar{u}_{i'})\\
&+\textstyle\sum_{j\in\mathscr{I}_F}\varepsilon_{ij}(x-\bar{x})^{2(k_i-1)}(u_i-\bar{u}_i)(u_j^*-\bar{u}_j^*)\\
&+\bar{q}_i\tfrac{\bar{x}^{2\bar{k}_i}}{2\bar{k}_i} + \bar{r}_i\tfrac{\bar{u}_i^{2\bar{k}_i}}{2\bar{k}_i} + \bar{c}_i\bar{x}^{2\bar{k}_i-1}\bar{u}_i\\
&+\textstyle\sum_{i'\in\mathscr{I}_L\backslash\{i\}}\bar{\varepsilon}_{ii'}\bar{x}^{2(\bar{k}_i-1)}\bar{u}_i\bar{u}_{i'}\\
&+\textstyle\sum_{j\in\mathscr{I}_F}\bar{\varepsilon}_{ij}\bar{x}^{2(\bar{k}_i-1)}\bar{u}_i\bar{u}_j^*\\
&+\left\{b_1(x-\bar{x}) + \textstyle\sum_{i'\in\mathscr{I}_L} b_{2i'}(u_{i'}-\bar{u}_{i'}) + \sum_{j\in\mathscr{I}_F} b_{2j}(u_j^*-\bar{u}_j^*)\right\}\hat{V}_{i,xm}\\
&+\{\bar{b}_1\bar{x} + \textstyle\sum_{i'\in\mathscr{I}_L}\bar{b}_{2i'}\bar{u}_{i'} + \sum_{j\in\mathscr{I}_F}\bar{b}_{2j}\bar{u}_j^*\}\hat{V}_{i,xm} + \tfrac{\sigma^2}{2}\hat{V}_{i,xxm} + J[\hat{V}_{i,m}].
\end{aligned}
$$

In view of the followers system, we have

$$
\left\{
\begin{aligned}
\frac{\partial(u_j^*-\bar{u}_j^*)}{\partial(u_i-\bar{u}_i)} &= -\frac{\varepsilon_{ji}}{(2k_j-1)r_j\eta_j^{2k_j-2}},\\
\frac{\partial\bar{u}_j^*}{\partial\bar{u}_i} &= -\frac{\bar{\varepsilon}_{ji}}{(2\bar{k}_j-1)\bar{r}_j\bar{\eta}_j^{2\bar{k}_j-2}}.
\end{aligned}
\right. \tag{11.49}
$$

The Stackelberg strategies of the leaders satisfy the following system:

$$
\begin{aligned}
0 &= r_i(u_i-\bar{u}_i)^{2k_i-1} + c_i(x-\bar{x})^{2k_i-1} + \textstyle\sum_{i'\in\mathscr{I}_L\backslash\{i\}}\varepsilon_{ii'}(x-\bar{x})^{2(k_i-1)}(u_{i'}-\bar{u}_{i'})\\
&+\textstyle\sum_{j\in\mathscr{I}_F}\varepsilon_{ij}(x-\bar{x})^{2(k_i-1)}(u_j^*-\bar{u}_j^*)\\
&-\textstyle\sum_{j\in\mathscr{I}_F}\varepsilon_{ij}(x-\bar{x})^{2(k_i-1)}(u_i-\bar{u}_i)\dfrac{\varepsilon_{ji}}{(2k_j-1)r_j\eta_j^{2k_j-2}}\\
&+\left[b_{2i} - \textstyle\sum_{j\in\mathscr{I}_F} b_{2j}\dfrac{\varepsilon_{ji}}{(2k_j-1)r_j\eta_j^{2k_j-2}}\right]\alpha_i(x-\bar{x})^{2k_i-1},\\
0 &= \bar{r}_i\bar{u}_i^{2\bar{k}_i-1} + \bar{c}_i\bar{x}^{2\bar{k}_i-1} + \textstyle\sum_{i'\in\mathscr{I}_L\backslash\{i\}}\bar{\varepsilon}_{ii'}\bar{x}^{2(\bar{k}_i-1)}\bar{u}_{i'}\\
&+\textstyle\sum_{j\in\mathscr{I}_F}\bar{\varepsilon}_{ij}\bar{x}^{2(\bar{k}_i-1)}\bar{u}_j^*\\
&-\textstyle\sum_{j\in\mathscr{I}_F}\bar{\varepsilon}_{ij}\bar{x}^{2(\bar{k}_i-1)}\bar{u}_i\dfrac{\bar{\varepsilon}_{ji}}{(2\bar{k}_j-1)\bar{r}_j\bar{\eta}_j^{2\bar{k}_j-2}}\\
&+\left[\bar{b}_{2i} - \textstyle\sum_{j\in\mathscr{I}_F}\bar{b}_{2j}\dfrac{\bar{\varepsilon}_{ji}}{(2\bar{k}_j-1)\bar{r}_j\bar{\eta}_j^{2\bar{k}_j-2}}\right]\bar{\alpha}_i\bar{x}^{2\bar{k}_i-1},
\end{aligned}
$$

whose solution provides the coefficients $(\eta_i^{ss}, \bar{\eta}_i^{ss})_{i\in\mathscr{L}}$. ■

Clearly, the mean-field-type Nash equilibrium (11.43) differs from the Stackelberg solution (11.48) when the ε_{ij} and $\bar{\varepsilon}_{ij}$ are nonzero.

11.4.2.1 Effect of the Total Number of Leaders on the Social Cost

We investigate the effect of the number of leaders on the global performance of the system. The total cost at the Stackelberg solution is

$$S(\mathcal{I}_L, m_0) = \sum_{i\in\mathcal{I}_L} \hat{V}_i(0, m_0) + \sum_{j\in\mathcal{I}_F} \hat{V}_j(0, m_0).$$

For $m_0 = \delta_{x_0}$ and $\bar{k}_i = \bar{k} \geq 1$, the total cost is

$$S(\mathcal{I}_L, m_0) = \left(\sum_{i\in\mathcal{I}_L} \bar{\alpha}_i(0) + \sum_{j\in\mathcal{I}_F} \bar{\alpha}_j(0)\right)\frac{x_0^{2\bar{k}}}{2\bar{k}}.$$

Uniform Coupling

When all other parameters are identical across the players except their role, that is, we have total symmetry at each layer, $S(\mathcal{I}_L, m_0)$ can be expressed as a function $|\mathcal{I}_L|$. It follows from (11.48) that

$$\begin{aligned}
&\chi := |\mathcal{I}_L|,\\
&0 = -\bar{r}(\bar{\eta}^{fo})^{2\bar{k}-1} - (|\mathcal{I}| - \chi - 1)\bar{\varepsilon}\bar{\eta}^{fo}\\
&-\chi\bar{\varepsilon}\bar{\eta}^{lead} + \bar{b}_2\bar{\alpha}^{fo} + c,\\
&0 = -\bar{r}(\bar{\eta}^{lead})^{2\bar{k}-1} - (\chi - 1)\bar{\varepsilon}\bar{\eta}^{lead}\\
&-(|\mathcal{I}| - \chi)\bar{\varepsilon}\bar{\eta}^{fo} + \bar{b}_2\bar{\alpha}^{lead} + \bar{c}\\
&+\frac{\bar{\varepsilon}(|\mathcal{I}|-\chi)(\bar{\varepsilon}\bar{\eta}^{lead}-\bar{\alpha}^{lead}\bar{b}_2)}{(2\bar{k}-1)\bar{r}(\bar{\eta}^{fo})^{2\bar{k}-2}},\\
&\bar{\alpha}^{lead}(0) = \bar{q}_T + \int_0^T \Big\{\bar{q} + \bar{r}(\bar{\eta}^{lead})^{2\bar{k}} - 2\bar{k}\bar{c}\bar{\eta}^{lead}\\
&+2\bar{k}\bar{\varepsilon}\bar{\eta}^{lead}[(\chi - 1)\bar{\eta}^{lead} + (|\mathcal{I}| - \chi)\bar{\eta}^{fo}]\\
&+2\bar{k}\bar{\alpha}^{lead}[\bar{b}_1 - \bar{b}_2\bar{\eta}^{lead}\chi - \bar{b}_2\bar{\eta}^{fo}(|\mathcal{I}| - \chi)]\Big\}dt\\
&\bar{\alpha}^{fo}(0) = \bar{q}_T + \int_0^T \Big\{\bar{q} + \bar{r}(\bar{\eta}^{fo})^{2\bar{k}} - 2\bar{k}\bar{c}\bar{\eta}^{fo}\\
&+2\bar{k}\bar{\varepsilon}\bar{\eta}^{fo}[(|\mathcal{I}| - \chi - 1)\bar{\eta}^{fo} + \chi\bar{\eta}^{lead}]\\
&+2\bar{k}\bar{\alpha}^{fo}[\bar{b}_1 - \bar{b}_2\bar{\eta}^{lead}\chi - \bar{b}_2\bar{\eta}^{fo}(|\mathcal{I}| - \chi)]\Big\}dt.
\end{aligned}$$

The optimal number of leaders is given by

$$|\mathcal{I}_L| \in \arg\min_{\chi}[\chi\bar{\alpha}^{lead}(0) + (|\mathcal{I}| - \chi)\bar{\alpha}^{fo}(0)],$$

Table 11.2 Parameters for the symmetric six-player MFTG

Parameter	Value
$\bar{c}_i = \bar{c}, \forall i \in \mathcal{I}$	0
$\bar{k}_i = \bar{k}, \forall i \in \mathcal{I}$	1
$\bar{\varepsilon}_i = \bar{\varepsilon}, \forall i \in \mathcal{I}$	1
$b_{2i} = b_2, \forall i \in \mathcal{I}$	0.1
$\bar{b}_{2i} = \bar{b}_2, \forall i \in \mathcal{I}$	0.5
$\bar{r}_i = \bar{r}, \forall i \in \mathcal{I}$	2
$\bar{q}_{iT} = \bar{q}_T, \forall i \in \mathcal{I}$	2

where $\bar{\alpha}$ depends on χ as well. We observe that the latter function is not necessarily monotone in $\chi = |\mathcal{I}_L|$. This means that increasing the number of leaders in the interaction does not necessarily improve the global performance of the system.

Impact of the Number of Leaders: Symmetric Case
We numerically investigate $S(|\mathcal{I}_L|, \delta_{x_0})$ as a function of $\chi = |\mathcal{I}_L|$ for $|\mathcal{I}| = 6$. Let us consider a symmetric six-player game problem involving the parameters presented in Table 11.2.

Figure 11.2 presents the evolution of both $\dot{\bar{\alpha}}^{\text{leader}}$ and $\dot{\bar{\alpha}}^{\text{follower}}$ for different numbers of leaders $|\mathcal{I}_L|$. Notice that the initial values $\bar{\alpha}_0^{\text{leader}}$ and $\bar{\alpha}_0^{\text{follower}}$ determine the optimal cost considering that

$$\bar{x}_0 = \bar{x}(0) = \int ym(0, dy) = \int y\delta_{x_0}(dy) = x_0.$$

Figure 11.2 and Table 11.3 also show that, under the given parametric choices, the lowest social cost is obtained when $|\mathcal{I}_L| = 2$, corresponding to a cost $S(|\mathcal{I}_L|, \delta_{x_0}) = 7.911$. These results offer insight into game design for the sake of either individual or global costs.

11.4.3 Multi-Layer Fully Hierarchical Games

A fully hierarchical game is one where there is one decision-maker per layer, i.e., the number of layers is exactly the number of decision-makers. This section examines fully hierarchical game with I decision-makers, where decision-maker i is at layer i. In this context, not only the set of decision-makers per level matters but also the number of hierarchical levels plays a key role in the global performance of the system. We also identify a specific range of parameters for which the Nash equilibrium coincides with the hierarchical solution independently of the number of layers and the order of play. In the previous sections, we had only bi-level problems. Here, we make as many levels as the number of decision-makers. There are $|\mathcal{I}|$ hierarchical levels. Each layer i, decision-maker i chooses a control strategy u_i

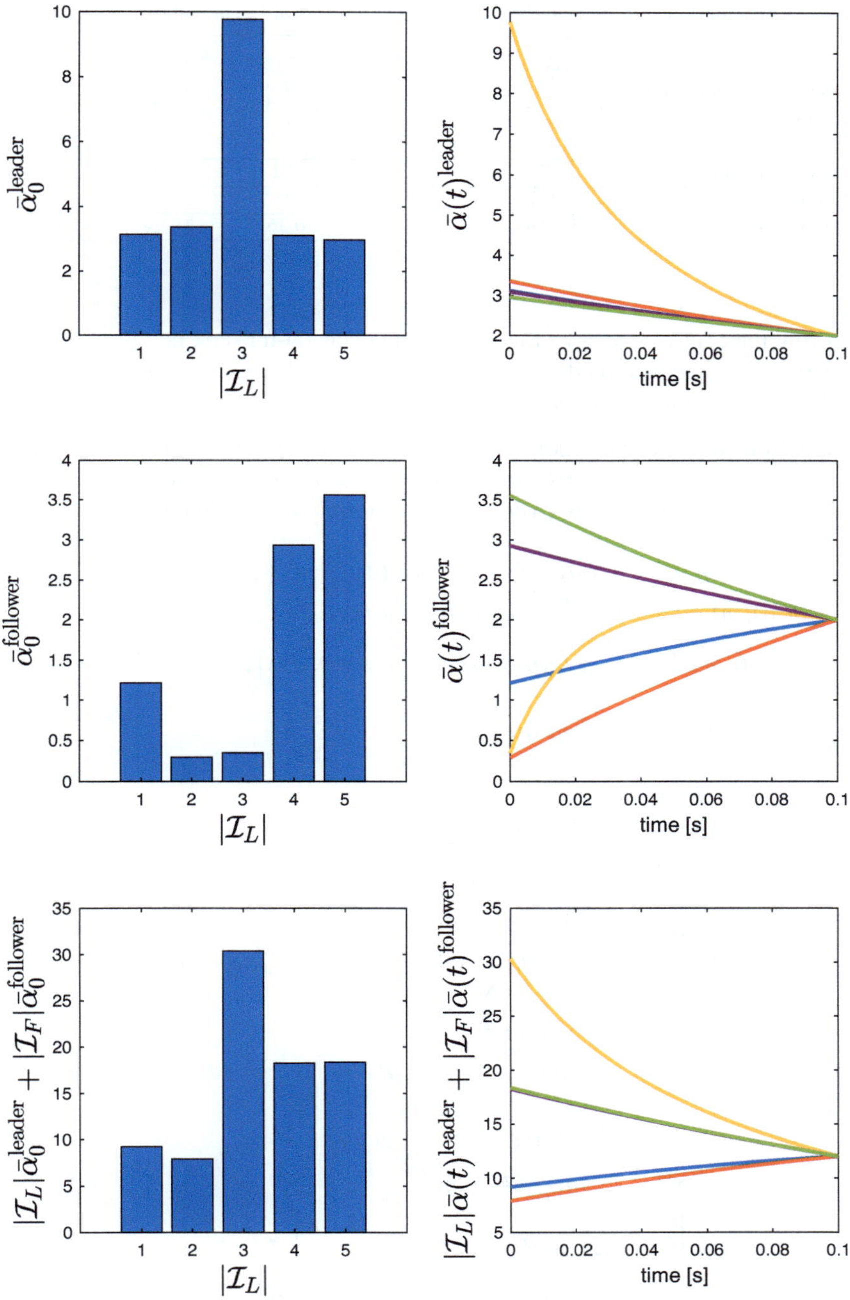

Fig. 11.2 Evolution of $\dot{\bar{\alpha}}^{\text{leader/follower}}$ and the corresponding initial values for different numbers of leaders

Table 11.3 Summary of $\bar{\alpha}_0^{\text{leader}}$, $\bar{\alpha}_0^{\text{follower}}$, and $S(|\mathcal{I}_L|, \delta_{x_0})$ for different numbers of leaders

Leader(s)-Follower(s) Structure	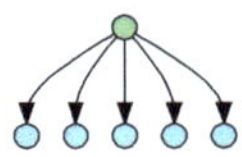	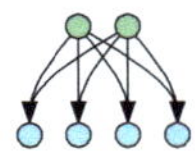	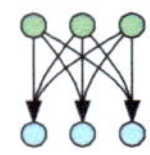	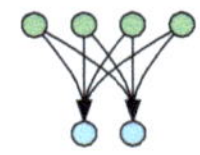	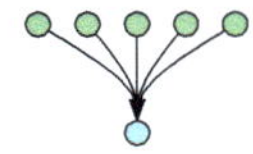
Individual leader cost	3.132	3.37	9.772	3.107	**2.968**
Individual follower cost	1.217	**0.2931**	0.3481	2.933	3.562
Global cost	9.219	**7.911**	30.36	18.29	18.4

knowing the control strategy of the preceding decision-makers, i.e., $\{i-1,\dots,1\}$. This becomes a sequential decision-making problem. We use a backward induction method to solve the hierarchical game problem.

Proposition 11.5 *Assume that all decision-makers in* $\{1,2,\dots,I-1\}$ *use linear-state-and-mean-field feedback strategies.The risk-neutral hierarchical mean-field-type solution with* I *levels is given in a semi-explicit form as follows:*

$$
\begin{array}{l}
u_i^{hs} = -\eta_i\left(x-\int ym(dy)\right) - \bar{\eta}_i\int ym(dy),\, i\in\mathcal{I},\\
\hat{V}_i(0,m) = \alpha_i(0)\int_x \frac{(x-\int ym_0(dy))^{2k_i}}{2k_i} m_0(dx)\\
+\bar{\alpha}_i(0)\frac{(\int ym_0(dy))^{2\bar{k}_i}}{2\bar{k}_i},\ \int ym(t,dy) = \left[\int ym(0,dy)\right] e^{\int_0^t\left[\bar{b}_1-\sum_j \bar{b}_{2j}\bar{\eta}_j\right]dt'},
\end{array}
$$

where the coefficient functions are given by

$$
\begin{array}{l}
\textit{Level } 1:\\
0 = -r_1\eta_1^{2k_1-1} + c_1 - \sum_{j=2}^I \varepsilon_{1,j}\eta_j\\
+\sum_{j=2}^I \varepsilon_{1,j}\eta_1\frac{\varepsilon_{j1}}{(2k_j-1)r_j}\eta_j^{-2(k_j-1)}\\
+\left\{b_{2,1}-\sum_{j=2}^I b_{2j}\frac{\varepsilon_{j1}}{(2k_j-1)r_j}\eta_j^{-2(k_j-1)}\right\}\alpha_1,\\
0 = \dot{\alpha}_1 + q_1 + r_1\eta_1^{2k_1} - 2k_1c_1\eta_1 + 2k_1\sum_{j=2}^I \varepsilon_{1j}\eta_1\eta_j\\
+2k_1\{b_1 - b_{21}\eta_1 - \sum_{j=2}^I b_{2j}\eta_j\}\alpha_1\\
+2k_1(2k_1-1)\alpha_1\frac{1}{2}\tilde{\sigma}^2\\
+\alpha_1\int_\Theta[(1+\tilde{\mu})^{2k_1} - 1 - 2k_1\tilde{\mu}]\nu(d\theta),\\
\alpha_1(T) = q_{1T},\\
0 = -\bar{r}_1\bar{\eta}_1^{2\bar{k}_1-1} + \bar{c}_1 - \sum_{j=2}^I \bar{\varepsilon}_{1,j}\bar{\eta}_j\\
+\sum_{j=2}^I \bar{\varepsilon}_{1,j}\bar{\eta}_1\frac{\bar{\varepsilon}_{j1}}{(2\bar{k}_j-1)\bar{r}_j}\bar{\eta}_j^{-2(\bar{k}_j-1)}\\
+\left\{\bar{b}_{21}-\sum_{j=2}^I \bar{b}_{2j}\frac{\bar{\varepsilon}_{j1}}{(2\bar{k}_j-1)\bar{r}_j}\bar{\eta}_j^{-2(\bar{k}_j-1)}\right\}\bar{\alpha}_1,\\
0 = \dot{\bar{\alpha}}_1 + \bar{q}_1 + \bar{r}_1\bar{\eta}_1^{2\bar{k}_1} - 2\bar{k}_1\bar{c}_1\bar{\eta}_1 + 2\bar{k}_1\sum_{j=2}^I \bar{\varepsilon}_{1j}\bar{\eta}_i\bar{\eta}_j\\
+2\bar{k}_1\{\bar{b}_1 - \bar{b}_{21}\bar{\eta}_1 - \sum_{j=2}^I \bar{b}_{2j}\bar{\eta}_j\}\bar{\alpha}_1,\\
\bar{\alpha}_1(T) = \bar{q}_{1T},
\end{array}
$$

Level i:

$$
\begin{aligned}
&0 = -r_i \eta_i^{2k_i-1} + c_i - \sum_{i'=1}^{i-1} \varepsilon_{I-1,i'}\eta_{i'} - \sum_{j=i+1}^{I} \varepsilon_{i,j}\eta_j \\
&+\sum_{j=i+1}^{I} \varepsilon_{i,j}\eta_i \frac{\varepsilon_{ji}}{(2k_j-1)r_j}\eta_j^{-2(k_j-1)} \\
&+\left[b_{2i} - \sum_{j=i+1}^{I} b_{2j}\frac{\varepsilon_{ji}}{(2k_j-1)r_j}\eta_j^{-2(k_j-1)}\right]\alpha_i, \\
&0 = \dot{\alpha}_i + q_i + r_i\eta_i^{2k_i} - 2k_i c_i \eta_i + 2k_i \sum_{i'=1}^{i-1} \varepsilon_{ii'}\eta_i\eta_{i'} \\
&+2k_i \sum_{j=i+1}^{I} \varepsilon_{ij}\eta_i\eta_j \\
&+2k_i\{b_1 - \sum_{i'=1}^{i-1} b_{2i'}\eta_{i'} - b_{2i}\eta_i - \sum_{j=i+1}^{I} b_{2j}\eta_j\}\alpha_i \\
&+2k_i(2k_i-1)\alpha_i \tfrac{1}{2}\tilde{\sigma}^2 \\
&+\alpha_i \int_{\Theta}[(1+\tilde{\mu})^{2k_i} - 1 - 2k_i\tilde{\mu}]\nu(d\theta), \\
&\alpha_i(T) = q_{iT}, \\
&0 = -\bar{r}_i \bar{\eta}_i^{2\bar{k}_i-1} + \bar{c}_i - \sum_{i'=1}^{i-1} \bar{\varepsilon}_{I-1,i'}\bar{\eta}_{i'} - \sum_{j=i+1}^{I} \bar{\varepsilon}_{i,j}\bar{\eta}_j \\
&+\sum_{j=i+1}^{I} \bar{\varepsilon}_{i,j}\bar{\eta}_i \frac{\bar{\varepsilon}_{ji}}{(2\bar{k}_j-1)\bar{r}_j}\bar{\eta}_j^{-2(\bar{k}_j-1)} \\
&+\left[\bar{b}_{2i} - \sum_{j=i+1}^{I} \bar{b}_{2j}\frac{\bar{\varepsilon}_{ji}}{(2\bar{k}_j-1)\bar{r}_j}\bar{\eta}_j^{-2(\bar{k}_j-1)}\right]\bar{\alpha}_i, \\
&0 = \dot{\bar{\alpha}}_i + \bar{q}_i + \bar{r}_i\bar{\eta}_i^{2\bar{k}_i} - 2\bar{k}_i\bar{c}_i\bar{\eta}_i + 2\bar{k}_i \sum_{i'=1}^{i-1} \bar{\varepsilon}_{ii'}\bar{\eta}_i\bar{\eta}_{i'} \\
&+2\bar{k}_i \sum_{j=i+1}^{I} \bar{\varepsilon}_{ij}\bar{\eta}_i\bar{\eta}_j \\
&+2\bar{k}_i\left[\bar{b}_1 - \sum_{i'=1}^{i-1} \bar{b}_{2i'}\bar{\eta}_{i'} - \bar{b}_{2i}\bar{\eta}_i - \sum_{j=i+1}^{I} \bar{b}_{2j}\bar{\eta}_j\right]\bar{\alpha}_i, \\
&\bar{\alpha}_i(T) = \bar{q}_{iT},
\end{aligned}
$$

Level I:

$$
\begin{aligned}
&\eta_I = \left(\frac{-\sum_{j=1}^{I-1} \varepsilon_{I,j}\eta_j + b_{2I}\alpha_I + c_I}{r_I}\right)^{\frac{1}{2k_I-1}}, \\
&0 = \dot{\alpha}_I + q_I + r_I\eta_I^{2k_I} - 2k_I c_I \eta_I + 2k_I \sum_{i'=1}^{I-1} \varepsilon_{Ii'}\eta_I\eta_{i'} \\
&+2k_I\{b_1 - \sum_{i'=1}^{I-1} b_{2i'}\eta_{i'} - b_{2I}\eta_I\}\alpha_I \\
&+2k_I(2k_I-1)\alpha_I \tfrac{1}{2}\tilde{\sigma}^2 \\
&+\alpha_I \int_{\Theta}[(1+\tilde{\mu})^{2k_I} - 1 - 2k_I\tilde{\mu}]\nu(d\theta), \\
&\alpha_I(T) = q_{IT}, \\
&\bar{\eta}_I = \left(\frac{-\sum_{j=1}^{I-1} \bar{\varepsilon}_{I,j}\bar{\eta}_j + \bar{b}_{2I}\bar{\alpha}_I + \bar{c}_I}{\bar{r}_I}\right)^{\frac{1}{2\bar{k}_I-1}}, \\
&0 = \dot{\bar{\alpha}}_I + \bar{q}_I + \bar{r}_I\bar{\eta}_I^{2\bar{k}_I} - 2\bar{k}_I\bar{c}_I\bar{\eta}_I + 2\bar{k}_I \sum_{i'=1}^{I-1} \bar{\varepsilon}_{Ii'}\bar{\eta}_I\bar{\eta}_{i'} \\
&+2\bar{k}_I\{\bar{b}_1 - \sum_{i'=1}^{i-1} \bar{b}_{2i'}\bar{\eta}_{i'} - \bar{b}_{2I}\bar{\eta}_I\}\bar{\alpha}_I, \\
&\bar{\alpha}_I(T) = \bar{q}_{IT},
\end{aligned}
$$

whenever these equations admit solutions.

Proof of Proposition 11.5 We use a backward induction process to prove the desired result.

I-th Hierarchical Level

Here, the preceding decision-makers have already chosen their strategies, and they are known to decision-maker I. Hence, the integrand Hamiltonian of I is

$$\begin{aligned}
H_I &= \inf_{u_I\in U_I}\{l_I + b\,\hat{V}_{I,xm}\} + \tfrac{\sigma^2}{2}\hat{V}_{I,xxm} + J[\hat{V}_{I,m}] \\
&= \inf_{u_I\in U_I} q_I\tfrac{(x-\bar{x})^{2k_I}}{2k_I} + r_I\tfrac{(u_I-\bar{u}_I)^{2k_I}}{2k_I} \\
&+c_I(x-\bar{x})^{2k_I-1}(u_I-\bar{u}_I) \\
&+\textstyle\sum_{i'=1}^{I-1}\varepsilon_{I,i'}(x-\bar{x})^{2(k_I-1)}(u_I-\bar{u}_I)(u_{i'}-\bar{u}_{i'}) \\
&+\Big[b_1(x-\bar{x})+\textstyle\sum_{i'=1}^{I-1}b_{2i'}(u_{i'}-\bar{u}_{i'})+b_{2,I}(u_I-\bar{u}_I)\Big]\hat{V}_{I,xm} \\
&+\bar{q}_I\tfrac{\bar{x}^{2\bar{k}_I}}{2\bar{k}_I}+\bar{r}_I\tfrac{\bar{u}_I^{2\bar{k}_I}}{2\bar{k}_I}+\bar{c}_I\bar{x}^{2\bar{k}_I-1}\bar{u}_I+\textstyle\sum_{i'=1}^{I-1}\bar{\varepsilon}_{I,i'}\bar{x}^{2(\bar{k}_I-1)}\bar{u}_I\bar{u}_{i'} \\
&+\Big[\bar{b}_1\bar{x}+\textstyle\sum_{i'=1}^{I-1}\bar{b}_{2i'}\bar{u}_{i'}+\bar{b}_{2,I}\bar{u}_I\Big]\hat{V}_{I-1,xm} \\
&+\tfrac{\sigma^2}{2}\hat{V}_{I,xxm}+J[\hat{V}_{I,m}].
\end{aligned} \tag{11.50}$$

It follows from strictly convex optimization above that the best-response strategy can be expressed as

$$\begin{aligned}
u_I^* - \bar{u}_I^* &= -\xi_1^{\frac{1}{2k_I-1}}, \\
\bar{u}_I^* &= -\xi_2^{\frac{1}{2\bar{k}_I-1}},
\end{aligned} \tag{11.51}$$

where

$$\begin{aligned}
\xi_1 &= \tfrac{1}{r_I}\Big(\textstyle\sum_{i'=1}^{I-1}\varepsilon_{I,i'}(x-\bar{x})^{2(k_I-1)}(u_{i'}-\bar{u}_{i'}) \\
&+b_{2,I}\hat{V}_{I,xm}+c_I(x-\bar{x})^{2k_I-1}\Big), \\
\xi_2 &= \tfrac{1}{\bar{r}_I}\Big(\textstyle\sum_{i'=1}^{I-1}\bar{\varepsilon}_{I,i'}\bar{x}^{2(\bar{k}_I-1)}\bar{u}_{i'}+\bar{b}_{2,I}\hat{V}_{I,xm}+\bar{c}_I\bar{x}^{2\bar{k}_I-1}.\Big)
\end{aligned} \tag{11.52}$$

In particular

$$\begin{aligned}
& i \le I-1: \\
& \tfrac{\partial(u_I^*-\bar{u}_I^*)}{\partial(u_i-\bar{u}_i)} = \tfrac{\varepsilon_{I,i}}{(2k_I-1)r_I}(x-\bar{x})^{2(k_I-1)}(u_I^*-\bar{u}_I^*)^{-2(k_I-1)}, \\
& \tfrac{\partial\bar{u}_I^*}{\partial\bar{u}_i} = \tfrac{\bar{\varepsilon}_{I,i}}{(2\bar{k}_I-1)\bar{r}_I}\bar{x}^{2(\bar{k}_I-1)}(\bar{u}_I^*)^{-2(\bar{k}_I-1)}.
\end{aligned}$$

If the preceding decision-makers $\{1, 2, \ldots, I-1\}$ have all used linear-state-and-mean-field feedback strategies, then the reaction of the I-th decision-maker who is at the I-th level of hierarchy can be rewritten as

$$
\begin{aligned}
u_I^{seq} &= -\eta_I \left(x - \int y m(dy)\right) - \bar{\eta}_I \int y m(dy),\\
\eta_I &= \left(\frac{-\sum_{j=1}^{I-1} \varepsilon_{I,j}\eta_j + b_{2I}\alpha_I + c_I}{r_I}\right)^{\frac{1}{2k_I-1}},\\
\bar{\eta}_I &= \left(\frac{-\sum_{j=1}^{I-1} \bar{\varepsilon}_{I,j}\bar{\eta}_j + \bar{b}_{2I}\bar{\alpha}_I + \bar{c}_I}{\bar{r}_I}\right)^{\frac{1}{2\bar{k}_I-1}}.
\end{aligned}
\tag{11.53}
$$

$(I-1)$-th Level

At the hierarchical level $I-1$, the preceding levels are $\{1, 2, \ldots, I-2\}$, and the succeeding level is I. Having the expression of the optimal response strategies of the last layer I, we can move to the preceding layer, i.e., $I-1$. Decision-maker $I-1$ has $u_1, \ldots, u_{I-2}$ and the reaction u_I^* of decision-maker I. Therefore, the integrand Hamiltonian of $I-1$ is given by

$$
\begin{aligned}
&H_{I-1}^r = \inf_{u_{I-1}\in U_{I-1}}\{l_i + b\, \hat{V}_{I-1,xm}\} + \tfrac{\sigma^2}{2}\hat{V}_{I-1,xxm} + J[\hat{V}_{I-1,m}]\\
&= \inf_{u_{I-1}\in U_{I-1}} q_{I-1}\frac{(x-\bar{x})^{2k_{I-1}}}{2k_{I-1}} + r_{I-1}\frac{(u_{I-1}-\bar{u}_{I-1})^{2k_{I-1}}}{2k_{I-1}}\\
&+c_{I-1}(x-\bar{x})^{2k_{I-1}-1}(u_{I-1}-\bar{u}_{I-1})\\
&+\sum_{i'=1}^{I-2}\varepsilon_{I-1,i'}(x-\bar{x})^{2(k_{I-1}-1)}(u_{I-1}-\bar{u}_{I-1})(u_{i'}-\bar{u}_{i'})\\
&+\varepsilon_{I-1,I}(x-\bar{x})^{2(k_{I-1}-1)}(u_{I-1}-\bar{u}_{I-1})(u_I^*-\bar{u}_I^*)\\
&+\Bigg[b_1(x-\bar{x}) + \sum_{i'=1}^{I-2} b_{2i'}(u_{i'}-\bar{u}_{i'})\\
&+b_{2,I-1}(u_{I-1}-\bar{u}_{I-1}) + b_{2I}(u_I^*-\bar{u}_I^*)\Bigg]\hat{V}_{I-1,xm}\\
&+\bar{q}_{I-1}\frac{\bar{x}^{2\bar{k}_{I-1}}}{2\bar{k}_{I-1}} + \bar{r}_{I-1}\frac{\bar{u}_{I-1}^{2\bar{k}_{I-1}}}{2\bar{k}_{I-1}}\\
&+\bar{c}_{I-1}\bar{x}^{2\bar{k}_{I-1}-1}\bar{u}_{I-1}\\
&+\sum_{i'=1}^{I-2}\bar{\varepsilon}_{I-1,i'}\bar{x}^{2(\bar{k}_{I-1}-1)}\bar{u}_{I-1}\bar{u}_{i'}\\
&+\bar{\varepsilon}_{I-1,I}\bar{x}^{2(\bar{k}_{I-1}-1)}\bar{u}_{I-1}\bar{u}_I^*\\
&+\{\bar{b}_1\bar{x} + \sum_{i'=1}^{I-2}\bar{b}_{2i'}\bar{u}_{i'} + \bar{b}_{2,I-1}\bar{u}_{I-1} + \bar{b}_{2I}\bar{u}_I^*\}\hat{V}_{I-1,xm}\\
&+\tfrac{\sigma^2}{2}\hat{V}_{I-1,xxm} + J[\hat{V}_{I-1,m}].
\end{aligned}
\tag{11.54}
$$

In view of (11.53), the terms with $\bar{u}_I^*$ depend on $\bar{u}_{I-1}, \bar{u}_{I-2}, \ldots, \bar{u}_1$. The first-order optimality condition for u_{I-1}^* yields

$$
\begin{aligned}
&0 = -r_{I-1}\eta_{I-1}^{2k_{I-1}-1} + c_{I-1} - \sum_{i'=1}^{I-2}\varepsilon_{I-1,i'}\eta_{i'} - \varepsilon_{I-1,I}\eta_I\\
&+\varepsilon_{I-1,I}\eta_{I-1}\frac{\varepsilon_{I,I-1}}{(2k_I-1)r_I}\eta_I^{-2(k_I-1)}\\
&+\left\{b_{2,I-1} - b_{2I}\frac{\varepsilon_{I,I-1}}{(2k_I-1)r_I}\eta_I^{-2(k_I-1)}\right\}\alpha_{I-1},\\
&0 = -\bar{r}_{I-1}\bar{\eta}_{I-1}^{2\bar{k}_{I-1}-1} + \bar{c}_{I-1} - \sum_{i'=1}^{I-2}\bar{\varepsilon}_{I-1,i'}\bar{\eta}_{i'} - \bar{\varepsilon}_{I-1,I}\bar{\eta}_I\\
&+\bar{\varepsilon}_{I-1,I}\bar{\eta}_{I-1}\frac{\bar{\varepsilon}_{I,I-1}}{(2\bar{k}_I-1)\bar{r}_I}\bar{\eta}_I^{-2(\bar{k}_I-1)}\\
&+\left\{\bar{b}_{2,I-1} - \bar{b}_{2,I}\frac{\bar{\varepsilon}_{I,I-1}}{(2\bar{k}_I-1)\bar{r}_I}\bar{\eta}_I^{-2(\bar{k}_I-1)}\right\}\bar{\alpha}_{I-1},
\end{aligned}
\tag{11.55}
$$

where we have used (11.53) for $i = I - 1$.

$$u_{I-1}^{seq} = -\eta_{I-1}\left(x - \int ym(dy)\right) - \bar{\eta}_{I-1}\int ym(dy). \tag{11.56}$$

i-th Level
For $i \in \{2, \dots, I-2\}$

$$\begin{array}{l}
H_i^r = \inf_{u_i\in U_i} q_i \frac{(x-\bar{x})^{2k_i}}{2k_i} + r_i \frac{(u_i-\bar{u}_i)^{2k_i}}{2k_i} + c_i(x-\bar{x})^{2k_i-1}(u_i-\bar{u}_i)\\
+\sum_{i'=1}^{i-1}\varepsilon_{ii'}(x-\bar{x})^{2(k_i-1)}(u_i-\bar{u}_i)(u_{i'}-\bar{u}_{i'})\\
+\sum_{j=i+1}^{I}\varepsilon_{ij}(x-\bar{x})^{2(k_i-1)}(u_i-\bar{u}_i)(u_j^*-\bar{u}_j^*)\\
+\left[b_1(x-\bar{x})+\sum_{i'=1}^{i-1} b_{2i'}(u_{i'}-\bar{u}_{i'})+b_{2i}(u_i-\bar{u}_i)+\sum_{j=i+1}^{I} b_{2j}(u_j^*-\bar{u}_j^*)\right]\hat{V}_{i,xm}\\
+\bar{q}_i\frac{\bar{x}^{2\bar{k}_i}}{2\bar{k}_i} + \bar{r}_i\frac{\bar{u}_i^{2\bar{k}_i}}{2\bar{k}_i} + \bar{c}_i\bar{x}^{2\bar{k}_i-1}\bar{u}_i + \sum_{i'=1}^{i-1}\bar{\varepsilon}_{ii'}\bar{x}^{2(\bar{k}_i-1)}\bar{u}_i\bar{u}_{i'}\\
+\sum_{j=i+1}^{I}\bar{\varepsilon}_{ij}\bar{x}^{2(\bar{k}_i-1)}\bar{u}_i\bar{u}_j^*\\
+\left\{\bar{b}_1\bar{x}+\sum_{i'=1}^{i-1}\bar{b}_{2i'}\bar{u}_{i'}+\bar{b}_{2i}\bar{u}_i+\sum_{j=i+1}^{I}\bar{b}_{2j}\bar{u}_j^*\right\}\hat{V}_{i,xm}\\
+\frac{\sigma^2}{2}\hat{V}_{i,xxm}+J[\hat{V}_{i,m}].
\end{array} \tag{11.57}$$

Via matching, from the first-order optimality condition, the coefficient functions $\eta_i, \bar{\eta}_i$ satisfy the following equations:

$$\begin{array}{l}
0 = -r_i\eta_i^{2k_i-1} + c_i - \sum_{i'=1}^{i-1}\varepsilon_{I-1,i'}\eta_{i'} - \sum_{j=i+1}^{I}\varepsilon_{i,j}\eta_j\\
+\sum_{j=i+1}^{I}\varepsilon_{i,j}\eta_i\frac{\varepsilon_{ji}}{(2k_j-1)r_j}\eta_j^{-2(k_j-1)}\\
+\left\{b_{2i}-\sum_{j=i+1}^{I} b_{2j}\frac{\varepsilon_{ji}}{(2k_j-1)r_j}\eta_j^{-2(k_j-1)}\right\}\alpha_i,\\
0 = -\bar{r}_i\bar{\eta}_i^{2\bar{k}_i-1} + \bar{c}_i - \sum_{i'=1}^{i-1}\bar{\varepsilon}_{I-1,i'}\bar{\eta}_{i'} - \sum_{j=i+1}^{I}\bar{\varepsilon}_{i,j}\bar{\eta}_j\\
+\sum_{j=i+1}^{I}\bar{\varepsilon}_{i,j}\bar{\eta}_i\frac{\bar{\varepsilon}_{ji}}{(2\bar{k}_j-1)\bar{r}_j}\bar{\eta}_j^{-2(\bar{k}_j-1)}\\
+\left\{\bar{b}_{2i}-\sum_{j=i+1}^{I}\bar{b}_{2j}\frac{\bar{\varepsilon}_{ji}}{(2\bar{k}_j-1)\bar{r}_j}\bar{\eta}_j^{-2(\bar{k}_j-1)}\right\}\bar{\alpha}_i.
\end{array} \tag{11.58}$$

The strategies are $u_i^{seq} = -\eta_i\left(x - \int ym(dy)\right) - \bar{\eta}_i\int ym(dy)$.

1-st Hierarchical Level
We now examine the first level of the hierarchy. The integrand Hamiltonian of decision-maker 1 is

$$
\begin{aligned}
H_1^r &= \inf_{u_1\in U_1} q_1\frac{(x-\bar{x})^{2k_1}}{2k_1} + r_1\frac{(u_1-\bar{u}_1)^{2k_1}}{2k_1} + c_1(x-\bar{x})^{2k_1-1}(u_1-\bar{u}_1)\\
&+\sum_{j=2}^{I}\varepsilon_{1j}(x-\bar{x})^{2(k_1-1)}(u_1-\bar{u}_1)(u_j^*-\bar{u}_j^*)\\
&+\left\{b_1(x-\bar{x})+b_{21}(u_1-\bar{u}_1)+\sum_{j=2}^{I}b_{2j}(u_j^*-\bar{u}_j^*)\right\}\hat{V}_{1,xm}\\
&+\bar{q}_1\frac{\bar{x}^{2\bar{k}_1}}{2\bar{k}_1}+\bar{r}_1\frac{\bar{u}_1^{2\bar{k}_1}}{2\bar{k}_1}+\bar{c}_1\bar{x}^{2\bar{k}_1-1}\bar{u}_1+\sum_{j=2}^{I}\bar{\varepsilon}_{1j}\bar{x}^{2(\bar{k}_1-1)}\bar{u}_1\bar{u}_j^*\\
&+\left\{\bar{b}_1\bar{x}+\bar{b}_{21}\bar{u}_1+\sum_{j=2}^{I}\bar{b}_{2j}\bar{u}_j^*\right\}\hat{V}_{1,xm}\\
&+\frac{\sigma^2}{2}\hat{V}_{1,xxm}+J[\hat{V}_{1,m}].
\end{aligned}
\tag{11.59}
$$

Via matching, from the first-order optimality condition, the coefficient functions $\eta_1, \bar{\eta}_1$ satisfy the following equations:

$$
\begin{aligned}
0 &= -r_1\eta_1^{2k_1-1}+c_1-\sum_{j=2}^{I}\varepsilon_{1j}\eta_j+\sum_{j=2}^{I}\varepsilon_{1j}\eta_1\frac{\varepsilon_{j1}}{(2k_j-1)r_j}\eta_j^{-2(k_j-1)}\\
&+\left\{b_{21}-\sum_{j=2}^{I}b_{2j}\frac{\varepsilon_{j1}}{(2k_j-1)r_j}\eta_j^{-2(k_j-1)}\right\}\alpha_1,\\
0 &= -\bar{r}_1\bar{\eta}_1^{2\bar{k}_1-1}+\bar{c}_1-\sum_{j=2}^{I}\bar{\varepsilon}_{1j}\bar{\eta}_j+\sum_{j=2}^{I}\bar{\varepsilon}_{1j}\bar{\eta}_1\frac{\bar{\varepsilon}_{j1}}{(2\bar{k}_j-1)\bar{r}_j}\bar{\eta}_j^{-2(\bar{k}_j-1)}\\
&+\left\{\bar{b}_{21}-\sum_{j=2}^{I}\bar{b}_{2j}\frac{\bar{\varepsilon}_{j1}}{(2\bar{k}_j-1)\bar{r}_j}\bar{\eta}_j^{-2(\bar{k}_j-1)}\right\}\bar{\alpha}_1.
\end{aligned}
\tag{11.60}
$$

The response strategy is $u_1^{seq} = -\eta_1\left(x-\int ym(dy)\right)-\bar{\eta}_1\int ym(dy)$.
Putting all together, we arrive at the desired result, which completes the proof. ∎

11.4.3.1 Design of the Order

This subsection examines the dependence of the social cost on the design of the order of hierarchy. In the heteregeneous case, the permutation of the players does affect significantly the social cost. Therefore, it is important to know who should be the leader and who should be the i-th follower so that the social cost is minimized. Toward such an analysis, we consider the following parameter values:

$$
\begin{aligned}
&b_{21}=0.1,\ \ b_{22}=0.2,\ \ b_{23}=0.3,\\
&\bar{b}_{21}=0.4,\ \ \bar{b}_{22}=0.5,\ \ \bar{b}_{23}=0.6,\\
&\bar{r}_1=1,\ \bar{r}_2=2,\ \bar{r}_3=3,\\
&\bar{q}_{1T}=2.1,\ \ \bar{q}_{2T}=2.2,\ \ \bar{q}_{3T}=2.3,\\
&\bar{b}_1=2,\ T=0.1, \qquad\qquad \bar{k}_i=\bar{k}=1,\ i\in\mathcal{I},
\end{aligned}
\tag{11.61}
$$

$$
\bar{\varepsilon}=\begin{pmatrix}1 & 1.2 & 1.1\\ 1.5 & 1 & 1.6\\ 1.3 & 1.4 & 1\end{pmatrix}.
\tag{11.62}
$$

Table 11.4 Social cost $S(\mathscr{I}_L, \delta_{x_0})$ for different hierarchical orders in a three-player case

Hierarchical Structure

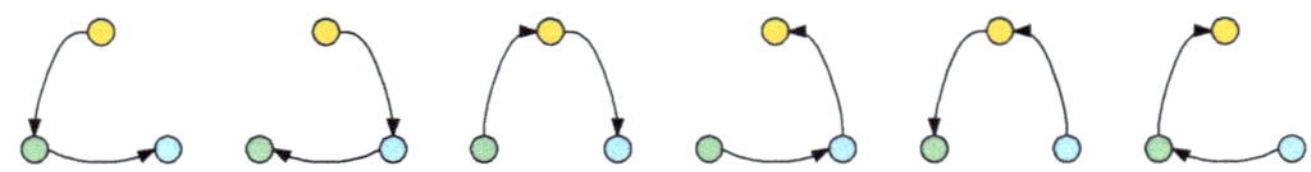

Combination label	1	2	3	4	5	6
Hierarchical order	{1}{2}{3}	{1}{3}{2}	{2}{1}{3}	{2}{3}{1}	{3}{1}{2}	{3}{2}{1}
Global cost	6.124	7.464	**5.864**	8.757	6.894	8.433

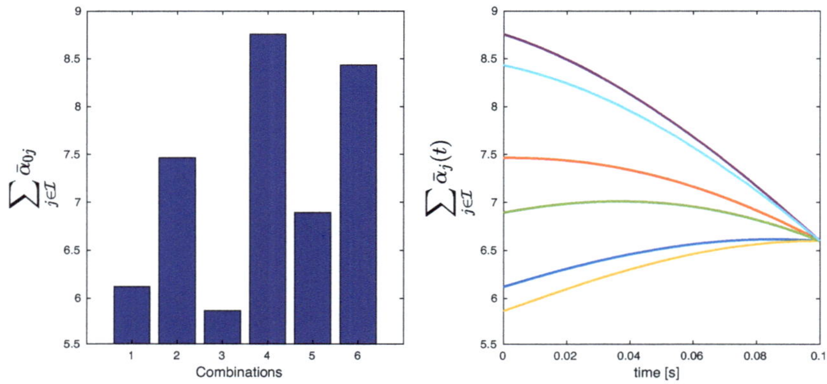

Fig. 11.3 Evolution of $\sum_{j\in\mathscr{I}} \dot{\bar{\alpha}}_j(t)$ and the corresponding initial values for different hierarchical structures

Table 11.5 Global cost for different hierarchy and leadership structures in the **heterogeneous** scenario

Leader(s)-Follower(s) Structure

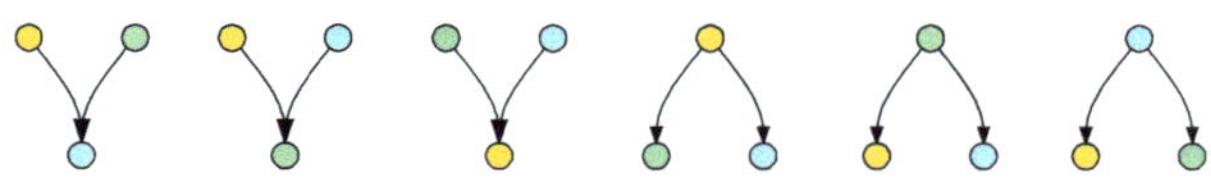

Leaders	{12}	{13}	{23}	{1}	{2}	{3}
Followers	{3}	{2}	{1}	{23}	{13}	{12}
Global cost	17.14	16.96	16.99	17.04	17.13	**16.92**

Table 11.4 presents a summary of the global costs for the six different possible hierarchical orders. It can be seen that the second configuration is the best in terms of leading to the lowest social cost. Moreover, Fig. 11.3 presents the evolution of $\sum_{j\in\mathscr{I}} \dot{\bar{\alpha}}_j(t)$ for all the possible hierarchical structures.

Table 11.5 presents the social cost for all hierarchical structures when there are three decision-makers and two levels.

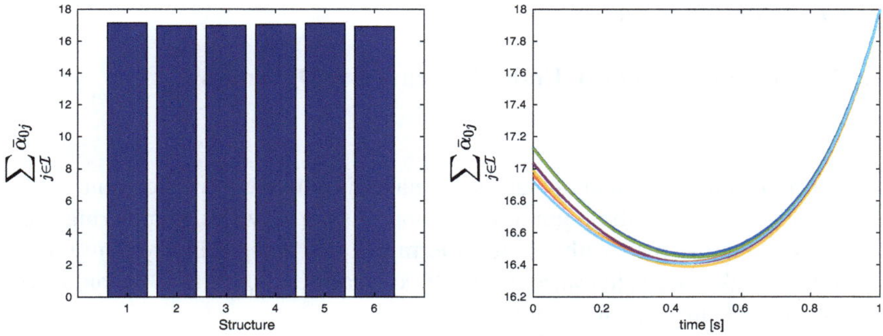

Fig. 11.4 Evolution of the sum of differential equations and the corresponding social cost for different hierarchical structures of the heterogeneous scenario

We observe from Table 11.5 and Fig. 11.4 that the structure where decision-maker 3 is the leader and decision-makers 1 and 2 are followers leads to a minimum social cost in the hierarchical solution.

11.5 Distributed Planning

This section examines probability distribution steering problems in a MFTG framework by exploring the role of decentralized strategies. When steering an initial distribution to a terminal one, one aims to minimize a certain cost functional of mean-field type. In particular, we are interested in how to design an add-on terminal cost so as to provide incentives for decision-makers to move collectively as specified and meet the initial-terminal conditions. More specifically

- We introduce distributed planning problems in MFTGs. We show that existence can be re-stored in a wide range of cases if the terminal function can be freely designed.
- We then establish relationships between the proposed scenario and distributed optimal transport theory with a more general cost functional.
- We derive particular optimality systems in a semi-explicit way for the case with moment constraints at the final time and non-quadratic cost.

Section 11.5.1 provides a background material on the planning problem in a deterministic setting. The framework is extended to distributed planning in Sect. 11.5.1.2. Section 11.5.2 presents the main problem statement and derives an optimality system. Solutions to the distributed planning problem under moment constraints are derived in Sect. 11.5.3.

11.5.1 Background

11.5.1.1 Planning Problem Under Deterministic Dynamics

A basic planning problem of a decision-maker is to move from the point $\bar{x}(t_0)$ at time instant t_0 to the point $\bar{x}(t_1)$ at time instant t_1 while minimizing the energy along the path with a careful terminal cost design. Equivalently, the decision-maker seeks to transport the Dirac measure $\delta_{\bar{x}(t_0)}$ at time instant t_0 to another Dirac measure $\delta_{\bar{x}(t_1)}$ at time instant t_1.

The basic planning problem can be solved using a well-designed dynamic optimization problem where the goal of the decision-maker is to choose the terminal objective function $\bar{\alpha}(t_1)\bar{x}^2(t_1)$ and a strategy $\bar{u}$ such that

$$\begin{aligned}&C(\bar{x}(t_0), \bar{x}(t_1)) := \inf_{\bar{u}}\{\bar{\alpha}(t_1)\bar{x}^2(t_1) + \textstyle\int_{t_0}^{t_1} \bar{u}^2(t)dt\}\\ &\text{subject to } d\bar{x} = \bar{u}dt,\\ &\bar{x}(t_0),\ \bar{x}(t_1) \text{ are given,}\end{aligned} \tag{11.63}$$

where $\bar{x}(t_0)$ and $\bar{x}(t_1)$ are two real numbers with the same sign and nonzero.

In the planning problem above, the optimal cost is $C(\bar{x}(t_0), \bar{x}(t_1))$. If we exchange the initial point $\bar{x}(t_0)$ and the terminal point $\bar{x}(t_1)$, we obtain $C(\bar{x}(t_1), \bar{x}(t_0))$ as the optimal cost, which is a similar problem.

To solve this optimal control under the given pair of initial-terminal point conditions, the decision-maker can use a backward induction method in two steps:

Optimization Step

Given $\bar{\alpha}(t_1)$, the decision-maker first solves the optimal control problem given by

$$\begin{aligned}&L := \bar{\alpha}(t_1)\bar{x}^2(t_1) + \textstyle\int_{t_0}^{t_1} \bar{u}^2(t)dt\\ &\inf_{\bar{u}} L\\ &\text{subject to } \dot{\bar{x}} = \bar{u},\ \bar{x}(t_0) = \bar{x}_0 \in \mathbb{R},\\ &\bar{u}(.) \in U = \mathbb{R} \text{ unconstrained.}\end{aligned}$$

This optimal control can be easily solved using a direct method.

It follows that

$$\begin{aligned}&\inf_{\bar{u}\in\mathscr{U}} L = \bar{\alpha}(t_0)\bar{x}^2(t_0),\\ &\text{state-feedback: } \bar{u}^* = -\bar{\alpha}\bar{x},\\ &\text{Riccati: } \dot{\bar{\alpha}} - \bar{\alpha}^2 = 0,\ \bar{\alpha}(t_1) \text{ fixed,}\\ &\implies \bar{\alpha}(t) = \frac{1}{\frac{1}{\bar{\alpha}(t_1)}+t_1-t},\\ &\bar{x}(t) = \bar{x}(t_0)e^{-\int_{t_0}^{t}\bar{\alpha}(t')dt'} = \bar{x}(t_0)\left(1 - \frac{t-t_0}{\frac{1}{\bar{\alpha}(t_1)}+t_1-t_0}\right).\end{aligned}$$

Matching Step

In order to design $\bar{\alpha}(t_1)$, one can match with the required conditions. The planning problem of the decision-maker is then reduced to the problem of finding $\bar{\alpha}(t_1)$ such that

$$\bar{x}(t_0)\left(1-\frac{t_1-t_0}{\frac{1}{\bar{\alpha}(t_1)}+t_1-t_0}\right)=\bar{x}(t_1),$$

which means, for $\bar{x}(0)\neq 0$ that

$$\frac{t_1-t_0}{\frac{1}{\bar{\alpha}(t_1)}+t_1-t_0}=\left[1-\frac{\bar{x}(t_1)}{\bar{x}(t_0)}\right].$$

It follows that

$$\bar{\alpha}(t_1)=\frac{\bar{x}(t_0)-\bar{x}(t_1)}{\bar{x}(t_1)(t_1-t_0)},$$

satisfies the requirement.

As expected, the optimal velocity has the direction of $\bar{x}(t_1)-\bar{x}(t_0)$ with a proper scaling. To connect this result with the euclidean distance between $\bar{x}(t_1)$ and $\bar{x}(t_0)$, we impose the terminal cost to be zero by adjusting the system as follows:

$$\begin{aligned}
&\inf_{\bar{u}} \int_{t_0}^{t_1} \bar{u}^2(t)dt\\
&\text{subject to}\\
&dy=\bar{u}dt,\\
&y(t_0)=\bar{x}(t_0)-\bar{x}(t_1),
\end{aligned}$$

where we set $y(t)=\bar{x}(t)-\bar{x}(t_1)$. The optimal cost is

$$\alpha(t_0)y^2(t_0)=\alpha(t_0)[\bar{x}(t_0)-\bar{x}(t_1)]^2,$$

Next, we present the deterministic distributed planning problem.

11.5.1.2 Distributed Planning Problem

The horizon of the interaction is $[t_0,t_1],\ \ t_0<t_1$. There are $I\geq 2$ decision-makers. The set of decision-makers is denoted by $\mathcal{I}=\{1,\ldots,I\}$. Decision-maker i has a control action $\bar{u}_i\in U_i=\mathbb{R}$. The state $\bar{x}$ is driven by an ordinary differential equation with a two-point boundary constraint given by

$$d\bar{x}=b(t,\bar{x},\bar{u}_1,\ldots,\bar{u}_I)dt,\ \ \bar{x}(t_0),\ \ \bar{x}(t_1),$$

where

$$\begin{aligned}&\text{Drift: } b : [t_0, t_1] \times \mathbb{R} \times \prod_{i\in\mathscr{I}} U_i \to \mathbb{R},\\ &b(\bar{x}, \bar{u}_1, \ldots, \bar{u}_I) = \bar{b}_1\bar{x} + \sum_{j\in\mathscr{I}} \bar{b}_{2j}\bar{u}_j,\\ &\bar{x}(t_1)\bar{x}(t_0) > 0.\end{aligned}$$

The cost functional of decision-maker i is

$$L_i(\bar{u}_1, \ldots, \bar{u}_I, \bar{x}(t_0), \bar{x}(t_1)) = h_i(\bar{x}(t_1)) + \int_{t_0}^{t_1} l_i(t, \bar{x}, \bar{u}_1, \ldots, \bar{u}_I)dt,$$

where

$$\begin{aligned}&l_i(t, \bar{x}, \bar{u}_1, \ldots, \bar{u}_I) = \bar{q}_i\frac{\bar{x}^{2\bar{k}_i}}{2\bar{k}_i} + \bar{r}_i\frac{\bar{u}_i^{2\bar{k}_i}}{2\bar{k}_i} + \bar{c}_i\bar{x}^{2\bar{k}_i-1}\bar{u}_i\\ &+\sum_{j\neq i} \bar{\varepsilon}_{ij}\bar{x}^{2(\bar{k}_i-1)}\bar{u}_i\bar{u}_j,\\ &h_i(\bar{x}) = \bar{\alpha}_i(t_1)\frac{\bar{x}^{2\bar{k}_i}}{2\bar{k}_i},\end{aligned}$$

where $\bar{k}_i \geq 1$ are natural numbers and the coefficients are time-dependent. The coefficient functions $\bar{q}_i, \bar{r}_i$ are nonnegative functions.

The distributed planning problem involves $I-$decision-makers and can be formulated as a differential game played over the horizon $[t_0, t_1]$. Given the strategies of the other decision-makers $(\bar{u}_1, \ldots, \bar{u}_{i-1}, \bar{u}_{i+1}, \ldots, \bar{u}_I)$, decision-maker $i \in \mathscr{I}$ chooses the terminal objective function h_i and a strategy $\bar{u}_i$ such that

$$\begin{aligned}&\inf_{u_i} L_i(\bar{u}_1, \ldots, \bar{u}_I, \bar{x}(t_0), \bar{x}(t_1))\\ &\text{subject to}\\ &d\bar{x} = b(t, \bar{x}, \bar{u}_1, \ldots, \bar{u}_I)dt,\\ &\bar{x}(t_0) \text{ fixed}, \bar{x}(t_1) \text{ fixed}.\end{aligned}$$

The two-point boundary conditions are, in general, challenging to meet in a first-order ordinary differential equation. However, here, each decision-maker has the freedom to design also its terminal cost h_i, which makes the problem feasible for a wider range of parameters. Notice that we are not optimizing over the best design of the function h_i,, but we are optimizing over the strategy u_i once the function h_i is chosen.

We solve the distributed planning problem as follows: We first solve the following system:

$$\begin{aligned}&i \in \mathscr{I},\\ &\inf_{\bar{u}_i} L_i(\bar{u}, \bar{x}(t_0), \bar{x}(t_1))\\ &\text{subject to}\\ &d\bar{x} = bdt,\\ &\bar{x}(t_0) \text{ fixed}.\end{aligned}$$

Its solution, if any, is given by

$$
\begin{aligned}
& i \in \mathcal{I}, \\
& \bar{u}_i = -\bar{\eta}_i \bar{x}, \\
& 0 = -\bar{r}_i \bar{\eta}_i^{2\bar{k}_i - 1} - \sum_{j \neq i} \bar{\varepsilon}_{ij} \bar{\eta}_j + \bar{b}_{2i} \bar{\alpha}_i + \bar{c}_i, \\
& v_i(t, \bar{x}) = \bar{\alpha}_i(t) \frac{\bar{x}^{2\bar{k}_i}}{2\bar{k}_i}, \\
& 0 = \dot{\bar{\alpha}}_i + \bar{q}_i + \bar{r}_i \bar{\eta}_i^{2\bar{k}_i} - 2\bar{k}_i \bar{c}_i \bar{\eta}_i + 2\bar{k}_i \sum_{j \neq i} \bar{\varepsilon}_{ij} \bar{\eta}_i \bar{\eta}_j \\
& +2\bar{k}_i \bar{\alpha}_i [\bar{b}_1 - \sum_j \bar{b}_{2j} \bar{\eta}_j], \\
& \bar{\alpha}_i(t_1) \text{ fixed}, \\
& \bar{x}(t) = [\bar{x}(t_0)]\, e^{\int_{t_0}^{t} \left[\bar{b}_1 - \sum_j \bar{b}_{2j} \bar{\eta}_j\right] dt'},
\end{aligned}
$$

whenever the above coefficient system admits a unique solution.

- Then, we compute the terminal point $\bar{x}(t_1)$ as a function of $\bar{\alpha}_i(t_1),\ i \in \mathcal{I}$ via the coefficients $\eta_i(t_1),\ i \in \mathcal{I}$.
- The distributed planning problem is then to design $\bar{\alpha}_i(t_1)$ as degrees of freedom such that the constraints are met at the specified times:

$$
\bar{x}(t_1) = \bar{x}(t_0) e^{\int_{t_0}^{t_1} \left[\bar{b}_1 - \sum_j \bar{b}_{2j} \bar{\eta}_j\right] dt}.
$$

The later equation offers several choices for the vector $\bar{\alpha}(t_1)$.

11.5.2 Planning in Mean-Field-Type Games

The horizon of the interaction is $[t_0, t_1],\quad t_0 < t_1$. There are $I \geq 2$ number of decision-makers. The set of decision-makers is denoted by $\mathcal{I} = \{1, 2, \ldots, I\}$. Decision-maker $i \in \mathcal{I}$ has a control action $u_i \in U_i = \mathbb{R}$. The state x is driven by drift-jump-diffusion of mean-field type given by

$$
\begin{aligned}
dx &= b(t, x, m, u_1, \ldots, u_I)dt + \sigma(t, x, m, u_1, \ldots, u_I)dB \\
&\quad + \int_{\Theta} \mu(t, x, m, u_1, \ldots, u_I), \theta)\tilde{N}(dt, d\theta), \\
&\quad x(t_0) \sim m(t_0, .),\ x(t_1) \sim m(t_1, .),
\end{aligned}
$$

where

Drift: b,
Diffusion: Brownian motion B,
Jump process: $N(dt, d\theta)$,
Compensated Jump: $\tilde{N}(dt, d\theta) = N(dt, d\theta) - \nu(d\theta)dt$,
$b, \sigma, \mu(., \theta), : \; [t_0, t_1] \times \mathbb{R} \times \mathscr{P}(\mathbb{R}) \times \prod_{j=1}^{I} U_j \to \mathbb{R}$.

The performance functional of decision-maker i is

$$L_i(u_1, \ldots, u_I, m_0) = h_i(x(t_1), m(t_1)) + \int_{t_0}^{t_1} l_i(t, x, m, u_1, \ldots, u_I)dt,$$

where $m(t, dy) = \mathbb{P}_{x(t)}(dy)$ is the probability measure of the state $x(t)$ at time t.

We assume the following information structure:

- perfect state measurement, i.e., all the decision-makers observe the state x,
- perfect knowledge of the model, i.e., the model above is known by all decision-makers.

In addition, each decision-maker is assumed to have the computational capability to be able to compute the aggregative term m from the model. Let $\mathscr{U}_i$ be the set of control strategies of decision-maker i that are progressively measurable with respect to the filtration generated by the unions of events in $\{B, N\}$.

The risk-neutral MFTG is given by $(\mathscr{I}, U_i, \mathscr{U}_i, E[L_i])_{i\in\mathscr{I}}$. We denote by $u^* = (u_1^*, \ldots, u_{i-1}^*, u_i^*, u_{i+1}^*, \ldots, u_I^*)$. A risk-neutral mean-field-type Nash equilibrium is a solution of the following fixed-point problem:

$$\begin{aligned} & i \in \mathscr{I}, \\ & E[L_i(u^*, m_0)] = \inf_{u_i \in \mathscr{U}_i} E[L_i(u_1^*, \ldots, u_{i-1}^*, u_i, u_{i+1}^*, \ldots, u_I^*, m_0)]. \end{aligned} \tag{11.64}$$

Let $\hat{V}_i(t, m)$ be the optimal cost-to-go from m at time $t \in (t_0, t_1)$ for decision-maker i given the strategies of the others:

$$\begin{aligned} \hat{V}_i(t, m) = \inf_{u_i \in \mathscr{U}_i} E[h_i(x(t_1), m(t_1)) + \int_t^{t_1} l_i(t, x, u_1^*, \ldots, u_{i-1}^*, u_i, u_{i+1}^*, \ldots, \\ u_I^*, m)dt' | m(t) = m]. \end{aligned} \tag{11.65}$$

Let $\hat{V}_{i,m}$ be the Gâteaux derivative of $\hat{V}_i(t, .)$ with respect to the measure m. Introduce the integrand Hamiltonian as

$$\begin{aligned} H_i(x, m, (\hat{V}_{j,m}, \hat{V}_{j,xm}, \hat{V}_{j,xxm})_j) = \inf_{u_i \in U_i} \Bigg\{ l_i + b\, \hat{V}_{i,xm} + \frac{\sigma^2}{2} \hat{V}_{i,xxm} \\ + \int_{\Theta} [\hat{V}_{i,m}(t_-, x + \mu) - \hat{V}_{i,m} - \mu \hat{V}_{i,xm}] \nu(d\theta) \Bigg\}. \end{aligned} \tag{11.66}$$

Denote the jump operator J as

$$J[\phi_i] := \int_{\Theta} [\phi_{i,m}(t_-, x+\mu) - \phi_{i,m} - \mu\phi_{i,xm}]\nu(d\theta),$$

and let J^* be the adjoint operator of J:

$$\langle J[\phi], m\rangle = \langle \phi, J^*[m]\rangle.$$

A sufficient condition for a risk-neutral Nash equilibrium system is given by the following (backward-forward) partial integrodifferential system:

$$\begin{array}{l} i \in \mathscr{I}, \\ 0 = \hat{V}_{i,t}(t,m) + \int_x H_i(x, m, (\hat{V}_{j,m}, \hat{V}_{j,xm}, \hat{V}_{j,xxm})_{j\in\mathscr{I}})m(t,dx), \\ \hat{V}_i(t_1, m) = \int m(dy)h_i(y,m), \\ m_t = -(mb)_x + \frac{1}{2}(m\sigma^2)_{xx} + J^*[m], \\ m(t_0, dy) = m_0(dy). \end{array} \tag{11.67}$$

The system (11.67) is an infinite PIDE system in m, and it provides the Nash equilibrium values of the MFTG. We use (11.67) to find risk-neutral mean-field-type Nash equilibria.

Next, we add the terminal constraint: Choose parameters of h_i to match the measure $m(t_1, .)$ at time t_1.

11.5.2.1 Steering Distributions

Given the strategies of the other decision-makers $(u_1^*, \ldots, u_{i-1}^*, u_{i+1}^*, \ldots, u_I^*)$, decision-maker i chooses parameters of the terminal objective function h_i and a strategy u_i such that

$$\begin{array}{l} \inf_{u_i} EL_i(u, m(t_0), m(t_1)) \\ \text{subject to} \\ dx = bdt + \sigma dB + \int_{\Theta} \mu(., \theta)\tilde{N}(dt, d\theta), \\ x(t_0) \sim m(t_0, .), \text{ fixed} \\ x(t_1) \sim m(t_1, .) \text{ fixed.} \end{array} \tag{11.68}$$

The distributed planning problem (11.68) extends the classical optimal transport problem and also the classical Schrödinger bridge problem (see [96, 97] and the references therein). That is, for a distribution of mass $m(t_0, dy)$, we wish to transport the mass in such a way that it is transformed into the distribution $m(t_1, dy)$ on the same space. The optimal transport theory is the study of optimal transportation and allocation between measures. The optimal transport problem was first introduced by [249] and formalized by [211], leading to the so-called Monge-Kantorovitch transportation problem.

11.5.2.2 Optimality System for Steering Distributions

In the distributed planning of mean-field type, the cost function is not necessarily Eu_i^2, and the terminal value of decision-maker i must satisfy $\hat{V}_i(T, m) = \int m(t_1, dy)h_i(y, m(t_1))$, where the equality is in a functional sense. The distributed planning problem is then to choose h_i such that the terminal distribution matches the prescribed distribution $m(t_1, dy)$:

$$\begin{array}{l} i \in \mathcal{I}, \\ 0 = \hat{V}_{i,t}(t, m) \\ + \int_x H_i(x, m, (\hat{V}_{j,m}, \hat{V}_{j,xm}, \hat{V}_{j,xxm})_{j\in\mathcal{I}})m(t, dx), \\ \hat{V}_i(t_1, m) = \int m(t_1, dy)h_i(y, m(t_1)), \\ m_t = -(mb)_x + \frac{1}{2}(m\sigma^2)_{xx} + J^*[m], \\ m(t_0, dy) = m_0(dy), h_i \text{ chosen to match a given } m(t_1, dy). \end{array}$$

In some problems, the planning is to satisfy some variance or moment constraints. This is illustrated in the next section.

11.5.3 Games with Moment Constraints

We will investigate a MFTG with the following data:

$$\begin{array}{l} l_i(t, x, u, m) = q_i \frac{(x-\bar{x})^{2k_i}}{2k_i} + r_i \frac{(u_i-\bar{u}_i)^{2k_i}}{2k_i} + c_i(x-\bar{x})^{2k_i-1}(u_i-\bar{u}_i) \\ + \sum_{j\in\mathcal{I}\setminus\{i\}} \varepsilon_{ij}(x-\bar{x})^{2(k_i-1)}(u_i-\bar{u}_i)(u_j-\bar{u}_j) \\ + \bar{q}_i \frac{\bar{x}^{2\bar{k}_i}}{2\bar{k}_i} + \bar{r}_i \frac{\bar{u}_i^{2\bar{k}_i}}{2\bar{k}_j} + \bar{c}_i \bar{x}^{2\bar{k}_i-1}\bar{u}_i \\ + \sum_{j\neq i} \bar{\varepsilon}_{ij} \bar{x}^{2(\bar{k}_i-1)} \bar{u}_i \bar{u}_j, \\ h_i(x, m) = \alpha_i(T) \frac{(x-\bar{x}(T))^{2k_i}}{2k_i} + \bar{\alpha}_i(T) \frac{\bar{x}(T)^{2\bar{k}_i}}{2\bar{k}_i}, \\ b(t, x, u, m) = b_1(x-\bar{x}) + \bar{b}_1\bar{x} \\ + \sum_{j\in\mathcal{I}} \left[b_{2j}(u_j-\bar{u}_j) + \bar{b}_{2j}\bar{u}_j\right], \\ \sigma(t, x, u, m) = (x-\bar{x})\tilde{\sigma}, \\ \mu(t, x, u, m, \theta) = (x-\bar{x})\tilde{\mu}(., \theta), \\ \bar{x}(t) = \int y m(t, dy), \\ \bar{u}(t) = \int u(t, y, m)m(t, dy), \end{array}$$

where $k_i \geq 1,\ \bar{k}_i \geq 1$ are natural numbers and the coefficients are time-dependent. The coefficient functions $q_i, r_i, \bar{q}_i, \bar{r}_i$ are nonnegative functions. $\alpha_i(T), \bar{\alpha}_i(T)$ are to be determined below.

In the distributed planning problem, the goal of decision-maker i is to choose the terminal weights $\alpha_i(T), \bar{\alpha}_i(T)$ and a strategy u_i such that

$$\begin{array}{l} \inf_{u_i} EL_i(u, m(t_0), d, \bar{x}(t_1)) \\ \text{subject to} \\ dx = bdt + \sigma dB + \int_{\Theta} \mu(., \theta)\tilde{N}(dt, d\theta), \\ x(t_0) \sim m(t_0, .),\ \text{fixed} \\ E[(x(t_1) - \int ym(t_1, dy))^{2k_i}] = \bar{c}_i,\ \text{fixed} \\ \int ym(t_1, dy) = \bar{x}(t_1),\ \text{fixed.} \end{array} \tag{11.69}$$

Proposition 11.6 *A risk-neutral mean-field-type Nash equilibrium is given in a semi-explicit way as follows:*

$$\begin{array}{l} u_i^{ne} = -\eta_i \left(x - \int ym(dy)\right) - \bar{\eta}_i \int ym(dy), \\ 0 = -r_i \eta_i^{2k_i - 1} - \sum_{j \neq i} \varepsilon_{ij} \eta_j + b_{2i}\alpha_i + c_i, \\ 0 = -\bar{r}_i \bar{\eta}_i^{2\bar{k}_i - 1} - \sum_{j \neq i} \bar{\varepsilon}_{ij} \bar{\eta}_j + \bar{b}_{2i}\bar{\alpha}_i + \bar{c}_i, \\ \hat{V}_i(t, m) = \alpha_i \int_x \frac{(x - \int ym(dy))^{2k_i}}{2k_i} m(dx) + \bar{\alpha}_i \frac{(\int ym(dy))^{2\bar{k}_i}}{2\bar{k}_i}, \\ 0 = \dot{\alpha}_i + q_i + r_i \eta_i^{2k_i} - 2k_i c_i \eta_i + 2k_i \sum_{j \neq i} \varepsilon_{ij} \eta_i \eta_j \\ +2k_i \alpha_i [b_1 - \sum_{j \in \mathcal{I}} b_{2j} \eta_j] + 2k_i (2k_i - 1)\alpha_i \frac{1}{2}\tilde{\sigma}^2 \\ +\alpha_i \int_{\Theta} [(1 + \tilde{\mu})^{2k_i} - 1 - 2k_i \tilde{\mu}] \nu(d\theta), \\ \alpha_i(T) \text{ to be determined}, \\ 0 = \dot{\bar{\alpha}}_i + \bar{q}_i + \bar{r}_i \bar{\eta}_i^{2\bar{k}_i} - 2\bar{k}_i \bar{c}_i \bar{\eta}_i + 2\bar{k}_i \sum_{j \neq i} \bar{\varepsilon}_{ij} \bar{\eta}_i \bar{\eta}_j \\ +2\bar{k}_i \bar{\alpha}_i [\bar{b}_1 - \sum_j \bar{b}_{2j} \bar{\eta}_j], \\ \bar{\alpha}_i(t_1) \text{ to be determined}, \\ \int ym(t, dy) = \left[\int ym(t_0, dy)\right] e^{\int_{t_0}^{t} \left[\bar{b}_1 - \sum_j \bar{b}_{2j} \bar{\eta}_j\right] dt'}, \end{array}$$

whenever the above coefficient system admits a unique solution integrable within the horizon.

The uniqueness of the coefficient system (11.70) in η requires a strong condition. For example, for $k = 1$, the determinant of the matrices Γ with entries $\gamma_{ii} = r_i,\ \gamma_{ij} = \varepsilon_{ij},\ i \neq j$ and $\bar{\Gamma}$ with entries $\bar{\gamma}_{ii} = \bar{r}_i,\ \bar{\gamma}_{ij} = \bar{\varepsilon}_{ij},\ i \neq j$ must be nonzero. When the determinant of the matrix Γ is zero, the resulting control strategies become non-admissible and the costs become infinite.

To solve the distributed planning problem, it remains to determine $\alpha_i(t_1), \bar{\alpha}_i(t_1), i \in \mathcal{I}$. $\bar{\alpha}_i(t_1)$ is determined from the evolution equation of the expected value and the constraint $\int ym(t_1, dy) = \bar{x}(t_1)$. The function $z_1(t) = \int ym(t, dy)$ satisfies

$$\begin{aligned} &\dot{z}_1 = z_1 \left\{ \bar{b}_1 - \sum_{j\in\mathscr{I}} \bar{b}_{2j}\bar{\eta}_j) \right\} \\ &z_1(t_1) = \bar{x}(t_1), \text{ fixed,} \end{aligned} \tag{11.70}$$

from which we deduce a system satisfied by $\bar{\alpha}_i(t_1), i \in \mathscr{I}$.

It remains to determine $\alpha_i(t_1)$. To do so, we establish the dynamics of $E[(x(t) - \int ym(t,dy))^{2k_i}]$ as time varies. Let $y_{k_i}(t) = E[(x(t) - \int ym(t,dy))^{2k_i}]$:

$$\begin{aligned} &i \in \mathscr{I}, \\ &\dot{y}_{k_i} = y_{k_i} \left\{ 2k_i(b_1 - \sum_{j\in\mathscr{I}} b_{2j}\eta_j) \right. \\ &\left. +k_i(2k_i-1)\tilde{\sigma}^2 + \int_\Theta \nu(d\theta)[(1+\tilde{\mu})^{2k_i} - 1 - 2k_i\tilde{\mu}] \right\} \\ &y_{k_i}(t_1) = \bar{c}_i \geq 0 \text{ fixed,} \end{aligned} \tag{11.71}$$

from which we deduce the elements $\alpha_i(t_1), i \in \mathscr{I}$. This system is underdetermined when the size of the ranges of k_i is smaller than the cardinality of the number of decision-makers. Another way is to consider several moment constraints.

11.6 Multi-Marginal Distributed Planning

The methodology presented here can be used to plan a path at multiple time instants $t_0 < t_{\frac{1}{n}} < t_{\frac{2}{n}} < \ldots < t_{\frac{n}{n}} = t_1$ such that $x(t_{\frac{k}{n}}) \sim m(t_{\frac{k}{n}}, .)$, where the $n+1$ marginal measures $m(t_{\frac{k}{n}}, .)$ are prescribed. The multi-marginal distributed planning problem becomes the design of a terminal objective function h_i and a strategy u_i such that

$$\begin{aligned} &i \in \mathscr{I}, \\ &\inf_{u_i} EL_i(u, (m(t_{\frac{k}{n}}, .))_k) \\ &\text{subject to} \\ &dx = bdt + \sigma dB + \int_\Theta \mu(.,\theta)\tilde{N}(dt, d\theta), \\ &x(t_0) \sim m(t_0, .), \text{ fixed} \\ &x(t_{\frac{k}{n}}) \sim m(t_{\frac{k}{n}}, .), \quad k \in \{1, \ldots, n-1\}, \text{ fixed} \\ &x(t_1) \sim m(t_1, .) \text{ fixed.} \end{aligned} \tag{11.72}$$

We solve the multi-marginal problem, and we decompose t into subproblems as follows:

$$\begin{aligned} &\inf_{u_i} Eh_i(\bar{x}(t_{\frac{k}{n}})) + \int_{t_{\frac{k-1}{n}}}^{t_{\frac{k}{n}}} l_i(t, \bar{x}, \bar{u})dt, \\ &\text{subject to} \\ &dx = bdt + \sigma dB + \int_\Theta \mu(.,\theta)\tilde{N}(dt, d\theta), \quad t \in (t_{\frac{k-1}{n}}, t_{\frac{k}{n}}) \\ &x(t_{\frac{k-1}{n}}) \sim m(t_{\frac{k-1}{n}}, .), \text{ fixed} \\ &x(t_{\frac{k}{n}}) \sim m(t_{\frac{k}{n}}, .), \text{ fixed.} \end{aligned} \tag{11.73}$$

Each of the subproblems is solved using the same method as above. We then combine the $n + 1$ systems together to obtain the desired result.

In this section, we have presented design and planning problems in MFTGs with perfect and complete information. We have provided semi-explicit solutions for a class of MFTGs with polynomial cost and moment constraints.

11.7 Summary

The Stackelberg solution can be determined in a semi-explicit form by means of a direct method for a broader class of non-quadratic hierarchical MFTG problems with possibly multiple leaders, multiple followers, multiple layers, and nonlinear distribution-dependent payoffs where the state dynamics is driven by conditional expected values of states, control actions, Brownian motions, Gauss-Volterra processes, jumps, and regime switching. In this chapter, we have restricted the analysis to linear-state-and-mean-field-type feedback strategies. This restriction has simplified the response functions.

11.8 Notes

For general Stackelberg games, we refer to Chapter 7 of Başar-Olsder's 1999 book. See also [19–21, 252, 290]. The work in [231] proposed a planning problem in mean-field games, in which a central planner would like to steer a population to a predetermined final configuration while still allowing individuals to choose their own strategies. This is a distributed planning problem from the perspective of the agents. Since then, several studies (see [1, 265, 268, 269] and the references therein) have been conducted in the context of mean-field games and optimal transport theory (see [96, 97]). The work in [3] examines optimal transport for a class of linear quadratic differential games. However, it has not been addressed in the context of MFTGs [102, 146, 330, 331, 333] in which the expected objective functional has a nonlinear dependence with respect to the measure of the state. Berge's solution concept has been examined in the context of MFTG in [350].

Chapter 12
Mean-Field-Type Games Driven by Rosenblatt Noises

Rosenblatt processes are non-Gaussian, non-Poisson, and non-Markov with long-range dependence. This chapter examines a class of mean-field-type games driven by Rosenblatt processes. We provide expressions for equilibrium strategies and corresponding equilibrium costs in linear state-and-mean-field-type feedback form for all decision-makers. The cost functional is non-quadratic and includes a fractional-integral of high-order polynomial. We show that the Ito's formula for state dynamics of mean-field-type-driven Rosenblatt process involves the third (or higher) derivative of the cost function in the general setting.

Interestingly, the equilibrium strategies obtained for the noise-free case (purely deterministic case) no longer provide an equilibrium when the noise is driven by a Rosenblatt process. The equilibrium strategies obtained for the Brownian motion, multi-fractional Brownian, Gauss-Volterra or a pure jump, and regime switching processes no longer provide an equilibrium when the noise is driven by a Rosenblatt process. In particular, the equilibrium strategies for the case when the state is driven by Brownian motion are no longer in equilibrium even in the quadratic cost case.

The chapter is organized as follows. Section 12.1 presents some evidence of evidence of non-Gaussianity in real data. Section 12.2 presents noise modeling in MFTG. Section 12.3 presents a class of MFTGs with a common state driven by a Rosenblatt noise and provides a direct method to solve the problem. Section 12.4 presents results of the cooperative case. Mixture of five types of noises is presented in Sect. 12.5. Section 12.6 examines long-run cost functional as the horizon length goes to infinity.

12.1 Evidence of Non-Gaussianity in Real Data

A Gaussian distribution, also known as the normal distribution, is a continuous probability distribution characterized by its bell-shaped curve, with well-defined

T. Başar et al., *Mean-Field-Type Game Theory I*, Static & Dynamic Game Theory: Foundations & Applications, https://doi.org/10.1007/978-3-032-07027-2_12

moments (mean and variance), where the tails decay exponentially at a rate determined by the standard deviation, no skew and is symmetric about its mean. Recent empirical studies across diverse sectors have demonstrated that classical Gaussian assumptions are often inadequate for describing stochastic phenomena encountered in engineered and socioeconomic systems. Traditional models in systems theory, control, and game theory typically rely on the central limit theorem, presuming that uncertainties and fluctuations are normally distributed. However, mounting evidence from various fields indicates that many systems exhibit noncentral limit, skews, heavy-tailed distributions, multimodal behaviors, and long-range dependencies that contradict these assumptions. For instance, in the e-commerce sector, the Whale Effect is characterized by the disproportionate impact of the top 5% of customers, who contribute up to 50% of total revenue. This phenomenon suggests that revenue distributions are heavily skewed and possess significant kurtosis, challenging the symmetry inherent in Gaussian models. In power grid operations, the interplay between solar surplus and evening demand produces dual net load peaks—features that standard Gaussian forecasts fail to capture. It is not a direct mixture of two Gaussian distribution as right skews are observed empirically. Similarly, agricultural supply chains, as observed by the startup Guinaga in mango production across Mali and Burkina Faso, experience dual harvest seasons that result in approximately 40% waste due to intervening periods of scarcity. Such dual modality in supply dynamics further emphasizes the limitations of Gaussian approximations. Additional evidence of non-Gaussian characteristics is evident in the temporal patterns of hardware demand cycles. The distinct adoption phases for central processing unit (CPU), graphics processing units (GPU), and tensor processing units (TPU) resources, driven by episodic shocks from cryptocurrency and machine intelligence sectors, result in multi-peaked demand curves. In the financial domain, the cyclic nature of machine intelligence funding—marked by rapid expansion phases followed by sharp contractions—deviates markedly from the unimodal trends predicted by normal distributions. Moreover, the volatile, sawtooth pattern of GPU pricing, influenced by abrupt surges in demand linked to crypto mining and large learning model developments, illustrates pricing dynamics with abrupt discontinuities and heavy tails. Other applications, such as underwater wireless communication and seasonal commodity pricing—as exemplified by onion prices in regions like Dogon Country and Ansongo—further underscore the prevalence of non-Gaussian fluctuations in real-world data. The inadequacy of Gaussian models in these settings necessitates the development of alternative mathematical frameworks. The Rosenblatt process, a prototypical example of a non-Gaussian self-similar process with long-range dependence, offers a rigorous tool for capturing higher-order dependencies and memory effects. Unlike Gaussian processes, which are completely determined by their first two moments, the Rosenblatt process incorporates cumulants of higher orders, thereby providing a more faithful representation of the stochastic properties observed in the aforementioned applications. Integrating the Rosenblatt process into systems and control theory enables the formulation of more accurate models for uncertainty propagation and feedback control. In particular, when extended to mean-field type game theory, this approach facilitates the study of strategic interactions among multiple agents (including machine intelligence agents) whose behaviors are influenced by non-Gaussian noise. Direct methods used to solve mean-field-type games allow for the reduction of high-dimensional agent-based models to tractable convex completion blocks, while the incorporation of non-Gaussian statistics ensures that these models remain robust in the presence of extreme events and irregular fluctuations.

12.1.1 e-Commerce: The Whale Effect in Customer Spending

In e-commerce, purchasing behavior rarely follows a Gaussian distribution. Instead, customer spending often adheres to a power-law, beta, gamma, Rosenblatt shape distribution, where a small fraction of users ("whales") account for a disproportionately large share of revenue. This phenomenon, known as the "whale effect," is a quintessential example of non-Gaussianity in action.

Why is it non-Gaussian? A Gaussian distribution assumes most data points cluster around the mean, with no skew, symmetrical tails. However, in e-commerce:

- Most users spend little to nothing: The majority of customers make infrequent or low-value purchases (e.g., one-time buyers).
- A tiny fraction drives revenue: The top 1–5% of customers ("whales") may generate 30–50% of total sales. These high-value customers buy luxury goods, bulk orders, or premium subscriptions repeatedly.

For instance, a luxury fashion platform might see 95% of users spend under $500 annually, while 5% spend over $10,000. This creates a long right tail in the spending histogram, starkly contrasting the bell-shaped Gaussian curve. There are several business implications of this situation. A heavy reliance on high-spending customers exposes businesses to revenue volatility, as the loss of just a few top spenders can significantly impact financial performance. To mitigate this risk, companies must deploy highly personalized marketing strategies, such as exclusive VIP programs, to maintain engagement and loyalty among their most valuable customers. Additionally, inventory decisions become increasingly strategic, with a focus on high-margin, niche products that cater to the preferences of these top-tier buyers, ensuring sustained profitability and competitive advantage.

Figure 12.1 presents a histogram mapping annual customer spending, with the x-axis spanning $0 to $50,000 and the y-axis indicating customer count. The distribution is strikingly uneven: A sharp spike on the left reflects a large base of low spenders, followed by a gradual decline through the midrange, and a long, thin tail extending far to the right, where high-value customers—those spending $10,000 or more—reside. This fits with a Rosenblatt distribution as will see below.

Overlaying a Gaussian curve underscores the mismatch: the real-world distribution is highly skewed, lacking the symmetry assumed in standard models. This "whale effect" highlights the fundamental non-Gaussian nature of e-commerce spending, reinforcing that businesses cannot rely on simple averages. Long-term

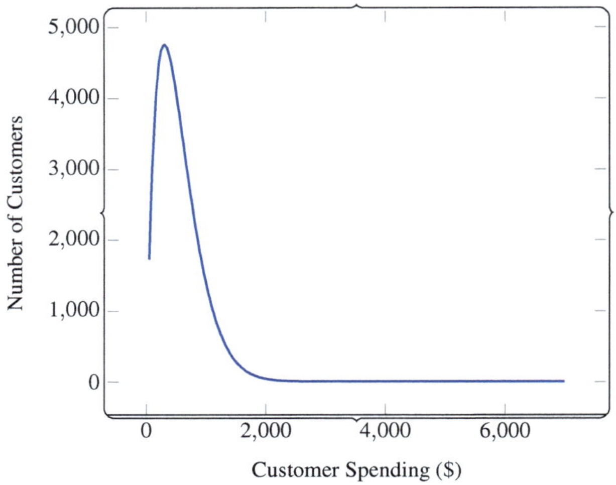

Fig. 12.1 Top 5% "whales" dominate revenue. Gaussian models underestimate tail risk by 60%

success depends on identifying and strategically catering to the outliers—the whales—who drive disproportionate value.

E-commerce businesses capitalize on the inherent asymmetry in customer spending by introducing low-budget products, transforming a vast base of minimal spenders into a meaningful revenue stream. This approach reinforces the non-Gaussian nature of spending behavior, where a small number of high-value customers drive disproportionate revenue, while the majority contribute through sheer volume.

By offering low-cost products, businesses create a scalable model that thrives on numbers. While individual low-spenders generate little revenue on their own, their collective purchasing power adds up. Selling a $5 accessory to 100,000 users, for example, yields $500,000—the same as a single whale spending half a million but with significantly lower risk. Lower price points also expand the customer base, drawing in price-sensitive users who might otherwise never transact. A streaming platform introducing a $3/month ad-supported tier, for instance, converts nonpaying users into revenue contributors, creating a sustainable pipeline of engaged customers.

The influx of low-cost purchases accentuates the structure of the spending distribution. More users cluster in the $0–$50 range, forming a pronounced left-side spike in the histogram, while high-value spenders on the right tail remain largely unchanged. This widens the gap between the mass market and the outliers, deepening the divide that defines non-Gaussianity in e-commerce.

Before introducing low-budget products, spending distribution is already skewed: 90% of users spend between $0 and $100, while only 2% surpass the $1,000 threshold. After introducing budget-friendly offerings, the left spike grows taller, with 95% of users now spending between $5 and $100. Meanwhile, whales

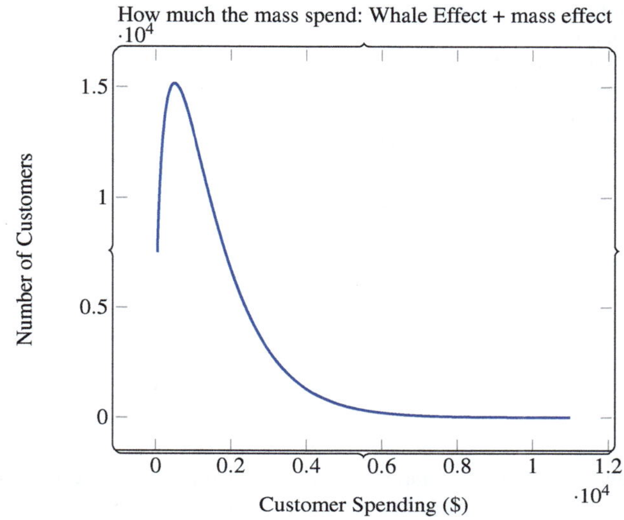

Fig. 12.2 The influx of low-cost purchases accentuates the structure of the spending distribution making it more skewed

continue their high-spending habits, leaving the right tail intact. The result is an even more pronounced power-law, beta, chi-2, gamma, Rosenblatt distribution, where the majority shift slightly from zero-spenders to low-spenders, but the overall shape of the curve grows more extreme—further defying Gaussian assumptions. Introducing low-budget products does not "normalize" spending, it amplifies non-Gaussianity by deepening the divide between the mass of small contributors and the elite few (see Fig. 12.2). This strategy acknowledges that e-commerce success lies not in chasing averages, but in catering to extremes: the volume of the many and the value of the few.

12.1.2 Distributed Power Networks: Prosumer-Driven Net Load

In modern distributed power networks—where consumers, prosumers (those who both consume and generate energy), and producers interact—the traditional assumption of Gaussian demand and supply patterns collapses. Instead, net load (grid demand minus localized renewable generation) follows a two-mode Rosenblatt shape, a clear departure from normality with profound implications for grid stability and energy markets.

The two-mode Rosenblatt shape, known for capturing skewness, bimodality and fat tails, is particularly suited for energy systems where physical constraints, such as

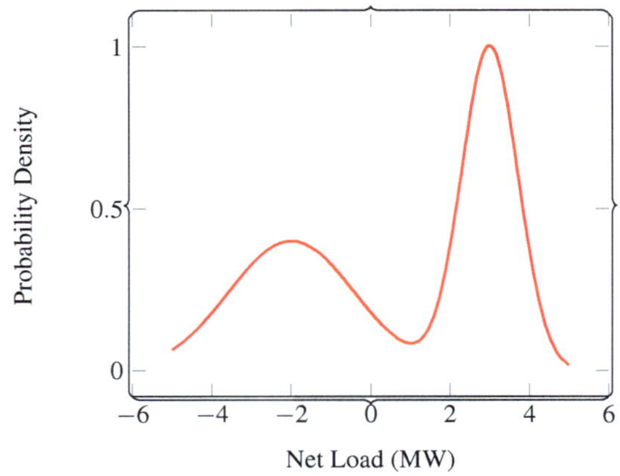

Fig. 12.3 Bimodal peaks (solar surplus vs. evening demand) make Gaussian load forecasting error-prone. Net load =grid demand-local renewable energy generation by prosumers

renewable generation limits, shape outcomes. Unlike two-mode Gaussian models, which assume symmetric variability around each mode, this framework reflects real-world fluctuations driven by weather patterns, time-of-day effects, and localized energy storage dynamics.

Net load asymmetry is largely driven by renewable generation. Solar panels and wind turbines only produce energy under specific conditions, creating sharp spikes and deep troughs in supply. On sunny days, prosumers inject excess solar power into the grid, often pushing net load toward zero or negative values. At night or on overcast days, they shift to heavy grid dependence, amplifying volatility. This effect is magnified by correlated behavior: prosumers in a given region experience similar weather patterns, causing synchronized surges and drop-offs in generation.

Battery storage further complicates the management. Prosumers with home storage charge their batteries during surplus periods and discharge them during deficits, introducing nonlinear shifts in net load. These abrupt transitions create structural breaks in grid demand, deviating sharply from the smooth fluctuations predicted by single-mode Gaussian models.

In Figs. 12.3 and 12.4, we consider a neighborhood with 1000 prosumers, each equipped with rooftop solar (max output = 5 kW). Over a single day, aggregated net load follows a two-mode Rosenblatt distribution. The net load cannot exceed grid demand or drop below negative thresholds, where excess energy is either curtailed or sold back under feed-in tariffs. The first peak occurs around midday, when solar generation surges and net load approaches its lower bound. The second peak emerges in the evening, when demand spikes and renewables contribute little. Extreme events, such as grid overload during heat waves when wind turbines underperform, occur far more frequently than a Gaussian model would predict. Overlaying a Gaussian curve on this distribution reveals its fundamental shortcomings. The normal assumption smooths over the dual peaks, ignores the hard boundaries imposed by generation constraints, and drastically underestimates the likelihood of extreme deviations. The real-world implications are significant. Traditional grid planning models, built on Gaussian assumptions, struggle to

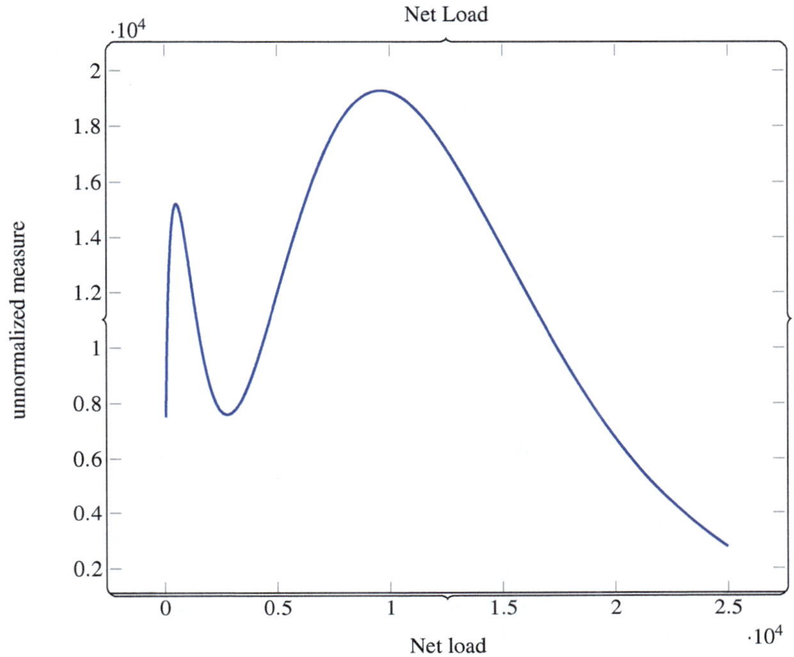

Fig. 12.4 Net load: change in the total demand due to renewable energy sources during the sunny days and the windy days modeled using two Rosenblatt shapes

anticipate blackouts triggered by synchronized solar drop-offs. In 2020, for instance, California's rolling blackouts were exacerbated by miscalculations in solar ramp forecasting, an oversight directly linked to an underappreciation of non-Gaussian dynamics.

Beyond grid stability, these structural distortions fuel market volatility. Electricity prices swing unpredictably as net load oscillates between extremes, sometimes dipping into negative territory when solar oversupply forces utilities to pay users to consume power. Storage systems designed for Gaussian expectations degrade prematurely, as unaccounted charge-discharge cycles stress batteries beyond intended use.

Empirical studies validate this shift (see Fig. 12.5). A 2023 and 2024 analysis of Germany's power grid found that net load in solar-dense regions followed a two-mode Rosenblatt distribution, exhibiting bimodal peaks at midday and evening. Similar observation was reported in the Sicily area in Italy where the price can be negative at certain days. In Spain, abundant solar capacity drives price pressures during sunny days when solar power generation peaks. Similarly, Finland's substantial wind energy capacity, particularly in coastal regions, causes price dips during periods of strong winds. In both cases, high renewable energy output can saturate the grid when demand is not high enough to absorb the excess supply. Negative

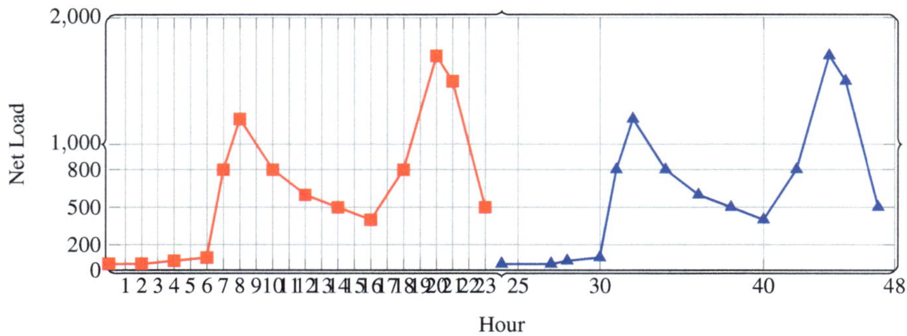

Fig. 12.5 Daily net load during the summer

electricity prices have become a regular feature of Europe's energy markets. As renewable energy capacity surges, particularly solar and wind, oversupply during low-demand periods leads to an increasing number of zero or negative price hour.

Traditional Gaussian-based models mispredicted grid stress events by 40%, while Rosenblatt-driven approaches significantly improved forecasting accuracy.

Addressing these challenges requires a fundamental rethinking of grid strategy. Non-Gaussian forecasting methods—such as Rosenblatt or mixture models—offer superior predictive power, enabling more resilient infrastructure planning. Dynamic pricing mechanisms, which penalize or reward users based on extreme net load fluctuations, incentivize prosumers to engage in load balancing. Micro-grids, designed to operate independently during demand troughs, provide localized stability, preventing wider grid disruptions.

The implications are clear. As renewable penetration accelerates, grid operators must abandon legacy assumptions and adapt to a system where extremes, not averages, dictate success or failure. The Rosenblatt distribution (or mixtures) encapsulates this reality, redefining energy networks as complex, constrained systems where ignoring statistical nuance is not just inefficient—it is catastrophic.

12.1.3 Agricultural Water Usage: Punjab's Trimodal Water Crisis

In Punjab, India—one of the world's most water-stressed agricultural regions—groundwater usage follows a three-mode Rosenblatt distribution, not the smooth Gaussian curve assumed by traditional models. This stark deviation from normality fuels widespread economic and environmental failures, as infrastructure and policy built on flawed assumptions struggle to keep pace with reality.

Water demand in Punjab surges at three distinct points in the agricultural cycle [119]. During pre-monsoon sowing (April–May), intensive irrigation is required

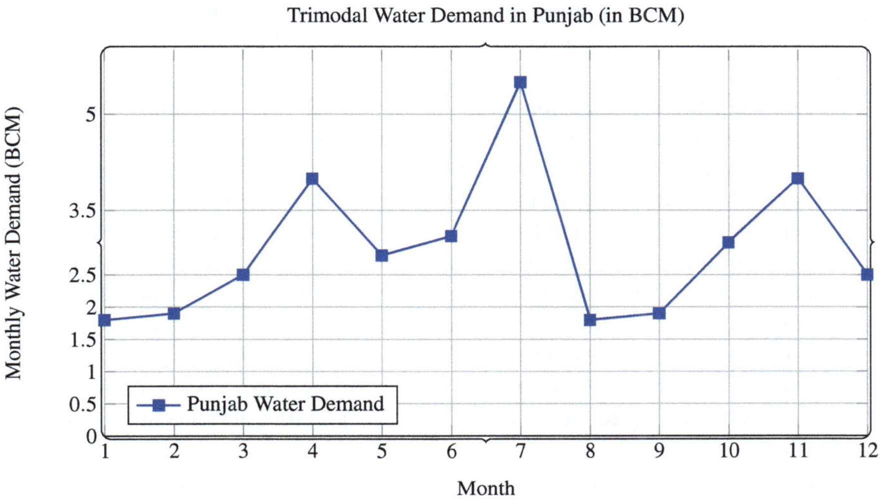

Fig. 12.6 Trimodal water demand in Punjab (averaged annual total: 30 BCM)

to prepare fields for rice, leading to the first peak in groundwater extraction. If the monsoon arrives late (June–July), farmers rely heavily on pumping to prevent seedling failure, creating a second, even larger peak. After the monsoon season (October–November), wheat planting begins, but depleted reservoirs force reliance on groundwater, driving the third peak. These three surges form a trimodal pattern, a defining feature of irrigation demand that contrasts sharply with the smooth, single-peak distribution assumed in Gaussian models.

Traditional forecasts fail to capture these abrupt shifts, leading to persistent shortages. Punjab's groundwater table declines by approximately 1 meter annually, driving $1.2 billion in crop losses due to irrigation failures. With 78% of aquifers overexploited and another 15% at critical levels, the consequences of mischaracterizing demand are severe. Gaussian-based models predict a stable annual demand of 20 billion cubic meters (BCM), yet actual demand regularly surges to 35 BCM, straining infrastructure and leaving millions of farmers vulnerable.

A Gaussian overlay on Punjab's actual water usage reveals its fundamental miscalculation. The traditional model assumes a smooth, single-peak curve centered on monsoon months, ignoring the sharp pre-monsoon and post-monsoon spikes. The real-world three-mode Rosenblatt shape (see Fig. 12.6), by contrast, exhibits three well-defined peaks, reflecting abrupt transitions in demand that render Gaussian-based infrastructure planning inadequate.

Ignoring this trimodal pattern has led to widespread failures. Canals and pump systems designed for "average" flows frequently collapse under peak stress, as seen in 2021 when 12,000 farmers lost entire crops after irrigation networks failed during the October surge. Agricultural policies continue to prioritize monsoon-season readiness while neglecting the far more dangerous pre- and post-monsoon crises. Supply chain disruptions compound the problem, with agrochemical and equipment suppliers routinely misallocating resources, leading to $500 million in losses in 2022 alone.

Recognizing the trimodal non-Gaussian nature of water demand is not merely about crisis management—it opens the door to solutions. AI-driven irrigation models, designed to optimize water use based on real demand cycles, have already shown promise in reducing peak stress. Programs that recharge aquifers during monsoon surpluses provide a buffer against future shortages. Shifts toward drought-resistant crop varieties aim to reduce dependence on unsustainable extraction.

Punjab's 2022 water crisis demonstrated the high cost of Gaussian assumptions. A forecasted 22 BCM demand underestimated the real 28 BCM requirement, triggered by a 6-week monsoon delay. The resulting outcome is widespread crop abandonment, financial losses, and an urgent policy shift toward peak-sensitive interventions, including restrictions on early rice planting and incentives for alternative crops.

Punjab's water crisis is not an anomaly but a case study in the dangers of misapplying Gaussian models to non-Gaussian systems. In resource-stressed environments, peaks, not averages, govern survival. Flattening these extremes leads to systemic failure, while embracing three-mode Rosenblatt dynamics enables more resilient infrastructure, policies, and economic strategies.

12.1.4 Mango Supply Chains in Mali and Burkina Faso

The mango supply chains of Mali and Burkina Faso exhibit a two-mode Rosenblatt shape, with two sharply distinct seasonal peaks. However, industry planning and infrastructure remain anchored in single-peak Gaussian models, leading to systemic waste, volatile pricing, and postharvest scarcity. Field research by the Malian startup Guinaga highlights how embracing these dynamics, rather than forcing flawed assumptions, can revolutionize West African agribusiness.

In February–March, mangoes from Siby in southwestern Mali dominate the market, supplying Bamako the capital city. This supply vanishes by late March, creating a scarcity gap before the second peak. By May and June, mangoes from Sikasso (Mali), Orodara, and Bobo-Dioulasso (Burkina Faso) flood the market, peaking in the third week of May (see Fig. 12.7). Poor processing capacity leads to severe postharvest loss, with thousands of tons rotting due to oversupply. After June, no fresh mangoes are available for drying, juicing, or export, despite strong market demand.

Traditional Gaussian models assume a single peak centered around the mean harvest period, smoothing out the supply curve. This misrepresentation leads to planning failures. While Gaussian assumptions predict a continuous and balanced supply, the reality follows a two-mode Rosenblatt shape with distinct peaks separated by a no-supply gap.

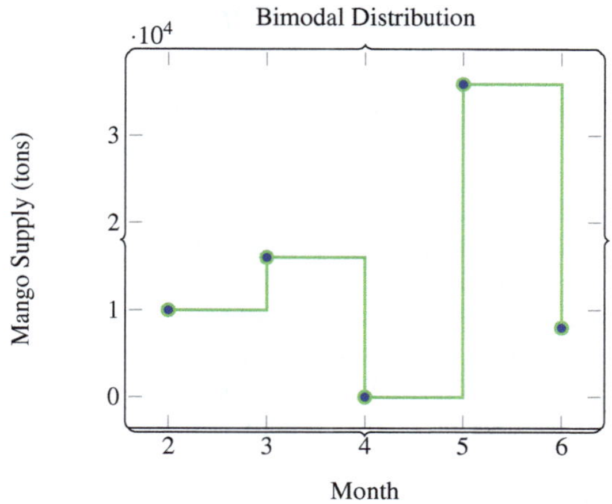

Fig. 12.7 Bimodal Mango Supply Dual harvest seasons create 2-month scarcity gap—invisible to single-peak models. Over an average of 75,000 tons were estimated by the Guinaga platform for the 2019–2024 period

Postharvest waste becomes unmanageable when market surges overwhelm storage and processing capacity. Without infrastructure to handle the supply extremes, large portions of the harvest are lost. Revenue opportunities disappear when dried mango and juice production shut down between July and January due to a lack of raw materials. Exporters and processors struggle with inconsistent supply, even as demand remains high. Price volatility intensifies as markets experience extreme fluctuations, with crashes during peak harvests and sharp increases post-June when mangoes become scarce.

The Malian agtech startup Guinaga conducted supply chain mapping across mango-growing regions, revealing critical inefficiencies in handling the two harvest peaks. Market dynamics remain unpredictable because traders and processors lack tools to navigate the disjointed seasons. Solutions include developing cold storage hubs to extend mango availability postharvest, establishing small-scale drying and juicing units near farms to absorb surplus, and deploying digital marketplaces to connect farmers with pre-committed buyers before harvest. By incorporating two-mode Rosenblatt shape models, Guinaga helps reduce losses, stabilize supply, and increase earnings for farmers and traders.

Mali and Burkina Faso's mango sector is a case study in how a two-mode Rosenblatt shape drives real-world agricultural dynamics. Ignoring these dual peaks leads to recurring failures—waste, price instability, and supply shortages. By shifting away from Gaussian assumptions and adopting new strategies, policy-makers and agribusinesses can turn volatility into opportunity, ensuring sustainable profits for farmers while meeting year-round consumer demand.

12.1.5 Dynamics of CPU, GPU, and TPU Adoption in Business

The evolution of computing hardware: central processing unit (CPUs), graphics processing units (GPUs), and tensor processing units (TPUs), in the business sector demonstrates clear deviations from Gaussian assumptions across demand, price, usage efficiency, and usage over time. These trends follow multi-peaked, Rosenblatt-like, or discontinuous patterns, shaped by technological innovations, market disruptions, and shifting workloads. The following analysis showcases this non-Gaussian behavior across key metrics:

Demand trends show a multi-peaked distribution, where CPU demand plateaus post-2010, GPU demand spikes sharply between 2015 and 2020, and TPU demand accelerates post-2020. There is no singular "mean" demand period, with spikes driven by discontinuous innovations, such as the release of GPT to the public in 2022. The period from 2000 to 2010 saw CPUs dominating with enterprise servers and PCs, driven by linear demand growth in conjunction with information technology expansion. Between 2010 and 2017, GPUs surged due to machine intelligence and machine learning adoption, with technologies like NVIDIA CUDA and deep learning. Bitcoin mining (2013–2017) caused a secondary spike in demand. From 2017 to 2025, TPUs and specialized machine intelligence chips, like AWS Inferentia, disrupted the dominance of GPUs, especially in cloud machine intelligence environments.

Regarding price trends, CPUs experienced a decline in price per core/thread following Moore's like Law until 2010, after which stagnation occurred due to physical limits. GPUs saw significant price surges during crypto growths (2017–2018, 2020–2021, 2024) and the machine intelligence growth cycle (2023), followed by price crashes during crypto winters. Meanwhile, TPUs saw consistent price drops as Google scaled cloud TPU offerings, benefiting from economies of scale, though their prices remained premium compared to GPUs. Price trends for GPUs exhibit a bimodal distribution, where speculative crypto and machine intelligence bubbles create volatile price peaks, while CPUs show a negative skew, reflecting long-term price decay with occasional plateaus.

Usage efficiency metrics reveal that CPUs saw flatlining of efficiency improvements post-2010 due to diminishing returns on clock speeds. GPUs initially showed significant efficiency improvements (FLOPS/$) until the inflationary effects of crypto mining (2017–2021). TPUs displayed exponential efficiency gains post-2018, optimized for machine intelligence inference, although their high upfront costs limited widespread adoption. The efficiency-to-price ratio for GPUs resembles a sawtooth wave, reflecting technological advances contrasted with speculative price inflation. Meanwhile, TPUs exhibited an exponential efficiency improvement, characterized by a nonlinear, power-law curve.

In terms of usage over time, CPUs were primarily used for general computing tasks like ERP and databases from 2000 to 2010. Between 2010 and 2020, GPUs dominated in fields like rendering, scientific computing, and machine intelligence training. From 2020 to 2025, TPUs and specialized machine intelligence chips

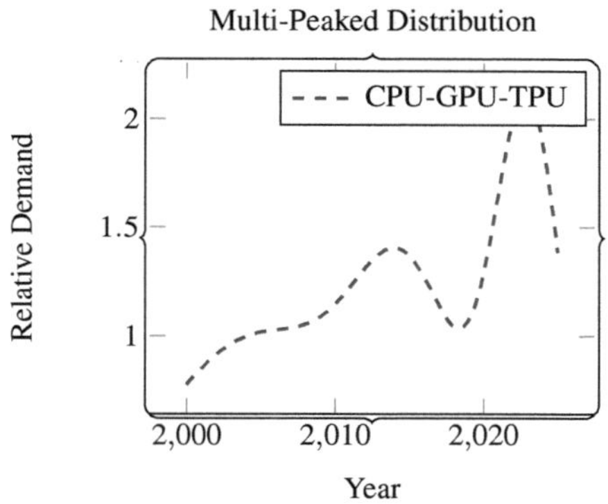

Fig. 12.8 Technology: CPU/GPU/TPU Adoption. Demand spikes driven by crypto (2017, 2021, 2024) and LLMs (2023) violate Gaussian stationarity

began to take over cloud machine intelligence inference tasks such as Google Search and GPT. These trends follow a multimodal distribution, with each hardware type having its own distinct adoption and obsolescence cycle. Legacy CPUs still exhibit long tails of usage, continuing to serve in certain environments despite being obsolete by newer standards.

Gaussian models fail to capture the complexities of the market due to factors like demand volatility from crypto bubbles, machine intelligence hype cycles, and supply chain shocks such as the 2021 chip shortage. Speculative markets, as seen with GPUs, and monopolistic forces, as seen with TPUs, create significant asymmetry and outliers in price distribution. Efficiency gains in hardware are often punctuated, as seen with TPU breakthroughs, or reversed, as seen with GPU price inflation during crypto booms. Additionally, legacy systems (e.g., CPUs) persist in usage far beyond their expected decay under Gaussian assumptions.

In Fig. 12.8, a real-world example of GPU demand from 2016 to 2025 illustrates the non-Gaussian trends. In 2016–2017, machine intelligence labs began purchasing GPUs for deep learning tasks. From 2017 to 2018, crypto miners hoarded GPUs, spiking prices by 300%. Between 2020 and 2021, COVID-driven remote work boosted demand for consumer GPUs, and in 2022–2023, machine intelligence startups like Stability AI and OpenAI triggered another GPU demand rush. It was followed by Llama of Meta, Grok of xAI, Gemini of Google, Qwen, DeepSeek, Kimi, etc. These fluctuating events create a multi-peaked demand curve, further underscoring the non-Gaussian nature of this market.

The evolution of computing hardware, driven by crypto-induced GPU demand, AI-powered TPU adoption, and technological breakthroughs, follows non-Gaussian patterns. Businesses relying on average forecasts are unprepared for market volatility, legacy system persistence, or pricing asymmetries. Understanding power-law dynamics, multimodal peaks, and unpredictable market events is crucial for thriving in the computing hardware sector (see price evolution in Fig. 12.9).

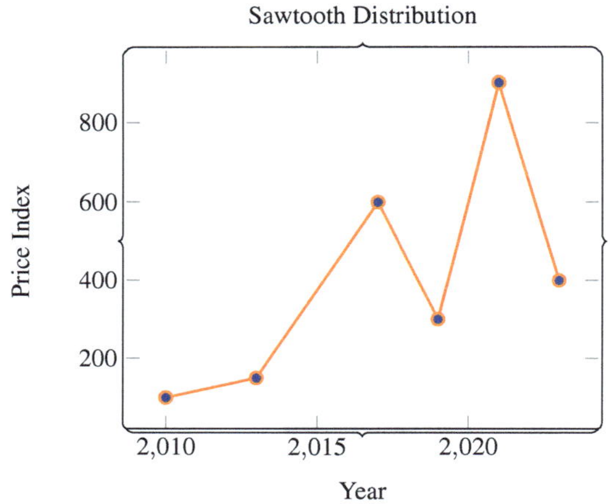

Fig. 12.9 Price index over the years. Crypto peaks (2017, 2021, 2024) and LLM peak (2023) create non-Gaussian price shocks

12.1.6 AI-Related Funding Dynamics

The evolution of machine intelligence funding, spanning startups, academia, and industry, exhibits multi-peaked, Rosenblatt-like, and discontinuous trends, starkly contrasting Gaussian assumptions. Technological breakthroughs, hypergrowth cycles, and market shocks shape this non-Gaussian landscape.

Demand trends reveal a multi-peaked distribution, with peaks in the 1960s (symbolic machine intelligence), 2010s (big data), and 2020s (generative machine intelligence). Legacy academic funding persists despite the dominance of industry. From the 1940s–1950s, machine intelligence funding was minimal, mainly driven by academia and military projects like ENIAC (Electronic Numerical Integrator and Computer). The first machine intelligence growth occurred in the 1960s–1970s, fueled by DARPA funding for symbolic machine intelligence projects like SHRDLU (an early natural-language understanding computer program that was developed by Terry Winograd at MIT in 1968–1970), with demand peaking in the mid-1960s. The machine intelligence winters of the 1980s–1990s followed, as funding collapsed due to unmet expectations, including the failure of expert systems. The 2000s–2010s saw a resurgence through machine learning, with institutions like IBM Watson and Google Brain reigniting demand. From 2012 to 2025, the deep learning explosion, driven by venture capital, saw a surge in demand, particularly post-2017 with the advent of transformer models and GPT-3.5-4-o.

In terms of price, the 1940s–1980s saw low "costs" due to limited compute power, with mainframes costing up to $1M in the 1960s. Prices rose in the 1990s–2010s as clusters and GPUs became prevalent, with training AlexNet in 2012 costing around $1M. By 2020–2025, exponential price growth became evident, with the cost of training GPT-4 reaching $100M and hyperscalers like AWS and Azure dominating infrastructure spending. Price growth post-2010 follows a power-law

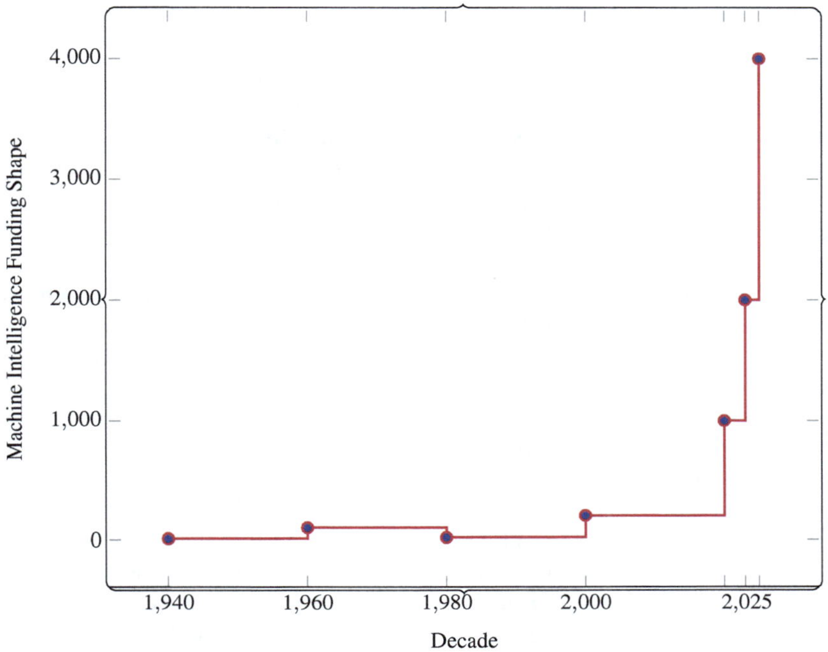

Fig. 12.10 Machine intelligence funding: Several peaks/waves/cycles-Machine Intelligence winters (1980s) vs. GPT-era hypergrowth defy smooth Gaussian projections

distribution, with costs growing 10x per decade, driven by compute and data scaling laws. Outliers, such as GPT-4o-5, skew averages.

Regarding usage efficiency, the 1940s–2000s saw low efficiency, with symbolic machine intelligence systems in the 1960s costing millions for narrow tasks. In the 2010s, efficiency spiked with GPUs (via NVIDIA CUDA), and models like ResNet delivered breakthroughs in deep learning per dollar spent. However, by the 2020s, diminishing returns were evident, with larger models like GPT-4 requiring 1000x more compute for marginal gains. Efficiency trends show an inverse U-shape, peaking between 2015 and 2020, then declining as models continued to grow. The efficiency-to-price ratio resembles a nonlinear hockey stick, with a steep rise followed by a plateau.

As illustrated in Fig. 12.10, usage over time follows an S-curve, with adoption initially slow before rapidly accelerating from 2010 to 2025. In the 1940s–1980s, machine intelligence was confined to academic labs like Stanford and MIT. By the 1990s–2010s, industry began adopting narrow machine intelligence applications (e.g., search algorithms, fraud detection). By 2020–2025, machine intelligence became ubiquitous, with deployments in chatbots, self-driving cars, and drug discovery. Startups scaled rapidly via cloud APIs. Adoption was marked by step-function jumps, with breakthroughs like AlphaGo in 2016 and ChatGPT in 2022 driving significant adoption spikes.

Gaussian models fail to account for the multi-peaked demand, exponential price growth, diminishing efficiency, and step-function adoption of machine intelligence technologies. Multiple disjointed peaks in demand (1960s, 2010s, 2020s) contradict a single mean/variance. Exponential cost growth and outliers like GPT-4-o-4.5 violate Gaussian symmetry, while efficiency's rise and fall creates an asymmetric, non-bell curve. Adoption follows a rapid S-curve with disruptive steps rather than smooth Gaussian growth.

A real-world example contrasts machine intelligence Winters and the generative machine intelligence growth. In the 1970s–1980s, funding dropped by 80% after the Lighthill Report, which criticized AI's failures. From 2020 to 2025, generative machine intelligence attracted over $300B in venture capital and corporate R&D, marking a stark contrast. This bimodal distribution (AI collapse vs. hypergrowth) cannot be modeled by a single Gaussian. AI funding dynamics are non-Gaussian due to hyper-growth/decline cycles (booms and busts), exponential compute costs (power-law scaling), and step-function adoption such as GPT. Entities like DARPA (1960s), NVIDIA (2010s), and OpenAI (2020s) thrived by adapting to these extremes. Traditional Gaussian models, assuming smooth and symmetrical trends, fail to predict events like machine intelligence winters or hypergrowth in generative AI. The future belongs to those who model multimodality, fat tails, and discontinuities—not averages.

12.1.7 Shallot/Onion Price in Dogon Country and in Ansongo

In a groundbreaking analysis conducted by the Malian startup Guinaga, the evolution of shallot and onion prices across two key regions, Dogon Country and Ansongo, has revealed a striking divergence from the expected Gaussian distribution commonly assumed in agricultural markets. The data, gathered through meticulous fieldwork over the past years (2019–2024), demonstrates the skewed nature of pricing dynamics in these regions, underscoring the significant role of seasonality, supply chain bottlenecks, and external shocks.

A closer look at the shallot price data (see Figs. 12.11 and 12.12) for Dogon Country, particularly, illustrates the sharp volatility that can only be described as non-Gaussian as it is highly skewed. The price trajectory follows a jagged pattern, with nonsymmetric shapes: one from June to November and another from December to April. This skewed shape, with abrupt increases during the raining season and crashes during the new local shallot season, starkly contrasts the smooth, bell-shaped curve typically associated with Gaussian processes. The shallot market, much like many other agricultural goods in Mali, is governed by more than just simple supply and demand; it is shaped by unpredictable factors such as sudden climate changes, transportation delays, and local demand surges during festive seasons.

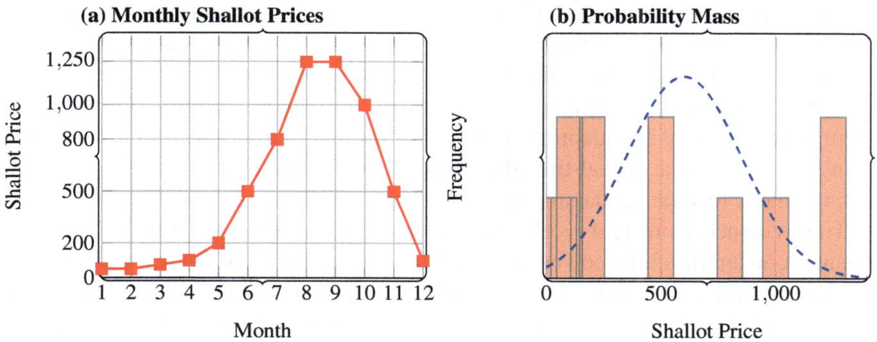

Fig. 12.11 The Gaussian fit does not capture the skewed and heavy-tailed behavior of the actual data

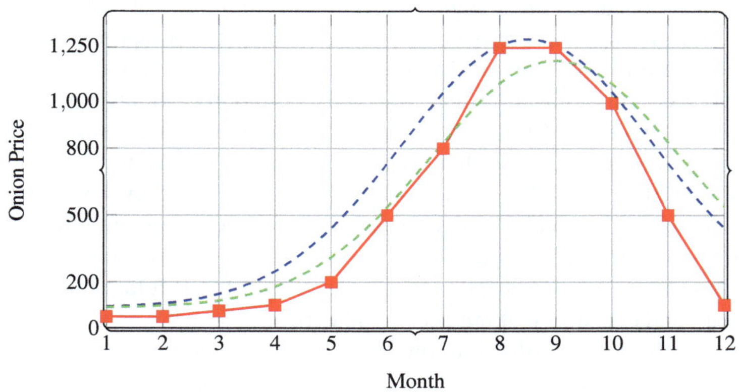

Fig. 12.12 Non-Gaussian fluctuations in shallot and onion prices over time

In Dogon Country, the price of shallots begins at a relatively low level (around 50 FCFA per kg) in January, steadily climbs as demand rises in the spring, and hits a staggering 1,250 FCFA per kg by August, only to plummet sharply toward the year-end. The rise and fall of prices are neither smooth nor predictable, defying standard economic models that rely on continuous, normally distributed data.

Similarly, the pricing behavior in Ansongo follows a comparable non-Gaussian trajectory, marked by extreme fluctuations that reflect the region's dependence on fluctuating water supply levels and market access, both of which are prone to abrupt changes. The typical assumption in standard market economics—that prices will gradually increase or decrease based on predictable trends—fails to account for the chaotic and unpredictable nature of agricultural production in these areas.

These findings are vital for policy-makers, traders, and agricultural stakeholders who have traditionally relied on Gaussian-based models to predict and stabilize market prices. Guinaga's research advocates for a new approach—one that embraces the inherent nonlinearity and unpredictability of local agricultural markets. By adopting data-driven, multimodal models that account for these peaks and valleys, stakeholders can mitigate the risk of overproduction and avoid costly shortages.

The non-Gaussian nature of shallot prices in Dogon Country and Ansongo exemplifies the need for adaptive strategies in rural agricultural economies, emphasizing the importance of real-time data, flexible pricing mechanisms, and responsive supply chains to weather the unpredictable ebbs and flows of these crucial markets.

The evidence from real-world data across diverse domains, from e-commerce revenue concentration to power grid fluctuations, agricultural cycles, and machine intelligence funding dynamics, demonstrates the failure of Gaussian models in capturing underlying patterns. Instead, many of these datasets exhibit multimode Rosenblatt-shaped distributions, characterized by multi-peaked, heavy-tailed, and discontinuous structures that defy the smooth, symmetric assumptions of the standard Gaussian framework. Whether it is the whale effect in e-commerce, the dual-seasonal dynamics in mango supply, or the multimodal volatility of GPU demand and pricing, the presence of distinct, dominant peaks and abrupt shifts challenges conventional statistical tools. Similarly, empirical evidence from underwater wireless communication and seasonal commodity pricing (as observed in onion markets in Dogon Country and Ansongo) further underscores how traditional Gaussian approaches fail to predict extreme variations. The trimodal water demand in Punjab and the multiphase machine intelligence funding cycles reinforce this pattern, showing that real-world processes often result from mixtures of independent, asymmetric drivers rather than a single, smooth probability distribution. Recognizing these non-Gaussian structures is crucial for accurate forecasting, risk assessment, and decision-making across industries.

Note that the Cauchy distribution (Fig. 12.13) belongs also to the family of continuous probability distribution and exhibits a bell shape but differs significantly from Gaussian distribution. The Cauchy distribution has heavy tails that decay at a much slower rate like inverse square which means it has no finite variance or mean. This heavy-tailed behavior makes the Cauchy distribution highly sensitive to outliers, and in practice, it is often used to model phenomena where extreme values are more likely than in a Gaussian distribution. The Cauchy distribution is used for modeling situations where extreme events dominate and the central measure (mean) is not meaningful.

12.1.8 Underwater Wireless Channel State Measurements

Long-range underwater acoustic communication (LR-UWAC), defined as the transmission of acoustic signals over distances spanning tens to hundreds of kilometers, is pivotal for sustained maritime operations. Applications range from marine environmental monitoring and geological surveying to biological data collection and acoustic tomography, which entails the remote measurement of sound speed and water temperature. The optimization of such communication systems hinges on

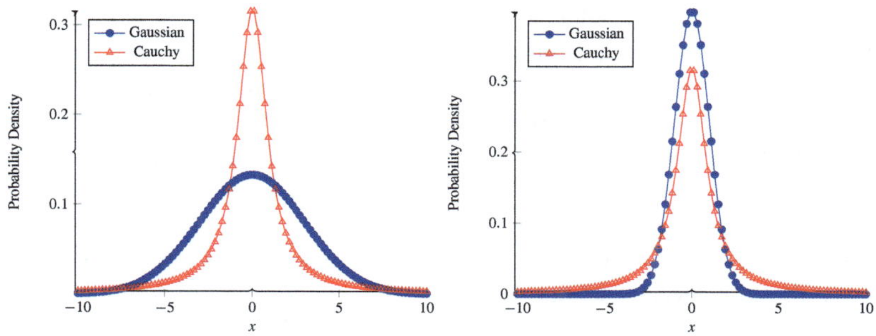

Fig. 12.13 Symmetry does not necessarily imply Gaussianity

precise channel state measurements, achieved through techniques including channel sounding, channel state information (CSI) estimation, Doppler spectrum analysis, and extensive large-scale modeling.

In channel sounding, both time-domain methods—employing wideband probe signals such as chirps or pseudorandom sequences—and frequency-domain approaches based on orthogonal frequency-division multiplexing (OFDM) are utilized to estimate the impulse and frequency responses of the channel. These measurements yield critical parameters such as delay spread and coherence bandwidth. CSI estimation, in turn, provides a quantitative assessment of the fading conditions inherent to the underwater environment. Complementary methodologies such as Doppler analysis, which captures mobility-induced frequency shifts, and random field models that quantify fading severity, are integrated with large-scale measurement campaigns. These campaigns, leveraging drive testing, massive multiple-input multiple-output (MIMO) arrays, and machine learning-enhanced ray tracing, elucidate spatial correlation and propagation trends across diverse aquatic environments, thereby informing the design of next-generation networks (5G, 6G, and beyond).

Accurate quantification of fading in underwater channels necessitates an examination of signal fluctuations induced by multipath propagation, absorption, scattering, and hydrodynamic effects. Traditional models based on Rayleigh, Rician, and Nakagami-m distributions, where the shape parameter m serves as an index of fading severity, are frequently employed. Empirical studies involve the deployment of hydrophones and transducers to capture variations in amplitude and phase over time. Spectral broadening techniques facilitate Doppler spread analysis, revealing motion-induced distortions attributable to water currents and marine life, while coherence time estimates offer insights into channel stability. The extraction of power delay profiles further distinguishes line of sight from multipath components, thereby refining the probabilistic characterization of fading.

In scenarios where conventional fading models inadequately capture memory effects and non-Gaussian statistics, the non-centered Rosenblatt process provides a robust alternative. Underwater acoustic channels, influenced by multipath reflections, turbulence, and Doppler shifts from water currents, often exhibit temporal correlations that defy simple Gaussian assumptions. By estimating the Hurst parameter, which quantifies long-range dependence, researchers can adjust the Rosenblatt model to accurately represent observed power variations. Empirical fitting typically involves the collection of time-series data from hydrophone arrays, the application of fractional-order spectral analysis, and the optimization of model parameters via likelihood estimation.

Since the early 1990s, a series of LR-UWAC experiments have been conducted across multiple international locales. In 1993, adaptive multichannel combining and equalization were demonstrated over a $203km$ link using a 12-element vertical array, achieving error-free communications at 660 bps [304]. Subsequent trials, such as those reported in 2001 [156], extended the communication range to 3250 km across the North Pacific Ocean from Southern California to the Hawaiian Islands, by coherently combining signals from 20 hydrophones, at a reduced data rate of 37.5 bps.

Experimental configurations typically involve a transmitter operating at variable water depths (e.g., 20, 35, and 50m) and multiple receivers deployed in both shallow and deep-water environments. One receiver, situated $100km$ northwest of the transmitter at approximately 1800 m depth, utilized a custom acoustic recorder based on the TASCAM DR100mk and Geospectrum hydrophones. A second receiver, deployed 50 km south at a comparable depth (roughly 60 m), allowed for concurrent data acquisition without the need for synchronization among vessels. Bathymetric profiles and sound speed profiles (SSP) were recorded, with negligible variation in sound speed (less than $1m/s$) observed between sites.

A variety of signal types, including chirp, JANUS, BPSK, QPSK, and OFDM, were transmitted. The chirp signal facilitated channel measurement, while the remaining signals supported data communication. Frame sizes ranged from 288 bits for JANUS to 4608 bits for QPSK, with a total of 43 frames received, 28 from the shallow site and 15 from the deep site. Variability in the received signal-to-noise ratio (SNR) was attributed to receiver drift and the time-varying nature of the channel, as exemplified by the amplitude attenuation profile in Fig. 12.14.

Timadie, a Mali-based startup focused on underwater communication, has conducted field experiments using remotely operated vehicles (ROVs) to explore the dynamics of channel states in submerged environments. The results revealed a Rosenblatt distribution rather than the traditionally assumed Gaussian shape, in the communication channel's state over time. This non-Gaussian distribution stems from the inherent characteristics of underwater acoustics, where factors like water salinity, temperature, and turbulence create highly variable and unpredictable signal attenuation. The illustration was done in the Niger River in West Africa. Operating at depths between 5 and 20 m, these ROVs communicate via acoustic modems that are particularly susceptible to multipath interference from riverbed reflections, water currents, and biological activity. Probe signals transmitted by each ROV allow for the continuous recording of received signal strength (RSSI) and signal-to-noise ratio (SNR) at 1-second intervals over a 10-minute period. The resultant power fluctuations exhibit persistent behavior, with an estimated Hurst parameter $H \approx 0.8$,

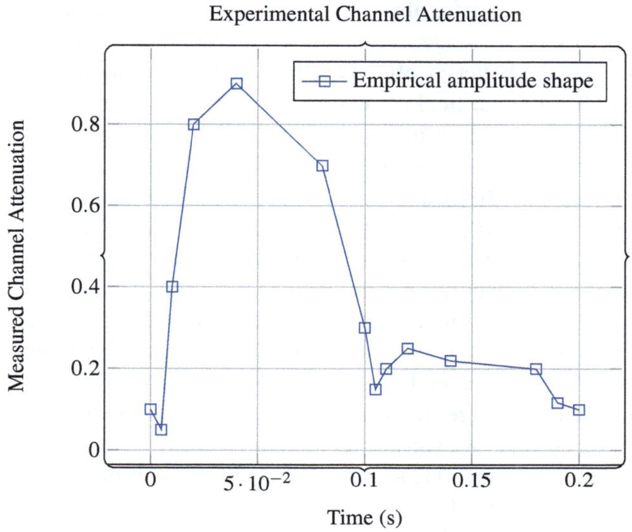

Fig. 12.14 Measured channel impulse response recorded under deep-water conditions at a transmitter depth of 20m

indicative of long-range dependence. The empirical probability density function of the received power reveals heavy-tailed, non-Gaussian characteristics that align well with the predictions of the Rosenblatt process—yielding a Wasserstein error on the order of 10^{-6}. In this dynamic environment, the inter-ROV distance, $d(t)$, influences the received signal power, modeled by the power-law decay:

$$P_r(t) = P_t - \eta \log_{10}\Big(1 + d(t)\Big) + 2\log_{10}\Big(|r^H(t)|\Big), \tag{12.1}$$

where P_t denotes the transmitted power, η the path loss exponent, and $r^H(t)$ encapsulates the long-range correlated fading component with Rosenblatt process. By integrating sophisticated measurement techniques with propagation module, this experiment predicts the behavior of underwater acoustic channels. The deployment of models such as the noncentered Rosenblatt process not only captures the inherent long-range dependence and non-Gaussian statistics of underwater fading but also informs the design of robust communication systems capable of operating in marine environments.

12.1.9 State of Health of Batteries of Electric Cabs

Accurately predicting electric vehicle battery health and remaining useful life (RUL) is a predictive maintenance challenge. The State of Health (SOH) assesses the battery's degradation over time, reflecting its ability to store and deliver energy compared to its pristine state, also typically represented as a percentage. SOH is affected by factors such as age, temperature, charging patterns, and usage, and it directly influences the battery's effective range and remaining life span. Both State of Charges and SOH are critical parameters monitored by the Battery Management System (BMS) for accurate range estimation and battery health assessment. The performance of batteries in electric vehicles is affected by both time-dependent degradation (calendar aging) and usage-dependent degradation (cycle aging). In regions with significant seasonal temperature variations, these effects differ between seasons. The RUL is the remaining cycles until capacity faded to $\alpha = 80\%$ of nominal (end-of-life threshold). A combined analysis that captures these differences over full 5 years, for predicting the RUL of batteries in electric vehicles operating in regions characterized by four distinct seasons. Battery capacity degradation results from both calendar aging, which follows an Arrhenius behavior, and cycle aging, which is dependent on usage intensity. Based on battery degradation data analysis with voltage (V), current (A), surface temperature (°C, via thermocouples), capacity (Ah), and internal resistance (Ω)—sampled at 1 Hz using programmable testbeds (Arbin BT2000), alongside derived metrics such as SOC and state of health (SOH); cycling protocols included CC-CV charging (1C rate, CC-CV charging is a typical method of charging rechargeable batteries such as li-ion, the operation switches between CC charging, which charges with a constant current, and CV that charges at a constant voltage, depending on the voltage of the rechargeable battery), dynamic discharge profiles (UDDS: Urban Dynamometer Driving Schedule), and calendar aging tests (20–80% SOC, 25–45°C) to emulate diverse operational stresses, battery chemistries, and characteristics impact the RUL. EV battery types include lithium nickel manganese cobalt oxide (NMC) batteries, lithium nickel cobalt aluminum oxide (NCA) batteries, lithium iron phosphate (LFP) batteries, lithium manganese oxide (LMO) batteries, lithium titanate (LTO) batteries, lithium polymer (LiPo) batteries, solid-state lithium-ion batteries, nickel-metal hydride (NiMH) batteries, lead-acid batteries, and lithium sulfur (Li-S) batteries.

The long-term degradation of electric vehicle (EV) batteries in high-utilization fleet applications, such as electric cabs, exhibits a stochastic profile that is well captured by a shape of a Rosenblatt distribution. The analysis of SoH in cabs older than 5 years reveals a departure from conventional Gaussian, instead of following a heavy-tailed structure with a varying Hurst parameter indicating different degrees of long-memory dependence in the degradation process. The presence of a wide spectrum of H values, spanning from 0.5 to 0.9, highlights the self-similar and fractal nature of battery wear, where certain conditions—such as operational stress, nonuniform charge cycles, and environmental variability—drive long-range correlations in SoH decline. This challenges traditional SoH prognostics and supports the need for fractional-order models to accurately predict remaining battery life in aged EV fleets. The probability density function of SoH, as illustrated in Fig. 12.15, underscores this non-Gaussian behavior, demonstrating asymmetry and heavy tails that signify the persistence of degradation modes. The Rosenblatt framework enables more precise risk-aware energy management strategies for second-life applications. The wide range of Hurst parameters observed in the data

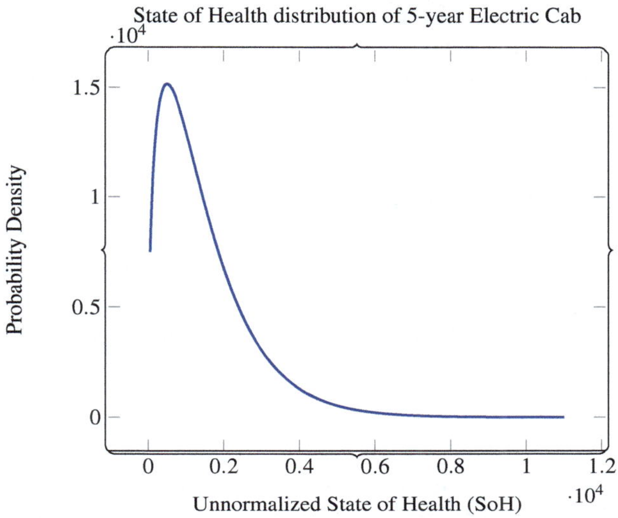

Fig. 12.15 Probability density function of SoH for EV batteries in 5+-year-old electric cabs, exhibiting a Rosenblatt-like heavy-tailed structure

suggests that the SoH of batteries in electric vehicles is affected by a variety of factors, including the type of battery, the driving habits of the driver, and the environment in which the vehicle is driven.

12.2 Noise Modeling in Mean-Field-Type Games

Different noise models have been proposed in mean-field-type game (MFTG) theory:

- Brownian motion noises: MFTGs with state driven by Brownian motions have been analyzed in Sect. 7.2. Note that the noise is a Brownian motion, but the state itself is not necessarily Markovian due to the presence of mean-field-type terms.
- Markovian noise: Beyond Brownian motion noises, MFTGs with state driven by Poisson jump and regime switching processes can be found in Sect. 14.2. Note that in this case, the noise is a jump process, but the state alone itself is not necessarily Markovian due to the presence of mean-field-type terms.
- Gauss-Volterra noises: Gauss-Volterra processes are first-order integral of a suitable kernel with Brownian noise. They do not satisfy the Markov property, and they are not martingales. These include fractional Brownian motions and multi-fractional Brownian motions. MFTG with state driven by Gauss-Volterra

processes have been discussed in Sects. 10 and 7.6 where semi-explicit solutions are provided for class of non-quadratic functions.

- MFTGs driven by non-Gaussian non-Poisson non-Markov noises, which are covered in this case

Consider the Wiener-Itô integral given by

$$Z^{\delta_1,\ldots,\delta_k}(t) = c(\delta) \int_{\mathbb{R}^k \setminus D_k} [\int_0^t \prod_{m=1}^k (t' - x_m)_+^{\delta_m} dt'] \prod_{l=1}^k dB(x_l)\tag{12.2}$$

where $D_k = \{(x_1, \ldots, x_k),\ x_1 = \ldots = x_k\}$ is the diagonal of $\mathbb{R}^k$, $c(\delta) \neq 0$, $\sum_{m=1}^k \delta_m \in (-\frac{k}{2} - \frac{1}{2}, -\frac{k}{2})$.

For $k = 2$, we have the second-order Hermite process, which is a Rosenblatt process with parameters (δ_1, δ_2). This is a process of interest, because it is one of the simplest non-Gaussian processes. This is a non-Markovian process and involves memory. We will study state dynamics that are driven by the process $r^H(t) := Z^{\delta,\delta}(t)$, where $\delta = \frac{H}{2} - 1$, with H a fractional index,

$$c(\frac{H}{2} - 1, \frac{H}{2} - 1) = \frac{\sqrt{2H(2H-1)}}{2B(1-H, \frac{H}{2})},$$

where $Beta(z_1, z_2) = \int_0^1 t^{z_1-1}(1-t)^{z_2-1}\, dt$ for complex number inputs z_1, z_2 such that Re (z_1), Re $(z_2) > 0$ is the Beta function and $\Gamma(z) = \int_0^\infty t^{z-1}e^{-t}\, dt$ is the gamma function with Re $(z) > 0$

A Rosenblatt process is a non-Gaussian non-Poisson non-Markovian second-order Hermite process $r^H(t) := Z^{\frac{H}{2}-1,\frac{H}{2}-1}(t)$.

12.3 Mean-Field-Type-Games Under Rosenblatt Noise

This section presents a class of noncooperative MFTGs with *nonquadratic cost* and a linear state driven by a Rosenblatt process. We start with the one-dimensional state case. Let $[0, T],\ T > 0$ be a fixed time horizon and $(\Omega, \mathcal{F}, \mathbb{F}^{r^H}, \mathbb{P})$ be a given filtered probability space. We consider several decision-makers in $\mathcal{I},\ 1 \leq |\mathcal{I}| < +\infty$ interacting during the time horizon $[0, T]$. An admissible control strategy u_i of the decision-maker $i \in \mathcal{I}$ is an $\mathbb{F}^{r^H}$-adapted and square integrable process with finite $2\bar{k}_i$-th moment with values in $\mathbb{R}$, and within the class of linear state-and-mean-field type feedback forms. We denote the set of all admissible controls by $\mathscr{U}_i$. A scalar linear-*non-quadratic* MFTG problem for a stochastic equation driven by a Rosenblatt process using a family of linear feedback control actions is formulated. Given the strategies $(u_j)_{j\neq i}$ of the decision-makers other than i,, the best-response problem of decision-maker $i \in \mathscr{I}$ under Rosenblatt process within the class of linear state-and-mean-field type feedback strategies is formulated as follows:

A MFTG with Rosenblatt Noise

Given $(u_j,\ j \in \mathscr{I}\backslash\{i\})$ in linear-and-mean-field-type feedback form, the best-response problem of the decision-maker i is

$$\begin{array}{l} \inf_{u_i} E[L_i(x,u)\,|x(0)=x_0], \\ \text{subject to} \\ x_t = x_0 + \int_0^t [b_1(x-\bar{x}) + \sum_{j\in\mathscr{I}} b_{2j}(u_j - \bar{u}_j)]dt' \\ +\int_0^t [(b_1+\bar{b}_1)\bar{x} + \sum_{j\in\mathscr{I}} (b_{2j}+\bar{b}_{2j})\bar{u}_j]dt' + r^H(t), \end{array} \tag{12.3}$$

where the cost functional of decision-maker i is given by

$$\begin{array}{l} L_i(x,u) = q_{iT}(x_T - \bar{x}_T)^2 + \bar{q}_{iT}\frac{\bar{x}_T^{2\bar{k}_i}}{2\bar{k}_i} \\ +\int_0^T \left[q_i(x-\bar{x})^2 + r_i(u_i-\bar{u}_i)^2 + \bar{q}_i\frac{\bar{x}^{2\bar{k}_i}}{2\bar{k}_i} + \bar{r}_i\frac{\bar{u}_i^{2\bar{k}_i}}{2\bar{k}_i} \right] dt \end{array} \tag{12.4}$$

with the (deterministic) coefficients

$$\begin{array}{l} q_i, \bar{q}_i\ :\ [0,T] \to \mathbb{R}_+,\ i \in \mathscr{I}, \\ r_i, \bar{r}_i > \epsilon > 0 :\ [0,T]\ \to \mathbb{R}_+,\ i \in \mathscr{I}, \\ b_1, \bar{b}_1, b_{2j}, \bar{b}_{2j}\ :\ [0,T] \to \mathbb{R},\ j \in \mathscr{I}, \end{array} \tag{12.5}$$

The functions $b_{2j}, \bar{b}_{2j}$ are not identically zero.

The notation $\bar{x}_t = E[x_t]$ represents the expected value of the state x_t at time t. $\bar{u}_i = E[u_i]$ is the expectation of the control action of decision-maker i. For $\bar{k}_i > 1$,, the running cost is *non-quadratic* in $\bar{x}, \bar{u}_i$. Notice that the expectation of the cost involves the variance of the state, $var[x]$, and the variance of the control action, $var[u_i]$. This makes it a *variance-aware game*. The presence of the variance term makes introduces nonlinearity into the expected cost, $E[L_i]$, for the probability measure of the state $\mathbb{P}_{x_t}$. Also the terms $\bar{x}^{2\bar{k}_i}$ and $\bar{u}_i^{2\bar{k}_i}$ are nonlinear in $\mathbb{P}_{x_t}$. Thus, we have a MFTG between the decision-makers in $\mathscr{I}$.

Solution Concepts
We recall here relevant definitions of the MFTG and their respective solution concepts.

Definition 12.1 (Mean-Field-Type Best-Response) Given the strategies $(u_j,\ j \neq i)$, of decision-makers other than i,, a best response strategy of decision-maker i is a strategy that solves the following problem of mean-field-type $\inf_{u_i} EL_i$ subject to (12.3). The set of best responses of i is denoted by $\mathrm{BR}_i((u_j)_{j\neq i})$.

Based on the best-response function $BR = (BR_i)_{i\in\mathscr{I}}$,, we define a Nash equilibrium of the MFTG.

Definition 12.2 (Mean-Field-Type Nash Equilibrium) A (mean-field-type) Nash equilibrium is a strategy profile $(u_j^*,\ j \in \mathscr{I})$, of all decision-makers such that for every decision-maker i, $u_i^* \in \mathrm{BR}_i((u_j^*)_{j\neq i})$.

Since a Rosenblatt process r^H is non-Gaussian, non-Markov, and non-Poisson, it requires a new Itô's formula. To date, there is no dynamic programming principle for the Rosenblatt process. There is no stochastic maximum principle available either for the Rosenblatt process. One way to solve MFTGs is by using a direct method, which we proceed with; the direct method consists of five elementary steps. The first step starts by setting the mean-field terms of the problem. The second step consists of identification of a partial guess functional where the coefficient functionals are time-dependent. The third step uses the stochastic integration formula. The fourth step uses a completion of terms in a one-shot game for both control actions and measure of the state and or control actions of all decision-makers. The approach used here combines some stochastic analysis for Rosenblatt processes, the method of completion of terms and some (component-wise) monotone operator game analysis. The fifth and last step uses an algebraic basis of linearly independent processes to identify the coefficients. The identification leads to a (possibly stochastic) differential system of equations, providing a semi-explicit representation of the solution.

The main result of this chapter characterizes the mean-field-type Nash equilibrium in linear state-and-mean-field-type feedback form.

Proposition 12.1 *Let $q_i > 0, \bar{q}_i > 0,\ r_i, \bar{r}_i > \epsilon > 0$. A Nash equilibrium of the MFTG (12.3) with a Rosenblatt process, having a linear state-and-mean-field type feedback form is given by*

$$\begin{aligned}
u_i^* &= -\eta_i(x-\bar{x}) - \bar{\eta}_i\bar{x},\\
\bar{\eta}_i &= \left(\frac{\bar{b}_{2i}\bar{\lambda}_i}{\bar{r}_i}\right)^{\frac{1}{2\bar{k}_i-1}},\\
EL_i &= \lambda_i(0)E(x_0-\bar{x}_0)^2 + \bar{\lambda}_i(0)\frac{1}{2\bar{k}_i}\bar{x}_0^{2\bar{k}_i} + \gamma_i(0),
\end{aligned} \tag{12.6}$$

whenever the following system of differential equations admits a positive solution which does not blow up (no singularity) within $[0, T]$:

$$
\begin{aligned}
&-\dot{\lambda}_i = q_i + 2b_1\lambda_i - \tfrac{b_{2i}^2}{r_i}\lambda_i^2,\\
&\lambda_i(T) = q_i(T),\\
&-\tfrac{d}{dt}\bar{\lambda}_i = \bar{q}_i + 2\bar{k}_i(b_1+\bar{b}_1)\bar{\lambda}_i - 2\bar{k}_i\bar{\lambda}_i \textstyle\sum_{j\neq i}(b_{2j}+\bar{b}_{2j})\left(\tfrac{(b_{2j}+\bar{b}_{2j})\bar{\lambda}_j}{\bar{r}_j}\right)^{\frac{1}{2\bar{k}_j-1}}\\
&-(2\bar{k}_i-1)\bar{r}_i\left(\tfrac{(b_{2i}+\bar{b}_{2i})\bar{\lambda}_i}{\bar{r}_i}\right)^{\frac{2\bar{k}_i}{2\bar{k}_i-1}},\\
&\bar{\lambda}_i(T) = \bar{q}_i(T),\\
&-\dot{\gamma}_i = r_i(\eta_i - \lambda_i\tfrac{b_{2i}}{r_i})^2 v_2(t) - 2\lambda_i \textstyle\sum_{j\in\mathscr{I}\setminus\{i\}} b_{2j}\eta_j v_2(t) + 2c\lambda_i o(t),\\
&\gamma_i(T) = 0,
\end{aligned}
\tag{12.7}
$$

where η_i *minimizes* $f_i(\eta)$:

$$
f_i(\eta) = r_i(\eta_i - \lambda_i\tfrac{b_{2i}}{r_i})^2 v_2(t) - 2\lambda_i \textstyle\sum_{j\in\mathscr{I}\setminus\{i\}} b_{2j}\eta_j v_2(t) + 2c\lambda_i o(t), \tag{12.8}
$$

and the vector η *is a fixed point of* $(\arg\min_{z_i} f_i(z_i, \eta_{-i}))_{i\in\mathscr{I}}$,

$$
\begin{aligned}
&b = [b_1 - \textstyle\sum_j b_{2j}\eta_j],\\
&o(t) = \left[\int_0^t c_3[t-s]^{2H-2} e^{-\int_0^s b(t'')dt''} ds\right] e^{\int_0^t b(t')\,dt'},\\
&v_2(t) = \left[v_2(0) + \int_0^t 2c\, o(s) e^{-\int_0^s 2b(t'')dt''} ds\right] e^{\int_0^t 2b(t')\,dt'}.
\end{aligned}
\tag{12.9}
$$

Sub-optimality of Strategies When Induced by Other Noises
We can derive several conclusions from Proposition 12.1:

- The equilibrium strategies obtained for the noise-free case (purely deterministic case) no longer provide an equilibrium when the noise is driven by a Rosenblatt process.
- The equilibrium strategies obtained for the Brownian motion, multifractional Brownian, Gauss-Volterra or a pure jump, and regime switching processes no longer provide an equilibrium when the noise is driven by a Rosenblatt process.
- In particular, the equilibrium strategies for the case when the state is driven by Brownian motion are no longer in equilibrium even in the quadratic cost case $\bar{k}_i = 1$ for which the third derivatives are zero.

We let $B^H(t) = \tilde{c}_2\int_{\mathbb{R}}(\int_0^t (t'-x)_+^{H-\frac{3}{2}} dt')dW(x)$, $H \in (\frac{1}{2}, 1)$, $\tilde{c}_2 = \sqrt{\frac{H(2H-1)}{B(2-2H, H-\frac{1}{2})}}$.

Integration Formula for the Rosenblatt Process

Lemma 12.1 *Let*

$$x_t = x_0 + \int_0^t d_1\, dt' + 2\tilde{c}\int_0^t d_2 dB^{\frac{H+1}{2}}(t') + \int_0^t d_3 dr^H(t')$$

for given functions (d_1, d_2, d_3). Then, the process y_t defined by $y_t = f(t, x_t)$ satisfies

$$y_t = y_0 + \int_0^t \tilde{d}_1\, dt' + 2\tilde{c}\int_0^t \tilde{d}_2 dB^{\frac{H+1}{2}}(t') + \int_0^t \tilde{d}_3 dr^H(t')$$

where

$$\begin{aligned}
\tilde{d}_1(t') &= f_{t'}(t', x_{t'}) + f_x(t', x_{t'})d_1 + 2cf_{xx}(t', x_{t'})(\nabla^{\frac{H}{2}} x_{t'})(t')d_2 \\
&+cf_{xx}(t', x_{t'})(\nabla^{\frac{H}{2},\frac{H}{2}} x_{t'})(t', t')d_3 + cf_{xxx}(t', x_{t'})[(\nabla^{\frac{H}{2}} x_{t'})(t')]^2 d_3, \\
\tilde{d}_2(t') &= f_x(t', x_{t'})d_2 + f_{xx}(t', x_{t'})(\nabla^{\frac{H}{2}} x_{t'})(t')d_3, \\
\tilde{d}_3(t') &= f_x(t', x_{t'})d_3,
\end{aligned} \tag{12.10}$$

with

$$\begin{aligned}
&c = c(\delta)\Gamma^2(\tfrac{H}{2}), \\
&\tilde{c} = \sqrt{\tfrac{(2H-1)\Gamma(1-\frac{H}{2})\Gamma(\frac{H}{2})}{(H+1)\Gamma(1-H)}}, \\
&\nabla^{\frac{H}{2}} = I_+^{\frac{H}{2}} \circ D, \quad \nabla^{\frac{H}{2},\frac{H}{2}} = I_{+,+}^{\frac{H}{2},\frac{H}{2}} \circ D^2, \\
&f : \mathbb{R} \to \mathbb{R}, \\
&(I_+^{\alpha} f)(x) = \tfrac{1}{\Gamma(\alpha)} \int_{-\infty}^x f(u)(x-u)^{\alpha-1} du, \\
&(\alpha_1, \alpha_2) \in (0,1)^2, \ f : \mathbb{R}^2 \to \mathbb{R}, \\
&(I_{+,+}^{\alpha_1,\alpha_2} f)(x_1, x_2) = \tfrac{1}{\Gamma(\alpha_1)\Gamma(\alpha_2)} \int_{-\infty}^{x_1}\int_{-\infty}^{x_2} f(u,v)(x_1-u)^{\alpha_1-1}(x_2-v)^{\alpha_2-1} du dv,
\end{aligned} \tag{12.11}$$

We will use the Itô's formula for Rosenblatt processes provided in Lemma 12.1. The Itô's formula for this type of non-Gaussian process involves the third-order derivatives in general.

Lemma 12.2 *Let $y_t = y_0 + \int_0^t drift\ dt' + r^H(t)$ for a certain drift functional. For $\lambda(t)f(y_t)$, with a smooth function such that $f_{yyy} = 0$ and a continuously differentiable function λ, the stochastic integration formula yields*

$$
\begin{aligned}
&\lambda(T)f(y_T) - \lambda(0)f(y_0) \\
&= \int_0^T [\dot{\lambda} f(y) + \lambda f_y.drift \, + c\lambda f_{yy}(y_t)(\nabla^{H/2,H/2}\, y_t)(t,t)]dt \\
&+2\tilde{c}\int_0^T [\lambda f_{yy}(y_t)(\nabla^{H/2} y_t)(t)]dB_t^{\frac{H+1}{2}} \\
&+\int_0^T \lambda f_y(y_t) dr^H(t),
\end{aligned} \tag{12.12}
$$

Taking expectation on both sides yields

$$
E[\lambda(T)f(y_T)-\lambda(0)f(y_0)]=E\int_0^T [\dot{\lambda} f(y)+\lambda f_y.drift \, +c\lambda f_{yy}(y_t)(\nabla^{\frac{H}{2},\frac{H}{2}}\, y_t)(t,t)]dt, \tag{12.13}
$$

The next Lemma provides the $\nabla^{\frac{H}{2},\frac{H}{2}}$ term as a function of the control actions.

Lemma 12.3 *Let the state dynamics be*

$$
\begin{aligned}
&\bar{x}_t = \bar{x}_0 + \int_0^t [(b_1+\bar{b}_1)\bar{x} + \sum_{j\in\mathcal{J}}(b_{2j}+\bar{b}_{2j})\bar{u}_j]dt', \\
&x_t - \bar{x}_t = x_0 - \bar{x}_0 \\
&+\int_0^t [b_1(x-\bar{x}) + \sum_{j\in\mathcal{J}} b_{2j}(u_j-\bar{u}_j)]dt' + r^H(t).
\end{aligned} \tag{12.14}
$$

Then, $(\nabla^{H/2,H/2}\,(x_t-\bar{x}_t))(t,t) =: O[x-\bar{x}](t) =: o(t)$ *satisfies the following ordinary differential equation*

$$
\begin{aligned}
&(\nabla^{H/2,H/2}\, y_t)(t,t) = o(t) = \int_0^t [b_1 - \sum_j b_{2j}\eta_j]o(t')dt' + g[r^H](t), \\
&g[r^H](t) = c_3 \int_0^t [t-t']^{2H-2}dt', \\
&o(0) = 0, \\
&c_3 = 2c(\delta)\frac{B^2(\frac{H}{2},1-H)}{\Gamma^2(\frac{H}{2})}.
\end{aligned} \tag{12.15}
$$

It is important to notice that $(\nabla^{H/2,H/2}\, y_t)(t,t)$ depends on the control η_i.

Lemma 12.4 *Let* $b = [b_1 - \sum_j b_{2j}\eta_j]$ *and* $(\eta_j)_{j\neq i}$ *be associated with the control actions of all decision-makers except the* i*-th one. Then,*

$$
\begin{aligned}
&o(t) = \left[\int_0^t c_3[t-s]^{2H-2}e^{-\int_0^s b(t'')dt''}ds\right]e^{\int_0^t b(t')\,dt'} \geq 0, \\
&o(o) = 0.
\end{aligned} \tag{12.16}
$$

Moreover, the function $\eta_i \mapsto o(t)$ *is strictly convex given* $(\eta_j)_{j\neq i}$.

Proof For $t > s \geq 0$,, the function $\eta \mapsto e^{\int_s^t \langle b_2,\eta\rangle dt''}$ is strictly convex for $b_{2k} \neq 0$, $\forall k \in \{1,\ldots,n\}$. In particular for $b_{2i} \neq 0$ the mapping $\eta_i \mapsto e^{\int_s^t \langle b_2,\eta\rangle dt''}$ is strictly convex for given $(\eta_j)_{j\neq i}$. ■

Lemma 12.5 *The dynamics of the variance* $v_2(t) = E(x_t - \bar{x}_t)^2$ *is given by*

$$
v_2(t) - v_2(0) = E[(x_t-\bar{x}_t)^2 - (x_0-\bar{x}_0)^2] = \int_0^t [2b(t')v_2 + 2c\, o(t')]dt' \tag{12.17}
$$

which can be expressed explicitly in terms of $v_2(0), b, o$ *as*

$$v_2(t) = \left[v_2(0) + \int_0^t 2c\, o(s) e^{-\int_0^s 2b(t'')dt''} ds \right] e^{\int_0^t 2b(t')\, dt'}.$$

The variance of the equilibrium state $var[x]$ *is independent of* $(\bar{\eta}_j, \bar{b}_1, \bar{b}_{2j}, \bar{r}_j, \bar{q}_j)_{j\in\mathcal{I}}$.

Proof Apply the stochastic integration formula with $\lambda(t) = 1$ and $f(y) = y^2$ in Lemma 12.2 and the double integration expression in Lemma 12.3. ■

Lemma 12.6 *Given* $(\eta_j)_{j\neq i}$ *of all the decision-makers except the* i*th one, the function* $\eta_i \mapsto v_2(t)$ *is strictly convex.*

Proof With $(\eta_j)_{j\neq i}$ given and fixed, from Lemma 12.3, $\eta_i \mapsto o(t)$ has exponential shape in b. Since $e^{-x\theta}$ is a strictly convex function, the same property follows for o in η_i. Now, let us look at the variance v_2 as a function of η_i. The map $\eta_i \mapsto v_2$ is strictly convex in η_i as $exp\ (linearmap)$ is strictly convex. This completes the proof. ■

Next we focus on the expected value of the state.

Lemma 12.7 *The expectation of the state at equilibrium is uniquely determined, and it is explicitly given by*

$$\begin{aligned} \bar{x}_t &= \bar{x}_0 + \textstyle\int_0^t [(b_1+\bar{b}_1)\bar{x} + \sum_{j\in\mathcal{I}} (b_{2j}+\bar{b}_{2j})\bar{u}_j] dt' = \bar{x}_0 e^{\int_0^t \bar{b}(t')dt'}, \\ \bar{b} &:= (b_1+\bar{b}_1) - \textstyle\sum_{i\in\mathcal{I}} (b_{2i}+\bar{b}_{2i}) \left(\frac{(b_{2i}+\bar{b}_{2i})\bar{\lambda}_i}{\bar{r}_i} \right)^{\frac{1}{2k_i-1}}. \end{aligned} \tag{12.18}$$

where $\bar{\lambda}_i$ *is given in Proposition 12.1 . The equilibrium mean state* $\bar{x}$ *is independent of the family* $(\eta_j, r_j, q_j)_{j\in\mathcal{I}}$.

Proof We use the expectation of the state obtained in Lemma 12.3 to derive the ordinary differential equation, which is linear in $\bar{x}$. The desired result follows. ■

Solving the One-Shot Game

The coefficients $(\eta_i)_i$ involved in the equilibrium strategies need to be identified. We propose to connect them to the one-shot game $\mathcal{G}(t) = (\mathcal{I}, (\mathbb{R})_{i\in\mathcal{I}}, (f_i)_{i\in\mathcal{I}})$.

Lemma 12.8 *For any time* $t > 0$,, *the one-shot game* $\mathcal{G}(t)$ *admits at least one Nash equilibrium.*

Proof Given the strategies $\eta_j,\ j \neq i$ of all decision-makers other than i,, the function $z_i \mapsto f_i(z_i, \eta_{-i})$ is continuous, and coercive on $\mathbb{R}$. Thus, $\mathcal{G}$ has a best-response correspondence BR_i such that $\eta_i \in \arg\min_{z_i} f_i(z_i, \eta_{-i}),\ i \in \mathcal{I}$. Given the strategies $\eta_j,\ j \neq i$ of all decision-makers other than i, the best response $BR_i = arg\min_{z_i} f_i(z_i, \eta_{-i})$ is unique from Lemma 12.6. An equilibrium of $\mathcal{G}(t)$

is a vector η that is a fixed point of $(\arg\min_{z_i} f_i(z_i, \eta_{-i}))_{i\in\mathcal{I}}$. By coercivity of the functions, the best responses can be restricted to compact convex sets. Then, the existence of equilibrium in $\mathcal{G}(t)$ follows for any time $t > 0$. ■

Note that a basic proximal algorithm or Bregman algorithm [167, 322] can be used to approximate the equilibrium.

We are now ready to prove the main result of this chapter.

Proof We combine the Lemmas 12.2, 12.3, and 12.5. We use the integration formula from Lemma 12.2 for $\lambda(x-\bar{x})^2$ and take its expectation. Given the strategies of the other decision-makers, $u_j - \bar{u}_j = -\eta_j(x-\bar{x}),\ j \neq i$, one has

$$\begin{array}{l} E[\lambda_i(T)(x(T)-\bar{x}(T))^2 - \lambda_i(0)(x_0-\bar{x}_0)^2] \\ = E\int_0^T [\dot{\lambda}_i + 2\lambda_i(b_1 - \sum_{j\in\mathcal{I}\setminus\{i\}} b_{2j}\eta_j)](x-\bar{x})^2 dt \\ -2E\int_0^T \lambda_i b_{2i}\eta_i(x-\bar{x})^2 dt \\ +2cE\int_0^T [\lambda_i(\nabla^{H/2,H/2}(x_t-\bar{x}_t))(t,t)]dt \end{array} \tag{12.19}$$

Let $\bar{k}_i > 1$ be an integer. We apply the standard integration formula to $\bar{x}_t = \bar{x}_0 + \int_0^t [(b_1+\bar{b}_1)\bar{x} + \sum_{j\in\mathcal{I}}(b_{2j}+\bar{b}_{2j})\bar{u}_j]dt'$, to arrive at

$$\begin{array}{l} \frac{\bar{\lambda}_i(T)}{2\bar{k}_i}\bar{x}_T^{2\bar{k}_i} - \frac{\bar{\lambda}_i(0)}{2\bar{k}_i}\bar{x}_0^{2\bar{k}_i} \\ = \int_0^T \Big[\frac{d}{dt}\bar{\lambda}_i\frac{1}{2\bar{k}_i}\bar{x}^{2\bar{k}_i} + 2\bar{k}_i\bar{\lambda}_i(b_1+\bar{b}_1)\frac{\bar{x}^{2\bar{k}_i}}{2\bar{k}_i} \\ +2\bar{k}_i\bar{\lambda}_i\sum_{j\in\mathcal{I}\setminus\{i\}}(b_{2j}+\bar{b}_{2j})\bar{u}_j\frac{1}{2\bar{k}_i}\bar{x}^{2\bar{k}_i-1}\Big]dt \\ +\int_0^T \bar{\lambda}_i(b_{2i}+\bar{b}_{2i})\bar{u}_i\bar{x}^{2\bar{k}_i-1}dt. \end{array} \tag{12.20}$$

Then,

$$\begin{array}{l} E[L_i(x,u) - \lambda_i(0)(x_0-\bar{x}_0)^2 - \bar{\lambda}_i(0)\frac{\bar{x}_0^{2\bar{k}_i}}{2\bar{k}_i}] \\ = E[q_{iT}(x_T-\bar{x}_T)^2 - \lambda_i(T)(x_T-\bar{x}_T)^2] \\ +E\int_0^T [\dot{\lambda}_i + q_i + 2\lambda_i(b_1 - \sum_{j\in\mathcal{I}\setminus\{i\}} b_{2j}\eta_j) - \lambda_i^2\frac{b_{2i}^2}{r_i}](x-\bar{x})^2 dt \\ +E\int_0^T r_i(\eta_i - \lambda_i\frac{b_{2i}}{r_i})^2(x-\bar{x})^2\, dt \\ +2cE\int_0^T [\lambda_i(\nabla^{H/2,H/2}(x_t-\bar{x}_t))(t,t)]dt \\ +\bar{q}_{iT}\frac{\bar{x}_T^{2\bar{k}_i}}{2\bar{k}_i} - \bar{\lambda}_i(T)\frac{\bar{x}_T^{2\bar{k}_i}}{2\bar{k}_i} \\ +\int_0^T \Big[\frac{d}{dt}\bar{\lambda}_i + \bar{q}_i + 2\bar{k}_i\bar{\lambda}_i(b_1+\bar{b}_1) \\ -2\bar{k}_i\bar{\lambda}_i\sum_{j\in\mathcal{I}\setminus\{i\}}(b_{2j}+\bar{b}_{2j})\bar{\eta}_j\Big]\frac{\bar{x}^{2\bar{k}_i}}{2\bar{k}_i}dt \\ +\int_0^T \bar{r}_i\frac{\bar{u}_i^{2\bar{k}_i}}{2\bar{k}_i} + \bar{\lambda}_i(b_{2i}+\bar{b}_{2i})\bar{u}_i\bar{x}^{2\bar{k}_i-1}dt \end{array} \tag{12.21}$$

We proceed with a convex completion by noting that

$$\bar{r}_i \frac{\bar{u}_i^{2\bar{k}}}{2\bar{k}_i} + \bar{\lambda}_i \bar{x}^{2\bar{k}_i-1}(b_{2i} + \bar{b}_{2i})\bar{u}_i + (2\bar{k}_i - 1)\bar{r}_i \left(\frac{(b_{2i}+\bar{b}_{2i})\bar{\lambda}_i}{\bar{r}_i}\right)^{\frac{2\bar{k}_i}{2\bar{k}_i-1}} \frac{\bar{x}^{2\bar{k}_i}}{2\bar{k}_i} \geq 0, \tag{12.22}$$

with equalities in (12.22) if, and only if, $\bar{u}_i = -\left(\frac{(b_{2i}+\bar{b}_{2i})\bar{\lambda}_i}{\bar{r}_i}\right)^{\frac{1}{2\bar{k}_i-1}} \bar{x}$, which means that $\bar{\eta}_i := \left(\frac{(b_{2i}+\bar{b}_{2i})\bar{\lambda}_i}{\bar{r}_i}\right)^{\frac{1}{2\bar{k}_i-1}}$.

The difference term in the cost functionals is therefore

$$\begin{aligned}
&E[L_i(x,u) - \lambda_i(0)(x_0 - \bar{x}_0)^2 - \bar{\lambda}_i(0)\frac{\bar{x}_0^{2\bar{k}_i}}{2\bar{k}_i}] \\
&= E[q_{iT}(x_T - \bar{x}_T)^2 - \lambda_i(T)(x_T - \bar{x}_T)^2] \\
&+ \int_0^T [\dot{\lambda}_i + q_i + 2\lambda_i b_1 - \lambda_i^2 \frac{b_{2i}^2}{r_i}]E(x - \bar{x})^2 dt \\
&+ \int_0^T \Big[r_i(\eta_i - \lambda_i \frac{b_{2i}}{r_i})^2 v_2(t) - 2\lambda_i \sum_{j\in\mathscr{I}\setminus\{i\}} b_{2j}\eta_j v_2(t) \\
&+2c\lambda_i(\nabla^{H/2,H/2}(x_t - \bar{x}_t))(t,t)\Big] dt + \bar{q}_{iT}\frac{\bar{x}_T^{2\bar{k}_i}}{2\bar{k}_i} - \bar{\lambda}_i(T)\frac{\bar{x}_T^{2\bar{k}_i}}{2\bar{k}_i} \\
&+ \int_0^T \Big[\frac{d}{dt}\bar{\lambda}_i + \bar{q}_i + 2\bar{k}_i\bar{\lambda}_i(b_1 + \bar{b}_1) - 2\bar{k}_i\bar{\lambda}_i \sum_{j\in\mathscr{I}\setminus\{i\}}(b_{2j} + \bar{b}_{2j})\bar{\eta}_j \\
&-(2\bar{k}_i - 1)\bar{r}_i \left(\frac{(b_{2i}+\bar{b}_{2i})\bar{\lambda}_i}{\bar{r}_i}\right)^{\frac{2\bar{k}_i}{2\bar{k}_i-1}}\Big] \frac{\bar{x}^{2\bar{k}_i}}{2\bar{k}_i} dt \\
&+ \int_0^T \Bigg[\bar{r}_i \frac{\bar{u}_i^{2\bar{k}_i}}{2\bar{k}_i} + \bar{\lambda}_i(b_{2i} + \bar{b}_{2i})\bar{u}_i \bar{x}^{2\bar{k}_i-1} \\
&+(2\bar{k}_i - 1)\bar{r}_i \left(\frac{(b_{2i}+\bar{b}_{2i})\bar{\lambda}_i}{\bar{r}_i}\right)^{\frac{2\bar{k}_i}{2\bar{k}_i-1}} \frac{\bar{x}^{2\bar{k}_i}}{2\bar{k}_i}\Bigg] dt
\end{aligned} \tag{12.23}$$

By setting

$$\begin{aligned}
&\lambda_i(T) = q_{iT}, \\
&0 = \dot{\lambda}_i + q_i + 2\lambda_i b_1 - \lambda_i^2 \frac{b_{2i}^2}{r_i}, \\
&\bar{\lambda}_i(T) = \bar{q}_{iT}, \\
&0 = \frac{d}{dt}\bar{\lambda}_i + \bar{q}_i + 2\bar{k}_i\bar{\lambda}_i(b_1 + \bar{b}_1) - 2\bar{k}_i\bar{\lambda}_i \sum_{j\in\mathscr{I}\setminus\{i\}}(b_{2j} + \bar{b}_{2j})\bar{\eta}_j \\
&-(2\bar{k}_i - 1)\bar{r}_i \left(\frac{(b_{2i}+\bar{b}_{2i})\bar{\lambda}_i}{\bar{r}_i}\right)^{\frac{2\bar{k}_i}{2\bar{k}_i-1}},
\end{aligned} \tag{12.24}$$

it follows that

$$\begin{aligned}
&E[L_i(x,u) - \lambda_i(0)(x_0 - \bar{x}_0)^2 - \bar{\lambda}_i(0)\frac{\bar{x}_0^{2\bar{k}_i}}{2\bar{k}_i}] \\
&= \int_0^T \Big[r_i(\eta_i - \lambda_i \frac{b_{2i}}{r_i})^2 v_2(t) - 2\lambda_i \sum_{j\in\mathscr{I}\setminus\{i\}} b_{2j}\eta_j v_2(t) \\
&+2c\lambda_i(\nabla^{H/2,H/2}(x_t - \bar{x}_t))(t,t)\Big] dt,
\end{aligned} \tag{12.25}$$

which completes the proof. ■

Special Cases

Variance-Aware Linear-Quadratic MFTG
Consider a special case of (12.3) with

$$
\begin{aligned}
& i \in \mathscr{I}, \\
& \inf_{u_i} q_{iT} E(x_T - \bar{x}_T)^2 + \bar{q}_{iT} \tfrac{\bar{x}_T^2}{2} \\
& + E \int_0^T \left[q_i (x - \bar{x})^2 + r_i (u_i - \bar{u}_i)^2 + \bar{q}_i \tfrac{\bar{x}^2}{2} + \bar{r}_i \tfrac{\bar{u}_i^2}{2} \right] dt \\
& \text{subject to} \\
& x_t = x_0 + \int_0^t [b_1 (x - \bar{x}) + \textstyle\sum_{j \in \mathscr{I}} b_{2j} (u_j - \bar{u}_j)] dt' \\
& + \int_0^t [(b_1 + \bar{b}_1)\bar{x} + \textstyle\sum_{j \in \mathscr{I}} (b_{2j} + \bar{b}_{2j}) \bar{u}_j] dt' + r^H(t),
\end{aligned}
\tag{12.26}
$$

Let $q_i > 0, \bar{q}_i > 0, r_i, \bar{r}_i > \epsilon > 0$. A Nash equilibrium of the MFTG (12.26) with a Rosenblatt process having linear state-and-mean-field-type feedback form is given by

$$
\begin{aligned}
& u_i^* = -\eta_i (x - \bar{x}) - \bar{\eta}_i \bar{x}, \\
& \bar{\eta}_i = \tfrac{\bar{b}_{2i} \bar{\lambda}_i}{\bar{r}_i}, \\
& E[L_i] = \lambda_i(0) E(x_0 - \bar{x}_0)^2 + \bar{\lambda}_i(0) \tfrac{1}{2\bar{k}_i} \bar{x}_0^{2\bar{k}_i} + \gamma_i(0),
\end{aligned}
\tag{12.27}
$$

whenever the following system of differential equations admits a positive solution which does not blow up (no singularity) within $[0, T]$.

$$
\begin{aligned}
& -\dot{\lambda}_i = q_i + 2 b_1 \lambda_i - \tfrac{b_{2i}^2}{r_i} \lambda_i^2, \\
& \lambda_i(T) = q_i(T), \\
& -\tfrac{d}{dt} \bar{\lambda}_i = \bar{q}_i + 2(b_1 + \bar{b}_1) \bar{\lambda}_i - 2 \bar{\lambda}_i \textstyle\sum_{j \neq i} \left(\tfrac{(b_{2j} + \bar{b}_{2j})^2}{\bar{r}_j} \bar{\lambda}_j \right) - \tfrac{(b_{2i} + \bar{b}_{2i})^2}{\bar{r}_i} \bar{\lambda}_i^2, \\
& \bar{\lambda}_i(T) = \bar{q}_i(T), \\
& -\dot{\gamma}_i = r_i (\eta_i - \lambda_i \tfrac{b_{2i}}{r_i})^2 v_2(t) - 2 \lambda_i \textstyle\sum_{j \in \mathscr{I} \setminus \{i\}} b_{2j} \eta_j v_2(t) + 2c \lambda_i o(t), \\
& \gamma_i(T) = 0,
\end{aligned}
\tag{12.28}
$$

where η_i minimizes $f_i(\eta)$:

$$
f_i(\eta) = r_i (\eta_i - \lambda_i \tfrac{b_{2i}}{r_i})^2 v_2(t) - 2 \lambda_i \textstyle\sum_{j \in \mathscr{I} \setminus \{i\}} b_{2j} \eta_j v_2(t) + 2c \lambda_i o(t).
\tag{12.29}
$$

The vector η is a fixed point of $(\arg\min_{z_i} f_i(z_i, \eta_{-i}))_{i \in \mathscr{I}}$,

(continued)

$$
\begin{aligned}
& b = [b_1 - \sum_j b_{2j}\eta_j], \\
& o(t) = \left[\int_0^t c_3[t-s]^{2H-2} e^{-\int_0^s b(t'')dt''} ds\right] e^{\int_0^t b(t')\,dt'}, \\
& v_2(t) = \left[v_2(0) + \int_0^t 2c\, o(s) e^{-\int_0^s 2b(t'')dt''} ds\right] e^{\int_0^t 2b(t')\,dt'}.
\end{aligned}
\tag{12.30}
$$

12.4 Weighted Cooperative MFTG Under Rosenblatt Noise

We now consider the cooperative case with the weights $\omega_j > 0,\ j \in \mathcal{I}$ and the power exponents $\bar{k}_i = \bar{k} \geq 1,\ \forall i \in \mathcal{I}$.

Cooperative MFTG with Rosenblatt Noise

$$
\begin{aligned}
& \inf_{(u_i)_{i\in\mathcal{I}}} E[\sum_{j\in\mathcal{I}} \omega_j L_j(x,u)], \\
& \text{subject to} \\
& x_t = x_0 + \int_0^t [b_1(x-\bar{x}) + \sum_{j\in\mathcal{I}} b_{2j}(u_j - \bar{u}_j)]dt' \\
& + \int_0^t [(b_1+\bar{b}_1)\bar{x} + \sum_{j\in\mathcal{I}} (b_{2j}+\bar{b}_{2j})\bar{u}_j]dt' + r^H(t),
\end{aligned}
\tag{12.31}
$$

Proposition 12.2 *Consider (12.31). Then, the optimal weighted cooperative cost is*

$$
\begin{aligned}
& [\sum_{i\in\mathcal{I}} \omega_i E L_i - \lambda(0)E(x_0-\bar{x}_0)^2 - \bar{\lambda}(0)\frac{\bar{x}_0^{2\bar{k}}}{2\bar{k}}] \\
& = \int_0^T \sum_{i\in\mathcal{I}} \omega_i r_i (\eta_i - \frac{\lambda b_{2i}}{\omega_i r_i})^2 v_2 + 2c\lambda o(t)\, dt, \\
& \text{the coefficients } \lambda, \bar{\lambda} \text{ solve} \\
& \lambda(T) = \sum_i \omega_i q_{iT}, \\
& 0 = \dot{\lambda} + \sum_i \omega_i q_i + 2\lambda b_1 - \sum_{i\in\mathcal{I}} \frac{b_{2i}^2}{\omega_i r_i}\lambda^2, \\
& \bar{\lambda}(T) = \sum_{i\in\mathcal{I}} \omega_i \bar{q}_{iT}, \\
& \frac{d}{dt}\bar{\lambda} + \sum_{i\in\mathcal{I}} \omega_i \bar{q}_i + 2\bar{k}\bar{\lambda}(b_1+\bar{b}_1) \\
& -(2\bar{k}-1)\sum_{i\in\mathcal{I}} \omega_i \bar{r}_i \left(\frac{(b_{2i}+\bar{b}_{2i})\bar{\lambda}}{\omega_i \bar{r}_i}\right)^{\frac{2\bar{k}}{2\bar{k}-1}} = 0, \\
& \text{and } \eta, \bar{\eta} \text{ are given by} \\
& \bar{\eta}_i^{coop} = \left(\frac{(b_{2i}+\bar{b}_{2i})\bar{\lambda}}{\omega_i \bar{r}_i}\right)^{\frac{1}{2\bar{k}-1}},\ \bar{u}_i^{coop} = -\bar{\eta}_i^{coop}\bar{x}, \\
& (\eta_i)_{i\in\mathcal{I}} \text{ minimizes } \sum_{i\in\mathcal{I}} \omega_i r_i (\eta_i - \frac{\lambda b_{2i}}{\omega_i r_i})^2 v_2 + 2c\lambda o(t), \\
& u_i^{coop} = -\eta_i(x-\bar{x}) - \bar{\eta}_i \bar{x},
\end{aligned}
\tag{12.32}
$$

Proof First note that

$$
\begin{aligned}
& E[\sum_i \omega_i L_i(x,u) - \lambda(0)(x_0-\bar{x}_0)^2 - \bar{\lambda}(0)\frac{\bar{x}_0^{2\bar{k}}}{2\bar{k}}] \\
&= E[\sum_i \omega_i q_{iT} - \lambda(T)](x_T-\bar{x}_T)^2 \\
&+ E\int_0^T [\dot{\lambda} + \sum_i \omega_i q_i + 2\lambda b_1 - \sum_{i\in\mathscr{I}} \frac{b_{2i}^2}{\omega_i r_i}\lambda^2](x-\bar{x})^2 dt \\
&+ E\int_0^T \sum_{i\in\mathscr{I}} \omega_i r_i(\eta_i - \frac{\lambda b_{2i}}{\omega_i r_i})^2 (x-\bar{x})^2\, dt \\
&+ 2cE\int_0^T \lambda o(t)dt + [\sum_{i\in\mathscr{I}} \omega_i \bar{q}_{iT} - \bar{\lambda}(T)]\frac{\bar{x}_T^{2\bar{k}}}{2\bar{k}} \\
&+ \int_0^T \left[\frac{d}{dt}\bar{\lambda} + \sum_{i\in\mathscr{I}} \omega_i \bar{q}_i + 2\bar{k}\bar{\lambda}(b_1+\bar{b}_1)\right]\frac{\bar{x}^{2\bar{k}}}{2\bar{k}} dt \\
&+ \int_0^T [\sum_{i\in\mathscr{I}} \omega_i \bar{r}_i \bar{\eta}_i^{2\bar{k}} - 2\bar{k}\bar{\lambda}\sum_{j\in\mathscr{I}}(b_{2j}+\bar{b}_{2j})\bar{\eta}_j]\frac{\bar{x}^{2\bar{k}}}{2\bar{k}} dt
\end{aligned}
\tag{12.33}
$$

We proceed with a convex completion by noting that

$$
\omega_i \bar{r}_i \bar{\eta}_i^{2\bar{k}} - 2\bar{k}\bar{\lambda}(b_{2i}+\bar{b}_{2i})\bar{\eta}_i + (2\bar{k}-1)\omega_i \bar{r}_i \left(\frac{(b_{2i}+\bar{b}_{2i})\bar{\lambda}}{\omega_i \bar{r}_i}\right)^{\frac{2\bar{k}}{2\bar{k}-1}} \geq 0, \tag{12.34}
$$

with equalities in (12.34) if, and only, if $\bar{\eta}_i := \left(\frac{(b_{2i}+\bar{b}_{2i})\bar{\lambda}}{\omega_i \bar{r}_i}\right)^{\frac{1}{2\bar{k}-1}}$.

The cost functional difference term is rewritten as

$$
\begin{aligned}
& E[\sum_i \omega_i L_i(x,u) - \lambda(0)(x_0-\bar{x}_0)^2 - \bar{\lambda}(0)\frac{\bar{x}_0^{2\bar{k}}}{2\bar{k}}] \\
&= E[\sum_i \omega_i q_{iT} - \lambda(T)](x_T-\bar{x}_T)^2 \\
&+ E\int_0^T [\dot{\lambda} + \sum_i \omega_i q_i + 2\lambda b_1 - \sum_{i\in\mathscr{I}} \frac{b_{2i}^2}{\omega_i r_i}\lambda^2](x-\bar{x})^2 dt \\
&+ \int_0^T \sum_{i\in\mathscr{I}} \omega_i r_i(\eta_i - \frac{\lambda b_{2i}}{\omega_i r_i})^2 v_2 + 2c\lambda o(t) dt \\
&+ [\sum_{i\in\mathscr{I}} \omega_i \bar{q}_{iT} - \bar{\lambda}(T)]\frac{\bar{x}_T^{2\bar{k}}}{2\bar{k}} \\
&+ \int_0^T \left[\frac{d}{dt}\bar{\lambda} + \sum_{i\in\mathscr{I}} \omega_i \bar{q}_i + 2\bar{k}\bar{\lambda}(b_1+\bar{b}_1) \right. \\
&\left. -(2\bar{k}-1)\sum_{i\in\mathscr{I}} \omega_i \bar{r}_i \left(\frac{(b_{2i}+\bar{b}_{2i})\bar{\lambda}}{\omega_i \bar{r}_i}\right)^{\frac{2\bar{k}}{2\bar{k}-1}} \right]\frac{\bar{x}^{2\bar{k}}}{2\bar{k}} dt \\
&+ \int_0^T \sum_{i\in\mathscr{I}} [\omega_i \bar{r}_i \bar{\eta}_i^{2\bar{k}} - 2\bar{k}\bar{\lambda}(b_{2i}+\bar{b}_{2i})\bar{\eta}_i]\frac{\bar{x}^{2\bar{k}}}{2\bar{k}} dt \\
&+ \int_0^T (2\bar{k}-1)\sum_{i\in\mathscr{I}} \omega_i \bar{r}_i \left(\frac{(b_{2i}+\bar{b}_{2i})\bar{\lambda}}{\omega_i \bar{r}_i}\right)^{\frac{2\bar{k}}{2\bar{k}-1}} \frac{\bar{x}^{2\bar{k}}}{2\bar{k}} dt
\end{aligned}
\tag{12.35}
$$

This completes the proof for the cooperative case. ■

12.5 Five Noises

In this section, we look at a mixture of five different noises in a particular class of mean-field-type games.

Table 12.1 Noises examined in MFTGs

Noises	Gaussian	Non-Gaussian
Markovian noise	Brownian motion	Poisson Jump, Regime Switching
non-Markovian noise	Fractional Brownian motion, Gauss-Volterra	Rosenblatt noise

$$\begin{array}{l} L_i = q_{iT}\frac{(x_T-\bar{x}_T)^2}{2} + \bar{q}_{iT}\frac{\bar{x}_T^{2\bar{k}_i}}{2\bar{k}_i} \\ + \int_0^T q_i\frac{(x-\bar{x})^2}{2} + r_i\frac{(u_i-\bar{u}_i)^2}{2} + \bar{q}_i\frac{\bar{x}^{2\bar{k}_i}}{2\bar{k}_i} + \bar{r}_i\frac{\bar{u}_i^{2\bar{k}_i}}{2\bar{k}_i}\, dt \\ + \int_0^T \bar{c}_i\bar{x}^{2\bar{k}_i-1}\bar{u}_i + \sum_{j\in\mathscr{I}\backslash\{i\}}\bar{\epsilon}_{ij}\bar{x}^{2(\bar{k}_i-1)}\bar{u}_i\bar{u}_j\, dt, \\ \text{subject to} \\ x(t) = x_0 + \int_0^t b_1 x + \bar{b}_1\bar{x} + \sum_{j\in\mathscr{I}}\left[b_{2j}u_j + \bar{b}_{2j}\bar{u}_j\right]\, dt' \\ + x_{noise}(t) \\ \\ \mathbb{P}(s(t+\delta) = s'|s) = \int_t^{t+\delta}\tilde{q}_{ss'}dt' + o(\delta),\ s'\neq s, \\ s(0) = s_0, \end{array} \tag{12.36}$$

The regime switching process $s(t)$ appears in the coefficient functions. We introduce five different problems, one for each noise structure and one mixed noise structure that may involve any combination of these five noises (Table 12.1).

State Driven by Brownian Motion

$$(PBm)\begin{cases} \text{Find } (u_1^*, \ldots, u_I^*) : \\ i \in \mathscr{I}, \\ E[L_i(u^*, m_0)] \\ = \inf_{u_i\in\mathscr{U}_i} E[L_i(u_1^*, \ldots, u_{i-1}^*, u_i, u_{i+1}^*, \ldots, u_I^*, m_0)], \\ x_{noise}(t) = \int_0^t \sigma_b dB(t'), \\ \mathscr{S} = \{s_0\} \end{cases} \tag{12.37}$$

State Driven by Regime Switching

$$(PRS)\begin{cases} \text{Find } (u_1^*, \ldots, u_I^*) : \\ i \in \mathscr{I}, \\ E[L_i(u^*, m_0)] \\ = \inf_{u_i\in\mathscr{U}_i} E[L_i(u_1^*, \ldots, u_{i-1}^*, u_i, u_{i+1}^*, \ldots, u_I^*, m_0)], \\ x_{noise}(t) = 0, \\ |\mathscr{S}| \geq 2 \end{cases} \tag{12.38}$$

State Driven by Poisson Jump

$$(PPJ)\begin{cases}\text{Find } (u_1^*,\ldots,u_I^*): \\ i\in\mathscr{I}, \\ E[L_i(u^*,m_0)] \\ =\inf_{u_i\in\mathscr{U}_i} E[L_i(u_1^*,\ldots,u_{i-1}^*,u_i,u_{i+1}^*,\ldots,u_I^*,m_0)], \\ x_{noise}(t)=\int_0^t\int_\theta \sigma_n(t',s,\theta)\tilde{N}(dt',d\theta), \\ \mathscr{S}=\{s_0\}\end{cases} \tag{12.39}$$

State Driven by Gauss-Volterra

$$(PGV)\begin{cases}\text{Find } (u_1^*,\ldots,u_I^*): \\ i\in\mathscr{I}, \\ E[L_i(u^*,m_0)] \\ =\inf_{u_i\in\mathscr{U}_i} E[L_i(u_1^*,\ldots,u_{i-1}^*,u_i,u_{i+1}^*,\ldots,u_I^*,m_0)], \\ x_{noise}(t)=\int_0^t \sigma_{gv} dB_{gv}(t'), \\ \mathscr{S}=\{s_0\}\end{cases} \tag{12.40}$$

State Driven by Rosenblatt Process

$$(PRh)\begin{cases}\text{Find } (u_1^*,\ldots,u_I^*): \\ i\in\mathscr{I}, \\ E[L_i(u^*,m_0)] \\ =\inf_{u_i\in\mathscr{U}_i} E[L_i(u_1^*,\ldots,u_{i-1}^*,u_i,u_{i+1}^*,\ldots,u_I^*,m_0)], \\ x_{noise}(t)=\int_0^t \sigma_r dr^H(t'), \\ \mathscr{S}=\{s_0\}\end{cases} \tag{12.41}$$

State Driven by Five Mixed Noises

$$(P5mixed)\begin{cases}\text{Find } (u_1^*,\ldots,u_I^*): \\ i\in\mathscr{I}, \\ E[L_i(u^*,m_0)] \\ =\inf_{u_i\in\mathscr{U}_i} E[L_i(u_1^*,\ldots,u_{i-1}^*,u_i,u_{i+1}^*,\ldots,u_I^*,m_0)], \\ x_{noise}(t)=\int_0^t \sigma_b dB(t')+\sigma_{gv} dB_{gv}(t') \\ +\int_0^t\int_\theta \sigma_n(t',s,\theta)\tilde{N}(dt',d\theta)+\int_0^t \sigma_r dr^H(t'), \\ |\mathscr{S}|\geq 2\end{cases} \tag{12.42}$$

Beyond mean-field-type Nash equilibrium, we are interested in multilayer Stackelberg solution when restricted to linear state-and-mean-field-type feedback strategies. In MFTG, multi-hierarchy setups elevate decision-making [157]. Envision distinct layers akin to game layers, each layer featuring one or several decision-

Table 12.2 Noise effects examined in MFTGs

MFTG	Without Rosenblatt	With Rosenblatt
1 layer	Explicit Nash MFTG	Semi-explicit
2 layers	Stackelberg MFTG	Semi-explicit
Fully hierarchical	Fully hierarchical MFTG	Semi-explicit

makers responsible for strategic choices. It resembles entities passing the decision baton from one layer to the next, with choices rippling across layers, emphasizing the need to grasp the overall impact. The game initiates at layer 1. Notably, the layer 1 decision-makers determine initial choices, influencing subsequent responses at layer 2. If multiple decision-makers exist at layer 1, they engage in a strategic layer game. First-layer decisions are observable by second-layer decision-makers, yet adjustments are unavailable once made.

Decision-makers at layer h base choices on preceding layers 1 to $h-1$, interacting with any other decision-maker at the same layer. Subsequently, layer $h+1$ decision-makers respond to decisions from layers 1 to h. While decisions up to the layer h are observable by $(h+1)$-th layer decision-makers, adjustments by h-th layer decision-makers post-choice are restricted.

In solving sequential decision-making, backward induction aids bottom-layer H decision-makers. They react with insight into preceding layers' choices, engaging with peers at the same layer. The layer $H-1$ decision-makers anticipate H-th decisions, optimizing actions based on layers 1 to $H-2$ choices.

For layer h decision-makers, anticipation extends to $h+1$ to H-th decision-makers, optimizing actions according to choices in layers 1 to $h-1$ and interacting with any other decision-maker at layer h.

Note that, in general, the layer game played at each layer can be different: It could be, for example, a team, self-interest, or Berge game between the decision-makers at the same layer (Table 12.2).

In terms of design, one could tackle the problem using the *Stackelberg, reverse Stackelberg, inverse Stackelberg, double Stackelberg, or conjectural variation approach.*

- Fractional Brownian motion-, Gauss-Volterra processes-, and Rosenblatt processes-based games cannot be solved with standard dynamic programming, because these are not Markovian and involve memory (long-range path dependence). To tackle them, the formulation needs to be in the space of measures, which is of infinite dimensions, which is suitable for MFTG.
- We provide a generic structure of multilevel hierarchical equilibria under mixture of five types of noises that include Rosenblatt process.
- In the absence of Rosenblatt process, we provide semi-explicit expression of multilevel hierarchical equilibria under mixture of the four other types of noises.
- We show that the equilibrium strategies obtained under the mixture of Brownian motions, Fractional Brownian motions, Gauss-Volterra, and Poisson processes no longer provide equilibrium when replaced by a new mixture of noises that involves the Rosenblatt noise process.

Table 12.3 Effects of Rosenblatt noises in MFTG equilibria

MFTG	Only with Brownian motion	If a Rosenblatt is added to
Nash	Equilibrium	No longer equilibrium
Stackelberg	Equilibrium	No longer equilibrium
Adversarial	Saddle point	No longer saddle point
Cooperation	Social optimum	No longer optimal

As particular cases, one has the following important observations (Table 12.3):

- the best-response strategy obtained under Brownian motions no longer provide a best-response strategy when the Brownian motion is replaced by a Rosenblatt noise.
- the best-response strategy obtained under Fractional Brownian motions no longer provide best-response strategy when the Fractional Brownian motion is replaced by a Rosenblatt noise.
- the best-response strategy obtained under Gauss-Volterra process no longer provide best-response strategy when the Gauss-Volterra is replaced by a Rosenblatt noise.
- the best-response strategy obtained under deterministic state setting no longer provide best-response strategy when replaced by an additive noise that is a Rosenblatt process.

This is a significant result as it says that if we make a modeling error by replacing a Rosenblatt process by a Gaussian process in the state dynamics, the resulting strategies become suboptimal, and the resulting outcomes are no longer equilibria. In other words, the structure of the noises plays a crucial role.

12.5.1 Mixture of Five Noises

We consider a class of conditional variance-aware games with $I \geq 2$ decision-makers interacting over the time horizon $[0, T]$ with $T > 0$. Let the cost functional of decision-maker $i \in \mathcal{I}$ be

$$\begin{aligned} L_i &= q_{iT}\frac{(x_T-\bar{x}_T)^2}{2} + \bar{q}_{iT}\frac{\bar{x}_T^{2\bar{k}_i}}{2\bar{k}_i} \\ &+ \int_0^T q_i\frac{(x-\bar{x})^2}{2} + r_i\frac{(u_i-\bar{u}_i)^2}{2} + \bar{q}_i\frac{\bar{x}^{2\bar{k}_i}}{2\bar{k}_i} + \bar{r}_i\frac{\bar{u}_i^{2\bar{k}_i}}{2\bar{k}_i}\, dt \\ &+ \int_0^T \bar{c}_i\bar{x}^{2\bar{k}_i-1}\bar{u}_i + \sum_{j\in\mathcal{I}\setminus\{i\}} \bar{\varepsilon}_{ij}\bar{x}^{2(\bar{k}_i-1)}\bar{u}_i\bar{u}_j\, dt, \end{aligned} \tag{12.43}$$

where

$$\begin{aligned} &\bar{k}_i \geq 1,\ m(t, dy) = \mathbb{P}_{x(t)\mid s(t)}(dy), \\ &\bar{x}(t) = \int y m(t, dy),\ \bar{u}_i(t) = \int u_i(t, y, s, m)m(t, dy), \\ &q_i(t, s), r_i(t, s), \bar{q}_i(t, s), \bar{r}_i(t, s) :\ [0, T]\times\mathcal{S} \to [\varepsilon_i, \infty), \varepsilon_i > 0, \\ &\bar{c}_i(t, s), \bar{\varepsilon}_{ij}(t, s) :\ [0, T]\times\mathcal{S} \to \mathbb{R}, \\ &u_i(t, x, s, m) \in U_i = \mathbb{R} \end{aligned} \tag{12.44}$$

Let the state dynamics x be real-valued, driven by five types of processes: Brownian motion, Gauss-Volterra, Poisson jump process, and Rosenblatt process and a regime switching process.

$$
\begin{aligned}
x(t) &= x_0 + \int_0^t b_1 x + \bar{b}_1 \bar{x} + \sum_{j\in\mathcal{I}} \left[b_{2j}u_j + \bar{b}_{2j}\bar{u}_j\right] dt' \\
&+ \int_0^t \left[\sigma_b dB(t') + \sigma_{gv} dB_{gv}(t') + \int_\theta \sigma_n(t', s, \theta)\tilde{N}(dt', d\theta)\right] \\
&+ \int_0^t \sigma_r dr^H(t'),
\end{aligned} \tag{12.45}
$$

$$
\mathbb{P}(s(t+\delta) = s'|s) = \int_t^{t+\delta} \tilde{q}_{ss'} dt' + o(\delta), \; s' \neq s,
$$

with

Drift coefficients:
$b_1(t,s), \bar{b}_1(t,s), b_{2j}(t,s), \bar{b}_{2j}(t,s) : \; [0,T]\times\mathcal{S} \to \mathbb{R}$,
Diffusion coefficient: $\sigma_b(t,s)$, Brownian motion B,
Gauss-Volterra coefficient: $\sigma_{gv}(t,s) : \; [0,T]\times\mathcal{S} \to \mathbb{R}$,
Gauss-Volterra B_{gv},
The variance of the process is $\int_0^t \sigma_{gv}(t',s)dB_{gv}(t')$
$\sigma^2_{cogv} =$
$\frac{d}{dt}\left[\int_0^t \left\{K(t'_+, t')\sigma_{gv}(t') + \int_{t'}^t \sigma_{gv}(t'')K(t',t'')dt''\right\}^2 dt'\right]$
for a certain kernel K.
Set of jump size Θ,
jump rate $\sigma_n(t,s,\theta) : \; [0,T]\times\mathcal{S}\times\Theta \to \mathbb{R}$,
Jump process: $N(dt, d\theta)$,
Compensated jump : $\tilde{N}(dt,d\theta) = N(dt,d\theta) - \nu(d\theta)dt$,
Regime switching: $s(t) \in \mathcal{S}$,
switching rate:$\tilde{q}_{ss'} > 0, s \neq s', \; \tilde{q}_{ss} = -\sum_{s'\neq s}\tilde{q}_{ss'}, \; s \in \mathcal{S}$
$\mathcal{S}$: a finite set of switching regimes,
Wiener-Itô integral: $Z^{\delta_1,\ldots,\delta_k}(t)$
$= c(\delta)\int_{\mathbb{R}^k\setminus D_k}[\int_0^t \prod_{m=1}^k (t'-x_m)_+^{\delta_m} dt'] \prod_{l=1}^k dB(x_l)$,
$D_k = \{(x_1,\ldots,x_k), \; x_1 = \ldots = x_k\} \subset \mathbb{R}^k$,
$c(\delta) \neq 0, \; \sum_{m=1}^k \delta_m \in (-\frac{k}{2}-\frac{1}{2}, -\frac{k}{2})$,
Rosenblatt coefficient: $\sigma_r(t,s) : \; [0,T]\times\mathcal{S} \to \mathbb{R}$,
Rosenblatt process : $r^H(t) := Z^{\delta,\delta}(t), \delta = \frac{H}{2} - 1$,
$c(\frac{H}{2}-1, \frac{H}{2}-1) = \frac{\sqrt{2H(2H-1)}}{2B(1-H,\frac{H}{2})}\Gamma^2(\frac{H}{2})$,
Beta function: $Beta(z_1,z_2) = \int_0^1 t^{z_1-1}(1-t)^{z_2-1}\,dt$,
$(z_1,z_2) : \mathrm{Re}\,(z_1), \mathrm{Re}\,(z_2) > 0$,
Gamma function: $\Gamma(z) = \int_0^\infty t^{z-1}e^{-t}\,dt, \; \mathrm{Re}\,(z) > 0$.

We assume that $s, x(0), B, N$ and r^H, are mutually independent.

Let $u_j,\ j \in \mathcal{I}$ and all the coefficient functions $b_1, \bar{b}_2, b_{2j}, \bar{b}_{2j}, \sigma_b^2, \sigma_{cogv}^2, \sigma_n^2, \sigma_r$ be integrable over $[0, T]$ for every regime switching $s \in \mathcal{S}$. Then, (12.45) admits a unique solution x. If in addition the random variable x_0 is square integrable over $\mathbb{R}$ then x is square integrable $[0, T] \times \mathbb{R}$.

Let $\mathcal{U}_i$ be the set of linear state and conditional mean-field-type feedback strategies of decision-maker i that are progressively measurable with respect to the filtration generated by the unions of events in $\{s, B, N, B_{gv}, r^H\}$. This means that we will restrict our analysis to situations where the control strategies have the following form $\alpha_i(t, s(t))x(t) + \bar{\alpha}_i(t, s(t))\bar{x}(t)$.

12.5.1.1 One Layer: Mean-Field-Type Nash Equilibrium

The risk-neutral MFTG is given by $(\mathcal{I}, U_i, \mathcal{U}_i, E[L_i])_{i\in\mathcal{I}}$. A risk-neutral mean-field-type Nash equilibrium is a strategy profile $(u_1^*, \dots, u_I^*)$ satisfying

$$\begin{aligned} & i \in \mathcal{I}, \\ & E[L_i(u^*, m_0)] \\ & = \inf_{u_i\in\mathcal{U}_i} E[L_i(u_1^*, \dots, u_{i-1}^*, u_i, u_{i+1}^*, \dots, u_I^*, m_0)]. \end{aligned} \tag{12.46}$$

12.5.1.2 Two Layers: One Leader and Many Followers

Given $u_l \in \mathcal{U}_l$, and $(u_j)_{j\in\mathcal{I}\backslash\{l,i\}}$, a risk-neutral best response strategy of follower i is a strategy that solves $\inf_{\mathcal{U}_i} E[L_i]$ subject to (12.45). The set of risk-neutral best responses of follower i is denoted by $\text{rnBR}_i(u_l; (u_j)_{j\in\mathcal{I}\backslash\{l,i\}})$.

In view of (12.43)–(12.45), the best response set here is a **singleton set**. This means that there is no ambiguity in the best response, and hence it will not require to define pessimistic Stackelberg, optimistic Stackelberg, mixed Stackelberg, etc.

A mean-field-type risk-neutral Nash equilibrium between the followers given the leader strategy u_l, is a strategy profile $(u_i^{\text{rn}},\ i \in \mathcal{I}\backslash\{l\})$ of all followers such that for each decision-maker $i \in \mathcal{I}\backslash\{l\}$, $u_i^{\text{rn}} \in \text{rnBR}_i(u_l; (u_j^{\text{rn}})_{j\in\mathcal{I}\backslash\{l,i\}})$.

A mean-field-type risk-neutral Stackelberg solution between the leader and multiple followers is a strategy profile $(u_l, (u_i^{\text{rn}},\ i \in \mathcal{I}\backslash\{l\}))$ of all decision-makers such that

$$\begin{aligned} & \text{for the leader:} \\ & u_l^{\text{rn}} \in \arg\min_{u_l\in\mathcal{U}_l}\{EL_l(x, u_l, (u_i^{\text{rn}})_{i\in\mathcal{I}\backslash\{l\}}) : \\ & u_i^{\text{rn}} \in \text{rnBR}_i(u_l; (u_j^{\text{rn}})_{j\in\mathcal{I}\backslash\{l,i\}}), i \in \mathcal{I}\backslash\{l\}\}, \\ & \\ & \text{and for every follower } i \in \mathcal{I}\backslash\{l\} : \\ & u_i^{\text{rn}} \in \text{rnBR}_i(u_l; (u_j^{\text{rn}})_{j\in\mathcal{I}\backslash\{l,i\}}). \end{aligned} \tag{12.47}$$

12.5.1.3 Two Layers: Multiple Leaders and Multiple Followers

A mean-field-type risk-neutral Stackelberg solution between multiple leaders and multiple followers is a strategy $((u_i^{ss})_{i\in\mathscr{I}_L}, (u_j^{ss})_{j\in\mathscr{I}_F})$, with $\mathscr{I} = \mathscr{I}_L \cup \mathscr{I}_F$. satisfying

$$\begin{aligned}
&\text{for the leaders:}\\
&i \in \mathscr{I}_L,\\
&u_i^{ss} \in \arg\min_{u_i\in\mathscr{U}_i}\Big\{EL_i(x, u_i, (u_{i'}^{ss})_{i\in\mathscr{I}_L\backslash\{i\}}, (u_j^{ss})_{j\in\mathscr{I}_F}) :\\
&u_j^{ss} \in \text{rnBR}_j((u_i^{ss})_{i\in\mathscr{I}_L}; (u_{j'}^{ss})_{j'\in\mathscr{I}_F\backslash\{j\}}),\ j \in \mathscr{I}_F\Big\},\\
&\text{and for every follower } j \in \mathscr{I}_F :\\
&u_j^{ss} \in \text{rnBR}_j((u_i^{ss})_{i\in\mathscr{I}_L}; (u_{j'}^{ss})_{j'\in\mathscr{I}_F\backslash\{j\}}).
\end{aligned} \tag{12.48}$$

12.5.1.4 Multilayer Fully Hierarchical Solution

We now examine the fully hierarchical game with I decision-makers. A fully hierarchical game is one where there is one decision-maker per layer, i.e., the number of layers is exactly the number of decision-makers. Since all permutations of the decision-makers are possible per layer, there are $I!$ configurations. We examine the case where decision-maker i is at i-th hierarchical layer. At each layer i,, decision-maker i chooses a control strategy u_i knowing the control strategy of the preceding decision-makers, i.e., $\{i-1,\ldots,1\}$.

At level k,, all decision-makers in $\{1, 2, \ldots, k-1\}$ used linear state-and-mean-field feedback strategies, and these are known to decision-maker k.

$$\begin{aligned}
&\text{Level 1:}\\
&u_1^{ss} \in \arg\min_{u_1\in\mathscr{U}_1}\Big\{EL_1(x, u_1, (u_j^{ss})_{j\in\mathscr{I}\backslash\{1\}}) :\\
&u_j^{ss} \in \text{rnBR}_j(u_1^{ss}; (u_{j'}^{ss})_{j'\in\mathscr{I}\backslash\{j,1\}}),\ \ j \in \mathscr{I}\backslash\{1\}\Big\},\\
&\text{Level } I-1 \geq k \geq 2 :\\
&u_k^{ss} \in \arg\min_{u_k\in\mathscr{U}_k}\Big\{\\
&EL_k(x, (u_i^{ss})_{i\in\{1,\ldots,k-1\}}; u_k, (u_{j'}^{ss})_{j'\in\{k+1,\ldots,I\}}) :\\
&u_j^{ss} \in \text{rnBR}_j((u_i^{ss})_{i\in\{1,\ldots,k\}}; (u_{j'}^{ss})_{j'\in\{k+1,\ldots,I\}}),\\
&j \in \{k+1,\ldots,I\}\Big\},\\
&\text{Level } I :\ u_I^{ss} \in \text{rnBR}_I((u_i^{ss})_{i\in\{1,\ldots,I-1\}}).
\end{aligned} \tag{12.49}$$

12.5.2 Semi-explicit Solution

We now provide the solution to the problems in (12.46), (12.47), (12.48), and (12.49) formulated above.

12.5.2.1 One-Layer Case: $H = 1$

Proposition 12.3 *Assume $q_i(t,s) > 0, \bar{q}_i(t,s) > 0,\ r_i(t,s), \bar{r}_i(t,s) > \epsilon_i > 0$. A Nash equilibrium of the MFTG (12.46) with a Rosenblatt process having linear state-and-mean-field-type feedback form is given by*

$$
\begin{array}{l}
i \in \mathcal{I}, \\
u_i^*(t) = -\sum_{s\in\mathcal{S}}[\eta_i(t,s)(x-\bar{x}) + \bar{\eta}_i(t,s)\bar{x}]\mathbb{I}_{\{s(t)=s\}}, \\
E[L_i] = E\left[\lambda_i(0,s_0)(x_0-\bar{x}_0)^2 + \bar{\lambda}_i(0,s_0)\frac{1}{2\bar{k}_i}\bar{x}_0^{2\bar{k}_i} + \gamma_i(0,s_0)\right],
\end{array}
\tag{12.50}
$$

whenever the following system of differential equations admits a positive solution which does not blow up (no singularity) within $[0, T]$.

$$
\begin{array}{l}
i \in \mathcal{I}, \\
-\dot{\lambda}_i(t,s) = q_i(t,s) + 2\lambda_i(t,s)b_1(t,s) - \frac{b_{2i}^2(t,s)}{r_i(t,s)}\lambda_i^2(t,s) \\
+\sum_{s'}[\lambda_i(t,s') - \lambda_i(t,s)]\tilde{q}_{ss'}, \\
\lambda_i(T,s) = q_i(T,s), \\
\\
\bar{r}_i\bar{\eta}_i^{2\bar{k}_i-1} + \sum_{j\in\mathcal{I}\backslash\{i\}}\bar{\epsilon}_{ij}\bar{\eta}_j = \bar{\lambda}_i(b_{2i}+\bar{b}_{2i}) + \bar{c}_i, \\
-\frac{d}{dt}\bar{\lambda}_i = \bar{q}_i + 2\bar{k}_i(b_1+\bar{b}_1)\bar{\lambda}_i + \sum_{s'}[\bar{\lambda}_i(t,s') - \bar{\lambda}_i(t,s)]\tilde{q}_{ss'} \\
-2\bar{k}_i\bar{\lambda}_i\sum_{j\neq i}(b_{2j}+\bar{b}_{2j})\bar{\eta}_j \\
-(2\bar{k}_i-1)\bar{r}_i\left(\frac{\bar{\lambda}_i(b_{2i}+\bar{b}_{2i})+\bar{c}_i-\sum_{j\in\mathcal{I}\backslash\{i\}}\bar{\epsilon}_{ij}\bar{\eta}_j}{\bar{r}_i}\right)^{\frac{2\bar{k}_i}{(2\bar{k}_i-1)}}, \\
\bar{\lambda}_i(T,s) = \bar{q}_i(T,s), \\
\\
-\dot{\gamma}_i(t,s) = g_i(\eta) + \frac{1}{2}\lambda_i(t,s(t))(\sigma_r^2(t,s(t)) + \sigma_{cogv}^2(t,s(t))) \\
+\frac{1}{2}\int\lambda_i(t,s(t))\sigma_n^2(t,s(t),\theta)\nu(d\theta) + \sum_{s'}[\gamma_i(t,s') - \gamma_i(t,s)]\tilde{q}_{ss'}, \\
\gamma_i(T,s) = 0,
\end{array}
\tag{12.51}
$$

where η_i minimizes $g_i(\eta)$:

$$
g_i(\eta) = r_i(\eta_i - \frac{\lambda_i b_{2i}}{r_i})^2\frac{1}{2}v_2 - \lambda_i\sum_{j\in\mathcal{I}\backslash\{i\}}b_{2j}\eta_j v_2 + c\lambda_i\sigma_r o,
\tag{12.52}
$$

the vector $(\eta_i)_{i\in\mathcal{I}}$ is a fixed point of $(\arg\min_{z_i} g_i(z_i, \eta_{-i}))_{i\in\mathcal{I}}$,

$$
\begin{aligned}
&o(t) = \int_0^t [b_1 - \sum_j b_{2j}\eta_j]o(t') + \sigma_{n2}(t')dt',\\
&o(0) = 0,\\
&\\
&v_2(t) - v_2(0)\\
&= \int_0^t 2[b_1 - \sum_j b_{2j}\eta_j]v_2(t') + \sigma_r^2(t', s(t')) + \sigma_{cogv}^2(t', s(t'))\\
&+ \int \sigma_n^2(t', s(t'), \theta)\nu(d\theta) + 2c\sigma_r(t', s(t'))o(t')dt'.
\end{aligned}
\tag{12.53}
$$

Remark 12.1 Note that these equations differ from those obtained in [157] for the case without Rosenblatt process.

- with only of the following processes—regime switching, Brownian motion, Gauss-Volterra, Poisson jump process, the minimization term is explicit: $\eta_i = \frac{\lambda_i b_{2i}}{r_i}$. When $\sigma_r = 0$, the λ system becomes

$$
\begin{aligned}
&i \in \mathcal{I},\ t \in [0, T],\\
&-\dot{\lambda}_i(t, s) = q_i(t, s) + 2\lambda_i(t, s)b_1(t, s) - \frac{b_{2i}^2(t,s)}{r_i(t,s)}\lambda_i^2(t, s)\\
&-2\lambda_i(t, s)\sum_{j\in\mathcal{I}\setminus\{i\}} \frac{b_{2j}^2(t,s)}{r_j(t,s)}\lambda_j(t, s)\\
&+\sum_{s'}[\lambda_i(t, s') - \lambda_i(t, s)]\tilde{q}_{ss'},\\
&\\
&\lambda_i(T, s) = q_i(T, s),\ s \in \mathcal{S},
\end{aligned}
\tag{12.54}
$$

- with only a Rosenblatt process: The minimization term is η_i is non-explicit due to the non-Gaussinity of r^H,, i.e., $\sigma_r \neq 0$.
- mixed noise with vs. without Rosenblatt process: The $\eta_i = \eta_i^{norh}$ obtained in the absence of Rosenblatt process does not correspond to the one obtained from g_i. As g_i is strictly convex in η_i,, this means that the equilibrium obtained under regime switching, Brownian motion, Gauss-Volterra, and Poisson jump process no longer provides equilibrium in the presence of Rosenblatt processes.
- A sufficient condition for the contraction of the best response is $\sum_{i,l} \frac{\bar{\epsilon}_{il}^2 \mathbb{I}_{l\neq i}}{(2\bar{k}_i-1)^2\bar{r}_i^2}$ $(\bar{\eta}_i^*)^{4-4\bar{k}_i} < 1$ which is satisfied when all the $\bar{\epsilon}_{il}^2$ are small enough. In particular, the condition is met when $\bar{\epsilon}_{il} = 0,\ \forall i, l$.

12.5.2.2 Two Layers: One Leader and Many Followers

We now provide the solution to the Problem in (12.47).

Proposition 12.4 *Let consider l to be leader and all the other decision-makers be followers as in (12.47). The Stackelberg solution is by*

$$
(\bar{\eta}_1^*(\bar{\eta}_l^*), \ldots, \bar{\eta}_{l-1}^*(\bar{\eta}_l^*), \bar{\eta}_l^*, \bar{\eta}_{l+1}^*(\bar{\eta}_l^*), \ldots, \bar{\eta}_I^*(\bar{\eta}_l^*))
$$

where

$$
\begin{aligned}
&\textit{For the followers: } i \in \mathscr{I}\backslash\{l\},\\
&\bar{\eta}_i(.|\bar{\eta}_l) = \left(\frac{\bar{\lambda}_i(b_{2i}+\bar{b}_{2i})+\bar{c}_i-\bar{\varepsilon}_{il}\bar{\eta}_l-\sum_{j\in\mathscr{I}\backslash\{i,l\}}\bar{\varepsilon}_{ij}\bar{\eta}_j}{\bar{r}_i}\right)^{\frac{1}{(2\bar{k}_i-1)}},\\
&\textit{For the leader}: l \in \mathscr{I}:\\
&\bar{\eta}_l^* \in \arg\min_{\bar{\eta}_l} O_l := \bar{r}_l\frac{\bar{\eta}_l^{2\bar{k}_l}}{2\bar{k}_l} - (\bar{\lambda}_l(b_{2l}+\bar{b}_{2l})+\bar{c}_l)\bar{\eta}_l\\
&-\bar{\lambda}_l\sum_{j\in\mathscr{I}\backslash\{l\}}(b_{2j}+\bar{b}_{2j})\bar{\eta}_j^* + \sum_{j\in\mathscr{I}\backslash\{l\}}\bar{\varepsilon}_{lj}\bar{\eta}_l\bar{\eta}_j^*
\end{aligned}
\tag{12.55}
$$

where

$$
\begin{aligned}
&-\tfrac{d}{dt}\bar{\lambda}_l = \bar{q}_l + 2\bar{k}_l(b_1+\bar{b}_1)\bar{\lambda}_i + \sum_{s'}[\bar{\lambda}_l(t,s')-\bar{\lambda}_l(t,s)]\tilde{q}_{ss'}\\
&+2\bar{k}_l O_l(\bar{\eta}_1^*(\bar{\eta}_l^*),\ldots,\bar{\eta}_{l-1}^*(\bar{\eta}_l^*),\bar{\eta}_l^*,\bar{\eta}_{l+1}^*(\bar{\eta}_l^*),\ldots,\bar{\eta}_I^*(\bar{\eta}_l^*)),\\
&\bar{\lambda}_l(T,s) = \bar{q}_l(T,s).
\end{aligned}
\tag{12.56}
$$

12.5.2.3 Two Layers: Multiple Leaders and Multiple Followers

We now provide the solution to the Problem in (12.48).

Proposition 12.5 *Let's consider (12.48) with the leaders set $\mathscr{I}_L$, and the followers set $\mathscr{I}_F$, disjoint $\mathscr{I}_L$ from such that*

$$
\mathscr{I}_L \cup \mathscr{I}_L = \mathscr{I},\ |\mathscr{I}_L| \geq 2.
$$

The Stackelberg solution is by

$$
((\bar{\eta}_l^*)_{l\in\mathscr{I}_L}, (\bar{\eta}_f^*((\bar{\eta}_l^*)_{l\in\mathscr{I}_L}),\ f\in\mathscr{I}_F)) = (\bar{\eta}_L, \bar{\eta}_F^*(\bar{\eta}_L)),
$$

where

$$
\begin{aligned}
&\textit{For the followers: } i \in \mathscr{I}_F,\\
&\bar{\eta}_i(.|\bar{\eta}_L) = \left(\frac{\bar{\lambda}_i(b_{2i}+\bar{b}_{2i})+\bar{c}_i-\sum_{l\in\mathscr{I}_L}\bar{\varepsilon}_{il}\bar{\eta}_l-\sum_{j\in\mathscr{I}_F\backslash\{i\}}\bar{\varepsilon}_{ij}\bar{\eta}_j}{\bar{r}_i}\right)^{\frac{1}{(2\bar{k}_i-1)}},\\
&\textit{For the leaders}: l \in \mathscr{I}:\\
&\bar{\eta}_l^* \in \arg\min_{\bar{\eta}_l} O_l := \bar{r}_l\frac{\bar{\eta}_l^{2\bar{k}_l}}{2\bar{k}_l} - (\bar{\lambda}_l(b_{2l}+\bar{b}_{2l})+\bar{c}_l)\bar{\eta}_l\\
&-\bar{\lambda}_l\sum_{j\in\mathscr{I}_L\backslash\{l\}}(b_{2j}+\bar{b}_{2j})\bar{\eta}_j - \bar{\lambda}_l\sum_{j\in\mathscr{I}_F}(b_{2j}+\bar{b}_{2j})\bar{\eta}_j^*\\
&+\sum_{j\in\mathscr{I}_L\backslash\{l\}}\bar{\varepsilon}_{lj}\bar{\eta}_l\bar{\eta}_j + \sum_{j\in\mathscr{I}_F}\bar{\varepsilon}_{lj}\bar{\eta}_l\bar{\eta}_j^*,
\end{aligned}
\tag{12.57}
$$

where

$$
\begin{array}{l}
-\frac{d}{dt}\bar{\lambda}_l = \bar{q}_l + 2\bar{k}_l(b_1+\bar{b}_1)\bar{\lambda}_i + \sum_{s'}[\bar{\lambda}_l(t,s') - \bar{\lambda}_l(t,s)]\tilde{q}_{ss'} \\
+2\bar{k}_l O_l(\bar{\eta}_L^*, \bar{\eta}_F^*(\bar{\eta}_L^*)), \\
\bar{\lambda}_l(T,s) = \bar{q}_l(T,s).
\end{array} \tag{12.58}
$$

12.5.2.4 One Decision-Maker per Layer

We now provide the solution to the Problem in (12.49).

Proposition 12.6 *Let's consider (12.49), and let*

$$
(\bar{\eta}_1^*, \bar{\eta}_2^*(\bar{\eta}_1^*), \bar{\eta}_3^*(\bar{\eta}_1^*, \bar{\eta}_2^*), \ldots, \bar{\eta}_I^*(\bar{\eta}_1^*, \bar{\eta}_2^*, \ldots, \bar{\eta}_{I-1}^*))
$$

with

$$
\begin{array}{l}
\textit{Layer } I: \\
\bar{\eta}_I^*(.|\bar{\eta}_1, \ldots, \bar{\eta}_{I-1}) = \left(\frac{\bar{\lambda}_I(b_{2I}+\bar{b}_{2I})+\bar{c}_I - \sum_{j=1}^{I-1}\bar{\varepsilon}_{Ij}\bar{\eta}_j}{\bar{r}_I}\right)^{\frac{1}{(2\bar{k}_I-1)}},
\end{array} \tag{12.59}
$$

At layer $I-1$: Given $(\bar{\eta}_1, \bar{\eta}_2, \ldots, \bar{\eta}_{I-2})$ and the reaction $\bar{\eta}_I^*(.|\bar{\eta}_1, \ldots, \bar{\eta}_{I-1})$, we compute $\bar{\eta}_{I-1}^*(\bar{\eta}_1, \bar{\eta}_2, \ldots, \bar{\eta}_{I-2})$

$$
\begin{array}{l}
\bar{\eta}_{I-1}^* \in \arg\min_{\bar{\eta}_{I-1}} O_{I-1} := \\
\bar{r}_{I-1}\frac{\bar{\eta}_{I-1}^{2\bar{k}_{I-1}}}{2\bar{k}_{I-1}} - (\bar{\lambda}_{I-1}(b_{2(I-1)} + \bar{b}_{2(I-1)}) + \bar{c}_{I-1})\bar{\eta}_{I-1} \\
-\bar{\lambda}_{I-1}\sum_{j=1}^{I-2}(b_{2j}+\bar{b}_{2j})\bar{\eta}_j - \bar{\lambda}_{I-1}(b_{2I}+\bar{b}_{2I})\bar{\eta}_I^* \\
+\sum_{j=1}^{I-2}\bar{\varepsilon}_{(I-1)j}\bar{\eta}_{I-1}\bar{\eta}_j + \bar{\varepsilon}_{(I-1)I}\bar{\eta}_{I-1}\bar{\eta}_I^*,
\end{array} \tag{12.60}
$$

At layer i :

$$
\begin{array}{l}
\textit{Given } (\bar{\eta}_1, \bar{\eta}_2, \ldots, \bar{\eta}_{i-1}) \textit{ and the reactions :} \\
\bar{\eta}_I^*(.|\bar{\eta}_1, \ldots, \bar{\eta}_i, \ldots, \bar{\eta}_{I-1}), \\
\bar{\eta}_{I-1}^*(\bar{\eta}_1, \bar{\eta}_2, \ldots, \bar{\eta}_i, \ldots, \bar{\eta}_{I-2}) \\
\ldots, \\
\bar{\eta}_{i+1}^*(.|\bar{\eta}_1, \ldots, \bar{\eta}_i),
\end{array} \tag{12.61}
$$

we recursively compute $\bar{\eta}_i^*(\bar{\eta}_1, \bar{\eta}_2, \ldots, \bar{\eta}_{i-1})$

$$
\begin{array}{l}
\bar{\eta}_i^* \in \arg\min_{\bar{\eta}_i} O_i := \\
\bar{r}_i\frac{\bar{\eta}_i^{2\bar{k}_i}}{2\bar{k}_i} - (\bar{\lambda}_i(b_{2i}+\bar{b}_{2i}) + \bar{c}_i)\bar{\eta}_i \\
-\bar{\lambda}_i\sum_{j=1}^{i-1}(b_{2j}+\bar{b}_{2j})\bar{\eta}_j - \bar{\lambda}_i\sum_{j=i+1}^{I}(b_{2j}+\bar{b}_{2j})\bar{\eta}_j^* \\
+\sum_{j=1}^{i-1}\bar{\varepsilon}_{ij}\bar{\eta}_i\bar{\eta}_j + \sum_{j=i+1}^{I}\bar{\varepsilon}_{ij}\bar{\eta}_i\bar{\eta}_j^*,
\end{array} \tag{12.62}
$$

with

$$
\begin{aligned}
&-\frac{d}{dt}\bar{\lambda}_i = \\
&\bar{q}_i + 2\bar{k}_i(b_1+\bar{b}_1)\bar{\lambda}_i + \sum_{s'}[\bar{\lambda}_i(t,s') - \bar{\lambda}_i(t,s)]\tilde{q}_{ss'} + 2\bar{k}_i O_i(\bar{\eta}^{ss}), \\
&\bar{\lambda}_i(T,s) = \bar{q}_i(T,s).
\end{aligned} \tag{12.63}
$$

Let $\bar{\eta}^{ss} = (\bar{\eta}_1^{ss}, \ldots, \bar{\eta}_I^{ss})$ *be*

$$
\begin{aligned}
&\bar{\eta}_I^{ss} = \bar{\eta}_I^*(.|\bar{\eta}_1^*, \ldots, \bar{\eta}_i^*, \ldots, \bar{\eta}_{I-1}^*), \\
&\bar{\eta}_{I-1}^{ss} = \bar{\eta}_{I-1}^*(\bar{\eta}_1^*, \bar{\eta}_2^*, \ldots, \bar{\eta}_i^*, \ldots, \bar{\eta}_{I-2}^*), \\
&\ldots, \\
&\bar{\eta}_{i+1}^{ss} = \bar{\eta}_{i+1}^*(.|\bar{\eta}_1^*, \ldots, \bar{\eta}_i^*), \\
&\bar{\eta}_i^{ss} = \bar{\eta}_i^*(\bar{\eta}_1^*, \bar{\eta}_2^*, \ldots, \bar{\eta}_{i-1}^*), \\
&\ldots, \\
&\bar{\eta}_2^{ss} = \bar{\eta}_2^*(\bar{\eta}_1^*), \\
&\bar{\eta}_1^{ss} = \bar{\eta}_1^*.
\end{aligned} \tag{12.64}
$$

If there is a solution to the above systems, then $(\eta, \bar{\eta}^{ss})$ *provides a Stackelberg solution.*

Remark 12.2 The existence of solution to the multilayer hierarchical MFTG are reduced to the existence of solution of the auxilliary stage games given by O_i. If there are H layers with $H \leq I$, and a partition $\mathscr{I}_1 \cup \mathscr{I}_2 \ldots \cup \mathscr{I}_H = \mathscr{I}$, Then auxiliary game is given by $(\mathscr{I}, U_i, O_{i,h},\ i \in \mathscr{I}_h,\ h \in \mathscr{H} = \{1, 2, \ldots, H\})$.

$$
\begin{aligned}
&i \in \mathscr{I}_h, \\
&O_{i,h} = \bar{r}_i \frac{\bar{\eta}_i^{2\bar{k}_i}}{2\bar{k}_i} - (\bar{\lambda}_i(b_{2i} + \bar{b}_{2i}) + \bar{c}_i)\bar{\eta}_i \\
&-\bar{\lambda}_i \sum_{j \in \mathscr{I}_1 \cup \ldots \cup \mathscr{I}_{h-1}} (b_{2j} + \bar{b}_{2j})\bar{\eta}_j \\
&-\bar{\lambda}_i \sum_{j \in \mathscr{I}_h \setminus \{i\}} (b_{2j} + \bar{b}_{2j})\bar{\eta}_j \\
&-\bar{\lambda}_i \sum_{j \in \mathscr{I}_{h+1} \cup \ldots \cup \mathscr{I}_H} (b_{2j} + \bar{b}_{2j})\bar{\eta}_j^* \\
&+\sum_{j \in \mathscr{I}_1 \cup \ldots \cup \mathscr{I}_{h-1}} \bar{\varepsilon}_{ij}\bar{\eta}_i\bar{\eta}_j \\
&+\sum_{j \in \mathscr{I}_h \setminus \{i\}} \bar{\varepsilon}_{ij}\bar{\eta}_i\bar{\eta}_j + \sum_{j \in \mathscr{I}_{h+1} \cup \ldots \cup \mathscr{I}_H} \bar{\varepsilon}_{ij}\bar{\eta}_i\bar{\eta}_j^*
\end{aligned} \tag{12.65}
$$

12.5.2.5 Proofs

We let $B^H(t) = \tilde{c}_2 \int_{\mathbb{R}} (\int_0^t (t' - x)_+^{H-\frac{3}{2}} dt') dW(x),\ H \in (\frac{1}{2}, 1),\ \tilde{c}_2 = \sqrt{\frac{H(2H-1)}{B(2-2H, H-\frac{1}{2})}}$.

Let

$$
\begin{aligned}
x(t) &= x_0 + \int_0^t d_1\, dt' + \int_0^t d_2 dB(t') + \int_0^t d_3 dB_{gv}(t') \\
&+ \int_0^t d_4 \tilde{N}(dt', d\theta) + 2\tilde{c} \int_0^t d_5 dB^{\frac{H+1}{2}}(t') + \int_0^t d_6 dr^H(t').
\end{aligned} \tag{12.66}
$$

for given functions $(d_1, d_2, d_3, d_4, d_5, d_6)$. Let $f(., s, .) \in C^{1,3}$ in (t, x) for any regime switching component s. Then, the process y defined by $y(t) = f(t, x(t), s(t))$ satisfies

$$\begin{array}{l} y(t) = y_0 + \int_0^t \tilde{d}_1 \, dt' + \int_0^t \tilde{d}_2 dB(t') + \int_0^t \tilde{d}_3 dB_{gv}(t') \\ + \int_0^t \tilde{d}_4 \tilde{N}(dt', d\theta) + 2\tilde{c} \int_0^t \tilde{d}_5 dB^{\frac{H+1}{2}}(t') + \int_0^t \tilde{d}_6 dr^H(t'), \end{array} \tag{12.67}$$

where

$$\begin{array}{l} \tilde{d}_1(t') = f_{t'}(t', x(t'), s(t')) + f_x(t', x(t'), s(t'))d_1 \\ + f_{xx}(t', x(t'), s(t'))\frac{1}{2}(d_2^2 + d_{3,cogv}^2) \\ + \int [f(t', x(t') + d_4, s(t')) - f(t', x(t'), s(t')) \\ - d_4 f_x(t', x(t'), s(t'))]\nu(d\theta) \\ + \sum_{s'}[f(t', x(t'), s') - f(t', x(t'), s)]\tilde{q}_{s's} \\ + 2c f_{xx}(t', x(t'))(\nabla^{\frac{H}{2}} x(t'))(t')d_5 \\ + c f_{xx}(t', x(t'), s(t'))(\nabla^{\frac{H}{2}, \frac{H}{2}} x)(t', t')d_6 \\ + c f_{xxx}(t', x(t'), s(t'))[(\nabla^{\frac{H}{2}} x(t'))(t')]^2 d_6, \\ \\ \tilde{d}_2(t') = f_x(t', x(t'), s(t'))d_2, \\ \\ \tilde{d}_3(t') = f_x(t', x(t'), s(t'))d_3, \\ \\ \tilde{d}_4(t') = [f(t', x(t', s(t')) + d_4, s(t')) - f(t', x(t'), s(t'))], \\ \\ \tilde{d}_5(t') = f_x(t', x(t'), s(t'))d_5 + f_{xx}(t', x(t'))(\nabla^{\frac{H}{2}} x)(t')d_6, \\ \\ \tilde{d}_6(t') = f_x(t', x(t'), s(t'))d_6, \end{array} \tag{12.68}$$

with

$$\begin{array}{l} c = c(\delta)\Gamma^2(\frac{H}{2}), \\ \tilde{c} = \sqrt{\frac{(2H-1)\Gamma(1-\frac{H}{2})\Gamma(\frac{H}{2})}{(H+1)\Gamma(1-H)}}, \\ \nabla^{\frac{H}{2}} = I_+^{\frac{H}{2}} \circ D, \quad \nabla^{\frac{H}{2}, \frac{H}{2}} = I_{+,+}^{\frac{H}{2}, \frac{H}{2}} \circ D^2, \\ f : \mathbb{R} \to \mathbb{R}, \\ (I_+^{\alpha} f)(t) = \frac{1}{\Gamma(\alpha)} \int_{-\infty}^t f(u)(t-u)^{\alpha-1} du, \\ (\alpha_1, \alpha_2) \in (0, 1)^2, \; f : \mathbb{R}^2 \to \mathbb{R}, \\ (I_{+,+}^{\alpha_1, \alpha_2} f)(t_1, t_2) = \\ \frac{1}{\Gamma(\alpha_1)\Gamma(\alpha_2)} \int_{-\infty}^{t_1} \int_{-\infty}^{t_2} f(u, v)(t_1 - u)^{\alpha_1 - 1}(t_2 - v)^{\alpha_2 - 1} du dv, \\ d_{3,cogv}^2 = \\ \frac{d}{dt}\left[\int_0^t \left\{ K(t'_+, t')d_3(t') + \int_{t'}^t d_3(t'')K(t', t'')dt'' \right\}^2 dt' \right] \end{array} \tag{12.69}$$

A proof of (12.75) can be easily obtained by combining Lemma 12.1 and [62]. ∎
The state dynamics (12.45) can be rewritten as

$$
\begin{array}{l}
x(t) = \bar{x}_0 + \int_0^t (b_1 + \bar{b}_1)\bar{x} + \sum_{j\in\mathcal{I}} (b_{2j} + \bar{b}_{2j})\bar{u}_j \, dt' \\
+x_0 - \bar{x}_0 + \int_0^t b_1(x - \bar{x}) + \sum_{j\in\mathcal{I}} b_{2j}(u_j - \bar{u}_j) \, dt' \\
+ \int_0^t \sigma_b dB(t') + \sigma_{gv} dB_{gv}(t') + \int_\theta \sigma_n(t', s, \theta)\tilde{N}(dt', d\theta) \\
+ \int_0^t \sigma_r dr^H(t')
\end{array}
\tag{12.70}
$$

which is decomposed as $x(t) = x(t) - \bar{x}(t) + \bar{x}(t)$ with

$$
\begin{array}{l}
\bar{x}(t) = \bar{x}_0 + \int_0^t (b_1 + \bar{b}_1)\bar{x} + \sum_{j\in\mathcal{I}} (b_{2j} + \bar{b}_{2j})\bar{u}_j \, dt', \\
\\
x(t) - \bar{x}(t) \\
= x_0 - \bar{x}_0 + \int_0^t b_1(x - \bar{x}) + \sum_{j\in\mathcal{I}} b_{2j}(u_j - \bar{u}_j) \, dt' \\
+ \int_0^t \sigma_b dB(t') + \sigma_{gv} dB_{gv}(t') + \int_\theta \sigma_n(t', s, \theta)\tilde{N}(dt', d\theta) \\
+ \int_0^t \sigma_r dr^H(t').
\end{array}
\tag{12.71}
$$

Equipped with the state decomposition (12.71), the cost functional of decision-maker i can be decomposed as $L_i = \hat{L}_i + \bar{L}_i$, $i \in \mathcal{I}$, where

$$
\begin{array}{l}
i \in \mathcal{I}, \\
\hat{L}_i = q_{iT}\frac{(x_T - \bar{x}_T)^2}{2} + \int_0^T q_i \frac{(x-\bar{x})^2}{2} + r_i \frac{(u_i - \bar{u}_i)^2}{2} \, dt, \\
\bar{L}_i = \bar{q}_{iT}\frac{\bar{x}_T^{2\bar{k}_i}}{2\bar{k}_i} + \int_0^T \bar{q}_i \frac{\bar{x}^{2\bar{k}_i}}{2\bar{k}_i} + \bar{r}_i \frac{\bar{u}_i^{2\bar{k}_i}}{2\bar{k}_i} \, dt \\
+ \int_0^T \bar{c}_i \bar{x}^{2\bar{k}_i - 1}\bar{u}_i + \sum_{j\in\mathcal{I}\backslash\{i\}} \bar{\varepsilon}_{ij}\bar{x}^{2(\bar{k}_i - 1)}\bar{u}_i\bar{u}_j \, dt,
\end{array}
\tag{12.72}
$$

Two subproblems are then formulated

$$
\begin{array}{l}
i \in \mathcal{I}, \\
\hat{L}_i = q_{iT}\frac{(x_T - \bar{x}_T)^2}{2} + \int_0^T q_i \frac{(x-\bar{x})^2}{2} + r_i \frac{(u_i - \bar{u}_i)^2}{2} \, dt, \\
\text{subject to} \\
x(t) - \bar{x}(t) \\
= x_0 - \bar{x}_0 + \int_0^t b_1(x - \bar{x}) + \sum_{j\in\mathcal{I}} b_{2j}(u_j - \bar{u}_j) \, dt' \\
+ \int_0^t \sigma_b dB(t') + \sigma_{gv} dB_{gv}(t') + \int_\theta \sigma_n(t', s, \theta)\tilde{N}(dt', d\theta) \\
+ \int_0^t \sigma_r dr^H(t'), \\
\mathbb{P}(s(t+\delta) = s'|s) = \int_t^{t+\delta} \tilde{q}_{ss'} dt' + o(\delta), \ s' \neq s, \\
s(0) = s_0 \in \mathcal{S},
\end{array}
\tag{12.73}
$$

and

$$
\begin{aligned}
& i \in \mathcal{I}, \\
& \bar{L}_i = \bar{q}_{iT} \frac{\bar{x}_T^{2\bar{k}_i}}{2\bar{k}_i} + \int_0^T \bar{q}_i \frac{\bar{x}^{2\bar{k}_i}}{2\bar{k}_i} + \bar{r}_i \frac{\bar{u}_i^{2\bar{k}_i}}{2\bar{k}_i}\, dt \\
& + \int_0^T \bar{c}_i \bar{x}^{2\bar{k}_i - 1} \bar{u}_i + \sum_{j \in \mathcal{I}\setminus\{i\}} \bar{\varepsilon}_{ij} \bar{x}^{2(\bar{k}_i - 1)} \bar{u}_i \bar{u}_j \, dt, \\
& \text{subject to} \\
& \bar{x}(t) = \bar{x}_0 + \int_0^t (b_1 + \bar{b}_1)\bar{x} + \sum_{j \in \mathcal{I}} (b_{2j} + \bar{b}_{2j}) \bar{u}_j \, dt', \\
& \mathbb{P}(s(t+\delta) = s' | s) = \int_t^{t+\delta} \tilde{q}_{ss'} dt' + o(\delta), \ s' \neq s, \\
& s(0) = s_0 \in \mathcal{S},
\end{aligned}
\tag{12.74}
$$

For the first subproblem, we use (12.75) with $\hat{f}_i(t, y(t), s(t)) = \lambda_i(t, s(t)) y^2(t)$, and $y(t) = x(t) - \bar{x}(t)$. For the second subproblem, we use (12.75) $\bar{f}_i(t, \bar{x}(t), s(t)) = \bar{\lambda}_i(t, s(t)) \frac{\bar{x}(t)^{2\bar{k}_i}}{2\bar{k}_i}$. Note that the subproblem (12.74) involves only the regime switching process $s(t)$ as stochastic element of the dynamics.

$$
E[\hat{f}_i(T, y(T), s(T)) - \hat{f}_i(0, y(0), s(0))] = E \int_0^T \tilde{d}_1 \, dt \tag{12.75}
$$

where

$$
\begin{aligned}
& \tilde{d}_1(t) = \\
& [\dot{\lambda}_i(t, s) + 2\lambda_i(t, s) b_1 + \sum_{s'} [\lambda_i(t, s') - \lambda_i(t, s)] \tilde{q}_{ss'}] y^2(t) \\
& + 2\lambda_i(t, s(t)) y(t) \sum_{j \in \mathcal{I}} b_{2j}(u_j - \bar{u}_j) \\
& + \lambda_i(t, s(t)) (\sigma_r^2(t, s(t)) + \sigma_{cogv}^2)(t, s(t)) \\
& + \int \lambda_i(t, s(t)) \sigma_n^2(t, s(t), \theta) \nu(d\theta) \\
& + 2c\lambda_i(t, s(t)) \sigma_r(t, s(t)) (\nabla^{\frac{H}{2}, \frac{H}{2}} y)(t, t),
\end{aligned}
\tag{12.76}
$$

$$
\begin{aligned}
& (\nabla^{\frac{H}{2}, \frac{H}{2}} y)(t, t) = I_{+,+}^{\frac{H}{2}, \frac{H}{2}} (D^2 y)(t, t) = o(t) \\
& = \int_0^t [b_1 - \sum_j b_{2j} \eta_j] o(t') + \sigma_{n2}(t') dt', \\
& o(0) = 0,
\end{aligned}
$$

One has $o(t) = \left[\int_0^t \sigma_{n2}(t') e^{-\int_0^{t'} b dt''} dt' \right] e^{\int_0^t b dt'}$ with $b = [b_1 - \sum_j b_{2j} \eta_j]$.

$$
\begin{aligned}
& \sigma_{n2}(t) = (\nabla^{\frac{H}{2}, \frac{H}{2}} z)(t, t), \\
& z(t) = \int_0^t \sigma_b dB(t') + \sigma_{gv} dB_{gv}(t') \\
& + \int_\theta \sigma_n(t', s, \theta) \tilde{N}(dt', d\theta) + \sigma_r dr^H(t'),
\end{aligned}
\tag{12.77}
$$

Note that σ_{n2} is a function of time and regime switching and does not depend on η.

$$
\begin{array}{l}
\bar{f}_i(T, \bar{x}(T), s(T)) - \bar{f}_i(0, \bar{x}(0), s(0)) = \\
\frac{\bar{\lambda}_i(T, s(T))}{2\bar{k}_i}\bar{x}^{2\bar{k}_i}(T) - \frac{\bar{\lambda}_i(0, s_0)}{2\bar{k}_i}\bar{x}_0^{2\bar{k}_i} \\
+\int_0^T [\frac{d}{dt}\bar{\lambda}_i(t, s) + 2\bar{k}_i\bar{\lambda}_i(t, s)(b_1 + \bar{b}_1) \\
+\sum_{s'}[\bar{\lambda}_i(t, s') - \bar{\lambda}_i(t, s)]\tilde{q}_{ss'}]\frac{1}{2\bar{k}_i}\bar{x}^{2\bar{k}_i} \\
+2\bar{k}_i\bar{\lambda}_i(t, s)\sum_{j\in\mathscr{I}\setminus\{i\}}(b_{2j} + \bar{b}_{2j})\bar{u}_j\frac{1}{2\bar{k}_i}\bar{x}^{2\bar{k}_i-1}\Big] \\
+\bar{\lambda}_i(t, s)(b_{2i} + \bar{b}_{2i})\bar{u}_i\bar{x}^{2\bar{k}_i-1}dt,
\end{array} \tag{12.78}
$$

$$
\begin{array}{l}
i \in \mathscr{I}, \\
E[\hat{L}_i - \hat{f}_i(0, (x - \bar{x})(0), s(0))] = \\
E[(q_{iT} - \lambda_i(T, s(T)))\frac{(x_T - \bar{x}_T)^2}{2}] \\
+E\int_0^T r_i\frac{(u_i - \bar{u}_i)^2}{2} + \lambda_i(t, s(t))(x - \bar{x})(t)b_{2i}(u_i - \bar{u}_i) \\
+[\dot{\lambda}_i(t, s) + q_i + 2\lambda_i(t, s)b_1 + \sum_{s'}[\lambda_i(t, s') - \lambda_i(t, s)]\tilde{q}_{ss'}]\frac{(x-\bar{x})^2}{2} \\
+\lambda_i(t, s(t))(x - \bar{x})(t)\sum_{j\in\mathscr{I}\setminus\{i\}} b_{2j}(u_j - \bar{u}_j) \\
+\frac{1}{2}\lambda_i(t, s(t))(\sigma_r^2(t, s(t)) + \sigma_{cogv}^2(t, s(t))) \\
+\frac{1}{2}\int \lambda_i(t, s(t))\sigma_n^2(t, s(t), \theta)\nu(d\theta) \\
+c\lambda_i(t, s(t))\sigma_r(t, s(t))o(t)dt,
\end{array} \tag{12.79}
$$

$$
\begin{array}{l}
i \in \mathscr{I}, \\
E[\bar{L}_i - \bar{f}_i(0, \bar{x}(0), s(0))] \\
= E[(\bar{q}_i(T, s(T)) - \bar{\lambda}_i(T, s(T)))\frac{\bar{x}(T)^{2\bar{k}_i}}{2\bar{k}_i}] \\
+E\int_0^T \bar{r}_i\frac{\bar{u}_i^{2\bar{k}_i}}{2\bar{k}_i} + \bar{\lambda}_i(t, s)(b_{2i} + \bar{b}_{2i})\bar{u}_i\bar{x}^{2\bar{k}_i-1} \\
+\bar{c}_i\bar{x}^{2\bar{k}_i-1}\bar{u}_i + \sum_{j\in\mathscr{I}\setminus\{i\}} \bar{\varepsilon}_{ij}\bar{x}^{2(\bar{k}_i-1)}\bar{u}_i\bar{u}_j \\
+[\frac{d}{dt}\bar{\lambda}_i(t, s) + \bar{q}_i + 2\bar{k}_i\bar{\lambda}_i(t, s)(b_1 + \bar{b}_1) \\
+\sum_{s'}[\bar{\lambda}_i(t, s') - \bar{\lambda}_i(t, s)]\tilde{q}_{ss'}]\frac{1}{2\bar{k}_i}\bar{x}^{2\bar{k}_i} \\
+2\bar{k}_i\bar{\lambda}_i(t, s)\sum_{j\in\mathscr{I}\setminus\{i\}}(b_{2j} + \bar{b}_{2j})\bar{u}_j\frac{1}{2\bar{k}_i}\bar{x}^{2\bar{k}_i-1}dt,
\end{array} \tag{12.80}
$$

Proof of Theorem 12.3 We proceed with a convex completion as follows. Recall that for smooth and strictly convex function f,, the Legendre-Fenchel transform is given by $f^*(x) = \sup_u\{xu - f(u)\}$. This means

$$f^*(x) + f(u) - xu \geq 0$$

for all (x, u). In addition, there is an equality if and only if $u^* = (f^*)'(x)$. The idea of convex completion is that if we have $f(u) - xu$, we will add an extra term $f^*(x)$ to it, so that it becomes nonnegative, and it is equal to 0 when $u = (f^*)'(x)$.

We apply the convex completion method to the following term:

$$\begin{aligned}&\bar{r}_i\frac{\bar{u}_i^{2\bar{k}_i}}{2\bar{k}_i}+\bar{u}_i\bar{x}^{2\bar{k}_i-1}[\bar{\lambda}_i(b_{2i}+\bar{b}_{2i})+\bar{c}_i-\sum_{j\in\mathcal{I}\backslash\{i\}}\bar{\varepsilon}_{ij}\bar{\eta}_j]\\&+(2\bar{k}_i-1)\bar{r}_i(\frac{[\bar{\lambda}_i(b_{2i}+\bar{b}_{2i})+\bar{c}_i-\sum_{j\in\mathcal{I}\backslash\{i\}}\bar{\varepsilon}_{ij}\bar{\eta}_j]}{\bar{r}_i})^{\frac{2\bar{k}_i}{(2\bar{k}_i-1)}}\frac{\bar{x}^{2\bar{k}_i}}{2\bar{k}_i}\geq 0,\end{aligned} \tag{12.81}$$

with equality in (12.81) if and only if
$\bar{u}_i(t)=-\bar{\eta}_i(t,s(t))\bar{x}(t)$ which means that

$$\bar{\eta}_i:=\left(\frac{\bar{\lambda}_i(b_{2i}+\bar{b}_{2i})+\bar{c}_i-\sum_{j\in\mathcal{I}\backslash\{i\}}\bar{\varepsilon}_{ij}\bar{\eta}_j}{\bar{r}_i}\right)^{\frac{1}{(2\bar{k}_i-1)}}.$$

On the other hand,

$$\begin{aligned}&r_i\frac{(u_i-\bar{u}_i)^2}{2}+\lambda_i(x-\bar{x})(t)b_{2i}(u_i-\bar{u}_i)\\&=[r_i(\eta_i-\frac{\lambda_i b_{2i}}{r_i})^2-\frac{\lambda_i^2 b_{2i}^2}{r_i}]\frac{1}{2}(x-\bar{x})^2(t)\end{aligned}$$

Therefore, we look for $\eta_i\in\arg\min g_i(\eta)$ with

$$g_i:=r_i(\eta_i-\frac{\lambda_i b_{2i}}{r_i})^2\frac{1}{2}v_2(t)-\lambda_i\sum_{j\in\mathcal{I}\backslash\{i\}}b_{2j}\eta_j v_2(t)+c\lambda_i\sigma_r o(t).$$

Note that η_i appears in the conditional variance $v_2(t)$ as well as in $o(t)$. By setting $\lambda_i(t,s(t))$ to 1 in (12.75), the dynamics of the conditional variance $v_2(t)=E[(x-\bar{x})^2(t)|s]$ is given by

$$\begin{aligned}&v_2(t)-v_2(0)\\&=\int_0^t 2b(t')v_2(t')+\sigma_r^2(t',s(t'))+\sigma_{cogv}^2(t',s(t'))\\&+\int\sigma_n^2(t',s(t'),\theta)\nu(d\theta)+2c\sigma_r(t',s(t'))o(t')dt'.\end{aligned} \tag{12.82}$$

Putting all together, we arrive at the problem formulated in (12.46) completing the proof of Theorem 12.3. ■

Proof of Proposition 12.4 We now examine the problem formulated in (12.47) and prove Proposition 12.4. Here decision-maker l is the leader and is alone at layer 1. By backward induction, we first solve the MFTG among the followers in $\mathcal{I}\backslash\{l\}$ which are given in Theorem 12.3 given η_l the control strategy of l.

Hence for $i\in\mathcal{I}\setminus\{l\}$, one has

$$\bar{\eta}_i(.|\bar{\eta}_l)=\left(\frac{\bar{\lambda}_i(b_{2i}+\bar{b}_{2i})+\bar{c}_i-\bar{\varepsilon}_{il}\bar{\eta}_l-\sum_{j\in\mathcal{I}\backslash\{i,l\}}\bar{\varepsilon}_{ij}\bar{\eta}_j}{\bar{r}_i}\right)^{\frac{1}{(2\bar{k}_i-1)}}.$$

Hence, the leader l modifies the cost functional L_l by taking into account the reaction of the followers and obtains the following cost functional

$$\begin{array}{l}\bar{L}_l = \bar{q}_{lT}\frac{\bar{x}_T^{2\bar{k}_l}}{2\bar{k}_l} + \int_0^T \bar{q}_l\frac{\bar{x}^{2\bar{k}_l}}{2\bar{k}_l} + \bar{r}_l\frac{\bar{u}_l^{2\bar{k}_l}}{2\bar{k}_l}\,dt\\ + \int_0^T \bar{c}_l\bar{x}^{2\bar{k}_l-1}\bar{u}_l + \sum_{j\in\mathscr{I}\backslash\{l\}}\bar{\varepsilon}_{lj}\bar{x}^{2(\bar{k}_l-1)}\bar{u}_l\bar{u}_j^*(\bar{u}_l)\,dt,\\ \text{subject to}\\ \bar{x}(t) = \bar{x}_0 + \int_0^t(b_1+\bar{b}_1)\bar{x} + (b_{2l}+\bar{b}_{2l})\bar{u}_l\\ +\sum_{j\in\mathscr{I}\backslash\{l\}}(b_{2j}+\bar{b}_{2j})\bar{u}_j^*(\bar{u}_l)\,dt',\\ \mathbb{P}(s(t+\delta)=s'|s) = \int_t^{t+\delta}\tilde{q}_{ss'}dt' + o(\delta),\ s'\neq s,\\ s(0)=s_0\in\mathscr{S},\end{array} \tag{12.83}$$

and hence one has

$$\begin{array}{l}\bar{r}_l\frac{\bar{u}_l^{2\bar{k}_l}}{2\bar{k}_l} + \bar{\lambda}_l(b_{2l}+\bar{b}_{2l})\bar{u}_l\bar{x}^{2\bar{k}_l-1}\\ +\bar{\lambda}_l\sum_{j\in\mathscr{I}\backslash\{l\}}(b_{2j}+\bar{b}_{2j})\bar{u}_j^*\bar{x}^{2\bar{k}_l-1}\\ +\bar{c}_l\bar{x}^{2\bar{k}_l-1}\bar{u}_l + \sum_{j\in\mathscr{I}\backslash\{l\}}\bar{\varepsilon}_{lj}\bar{x}^{2(\bar{k}_l-1)}\bar{u}_l\bar{u}_j^*\\ = \left\{\bar{r}_l\frac{\bar{\eta}_l^{2\bar{k}_l}}{2\bar{k}_l} - (\bar{\lambda}_l(b_{2l}+\bar{b}_{2l})+\bar{c}_l)\bar{\eta}_l\right.\\ \left.-\bar{\lambda}_l\sum_{j\in\mathscr{I}\backslash\{l\}}(b_{2j}+\bar{b}_{2j})\bar{\eta}_j^* + \sum_{j\in\mathscr{I}\backslash\{l\}}\bar{\varepsilon}_{lj}\bar{\eta}_l\bar{\eta}_j^*\ \right\}\bar{x}^{2\bar{k}_l} =: O_l\bar{x}^{2\bar{k}_l}.\end{array} \tag{12.84}$$

Note that

$$\begin{array}{l}\text{for } i\neq l,\\ \partial_{\bar{\eta}_l}\bar{\eta}_i^* = \frac{-\bar{\varepsilon}_{il}}{(2\bar{k}_i-1)\bar{r}_i}\left(\frac{\bar{\lambda}_i(b_{2i}+\bar{b}_{2i})+\bar{c}_i-\bar{\varepsilon}_{il}\bar{\eta}_l-\sum_{k\in\mathscr{I}\backslash\{i,l\}}\bar{\varepsilon}_{ik}\bar{\eta}_k}{\bar{r}_i}\right)^{\frac{1}{(2\bar{k}_i-1)}-1}\\ = \frac{-\bar{\varepsilon}_{il}}{(2\bar{k}_i-1)\bar{r}_i}(\bar{\eta}_i^*)^{2-2\bar{k}_i}\end{array} \tag{12.85}$$

Combining (12.84) and (12.85), we arrive at the following first-order optimality condition

$$\begin{array}{l}0 = \partial_{\bar{\eta}_l}O_l(\bar{\eta}_1^*(\bar{\eta}_l),\ldots,\bar{\eta}_{l-1}^*(\bar{\eta}_l),\bar{\eta}_l,\bar{\eta}_{l+1}^*(\bar{\eta}_l),\ldots,\bar{\eta}_I^*(\bar{\eta}_l))\\ = \left\{\bar{r}_l\bar{\eta}_l^{2\bar{k}_l-1} - (\bar{\lambda}_l(b_{2l}+\bar{b}_{2l})+\bar{c}_l)\right.\\ +\bar{\lambda}_l\sum_{j\in\mathscr{I}\backslash\{l\}}(b_{2j}+\bar{b}_{2j})\frac{\varepsilon_{jl}}{(2\bar{k}_j-1)\bar{r}_j}(\bar{\eta}_j^*)^{2-2\bar{k}_j}\\ \left.+\sum_{j\in\mathscr{I}\backslash\{l\}}\bar{\varepsilon}_{lj}\bar{\eta}_j^* - \sum_{j\in\mathscr{I}\backslash\{l\}}\bar{\eta}_l\frac{\bar{\varepsilon}_{lj}\bar{\varepsilon}_{jl}}{(2\bar{k}_j-1)\bar{r}_j}(\bar{\eta}_j^*)^{2-2\bar{k}_j}\right\}\end{array} \tag{12.86}$$

From the power differentiation property, we have

$$\begin{aligned}
&\text{for } j \neq l,\\
&\partial_{\bar{\eta}_l}\left\{(\bar{\eta}_j^*)^{2-2\bar{k}_j}\right\} = (2-2\bar{k}_j)\frac{-\bar{\varepsilon}_{jl}}{(2\bar{k}_j-1)\bar{r}_j}(\bar{\eta}_j^*)^{2-2\bar{k}_j}(\bar{\eta}_j^*)^{1-2\bar{k}_j}\\
&= \frac{(2\bar{k}_j-2)}{(2\bar{k}_j-1)}\frac{\bar{\varepsilon}_{jl}}{\bar{r}_j}(\bar{\eta}_j^*)^{3-4\bar{k}_j}
\end{aligned} \tag{12.87}$$

The second-order condition yields

$$\begin{aligned}
&\text{for } i \neq l,\ 0 < \partial^2_{\bar{\eta}_l\bar{\eta}_l} O_l = \Big\{(2\bar{k}_l-1)\bar{r}_l\bar{\eta}_l^{2\bar{k}_l-2}\\
&+\bar{\lambda}_l\textstyle\sum_{j\in\mathscr{I}\setminus\{l\}}(b_{2j}+\bar{b}_{2j})\frac{\bar{\varepsilon}_{jl}}{(2\bar{k}_j-1)\bar{r}_j}\frac{(2\bar{k}_j-2)}{(2\bar{k}_j-1)}\frac{\bar{\varepsilon}_{jl}}{\bar{r}_j}(\bar{\eta}_j^*)^{3-4\bar{k}_j}\\
&-2\textstyle\sum_{j\in\mathscr{I}\setminus\{l\}}\frac{\bar{\varepsilon}_{lj}\bar{\varepsilon}_{jl}}{(2\bar{k}_j-1)\bar{r}_j}(\bar{\eta}_j^*)^{2-2\bar{k}_j}\\
&-\textstyle\sum_{j\in\mathscr{I}\setminus\{l\}}\bar{\eta}_l\frac{\bar{\varepsilon}_{lj}\bar{\varepsilon}_{jl}}{(2\bar{k}_j-1)\bar{r}_j}\frac{(2\bar{k}_j-2)}{(2\bar{k}_j-1)}\frac{\bar{\varepsilon}_{jl}}{\bar{r}_j}(\bar{\eta}_j^*)^{3-4\bar{k}_j}\Big\}
\end{aligned} \tag{12.88}$$

Note that this condition reduces to

$$0 < \bar{r}_l - 2\sum_{j\in\mathscr{I}\setminus\{l\}}\frac{\bar{\varepsilon}_{lj}\bar{\varepsilon}_{jl}}{(2\bar{k}_j-1)\bar{r}_j}$$

when $\bar{k}_j = \bar{k}_l = 1,\ \forall j$, which is satisfied, for example, when $\bar{\varepsilon}_{lj}\bar{\varepsilon}_{jl} \leq 0,\ j \neq l.$
The $\bar{\lambda}_l$ is then modified to be

$$\begin{aligned}
&-\tfrac{d}{dt}\bar{\lambda}_l = \bar{q}_l + 2\bar{k}_l(b_1+\bar{b}_1)\bar{\lambda}_i + \textstyle\sum_{s'}[\bar{\lambda}_l(t,s') - \bar{\lambda}_l(t,s)]\tilde{q}_{ss'}\\
&+2\bar{k}_l O_l(\bar{\eta}_1^*(\bar{\eta}_l^*),\ldots,\bar{\eta}_{l-1}^*(\bar{\eta}_l^*),\bar{\eta}_l^*,\bar{\eta}_{l+1}^*(\bar{\eta}_l^*),\ldots,\bar{\eta}_I^*(\bar{\eta}_l^*)),\\
&\bar{\lambda}_l(T,s) = \bar{q}_l(T,s).
\end{aligned} \tag{12.89}$$

This completes the proof of Proposition 12.4. ■

Proof of Proposition 12.5 We now examine the problem formulated in (12.48) and prove Proposition 12.5. It follows similar steps as in the proof of Proposition 12.4; the only difference is that there are several leaders and therefore there will be a Nash game between leaders. By splitting the set $\mathscr{I}$, we arrive at the desired system provided in Proposition 12.5. ■

Proof of Proposition 12.6 To solve the Problem formulated in (12.49) and prove Proposition 12.6, we apply recursively the method layer by layer as in Proposition 12.5. ■

We studied the effect of noises in a class of multilayer MFTGs. The analysis here is limited to linear state-and-mean-field-type feedback strategies. This restriction has simplified the analysis of the Stackelberg solution. It would be interesting to investigate the effect of noises under more general strategies. In discrete time, MFTG framework can be tacked under more general noises including non-Gaussianity as the telescopic sum can be used. In continuous time however, it requires an integration formula for the associated stochastic process. The stochastic integration formula differs from Brownian motion to Gauss-Volterra to Rosenblatt process. Each new (nonstandard) noise would require another integration formula making the analysis more involved. It would be interesting to tackle the continuous time MFTGs with state driven by general square integrable noises. It would be interesting to investigate also the inverse or reverse Stackelberg solution by modifying the methodology developed here with $\bar{\eta}^{iss} = (\bar{\eta}_1^{iss}, \ldots, \bar{\eta}_I^{iss})$ where

$$\begin{aligned}
\bar{\eta}_I^{iss} &= \bar{\eta}_I^*, \\
\bar{\eta}_{I-1}^{iss} &= \bar{\eta}_{I-1}^*(\bar{\eta}_I^*), \\
&\ldots, \\
\bar{\eta}_i^{iss} &= \bar{\eta}_i^*(\bar{\eta}_{i+1}^*, \bar{\eta}_2^*, \ldots, \bar{\eta}_I^*), \\
&\ldots, \\
\bar{\eta}_2^{iss} &= \bar{\eta}_2^*(\bar{\eta}_3^*, \ldots, \bar{\eta}_I^*), \\
\bar{\eta}_1^{iss} &= \bar{\eta}_1^*(\bar{\eta}_2^*, \ldots, \bar{\eta}_I^*).
\end{aligned} \tag{12.90}$$

12.6 Long-Run Behavior Under Rosenblatt Processes

In this section, we examine the average cost when the horizon length gets large. The proofs of the results presented in this section can be derived from the previous section.

12.6.1 Control of Linear Systems Driven by Rosenblatt Noise

Optimal control problems for systems driven by Rosenblatt noise are a new area of research with significant implications for both theoretical advancements and practical applications. Traditional control theory often relies on Gaussian, Poissonian, and Markovian noise assumptions, which fail to capture the many shapes observed in real-world data. Empirical evidence from various domains, such as e-commerce, power grid operations, agricultural supply chains, and hardware demand cycles, consistently demonstrates non-Gaussian characteristics, including skewed, heavy tails, multimodal behaviors, and strong long-range dependencies. The Rosenblatt process can capture some of these higher-order dependencies and memory effects. By integrating Rosenblatt noise into control theory, we aim to obtain more accurate models for uncertainty propagation and feedback control, which are crucial for systems characterized by extreme events and noncentral limit fluctuations. This approach not only enhances predictive accuracy but also improves control performance in systems where traditional Gaussian models fall short.

We start with the infinite horizon stochastic optimal control with long-time average cost and linear state dynamics driven by Rosenblatt noise process.

$$
\begin{aligned}
&L_\infty(u) = \limsup_{T\to\infty} \tfrac{1}{T} E \int_0^T (qx^2(t) + ru^2(t))dt,\\
&x(t) = x_0 + \int_0^t (b_1 x(t') + b_2 u(t'))dt' + r^H(t),\\
&(b_1, b_2, x_0) \in \mathbb{R}^3, b_2 \neq 0,\ q > 0, r > 0, H \in (\tfrac{1}{2}, 1),\\
&\mathcal{U} := \{(u(t))_{t\geq 0} : u(t) = Kx(t) \text{ with } K \in \mathbb{R}\}
\end{aligned}
$$

The control of the ergodic cost under Rosenblatt is given by $\inf_{u\in\mathcal{U}} L_\infty(u)$ subject to the state dynamics driven by Rosenblatt noise.

We use a direct method to find linear state-feedback optimal control.

Proposition 12.7 *The optimal gain in the family of admissible feedbacks is given by*

$$
\hat{K} = -\frac{b_1 + \sqrt{b_1^2 + 4H(1-H)\frac{b_2^2 q}{r}}}{2b_2(1-H)},
$$

and the optimal cost is given by

$$
L_\infty(\hat{K}) = \frac{\Gamma(2H)}{[-(b_1+b_2\hat{K})]^{2H-1}}\left(-\frac{r\hat{K}}{b_2}\right) = \frac{\Gamma(2H+1)}{2[-(b_1+b_2\hat{K})]^{2H}}\left(q + r\hat{K}^2\right).
$$

Proposition 12.7 establishes the optimal gain for a linear-quadratic control problem driven by a Rosenblatt process by an explicit representation from a variation-of-constants formula. It is combined with a stochastic calculus formula for the Rosenblatt process and is applied to the squared state variable. By taking expectations and analyzing the asymptotic behavior of the integral terms, the proof derives an integral equation for the expected value of the squared state variable in the long run. The optimal gain is found by minimizing the cost functional, leading to a quadratic equation whose solution provides the optimal gain and cost.

Proof The state equation has a solution that can be explicitly written using a variation-of-constants formula or integration by parts, that is,

$$
x(t) = e^{(b_1+b_2K)t}x_0 + \int_0^t e^{(b_1+b_2K)(t-t')}dr^H(t').
$$

Using this explicit form of this solution, we set for $y_t = X_t$ and $f(t,x) = x^2$, and we use the stochastic calculus formula to obtain

$$
\begin{aligned}
x^2(T) = x_0^2 + \int_0^T &\left[2(b_1 + b_2K)x^2(t') + 2c\nabla^{\frac{H}{2},\frac{H}{2}} x_{t'}(t',t')\right]dt'\\
&+2\tilde{c}\int_0^T 2\nabla^{\frac{H}{2}} x(t')dB^{\frac{H}{2}+\frac{1}{2}}(t') + \int_0^T 2x(t')dr^H(t'),
\end{aligned}
$$

where $T > 0$. By taking the expectation of both sides of the above equation, it follows that

$$Ex(T)^2 = x_0^2 + 2(b_1 + b_2K)E\int_0^T x(t)^2dt + 2cE\int_0^T \nabla^{\frac{H}{2},\frac{H}{2}} x_{t'}(t',t')dt',$$

because the stochastic integrals have zero expectation. The second-order fractional stochastic derivative can be computed as follows. Denote

$$o_t(t') \triangleq \nabla^{\frac{H}{2},\frac{H}{2}} x_t(t',t').$$

Applying the operator $\nabla^{\frac{H}{2},\frac{H}{2}}$ to equation $x(t) = x_0 + \int_0^t (b_1x(t') + b_2u(t'))dt' + r^H(t)$, , it follows that $o_t(u)$ satisfies the nonhomogeneous linear differential equation

$$o_t(t') = (b_1 + b_2K)\int_0^t o_s(t')ds + \nabla^{\frac{H}{2},\frac{H}{2}} R_t(t',t')$$

for almost every $t' \in \mathbb{R}$. We use

$$\nabla^{\frac{H}{2},\frac{H}{2}} r_t^H(t',t') = \tilde{C}_H\int_0^t |t'-r|^{2H-2}dr,$$

where the constant is given by

$$\tilde{C}_H \triangleq 2c(\delta)\frac{B^2\left(\frac{H}{2},1-\frac{H}{2}\right)}{\Gamma^2(\frac{H}{2})}$$

By solving the integral equation, it follows that

$$o_t(t') = \tilde{C}_H\int_0^t e^{(b_1+b_2K)(t-r)}|t'-r|^{2H-2}dr,$$

which by a change of variables yields

$$o_t(t) = \nabla^{\frac{H}{2},\frac{H}{2}} X_t(t,t) = \tilde{C}_H\int_0^t e^{(b_1+b_2K)r}r^{2H-2}dr.$$

By substituting this last expression into the equation, simplifying the constants, and dividing both sides by T, it follows that

$$\frac{1}{T}Ex(T)^2 = \frac{1}{T}x_0^2 + 2(b_1+b_2K)\frac{1}{T}E\int_0^T x(t')^2dt' + 2H(2H-1)\frac{1}{T}$$
$$\int_0^T\int_0^t e^{(b_1+b_2K)r}r^{2H-2}drdt.$$

The convergence of the separate terms in the above equation is now studied. Without loss of generality, it can be assumed that $b_1 + b_2 K < 0$ because, otherwise, the cost $L_\infty(K)$ would be infinite. Consequently, by the definition of the Gamma function, there is the convergence

$$\int_0^t e^{(b_1+b_2K)r} r^{2H-2} dr \to \frac{\Gamma(2H-1)}{[-(b_1+b_2K)]^{2H-1}}$$

which assures that also the average

$$\frac{1}{T}\int_0^T \int_0^s e^{(b_1+b_2K)r} r^{2H-2} drds$$

converges to the same constant as $T \to \infty$. This convergence can also be used to show that

$$\lim_{T\to\infty} \frac{1}{T} Ex(T)^2 = 0$$

because $Ex(T)^2$ can be computed explicitly using formula. Hence, by letting $T \to \infty$ in the equation and using the definition of the cost functional, it follows that

$$0 = \frac{2(b_1+b_2K)}{q+rK^2} L_\infty(K) + \frac{\Gamma(2H+1)}{[-(b_1+b_2K)]^{2H-1}}$$

from which $L_\infty(K)$ can be isolated:

$$L_\infty(K) = \frac{(2H+1)(q+rK^2)}{2[-(b_1+b_2K)]^{2H}}.$$

Minimizing the above formula, it follows that the optimal gain $\hat{K}$ must satisfy the quadratic equation

$$(1-H)b\hat{K}^2 + b_1\hat{K} - H\frac{b_2 q}{r} = 0$$

from which the formula follows. Moreover, note that

$$\frac{H(q+r\hat{K}^2)}{-(b_1+b_2\hat{K})} = -\frac{r}{b_2}\hat{K}$$

which, together with the equation, yields the optimal cost. This completes the proof. ■

We now link the optimal gain to the positive solution of an algebraic Riccati equation, connecting the control problem to classical control theory results.

It follows from the quadratic equation $(1-H)b_2\hat{K}^2 + b_1\hat{K} - H\frac{b_2 q}{r} = 0$ that if P is the positive solution to the following algebraic Riccati equation

$$(1-H)\frac{b_2^2}{r}P^2 + b_1 P - Hq = 0,$$

then the optimal gain can be written in the form

$$\hat{K} = \frac{b_2}{r}P.$$

What If We Replace the Rosenblatt Noise by a Standard One and Optimize?

$$\begin{aligned}
&\inf_{u\in\mathcal{U}} \limsup_{T\to\infty} \tfrac{1}{T} E\int_0^T (qx^2(t) + ru^2(t))dt,\\
&x(t) = x_0 + \int_0^t (b_1 x(t') + b_2 u(t'))dt' + \cancel{r^H(t)}\\
&\qquad\qquad\qquad\qquad\qquad\qquad\qquad \approx B(t)\\
&\qquad\qquad\qquad\qquad\qquad\qquad\qquad \approx B^H(t)\\
&\qquad\qquad\qquad\qquad\qquad\qquad\qquad \approx \tilde{N}(t,.)\\
&\qquad\qquad\qquad\qquad\qquad\qquad\qquad \approx 0.
\end{aligned}$$

This operation leads to suboptimality. Approximating the Rosenblatt noise by another one such as Brownian motion, fractional Brownian, Gauss-Volterra noise, compensated Poisson jump process leads to suboptimal control. Figures 12.16, 12.17, and 12.18 illustrate failures of noise approximation-optimization cascade.

12.6.2 Variance-Aware Control Driven by Rosenblatt Noise

We examine the infinite horizon variance-aware optimal control with long-time average cost and linear state dynamics driven by Rosenblatt noise process.

$$\begin{aligned}
&L_\infty(u) = \limsup_{T\to\infty} \tfrac{1}{T}\int_0^T (q\ var(x(t)) + r\ var(u(t)) + \bar{q}\bar{x}^2(t) + \bar{r}\bar{u}^2(t))dt,\\
&x(t) = x_0 + \int_0^t (b_1 x(t') + b_2 u(t'))dt' + r^H(t) + \int_0^t (\bar{b}_0 + \bar{b}_1\bar{x}(t') + \bar{b}_2\bar{u}(t'))dt',\\
&(b_1, b_2, x_0) \in \mathbb{R}^3, b_2 \neq 0,\ q, \bar{q} > 0, r, \bar{r} > 0, H \in (\tfrac{1}{2}, 1),\\
&(\bar{b}_0, \bar{b}_1, \bar{b}_2) \in \mathbb{R}^3, \bar{b}_1 \neq 0\\
&\mathcal{U} := \big\{(u(t))_{t\geq 0} : u(t) = K(x(t) - \bar{x}(t)) + \bar{K}\bar{x}(t) \text{ with } (K, \bar{K}) \in \mathbb{R}^2,\\
&\qquad b_1 + Kb_2 < 0, b_1 + \bar{b}_1 + \bar{K}(b_2 + \bar{b}_2) < 0\big\}
\end{aligned}$$

The variance-aware control of the ergodic cost under Rosenblatt is given by $\inf_{u\in\mathcal{U}} L_\infty(u)$ subject to the McKean-Vlasov linear state dynamics driven by Rosenblatt noise.

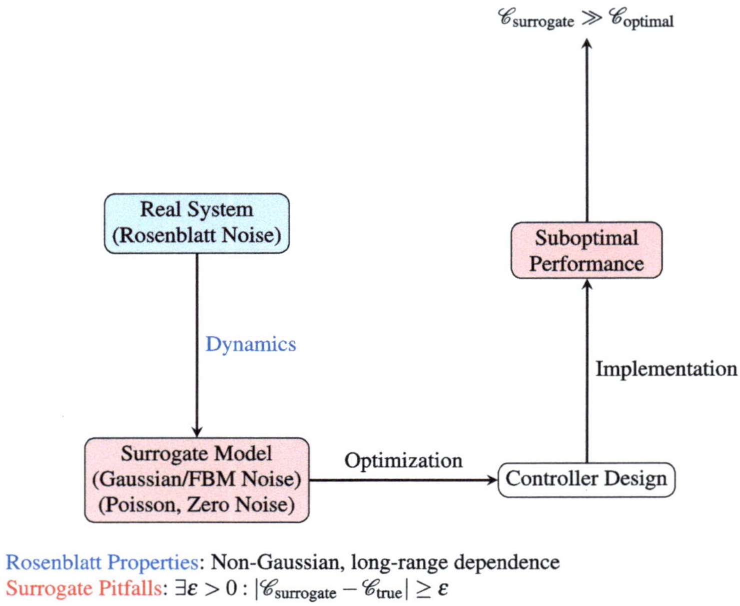

Fig. 12.16 Noise approximation-optimization cascade leads to suboptimality

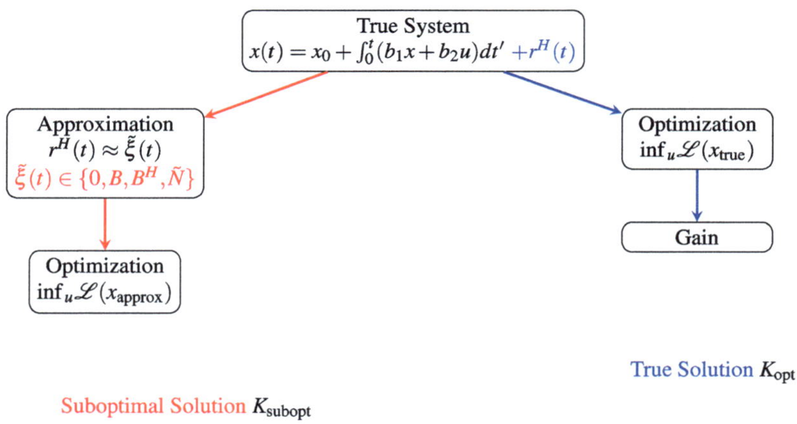

Fig. 12.17 Replacing Rosenblatt noise by another noise approximation-optimization cascade leads to suboptimality

We use a direct method to find linear state-and-mean-field-type feedback optimal control.

Proposition 12.8 *The optimal gains in the family of admissible state-mean-field-type feedbacks are given by*

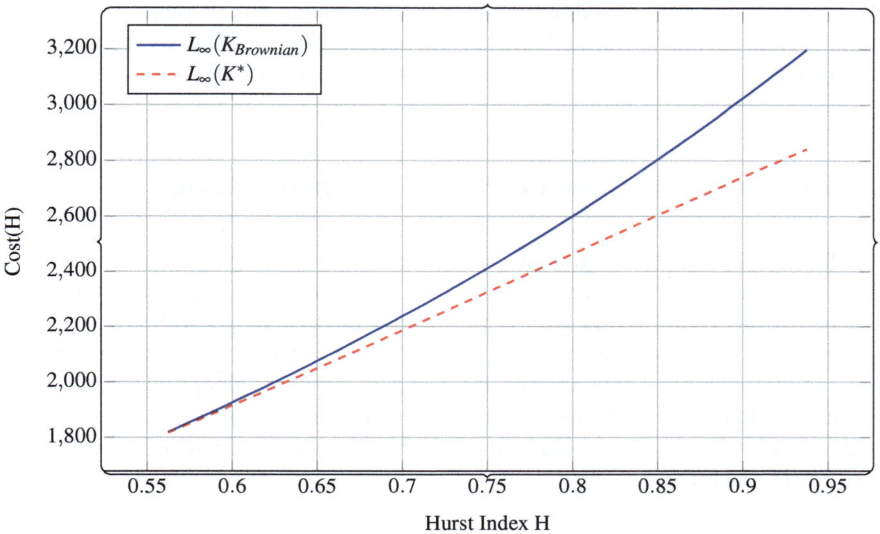

Fig. 12.18 The cost of simplification grows without bound. The price simplicity is infinite

$$K^* = -\frac{b_1 + \sqrt{b_1^2 + 4H(1-H)\frac{b_2^2 q}{r}}}{2b_2(1-H)},$$

$$\bar{K}^* = \frac{(b_2+\bar{b}_2)\bar{q}}{\bar{r}(b_1+\bar{b}_1)},\ b_1+\bar{b}_1 \neq 0,$$

and the optimal cost is given by

$$L_\infty(K^*, \bar{K}^*) = \frac{\Gamma(2H)}{[-(b_1+b_2\hat{K})]^{2H-1}}\left(-\frac{r\hat{K}}{b_2}\right) + \frac{\bar{b}_0^2\bar{q}\bar{r}}{(\bar{b}_1+\bar{b}_1)^2\bar{r}+(b_2+\bar{b}_2)^2\bar{q}}$$

What If We Replace the Rosenblatt by a Standard One and Minimize the Variance?

$$\begin{aligned}
&\inf_{u\in\mathcal{U}} \limsup_{T\to\infty} \tfrac{1}{T}\int_0^T (q\ var(x(t)) + r\ var(u(t)) + \bar{q}\bar{x}^2(t) + \bar{r}\bar{u}^2(t))dt,\\
&x(t) = x_0 - \bar{x}_0 + \int_0^t (b_1(x(t') - \bar{x}(t')) + b_2(u(t') - \bar{u}(t')))dt'\\
&+\bar{x}_0 + \int_0^t (\bar{b}_0 + (\bar{b}_1+\bar{b}_1)\bar{x}(t') + (b_2+\bar{b}_2)\bar{u}(t'))dt' + r^H(t),\\
&\qquad\qquad\qquad\qquad\qquad\qquad\qquad\qquad \approx B(t)\\
&\qquad\qquad\qquad\qquad\qquad\qquad\qquad\qquad \approx B^H(t)\\
&\qquad\qquad\qquad\qquad\qquad\qquad\qquad\qquad \approx \tilde{N}(t,.)\\
&\qquad\qquad\qquad\qquad\qquad\qquad\qquad\qquad \approx 0
\end{aligned}$$

This operation leads to suboptimality. Approximating the Rosenblatt noise by another one such as Brownian motion, fractional Brownian, Gauss-Volterra noise, compensated Poisson jump process leads to suboptimal control.

12.6.3 Zero-Sum Games Driven by Rosenblatt Noise

We formulate and solve zero-sum games where the state dynamics are driven by Rosenblatt noise. These games are particularly relevant in adversarial settings, where two decision-makers with opposing objectives interact within a stochastic environment. The key result is the establishment of saddle-point equilibria, providing conditions under which each decision-maker can adopt an optimal strategy that minimizes their maximum possible loss. This is a crucial development, as it extends classical game theory to more accurately reflect the stochastic nature of real-world adversarial interactions.

Let

$$
\begin{aligned}
L_T(u, v) &= \tfrac{1}{T} E\left[\int_0^T \left(qx^2(t) + ru^2(t) - sv^2(t)\right) dt\right],\\
L_{1,2,\infty}(u, v) &= \liminf_{u, T\to\infty} \limsup_{v, T\to\infty} L_T(u, v)\\
L_{2,1,\infty}(u, v) &= \limsup_{v, T\to\infty} \liminf_{u, T\to\infty} L_T(u, v)
\end{aligned}
$$

$$
x(t) = x_0 + \int_0^t [b_1 x(t') + b_2 u(t') + b_3 v(t')]dt' + r^H(t),
$$

$$
(b_1, b_2, b_3, x_0) \in \mathbb{R}^4,\ b_2, b_3 \neq 0,\ q, r, s > 0, H \in (\tfrac{1}{2}, 1),
$$

$$
\begin{aligned}
&\limsup_{T\to\infty} \tfrac{1}{T} E\left[\int_0^T \left(qx^2(t) + ru^2(t) + sv^2(t)\right) dt\right] < \infty\\
&s^2 b_1^2 b_2^2 \geq 4\left(-\tfrac{b_2^3 s^2}{b_3 r} + b_2 s b_3 + \tfrac{H b_2^3 s^2}{b_3^2} - H b_2 s\right)(s b_1 b_2 + H b_2 q)\\
&\mathcal{U} := \{(u(t))_{t\geq 0} : u_t = Kx(t) \text{ with } K \in \mathbb{R}\}\\
&\mathcal{V} := \{(v(t))_{t\geq 0} : v(t) = Lx(t) \text{ with } L \in \mathbb{R}\}
\end{aligned}
$$

Zero-sum game with ergodic cost under Rosenblatt noise is $\inf_{u\in\mathcal{U}} \sup_{v\in\mathcal{V}} L_\infty(u, v)$ subject to the state dynamics driven by Rosenblatt noise.

Proposition 12.9 *MiniMax=Maximin: The zero-sum game under Rosenblatt has a value, and each decision-maker has an optimal strategy.*

Proposition 12.10 *Explicit Solution of MiniMax: The saddle-point gains in the family of admissible linear state feedbacks are given by*

$$
K = -\frac{b_2 s}{b_3 r} L, \quad (-\frac{b_2^3 s^2}{b_3 r} + b_2 s b_3 + \frac{H b_2^3 s^2}{b_3^2} - H b_2 s)L^2 + s b_1 b_2 L + H b_2 q = 0,
$$

and the equilibrium cost (value) is given by $L_\infty(K, L) = \frac{\Gamma(2H+1)(q+rK^2-sL^2)}{2[-(b_1+b_2K+b_3L)]^{2H}}$.

What If We Replace the Rosenblatt by a Standard One and Find Saddle Point

$$
\begin{aligned}
&\inf_u \sup_v \limsup_{T\to\infty} \\
&\tfrac{1}{T} E\left[\int_0^T \left(qx^2(t) + qu^2(t) - sv^2(t)\right) dt\right], \\
&x(t) = x_0 + \int_0^t (b_1 x(t') + b_2 u(t') + b_3 v(t'))dt' + r^H(t) \\
&\qquad\qquad\qquad\qquad\qquad\qquad\qquad\qquad \approx B(t) \\
&\qquad\qquad\qquad\qquad\qquad\qquad\qquad\qquad \approx B^H(t) \\
&\qquad\qquad\qquad\qquad\qquad\qquad\qquad\qquad \approx \tilde{N}(t) \\
&\qquad\qquad\qquad\qquad\qquad\qquad\qquad\qquad \approx 0.
\end{aligned}
$$

This operation leads to nonequilibrium. Approximating the Rosenblatt noise by another one such as Brownian motion, fractional Brownian, and Gauss-Volterra noise compensated Poisson jump process does not provide the saddle point.

12.6.4 Nonzero-Sum Games Driven by Rosenblatt Noise

We now consider nonzero-sum games under Rosenblatt noise where the interests of the decision-makers are not strictly opposed but rather involve a mix of cooperation and competition. Unlike zero-sum games, where one decision-maker's gain is exactly balanced by the other's loss, nonzero-sum games allow for the possibility of mutual benefit or shared losses. This is crucial, because it extends the analysis of game theory to more realistic scenarios where decisoin-makers' objectives are interdependent and influenced by non-Gaussian noise.

$$
\begin{aligned}
&L_\infty(u) = \limsup_{T\to\infty} \tfrac{1}{T} E \int_0^T (q_i x^2(t) + r_i u_i^2(t))dt, \\
&x(t) = x_0 + \int_0^t (b_1 x(t') + \textstyle\sum_{j\in\mathcal{J}} b_{2j} u_j(t'))dt' + r^H(t), \\
&(b_1, b_{2j}, x_0) \in \mathbb{R}^3, b_{2j} \neq 0,\ q_j > 0, r_j > 0,\ H \in (\tfrac{1}{2}, 1), \\
&\mathcal{U}_i := \{(u_i(t))_{t\geq 0} : u_i(t) = K_i x(t) \text{ with } K_i \in \mathbb{R}, (b_1 + \textstyle\sum_{j\in\mathcal{J}} b_{2j} K_j) < 0\}
\end{aligned}
$$

The control of the ergodic cost under Rosenblatt is given by $\inf_{u_i \in \mathcal{U}_i} L_{i,\infty}(u)$ subject to the state dynamics driven by Rosenblatt noise.

We use a direct method to find linear state-feedback best-response control strategy of agent i.

Proposition 12.11 *Given the admissible strategies of the other agents* $(u_j \in \mathcal{U}_j,\ j \neq i)$, *the best-response gain of agent i in the family of admissible feedbacks* *is given by* $K_i^* = -\frac{a+\sqrt{a^2+4H(1-H)\frac{b_{2i}^2 q_i}{r_i}}}{2b_{2i}(1-H)}$, *with* $a := b_1 + \sum_{j\in\mathscr{I}\backslash\{i\}} b_{2j}K_j$, *and the best-response cost is given by* $L_{i,\infty}(K^*) = \frac{\Gamma(2H)}{[-(b_1+\sum_{j\in\mathscr{I}} b_{2j}K_j)]^{2H-1}}\left(-\frac{r_i K_i^*}{b_{2i}}\right)$

Proposition 12.12 *If there exists a solution to the following system:*

$$\begin{aligned} & i \in \mathscr{I}, \\ & K_i = -\frac{b_1+\sum_{j\in\mathscr{I}\backslash\{i\}} b_{2j}K_j + \sqrt{(b_1+\sum_{j\in\mathscr{I}\backslash\{i\}} b_{2j}K_j)^2+4H(1-H)\frac{b_{2i}^2 q_i}{r_i}}}{2b_{2i}(1-H)}, \\ & (b_1 + \textstyle\sum_{j\in\mathscr{I}} b_{2j}K_j) < 0, \end{aligned}$$

then, the nonzero-sum stochastic differential game under Rosenblatt has an equilibrium in stationary linear state-feedback strategies.

Note that this result is restricted to the class of linear state-feedback strategies.

What If We Replace the Rosenblatt by a Standard Brownian Motion and Solve the Noncooperative Game?

$$\begin{aligned} & \inf_{u\in\mathscr{U}} \limsup_{T\to\infty} \tfrac{1}{T} E \int_0^T (q_i x^2(t) + r_i u_i^2(t))dt, \\ & x(t) = x_0 + \int_0^t (b_1 x(t') + \textstyle\sum_{j\in\mathscr{I}} b_{2j}u_j(t'))dt' + r^H(t), \\ & \qquad \approx B(t) \\ & \qquad \approx B^H(t) \\ & \qquad \approx \tilde{N}(t,.) \\ & \qquad \approx 0 \end{aligned}$$

This operation leads to nonequilibrium. Approximating the Rosenblatt noise by another one such as Brownian motion, fractional Brownian, and Gauss-Volterra noise, compensated Poisson jump process no longer provides an equilibrium strategy.

Notes

Empirical evidence from various domains, including water demand, e-commerce, power grid operations, wireless channels, and agricultural supply chains, demonstrates the prevalence of non-Gaussian characteristics such as skews, heavy tails, and strong long-range dependencies. The Rosenblatt process, a non-Gaussian non-Markovian, self-similar process, offers a baseline framework for capturing some of the behaviors observed in real data. MFTGs driven by non-Gaussian, non-Poisson, and non-Markov noise have been examined in [136–139].

Chapter 13
Mean-Field-Type Games with Asymmetric Information

In the study of the mean-field-type game (MFTG) theory, understanding how agents interact and make choices in asymmetric information environments has always been a central challenge. Mean-field-type games with asymmetric information explore scenarios where individual decision-makers operate under conditions of partial observation and varying levels of information. This chapter presents MFTGs in which the decision-makers have different information structures and different perceptions of the state process. The observations of the decision-makers are correlated. The payoff functionals involve couplings between the controls of the decision-makers and a covariance between the state and the control action of the decision-maker. The state dynamics is driven by a Gauss-Volterra jump-diffusion process, which includes fractional Brownian motions and multi-fractional Brownian motions. The state dynamics are of a conditional McKean-Vlasov type as its coefficients depend on the expected value of the state and the expected value of the control actions conditioned on the switching regime.

When decision-makers have different perceptions about the state, a higher-order conditioning occurs. The conditioning of the other decision-makers' conditioning plays a crucial role in determining outcomes of the dynamic game. For consistent filters, the decision-makers must have filters that others will coordinate with them. The question we raise is whether decision-makers consider these higher-order conditioning (the conditionings of the conditional expectations with respect to individual observations) when making decisions in the dynamic MFTG.

We identify situations when infinite regress occurs as well as when it does not. Several semi-explicitly solvable examples are provided by exploiting scenarios

T. Başar et al., *Mean-Field-Type Game Theory I*, Static & Dynamic Game Theory: Foundations & Applications, https://doi.org/10.1007/978-3-032-07027-2_13

where infinite regress of beliefs/estimates can be avoided by finding a basis-preserving functional.

13.1 Partial Observation of the State Under Correlated Noise

Consider $I \geq 2$ decision-makers interacting in a dynamic environment with a state $x \in \mathbb{R}^d$, $d \geq 1$ and different switching regimes represented by s. The complete scope of the state x is unobservable by a single decision-maker. Decision-maker i observes its own version of x which is y_i. The signal-observation process is therefore $(x, s, (y_i)_{i\in\mathcal{I}})$:

$$\begin{aligned} & dx = bdt + \sigma dB + \int_\theta \mu d\tilde{N} + \sigma_{gv} dB_{gv} + \sum_{j\in\mathcal{I}} \sigma_{o,j} dB_{o,j}, \\ & x(0) = x_0, \\ & dy_i = [h_{1i}x + \bar{h}_{1i}\bar{x} + \sum_l \bar{h}_{2il}\bar{u}_l]dt + dB_{o,i}, \ \ i \in \mathcal{I}, \\ & y_i(0) = 0, \end{aligned} \tag{13.1}$$

where the coefficients are given by

$$b = b_1 x + \bar{b}_1 \bar{x} + \sum_j [b_{2j} u_j + \bar{b}_{2j} \bar{u}_j],$$

where $b_k(t, s), \bar{b}_k(t, s)$ are matrices and $\phi = \bar{\phi}_1\bar{x} + \sum_{i\in\mathcal{I}} \bar{\phi}_{2i}\bar{u}_i$, with $\phi \in \{\sigma, \sigma_{gv}\}$, $\phi_k = \phi_k(t, s) \in \mathbb{R}^{d\times d}$ depends on time and regime switching, $\mu_k = \mu_k(t, s, \theta)$, where θ is the jump size, $\bar{x} = E[x|\mathcal{F}^s]$, and $\sigma_{o,j}(t)$ is state-independent.

The noises in (13.1) are described as follows:

- s is a regime switching process over a finite set $\mathcal{S} = \{1, 2, \ldots, S\}$ with transition rate $q_{ss'} > 0$, $s' \neq s$, and $q_{ss} := -\sum_{s'\neq s} q_{ss'}$.
- B is a Brownian motion, and σdB is of compatible dimension with x.
- a jump process denoted by N and compensated jump process denoted by $\tilde{N}(dt, d\theta) = N(dt, d\theta) - \nu(d\theta)dt$, where ν is a sigma-finite measure over the set of jump sizes Θ.
- B_{gv} denotes a Gauss-Volterra process. The Gauss-Volterra processes are singular integrals of a standard Brownian motion and include (i) fractional Brownian motions, (ii) Liouville fractional Brownian motions, and (iii) multi-fractional Brownian motions. The difficulty of finding a semi-explicit solution is significantly increased if the noise process, and thereby the state process, is driven by non-Markov processes or non-martingales. The Gauss-Volterra process B_{gv} has zero mean and covariance given by

$$c(t, t') = E[B_{gv}(t)B_{gv}(t')] = \int_0^{\min(t,t')} K(t, t'')K(t', t'')dt'', \tag{13.2}$$

where K is a suitable integrable kernel. The variance of the process $\int_0^t \sigma_{gv}(t', s)dB_{gv}(t')$ is given by

$$\sigma^2_{cogv} = \frac{d}{dt}\left[\int_0^t \left\{K(t'_+, t')\sigma_{gv}(t') + \int_{t'}^t \sigma_{gv}(t'')K(t', t'')dt''\right\}^2 dt'\right].$$

The observation process of decision-maker i is (y_i, s). The observation coefficient functions $h_{1i}(t), \bar{h}_{1i}(t), \bar{h}_{2il}(t) \in \mathbb{R}^{d\times d}$. Based on its own partially observed state y_i and switching regime, decision-maker i wants to optimize its long-term payoff functional $E[L_i(x, u, s)]$. An admissible pure strategy u_i of decision-maker i is a $\mathscr{F}^{y_i,s}$-adapted process with values in $\mathbb{R}^d$. We identify a structure in which there is no infinite regress in the context of MFTGs. The matrix $\bar{b}_2(t, s)$ is not necessarily diagonal. Therefore, there is still a coupling between the decision-makers' conditional control actions via $\{E[u_i|\mathscr{F}^s]\}_i$. We will show below that with the control-coupling through the terms $\{E[u_i|\mathscr{F}^s]\}_i$, there is no infinite regress.

13.2 Mean-Field-Type Filter

One important element of the conditional MFTG is to determine and characterize the mean-field-type filters of decision-maker i, i.e., the distribution of state given the observations, the conditional expectation of the state, and the conditional expectation of the conditional expectation of other decision-makers' observations.

For the mean-field-type filter $\hat{x}$, we follow the stochastic filtering theory procedure by inserting the expression of the innovation process in the dynamics of x and obtain

$$\begin{aligned} d\hat{x} &= \{\hat{b} - \textstyle\sum_{j\in\mathscr{I}} \sigma_{o,j}[h_{1j}\hat{x} + \bar{h}_{1j}\bar{x} + \sum_l \bar{h}_{2jl}\bar{u}_l]\}dt \\ &+\sigma dB + \textstyle\int_\theta \mu d\tilde{N} + \sigma_{gv}dB_{gv} + \sum_{j\in\mathscr{I}} \sigma_{o,j}dy_j, \\ \hat{x}(0) &= \hat{x}_0. \end{aligned} \tag{13.3}$$

In this setting, decision-makers have different perceptions of the same state. The matrices $b_1, h_1, \sigma_o, \bar{b}_1, \bar{b}_2, \bar{h}_1, \bar{h}_2, \sigma, \mu, \sigma_{gv}$ are not necessarily diagonal. The state equation becomes

$$\begin{aligned} d\hat{x} = &\big\{b_1\hat{x} + \bar{b}_1\bar{x} + \textstyle\sum_k b_{2k}u_k + \sum_k \bar{b}_{2k}\bar{u}_k \\ &-[\textstyle\sum_{m\in\mathscr{I}} \sigma_{om}h_{1m}]\hat{x} - [\sum_{m\in\mathscr{I}} \sigma_{om}\bar{h}_{1m}]\bar{x} \\ &- \textstyle\sum_j(\sum_{m\in\mathscr{I}} \sigma_{om}\bar{h}_{2mj})\bar{u}_j \big\}dt \\ &+\sigma dB + \textstyle\int_\theta \mu d\tilde{N} + \sigma_{gv}dB_{gv} + \sum_{j\in\mathscr{I}} \sigma_{o,j}dy_j. \end{aligned} \tag{13.4}$$

The basic state-feedback reasoning of this linear estimation will require the vector obtained by concatenating the following elements:

$$(\bar{x}, \hat{x}, (\hat{x}_{|i}, (\hat{x}_{|ji})_{j\in\mathcal{I}\backslash\{i\}}, \quad i \in \mathcal{I})).$$

We introduce the following notation:

$$\begin{array}{l} \hat{x} = E[x|\mathcal{F}^{y,s}], \hat{x}_{|j} = E[x|\mathcal{F}^{y_j,s}], \\ \hat{x}_{|ji} = E[\hat{x}_{|j}|\mathcal{F}^{y_i}], \hat{u}_{|ji} = E[\hat{u}_{|j}|\mathcal{F}^{y_i}]. \end{array} \tag{13.5}$$

Next, we discuss how to use these filters in a class of MFTGs.

13.3 Risk-Neutral Decision-Makers with Asymmetric Information

The information structure of decision-maker i is $\mathcal{F}^{y_i,s}$. The set of admissible strategies of decision-maker i is

$$\mathcal{U}_i = \{u_i \mid u_i \text{ is } \mathcal{F}^{y_i,s} - \text{adapted}, \ u_i(.) \in U_i, a.e, t \in [0, T], \ \mathbb{P} - a.s\}. \tag{13.6}$$

Let the cost functional of decision-maker i be

$$\begin{array}{l} \tilde{L}_i = \langle \hat{q}_{iT}(\hat{x} - \hat{x}_{|i}), (\hat{x} - \hat{x}_{|i})\rangle + \langle (\hat{q}_{iT} + q_{iT})(\hat{x}_{|i} - \bar{x}), (\hat{x}_{|i} - \bar{x})\rangle \\ +\langle (q_{iT} + \bar{q}_{iT})\bar{x}, \bar{x}\rangle \\ + \int_0^T \langle \hat{q}_i(\hat{x} - \hat{x}_{|i}), (\hat{x} - \hat{x}_{|i})\rangle + \langle (\hat{q}_i + q_i)(\hat{x}_{|i} - \bar{x}), (\hat{x}_{|i} - \bar{x})\rangle + \langle (q_i + \bar{q}_i)\bar{x}, \bar{x}\rangle dt \\ + \int_0^T \langle r_i(u_i - \bar{u}_i), (u_i - \bar{u}_i)\rangle + \sum_{j\in\mathcal{I}} \langle \bar{r}_{ij}\bar{u}_j, \bar{u}_j\rangle dt. \end{array} \tag{13.7}$$

Decision-maker i seeks to solve the following problem:

$$\begin{cases} \inf_{u_i\in\mathcal{U}_i} E[\tilde{L}_i] \text{ subject to} \\ (13.4). \end{cases} \tag{13.8}$$

The asymmetric information game is

$$(\mathcal{I}, (U_i, (\mathcal{F}^{y_i,s}, \mathcal{U}_i, E\tilde{L}_i)_{i\in\mathcal{I}}).$$

We use the following decomposition:

$$(\hat{x} - \hat{x}_{|i}) = (\hat{x} - \bar{x}) + (\bar{x} - \hat{x}_{|i}),$$

and expand the quadratic term

$$\begin{cases} \langle \alpha_i(\hat{x} - \hat{x}_{|i}), (\hat{x} - \hat{x}_{|i})\rangle \\ = \langle \alpha_i(\hat{x} - \bar{x}), (\hat{x} - \bar{x})\rangle + \langle \alpha_i(\bar{x} - \hat{x}_{|i}), (\bar{x} - \hat{x}_{|i})\rangle \\ -\langle \alpha_i(\hat{x} - \bar{x}), (\hat{x}_{|i} - \bar{x})\rangle - \langle \alpha_i(\hat{x}_{|i} - \bar{x}), (\hat{x} - \bar{x})\rangle. \end{cases} \tag{13.9}$$

We compute the weighted conditional variance as

$$\begin{cases} E[\langle \alpha_i(\hat{x} - \hat{x}_{|i}), (\hat{x} - \hat{x}_{|i})\rangle | \mathscr{F}_i] \\ = E[\langle \alpha_i(\hat{x} - \bar{x}), (\hat{x} - \bar{x})\rangle | \mathscr{F}_i] + E[\langle \alpha_i(\hat{x}_{|i} - \bar{x}), (\hat{x}_{|i} - \bar{x})\rangle | \mathscr{F}_i]. \end{cases} \tag{13.10}$$

The cost function is then rewritten as

$$\begin{aligned} E[\tilde{L}_i] &= E\langle \hat{q}_{iT}(\hat{x} - \bar{x}), (\hat{x} - \bar{x})\rangle + E\langle (\hat{q}_{iT} + q_{iT})(\hat{x}_{|i} - \bar{x}), (\hat{x}_{|i} - \bar{x})\rangle \\ &+ E\langle (q_{iT} + \bar{q}_{iT})\bar{x}, \bar{x}\rangle \\ &+ E\int_0^T \langle \hat{q}_i(\hat{x} - \hat{x}_{|i}), (\hat{x} - \hat{x}_{|i})\rangle + \langle (\hat{q}_i + q_i)(\hat{x}_{|i} - \bar{x}), (\hat{x}_{|i} - \bar{x})\rangle \\ &+ \langle (q_i + \bar{q}_i)\bar{x}, \bar{x}\rangle dt \\ &+ E\int_0^T \langle r_i(u_i - \bar{u}_i), (u_i - \bar{u}_i)\rangle + \sum_{j\in\mathscr{I}} \langle \bar{r}_{ij}\bar{u}_j, \bar{u}_j\rangle dt. \end{aligned} \tag{13.11}$$

13.3.1 When Infinite Regress Occurs

When seeking state-and-mean-field-type feedback equilibrium strategies, the following states appear naturally:

- y_i, s are observed by decision-maker i.
- $\hat{x}_{|i}$ is computed by decision-maker i from $\hat{x}$. Since the dynamics of $\hat{x}$ involves $[\hat{x}, \bar{x}, (y_j, u_j, \bar{u}_j)_{j\in\mathscr{I}}]$, the dynamics of $\hat{x}_{|i}$ involves the filters $\hat{x}_{|i}, \bar{x}, (y_{j|i}, u_{j|i}, \bar{u}_j)_{j\in\mathscr{I}}$.
- This requires the computation of the term $(u_{j|i})_{j\in\mathscr{I}}$ in the state dynamics. If u_j involves $\hat{x}_{|j}$ as in state-feedback, then $u_{j|i}$ will bring $\hat{x}_{|ji}$. Decision-maker i is now facing the state of $\hat{x}_{|ji}$.
- From decision-maker i's cost functional perspective, Itô's formula for jump-diffusion-regime-switching and Gauss-Volterra processes, the product terms $\hat{x}_{|i}\hat{x}_{|ji}$ and $[\hat{x}\hat{x}_{|j}]_{|i}$ will appear if the cost involves the quadratic terms $\hat{x}^2$ and $\hat{x}_i^2$.
- This may require the dynamics of $\hat{x}_{|ji}$ which involves $[\hat{x}_{|ji}, \bar{x}, (y_{k|ji}, u_{k|ji}, \bar{u}_k)_{k\in\mathscr{I}}]$. The iteration of the procedure leads to $\hat{x}_{|kji}$ and so on.

The reasoning above leads clearly to an infinite-dimensional fixed-point problem. In order to derive a finite-dimensional semi-explicit solution, we will identify below particular settings in which the infinite regress does not occur. No sophisticated fixed-point theory will be required. The solvable examples are obtained from the fact that the mean-field-type terms $\bar{x}$ and $(\bar{u}_j)_j$ are not affected by the higher-order conditioning.

13.3.2 Change of Probability Measure

We change the probability measure $dP^u = [\frac{dP^u}{dP}]dP = \rho(T)dP$ with $\frac{dP^u}{dP} = \rho(T) = \prod_{i\in\mathcal{I}} \rho_i$ with

$$\begin{array}{l} i \in \mathcal{I}, \\ d\rho_i = \rho_i \langle [h_{1i}x + \bar{h}_{1i}\bar{x} + \sum_l h_{2il}\bar{u}_l], dy_i\rangle, \\ \rho(0) = 1. \end{array} \tag{13.12}$$

Thanks to this change of measure and by the Girsanov theorem, the trend of the observation process y can be handled:

$$e^{\sum_i \left[-\int_0^t [h_{1i}x+\bar{h}_{1i}\bar{x}+\sum_l h_{2il}\bar{u}_l]^* dy_i - \frac{1}{2}\int_0^t \left| h_{1i}x+\bar{h}_{1i}\bar{x}+\sum_l h_{2il}\bar{u}_l \right|^2 dt' \right]},$$

and the observation process becomes a $\mathbb{P}^u$-Brownian motion.

13.3.3 Reformulated Mean-Field-Type Game

The cost functional of decision-maker i is

$$\begin{array}{l} L_i := \langle \hat{q}_{iT}(\hat{x} - \bar{x}), (\hat{x} - \bar{x})\rangle + \langle (\hat{q}_{iT} + q_{iT})(\hat{x}_{|i} - \bar{x}), (\hat{x}_{|i} - \bar{x})\rangle \\ +\langle (q_{iT} + \bar{q}_{iT})\bar{x}, \bar{x}\rangle \\ + \int_0^T \langle \hat{q}_i(\hat{x} - \hat{x}_{|i}), (\hat{x} - \hat{x}_{|i})\rangle + \langle (\hat{q}_i + q_i)(\hat{x}_{|i} - \bar{x}), (\hat{x}_{|i} - \bar{x})\rangle + \langle (q_i + \bar{q}_i)\bar{x}, \bar{x}\rangle dt \\ + \int_0^T \langle r_i(u_i - \bar{u}_i), (u_i - \bar{u}_i)\rangle + \sum_{j\in\mathcal{I}} \langle \bar{r}_{ij}\bar{u}_j, \bar{u}_j\rangle dt. \end{array} \tag{13.13}$$

The best-response problem of decision-maker i is

$$\begin{array}{l} \inf_{u_i} E[L_i] \text{ subject to} \\ d\hat{x} = \left\{ [b_1 - \sum_{m\in\mathcal{I}} \sigma_{om} h_{1m}]\hat{x} + [\bar{b}_1 - \sum_{m\in\mathcal{I}} \sigma_{om}\bar{h}_{1m}]\bar{x} \right. \\ \left. + \sum_k b_{2k}u_k + \sum_j [\bar{b}_{2j} - (\sum_{m\in\mathcal{I}} \sigma_{om}\bar{h}_{2mj})]\bar{u}_j \right\} dt \\ +\sigma dB + \int_\theta \mu d\tilde{N} + \sigma_{gv} dB_{gv} + \sum_{j\in\mathcal{I}} \sigma_{o,j} dy_j, \\ \hat{x}(0) \text{ given}, \\ d\rho = \rho \sum_i \langle [h_{1i}x + \bar{h}_{1i}\bar{x} + \sum_j \bar{h}_{2ij}\bar{u}_j)], dy_i\rangle, \\ \rho(0) = 1. \end{array} \tag{13.14}$$

Sufficient conditions for optimality of the above problem can be derived from Chap. 8. In particular

- The stochastic maximum principle (of mean-field type) provides a representation of the optimal control strategies as a functional filter of the adjoint process (p_1, p_2) associated with $(\hat{x}, \rho)$:

$$u_i = -\frac{1}{2}\sum_s r_i^{-1} b_{2i} E[p_1|\mathscr{F}^{y_i,s}]\mathbb{1}_{\{s(t)=s\}}$$
$$-\frac{1}{2}\sum_s \bar{r}_{ii}^{-1}(b_{2i}+\bar{b}_{2i})E[p_1|\mathscr{F}^s]\mathbb{1}_{\{s(t)=s\}}. \quad (13.15)$$

Note that this is not an explicit representation in a conditional state-and-mean-field-type feedback form. For explicit representation, the filter $E[p_1|\mathscr{F}^{y_i,s}]$ needs to be computed.
- The dynamic programming system (of mean-field type) yields a representation in infinite dimensions.

Based on the discussion above, we choose a specific form that can be solved in a conditional state-and-mean-field-type feedback form, in finite dimensions.

13.4 Individual State Observation

Assume that b_1, h_1, σ_o are diagonal matrices, but $\bar{b}_1, \bar{b}_2, \bar{h}_1, \bar{h}_2, \sigma, \mu, \sigma_{gv}$ are not necessarily diagonal. The state equation of decision-maker i is given by

$$\begin{cases} i \in \mathscr{I}, \\ d\hat{x}_i = \Big\{b_{1ii}\hat{x}_i + b_{2ii}u_i + \sum_{j=1}\bar{b}_{1ij}\bar{x}_j + \sum_{j=1}\bar{b}_{2ij}\bar{u}_j \\ -\sigma_{o,i}[h_{1ii}\hat{x}_i + \sum_j(\bar{h}_{1ij}\bar{x}_j + \bar{h}_{2ij}\bar{u}_j)]\Big\}\, dt \\ +\sum_{j\in\mathscr{I}}\sigma_{ij}dB_j + \int_\theta \sum_j \mu_{ij}d\tilde{N}_j + \sum_j \sigma_{gv,ij}dB_{gv,j} + \sigma_{o,i}dy_i, \\ \hat{x}_i(0) \text{ given}, \\ d\rho_i = \rho_i\langle [h_{1ii}\hat{x}_i + \sum_j(\bar{h}_{1ij}\bar{x}_j + \bar{h}_{2ij}\bar{u}_j)], dy_i\rangle\,, \\ \rho(0) = 1. \end{cases} \quad (13.16)$$

In the individual state dynamics (13.16), the decision-makers are coupled through the conditional expectations of their individual states and control actions with respect to the regime switching.

The cost functional is

$$\begin{aligned} E[L_i] &= E\langle \hat{q}_{iT}(\hat{x}_i - \bar{x}_i), (\hat{x}_i - \bar{x}_i)\rangle + E\langle (\hat{q}_{iT} + q_{iT})(\hat{x}_{i|i} - \bar{x}_i), (\hat{x}_{i|i} - \bar{x}_i)\rangle \\ &+E\langle (q_{iT} + \bar{q}_{iT})\bar{x}_i, \bar{x}_i\rangle \\ &+E\int_0^T \langle \hat{q}_i(\hat{x}_i - \bar{x}_i), (\hat{x}_i - \bar{x}_i)\rangle + \langle (\hat{q}_i + q_i)(\hat{x}_{i|i} - \bar{x}_i), (\hat{x}_{i|i} - \bar{x}_i)\rangle \\ &+\langle (q_i + \bar{q}_i)\bar{x}_i, \bar{x}_i\rangle dt \\ &+E\int_0^T \langle r_i(u_i - \bar{u}_i), (u_i - \bar{u}_i)\rangle + \sum_{j\in\mathscr{I}}\langle \bar{r}_{ij}\bar{u}_j, \bar{u}_j\rangle dt. \end{aligned} \quad (13.17)$$

In the cost functional (13.17), the decision-makers are coupled through the conditional expectations of their individual control actions with respect to the regime switching.

By introducing the following augmented state $[z_i, \bar{x}]$

$$z_i = (\hat{x}_i, \hat{x}_{i|i}), \bar{x} = (\bar{x}_j)_j,$$

the dynamics of $[z_i, \bar{x}]$ can be expressed as a linear functional of augmented $[z_i - \bar{z}_i, \bar{x}]$, with $\bar{z}_i = (\bar{x}_i, \bar{x}_i)$, and augmented control actions $[u_i - \bar{u}_i, \bar{u}]$,

and the cost functional has the following structure:

$$\begin{aligned} &\langle q_i(z_i - \bar{z}_i), (z_i - \bar{z}_i)\rangle + \langle \bar{q}_i \bar{x}, \bar{x}\rangle \\ &+\langle r_i(u_i - \bar{u}_i), (u_i - \bar{u}_i)\rangle + \textstyle\sum_{j\in\mathcal{I}} \langle \bar{r}_{ij}\bar{u}_j, \bar{u}_j\rangle. \end{aligned} \tag{13.18}$$

This is now a standard LQ-MFTG with augmented state $[z_i, \bar{x}]$, $i \in \mathcal{I}$ whose solution can be obtained following a method similar to that in Sect. 7.10 with guess functional:

$$f_i(t, \hat{x}, s) = \langle \alpha_i(z_i - \bar{z}_i), (z_i - \bar{z}_i)\rangle + \langle \bar{\alpha}_i \bar{x}, \bar{x}\rangle,$$

where $\alpha_i, \bar{\alpha}_i$ are matrices of compatible dimensions. The equilibrium control strategy has the following structure:

$$-\eta_i(\hat{x}_{i|i} - \bar{x}_i) - \sum_j \bar{\eta}_{ij}\bar{x}_j,$$

where η satisfies a deterministic Riccati equation and $\bar{\eta}$ is a matrix depending on the matrix $\bar{\alpha}_i$.

This chapter has provided foundational elements of MFTGs with asymmetric information [122, 132], shedding light on the filtered dynamics of decision-making in environments characterized by partial observation and varying levels of information. We have presented MFTGs under Gauss-Volterra processes and asymmetric state information. We have examined how decision-makers filter and interpret information, the implications of asymmetric information for variance-aware decision-makers, and the role of individual state observation in shaping equilibrium outcomes.

13.5 Notes

MFTGs with asymmetric information involve the mean-field-type filters $m_i(t, .) = \mathbb{P}_{x(t)|\mathcal{F}_t^{y_i}}$, $i \in \mathcal{I}$ the conditional distribution of the state $x(t)$ given $(y_i(t'))_{t'\leq t}$ the observation process of decision-maker i up to time t. As m_i, $i \in \mathcal{I}$ are processes, decision-maker j may need to condition it on $\mathcal{F}_t^{y_j}$ at time t. This leads to $m_i(t, .)|\mathcal{F}_t^{y_j}$ and so on. We refer the reader to [121, 129, 131, 166, 168, 310].

Chapter 14
Difference Games of Mean-Field Type

This chapter covers foundations of mean-field-type games in discrete time with scalar states via illustrative variance-aware examples. We start with linear-quadratic discrete-time mean-field games in Sect. 14.1. Section 14.1.2 collects several results on discrete-time risk-neutral linear-quadratic mean-field games. In Sect. 14.1.3, we formulate risk-sensitive discrete-time game problem, carefully explaining the difference between risk-sensitive and risk-neutral Riccati equations. Section 14.1.4 is devoted to the study of robust mean-field difference games. Section 14.1.5 focuses on social optima of linear-quadratic mean-field difference games using mean-field-type control. Section 14.2 deals with linear-quadratic discrete-time mean-field-type games (LQ-DT-MFTGs). Section 14.2.1 formulates the basic LQ-DT-MFTGs with risk-neutral cost functionals. Section 14.2.3 discusses a LQ-DT-MFTGs driven by combined additive and multiplicative noises. Section 14.2.5 explains to transition from discrete to continuous time. Section 14.5 is on discrete-time adversarial MFTGs. Section 14.5.1 presents a robust LQ-DT-MFTGs. Section 14.5.3 focuses on the solvability of the saddle-point problem by carefully explaining the difference between nonzero-sum and zero-sum game strategies. Section 14.6 discusses more general discrete-time MFTG. Both discrete-time DPP and SMP are examined in the space of probability measures. The chapter ends with some notes and discussion of related literature in Sect. 14.7.

14.1 Linear-Quadratic Discrete-Time Mean-Field Games

This section considers discrete-time linear-quadratic scalar mean-field games. We present linear-quadratic problem classes with a focus on problems with continuous state variables in discrete time that are driven by standard white noises. A simple and basic method is proposed for finding both equilibrium strategies and equilibrium

T. Başar et al., *Mean-Field-Type Game Theory I*, Static & Dynamic Game Theory: Foundations & Applications, https://doi.org/10.1007/978-3-032-07027-2_14

cost. The continuous time analogs of the problems studied here have been already analyzed in Chap. 7.

14.1.1 Discrete-Time Controlled Stochastic Process

Let $K \geq 1$ be a fixed time and $(\Omega, \mathscr{F}, \mathbb{F}, \mathbb{P})$ be a given filtered probability space on which a one-dimensional standard white noise $w = \{w_k\}_{k\geq 0}$ is defined; we have the filtration $\mathbb{F} = \{\mathscr{F}_k,\ 0 \leq k \leq K-1\}$ as the natural filtration of w augmented by $\mathbb{P}$-null sets of $\mathscr{F}$. A set S is $\mathbb{P}$-null (or zero measure) if its measure is zero under $\mathbb{P}$.

We introduce the following notations:

- Let $i \geq 1$. $l^i(0, K; \mathbb{R})$ is the set of sequences $(f_k)_{k\in\{0,\ldots,K-1\}} \in \mathbb{R}^K$ such that $\sum_{k=0}^{K-1} |f_k|^i dt < \infty$.
- $l^i_{\mathbb{F}}(0, K; \mathbb{R})$ is the set of $\mathbb{F}$-adapted $\mathbb{R}$-valued processes X such that $E\left[\sum_{k=0}^{K-1} |X_k|^i\right] < \infty$.
- $|\phi|_K := \sup_{0\leq k\leq K} |\phi_k|$

We have a discrete-time controlled stochastic process described by the following model

$$x_{k+1} = [ax_k + \bar{a}m_k + bu_k] + \sigma w_k, \quad x_0 \in l^2(\mathbb{R}) \tag{14.1}$$

where k stands for the discrete-time instant and $a, \bar{a}, b, \sigma$ are real numbers, x_k represents the state of a generic decision-maker, m_k is the mean of the states of all the other decision-makers to be specified, and m is a deterministic function such that $|m|_K < \infty$. For simplicity, we consider only the situation where w_k is a one-dimensional martingale difference sequence with

$$E[w_{k+1} \mid \mathscr{F}_k] = 0, \ E[w_{k+1}^2 \mid \mathscr{F}_k] = 1,$$

where $\mathscr{F}_k$ is the σ-algebra generated by $(x_0, \{w_k\}_{0\leq k\leq K-1})$. The initial value x_0 and the disturbance $\{w_k\}_{0\leq k\leq K-1}$ are assumed to be independent of each other. A typical example of process w would be $w_{k+1} = B_{k+1} - B_k$, with B_k is a Brownian motion. In the discrete-time Liouville dynamics (14.1), the process $\{x_k\}_{0\leq k\leq K}$ is the state process and $\{u_k\}_{0\leq k\leq K-1}$ is the control input of the generic decision-maker. The process $\{w_k\}_{0\leq k\leq K-1}$ represents a stochastic disturbance to the state and is defined on a probability space $(\Omega, \mathscr{F}, \mathbb{P})$.

We next introduce performance index (criterion) of the following form for each generic decision-maker over the horizon $\{0, \ldots, K-1\}$.

$$L(u, m) = q_K x_K^2 + \bar{q}_K (x_K - m_K)^2 + \sum_{k=0}^{K-1} q_k x_k^2 + \bar{q}_k (x_k - m_k)^2 + r_k u_k^2, \tag{14.2}$$

where $q_k, \bar{q}_k$ are scalar positive weighting coefficients of the state and the difference between the state and the mean-field at the time step k, and the control input of a generic decision-maker is u_k over the time horizon $0 \leq k \leq K-1$.[1]

Definition 14.1 (Admissible Control Strategy) The set of admissible control strategies for a generic decision-maker is

$$\mathscr{U} = \left\{ \begin{array}{l} u = (u_0, \ldots, u_{K-1}) \mid u_k : \Omega \to \mathbb{R}, \;\; u_k \text{ is } \mathscr{F}_k\text{-measurable}, \\ E[u_k^2] < +\infty, \;\; k \in \{0, \ldots, K-1\}, \\ u_k \in U \; a.e. \; k \in \{0, \ldots, K-1\}, \; \mathbb{P}\text{-}a.s. \end{array} \right\} \tag{14.3}$$

Lemma 14.1 *Let m such that $|m|_K < +\infty$. For any initial random variable X_0 such that $E[X_0^2] < +\infty$, and any strategy profile $u \in \mathscr{U}$, the state process is well defined. Moreover, for any $1 \leq k \leq K$, the first and the second moments of x_k are finite. $E[|x_k|^l] < +\infty$, $l \in \{1, 2\}$.*

The proof is elementary and follows from the recursive deterministic equation

$$E[x_{k+1}] = aE[x_k] + \bar{a}m_k + bE[u_k] \tag{14.4}$$

A similar recursive equation can be established for the second moment.

14.1.2 Linear-Quadratic Discrete-Time Mean-Field Game

In this section, we provide the solution to the class of discrete-time risk-neutral mean-field games described by a linear stochastic difference equation (14.1) and quadratic cost functional (14.2) for each generic decision-maker.

Consider a very large number of risk-neutral decision-makers. Given m,, the best-response optimal control of a generic decision-maker is the following risk-neutral linear-quadratic discrete-time mean-field problem where the mean-field term is through the mean of the states of all the decision-makers (which is implicit here):

$$\left\{ \begin{array}{l} \inf_{u(\cdot)\in\mathscr{U}} E\left[L(u, m)\right], \\ \text{subject to} \quad \text{state satisfying (14.1) where } \; L \; \text{ is defined in (14.2)} \end{array} \right. \tag{14.5}$$

where m_k is the mean state trajectory created by all decision-makers at an equilibrium (if it exists).

[1] Taking both q_k and $\bar{q}_k$ to be positive captures most realistic scenarios where the generic agent wants his state to be as close to the mean field as possible while also wanting it to be close to zero. The positivity on $\bar{q}_k$ could in fact be relaxed to $q_k + \bar{q}_k > 0$, which would also reflect a meaningful model, this time allowing for $\bar{q}_k$ to be negative, capturing the scenario where the agent wants his state x_k to be as far as possible from the mean field m_k, while still wanting it to be close to zero.

Definition 14.2 (Best Response to Mean-Field) Any $\bar{u}(\cdot) \in \mathscr{U}$ satisfying the infimum in (14.5) is called a risk-neutral best-response strategy of a generic decision-maker to the mean-field term $\{m_k\}_{k\in\{0,\ldots,K\}}$. The corresponding state process, solution of (14.5), is denoted by $(\bar{x}_k)_k := (x_k^{\bar{u}}[m, x_0])_k$.

The risk-neutral mean-field equilibrium problem we are concerned with is to characterize the triplet $(\bar{x}, \bar{u}, m)$ solution of the problem (14.5) and the mean state created by all the decision-makers coincide with m,, i.e.,

$$m_k = E[x_k^{\bar{u}}[m]],$$

which is a fixed-point equation in m.

Definition 14.3 A risk-neutral mean-field equilibrium is a collection (x^*, u^*, m^*) such that $\bar{u}$ minimizes (14.5) and $E[x_k^*] = m_k^*$ for $k \in \{0, \ldots, K\}$.

Note that, in this particular case, we define a risk-neutral mean-field equilibrium through the mean state and not the distribution itself, since high-order moments do not enter into the solution process.

Proposition 14.1 *For the best response problem (14.5), under the standing assumptions that for $k \in \{0, \ldots, K-1\}$, $q_k > 0$, $(q_k + \bar{q}_k) > 0$, $r_k > 0$, and given m such that $|m|_K < +\infty$,, there is a unique best-response strategy of a generic decision-maker, and it is given by*

$$\begin{cases} \bar{u} = -\frac{b}{r_k+b^2\beta_{k+1}}[a\beta_{k+1}(x_k - m_k) + (a+\bar{a})\beta_{k+1}m_k + \eta_{k+1}m_{k+1}], \\ \beta_k = a^2\beta_{k+1} + q_k + \bar{q}_k - \frac{b^2a^2\beta_{k+1}^2}{r_k+b^2\beta_{k+1}}, \\ \beta_K = q_K + \bar{q}_K, \\ \eta_k m_k = a\bar{a}\beta_{k+1}m_k + a\eta_{k+1}m_{k+1} - \bar{q}_k m_k - \frac{b^2}{r_k+b^2\beta_{k+1}}(\bar{a}\beta_{k+1}m_k + \eta_{k+1}m_{k+1})\beta_{k+1}a, \\ \eta_K = -\bar{q}_K, \\ \gamma_k m_k^2 = \bar{a}^2\beta_{k+1}m_k^2 + 2\eta_{k+1}m_{k+1}[\bar{a}m_k] + \gamma_{k+1}m_{k+1}^2 + \bar{q}_k m_k^2 \\ -\frac{b^2}{r_k+b^2\beta_{k+1}}(\bar{a}\beta_{k+1}m_k + \eta_{k+1}m_{k+1})^2, \\ \gamma_K = \bar{q}_K, \\ E[L(\bar{u}, m)] = \beta_0 x_0^2 + 2\eta_0 x_0 m_0 + \gamma_0 m_0^2 + \sum_{k=0}^{K-1}\sigma^2\beta_{k+1}. \end{cases} \tag{14.6}$$

14.1.2.1 Risk-Neutral Mean-Field Equilibrium

We now look for an equilibrium of the risk-neutral linear-quadratic discrete-time mean-field game in the infinite population regime, which we call risk-neutral mean-field equilibrium. If $(\bar{x}, \bar{u}, m)$ is a mean-field equilibrium, then m solves the following fixed-point equation

$$m_{k+1} = (a+\bar{a})m_k - \frac{b^2}{r_k + b^2\beta_{k+1}}[(a+\bar{a})\beta_{k+1}m_k + \eta_{k+1}m_{k+1}] \tag{14.7}$$

If $r_k + b^2(\beta_{k+1} + \eta_{k+1})$ is nonzero, one can invert (14.7) to get m_{k+1}.

$$m_{k+1} = \left[1 + \frac{b^2\eta_{k+1}}{r_k + b^2\beta_{k+1}}\right]^{-1}\left[(a+\bar{a}) - \frac{b^2(a+\bar{a})\beta_{k+1}}{r_k + b^2\beta_{k+1}}\right]m_k \tag{14.8}$$

$$= \frac{(a+\bar{a})r_k}{r_k + b^2(\beta_{k+1} + \eta_{k+1})}m_k =: \epsilon_k m_k \tag{14.9}$$

where $\epsilon_k = \frac{(a+\bar{a})r_k}{r_k + b^2(\beta_{k+1}+\eta_{k+1})}$. It follows that

$$\eta_k = a\bar{a}\beta_{k+1} + a\eta_{k+1}\epsilon_k - \bar{q}_k - \frac{b^2}{r_k + b^2\beta_{k+1}}(\bar{a}\beta_{k+1} + \eta_{k+1}\epsilon_k)\beta_{k+1}a.$$

$$\eta_k = -q_k + \frac{ar_k}{r_k + b^2\beta_{k+1}}[\bar{a}\beta_{k+1} + \eta_{k+1}\epsilon_k],$$

and

$$\gamma_k = \bar{a}^2\beta_{k+1} + 2\eta_{k+1}\epsilon_k\bar{a} + \gamma_{k+1}\epsilon_k^2 + \bar{q}_k - \frac{b^2}{r_k + b^2\beta_{k+1}}(\bar{a}\beta_{k+1} + \eta_{k+1}\epsilon_k)^2.$$

Proposition 14.2 *Consider the equilibrium problem associated with (14.5). Suppose that for* $k \in \{0, \ldots, K-1\}$ $q_k > 0$, $(q_k + \bar{q}_k) > 0$, $r_k > 0$, *and* $r_k + b^2(\beta_{k+1} + \eta_{k+1}) \neq 0$. *Then, there is a unique risk-neutral mean-field equilibrium, and it is given by*

$$\begin{cases} \bar{u} = -\frac{b}{r_k+b^2\beta_{k+1}}a\beta_{k+1}(x_k - m_k) - \frac{b}{r_k+b^2\beta_{k+1}}\{(a+\bar{a})\beta_{k+1} + \eta_{k+1}\epsilon_k\}m_k, \\ \beta_k = a^2\beta_{k+1} + q_k + \bar{q}_k - \frac{b^2a^2\beta_{k+1}^2}{r_k+b^2\beta_{k+1}}, \\ \beta_K = q_K + \bar{q}_K, \\ \eta_k = a\bar{a}\beta_{k+1} + a\eta_{k+1}\epsilon_k - \bar{q}_k - \frac{b^2}{r_k+b^2\beta_{k+1}}(\bar{a}\beta_{k+1} + \eta_{k+1}\epsilon_k)\beta_{k+1}a \\ = -q_k + \frac{ar_k}{r_k+b^2\beta_{k+1}}[\bar{a}\beta_{k+1} + \eta_{k+1}\epsilon_k], \\ \eta_K = -\bar{q}_K, \\ \gamma_k = \bar{a}^2\beta_{k+1} + 2\eta_{k+1}\epsilon_k\bar{a} + \gamma_{k+1}\epsilon_k^2 + \bar{q}_k - \frac{b^2}{r_k+b^2\beta_{k+1}}(\bar{a}\beta_{k+1} + \eta_{k+1}\epsilon_k)^2, \\ \gamma_K = \bar{q}_K, \\ \epsilon_k = \frac{(a+\bar{a})r_k}{r_k+b^2(\beta_{k+1}+\eta_{k+1})}. \end{cases} \tag{14.10}$$

14.1.3 Linear-Quadratic Mean-Field Games: Risk-Sensitive Case

We now consider the case where all the decision-makers are risk-sensitive. The best response of a decision-maker is the following discrete-time risk-sensitive linear-quadratic optimal control where the mean-field term is through the mean of the states of all the decision-makers (which is again implicit in the formulation):

$$\begin{cases} \inf_{u(\cdot)\in\mathcal{U}} E\left[e^{\theta L(u,m)}\right], \\ \text{subject to} \quad \text{state satisfying (14.1) where } L \text{ is defined in (14.2),} \\ \text{and } \theta \text{ is the risk-sensitivity parameter.} \end{cases} \tag{14.11}$$

As in the risk-neutral case, any $\bar{u}(\cdot) \in \mathcal{U}$ satisfying the infimum in (14.11) is called a risk-sensitive best response of a generic decision-maker to the mean-field term $m = (m_k)_k$. The following proposition provides the unique solution to this optimal control problem which is standard.

Proposition 14.3 *For the risk-sensitive best response problem (14.11), under the governing assumptions that for $k \in \{0, \ldots, K-1\}$, $q_k > 0$, $(q_k+\bar{q}_k) > 0$, $r_k > 0$, $\tilde{\beta} > 0$ and given m such that $|m|_K < +\infty$,, there exists a unique risk-sensitive best-response strategy of a generic decision-maker, which is given by*

$$\begin{cases} \bar{u} = -b.\frac{\tilde{\beta}_{k+1}(ax_k+\bar{a}m_k)+(1+2\theta\sigma^2\beta_{k+1})\eta_{k+1}m_{k+1}}{r_k+b^2\tilde{\beta}_{k+1}} \\ = -\frac{b}{r_k+b^2\tilde{\beta}_{k+1}}[a\tilde{\beta}_{k+1}(x_k-m_k)+(a+\bar{a})\tilde{\beta}_{k+1}m_k+(1+2\theta\sigma^2\beta_{k+1})\eta_{k+1}m_{k+1}], \\ \beta_k = a^2\tilde{\beta}_{k+1}+q_k+\bar{q}_k-\frac{b^2a^2\tilde{\beta}_{k+1}^2}{r_k+b^2\tilde{\beta}_{k+1}}, \\ \tilde{\beta}_k = \beta_k(1+2\theta\sigma^2\beta_k), \\ \beta_K = q_K+\bar{q}_K, \\ \\ E[e^{L(\bar{u},m)}] = e^{\theta[\beta_0x_0^2+2\eta_0x_0m_0+\gamma_0m_0^2+\sum_{k=0}^{K-1}\sigma^2\beta_{k+1}w_k^2]}. \end{cases} \tag{14.12}$$

We note that the presence of the term $\tilde{\beta}_k = \beta_k(1+2\theta\sigma^2\beta_k)$ in the risk-sensitive discrete-time Riccati equation comes from the completion of the exponential martingale in the derivation of the solution. Note also that we recover the risk-neutral discrete-time Riccati equation of Proposition 14.2 as θ goes to 0.

14.1.4 Robust Discrete-Time Mean-Field Games

We now consider a variation on discrete time LQ- MFG with an additional malicious/disturbance term v_k at time k in the state equation. We assume that the disturbance term $v = (v_k)_k$ takes values in $V = \mathbb{R}$ and define the set of disturbance strategies, $\mathcal{V}$, in a similar fashion as $\mathcal{U}$.

$$\mathscr{V} = \left\{ \begin{array}{l} v = (v_0, \ldots, v_{K-1}) \mid v_k : \Omega \to \mathbb{R}, \ \ v_k \text{ is } \mathscr{F}_k\text{-measurable}, \\ E[v_k^2] < +\infty, \ \ k \in \{0, \ldots, K-1\}, \\ v_k \in V \ a.e. \ k \in \{0, \ldots, K-1\}, \ \ \mathbb{P} - a.s. \end{array} \right\} \tag{14.13}$$

The state, therefore, now evolves according to

$$x_{k+1} = [ax_k + \bar{a}m_k + bu_k + cv_k] + \sigma w_k, \ x(0) = X_0,$$

The cost function is accordingly modified to

$$L(u, v, m) = q_K x_K^2 + \bar{q}_K (x_K - m_K)^2 + \sum_{k=0}^{K-1} q_k x_k^2 + \bar{q}_k (x_k - m_k)^2$$
$$+ r_k u_k^2 - s_k v_k^2, \tag{14.14}$$

with $s_k > 0$ for all k, which places a positive cost on $\{v_k\}$, the action of the maximizer, where $\{u_k\}$ is still the minimizing variable. As before, we take $r_k > 0, q_k > 0, \ q_k + \bar{q}_k > 0, \forall k \in \{0, \ldots, K-1\}$. The resulting game is called *robust LQ MFG in discrete time*.

be the cost functional of a generic decision-maker with strategy u under disturbance v when the mean-field process is m. The best response of a generic decision-maker is obtained for a given m by solving the following minimax problem

$$\left\{ \begin{array}{l} \inf_{u(\cdot)\in\mathscr{U}} \sup_{v(\cdot)\in\mathscr{V}} E[L_2(u, v, m)], \\ \text{subject to} \\ x_{k+1} = [ax_k + \bar{a}m_k + bu_k + cv_k] + \sigma w_k, \ x(0) = X_0. \end{array} \right. \tag{14.15}$$

This is a dynamic discrete-time linear-quadratic zero-sum game, which admits a unique saddle-point solution as given below. See also [22], Chapter 6.

Proposition 14.4 *Let* $r_k > 0, q_k > 0, \ q_k + \bar{q}_k > 0, \ s_k > 0, \ (\frac{b^2}{r_k} - \frac{c^2}{s_k}) > 0,$ $\forall k \in \{0, \ldots, K-1\}$.

Then the best response strategies are

$$\begin{array}{l} v_k = \frac{r_k c}{r_k s_k + \beta_{k+1}(s_k b^2 - r_k c^2)} \{\beta_{k+1}(ax_k + \bar{a}m_k) + \eta_{k+1} m_{k+1}\}, \\ u_k = -\frac{s_k b}{r_k s_k + \beta_{k+1}(s_k b^2 - r_k c^2)} \{\beta_{k+1}(ax_k + \bar{a}m_k) + \eta_{k+1} m_{k+1}\} \\ \beta_k = a^2 \beta_{k+1} + (q_k + \bar{q}_k) + \beta_{k+1}^2 a^2 \left[\frac{s_k b^2 - r_k c^2}{r_k s_k + \beta_{k+1}(s_k b^2 - r_k c^2)} \right], \\ \beta_K = q_K + \bar{q}_K, \\ \eta_k = (a\bar{a} - \bar{q}_k) + \eta_{k+1}\mu_{k+1} a \\ +\beta_{k+1} a [\beta_{k+1}\bar{a} + \eta_{k+1}\mu_{k+1}] \left[\frac{s_k b^2 - r_k c^2}{r_k s_k + \beta_{k+1}(s_k b^2 - r_k c^2)} \right], \\ \eta_K = -\bar{q}_K, \end{array} \tag{14.16}$$

Proposition 14.5 *The equilibrium cost is given by*

$$\begin{array}{l} E[L(u,v,m)] = \ \beta_0 x_0^2 + 2\eta_0 x_0 m_0 + \gamma_0 m_0^2 + E\sum_{k=0}^{K-1} \beta_{k+1}\sigma^2 \\ \gamma_k = \bar{a}^2 + \bar{q}_k + 2\eta_{k+1}\mu_{k+1}\bar{a} + \gamma_{k+1}\mu_{k+1}^2 \\ +[\beta_{k+1}\bar{a} + \eta_{k+1}\mu_{k+1}]^2 \left[\frac{s_k b^2 - r_k c^2}{r_k s_k + \beta_{k+1}(s_k b^2 - r_k c^2)}\right], \\ \gamma_K = \bar{q}_K, \\ m_{k+1} = m_0 \prod_{i=0}^{k} \mu_{i+1}, \\ \mu_{k+1} = \frac{r_k s_k (a+\bar{a})}{r_k s_k + (\beta_{k+1} - \eta_{k+1})(s_k b^2 - r_k c^2)}. \end{array} \tag{14.17}$$

14.1.5 Linear-Quadratic Single Decision-Maker Mean-Field-Type Problem

In this subsection, we consider a single decision-maker with the following linear-quadratic discrete-time optimization problem of the mean-field type.

The state still evolves according to (14.1) with m_k replaced by the expected value of x_k, i.e., $E[x_k]$. A similar substitution is made for m_k in the cost function (14.2), that is,

$$\hat{L}(u) := q_K x_K^2 + \bar{q}_K (x_K - E[x_K])^2 + \sum_{0}^{K-1} q_k x_k^2 + \bar{q}_k (x_k - E[x_k])^2 + r_k u_k^2.$$

$$\begin{cases} \inf_{u(\cdot)\in\mathscr{U}} E[\hat{L}(u)] \\ \text{subject to} \\ x_{k+1} = \{a x_k + \bar{a} E[x_k] + b u_k\} + \sigma w_k, \\ x_0 \sim \mathscr{L}(X_0), \qquad E[X_0] = m_0, \end{cases} \tag{14.18}$$

whose solution is given in the following proposition.

Proposition 14.6 *Consider the problem (14.18) under the assumption that for* $k \in \{0, \ldots, K-1\}$ $q_k > 0,\ (q_k + \bar{q}_k) > 0,\ r_k > 0$. *Then, there exists a unique global optimum, and it is given by*

$$\begin{cases} \hat{\beta}_k = a^2 \hat{\beta}_{k+1} + q_k + \bar{q}_k - \frac{b^2 a^2 \hat{\beta}_{k+1}^2}{r_k + \hat{\beta}_{k+1} b^2} = q_k + \bar{q}_k + \frac{a^2 \hat{\beta}_{k+1} r_k}{r_k + \hat{\beta}_{k+1} b^2} \geq 0 \\ \hat{\beta}_K = q_K + \bar{q}_K > 0 \\ \hat{\gamma}_k = (a+\bar{a})^2 \hat{\gamma}_{k+1} + q_k - \frac{b^2 (a+\bar{a})^2 \hat{\gamma}_{k+1}^2}{r_k + \hat{\gamma}_{k+1} b^2} = q_k + \frac{(a+\bar{a})^2 \hat{\gamma}_{k+1} r_k}{r_k + \hat{\gamma}_{k+1} b^2} \\ \hat{\gamma}_K = q_K > 0 \\ \hat{u}_k = -\frac{b a \hat{\beta}_{k+1}}{r_k + \hat{\beta}_{k+1} b^2}(x_k - E x_k) - \frac{b(a+\bar{a})\hat{\gamma}_{k+1}}{r_k + \hat{\gamma}_{k+1} b^2} E x_k \\ E[\hat{L}(\hat{u}, \hat{m})] = \hat{\beta}_0 (x_0 - E x_0)^2 + \hat{\gamma}_0 (E x_0)^2 + \sum_{k=0}^{K-1} \hat{\beta}_{k+1} \sigma^2. \end{cases} \tag{14.19}$$

Remark 14.1 It is important to notice that even in the linear-quadratic setup, the optimality system of the equilibrium problem, that is given in Proposition 14.2, is different from the solution to the global optimum problem given in Proposition 14.6.

We now prove Proposition 14.6. We write

$$\begin{cases} E[q_k x_k^2 + \bar{q}_k(x_k - E[x_k])^2] = (q + \bar{q}_k)(x_k - E[x_k])^2 + q_k(E[x_k])^2 \\ = (q + \bar{q}_k)var(x_k - E[x_k]) + q_k(E[x_k])^2. \end{cases} \tag{14.20}$$

$$\begin{cases} E[\hat{L}(u)] := E\{(q_K + \bar{q}_K)(x_K - E[x_K])^2 + q_K(E[x_K])^2 \\ + \sum_{k=0}^{K-1}(q_k + \bar{q}_k)(x_k - E[x_k])^2 + q_k(E[x_k])^2 + r_k u_k^2\}. \end{cases} \tag{14.21}$$

We try a candidate optimal cost structure in the following form $\hat{\beta}_k(x_k - Ex_k)^2 + \hat{\gamma}_k(Ex_k)^2$. We compute the difference between the expected global cost and the one obtained by unknown sequences $\hat{\beta}$, $\hat{\gamma}$ to be determined.

$$\begin{aligned} & E\{\hat{L}(u,m) - \hat{\beta}_0(x_0 - Ex_0)^2 - \hat{\gamma}_0(Ex_0)^2\} \\ & = [(q_K + \bar{q}_K) - \hat{\beta}_K]E(x_K - Ex_K)^2 + [q_K - \hat{\gamma}_K](Ex_K)^2 + [q_K - \hat{\gamma}_K](Ex_K)^2 \\ & + \sum_{k=0}^{K-1}[\hat{\beta}_{k+1}(x_{k+1} - Ex_{k+1})^2 - \hat{\beta}_k(x_k - Ex_k)^2] + [\hat{\gamma}_{k+1}(Ex_{k+1})^2 - \hat{\gamma}_k(Ex_k)^2] \\ & + \sum_{k=0}^{K-1}\{(q_k + \bar{q}_k)var(x_k) + q_k(E[x_k])^2 + r_k E[u_k - Eu_k]^2 + r_k(Eu_k)^2\} \end{aligned} \tag{14.22}$$

We will express the terms $(x_{k+1} - Ex_{k+1})^2$ and $(Ex_{k+1})^2$ as functions of $(x_k - Ex_k)$, $(u_{ik} - Eu_{ik})$, (Ex_k), (Eu_{ik}).

$$\hat{\beta}_{k+1}E(x_{k+1} - Ex_{k+1})^2 = \hat{\beta}_{k+1}E[a(x_k - Ex_k) + b(u_k - Eu_k) + \sigma W_k]^2 =$$

$$Ea^2\hat{\beta}_{k+1}(x_k - Ex_k)^2 + 2ab\hat{\beta}_{k+1}(x_k - Ex_k)(u_k - Eu_k) + \hat{\beta}_{k+1}b^2(u_k - Eu_k)^2 + \hat{\beta}_{k+1}\sigma^2.$$

We now compute $\hat{\gamma}_{k+1}(Ex_{k+1})^2$:

$$\begin{aligned} & \hat{\gamma}_{k+1}(Ex_{k+1})^2 = \hat{\gamma}_{k+1}[(a + \bar{a})Ex_k + bEu_k]^2 \\ & = (a + \bar{a})^2\hat{\gamma}_{k+1}(Ex_k)^2 + 2(a + \bar{a})b\hat{\gamma}_{k+1}Ex_k Eu_k + \hat{\gamma}_{k+1}b^2[Eu_k]^2. \end{aligned} \tag{14.23}$$

$$
\begin{aligned}
&E\{\hat{L}(u,m) - \hat{\beta}_0(x_0 - Ex_0)^2 - \hat{\gamma}_0(Ex_0)^2\} \\
&= \sum_{k=0}^{K-1}[a^2\hat{\beta}_{k+1} - \hat{\beta}_k + q_k + \bar{q}_k](x_k - Ex_k)^2 \\
&+ \sum_{k=0}^{K-1}[(a+\bar{a})^2\hat{\gamma}_{k+1} - \hat{\gamma}_k + q_k](Ex_k)^2 \\
&+ \sum_{k=0}^{K-1} 2ab\hat{\beta}_{k+1}(x_k - Ex_k)(u_k - Eu_k) + (r_k + \hat{\beta}_{k+1}b^2)(u_k - Eu_k)^2 \\
&+ \sum_{k=0}^{K-1} 2(a+\bar{a})b\hat{\gamma}_{k+1}(Ex_k)Eu_k + (r_k + \hat{\gamma}_{k+1}b^2)[Eu_k]^2 + \sum_{k=0}^{K-1} \hat{\beta}_{k+1}\sigma^2 \\
&= \sum_{k=0}^{K-1}[a^2\hat{\beta}_{k+1} - \hat{\beta}_k + q_k + \bar{q}_k](x_k - Ex_k)^2 \\
&+ \sum_{k=0}^{K-1}[(a+\bar{a})^2\hat{\gamma}_{k+1} - \hat{\gamma}_k + q_k](Ex_k)^2 \\
&+ \sum_{k=0}^{K-1}(r_k + \hat{\beta}_{k+1}b^2)\{u_k - Eu_k + \tfrac{ab\hat{\beta}_{k+1}}{r_k+\hat{\beta}_{k+1}b^2}(x_k - Ex_k)\}^2 \\
&- \sum_{k=0}^{K-1} \tfrac{a^2b^2\hat{\beta}_{k+1}^2}{r_k+\hat{\beta}_{k+1}b^2}(x_k - Ex_k)^2 \\
&+ \sum_{k=0}^{K-1}(r_k + \hat{\gamma}_{k+1}b^2)\{Eu_k + \tfrac{(a+\bar{a})b\hat{\gamma}_{k+1}}{r_k+\hat{\gamma}_{k+1}b^2}Ex_k\}^2 \\
&\sum_{k=0}^{K-1} -\tfrac{(a+\bar{a})^2b^2\hat{\gamma}_{k+1}^2}{r_k+\hat{\gamma}_{k+1}b^2}(Ex_k)^2 + \sum_{k=0}^{K-1}\hat{\beta}_{k+1}\sigma^2 \\
&= \sum_{k=0}^{K-1}[a^2\hat{\beta}_{k+1} - \hat{\beta}_k + q_k + \bar{q}_k - \tfrac{a^2b^2\hat{\beta}_{k+1}^2}{r_k+\hat{\beta}_{k+1}b^2}](x_k - Ex_k)^2 \\
&+ \sum_{k=0}^{K-1}[(a+\bar{a})^2\hat{\gamma}_{k+1} - \hat{\gamma}_k + q_k - \tfrac{(a+\bar{a})^2b^2\hat{\gamma}_{k+1}^2}{r_k+\hat{\gamma}_{k+1}b^2}](Ex_k)^2 \\
&+ \sum_{k=0}^{K-1}(r_k + \hat{\beta}_{k+1}b^2)\{u_k - Eu_k + \tfrac{ab\hat{\beta}_{k+1}}{r_k+\hat{\beta}_{k+1}b^2}(x_k - Ex_k)\}^2 \\
&+ \sum_{k=0}^{K-1}(r_k + \hat{\gamma}_{k+1}b^2)\{Eu_k + \tfrac{(a+\bar{a})b\hat{\gamma}_{k+1}}{r_k+\hat{\gamma}_{k+1}b^2}Ex_k\}^2 \\
&+ \sum_{k=0}^{K-1}\hat{\beta}_{k+1}\sigma^2.
\end{aligned}
\tag{14.24}
$$

Taking the minimum over u yields the desired result.

14.2 Linear-Quadratic Discrete-Time Mean-Field-Type Games

This section deals with linear-quadratic discrete-time mean-field-type games (LQ DT MFTGs), that is, MFTGs where the state process evolves according to a linear stochastic difference equation and the cost functional for each decision-maker is quadratic.

In the risk-neutral LQ DT MFTG, we show that there exists a unique strategy that is best response to the expected values of the state and provide a simple sufficient condition of existence and uniqueness of mean-field equilibrium. We also provide a closed-form expression of the global optimum as a state-feedback strategy.

Next, we explore how randomness and uncertainty affect the game, both when noises are additive and when they're multiplicative. We'll also use some method to the methodology easier to understand. Finally, we study the transition from small time steps to a continuous flow of time.

14.2.1 Formulation

We consider a simple class of LQ DT MFTGs between I decision-makers. The system equation is the following linear stochastic difference equation of mean-field type with $k \in \{0, 1, \ldots, K-1\}$, $K \geq 1$,

$$\begin{aligned} x_{k+1} &= \left\{ a x_k + \bar{a} E[x_k] + \textstyle\sum_{j=1}^{I} b_j u_{jk} \right\} + \sigma w_k, \\ x_0 &\sim \mathscr{L}(X_0), \qquad E[X_0] = m_0, \end{aligned} \tag{14.25}$$

representing a common state of the interactive system between the I decision-makers and $a, \bar{a}, b_i, \sigma, m_0$ are real numbers. In the McKean-Vlasov discrete time dynamics (14.25), the process $\{x_k\}_{0 \leq k \leq K}$ is the state process and $\{u_{ik}, \}_{0 \leq k \leq K-1}$ is the control process of the decision-maker i. The process $\{w_k\}_{0 \leq k \leq K-1}$ represents a stochastic disturbance to the state and is defined on a probability space $(\Omega, \mathscr{F}, \mathbb{P})$. For simplicity, we consider only the situation where $\{w_k\}_k$ is a one-dimensional martingale difference sequence with

$$E[w_{k+1} \mid \mathscr{F}_k] = 0, \;\; E[w_{k+1}^2 \mid \mathscr{F}_k] = 1,$$

where $\mathscr{F}_k$ is the σ-algebra generated by $(x_0, \{w_k\}_{0 \leq k \leq K-1})$. The initial value x_0 and the disturbance $\{w_k\}_{0 \leq k \leq K-1}$ are assumed to be independent of each other. A typical example of process w would be $w_{k+1} = B_{k+1} - B_k$, with B_k a Brownian motion. We introduce the set of admissible control strategies for decision-maker i as

$$\begin{aligned} \mathscr{U}_i = \{ & u_i = (u_{i0}, \ldots, u_{i,K-1}) \mid u_{i,k} : \Omega \to \mathbb{R}, \;\; u_{ik} \text{ is } \mathscr{F}_k\text{-measurable}, \\ & E[u_{ik}^2] < +\infty \} \end{aligned}$$

Lemma 14.2 *For any initial random variable X_0 such that $E[X_0^2] < +\infty$, and any strategy profile $u = (u_1, \ldots, u_I) \in \prod_{j=1}^{I} \mathscr{U}_j$,, the state process is well defined. Moreover, for any $1 \leq k \leq K$,, the first and second moments of x_k are finite. $E[|x_k|^l] < +\infty$, $l \in \{1, 2\}$.*

The proof is elementary and follows from the equalities below

$$\begin{aligned} E x_{k+1} &= (a + \bar{a}) E x_k + \textstyle\sum_{i=1}^{I} b_i E[u_{ik}] \\ &= (a + \bar{a})^{k+1} E x_0 + \textstyle\sum_{l=0}^{k} (a + \bar{a})^l \sum_{i=1}^{I} b_i E[u_{i,k-l}] \end{aligned} \tag{14.26}$$

A similar recursive equation can be established for the second moment.

We take the cost functional for decision-maker i associated with (14.25) to be

$$L_i(u) = q_{iK} x_K^2 + \bar{q}_{iK} (E[x_K])^2 + \textstyle\sum_0^{K-1} q_{ik} x_k^2 + \bar{q}_{ik} (E[x_k])^2 + r_{ik} u_{ik}^2. \tag{14.27}$$

which is composed of the terminal cost $q_{iK}x_K^2 + \bar{q}_{iK}(E[x_K])^2$ and a running cost of $q_{ik}x_k^2 + \bar{q}_{ik}(E[x_k])^2 + r_{ik}u_{ik}^2$. The coefficients $q_i, \bar{q}_i, r_{ik}$ are assumed to be positive real numbers.

The decision-makers interact through the common state x which influences the cost function. This is a dynamic game where the decision-makers are not necessarily indistinguishable because the coefficients $q_i, \bar{q}_i, r_i, b_i$, may be different across different decision-makers. Moreover, each decision-maker i influences the mean-field term $E[x]$ through its control u_i. In this model, the contribution of a single decision-maker (say i) in the mean-field term $E[x]$ cannot be neglected. This provides novel features to the mean-field framework.

The expected value of cost L_i of decision-maker i can be written as

$$\begin{aligned} &E[L_i(u)] = [q_{iK}var(x_K) + (q_{iK} + \bar{q}_{iK})(E[x_K])^2] \\ &+ \sum_{k=0}^{K-1} q_{ik}var(x_k) + (q_{ik} + \bar{q}_{ik})(E[x_k])^2 + \sum_{k=0}^{K-1} r_{ik}var(u_{ik}) + r_{ik}(Eu_{ik})^2 \end{aligned} \tag{14.28}$$

Under the assumptions of Lemma 14.2, the risk-neutral cost functional $E[L_i(u)]$ is well defined.

14.2.1.1 Best-Response of a Decision-Maker and Mean-Field-Type Equilibrium

We present necessary and sufficient conditions for the solvability of these basic LQ DT MFTGs. The methodology we employ in obtaining the best response of a decision-maker is based on a simple completion of squares. Given $(u_j \in \mathcal{U}_j,\ j \neq i)$, the best-response problem of decision-maker i is given by

Best-response problem of decision-maker i.

$$\begin{cases} \inf_{u_i\in\mathcal{U}_i} E[L_i(u_1,\ldots,u_I)] \\ \text{subject to } \ x_{k+1} = \left\{ax_k + \bar{a}Ex_k + \sum_{i=1}^{I} b_i u_{ik}\right\} + \sigma w_k, \\ x_0 \sim \mathcal{L}(X_0), \qquad E[X_0] = m_0. \end{cases} \tag{14.29}$$

Definition 14.4 (BR$_i$: Best Response of Decision-Maker i) Any $\bar{u}_i(\cdot) \in \mathcal{U}_i$ achieving the infimum in (14.29) is called a risk-neutral best-response strategy of decision-maker i to the other decision-makers' strategies $u_{-i} \in \prod_{j\neq i} \mathcal{U}_j$.

Solution to the best-response problem (14.29).

For problem (14.29), under the assumption that for $k \in \{0, \dots, K-1\}$, $q_{jk} > 0,\ (q_{jk}+\bar{q}_{jk}) > 0,\ r_{jk} > 0$, and given linear feedback state of the other decision-makers, there is a unique best response of decision-maker i and a unique Nash equilibrium given by

$$\begin{cases} u_{i,k} = \eta_{ik}(x_k - Ex_k) + \bar{\eta}_{ik} Ex_k, \\ \eta_{ik} = -\frac{[ab_i\beta_{i,k+1} + b_i\beta_{i,k+1}\sum_{j\neq i} b_j\eta_{jk}]}{r_{ik}+b_i^2\beta_{i,k+1}}, \\ \bar{\eta}_{ik} = -\frac{b_i\gamma_{i,k+1}(a+\bar{a}+\sum_{j\neq i} b_j\bar{\eta}_{j,k})}{r_{ik}+b_i^2\gamma_{i,k+1}}, \\ \beta_{ik} = q_{ik} + \beta_{i,k+1}\{a^2 + 2a\sum_{j\neq i} b_j\eta_{jk} + [\sum_{j\neq i} b_j\eta_{jk}]^2\} \\ -\frac{[ab_i\beta_{i,k+1} + b_i\beta_{i,k+1}\sum_{j\neq i} b_j\eta_{jk}]^2}{r_{ik}+b_i^2\beta_{i,k+1}} \\ \beta_{iK} = q_{iK} > 0 \\ \gamma_{ik} = (q_{ik}+\bar{q}_{ik}) + \gamma_{i,k+1}(a+\bar{a}+\sum_{j\neq i} b_j\bar{\eta}_{j,k})^2 - \frac{(b_i\gamma_{i,k+1}(a+\bar{a}+\sum_{j\neq i} b_j\bar{\eta}_{j,k}))^2}{r_{ik}+b_i^2\gamma_{i,k+1}} \\ \gamma_{iK} = q_{iK} + \bar{q}_{iK} > 0 \end{cases} \tag{14.30}$$

and the Nash equilibrium cost of decision-maker i is

$$E[L_i(\hat{u})] = E\beta_{i0}(x_0 - Ex_0)^2 + \gamma_{i0}(Ex_0)^2 + \sum_{k=0}^{K-1} \beta_{i,k+1}\sigma^2.$$

The expected value of the control strategy is $Eu_{i,k} = \bar{\eta}_{ik}Ex_k$, and the evolution of the expected value of the state is $Ex_{k+1} = [a + \bar{a} + \sum_{i=1}^{I} \bar{\eta}_{ik}]Ex_k$. Thus, by induction,

$$Ex_{k+1} = \left[a + \bar{a} + \sum_{i=1}^{I} \bar{\eta}_{ik}\right] Ex_k = (Ex_0)\prod_{l=0}^{k}\left[a + \bar{a} + \sum_{i=1}^{I} \bar{\eta}_{i,l}\right].$$

Plugging it into the expression of the best response, we obtain the dependence of the optimal strategy of decision-maker i as function of the initial term Ex_0 which may create a time inconsistency.

We now prove the result (14.30). We compute the difference between the expected cost and the one obtained by unknown sequences β, γ yet to be determined.

$$
\begin{aligned}
&E\{L_i(u) + \beta_{i,K}(x_K - Ex_K)^2 - \beta_{i,0}(x_0 - Ex_0)^2 + \gamma_{i,K}(Ex_K)^2 - \gamma_{i,0}(Ex_0)^2\} \\
&= q_{iK} var(x_K) + (q_{iK} + \bar{q}_{iK})(E[x_K])^2 \\
&+ \sum_{k=0}^{K-1}\{q_{ik} var(x_k) + (q_{ik} + \bar{q}_{ik})(E[x_k])^2 \\
&+ r_{ik} E[u_{ik} - Eu_{ik}]^2 + r_{ik}(Eu_{ik})^2\} \\
&+ E\sum_{k=0}^{K-1} \beta_{i,k+1}(x_{k+1} - Ex_{k+1})^2 - \beta_{i,k}(x_k - Ex_k)^2 \\
&+ \gamma_{i,k+1}(Ex_{k+1})^2 - \gamma_{i,k}(Ex_k)^2
\end{aligned} \tag{14.31}
$$

We will express the terms $(x_{k+1} - Ex_{k+1})^2$ and $(Ex_{k+1})^2$ as functions of $(x_k - Ex_k), (u_{ik} - Eu_{ik}), (Ex_k), (Eu_{ik})$.

$$
\begin{aligned}
&\beta_{i,k+1} E(x_{k+1} - Ex_{k+1})^2 = \beta_{i,k+1}\{a^2 + 2a\sum_{j\neq i} b_j \eta_{jk} \\
&+ [\sum_{j\neq i} b_j \eta_{jk}]^2\}(x_k - Ex_k)^2 \\
&+ \beta_{i,k+1}\{2ab_i(x_k - Ex_k)(u_{ik} - Eu_{ik}) + 2b_i(u_{ik} - Eu_{ik})(x_k - Ex_k)\sum_{j\neq i} b_j \eta_{jk} \\
&+ b_i^2(u_{ik} - Eu_{ik})^2\} + \beta_{i,k+1}\sigma^2,
\end{aligned} \tag{14.32}
$$

where $u_{jk} = \eta_{jk}(x_k - Ex_k) + \bar{\eta}_{j,k} Ex_k,\ j \neq i$. We now compute $\gamma_{i,k+1}(Ex_{k+1})^2$:

$$
\begin{aligned}
&\gamma_{i,k+1}(Ex_{k+1})^2 = \gamma_{i,k+1}[(a + \bar{a})Ex_k + b_i Eu_i + Ex_k \sum_{j\neq i} b_j \bar{\eta}_{j,k}]^2 \\
&= \gamma_{i,k+1}[(a + \bar{a} + \sum_{j\neq i} b_j \bar{\eta}_{j,k})Ex_k + b_i Eu_i]^2 \\
&= \gamma_{i,k+1}(a + \bar{a} + \sum_{j\neq i} b_j \bar{\eta}_{j,k})^2 (Ex_k)^2 \\
&+ b_i^2 \gamma_{i,k+1}(Eu_i)^2 + 2b_i \gamma_{i,k+1}(a + \bar{a} + \sum_{j\neq i} b_j \bar{\eta}_{j,k}) Ex_k Eu_i.
\end{aligned} \tag{14.33}
$$

The difference between the cost and the guess functional is

$$
\begin{aligned}
&E\{L_i(u) + \beta_{i,K}(x_K - Ex_K)^2 - \beta_{i,0}(x_0 - Ex_0)^2 + \gamma_{i,K}(Ex_K)^2 - \gamma_{i,0}(Ex_0)^2\} \\
&= [q_{iK} var(x_K) + (q_{iK} + \bar{q}_{iK})(E[x_K])^2] \\
&+ \sum_{k=1}^{K-1}[q_{ik} + \beta_{i,k+1}\{a^2 + 2a\sum_{j\neq i} b_j \eta_{jk} + [\sum_{j\neq i} b_j \eta_{jk}]^2\} - \beta_{i,k}] var(x_k) \\
&+ \sum_{k=1}^{K-1}[(q_{ik} + \bar{q}_{ik}) + \gamma_{i,k+1}(a + \bar{a} + \sum_{j\neq i} b_j \bar{\eta}_{j,k})^2 - \gamma_{i,k}](Ex_k)^2 \\
&+ \sum_{k=1}^{K-1}(r_{ik} + b_i^2 \beta_{i,k+1}) E[u_{ik} - Eu_{ik}]^2 + \\
&+ \sum_{k=1}^{K-1}[2ab_i\beta_{i,k+1} + 2b_i\beta_{i,k+1}\sum_{j\neq i} b_j \eta_{jk}](x_k - Ex_k)(u_{ik} - Eu_{ik}) \\
&+ \sum_{k=1}^{K-1}(r_{ik} + b_i^2 \gamma_{i,k+1})(Eu_{ik})^2 + 2b_i\gamma_{i,k+1}(a + \bar{a} + \sum_{j\neq i} b_j \bar{\eta}_{j,k}) Ex_k Eu_i \\
&+ \sum_{k=1}^{K-1} \beta_{i,k+1}\sigma^2
\end{aligned} \tag{14.34}
$$

By completing the squares $u_{ik} - Eu_{ik}$ and Eu_{ik}, one obtains

$$
\begin{aligned}
&E\{L_i + \beta_{i,K}(x_K - Ex_K)^2 - \beta_{i,0}(x_0 - Ex_0)^2 + \gamma_{i,K}(Ex_K)^2 - \gamma_{i,0}(Ex_0)^2\}\\
&= q_{iK} var(x_K) + (q_{iK} + \bar{q}_{iK})(E[x_K])^2\\
&+ \sum_{k=1}^{K-1}[q_{ik} + \beta_{i,k+1}\{a^2 + 2a\sum_{j\neq i} b_j\eta_{jk} + [\sum_{j\neq i} b_j\eta_{jk}]^2\} - \beta_{i,k}]var(x_k)\\
&+ \sum_{k=1}^{K-1}[(q_{ik} + \bar{q}_{ik}) + \gamma_{i,k+1}(a + \bar{a} + \sum_{j\neq i} b_j\bar{\eta}_{j,k})^2 - \gamma_{i,k}](Ex_k)^2\\
&+ \sum_{k=1}^{K-1}(r_{ik} + b_i^2\beta_{i,k+1})\left\{u_{ik} - Eu_{ik} + \frac{[ab_i\beta_{i,k+1} + b_i\beta_{i,k+1}\sum_{j\neq i} b_j\eta_{jk}](x_k - Ex_k)}{r_{ik} + b_i^2\beta_{i,k+1}}\right\}^2\\
&- \sum_{k=1}^{K-1}\frac{[ab_i\beta_{i,k+1} + b_i\beta_{i,k+1}\sum_{j\neq i} b_j\eta_{jk}]^2(x_k - Ex_k)^2}{r_{ik} + b_i^2\beta_{i,k+1}}\\
&+ \sum_{k=1}^{K-1}(r_{ik} + b_i^2\gamma_{i,k+1})\left\{Eu_{ik} + \frac{b_i\gamma_{i,k+1}(a + \bar{a} + \sum_{j\neq i} b_j\bar{\eta}_{j,k})Ex_k}{r_{ik} + b_i^2\gamma_{i,k+1}}\right\}^2\\
&- \sum_{k=1}^{K-1}\frac{(b_i\gamma_{i,k+1}(a + \bar{a} + \sum_{j\neq i} b_j\bar{\eta}_{jk}))^2(Ex_k)^2}{r_{ik} + b_i^2\gamma_{i,k+1}} + \sum_{k=1}^{K-1}\beta_{i,k+1}\sigma^2.
\end{aligned}
\tag{14.35}
$$

It turns out that

$$
-Eu_{ik} = \frac{b_i\gamma_{i,k+1}(a + \bar{a} + \sum_{j\neq i} b_j\bar{\eta}_{j,k})Ex_k}{r_{ik} + b_i^2\gamma_{i,k+1}},
$$

and hence the best response strategy at time k is

$$
\begin{aligned}
u_{ik} &= Eu_{ik} - \frac{[ab_i\beta_{i,k+1} + b_i\beta_{i,k+1}\sum_{j\neq i} b_j\eta_{jk}](x_k - Ex_k)}{r_{ik} + b_i^2\beta_{i,k+1}}\\
&= -\frac{b_i\gamma_{i,k+1}(a + \bar{a} + \sum_{j\neq i} b_j\bar{\eta}_{j,k})}{r_{ik} + b_i^2\gamma_{i,k+1}}Ex_k - \frac{[ab_i\beta_{i,k+1} + b_i\beta_{i,k+1}\sum_{j\neq i} b_j\eta_{jk}]}{r_{ik} + b_i^2\beta_{i,k+1}}(x_k - Ex_k)
\end{aligned}
\tag{14.36}
$$

where β and γ are positive solutions of the following system

$$
\begin{aligned}
&\beta_{i,k} = q_{ik} + \beta_{i,k+1}\{a^2 + 2a\sum_{j\neq i} b_j\eta_{jk} + (\sum_{j\neq i} b_j\eta_{jk})^2\}\\
&\quad - \frac{[ab_i\beta_{i,k+1} + b_i\beta_{i,k+1}\sum_{j\neq i} b_j\eta_{jk}]^2}{r_{ik} + b_i^2\beta_{i,k+1}}\\
&\beta_{i,K} = q_{iK} > 0,\\
&\gamma_{i,k} = (q_{ik} + \bar{q}_{ik}) + \gamma_{i,k+1}(a + \bar{a} + \sum_{j\neq i} b_j\bar{\eta}_{j,k})^2\\
&\quad - \frac{(b_i\gamma_{i,k+1}(a + \bar{a} + \sum_{j\neq i} b_j\bar{\eta}_{j,k}))^2}{r_{ik} + b_i^2\gamma_{i,k+1}}\\
&\gamma_{i,K} = (q_{iK} + \bar{q}_{iK}) > 0
\end{aligned}
\tag{14.37}
$$

This means that the best response strategy of decision-maker i is

$$
\begin{aligned}
u_{ik} &= \eta_{ik}(x_k - Ex_k) + \bar{\eta}_{ik}Ex_k,\\
\eta_{ik} &= -\frac{b_i\beta_{i,k+1}[a + \sum_{j\neq i} b_j\eta_{jk}]}{r_{ik} + b_i^2\beta_{i,k+1}},\\
\bar{\eta}_{ik} &= -\frac{b_i\gamma_{i,k+1}(a + \bar{a} + \sum_{j\neq i} b_j\bar{\eta}_{j,k})}{r_{ik} + b_i^2\gamma_{i,k+1}}.
\end{aligned}
\tag{14.38}
$$

The risk-neutral mean-field equilibrium problem we are concerned with is to characterize the triplet $(x, u, E[x])$ such that for every decision-maker i, u_i is an optimizer of the best response problem (14.29), and the expected value of the resulting state created by all the decision-makers is $E[x]$. This means that an equilibrium is a fixed point of the best response correspondence $BR = (BR_1, \ldots, BR_I)$ where $BR_i : \prod_{j\neq i} \mathcal{U}_j \to \mathcal{U}_i$ is the best-response set of decision-maker i.

In accordance to the above, determine the system satisfied by an equilibrium strategy. To do so, we need to write down the fixed-point system. From the system of η, we deduce that

$$b_i\beta_{i,k+1}\sum_{j\neq i} b_j\eta_{jk} = -\eta_{ik}[r_{ik} + b_i^2\beta_{i,k+1}] - ab_i\beta_{i,k+1},$$

which means that if $(\beta_{i,k+1})_i$ is given, then $\eta_k = (\eta_{1k}, \ldots, \eta_{Ik})$ is $\eta_k = -am^{-1}\tilde{c}$ where $\tilde{c}_i = b_i\beta_{i,k+1}$ and $m_{ij} = b_i\beta_{i,k+1}b_j$, $m_{ii} = [r_{ik} + b_i^2\beta_{i,k+1}]$. In terms of information structure, a decision-maker may require to compute the real numbers $\eta_{j,k}$ of the other decision-makers at time k in order to get its best response strategy at time k, and this is in addition to the common state observation x_k. We can find $\bar{\eta}$ in a similar way from the vector $(\gamma_{i,k+1})_i$.

The system of β can be rewritten as

$$\beta_{i,k} = q_{ik} + r_{ik}\frac{\eta_{ik}^2[r_{ik} + b_i^2\beta_{i,k+1}]}{b_i^2\beta_{i,k+1}}.$$

The system of γ can be rewritten as

$$\gamma_{i,k} = (q_{ik} + \bar{q}_{ik}) + r_{ik}\frac{\bar{\eta}_{ik}^2(r_{ik} + b_i^2\gamma_{i,k+1})}{b_i^2\gamma_{i,k+1}}.$$

Since the equation β is nonlinear in η, we have to deal with nonlinear fixed-point equation, whose admissible solution (if any) provides an equilibrium strategy.

Proposition 14.7 *If there exist η, $\bar{\eta}$, β, γ such that*

$$\begin{cases} \eta_k = (\eta_{1k}, \ldots, \eta_{Ik}), \tilde{c}_i = b_i\beta_{i,k+1},\ m_{ij} = b_i\beta_{i,k+1}b_j, m_{ii} = [r_{ik} + b_i^2\beta_{i,k+1}],\\ \eta_k = -am^{-1}\tilde{c},\\ \bar{c}_i = b_i\gamma_{i,k+1},\ \bar{m}_{ij} = b_i\gamma_{i,k+1}b_j,\ \bar{m}_{ii} = [r_{ik} + b_i^2\gamma_{i,k+1}],\\ \bar{\eta}_k = -(a+\bar{a})\bar{m}^{-1}\bar{c},\\ \beta_{i,k} = q_{ik} + r_{ik}\frac{\eta_{ik}^2[r_{ik}+b_i^2\beta_{i,k+1}]}{b_i^2\beta_{i,k+1}},\\ \gamma_{i,k} = (q_{ik} + \bar{q}_{ik}) + r_{ik}\frac{\bar{\eta}_{ik}^2(r_{ik}+b_i^2\gamma_{i,k+1})}{b_i^2\gamma_{i,k+1}}, \end{cases} \tag{14.39}$$

then the mean-field-type Nash equilibrium of the LQ DT MFTG is given by

$$\begin{cases} u_{ik} = \eta_{ik}(x_k - Ex_k) + \bar{\eta}_{ik}Ex_k,\\ Ex_{k+1} = (Ex_0)\prod_{l=0}^{k}[a + \bar{a} + \sum_{i=1}^{I}\bar{\eta}_{i,l}]. \end{cases} \tag{14.40}$$

14.2.2 Risk-Neutral Global Optimum

We now analyze at the risk-neutral global optimum of the interactive system where the decision-makers collectively minimize the sum of individual costs. The problem is formulated as follows, with its solution given in Proposition 14.8.

$$\begin{cases} i \in \mathcal{I} \\ \inf_{u(\cdot)\in\mathcal{U}:=\prod_j \mathcal{U}_j} E[\sum_{i=1}^{I} L_i(u_1,\dots,u_I)] \\ \text{subject to } \ x_{k+1} = \left\{ax_k + \bar{a}Ex_k + \sum_{i=1}^{I} b_i u_{ik}\right\} + \sigma w_k, \\ x_0 \sim \mathcal{L}(X_0), \qquad E[X_0] = m(0) \end{cases} \tag{14.41}$$

Let $b = (b_1,\dots,b_I)$, $u_k = (u_{1k},\dots,u_{Ik})$. and $R_k = diag(r_{1k},\dots,r_{Ik})$. Denote by b' the transpose vector of b.

Proposition 14.8 *For problem (14.41), under the assumption that for* $k \in \{0,\dots,K-1\}$ $\sum_{j=1}^{I} q_{jk} > 0$, $\sum_{j=1}^{I}(q_{jk} + \bar{q}_{jk}) > 0$, $r_{ik} > 0$,, *there is a unique global optimum, and it is given by*

$$\begin{cases} \hat{u}_k = -[R_k + b'\hat{\beta}_{k+1}b]^{-1}b'a\hat{\beta}_{k+1}(x_k - Ex_k) - [R_k + b'\hat{\gamma}_{k+1}b]^{-1}b'(a+\bar{a})\hat{\gamma}_{k+1}Ex_k \\ \hat{\beta}_k = a^2\hat{\beta}_{k+1} + Q_k - \{[R_k + b'\hat{\beta}_{k+1}b]^{-1}b'\}'ba^2\hat{\beta}_{k+1}^2 \\ \hat{\beta}_K = Q_K := \sum_{i=1}^{I} q_{iK} > 0 \\ \hat{\gamma}_k = (a+\bar{a})^2\hat{\gamma}_{k+1} + (Q_k + \bar{Q}_k) - \{[R_k + b'\hat{\gamma}_{k+1}b]^{-1}b'\}'b(a+\bar{a})^2\hat{\gamma}_{k+1}^2 \\ \hat{\gamma}_K = Q_K + \bar{Q}_K := \sum_{i=1}^{I} q_{iK} + \bar{q}_{iK} > 0 \end{cases} \tag{14.42}$$

Furthermore, $\hat{\beta}$ *and* $\hat{\gamma}$ *satisfy with the following property* $\hat{\beta}_k > 0$ *and* $\hat{\gamma}_k > 0$, *and the globally optimum cost is*

$$E\left[\sum_i L_i(\hat{u})\right] = E\hat{\beta}_0(x_0 - Ex_0)^2 + \hat{\gamma}_0(Ex_0)^2 + \sum_{k=0}^{K-1}\hat{\beta}_{k+1}\sigma^2.$$

Note that Proposition 14.8 tells us that the global optimum of the MFTG (14.41) is a linear state-and-mean-field-type feedback in the form $\alpha_k x_k + \bar{\alpha}_k Ex_k$.

Proof of Proposition 14.8 We first compute the difference between the expected global cost and the one obtained by unknown sequences $\hat{\beta}, \hat{\gamma}$ yet to be determined.

$$
\begin{aligned}
&E\{\sum_i L_i - \hat{\beta}_0(x_0 - Ex_0)^2 - \hat{\gamma}_0(Ex_0)^2\} \\
&= E\sum_{k=0}^{K-1}[a^2\hat{\beta}_{k+1} - \hat{\beta}_k + Q_k - \{[R_k + \hat{\beta}_{k+1}b'b]^{-1}b'\}'ba^2\hat{\beta}_{k+1}^2]var(x_k) \\
&+E\sum_{k=0}^{K-1}\{u_k - Eu_k + [R_k + b'\hat{\beta}_{k+1}b]^{-1}a\hat{\beta}_{k+1}b'(x_k - Ex_k)\}'[R_k + \hat{\beta}_{k+1}b'b] \\
&\{u_k - Eu_k + [R_k + \hat{\beta}_{k+1}b'b]^{-1}a\hat{\beta}_{k+1}b'(x_k - Ex_k)\} \\
&+\sum_{k=0}^{K-1}\hat{\beta}_{k+1}\sigma^2 \\
&+E\sum_{k=0}^{K-1}[(a+\bar{a})^2\hat{\gamma}_{k+1} - \hat{\gamma}_k + (Q_k + \bar{Q}_k) - \{[R_k + b'\hat{\gamma}_{k+1}b]^{-1}b'\}'b(a+\bar{a})^2\hat{\gamma}_{k+1}^2](E[x_k])^2 \\
&+\{Eu_k + [R_k + b'\hat{\gamma}_{k+1}b]^{-1}(a+\bar{a})\hat{\gamma}_{k+1}b'Ex_k\}'[R_k + \hat{\gamma}_{k+1}b'b]\{Eu_k + [R_k + b'\hat{\gamma}_{k+1}b]^{-1} \\
&(a+\bar{a})\hat{\gamma}_{k+1}b'Ex_k\}
\end{aligned} \tag{14.43}
$$

It follows that

$$
\begin{cases}
\hat{\beta}_k = a^2\hat{\beta}_{k+1} + Q_k - \{[R_k + b'\hat{\beta}_{k+1}b]^{-1}b'\}'ba^2\hat{\beta}_{k+1}^2 \\
\hat{\beta}_K = Q_K = \sum_{i=1}^{I} q_{iK} > 0 \\
\hat{\gamma}_k = (a+\bar{a})^2\hat{\gamma}_{k+1} + (Q_k + \bar{Q}_k) - \{[R_k + b'\hat{\gamma}_{k+1}b]^{-1}b'\}'b(a+\bar{a})^2\hat{\gamma}_{k+1}^2 \\
\hat{\gamma}_K = Q_K + \bar{Q}_K = \sum_{i=1}^{I} q_{iK} + \bar{q}_{iK} > 0 \\
\hat{u}_k = -[R_k + b'\hat{\beta}_{k+1}b]^{-1}b'a\hat{\beta}_{k+1}(x_k - Ex_k) + E\hat{u}_k \\
E\hat{u}_k = -[R_k + b'\hat{\gamma}_{k+1}b'b]^{-1}b'(a+\bar{a})\hat{\gamma}_{k+1}Ex_k \\
E[\sum_i L_i(\hat{u})] = \hat{\beta}_0(x_0 - Ex_0)^2 + \hat{\gamma}_0(Ex_0)^2 + \sum_{k=0}^{K-1}\hat{\beta}_{k+1}\sigma^2
\end{cases} \tag{14.44}
$$

By backward induction, one can easily show that since the coefficients q_k, $\bar{q}_k + q_k$ are strictly positive and $r_{ik} > 0$,, the solutions $\hat{\beta}$ and $\hat{\gamma}$ are positive sequences. It is important to notice that $[R_k + \hat{\beta}_{k+1}b'b]$ is invertible, because it is a symmetric positive definite matrix. Therefore, there is no additional condition for the existence of a global optimal strategy, because it does not involve a fixed-point system.

It is easy that for a single decision-maker mean-field-type optimization, i.e., $I = 1$, the result of Proposition 14.18 is recovered. ■

14.2.3 *Difference Games with Additive and Multiplicative Noises*

We consider in this subsection a variation on the LQ DT MFTG discussed where now the mean-field terms are

- the expectation of the state in the drift and noise coefficient functions.
- the expectation of the control actions, in the drift and noise coefficient functions.

The state process is still scalar and the running costs of the decision-makers The instant cost function includes

- the expectation of the state, $E[x]$,
- the expectation of the control strategy, $E[u_j]$, j,
- the variance of the state, $\mathrm{var}(x)$
- the square of the conditional expected value of the state, $\{E[x]\}^2$
- the variance of the control-action process, $\mathrm{var}(u_j)$
- the square of the expected value of the control-action process, $\{E[u_j]\}^2$
- A cross-term such as covariance between the state and the control $\mathrm{cov}(x, u_j)$
- A product term between the expected value of the control-action and the expected value of the state, $E[u_j].E[x]$

The following boxed best response problem

Best-Response Problem with Additive and Multiplicative Noises.

$$\begin{cases} i \in \{1, 2, \ldots, I\}, \\ L_i = q_{iK}x_K^2 + \bar{q}_{iK}(E[x_K])^2 + \epsilon_{i3K}E[x_K] \\ +\sum_{k=0}^{K-1} q_{ik}x_k^2 + \bar{q}_{ik}(E[x_k])^2 + r_{ik}u_{ik}^2 + \bar{r}_{ik}(E[u_{ik}])^2 \\ +\sum_{k=0}^{K-1} 2\epsilon_{i1k}x_k u_{ik} + 2\bar{\epsilon}_{i1k}(E[x_k])E[u_{ik}] + \epsilon_{2ik}E[u_{ik}] + \epsilon_{i3k}E[x_k], \\ \\ \inf_{u_i \in \mathcal{U}_i} E[L_i], \\ \text{subject to} \\ x_{k+1} = \left(b_{0k} + b_{1k}x_k + \bar{b}_{1k}E[x_k] + \sum_{j=1}^{I} b_{2jk}u_{jk} + \sum_{j=1}^{I} \bar{b}_{2jk}E[u_{jk}]\right) \\ + \left(\sigma_{0k} + \sigma_{1k}x_k + \bar{\sigma}_{1k}E[x_k] + \sum_{j=1}^{I} \sigma_{2jk}u_{jk} + \sum_{j=1}^{I} \bar{\sigma}_{2jk}E[u_{jk}]\right) w_{k+1}, \\ x(0) = x_0, \end{cases} \tag{14.45}$$

where

$$\begin{cases} q_{ik} > 0, q_{ik} + \bar{q}_{ik} > 0, \\ r_{ik} > 0, r_{ik} + \bar{r}_{ik} > 0, \\ \epsilon_{i1k}, \bar{\epsilon}_{i1k}, \epsilon_{2ik}, \epsilon_{i3k} \in \mathbb{R}, \\ b_{0k}, b_{1k}, b_{2jk}, \bar{b}_{1k}, \bar{b}_{2jk} \in \mathbb{R}, \\ \sigma_{0k}, \sigma_{1k}, \sigma_{2jk}, \bar{\sigma}_{1k}, \bar{\sigma}_{2jk} \in \mathbb{R}, \\ w_{k+1} \text{ unit Gaussian} \end{cases} \tag{14.46}$$

By setting the coefficients $\sigma_1, \bar{\sigma}_1, \sigma_{2j}, \bar{\sigma}_{2j}$ equal to zero, one obtains an additive noise model where the noise is uncorrelated with the state process. When one of these coefficients is nonzero, for example, when $\sigma_1 \neq 0$,, one obtains a multiplicative noise model with $x_k w_{k+1}$. The model above includes both additive and multiplicative noises with control-and-mean-field dependence.

14.2.4 Reformulation with Covariance and Variance

We now provide below an equivalent formulation of Problem (14.45) using the same technique as in Sect. 7.6, followed by the solution to the best response problem.

Equivalent formulation of Problem (14.45).

$$
\begin{aligned}
& i \in \{1, 2, \ldots, I\}, \\
& \hat{L}_i = q_{iK}\ var(x_K) + (q_{iK} + \bar{q}_{iK})(E[x_K])^2 + \epsilon_{i3K} E[x_K] \\
& + \textstyle\sum_{k=0}^{K-1} q_{ik}\ var(x_k) + (q_{ik} + \bar{q}_{ik})(E[x_k])^2 + r_{ik}\ var(u_{ik}) + (r_{ik} + \bar{r}_{ik})(E[u_{ik}])^2 \\
& + \textstyle\sum_{k=0}^{K-1} 2\epsilon_{i1k}\ cov(x_k, u_{ik}) + 2(\epsilon_{i1k} + \bar{\epsilon}_{i1k})(E[x_k])E[u_{ik}] + \epsilon_{2ik} E[u_{ik}] + \epsilon_{i3k} E[x_k],
\end{aligned}
$$

$$
\begin{aligned}
& \inf_{u_i \in \mathcal{U}_i} E[\hat{L}_i], \\
& \text{subject to} \\
& x_{k+1} = \Big\{ b_{0k} + b_{1k}(x_k - E[x_k]) + (b_{1k} + \bar{b}_{1k}) E[x_k] \\
& + \textstyle\sum_{j=1}^{I} b_{2jk}(u_{jk} - E[u_{jk}]) + \sum_{j=1}^{I} (b_{2jk} + \bar{b}_{2jk}) E[u_{jk}] \Big\} \\
& + \{ \sigma_{0k} + \sigma_{1k}(x_k - E[x_k]) + (\sigma_{1k} + \bar{\sigma}_{1k}) E[x_k] \\
& + \textstyle\sum_{j=1}^{I} \sigma_{2jk}(u_{jk} - E[u_{jk}]) + \sum_{j=1}^{I} (\sigma_{2jk} + \bar{\sigma}_{2jk}) E[u_{jk}] \Big\}\ w_{k+1}, \\
& x(0) = x_0,
\end{aligned}
\tag{14.47}
$$

Solution to the Best-Response Problem (14.45).

For problem (14.45), under the positivity and invertibility conditions, the mean-field-type Nash equilibrium strategy is

$$\left\{\begin{array}{l} i \in \{1, 2, \ldots, I\}, \\ u_{ik} = -\eta_{ik}(x_k - Ex_k) - \bar{\eta}_{ik} Ex_k - \hat{\eta}_{ik}, \\ \eta_{ik} = -\frac{1}{c_{ik}}\alpha_{i,k+1}\sum_{j\neq i} b_{2ik}b_{2jk}\eta_{jk} - \frac{1}{c_{ik}}\alpha_{i,k+1}\sigma_{2ik}\sum_{j\neq i}\sigma_{2jk}\eta_{jk} \\ +\frac{1}{c_{ik}}\alpha_{i,k+1}b_{1k}b_{2ik} + \frac{1}{c_{ik}}\alpha_{i,k+1}\sigma_{1k}\sigma_{2ik} + \frac{\epsilon_{i1k}}{c_{ik}}, \\ c_{ik} = r_{ik} + \alpha_{i,k+1}(b_{2ik}^2 + \sigma_{2ik}^2), \\ \bar{\eta}_{ik} = -\frac{1}{\bar{c}_{ik}}\alpha_{i,k+1}(\sigma_{2ik} + \bar{\sigma}_{2ik})\sum_{j\neq i}(\sigma_{2jk} + \bar{\sigma}_{2jk})\bar{\eta}_{jk} \\ -\frac{1}{\bar{c}_{ik}}\beta_{i,k+1}(b_{2ik} + \bar{b}_{2ik})\sum_{j\neq i}(b_{2jk} + \bar{b}_{2jk})\bar{\eta}_{jk} \\ +\frac{1}{\bar{c}_{ik}}\alpha_{i,k+1}(\sigma_{1k} + \bar{\sigma}_{1k})(\sigma_{2ik} + \bar{\sigma}_{2ik}) \\ +\frac{1}{\bar{c}_{ik}}\beta_{i,k+1}(b_{1k} + \bar{b}_{1k})(b_{2ik} + \bar{b}_{2ik}) \\ +\frac{1}{\bar{c}_{ik}}(\epsilon_{i1k} + \bar{\epsilon}_{i1k}), \\ \bar{c}_{ik} = (r_{ik} + \bar{r}_{ik}) + \alpha_{i,k+1}(\sigma_{2ik} + \bar{\sigma}_{2ik})^2 + \beta_{i,k+1}(b_{2ik} + \bar{b}_{2ik})^2, \\ \hat{\eta}_{ik} = -\frac{1}{\bar{c}_{ik}}\alpha_{i,k+1}(\sigma_{2ik} + \bar{\sigma}_{2ik})\sum_{j\neq i}(\sigma_{2jk} + \bar{\sigma}_{2jk})\hat{\eta}_{jk} \\ -\frac{1}{\bar{c}_{ik}}\beta_{i,k+1}(b_{2ik} + \bar{b}_{2ik})\sum_{j\neq i}(b_{2jk} + \bar{b}_{2jk})\hat{\eta}_{jk} \\ +\frac{1}{\bar{c}_{ik}}\alpha_{i,k+1}\sigma_{0k}(\sigma_{2ik} + \bar{\sigma}_{2ik}) \\ +\frac{1}{\bar{c}_{ik}}\beta_{i,k+1}b_{0k}(b_{2ik} + \bar{b}_{2ik}) + \frac{1}{2\bar{c}_{ik}}(\gamma_{i,k+1}(b_{2ik} + \bar{b}_{2ik}) + \epsilon_{2ik}), \end{array}\right. \tag{14.48}$$

and the mean-field-type Nash equilibrium cost of decision-maker i is

$$E[L_i^*] = \alpha_{i0}(x_0 - Ex_0)^2 + \beta_{i0}(Ex_0)^2 + \gamma_{i0}(Ex_0) + \delta_{i0}.$$

Coefficients of Solution to (14.45).

$$\begin{aligned}
&\alpha_{ik} = q_{ik} - \eta_{ik}^2 c_{ik} + \alpha_{i,k+1}[b_{1k} - \textstyle\sum_{j\neq i} b_{2jk}\eta_{jk}]^2\\
&+\alpha_{i,k+1}[\sigma_{1k} - \textstyle\sum_{j\neq i} \sigma_{2jk}\eta_{jk}]^2,\\
&\alpha_{iK} = q_{iK},\\
&\beta_{ik} = (q_{ik} + \bar{q}_{ik}) - \bar{c}_{ik}\bar{\eta}_{ik}^2\\
&+\beta_{i,k+1}[(b_{1k} + \bar{b}_{1k}) - \textstyle\sum_{j\neq i}(b_{2jk} + \bar{b}_{2jk})\bar{\eta}_{jk}]^2\\
&+\alpha_{i,k+1}[(\sigma_{1k} + \bar{\sigma}_{1k}) - \textstyle\sum_{j\neq i}(\sigma_{2jk} + \bar{\sigma}_{2jk})\bar{\eta}_{jk}]^2,\\
&\beta_{iK} = q_{iK} + \bar{q}_{iK},\\
&\gamma_{ik} = \gamma_{i,k+1}(b_{1k} + \bar{b}_{1k}) + \epsilon_{i3k} - 2\bar{c}_{ik}\bar{\eta}_{ik}\hat{\eta}_{ik}\\
&+2\alpha_{i,k+1}\sigma_{0k}(\sigma_{1k} + \bar{\sigma}_{1k}) + 2\beta_{i,k+1}b_{0k}(b_{1k} + \bar{b}_{1k})\\
&+2\alpha_{i,k+1}[\textstyle\sum_{j\neq i}(\sigma_{2jk} + \bar{\sigma}_{2jk})\bar{\eta}_{jk}][\textstyle\sum_{j\neq i}(\sigma_{2jk} + \bar{\sigma}_{2jk})\hat{\eta}_{jk}]\\
&+2\beta_{i,k+1}[\textstyle\sum_{j\neq i}(b_{2jk} + \bar{b}_{2jk})\bar{\eta}_{jk}][\textstyle\sum_{j\neq i}(b_{2jk} + \bar{b}_{2jk})\hat{\eta}_{jk}]\\
&-2\alpha_{i,k+1}(\sigma_{1k} + \bar{\sigma}_{1k})\textstyle\sum_{j\neq i}(\sigma_{2jk} + \bar{\sigma}_{2jk})\hat{\eta}_{jk}\\
&-2\beta_{i,k+1}\textstyle\sum_{j\neq i}(b_{1k} + \bar{b}_{1k})(b_{2jk} + \bar{b}_{2jk})\hat{\eta}_{jk}\\
&-2\alpha_{i,k+1}\sigma_{0k}\textstyle\sum_{j\neq i}(\sigma_{2jk} + \bar{\sigma}_{2jk})\bar{\eta}_{jk}\\
&-2\beta_{i,k+1}\textstyle\sum_{\neq i}^{I} b_{0k}(b_{2jk} + \bar{b}_{2jk})\bar{\eta}_{jk} - \gamma_{i,k+1}\textstyle\sum_{j\neq i}(b_{2jk} + \bar{b}_{2jk})\bar{\eta}_{jk},\\
&\gamma_{iK} = \epsilon_{i3K},\\
&\delta_{ik} = \delta_{i,k+1} - \bar{c}_{ik}\hat{\eta}_{ik}^2 + \alpha_{i,k+1}\sigma_{0k}^2 + \beta_{i,k+1}b_{0k}^2 + \gamma_{i,k+1}b_{0k}\\
&+\alpha_{i,k+1}\{\textstyle\sum_{j\neq i}(\sigma_{2jk} + \bar{\sigma}_{2jk})\hat{\eta}_{jk}\}^2 + \beta_{i,k+1}[\textstyle\sum_{j\neq i}(b_{2jk} + \bar{b}_{2jk})\hat{\eta}_{jk}]^2\\
&-2\alpha_{i,k+1}\sigma_{0k}\textstyle\sum_{j\neq i}(\sigma_{2jk} + \bar{\sigma}_{2jk})\hat{\eta}_{jk} - 2\beta_{i,k+1}\textstyle\sum_{\neq i}^{I} b_{0k}(b_{2jk} + \bar{b}_{2jk})\hat{\eta}_{jk}\\
&-\gamma_{i,k+1}\textstyle\sum_{j\neq i}(b_{2jk} + \bar{b}_{2jk})\hat{\eta}_{jk},\\
&\delta_{iK} = 0.
\end{aligned} \tag{14.49}$$

14.2.4.1 Proof of Nash Equilibrium 14.62

Based on the structure of the terminal cost in (14.47), we start with a guess functional for decision-maker i tin the following form

$$\alpha_{ik}(x_k - Ex_k)^2 + \beta_{ik}(Ex_k)^2 + \gamma_{ik}(Ex_k) + \delta_{ik},$$

where $\alpha_{ik}, \beta_{ik}, \gamma_{ik}, \delta_{ik}$ are time-varying coefficients to be determined.

We use the telescopic sum

$$f_K = f_0 + \sum_{k=0}^{K-1} (f_{k+1} - f_k),$$

to derive the difference between the guess (ansatz) and the cost functionals.

$$\begin{aligned}
&E\{L_i - \alpha_{i0}(x_0 - Ex_0)^2 - \beta_{i0}(Ex_0)^2 - \gamma_{i0}(Ex_0) - \delta_{i0}\} \\
&= [q_{iK} - \alpha_{iK}]var(x_K) + [\beta_{iK} - q_{iK} - \bar{q}_{iK}](E[x_K])^2 \\
&+(\epsilon_{i3K} - \gamma_{iK})E[x_K] + (0 - \delta_{iK}) \\
&+ \textstyle\sum_{k=0}^{K-1} \{q_{ik} var(x_k) + (q_{ik} + \bar{q}_{ik})(E[x_k])^2 + r_{ik} E[u_{ik} - Eu_{ik}]^2 \\
&+(r_{ik} + \bar{r}_{ik})(Eu_{ik})^2\} \\
&+E \textstyle\sum_{k=0}^{K-1} \{\alpha_{i,k+1}(x_{k+1} - Ex_{k+1})^2 - \alpha_{ik}(x_k - Ex_k)^2\} \\
&+ \textstyle\sum_{k=0}^{K-1} \{\beta_{i,k+1}(Ex_{k+1})^2 - \beta_{i,k}(Ex_k)^2\} \\
&+ \textstyle\sum_{k=0}^{K-1} \{\gamma_{i,k+1}(Ex_{k+1}) - \gamma_{ik}(Ex_k)\} + \textstyle\sum_{k=0}^{K-1} \{\delta_{i,k+1} - \delta_{ik}\} \\
&+ \textstyle\sum_{k=0}^{K-1} 2\epsilon_{i1k}\, cov(x_k, u_{ik}) + 2(\epsilon_{i1k} + \bar{\epsilon}_{i1k})(E[x_k])E[u_{ik}] + \epsilon_{2ik} E[u_{ik}] \\
&+\epsilon_{i3k} E[x_k].
\end{aligned} \tag{14.50}$$

We next express the terms $(x_{k+1} - Ex_{k+1})^2$ and $(Ex_{k+1})^2$ as functions of $(x_k - Ex_k)$, $(u_{ik} - Eu_{ik})$, (Ex_k), (Eu_{ik}).

$$
\begin{aligned}
&E\{L_i-\alpha_{i0}(x_0-Ex_0)^2-\beta_{i0}(Ex_0)^2-\gamma_{i0}(Ex_0)-\delta_{i0}\}\\
&=[q_{iK}-\alpha_{iK}]var(x_K)+[\beta_{iK}-q_{iK}-\bar{q}_{iK}](E[x_K])^2\\
&+(\epsilon_{i3K}-\gamma_{iK})E[x_K]+(0-\delta_{iK})\\
&+E\sum_{k=0}^{K-1}\{r_{ik}+\alpha_{i,k+1}(b_{2ik}^2+\sigma_{2ik}^2)\}(u_{ik}-E[u_{ik}])^2\\
&+E\sum_{k=0}^{K-1}2\alpha_{i,k+1}b_{2ik}(u_{ik}-E[u_{ik}])\sum_{j\neq i}b_{2jk}(u_{jk}-E[u_{jk}])\\
&+E\sum_{k=0}^{K-1}2\alpha_{i,k+1}b_{1k}b_{2ik}(u_{ik}-E[u_{ik}])(x_k-E[x_k])\\
&+E\sum_{k=0}^{K-1}2\alpha_{i,k+1}w_{k+1}^2\sigma_{2ik}(u_{ik}-E[u_{ik}])\sum_{j\neq i}\sigma_{2jk}(u_{jk}-E[u_{jk}])\\
&+E\sum_{k=0}^{K-1}2\alpha_{i,k+1}w_{k+1}^2\sigma_{1k}\sigma_{2ik}(u_{ik}-E[u_{ik}])(x_k-E[x_k])\\
&+\sum_{k=0}^{K-1}2\epsilon_{i1k}\,cov(x_k,u_{ik})\\
&+E\sum_{k=0}^{K-1}\{(r_{ik}+\bar{r}_{ik})+\alpha_{i,k+1}(\sigma_{2ik}+\bar{\sigma}_{2ik})^2+\beta_{i,k+1}(b_{2ik}+\bar{b}_{2ik})^2\}(Eu_{ik})^2\\
&+E\sum_{k=0}^{K-1}2\alpha_{i,k+1}w_{k+1}^2(\sigma_{2ik}+\bar{\sigma}_{2ik})(E[u_{ik}])\sum_{j\neq i}(\sigma_{2jk}+\bar{\sigma}_{2jk})E[u_{jk}]\\
&+E\sum_{k=0}^{K-1}2\alpha_{i,k+1}w_{k+1}^2\sigma_{0k}(\sigma_{2ik}+\bar{\sigma}_{2ik})E[u_{ik}]\\
&+E\sum_{k=0}^{K-1}2\alpha_{i,k+1}w_{k+1}^2(\sigma_{1k}+\bar{\sigma}_{1k})(\sigma_{2ik}+\bar{\sigma}_{2ik})E[u_{ik}]E[x_k]\\
&+E\sum_{k=0}^{K-1}2\beta_{i,k+1}E[u_{ik}](b_{2ik}+\bar{b}_{2ik})\sum_{j\neq i}(b_{2jk}+\bar{b}_{2jk})E[u_{jk}]\\
&+E\sum_{k=0}^{K-1}2\beta_{i,k+1}b_{0k}(b_{2ik}+\bar{b}_{2ik})E[u_{ik}]\\
&+E\sum_{k=0}^{K-1}2\beta_{i,k+1}(b_{1k}+\bar{b}_{1k})(b_{2ik}+\bar{b}_{2ik})E[u_{ik}]E[x_k]\\
&+\sum_{k=0}^{K-1}2(\epsilon_{i1k}+\bar{\epsilon}_{i1k})(E[x_k])E[u_{ik}]\\
&+E\sum_{k=0}^{K-1}\gamma_{i,k+1}(b_{2ik}+\bar{b}_{2ik})E[u_{ik}]+\sum_{k=0}^{K-1}\epsilon_{2ik}E[u_{ik}]\\
&+E\sum_{k=0}^{K-1}\{\alpha_{i,k+1}b_{1k}^2-\alpha_{ik}+q_{ik}\}(x_k-E[x_k])^2\\
&+E\sum_{k=0}^{K-1}\alpha_{i,k+1}w_{k+1}^2\sigma_{1k}^2(x_k-E[x_k])^2\\
&+E\sum_{k=0}^{K-1}\{\beta_{i,k+1}(b_{1k}+\bar{b}_{1k})^2-\beta_{ik}+(q_{ik}+\bar{q}_{ik})\}(E[x_k])^2\\
&+E\sum_{k=0}^{K-1}\alpha_{i,k+1}w_{k+1}^2(\sigma_{1k}+\bar{\sigma}_{1k})^2(E[x_k])^2\\
&+E\sum_{k=0}^{K-1}2\alpha_{i,k+1}w_{k+1}^2\sigma_{0k}(\sigma_{1k}+\bar{\sigma}_{1k})E[x_k]\\
&+E\sum_{k=0}^{K-1}2\beta_{i,k+1}b_{0k}(b_{1k}+\bar{b}_{1k})E[x_k]\\
&+E\sum_{k=0}^{K-1}\{\gamma_{i,k+1}(b_{1k}+\bar{b}_{1k})-\gamma_{ik}+\epsilon_{i3k}\}E[x_k]\\
&+E\sum_{k=0}^{K-1}\alpha_{i,k+1}\sigma_{0k}^2w_{k+1}^2+E\sum_{k=0}^{K-1}\beta_{i,k+1}b_{0k}^2\\
&+E\sum_{k=0}^{K-1}\gamma_{i,k+1}b_{0k}+E\sum_{k=0}^{K-1}\{\delta_{i,k+1}-\delta_{ik}\}\\
&+E\sum_{k=0}^{K-1}\alpha_{i,k+1}[\sum_{j\neq i}b_{2jk}(u_{jk}-E[u_{jk}])]^2\\
&+E\sum_{k=0}^{K-1}\alpha_{i,k+1}w_{k+1}^2[\sum_{j\neq i}\sigma_{2jk}(u_{jk}-E[u_{jk}])]^2\\
&+E\sum_{k=0}^{K-1}2\alpha_{i,k+1}\sum_{j\neq i}b_{1k}b_{2jk}(u_{jk}-E[u_{jk}])(x_k-E[x_k])\\
&+E\sum_{k=0}^{K-1}2\alpha_{i,k+1}w_{k+1}^2\sigma_{1k}\sum_{j\neq i}\sigma_{2jk}(u_{jk}-E[u_{jk}])(x_k-E[x_k])\\
&+E\sum_{k=0}^{K-1}\alpha_{i,k+1}w_{k+1}^2\{\sum_{j\neq i}(\sigma_{2jk}+\bar{\sigma}_{2jk})E[u_{jk}]\}^2\\
&+E\sum_{k=0}^{K-1}\beta_{i,k+1}\{\sum_{j\neq i}(b_{2jk}+\bar{b}_{2jk})E[u_{jk}]\}^2\\
&+E\sum_{k=0}^{K-1}2\alpha_{i,k+1}w_{k+1}^2(\sigma_{1k}+\bar{\sigma}_{1k})\sum_{j\neq i}(\sigma_{2jk}+\bar{\sigma}_{2jk})E[u_{jk}]E[x_k]\\
&+E\sum_{k=0}^{K-1}2\beta_{i,k+1}\sum_{j\neq i}(b_{1k}+\bar{b}_{1k})(b_{2jk}+\bar{b}_{2jk})E[u_{jk}]E[x_k]\\
&+E\sum_{k=0}^{K-1}2\alpha_{i,k+1}w_{k+1}^2\sigma_{0k}\sum_{j\neq i}(\sigma_{2jk}+\bar{\sigma}_{2jk})E[u_{jk}]\\
&+E\sum_{k=0}^{K-1}2\beta_{i,k+1}\sum_{\neq i}^{I}b_{0k}(b_{2jk}+\bar{b}_{2jk})E[u_{jk}]\\
&+E\sum_{k=0}^{K-1}\gamma_{i,k+1}\sum_{j\neq i}(b_{2jk}+\bar{b}_{2jk})E[u_{jk}]
\end{aligned}
\tag{14.51}
$$

We then complete squares to arrive at

$$\begin{aligned}
&E\sum_{k=0}^{K-1}\{r_{ik}+\alpha_{i,k+1}(b_{2ik}^2+\sigma_{2ik}^2)\}(u_{ik}-E[u_{ik}])^2\\
&+E\sum_{k=0}^{K-1}2\alpha_{i,k+1}b_{2ik}(u_{ik}-E[u_{ik}])\sum_{j\neq i}b_{2jk}(u_{jk}-E[u_{jk}])\\
&+E\sum_{k=0}^{K-1}2\alpha_{i,k+1}b_{1k}b_{2ik}(u_{ik}-E[u_{ik}])(x_k-E[x_k])\\
&+E\sum_{k=0}^{K-1}2\alpha_{i,k+1}w_{k+1}^2\sigma_{2ik}(u_{ik}-E[u_{ik}])\sum_{j\neq i}\sigma_{2jk}(u_{jk}-E[u_{jk}])\\
&+E\sum_{k=0}^{K-1}2\alpha_{i,k+1}w_{k+1}^2\sigma_{1k}\sigma_{2ik}(u_{ik}-E[u_{ik}])(x_k-E[x_k])\\
&+\sum_{k=0}^{K-1}2\epsilon_{i1k}\,cov(x_k,u_{ik})\\
&=E\sum_{k=0}^{K-1}c_{ik}\{u_{ik}-E[u_{ik}]+\eta_{ik}(x_k-E[x_k])\}^2-\eta_{ik}^2c_{ik}(x_k-E[x_k])^2,
\end{aligned}\tag{14.52}$$

where

$$\begin{aligned}
&c_{ik}=r_{ik}+\alpha_{i,k+1}(b_{2ik}^2+\sigma_{2ik}^2),\\
&\eta_{ik}=-\frac{1}{c_{ik}}\alpha_{i,k+1}b_{2ik}\sum_{j\neq i}b_{2jk}\eta_{jk}-\frac{1}{c_{ik}}\alpha_{i,k+1}\sigma_{2ik}\sum_{j\neq i}\sigma_{2jk}\eta_{jk}\\
&+\frac{1}{c_{ik}}\alpha_{i,k+1}b_{1k}b_{2ik}+\frac{1}{c_{ik}}\alpha_{i,k+1}\sigma_{1k}\sigma_{2ik}+\frac{\epsilon_{i1k}}{c_{ik}}.
\end{aligned}\tag{14.53}$$

Similarly,

$$\begin{aligned}
&E\sum_{k=0}^{K-1}\{(r_{ik}+\bar r_{ik})+\alpha_{i,k+1}(\sigma_{2ik}+\bar\sigma_{2ik})^2+\beta_{i,k+1}(b_{2ik}+\bar b_{2ik})^2\}(Eu_{ik})^2\\
&+E\sum_{k=0}^{K-1}2\alpha_{i,k+1}w_{k+1}^2(\sigma_{2ik}+\bar\sigma_{2ik})(E[u_{ik}])\sum_{j\neq i}(\sigma_{2jk}+\bar\sigma_{2jk})E[u_{jk}]\\
&+E\sum_{k=0}^{K-1}2\beta_{i,k+1}E[u_{ik}](b_{2ik}+\bar b_{2ik})\sum_{j\neq i}(b_{2jk}+\bar b_{2jk})E[u_{jk}]\\
&+E\sum_{k=0}^{K-1}2\alpha_{i,k+1}w_{k+1}^2(\sigma_{1k}+\bar\sigma_{1k})(\sigma_{2ik}+\bar\sigma_{2ik})E[u_{ik}]E[x_k]\\
&+E\sum_{k=0}^{K-1}2\beta_{i,k+1}(b_{1k}+\bar b_{1k})(b_{2ik}+\bar b_{2ik})E[u_{ik}]E[x_k]\\
&+\sum_{k=0}^{K-1}2(\epsilon_{i1k}+\bar\epsilon_{i1k})(E[x_k])E[u_{ik}]\\
&+E\sum_{k=0}^{K-1}2\alpha_{i,k+1}w_{k+1}^2\sigma_{0k}(\sigma_{2ik}+\bar\sigma_{2ik})E[u_{ik}]\\
&+E\sum_{k=0}^{K-1}2\beta_{i,k+1}b_{0k}(b_{2ik}+\bar b_{2ik})E[u_{ik}]\\
&+E\sum_{k=0}^{K-1}\gamma_{i,k+1}(b_{2ik}+\bar b_{2ik})E[u_{ik}]\\
&+\sum_{k=0}^{K-1}\epsilon_{2ik}E[u_{ik}]\\
&=E\sum_{k=0}^{K-1}\bar c_{ik}\{Eu_{ik}+\eta_{ik}E[x_k]+\hat\eta_{ik}\}^2-\bar c_{ik}[\bar\eta_{ik}E[x_k]+\hat\eta_{ik}]^2
\end{aligned}\tag{14.54}$$

with

$$\begin{aligned}
&\bar c_{ik}=(r_{ik}+\bar r_{ik})+\alpha_{i,k+1}(\sigma_{2ik}+\bar\sigma_{2ik})^2+\beta_{i,k+1}(b_{2ik}+\bar b_{2ik})^2,\\
&\bar\eta_{ik}E[x_k]+\hat\eta_{ik}=\frac{1}{\bar c_{ik}}\alpha_{i,k+1}(\sigma_{2ik}+\bar\sigma_{2ik})\sum_{j\neq i}(\sigma_{2jk}+\bar\sigma_{2jk})E[u_{jk}]\\
&+\frac{1}{\bar c_{ik}}\beta_{i,k+1}(b_{2ik}+\bar b_{2ik})\sum_{j\neq i}(b_{2jk}+\bar b_{2jk})E[u_{jk}]\\
&+\frac{1}{\bar c_{ik}}\alpha_{i,k+1}(\sigma_{1k}+\bar\sigma_{1k})(\sigma_{2ik}+\bar\sigma_{2ik})E[x_k]\\
&+\frac{1}{\bar c_{ik}}\beta_{i,k+1}(b_{1k}+\bar b_{1k})(b_{2ik}+\bar b_{2ik})E[x_k]\\
&+\frac{1}{\bar c_{ik}}(\epsilon_{i1k}+\bar\epsilon_{i1k})(E[x_k])+\frac{1}{\bar c_{ik}}\alpha_{i,k+1}\sigma_{0k}(\sigma_{2ik}+\bar\sigma_{2ik})\\
&+\frac{1}{\bar c_{ik}}\beta_{i,k+1}b_{0k}(b_{2ik}+\bar b_{2ik})+\frac{1}{2\bar c_{ik}}(\gamma_{i,k+1}(b_{2ik}+\bar b_{2ik})+\epsilon_{2ik}).
\end{aligned}\tag{14.55}$$

This leads to

$$
\begin{aligned}
&\bar{\eta}_{ik} = -\tfrac{1}{\bar{c}_{ik}}\alpha_{i,k+1}(\sigma_{2ik}+\bar{\sigma}_{2ik})\textstyle\sum_{j\neq i}(\sigma_{2jk}+\bar{\sigma}_{2jk})\bar{\eta}_{jk}\\
&-\tfrac{1}{\bar{c}_{ik}}\beta_{i,k+1}(b_{2ik}+\bar{b}_{2ik})\textstyle\sum_{j\neq i}(b_{2jk}+\bar{b}_{2jk})\bar{\eta}_{jk}\\
&+\tfrac{1}{\bar{c}_{ik}}\alpha_{i,k+1}(\sigma_{1k}+\bar{\sigma}_{1k})(\sigma_{2ik}+\bar{\sigma}_{2ik})\\
&+\tfrac{1}{\bar{c}_{ik}}\beta_{i,k+1}(b_{1k}+\bar{b}_{1k})(b_{2ik}+\bar{b}_{2ik})+\tfrac{1}{\bar{c}_{ik}}(\epsilon_{i1k}+\bar{\epsilon}_{i1k}),\\
&\hat{\eta}_{ik} = -\tfrac{1}{\bar{c}_{ik}}\alpha_{i,k+1}(\sigma_{2ik}+\bar{\sigma}_{2ik})\textstyle\sum_{j\neq i}(\sigma_{2jk}+\bar{\sigma}_{2jk})\hat{\eta}_{jk}\\
&-\tfrac{1}{\bar{c}_{ik}}\beta_{i,k+1}(b_{2ik}+\bar{b}_{2ik})\textstyle\sum_{j\neq i}(b_{2jk}+\bar{b}_{2jk})\hat{\eta}_{jk}\\
&+\tfrac{1}{\bar{c}_{ik}}\alpha_{i,k+1}\sigma_{0k}(\sigma_{2ik}+\bar{\sigma}_{2ik})\\
&+\tfrac{1}{\bar{c}_{ik}}\beta_{i,k+1}b_{0k}(b_{2ik}+\bar{b}_{2ik})+\tfrac{1}{2\bar{c}_{ik}}(\gamma_{i,k+1}(b_{2ik}+\bar{b}_{2ik})+\epsilon_{2ik}).
\end{aligned}
\tag{14.56}
$$

Now collecting all together, we arrive at

$$
\begin{aligned}
&E\{L_i-\alpha_{i0}(x_0-Ex_0)^2-\beta_{i0}(Ex_0)^2-\gamma_{i0}(Ex_0)-\delta_{i0}\}\\
&=[q_{iK}-\alpha_{iK}]var(x_K)+[\beta_{iK}-q_{iK}-\bar{q}_{iK}](E[x_K])^2\\
&+(\epsilon_{i3K}-\gamma_{iK})E[x_K]+(0-\delta_{iK})\\
&+E\textstyle\sum_{k=0}^{K-1}c_{ik}\{u_{ik}-E[u_{ik}]+\eta_{ik}(x_k-E[x_k])\}^2\\
&+E\textstyle\sum_{k=0}^{K-1}\Big\{\alpha_{i,k+1}b_{1k}^2-\alpha_{ik}+q_{ik}-\eta_{ik}^2c_{ik}\\
&+\alpha_{i,k+1}\sigma_{1k}^2+\alpha_{i,k+1}[\textstyle\sum_{j\neq i}b_{2jk}\eta_{jk}]^2+\alpha_{i,k+1}[\sum_{j\neq i}\sigma_{2jk}\eta_{jk}]^2\\
&-2\alpha_{i,k+1}\textstyle\sum_{j\neq i}b_{1k}b_{2jk}\eta_{jk}-2\alpha_{i,k+1}\sigma_{1k}\sum_{j\neq i}\sigma_{2jk}\eta_{jk}\Big\}(x_k-E[x_k])^2\\
&+E\textstyle\sum_{k=0}^{K-1}\bar{c}_{ik}\{Eu_{ik}+(\eta_{ik}E[x_k]+\hat{\eta}_{ik})\}^2\\
&+E\textstyle\sum_{k=0}^{K-1}\Big\{\beta_{i,k+1}(b_{1k}+\bar{b}_{1k})^2-\beta_{ik}+(q_{ik}+\bar{q}_{ik})-\bar{c}_{ik}\bar{\eta}_{ik}^2\\
&+\alpha_{i,k+1}(\sigma_{1k}+\bar{\sigma}_{1k})^2+\alpha_{i,k+1}[\textstyle\sum_{j\neq i}(\sigma_{2jk}+\bar{\sigma}_{2jk})\bar{\eta}_{jk}]^2\\
&+\beta_{i,k+1}[\textstyle\sum_{j\neq i}(b_{2jk}+\bar{b}_{2jk})\bar{\eta}_{jk}]^2\\
&-2\alpha_{i,k+1}(\sigma_{1k}+\bar{\sigma}_{1k})\textstyle\sum_{j\neq i}(\sigma_{2jk}+\bar{\sigma}_{2jk})\bar{\eta}_{jk}\\
&-2\beta_{i,k+1}\textstyle\sum_{j\neq i}(b_{1k}+\bar{b}_{1k})(b_{2jk}+\bar{b}_{2jk})\bar{\eta}_{jk}\Big\}(E[x_k])^2\\
&+E\textstyle\sum_{k=0}^{K-1}\Big\{\gamma_{i,k+1}(b_{1k}+\bar{b}_{1k})-\gamma_{ik}+\epsilon_{i3k}-2\bar{c}_{ik}\bar{\eta}_{ik}\hat{\eta}_{ik}\\
&+2\alpha_{i,k+1}\sigma_{0k}(\sigma_{1k}+\bar{\sigma}_{1k})+2\beta_{i,k+1}b_{0k}(b_{1k}+\bar{b}_{1k})\\
&+2\alpha_{i,k+1}[\textstyle\sum_{j\neq i}(\sigma_{2jk}+\bar{\sigma}_{2jk})\bar{\eta}_{jk}][\sum_{j\neq i}(\sigma_{2jk}+\bar{\sigma}_{2jk})\hat{\eta}_{jk})]\\
&+2\beta_{i,k+1}[\textstyle\sum_{j\neq i}(b_{2jk}+\bar{b}_{2jk})\bar{\eta}_{jk}][\sum_{j\neq i}(b_{2jk}+\bar{b}_{2jk})\hat{\eta}_{jk}]\\
&-2\alpha_{i,k+1}(\sigma_{1k}+\bar{\sigma}_{1k})\textstyle\sum_{j\neq i}(\sigma_{2jk}+\bar{\sigma}_{2jk})\hat{\eta}_{jk}\\
&-2\beta_{i,k+1}\textstyle\sum_{j\neq i}(b_{1k}+\bar{b}_{1k})(b_{2jk}+\bar{b}_{2jk})\hat{\eta}_{jk}\\
&-2\alpha_{i,k+1}\sigma_{0k}\textstyle\sum_{j\neq i}(\sigma_{2jk}+\bar{\sigma}_{2jk})\bar{\eta}_{jk}\\
&-2\beta_{i,k+1}\textstyle\sum_{\neq i}^{I}b_{0k}(b_{2jk}+\bar{b}_{2jk})\bar{\eta}_{jk}-\gamma_{i,k+1}\sum_{j\neq i}(b_{2jk}+\bar{b}_{2jk})\bar{\eta}_{jk}\Big\}Ex_k\\
&+\textstyle\sum_{k=0}^{K-1}\Big\{\delta_{i,k+1}-\delta_{ik}-\bar{c}_{ik}\hat{\eta}_{ik}^2+\alpha_{i,k+1}\sigma_{0k}^2+\beta_{i,k+1}b_{0k}^2+\gamma_{i,k+1}b_{0k}\\
&+\alpha_{i,k+1}\{\textstyle\sum_{j\neq i}(\sigma_{2jk}+\bar{\sigma}_{2jk})\hat{\eta}_{jk}\}^2\\
&+\beta_{i,k+1}[\textstyle\sum_{j\neq i}(b_{2jk}+\bar{b}_{2jk})\hat{\eta}_{jk}]^2-2\alpha_{i,k+1}\sigma_{0k}\sum_{j\neq i}(\sigma_{2jk}+\bar{\sigma}_{2jk})\hat{\eta}_{jk}\\
&-2\beta_{i,k+1}\textstyle\sum_{\neq i}^{I}b_{0k}(b_{2jk}+\bar{b}_{2jk})\hat{\eta}_{jk}-\gamma_{i,k+1}\sum_{j\neq i}(b_{2jk}+\bar{b}_{2jk})\hat{\eta}_{jk}\Big\}
\end{aligned}
\tag{14.57}
$$

Through proper matching, one obtains the desired result. ∎

14.2.5 From Discrete to Continuous Time

We study in this subsection the realtonship between continuous time LQ MFTG and their counterparts in discrete time, obtaining through discretization of the time interval. Toward that end, let the discretization parameter (that is the step size) be $\Delta > 0$. Further, let

$$\begin{aligned}
&\tilde{b}_{0k} = b_{0k}\Delta,\ \tilde{b}_{1k} = 1 + b_{1k}\Delta,\\
&\tilde{b}_{2jk} = b_{2jk}\Delta, \tilde{\bar{b}}_{2jk} = \bar{b}_{2jk}\Delta,\\
&\tilde{\sigma} = \sigma\sqrt{\Delta},\ \tilde{\bar{\sigma}} = \bar{\sigma}\sqrt{\Delta},\\
&\tilde{q}_{ik} = q_{ik}\Delta,\ \tilde{\bar{q}}_{ik} = \bar{q}_{ik}\Delta,\\
&\tilde{r}_{ik} = r_{ik}\Delta,\ \tilde{\bar{r}}_{ik} = \bar{r}_{ik}\Delta,\\
&\tilde{\epsilon}_{2ik} = \epsilon_{2ik}\Delta\ ,\ \tilde{\epsilon}_{i3k} = \epsilon_{i3k}\Delta,\\
&w_{k+1} = \frac{1}{\sqrt{\Delta}}(B_{(k+1)\Delta} - B_{k\Delta}),\ K = \left[\frac{T}{\Delta}\right].
\end{aligned} \tag{14.58}$$

Now, in the original formulation of the LQ MFTG, we replace he original parameters with the corresponding tilded ones above. It follows from the scaling parameters above that

$$\begin{cases}
\frac{\tilde{b}_{0k}\tilde{b}_{2jk}}{\Delta} \to 0,\\
\frac{(b_{2i}+\bar{b}_{2i})(b_{2j}+\bar{b}_{2j})}{\Delta} \to 0,\\
\{\tilde{b}_{1k} - \sum_{j\neq i}\tilde{b}_{2jk}\eta_{jk}\}^2 = 1 + \Delta[2b_1 - 2\sum_{j\neq i} b_{2jk}\eta_{jk}] + O(\Delta^2),\\
w_{k+1} \to dB(t),
\end{cases} \tag{14.59}$$

as the step size Δ goes to zero,

14.2.5.1 Limiting Cost as the Time-Scale Vanishes

We now look at the limiting game as Δ goes to zero. The cost functional and the state dynamics become

$$\begin{cases}
i \in \{1, 2, \ldots, I\},\\
\hat{L}_i = q_i(T)\ var(x(T)) + (q_i(T) + \bar{q}_i(T))(E[x(T)])^2 + \epsilon_{i3}(T)E[x(T)]\\
+\int_0^T q_i\ var(x_k) + (q_i + \bar{q}_i)(E[x])^2 + r_i\ var(u_i) + (r_i + \bar{r}_i)(E[u_i])^2\, dt\\
+\int_0^T 2\epsilon_{i1}\ cov(x, u_i) + 2(\epsilon_{i1} + \bar{\epsilon}_{i1})(E[x])E[u_i] + \epsilon_{2i}E[u_i] + \epsilon_{i3}E[x]\, dt,\\
\\
dx = \big\{b_0 + b_1(x - E[x]) + (b_1 + \bar{b}_1)E[x]\\
+\sum_{j=1}^I b_{2j}(u_j - E[u_j]) + \sum_{j=1}^I (b_{2j} + \bar{b}_{2j})E[u_j]\big\}\\
+\{\sigma_0 + \sigma_1(x - E[x]) + (\sigma_1 + \bar{\sigma}_1)E[x]\\
+\sum_{j=1}^I \sigma_{2j}(u_j - E[u_j]) + \sum_{j=1}^I (\sigma_{2j} + \bar{\sigma}_{2j})E[u_j]\big\}\, dB(t),\\
x(0) = X_0,
\end{cases} \tag{14.60}$$

This is a continuous time LQ MFTG.

14.2.5.2 Limiting Equilibrium Strategies and Costs as the Time Scale Vanishes

Closed-form expressions for the mean-field equilibrium and the corresponding costs for the decision-makers of the game problem (14.60) were provided in Sect. 7.6.

We verify here that the discrete time solution converges to the continuous-time solution as Δ goes to zero. Note that the denominator coefficients become

$$\begin{cases} \frac{\tilde{c}_{ik}}{\Delta} \to c_i := r_i + \alpha_i \sigma_{2i}^2, \\ \frac{\bar{\tilde{c}}_{ik}}{\Delta} \to \bar{c}_i := (r_i + \bar{r}_i) + \alpha_i(\sigma_{2i} + \bar{\sigma}_{2i})^2, \end{cases} \tag{14.61}$$

$$\begin{cases} u_i = -\eta_i(x - Ex) - \bar{\eta}_i Ex - \hat{\eta}_i, \\ \eta_i = -\frac{1}{c_i}\alpha_i\sigma_{2i}\sum_{j\neq i}\sigma_{2j}\eta_j + \frac{1}{c_i}\alpha_i b_{2i} + \frac{1}{c_i}\alpha_i\sigma_1\sigma_{2i} + \frac{\epsilon_{i1}}{c_i}, \\ \\ \bar{\eta}_i = -\frac{1}{\bar{c}_i}\alpha_i(\sigma_{2i} + \bar{\sigma}_{2i})\sum_{j\neq i}(\sigma_{2j} + \bar{\sigma}_{2j})\bar{\eta}_j \\ +\frac{1}{\bar{c}_i}\alpha_i(\sigma_1 + \bar{\sigma}_1)(\sigma_{2i} + \bar{\sigma}_{2i}) + \frac{1}{\bar{c}_i}(\epsilon_{i1} + \bar{\epsilon}_{i1}), \\ \hat{\eta}_i = -\frac{1}{\bar{c}_i}\alpha_i(\sigma_{2i} + \bar{\sigma}_{2i})\sum_{j\neq i}(\sigma_{2j} + \bar{\sigma}_{2j})\hat{\eta}_j \\ +\frac{1}{\bar{c}_i}\alpha_i\sigma_0(\sigma_{2i} + \bar{\sigma}_{2i}) + \frac{1}{2\bar{c}_i}(\gamma_i(b_{2i} + \bar{b}_{2i}) + \epsilon_{2i}), \end{cases} \tag{14.62}$$

and the equilibrium cost of decision-maker i is

$$E[\hat{L}_i] = \alpha_i(0)(x_0 - Ex_0)^2 + \beta_i(0)(Ex_0)^2 + \gamma_i(0)(Ex_0) + \delta_i(0).$$

The scaling $\frac{\alpha_{i,k+1}-\alpha_{ik}}{\Delta}$ converges to the time derivative of α_i as Δ goes to zero.

$$\frac{\alpha_{i,k+1} - \alpha_{ik}}{\Delta} \to \dot{\alpha}_i.$$

The limiting coefficient α_i solves the set of ordinary differential equations:

$$\begin{cases} \dot{\alpha}_i = -q_i - 2\alpha_i b_1 - \alpha_i\sigma_1^2 + \eta_i^2 c_i + 2\alpha_i\sum_{j\neq i} b_{2j}\eta_j \\ +2\alpha_i\sigma_1\sum_{j\neq i}\sigma_{2j}\eta_{jk} - \alpha_i[\sum_{j\neq i}\sigma_{2j}\eta_j]^2, \\ \alpha_i(T) = q_i(T). \end{cases} \tag{14.63}$$

The scaling $\frac{\beta_{i,k+1}-\beta_{ik}}{\Delta}$ converges to the time derivative of β_i as Δ goes to zero.

$$\frac{\beta_{i,k+1} - \beta_{ik}}{\Delta} \to \dot{\beta}_i.$$

The limiting coefficient β_i solves the set of ordinary differential equations:

$$\begin{cases} \dot{\beta}_i = -(q_i + \bar{q}_i) - 2\beta_i(b_1 + \bar{b}_1) + \bar{c}_i \bar{\eta}_i^2 + 2\beta_i \sum_{j\neq i}(b_{2j} + \bar{b}_{2j})\bar{\eta}_j \\ -\beta_i[(\sigma_1 + \bar{\sigma}_1) - \sum_{j\neq i}(\sigma_{2j} + \bar{\sigma}_{2j})\bar{\eta}_j]^2, \\ \beta_i(T) = q_i(T) + \bar{q}_i(T), \end{cases} \tag{14.64}$$

The scaling $\frac{\gamma_{i,k+1}-\gamma_{ik}}{\Delta}$ converges to the time derivative of γ_i as Δ goes to zero.

$$\frac{\gamma_{i,k+1} - \gamma_{ik}}{\Delta} \to \dot{\gamma}_i,$$

The limiting coefficient γ_i solves the set of ordinary differential equations:

$$\begin{cases} \dot{\gamma}_i = -2\beta_i b_0 - \gamma_i(b_1 + \bar{b}_1) - \epsilon_{i3} + 2\bar{c}_i \bar{\eta}_{ik}\hat{\eta}_{ik} \ - 2\alpha_i\sigma_0(\sigma_1 + \bar{\sigma}_1) \\ -2\alpha_i[\sum_{j\neq i}(\sigma_{2j} + \bar{\sigma}_{2j})\bar{\eta}_j][\sum_{j\neq i}(\sigma_{2j} + \bar{\sigma}_{2j})\hat{\eta}_j)] \\ +2\alpha_i(\sigma_1 + \bar{\sigma}_1)\sum_{j\neq i}(\sigma_{2j} + \bar{\sigma}_{2j})\hat{\eta}_j + 2\alpha_i\sigma_0\sum_{j\neq i}(\sigma_{2j} + \bar{\sigma}_{2j})\bar{\eta}_j \\ +2\beta_i\sum_{j\neq i}(b_{2j} + \bar{b}_{2j})\hat{\eta}_j + \gamma_i\sum_{j\neq i}(b_{2j} + \bar{b}_{2j})\bar{\eta}_j, \\ \gamma_i(T) = \epsilon_{i3}(T), \end{cases} \tag{14.65}$$

Similarly, $\frac{\delta_{i,k+1}-\delta_{ik}}{\Delta} \to \dot{\delta}_i$, and the limiting coefficient δ_i solve the set of ordinary differential equations:

$$\begin{cases} \dot{\delta}_i = -\gamma_i b_0 - \alpha_i\sigma_0^2 + \bar{c}_i\hat{\eta}_i^2 - \alpha_i\{\sum_{j\neq i}(\sigma_{2j} + \bar{\sigma}_{2j})\hat{\eta}_j\}^2 + 2\alpha_i\sigma_0\sum_{j\neq i}(\sigma_{2j} + \bar{\sigma}_{2j})\hat{\eta}_j \\ +\gamma_i\sum_{j\neq i}(b_{2j} + \bar{b}_{2j})\hat{\eta}_j, \\ \delta_i(T) = 0. \end{cases} \tag{14.66}$$

Thus, we retrieve the results of Sect. 7.6 in the absence of jump terms as Δ vanishes.

14.3 Higher-Order Cost

This section presents a unified framework for solving discrete-time optimal control and game problems with higher-order state and control costs of power-law form. By building convex-completion techniques, we derive semi-explicit expressions for equilibrium strategies, cost-to-go functions, and recursive coefficient dynamics across deterministic, stochastic, and multi-agent system settings. Key contributions include variance-aware solutions under additive and multiplicative noise, extensions to mean-field-type-dependent dynamics, and conditions that ensure the positivity of recursive coefficients. In particular, we establish that higher-order costs induce less aggressive control policies compared to quadratic formulations, a finding that is validated through numerical analyses. The results provide a foundational methodology for analyzing nonlinear systems under higher-order penalization, bridging classical optimal control theory, game theory, and mean-field-type game theory with modern computational tools for applications ranging from engineering to multi-agent machine intelligence systems. The key contributions of this section are summarized as follows:

Deterministic Control with Higher-Order Costs:

- We start with a basic framework for solving discrete-time optimal control problems with non-quadratic cost functions, specifically the quartic terms $\bar{x}^4$ in state and quartic in control $\bar{u}^4$
- We then develop a convex-completion approach to derive semi-explicit solutions for the optimal control and the associated cost-to-go function.
- We establish recursive equations for cost coefficients and prove conditions ensuring their positivity, guaranteeing well-posed solutions.
- We extend the method to generalized even-order cost functions of the form $\bar{x}^{2p}$ and $\bar{u}^{2p}$.
- We present numerical examples illustrating the impact of increasing p on optimal state trajectories, control actions, and optimal cost to go.

Variance-Aware Control with Additive Noise:

- We extend the deterministic control framework to stochastic systems with additive noise, incorporating variance-aware cost functions.
- We then derive semi-explicit optimal control solutions for systems with quartic $var(x) + \bar{x}^4$ and $var(u) + \bar{u}^4$ and higher-order penalties $var(x) + \bar{x}^{2p}$ and $var(u) + \bar{u}^{2p}$ under stochastic perturbations.
- We establish recursive conditions for cost coefficients and ensure their positivity and stability and demonstrate through semi-explicit solutions that higher-order variance-aware control leads to less aggressive control actions.

Risk-Aware Stochastic Control with Multiplicative Noise

- We generalize the variance-aware control framework to stochastic systems with multiplicative noise, capturing uncertainty-dependent dynamics.
- We develop recursive cost-to-go formulations accounting for both state variance and higher-order cost penalties.
- We derive a semi-explicit optimal control strategy that balances risk sensitivity and nonlinear cost minimization.
- We extend the analysis to general even-order cost functions and solve it semi-explicitly.

Selfish Agents with Higher-Order Costs

- We then introduce a multi-agent game framework where each agent optimizes a higher-order cost function.
- We develop semi-explicit equilibrium strategies and equilibrium costs for selfish agents interacting through a shared system dynamics model.
- We establish recursive conditions for the existence of state-feedback Nash equilibrium strategies with quartic and higher-order costs.

- We extend the approach to general even-order cost functions and discusses the feasibility of solving such games using state-of-the-art neural networks.

Selfish Agents with Variance-Aware Higher-Order Costs

- We extend the game-theoretic model to stochastic environments where agents account for both mean-state deviations and variance-aware penalties $var(x) + (E[x])^{2p}$, $var(u_{ik}) + (E[u_{ik}])^{2p}$ per agent, x being a common state to all agents and u_{ik} being the control action picked by agent i at time step k.
- We derive recursive state-and-mean-field-type feedback Nash equilibrium strategies that incorporate both risk sensitivity and higher-order cost minimization.
- We establish conditions for the well-posedness of the variance-aware game and present validation of equilibrium strategies.

Risk-Aware Games Under Multiplicative Noise

- We finally develop a stochastic game framework where agents operate in environments with multiplicative noise that is state-control action-and-mean-field-type dependent.
- We establish semi-explicit state-and-mean-field-type feedback Nash equilibrium strategies and equilibrium costs for risk-sensitive agents with higher-order cost penalties $(x - E[x])^{2o} + (E[x])^{2p}$, $(u_{ik} - E[u_{ik}])^{2o} + (E[u_{ik}])^{2p}$, $o, p \geq 1$.
- We extend the recursive solution framework to general even-order cost functions, ensuring applicability to a broad class of nonlinear game settings.
- We provide a verification method using a measure-space formulation, confirming the robustness of the proposed equilibrium solutions (Table 14.1).

Table 14.1 Summary of contributions across different problem types

Control type	Optimal state	Optimal control	Optimal cost-to-go
Deterministic control with higher-order costs	(✓)	(✓)	(✓)
Variance-aware control with additive noise	(✓)	(✓)	(✓)
Risk-aware stochastic control with multiplicative noise	(✓)	(✓)	(✓)
Game type	Equilibrium state	Equilibrium strategy (per agent)	Equilibrium cost (per agent)
Selfish agents with higher-order costs	(✓)	(✓)	(✓)
Selfish agents with variance-awareness	(✓)	(✓)	(✓)
Risk-awareness under multiplicative noise	(✓)	(✓)	(✓)

14.3.1 How Does the Convex Completion Works?

As it was presented before in the literature review, there are many works on the computation of semi-explicit solutions to mean-field-type game problems by following the so-called direct method. One of the key elements in this method, in the framework of linear-quadratic settings, is the square completion in order to optimize the Hamiltonian over the control actions. This work uses convex completion as briefly introduced next.

Let X be a real vector space and X^* be its dual space. Let $f : X \to \mathbb{R} \cup \{+\infty\}$ be a proper, convex, and lower semicontinuous function. The Legendre-Fenchel conjugate (or transform) of f is defined as

$$f^*(y) = \sup_{x \in X}\{\langle y, x\rangle - f(x)\},$$

where $\langle y, x\rangle$ denotes the dual pairing between $y \in X^*$ and $x \in X$. The convexity inequality, also known as Fenchel's inequality or Young's inequality, states that for any $x \in X$ and $y \in X^*$:

$$f(x) + f^*(y) \geq \langle y, x\rangle.$$

The convex completion method uses this inequality as follows. After making the difference between the candidate cost and game cost functional of the agent, one gets a function of the control of that agent, say $f_i(u_i)$. We complete this cost by adding and removing $f_i^*(v_i) - \langle v_i, u_i\rangle$.

$$f_i(u_i) + f_i^*(v_i) - \langle v_i, u_i\rangle \geq 0$$

for control actions (u_i, v_i). The equality holds if and only if $v_i \in \partial f_i(u_i)$, where $\partial f_i(u_i)$ is the sub-differential of f_i at u_i. The sub-differential is a generalization of the gradient for non-differentiable functions. This extra term added to the difference cost goes to the coefficient matching terms making the solution semi-explicit.

14.3.2 Convexity of Higher-Order Terms

Next, Lemma 14.3 shows a convexity property that the development of the section relies on.

Lemma 14.3 *Let $p \geq 1, a \neq 0, b \neq 0$. Then the mapping $z \mapsto f(z) = z^{2p} + (az + b)^{2p}$ is strictly convex.* ∎

Proof The function f is twice differentiable as it is polynomial. It is strictly convex if its second derivative $f''(z) > 0$, for all z. We differentiate twice to obtain

$$f''(z) = 2p(2p-1)z2p-2 + 2p(2p-1)a^2(az+b)^{2p-2}.$$

We observe the following:

- When $p \geq 1$, the term $2p(2p-1)z^{2p-2}$ is *positive* for all $z \neq 0$.
- When $p \geq 1, a, b \neq 0$, the term $2p(2p-1)a^2(az+b)^{2p-2}$ is always *positive* for $z \neq -b/a \neq 0$.

Since both terms are nonnegative and do not vanish simultaneously, the function is strictly convex under the assumption that $p \geq 1, a, b \neq 0$. This complete the proof. ∎

14.3.3 Deterministic Control with Higher-Order Costs

In this section, we explicitly solve the discrete-time optimal control problem with higher-order costs. We derive closed-form solutions for the optimal control, the optimal cost, and the recursive coefficients for the cost-to-go function. Additionally, we extend the methodology to address general higher-order costs of the form $\bar{u}^{2p}$ for arbitrary $p \geq 1$, where $\bar{u}$ denotes the expected control input.

14.3.3.1 Optimal Control with Quartic Cost

We consider the following discrete-time dynamical system:

$$\bar{x}_{k+1} = \bar{a}_k \bar{x}_k + \bar{b}_k \bar{u}_k, \quad k \in \{0, \ldots, N-1\}, \tag{14.67}$$

where $\bar{x}_k \in \mathbb{R}$ is the state, $\bar{u}_k \in \mathbb{R}$ is the control input, and $\bar{a}_k, \bar{b}_k \in \mathbb{R}$ are system parameters. The dynamics in (14.67) evolves within a discrete-time horizon $[0 \ldots N] := [0, N] \cap \mathbb{Z}$ with $N \in \mathbb{Z}_+$. The initial state $\bar{x}_0$ is given. The goal is to minimize the cost function

$$L(x_0, \bar{u}) = \bar{q}_N \bar{x}_N^4 + \sum_{k=0}^{N-1} \left(\bar{q}_k \bar{x}_k^4 + \bar{r}_k \bar{u}_k^4 \right), \tag{14.68}$$

where $\bar{q}_k, \bar{q}_N, \bar{r}_k > 0$ are weights penalizing the state and control deviations. Therefore, the optimal control problem is

$$\min_{\bar{u} \in \mathscr{U}_k} L(x_0, \bar{u}), \text{ s.t. (14.67)}, \ x_0 \text{ given.} \tag{14.69}$$

In order to solve the problem in (14.69) we need to specify the information structure and strategy available to the agent who choose the control action. The set of control actions at time step k is

$$\bar{U}_k = \{\bar{u}_k \in \mathbb{R} |\ \bar{u}_k^4 < +\infty\} = \mathbb{R},$$

and the set

$$\bar{\mathcal{U}}_k = \{(\bar{u}_k)_{k\in\{0,\dots,N-1\}} \mid \bar{u}_k(.) \in \bar{U}_k, \bar{u}_k \text{ Lebesgue measurable}\}.$$

We restrict our attention to state-feedback control strategies, i.e., $\bar{u}_k(\bar{x}_k)$. This means that instead of considering the entire past history $(\bar{x}_0, \bar{x}_1, \dots, \bar{x}_{k-1}, \bar{x}_k)$, we choose to work only with the current state $\bar{x}_k$.

Note that for any $\bar{u} \in \bar{\mathcal{U}}$, the system (14.67) admits unique solution denoted by $\bar{x} := \bar{x}^{\bar{x}_0,\bar{u}}$ meaning it is determined by the initial state and the control strategy employed over the path. In this particular case, the solution is explicitly obtained by recursion of linear (affine) operators as shown next in Proposition 14.9.

Proposition 14.9 *The optimal control input for the problem in* (14.69) *with system dynamics in* (14.67) *and cost functional* (14.68) *is given by*

$$\bar{u}_k^* = -\frac{\left(\frac{\bar{\alpha}_{k+1}\bar{b}_k}{\bar{r}_k}\right)^{1/3}\bar{a}_k}{1+\left(\frac{\bar{\alpha}_{k+1}\bar{b}_k}{\bar{r}_k}\right)^{1/3}\bar{b}_k}\bar{x}_k. \tag{14.70}$$

where α solves the following backward recursive equation:

$$\bar{\alpha}_k = \bar{q}_k + \bar{r}_k\left(\frac{\left(\frac{\bar{\alpha}_{k+1}\bar{b}_k}{\bar{r}_k}\right)^{1/3}\bar{a}_k}{1+\left(\frac{\bar{\alpha}_{k+1}\bar{b}_k}{\bar{r}_k}\right)^{1/3}\bar{b}_k}\right)^4 + \bar{\alpha}_{k+1}\left(\bar{a}_k - \frac{\left(\frac{\bar{\alpha}_{k+1}\bar{b}_k}{\bar{r}_k}\right)^{1/3}\bar{a}_k\bar{b}_k}{1+\left(\frac{\bar{\alpha}_{k+1}\bar{b}_k}{\bar{r}_k}\right)^{1/3}\bar{b}_k}\right)^4, \tag{14.71}$$

with terminal condition $\bar{\alpha}_N = \bar{q}_N$ depending on the cost weight. ∎

Proof We propose a candidate for the optimal cost function at time k to be

$$f_k(\bar{x}_k) = \bar{\alpha}_k\bar{x}_k^4, \tag{14.72}$$

where $\bar{\alpha}_k > 0$ is a scalar coefficient to be determined recursively. Then, we use telescopic sum, i.e.,

$$f_N(\bar{x}_N) = f_0(\bar{x}_0) + \sum_{k=0}^{N-1} f_{k+1}(\bar{x}_{k+1}) - f_k(\bar{x}_k).$$

By taking the difference $L(x_0, u) - f_0(\bar{x}_0)$, we obtain

$$L(x_0, \bar{u}) - f_0(\bar{x}_0) = \bar{q}_N \bar{x}_N^4 - f_N(\bar{x}_N) + \sum_{k=0}^{N-1} \left(\bar{q}_k \bar{x}_k^4 + \bar{r}_k \bar{u}_k^4 \right)$$
$$+ f_{k+1}(\bar{x}_{k+1}) - f_k(\bar{x}_k).$$

Now, let us introduce the following term:

$$C_k = \min_{\bar{u}_k} \left(\bar{r}_k \bar{u}_k^4 + \bar{\alpha}_{k+1} (\bar{a}_k \bar{x}_k + \bar{b}_k \bar{u}_k)^4 \right),$$

and it yields

$$L(x_0, \bar{u}) - f_0(\bar{x}_0) = \bar{q}_N \bar{x}_N^4 - f_N(\bar{x}_N)$$
$$+ \sum_{k=0}^{N-1} \left(\bar{r}_k \bar{u}_k^4 + f_{k+1}(\bar{x}_{k+1}) \right) - C_k + \left(C_k + \bar{q}_k \bar{x}_k^4 - f_k(\bar{x}_k) \right).$$

Note here that we are not using the completing the squares technique but a completion of convex terms. This generalization to convex completion is very important as we will see in the sequel to handle higher-order cost. The advantage of convex completion is that it covers more general convex functionals that are invariant by the structure of the control actions used by the agent. Here, the running cost function is convex $(\bar{x}_k, \bar{u}_k)$. To find the optimal control, we differentiate the cost function with respect to $\bar{u}_k$:

$$4\bar{r}_k \bar{u}_k^3 + 4\bar{\alpha}_{k+1} (\bar{a}_k \bar{x}_k + \bar{b}_k \bar{u}_k)^3 \bar{b}_k = 0.$$

Simplify and isolate $(\bar{a}_k \bar{x}_k + \bar{b}_k \bar{u}_k)^3$:

$$\bar{u}_k^3 = -\frac{\bar{\alpha}_{k+1} \bar{b}_k}{\bar{r}_k} (\bar{a}_k \bar{x}_k + \bar{b}_k \bar{u}_k)^3,$$

and taking the cube root one obtains the announced optimal control input $\bar{u}_k^*$ as in (14.70). Now, let us compute the recursive equation for $\bar{\alpha}_k$. Substitute $\bar{u}_k^*$ back into the difference equation, and collect terms proportional to $\bar{x}_k^4$ obtaining (14.71), completing the proof. ■

Considering quartic cost functional implies that the cost can only get nonnegative values as $\bar{q}_k, \bar{q}_N, \bar{r}_k > 0$. Therefore, according to the ansatz for the value function in (14.72), the recursive solution for $\bar{\alpha}$ in (14.71) should be positive. Remark 14.2 shows the conditions under which there is existence of positive $\bar{\alpha}_k$.

Remark 14.2 The coefficients $\bar{\alpha}_k$ remain positive in (14.71) if

- $\bar{q}_k, \bar{q}_N, \bar{r}_k > 0$,
- $\bar{a}_k, \bar{b}_k$ are bounded,
- The denominator in $\bar{u}_k^*$ satisfies $1 + \left(\frac{\bar{\alpha}_{k+1} \bar{b}_k}{\bar{r}_k} \right)^{1/3} \bar{b}_k > 0$.

These conditions are trivially satisfied in most systems, ensuring $\bar{\alpha}_k > 0$ for all $k \in \{0, 1, \ldots, N\}$. ▮

14.3.3.2 Extension to $\bar{x}^{2p}$ Costs

We now provide for discrete-time optimal control problems with costs of the form $\bar{x}_k^{2p}$ and $\bar{u}_k^{2p}$, where $p \geq 1$. The solutions include the optimal control law, the recursive relation for the cost-to-go coefficient $\bar{\alpha}_k$, and the expression for the optimal cost. We consider the same discrete-time system as in (14.67). The objective is to minimize the 2p-order cost function

$$L(x_0, u) = \bar{q}_N \bar{x}_N^{2p} + \sum_{k=0}^{N-1} \left(\bar{q}_k \bar{x}_k^{2p} + \bar{r}_k \bar{u}_k^{2p} \right), \tag{14.73}$$

where $\bar{q}_k, \bar{q}_N, \bar{r}_k > 0$ are weights.

Proposition 14.10 *The optimal control input for the problem in* (14.69) *with system dynamics in* (14.67) *and cost functional* (14.73) *is given by*

$$\bar{u}_k^* = -\frac{\left(\frac{\bar{\alpha}_{k+1}\bar{b}_k}{\bar{r}_k}\right)^{1/(2p-1)} \bar{a}_k}{1 + \left(\frac{\bar{\alpha}_{k+1}\bar{b}_k}{\bar{r}_k}\right)^{1/(2p-1)} \bar{b}_k} \bar{x}_k, \tag{14.74}$$

where α solves the following backward recursive equation

$$\begin{aligned} \bar{\alpha}_k &= \bar{q}_k + \bar{r}_k \left(\frac{\left(\frac{\bar{\alpha}_{k+1}\bar{b}_k}{\bar{r}_k}\right)^{1/(2p-1)} \bar{a}_k}{1 + \left(\frac{\bar{\alpha}_{k+1}\bar{b}_k}{\bar{r}_k}\right)^{1/(2p-1)} \bar{b}_k} \right)^{2p} \\ &+ \bar{\alpha}_{k+1} \left(\bar{a}_k - \bar{b}_k \frac{\left(\frac{\bar{\alpha}_{k+1}\bar{b}_k}{\bar{r}_k}\right)^{1/(2p-1)} \bar{a}_k}{1 + \left(\frac{\bar{\alpha}_{k+1}\bar{b}_k}{\bar{r}_k}\right)^{1/(2p-1)} \bar{b}_k} \right)^{2p}, \end{aligned} \tag{14.75}$$

with terminal condition $\bar{\alpha}_N = \bar{q}_N$ depending on the cost weight. ▮

Proof The candidate cost is assumed to have the form

$$f_k(\bar{x}_k) = \bar{\alpha}_k \bar{x}_k^{2p},$$

where $\bar{\alpha}_k > 0$ is a coefficient to be determined recursively, i.e., the optimal cost at each time step is proportional to $\bar{x}_k^{2p}$. The difference term now has

$$C_k = \min_{\bar{u}_k} \left(\bar{r}_k \bar{u}_k^{2p} + \bar{\alpha}_{k+1} (\bar{a}_k \bar{x}_k + \bar{b}_k \bar{u}_k)^{2p} \right).$$

Differentiating with respect to $\bar{u}_k$ gives

$$2p\, \bar{r}_k \bar{u}_k^{2p-1} + 2p\, \bar{\alpha}_{k+1} \bar{b}_k (\bar{a}_k \bar{x}_k + \bar{b}_k \bar{u}_k)^{2p-1} = 0.$$

Rearrange terms yields

$$\bar{r}_k \bar{u}_k^{2p-1} + \bar{\alpha}_{k+1} \bar{b}_k (\bar{a}_k \bar{x}_k + \bar{b}_k \bar{u}_k)^{2p-1} = 0.$$

Isolate $(\bar{a}_k \bar{x}_k + \bar{b}_k \bar{u}_k)^{2p-1}$, i.e.,

$$\bar{u}_k^{2p-1} = -\frac{\bar{\alpha}_{k+1} \bar{b}_k}{\bar{r}_k} (\bar{a}_k \bar{x}_k + \bar{b}_k \bar{u}_k)^{2p-1}.$$

Now, by taking the $(2p - 1)$-th root on both sides, one obtains the optimal control input as in (14.74). Finally, by substituting the optimal control $\bar{u}_k^*$ back into the candidate cost yields the recursive equation announced in (14.75), completing the proof. ∎

The result presented in Proposition 14.10 is more general than the quadratic cost that has been reported in the literature. Hence, one can easily retrieve the quadratic solution by adjusting the p-value as shown in the remark below.

Remark 14.3 For $p = 1$, the latter equation becomes

$$\bar{\alpha}_k = \bar{q}_k + \frac{\bar{\alpha}_{k+1} \bar{a}_k^2 \bar{r}_k}{\bar{r}_k + \bar{\alpha}_{k+1} \bar{b}_k^2},$$

with terminal condition $\bar{\alpha}_N = \bar{q}_N$, which is the Riccati equation for the discrete-time linear-quadratic regulator (LQR) case. ▮

14.3.4 Variance-Aware Control with Additive Noise

Non-quadratic costs, such as $(Ex)^4$ or $(Eu)^4$, introduce another class of realistic representation of these nonlinearities, emphasizing large deviations in state and control. We solve the variance-aware discrete-time stochastic optimal control problem with higher-order costs. We derive semi-explicit solutions for the variance-aware stochastic optimal control, the optimal variance-aware cost, and the recursive coefficients for the variance-aware cost-to-go function. Additionally, we extend the methodology to address general higher-order costs $(Ex)^{2p}$ for arbitrary $p \geq 1$.

14.3.4.1 Variance-Aware Quartic Cost with Additive Noise

We consider the following discrete-time dynamical system:

$$x_{k+1} = \bar{a}_k x_k + \bar{b}_k u_k + \epsilon_{k+1}, \quad k \in \{0, \ldots, N-1\} \tag{14.76}$$

where $x_k \in \mathbb{R}$ is the state, $u_k \in \mathbb{R}$ is the control input, and $\bar{a}_k, \bar{b}_k \in \mathbb{R}$ are system parameters. The initial state x_0 is a given random variable that is independent of ϵ. $\{\epsilon_k\}_k$ is a zero-mean random process, with $\epsilon_0 = 0$. The state dynamics in (14.76) can be rewritten as

$$x_{k+1} = \bar{a}_k \bar{x}_k + \bar{b}_k \bar{u}_k + \bar{a}_k(x_k - \bar{x}_k) + \bar{b}_k(u_k - \bar{u}_k) + \epsilon_{k+1}, \quad k \in \{0, \ldots, N-1\},$$

The goal is to minimize the expected value of the following cost function:

$$L(x_0, u) = q_k var(x_N) + \bar{q}_N \bar{x}_N^4 + \sum_{k=0}^{N-1} \left(q_k var(x_k) + \bar{q}_k \bar{x}_k^4 + r_k var(u_k) + \bar{r}_k \bar{u}_k^4 \right),$$

where $q_k, q_N, r_k, \bar{q}_k, \bar{q}_N, \bar{r}_k > 0$ are weights penalizing the state and control deviations. The variance of the system state and control input are given by $var(x) = E[(x - \bar{x})^2]$, and $var(u) = E[(u - \bar{u})^2]$, respectively. Therefore the variance-aware control problem becomes

$$\min_{\bar{u} \in \bar{\mathscr{U}}_k} E[L(x_0, \bar{u})], \text{ s.t. (14.76)}, \ x_0 \text{ given.} \tag{14.77}$$

Proposition 14.11 *The optimal control input for the variance-aware problem in (14.77) is given by*

$$u_k^* = -\frac{\alpha_{k+1} \bar{b}_k \bar{a}_k}{(r_k + \alpha_{k+1} \bar{b}_k^2)}(x_k - \bar{x}_k) - \frac{\left(\frac{\bar{\alpha}_{k+1} \bar{b}_k}{\bar{r}_k}\right)^{1/3} \bar{a}_k}{1 + \left(\frac{\bar{\alpha}_{k+1} \bar{b}_k}{\bar{r}_k}\right)^{1/3} \bar{b}_k} \bar{x}_k, \tag{14.78}$$

where $\bar{\alpha}_k$, $\bar{\gamma}_k$, and α_k solve the following recursive equations:

$$\bar{\alpha}_k = \bar{q}_k + \bar{r}_k \left(\frac{\left(\frac{\bar{\alpha}_{k+1} \bar{b}_k}{\bar{r}_k}\right)^{1/3} \bar{a}_k}{1 + \left(\frac{\bar{\alpha}_{k+1} \bar{b}_k}{\bar{r}_k}\right)^{1/3} \bar{b}_k} \right)^4 + \bar{\alpha}_{k+1} \left(\bar{a}_k - \frac{\left(\frac{\bar{\alpha}_{k+1} \bar{b}_k}{\bar{r}_k}\right)^{1/3} \bar{a}_k \bar{b}_k}{1 + \left(\frac{\bar{\alpha}_{k+1} \bar{b}_k}{\bar{r}_k}\right)^{1/3} \bar{b}_k} \right)^4, \tag{14.79}$$

$$\bar{\gamma}_k = \bar{\gamma}_{k+1} + E\left(\alpha_{k+1} \epsilon_{k+1}^2\right), \tag{14.80}$$

$$\alpha_k = q_k + r_k \left(\frac{\alpha_{k+1} \bar{b}_k \bar{a}_k}{(r_k + \alpha_{k+1} \bar{b}_k^2)} \right)^2 + \alpha_{k+1} \left(\bar{a}_k - \frac{\alpha_{k+1} \bar{b}_k^2 \bar{a}_k}{(r_k + \alpha_{k+1} \bar{b}_k^2)} \right)^2. \tag{14.81}$$

with $\bar{\alpha}_N = \bar{q}_N$, $\bar{\gamma}_N = 0$, *and* $\alpha_N = q_N$ *and terminal conditions.* ∎

Proof We propose a candidate cost function at time k to be

$$f_k(x_k, \bar{x}_k) = \alpha_k var(x_k) + \bar{\alpha}_k \bar{x}_k^4 + \bar{\gamma}_k,$$

where $\alpha_k, \bar{\alpha}_k \geq 0$ are scalar coefficients to be determined recursively. Now, applying telescopic sum yields

$$f_N(x_N) = f_0(x_0) + \sum_{k=0}^{N-1} f_{k+1}(x_{k+1}) - f_k(x_k).$$

By taking the difference $L - f_0(x_0)$, we obtain

$$L(x_0, u) - f_0(x_0) = q_N var(x_N) + \bar{q}_N \bar{x}_N^4 - f_N(x_N)$$
$$+ \sum_{k=0}^{N-1} \left(q_k var(x_k) + \bar{q}_k \bar{x}_k^4 + r_k var(u_k) + \bar{r}_k \bar{u}_k^4\right) + f_{k+1}(x_{k+1}) - f_k(x_k).$$

Let us now consider the term

$$C_k = \min_{u_k} \left(r_k var(u_k) + \alpha_{k+1} var(x_{k+1}) + \bar{r}_k \bar{u}_k^4 + \bar{\alpha}_{k+1}(\bar{a}_k \bar{x}_k + \bar{b}_k \bar{u}_k)^4\right),$$

and

$$\begin{aligned} L(x_0, u) - f_0(x_0) = {} & q_N var(x_N) + \bar{q}_N \bar{x}_N^4 - f_N(x_N) \\ & + \sum_{k=0}^{N-1} r_k var(u_k) + \bar{r}_k \bar{u}_k^4 + f_{k+1}(x_{k+1}) - C_k \\ & + (C_k + q_k var(x_k) + \bar{q}_k \bar{x}_k^4 - f_k(x_k)). \end{aligned} \tag{14.82}$$

Now, the following term can be rewritten as follows:

$$\begin{aligned} r_k var(u_k) + \alpha_{k+1} var(x_{k+1}) = {} & E(r_k(u_k - \bar{u}_k)^2 + \alpha_{k+1}(\bar{a}_k(x_k - \bar{x}_k) \\ & + \bar{b}_k(u_k - \bar{u}_k) + \epsilon_{k+1})^2) \end{aligned}$$

which means that

$$(u_k - \bar{u}_k) = -\frac{\alpha_{k+1} \bar{b}_k \bar{a}_k}{(r_k + \alpha_{k+1} \bar{b}_k^2)} (x_k - \bar{x}_k)$$

To find the deterministic part of optimal control $\bar{u}_k^*$, we differentiate the cost function with respect to u_k, i.e.,

$$4\bar{r}_k\bar{u}_k^3 + 4\bar{\alpha}_{k+1}(\bar{a}_k\bar{x}_k + \bar{b}_k\bar{u}_k)^3\bar{b}_k = 0.$$

Simplify and isolate $(\bar{a}_k\bar{x}_k + \bar{b}_k\bar{u}_k)^3$:

$$\bar{u}_k^3 = -\frac{\bar{\alpha}_{k+1}\bar{b}_k}{\bar{r}_k}(\bar{a}_k\bar{x}_k + \bar{b}_k\bar{u}_k)^3.$$

Taking the cube root:

$$\bar{u}_k^* = -\frac{\left(\frac{\bar{\alpha}_{k+1}\bar{b}_k}{\bar{r}_k}\right)^{1/3}\bar{a}_k\bar{x}_k}{1+\left(\frac{\bar{\alpha}_{k+1}\bar{b}_k}{\bar{r}_k}\right)^{1/3}\bar{b}_k}.$$

Therefore, the optimal control is the announced one in (14.78). Now, substituting u_k^* back into the difference equation in (14.82) and collecting terms proportional to x_k^4, one obtains the announced recursive equations for $\bar{\alpha}_k$, $\bar{\gamma}_k$, and α_k, completing the proof. ■

Remark 14.4 The coefficients $\alpha_k, \bar{\alpha}_k$ remain positive if

1. $q_k, q_N, r_k, \bar{q}_k, \bar{q}_N, \bar{r}_k > 0$,
2. $\bar{a}_k, \bar{b}_k$ are bounded,
3. The denominator in $\bar{u}_k^*$ satisfies $1 + \left(\frac{\bar{\alpha}_{k+1}\bar{b}_k}{\bar{r}_k}\right)^{1/3}\bar{b}_k > 0$.
4. The denominator in $u_k - \bar{u}_k^*$ satisfies $(r_k + \alpha_{k+1}\bar{b}_k^2) > 0$.

These conditions are trivially satisfied in most systems, ensuring $\alpha_k > 0$ and $\bar{\alpha}_k$ for all $k \in \{0, 1, \ldots, N\}$. ▮

14.3.4.2 Extension to $\bar{x}^{2p}$ Costs

We now provide for discrete-time optimal control problems with costs of the form x_k^{2p} and u_k^{2p}, where $p \geq 1$. The solutions include the optimal control law, the recursive relation for the cost-to-go coefficient α_k, and the expression for the optimal cost. We consider the same discrete-time system as in (14.76). The objective is to minimize the variance-aware $2p$-order cost function

$$\begin{aligned} L(x_0, u) &= q_N var(x_N) + \bar{q}_N\bar{x}_N^{2p} \\ &+ \textstyle\sum_{k=0}^{N-1}\left(q_k var(x_k) + \bar{q}_k\bar{x}_k^{2p} + r_k var(u_k) + \bar{r}_k\bar{u}_k^{2p}\right), \end{aligned} \tag{14.83}$$

where $q_k, q_N, r_k, \bar{q}_k, \bar{q}_N, \bar{r}_k > 0$ are weights.

Proposition 14.12 *The optimal control input for the variance-aware problem in* (14.77) *with dynamics in* (14.76) *and cost functional* (14.83) *is given by*

$$u_k^* = -\frac{\alpha_{k+1}\bar{b}_k\bar{a}_k}{(r_k + \alpha_{k+1}\bar{b}_k^2)}(x_k - \bar{x}_k) - \frac{\left(\frac{\bar{\alpha}_{k+1}\bar{b}_k}{\bar{r}_k}\right)^{1/(2p-1)} \bar{a}_k\bar{x}_k}{1 + \left(\frac{\bar{\alpha}_{k+1}\bar{b}_k}{\bar{r}_k}\right)^{1/(2p-1)} \bar{b}_k}, \tag{14.84}$$

where $\bar{\alpha}_k$, $\bar{\gamma}_k$, *and* α_k *solve the following recursive equations:*

$$\begin{aligned}
\bar{\alpha}_k \quad &= \bar{q}_k + \bar{r}_k \left(\frac{\left(\frac{\bar{\alpha}_{k+1}\bar{b}_k}{\bar{r}_k}\right)^{1/(2p-1)} \bar{a}_k}{1+\left(\frac{\bar{\alpha}_{k+1}\bar{b}_k}{\bar{r}_k}\right)^{1/(2p-1)} \bar{b}_k} \right)^{2p} \\
&+\bar{\alpha}_{k+1} \left(\bar{a}_k - \bar{b}_k \frac{\left(\frac{\bar{\alpha}_{k+1}\bar{b}_k}{\bar{r}_k}\right)^{1/(2p-1)} \bar{a}_k}{1+\left(\frac{\bar{\alpha}_{k+1}\bar{b}_k}{\bar{r}_k}\right)^{1/(2p-1)} \bar{b}_k} \right)^{2p}, \qquad (14.85) \\
\bar{\gamma}_k \quad &= \bar{\gamma}_{k+1} + E\left(\alpha_{k+1}\epsilon_{k+1}^2\right), \qquad (14.86)
\end{aligned}$$

$$\alpha_k = q_k + r_k \left(\frac{\alpha_{k+1}\bar{b}_k\bar{a}_k}{(r_k+\alpha_{k+1}\bar{b}_k^2)} \right)^2 + \alpha_{k+1} \left(\bar{a}_k - \frac{\alpha_{k+1}\bar{b}_k^2\bar{a}_k}{(r_k+\alpha_{k+1}\bar{b}_k^2)} \right)^2, \tag{14.87}$$

with $\bar{\alpha}_N = \bar{q}_N$, $\bar{\gamma}_N = 0$, *and* $\alpha_N = q_N$ *and terminal conditions.* ∎

Proof The candidate cost is assumed to have the form

$$f_k(x_k, \bar{x}_k) = \alpha_k var(x_k) + \bar{\alpha}_k \bar{x}_k^{2p} + \bar{\gamma}_k,$$

where $\alpha_k \geq 0, \bar{\alpha}_k \geq 0$ is a coefficient to be determined recursively. The expected value of difference term now has

$$C_k = \min_{u_k} \left(r_k var(u_k) + \alpha_{k+1} var(x_{k+1}) + \bar{r}_k \bar{u}_k^{2p} + \bar{\alpha}_{k+1}(\bar{a}_k\bar{x}_k + \bar{b}_k\bar{u}_k)^{2p} \right).$$

The stochastic part of the optimal control is determined by

$$\begin{aligned}
&r_k var(u_k) + \alpha_{k+1} var(x_{k+1}) \\
&= E(r_k(u_k - \bar{u}_k)^2 + \alpha_{k+1}(\bar{a}_k(x_k - \bar{x}_k) + \bar{b}_k(u_k - \bar{u}_k) + \epsilon_{k+1})^2)
\end{aligned}$$

which means that

$$(u_k - \bar{u}_k) = -\frac{\alpha_{k+1}\bar{b}_k\bar{a}_k}{(r_k + \alpha_{k+1}\bar{b}_k^2)}(x_k - \bar{x}_k).$$

The deterministic part of the optimal control is obtained as follows. Differentiating with respect to $\bar{u}_k$ gives

$$2p\,\bar{r}_k\bar{u}_k^{2p-1} + 2p\,\bar{\alpha}_{k+1}\bar{b}_k(\bar{a}_k\bar{x}_k + \bar{b}_k\bar{u}_k)^{2p-1} = 0.$$

Rearrange terms

$$\bar{r}_k\bar{u}_k^{2p-1} + \bar{\alpha}_{k+1}\bar{b}_k(\bar{a}_k\bar{x}_k + \bar{b}_k\bar{u}_k)^{2p-1} = 0.$$

Isolate $(\bar{a}_k\bar{x}_k + \bar{b}_k\bar{u}_k)^{2p-1}$:

$$\bar{u}_k^{2p-1} = -\frac{\bar{\alpha}_{k+1}\bar{b}_k}{\bar{r}_k}(\bar{a}_k\bar{x}_k + \bar{b}_k\bar{u}_k)^{2p-1}.$$

Take the $(2p-1)$-th root on both sides:

$$\bar{u}_k = -\frac{\left(\frac{\bar{\alpha}_{k+1}\bar{b}_k}{\bar{r}_k}\right)^{1/(2p-1)}\bar{a}_k\bar{x}_k}{1 + \left(\frac{\bar{\alpha}_{k+1}\bar{b}_k}{\bar{r}_k}\right)^{1/(2p-1)}\bar{b}_k}.$$

which leads to concluding that the optimal control input is given by the announced expression in (14.84). Substituting the optimal control u_k^* back into the candidate cost yields the announced recursive equations for $\bar{\alpha}_k$, $\bar{\gamma}_k$, and α_k, where we have used that

$$(X + \epsilon_{k+1})^2 = X^2 + \epsilon_{k+1}^2 + 2X\epsilon_{k+1}$$

completing the proof. ■

Remark 14.5 Note that for $p = 1$, the latter equation becomes

$$\bar{\alpha}_k = \bar{q}_k + \frac{\bar{\alpha}_{k+1}\bar{a}_k^2\bar{r}_k}{\bar{r}_k + \bar{\alpha}_{k+1}\bar{b}_k^2},$$
$$\bar{\alpha}_N = \bar{q}_N$$

which is the Riccati equation for discrete-time linear-quadratic regulator (LQR) case, and

$$f_0(x_0, \bar{x}_0) = \alpha_0 var(x_0) + \bar{\alpha}_0\bar{x}_0^{2p} + \bar{\gamma}_0.$$

is the optimal cost. ▮

14.3.5 Variance-Aware Stochastic Cost with Multiplicative Noise

14.3.5.1 Variance-Aware Quartic Cost with Multiplicative Noise

We consider the following discrete-time dynamical system with multiplicative state-and-mean state dependent noise

$$x_{k+1} = \bar{a}_k x_k + \bar{b}_k u_k + (x_k - \bar{x}_k)\epsilon_{k+1}, \quad k \in \{0, \dots, N-1\} \tag{14.88}$$

where $x_k \in \mathbb{R}$ is the state, $u_k \in \mathbb{R}$ is the control input, and $\bar{a}_k, \bar{b}_k \in \mathbb{R}$ are system parameters. The initial state x_0 is a given random variable that is independent of ϵ. $\{\epsilon_k\}_k$ is a zero-mean random process, with $\epsilon_0 = 0$. The state dynamics can be rewritten as

$$\begin{aligned} x_{k+1} &= \bar{a}_k \bar{x}_k + \bar{b}_k \bar{u}_k + \bar{a}_k(x_k - \bar{x}_k) + \bar{b}_k(u_k - \bar{u}_k) + (x_k - \bar{x}_k)\epsilon_{k+1}, \\ & k \in \{0, \dots, N-1\}, \end{aligned}$$

The goal is to minimize the expected value of the cost function

$$L(x_0, u) = q_k var(x_N) + \bar{q}_N \bar{x}_N^4 + \sum_{k=0}^{N-1} \left(q_k var(x_k) + \bar{q}_k \bar{x}_k^4 + r_k var(u_k) + \bar{r}_k \bar{u}_k^4 \right), \tag{14.89}$$

where $q_k, q_N, r_k, \bar{q}_k, \bar{q}_N, \bar{r}_k > 0$ are weights penalizing the state and control deviations. Therefore, the control problem is given by

$$\min_{\bar{u} \in \bar{\mathscr{U}}_k} E[L(x_0, \bar{u})], \text{ s.t. } (14.88),\ x_0 \text{ given.} \tag{14.90}$$

Proposition 14.13 *The optimal control input for the variance-aware problem with multiplicative noise in* (14.90) *with dynamics in* (14.88) *and cost functional* (14.89) *is given by*

$$u_k^* = -\frac{\alpha_{k+1} \bar{b}_k \bar{a}_k}{(r_k + \alpha_{k+1} \bar{b}_k^2)}(x_k - \bar{x}_k) - \frac{\left(\frac{\bar{\alpha}_{k+1} \bar{b}_k}{\bar{r}_k}\right)^{1/3} \bar{a}_k}{1 + \left(\frac{\bar{\alpha}_{k+1} \bar{b}_k}{\bar{r}_k}\right)^{1/3} \bar{b}_k} \bar{x}_k. \tag{14.91}$$

where $\bar{\alpha}_k$ and $\bar{\gamma}_k$ solve the following recursive equations

$$\bar{\alpha}_k = \bar{q}_k + \bar{r}_k \left(\frac{\left(\frac{\bar{\alpha}_{k+1} \bar{b}_k}{\bar{r}_k}\right)^{1/3} \bar{a}_k}{1 + \left(\frac{\bar{\alpha}_{k+1} \bar{b}_k}{\bar{r}_k}\right)^{1/3} \bar{b}_k} \right)^4 + \bar{\alpha}_{k+1} \left(\bar{a}_k - \frac{\left(\frac{\bar{\alpha}_{k+1} \bar{b}_k}{\bar{r}_k}\right)^{1/3} \bar{a}_k \bar{b}_k}{1 + \left(\frac{\bar{\alpha}_{k+1} \bar{b}_k}{\bar{r}_k}\right)^{1/3} \bar{b}_k} \right)^4, \tag{14.92}$$

$$\alpha_k = q_k + r_k\left(\frac{\alpha_{k+1}\bar{b}_k\bar{a}_k}{(r_k+\alpha_{k+1}\bar{b}_k^2)}\right)^2 + \alpha_{k+1}\left(\bar{a}_k - \frac{\alpha_{k+1}\bar{b}_k^2\bar{a}_k}{(r_k+\alpha_{k+1}\bar{b}_k^2)}\right)^2 + \alpha_{k+1}E\left(\epsilon_{k+1}^2\right), \quad (14.93)$$

with $\bar{\alpha}_N = \bar{q}_N$ *and* $\alpha_N = q_N$ *as terminal conditions.* ∎

Proof We propose a candidate cost function at time k to be

$$f_k(x_k, \bar{x}_k) = \alpha_k var(x_k) + \bar{\alpha}_k \bar{x}_k^4,$$

where $\alpha_k, \bar{\alpha}_k \geq 0$ are scalar coefficients to be determined recursively. We proceed now with the telescopic sum, i.e.,

$$f_N(x_N) = f_0(x_0) + \sum_{k=0}^{N-1} f_{k+1}(x_{k+1}) - f_k(x_k).$$

By taking the difference $L(x_0, u) - f_0(x_0)$, we obtain

$$\begin{aligned} &L(x_0, u) - f_0(x_0) = q_N var(x_N) + \bar{q}_N \bar{x}_N^4 - f_N(x_N) \\ &+ \textstyle\sum_{k=0}^{N-1} \left(q_k var(x_k) + \bar{q}_k \bar{x}_k^4 + r_k var(u_k) + \bar{r}_k \bar{u}_k^4\right) + f_{k+1}(x_{k+1}) - f_k(x_k). \end{aligned}$$

Let us introduce the following term:

$$C_k = \min_{u_k}\left(r_k var(u_k) + \alpha_{k+1} var(x_{k+1}) + \bar{r}_k \bar{u}_k^4 + \bar{\alpha}_{k+1}(\bar{a}_k\bar{x}_k + \bar{b}_k\bar{u}_k)^4\right).$$

Then, the difference $L(x_0, u) - f_0(x_0)$ is rewritten as

$$\begin{aligned} L(x_0, u) - f_0(x_0) &= q_N var(x_N) + \bar{q}_N \bar{x}_N^4 - f_N(x_N) \\ &\quad + \sum_{k=0}^{N-1}\left(r_k var(u_k) + \bar{r}_k \bar{u}_k^4 + f_{k+1}(x_{k+1})\right) - C_k \\ &\quad + \left(C_k + q_k var(x_k) + \bar{q}_k \bar{x}_k^4 - f_k(x_k)\right). \end{aligned}$$

where

$$\begin{aligned} r_k var(u_k) + \alpha_{k+1} var(x_{k+1}) &= E(r_k(u_k - \bar{u}_k)^2 + \alpha_{k+1}(\bar{a}_k(x_k - \bar{x}_k) \\ &\quad + \bar{b}_k(u_k - \bar{u}_k) + (x_k - \bar{x}_k)\epsilon_{k+1})^2) \end{aligned}$$

which means that

$$(u_k - \bar{u}_k) = -\frac{\alpha_{k+1}\bar{b}_k\bar{a}_k}{(r_k + \alpha_{k+1}\bar{b}_k^2)}(x_k - \bar{x}_k).$$

To find the deterministic part of optimal control $\bar{u}_k^*$, we differentiate the cost function with respect to $\bar{u}_k$, i.e.,

$$4\bar{r}_k\bar{u}_k^3 + 4\bar{\alpha}_{k+1}(\bar{a}_k\bar{x}_k + \bar{b}_k\bar{u}_k)^3\bar{b}_k = 0.$$

Simplify and isolate $(\bar{a}_k\bar{x}_k + \bar{b}_k\bar{u}_k)^3$ resulting in

$$\bar{u}_k^3 = -\frac{\bar{\alpha}_{k+1}\bar{b}_k}{\bar{r}_k}(\bar{a}_k\bar{x}_k + \bar{b}_k\bar{u}_k)^3.$$

Taking the cube root

$$\bar{u}_k^* = -\frac{\left(\frac{\bar{\alpha}_{k+1}\bar{b}_k}{\bar{r}_k}\right)^{1/3}\bar{a}_k\bar{x}_k}{1 + \left(\frac{\bar{\alpha}_{k+1}\bar{b}_k}{\bar{r}_k}\right)^{1/3}\bar{b}_k}.$$

Then, the optimal control is obtained corresponding to (14.91). Now, by substituting u_k^* back into the difference equation and collecting terms proportional to x_k^4, one obtains the recursive equation for α_k and $\bar{\alpha}_k$ as in (14.92), completing the proof. ■

Remark 14.6 The coefficients $\alpha_k, \bar{\alpha}_k$ remain positive if

1. $q_k, q_N, r_k, \bar{q}_k, \bar{q}_N, \bar{r}_k > 0$,
2. $\bar{a}_k, \bar{b}_k$ are bounded,
3. The denominator in $\bar{u}_k^*$ satisfies $1 + \left(\frac{\bar{\alpha}_{k+1}\bar{b}_k}{\bar{r}_k}\right)^{1/3}\bar{b}_k > 0$.
4. The denominator in $u_k - \bar{u}_k^*$ satisfies $(r_k + \alpha_{k+1}\bar{b}_k^2) > 0$. ▮

These conditions are trivially satisfied in most systems, ensuring $\alpha_k > 0$ and $\bar{\alpha}_k$ for all $k \in \{0, 1, \ldots, N\}$

14.3.5.2 Extension to $\bar{x}^{2p}$ Costs

We now provide for discrete-time optimal control problems with costs of the form x_k^{2p} and u_k^{2p}, where $p \geq 1$. The solutions include the optimal control law, the recursive relation for the cost-to-go coefficient α_k,, and the expression for the optimal cost. We consider the same discrete-time system as in (14.88). Now, the objective is to minimize the variance-aware 2p-order cost function

$$L = q_N var(x_N) + \bar{q}_N\bar{x}_N^{2p} + \sum_{k=0}^{N-1}\left(q_k var(x_k) + \bar{q}_k\bar{x}_k^{2p} + r_k var(u_k) + \bar{r}_k\bar{u}_k^{2p}\right), \tag{14.94}$$

where $q_k, q_N, r_k, \bar{q}_k, \bar{q}_N, \bar{r}_k > 0$ are weights.

Proposition 14.14 *The optimal control input for the variance-aware problem with multiplicative noise in* (14.90) *with dynamics in* (14.88) *and cost functional* (14.94) *is given by*

$$u_k^* = -\frac{\alpha_{k+1}\bar{b}_k\bar{a}_k}{(r_k + \alpha_{k+1}\bar{b}_k^2)}(x_k - \bar{x}_k) - \frac{\left(\frac{\bar{\alpha}_{k+1}\bar{b}_k}{\bar{r}_k}\right)^{1/(2p-1)}\bar{a}_k\bar{x}_k}{1 + \left(\frac{\bar{\alpha}_{k+1}\bar{b}_k}{\bar{r}_k}\right)^{1/(2p-1)}\bar{b}_k}.$$

where $\bar{\alpha}_k$ *and* $\bar{\alpha}_k$ *solve the following recursive equations:*

$$\begin{aligned}
\bar{\alpha}_k &= \bar{q}_k + \bar{r}_k\left(\frac{\left(\frac{\bar{\alpha}_{k+1}\bar{b}_k}{\bar{r}_k}\right)^{1/(2p-1)}\bar{a}_k}{1+\left(\frac{\bar{\alpha}_{k+1}\bar{b}_k}{\bar{r}_k}\right)^{1/(2p-1)}\bar{b}_k}\right)^{2p} \\
&+\bar{\alpha}_{k+1}\left(\bar{a}_k - \bar{b}_k\frac{\left(\frac{\bar{\alpha}_{k+1}\bar{b}_k}{\bar{r}_k}\right)^{1/(2p-1)}\bar{a}_k}{1+\left(\frac{\bar{\alpha}_{k+1}\bar{b}_k}{\bar{r}_k}\right)^{1/(2p-1)}\bar{b}_k}\right)^{2p}, \\
\alpha_k &= q_k + r_k\left(\frac{\alpha_{k+1}\bar{b}_k\bar{a}_k}{(r_k+\alpha_{k+1}\bar{b}_k^2)}\right)^2 + \alpha_{k+1}\left(\bar{a}_k - \frac{\alpha_{k+1}\bar{b}_k^2\bar{a}_k}{(r_k+\alpha_{k+1}\bar{b}_k^2)}\right)^2 + \alpha_{k+1}E\left(\epsilon_{k+1}^2\right),
\end{aligned}$$

with $\bar{\alpha}_N = \bar{q}_N$ *and* $\alpha_N = q_N$ *as terminal conditions.* ∎

Proof The candidate cost is assumed to have the form

$$f_k(x_k, \bar{x}_k) = \alpha_k var(x_k) + \bar{\alpha}_k\bar{x}_k^{2p},$$

where $\alpha_k \geq 0, \bar{\alpha}_k \geq 0$ is a coefficient to be determined recursively. The expected value of difference term now has

$$\begin{aligned}
C = \min_{u_k}\Big(&r_k var(u_k) + \alpha_{k+1}var(x_{k+1}) + \bar{r}_k\bar{u}_k^{2p} \\
&+\bar{\alpha}_{k+1}(\bar{a}_k\bar{x}_k + \bar{b}_k\bar{u}_k)^{2p}\Big).
\end{aligned}$$

The stochastic part of the optimal control is determined by

$$\begin{aligned}
&r_k var(u_k) + \alpha_{k+1}var(x_{k+1}) \\
&= E(r_k(u_k - \bar{u}_k)^2 + \alpha_{k+1}(\bar{a}_k(x_k - \bar{x}_k) + \bar{b}_k(u_k - \bar{u}_k) + (x_k - \bar{x}_k)\epsilon_{k+1})^2)
\end{aligned}$$

which means that

$$(u_k - \bar{u}_k) = -\frac{\alpha_{k+1}\bar{b}_k\bar{a}_k}{(r_k + \alpha_{k+1}\bar{b}_k^2)}(x_k - \bar{x}_k).$$

The deterministic part of the optimal control is obtained as follows. Differentiating with respect to $\bar{u}_k$ gives

$$2p\,\bar{r}_k\bar{u}_k^{2p-1} + 2p\,\bar{\alpha}_{k+1}\bar{b}_k(\bar{a}_k\bar{x}_k + \bar{b}_k\bar{u}_k)^{2p-1} = 0.$$

Rearrange terms

$$\bar{r}_k\bar{u}_k^{2p-1} + \bar{\alpha}_{k+1}\bar{b}_k(\bar{a}_k\bar{x}_k + \bar{b}_k\bar{u}_k)^{2p-1} = 0.$$

Isolate $(\bar{a}_k\bar{x}_k + \bar{b}_k\bar{u}_k)^{2p-1}$:

$$\bar{u}_k^{2p-1} = -\frac{\bar{\alpha}_{k+1}\bar{b}_k}{\bar{r}_k}(\bar{a}_k\bar{x}_k + \bar{b}_k\bar{u}_k)^{2p-1}.$$

Take the $(2p-1)$-th root on both sides

$$\bar{u}_k = -\frac{\left(\frac{\bar{\alpha}_{k+1}\bar{b}_k}{\bar{r}_k}\right)^{1/(2p-1)}\bar{a}_k\bar{x}_k}{1 + \left(\frac{\bar{\alpha}_{k+1}\bar{b}_k}{\bar{r}_k}\right)^{1/(2p-1)}\bar{b}_k},$$

and we obtain the announced optimal control input u_k^* in (14.95). Substituting the optimal control u_k^* back into the candidate cost yields the recursive equations for $\bar{\alpha}_k$ and $\bar{\alpha}_k$ and in (14.95). where we have used that

$$(aY + Y\epsilon_{k+1})^2 = Y^2(a^2 + \epsilon_{k+1}^2 + 2a\epsilon_{k+1}),$$

completing the proof. ■

Remark 14.7 The optimal cost is $f_0(x_0, \bar{x}_0) = \alpha_0 var(x_0) + \bar{\alpha}_0\bar{x}_0^{2p}$ meaning that as the variance of noises $E\left(\epsilon_{k+1}^2\right)$ vanishes, we retrieve the solution of the deterministic optimal control. ▮

14.3.6 Higher-Order Moment Costs with Multiplicative State-Control and Mean-Field-Type-Dependent Noise

We now provide for discrete-time optimal control problems with $2o$-moment costs of the form $E[(x_k - \bar{x}_k)^{2o}]$ and $E[(u_k - \bar{u}_k)^{2o}]$, where $o \geq 1$. The solutions include the optimal control law, the recursive relation for the cost-to-go coefficient α_k,, and the expression for the optimal cost. We consider the following discrete-time system of mean-field type

$$x_{k+1} = \left(\bar{a}_k\bar{x}_k + \bar{b}_k\bar{u}_k\right) + (a_k(x_k - \bar{x}_k) + b_k(u_k - \bar{u}_k))\,\epsilon_{k+1}, \quad k \in \{0, \ldots, N-1\},$$

where $x_k \in \mathbb{R}$ is the state, $u_k \in \mathbb{R}$ is the control input, and $\bar{a}_k, \bar{b}_k \in \mathbb{R}$ are system parameters. The objective is to minimize the $2o$-th moment-dependent cost function

$$\begin{aligned} L &= q_N E[(x_N - \bar{x}_N)^{2o}] + \bar{q}_N \bar{x}_N^{2p} \\ &+ \sum_{k=0}^{N-1} \left(q_k E[(x_k - \bar{x}_k)^{2o}] + \bar{q}_k \bar{x}_k^{2p} + r_k E[(u_k - \bar{u}_k)^{2o}] + \bar{r}_k \bar{u}_k^{2p} \right), \end{aligned} \tag{14.95}$$

where $q_k, q_N, r_k, \bar{q}_k, \bar{q}_N, \bar{r}_k > 0$ are weights. The control problem is as follows:

$$\min_{\bar{u} \in \bar{\mathscr{U}}_k} E[L(x_0, \bar{u})], \text{ s.t. (14.95)}, \ x_0 \text{ given.} \tag{14.96}$$

Proposition 14.15 *The optimal control input for the higher-order moment problem in* (14.96) *with dynamics in* (14.95) *and cost functional* (14.95), *is given by*

$$u_k^* = -\frac{\left(\frac{\alpha_{k+1} b_k m_{k+1,2o}}{r_k}\right)^{1/(2o-1)} a_k}{1+\left(\frac{\alpha_{k+1} b_k m_{k+1,2o}}{r_k}\right)^{1/(2o-1)} b_k}(x_k - \bar{x}_k) - \frac{\left(\frac{\bar{\alpha}_{k+1} \bar{b}_k}{\bar{r}_k}\right)^{1/(2p-1)} \bar{a}_k}{1+\left(\frac{\bar{\alpha}_{k+1} \bar{b}_k}{\bar{r}_k}\right)^{1/(2p-1)} \bar{b}_k}\bar{x}_k. \tag{14.97}$$

where $\bar{\alpha}_k$ and α_k solve the following recursive equations

$$\begin{aligned} \bar{\alpha}_k &= \bar{q}_k + \bar{r}_k \left(\frac{\left(\frac{\bar{\alpha}_{k+1} \bar{b}_k}{\bar{r}_k}\right)^{1/(2p-1)} \bar{a}_k}{1+\left(\frac{\bar{\alpha}_{k+1} \bar{b}_k}{\bar{r}_k}\right)^{1/(2p-1)} \bar{b}_k} \right)^{2p} \\ &+ \bar{\alpha}_{k+1} \left(\bar{a}_k - \bar{b}_k \frac{\left(\frac{\bar{\alpha}_{k+1} \bar{b}_k}{\bar{r}_k}\right)^{1/(2p-1)} \bar{a}_k}{1+\left(\frac{\bar{\alpha}_{k+1} \bar{b}_k}{\bar{r}_k}\right)^{1/(2p-1)} \bar{b}_k} \right)^{2p}, \\ \alpha_k &= q_k + r_k \left(\frac{\left(\frac{\alpha_{k+1} b_k m_{k+1,2o}}{r_k}\right)^{1/(2o-1)} a_k}{1+\left(\frac{\alpha_{k+1} b_k m_{k+1,2o}}{r_k}\right)^{1/(2o-1)} b_k} \right)^{2o} \\ &+ \alpha_{k+1} \left(a_k - b_k \frac{\left(\frac{\alpha_{k+1} b_k m_{k+1,2o}}{r_k}\right)^{1/(2o-1)} a_k}{1+\left(\frac{\alpha_{k+1} b_k m_{k+1,2o}}{r_k}\right)^{1/(2o-1)} b_k} \right)^{2o}, \end{aligned} \tag{14.98}$$

with $\bar{\alpha}_N = \bar{q}_N$ and $\alpha_N = q_N$ as terminal conditions and $m_{k+1,2o} = E[\epsilon_{k+1}^{2o}]$. ∎

Proof The candidate cost is assumed to have the form

$$f_k(x_k, \bar{x}_k) = \alpha_k E[(x_k - \bar{x}_k)^{2o}] + \bar{\alpha}_k \bar{x}_k^{2p},$$

where $\alpha_k \geq 0$, $\bar{\alpha}_k \geq 0$ are coefficients to be determined recursively. The expected value of difference term now has

$$\begin{aligned} C_k = \min_{u_k} \Big(& r_k E[(u_k - \bar{u}_k)^{2o}] + \alpha_{k+1} E[(x_{k+1} - \bar{x}_{k+1})^{2o}] + \bar{r}_k \bar{u}_k^{2p} \\ & + \bar{\alpha}_{k+1} (\bar{a}_k \bar{x}_k + \bar{b}_k \bar{u}_k)^{2p} \Big) \end{aligned}$$

This minimization is subject to $\mathscr{F}_k$ adaptivity of optimal control. We compute

$$
\begin{aligned}
E[(x_{k+1}-\bar{x}_{k+1})^{2o}] &= E[(a_k(x_k-\bar{x}_k)+b_k(u_k-\bar{u}_k))\epsilon_{k+1})^{2o}] \\
&= E[(a_k(x_k-\bar{x}_k)+b_k(u_k-\bar{u}_k))^{2o}]m_{k+1,2o}
\end{aligned}
$$

The stochastic part of the control yields

$$
u_k-\bar{u}_k = -\frac{\left(\frac{\alpha_{k+1}b_k m_{k+1,2o}}{r_k}\right)^{1/(2o-1)} a_k}{1+\left(\frac{\alpha_{k+1}b_k m_{k+1,2o}}{r_k}\right)^{1/(2o-1)} b_k}(x_k-\bar{x}_k)
$$

The deterministic part of the control yields

$$
\bar{u}_k = -\frac{\left(\frac{\bar{\alpha}_{k+1}\bar{b}_k}{\bar{r}_k}\right)^{1/(2p-1)} \bar{a}_k\bar{x}_k}{1+\left(\frac{\bar{\alpha}_{k+1}\bar{b}_k}{\bar{r}_k}\right)^{1/(2p-1)} \bar{b}_k}.
$$

We can easily check that the obtained optimal control is the announced in (14.97), which is adapted to the filtration generated by $\{x_0,\ldots,x_k\}$. Substituting the optimal control u_k^* back into the candidate cost yields (14.98), and the optimal cost is

$$
f_0(x_0,\bar{x}_0) = \alpha_0 E[(x_0-\bar{x}_0)^{2o}] + \bar{\alpha}_0\bar{x}_0^{2p},
$$

completing the proof. ■

14.3.6.1 Verification Method in the Space of Measures

The above analysis covers noise beyond the Markovian noise. We now verify the proposed solution by checking the case where the noise process is Markovian. We do, so we work directly on the space of probability measures associated with x_k and denote by $\mu_k = \mathbb{P}_{x_k}$ its probability distribution. Then μ_{k+1} can be obtained $\mathbb{P}_{\epsilon_{k+1}}, \mu_k$, and x_k, Ex_k, u_k, Eu_k. Note here that the mean-field type is involved in the transition kernel. The cost can also be written in terms of μ_k and the functional u_k of mean-field type. Our new augmented state becomes μ, and the system becomes a big deterministic system in the space of measures. The advantage now is that one can write a dynamic programming principle of this new augmented state.

$$
V_k(\mu_k) = \inf_{u_k}\left(q_k\int\left[\left(z-\int y\mu_k(dy)\right)^{2o}\right]\mu_k(dz) + \bar{q}_k\left(\int y\mu_k(dy)\right)^{2p}\right.
$$

$$+ r_k \int \left[\left(u_k(z, \mu_k) - \int u_k(y, \mu_k)\mu_k(dy) \right)^{2o} \right] \mu_k(dz)$$

$$+ \bar{r}_k \left(\int u(y, \mu_k)\mu_k(dy) \right)^{2p} + V_{k+1}(\mu_{k+1}) \Big)$$

and the terminal value is

$$V_N(\mu_N) = q_N \int \left[\left(z - \int y\mu_N(dy) \right)^{2o} \right] \mu_N(dz) + \bar{q}_N \left(\int y\mu_N(dy) \right)^{2p},$$

where

$$\mu_{k+1}(dy) = \begin{cases} \mathbb{P}_{\epsilon_{k+1}}((a_k(x_k - Ex_k) + b_k(u_k - Eu_k))^{-1} [(dy - (\bar{a}_k Ex_k + \bar{b}_k Eu_k))]) \text{ if} \\ (a_k(x_k - Ex_k) + b_k(u_k - Eu_k) \neq 0, \\ \delta_{\bar{a}_k Ex_k + \bar{b}_k Eu_k}(dy) \text{ otherwise.} \end{cases}$$

We seek for $V_k(\mu_k)$ in the form

$$\alpha_k \int \left[\left(z - \int y\mu_k(dy) \right)^{2o} \right] \mu_k(dz) + \bar{\alpha}_k \left(\int y\mu_k(dy) \right)^{2p}.$$

Inserting the above yields the same dynamics for α_k and $\bar{\alpha}_k$ as well as the optimal control structure

$$u_k(x, \mu_k) = -\frac{\left(\frac{\alpha_{k+1} b_k m_{k+1,2o}}{r_k}\right)^{1/(2o-1)} a_k}{1 + \left(\frac{\alpha_{k+1} b_k m_{k+1,2o}}{r_k}\right)^{1/(2o-1)} b_k} \left(x - \int y\mu_k(dy) \right)$$

$$- \frac{\left(\frac{\bar{\alpha}_{k+1} \bar{b}_k}{\bar{r}_k}\right)^{1/(2p-1)} \bar{a}_k}{1 + \left(\frac{\bar{\alpha}_{k+1} \bar{b}_k}{\bar{r}_k}\right)^{1/(2p-1)} \bar{b}_k} \left(\int y\mu_k(dy) \right).$$

14.3.7 Selfish Agents With Higher-Order Costs

In this section, we provide semi-explicit solution of discrete-time game problems with non-quadratic cost functions, specifically focusing on the minimization of a fourth-order state and control cost of each agent. By using convex-completion techniques, we derive semi-explicit expressions for the equilibrium strategies, the associated equilibrium cost-to-go function, and a recursive relation for the cost coefficients. Furthermore, we establish conditions for the existence of positive coefficients in the cost recursion and extend the method to generalized higher-order costs of the power-law form of arbitrary even order. The findings provide a foundation for analyzing discrete time games with nonquadratic cost.

14.3.7.1 Selfish Agents with Quartic Costs

We consider the following discrete-time dynamical system influenced by the decision of $I \geq 2$ agents:

$$\bar{x}_{k+1} = \bar{a}_k \bar{x}_k + \sum_{i \in \mathscr{I}} \bar{b}_{ik} \bar{u}_{ik}, \quad k \in \{0, \ldots, N-1\}, \tag{14.99}$$

where $\mathscr{I} = \{1, \ldots, I\}$, is the set of agents, $\bar{x}_k \in \mathbb{R}$ is the state, $\bar{u}_{ik} \in \mathbb{R}$ is the control input of agent i, and $\bar{a}_k, \bar{b}_k \in \mathbb{R}$ are system parameters at time step k. The initial state $\bar{x}_0$ is given. The goal is to minimize the cost function

$$L_i(x_0, u) = \bar{q}_{iN} \bar{x}_N^4 + \sum_{k=0}^{N-1} \left(\bar{q}_{ik} \bar{x}_k^4 + \bar{r}_{ik} \bar{u}_{ik}^4 \right), \tag{14.100}$$

where $\bar{q}_{ik}, \bar{q}_{iN}, \bar{r}_{ik} > 0$ are weights penalizing the state and control deviations. The game problem is given by

$$\forall i \in \mathscr{I}: \min_{\bar{u}_i \in \mathscr{U}_{ik}} E[L_i(x_0, \bar{u})], \text{ s.t. (14.99)}, \; x_0 \text{ given.} \tag{14.101}$$

Proposition 14.16 *The equilibrium strategy for the game problem in* (14.101) *with dynamics in* (14.99) *and cost functional in* (14.100) *is given by*

$$\bar{u}_{ik}^* = -\frac{\left(\frac{\bar{\alpha}_{i,k+1}\bar{b}_{ik}}{\bar{r}_{ik}}\right)^{1/3}}{1 + \left(\frac{\bar{\alpha}_{i,k+1}\bar{b}_{ik}}{\bar{r}_{ik}}\right)^{1/3} \bar{b}_{ik}} \left(\bar{a}_k \bar{x}_k + \sum_{j \in \mathscr{I} \setminus \{i\}} \bar{b}_{jk} \bar{u}_{jk} \right). \tag{14.102}$$

where $\bar{\alpha}_{ik}$ solves the following recursive equation

$$\bar{\alpha}_{ik} = \bar{q}_{ik} + \bar{r}_{ik} \left((\bar{G}^{-1}\bar{c})_i \bar{a}_k \right)^4 + \bar{\alpha}_{i,k+1} \left(\bar{a}_k - \sum_{j \in \mathscr{I}} (\bar{E}^{-1}\bar{c})_j \bar{a}_k \bar{b}_{jk} \right)^4, \tag{14.103}$$

where $\bar{G}_k$ is the square matrix with $\bar{g}_{ii,k} = 1$, $\bar{g}_{ij,k} = \bar{c}_{ik} \bar{b}_{jk}$,

$$\bar{c}_{ik} = \frac{\left(\frac{\bar{\alpha}_{i,k+1}\bar{b}_{ik}}{\bar{r}_{ik}}\right)^{1/3}}{1 + \left(\frac{\bar{\alpha}_{i,k+1}\bar{b}_{ik}}{\bar{r}_{ik}}\right)^{1/3} \bar{b}_{ik}}$$

and $\bar{\alpha}_{iN} = \bar{q}_{iN}$ *as terminal conditions.* ∎

Proof We propose a candidate cost function at time k to be

$$f_{ik}(\bar{x}_k) = \bar{\alpha}_{ik}\bar{x}_k^4,$$

where $\bar{\alpha}_{ik} > 0$ is a scalar coefficient to be determined recursively. Using the telescopic sum yields

$$f_{i,N}(\bar{x}_N) = f_{i,0}(\bar{x}_0) + \sum_{k=0}^{N-1} f_{i,k+1}(\bar{x}_{k+1}) - f_{i,k}(\bar{x}_k).$$

By taking the difference $L_i(x_0, u) - f_{i,0}(\bar{x}_0)$, we obtain

$$L_i(x_0, u) - f_{i,0}(\bar{x}_0) = \bar{q}_{iN}\bar{x}_N^4 - f_{iN}(\bar{x}_N) +$$
$$\textstyle\sum_{k=0}^{N-1} \left(\bar{q}_{ik}\bar{x}_k^4 + \bar{r}_{ik}\bar{u}_{ik}^4\right) + f_{i,k+1}(\bar{x}_{k+1}) - f_{i,k}(\bar{x}_k).$$

Introducing the term

$$C_{ik} = \min_{\bar{u}_{ik}} \left(\bar{r}_{ik}\bar{u}_k^4 + \bar{\alpha}_{i,k+1} \left(\bar{a}_k\bar{x}_k + \sum_{i\in\mathscr{I}} \bar{b}_{ik}\bar{u}_{ik} \right)^4 \right). \tag{14.104}$$

and replacing back in the difference equation yield

$$L_i(x_0, u) - f_{i,0}(\bar{x}_0) = \bar{q}_{iN}\bar{x}_N^4 - f_{i,N}(\bar{x}_N) +$$
$$\textstyle\sum_{k=0}^{N-1} \left(\bar{r}_{ik}\bar{u}_{ik}^4 + f_{i,k+1}(\bar{x}_{k+1})\right) - C_{ik} + \left(C_{ik} + \bar{q}_{ik}\bar{x}_k^4 - f_{i,k}(\bar{x}_k)\right).$$

To find the best response strategy, we differentiate the cost function with respect to $\bar{u}_k$:

$$4\bar{r}_{ik}\bar{u}_{ik}^3 + 4\bar{\alpha}_{i,k+1} \left(\bar{a}_k\bar{x}_k + \sum_{j\in\mathscr{I}} \bar{b}_{jk}\bar{u}_{jk} \right)^3 \bar{b}_{ik} = 0. \tag{14.105}$$

As the terms are continuously differentiable and convex, the first-order condition (Fermat's rule) is also sufficient. We simplify and isolate $(\bar{a}_k\bar{x}_k + \sum_{i\in\mathscr{I}} \bar{b}_{ik}\bar{u}_{ik})^3$:

$$\bar{u}_{ik}^3 = -\frac{\bar{\alpha}_{i,k+1}\bar{b}_{ik}}{\bar{r}_{ik}} \left(\bar{a}_k\bar{x}_k + \sum_{j\in\mathscr{I}} \bar{b}_{jk}\bar{u}_{jk} \right)^3. \tag{14.106}$$

Taking the cube root, one obtains the announced optimal strategy $\bar{u}_{ik}^*$ as in (14.102). This means that $\bar{G}_k\bar{u}_k = -\bar{c}_k(\bar{a}_k\bar{x}_k)$. In addition, if the matrix G is invertible, then it follows that the equilibrium strategy of agent i is given by

$$\bar{u}_{ik}^* = -(\bar{G}_k^{-1}\bar{c}_k)_i(\bar{a}_k\bar{x}_k).$$

Substitute $\bar{u}_{ik}^*$ back into the difference equation, and collect terms proportional to $\bar{x}_k^4$; one obtains the recursive equation for $\bar{\alpha}_{ik}$, completing the proof. ■

Remark 14.8 Assume that $\bar{G}$ is invertible, the coefficients $\bar{\alpha}_k$ remain positive if

1. $\bar{q}_{ik}, \bar{q}_{iN}, \bar{r}_{ik} > 0$,
2. $\bar{a}_k, \bar{b}_{ik}$ are bounded,
3. The denominator in $\bar{u}_{ik}^*$ satisfies $1 + \left(\frac{\bar{\alpha}_{i,k+1}\bar{b}_{ik}}{\bar{r}_{ik}}\right)^{1/3}\bar{b}_{ik} > 0$.

These conditions are trivially satisfied in most physical systems, ensuring $\bar{\alpha}_{ik} > 0$ for all $k \in \{0, 1, \dots, N\}$. ■

14.3.7.2 Extension to $\bar{x}^{2p}$ Costs

We now provide for discrete-time games with individual costs of the form $\bar{x}_k^{2p}$ and $\bar{u}_{ik}^{2p}$, where $p \geq 1$. The solutions include the optimal control law, the recursive relation for the cost-to-go coefficient $\bar{\alpha}_{ik}$,, and the expression for the optimal cost. We consider the same discrete-time system as in (14.99). The objective is to minimize the $2p$-order cost function of agent i:

$$L_i(x_0, u) = \bar{q}_{iN}\bar{x}_N^{2p} + \sum_{k=0}^{N-1} \bar{q}_{ik}\bar{x}_k^{2p} + \bar{r}_{ik}\bar{u}_{ik}^{2p}, \tag{14.107}$$

where $\bar{q}_{ik}, \bar{q}_{iN}, \bar{r}_{ik} > 0$ are weights.

Proposition 14.17 *The optimal strategy for the game problem in* (14.101) *with dynamics in* (14.99) *and cost functional in* (14.113) *is given by*

$$\bar{u}_{ik}^* = -\frac{\left(\frac{\bar{\alpha}_{i,k+1}\bar{b}_{ik}}{\bar{r}_{ik}}\right)^{1/(2p-1)}}{1 + \left(\frac{\bar{\alpha}_{i,k+1}\bar{b}_{ik}}{\bar{r}_{ik}}\right)^{1/(2p-1)}\bar{b}_{ik}} \left(\bar{a}_k\bar{x}_k + \sum_{j\in\mathcal{I}\setminus\{i\}} \bar{b}_{jk}\bar{u}_{jk}\right), \tag{14.108}$$

where $\bar{\alpha}_{i,k}$ solves the following recursive equation

$$\bar{\alpha}_{i,k} = \bar{q}_{ik} + \bar{r}_{ik}\left((\bar{G}_k^{-1}\bar{c}_k)_i\bar{a}_k\right)^{2p} + \bar{\alpha}_{i,k+1}\left(\bar{a}_k - \sum_{j\in\mathcal{I}}(\bar{G}_k^{-1}\bar{c}_k)_j\bar{a}_k\bar{b}_{jk}\right)^{2p}, \tag{14.109}$$

with $\bar{G}$ being a square matrix with $\bar{g}_{ii} = 1$, $\bar{g}_{ij} = \bar{c}_i\bar{b}_{jk}$ and

$$\bar{c}_i = \frac{\left(\frac{\bar{\alpha}_{i,k+1}\bar{b}_{ik}}{\bar{r}_{ik}}\right)^{1/(2p-1)}}{1+\left(\frac{\bar{\alpha}_{i,k+1}\bar{b}_{ik}}{\bar{r}_{ik}}\right)^{1/(2p-1)}\bar{b}_{ik}},$$

and $\bar{\alpha}_{iN} = \bar{q}_{iN}$ *as terminal condition.* ∎

Proof The candidate cost of agent i is assumed to have the following form

$$f_{ik}(\bar{x}_k) = \bar{\alpha}_{ik}\bar{x}_k^{2p},$$

where $\bar{\alpha}_{ik} > 0$ is a coefficient to be determined recursively. The difference term now has

$$C_{ik} = \min_{\bar{u}_{ik}}\left(\bar{r}_{ik}\bar{u}_{ik}^{2p} + \bar{\alpha}_{i,k+1}(\bar{a}_k\bar{x}_k + \sum_{j\in\mathscr{I}}\bar{b}_{jk}\bar{u}_{jk})^{2p}\right).$$

Differentiating with respect to $\bar{u}_{ik}$ gives

$$2p\,\bar{r}_{ik}\bar{u}_{ik}^{2p-1} + 2p\,\bar{\alpha}_{i,k+1}\bar{b}_{ik}\left(\bar{a}_k\bar{x}_k + \sum_{j\in\mathscr{I}}\bar{b}_{jk}\bar{u}_{jk}\right)^{2p-1} = 0.$$

Rearrange terms

$$\bar{r}_{ik}\bar{u}_{ik}^{2p-1} + \bar{\alpha}_{i,k+1}\bar{b}_{ik}\left(\bar{a}_k\bar{x}_k + \sum_{j\in\mathscr{I}}\bar{b}_{jk}\bar{u}_{jk}\right)^{2p-1} = 0.$$

We isolate $\left(\bar{a}_k\bar{x}_k + \sum_{j\in\mathscr{I}}\bar{b}_{jk}\bar{u}_{jk}\right)^{2p-1}$:

$$\bar{u}_{ik}^{2p-1} = -\frac{\bar{\alpha}_{i,k+1}\bar{b}_{ik}}{\bar{r}_{ik}}\left(\bar{a}_k\bar{x}_k + \sum_{j\in\mathscr{I}}\bar{b}_{jk}\bar{u}_{jk}\right)^{2p-1}.$$

Take the $(2p-1)$-th root on both sides, one gets the optimal strategy in (14.108). Meaning that $\bar{G}_k\bar{u}_k = -\bar{c}_k(\bar{a}_k\bar{x}_k)$ where $\bar{G}$ is the announced square matrix. If the matrix G_k is invertible, then it follows that the equilibrium strategy of agent i is given by

$$\bar{u}_{ik}^{*} = -\left(\bar{G}_k^{-1}\bar{c}_k\right)_i(\bar{a}_k\bar{x}_k)$$

Substituting the best response strategy $\bar{u}_{ik}^*$ back into the candidate cost yields the recursive equation for $\bar{\alpha}_{i,k}$ in (14.109). The equilibrium cost at each time step is proportional to $\bar{x}_k^{2p}$:

$$f_{ik}(\bar{x}_k) = \bar{\alpha}_{ik}\bar{x}_k^{2p}, \tag{14.110}$$

completing the proof. ■

14.3.8 Selfish Agents with Variance-Aware Higher-Order Costs

We now focus on variance-aware discrete-time games with finitely many agents with non-quadratic cost functions, specifically focusing on the minimization of a fourth-order state and control cost per agent. By using convex-completion techniques, we derive semi-explicit expressions for the equilibrium strategies and the associated variance-aware cost-to-go function, and a recursive relation for the cost coefficients. Furthermore, we establish conditions for the existence of positive coefficients in the variance-aware cost recursion and extend the method to generalized higher-order costs of the even power-law form.

14.3.8.1 Selfish Agents with Variance-Aware Quartic Costs

We consider the following discrete-time dynamical system

$$x_{k+1} = \bar{a}_k x_k + \sum_{i\in\mathcal{I}} \bar{b}_{ik} u_{ik} + \epsilon_{k+1}, \quad k \in \{0, \ldots, N-1\} \tag{14.111}$$

where $\mathcal{I} = \{1, \ldots, I\}$, is the set of agents, $\bar{x}_k \in \mathbb{R}$ is the state, $\bar{u}_{ik} \in \mathbb{R}$ is the control input of agent i, and $\bar{a}_k, \bar{b}_k \in \mathbb{R}$ are system parameters at time step k. The initial state x_0 is a given random variable that is independent of ϵ. $\{\epsilon_k\}_k$ is a zero-mean random process, with $\epsilon_0 = 0$. The state dynamics can be rewritten as

$$\begin{aligned} x_{k+1} = {} & \bar{a}_k \bar{x}_k + \sum_{i\in\mathcal{I}} \bar{b}_{ik}\bar{u}_{ik} + \bar{a}_k(x_k - \bar{x}_k) + \sum_{i\in\mathcal{I}} \bar{b}_{ik}(u_{ik} - \bar{u}_{ik}) + \epsilon_{k+1}, \\ & k \in \{0, \ldots, N-1\}, \end{aligned} \tag{14.112}$$

The goal is to minimize the expected value of the cost function

$$L_{ik}(x_0, u) = q_{iN} var(x_N) + \bar{q}_{iN}\bar{x}_N^4$$

$$+\sum_{k=0}^{N-1}\left(q_{ik}var(x_k)+\bar{q}_{ik}\bar{x}_k^4+r_{ik}var(u_k)+\bar{r}_{ik}\bar{u}_k^4\right), \tag{14.113}$$

where $q_{ik}, q_{iN}, r_{ik}, \bar{q}_{ik}, \bar{q}_{iN}, \bar{r}_{ik} > 0$ are weights penalizing the state and control deviations. The game theoretical problem is

$$\forall i \in \mathcal{I}: \min_{\bar{u}_i \in \bar{\mathcal{U}}_{ik}} E[L_i(x_0, \bar{u})], \text{ s.t. (14.111)}, \ x_0 \text{ given.} \tag{14.114}$$

Proposition 14.18 *The best-response strategy solution for problem in* (14.114) *with system dynamics as in* (14.111) *and cost functional in* (14.113) *is given by*

$$u_{i,k}^* = -(\bar{G}_k^{-1}\bar{c}_k)_i(\bar{a}_k\bar{x}_k) - (G_k^{-1}c_k)_i[\bar{a}_k(x_k-\bar{x}_k)], \tag{14.115}$$

with G_k is the matrix with $e_{ii,k}=1$ and $e_{ij,k}=c_{ik}\bar{b}_{jk}$ and

$$c_{ik} = \frac{\alpha_{i,k+1}\bar{b}_{ik}}{(r_{ik}+\alpha_{i,k+1}\bar{b}_{ik}^2)},$$

and $\bar{G}_k$ is the square matrix with $\bar{g}_{ii,k}=1$, $\bar{g}_{ij,k}=\bar{c}_{ik}\bar{b}_{jk}$ and

$$\bar{c}_{ik} = \frac{\left(\frac{\bar{\alpha}_{i,k+1}\bar{b}_{ik}}{\bar{r}_{ik}}\right)^{1/3}}{1+\left(\frac{\bar{\alpha}_{i,k+1}\bar{b}_{ik}}{\bar{r}_{ik}}\right)^{1/3}\bar{b}_{ik}},$$

where α_{ik} and $\bar{\alpha}_{ik}$ solve the following recursive equations

$$\bar{\alpha}_{ik}=\bar{q}_{ik}+\bar{r}_{ik}\left((\bar{G}_k^{-1}\bar{c}_k)_i\bar{a}_k\right)^4+\bar{\alpha}_{k+1}\left(\bar{a}_k-\sum_{j\in\mathcal{I}}(\bar{G}_k^{-1}\bar{c}_k)_j\bar{a}_k\bar{b}_{jk}\right)^4, \tag{14.116}$$

$$\alpha_{ik}=q_{ik}+r_{ik}\left((G_k^{-1}c_k)_i\bar{a}_k\right)^2+\alpha_{i,k+1}\left(\bar{a}_k-\sum_{j\in\mathcal{I}}(G_k^{-1}c_k)_j\bar{a}_k\bar{b}_{jk}\right)^2, \tag{14.117}$$

with $\bar{\alpha}_{iN}=\bar{q}_{iN}$ and $\alpha_{iN}=q_{iN}$ as terminal conditions. ∎

Proof We propose a candidate cost function at time k to be

$$f_{ik}(x_k,\bar{x}_k) = \alpha_{ik}var(x_k)+\bar{\alpha}_{ik}\bar{x}_k^4+\bar{\gamma}_{ik},$$

where $\alpha_{ik}, \bar{\alpha}_{ik}, \bar{\gamma}_{ik} \geq 0$ are scalar coefficients to be determined recursively. Computing by telescopic sum yields

$$f_{iN}(x_N) = f_{i0}(x_0) + \sum_{k=0}^{N-1} f_{i,k+1}(x_{k+1}) - f_{i,k}(x_k).$$

By taking the difference $L_i(x_0, u) - f_{i0}(x_0)$, we obtain

$$L_i(x_0, u) - f_{i0}(x_0) = q_{iN} var(x_N) + \bar{q}_{iN}\bar{x}_N^4 - f_N(x_N)$$
$$+ \sum_{k=0}^{N-1} \left(q_{ik} var(x_k) + \bar{q}_{ik}\bar{x}_k^4 + r_{ik} var(u_{ik}) + \bar{r}_{ik}\bar{u}_{ik}^4\right) + f_{i,k+1}(x_{k+1}) - f_{i,k}(x_k).$$

Let us introduce the following term

$$C_{ik} = \min_{u_{ik}} \Big(r_{ik} var(u_{ik}) + \alpha_{i,k+1} var(x_{k+1}) + \bar{r}_{ik}\bar{u}_{ik}^4$$
$$+\bar{\alpha}_{i,k+1}(\bar{a}_k\bar{x}_k + \sum_{j\in\mathcal{I}} \bar{b}_{jk}\bar{u}_{jk})^4\Big). \tag{14.118}$$

and replacing back in the difference equation, one arrives at

$$L_i(x_0, u) - f_{i0}(x_0) = q_{iN} var(x_N) + \bar{q}_{iN}\bar{x}_N^4 - f_{iN}(x_N)$$
$$+ \sum_{k=0}^{N-1} \left(r_{ik} var(u_{ik}) + \bar{r}_{ik}\bar{u}_{ik}^4 + f_{i,k+1}(x_{k+1})\right) - C_{ik}$$
$$+ \left(C_{ik} + q_{ik} var(x_k) + \bar{q}_{ik}\bar{x}_k^4 - f_{i,k}(x_k)\right).$$

and we have that

$$r_{ik} var(u_{ik}) + \alpha_{i,k+1} var(x_{k+1})$$
$$= E\left[r_{ik}(u_{ik} - \bar{u}_{ik})^2 + \alpha_{i,k+1}\left(\bar{a}_k(x_k - \bar{x}_k) + \sum_{j\in\mathcal{I}} \bar{b}_{jk}(u_{jk} - \bar{u}_{jk}) + \epsilon_{k+1}\right)^2\right],$$

which means that

$$(u_{ik} - \bar{u}_{ik}) = -\frac{\alpha_{i,k+1}\bar{b}_k}{(r_k + \alpha_{i,k+1}\bar{b}_k^2)}\left(\bar{a}_k(x_k - \bar{x}_k) + \sum_{j\in\mathcal{I}\setminus\{i\}} \bar{b}_{jk}(u_{jk} - \bar{u}_{jk})\right), \tag{14.119}$$

and using G_k matrix, we have $(u_{ik} - \bar{u}_{ik}) = -G_k^{-1}c_k[\bar{a}_k(x_k - \bar{x}_k)]$ whenever G_k is invertible. To find the deterministic part of best response strategy $\bar{u}_{ik}^*$ of agent i,, we differentiate the cost function with respect to u_{ik}, i.e.,

$$4\bar{r}_{ik}\bar{u}_{ik}^3 + 4\bar{\alpha}_{i,k+1}\left(\bar{a}_k\bar{x}_k + \sum_{j\in\mathcal{I}} \bar{b}_{jk}\bar{u}_{jk}\right)^3 \bar{b}_{ik} = 0.$$

We simplify and isolate $\left(\bar{a}_k\bar{x}_k + \sum_{j\in\mathcal{I}} \bar{b}_{jk}\bar{u}_{jk}\right)^3$:

$$\bar{u}_{ik}^3 = -\frac{\bar{\alpha}_{i,k+1}\bar{b}_{ik}}{\bar{r}_{ik}}\left(\bar{a}_k\bar{x}_k + \sum_{j\in\mathcal{I}} \bar{b}_{jk}\bar{u}_{jk}\right)^3.$$

Taking the cube root

$$\bar{u}_{ik}^* = -\frac{\left(\frac{\bar{\alpha}_{i,k+1}\bar{b}_{ik}}{\bar{r}_{ik}}\right)^{1/3}}{1+\left(\frac{\bar{\alpha}_{i,k+1}\bar{b}_{ik}}{\bar{r}_{ik}}\right)^{1/3}\bar{b}_{ik}}\left(\bar{a}_k\bar{x}_k + \sum_{j\in\mathcal{I}\setminus\{i\}} \bar{b}_{jk}\bar{u}_{jk}\right) \tag{14.120}$$

This means that $\bar{G}_k\bar{u}_k = -\bar{c}_k(\bar{a}_k\bar{x}_k)$ where $\bar{G}_k$ is the aforementioned square matrix. If the matrix $\bar{G}_k$ is invertible, then it follows that the equilibrium strategy of agent i is given by

$$\bar{u}_{ik}^* = -(\bar{G}_k^{-1}\bar{c}_k)_i(\bar{a}_k\bar{x}_k). \tag{14.121}$$

Now, substitute $u_{ik}^* - \bar{u}_{ik}^*$ and $\bar{u}_{ik}^*$ back into the difference equation and collect terms proportional to $(x_k - \bar{x}_k)^4$, $\bar{x}_k^4$. Thus, one obtains the announced recursive equations for α_{ik} and $\bar{\alpha}_{ik}$, completing the proof. ■

Remark 14.9 If G_k and $\bar{G}_k$ are invertible, then the coefficients $\alpha_{ik}, \bar{\alpha}_{ik}$ remain positive if

1. $q_{ik}, q_{iN}, r_{ik}, \bar{q}_{ik}, \bar{q}_{iN}, \bar{r}_{ik} > 0$,
2. $\bar{a}_k, \bar{b}_{ik}$ are bounded,
3. The denominator in $\bar{u}_{ik}^*$ satisfies $1+\left(\frac{\bar{\alpha}_{i,k+1}\bar{b}_{ik}}{\bar{r}_{ik}}\right)^{1/3}\bar{b}_{ik} > 0$.
4. The denominator in $u_{ik} - \bar{u}_{ik}^*$ satisfies $(r_{ik} + \alpha_{i,k+1}\bar{b}_{ik}^2) > 0$.

Then $\alpha_{ik} > 0$ and $\bar{\alpha}_{ik}$ for all $k \in \{0, 1, \ldots, N\}$. ▮

14.3.8.2 Extension to $\bar{x}^{2p}$ Costs

We now provide for discrete-time games with costs of the form x_k^{2p} and u_{ik}^{2p}, where $p \geq 1$. The solutions include the equilibrium strategy law, the recursive relation for the cost-to-go coefficient α_k,, and the expression for the equilibrium cost. We consider the same discrete-time system as in (14.111). The objective is to minimize the variance-aware $2p$-order cost function

$$L_i(x_0, u) = q_{iN} var(x_N) + \bar{q}_{iN}\bar{x}_N^{2p} + \sum_{k=0}^{N-1}\left(q_{ik} var(x_k) + \bar{q}_{ik}\bar{x}_k^{2p} + r_{ik} var(u_{ik}) + \bar{r}_{ik}\bar{u}_{ik}^{2p}\right), \tag{14.122}$$

where $q_{ik}, q_{iN}, r_{ik}, \bar{q}_{ik}, \bar{q}_{iN}, \bar{r}_{ik} > 0$ are weights.

Proposition 14.19 *The optimal strategy solution for problem in* (14.114) *with system dynamics as in* (14.111) *and new cost functional in* (14.122) *is given by*

$$u_{i,k}^* = -(\bar{G}_k^{-1}\bar{c}_k)_i(\bar{a}_k\bar{x}_k) - (G_k^{-1}c_k)_i\bar{a}_k(x_k - \bar{x}_k) \tag{14.123}$$

with G_k *is a matrix with* $e_{ii,k} = 1, e_{ij,k} = c_{ik}\bar{b}_{jk}$,

$$c_{ik} = \frac{\alpha_{i,k+1}\bar{b}_{ik}}{(r_{ik} + \alpha_{i,k+1}\bar{b}_{ik}^2)};$$

and $\bar{G}_k$ *is the square matrix with* $\bar{g}_{ii,k} = 1$, $\bar{g}_{ij,k} = \bar{c}_{ik}\bar{b}_{jk}$ *and*

$$\bar{c}_{ik} = \frac{\left(\frac{\bar{\alpha}_{i,k+1}\bar{b}_{ik}}{\bar{r}_{ik}}\right)^{1/(2p-1)}}{1 + \left(\frac{\bar{\alpha}_{i,k+1}\bar{b}_{ik}}{\bar{r}_{ik}}\right)^{1/(2p-1)}\bar{b}_{ik}},$$

and with the following recursive equations

$$\begin{aligned}
\bar{\alpha}_{i,k} &= \bar{q}_{ik} + \bar{r}_{ik}\left((\bar{G}_k^{-1}\bar{c}_k)_i\bar{a}_k\right)^{2p} + \bar{\alpha}_{i,k+1}\left(\bar{a}_k - \sum_{j\in\mathcal{I}}(\bar{G}^{-1}\bar{c})_j\bar{a}_k\bar{b}_{jk}\right)^{2p}, \\
\bar{\gamma}_{i,k} &= \bar{\gamma}_{i,k+1} + E\left(\alpha_{i,k+1}\epsilon_{k+1}^2\right), \\
\alpha_{ik} &= q_{ik} + r_{ik}\left((G_k^{-1}c_k)_i\bar{a}_k\right)^2 + \alpha_{k+1}\left(\bar{a}_k - \sum_j(G_k^{-1}c_k)_j\bar{a}_k b_{jk}\right)^2,
\end{aligned} \tag{14.124}$$

and with $\bar{\alpha}_{iN} = \bar{q}_{iN}$, $\bar{\gamma}_{iN} = 0$, *and* $\alpha_{iN} = q_{iN}$ *as terminal conditions.* ∎

Proof The candidate cost is assumed to have the form

$$f_{ik}(x_k, \bar{x}_k) = \alpha_{ik} var(x_k) + \bar{\alpha}_{ik}\bar{x}_k^{2p} + \bar{\gamma}_{ik},$$

where $\alpha_{ik}, \bar{\alpha}_{ik}, \bar{\gamma}_{ik} \geq 0$ is a coefficient to be determined recursively. The expected value of difference term now has

$$C_{ik} = \min_{u_{ik}}\left(r_{ik} var(u_{ik}) + \alpha_{i,k+1} var(x_{k+1}) + \bar{r}_{ik}\bar{u}_{ik}^{2p} + \bar{\alpha}_{i,k+1}(\bar{a}_k\bar{x}_k + \sum_{j\in\mathcal{I}}\bar{b}_{jk}\bar{u}_{jk})^{2p}\right).$$

The stochastic part of the best response strategy is determined by

$$
\begin{aligned}
& r_{ik} var(u_{ik}) + \alpha_{i,k+1} var(x_{k+1}) \\
& = E(r_{ik}(u_{ik} - \bar{u}_{ik})^2 + \alpha_{i,k+1}(\bar{a}_k(x_k - \bar{x}_k) + \sum_{j\in\mathcal{I}} \bar{b}_{jk}(u_{jk} - \bar{u}_{jk}) + \epsilon_{k+1})^2)
\end{aligned} \tag{14.125}
$$

which means that

$$
(u_k - \bar{u}_k) = -G_k^{-1} c_k \bar{a}_k (x_k - \bar{x}_k)
$$

with the announced matrix G_k. The deterministic part of the best response strategy is obtained as follows:

$$
\bar{u}_{ik} = -\frac{\left(\frac{\bar{\alpha}_{i,k+1}\bar{b}_{ik}}{\bar{r}_{ik}}\right)^{1/(2p-1)}}{1 + \left(\frac{\bar{\alpha}_{i,k+1}\bar{b}_{ik}}{\bar{r}_{ik}}\right)^{1/(2p-1)} \bar{b}_{ik}} \left(\bar{a}_k \bar{x}_k + \sum_{j\in\mathcal{I}\setminus\{i\}} \bar{b}_{jk}\bar{u}_{jk}\right). \tag{14.126}
$$

This means that $\bar{G}_k \bar{u}_k = -\bar{c}_k(\bar{a}_k \bar{x}_k)$ where $\bar{G}_k$ is the announced square matrix. If the matrix $\bar{G}_k$ is invertible, then it follows that the equilibrium strategy of agent i is given by

$$
\bar{u}_{ik}^* = -(\bar{G}_k^{-1}\bar{c}_k)_i(\bar{a}_k \bar{x}_k) \tag{14.127}
$$

and the announced optimal strategy (14.123) is obtained. Substituting the equilibrium strategy u_{ik}^* back into the candidate cost yields one arrives at the announced recursive equations, where we have used that

$$
(X + \epsilon_{k+1})^2 = X^2 + \epsilon_{k+1}^2 + 2X\epsilon_{k+1}.
$$

Finally, the equilibrium cost of agent i is

$$
f_{i0}(x_0, \bar{x}_0) = \alpha_{i0} var(x_0) + \bar{\alpha}_{i0}\bar{x}_0^{2p} + \bar{\gamma}_{i0}.
$$

completing the proof. ■

14.3.9 Risk-Aware Games Under Multiplicative Noise

Non-quadratic costs, such as $\bar{x}^4$, $(x-\bar{x})^4$, $\bar{u}^4$ or $(x-\bar{u})^4$, introduce another class of realistic representation of these nonlinearities, emphasizing large deviations in state and control.

We now explicitly solve the variance-aware discrete-time games with higher-order costs. We derive semi-explicit solutions for the variance-aware equilibrium, the variance-aware equilibrium cost, and the recursive coefficients for the variance-aware cost-to-go function. Additionally, we extend the methodology to address general higher-order costs $(x_k - Ex_k)^{2o}$ for arbitrary $o \geq 1$.

14.3.9.1 Variance-Aware Quartic Cost Under Multiplicative Noise

We consider the following discrete-time dynamical system with multiplicative state-and-mean-state-dependent noise

$$x_{k+1} = \bar{a}_k x_k + \sum_{i\in\mathcal{I}} \bar{b}_{ik} u_{ik} + (x_k - \bar{x}_k)\epsilon_{k+1}, \quad k \in \{0, \ldots, N-1\} \tag{14.128}$$

where $\mathcal{I} = \{1, \ldots, I\}$, is the set of agents, $\bar{x}_k \in \mathbb{R}$ is the state, $\bar{u}_{ik} \in \mathbb{R}$ is the control input of agent i, $\bar{a}_k, \bar{b}_k \in \mathbb{R}$ are system parameters at time step k. The initial state x_0 is a given random variable that is independent of ϵ. $\{\epsilon_k\}_k$ is a zero-mean random process, with $\epsilon_0 = 0$. The state dynamics can be rewritten as

$$x_{k+1} = \bar{a}_k \bar{x}_k + \sum_{i\in\mathcal{I}} \bar{b}_{ik}\bar{u}_{ik} + \bar{a}_k(x_k - \bar{x}_k) + \sum_{i\in\mathcal{I}} \bar{b}_{ik}(u_{ik} - \bar{u}_{ik}) + (x_k - \bar{x}_k)\epsilon_{k+1},$$
$$k \in \{0, \ldots, N-1\},$$

The goal is to minimize the expected value of the cost function

$$L_i(x_0, u) = q_{iN} var(x_N) + \bar{q}_{iN}\bar{x}_N^4 + \sum_{k=0}^{N-1} \left(q_{ik} var(x_k) + \bar{q}_{ik}\bar{x}_k^4 + r_{ik} var(u_{ik}) + \bar{r}_{ik}\bar{u}_{ik}^4 \right), \tag{14.129}$$

where $q_{ik}, q_{iN}, r_{ik}, \bar{q}_{ik}, \bar{q}_{iN}, \bar{r}_{ik} > 0$ are weights penalizing the state and control deviations. The game theoretical problem is

$$\forall i \in \mathcal{I}: \min_{\bar{u}_i \in \mathcal{U}_{ik}} E[L_i(x_0, \bar{u})], \text{ s.t. (14.128)}, \ x_0 \text{ given.} \tag{14.130}$$

Proposition 14.20 *The optimal strategy for the problem in* (14.130) *is given by*

$$u_{ik}^* = -(G_k^{-1} c_k)_i [\bar{a}_k(x_k - \bar{x}_k)] - (\bar{G}_k^{-1}\bar{c}_k)_i[\bar{a}_k \bar{x}_k] \tag{14.131}$$

where G_k the matrix with $e_{ii,k} = 1$ and $e_{ij,k} = c_{ik}\bar{b}_{jk}$ and

$$c_{ik} = \frac{\alpha_{i,k+1}\bar{b}_{ik}}{(r_{ik} + \alpha_{i,k+1}\bar{b}_{ik}^2)};$$

and $\bar{G}_k$ is the square matrix with $\bar{g}_{ii,k} = 1$, $\bar{g}_{ij,k} = \bar{c}_{ik}\bar{b}_{jk}$ and

$$\bar{c}_{ik} = \frac{\left(\frac{\bar{\alpha}_{i,k+1}\bar{b}_{ik}}{\bar{r}_{ik}}\right)^{1/3}}{1 + \left(\frac{\bar{\alpha}_{i,k+1}\bar{b}_{ik}}{\bar{r}_{ik}}\right)^{1/3}\bar{b}_{ik}}.$$

The terms α_{ik} $\bar{\alpha}_{ik}$ solve the following recursive equations:

$$\bar{\alpha}_{ik} = \bar{q}_{ik} + \bar{r}_{ik}\left((\bar{G}_k^{-1}\bar{c}_k)_i\bar{a}_k\right)^4 + \bar{\alpha}_{i,k+1}\left(\bar{a}_k - \sum_{j\in\mathcal{I}}(\bar{G}_k^{-1}\bar{c}_k)_j\bar{a}_k\bar{b}_{jk}\right)^4, \tag{14.132}$$

$$\alpha_{ik} = q_{ik} + r_{ik}\left((G_k^{-1}c_k)_i\bar{a}_k\right)^2 + \alpha_{i,k+1}\left(\bar{a}_k - \sum_{j\in\mathcal{I}}(G_k^{-1}c_k)_j\bar{a}_k\bar{b}_{jk}\right)^2$$
$$+ \alpha_{i,k+1}E\left(\epsilon_{k+1}^2\right), \tag{14.133}$$

with terminal conditions $\bar{\alpha}_{iN} = \bar{q}_{iN}$ and $\alpha_{iN} = q_{iN}$. ∎

Proof We propose a candidate cost function at time k to be

$$f_{ik}(x_k, \bar{x}_k) = \alpha_{ik}var(x_k) + \bar{\alpha}_{ik}\bar{x}_k^4,$$

where $\alpha_{ik}, \bar{\alpha}_{ik} \geq 0$ are scalar coefficients to be determined recursively. Using the telescopic sum yields

$$f_{iN}(x_N) = f_{i0}(x_0) + \sum_{k=0}^{N-1} f_{i,k+1}(x_{k+1}) - f_{i,k}(x_k).$$

By taking the difference $L_i(x_0, u) - f_{i0}(x_0)$, we obtain

$$L_i(x_0, u) - f_{i0}(x_0) = q_{iN}var(x_N) + \bar{q}_{iN}\bar{x}_N^4 - f_{iN}(x_N)$$
$$+ \sum_{k=0}^{N-1}\left(q_{ik}var(x_k) + \bar{q}_{ik}\bar{x}_k^4 + r_{ik}var(u_k) + \bar{r}_{ik}\bar{u}_k^4\right) + f_{i,k+1}(x_{k+1}) - f_{i,k}(x_k).$$

Considering the following term

$$C_{ik} = \min_{u_{ik}}\Big(r_{ik}var(u_{ik}) + \alpha_{i,k+1}var(x_{k+1}) + \bar{r}_{ik}\bar{u}_{ik}^4$$

$$+\bar{\alpha}_{i,k+1}\left(\bar{a}_k\bar{x}_k+\sum_{j\in\mathcal{I}}\bar{b}_{jk}\bar{u}_{jk}\right)^4\Bigg),$$

the difference equation can be rewritten as

$$\begin{aligned}L_i(x_0,u)-f_{i0}(x_0)&=q_{iN}var(x_N)+\bar{q}_{iN}\bar{x}_N^4-f_N(x_N)\\&+\sum_{k=0}^{N-1}\Big(r_{ik}var(u_{ik})+\bar{r}_{ik}\bar{u}_{ik}^4+f_{i,k+1}(x_{k+1})\Big)-C_{ik}\\&+\Big(C_{ik}+q_{ik}var(x_k)+\bar{q}_{ik}\bar{x}_k^4-f_{i,k}(x_k)\Big).\end{aligned}$$

We have that

$$\begin{aligned}&r_{ik}var(u_{ik})+\alpha_{i,k+1}var(x_{k+1})\\&=E\left[r_{ik}(u_{ik}-\bar{u}_{ik})^2+\alpha_{i,k+1}\left(\bar{a}_k(x_k-\bar{x}_k)+\sum_{j\in\mathcal{I}}\bar{b}_{jk}(u_{jk}-\bar{u}_{jk})\right.\right.\\&\left.\left.+(x_k-\bar{x}_k)\epsilon_{k+1}\right)^2\right]\end{aligned}$$

which means that

$$u_{ik}-\bar{u}_{ik}=-\frac{\alpha_{i,k+1}\bar{b}_k}{(r_k+\alpha_{i,k+1}\bar{b}_k^2)}\left(\bar{a}_k(x_k-\bar{x}_k)+\sum_{j\in\mathcal{I}\backslash\{i\}}\bar{b}_{jk}(u_{jk}-\bar{u}_{jk})\right).$$

Using the introduced matrix G_k, it follows that $(u_k-\bar{u}_k)=-G_k^{-1}c_k[\bar{a}_k(x_k-\bar{x}_k)]$ whenever G_k is invertible. To find the deterministic part of best response strategies $\bar{u}_{ik}^*$, we differentiate the cost function with respect to u_{ik} :

$$4\bar{r}_{ik}\bar{u}_{ik}^3+4\bar{\alpha}_{i,k+1}\left(\bar{a}_k\bar{x}_k+\sum_{j\in\mathcal{I}}\bar{b}_{jk}\bar{u}_{jk}\right)^3\bar{b}_{ik}=0.$$

We simplify and isolate $\left(\bar{a}_k\bar{x}_k+\sum_{j\in\mathcal{I}}\bar{b}_{jk}\bar{u}_{jk}\right)^3$:

$$\bar{u}_{ik}^3=-\frac{\bar{\alpha}_{i,k+1}\bar{b}_{ik}}{\bar{r}_{ik}}\left(\bar{a}_k\bar{x}_k+\sum_{j\in\mathcal{I}}\bar{b}_{jk}\bar{u}_{jk}\right)^3.$$

Taking the cube root:

$$\bar{u}_{ik}^{*} = -\frac{\left(\frac{\bar{\alpha}_{i,k+1}\bar{b}_{ik}}{\bar{r}_{ik}}\right)^{1/3}}{1+\left(\frac{\bar{\alpha}_{i,k+1}\bar{b}_{ik}}{\bar{r}_{ik}}\right)^{1/3}\bar{b}_{ik}}\left(\bar{a}_k\bar{x}_k + \sum_{j\in\mathscr{I}\setminus\{i\}} \bar{b}_{jk}\bar{u}_{jk}\right).$$

This means that $\bar{G}_k\bar{u}_k = -\bar{c}_k(\bar{a}_k\bar{x}_k)$ where $\bar{G}_k$ is the announced square matrix. If the matrix $\bar{G}_k$ is invertible, then it follows that the equilibrium strategy of agent i is given by

$$\bar{u}_{ik}^{*} = -(\bar{G}_k^{-1}\bar{c}_k)_i(\bar{a}_k\bar{x}_k).$$

Then, the announced equilibrium strategy in 14.131 of agent i is obtained. Substitute u_{ik}^{*} back into the difference equation, and collect terms proportional to $(x_k - \bar{x}_k)^4$, and the recursive equations in (14.132) are obtained, completing the proof. ■

Remark 14.10 If G_k and $\bar{G}_k$ are invertible, then the coefficients $\alpha_k, \bar{\alpha}_k$ remain positive if

1. $q_{ik}, q_{iN}, r_{ik}, \bar{q}_{ik}, \bar{q}_{iN}, \bar{r}_{ik} > 0$,
2. $\bar{a}_k, \bar{b}_{ik}$ are bounded,
3. The denominator in $\bar{u}_{ik}^{*}$ satisfies $1+\left(\frac{\bar{\alpha}_{i,k+1}\bar{b}_{ik}}{\bar{r}_{ik}}\right)^{1/3}\bar{b}_{ik} > 0$.
4. The denominator in $u_{ik} - \bar{u}_{ik}^{*}$ satisfies $(r_k + \alpha_{i,k+1}\bar{b}_{ik}^{2}) > 0$.

Then $\alpha_{ik} > 0$ and $\bar{\alpha}_{ik} > 0$ for all $k \in \{0, 1, \ldots, N\}$.

14.3.9.2 Extension to $\bar{x}^{2p}$ Costs

We now provide for discrete-time optimal control problems with costs of the form $\bar{x}_k^{2p}$ and $\bar{u}_{ik}^{2p}$, where $p \geq 1$. The solutions include the optimal control law, the recursive relation for the cost-to-go coefficient α_{ik}, and the expression for the optimal cost. We consider the discrete-time system in (14.128). The new objective is to minimize the variance-aware $2p$-order cost function

$$L_i = q_{iN}var(x_N) + \bar{q}_{iN}\bar{x}_N^{2p} + \sum_{k=0}^{N-1} q_{ik}var(x_k) + \bar{q}_{ik}\bar{x}_k^{2p} + r_{ik}var(u_{ik}) + \bar{r}_{ik}\bar{u}_{ik}^{2p}, \tag{14.134}$$

where $q_{ik}, q_{iN}, r_{ik}, \bar{q}_{ik}, \bar{q}_{iN}, \bar{r}_{ik} > 0$ are weights.

Proposition 14.21 *The equilibrium strategy for the problem in* (14.130) *with the cost functional in* (14.134) *is*

$$u_{ik}^{*} = -\left(G_k^{-1}c_k\right)_i \bar{a}_k(x_k - \bar{x}_k) - \left(\bar{G}_k^{-1}\bar{c}_k\right)_i \bar{a}_k\bar{x}_k \tag{14.135}$$

where G_k is a matrix with $e_{ii,k} = 1$, $e_{ij,k} = c_{ik}\bar{b}_{jk}$,

$$c_{ik} = \frac{\alpha_{i,k+1}\bar{b}_{ik}}{(r_{ik} + \alpha_{i,k+1}\bar{b}_{ik}^2)};$$

and $\bar{G}$ is the square matrix with $\bar{g}_{ii} = 1$, $\bar{g}_{ij,k} = \bar{c}_{ik}\bar{b}_{jk}$ and

$$\bar{c}_{ik} = \frac{\left(\frac{\bar{\alpha}_{i,k+1}\bar{b}_{ik}}{\bar{r}_{ik}}\right)^{1/(2p-1)}}{1 + \left(\frac{\bar{\alpha}_{i,k+1}\bar{b}_{ik}}{\bar{r}_{ik}}\right)^{1/(2p-1)}\bar{b}_{ik}}.$$

The terms $\bar{\alpha}_{ik}$ and α_{ik} solve the following recursive equations

$$\bar{\alpha}_{ik} = \bar{q}_{ik} + \bar{r}_{ik}\left((\bar{G}_k^{-1}\bar{c}_k)_i\bar{a}_k\right)^{2p} + \bar{\alpha}_{k+1}\left(\bar{a}_k - \sum_{j\in\mathcal{I}}(\bar{G}_k^{-1}\bar{c}_k)_j\bar{a}_k\bar{b}_{jk}\right)^{2p},$$

$$\alpha_{ik} = q_{ik} + r_{ik}\left((G_k^{-1}c_k)_i\bar{a}_k\right)^2 + \alpha_{i,k+1}\left(\bar{a}_k - \sum_{j\in\mathcal{I}}(G_k^{-1}c_k)_j\bar{a}_k\bar{b}_{jk}\right)^2$$
$$+ \alpha_{i,k+1}E\left(\epsilon_{k+1}^2\right). \tag{14.136}$$

with terminal conditions $\bar{\alpha}_{iN} = \bar{q}_{iN}$ and $\alpha_{iN} = q_{iN}$. ∎

Proof The candidate cost of agent i is assumed to have the form

$$f_{ik}(x_k, \bar{x}_k) = \alpha_{ik} var(x_k) + \bar{\alpha}_{ik}\bar{x}_k^{2p},$$

where $\alpha_{ik}, \bar{\alpha}_{ik}, \bar{\gamma}_{ik} \geq 0$ is a coefficient to be determined recursively. The expected value of difference term now has

$$C_{ik} = \min_{u_{ik}}\left(r_{ik}var(u_{ik}) + \alpha_{i,k+1}var(x_{k+1}) + \bar{r}_{ik}\bar{u}_{ik}^{2p}\right.$$
$$\left.+\bar{\alpha}_{i,k+1}(\bar{a}_k\bar{x}_k + \sum_{j\in\mathcal{I}}\bar{b}_{jk}\bar{u}_{jk})^{2p}\right).$$

The stochastic part of the best response strategy is determined by

$$r_{ik}var(u_{ik}) + \alpha_{i,k+1}var(x_{k+1}) = E(r_{ik}(u_{ik} - \bar{u}_{ik})^2 + \alpha_{i,k+1}(\bar{a}_k(x_k - \bar{x}_k)$$
$$+ \sum_{j\in\mathcal{I}}\bar{b}_{jk}(u_{jk} - \bar{u}_{jk}) + (x_k - \bar{x}_k)\epsilon_{k+1})^2)$$

which means that $(u_k - \bar{u}_k) = -G_k^{-1} c_k[\bar{a}_k(x_k - \bar{x}_k)]$ with G_k being the announced matrix. The deterministic part of the best response strategy of agent i is obtained as follows:

$$\bar{u}_{ik} = -\frac{\left(\frac{\bar{\alpha}_{i,k+1}\bar{b}_{ik}}{\bar{r}_{ik}}\right)^{1/(2p-1)}}{1 + \left(\frac{\bar{\alpha}_{i,k+1}\bar{b}_{ik}}{\bar{r}_{ik}}\right)^{1/(2p-1)} \bar{b}_{ik}} \left(\bar{a}_k \bar{x}_k + \sum_{j \in \mathcal{I}\backslash\{i\}} \bar{b}_{jk}\bar{u}_{jk}\right).$$

This means that $\bar{G}_k \bar{u}_k = -\bar{c}_k(\bar{a}_k \bar{x}_k)$ where $\bar{G}$ is the announced square matrix. If the matrix $\bar{G}_k$ is invertible, then it follows that the equilibrium strategy of agent i is given by

$$\bar{u}_{ik}^* = -(\bar{G}_k^{-1}\bar{c}_k)_i(\bar{a}_k \bar{x}_k) \tag{14.137}$$

Combining together the equilibrium strategy of agent i yields the optimal strategy in (14.135). Substituting the equilibrium strategy u_{ik}^* back into the candidate cost yields the recursive equations in (14.136), where we have used that

$$(aY + Y\epsilon_{k+1})^2 = Y^2(a^2 + \epsilon_{k+1}^2 + 2a\epsilon_{k+1}).$$

The equilibrium cost of i is

$$f_{i0}(x_0, \bar{x}_0) = \alpha_{i0} var(x_0) + \bar{\alpha}_{i0}\bar{x}_0^{2p}.$$

As the variance of the noises $E\left(\epsilon_{k+1}^2\right)$ vanish, we retrieve the solution of deterministic game. ■

14.3.9.3 Extension to $E[(x - Ex)^{2o}]$ Costs With Multiplicative State-Control and Mean-Field-Type-Dependent Noise

We now provide for discrete-time games with $2o$-moment costs of the form $E[(x_k - Ex_k)^{2o}]$ and $E[(u_{ik} - Eu_{ik})^{2o}]$, where $o \geq 1$. The solutions include the optimal control law, the recursive relation for the cost-to-go coefficient α_{ik},, and the expression for the equilibrium costs. We consider the following discrete-time system of mean-field type. For $k \in \{0, \ldots, N-1\}$,

$$x_{k+1} = \left(\bar{a}_k Ex_k + \sum_{i \in \mathcal{I}} \bar{b}_{ik} Eu_{ik}\right) + \left(a_k(x_k - Ex_k) + \sum_{i \in \mathcal{I}} b_{ik}(u_{ik} - Eu_{ik})\right)\epsilon_{k+1}, \tag{14.138}$$

where $x_k \in \mathbb{R}$ is the state, $u_{ik} \in \mathbb{R}$ is the control input, and $\bar{a}_k, \bar{b}_{ik} \in \mathbb{R}$ are system parameters. The objective is to minimize the $2o$-th moment-dependent cost function

$$
\begin{aligned}
L_i(x_0, u) &= q_{iN} E[(x_N - \bar{x}_N)^{2o}] + \bar{q}_{iN}(Ex_N)^{2p} \\
&\quad + \sum_{k=0}^{N-1} q_{ik} E[(x_k - \bar{x}_k)^{2o}] + \bar{q}_{ik}(Ex_k)^{2p} \\
&\quad + \sum_{k=0}^{N-1} r_{ik} E[(u_{ik} - \bar{u}_{ik})^{2o}] + \bar{r}_{ik}(\bar{u}_{ik})^{2p},
\end{aligned}
\tag{14.139}
$$

where $q_{ik}, q_{iN}, r_{ik}, \bar{q}_{ik}, \bar{q}_{iN}, \bar{r}_{ik} > 0$ are weights. The equilibrium problem is stated as follows:

$$
\forall i \in \mathcal{I}: \quad \min_{\bar{u}_i \in \bar{\mathcal{U}}_{ik}} E[L_i(x_0, \bar{u})], \text{ s.t. (14.138)}, \ x_0 \text{ given.} \tag{14.140}
$$

Proposition 14.22 *An equilibrium strategy for the problem in* (14.140) *with the cost functional in* (14.139) *is*

$$
u_{ik}^* = -(\tilde{G}_k^{-1}\tilde{c}_k)_i a_k (x_k - \bar{x}_k) - (\bar{G}_k^{-1}\bar{c}_k)_i \bar{a}_k \bar{x}_k, \tag{14.141}
$$

where $\tilde{G}_k$ is a matrix with $\tilde{g}_{ii,k} = 1$, $\tilde{g}_{ij,k} = \tilde{c}_{ik} b_{jk}$, and

$$
\tilde{c}_{ik} = \frac{\left(\frac{\alpha_{i,k+1} b_{ik} m_{k+1,2o}}{r_{ik}}\right)^{1/(2o-1)}}{1 + \left(\frac{\alpha_{i,k+1} b_{ik} m_{k+1,2o}}{r_{ik}}\right)^{1/(2o-1)} b_{ik}};
$$

and $\bar{G}$ is the square matrix with $\bar{g}_{ii} = 1$, $\bar{g}_{ij,k} = \bar{c}_{ik} \bar{b}_{jk}$ and

$$
\bar{c}_{ik} = \frac{\left(\frac{\bar{\alpha}_{i,k+1} \bar{b}_{ik}}{\bar{r}_{ik}}\right)^{1/(2p-1)}}{1 + \left(\frac{\bar{\alpha}_{i,k+1} \bar{b}_{ik}}{\bar{r}_{ik}}\right)^{1/(2p-1)} \bar{b}_{ik}}.
$$

The terms $\alpha_{i,k}$ and $\bar{\alpha}_{i,k}$ solve the following recursive equations

$$
\bar{\alpha}_{ik} = \bar{q}_{ik} + \bar{r}_{ik}\left((\bar{G}_k^{-1}\bar{c}_k)_i \bar{a}_k\right)^{2p} + \bar{\alpha}_{k+1}\left(\bar{a}_k - \sum_{j \in \mathcal{I}} (\bar{G}_k^{-1}\bar{c}_k)_j \bar{a}_k \bar{b}_{jk}\right)^{2p}, \tag{14.142}
$$

$$
\alpha_{ik} = q_{ik} + r_{ik}\left((\tilde{G}_k^{-1}\tilde{c}_k)_i a_k\right)^{2o} + \alpha_{i,k+1}\left(a_k - \sum_{j \in \mathcal{I}} (\tilde{G}_k^{-1}\tilde{c}_k)_j a_k b_{jk}\right)^{2o}, \tag{14.143}
$$

with terminal conditions $\bar{\alpha}_{iN} = \bar{q}_{iN}$ and $\alpha_{iN} = q_{iN}$; and $m_{k+1,2o} = E[\epsilon_{k+1}^{2o}]$. ∎

Proof The candidate cost is assumed to have the form

$$f_{ik}(x_k, \bar{x}_k) = \alpha_{ik} E[(x_k - \bar{x}_k)^{2o}] + \bar{\alpha}_{ik}(\bar{x}_k)^{2p},$$

where $\alpha_{ik} \geq 0, \bar{\alpha}_{ik} \geq 0$ is a coefficient to be determined recursively. The expected value of difference term now has

$$C_{ik} = \min_{u_{ik}} \Big(r_{ik} E[(u_{ik} - \bar{u}_{ik})^{2o}] + \alpha_{i,k+1} E[(x_{k+1} - \bar{x}_{k+1})^{2o}] \quad (14.144)$$

$$+ \bar{r}_{ik} \bar{u}_{ik}^{2p} + \bar{\alpha}_{i,k+1} \left(\bar{a}_k \bar{x}_k + \sum_{j \in \mathscr{I}} \bar{b}_{jk} \bar{u}_{jk} \right)^{2p} \Bigg). \quad (14.145)$$

This minimization is subject to $\mathscr{F}_k$ adaptivity of equilibrium strategy. We compute

$$E\left[(x_{k+1} - \bar{x}_{k+1})^{2o}\right] = E\left[(a_k(x_k - \bar{x}_k) + \sum_{i \in \mathscr{I}} b_{ik}(u_{ik} - \bar{u}_{ik}))\epsilon_{k+1})^{2o} \right]$$

$$= E\left[(a_k(x_k - \bar{x}_k) + \sum_{i \in \mathscr{I}} b_{ik}(u_{ik} - \bar{u}_{ik}))^{2o} \right] m_{k+1,2o}.$$

The stochastic part of the best response strategy of agent i yields

$$u_{ik} - \bar{u}_{ik} = -\frac{\left(\frac{\alpha_{i,k+1} b_{ik} m_{k+1,2o}}{r_{ik}}\right)^{1/(2o-1)}}{1 + \left(\frac{\alpha_{i,k+1} b_{ik} m_{k+1,2o}}{r_{ik}}\right)^{1/(2o-1)} b_{ik}}$$

$$\left(a_k(x_k - \bar{x}_k) + \sum_{j \in \mathscr{I} \setminus \{i\}} b_{jk}(u_{jk} - \bar{u}_{jk}) \right).$$

Then

$$u_k - E u_k = \tilde{G}_k^{-1} \tilde{c}_k [a_k(x_k - \bar{x}_k)]$$

with the announced matrix $\tilde{G}$. The deterministic part of the control yields

$$\bar{u}_{ik} = -\frac{\left(\frac{\bar{\alpha}_{i,k+1} \bar{b}_k}{\bar{r}_k}\right)^{1/(2p-1)}}{1 + \left(\frac{\bar{\alpha}_{i,k+1} \bar{b}_k}{\bar{r}_k}\right)^{1/(2p-1)} \bar{b}_k} \left(\bar{a}_k \bar{x}_k + \sum_{j \in \mathscr{I} \setminus \{i\}} \bar{b}_{jk} E u_{jk} \right).$$

We can easily check that the obtained equilibrium strategy of agent i is the presented in (14.141), and it is adapted to the filtration generated by $\{x_0, \bar{x}_0, \ldots, x_k, \bar{x}_k\}$.

Substituting the equilibrium strategy u_{ik}^* back into the candidate cost of i yields the recursive equations in (14.142).

Note that the α_{ik} and α_{jk} are coupled through $\tilde{G}$ and $\tilde{c}$. The $\bar{\alpha}_{ik}$ and $\bar{\alpha}_{jk}$ are coupled through $\bar{G}$ and $\bar{c}$. Finally, the equilibrium cost of agent $i \in \mathscr{I} = \{1, \dots, I\}$, is

$$f_{i0}(x_0, \bar{x}_0) = \alpha_{i0} E[(x_0 - \bar{x}_0)^{2o}] + \bar{\alpha}_{i0}\bar{x}_0^{2p}, \tag{14.146}$$

completing the proof. ■

14.3.9.4 Verification Method in the Space of Measures

The above analysis covers noise beyond the Markovian noise. We now verify the proposed solution by checking the case where the noise process is Markovian. To do so, we work directly on the space of probability measures associated with x_k and denote by $\mu_k = \mathbb{P}_{x_k}$ its probability distribution. Then μ_{k+1} can be obtained $\mathbb{P}_{\epsilon_{k+1}}$, μ_k and $x_k, \bar{x}_k, u_{ik}, \bar{u}_{ik}$. Note here that the mean-field type is involved in the transition kernel. The cost can also be written in terms of μ_k and the functional u_{ik}, of mean-field type. Our new augmented state becomes μ, and the system becomes a big deterministic system in the space of measures. The advantage now is that one can write a dynamic programming principle of this new augmented state.

$$i \in \mathscr{I} = \{1, \dots, I\},$$

$$\begin{aligned} V_{ik}(\mu_k) = \inf_{u_{ik}} \Bigg(& q_{ik} \int \left[(z - \int y\mu_k(dy))^{2o} \right] \mu_k(dz) + \bar{q}_{ik} \left(\int y\mu_k(dy) \right)^{2p} \\ & + r_{ik} \int \left[\left(u_{ik}(z, \mu_k) - \int u_{ik}(y, \mu_k)\mu_k(dy) \right)^{2o} \right] \mu_k(dz)) \\ & + \bar{r}_{ik} \left(\int u_{ik}(y, \mu_k)\mu_k(dy) \right)^{2p} + V_{i,k+1}(\mu_{k+1}) \Bigg) \end{aligned}$$

and the terminal cost of agent i is

$$V_{i,N}(\mu_N) = q_{iN} \int \left[\left(z - \int y\mu_N(dy) \right)^{2o} \right] \mu_N(dz) + \bar{q}_{iN} \left(\int y\mu_N(dy) \right)^{2p}.$$

where

$$\mu_{k+1}(dy) = \begin{cases} \mathbb{P}_{\epsilon_{k+1}}((a_k(x_k - Ex_k) + \sum_{i \in \mathscr{I}} b_{ik}(u_{ik} - Eu_{ik}))^{-1} [(dy - (\bar{a}_k Ex_k + \sum_{i \in \mathscr{I}} \bar{b}_{ik} Eu_{ik}))]) \text{ if} \\ (a_k(x_k - Ex_k) + \sum_{i \in \mathscr{I}} b_{ik}(u_{ik} - Eu_{ik}) \neq 0 \\ \delta_{\bar{a}_k Ex_k + \sum_{i \in \mathscr{I}} \bar{b}_{ik} Eu_{ik}}(dy) \text{ otherwise} \end{cases}$$

We seek for $V_{ik}(\mu_k)$ in the form

$$\alpha_{ik}\int\left[\left(z-\int y\mu_k(dy)\right)^{2o}\right]\mu_k(dz)+\bar{\alpha}_{ik}\left(\int y\mu_k(dy)\right)^{2p}.$$

Inserting the above yields the same dynamics for α_{ik} and $\bar{\alpha}_{ik}$ as well as the equilibrium structure

$$u_{ik}^*(x,\mu_k)=-(\tilde{G}_k^{-1}\tilde{c}_k)_i a_k\left(x-\int y\mu_k(dy)\right)-(\bar{G}_k^{-1}\bar{c}_k)_i\bar{a}_k\left(\int y\mu_k(dy)\right).$$

14.4 Summary

We investigated higher-order costs in the context of discrete time mean-field-type optimal control and game theory, addressing limitations in traditional quadratic cost models and offering solutions for various scenarios. We presented semi-explicit solutions for different problem types, including deterministic control with higher-order costs, variance-aware control with additive noise, risk-aware stochastic control with multiplicative noise, selfish agents with higher-order costs, selfish agents with variance-aware higher-order costs, and risk-aware games under multiplicative noise. These solutions provide a valuable tool for understanding and controlling multi-agent systems that involve nonlinear penalties and risk-sensitive behavior. The semi-explicit solutions derived in this section offer a significant advantage for training, verifying, and validating transformer-based deep neural networks in the context of higher-order cost minimization and state-and-mean-field-type feedback equilibrium computation. This is primarily because these solutions provide a ground truth or baseline against which the performance of the neural networks can be directly compared. This is a notable advancement compared to previous approaches where the absence of such solutions made it challenging to assess the accuracy and reliability of neural network-based approximations, especially for higher-order cost functions that exhibit sensitive dependencies on system parameters. The semi-explicit solutions enable training with supervision as the solutions can be used to generate training data with known optimal control strategies or equilibrium solutions, allowing for supervised learning of the deep neural networks. This can lead to more efficient and accurate training as there is a reference to compare with. The solution provided here can be also used for verification of network of network (NoN) outputs including multi-agent unsupervised or reinforcement learning methods, which are often used in the absence of ground truth. The semi-explicit solutions provide a means to directly verify the outputs of both shallow and deep neural networks. By comparing the network's predicted control actions or equilibrium states with the semi-explicit solutions, one can quantitatively assess the network's accuracy and identify potential errors or biases. The semi-explicit solutions can be used to validate the generalization ability of the shallow and deep neural networks. By evaluating the network's performance on a set of test cases with known semi-explicit solutions, one can assess how well the network generalizes to unseen scenarios and system parameters. The availability of semi-explicit solutions for MFTGs with non-quadratic cost functions marks a breakthrough in the field. It enables a more rigorous and reliable approach to developing and deploying deep learning algorithms for mean-field-type control and game-theoretic problems. The semi-explicit solutions derived here offer also a transformative approach to machine intelligence risk quantification, particularly in scenarios with higher-order risk measures, which are essential for capturing nonlinear risk phenomena, such as those associated with cascading failures, tipping points, or existential threats. These semi-explicit solutions provide a means to quantify these risks more accurately than traditional methods, enabling an understanding of potential machine intelligence harms and facilitating the evaluation of risk mitigation strategies by assessing their impact on the overall risk profile. Furthermore, the insights gained from our semi-explicit solutions can inform machine intelligence safety research by highlighting critical areas where further investigation is needed, guiding the development of targeted safety mechanisms. The semi-explicit nature of our solutions also offers a degree of transparency and explainability that is often lacking in purely numerical or simulation-based approaches, facilitating informed decision-making by policy-makers. The ability to quantify machine intelligence risks accurately and reliably is essential for ensuring the safe and beneficial development of advanced machine intelligence systems, and our findings provide a valuable tool for achieving this goal, paving the way for more robust and comprehensive machine intelligence risk quantification and management frameworks.

14.5 Discrete Time Adversarial Mean-Field-Type Games

In this section, we formulate and solve a class of discrete time linear-quadratic adversarial MFTGs. The state dynamics are described by a linear stochastic difference equation having mean-field terms such as the expected value of the state and the expected value of the control action. It involves both additive and multiplicative noises as well as state, control, and mean-field-dependent noises. The cost functionals include the variance of the state, the variances of the control actions, the covariances between state and controls, the expected values of the state, and control action of both decision-makers.

The optimal strategies and equilibrium cost for the decision-makers are obtained in a semi-explicit way using a simple direct method.

Specifics of the *adversarial* MFTGs are as follows. A defender and an attacker interact over the discrete time horizon $\{0, \ldots, K-1\}$, $K \geq 2$. The mean-field terms involved are

- the expectation of the state in the drift and noise coefficient functions.
- the expectation of the control actions in the drift and noise coefficient functions.

The state process is scalar, and running cost terms include the following:

- the expectation of the state, $E[x]$,
- the expectation of the control strategy, $E[u_1]$, $E[u_2]$
- the variance of the state, $\text{var}(x)$
- the square of the conditional expected value of the state, $\{E[x]\}^2$
- the variances of the control-action processes, $\text{var}(u_1)$, $\text{var}(u_2)$
- the square of the expected values of the control-action processes $\{E[u_1]\}^2$, $\{E[u_2]\}^2$
- Two types of cross-terms: the covariances between the state and the controls, $\text{cov}(x, u_1)$, $\text{cov}(x, u_2)$, and the covariance between the controls of 1 and 2, $\text{cov}(u_1, u_2)$
- A product term between the expected value of the control action and the expected value of the state, $E[u_1].E[x]$, $E[u_2]E[x]$.
- A product term between the expected values of the control actions, $E[u_1]E[u_2]$.

To solve the aforementioned problem in a semi-explicit way, we follow a direct method in four steps. The first step consists and reformulates an equivalent adversarial problem with orthogonal decomposition. The second step uses a guess functional and its telescopic sum. The third step uses a completion of squares for $u_1 - E[u_1]$, $E[u_1]$, $u_2 - E[u_2]$, and $E[u_2]$.The fourth step uses a matching technique to derive explicitly difference equations.

14.5.1 Adversarial Game

As formulated in terms above, we have a discrete time MFTG between defender DM1 and attacker DM2. The cost functional to be minimized by the defender and maximized by the attacker and the state dynamics are interactive with mean-field coupling, and the state dynamics is subject to both additive and multiplicative noises. The cost function L and the state dynamics are given as follows:

$$\begin{cases} L = q_K x_K^2 + \bar{q}_K (E[x_K])^2 + \phi_{3K} E[x_K] \\ + \sum_{k=0}^{K-1} q_k x_k^2 + \bar{q}_k (E[x_k])^2 + r_{1k} u_{1k}^2 + \bar{r}_{1k} (E[u_{1k}])^2 \\ + \sum_{k=0}^{K-1} 2\epsilon_{1k} x_k u_{1k} + 2\bar{\epsilon}_{1k} (E[x_k]) E[u_{1k}] + 2\phi_{1k} E[u_{1k}] + 2\phi_{3k} E[x_k] \\ + \sum_{k=0}^{K-1} r_{2k} u_{2k}^2 + \bar{r}_{2k} (E[u_{2k}])^2 \\ + \sum_{k=0}^{K-1} 2\epsilon_{2k} x_k u_{2k} + 2\bar{\epsilon}_{2k} (E[x_k]) E[u_{2k}] + 2\phi_{2k} E[u_{2k}] \\ + \sum_{k=0}^{K-1} 2 s_k u_{1k} u_{2k} + 2\bar{s}_k (E[u_{1k}])(E[u_{2k}]), \\ \\ x_{k+1} = \left(b_{0k} + b_{1k} x_k + \bar{b}_{1k} E[x_k] + \sum_{m \in \{1,2\}} b_{m2k} u_{mk} + \sum_{m \in \{1,2\}} \bar{b}_{m2k} E[u_{mk}] \right) \\ + \left(\sigma_{0k} + \sigma_{1k} x_k + \bar{\sigma}_{1k} E[x_k] + \sum_{m \in \{1,2\}} \sigma_{m2k} u_{mk} + \sum_{m \in \{1,2\}} \bar{\sigma}_{m2k} E[u_{mk}] \right) w_{k+1}, \\ x(0) = X_0, \end{cases} \tag{14.147}$$

where

$$\begin{cases} q_k > 0, q_k + \bar{q}_k > 0, \\ r_{1k} > 0, r_{1k} + \bar{r}_{1k} > 0, \\ r_{2k} < 0, r_{2k} + \bar{r}_{2k} < 0, \\ \epsilon_{1k}, \bar{\epsilon}_{1k}, \phi_{1k}, \in \mathbb{R}, \\ b_{0k}, b_{1k}, b_{22k}, \bar{b}_{1k}, \bar{b}_{22k} \in \mathbb{R}, \\ \sigma_{0k}, \sigma_{1k}, \sigma_{22k}, \bar{\sigma}_{1k}, \bar{\sigma}_{22k} \in \mathbb{R}, \\ w_{k+1} \perp X_0, \end{cases} \tag{14.148}$$

In the discrete time dynamics (14.147), the process $\{x_k\}_{0\leq k\leq K}$ is the state process and $\{u_{mk}, \}_{0\leq k\leq K-1}$ is the control process of the decision-maker $m \in \{1, 2\}$. The process $\{w_k\}_{0\leq k\leq K-1}$ represents a stochastic disturbance to the state and is defined on a probability space $(\Omega, \mathscr{F}, \mathbb{P})$. For simplicity, we consider only the situation where w_k is a one-dimensional martingale difference sequence with

$$E[w_{k+1} \mid \mathscr{F}_k] = 0, \;\; E[w_{k+1}^2 \mid \mathscr{F}_k] = 1,$$

where $\mathscr{F}_k$ is the σ-algebra generated by $(x_0, \{w_k\}_{0\leq k\leq K-1})$. The initial value x_0 and the disturbance $\{w_k\}_{0\leq k\leq K-1}$ are assumed to be independent of each other. It is important to note that the stochastic process $\{w_k\}$ is relatively general here, and it does not need to be a Gaussian.

We introduce the following set of admissible control strategies of both decision-maker $m \in \{1, 2\}$ as

$$\mathscr{U}_m = \left\{ \begin{array}{l} u_m = (u_{m0}, \ldots, u_{m,K-1}) \mid u_{mk} : \Omega \to \mathbb{R}, \\ u_{mk} \text{ is } \mathscr{F}_k - \text{measurable}, \;\; E[u_{mk}^2] < +\infty \end{array} \right\} \tag{14.149}$$

For any initial random variable X_0 such that $E[X_0^2] < +\infty$, and any strategy profile $u = (u_1, u_2) \in \prod_{m\in\{1,2\}} \mathscr{U}_m$,, the state process is well defined. Moreover, for any $1 \leq k \leq K$,, the first and the second moments of x_k are finite.

$$E[|x_k|^l] < +\infty, \; l \in \{1, 2\}. \tag{14.150}$$

Note that what we have here is a mean-field-type zero-sum game with the expected cost $E[L(u_1, u_2)]$, to be minimized by $u_1 \in \mathscr{U}_1$ and $u_2 \in \mathscr{U}_2$. Hence, what we are looking for is a saddle point solution, which is called mean-field-type saddle point solution, because of the presence of the mean-field terms. The corresponding value of $E[L(u_1, u_2)]$, is known as the mean-field value of zero-sum MFTG.

14.5.2 Reformulation With Covariance and Variance

We now provide an equivalent formulation of Problem (14.147) using covariance and variance. To do so, we use the following decompositions

$$x = x - E[x] + E[x], \qquad u_m = u_m - E[u_m] + E[u_m]. \tag{14.151}$$

Moreover, these decompositions are orthogonal as

$$E\{(x - E[x])E[x]\} = 0, \quad E\{(u_m - E[u_m])E[u_m]\} = 0. \tag{14.152}$$

As we will see below, these orthogonal decompositions simplify the analysis in the expansion.

With this transformation, and with insertion of inf-sup operation, the game with cost function and state dynamics as given Problem (14.147) becomes equivalent to

$$\begin{cases} \hat{L} = q_K\ var(x_K) + (q_K + \bar{q}_K)(E[x_K])^2 + \phi_{3K} E[x_K] \\ + \sum_{k=0}^{K-1} q_k\ var(x_k) + (q_k + \bar{q}_k)(E[x_k])^2 \\ + \sum_{k=0}^{K-1} r_{1k}\ var(u_{1k}) + (r_{1k} + \bar{r}_{1k})(E[u_{1k}])^2 \\ + \sum_{k=0}^{K-1} 2\epsilon_{1k}\ cov(x_k, u_{1k}) + 2(\epsilon_{1k} + \bar{\epsilon}_{1k})(E[x_k])E[u_{1k}] + 2\phi_{1k}E[u_{1k}] + 2\phi_{3k}E[x_k] \\ + \sum_{k=0}^{K-1} r_{2k}\ var(u_{2k}) + (r_{2k} + \bar{r}_{2k})(E[u_{2k}])^2 \\ + \sum_{k=0}^{K-1} 2\epsilon_{2k}\ cov(x_k, u_{2k}) + 2(\epsilon_{2k} + \bar{\epsilon}_{2k})(E[x_k])E[u_{2k}] + 2\phi_{2k}E[u_{2k}] \\ + \sum_{k=0}^{K-1} 2s_k\ cov(u_{1k}, u_{2k}) + 2(s_k + \bar{s}_k)(E[u_{1k}])(E[u_{2k}]), \\ \\ \inf_{u_1 \in \mathcal{U}_1} \sup_{u_2 \in \mathcal{U}_2} E[\hat{L}_2(u_1, u_2)], \\ \text{subject to} \\ x_{k+1} = \Big\{ b_{0k} + b_{1k}(x_k - E[x_k]) + (b_{1k} + \bar{b}_{1k})E[x_k] \\ + \sum_{m \in \{1,2\}} b_{m2k}(u_{mk} - E[u_{mk}]) + \sum_{m \in \{1,2\}} (b_{m2k} + \bar{b}_{m2k})E[u_{mk}] \Big\} \\ + \{ \sigma_{0k} + \sigma_{1k}(x_k - E[x_k]) + (\sigma_{1k} + \bar{\sigma}_{1k})E[x_k] \\ + \sum_{m \in \{1,2\}} \sigma_{m2k}(u_{mk} - E[u_{mk}]) + \sum_{m \in \{1,2\}} (\sigma_{m2k} + \bar{\sigma}_{m2k})E[u_{mk}] \Big\} w_{k+1}, \\ x(0) = x_0, \end{cases} \tag{14.153}$$

where $var(x_k) = E[(x_k - Ex_k)^2]$, is the variance of the state x_k and $cov(x_k, u_{ik}) = E[(x_k - Ex_k)(u_{ik} - Eu_{ik})]$, is the covariance between the state x_k and the control action u_{ik} for $i \in \{1, 2\}$. The proof of (14.153) follows immediately from (14.151) and (14.152).

14.5.3 Semi-explicit Solution

We provide a generic solution to Problem (14.153) with both additive and multiplicative noises. The solution is presented in a semi-explicitly way using coefficients satisfying a deterministic discrete time system.

The optimal strategy of the decision-maker 1 (the defender), who minimizes the cost, is

$$\begin{cases} u_{1k} = -\eta_{1k}(x_k - Ex_k) - \bar{\eta}_{1k} Ex_k - \hat{\eta}_{1k}, \\ \eta_{1k} = -\frac{\eta_{2k}}{c_{1k}}\{\alpha_{k+1}[b_{12k}b_{22k} + \sigma_{12k}\sigma_{22k}] + s_k\} + \frac{1}{c_{1k}}\alpha_{k+1}\sigma_{1k}\sigma_{12k} + \frac{1}{c_{1k}}\alpha_{k+1}b_{1k}b_{12k} + \frac{1}{c_{1k}}\epsilon_{1k}, \\ c_{1k} = r_{1k} + \alpha_{1,k+1}(b_{12k}^2 + \sigma_{12k}^2), \\ \bar{\eta}_{1k} = -\frac{1}{\bar{c}_{1k}}\{\alpha_{k+1}(\sigma_{12k} + \bar{\sigma}_{12k})(\sigma_{22k} + \bar{\sigma}_{22k}) + \beta_{k+1}(b_{12k} + \bar{b}_{12k})(b_{22k} + \bar{b}_{22k}) + (s_k + \bar{s}_k)\}\bar{\eta}_{2k} \\ +\frac{1}{\bar{c}_{1k}}\{\alpha_{k+1}(\sigma_{1k} + \bar{\sigma}_{1k})(\sigma_{12k} + \bar{\sigma}_{12k}) + \beta_{k+1}(b_{1k} + \bar{b}_{1k})(b_{12k} + \bar{b}_{12k}) + (\epsilon_{1k} + \bar{\epsilon}_{1k})\}, \\ \bar{c}_{1k} = (r_{1k} + \bar{r}_{1k}) + \alpha_{k+1}(\sigma_{12k} + \bar{\sigma}_{12k})^2 + \beta_{k+1}(b_{12k} + \bar{b}_{12k})^2, \\ \hat{\eta}_{1k} = -\frac{1}{\bar{c}_{1k}}\{\alpha_{k+1}(\sigma_{12k} + \bar{\sigma}_{12k})(\sigma_{22k} + \bar{\sigma}_{22k}) + \beta_{k+1}(b_{12k} + \bar{b}_{12k})(b_{22k} + \bar{b}_{22k}) + (s_k + \bar{s}_k)\}\bar{\eta}_{2k} \\ +\frac{1}{\bar{c}_{1k}}[\alpha_{k+1}\sigma_{0k}(\sigma_{12k} + \bar{\sigma}_{12k}) + \beta_{k+1}b_{0k}(b_{12k} + \bar{b}_{12k}) + \gamma_{k+1}(b_{12k} + \bar{b}_{12k}) + \phi_{1k}, \end{cases} \tag{14.154}$$

The optimal strategy of the decision-maker 2 who maximizes the cost is

$$\begin{cases} u_{2k} = -\eta_{2k}(x_k - Ex_k) - \bar{\eta}_{2k} Ex_k - \hat{\eta}_{2k}, \\ \eta_{2k} = \frac{1}{c_{2k}}[\alpha_{k+1}\sigma_{1k}\sigma_{22k} + \alpha_{k+1}b_{1k}b_{22k} + \epsilon_{2k}] \\ -\frac{1}{c_{2k}}\frac{1}{c_{1k}}\{\alpha_{k+1}[b_{12k}b_{22k} + \sigma_{12k}\sigma_{22k}] + s_k\}\{\alpha_{k+1}[\sigma_{1k}\sigma_{12k} + b_{1k}b_{12k}] + \epsilon_{1k}\}, \\ c_{2k} = r_{2k} + \alpha_{k+1}(b_{22k}^2 + \sigma_{22k}^2) - \frac{1}{c_{1k}}\{\alpha_{k+1}[b_{12k}b_{22k} + \sigma_{12k}\sigma_{22k}] + s_k\}^2, \\ \bar{\eta}_{2k} = \frac{\kappa_{22}}{\bar{c}_{2k}}, \\ \bar{c}_{2k} = (r_{2k} + \bar{r}_{2k}) + \alpha_{k+1}(\sigma_{22k} + \bar{\sigma}_{22k})^2 + \beta_{k+1}(b_{22k} + \bar{b}_{22k})^2 \\ -\frac{1}{\bar{c}_{1k}}\{\alpha_{k+1}(\sigma_{12k} + \bar{\sigma}_{12k})(\sigma_{22k} + \bar{\sigma}_{22k}) + \beta_{k+1}(b_{12k} + \bar{b}_{12k})(b_{22k} + \bar{b}_{22k}) + (s_k + \bar{s}_k)\}^2, \\ \kappa_{22} := \beta_{k+1}(b_{1k} + \bar{b}_{1k})(b_{22k} + \bar{b}_{22k}) + \alpha_{k+1}(\sigma_{1k} + \bar{\sigma}_{1k})(\sigma_{22k} + \bar{\sigma}_{22k}) + (\epsilon_{2k} + \bar{\epsilon}_{2k}) - \frac{1}{\bar{c}_{1k}}\kappa_{11}\kappa_{12}, \\ \hat{\eta}_{2k} = \frac{\kappa_{23}}{\bar{c}_{2k}}, \\ \kappa_{23} := \alpha_{k+1}\sigma_{0k}(\sigma_{22k} + \bar{\sigma}_{22k}) \\ +\beta_{k+1}\, b_{0k}(b_{22k} + \bar{b}_{22k}) + \gamma_{k+1}(b_{22k} + \bar{b}_{22k}) + \phi_{2k} - \frac{1}{\bar{c}_{1k}}\kappa_{11}\kappa_{13}, \\ \kappa_{11} := \{\alpha_{k+1}(\sigma_{12k} + \bar{\sigma}_{12k})(\sigma_{22k} + \bar{\sigma}_{22k}) + \beta_{k+1}(b_{12k} + \bar{b}_{12k})(b_{22k} + \bar{b}_{22k}) + (s_k + \bar{s}_k)\}, \\ \kappa_{12} := \{\alpha_{k+1}(\sigma_{1k} + \bar{\sigma}_{1k})(\sigma_{12k} + \bar{\sigma}_{12k}) + \beta_{k+1}(b_{1k} + \bar{b}_{1k})(b_{12k} + \bar{b}_{12k}) + (\epsilon_{1k} + \bar{\epsilon}_{1k})\}, \\ \kappa_{13} := \alpha_{k+1}\sigma_{0k}(\sigma_{12k} + \bar{\sigma}_{12k}) + \beta_{k+1}b_{0k}(b_{12k} + \bar{b}_{12k}) + \gamma_{k+1}(b_{12k} + \bar{b}_{12k}) + \phi_{1k}, \end{cases} \tag{14.155}$$

and the value of the MFTG is

$$E[L^*] = \alpha_0(x_0 - Ex_0)^2 + \beta_0(Ex_0)^2 + 2\gamma_0(Ex_0) + \delta_0.$$

The coefficient functions here are given by

$$
\begin{aligned}
&\alpha_k = \alpha_{k+1}(b_{1k}^2 + \sigma_{1k}^2) + q_k - c_{2k}\eta_{2k}^2 - \tfrac{1}{c_{1k}}\{\alpha_{k+1}[\sigma_{1k}\sigma_{12k} + b_{1k}b_{12k}] + \epsilon_{1k}\}^2,\\
&\alpha_K = q_K,\\
&\beta_k = \beta_{k+1}(b_{1k} + \bar{b}_{1k})^2 + (q_k + \bar{q}_k) + \alpha_{k+1}(\sigma_{1k} + \bar{\sigma}_{1k})^2 - \tfrac{1}{\bar{c}_{1k}}\kappa_{12}^2 - \tfrac{1}{\bar{c}_{2k}}\kappa_{22}^2,\\
&\beta_K = q_K + \bar{q}_K,\\
&\gamma_{1k} = \gamma_{k+1}(b_{1k} + \bar{b}_{1k}) + \phi_{3k} + \alpha_{k+1}\sigma_{0k}(\sigma_{1k} + \bar{\sigma}_{1k})\\
&+\beta_{k+1}b_{0k}(b_{1k} + \bar{b}_{1k}) - \tfrac{1}{\bar{c}_{1k}}\kappa_{12}\kappa_{13} - \tfrac{1}{\bar{c}_{2k}}\kappa_{22}\kappa_{23},\\
&\gamma_K = \epsilon_{3K},\\
&\delta_k = \delta_{k+1} + \alpha_{k+1}\sigma_{0k}^2 + \beta_{k+1}b_{0k}^2 + 2\gamma_{k+1}b_{0k} - \tfrac{1}{\bar{c}_{1k}}\kappa_{13}^2 - \tfrac{1}{\bar{c}_{2k}}\kappa_{23}^2,\\
&\delta_K = 0.
\end{aligned}
\tag{14.156}
$$

We carry out the proof in a few steps. The first step of the proof uses a guess functional and its telescopic sum. The second step uses a square completion for $u_m - E[u_m]$, and $E[u_m]$. The third step uses a matching technique.

14.5.3.1 Guess Functional Telescopic Sum

Given a generic functional f_k, we use the telescopic sum

$$
f_K = f_0 + \sum_{k=0}^{K-1}(f_{k+1} - f_k), \tag{14.157}
$$

to derive the difference between the guess (ansatz) functions and and the cost functional. Applying telescopic sum to

$$
f_k = \alpha_k(x_k - Ex_k)^2 + \beta_k(Ex_k)^2 + 2\gamma_k(Ex_k) + \delta_k,
$$

where $\alpha, \beta, \gamma, \delta$ are deterministic sequences yet to be determined yields a difference equation with several mean-field terms.

The computation of $E[L - \alpha_0(x_0 - Ex_0)^2 - \beta_0(Ex_0)^2 - 2\gamma_0(Ex_0) - \delta_0]$ yields

$$
\begin{aligned}
&E\{\hat{L} - \alpha_0(x_0 - Ex_0)^2 - \beta_0(Ex_0)^2 - 2\gamma_0(Ex_0) - \delta_0\}\\
&= [q_K - \alpha_K]var(x_K) + [q_K + \bar{q}_K - \beta_K](E[x_K])^2\\
&+2(\phi_{3K} - \gamma_K)E[x_K] + (0 - \delta_K)\\
&+\sum_{k=0}^{K-1}\Big[\{r_{1k} + \alpha_{k+1}(b_{12k}^2 + \sigma_{12k}^2)\}var(u_{1k})\\
&+2\alpha_{k+1}b_{12k}b_{22k}(u_{1k} - E[u_{1k}])(u_{2k} - E[u_{2k}])\\
&+2\alpha_{k+1}\sigma_{12k}\sigma_{22k}(u_{1k} - E[u_{1k}])(u_{2k} - E[u_{2k}]) + 2s_k\, cov(u_{1k}, u_{2k})\\
&+2\alpha_{k+1}\sigma_{1k}\sigma_{12k}(u_{1k} - E[u_{1k}])(x_k - E[x_k])\\
&+2\alpha_{k+1}b_{1k}b_{12k}(u_{1k} - E[u_{1k}])(x_k - E[x_k]) + 2\epsilon_{1k}\, cov(x_k, u_{1k})\\
&+\{(r_{1k} + \bar{r}_{1k}) + \alpha_{k+1}(\sigma_{12k} + \bar{\sigma}_{12k})^2 + \beta_{k+1}(b_{12k} + \bar{b}_{12k})^2\}(E[u_{1k}])^2\\
&+2\alpha_{k+1}(\sigma_{12k} + \bar{\sigma}_{12k})(\sigma_{22k} + \bar{\sigma}_{22k})(E[u_{1k}])E[u_{2k}]\\
&+2\beta_{k+1}(b_{12k} + \bar{b}_{12k})(b_{22k} + \bar{b}_{22k})E[u_{1k}]E[u_{2k}] + 2(s_k + \bar{s}_k)(E[u_{1k}])(E[u_{2k}])\\
&+2\alpha_{k+1}(\sigma_{1k} + \bar{\sigma}_{1k})(\sigma_{12k} + \bar{\sigma}_{12k})E[u_{1k}]E[x_k]\\
&+2\beta_{k+1}(b_{1k} + \bar{b}_{1k})(b_{12k} + \bar{b}_{12k})E[u_{1k}]E[x_k]\\
&+2(\epsilon_{1k} + \bar{\epsilon}_{1k})(E[x_k])E[u_{1k}]\\
&+2\alpha_{k+1}\sigma_{0k}(\sigma_{12k} + \bar{\sigma}_{12k})E[u_{1k}]\\
&+2\beta_{k+1}b_{0k}(b_{12k} + \bar{b}_{12k})E[u_{1k}]\\
&+2\gamma_{k+1}(b_{12k} + \bar{b}_{12k})E[u_{1k}] + 2\phi_{1k}E[u_{1k}]\\
&+\{r_{2k} + \alpha_{k+1}(b_{22k}^2 + \sigma_{22k}^2)\}\, var(u_{2k})\\
&+2\alpha_{k+1}\sigma_{1k}\sigma_{22k}(u_{2k} - E[u_{2k}])(x_k - E[x_k])\\
&+2\alpha_{k+1}b_{1k}b_{22k}(u_{2k} - E[u_{2k}])(x_k - E[x_k]) + 2\epsilon_{2k}\, cov(x_k, u_{2k})\\
&+\{(r_{2k} + \bar{r}_{2k}) + \alpha_{k+1}(\sigma_{22k} + \bar{\sigma}_{22k})^2 + \beta_{k+1}(b_{22k} + \bar{b}_{22k})^2\}(E[u_{2k}])^2\\
&+2\beta_{k+1}(b_{1k} + \bar{b}_{1k})(b_{22k} + \bar{b}_{22k})E[u_{2k}]E[x_k]\\
&+2\alpha_{k+1}(\sigma_{1k} + \bar{\sigma}_{1k})(\sigma_{22k} + \bar{\sigma}_{22k})E[u_{2k}]E[x_k]\\
&+2(\epsilon_{2k} + \bar{\epsilon}_{2k})E[u_{2k}](E[x_k])\\
&+2\alpha_{k+1}\sigma_{0k}(\sigma_{22k} + \bar{\sigma}_{22k})E[u_{2k}]\\
&+2\beta_{k+1}\, b_{0k}(b_{22k} + \bar{b}_{22k})E[u_{2k}]\\
&+2\gamma_{k+1}(b_{22k} + \bar{b}_{22k})E[u_{2k}] + 2\phi_{2k}E[u_{2k}]\\
&+\{\alpha_{k+1}(b_{1k}^2 + \sigma_{1k}^2) - \alpha_k + q_k\}(x_k - E[x_k])^2\\
&+\{\beta_{k+1}(b_{1k} + \bar{b}_{1k})^2 - \beta_k + (q_k + \bar{q}_k) + \alpha_{k+1}(\sigma_{1k} + \bar{\sigma}_{1k})^2\}(E[x_k])^2\\
&+\{2\gamma_{k+1}(b_{1k} + \bar{b}_{1k}) - 2\gamma_k + 2\phi_{3k} + 2\alpha_{k+1}\sigma_{0k}(\sigma_{1k} + \bar{\sigma}_{1k})\\
&+2\beta_{k+1}b_{0k}(b_{1k} + \bar{b}_{1k})\}E[x_k]\\
&+\delta_{k+1} - \delta_k + \alpha_{k+1}\sigma_{0k}^2 + \beta_{k+1}b_{0k}^2 + 2\gamma_{k+1}b_{0k}\Big]
\end{aligned}
\tag{14.158}
$$

In order to carry out the infsup operation, we reorganize the optimization problem into a strictly convex-concave-like form by completing the squares. A sufficient condition for saddle point is simply obtained with strict convexity in u_1 and strict concavity in u_2. Since the running cost functional is quadratic, the convex-concave property can be checked using the Hessian matrix. Therefore, we use a completion of squares which is detailed in the next subsection.

14.5.3.2 Joint Completion of Squares and Matching

We complete the squares for the terms in $(u_{1k} - E[u_{1k}])$

$$
\begin{aligned}
&\{r_{1k} + \alpha_{k+1}(b_{12k}^2 + \sigma_{12k}^2)\}var(u_{1k}) + 2\alpha_{k+1}b_{12k}b_{22k}(u_{1k} - E[u_{1k}])(u_{2k} - E[u_{2k}])\\
&+2\alpha_{k+1}\sigma_{12k}\sigma_{22k}(u_{1k} - E[u_{1k}])(u_{2k} - E[u_{2k}]) + 2s_k\, cov(u_{1k}, u_{2k})\\
&+2\alpha_{k+1}\sigma_{1k}\sigma_{12k}(u_{1k} - E[u_{1k}])(x_k - E[x_k])\\
&+2\alpha_{k+1}b_{1k}b_{12k}(u_{1k} - E[u_{1k}])(x_k - E[x_k]) + 2\epsilon_{1k}\, cov(x_k, u_{1k})\\
&= c_{1k}\{u_{1k} - E[u_{1k}] + \eta_{1k}(x_k - E[x_k])\}^2\\
&-\tfrac{1}{c_{1k}}\{\alpha_{k+1}[b_{12k}b_{22k} + \sigma_{12k}\sigma_{22k}] + s_k\}^2(u_{2k} - E[u_{2k}])^2\\
&-\tfrac{1}{c_{1k}}\{\alpha_{k+1}[b_{12k}b_{22k} + \sigma_{12k}\sigma_{22k}] + s_k\}\{\alpha_{k+1}[\sigma_{1k}\sigma_{12k} + b_{1k}b_{12k}] + \epsilon_{1k}\}\\
&(x_k - E[x_k])(u_{2k} - E[u_{2k}])\\
&-\tfrac{1}{c_{1k}}\{\alpha_{k+1}[\sigma_{1k}\sigma_{12k} + b_{1k}b_{12k}] + \epsilon_{1k}\}^2(x_k - E[x_k])^2,
\end{aligned}
\tag{14.159}
$$

where

$$
\begin{aligned}
&c_{1k} = r_{1k} + \alpha_{1,k+1}(b_{12k}^2 + \sigma_{12k}^2),\\
&\eta_{1k} = -\tfrac{\eta_{2k}}{c_{1k}}\{\alpha_{k+1}[b_{12k}b_{22k} + \sigma_{12k}\sigma_{22k}] + s_k\}\\
&+\tfrac{1}{c_{1k}}\alpha_{k+1}\sigma_{1k}\sigma_{12k} + \tfrac{1}{c_{1k}}\alpha_{k+1}b_{1k}b_{12k} + \tfrac{1}{c_{1k}}\epsilon_{1k},
\end{aligned}
\tag{14.160}
$$

Similarly, we complete the squares for the terms in $E[u_{1k}]$:

$$
\begin{aligned}
&\{(r_{1k} + \bar{r}_{1k}) + \alpha_{k+1}(\sigma_{12k} + \bar{\sigma}_{12k})^2 + \beta_{k+1}(b_{12k} + \bar{b}_{12k})^2\}(E[u_{1k}])^2\\
&+2\alpha_{k+1}(\sigma_{12k} + \bar{\sigma}_{12k})(\sigma_{22k} + \bar{\sigma}_{22k})(E[u_{1k}])E[u_{2k}]\\
&+2\beta_{k+1}(b_{12k} + \bar{b}_{12k})(b_{22k} + \bar{b}_{22k})E[u_{1k}]E[u_{2k}] + 2(s_k + \bar{s}_k)(E[u_{1k}])(E[u_{2k}])\\
&+2\alpha_{k+1}(\sigma_{1k} + \bar{\sigma}_{1k})(\sigma_{12k} + \bar{\sigma}_{12k})E[u_{1k}]E[x_k]\\
&+2\beta_{k+1}(b_{1k} + \bar{b}_{1k})(b_{12k} + \bar{b}_{12k})E[u_{1k}]E[x_k] + 2(\epsilon_{1k} + \bar{\epsilon}_{1k})(E[x_k])E[u_{1k}]\\
&+2\alpha_{k+1}\sigma_{0k}(\sigma_{12k} + \bar{\sigma}_{12k})E[u_{1k}]\\
&+2\beta_{k+1}b_{0k}(b_{12k} + \bar{b}_{12k})E[u_{1k}] + 2\gamma_{k+1}(b_{12k} + \bar{b}_{12k})E[u_{1k}] + 2\phi_{1k}E[u_{1k}]\\
&= \bar{c}_{1k}\{Eu_{1k} + \eta_{1k}E[x_k] + \hat{\eta}_{1k}\}^2 - \tfrac{1}{\bar{c}_{1k}}\kappa_{11}^2(E[u_{2k}])^2 - \tfrac{1}{\bar{c}_{1k}}\kappa_{12}^2(E[x_k])^2 - \tfrac{1}{\bar{c}_{1k}}\kappa_{13}^2\\
&-\tfrac{1}{\bar{c}_{1k}}2\kappa_{11}\kappa_{12}(E[u_{2k}])(E[x_k]) - \tfrac{1}{\bar{c}_{1k}}2\kappa_{11}\kappa_{13}(E[u_{2k}]) - \tfrac{1}{\bar{c}_{1k}}2\kappa_{12}\kappa_{13}(E[x_k]),
\end{aligned}
\tag{14.161}
$$

$$
\begin{aligned}
&\bar{c}_{1k} := (r_{1k}+\bar{r}_{1k}) + \alpha_{k+1}(\sigma_{12k}+\bar{\sigma}_{12k})^2 + \beta_{k+1}(b_{12k}+\bar{b}_{12k})^2,\\
&\bar{\eta}_{1k} = -\frac{1}{\bar{c}_{1k}}\{\alpha_{k+1}(\sigma_{12k}+\bar{\sigma}_{12k})(\sigma_{22k}+\bar{\sigma}_{22k}) + \beta_{k+1}(b_{12k}+\bar{b}_{12k})(b_{22k}+\bar{b}_{22k}) + (s_k+\bar{s}_k)\}\bar{\eta}_{2k}\\
&+\frac{1}{\bar{c}_{1k}}\{\alpha_{k+1}(\sigma_{1k}+\bar{\sigma}_{1k})(\sigma_{12k}+\bar{\sigma}_{12k}) + \beta_{k+1}(b_{1k}+\bar{b}_{1k})(b_{12k}+\bar{b}_{12k}) + (\epsilon_{1k}+\bar{\epsilon}_{1k})\},\\
&\hat{\eta}_{1k} = -\frac{1}{\bar{c}_{1k}}\{\alpha_{k+1}(\sigma_{12k}+\bar{\sigma}_{12k})(\sigma_{22k}+\bar{\sigma}_{22k}) + \beta_{k+1}(b_{12k}+\bar{b}_{12k})(b_{22k}+\bar{b}_{22k}) + (s_k+\bar{s}_k)\}\bar{\eta}_{2k}\\
&+\frac{1}{\bar{c}_{1k}}[\alpha_{k+1}\sigma_{0k}(\sigma_{12k}+\bar{\sigma}_{12k}) + \beta_{k+1}b_{0k}(b_{12k}+\bar{b}_{12k}) + \gamma_{k+1}(b_{12k}+\bar{b}_{12k}) + \phi_{1k}],\\
&\bar{\tau}_{1k} =\\
&\{\alpha_{k+1}(\sigma_{12k}+\bar{\sigma}_{12k})(\sigma_{22k}+\bar{\sigma}_{22k}) + \beta_{k+1}(b_{12k}+\bar{b}_{12k})(b_{22k}+\bar{b}_{22k}) + (s_k+\bar{s}_k)\}(E[u_{2k}])\\
&+\{\alpha_{k+1}(\sigma_{1k}+\bar{\sigma}_{1k})(\sigma_{12k}+\bar{\sigma}_{12k} + \beta_{k+1}(b_{1k}+\bar{b}_{1k})(b_{12k}+\bar{b}_{12k}) + (\epsilon_{1k}+\bar{\epsilon}_{1k})\}(E[x_k])\\
&+\alpha_{k+1}\sigma_{0k}(\sigma_{12k}+\bar{\sigma}_{12k}) + \beta_{k+1}b_{0k}(b_{12k}+\bar{b}_{12k}) + \gamma_{k+1}(b_{12k}+\bar{b}_{12k}) + \phi_{1k}\\
&=: \kappa_{11}(E[u_{2k}]) + \kappa_{12}(E[x_k]) + \kappa_{13},\\
&\kappa_{11} := \{\alpha_{k+1}(\sigma_{12k}+\bar{\sigma}_{12k})(\sigma_{22k}+\bar{\sigma}_{22k}) + \beta_{k+1}(b_{12k}+\bar{b}_{12k})(b_{22k}+\bar{b}_{22k}) + (s_k+\bar{s}_k)\},\\
&\kappa_{12} := \{\alpha_{k+1}(\sigma_{1k}+\bar{\sigma}_{1k})(\sigma_{12k}+\bar{\sigma}_{12k}) + \beta_{k+1}(b_{1k}+\bar{b}_{1k})(b_{12k}+\bar{b}_{12k}) + (\epsilon_{1k}+\bar{\epsilon}_{1k})\},\\
&\kappa_{13} : \alpha_{k+1}\sigma_{0k}(\sigma_{12k}+\bar{\sigma}_{12k}) + \beta_{k+1}b_{0k}(b_{12k}+\bar{b}_{12k}) + \gamma_{k+1}(b_{12k}+\bar{b}_{12k}) + \phi_{1k},\\
&\bar{\tau}_{1k}^2 = \kappa_{11}^2(E[u_{2k}])^2 + \kappa_{12}^2(E[x_k])^2 + \kappa_{13}^2\\
&+2\kappa_{11}\kappa_{12}(E[u_{2k}])(E[x_k]) + 2\kappa_{11}\kappa_{13}(E[u_{2k}]) + 2\kappa_{12}\kappa_{13}(E[x_k]),
\end{aligned}
\tag{14.162}
$$

By inserting (14.159) and (14.161), the cost difference (14.158) becomes

$$
\begin{aligned}
&E\{L - \alpha_0(x_0 - Ex_0)^2 - \beta_0(Ex_0)^2 - 2\gamma_0(Ex_0) - \delta_0\}\\
&= [q_K - \alpha_K]var(x_K) + [q_K + \bar{q}_K - \beta_K](E[x_K])^2 + 2(\phi_{3K} - \gamma_K)E[x_K] + (0 - \delta_K)\\
&+\sum_{k=0}^{K-1}\Big[c_{1k}\{u_{1k} - E[u_{1k}] + \eta_{1k}(x_k - E[x_k])\}^2\\
&+\bar{c}_{1k}\{Eu_{1k} + \eta_{1k}E[x_k] + \hat{\eta}_{1k}\}^2\\
&+\{r_{2k} + \alpha_{k+1}(b_{22k}^2 + \sigma_{22k}^2)\}\ var(u_{2k})\\
&-\frac{1}{c_{1k}}\{\alpha_{k+1}[b_{12k}b_{22k} + \sigma_{12k}\sigma_{22k}] + s_k\}^2(u_{2k} - E[u_{2k}])^2\\
&+2\alpha_{k+1}\sigma_{1k}\sigma_{22k}(u_{2k} - E[u_{2k}])(x_k - E[x_k])\\
&+2\alpha_{k+1}b_{1k}b_{22k}(u_{2k} - E[u_{2k}])(x_k - E[x_k]) + 2\epsilon_{2k}\ cov(x_k, u_{2k})\\
&-\frac{2}{c_{1k}}\{\alpha_{k+1}[b_{12k}b_{22k} + \sigma_{12k}\sigma_{22k}] + s_k\}\{\alpha_{k+1}[\sigma_{1k}\sigma_{12k} + b_{1k}b_{12k}] + \epsilon_{1k}\}cov(x_k, u_{2k})\\
&+\{(r_{2k}+\bar{r}_{2k}) + \alpha_{k+1}(\sigma_{22k}+\bar{\sigma}_{22k})^2 + \beta_{k+1}(b_{22k}+\bar{b}_{22k})^2\}(E[u_{2k}])^2 - \frac{1}{\bar{c}_{1k}}\kappa_{11}^2(E[u_{2k}])^2\\
&+2\beta_{k+1}(b_{1k}+\bar{b}_{1k})(b_{22k}+\bar{b}_{22k})E[u_{2k}]E[x_k]\\
&+2\alpha_{k+1}(\sigma_{1k}+\bar{\sigma}_{1k})(\sigma_{22k}+\bar{\sigma}_{22k})E[u_{2k}]E[x_k]\\
&+2(\epsilon_{2k}+\bar{\epsilon}_{2k})E[u_{2k}](E[x_k]) - \frac{1}{\bar{c}_{1k}}2\kappa_{11}\kappa_{12}(E[u_{2k}])(E[x_k])\\
&+2\alpha_{k+1}\sigma_{0k}(\sigma_{22k}+\bar{\sigma}_{22k})E[u_{2k}]\\
&+2\beta_{k+1}\ b_{0k}(b_{22k}+\bar{b}_{22k})E[u_{2k}] + 2\gamma_{k+1}(b_{22k}+\bar{b}_{22k})E[u_{2k}] + 2\phi_{2k}E[u_{2k}]\\
&-\frac{1}{\bar{c}_{1k}}2\kappa_{11}\kappa_{13}(E[u_{2k}])\\
&+\{\alpha_{k+1}(b_{1k}^2 + \sigma_{1k}^2) - \alpha_k + q_k\}(x_k - E[x_k])^2\\
&-\frac{1}{c_{1k}}\{\alpha_{k+1}[\sigma_{1k}\sigma_{12k} + b_{1k}b_{12k}] + \epsilon_{1k}\}^2(x_k - E[x_k])^2\\
&+\{\beta_{k+1}(b_{1k}+\bar{b}_{1k})^2 - \beta_k + (q_k+\bar{q}_k) + \alpha_{k+1}(\sigma_{1k}+\bar{\sigma}_{1k})^2\}(E[x_k])^2\\
&-\frac{1}{\bar{c}_{1k}}\kappa_{12}^2(E[x_k])^2\\
&+\{2\gamma_{k+1}(b_{1k}+\bar{b}_{1k}) - 2\gamma_k + 2\phi_{3k} + 2\alpha_{k+1}\sigma_{0k}(\sigma_{1k}+\bar{\sigma}_{1k}) + 2\beta_{k+1}b_{0k}(b_{1k}+\bar{b}_{1k})\}E[x_k]\\
&-\frac{1}{\bar{c}_{1k}}2\kappa_{12}\kappa_{13}(E[x_k]) + \delta_{k+1} - \delta_k + \alpha_{k+1}\sigma_{0k}^2 + \beta_{k+1}b_{0k}^2 + 2\gamma_{k+1}b_{0k} - \frac{1}{\bar{c}_{1k}}\kappa_{13}^2\Big]
\end{aligned}
\tag{14.163}
$$

Completion of squares for $u_2 - Eu_2$ yields

$$\begin{aligned}
&\{r_{2k} + \alpha_{k+1}(b_{22k}^2 + \sigma_{22k}^2)\}\ var(u_{2k})\\
&-\frac{1}{c_{1k}}\{\alpha_{k+1}[b_{12k}b_{22k} + \sigma_{12k}\sigma_{22k}] + s_k\ \}^2(u_{2k} - E[u_{2k}])^2\\
&+2\alpha_{k+1}\sigma_{1k}\sigma_{22k}(u_{2k} - E[u_{2k}])(x_k - E[x_k])\\
&+2\alpha_{k+1}b_{1k}b_{22k}(u_{2k} - E[u_{2k}])(x_k - E[x_k]) + 2\epsilon_{2k}\ cov(x_k, u_{2k})\\
&-\frac{2}{c_{1k}}\{\alpha_{k+1}[b_{12k}b_{22k} + \sigma_{12k}\sigma_{22k}] + s_k\ \}\{\alpha_{k+1}[\sigma_{1k}\sigma_{12k} + b_{1k}b_{12k}] + \epsilon_{1k}\}\\
&cov(x_k, u_{2k})\\
&= c_{2k}\{u_{2k} - E[u_{2k}] + \eta_{2k}(x_k - E[x_k])\}^2 - c_{2k}\eta_{2k}^2(x_k - E[x_k])^2
\end{aligned} \tag{14.164}$$

where c_{2k} and η_{2k} are given by

$$\begin{aligned}
&c_{2k} = r_{2k} + \alpha_{k+1}(b_{22k}^2 + \sigma_{22k}^2) - \frac{1}{c_{1k}}\{\alpha_{k+1}[b_{12k}b_{22k} + \sigma_{12k}\sigma_{22k}] + s_k\ \}^2,\\
&\eta_{2k} = \frac{1}{c_{2k}}[\alpha_{k+1}\sigma_{1k}\sigma_{22k} + \alpha_{k+1}b_{1k}b_{22k} + \epsilon_{2k}]\\
&-\frac{1}{c_{2k}}\frac{1}{c_{1k}}\{\alpha_{k+1}[b_{12k}b_{22k} + \sigma_{12k}\sigma_{22k}] + s_k\ \}\{\alpha_{k+1}[\sigma_{1k}\sigma_{12k} + b_{1k}b_{12k}] + \epsilon_{1k}\},
\end{aligned} \tag{14.165}$$

Similarly, completion of squares for Eu_2 yields

$$\begin{aligned}
&\{(r_{2k} + \bar{r}_{2k}) + \alpha_{k+1}(\sigma_{22k} + \bar{\sigma}_{22k})^2 + \beta_{k+1}(b_{22k} + \bar{b}_{22k})^2\}(E[u_{2k}])^2 - \frac{1}{\bar{c}_{1k}}\kappa_{11}^2(E[u_{2k}])^2\\
&+2\beta_{k+1}(b_{1k} + \bar{b}_{1k})(b_{22k} + \bar{b}_{22k})E[u_{2k}]E[x_k] + 2\alpha_{k+1}(\sigma_{1k} + \bar{\sigma}_{1k})(\sigma_{22k} + \bar{\sigma}_{22k})E[u_{2k}]E[x_k]\\
&+2(\epsilon_{2k} + \bar{\epsilon}_{2k})E[u_{2k}](E[x_k]) - \frac{1}{\bar{c}_{1k}}2\kappa_{11}\kappa_{12}(E[u_{2k}])(E[x_k]) + 2\alpha_{k+1}\sigma_{0k}(\sigma_{22k} + \bar{\sigma}_{22k})E[u_{2k}]\\
&+2\beta_{k+1}\ b_{0k}(b_{22k} + \bar{b}_{22k})E[u_{2k}] + 2\gamma_{k+1}(b_{22k} + \bar{b}_{22k})E[u_{2k}] + 2\phi_{2k}E[u_{2k}] - \frac{1}{\bar{c}_{1k}}2\kappa_{11}\kappa_{13}(E[u_{2k}])\\
&= \bar{c}_{2k}\{E[u_{2k}] + \bar{\eta}_{2k}E[x_k] + \hat{\eta}_{2k}\}^2 - \frac{1}{\bar{c}_{2k}}\kappa_{22}^2(E[x_k])^2 - \frac{2}{\bar{c}_{2k}}\kappa_{22}\kappa_{23}E[x_k] - \frac{1}{\bar{c}_{2k}}\kappa_{23}^2,
\end{aligned} \tag{14.166}$$

where

$$\begin{aligned}
&c_{2k} < 0,\\
&\bar{c}_{2k} = (r_{2k} + \bar{r}_{2k}) + \alpha_{k+1}(\sigma_{22k} + \bar{\sigma}_{22k})^2 + \beta_{k+1}(b_{22k} + \bar{b}_{22k})^2 - \frac{1}{\bar{c}_{1k}}\kappa_{11}^2\\
&= (r_{2k} + \bar{r}_{2k}) + \alpha_{k+1}(\sigma_{22k} + \bar{\sigma}_{22k})^2 + \beta_{k+1}(b_{22k} + \bar{b}_{22k})^2\\
&-\frac{1}{\bar{c}_{1k}}\{\alpha_{k+1}(\sigma_{12k} + \bar{\sigma}_{12k})(\sigma_{22k} + \bar{\sigma}_{22k}) + \beta_{k+1}(b_{12k} + \bar{b}_{12k})(b_{22k} + \bar{b}_{22k}) + (s_k + \bar{s}_k)\}^2,\\
&\tau_{2k} = \Big\{\beta_{k+1}(b_{1k} + \bar{b}_{1k})(b_{22k} + \bar{b}_{22k})\\
&+\alpha_{k+1}(\sigma_{1k} + \bar{\sigma}_{1k})(\sigma_{22k} + \bar{\sigma}_{22k}) + (\epsilon_{2k} + \bar{\epsilon}_{2k}) - \frac{1}{\bar{c}_{1k}}\kappa_{11}\kappa_{12}\Big\}(E[x_k])\\
&+\alpha_{k+1}\sigma_{0k}(\sigma_{22k} + \bar{\sigma}_{22k})\\
&+\beta_{k+1}\ b_{0k}(b_{22k} + \bar{b}_{22k}) + \gamma_{k+1}(b_{22k} + \bar{b}_{22k}) + \phi_{2k} - \frac{1}{\bar{c}_{1k}}\kappa_{11}\kappa_{13}\\
&=: \kappa_{22}E[x_k] + \kappa_{23},\\
&\kappa_{22} := \beta_{k+1}(b_{1k} + \bar{b}_{1k})(b_{22k} + \bar{b}_{22k}) + \alpha_{k+1}(\sigma_{1k} + \bar{\sigma}_{1k})(\sigma_{22k} + \bar{\sigma}_{22k}) + (\epsilon_{2k} + \bar{\epsilon}_{2k}) - \frac{1}{\bar{c}_{1k}}\kappa_{11}\kappa_{12},\\
&\kappa_{23} := \alpha_{k+1}\sigma_{0k}(\sigma_{22k} + \bar{\sigma}_{22k}) + \beta_{k+1}\ b_{0k}(b_{22k} + \bar{b}_{22k}) + \gamma_{k+1}(b_{22k} + \bar{b}_{22k}) + \phi_{2k} - \frac{1}{\bar{c}_{1k}}\kappa_{11}\kappa_{13},\\
&\bar{\eta}_{2k} = \frac{\kappa_{22}}{\bar{c}_{2k}},\\
&\hat{\eta}_{2k} = \frac{\kappa_{23}}{\bar{c}_{2k}},
\end{aligned} \tag{14.167}$$

Thus,

$$
\begin{aligned}
& E\{\hat{L} - \alpha_0(x_0 - Ex_0)^2 - \beta_0(Ex_0)^2 - 2\gamma_0(Ex_0) - \delta_0\} \\
& = [q_K - \alpha_K]var(x_K) + [q_K + \bar{q}_K - \beta_K](E[x_K])^2 + 2(\phi_{3K} - \gamma_K)E[x_K] + (0 - \delta_K) \\
& + \sum_{k=0}^{K-1} \big[c_{1k}\{u_{1k} - E[u_{1k}] + \eta_{1k}(x_k - E[x_k])\}^2 \\
& + \bar{c}_{1k}\{Eu_{1k} + \eta_{1k}E[x_k] + \hat{\eta}_{1k}\}^2 \\
& + c_{2k}\{u_{2k} - E[u_{2k}] + \eta_{2k}(x_k - E[x_k])\}^2 - c_{2k}\eta_{2k}^2(x_k - E[x_k])^2 \\
& + \bar{c}_{2k}\{E[u_{2k}] + \bar{\eta}_{2k}E[x_k] + \hat{\eta}_{2k}\}^2 - \tfrac{1}{\bar{c}_{2k}}\kappa_{22}^2(E[x_k])^2 - \tfrac{2}{\bar{c}_{2k}}\kappa_{22}\kappa_{23}E[x_k] - \tfrac{1}{\bar{c}_{2k}}\kappa_{23}^2 \\
& + \{\alpha_{k+1}(b_{1k}^2 + \sigma_{1k}^2) - \alpha_k + q_k\}(x_k - E[x_k])^2 \\
& - \tfrac{1}{c_{1k}}\{\alpha_{k+1}[\sigma_{1k}\sigma_{12k} + b_{1k}b_{12k}] + \epsilon_{1k}\}^2(x_k - E[x_k])^2 \\
& + \{\beta_{k+1}(b_{1k} + \bar{b}_{1k})^2 - \beta_k + (q_k + \bar{q}_k) + \alpha_{k+1}(\sigma_{1k} + \bar{\sigma}_{1k})^2\}(E[x_k])^2 \\
& - \tfrac{1}{\bar{c}_{1k}}\kappa_{12}^2(E[x_k])^2 \\
& + \{2\gamma_{k+1}(b_{1k} + \bar{b}_{1k}) - 2\gamma_k + 2\phi_{3k} + 2\alpha_{k+1}\sigma_{0k}(\sigma_{1k} + \bar{\sigma}_{1k}) + 2\beta_{k+1}b_{0k}(b_{1k} + \bar{b}_{1k})\}E[x_k] \\
& - \tfrac{1}{\bar{c}_{1k}}2\kappa_{12}\kappa_{13}(E[x_k]) + \delta_{k+1} - \delta_k + \alpha_{k+1}\sigma_{0k}^2 + \beta_{k+1}b_{0k}^2 + 2\gamma_{k+1}b_{0k} - \tfrac{1}{\bar{c}_{1k}}\kappa_{13}^2\big]
\end{aligned}
\tag{14.168}
$$

Based on the above calculations, the difference between the guess and cost functionals yields

$$
\begin{aligned}
& E\{\hat{L} - \alpha_0(x_0 - Ex_0)^2 - \beta_0(Ex_0)^2 - 2\gamma_0(Ex_0) - \delta_0\} \\
& = [q_K - \alpha_K]var(x_K) + [q_K + \bar{q}_K - \beta_K](E[x_K])^2 + 2(\phi_{3K} - \gamma_K)E[x_K] + (0 - \delta_K) \\
& + \sum_{k=0}^{K-1} \big[c_{1k}\{u_{1k} - E[u_{1k}] + \eta_{1k}(x_k - E[x_k])\}^2 + \bar{c}_{1k}\{Eu_{1k} + \eta_{1k}E[x_k] + \hat{\eta}_{1k}\}^2 \\
& + c_{2k}\{u_{2k} - E[u_{2k}] + \eta_{2k}(x_k - E[x_k])\}^2 + \bar{c}_{2k}\{E[u_{2k}] + \bar{\eta}_{2k}E[x_k] + \hat{\eta}_{2k}\}^2 \\
& + \Big\{\alpha_{k+1}(b_{1k}^2 + \sigma_{1k}^2) - \alpha_k + q_k - c_{2k}\eta_{2k}^2 \\
& - \tfrac{1}{c_{1k}}\{\alpha_{k+1}[\sigma_{1k}\sigma_{12k} + b_{1k}b_{12k}] + \epsilon_{1k}\}^2\Big\}(x_k - E[x_k])^2 \\
& + \Big\{\beta_{k+1}(b_{1k} + \bar{b}_{1k})^2 - \beta_k + (q_k + \bar{q}_k) + \alpha_{k+1}(\sigma_{1k} + \bar{\sigma}_{1k})^2 \\
& - \tfrac{1}{\bar{c}_{1k}}\kappa_{12}^2 - \tfrac{1}{\bar{c}_{2k}}\kappa_{22}^2\Big\}(E[x_k])^2 \\
& + 2\Big\{\gamma_{k+1}(b_{1k} + \bar{b}_{1k}) - \gamma_k + \phi_{3k} + \alpha_{k+1}\sigma_{0k}(\sigma_{1k} + \bar{\sigma}_{1k}) \\
& + \beta_{k+1}b_{0k}(b_{1k} + \bar{b}_{1k}) - \tfrac{1}{\bar{c}_{1k}}\kappa_{12}\kappa_{13} - \tfrac{1}{\bar{c}_{2k}}\kappa_{22}\kappa_{23}\Big\}E[x_k] \\
& + \delta_{k+1} - \delta_k + \alpha_{k+1}\sigma_{0k}^2 + \beta_{k+1}b_{0k}^2 + 2\gamma_{k+1}b_{0k} - \tfrac{1}{\bar{c}_{1k}}\kappa_{13}^2 - \tfrac{1}{\bar{c}_{2k}}\kappa_{23}^2\big]
\end{aligned}
\tag{14.169}
$$

By process of matching terms, one obtains the announced result provided that $c_{2k}, \bar{c}_{2k}$ is negative and $c_{1k}, \bar{c}_{1k}$ is positive.

In this section, we have shown that a saddle point can be determined in a semi-explicit way for the discrete time linear-quadratic adversarial game problem where both additive and multiplicative noises are involved. We have used a direct method to solve the problem. The method is basic and applies the telescopic formula.

14.6 Discrete Time Mean-Field-Type Games

We now consider a class of discrete-time MFTGs over the horizon $\mathcal{T} := \{0, 1, \ldots, T-1\}$, with $T \geq 1$. The decision-makers are $\mathcal{I} = \{1, 2, \ldots, I\}$ with $1 < I < +\infty$. We are given $I+1$ measurable spaces $(\mathcal{X}, \mathcal{B}(\mathcal{X}))$ and $(A_i, \mathcal{B}(A_i))_{i\in\mathcal{I}}$. We denote by $\mathcal{P}(A_i))$ the set of probability measures on A_i.

The state x_t of the interacting system has a probability measure denoted by m_t.

14.6.1 Feedback Strategies

We restrict our attention to state-mean-field-type feedback strategies of the decision-makers. A state-mean-field-type feedback strategy of the decision-makers is a measurable function of (x, m). The strategy profile a_t of all decision-makers has a probability measure denoted by $\tilde{m}_t$.

The Kolmogorov equation

$$\begin{array}{l} S \subseteq \mathcal{S}, \\ m_{t+1}(S) = \int_x P_t(x, a_t, m_t, \tilde{m}_t, S) m_t(dx), \ t \in \{0, 1, \ldots, T-1\} \\ m_0(dx), \end{array} \tag{14.170}$$

where S is a subset of the state space.

The Bellman system of the discrete-time MFTG is given by

$$\begin{array}{l} i \in \mathcal{I}, \\ V_{i,T}(m_T) = \int_x g_i(x, m_T) m_T(dx), \\ t \in \{1, 2, \ldots, T-1\} : \\ V_{i,t}(m_t) = \sup_{a_{i,t}\in A_i} \{\int_x r_{i,t}(x, a_t, m_t, \tilde{m}_t) m_t(dx) + V_{i,t+1}(m_{t+1})\}, \\ m_{t+1}(S) = \int_x P_t(x, a_t, m_t, \tilde{m}_t, S) m_t(dx), \\ m_0(dx). \end{array} \tag{14.171}$$

The mean-field-type equilibrium system yields

$$\begin{array}{l} i \in \mathcal{I}, \\ V_{i,T}(m_T) = \int_x g_i(x, m_T) m_T(dx), \\ t \in \{0, 1, \ldots, T-1\} : \\ V_{i,t}(m_t) = \sup_{a_i\in\mathcal{A}_i} \{\int_x g_i(x, m_T) m_T(dx) + \sum_{k=t}^{T-1} \int_x r_{i,k}(x, a_k, m_k, \tilde{m}_k) m_k(dx)\}, \\ m_{k+1}(S) = \int_x P_k(x, a_k, m_k, \tilde{m}_k, S) m_k(dx), \\ m_t(dx). \end{array} \tag{14.172}$$

The equilibrium payoff vector is $(V^*_{i,0}(m_0))_{i\in\mathscr{I}}$, and the equilibrium strategies are the ones achieving (14.171).

14.6.2 Adversarial Mean-Field-Type Games

We now consider a class of adversarial discrete-time MFTGs over the horizon $\mathscr{T} = \{0, 1, 2, \ldots, T-1\}$. The decision-makers are $\mathscr{I} = \{1, 2\}$.

The Bellman-Shapley maxmin system of discrete-time MFTG is given by

$$\begin{array}{l} \underline{V}_T(m_T) = \int_x g(x, m_T)m_T(dx), \\ t \in \{0, 1, 2, \ldots, T-1\}: \\ \underline{V}_t(m_t) = \sup_{a_{1,t}\in A_1}\inf_{a_{2,t}\in A_2}\{\int_x r_t(x, a_{1,t}, a_{2,t}, m_t, \tilde{m}_t)m_t(dx) + \underline{V}_{t+1}(m_{t+1})\}, \\ m_{t+1}(S) = \int_x P_t(x, a_t, m_t, \tilde{m}_t, S)m_t(dx), \\ m_0(dx). \end{array} \tag{14.173}$$

The Bellman-Shapley minmax system of discrete-time MFTG is given by

$$\begin{array}{l} \bar{V}_T(m_T) = \int_x g(x, m_T)m_T(dx), \\ t \in \{0, 1, 2, \ldots, T-1\}: \\ \bar{V}_t(m_t) = \inf_{a_{2,t}\in A_2}\sup_{a_{1,t}\in A_1}\{\int_x r_t(x, a_t, m_t, \tilde{m}_t)m_t(dx) + \bar{V}_{t+1}(m_{t+1})\}, \\ m_{t+1}(S) = \int_x P_t(x, a_t, m_t, \tilde{m}_t, S)m_t(dx), \\ m_0(dx). \end{array} \tag{14.174}$$

14.6.3 Equilibrium Q-Values

Consider $(m_0, m_1, \ldots, m_{T-1}, m_T)$. The value functions are

$$(V_{i,0}(m_0), V_{i,1}(m_1), \ldots, V_{i,T-1}(m_{T-1}), V_{i,T}(m_T)).$$

Let $Q_{i,t},\ i \in \mathscr{I},\ 0 \le t \le T-1$ be

$$\begin{array}{l} Q_{i,t}(m_t, a_t) \\ := \int r_{i,t}(x, a_t, m_t, \tilde{m}_t)m_t(dx) + V_{i,t+1}\left(\int_x P_t(x, a_t, m_t, \tilde{m}_t, .)m_t(dx)\right) \end{array} \tag{14.175}$$

Then,

$$\sup_{a_{i,t}\in A_i} Q_{i,t}(m_t, a_t) = V_{i,t}(m_t),\ i \in \mathscr{I},\ 0 \le t \le T-1. \tag{14.176}$$

Hence

$$Q_{i,t}(m_t, a_t) := \int r_{i,t}(x, a_t, m_t, \tilde{m}_t)m_t(dx) + \sup_{a_{i,t+1}\in A_i} Q_{i,t+1}(m_{t+1}, a_{t+1}) \tag{14.177}$$

Thus, we introduce the error of decision-maker i at time t as

$$\epsilon_{i,t} = \left\{ \int r_{i,t}(x, a_t, m_t, \tilde{m}_t)m_t(dx) + \sup_{a_{i,t+1}\in A_i} Q_{i,t+1}(m_{t+1}, a_{t+1}) - Q_{i,t}(m_t, a_t) \right\}.$$

Thus, error of decision-maker i over the entire horizon is $\epsilon_i = (\epsilon_{i,t})_{t\in\mathcal{T}}$.

As open-loop strategies are different from feedback strategies, we need to prepare the input data of open-loop strategies.

14.6.4 Open-Loop Strategies

Given $(m_0, a_0, a_1, \ldots, a_{T-1})$,, one can compute $(m_1, \ldots, m_T)$ using the Kolmogorov equation in (14.170). Note that m_0 is not a decision variable.

$$\begin{array}{l} i \in \mathcal{I}, \\ R_i(m_0, a, \lambda, \mu) = \int g_i(x, m_T)m_T(dx) + \sum_{k=0}^{T-1} \int_x r_{i,k}(x, a_k, m_k, \tilde{m}_k)m_k(dx) \\ -\sum_{k=0}^{T-1} \int \lambda_{i,k+1}(dy)\{m_{k+1}(y) - \int_x P_k(x, a_k, m_k, \tilde{m}_k, y)m_k(dx)\} \\ -\sum_{k=0}^{T} \mu_{1,k}(\int m_k(dz) - 1). \end{array} \tag{14.178}$$

The primal game

$$\sup_{a_i} \inf_{(\lambda_i,\mu_i)} R_i(m_0, a, \lambda, \mu),\ i \in \mathcal{I}$$

and the dual game

$$\inf_{(\lambda_i,\mu_i)} \sup_{a_i} R_i(m_0, a, \lambda, \mu),\ i \in \mathcal{I}.$$

$$\begin{array}{l} i \in \mathcal{I}, \\ \lambda_{i,T}(x) = g_i(x, m_T) + \int g_{i,m_T}(x, m_T)(z)m_T(dz), \\ k \in \{0, \ldots, T-1\} : \\ \lambda_{i,k}(x) = r_{i,k}(x, a_k, m_k, \tilde{m}_k) + \int_x r_{i,k,m_k}(x, a_k, m_k, \tilde{m}_k)(y)m_k(dy) \\ +\int \lambda_{i,k+1}(dy)P_k(x, a_k, m_k, \tilde{m}_k, y) \\ +\int\int \lambda_{i,k+1}(dy)P_{k,m_k}(x, a_k, m_k, \tilde{m}_k, y)(z)m_k(dz), \\ m_{k+1}(S) = \int_x P_k(x, a_k, m_k, \tilde{m}_k, S)m_k(dx), \\ \int m_k(dz) = 1, \\ a_{i,k}^* \in \arg\max_{a_{i,k}}\{r_{i,k}(x, a_k, m_k, \tilde{m}_k) + \int \lambda_{i,k+1}(y)P_k(x, a_k, m_k, \tilde{m}_k, dy)\}, \end{array} \tag{14.179}$$

By introducing $H_{i,k} = r_{i,k}(x, a_k, m_k, \tilde{m}_k) + \int \lambda_{i,k+1}(y) P_k(x, a_k, m_k, \tilde{m}_k, dy)$, the system (14.179) can be rewritten as

$$\begin{array}{l} i \in \mathscr{I}, \\ \lambda_{1,T}(x) = g_i(x, m_T) + \int g_{i,m_T}(x, m_T)(z) m_T(dz), \\ k \in \{0, \ldots, T-1\} : \\ \lambda_{i,k} = H_{i,k} + \int H_{i,k,m_k}(z) m_k(dz), \\ m_{k+1}(S) = \int P_k(x, a_k, m_k, \tilde{m}_k, S) m_k(dx), \\ \int m_k(dz) = 1, \\ a^*_{i,k} \in \arg\max_{a_{i,k}} H_{i,k}. \end{array} \tag{14.180}$$

14.6.5 Adversarial Mean-Field-Type Games

We consider a class of adversarial discrete-time MFTGs over the horizon $\mathscr{T} = \{0, 1, 2, \ldots, T-1\}$. The decision-makers are $\mathscr{I} = \{1, 2\}$.

The Bellman-Shapley maxmin system of discrete-time MFTG is given by

$$\begin{array}{l} \underline{V}_T(m_T) = \int_x g(x, m_T) m_T(dx), \\ t \in \{0, 1, 2, \ldots, T-1\} : \\ \underline{V}_t(m_t) = \sup_{a_{1,t} \in A_1} \inf_{a_{2,t} \in A_2} \{\int_x r_t(x, a_{1,t}, a_{2,t}, m_t, \tilde{m}_t) m_t(dx) + \underline{V}_{t+1}(m_{t+1})\}, \\ m_{t+1}(S) = \int_x P_t(x, a_t, m_t, \tilde{m}_t, S) m_t(dx), \\ m_0(dx). \end{array} \tag{14.181}$$

The Bellman-Shapley minmax system of discrete-time MFTG is given by

$$\begin{array}{l} \bar{V}_T(m_T) = \int_x g(x, m_T) m_T(dx), \\ t \in \{0, 1, 2, \ldots, T-1\} : \\ \bar{V}_t(m_t) = \inf_{a_{2,t} \in A_2} \sup_{a_{1,t} \in A_1} \{\int_x r_t(x, a_t, m_t, \tilde{m}_t) m_t(dx) + \bar{V}_{t+1}(m_{t+1})\}, \\ m_{t+1}(S) = \int_x P_t(x, a_t, m_t, \tilde{m}_t, S) m_t(dx), \\ m_0(dx). \end{array} \tag{14.182}$$

14.6.6 MinMaxQ-Values

Consider $(m_0, m_1, \ldots, m_{T-1}, m_T)$. The value functions are

$$(V_{1,0}(m_0), V_{1,1}(m_1), \ldots, V_{1,T-1}(m_{T-1}), V_{1,T}(m_T)).$$

Let $Q_t,\ 0 \leq t \leq T-1$ be

$$Q_t(m_t, a_t) \\ := \int r_t(x, a_t, m_t, \tilde{m}_t)m_t(dx) + V_{t+1}\left(\int_x P_t(x, a_t, m_t, \tilde{m}_t, .)m_t(dx)\right) \tag{14.183}$$

Then,

$$\sup_{a_{1,t}\in A_1} \inf_{a_{2,t}\in A_2} Q_t(m_t, a_t) = V_t(m_t), \quad 0 \leq t \leq T-1. \tag{14.184}$$

Hence

$$Q_t(m_t, a_t) := \int r_t(x, a_t, m_t, \tilde{m}_t)m_t(dx) + \sup_{a_1\in A_1} \inf_{a_2\in A_2} Q_{t+1}(m_{t+1}, a_{t+1}) \tag{14.185}$$

14.7 Notes

A class of discrete-time mean-field games was introduced in [209, 210] under the name of anonymous sequential games. Since then, both finite and infinite time-horizon have been analyzed in the literature [25, 284–286, 362]. See [44, 51, 51, 112, 121, 132, 132, 133, 294–296, 323, 333] for more details on discrete-time MFTGs.

The transition from discrete to continuous time as the scale moves to zero is important in variance-aware problems [40, 41, 43, 44, 166, 296, 350]. Relationship between risk-sensitive and robust mean field games can be found in [251].

In order to apply DPP and SMP, the problems are transformed into deterministic ones by writing the the payoffs and the state transitions in infinite dimensions. Learning and approximate mean-field Q-values can be found in [4, 23, 24, 239, 351, 352].

Part IV
Wrap-up

Chapter 15
Conclusions and New Directions

15.1 Conclusions

In "Mean-Field-Type Game Theory: Foundations and New Directions," we have presented key foundational elements. This concluding chapter of Volume I offers a reflective overview of our exploration and highlights the key takeaways from our journey through this emerging field. Throughout the book, we have presented various strategic interactions, decision-making, and dynamic systems. From one-shot MFTGs to complex hierarchical structures, from linear-quadratic-induced interactions to Rosenblatt noise-driven dynamics, we have traversed a wide spectrum of concepts and applications, all united under the banner of the MFTG theory. At the heart of MFTGs lies the idea that individual actions and probability distributions behind the collective strategies of decision-makers, when aggregated over the distributions, can give rise to strategic behaviors and emergent phenomena.

Our exploration has revealed several fundamental principles and tools that underpin the MFTG theory. The direct method, verification and validation method, Wiener chaos expansion, stochastic maximum principle, and dynamic programming principle have emerged as useful tools for analyzing and optimizing strategies in dynamic risk-aware environments.

We have also examined risk awareness in MFTGs, emphasizing the importance of considering risk and uncertainty when making strategic decisions. Linear, quadratic, and non-quadratic models have been explored, each offering unique insights into the dynamics of MFTGs.

Additionally, we have covered asymmetric information across decision-makers and adversarial interactions, further expanding the scope and applicability of MFTGs.

As we conclude our journey in this volume, it is clear that MFTGs are more than just theoretical constructs-they are powerful tools for understanding, modeling, and optimizing real-world interactive systems. More applications are provided in Volume II of the book: Mean-Field-Type Game Theory: Applications.

T. Başar et al., *Mean-Field-Type Game Theory I*, Static & Dynamic Game Theory: Foundations & Applications, https://doi.org/10.1007/978-3-032-07027-2_15

15.2 New Directions

Emerging directions in the MFTG theory continue to evolve, extending its reach into new and exciting directions. Some of the emerging areas and trends within the MFTG theory are

- Distributionally Robust MFTGs: Exploring distributional robust decision-making strategies in MFTGs under uncertainty, where the uncertainty is characterized by probability distributions rather than specific scenarios.
- MFTG Theory Driven by Mixed Noises: Investigating the impact of mixed noise sources (e.g., from modeling error, computational error, measurement error, perception error, nature) and combining different types of stochastic processes such as Brownian motion, jump process, regime switching, fractional Brownian motion, Gauss-Volterra processes, and Rosenblatt processes of different types on the dynamics and equilibria.
- Data-Driven MFTG Theory: Leveraging the power of data analytics and machine learning to inform strategies and decision-making in MFTGs, using real-world data to model and optimize interactions.
- MFTG Theory with Random Number of Interacting Decision-Makers: Exploring MFTGs where the number of decision-makers is itself a random variable, addressing scenarios with varying group sizes.
- MFTG Theory with Random Horizon Duration and Sleeping Mode: Examining MFTGs with variable time horizons and periods of inactivity (sleeping mode), which are prevalent in various dynamic systems.
- MFTG Theory Driven by Non-Gaussian, Non-Poissonian, Non-Markovian Processes: Extending the foundations of MFTGs to encompass more realistic stochastic processes beyond the classical Gaussian, Poisson, and Markovian models.
- MFTG Theory with Changing Empathy Structures: Investigating how changing empathy structures among decision-makers influence strategic interactions and cooperation within MFTGs.
- MFTG Theory for Risk-Aware Audio-Enabled Machine Intelligence: Integrating the MFTG theory into the co-construction of audio-enabled intelligent systems in 7163+ local languages and dialects, where autonomous decision-makers engage in strategic decision-making, leading to more robust, risk-aware, explainable, and adaptable machine intelligence.
- MFTG Theory for Strategic Machine Intelligence: Strategic machine intelligence is a paradigm of machine intelligence where multiple autonomous agents, each possessing its own machine intelligence, operate within a dynamic environment. These agents sense their surroundings, learn from data and experiences, and adapt their strategies accordingly, all while considering the influences of other agents' machine intelligence. This approach enables intelligent entities to make informed decisions, collaborate, compete, negotiate, or coordinate within complex multi-agent systems, with applications spanning autonomous vehicles, robotics, finance, supply chain management, and beyond, addressing the intricate

challenges of dynamic decision-making in interactive environments. In this context, one of the future directions is to integrate MFTG into the development of multiple agents, each having its own risk-aware machine intelligence.

- Strategic Machine Intelligence for MFTG Theory: Developing advanced strategic machine intelligence techniques to analyze and optimize MFTGs, enhancing our understanding of complex systems.
- MFTG Theory-Based Risk Awareness: Expanding the application of risk-aware MFTGs to various domains, including finance, supply chain management, and environmental planning.
- Variance-Aware MFTG Theory: Focusing on strategies that explicitly consider variance and its impact on decision-making within MFTGs, particularly in risk management.
- Quantile-Aware MFTG Theory: Exploring MFTGs that prioritize specific quantiles of the collective performance, catering to scenarios where extreme outcomes are of particular concern.
- Co-opetitive MFTG Theory: Investigating cooperative and competitive interactions within MFTGs, where decision-makers simultaneously collaborate and compete to achieve their objectives.
- Cooperative MFTG Theory: Advancing the theory and applications of cooperative MFTGs, with a focus on collective decision-making among decision-makers for mutual benefits.
- Coalitional MFTG Theory: Examining how coalitions and alliances among decision-makers in MFTGs influence the dynamics and equilibrium of the system, with applications in multi-agent systems and network science.

Appendix A
State Dynamics of Mean-Field Type

In this appendix, we collect together the results that have been used and referenced in different chapters of the book.

A.1 Inequalities

Lemma A.1 (Burkholder–Davis–Gundy Inequality) *For all $p > 0$, there exist positive constants c_p and C_p such that for all continuous local martingales $M = (M_t)_t$ and all $T > 0$, we have*

$$c_p E\left[\langle M\rangle_T^{p/2}\right] \leq E\left[\sup_{0\leq t\leq T} |M_t|^p\right] \leq C_p E\left[\langle M\rangle_T^{p/2}\right]. \tag{A.1}$$

See [270], Theorem 48, p. 195.

A.2 Martingale Representation Theorem

Lemma A.2 *On the filtered probability space $\left(\Omega, \mathscr{F}, (\mathscr{F}_t)_{0\leq t\leq T}, \mathbb{P}\right)$, let N be a continuous local martingale and M be any other local martingale. Then, there exists a progressively measurable process ξ and local martingale L such that*

$$M_t = M_0 + \int_0^t \xi(s)\, dN_s + L_t, \quad \langle L, N\rangle_t = 0, \quad 0 \leq t \leq T,$$

and

T. Başar et al., *Mean-Field-Type Game Theory I*, Static & Dynamic Game Theory: Foundations & Applications, https://doi.org/10.1007/978-3-032-07027-2

$$E\left[\int_0^T |\xi(s)|^2 d\langle N\rangle_s\right] < +\infty.$$

See [217], Proposition 4.1, for the Brownian motion case and [270], Corollary 2, p. 183, for the general case.

A.3 Backward Stochastic Differential Equations

A.3.1 Backward SDEs of Mean-Field Type

Let $W = (W_t)_{t\in[0,T]}$ be a standard d-dimensional Brownian motion on a filtered probability space $(\Omega, \mathscr{F}, P)$ where the filtration $\mathbb{F} = (\mathscr{F}_t)_{t\geq 0}$ is the natural filtration of W generated by the Brownian motion W and $T > 0$ is a fixed finite horizon.
We denote by $S^2_{\mathbb{F}}([0,T])$ the set of $\mathbb{R}^d$-valued progressively measurable processes Y such that

$$E\Big[\sup_{t\in[0,T]} |Y_t|^2\Big] < +\infty,$$

and by $L^2_{\mathbb{F}}([0,T];\mathbb{R}^{d\times d})$ the set of $\mathbb{R}^{d\times d}$-valued progressively measurable processes Z such that

$$E\Big[\int_0^T |Z_t|^2 dt\Big] < +\infty.$$

Consider a pair (ξ, g), called the terminal value and driver (or generator), satisfying

- (A1) $\xi \in L^2(\Omega, \mathscr{F}_T, \mathbb{P}; \mathbb{R}^d)$
- (A2) $g : \Omega \times [0,T] \times \mathbb{R}^d \times \mathbb{R}^d \times \mathbb{R}^{d\times d} \times \mathbb{R}^{d\times d} \longrightarrow \mathbb{R}^d$ such that
  - $g(t, y_1, y_2, z_1, z_2)$ is $\mathbb{F}$-progressively measurable for all (y_1, y_2, z_1, z_2).
  - $g(\cdot, 0, 0, 0, 0) \in L^2_{\mathbb{F}}([0,T];\mathbb{R}^d)$.
  - g is uniformly Lipschitz in (y_1, y_2, z_1, z_2), i.e. there is some constant $C_g \geq 0$ such that

$$|g(t, y_1, y_2, z_1, z_2) - g(t, \bar{y}_1, \bar{y}_2, \bar{z}_1, \bar{z}_2)| \leq C\sum_{i=1}^{2}(|y_i - \bar{y}_i| + |z_i - \bar{z}_i|),$$

for all $y_1, \bar{y}_1, y_2, \bar{y}_2,\ z_1, \bar{z}_1, z_2, \bar{z}_2,\ dt \otimes d\mathbb{P}$-a.e.

We consider the backward stochastic differential equation (BSDE) of mean-field type:

$$- dY_t = g(t, Y_t, E[Y_t], Z_t, E[Z_t])dt - ZdW_t, \quad Y_T = \xi. \tag{A.2}$$

Lemma A.3 *Given the pair* (ξ, g) *satisfying (A1) and (A2), there exists a unique solution* $(Y, Z) \in S^2_{\mathbb{F}}([0, T]) \times L^2_{\mathbb{F}}([0, T]; \mathbb{R}^{d\times d})$ *to the BSDE (A.2), i.e., the solution satisfies*

$$E\left[\sup_{t\in[0,T]} |Y_t|^2 + \int_0^T |Z_t|^2 dt\right] < +\infty.$$

See [77]. ■

A.3.2 Linear BSDEs

We consider a particular case of a standard BSDE where the driver g is linear in y and z. The linear BSDE takes the form

$$- dY_t = (A_t Y_t + B_t Z_t + C_t) - Z_t dW_t, \quad Y_T = \xi, \tag{A.3}$$

where A, B are bounded progressively measurable processes and C is a process in $L^2_{\mathbb{F}}([0, T]; \mathbb{R}^d)$.

Lemma A.4 *The unique solution* (Y, Z) *to the linear BSDE (A.3) is given by*

$$Y_t = E\left[\left[\int_t^T \Phi^T(s, t)C_s ds + \Phi^T(T, t)\xi \middle| \mathscr{F}_t\right]\right], \tag{A.4}$$

where $\Phi(s, t)$, $(s \geq t)$ *is the solution of the linear equation*

$$\begin{cases} d\Phi(s, t) = A_s\Phi(s, t)ds + B_s\Phi(s, t)dW_s, \\ \Phi(t, t) = I_d, \end{cases} \tag{A.5}$$

where Φ^* *is the transpose of the matrix* Φ.

Proof Apply Itô's formula to $\Phi(s, t)Y_s$. ■

A.4 Convergence of Probability Measures

Lemma A.5 (Skorokhod Selection Theorem) *Let (S, ρ) be a complete separable metric space and $\mathbb{P}$ and $\mathbb{P}_I$, $I \in \{1, 2, \ldots\}$ be probability measures on $(S, \mathscr{B}(S))$ such that $\mathbb{P}_I$ converges weakly to $\mathbb{P}$. Then, on a probability space $(\widetilde{\Omega}, \widetilde{\mathscr{F}}, \widetilde{\mathbb{P}})$, we can construct S-valued random variables X^I, $I \in \{1, 2, \ldots\}$ and X such that*

$$\begin{cases} \mathbb{P} = \widetilde{\mathbb{P}}_X, \\ \mathbb{P}_I = \widetilde{\mathbb{P}}_{X^I}, \quad I \in \{1, 2, \ldots\}, \\ X^I \longrightarrow X \ \ \widetilde{\mathbb{P}} \text{ a.s. } as\ I \to \infty. \end{cases} \tag{A.6}$$

See [256], page 9.

Lemma A.6 (Kolmogorov's Continuity Theorem) *Let $(X_t^I)_{I \geq 0}$ be a sequence of d-dimensional continuous processes satisfying the following two conditions:*

- *There exist positive constants m and γ such that $E\left[\left|X_0^I\right|^\gamma\right] \leq m, \ \ I \geq 0$.*
- *There exist positive constants α, β and m_k, $k = 1, 2, \ldots$, such that*

$$E\left[\left|X_t^I - X_s^I\right|^\alpha\right] \leq m_k \left|t - s\right|^{1+\beta}, \quad I \geq 0, \quad t, s \in [0, k], \ k = 1, 2, \ldots.$$

Then, there exist a subsequence $(n_k)_k$, a probability space $(\widetilde{\Omega}, \widetilde{\mathscr{F}}, \widetilde{\mathbb{P}})$ and d-dimensional continuous processes $\widetilde{X}^{n_k}, k \in \{1, 2, \ldots\}$, and $\widetilde{X}$ defined on it such that

- *The laws of $\widetilde{X}^{n_k}$ and X^{n_k} coincide.*
- *$\widetilde{X}_t^{n_k}$ converges to $\widetilde{X}_t$ uniformly on every finite time interval, $\widetilde{\mathbb{P}}$a.s.*

See [256], page 18. ■

Lemma A.7 *Let $C_{E'} = C\left([0, 1], E'\right)$ be the space of continuous mappings from $[0, 1]$ to E', the topological dual of some Frêchet space E. Let $(\mathbb{P}_I)_I$ to be a sequence of probability measures on $C_{E'}$. Suppose that for each ξ in E, the sequence $\left(\mathbb{P}_I \Pi_\xi^{-1}\right)$ is tight in C, where $\Pi_\xi : x \in C_{E'} \longrightarrow \langle x_., \xi\rangle \in C = C\left([0, 1], \mathbb{R}^d\right)$. Then, the sequence $(\mathbb{P}_I)$ itself is tight.*

See [246]. ■

Appendix B
Mean-Field Convergence

The extension of (i) the law of large numbers, (ii) the central limit theorem, and (iii) the large deviation principle, from independent random variables to sequences of indistinguishable random variables, have been widely studied since the appearance in Blum, Chernoff, and co-authors [71]. Below, we state some well-known results and explain how they can be used in the McKean-Vlasov context with the L^{α}-norm.

B.1 Indistinguishability

The notion of indistinguishability (or exchangeability or interchangeability) is introduced in order to discuss the existence of a limiting measure and mean-field convergence of the empirical measure of virtual particle states in the framework of de Finetti–Hewitt–Savage [5–7, 109–111, 190]. The indistinguishability property is called exchangeability in the probability literature.

Let $\mathscr{X}$ be a separable, complete, and metrizable topological space (Polish space).

Definition B.1 (Indistinguishability) A collection $(x_1, x_2, \ldots, x_I)$ of $\mathscr{X}$-valued random variables/processes is indistinguishable (or exchangeable) if the joint law is invariant by permutation over the index set $\{1, \ldots, I\}$, i.e., for any permutation σ over the set $\{1, 2, \ldots, I\}$, one has

$$\mathscr{L}(x_1, x_2, \ldots, x_I) = \mathscr{L}(x_{\sigma(1)}, \ldots, x_{\sigma(I)}), \tag{B.1}$$

where $\mathscr{L}(X)$ denotes the law of the random variable X. An infinite family of random variables/processes $(x_1, x_2, \ldots)$ is indistinguishable if for every finite I the family $(x_1, x_2, \ldots, x_I)$ is indistinguishable.

This says that the order (position) of the random variable in the family does not change the joint distribution. From (B.1), we also have that, for any measurable

T. Başar et al., *Mean-Field-Type Game Theory I*, Static & Dynamic Game Theory: Foundations & Applications, https://doi.org/10.1007/978-3-032-07027-2

operator O

$$\mathcal{L}\left(x_1, x_2, \ldots, x_I,\ O(\frac{1}{I}\sum_{i=1}^{I}\delta_{x_i})\right)$$

$$= \mathcal{L}\left(x_{\sigma(1)}, \ldots, x_{\sigma(I)},\ O(\frac{1}{I}\sum_{i=1}^{I}\delta_{x_{\sigma(i)}})\right), \tag{B.2}$$

where we do not permute the last component.

For indistinguishable random variables/processes, the convergence of the empirical measure $m_I := \frac{1}{I}\sum_{i=1}^{I}\delta_{x_i}$ has been widely studied. This sits at the intersection of group theory and probability theory. The symmetry group properties have been used to derive some properties of the distributions of the processes. The *de Finetti–Hewitt–Savage* theory provides the mean-field convergence of such a measure-valued process.

When studying convergence of measures, an important issue is the choice of the probability metric. In order to measure the gap between two probability measures, we introduce the Wasserstein (Vasershtein) metric (also called Monge-Kantorovich metric) d_α of order $\alpha \geq 1$.

Definition B.2 (Wasserstein) The following is called Wasserstein metric:

$$d_\alpha^\alpha(\mu, \nu) = \inf\left\{\int_{(x,y)} d_0(x, y)^\alpha \gamma(dx, dy);\ \gamma \in \mathcal{P}(\mathcal{X} \times \mathcal{X}),\ \gamma_x = \mu,\ \gamma_y = \nu\right\},$$

where γ_x denotes the marginal with respect to the x-component, where d_0 is a reference metric on $\mathcal{X}$ (such a metric exists because $\mathcal{X}$ is assumed to be metrizable).

The Kantorovich-Rubinstein 1958 theorem provides a dual representation of d_1 in terms of a Lipschitz-bounded metric:

$$d(\mu, \nu) := d_1(\mu, \nu) = \sup\left\{\int \phi d(\mu - \nu);\ \left|\phi\right|_{Lip} \leq 1\right\},$$

where $\left|\phi\right|_{Lip} = \left|\phi\right|_\infty + \sup_{x \neq y} \frac{|\phi(x)-\phi(y)|}{d_0(x,y)}$ is the Lipschitz norm of ϕ. It can be shown that d_α is a metric (a true distance in a topological sense), i.e., it satisfies the axioms of a metric. For Polish spaces $\mathcal{X}$, the Wasserstein distance d_1 is known to metrize the weak topology over $\mathcal{X}$. As stated in [360], the Wasserstein distance has the following properties: for any $1 \leq \alpha < +\infty$

$$\lim_I d_\alpha(m_I, m) = 0,$$

implies, in particular, that

- m_I converges to m in distribution (weak convergence of probability measures), i.e.

$$E_{m_I}[\phi] := \int \phi dm_I \to E_m[\phi],$$

as $I \to +\infty$, for any measurable bounded and Lipschitz functions ϕ.
- $\int d_0^\alpha(x, y) m_I(dy) < +\infty$ for some $x \in \mathscr{X}$.

Thanks to these nice properties, the Wasserstein distance d_α is an appropriate candidate for the convergence of the empirical measure in the weak sense.

Theorem B.1 (de Finetti–Hewitt–Savage) *Let $x_1, x_2, \ldots,$ be an indistinguishable sequence of $\mathscr{X}$-valued random variables, where $\mathscr{X}$ is a Polish space. Then, there is a $\mathscr{P}(\mathscr{X})$-valued random measure m such that*

$$m = \lim_{I \to \infty} \frac{1}{I} \sum_{i=1}^{I} \delta_{x_i}, \text{ almost surely,}$$

where $\mathscr{P}(\mathscr{X})$ denotes the space of probability measures on $\mathscr{X}$. Conditioned on m, the random variables $x_1, x_2, \ldots$ are i.i.d. with distribution m; that is, for each measurable bounded function ϕ

$$E\left(\phi(x_1, x_2, \ldots, x_k) \mid m\right) = \int \phi(y_1, \ldots, y_k) m(dy_1) \ldots m(dy_k).$$

In addition, if the moments of x_i are finite then

$$d_1\left(m, \frac{1}{I}\sum_{i=1}^{I} \delta_{x_i}\right) \leq \frac{C_1}{\sqrt{I}} = O\left(\frac{1}{\sqrt{I}}\right),$$

where $C_1 > 0$ and $d = d_1$ denotes the Wasserstein metric of order one.

Note that the convergence in Theorem B.1 is in the weak sense since the Monge-Kantorovich distance d_1 metrizes the weak topology. Theorem B.1 has been proved by de Finetti [111] for infinite binary sequences and has been extended by Hewitt and Savage [190] to continuous and compact state spaces. A simple and elegant proof can be found in Aldous [5], pp. 18–22, for the general state space. The rate of convergence for the Monge-Kantorovich distance is obtained following the line of the law of large numbers of interacting systems. Theorem B.1 was initially used for static (time-independent) maps. Then, several applications in mathematical physics and biology, with dynamical models, entered the picture. These are dynamically interacting particles, genes, molecules, or nodes. Theorem B.1 was then extended to the dynamical case in at least two ways: (i) pathwise (up to a certain time step T) and (ii) at each time step t.

B.2 Large Deviation Principle for m_I

We say that for any time t, the probability measure $(m_I(t))_{I\geq 0}$ on a topological space obeys a large deviation principle with rate functions $(I(t,.))$ and in the scale $(a_I)_I$ if $(a_I)_I$ is a real-valued sequence satisfying $a_I \to \infty$ and I is a nonnegative, lower semicontinuous function such that

$$- \inf_{x\in int(B)} I(t,x) \leq \underline{\lim}_I \frac{1}{a_I} \log m_I(t,B) \leq \overline{\lim}_I \frac{1}{a_I} \log m_I(t,B) \\ \leq -\inf_{x\in cl(B)} I(t,x),$$

for any measurable set B, whose interior is denoted by $int(B)$ and closure by $cl(B)$. If the level sets $\{x : I(t,x) \leq \beta\}$ are compact for every $\beta < +\infty$, $I(t,.)$ is called a *good rate function*. We introduce $\tilde{H}(.|.)$ as the relative entropy function (defined also above)

$$\tilde{H}(\mu|\,\nu) := \int \log(\frac{d\mu}{d\nu})\, d\mu,$$

if μ is absolutely continuous with respect to ν and $+\infty$ otherwise.

The main advantage of having this type of result is the decay of $m_I(t,B)$ as n gets large. Basically, when the two limits are identical, $m_I(t,B)$ is the order of $e^{-a_I R}$ where $R = \inf_{x\in B} I(t,x) > 0$. As a consequence, the weak convergence from Theorem B.1 and the central limit theorems can be derived from these inequalities.

Next, we provide a large deviation principle result [360].

Theorem B.2 *Assume that initially $m_I(0)$ follows a large deviation principle with rate $I(0,m(0))$ on the set of probability measures $\mathscr{P}(\mathscr{X})$. Then, $(m_I(t,.))_{t\in[0,T]}$ follows a large deviation principle on the set of cadlag functions from $[0,T]$ to $\mathscr{P}(\mathscr{X})$ with good rate*

$$I(t,m(t)) = \begin{cases} I(0,m(0)) + \int_0^1 \tilde{H}(\,\dot{m}(s) \mid m(0)) \text{ if } \dot{m}(s) \text{ a.c.} \\ +\infty \text{ otherwise,} \end{cases}$$

where a.c. means absolutely continuous.

B.3 Existence of a Solution to the State Equation

Definition B.3 Consider two processes $(x(t))_{t\in[0,T]}$ and $(y(t))_{t\in[0,T]}$. Define $D_{T,\alpha}(\mu,\nu)$ as

$$D_{T,\alpha}(\mu,\nu) := \inf \left\{ \left(E[\sup_{t\in[0,T]} d_0(x(t),y(t))^\alpha] \right)^{1/\alpha} ;\ \mathscr{L}(x) = \mu,\ \mathscr{L}(y) = \nu \right\}.$$

To prove the existence of a solution with respect to the Wasserstein distance, we adopt a contraction-type approach. Then, we construct a Cauchy sequence (L^{α} space which is a Polish space). Consequently, the solution is almost unique. Consider the SDE given by

$$x(t) = x(0) + \int_0^t \sigma(., s, x(s))dB(s) + \int_0^t \left[\int_y |b|^{\alpha}(., s, x(s), y, u)m(s, dy)\right]^{1/\alpha}$$

$$ds =: RH(x)[m],$$

which leads to a fixed-point stochastic equation in m: $m = \mathscr{L}(x(t)) = \mathscr{L}(RH(x)[m])$.

Consider two measures $m_1(s, dy)$ and $m_2(s, dy)$ such that $D_{T,\alpha}(m_i, m_0) < +\infty$. Then

$$D_{t,\alpha}\left(\mathscr{L}(RH(x)[m_1]), \mathscr{L}(RH(x)[m_2])\right) \leq 2Le^{2Lt}\int_0^t D_{s,\alpha}(m_1, m_2)\, ds,$$

for any $t \in [0, T]$. Let $c > 0$, $\Psi(f)(t) \leq c\int_0^t f(s)\, ds$, $t \in [0, T]$ for some operator Ψ and $M = \sup_{s\in[0,T]} f(s)$.

$$\Psi^2(f)(t) \leq c\int_0^t \Phi(f)(s)\, ds = c\int_0^t [cMs]\, ds = c^2\frac{t^2}{2}M.$$

Then, by induction, $\Phi^k(f)(t) \leq \left(c^k\frac{T^k}{k!}\right)M$, and

$$D_{t,\alpha}\left(\mathscr{L}^{k+1}(\mu_0), \mathscr{L}^k(\mu_0)\right) \leq C_T^k\frac{T^k}{k!}D_{T,\alpha}\left(\mathscr{L}(\mu_0), \mu_0\right) < +\infty,$$

where $C_T = 2Le^{2LT}$ and where we have used the Lipschitz continuity L of b and the Minkowski inequality for $p \geq 1$. By summing up over natural numbers k, one gets $\sum_{k\geq k_0} D_{t,\alpha}\left(\mathscr{L}^{k+1}(\mu_0), \mathscr{L}^k(\mu_0)\right) < +\infty$. Thus, $(\mathscr{L}^k(\mu_0))_{k\geq k_0}$ is a Cauchy sequence with respect to the metric $D_{T,\alpha}$, for $k_0 \geq 1$. Since we are in a complete metric space, this sequence converges to some fixed-point (say, m). Then, $m = \mathscr{L}(RH(x)[m]) = \mathscr{L}(m)$ is the unique fixed-point solution of the SDE.

B.3.1 Pathwise Mean-Field Convergence

Consider I independent random processes $x^*_{i,I}$ satisfying

$$x^*_{i,I}(t) = x^*_{i,I}(0) + \int_0^t \sigma(., s, x^*_{i,I}(s))dB(s)$$

$$+\int_0^t \left[\int_y |b|^\alpha(., s, x^*_{i,I}(s), y, u)m^*(s, dy)\right]^{1/\alpha} ds,$$

with $m^*(t, .) = \mathscr{L}(x^*(t))$ and the particle representation

$$x_{i,I}(t) = x_{i,I}(0) + \int_0^t \sigma(., s, x_{i,I}(s))dB(s)$$
$$+\int_0^t \left[\int_y |b|^\alpha(., s, x_{i,I}(s), y, u)m_I(s, dy)\right]^{1/\alpha} ds,$$

where $m^*_I(t, .) = \frac{1}{I}\sum_{i=1}^I \delta_{x^*_{i,I}(t)}$.

To prove the convergence of $x_{i,I}(t)$ as I becomes large, it suffices to prove the convergence for the coefficients with the mean-field terms. We take the difference between the two drift terms:

$$D = \int_0^t \left[\int_y |b|^\alpha(., s, x^*_{i,I}(s), y, u)m^*(s, dy)\right]^{1/\alpha} ds$$
$$-\int_0^t \left[\int_y |b|^\alpha(., s, x_{i,I}(s), y, u)m_I(s, dy)\right]^{1/\alpha} ds. \quad \text{(B.3)}$$

We decompose D into three separate terms as follows:

$$\begin{aligned}
D =& \int_0^t \left[\int_y |b|^\alpha(., s, x^*_{i,I}(s), y, u)m^*(s, dy)\right]^{1/\alpha} ds \\
&-\int_0^t \left[\int_y |b|^\alpha(., s, x^*_{i,I}(s), y, u)m^*_I(s, dy)\right]^{1/\alpha} ds \\
&+\int_0^t \left[\int_y |b|^\alpha(., s, x^*_{i,I}(s), y, u)m^*_I(s, dy)\right]^{1/\alpha} ds \\
&-\int_0^t \left[\int_y |b|^\alpha(., s, x^*_{i,I}(s), y, u)m_I(s, dy)\right]^{1/\alpha} ds \\
&+\int_0^t \left[\int_y |b|^\alpha(., s, x^*_{i,I}(s), y, u)m_I(s, dy)\right]^{1/\alpha} ds \\
&-\int_0^t \left[\int_y |b|^\alpha(., s, x_{i,I}(s), y, u)m_I(s, dy)\right]^{1/\alpha} ds, \\
D =& I_1 + I_2 + I_3.
\end{aligned} \quad \text{(B.4)}$$

The first term I_1 (see (B.4)) deals only with i.i.d. random variables. Therefore, the convergence for that part is classical. For the second term I_2, we use the triangule inequality for $\alpha \geq 1$. By Lipschitz continuity of b, we get $|I_2| \leq \frac{L}{I^{\frac{1}{\alpha}}} \int_0^t \left[\sum_{i=1}^I \left|x_{i,I}^*(s) - x_{i,I}(s)\right|^\alpha\right]^{1/\alpha} ds$. By Lipschitz continuity of b, $|I_3| \leq \int_0^t |x_{i,I}^*(s') - x_{i,I}(s)|\, ds$.

By Hölder inequality

$$\left\{E[\sup_{s\in[0,T]} |x_{i,I}^*(s) - x_{i,I}(s)|^\alpha]\right\}^{1/\alpha} \leq \int_0^T \left\{E\left|Y_{I,s}\right|^\alpha\right\}^{1/\alpha} ds \tag{B.5}$$

$$+ \frac{L}{I^{\frac{1}{\alpha}}} \int_0^T \left[E\sup\left|\sum_{j=1}^I |x_{j,I}^*(s) - x_{j,I}(s)|\right|^\alpha\right]^{1/\alpha} ds$$

$$+ L \int_0^T \{E[\sup_{s\in[0,t]} |x_{i,I}^*(s) - x_{i,I}(s)|^\alpha]\}^{1/\alpha} dt,$$

where

$$Y_{I,s} = \left[\int_y |b|^\alpha(., s, x_{i,I}^*(s), y, u) m^*(s, dy)\right]^{1/\alpha}$$

$$- \left[\int_y |b|^\alpha(., s, x_{i,I}^*(s), y, u) m_I^*(s, dy)\right]^{1/\alpha}.$$

Summing over i and using the Gronwall lemma yields

$$\{E[\sup_{s\in[0,T]} |x_{i,I}^*(s) - x_{i,I}(s)|^\alpha]\}^{1/\alpha} \leq 2Le^{2LT} \int_0^T \left\{E\left|Y_{I,s}\right|^\alpha\right\}^{1/\alpha} ds,$$

which provides a convergence rate of $\sqrt{I}$ since

$$\sup_I \sqrt{I} \left\{E \sup_{s\in[0,T]} \left|Y_{I,s}\right|^\alpha\right\}^{1/\alpha} < +\infty.$$

This result shows that when $\alpha \geq 1$ and when the initial distributions of the virtual particles are mutually independent, with the same distribution as $x(0)$, then the particle interaction model with fixed control u soon destroys that independence through the empirical measure m_I. But, for a given finite time t, when the number of particles becomes large, the mean-field convergence implies that the distributions become approximately independent again conditioning on $m(t, .)$, so that independence is still retained. This is called *propagation of chaos*.

Note that this result is limited to a finite horizon. For long-term behavior, one needs to study the asymptotics (infinite horizon in time) of the SDEs in order to derive the propagation (or non-propagation) of chaos property.

B.3.2 State-Action k-Wise Indistinguishability

In the context of game theory, one may need not just the dependence of (x_i, m_I) but also the dependence on (x_{i_1}, x_{i_2}, m_I), $(x_{i_1}, x_{i_2}, x_{i_3}, m_I)$, $\ldots$, $(x_{i_1}, x_{i_2}, x_{i_3}, \ldots, x_{i_k}, m_I)$ in the quantities of interests.These will require pairwise-indistinguishability, 3-wise, $\ldots$, k-wise indistinguishability, which corresponds to the indistinguishability of the family $\{(x_{i_1}, x_{i_2}) \mid (i_1, i_2) \in \mathscr{I}^2, i_1 \neq i_2\}$, $\{(x_{i_1}, x_{i_2}, x_{i_3}) \mid (i_1, i_2, i_3) \in \mathscr{I}^3, i_1 \neq i_2 \neq i_3\}$, $\{(x_{i_1}, x_{i_2}, x_{i_3}, \ldots, x_{i_k}) \mid (i_1, i_2, i_3, \ldots, i_k) \in \mathscr{I}^k, i_1 \neq i_2 \neq i_3 \ldots \neq i_k\}$, respectively. This leads to the introduction of the empirical measures:

$$\begin{cases} \frac{1}{I(I-1)} \sum_{(i_1,i_2)\in\mathscr{I}^2, i_1\neq i_2} \delta_{(x_{i_1},x_{i_2})} \\ \frac{1}{I(I-1)(I-2)} \sum_{(i_1,i_2,i_3)\in\mathscr{I}^3, i_1\neq i_2\neq i_3} \delta_{(x_{i_1},x_{i_2},x_{i_3})} \\ \frac{1}{I(I-1)(I-2)\ldots(I-k)} \sum_{(i_1,i_2,i_3,\ldots,i_k)\in\mathscr{I}^k, i_1\neq i_2\neq i_3\neq i_k} \delta_{(x_{i_1},x_{i_2},x_{i_3},\ldots,x_{i_k})}. \end{cases} \tag{B.6}$$

These higher order interactions occur when $k \geq 2$.

In game theory, indistinguishability refers to a situation where decision-makers cannot distinguish between strategic choices or outcomes based on available information. Therefore, the decisions of interacting decision-making should appear in the quantities of interests (state dynamics, payoffs), and hence, the process x_i of decision-maker i will now be replaced by the new process $y_i = (x_i, a_i)$.

This leads to the introduction of the following empirical measures:

$$\begin{cases} \frac{1}{I} \sum_{i\in\mathscr{I}} \delta_{y_i}, \\ \frac{1}{I(I-1)} \sum_{(i_1,i_2)\in\mathscr{I}^2, i_1\neq i_2} \delta_{(y_{i_1},y_{i_2})}, \\ \frac{1}{I(I-1)(I-2)} \sum_{(i_1,i_2,i_3)\in\mathscr{I}^3, i_1\neq i_2\neq i_3} \delta_{(y_{i_1},y_{i_2},y_{i_3})}, \\ \\ \frac{1}{I(I-1)(I-2)\ldots(I-k)} \sum_{(i_1,i_2,i_3,\ldots,i_k)\in\mathscr{I}^k, i_1\neq i_2\neq i_3\neq i_k} \delta_{(y_{i_1},y_{i_2},y_{i_3},\ldots,y_{i_k})}, \\ k \geq 3 \end{cases} \tag{B.7}$$

Pairwise indistinguishability in this context means that within the family, any two distinct state actions of the decision-makers are indistinguishable from each other when evaluated in the quantities of interests.

Similarly, k-wise indistinguishability in this context means that within the family, any k distinct state action elements of decision-makers are indistinguishable from each other when evaluated in the quantities of interests.

When the state action is indistinguishable, the mean-field convergence implies

$$
\begin{cases}
\frac{1}{I}\sum_{i\in\mathcal{I}}\delta_{y_i} \implies \mu,\\
\frac{1}{I(I-1)}\sum_{(i_1,i_2)\in\mathcal{I}^2, i_1\neq i_2}\delta_{(y_{i_1},y_{i_2})} \implies \mu\otimes\mu,\\
\frac{1}{I(I-1)(I-2)}\sum_{(i_1,i_2,i_3)\in\mathcal{I}^3, i_1\neq i_2\neq i_3}\delta_{(y_{i_1},y_{i_2},y_{i_3})} \implies \mu\otimes\mu\otimes\mu,\\
\\
\frac{1}{I(I-1)(I-2)\ldots(I-k)}\sum_{(i_1,i_2,i_3,\ldots,i_k)\in\mathcal{I}^k, i_1\neq i_2\neq i_3\neq i_k}\delta_{(y_{i_1},y_{i_2},y_{i_3},\ldots,y_{i_k})}\\
\implies \underbrace{\mu\otimes\mu\otimes\mu\ldots\otimes\mu,}_{k\ times}\\
k\geq 3.
\end{cases}
\tag{B.8}
$$

This last property is important when examining population games, evolutionary games, crowding games, and mean-field games with higher-order interactions. In particular, in the finite population regime, when one decision-maker deviates, the indistinguishability property may fail.

References

1. Y. Achdou, F. Camilli, and I. Capuzzo-Dolcetta. Mean field games: numerical methods for the planning problem. *SIAM Journal on Control and Optimization*, 50:77–109, 2012.
2. Yves Achdou, Manh-Khang Dao, Olivier Ley, and Nicoletta Tchou. A class of infinite horizon mean field games on networks. *arXiv preprint arXiv:1805.11290*, 2018.
3. D.O. Adu, T. Başar, and B. Gharesifard. Optimal transport for a class of linear quadratic differential games. *IEEE Transactions on Automatic Control, 67(11):6287–6294, November*, 2022.
4. S. Aggarwal, M.A. uz Zaman, and T. Başar. Linear quadratic mean-field games with communication constraints. *Proc. 2022 American Control Conference (ACC 2022), Atlanta, June 8–10, 2022, pp. 1323–1329.*, 2022.
5. D. J. Aldous. Exchangeability and related topics. *Ecole d'ete de Probabilites de Saint-Flour XIII - 1983 Lecture Notes in Mathematics, 1985, Volume 1117/1985, 1–198*, 1983.
6. David J Aldous. *Exchangeability and related topics*. pp. 1–198 in: Lecture Notes in Mathematics 117, Springer, 1985.
7. David J Aldous, Illdar A Ibragimov, Jean Jacod, and David J Aldous. *Exchangeability and related topics*. Springer, 1985.
8. N. Almayouf, E. Bachini, A. Chapouto, R. Ferreira, D.-A. Gomes, D. Jordao, D.-E. Junior, A. Karag-ulyan, J. Monasterio, L. Nurbekyan, G. Pagliar, M. Piccirilli, S. Pratapsi, M. Prazeres, J. Reis, A. Rodrigues, O. Romero, M. Sargsyan, T. Seneci, C. Song, K. Terai, R. Tomisaki, H. Velasco-Perez, V. Voskanyan, and X. Yang. Existence of positive solutions for an approximation of stationary mean-field games. *Journal of Mathematics*, 2015.
9. Eitan Altman, Yezekael Hayel, Hamidou Tembine, and Rachid El-Azouzi. Markov decision evolutionary games with expected average fitness. *Evolutionary Ecology Research*, 11(4):677–689, 2009.
10. David Ambrose and Alpár Mészáros. Well-posedness of mean field games master equations involving non-separable local hamiltonians. *Transactions of the American Mathematical Society*, 376(04):2481–2523, 2023.
11. Gholamali Aminian, Yixuan He, Gesine Reinert, Lukasz Szpruch, and Samuel N Cohen. Generalization error of graph neural networks in the mean-field regime. *arXiv preprint arXiv:2402.07025*, 2024.
12. Berkay Anahtarci, Can Deha Kariksiz, and Naci Saldi. Maximum causal entropy inverse reinforcement learning for mean-field games. *arXiv preprint arXiv:2401.06566*, 2024.
13. M. Arapostathis, A. Biswas, and J. Carroll. On solutions of mean field games with ergodic cost. *Commun. Partial Differ. Equ.*, 23(11–12):2187–2217, 2016.

T. Başar et al., *Mean-Field-Type Game Theory I*, Static & Dynamic Game Theory: Foundations & Applications, https://doi.org/10.1007/978-3-032-07027-2

14. Alexander Aurell. Mean-field type games between two players driven by backward stochastic differential equations. *Games*, 9(4):88, 2018.
15. Alexander Aurell and Boualem Djehiche. Mean-field type modeling of nonlocal crowd aversion in pedestrian crowd dynamics. *SIAM Journal on Control and Optimization*, 56(1):434–455, 2018.
16. Alexander Aurell and Boualem Djehiche. Modeling tagged pedestrian motion: A mean-field type game approach. *Transportation research part B: methodological*, 121:168–183, 2019.
17. Yurii Averboukh. Lattice approximations of the first-order mean field type differential games. *Nonlinear Differential Equations and Applications NoDEA*, 28:1–42, 2021.
18. S. Mèlèard B. Jourdain and W. Woyczynski. Nonlinear SDEs driven by lévy processes and related PDEs. *Alea*, 4:1–29, 2008.
19. T Başar. Performance bounds for hierarchical systems under partial dynamic information. *Journal of Optimization Theory and Applications, 39(1):67–87, January*, 1983.
20. T Başar. Affine incentive schemes for stochastic systems with dynamic information. *SIAM Journal on Control and Optimization, 22(2):199–210, March*, 1984.
21. T Başar. Stochastic incentive problems with partial dynamic information and multiple levels of hierarchy. *European J. Political Economy (special issue on "Economic Design"), V:203–217*, 1989.
22. T Başar and G J Olsder. Dynamic noncooperative game theory. *Society for Industrial and Applied Mathematics*, 1998.
23. T. Başar and G. Zaccour. editors. handbook of dynamic game theory, volume i (theory of dynamic games). *Springer International Publishing*, 2018.
24. T. Başar and G. Zaccour. editors. handbook of dynamic game theory, volume ii (applications of dynamic games). *Springer International Publishing*, 2018.
25. Tamer Başar. A general theory for discrete-time mean-field games. *MFG Workshop*, 2017.
26. Tamer Başar. A consensus problem in mean field setting with noisy measurements of target. In *Annual American Control Conference, ACC , Milwaukee, WI, USA, June 27–29, 2018*, pages 6521–6526, 2018.
27. Reginald A Banez, Lixin Li, Chungang Yang, Zhu Han, Reginald A Banez, Lixin Li, Chungang Yang, and Zhu Han. Mean-field-type game for multi-access edge computing networks. *Mean Field Game and its Applications in Wireless Networks*, pages 147–178, 2021.
28. Reginald A Banez, Lixin Li, Chungang Yang, Lingyang Song, and Zhu Han. A mean-field-type game approach to computation offloading in mobile edge computing networks. In *IEEE International Conference on Communications (ICC)*, pages 1–6. IEEE, 2019.
29. Reginald A Banez, Hamidou Tembine, Lixin Li, Chungang Yang, Lingyang Song, Zhu Han, and H Vincent Poor. Mean-field-type game-based computation offloading in multi-access edge computing networks. *IEEE Transactions on Wireless Communications*, 19(12):8366–8381, 2020.
30. M. Bardi. Explicit solutions of some linear-quadratic mean field games. *Workshop on Mean Field Games, Roma*, 2011.
31. Martino Bardi. Explicit solutions of some linear-quadratic mean field games. *Networks & Heterogeneous Media*, 7:243–261, 2012.
32. Martino Bardi and Fabio S. Priuli. Linear-quadratic n-person and mean-field games with ergodic cost. *CDC*, 2013.
33. J Barreiro-Gomez, Tyrone E Duncan, and Hamidou Tembine. Linear-quadratic mean-field-type games-based stochastic model predictive control: A microgrid energy storage application. In *2019 American Control Conference (ACC)*, pages 3224–3229. IEEE, 2019.
34. J Barreiro-Gomez and H Tembine. Mean-field-type model predictive control: An application to water distribution networks. *IEEE Access, 10.1109/ACCESS.2019.2940761*, 7, 2019.
35. Julian Barreiro-Gomez and Salah E Choutri. Data-driven stability of stochastic mean-field type games via noncooperative neural network adversarial training. *Asian Journal of Control*, 2023.
36. Julian Barreiro-Gomez, Salah Eddine Choutri, and Hamidou Tembine. Risk-awareness in multi-level building evacuation with smoke: Burj Khalifa case study. *Automatica*, 129:109625, 2021.

37. Julian Barreiro-Gomez, Boualem Djehiche, Tyrone E Duncan, Bozenna Pasik-Duncan, and Hamidou Tembine. Fractional mean-field-type games under non-quadratic costs: A direct method. In *2019 IEEE 58th Conference on Decision and Control (CDC)*, pages 293–298. IEEE, 2019.
38. Julian Barreiro-Gomez, Florian Dörfler, and Hamidou Tembine. Distributed robust population games with applications to optimal frequency control in power systems. In *2018 Annual American Control Conference (ACC)*, pages 5762–5767. IEEE, 2018.
39. Julian Barreiro-Gomez, Tyrone E Duncan, Bozenna Pasik-Duncan, and Hamidou Tembine. Semiexplicit solutions to some nonlinear nonquadratic mean-field-type games: a direct method. *IEEE Transactions on Automatic Control*, 65(6):2582–2597, 2019.
40. Julian Barreiro-Gomez, Tyrone E Duncan, and Hamidou Tembine. Co-opetitive linear-quadratic mean-field-type games. *IEEE Transactions on Cybernetics*, 50(12):5089–5098, 2019.
41. Julian Barreiro-Gomez, Tyrone E Duncan, and Hamidou Tembine. Linear–quadratic mean-field-type games: Jump–diffusion process with regime switching. *IEEE Transactions on Automatic Control*, 64(10):4329–4336, 2019.
42. Julian Barreiro-Gomez, Tyrone E Duncan, and Hamidou Tembine. Linear-quadratic mean-field-type games with multiple input constraints. *IEEE Control Systems Letters*, 3(3):511–516, 2019.
43. Julian Barreiro-Gomez, Tyrone E Duncan, and Hamidou Tembine. Matrix-valued mean-field-type games: Risk-sensitive, adversarial, and risk-neutral linear-quadratic case. *arXiv preprint arXiv:1904.11346*, 2019.
44. Julian Barreiro-Gomez, Tyrone E Duncan, and Hamidou Tembine. Discrete-time linear-quadratic mean-field-type repeated games: Perfect, incomplete, and imperfect information. *Automatica*, 112:108647, 2020.
45. Julian Barreiro-Gomez, Germán Obando, Andres Pantoja, and Hamidou Tembine. Heterogeneous multi-population evolutionary dynamics with migration constraints. *IFAC-PapersOnLine*, 53(2):16852–16857, 2020.
46. Julian Barreiro-Gomez and Hamidou Tembine. Blockchain token economics: A mean-field-type game perspective. *IEEE Access*, 7:64603–64613, 2019.
47. Julian Barreiro-Gomez and Hamidou Tembine. Constrained mean-field-type games: Stationary case. In *2019 IEEE 58th Conference on Decision and Control (CDC)*, pages 2208–2213. IEEE, 2019.
48. Julian Barreiro-Gomez and Hamidou Tembine. A MATLAB-based mean-field-type games toolbox: Continuous-time version. *IEEE Access*, 7:126500–126514, 2019.
49. Julian Barreiro-Gomez and Hamidou Tembine. Mean-field-type model predictive control: An application to water distribution networks. *IEEE Access*, 7:135332–135339, 2019.
50. Julian Barreiro-Gomez and Hamidou Tembine. A tutorial on mean-field-type games and risk-aware controllers. *Annual Reviews in Control*, 50:317–334, 2020.
51. Julian Barreiro-Gomez and Hamidou Tembine. *Mean-Field-Type Games for Engineers*. CRC Press, 2021.
52. Julian Barreiro-Gomez, Hamidou Tembine, Leonardo Stella, Dario Bauso, and Patrizio Colaneri. Risk-aware control and games in engineering. In *2020 59th IEEE Conference on Decision and Control (CDC)*, pages 3860–3870. IEEE, 2020.
53. Dario Bauso, Ben Mansour Dia, Boualem Djehiche, Hamidou Tembine, and Raul Tempone. Mean-field games for marriage. *PloS one*, 9(5):e94933, 2014.
54. Dario Bauso, Jian Gao, and Hamidou Tembine. Distributionally robust games: f-divergence and learning. In *Proceedings of the 11th EAI International Conference on Performance Evaluation Methodologies and Tools*, pages 148–155, 2017.
55. Dario Bauso, Hamidou Tembine, and Tamer Başar. Robust mean field games. *Dynamic Games and Applications*, 6(3):277–303, Sep 2016.
56. R. Bellman. On the theory of dynamic programming. *Proceedings of the National Academy of Sciences of the U.S.A*, 38:716–719, 1952.

57. Michel Benaim and Jean-Yves Le Boudec. A class of mean field interaction models for computer and communication systems. *Perform. Eval.*, 65(11–12):823–838, 2008.
58. V. E. Beneš. Existence of optimal stochastic control laws. *SIAM J. Control*, 9:446–472, 1971.
59. A. Bensoussan, B. Djehiche, H. Tembine, and S. C. P. Yam. Mean-field-type games with jump and regime switching. *Dynamic Games and Applications* https://doi.org/10.1007/s13235--019-00306-2, 2020.
60. A. Bensoussan, J. Frehse, and Yam S.C.P. Mean field games and mean field type control theory. *SpringerBriefs in Mathematics, Springer*, 2013.
61. A. Bensoussan, K. C. J. Sung, S. C. P. Yam, and S. P. Yung. Linear-quadratic mean field games. *Journal of Optimization Theory and Applications*, 169(2):496–529, May 2016.
62. Alain Bensoussan, Boualem Djehiche, Hamidou Tembine, and Sheung Chi Phillip Yam. Mean-field-type games with jump and regime switching. *Dynamic Games and Applications*, 10:19–57, 2020.
63. Alain Bensoussan, Jens Frehse, and Sheung Chi Phillip Yam. The master equation in mean field theory. *Journal de Mathématiques Pures et Appliquées*, 103(6):1441–1474, 2015.
64. Philippe Bergault and Leandro Sánchez-Betancourt. A mean field game between informed traders and a broker. *arXiv preprint arXiv:2401.05257*, 2024.
65. J. Bergin and D. Bernhardt. Anonymous sequential games with aggregate uncertainty. *Journal of Mathematical Economics*, 21:543–562, 1992.
66. Jules Berry, Olivier Ley, and Francisco J Silva. Approximation and perturbations of stable solutions to a stationary mean field game system. *arXiv preprint arXiv:2402.16377*, 2024.
67. Charles Bertucci. Monotone solutions for mean field games master equations: finite state space and optimal stopping. *Journal de l'Ecole Polytechnique Mathématiques*, 8:1099–1132, 2021.
68. Charles Bertucci, Jean-Michel Lasry, and Pierre-Louis Lions. On lipschitz solutions of mean field games master equations. *arXiv preprint arXiv:2302.05218*, 2023.
69. D. Blackwell. Discounted dynamic programming. *Ann. Math. Statist*, 36(1), 1965.
70. D. Blackwell and T. S. Fergusson. The big match. *Ann. Math. Statist.*, 39:159–163, 1968.
71. JR Blum, H Chernoff, M Rosenblatt, and H Teicher. Central limit theorems for interchangeable processes. *Canadian Journal of Mathematics*, 10:222–229, 1958.
72. Karl-Wilhelm Georg Bollweg and Thilo Meyer-Brandis. Mean-field SDEs driven by g-Brownian motion. *arXiv preprint arXiv:2401.09113*, 2024.
73. Ahmed Bonfoh and Cyril D. Enyi. Large time behavior of a conserved phase-field system. *Communications on Pure & Applied Analysis*, 15 (4):1077–1105, 2016.
74. Pierre Brèmaud. *Point Processes and Queues: Martingale Dynamics*, volume 101. Springer-Verlag, Berlin, 1981.
75. L. M. Briceno-Arias, D. Kalise, and F. J. Silva. Proximal methods for stationary mean field games with local couplings. *SIAM Journal on Control and Optimization*, 2018.
76. R. Buckdahn, B. Djehiche, and J. Li. A general stochastic maximum principle for SDEs of mean-field type. *Applied Math. and Optimization*, 64(2):197–216, 2011.
77. R. Buckdahn, J. Li, and S. Peng. Mean-field backward stochastic differential equations and related partial differential equations. *Stoch. Process. Appl.*, 119(10):3133–3154, 2009.
78. Rainer Buckdahn, Boualem Djehiche, and Juan Li. A general stochastic maximum principle for SDEs of mean-field type. *Applied Mathematics & Optimization*, 64(2):197–216, Oct 2011.
79. Martin Burger and Marco Di Francesco. Large time behavior of nonlocal aggregation models with nonlinear diffusion. *Networks & Heterogeneous Media*, 3(4):749–785, 2008.
80. Elisa Calzola, Elisabetta Carlini, and Francisco J Silva. A high-order scheme for mean field games. *Journal of Computational and Applied Mathematics*, page 115769, 2024.
81. P. Cardaliaguet. Long time average of first order mean field games and weak KAM theory. *Dynamic Games and Applications*, 3:473–488, December 2013.
82. P. Cardaliaguet, F. Delarue, J.-M. Lasry, and P.-L. Lions. The master equation and the convergence problem in mean field games. *Princeton University Press*, 2015.
83. P. Cardaliaguet, J. Lasry, P. Lions, and A. Porretta. Long time average of mean field games. *Networks & Heterogeneous Media*, 7 (2):279–301, 2012.

84. P. Cardaliaguet, J. Lasry, P. Lions, and A. Porretta. Long time average of mean field games with a nonlocal coupling. *SIAM Journal on Control and Optimization*, 51(5):3558–3591, 2013.
85. Pierre Cardaliaguet, Marco Cirant, and Alessio Porretta. Splitting methods and short time existence for the master equations in mean field games. *Journal of the European Mathematical Society*, 25(5):1823–1918, 2022.
86. Pierre Cardaliaguet, François Delarue, Jean-Michel Lasry, and Pierre-Louis Lions. *The master equation and the convergence problem in mean field games:(ams-201)*. Princeton University Press, 2019.
87. Pierre Cardaliaguet and Alessio Porretta. Long time behavior of the master equation in mean field game theory. *MFG Workshop*, 2018.
88. Pierre Cardaliaguet and Alessio Porretta. Long time behavior of the master equation in mean-field games. *Analysis & PDE*, 2019.
89. Pierre Cardaliaguet and Panagiotis Souganidis. Monotone solutions of the master equation for mean field games with idiosyncratic noise. *SIAM Journal on Mathematical Analysis*, 54(4):4198–4237, 2022.
90. E Carlini, FJ Silva, and A Zorkot. Newton schemes for mean field games. *Preprint*, 2024.
91. Elisabetta Carlini, Francisco J Silva, and Ahmad Zorkot. A Lagrange–Galerkin scheme for first order mean field game systems. *SIAM Journal on Numerical Analysis*, 62(1):167–198, 2024.
92. R. Carmona and F. Delarue. Probabilistic theory of mean field games with applications i-ii. *Springer, ISBN 978-3-319-59820-8*, 2018.
93. René Carmona and Francois Delarue. The master equation for large population equilibriums. In *Stochastic Analysis and Applications 2014: In Honour of Terry Lyons*, pages 77–128. Springer, 2014.
94. Alekos Cecchin and François Delarue. Weak solutions to the master equation of potential mean field games. *arXiv preprint arXiv:2204.04315*, 2022.
95. Alekos Cecchin and Guglielmo Pelino. Convergence, fluctuations and large deviations for finite state mean field games via the master equation. *Stochastic Processes and their Applications*, 129(11):4510–4555, 2019.
96. Y. Chen, T. T. Georgiou, and M. Pavon. On the relation between optimal transport and Schrödinger bridges: A stochastic control viewpoint. *Journal of Optimization Theory and Applications*, 169(2):671–691, 2016.
97. Y. Chen, T. T Georgiou, and M. Pavon. Steering the distribution of agents in mean-field games system. *Journal of Optimization Theory and Applications*, 179(1):332–357, 2018.
98. Salah Eddine Choutri, Boualem Djehiche, and Hamidou Tembine. Optimal control and zero-sum games for Markov chains of mean-field type. *Mathematical Control & Related Fields*, 9(2156-8472-2019-3-571):571, 2019.
99. Salah Eddine Choutri and Tembine Hamidou. A stochastic maximum principle for Markov chains of mean-field type. *Games*, 9(4):84, 2018.
100. M. Cirant. Stationary focusing mean-field games. *Commun. Partial Differ. Equ.*, 41(8):1324–1346, 2016.
101. Marco Cirant and Davide Francesco Redaelli. Some remarks on linear-quadratic closed-loop games with many players. *arXiv preprint arXiv:2401.06534*, 2024.
102. Abdoul Karim Cissé and Hamidou Tembine. Cooperative mean-field type games. *IFAC Proceedings Volumes*, 47(3):8995–9000, 2014.
103. R.J. Cohen, S.N. snf Elliott. Existence, uniqueness and comparisons for bsdes in general spaces. *Annals of Probability*, 40(5):2264–2297, 2012.
104. Marco Confortola, Fulvia snd Fuhrman and Jean Jacod. Backward stochastic differential equations driven by a marked point process: an elementary approach, with an application to optimal control. *The Annals of Applied Probability*, 26:1743–1773, 2016.
105. Fabio Coppini, Anna De Crescenzo, and Huyen Pham. Nonlinear Graphon mean-field systems. *arXiv preprint arXiv:2402.08628*, 2024.

106. Julio C Correa and Edgard A Pimentel. A hessian-dependent functional with free boundaries and applications to mean-field games. *The Journal of Geometric Analysis*, 34(3):1–21, 2024.
107. Andrea Cosso, Fausto Gozzi, Idris Kharroubi, Huyên Pham, and Mauro Rosestolato. Master Bellman equation in the Wasserstein space: Uniqueness of viscosity solutions. *Transactions of the American Mathematical Society*, 377(01):31–83, 2024.
108. Donald Dawson and Xiaogu Zheng. Law of large numbers and central limit theorem for unbounded jump mean-field models. *Advances in Applied Mathematics*, 12(3):293–326, 1991.
109. B. de Finetti. Funzione caratteristica di un fenomeno aleatorio. *Atti della R. Accademia Nazionale dei Lincei, Ser. 6, Memorie, Classe di Scienze Fisiche, Matematiche e Naturali*, 4:251–299, 1931.
110. B. de Finetti. La prevision: ses lois logiques, ses sources subjectives. *Annales de l'Institut Henri Poincare*, 7:1–68, 1937.
111. Bruno De Finetti. Funzione caratteristica di un fenomeno aleatorio. In *Atti del Congresso Internazionale dei Matematici: Bologna del 3 al 10 de settembre di 1928*, pages 179–190, 1929.
112. M. Debbah and H. Tembine. Reacting to the interference field, mechanisms and games for dynamic spectrum allocation. *Cambridge University Press*, 2013.
113. François Delarue, Daniel Lacker, and Kavita Ramanan. From the master equation to mean field game limit theory: a central limit theorem. *Electron. J. Probab.*, 24:1–54, 2019.
114. François Delarue, Daniel Lacker, and Kavita Ramanan. From the master equation to mean field game limit theory: Large deviations and concentration of measure. *Ann. Probab. 48(1): 211–263*, 2020.
115. François Delarue and Youssef Ouknine. Intrinsic regularization by noise for $1d$ mean field games. *arXiv preprint arXiv:2401.13844*, 2024.
116. François Delarue and Rinel Foguen Tchuendom. Selection of equilibria in a linear quadratic mean-field game. *Stochastic Processes and their Applications*, 130(2):1000–1040, 2020.
117. François Delarue and Athanasios Vasileiadis. Exploration noise for learning linear-quadratic mean field games. *arXiv preprint arXiv:2107.00839*, 2021.
118. Robert Denkert and Ulrich Horst. Extended mean-field games with multi-dimensional singular controls and non-linear jump impact. *arXiv preprint arXiv:2402.09317*, 2024.
119. N. Devineni, S. Perveen, and U. Lall. Solving groundwater depletion in india while achieving food security. *Nature Communications*, 13:3374, 2022.
120. Jodi Dianetti, Salvatore Federico, Giorgio Ferrari, and Giuseppe Floccari. Multiple equilibria in mean-field game models for large oligopolies with strategic complementarities. *arXiv preprint arXiv:2401.17034*, 2024.
121. B. Djehiche, A. Tcheukam, and H. Tembine. Mean-field-type games in engineering. *AIMS Electronics and Electrical Engineering*, 1(1):18–73, 2017.
122. B. Djehiche and H. Tembine. Risk-sensitive mean-field type control under partial observation. *Chapter Stochastics of Environmental and Financial Economics,vol. 138 of the series Springer Proceedings in Mathematics & Statistics*, pages 243–26, 2016.
123. B. Djehiche, H. Tembine, and Raul Tempone. A stochastic maximum principle for risk-sensitive mean-field type control. *IEEE Trans. Automat. Contr.*, 60(10):2640–2649, 2015.
124. Boualem Djehiche, Julian Barreiro-Gomez, and Hamidou Tembine. Electricity price dynamics in the smart grid: A mean-field-type game perspective. In *23rd International Symposium on Mathematical Theory of Networks and Systems Hong Kong University of Science and Technology, Hong Kong*, 2018.
125. Boualem Djehiche, Julian Barreiro-Gomez, and Hamidou Tembine. Mean-field-type games for blockchain-based distributed power networks. In *Beyond Traditional Probabilistic Methods in Economics 2*, pages 45–64. Springer, 2019.
126. Boualem Djehiche, Julian Barreiro-Gomez, and Hamidou Tembine. Mean-field-type games for blockchain-based distributed power networks. *Book chapter in Beyond Traditional Probabilistic Methods in Economics, Editors: Vladik Kreinovich, Nguyen Ngoc Thach,Nguyen Duc Trung, Dang Van Thanh, Pages 45–64, Part of the Studies in Computational Intelligence book series (SCI, volume 809), Springer*, 2019.

127. Boualem Djehiche, Julian Barreiro-Gomez, and Hamidou Tembine. Price dynamics for electricity in smart grid via mean-field-type games. *Dynamic Games and Applications*, 10:798–818, 2020.
128. Boualem Djehiche, Alain Tcheukam, and Hamidou Tembine. A mean-field game of evacuation in a multi-level building. *Special Session 118: Mean field games and applications, The 11th AIMS Conference on Dynamical Systems, Differential Equations and Applications, July 1–July 5, Orlando, Florida, USA*, 2016.
129. Boualem Djehiche, Alain Tcheukam, and Hamidou Tembine. Mean-field games for evacuation: Simulation results. *17th International Symposium on Dynamic Games and Applications, July 12–15, Urbino, Italy*, 2016.
130. Boualem Djehiche, Alain Tcheukam, and Hamidou Tembine. A mean-field game of evacuation in multilevel building. *IEEE Transactions on Automatic Control*, 62(10):5154–5169, 2017.
131. Boualem Djehiche, Alain Tcheukam, and Hamidou Tembine. A mean-field game of evacuation in multilevel building. *IEEE Trans. Automat. Contr.*, 62(10):5154–5169, 2017.
132. Boualem Djehiche, Alain Tcheukam, and Hamidou Tembine. Mean-field-type games in engineering. *AIMS Electronics and Electrical Engineering*, doi: https://doi.org/10.3934/ElectrEng.2017.1.18, 1:18–73, 2017.
133. Bozhang Dong, Tianyang Nie, and Zhen Wu. Maximum principle for discrete-time stochastic control problem of mean-field type. *Automatica*, 144:110497, 2022.
134. Federica Dragoni and Ermal Feleqi. Ergodic mean field games with hörmander diffusions. *Calculus of Variations and Partial Differential Equations*, 57(5):116, Jul 2018.
135. Roxana Dumitrescu, Marcos Leutscher, and Peter Tankov. Energy transition under scenario uncertainty: a mean-field game of stopping with common noise. *Mathematics and Financial Economics*, pages 1–42, 2024.
136. Tyrone E Duncan, Bozenna Pasik-Duncan, and Hamidou Tembine. Mean-field-type games driven by Rosenblatt process: non-quadratic cost. *Preprint*, 2023.
137. Tyrone E Duncan, Bozenna Pasik-Duncan, and Hamidou Tembine. Multi-layer mean-field-type games driven by Rosenblatt process: non-quadratic cost. *Preprint*, 2023.
138. Tyrone E Duncan, Bozenna Pasik-Duncan, and Hamidou Tembine. Mean-field-type games driven by Rosenblatt processes. *CODIT*, pages 982–987, 2024.
139. Tyrone E Duncan, Bozenna Pasik-Duncan, and Hamidou Tembine. Multi-layer mean-field-type games driven by five noises. *IEEE CDC*, pages 1850–1855, 2024.
140. P. K. Dutta. A folk theorem for stochastic games. *University of Rochester - Center for Economic Research (RCER)*, 1991.
141. P. K Dutta. A folk theorem for stochastic games. *Journal of Economic Theory*, 66:1–32, 1995.
142. Federgruen, A. On N-person stochastic games with denumerable state space, *Advances in Applied Probability*, 10(2):52–471, 1978.
143. Feinberg E. and Shwartz A. (Eds). Handbook of Markov decision processes: Methods and applications. *International Series of Operations Research and Management Science, 40, Kluwer Academic Publishers*, 2002.
144. Nicole El Karoui and Said Hamadène. BSDEs and risk-sensitive control, zero-sum and nonzero-sum game problems of stochastic functional differential equations. *Stochastic Processes and their Application*, 107:145–169, 2003.
145. Nicole El Karoui, Shige Peng, and Marie-Claire Quenez. Backward stochastic differential equations in finance. *Mathematical Finance*, 7(1):1–71, 1997.
146. Zahrate El Oula Frihi, Julian Barreiro-Gomez, Salah Eddine Choutri, and Hamidou Tembine. Hierarchical structures and leadership design in mean-field-type games with polynomial cost. *Games*, 11(3):30, 2020.
147. Stewart N Ethier and Thomas G Kurtz. *Markov processes: characterization and convergence*. John Wiley & Sons, 2009.
148. Salvatore Federico, Fausto Gozzi, and Daria Ghilli. Linear-quadratic mean field games in Hilbert spaces. *arXiv preprint arXiv:2402.14935*, 2024.

149. E. Feleqi. The derivation of ergodic mean field game equations for several populations of players. *Dyn. Games Appl. 3(4)*, pages 523–536, 2013.
150. Shui Feng. Large deviations for empirical process of mean-field interacting particle system with unbounded jumps. *The Annals of Probability*, pages 2122–2151, 1994.
151. Shui Feng and Xiaogu Zheng. Solutions of a class of nonlinear master equations. *Stochastic processes and their applications*, 43(1):65–84, 1992.
152. Christian Fiedler, Michael Herty, and Sebastian Trimpe. On kernel-based statistical learning theory in the mean field limit. *Advances in Neural Information Processing Systems*, 36, 2024.
153. J. Filar and K. Vrieze. Competitive Markov decision processes. *Springer-Verlag, New York*, 1997.
154. AM Fink. Equilibrium in a stochastic n-person game. *J Sci Hiroshima Univ*, 28:89–9, 1964.
155. J. Flesch, G. Schoenmakers, and K. Vrieze. Stochastic games on a product state space. *Math. Oper. Res.*, 33:403–420, 2008.
156. L. Freitag and M. Stojanovic. Basin-scale acoustic communication: A feasibility study using tomography m-sequences. In *Proceedings of MTS/IEEE OCEANS*, pages 2256–2261, Honolulu, HI, USA, Nov. 2001.
157. Zahrate El Oula Frihi, Julian Barreiro-Gomez, Salah Eddine Choutri, Boualem Djehiche, and Hamidou Tembine. Stackelberg mean-field-type games with polynomial cost. *IFAC-PapersOnLine*, 53(2):16920–16925, 2020.
158. Zahrate El Oula Frihi, Salah Eddine Choutri, Julian Barreiro-Gomez, and Hamidou Tembine. Hierarchical mean-field type control of price dynamics for electricity in smart grid. *Journal of Systems Science and Complexity*, 35(1):1–17, 2022.
159. Zuyue Fu, Zhuoran Yang, Yongxin Chen, and Zhaoran Wang. Actor-critic provably finds Nash equilibria of linear-quadratic mean-field games. *arXiv preprint arXiv:1910.07498*, 2019.
160. D. Fudenberg and Y. Yamamoto. The folk theorem for irreducible stochastic games with imperfect public monitoring. *Journal of Economic Theory*, 146:1664–1683, 2011.
161. Wilfrid Gangbo and Alpár R Mészáros. Global well-posedness of master equations for deterministic displacement convex potential mean field games. *Communications on Pure and Applied Mathematics*, 75(12):2685–2801, 2022.
162. Wilfrid Gangbo, Alpár R Mészáros, Chenchen Mou, and Jianfeng Zhang. Mean field games master equations with nonseparable Hamiltonians and displacement monotonicity. *The Annals of Probability*, 50(6):2178–2217, 2022.
163. Wilfrid Gangbo and Andrzej Święch. Existence of a solution to an equation arising from the theory of mean field games. *Journal of Differential equations*, 259(11):6573–6643, 2015.
164. J. Gao and H. Tembine. Distributionally robust games for deep generative learning. *IEEE World Congress on Computational Intelligence Windsor Convention Centre, Rio de Janeiro, Brazil 08–13 July*, 2018.
165. J. Gao and H. Tembine. Distributionally robust games: Wasserstein metric. *International Joint Conference on Neural Networks (IJCNN), Rio de Janeiro, Brazil, July*, 2018.
166. Jian Gao and Hamidou Tembine. Distributed mean-field-type filters for big data assimilation. In *2016 IEEE 18th International Conference on High Performance Computing and Communications; IEEE 14th International Conference on Smart City; IEEE 2nd International Conference on Data Science and Systems (HPCC/SmartCity/DSS)*, pages 1446–1453. IEEE, 2016.
167. Jian Gao and Hamidou Tembine. Bregman learning for generative adversarial networks. In *2018 Chinese Control And Decision Conference (CCDC)*, pages 82–89. IEEE, 2018.
168. Jian Gao and Hamidou Tembine. Distributed mean-field-type filters for traffic networks. *IEEE Transactions on Intelligent Transportation Systems*, 20(2):507–521, 2018.
169. Jian Gao and Hamidou Tembine. Distributionally robust games: Wasserstein metric. In *2018 International Joint Conference on Neural Networks (IJCNN)*, pages 1–8. IEEE, 2018.
170. Jian Gao, Yida Xu, Julian Barreiro-Gomez, Massa Ndong, Michalis Smyrnakis, and Hamidou Tembine. Distributionally robust optimization. *Book Optimization Algorithms-Examples*, page 1, 2018.

171. Yuan Gao, Wuchen Li, and Jian-Guo Liu. Master equations for finite state mean field games with nonlinear activations. *arXiv preprint arXiv:2212.05675*, 2022.
172. Yuan Gao and Di Qi. Mean field games for controlling coherent structures in nonlinear fluid systems. *arXiv preprint arXiv:2401.10356*, 2024.
173. R Gâteaux. Sur les fonctionnelles continues et les fonctionnelles analytiques. *Comptes rendus hebdomadaires des séances de l'Académie des sciences, Paris*, 157:325–327, 1913.
174. R Gâteaux. Fonctions d'une infinité de variables indépendantes. *Bulletin de la Société Mathématique de France*, 47:70–96, 1919.
175. D. Gillette. Stochastic games with zero stop probabilities. *In Contributions to the Theory of Games (ed.) M. Dresher et al., Princeton University Press, Princeton, N.J.*, pages 179–188, 1957.
176. D.-A. Gomes and H. Mitake. Existence for stationary mean-field games with congestion and quadratic hamiltonians. *NoDEA Nonlinear Differential Equations Appl.*, 22(6):1897–1910, 2015.
177. D.A. Gomes, L. Nurbekyan, and E. Pimentel. Economic models and mean-field games theory. *In: 30th Brazilian Mathematics Colloquium*, 2015.
178. D.A. Gomes, S. Patrizi, and V. Voskanyan. On the existence of classical solutions for stationary extended mean field games. *Nonlinear Anal. Theory Methods Appl.*, 99:49–79, 2014.
179. Diogo A. Gomes, Joana Mohr, and Rafael Rigão Souza. Discrete time, finite state space mean field games. *Springer Berlin Heidelberg*, pages 385–389, 2011.
180. Diogo A Gomes and Vardan K Voskanyan. Extended deterministic mean-field games. *SIAM Journal on Control and Optimization*, 54(2):1030–1055, 2016.
181. P Jameson Graber. Linear quadratic mean field type control and mean field games with common noise, with application to production of an exhaustible resource. *Applied Mathematics & Optimization*, 74:459–486, 2016.
182. P Jameson Graber and Ronnie Sircar. Master equation for Cournot mean field games of control with absorption. *Journal of Differential Equations*, 343:816–909, 2023.
183. Xin Guo, Anran Hu, and Junzi Zhang. MF-OMO: An optimization formulation of mean-field games. *SIAM Journal on Control and Optimization*, 62(1):243–270, 2024.
184. S. Hamadène and J. P. Lepeltier. Backward equations, stochastic control and zero-sum stochastic differential games. *Stochastics Stochastics Rep.*, 54(3–4):221–231, 1995.
185. Tembine Hamidou. Mean-field-type game theory. *Special issue on Nonasymptotic Mean-Field-Type Game Theory*, 2014.
186. Jiequn Han, Ruimeng Hu, and Jihao Long. Learning high-dimensional Mckean–Vlasov forward-backward stochastic differential equations with general distribution dependence. *SIAM Journal on Numerical Analysis*, 62(1):1–24, 2024.
187. Naveed Ul Hassan, Mohamad Assaad, and Hamidou Tembine. Distributed h-infinity-based power control in a dynamic wireless network environment. *IEEE Communications Letters*, 17(6):1124–1127, 2013.
188. Cong He and Hongjun Yu. Large time behavior of the solution to the landau equation with specular reflective boundary condition. *Kinetic & Related Models*, 6 (3):601–623, 2013.
189. Daniel Hernández-Hernández and Joshué Heli Ricalde-Guerrero. Mean-field games with common Poissonian noise: a maximum principle approach. *arXiv preprint arXiv:2401.10952*, 2024.
190. E. Hewitt and L. J. Savage. Symmetric measures on cartesian products. *Transactions of the American Mathematical Society*, 80:470–501, 1955.
191. E. Hopf. Generalized solutions of nonlinear equations of first order. *J. Math. Mech. 14*, pages 951–973, 1965.
192. Hörner, Johannes and Sugaya, Takuo and Takahashi, Satoru and Vieille, Nicolas, Belief-free equilibria in games with long-run monitoring, *Econometrica*, 77(6):1751–1801, 2009
193. J. Hörner, T. Sugaya, S. Takahashi, and N. Vieille. Recursive methods in discounted stochasitc games : An algorithm for $\delta \longrightarrow 0$ and a folk theorem. *Econometrica*, 79(4):1277–1318, 2011.

194. Yao Hu, Zhou Su, and Qichao Xu. Multi-UAV network logistics task allocation algorithm based on mean-field-type game. In *International Conference on Smart Objects and Technologies for Social Good*, pages 3–18. Springer, 2022.
195. Ying Hu, Jianhui Huang, and Xun Li. Linear quadratic mean field game with control input constraint. *ESAIM: Control, Optimisation and Calculus of Variations*, 24(2):901–919, 2018.
196. Han Huang and Rongjie Lai. Unsupervised solution operator learning for mean-field games via sampling-invariant parametrizations. *arXiv preprint arXiv:2401.15482*, 2024.
197. Jianhui Huang and Na Li. Linear–quadratic mean-field game for stochastic delayed systems. *IEEE Transactions on Automatic Control*, 63(8):2722–2729, 2018.
198. Jianhui Huang and Tinghan Xie. A class of mean-field games with optimal stopping and its applications. *Asian Journal of Control*, 26(1):56–69, 2024.
199. Jiawei Huang, Niao He, and Andreas Krause. Model-based RL for mean-field games is not statistically harder than single-agent RL. *arXiv preprint arXiv:2402.05724*, 2024.
200. M. Huang, P.E. Caines, and R.P. Malhamé. Large-population cost-coupled LQG problems with nonuniform agents: Individual-mass behavior and decentralized ε-Nash equilibria. *IEEE Transactions on Automatic Control*, 52(9):1560–1571, 2007.
201. M. Huang, P.E. Caines, and R.P. Malhamé. A locality generalization of the NCE (mean field) principle: agent specific cost interactions. *Proc. 47th IEEE CDC, Cancun, Mexico*, pages 5539–5544, December 2008.
202. Minyi Huang and Mengjie Zhou. Linear quadratic mean field games: Asymptotic solvability and relation to the fixed point approach. *IEEE Transactions on Automatic Control*, 65(4):1397–1412, 2019.
203. P.E. Huang, R. P. Malhame, and P.E Caines. Nash equilibria for large-population linear stochastic systems of weakly coupled agents. *In Analysis, Control and Optimization of Complex Dynamic Systems (GERAD 25th Annivesary Series), E. K. Boukas and R. P. Malhame Eds., Chapter 9, Springer, New York*, pages 215–252, 2005.
204. Pengyan Huang, Guangchen Wang, Shujun Wang, and Hua Xiao. A mean-field game for a forward-backward stochastic system with partial observation and common noise. *IEEE/CAA Journal of Automatica Sinica*, 11(3):746–759, 2024.
205. D. H. Jacobson and D. Q. Mayne. Differential dynamic programming. *American Elsevier Pub. Co., New York, NY*, 1970.
206. Espen Robstad Jakobsen and Artur Rutkowski. The master equation for mean field game systems with fractional and nonlocal diffusions. *arXiv preprint arXiv:2305.18867*, 2023.
207. Jiamin Jian, Peiyao Lai, Qingshuo Song, and Jiaxuan Ye. The convergence rate of the equilibrium measure for the hybrid LQG mean field game. *Nonlinear Analysis: Hybrid Systems*, 52:101454, 2024.
208. Sixian Jin and Qingshuo Song. Stochastic maximum principle for a time-changed mean field game. *Mathematical Control and Related Fields*, 14(1):191–198, 2024.
209. B. Jovanovic. Selection and the evolution of industry. *Econometrica*, 50:649–670, 1982.
210. B. Jovanovic and R. W. Rosenthal. Anonymous sequential games. *Journal of Mathematical Economics*, 17:77–87, 1988.
211. L. V. Kantorovich. On the translocation of masses. *Dokl. Akad. Nauk SSSR*, 37:199–201, 1942.
212. I. Karatzas and S.E. Shreve. Brownian motion and stochastic calculus. *2nd ed., Springer, New York*, 1991.
213. M. A. Khan and H. Tembine. Random matrix games in wireless networks. *IEEE Global High Tech Congress on Electronics (GHTCE 2012), November 18–20, 2012, Shenzhen, China*, 2012.
214. Michael V Klibanov and Yurii Averboukh. Lipschitz stability estimate and uniqueness in the retrospective analysis for the mean field games system via two Carleman estimates. *SIAM Journal on Mathematical Analysis*, 56(1):616–636, 2024.
215. Michael V Klibanov, Jingzhi Li, and Hongyu Liu. On the mean field games system with lateral cauchy data via Carleman estimates. *Journal of Inverse and Ill-posed Problems*, 2024.
216. Yunbum Kook, Matthew S Zhang, Sinho Chewi, Murat A Erdogdu, et al. Sampling from the mean-field stationary distribution. *arXiv preprint arXiv:2402.07355*, 2024.

217. H. Kunita and S. Watanabe. On square integrable martingales. *Nagoya Math. J. vol. 30, pp. 209–245.*, 1967.
218. Glicksberg I. L. A further generalization of the Kakutani's fixed-point theorem, with application to Nash equilibrium points. *Proceedings of the American Mathematical Society*, 3(1):170–174, 1952.
219. Razvan-Andrei Lascu, Mateusz B Majka, and Lukasz Szpruch. Mirror descent-ascent for mean-field min-max problems. *arXiv preprint arXiv:2402.08106*, 2024.
220. J.-M. Lasry and P.-L. Lions. Jeux à champ moyen. i. le cas stationnaire. *C. R. Math. Acad. Sci. Paris*, 343(9):619–625, 2006.
221. J.M. Lasry and P.L. Lions. Jeux à champ moyen i. le cas stationnaire. *C. R. Acad. Sci. Paris*, 343 (9), 2006.
222. J.M. Lasry and P.L. Lions. Jeux à champ moyen ii. horizon fini et controle optimal. *C. R. Acad. Sci. Paris*, 343(10), 2006.
223. J.M. Lasry and P.L. Lions. Mean field games. *Japan. J. Math.*, 2:229–260, 2007.
224. P. D. Lax. Hyperbolic systems of conservation laws ii. *Commun. Pure Appl. Math.*, 10:537–566, 1957.
225. Pengdeng Li, Runsheng Yu, Xinrun Wang, and Bo An. Transition-informed reinforcement learning for large-scale stackelberg mean-field games. *Preprint*, 2024.
226. Xiaochen Li and Kai Du. Sequential propagation of chaos for mean-field BSDE systems. *Chinese Annals of Mathematics, Series B*, 45(1):11–40, 2024.
227. Yong Liang, Bing-Chang Wang, and Huanshui Zhang. Discrete-time indefinite linear-quadratic mean field games and control: The finite-population case. *Automatica*, 162:111518, 2024.
228. Zongxia Liang and Keyu Zhang. A mean field game approach to relative investment-consumption games with habit formation. *arXiv preprint arXiv:2401.15659*, 2024.
229. Liggett, Thomas M. and Lippman, Steven A. Stochastic games with perfect information and no discount factor, *Siam Review*, 11(4):604-607, 1969.
230. Yiqing Lin and Kun Xu. Mean-field reflected bsdes driven by a marked point process. *arXiv preprint arXiv:2401.07723*, 2024.
231. P.-L. Lions. Mean-field games lectures. *Cours au College de France* www.college-de-france.fr, 2009.
232. Pierre-Louis Lions and Panagiotis E Souganidis. A convergence result for a local planning problem for mean field games and rigorous proof of a Freidlin–Ventchel-type large deviations principle for the $1 + 1$ KPZ equation. *arXiv preprint arXiv:2401.02024*, 2024.
233. Hongyu Liu and Shen Zhang. Inverse boundary problem for a mean field game system with probability density constraint. *arXiv preprint arXiv:2402.13274*, 2024.
234. Lu-ping Liu and Wen-sheng Jia. Well-posedness for mean field games with finite state and action space. *Journal of Optimization Theory and Applications*, pages 1–18, 2024.
235. Xikui Liu, Yapeng Teng, and Yan Li. A design proposal of finite-time h-infinity controller for stochastic mean-field systems. *Asian Journal of Control*, 2024.
236. Peng Luo and Ludovic Tangpi. Laplace principle for large population games with control interaction. *Stochastic Processes and their Applications*, page 104314, 2024.
237. X. Luo and H. Tembine. Evolutionary coalitional games in random access control. *32nd IEEE International Conference on Computer Communications INFOCOM miniconference, April 14–19, Turin, Italy*, 2013.
238. Yan Ma and Minyi Huang. Linear quadratic mean field games with a major player: The multi-scale approach. *Automatica*, 113:108774, 2020.
239. W. Mao, H. Qiu, C. Wang, H. Franke, Z. Kalbarczyk, R. Iyer, and T. Başar. A mean-field game approach to cloud resource management with function approximation. *Proc. 36th Conference on Neural Information Processing Systems (NeurIPS 2022; November 26-December 4, 2022), New Orleans, Louisiana.*, 2022.
240. Sergio Mayorga. Short time solution to the master equation of a first order mean field game. *Journal of Differential Equations*, 268(10):6251–6318, 2020.

241. P. Mehta and S. Meyn. Q-learning and Pontryagin's minimum principle. *in IEEE Proc. CDC*, 2009.
242. J.F. Mertens and T. Parthasarathy. Non zero sum stochastic games. *In Stochastic Games and Related Topics, Raghavan T.E.S. et al. (eds.), Kluwer Academic Publishers*, pages 145–148, 1991.
243. A. R Meszaros and Francisco J. Silva. On the variational formulation of some stationary second order mean field game systems. *SIAM Journal on Mathematical Analysis, Vol. 50, Iss. 1 (2018)10.1137/17M1125960*, 2017.
244. Alpár R Mészáros and Chenchen Mou. Mean field games systems under displacement monotonicity. *SIAM Journal on Mathematical Analysis*, 56(1):529–553, 2024.
245. Enzo Miller and Huyen Pham. Linear-quadratic Mckean–Vlasov stochastic differential games. *Modeling, Stochastic Control, Optimization, and Applications*, pages 451–481, 2019.
246. Itaru Mitoma. Tightness of probabilities on c ([0, 1]; y') and d ([0, 1]; y'). *The Annals of Probability*, pages 989–999, 1983.
247. Partha Sarathi Mohapatra and Puduru Viswanadha Reddy. Linear-quadratic mean-field-type difference games with coupled affine inequality constraints. *IEEE Control Systems Letters*, 2023.
248. D. Monderer and L. S. Shapley. Potential games. *Games and Economic Behavior*, 14:124–143, 1996.
249. G. Monge. Mémoire sur la théorie des déblais et des remblais. *In Histoire de l'Académie Royale des Sciences de Paris*, pages 666–704, 1781.
250. J. Moon and T. Başar. Linear quadratic risk-sensitive mean field games. *CDC*, 2014.
251. J. Moon and T. Başar. Linear quadratic risk-sensitive and robust mean-field games. *Preprint*, 2017.
252. J. Moon and T. Başar. Linear quadratic mean-field Stackelberg differential games. *Automatica, 97:200–213, November*, 2018.
253. Jun Moon and Tamer Başar. Linear quadratic risk-sensitive and robust mean field games. *IEEE Trans. Automat. Contr.*, 62(3):1062–1077, 2017.
254. Chenchen Mou and Jianfeng Zhang. Wellposedness of second order master equations for mean field games with nonsmooth data. *arXiv preprint arXiv:1903.09907*, 2019.
255. Chenchen Mou and Jianfeng Zhang. Mean field game master equations with anti-monotonicity conditions. *arXiv preprint arXiv:2201.10762*, 2022.
256. S. Watanabe N. Ikeda. Stochastic differential equations and diffusion processes. *North Holland, Amsterdam.*, 1981.
257. G Namah and J-M Roquejoffre. Remarks on the long time behaviour of the solutions of Hamilton–Jacobi equations. *Commun Partial Differ Equ*, 24(5–6):883–893, 1999.
258. Nash, John, Equilibrium points in n-person games, Proceedings of the National Academy of Sciences, 36(1):48–49, 1950
259. Nash, John, Non-cooperative games, *Annals of Mathematics*, 286–295, 1951
260. A. Neyman and S Sorin. Repeated games with public uncertain duration process. *Int J Game Theory*, 39:29–52, 2010.
261. Minh Nguyen. Finite and mean field games for optimal investment with HARA utility function and the presence of risk-seeking agents. *Preprint*, 2024.
262. Panrui Ni. Time periodic solutions of first order mean field games from the perspective of Mather theory. *arXiv preprint arXiv:2401.07155*, 2024.
263. O. A. Oleinik. Discontinuous solutions of non-linear differential equations. *Uspekhi Mat. Nauk, (in Russian), English transl. in Amer. Math. Soc. Transl. 26 (1963) 95–172*, 12:3–73, 1957.
264. O. A. Oleinik. Construction of a generalized solution of the cauchy problem for a quasi-linear equation of first order by the introduction of vanishing viscosity. *Uspekhi Mat. Nauk*, 14:159–164, 1959.
265. Carlo Orrieri, Alessio Porretta, and Giuseppe Savaré. A variational approach to the mean field planning problem. *Journal of Functional Analysis*, 277(6):1868 – 1957, 2019.

266. Yohance AP Osborne and Iain Smears. Analysis and numerical approximation of stationary second-order mean field game partial differential inclusions. *SIAM Journal on Numerical Analysis*, 62(1):138–166, 2024.
267. E. Pardoux and S. Peng. Adapted solution of a backward stochastic differential equation. *Systems and Control Letters*, 14:55–61, 1990.
268. A. Porretta. On the planning problem for a class of mean field games. *Comptes Rendus Mathematique*, 351(11–12):457–462, 2013.
269. A. Porretta. On the planning problem for the mean field games system. *Dyn. Games Appl.*, 4(2):231–256, 2014.
270. P. Protter. Stochastic differential equations. stochastic integration and differential equations. *(Second edition, Version 2.1), Springer Berlin Heidelberg New York.*, 2005.
271. Lakshmanan R. and Pichler A. Expectile in risk-sensitive averse stochastic programming and dynamic optimization. *Pure and applied functional analysis*, 2024.
272. J. Li R. Buckdahn and J. Ma. A stochastic maximum principle for general mean-field systems. *Applied Mathematics & Optimization*, 74(3):507–534, 2016.
273. Parthasarathy, T. and Tijs, S. H. and Vrieze, O. J., On stochastic games with finite state and action spaces, *International Journal of Game Theory*, 13(2):99–104, 1984.
274. Giorgia Ramponi, Pavel Kolev, Olivier Pietquin, Niao He, Mathieu Laurière, and Matthieu Geist. On imitation in mean-field games. *Advances in Neural Information Processing Systems*, 36, 2024.
275. Kui Ren, Nathan Soedjak, Kewei Wang, and Hongyu Zhai. Reconstructing a state-independent cost function in a mean-field game model. *arXiv preprint arXiv:2402.09297*, 2024.
276. Lu Ren, Yuxin Jin, Zijia Niu, Guofang Wang, Wang Yao, and Xiao Zhang. Optimal strategies for large-scale pursuers against one evader: A mean field game-based hierarchical control approach. *Systems & Control Letters*, 183:105697, 2024.
277. Espen Robstad Jakobsen and Artur Rutkowski. The master equation for mean field game systems with fractional and nonlocal diffusions. *arXiv e-prints*, pages arXiv–2305, 2023.
278. Rogers, P. D. Variations on games of perfect information PhD Thesis, University of California, Berkeley, 1969.
279. L Chris G Rogers and David Williams. *Diffusions, Markov processes and martingales: Volume 2, Itô calculus*, volume 2. Cambridge university press, 2000.
280. Esteban J Rolón Gutiérrez, Son Luu Nguyen, and George Yin. Markovian-switching systems: Backward and forward-backward stochastic differential equations, mean-field interactions, and nonzero-sum differential games. *Applied Mathematics & Optimization*, 89(2):33, 2024.
281. Giulia Rossi, Alain Tcheukam, and Hamidou Tembine. Empathy in one-shot prisoner dilemma. *arXiv preprint arXiv:1702.05361*, 2017.
282. Giulia Rossi, Alain Tcheukam, and Hamidou Tembine. How much does users' psychology matter in engineering mean-field-type games. *arXiv preprint arXiv:1702.05355*, 2017.
283. N. Saldi, T. Başar, and M. Raginsky. Markov–Nash equilibria in mean-field games with discounted cost. *SIAM Journal on Control and Optimization*, 56(6):4256–4287, 2018.
284. N. Saldi, T. Başar, and M. Raginsky. Approximate Nash equilibria in partially observed stochastic games with mean-field interactions. *Mathematics of Operations Research, 44(3):1006–1033*, 2019.
285. N. Saldi, T. Başar, and M. Raginsky. Approximate Markov–Nash equilibria for discrete-time risk-sensitive mean-field games. *Mathematics of Operations Research, 45(4):1596–1620, November*, 2020.
286. N. Saldi, T. Başar, and M. Raginsky. Partially observed discrete-time risk-sensitive mean-field games. *Dynamic Games and Applications, 13:929–960*, 2023.
287. Naci Saldi, Tamer Başar, and Maxim Raginsky. Markov–Nash equilibria in mean-field games with discounted cost. In *American Control Conference, ACC , Seattle, WA, USA, May 24–26*, pages 3676–3681, 2017.
288. Jinghai Shao, Taoran Tian, and Shen Wang. Conditional Mckean–Vlasov SDEs with jumps and Markovian regime-switching: wellposedness, propagation of chaos, averaging principle. *Journal of Mathematical Analysis and Applications*, 534(2):128080, 2024.

289. L. S. Shapley. Stochastic games. *Proceedings of the National Academy of Sciences of the United States of America*, 39(10):1095–1100, 1953.
290. H. Shen and T. Başar. Optimal nonlinear pricing for a monopolistic network service provider with complete and incomplete information. *IEEE J. on Selected Areas in Communications (JSAC) Special Issue: Non-Cooperative Behavior in Networking, 25(6):1216–1223, June*, 2007.
291. Eran Shmaya and Bruno Ziliotto. Bayesian learning in mean field games. *arXiv preprint arXiv:2401.17696*, 2024.
292. Joshua Sin, John W Bonnes, Luke C Brown, and David M Ambrose. Existence and computation of stationary solutions for congestion-type mean field games via bifurcation theory and forward-forward problems. *Journal of Dynamics and Games*, 11(1):48–62, 2024.
293. Alain Tcheukam Siwe and Hamidou Tembine. Mean-field-type games on airline networks and airport queues: Braess paradox, its negation, and crowd effect. In *2016 13th International Multi-Conference on Systems, Signals & Devices (SSD)*, pages 595–600. IEEE, 2016.
294. Alain Tcheukam Siwe and Hamidou Tembine. Network security as public good: A mean-field-type game theory approach. In *2016 13th International Multi-Conference on Systems, Signals & Devices (SSD)*, pages 601–606. IEEE, 2016.
295. Alain Tcheukam Siwe and Hamidou Tembine. Network security as public good: A mean-field-type game theory approach. *13th International Multi-Conference on Systems, Signals & Devices. Conference on Systems, Automation & Control. March 21–24 Leipzig, Germany, pp. 598–603*, 2016.
296. Alain Tcheukam Siwe and Hamidou Tembine. On the distributed mean-variance paradigm. In *2016 13th International Multi-Conference on Systems, Signals & Devices (SSD)*, pages 607–612. IEEE, 2016.
297. A. Vladimirovich Skorokhod. *Studies in the theory of random processes*, volume 7021. Courier Dover Publications, 1982.
298. M. Sobel. Noncooperative stochastic games. *Ann. Math. Statist.*, 42:1930–1935, 1971.
299. M. Sobel. Continuous stochastic games. *J. Appl. Prob.*, 10:597–604, 1973.
300. Alexander Sokol and Niels Richard Hansen. Exponential martingales and changes of measure for counting processes. *Stochastic analysis and applications*, 33:823–843, 2015.
301. Eilon Solan. Discounted stochastic games. *Mathematics of Operations Research*, 23:1010–1021, 1998.
302. Eilon Solan. Stochastic games. *Encyclopedia of Database Systems, Springer*, 2009.
303. S Sorin. A first course on zero-sum repeated games. *Mathématiques et Applications, Springer, 37, Springer, 212p.*, 2002.
304. M. Stojanovic, J. Catipovic, and J. G. Proakis. Adaptive multichannel combining and equalization for underwater acoustic communications. *Journal of the Acoustical Society of America*, 94(3):1621–1631, 1993.
305. D. W. Stroock and S. R. S. Varadhan. Multidimensional diffusion processes. *Springer Verlag, New York*, 1979.
306. Yabing Sun, Jie Yang, and Weidong Zhao. An accurate numerical scheme for mean-field forward and backward SDEs with jumps. *Preprint*, 2024.
307. Yan Sun, Lixin Li, Qianqian Cheng, Dawei Wang, Wei Liang, Xu Li, and Zhu Han. Joint trajectory and power optimization in multi-type UAVs network with mean field q-learning. In *2020 IEEE International Conference on Communications Workshops (ICC Workshops)*, pages 1–6. IEEE, 2020.
308. Yan Sun, Lixin Li, Kaiyuan Xue, Xu Li, Wei Liang, and Zhu Han. Inhomogeneous multi-UAV aerial base stations deployment: a mean-field-type game approach. In *2019 15th International Wireless Communications & Mobile Computing Conference (IWCMC)*, pages 1204–1208. IEEE, 2019.
309. M Takahashi. Stochastic games with infinitely many strategies. *J Sci Hiroshima Univ Ser A-I*, 26:123–134, 1962.
310. Alain Tcheukam, Boualem Djehiche, and Hamidou Tembine. Evacuation of multi-level building: Design, control and strategic flow. *35th Chinese Control Conference (CCC2016), July 26–28 Southwest China, Chengdu*, 2016.

311. Alain Tcheukam and Hamidou Tembine. Mean-field-type games for distributed power networks in presence of prosumers. In *2016 Chinese Control and Decision Conference (CCDC)*, pages 446–451. IEEE, 2016.
312. Rinel Foguen Tchuendom. Uniqueness for linear-quadratic mean field games with common noise. *Dynamic Games and Applications*, 8:199–210, 2018.
313. H. Tembine. Population games in large-scale networks: time delays, mean field and applications. *LAP Lambert Academic Publishing, ISBN 978-3-8383-6392-9, 250 pages*, December 2010.
314. H. Tembine. Hybrid mean field game dynamics in large populations. *American Control Conference, ACC 2011, San Francisco, California, US.*, 2011.
315. H. Tembine. Mean field stochastic games: convergence, q/h learning, optimality. *American Control Conference, ACC 2011, San Francisco, California, US*, 2011.
316. H. Tembine. Non-asymptotic mean-field games. *International Conference on Network Games, Control and Optimization (NETGCOOP), November 28–30, Avignon, France*, 2012.
317. H. Tembine. Fast distributed strategic learning for global optima in queueing access games. *19th World Congress of the International Federation of Automatic Control (IFAC), Cape Town, South Africa, 24–29 August*, 2014.
318. H. Tembine. Nonasymptotic mean-field games. *IEEE Transactions on Systems, Man, and Cybernetics, Part B: Cybernetics, vol. 44, number 12, December*, 2014.
319. H. Tembine. Risk-sensitive mean-field-type games with lp-norm drifts. *Automatica*, 59:224 – 237, 2015.
320. H Tembine. Mean-field-type game theory. *MFTG Tutorial notes*, 2017.
321. H. Tembine. Quantile-based mean-field games with common noise. *Preprint*, 2017.
322. H. Tembine. Deep learning meets game theory: Bregman-based algorithms for interactive deep generative adversarial networks. *IEEE Transactions on Cybernetics, 14 pages, volume 50 , Issue: 3 , March, pp. 1132–1145*, 2020.
323. H. Tembine and Minyi Huang. Mean field difference games. *CDC-ECC, 50th IEEE Conference on Decision and Control and European Control Conference, Orlando, Florida.*, December 12–15 2011.
324. H. Tembine, J. Y. Le Boudec, R. ElAzouzi, and E. Altman. Mean field asymptotic of Markov decision evolutionary games and teams. *in the Proc. of GameNets*, May 2009.
325. H. Tembine, Z. Quanyan, and T. Başar. Risk-sensitive mean field stochastic differential games. *IFAC World Congress*, August 2011.
326. H. Tembine, R. Tempone, and Pedro Vilanova. Mean field interaction in chemical reaction networks. *Allerton Conference on Communication, Control and Computing, 28–30 Sept. pp. 991–997, Print ISBN: 978-1-4577-1817-5*, 2011.
327. H. Tembine, Q. Zhu, and T. Başar. Risk-sensitive mean-field games. *IEEE Transactions on Automatic Control*, 59(4):835–850, April 2014.
328. Hamidou Tembine. Large-scale games in large-scale systems. *arXiv preprint arXiv:1111.2285*, 2011.
329. Hamidou Tembine. Distributed massive mimo network games: Risk and altruism. In *2015 54th IEEE Conference on Decision and Control (CDC)*, pages 3481–3486. IEEE, 2015.
330. Hamidou Tembine. Risk-sensitive mean-field-type games with lp-norm drifts. *Automatica*, 59:224–237, 2015.
331. Hamidou Tembine. Uncertainty quantification in mean-field-type teams and games. In *2015 54th IEEE Conference on Decision and Control (CDC)*, pages 4418–4423. IEEE, 2015.
332. Hamidou Tembine. Mean-field-type optimization for demand-supply management under operational constraints in smart grid. *Energy Systems*, 7(2):333–356, 2016.
333. Hamidou Tembine. Mean-field-type games. *AIMS Math*, 2(4):706–735, 2017.
334. Hamidou Tembine. Payoff measurement noise in risk-sensitive mean-field-type games. In *2017 29th Chinese Control And Decision Conference (CCDC)*, pages 3764–3769. IEEE, 2017.
335. Hamidou Tembine. *Distributed strategic learning for wireless engineers*. CRC Press, 2018.

336. Hamidou Tembine. Covid-19: data-driven mean-field-type game perspective. *Games*, 11(4):51, 2020.
337. Hamidou Tembine. Distributed planning in mean-field-type games. *IFAC-PapersOnLine*, 53(2):2183–2188, 2020.
338. Hamidou Tembine. Master adjoint systems in mean-field-type games. *Preprint*, 2020.
339. Hamidou Tembine. Master adjoint systems in mean-field-type games. *in Communications in Information and Systems, vol.21, 4, pp. 623–650*, 2021.
340. Hamidou Tembine. Master adjoint systems in mean-field-type games. *CoScience: Collective Science Platform, Self-Science Journal sjscience.org*, 2022.
341. Hamidou Tembine, Eitan Altman, Rachid ElAzouzi, and Yezekael Hayel. Stochastic population games with individual independent states and coupled constraints. In *Proceedings of the 3rd International Conference on Performance Evaluation Methodologies and Tools*, pages 1–10, 2008.
342. Hamidou Tembine, Massa NDong Issa Bamia, Bakary Coulibaly Oumar Issiaka Traore, Moussa Traore, Moussa Sanogo, Mamadou Eric Sangare, Salif Kante, Daryl Noupa Yongueng, Hafiz Tiomoko Ali, Malik Tiomoko, Frejus Laleye, Boualem Djehiche, Wesmanegda Elisee Dipama, Idris Baba Saje, Hammid Mohammed Ibrahim, Moumini Sanogo, Marie Coursel Nininahazwe, Abdul-Latif Siita, Haine Mhlongo, Teddy Nelvy Dieu Merci Kouka, Mariam Serine Jeridi, Mutiyamuogo Parfait Mupenge, Lekoueiry Dehah, Abdoul Aziz Bio Sidi D Bouko, Wilfried Franceslas Zokoue, Odette Richette Sambila, Alina RS Mbango, Mady Diagouraga, Oumarou Moussa Sanoussi, Gizachew Dessalegn, Mohamed Lamine Samoura, and Bintou Laetitia Audrey Coulibaly. Breaking the barriers of text-hungry and audio-deficient ai. *AI MALI*, 2025.
343. Hamidou Tembine, Abdellatif Kobbane, and Mohammed El Koutbi. Robust power allocation games under channel uncertainty and time delays. In *2010 IFIP Wireless Days*, pages 1–5. IEEE, 2010.
344. Hamidou Tembine and Ousmane Kodio. Reverse Ishikawa–Nesterov learning scheme for fractional mean-field games. *IFAC-PapersOnLine*, 50(1):8090–8096, 2017.
345. Hamidou Tembine, Raul Tempone, and Pedro Vilanova. Mean-field learning: a survey. *arXiv preprint arXiv:1210.4657*, 2012.
346. Hamidou Tembine, Raul Tempone, and Pedro Vilanova. Mean-field learning for satisfactory solutions. In *52nd IEEE Conference on Decision and Control*, pages 4871–4876. IEEE, 2013.
347. Joana Terra and Noemi Wolanski. Large time behavior for a nonlocal diffusion equation with absorption and bounded initial data. *Discrete & Continuous Dynamical Systems - A*, 31 (2):581–605, 2011.
348. Thuijsman, Frank, Optimality and equilibria in stochastic games, Centrum voor Wiskunde en Informatica, 82, 1992.
349. F Thuijsman and TES Raghavan. Perfect information stochastic games and related classes. *Int J Game Theory*, 1997.
350. Noureddine Toumi, Julian Barreiro-Gomez, Tyrone E Duncan, and Hamidou Tembine. Berge equilibrium in linear-quadratic mean-field-type games. *Journal of the Franklin Institute*, 357(15):10861–10885, 2020.
351. M. A. uz Zaman, S. Bhatt, A. Koppel, and T. Başar. Oracle-free reinforcement learning in mean-field games along a single sample path. *Proc. 25th International Conference on Artificial Intelligence and Statistics (AISTATS 2023), Valencia, Spain, April 25–27,*, 2023.
352. ***M.A. uz Zaman, S. Bhatt, and T. Başar. Adversarial linear-quadratic mean-field games over multigraphs. *Proc. 60th IEEE Conference on Decision and Control (CDC'21, Dec 13–15, 2021; Austin, Texas), pp. 209–214.202*, 2021.
353. Muhammad Aneeq uz Zaman, Kaiqing Zhang, Erik Miehling, and Tamer Başar. Reinforcement learning in non-stationary discrete-time linear-quadratic mean-field games. In *2020 59th IEEE Conference on Decision and Control (CDC)*, pages 2278–2284. IEEE, 2020.
354. Deepanshu Vasal and Randall Berry. Master equation for discrete-time Stackelberg mean field games with a single leader. In *2022 IEEE 61st Conference on Decision and Control (CDC)*, pages 5529–5535. IEEE, 2022.

355. N Vieille. Equilibrium in 2-person stochastic games i: A reduction. *Israel J Math*, 119:55–91, 2000.
356. N Vieille. Equilibrium in 2-person stochastic games ii: The case of recursive games. *Israel J Math*, 119:93–126, 2000.
357. N Vieille. Stochastic games: Recent results. *In: Aumann RJ, Hart S (eds) Handbook of game theory with economic applications*, 3:1833–1850, 2002.
358. P. Vilanova, H. Tembine, and M. Debbah. Noisy mean field stochastic games with network applications. *ICIAM - 7th International Congress on Industrial and Applied Mathematics, ICIAM Vancouver, MS403 Multiscale Approximations Of Kinetic Monte Carlo Simulations.*, 2011.
359. Juan Sebastián Villalba Torres. A mean field-type game approach for resilient synchronization in multi-agent systems. *Universidad de los Andes*, 2022.
360. Cédric Villani et al. *Optimal transport: old and new*, volume 338. Springer, 2009.
361. Von Neumann, John and Morgenstern, Oskar, Theory of games and economic behavior, Princeton University Press, 1944
362. X. Wang, N. Xiao, L. Xie, E. Frazzoli, and D. Rus. Discrete-time mean field games in multi-agent systems. In *2014 13th International Conference on Control Automation Robotics Vision (ICARCV)*, pages 711–716, Dec 2014.
363. Wardrop, John Glen, Some theoretical aspects of road traffic research, *Proceedings of the Institution of Civil Engineers*, 1(3):325–362, 1952
364. C.I.C.H. Watkins and P. Dayan. Q-learning. *Machine Learning*, 8:279–292, 1992.
365. Dongxuan Wu, Yaru Zhang, Liping Xu, and Zhi Li. Strong convergence of Euler–Maruyama schemes for doubly perturbed Mckean–Vlasov stochastic differential equations. *Communications in Nonlinear Science and Numerical Simulation*, page 107927, 2024.
366. Yida Xu and Hamidou Tembine. Risk-sensitive lowest unique bid auctions: Empirical and analytical investigations. *20th World Congress of the International Federation of Automatic Control (IFAC) , Toulouse, France, 9–14 July*, 2017.
367. Xiangjiang Yang, Xiaotian Zhou, Haixia Zhang, and Dongfeng Yuan. Co-opetitive mean field type game based task offloading strategy in multi-access edge computing networks. *IET Communications*, 16(16):1946–1956, 2022.
368. Batuhan Yardim, Artur Goldman, and Niao He. When is mean-field reinforcement learning tractable and relevant? *arXiv preprint arXiv:2402.05757*, 2024.
369. Jiajia Yu, Rongjie Lai, Wuchen Li, and Stanley Osher. A fast proximal gradient method and convergence analysis for dynamic mean field planning. *Mathematics of Computation*, 93(346):603–642, 2024.
370. Jiajia Yu, Quan Xiao, Tianyi Chen, and Rongjie Lai. A bilevel optimization method for inverse mean-field games. *arXiv preprint arXiv:2401.05539*, 2024.
371. Benjamin J Zhang, Siting Liu, Wuchen Li, Markos A Katsoulakis, and Stanley J Osher. Wasserstein proximal operators describe score-based generative models and resolve memorization. *arXiv preprint arXiv:2402.06162*, 2024.
372. Fengzhuo Zhang, Vincent Tan, Zhaoran Wang, and Zhuoran Yang. Learning regularized monotone Graphon mean-field games. *Advances in Neural Information Processing Systems*, 36, 2024.
373. Xin Zhang, Robert J Elliott, and Tak Kuen Siu. A stochastic maximum principle for a Markov regime-switching jump-diffusion model and its application to finance. *SIAM Journal on Control and Optimization*, 50(2):964–990, 2012.
374. Jianjun Zhou, Nizar Touzi, and Jianfeng Zhang. Viscosity solutions for HJB equations on the process space: Application to mean field control with common noise. *arXiv preprint arXiv:2401.04920*, 2024.
375. Mo Zhou, Stanley Osher, and Wuchen Li. A deep learning algorithm for computing mean field control problems via forward-backward score dynamics. *arXiv preprint arXiv:2401.09547*, 2024.
376. Quanyan Zhu, Hamidou Tembine, and Tamer Başar. Evolutionary games for hybrid additive white gaussian noise multiple access control. In *GLOBECOM 2009—2009 IEEE Global Telecommunications Conference*, pages 1–6. IEEE, 2009.

Index

T. Başar et al., *Mean-Field-Type Game Theory I*, Static & Dynamic Game Theory: Foundations & Applications, https://doi.org/10.1007/978-3-032-07027-2